ENGINEERING CHEMISTRY

For the Students of B.E., B.Tech., B.Sc. [Engg.] A.M.I.E., M.Sc. (Environmental Chemistry), M. Tech. (Environmental Engineering) and other Competitive Courses

S.S. DARA
M.Sc., Ph.D.
Former Professor & Head,
Department of Applied Chemistry
Visvesvaraya National Institute of Technology (Formerly VRCE),
Nagpur

Revised by

S.S. UMARE
M.Sc., Ph.D.
Professor & Head,
Department of Applied Chemistry
Visvesvaraya National Institute of Technology (Formerly VRCE),
Nagpur

S. CHAND PUBLISHING

S Chand And Company Limited
(ISO 9001 Certified Company)

S Chand And Company Limited

(ISO 9001 Certified Company)

Head Office: Block B-1, House No. D-1, Ground Floor, Mohan Co-operative Industrial Estate, New Delhi – 110 044 | Phone: 011-66672000

Registered Office: A-27, 2nd Floor, Mohan Co-operative Industrial Estate, New Delhi – 110 044
Phone: 011-49731800

www.schandpublishing.com; e-mail: info@schandpublishing.com

Branches

Chennai	:	Ph: 23632120; chennai@schandpublishing.com
Guwahati	:	Ph: 2738811, 2735640; guwahati@schandpublishing.com
Hyderabad	:	Ph: 40186018; hyderabad@schandpublishing.com
Jalandhar	:	Ph: 4645630; jalandhar@schandpublishing.com
Kolkata	:	Ph: 23357458, 23353914; kolkata@schandpublishing.com
Lucknow	:	Ph: 4003633; lucknow@schandpublishing.com
Mumbai	:	Ph: 25000297; mumbai@schandpublishing.com
Patna	:	Ph: 2260011; patna@schandpublishing.com

© S Chand And Company Limited, 1986

All rights reserved. No part of this publication may be reproduced or copied in any material form (including photocopying or storing it in any medium in form of graphics, electronic or mechanical means and whether or not transient or incidental to some other use of this publication) without written permission of the copyright owner. Any breach of this will entail legal action and prosecution without further notice.

Jurisdiction: All disputes with respect to this publication shall be subject to the jurisdiction of the Courts, Tribunals and Forums of New Delhi, India only.

First Edition 1986
Subsequent Editions and Reprint 1988, 90, 92, 94, 95, 96, 97, 99, 2000, 2001, 2003, 2006, 2007, 2008 (Twice), 2009 (Twice), 2010
Twelfth Edition 2010; Reprint 2013, 2014 (Thrice), 2017 (Twice), 2020, 2021 (Twice)

Reprint 2022 (Thrice)

ISBN: 978-81-219-0359-2   **Product Code:** H3ECH68CHEM10ENAL10O

PRINTED IN INDIA

By Vikas Publishing House Private Limited, Plot 20/4, Site-IV, Industrial Area Sahibabad, Ghaziabad – 201 010 and Published by S Chand And Company Limited, A-27, 2nd Floor, Mohan Co-operative Industrial Estate, New Delhi – 110 044

PREFACE TO THE TWELFTH EDITION

Any good text book, particularly that in the fast changing fields such as engineering and technology, is not only expected to cater to the current curricular requirements of various institutions but also should provide a glimplse towards the latest developments in the concerned subject and the relevant disciplines. It should guide the periodic review and updating of the curriculum. It is precisely with this spirit that new topics have been constantly added in every Edition of this book. This approach has been appreciated and encouraged by the students and the faculty of the various engineering institutions in the country as indicated by the phenomenal response received for this book over the past two decades. This new Twelfth Edition of the book is another effort in that direction.

In this edition, several chapters have been updated and revised keeping in view of the recent developments. In the chapter on "Water treatment" desalination of water and some additional numericals on water treatment have been included. In the chapter on "Fuel and Combustion" the new topics such as catalytic converter, LPG, CNG, power alcohol, biodiesel and some numbericals on combustion calculation of current interest have been included. In the chapter "Cement" topics on properties such as soundness, fineness of cement and use of fly ash as cementing material have been included. In the chapter on "Lubricants" viscosity index and re-refining of lubricating oil is included. In the chapter on "Corrosion" the problem on corrosion tendency of metal is included.

In the chapter on polymer serveral advance topics such as conducting polymers, biopolymers, low dielectric constant polymers, liquid crystal polymers have been added and lot of text in this chapter has been re-written for greater clarity and simplicity. New diagrams have been incorporated and a number of the old one have been improved upon.

The text in the chapter composite have been added. One detailed section on the magnetic material have been added in the chapter, structure of solids. The title of the chapter "Ceramics" have been change by "Glass and Ceramics" and the text on glass is included. Further, the printing mistakes have been corrected and several other chapters have been updated wherever possible.

I sincerely thank the student and teaching community of engineering and technology faculty all over the country. It is solely their encouragement, suggestions, feedback and constructive criticism that is responsible in carving out this book in the present form.

This twelfth revised, enlarged and enriched edition of the book is sincerely offered at the service of students and teaching fraternity associated with engineerng chemistry from the various engineering and technological institutions all over the country. It is hoped that this new edition of the book will be received with vibrant enthusiasm.

AUTHORS

PREFACE TO THE FIRST EDITION

This book is written exclusively for students of various branches of engineering, keeping in view their professional requirements, after entering into their practical life. Many new products of the chemical industries are finding increasing application in all the fields of engineering. The scope of their application is mostly dictated by their chemical behaviour under a given set of conditions. For instance, an ideal selection of an appropriate metal, metal, alloy, or combination of metals, the design of an equipment to minimize corrosion, the selection of a proper lubricating oil to minimize friction and wear, the selection of suitable additives for a special cement or the selection of the right type of a ceramic, plastic or rubber for satisfactory performance for a given purpose under a given set of conditions, can be made only on the basis of the chemical properties of materials, even more than on their physical properties because slight changes in chemical composition may alter the physical properties considerably. Inadequate knowledge of the chemical principles involved may lead to serious errors in the selection and application of the materials used in any field of engineering. It is for this reason that the engineering faculties of many foreign universities are insisting on a second course in Chemistry for their students. This book lends further support for their conviction that "the engineering graduate who knows the differences in chemical properties of alternative materials and who understands the general chemical principles on which their behaviour depends will prove to be a better and more successful engineer than one who does not."

This book embodies 12 chapters which are of basic importance in the curriculum of engineering students and provide a core course of engineering chemistry for all branches of engineering. Each chapter consists of a methodical introduction, historical background, discussion of basic physico-chemical principles involved and practical applications and significance. Chapters on Water and Fuels also contain systematic methods of solving problems on Water Treatment and Combustion Calculations followed by several worked out examples. Further, at the end, enquiring questions on all the chapters are given which also include typical objective questions and answers. A list of reference books has also been included at the end, under bibliography.

This book is written solely with a conviction to severe the academic and professional requirements of the students of all branches of engineering.

Any suggestions and constructive criticism towards this objective are welcome.

Jan. 1986

AUTHOR

ACKNOWLEDGEMENT

I wish to express my gratitude to late Prof. S. S. Dara (Prime author of this book) the former, Head Department of Chemistry of our Institution who was the source of inspiration for review of this book.

I sincerely acknowledge the encouragement, moral support and valuable suggestion of all of the following.

Prof. S. S. Gokhle, Director, VNIT Nagpur. Prof. M. C. Gupta, Prof. S. G. Viswanath of Nagpur University. All my collegues, teaching Engineering Chemistry in VNIT, Nagpur University, Amravati University, and other Universities in Maharashtra and India. Dr. B. M. Rao, Dr. J. D. Ekte, Dr. A. Kumar, Dr. R. K. Kowadkar, Dr. C. Das, Dr. R. T. Jadhav, Dr. S. J. Juneja, Dr. S. B. Gholse, Dr. M. K. N. Yenki., Prof. N. Sulochana NIT Tiruchirappalli, Prof. P. N. Rao NIT Warangal, Dr. A. C. Hegde NIT Suratkal, Dr. R. K, Patel NIT Rourkela, Dr. Masood Alam Jamia Milha Islamia University, New Delhi, Dr. V. K. Srivastava, Institute of Petroleum Technology, Gandhinagar, Dr. S. K. Singh, Institute of Technology, GGU Bilaspur, Dr. Y. Sharma, BHU Varanasi, Prof. A. K. Mishra Sagar University, Dr. Aswar Amravati University. I am duly bound to express my thanks to the authors and publishers of all the books which have been referred during the course of preparation of this revised book. Further, I wish to record my appreciation to Chitriv's Computers, Nagpur for typing the manuscript.

Last but not least, I wish to express my sincere appreciation to Mrs. Nirmala Gupta, Chairperson cum Managing Director, Shri Navin Joshi V.P. (Publishing) and Shri. Bhagirath Kaushik, General Manager (Sales) S.Chand & Company Ltd. and all the Branch Managers of S. Chand Company Ltd. for their whole hearted cooperation in all aspects of the revised publication and promotion of this book.

AUTHORS

5.22	Marketing and manufacture of lubricants in India	282
5.23	Conclusion	283
5.24	Electrical insulating oils	283
5.25	White oils	285

6. PORTLAND CEMENT — 287–312

6.1	Introduction	287
6.2	Raw materials	288
6.3	Important process parameters for manufacturing a good cement clinker.	289
6.4	Methods of manufacturing cements	289
6.5	Dry process Vs. wet process	290
6.6	Sequence of operations	291
6.7	Characteristics of constitutional compounds	293
6.8	Additives for cement	294
6.9	Properties of cement –	295
6.10	Testing of cement	297
6.11	General Composition of ordinary portland cement	298
6.12	Reactions taking place in the rotary kiln	298
6.13	Thermochemical changes taking place during cement formation	299
6.14	Action of some chemicals on concrete	299
6.15	Computation of the amount of constitutional compounds	299
6.16	Types of Portland cement and its derivatives	302
6.17	Other types of cement	306
6.18	Use of fly ash as cementing material	306
6.19	Mortars and concretes	307
6.20	Prestressed concrete	307
6.21	Post–tensioning	308
6.22	Curing	308
6.23	Overall scenario of cement industry	308

7. PHASE RULE — 313–334

7.1	Introduction	313
7.2	Gibb's Phase Rule	313
7.3	Application of Phase Rule to One-Component System	317
7.4	Two-Component System	323
7.5	Uses of Phase Rule	333
7.6	Limitations of Phase Rule	333

8. CHEMICAL BONDING — 335–364

8.1	Ionic or Electrovalent Bond	335
8.2	Covalent Bond	338
8.3	Exceptions to the Octet rule	340
8.4	Resonance	341
8.5	Variable Valency	342
8.6	Coordinate or Dative Bond	342
8.7	Complex Ions	344
8.8	Coordination Number or Ligancy	345

(xii)

8.9	Werner's Coordination Theory	345
8.10	Hydrogen Bond	346
8.11	Valence Bond Theory	349
8.12	Metallic Bond	357

9. POLYMERS — 365–444

9.1	Introduction	365
9.1.1	Classification of Polymers	366
9.1.2	Types of Polymerization	372
9.1.3	Mechanism of Chain Polymerization	373
9.1.4	Serio-specific Polymerization	377
9.1.5	Step Polymerization	378
9.1.6	Polymerizability of a Monomer	378
9.1.7	Thermodynamics of Polymerization Process	378
9.1.8	Practical Methods of Polymerization	379
9.1.9	Molecular Weight of Polymers	381
9.1.10	Engineering and Speciality	383
9.1.11	Electrically Conducting Polymers	384
9.1.12	Photoconductive Polymers	388
9.1.13	Structure Property Relationships in Polymers	390
9.2	Resins and Platics	932
9.2.1	Constituents of Plastics	393
9.2.2	Fabrication of Plastic Articles	394
9.2.3	Thermoplastic Resins	396
9.2.4	Thermoset Resins	407
9.2.5	Low Dielectric Constant Polymers	418
9.2.6	Biopolymers	420
9.2.7	Liquid Crystal Polymers	423
9.3	Rubbers	425
9.3.1	Natural Rubber	429
9.3.2	Synthetic Rubbers	431
9.4	Flow sheet for Producing Scheme Important Polymers	436

10. COMPOSITE MATERIALS — 445–457

10.1	Introduction	445
10.2	Constitution	445
10.3	Classification	448
10.4	(A) Particle-Reinforced Composites	448
10.4	(B) Fibre-Reinforced Composites	448
10.5	Fibre Glass - Reinforced Composites	451
10.6	Other Fibre-Reinforced Composites	451
10.7	Metal Matrix - Fiber Composites	451
10.8	Hybrid Composites	452
10.9	Processing of Fiber - Reinforced Composites	452
10.10	Structural Composites	455
10.11	Applications of Composite Materials	456

21. BATTERIES AND BATTERY TECHNOLOGY 710–752
- 21.1 Introduction — 711
- 21.2 Theoretical Principles — 712
- 21.3 Primary Cells — 717
- 21.4 Secondary Batteries — 729
- 21.5 Reserve Batteries — 735
- 21.6 Fuel Cells — 740
- 21.7 Solar Cells — 752

22. INSTRUMENTAL TECHNIQUES IN CHEMICAL ANALYSIS 753–797
- 22.1. Colorimetry and Visible Spectroseopy — 753
- 22.2. Ultraviolet Spectroscopy — 761
- 22.3. Infrared Spectrophotometry — 768
- 22.4. Chromatography — 778
- 22.5. Nuclear Magnetic Resonance (NMR) Spectroscopy — 788
- 22.6. Flame Photometry — 792
- 22.7. Atomic Absorption Spectrometry — 794

23. GREEN CHEMISTRY FOR CLEAN TECHNOLOGY 798-816
- 23.1 Introduction — 798
- 23.2 Goals of green chemistry — 798
- 23.3 Significance of green chemistry — 799
- 23.4 Basic components of green chemistry research — 799
- 23.5 Atom economy — 808
- 23.6 Functional group approaches to green chemistry — 809
- 23.7 Optimization of frameworks for the design of greener synthetic pathways. — 811
- 23.8 Industrial Applications of Green Chemistry — 812
- 23.9 Conclusion — 815

24. MECHANISM OF ORGANIC REACTIONS 817-855
- 24.1 Introduction — 817
- 24.2 Electron Displacement Effects — 817
- 24.3 Reaction Mechanism — 826
- 24.4 Energy requirements of a reaction — 835
- 24.5 Types of organic reactions and mechanism — 838
- 24.6 Mechanism of some reactions — 850

25. REACTION DYNAMICS & CATALYSIS 856-924
- 25.1 Introduction — 856
- 25.2 Rate of a reaction or reaction velocity — 856
- 25.3 Reaction rate and time — 857
- 25.4 Factors influencing the reaction rate — 857
- 25.5 Rate law (or rate equation) and rate constant — 858
- 25.6 Measurement of rate of reaction — 859
- 25.7 Order of a reaction — 859
- 25.8 Zero order reaction
- 25.9 Molecularity of a reaction
- 25.10 Pseudo-order reactions

(xvi)

25.11	Integrated rate equations	862
25.12	Reactions involving more than three molecules	874
25.13	Methods for determination of order of a reaction	874
25.14	Complex or simultaneous or composite reactions	877
25.15	Theories of reaction rates	883
25.16	Effect of temperature on rates of reaction Arrhenius equation	887
25.17	Activation energy and catalysis	888
25.18	Examples	889

CATALYSIS

25.19	Introduction	901
25.20	Action of a catalyst	901
25.21	Characteristics of catalytic reactions (or criteria of catalysis)	901
25.22	Types of catalysis	904
25.23	Catalytic promoters	906
25.24	Catalytic poisons	907
25.25	Negative catalysis and inhibition	908
25.26	Autocatalysis	910
25.27	Induced catalysis	911
25.28	Activation energy and catalysis	911
25.29	Theories of catalysis	912
25.30	Acid-base catalysis	918
25.31	Enzyme catalysis	921
25.32	Some industrial processes using catalysts important	922
25.33	Criterial for choosing a catalyst for industrial application	923

26. PHOTOCHEMISTRY   **925-947**

26.1	Introduction	925
26.2	Photochemical reactions	925
26.3	Laws of Photochemistry	926
26.4	Quantum efficiency	927
26.5	High and low quantum yields	931
26.6	Mechanism of some photochemical reactions	931
26.7	Photosynthesis	935
26.8	Types of photochemical reactions	936
26.9	Apparatus for photochemical studies	937
26.10	Applications of photochemistry in technology	938
26.11	Photochemistry of vision	938
26.12	Photosynthesis and Bioenergetics	940

QUESTION BANK	948–962
APPENDIX-1 : El Nino Phenomenon and Its Effects	963–964
APPENDIX-2 : Basic Principles of Green Chemistry	965–965
BIBLIOGRAPHY	966–968
INDEX	969–972

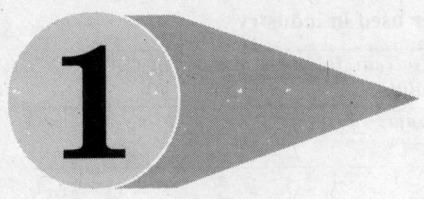

Water Treatment

"Water is one of the most abundant commodities in nature, but is also the most misused one."

1.1. INTRODUCTION

One of the basic necessities of life is water. Living things exist on the earth because this is the only planet that has the presence of water. Water is necessary for the survival of all living things be it plant or animal life.

Water is one of the most abundant commodities in nature but is also the most misused one. Although earth is a blue planet and 80% of its surface is covered by water, the hard fact of life is that about 97% of it is locked in the oceans, and sea which is too saline to drink and for direct use for agricultural or industrial purposes. 2.4 % is trapped in polar ice caps and giant glaciers, from which icebergs break off and slowly melt at sea. >1% water is used by man for various development, industrial, agricultural, steam generation domestic.

1·2. SOURCES OF WATER

Water is required for agricultural, municipal and industrial purposes. For industrial purposes, natural waters may be broadly divided into the following categories:
(1) Surface waters:
   (*a*) Flowing waters *e.g.*, streams and rivers (Moorland surface drainage)
   (*b*) Still waters *e.g.*, ponds, lakes and reservoirs (Lowland surface drainage)
(2) Underground water: Water from shallow and deep springs and wells
(3) Rain water
(4) Estuarine and sea water

From the point of view of industrial applications, it is not usually feasible to use rain water and sea water. Rain water is irregular in supply and generally expensive to collect. Estuarine and sea waters are too saline for most industrial uses except cooling. The three major sources of water for industrial use are
   (*a*) Moorland surface drainage.
   (*b*) Lowland surface drainage.
   (*c*) Deep well water.

The important properties of these three types of waters are given in Table – 1.

Table 1. Analyses of some types of water used in industry

Sr. No.	Type of water	Ionic constituents and their concentration in ppm	pH	Silica, ppm	Dissolved solids ppm	Hardness in terms of $CaCO_3$ in (in ppm) alkaline	non-alkaline	Total
1.	Moorland surface drainage	Na^+ — 8 Ca^{2+} — 7 Mg^{2+} — 6 HCO_3^- — 15 Cl^- — 10 SO_4^{2-} — 31 NO_3^- — Traces	6.7	8	77	12	30	42
2.	Low-land surface drainage	Na^+ — 25 Ca^2 — 63 Mg^{2+} — 18 HCO_3^- — 160 Cl^- — 43 SO_4^{2-} — 80 NO_3^- — 15	7.5	10	333	130	100	230
3.	Deep well water	Na^+ — 74 Ca^2 — 48 Mg^{2+} — 20 HCO_3^- — 350 Cl^- — 48 SO_4^{2-} — 10 NO_3^- — Traces	7.0	15	388	203	0	203

(a) **Moorland surface drainage**

Water from this source is fairly constant in composition. It is generally clear and cloured brown. It is slightly acidic due to the presence of dissolved carbon dioxide and of weak organic acids, which renders it corrosive. Although its hardness is low, it can cause scale formation in boilers unless it is suitably treated before use. It contains some strains of iron bacteria which must be removed by chlorination to prevent their deposition in the pipe-lines. It possesses a tendency to dissolve lead and copper. This fact should be considered if the water is used for drinking purposes.

(b) **Lowland surface drainage**

Water from this source vary widely in composition from place to place. It is not generally coloured but may contain fine mud in suspension, which does not easily settle unless with the help of coagulants. Its hardness is usually high and hence can cause serious scale formation in boilers, economizers and coolers, unless the water is properly treated before use. When the water is heated in boilers, the CO_2 produced from the bicarbonate ions passes off with the steam and dissolves in the condensate, forming carbonic acid which is corrosive to the mild steel. River and canal waters may get contaminated by sewage and industrial wastes, which may require preliminary treatment prior to softening.

(c) **Deep well waters**

This type of water is fairly constant in composition, unless contaminated by other waters percolating through faults in the surrounding strata. When freshly drawn, this is usually colourless, clear and devoid of finely divided suspended matter and hence sparkling. This type of water may develop a brown opalescence on exposure to air due to the presence of small amounts of ferrous iron, which gets converted into hydrated ferric oxide. Traces of manganese as well as H_2S may also be present.

In many deep well waters, the concentration of the bicarbonate is more than equivalent to the combined concentrations of Ca^{2+} and Mg^{2+} ions so that Na_2CO_3 may be considered to be

WATER TREATMENT

present (*vide* Table-1). The hardness is then entirely alkaline hardness. The sulphate content is often very low.

Water from shallow wells possess a composition similar to that of low-land surface drainage waters. The concentration of bicarbonate ions is less than the combined concentrations of Ca^{2+} and Mg^{2+} ions. Hence, the water contains non-alkaline hardness.

Some deep well waters contain considerable amounts of free CO_2. Because of the hardness present and because of the carbonate ions produced due to thermal decomposition of bicarbonate ions, deep well waters, like low-land waters, give rise to severe scale formation in boilers. The high silica content also contributes to the formation of hard scales in boilers. CO_2 formed by the decomposition of bicarbonate ions, similar to lowland waters, also results in the production of acid condensate leading to corrosion.

1·3. EFFECT OF WATER ON ROCKS AND MINERALS

1. **Dissolution.** Some mineral constituents of rocks such as NaCl and $CaSO_4 \cdot 2H_2O$ readily dissolve in water.

2. **Hydration.** Some minerals are easily hydrated with the consequent increase in volume leading to disintegration of the rocks in which these minerals are present. Example are:

(a) $CaSO_4 \xrightarrow{\text{Hydration}} CaSO_4 \cdot 2H_2O$ (accompanied by an expansion of 33%)
 anhydrate   gypsum

(b) $Mg_2SiO_4 \xrightarrow{\text{Hydration}} Mg_2SiO_4 \cdot XH_2O$
 Olivine   Serpentine

3. **Effect of dissolved oxygen.** This leads to oxidation and hydration

$Fe_3O_4 \xrightarrow{\text{oxidation}} Fe_2O_3 \xrightarrow{\text{hydration}} 3Fe_2O_3 \cdot 2H_2O$
Magnatite   haematite   Limonite

$2FeS_2 + 7O_2 + 2H_2O \longrightarrow 2FeSO_4 + 2H_2SO_4$
Marcesite

4. **Effect of dissolved CO_2**

(a) Water containing dissolved CO_2 convert the insoluble carbonates of Ca, Mg and Fe into their relatively soluble bicarbonates

$CaCO_{3(s)} + H_2CO_3 \longrightarrow Ca(HCO_3)_2$
$MgCO_{3(s)} + H_2CO_3 \longrightarrow Mg(HCO_3)_2$

(b) Rock forming minerals like silicates and alumino-silicates of Na, K, Ca and Fe are attacked by CO_2, producing soluble carbonates, bicarbonates and silica

$K_2O \cdot Al_2O_3 \cdot 6SiO_2 + 2H_2O + CO_2 \longrightarrow Al_2O_3 \cdot 2SiO_2 \cdot 2H_2O + K_2CO_3 + 4SiO_2$
orthoclase   kaolin

(c) Rocks containing felspar disintegrate and charge nearby river water with dissolved salts, fine clay and silica in suspension.

Thus, water collects impurities from the ground, rocks or soil with which it comes into contact. Contamination of water may also result from sewage or industrial wastes, either by actual contact when these are allowed to flow into the running water, or by percolation through the ground.

1·4. TYPES OF IMPURITIES PRESENT IN WATER

The impurities present in natural waters may be broadly classified as follows:

(1) **Dissolved impurities**

(a) Inorganic salts *e.g.*,
 (i) Cations : Ca^{2+}, Mg^{2+}, Na^+, K^+, Fe^{2+}, Al^{3+} and sometimes traces of Zn^{2+} and Cu^{2+}
 (ii) Anions : Cl^-, SO_4^{2-}, NO_3^-, HCO_3^-, and sometimes F^- and NO_2^-

(b) Gases e.g., CO_2, O_2, N_2, oxides of N_2 and sometimes NH_3, H_2S
(c) Organic salts

(2) **Suspended impurities**
(a) Inorganic e.g., clay and sand
(b) Organic e.g., oil globules, vegetable and animal matter.

Finely divided clay and silica, aluminium hydroxide, ferric hydroxide, organic waste products, humic acids, colouring matter, complex protein, aminoacids, which are generally classified as albunoid ammonia.

(3) **Colloidal impurities**
Clay and finly divided silica colloidal pertical of 10^{-4} - 10^{-6} mm size.

(4) **Micro-organisms**
Bacterias, fungi, algae other micro-organisms and other forms of animal and vegetable life.

1·5. EFFECTS OF IMPURITIES IN NATURAL WATERS

The various types of impurities present in natural waters impart some properties on the waters. From the point of view of industrial use, the characteristics and effects of the impurities on the water quality are discussed under the following headings:
(1) Colour
(2) Tastes and odours
(3) Turbidity and sediment
(4) Micro-organisms
(5) Dissolved mineral matter:
   (a) hardness,
   (b) alkalinity,
   (c) total solids, and
   (d) corrosion
(6) Dissolved gases
(7) Silica content and
(8) Oxidability.

1·5. (1) Colour. Colour is found mostly in surface waters, although water from some shallow wells, springs and deep wells may also be occasionally coloured. The colours of natural waters range from yellowish-brown to dark brown. The colours and the materials which cause it are often objectionable in which the water and the manufactured product come into contact, e.g., dyeing, scouring and laundering.

The colour of natural waters is mainly due to the presence of dissolved or colloidally dispersed organic matter. The measurement of colour is usually made with a tintometer and the result is expressed in "Hazen units" or "Standard units of colour". In determining the colour of water, it is the true colour as expressed in the standard units, that is of interest and not the apparent colour.

It is known that solutions of potassium chloroplatinate tinted with small amounts of cobalt chloride give colours similar to the colours of natural waters. The colour produced by 1 mg/litre (1 ppm) of platinum (used as K_2PtCl_6) is taken as the standard unit of colour. The colour determination is done only after the removal of suspended matter by centrifugation.

The colour standards are usually prepared by dissolving 1.2545 g of K_2PtCl_6 (containing 0.5 g of Pt) and 1 g of crystallized cobaltious chloride ($CoCl_2 . 6H_2O$) containing about 0.248 g of cobalt in water with 100 ml of concentrated HCl and diluting to 1 liter with distilled water. This solution is deemed to possess 500 units of colour, as per the American Public Health Association's books of standard methods of water analysis.

WATER TREATMENT

Removal or reduction of colour and organic matter is generally accomplished by coagulation, settling, adsorption, filtration and sometimes super chlorination.

1·5. (2) Tastes and odours. Most of the odours in natural waters, with the exception of H_2S, and iron are organic in nature. The odours and tastes observed in chlorinated waters are due to compounds formed by the reaction of chlorine on traces of organic matter present in the water. These organic tastes and odours are usually confined to surface waters and are either very low or totally absent in deep well waters.

Disagreeable odours and tastes are objectionable for various industrial processes such as beverages, food products, paper, pulp and textiles. Organic tastes and odours may be removed by means of activated carbon, aeration, or aeration followed by activated carbon treatment. The removal of inorganic odours and tastes due to H_2S or iron will have to be removed by chemical methods like oxidation, chlorination or precipitation.

1·5. (3) Turbidity and sediment. "Turbidity" is imparted to natural waters due to the presence of finely divided, insoluble impurities which remain suspended in water and reduce its clarity. These suspended impurities may be inorganic in nature (*e.g.,* clay, silt, silica, ferric hydroxide, calcium carbonate, sulphur, etc.) or organic (*e.g.,* finely divided vegetable or animal matter, oils, fats, greases, micro-organisms etc.).

Since the different materials that causes turbidity in natural water, an arbitary standard is used 1 mg SiO_2/L = 1 unit of turbidity. Standard suspensions of pure silica are used for measuring turbidity. However, the actual standards of turbidity are determined from the depths of liquid in a standard Jackson candle turbiditimeter, in which the flame of a candle disappears when viewed lengthwise through the tube. For instance, the depths of 72.9 cm, 21.5 cm, 4.5 cm and 2.3 cm correspond to the turbidities of 25 ppm, 100 ppm, 500 ppm and 1000 ppm respectively.

Nowadays, turbidity is conveniently measured with the help of more reliable, sensitive instrument photometers. A beam of light from a source produced by a standardized electric bulb is passed through a sample. The light emerging from the sample is then directed through a photometer which measures the light absorbed. The reading on the meter is calibrated in terms of turbidity. Formazin polymer suspension has now mostly replaced silica suspension as the standard because it provides more reproducible results. Therefore, the turbidity is now also expressed as *"formazin turbidity units"* (*FTU*).

Another method used for measurement of turbidity is by light scattering. The light falling on the sample is scattered because of the turbidity present. The scattered light is then measured by putting a photometer at right angles to the path of the incident light generated by the light source. This technique of measurement of light scattered at an angle of 90° is called *"nephelometry"*. Accordingly, the unit of turbidity is *"nephelometric turbidity unit"* (*NTU*).

Tolerances of turbidity for different industries depend upon the type of industry, the grade of the product being manufactured, the nature of the turbidity present and the type of wet processing being practised. Suspended silt and mud which cause turbidity may be objectionable in boilers and in cooling-water systems. Colloidal or dissolved organic matter causing turbidity may interfere with water-softening processes. For example, the Zeolites and cation-exchangers used in water-softening processes may be coated with coagulated organic or suspended matter which leads to reduction in their efficiency.

Turbidity of water may be removed by sedimentation, followed by (*a*) coagulation and filtration, (*b*) coagulation and settling, or (*c*) coagulation, settling and filtration.

1·5. (4) Micro-organisms. Micro-oganisms are more abundant in surface waters (since these come into contact with air, soil and vegetation in which the organisms originally existed), whereas in deep well waters, the bacterial count is often low or even absent. The growth of these organisms in water used for industrial purposes may cause serious problems and hence effective measures have to be taken to prevent the growth of these organisms. Organic growths generally take place most readily in water at temperatures ranging from 10°C — 35°C. Many of them form coatings in pipe lines, thus reducing their carrying capacity considerably. These coatings frequently break loose in large masses which may completely block the flow through valves, pumps, nozzles and other parts of the water distribution systems. In filters and water softeners employing granular media, the granules may become matted together by such organic growths, thus impairing their operation by lowering their flow rates and also resulting in channeling and overturning of the beds.

The commonest types of living organisms that are important from the point of view of treatment are algae, fungi and bacteria, all of which often form slime with consequent fouling and corrosion. The slime surrounding the organisms causes them to adhere to metal surfaces. This leads to difficulties such as reduced heat transfers and tube blockages. The growth of marine organisms, particularly mussels, in sea-water systems may lead to serious reduction in the carrying capacity of the pipe lines.

Control of algal, fungal and bacterial growths is usually achieved by chlorination. Water soluble solid sterilizing agents such as $CuSO_4$, sodium pentachlorophenate and organic mercurials are used in special circumstances. The growth of mussels may be prevented by chlorination.

Algae and other chlorophyl-containing plant need sunlight for their growth. Hence, the growth of these organisms can be prevented by storing the water in covered reservoirs. In many industrial plants, the organic growths are removed in the settling basin of the water-treatment plant. This is usually done with chlorine and coagulation, settling and filtration, to remove the remains. The use of chlorine in this way is called pre-chlorination, which is helpful in reducing the dosages of coagulant required. In many other cases, the bulk of the organic matter is first removed by coagulation, settling and filtration followed by chlorination. This method is called post-chlorination. Sometimes, both pre-chlorination and post-chlorination are used.

Iron and manganese bacterial growths, known as "Crenothrix" are best prevented by removal of these metals, followed by chlorination. In the case of sulphur waters, the H_2S should be first removed followed by chlorination to remove the last traces of H_2S and to kill any sulphate-reducing bacteria which may be present.

1·5. (5) Dissolved mineral matter. For most of the industrial uses, only the following mineral constituents are usually determined: Ca, Mg, Na, K, bicarbonate, carbonate, hydroxide, chloride, sulphate, nitrate, fluoride, silica, Fe, Mn and mineral acid. The most important quality of the dissolved mineral matter from the point of industrial application include hardness and alkalinity.

Hardness. Hardness was originally defined as the soap consuming capacity of a water sample. Soaps generally consists of the sodium salts of long-chain fatty acids such as oleic acid, palmetic acid and stearic acid. The soap consuming capacity of water is mainly due to the presence of calcium and magnesium ions. These ions react with the sodium salts of long-chain fatty acids present in the soap to form insoluble scums of calcium and magnesium soaps which do not possess any detergent value (cleaning tendency).

$$2\,C_{17}H_{35}COONa + CaCl_2 \longrightarrow (C_{17}H_{35}COO)_2Ca \downarrow + 2\,NaCl$$
Soap (soluble)   Calcium soap (insoluble ppt.)

Other metal ions like Fe^{2+}, Mn^{2+} and Al^{3+} also react with the soap in the same fashion, thus contributing to hardness but generally, these are present in natural waters only in traces. Further, acids such as carbonic acid can also cause free fatty acid to separate from soap solution and thus contribute to hardness. However, in practice, the hardness of a water sample is usually taken as a measure of its Ca^{2+} and Mg^{2+} content.

WATER TREATMENT

Temporary and permanent hardness

When natural water is boiled, the bicarbonate ions present are decomposed to form carbonate ions and carbon dioxide is set free. The hardness so precipitated was referred to as "temporary hardness", but this term is now referred to all the hardness associated with the bicarbonate content of the water (*i.e.,* that determinable by titration with acid). This is due to the fact that $CaCO_3$ and more particularly $MgCO_3$ have appreciable, though slight, solubility in water.

$$Ca(HCO_3)_2 \xrightarrow{\Delta} CaCO_3 \downarrow + H_2O + CO_2 \uparrow$$

$$Mg(HCO_3)_2 \xrightarrow{\Delta} \underset{\text{(Insoluble)}}{Mg(OH)_2} \downarrow + H_2O + CO_2 \uparrow$$

The difference between the temporary and total hardness is referred to as "permanent hardness", since this is not removed by boiling the water. The permanent hardness is regarded as comprising of the dissolved chlorides, sulphates and nitrates of calcium and magnesium.

Alkaline and non-alkaline hardness

The terms, temporary and permanent hardness, are gradually being replaced by the preferred terms alkaline and non-alkaline hardness.

"Alkaline hardness" is defined as the hardness due to the bicarbonates, carbonates and hydroxides of the hardness-producing metals. It is also called "carbonate hardness".

In a raw water, the alkaline hardness is almost always the hardness associated with the bicarbonates. However, a treated or boiler water may also contain hardness due to small quantities of $CaCO_3$ and $Mg(OH)_2$ in solution. The alkalinity as measured by titration with mineral acid using methyl orange as indicator is equal to the sum of the concentrations of the bicarbonates, carbonate and hydroxide expressed in equivalents. If this alkalinity is less than the total hardness also expressed in equivalents, then the alkaline hardness is equal to the alkalinity. Conversely, when the alkalinity to methyl orange is equal to or greater than the total hardness, the alkaline hardness is equal to the total hardness".

The "non-alkaline hardness" is obtained by subtracting the "alkaline hardness" from the "total hardness". This is also known as "non-carbonate hardness".

Estimation of hardness

Hardness is usually determined by the following two methods:

(1) Soap solution method

Soluble soaps consist of sodium or potassium salts of higher fatty acids, such as oleic acid, stearic acid and palmitic acid. These soaps give lather with hard water only after sufficient quantity of the soap is added to precipitate all the hardness causing metal ions present in water.

$$\underset{\substack{\text{Calcium or magnesium}\\\text{bicarbonate}}}{Ca \text{ or } Mg\,(HCO_3)_2} + \underset{\substack{\text{Sodium stearate}\\\text{(soluble soap with}\\\text{detergent value)}}}{2C_{17}H_{35}COONa} \longrightarrow \underset{\substack{\text{Ca or Mg stearate}\\\text{(insoluble soap with}\\\text{no detergent value)}}}{(C_{17}H_{35}COO)_2\,Ca \text{ or } Mg} + 2\,NaHCO_3$$

$$CaCl_2 \text{ or } MgCl_2 + 2C_{17}H_{35}COONa \longrightarrow (C_{17}H_{35}COO)_2 Ca \text{ or } Mg + 2NaCl$$

$$CaSO_4 \text{ or } MgSO_4 + 2C_{17}H_{35}COONa \longrightarrow (C_{17}H_{35}COO)_2 Ca \text{ or } Mg + Na_2SO_4.$$

Thus, after precipitation of all the hardness causing metal ions present in the hard water sample further addition of soap gives lather.

The total hardness of a water sample can be determined by litrating an aliquot of the sample against a standard soap solution in alcohol. The appearance of a stable lather persisting even after shaking for about 2 minutes marks end-point. If the water sample is boiled for 30 minutes to remove the temporary hardness and then is titrated with the standard soap solution as described above, the titre value corresponds to the permanent hardness of the sample. The difference between the two measurements correspond to the temporary hardness.

(2) EDTA method

This method gives more accurate results than the soap solution method.

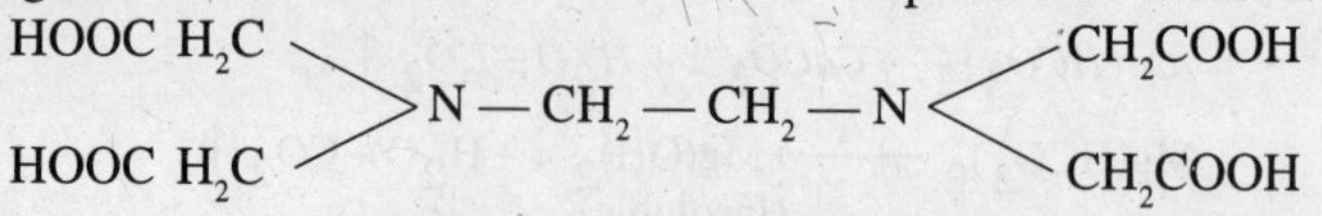

$$\begin{array}{c} HOOC\ H_2C \\ HOOC\ H_2C \end{array} \!\!\!>\!\! N-CH_2-CH_2-N \!\!<\!\!\! \begin{array}{c} CH_2COOH \\ CH_2COOH \end{array}$$

Fig. 1.1. Structure of EDTA

EDTA can be represented by H_4Y. H_4Y from is not used because of its limited solubility, again Na_4Y form also not used because of its extensive hydrolysis in solution which make the solution highly alkaline Na_2H_2Y form is mostly used in analytical work since it can be obtained in high state of purity.

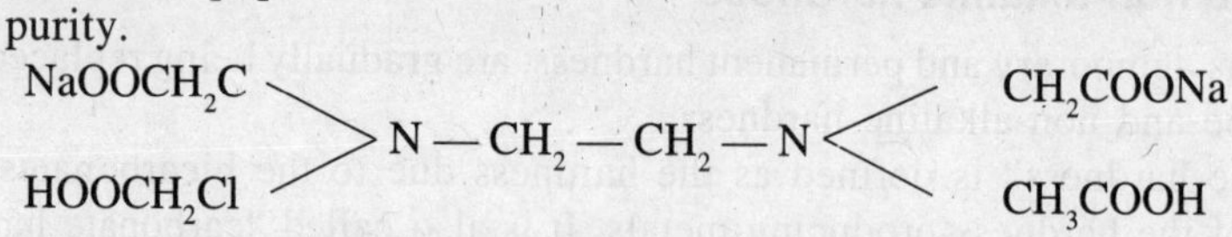

$$\begin{array}{c} NaOOCH_2C \\ HOOCH_2Cl \end{array} \!\!\!>\!\! N-CH_2-CH_2-N \!\!<\!\!\! \begin{array}{c} CH_2COONa \\ CH_3COOH \end{array}$$

Ethylene diamine tetra acetic acid (EDTA) (Fig. 1.1) forms complexes with Ca^{2+} and Mg^{2+}, as well as with many other metal cations, in aqueous solution. These complexes have the general formula given in Fig. 1.2 below:

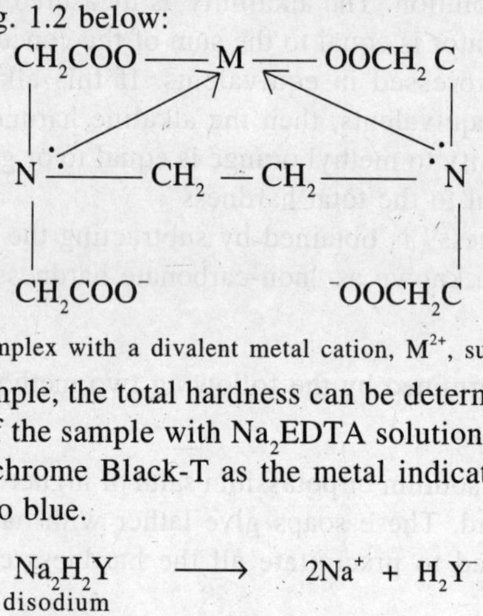

$$\begin{array}{c} CH_2COO\!\!\!-\!\!\!-M\!\!\!-\!\!\!-OOCH_2C \\ | \hspace{2cm} | \\ N\!\!\!-\!\!\!-CH_2-CH_2\!\!\!-\!\!\!-N \\ | \hspace{2cm} | \\ CH_2COO^- \hspace{1cm} {}^-OOCH_2C \end{array}$$

Fig. 1.2. EDTA complex with a divalent metal cation, M^{2+}, such as Ca^{2+}/Mg^{2+}, etc.

Thus in a hard water sample, the total hardness can be determined by titrating the Ca^{2+} and Mg^{2+} present in an aliquot of the sample with Na_2EDTA solution, using NH_4Cl-NH_4OH buffer solution of pH 10 and Eriochrome Black-T as the metal indicator. The colour change at the end-point is from wine red to blue.

$$Na_2H_2Y \longrightarrow 2Na^+ + H_2Y^{--}$$
disodium
EDTA
– solution

$$Mg^{2+} + HD^{2-} \longrightarrow MgD^- + H^+$$
(Indicator) \hspace{1cm} (Metal-indicator complex)
blue \hspace{3cm} wine red

$$(Mg\ or\ Ca)\ D^- + H_2Y^{--} \longrightarrow (Mg\ or\ Ca)\ Y^{--} + HD^{--} + H^+$$
Metal indicator \hspace{0.5cm} From \hspace{1cm} metal-EDTA \hspace{1cm} Free
complex (wine \hspace{0.3cm} disodium \hspace{0.8cm} complex \hspace{1.2cm} indicator
red colour) \hspace{0.8cm} EDTA \hspace{1cm} (colourless) \hspace{1cm} (blue colour)
\hspace{2.5cm} solution

*****Note.** For details regarding the various methods of determining different types of hardness, the students may refer to the Textbook on "Experiments and Calculations in Engineering Chemistry" by S.S. DARA.

WATER TREATMENT

Permanent hardness can be determined by precipitating the temporary hardness by prolonged boiling for about 30 minutes followed by titration with the Na_2EDTA solution as above. The difference in the titre values corresponds to the temporary hardness of the water sample.*

UNITS OF HARDNESS

(1) **Parts per million (ppm).** One part per million (ppm) is a unit weight of solute per million unit weights of solution. In dilute solutions of density ≈ 1, 1 ppm = 1 mg/liter. It is customary to express hardness in terms of equivalents of $CaCO_3$. Hence, all the hardness causing impurities are first converted in terms of their respective weights equivalent to $CaCO_3$ and the sum total of the same is expressed in parts per million.

$$\left. \begin{array}{l} \text{Equivalent of} \\ CaCO_3 \text{ for a hardness} \\ \text{causing substance} \end{array} \right\} = \frac{\text{weight of the substance causing hardness} \times 50}{\text{Equivalent weight of the substance causing hardness}}$$

(Since chemical equivalent weight of $CaCO_3$ = 50)

For instance, 136 parts by weight of $CaSO_4$ would react with the same amount of soap as 100 parts by weight of $CaCO_3$ (i.e., 2 equivalents of $CaCO_3$). Hence, in order to convert the weight of $CaSO_4$ present as its $CaCO_3$ equivalent, the weight of $CaSO_4$ should be multiplied by a factor of $\frac{100}{136}$ or $\frac{50}{68}$. Similarly, if 'a' gms of $CaCl_2$, 'b' gms of $MgSO_4$, 'c' gms of $MgCl_2$ 'd' gms of $Ca(HCO_3)_2$ and 'e' gms of $Mg(HCO_3)_2$ are present in a hard water sample, each of them can be converted in terms of their weight equivalent of O_3 by multiplying with $\frac{100}{111}, \frac{100}{120}, \frac{100}{95}, \frac{100}{162}$ and $\frac{100}{146}$ respectively. The sum total of the values so obtained expressed as parts per million represents the hardness of the water sample under consideration.

$$\begin{aligned} 100 \text{ g } CaCO_3 &\equiv 111 \text{ g } CaCl_2 \\ &\equiv 120 \text{ g } MgSO_4 \\ &\equiv 95 \text{ g } MgCl_2 \\ &\equiv 162 \text{ g } Ca(HCO_3)_2 \\ &\equiv 146 \text{ g } Mg(HCO_3)_2 \\ &\equiv 148 \text{ g } Mg(NO_3)_2 \\ &\equiv 44 \text{ g } CO_2 \\ &\equiv 136 \text{ g } CaSO_4 \end{aligned}$$

(2) **Equivalents per million (epm).** One equivalent per million is a unit chemical equivalent weight of solute per million weight units of solution. In dilute solutions of density not differing very much from unity, 1 epm = 1 milligram equivalent per litre; and in titrimetry, 1 epm is conventionally taken as equal to 1 ml of 1 N solution per litre.

Thus,
$$\begin{aligned} 1 \text{ epm of Mg} &\equiv 12 \text{ ppm of Mg} \\ &\equiv 50 \text{ ppm of } CaCO_3 \\ &\equiv 42 \text{ ppm of } MgCO_3 \\ &\equiv 73 \text{ ppm } Mg(HCO_3)_2 \\ &\equiv 81 \text{ ppm } Ca(HCO_3)_2 \\ &\equiv 68 \text{ ppm } CaSO_4 \\ &\equiv 47.5 \text{ ppm } MgCl_2 \\ &\equiv 55.5 \text{ ppm } CaCl_2 \\ &\equiv 60 \text{ ppm } MgSO_4 \text{ and so on.} \end{aligned}$$

It should be noted that for any dissolved substance, a concentration of 1 epm is equal to "50 ppm as $CaCO_3$".

(3) **Grains per imperial gallon (gpg).** In English system, hardness is expressed in terms of grains $\left(1 \text{ grain} = \dfrac{1}{7{,}000} \text{ lb}\right)$ per gallon (10 lbs); *i.e.* parts per 70,000 parts 1 grain per gallon is also called as degree Clark. Thus, 9 degrees Clark means that 9 grains in terms of $CaCO_3$ are present per gallon of water; or 9 parts are present per 70,000 parts of water. On ppm scale, it means that $9 \times \dfrac{10^6}{70{,}000} = 128.57$ ppm of hardness (as $CaCO_3$) is present in the water sample.

Inter-relationship between various units of hardness

$1 \text{ p.p.m.} = 1 \text{ mg/l} = 0.1°$ French $= 0.07°$ Clark
$\phantom{1 \text{ p.p.m.}} = 0.07$ grains per imperial gallon
$\phantom{1 \text{ p.p.m.}} = 0.0583$ grains per U.S. gallon
$\phantom{1 \text{ p.p.m.}} = 0.02$ epm as $CaCO_3$

$1°$ Clark $= 14.3$ ppm $= 1.43°$ French
$\phantom{1° \text{ Clark}} = 1$ grain per imperial gallon
$\phantom{1° \text{ Clark}} = 0.833$ grains per U.S. gallon

1 grain per U.S. gallon $= 17.1$ ppm $= 17.1$ mg/l
$\phantom{1 \text{ grain per U.S. gallon}} = 1.2$ grains per imperial gallon

$1°$ Fr $= 10$ ppm $= 10$ mg/l $= 0.7°$ Clark
$1°$ Russian $= 1$ part $Ca/10^6$ parts of water
$1°$ German $= 1$ part $Ca/10^5$ parts water.

10 ppm as CaO or 17.9 ppm as $CaCO_3$.

Water containing less than 150 ppm of hardness are classified generally as "good", those containing 150 to 300 ppm as fair and those exceeding 300 ppm as "bad" as per Bureau of Indian standard BIS.

Example 1. *Calculate the temporary and permanent hardness of a water sample, having the following analysis:*

$$Mg(HCO_3)_2 - 73 \text{ mg/l}$$
$$Ca(HCO_3)_2 - 162 \text{ mg/l}$$
$$CaSO_4 - 136 \text{ mg/l}$$
$$MgCl_2 - 95 \text{ mg/l}$$
$$CaCl_2 - 111 \text{ mg/l}$$
$$NaCl - 100 \text{ mg/l}$$

Solution.

Salt	$CaCO_3$ equivalent
$Mg(HCO_3)_2$	$73 \times \dfrac{100}{146} = 50$ mg/l
$Ca(HCO_3)_2$	$162 \times \dfrac{100}{162} = 50$ mg/l
$CaSO_4$	$136 \times \dfrac{100}{136} = 100$ mg/l
$MgCl_2$	$95 \times \dfrac{100}{95} = 100$ mg/l
$CaCl_2$	$111 \times \dfrac{100}{111} = 100$ mg/l
$NaCl$	Does not contribute to hardness and hence ignored.

WATER TREATMENT

$$\text{Temporary hardness} \equiv [Mg(HCO_3)_2] + [Ca(HCO_3)_2]$$
$$= 50 \text{ mg/l} + 100 \text{ mg/l}$$
$$= 150 \text{ mg/l or } 150 \text{ ppm}$$
$$= 150 \times 0.07° \text{ Clark} = \mathbf{10.5° \text{ Clark}}$$

$$\text{Permanent hardness} \equiv [CaSO_4] + [MgCl_2] + [CaCl_2]$$
$$= 100 \text{ mg/l} + 100 \text{ mg/l} + 100 \text{ mg/l}$$
$$= 300 \text{ mg/l or } 300 \text{ ppm}$$
$$= 300 \times 0.07° \text{ Clark} = \mathbf{21° \text{ Clark}}$$

$$\therefore \text{Total hardness} = 150 + 300 = 450 \text{ ppm}$$
$$= 450 \times 0.07° \text{ Clark}$$
$$= \mathbf{31.5° \text{ Clark}}$$

Alternative method of calculation

Equivalent weights of the different salts involved are as follows:

Salt		Equivalent weight
$Mg(HCO_3)_2$	—	$\frac{146}{2} = 73$
$Ca(HCO_3)_2$	—	$\frac{162}{2} = 81$
$CaSO_4$	—	$\frac{136}{2} = 68$
$MgCl_2$	—	$\frac{95}{2} = 47.5$
$CaCl_2$	—	$\frac{111}{2} = 55.5$
$CaCO_3$	—	$\frac{100}{2} = 50$

Also,

$$\text{Equivalents per million, e.p.m.} = \frac{\text{Weight in ppm or mg/l}}{\text{Equivalent weight}}$$

Now,

Salt		Equivalents per million
$Mg(HCO_3)_2$	—	$\frac{73}{73} = 1 \text{ epm}$
$Ca(HCO_3)_2$	—	$\frac{162}{81} = 2 \text{ epm}$
$CaSO_4$	—	$\frac{136}{68} = 2 \text{ epm}$
$MgCl_2$	—	$\frac{95}{47.5} = 2 \text{ epm}$
$CaCl_2$	—	$\frac{111}{55.5} = 2 \text{ epm}$
$NaCl$	—	Does not contribute to hardness and hence ignored.

Temporary hardness
$$= 50 \times \{[Mg(HCO_3)_2] + [Ca(HCO_3)_2]\}$$
$$= 50 \times \{1 + 2\} = \mathbf{150 \text{ ppm as } CaCO_3}$$

	$= 150 \times 0.07°$ Clark
	$= \mathbf{10.5°\ Clark}$
	(since 1 epm of each salt 50 ppm of $CaCO_3$)
Permanent hardness	$= 50 \times \{[CaSO_4] + [MgCl_2] + [CaCl_2]\}$
	$= 50 \times \{2+2+2\} = \mathbf{300\ ppm\ as\ CaCO_3}$
	$= 300 \times 0.07°$ Clark
	$= \mathbf{21°\ Clark}$
Total hardness	$=$ Temporary hardness + permanent hardness
	$= 150\ \text{ppm} + 300\ \text{ppm} = \mathbf{450\ ppm}$
	$= 450 \times 0.07°\ \text{Clark} = \mathbf{31.5°\ Clark.}$

Alkalinity. By alkalinity of water we mean the total content of those substances in water that cause an increased concentration of OH⁻ ions upon dissociation or due to hydrolysis. The alkalinity of natural waters is generally due to the presence in them of HCO_3^-, SiO_3^{2-}, $HSiO_3^-$, and sometimes CO_3^{2-} ions and also due to the presence of salts of some weak organic acids, known as humates, that bind H⁺ ions as a result of hydrolysis, thereby increasing the concentration of OH⁻ ions. In addition to the above, the alkalinity of boiler water is also conditioned by the presence of PO_4^{3-} and OH⁻ ions. Also, the presence of salts of weak acids such as silicates and borates induces buffer capacity in water and resists the lowering of pH. Surface waters containing algae and also water treated by lime-soda process may contain considerable quantities of alkalinity due to CO_3^{2-} and OH⁻.

Depending on the anion that is present in water (HCO_3^-, CO_3^{2-} or OH⁻), alkalinity is classified respectively as bicarbonate alkalinity, carbonate alkalinity or hydroxide alkalinity. The maximum contraminant level for alkalinity is 200 ppm for domastic purpose as per BIS.

Highly alkaline waters may lead to caustic embrittlement and also may cause deposition of precipitates and sludges in boiler tubes and pipes.

With respect to the constituents causing alkalinity in natural waters, the following situations may arise :

1. Hydroxides only
2. Carbonates only
3. Bicarbonates only
4. Hydroxides and carbonates
5. Carbonates and bicarbonates

(**Notes.** The possibility of hydroxides and bicarbonates existing together is ruled out because of the fact that they combine with each other as follows forming the carbonates:

$$OH^- + HCO_3^- = CO_3^- + H_2O$$

The types and extent of alkalinity present in a water sample may be conveniently determined by titrating an aliquot of the sample with a standard acid to phenolphthalein end-point, P, and continuing the titration to methyl orange end-point M. The reactions taking place may be represented by the following equations:

$$OH^- + H^+ \longrightarrow H_2O \qquad \ldots(1)$$

$$CO_3^{--} + H^+ \longrightarrow HCO_3^- \qquad \ldots(2)$$

$$HCO_3^- + H^+ \longrightarrow H_2CO_3 \longrightarrow H_2O + CO_2 \qquad \ldots(3)$$

The volume of acid run-down upto phenolphthalein end-point P corresponds to the completion of equations (1) and (2) given above, while the volume of acid run-down after P corresponds to the completion of equation (3). The total amount of acid used from beginning of the experiment, *i.e.*, the methyl orange end-point M, corresponds to the total alkalinity present which represents the completion of reactions (1) to (3).

WATER TREATMENT

The results may be summarized in the following Table 2, from which the amounts of hydroxides, carbonates and bicarbonates present in the water sample may be computed:

Table 2

Results of titrations to phenol-phthalein end point, P, and methyl orange end-point M.	Hydroxide OH^-	Carbonate CO_3^{2-}	Bicarbonate HCO_3^-
P = 0	Nil	Nil	M
P = M	M	Nil	Nil
P = $\frac{1}{2}$ M	Nil	2 P	Nil
P > $\frac{1}{2}$ M	[2 P − M]	2 [M − P]	Nil
P < $\frac{1}{2}$ M	Nil	2 P	[M − 2 P]

Alkalinity is generally expressed as parts per million (ppm) in terms of $CaCO_3$. 1000 ml of N/50 acid solution ≡ 1000 mg of $CaCO_3$. Hence

$$\text{Alkalinity} = \frac{\text{Vol. of N/50 acid} \times 1000}{\text{Vol. of sample taken for titration}} \text{ mg/l or ppm}$$

On the basis of the analysis of water with respect of alkalinity and total hardness, the amounts of carbonate hardness (temporary hardness or alkaline hardness) and non-carbonate hardness (non-alkaline hardness or permanent hardness) present in the water can be determined as follows:

1. If the methyl orange alkalinity of the water equals or exceeds the total hardness, all the hardness is present as carbonate hardness.
2. If the methyl organe alkalinity of water is less than the total hardness, the carbonate hardness equals the alkalinity.
3. The non-carbonate hardness, under conditions in (2) above, is equal to the total hardness minus the methyl organe alkalinity.

Further, on the basis of hardness and alkalinity, natural waters can be subdivided into two groups : Non-alkaline and alkaline. If the hardness is greater than alkalinity, the water is called non-alkaline. If the hardness of the water is lesser than alkalinity, the water is characterised as alkaline. Non-alkaline waters are more frequently encountered in nature, which are characterized by different kinds of hardness as follows:

$$H_t = (H_c + H_{nc}) = (H_{Ca} + H_{Mg})$$

where
H_t = total hardness,
H_c = carbonate hardness,
H_{nc} = non-carbonate hardness,
H_{Ca} = calcium hardness and
H_{Mg} = magnesium hardness.

Example 1. *100 ml of a raw water sample on titration with N/50 H_2SO_4 required 12.4 ml of the acid to phenolphthalein end-point and 15.2 ml of the acid to methyl orange end-point. Determine the type and extent of alkalinity present in the water sample.*

Solution.

P = 12.4 ml of N/50 H_2SO_4
M = 15.2 ml of N/50 H_2SO_4

Since $P > \frac{1}{2} M$, the water sample must contain only OH^- and CO_3^{--} alkalinities and there cannot be any HCO_3^- alkalinity (vide Table 2).

Further,
1. The volume of N/50 H_2SO_4 equivalent to OH^- present in 100 ml of the water sample
$$= [2P - M] = (2 \times 12.4) \text{ ml} - 15.2 \text{ ml}$$
$$= 24.8 \text{ ml} - 15.2 \text{ ml} = 9.6 \text{ ml. and}$$
2. The volume of N/50 H_2SO_4 equivalent to CO_3^{--} present in 100 ml of the water sample
$$= 2[M - P] = 2[15.2 - 12.4] \text{ ml}$$
$$= 2 \times 2.8 \text{ ml} = 5.6 \text{ ml}$$
3. Since the equivalent weight of $CaCO_3 = 50$
1 ml of 1 N $H_2SO_4 \equiv 50$ mg of $CaCO_3$

Alkalinity due to OH^-

Since 1 ml of 1N $H_2SO_4 \equiv 50$ mg $CaCO_3$

$$9.6 \text{ ml of N/50 } H_2SO_4 \equiv 50 \times 9.6 \times \frac{1}{50} \text{ mg } CaCO_3$$

$$= 9.6 \text{ mg } CaCO_3/100 \text{ ml of water sample}$$

∴ Amount of OH^- present in 1 litre of the water sample

$$= 9.6 \times \frac{1000}{100} = 96 \text{ mg/l as } CaCO_3$$

∴ Alkalinity of the water sample due to OH^-
$$= \textbf{96 ppm}$$

Alkalinity due to CO_3^{--}

Similarly,

$$5.6 \text{ ml of N/50 } H_2SO_4 \equiv 50 \times 5.6 \times \frac{1}{50} \text{ mg } CaCO_3$$

$$= 5.6 \text{ mg } CaCO_3/100 \text{ ml of water sample}$$

∴ Amount of CO_3^{--} present in 1 litre of the water sample

$$= 5.6 \times \frac{1000}{100} = 56 \text{ mg/l as } CaCO_3$$

∴ Alkalinity of the water sample due to CO_3^{--}
$$= \textbf{56 ppm}$$

Report

The given water sample contains:
OH^- alkalinity = 96 ppm
CO_3^{--} alkalinity = 56 ppm
Total alkalinity = 152 ppm

Example 2. *A water sample is not alkaline to phenolphthalein. However, 100 ml of the sample, on titration with N/50 HCl, required 16.9 ml to obtain the end-point, using methyl orange as indicator. What are the types and amount of alkalinity present in the sample ?*

Solution.
$$P = 0; \quad M = 16.9 \text{ ml.}$$

Hence, the alkalinity of the water sample is only due to HCO_3^- ions.

∴ Volume of N/50 HCl equivalent to the HCO_3^- present in 100 ml of the water sample = 16.9 ml
Since 1 ml of 1N HCl ≡ 50 mg $CaCO_3$

WATER TREATMENT

$$16.9 \text{ ml of N/50 HCl} \equiv 50 \times 16.9 \times \frac{1}{50} \text{ mg}$$
$$= 16.9 \text{ mg CaCO}_3/100 \text{ ml of water sample}$$

∴ Amount of HCO_3^- present in 1 litre of the water sample

$$\equiv 16.9 \times \frac{1000}{100} = 169 \text{ mg as CaCO}_3$$

∴ Alkalinity ≡ **169 ppm**

Report

The given water sample consists of only HCO_3^- alkalinity and it is 169 p.p.m.

Example 3. *A water sample is alkaline to both phenolphthalein as well as methyl orange. 100 ml of the water sample on titration with N/50 HCl required 4.7 ml of the acid to phenolphthalein end-point. When a few drops of methyl orange are added to the same solution and the litration further continued, the yellow colour of the solution just turned red after addition of another 10.5 ml of the acid solution. Elucidate on the type an extent of alkalinity present in the water sample.*

Solution.
$$P = 4.7 \text{ ml of N/50 HCl}$$
$$M = (4.7 + 10.5) = 15.2 \text{ ml of N/50 HCl}$$

Since $P < \frac{1}{2} M$, the water sample must contain CO_3^{--} alkalinities and HCO_3^- alkalinity only and not OH^- alkalinity.

Further,
(i) The volume of N/50 HCl equivalent to CO_3^{--} present in 100 ml of the water sample ≡ 2 [P] = 2 × 4.7 = 9.4 ml, and
(ii) The volume of N/50 HCl equivalent to HCO_3^- present in 100 ml of the water sample ≡ M − 2P = (15.2 − 9.4) ml = 5.8 ml

Alkalinity due to CO_3^{--}

Since 1 ml of 1N HCl ≡ 50 mg of $CaCO_3$

$$9.4 \text{ ml of N/50 HCl} \equiv 50 \times 9.4 \times \frac{1}{50} = 9.4 \text{ mg CaCO}_3/100 \text{ ml of water sample}$$

∴ Amount of CO_3^{--} present in 1 litre of the water sample

$$= 9.4 \times \frac{1000}{100} = 94 \text{ mg/l} = 94 \text{ ppm as CaCO}_3$$

∴ Alkalinity due to CO_3^{--} = **94 ppm**

Alkalinity due to HCO_3^-

$$5.6 \text{ ml of N/50 HCl} \equiv 50 \times 5.8 \times \frac{1}{50} = 5.8 \text{ mg of CaCO}_3/100 \text{ ml of water sample}$$

∴ Amount of HCO_3^- present in 1 litre of the water sample

$$= 5.8 \times \frac{1000}{100} = 58 \text{ mg/l as CaCO}_3$$

∴ Alkalinity due to HCO_3^- = **58 ppm**

Report

The given water sample contains:
CO_3 alkalinity = 94 ppm

HCO_3^- alkalinity = 58 ppm
Total alkalinity = 152 ppm

Example 4. *100 ml of a water sample, on titration with N/50 H_2SO_4, gave a titre value of 5.8 ml to phenolphthalein end-point and 11.6 ml to methyl orange end-point. Calculate the alkalinity of the water sample in terms of $CaCO_3$ and comment on the type of alkalinity present.*

Solution.

$$P = 5.8 \text{ ml}; \quad M = 11.6 \text{ ml.}$$

Since $P = \frac{1}{2} M$, it means that all the alkalinity present in the water sample is due to CO_3^{--} only: while OH^- and HCO_3^- are absent.

Further, the volume of N/50 H_2SO_4 equivalent to CO_3^{--} present in 100 ml of the water sample

$$= 2P = 2 \times 5.8 = 11.6 \text{ ml}$$

Since 1 ml of 1N HCl ≡ 50 mg of $CaCO_3$

$$11.6 \text{ ml of N/50 HCl} \equiv 50 \times \frac{11.6}{1} \times \frac{1}{50}$$

$$= 11.6 \text{ mg of } CaCO_3/100 \text{ ml of water sample}$$

∴ Strength of CO_3^{--} in terms of $CaCO_3$

$$= 11.6 \times \frac{1000}{100} = \mathbf{116 \text{ mg/l}}$$

$$= 116 \text{ ppm}$$

Report

The alkalinity of the water sample is 116 ppm, which is only due to CO_3^{--}.

Example 5. *100 ml of a water sample, on titration with N/50 H_2SO_4, using phenolphthalein as indicator, gave the end-point when 5.0 ml of acid were run down. Another aliquot of 100 ml of the sample also required 5.0 ml of the acid to obtain methylorange end-point. What is the type of alkalinity present in the sample and what is its magnitude?*

Solution.

$$P = 5.0 \text{ ml}; \quad M = 5.0 \text{ ml.}$$

Since P = M, it is obvious that the water sample contains only hydroxide alkalinity and it is not a natural water sample. Further, since 1 ml of 1N H_2SO_4 ≡ 50 mg of $CaCO_3$

$$5 \text{ ml of N/50 } H_2SO_4 \equiv 50 \times \frac{5}{1} \times \frac{1}{50}$$

$$= 5 \text{ mg of } CaCO_3/100 \text{ ml of water sample}$$

∴ The amount of OH^- present in 1 litre of the water sample

$$= 5 \times \frac{1000}{100} = 50 \text{ mg of } CaCO_3$$

∴ Alkalinity of the water sample = 50 mg/l or **50 ppm**

Chlorides

Chlorides are present in water generally as NaCl, $MgCl_2$ and $CaCl_2$. Although chlorides are not considered as harmful as such, their concentrations over 250 mg/L impart peculiar taste to the water which is objectionable or unacceptable for drinking purposes for most people from aesthetic point of view. Hence the secondary standard for chlorides is 250 mg/L. Further, presence of unusually high concentrations of chloride in water generally indicates pollution from domestic sewage or from industrial wastewaters, presence of chlorides is also undesirable in boiler feed water. Salts like $MgCl_2$ may undergo hydrolysis under the high pressure and temperature prevailing in the boiler, generating hydrochloric acid which causes corrosion of boiler parts.

Sulphates

Sulphates are among the major anions present in natural water. When sulphates are present in excessive amounts in drinking water, they may produce a laxative or cathartic effect on the people consuming such water. The secondary maximum contaminant level (SMCL) for sulphates is 250 mg/L.

Nitrates

Excessive concentrations of nitrates are objectionable particularly for infants. The maximum contaminant level (MCL) for nitrates is 10 mg/L.

In agricultural regions, ground water can have significant concentrations of nitrates from unused fertilizer leaching into the underlying aquifers.

Surface waters can be polluted by nitrates both from discharge of municipal wastewater and from drainage from agricultural lands.

Ingestion of excessive nitrates in drinking water by infants causes a disease known as *"methemoglobinemia"* (infant cyanosis or blue baby syndrome). In the intestines of infants (particularly those below 6 months of age), nitrates can be reduced to nitrites, which are absorbed into the blood, oxidizing the iron present in the blood, thereby resulting in cyanosis which causes a blue colour to the baby. That is why, this condition is known as *"Blue-baby syndrome"*. Infant methemoglobinemia can be readily diagnosed by medical doctors and is treated readily by injecting methylene blue into the infant's blood.

Nitrates can be effectively removed from water with the help of strongly basic anion exchange resins. However, high operating cost and the disposal of huge quantity of waste brine from regeneration of the resin pose major limitations for this treatment process.

Fluorides

Flouride is found in groundwater as a result of dissolution from geologic formations. It is particularly found in ground waters that come into contact with fluoride containing minerals such as fluorspar (CaF_2), fluorapatite [$Ca_{10}F_2(PO_4)_6$], cryolite (Na_3AlF_6) and igneous rocks containing fluosilicates.

Surface waters generally contain much smaller concentrations of fluoride, unless they are contaminated otherwise. Water pollution by fluorides may be caused by the contaminated domestic sewage and the run-off from agricultural lands where phosphatic fertilizers have been used. Phosphatic fertilizers may contain 0.5 to 4% of fluorine by weight as an impurity.

The fluoride concentration found in some water samples is 1.5 to 6 mg/L, and in extreme cases, it may be as high as 16 to 36 mg/L. Unsightly fluorosis may be caused when the fluoride level in drinking water exceeds 4 mg/L.

Optimum fluoride concentrations prescribed in public water supplies generally are in the range of 0.7 to 1.2 mg/L, depending on the annual average of maximum daily air temperature of the place, on which the consumption of the water by the people depends. Beneficial health effects have been observed where the fluoride levels are optimum. However, too low or too high concentrations of fluoride in drinking water are problematic, and such situations may have to be tackled by fluoridation (addition of fluoride) or defluoridation (removal of fluoride) respectively.

Absence or low concentration of fluoride in drinking water causes a high incidence of dental caries, particularly in children. On the other hand, excessive concentration of fluoride in drinking water causes *"fluorosis"* which is manifested in mottling of teeth, discoloration and at times, chipping of teeth.

Drinking water of some cities in U.K. is fluorine-deficient, whereas Nalgonda area of Andhra Pradesh, some villages near Chandrapur in Maharashtra, and some parts of Punjab, Rajasthan, Tamil Nadu, Karnataka and U.P. contain high levels of fluoride in drinking water. A few cases of *"crippling fluorosis"* were reported in China and India at places where the surface well waters contained fluoride levels of about 10 mg/L. This disease is characterized by the development of hyper-mineralisation of skeleton, body outgrowths (exostoses) and calcification of ligaments. Stiff painful joints and immobilization that follows, result in the crippling effects.

Defluoridation of drinking water containing high-fluoride levels can be achieved by
  (a) Precipitation using aluminium salts in alkaline media followed by settling or flotation followed by filtration.
  (b) Using strongly basic anion exchange resin, and
  (c) by adsorption on activated carbon.

Manganese

Manganese is objectionable in water supplies because it imparts brownish colour to laundered goods. It also affects the table of beverages such as coffee and tea by imparting a sort of medicinal taste. The SMCL (Secondary maximum contaminant level) for Mn is 0.05 mg/L.

Lead

Epidemiological, toxicological and chemical studies proved that exposure to lead can affect human health. The blood-forming system, the nervous system and the renal system of the human body are most sensitive to lead. Blood lead levely of 0.8 to 0.1 µg/L can inhibit enzymatic actions. Also, lead can alter physical and mental development in children, interfere with their growth, decrease attention span and hearing and also interfere with heme synthesis. In older people, lead can increase blood pressure. In drinking water, lead pollution may be caused from pipe solders.

Copper

Copper imparts undesirable taste to drinking water. Plumbing used to convey water in the household, water distribution system is the main source of copper in drinking water. In small amounts, it may not be detrimental to human health but at higher concentration it may cause stomach and internal distress. It may also cause Wilson's disease. A type of polyvinyl chloride called PVC is used to replace copper for household plumbing.

Iron

Iron is objectionable in water supplies because it imparts unacceptable brownish colour to laundered goods. It also affects the taste of beverages like tea and coffee. The secondary maximum contaminant level (SMCL) for Fe is 0.3 mg/L.

Arsenic

Arsenic is a metalloid and is also considered along with toxic heavy metals.

Arsenic is a common constituent of earth's crust and is an impurity in metallic ores and industrial materials.

Arsenic is well-known for its toxicity and even a very small dose can result in severe poisoning. Environmental interest in arsenic and its compounds arises from their extreme toxicity to humans on ingestion and inflammation caused by skin contact. In fresh-water, arsenic is normally present in the anionic form as either arsenate (As^{5+}) or arsenite (As^{3+}). Due to its extreme toxicity tentative limit of 0.01 and 0.05 mg/L has been recommended for arsenic in drinking water by WHO (World Health Organisation) and BIS respectively. WHO revised the recommended limit for arsenic in drinking water is 0.01 mg/L in 1993. Therefore, analytical

WATER TREATMENT

methods for detection and determination of arsenic must be capable of analysing it at nanogram levels present in water. One of the most commonly used method for analysis of arsenic depends on the generation of arsine (AsH_3), its reaction with silver diethyl dithiocarbamate (SDDC) which develops a colour and measurement of the colour intensity by spectrophotometry. A pre-concentration step may be needed to measure arsenic present in drinking water at trace levels.

Exposure to arsenic may come from natural sources or industrial sources or from self-administered sources (from arsenic containing insecticides, herbicides, or rodenticides unintentionally by children or deliberately in suicide attempts by adults). Occupational and environmental health problems can result from the exposure of arsenicals. Exposure to arsine gas is an environmental hazard. Arsine is a colourless, non-irritating gas that causes a rapid and unique destruction of red blood cells and may cause kidney failure and eventual death. Most cases of arsine gas poisoning occurred due to the use of acids on crude metals (or ores), of which one of them or both may contain arsenic as an impurity.

Arsenic is used in cotton industry, and in electronics especially in photocopying and high speed computers. Lead arsenate and sodium/calcium arsenite are used as pesticides. Monosodium arsenite and dimethyl arsenic acid are used as weed killers. Fluochrome arsenate phenoyl (FCAP) and chromated copper arsenate (CCA) are used as wood preservatives. Arsenic pollution of the environment also results from its release into air by the smelting of ores containing arsenic, combustion of coal and through use of arsenical pesticides. Arsenic compounds are also used in live stock and poultry-farming.

Trivalent arsenic is more toxic than its pentavalent form, because arsenites are strongly bound to –SH groups present in proteins and result in enzyme inhibition. However, arsenates inhibit ATP synthesis. Arsenic can cross placental membrane and is known to be teratogenic to animals. Arsenic induces skin lesions and may lead to the development of skin cancer. Inorganic arsenicals can cause lung-cancer in humans.

Arsenic poisoning due to consumption of groundwaters has been reported from West Bengal in India, Bangla Desh, Argentina, China, Chile, Mangolia, Mexico, Taiwan, Thailand, Vietnam and USA.

About 1.5 million people in West Bengal are either suffering from arsenic poisoning due to consumption of contaminated ground water or face the risk of arsenic poisoning.

Treatment with coagulants like alum and ferric chloride can effectively remove arsenic in water. Several new approaches are being investigated for more effective removal of arsenic from contaminated water, which include photocatalytic and adsorption processes.

Dissolved solids content. The dissolved solids content of a water sample is important in deciding whether the water is suitable for boiler feed purposes. It can be determined by evaporating an aliquot of the filtered sample to dryness and further heating the residue to constant weight at 110° ± 10°C. However, during this heating, the bicarbonate ions are decomposed to give carbonate ions and gaseous CO_2, according to the equation $2HCO_3^- = CO_3^{2-} + H_2O + CO_2$. The bicarbonate ions thus yield only carbonate ions.

The total solids content comprises of dissolved solids and suspended solids.

Corrosion. The corrosiveness of a raw water is determined by the nature of the impurities present in it. Thus the common constructional materials are attacked by water containing dissolved acids, oxygen and salts. Under boiler conditions, concentrated alkalies can also attack steel. The most objectionable consequences of corrosion are: (*a*) loss of mechanical strength arising from the general thinning of the metal or development of cracks, (*b*) pitting and perforation of the metal, (*c*) contamination of the water and (*d*) interference with heat transfer and the consequent fall in efficiency.

1·5 (6) Dissolved gases.
The dissolved gases occurring in various water supplies are:
   (i) Carbon dioxide (CO_2).
  (ii) Oxygen (O_2).
 (iii) Nitrogen (N_2)
  (iv) Hydrogen Sulphide (H_2S), and
   (v) Methane (CH_4).

(i) Carbon dioxide

Varying amounts of free CO_2 are present in natural waters. The amount of CO_2 picked up by rain water from the atmosphere is very small and ranges from 0.5 to 2 ppm. Most of the surface waters, if sampled at their surfaces, contain CO_2 in the range of 0 to 5 ppm. In lake waters, the CO_2 concentration in the surface samples range from 0 to 2 ppm and it increases with the depth. This is due to the fact that CO_2 is generated at the bottom of the lake, due to the decay of organic matter, while in the upper layers, microscopic plants consume CO_2 for photosynthesis and give out oxygen. Rivers containing considerable organic matter showed as much as 50 ppm of CO_2 content. Rivers receiving acid mine waters or acid wastes may also show high CO_2 content. Ground waters usually contain 1 to 50 ppm CO_2 while shallow wells, in areas overlain by peaty soils, may show 50 to over 300 ppm of CO_2.

Phenolphthalein alkalinity in natural waters

Freshly sampled natural waters usually contain some free CO_2. Some of them sometimes show an appreciable phenolphthalein alkalinity due to photosynthesis of the large or microscopic plants present, which under the influence of sunlight, breathe-in CO_2 and breathe-out oxygen. This process may continue even after the free CO_2 supply gets exhausted by deriving the half-bound CO_2 content from the bicarbonates present, thus forming the normal carbonate and imparting phenolphthalein alkalinity to the water.

Effect of CO_2 on pH values

When CO_2 dissolves in water, it forms weakly dissociated carbonic acid, H_2CO_3 to some extent. If the water is free from all traces of alkali and is saturated with CO_2 (about 1450 ppm at 77°F), then the pH of it is about 3.8. Such a low pH would not be found in natural waters (except for waters containing free mineral acidity) because of the presence of some bicarbonate alkalinity.

(Pure distilled water, in equilibrium with the CO_2 content of the atmosphere, will have a pH of about 5.7. However, even the slightest trace of alkalinity raises the pH, so that most distilled waters in glass vessels show a pH around 6.4).

If the water contains bicarbonate alkalinity also, then its pH value depends not only on the free CO_2 content, but instead, depends on the ratio of free CO_2 to the methyl orange alkalinity of the water. Hence, the pH value of any water having a free CO_2 may be calculated from the relative proportions of the free CO_2 and its bicarbonate alkalinity. Similarly, if the pH of the water and its bicarbonate alkalinity are known, the free CO_2 content may be calculated.

Dissolved carbon dioxide, if present in boiler-feed water, produces an acid solution which leads to a general corrosive attack on the metal. Further, CO_2 is also an accelerating factor in dissolved-oxygen corrosion. Hence, water which contains, in addition to dissolved O_2, a high CO_2 content in relation to its alkalinity, will be much more corrosive than a water containing a low CO_2 content in relation to its alkalinity. In other words, a water having a dissolved O_2 content is much more corrosive, if it has a low pH than if it has a higher pH.

Carbon dioxide in water may be removed or reduced by means of an aerator, degasifier or vacuum deaerator. Carbon dioxide in water may also be neutralized by the addition of lime or an alkali such as caustic soda, but these procedures are limited only for raw or treated waters containing relatively small amounts of CO_2. CO_2 may also be partially removed from water by filtration through a neutralizing filter, employing a bed of granulated calcite. Some of this dissolves in the water forming Ca $(HCO_3)_2$ while bringing the pH up to about 7.2. Such filters are widely used in the household field of water treatment and to some extent in industry also.

(ii) Oxygen

The solubility of pure oxygen at 32°F and atmospheric pressure is 48.89 ml per litre; while under the same conditions, the solubility of nitrogen is 23.54 ml per litre. Consequently, when air is dissolved in water, the two main components, namely N_2 and O_2, exist in different proportions in solution than that existing in the atmosphere.

According to Henry's law, the solubility of a gas is proportional to the absolute pressure. Thus, if pressure is increased, the amount of air or O_2, which can be held in solution at a given temperature, is also increased proportionally.

Dissolved oxygen is very corrosive to metals like iron, steel, galvanized iron and brass, which are widely used for making vessels for conducting and holding water. Low pH values and elevated temperatures accelerate the rate of this corrosion.

Inhibition or reduction of dissolved oxygen corrosion could be achieved by deaeration of the boiler feed water, vacuum deaeration, sodium sulphite treatment, treatment with sodium silicate plus caustic soda, chromate treatment and by cathodic protection.

Dissolve oxygen is needed for living orgenisms to maintain their biological process. As per BIS domastic water must have 4-6 ppm dissolve oxygen and 0.01-0.05 ppm for boiler water.

(iii) Nitrogen

Nitrogen is a rather inert gas which has no corrosive effects on metals. In analyzing waters, nitrogen is practically never determined, since it is inert and is relatively unimportant as far as water treatment is concerned.

(iv) Hydrogen sulphide

Waters which contain sulphides are known as "sulphur waters" which are characterized by offensive odours and their marked corrosiveness. Most sulphur waters are ground waters. H_2S, like CO_2, when dissolved in water is feebly ionized. Oxidation of sulphides by dissolved oxygen is apparently a rather slow process. Chlorine may also be used to oxidize H_2S, but the process is rather expensive on raw "sulphur waters" because it takes eight atoms of chlorine to oxidize one molecule of H_2S to form a sulphate.

$$H_2S + 4Cl_2 + 4H_2O \longrightarrow H_2SO_4 + 8HCl$$

(v) Methane

In general, well waters containing methane have been found in the glacial drift and in oil and gas well areas. While methane of itself, is apparently unobjectionable in drinking water, it would be advisable to aerate the water for either industrial or household use, so as to eliminate fire and explosion hazard.

1·5 (7) Silica Content. Silica content means the concentration of silicic acid, H_2SiO_3 (expressed as SiO_2) present in water. The presence of SiO_2 in boiler feed water, especially the high pressure boilers, gives rise to many difficulties in operating the power generation equipment because of the formation of silicate scales of low thermal conductivity and similar heavy deposits on the blades and nozzles of turbines. This is the reason why the water-treatment technology always includes a process ensuring partial or total desilication of raw water. In modern practice, silica is removed from raw water by using ionite exchangers.

The concentration of SiO_2 in natural water varies over wide limits, from 5 to 90 mg/l. It decreases with increasing mineralization (salt content) of water. Natural waters contain silicic acid both in an ionic ($HSiO_3^-$) and in a colloidal state. This complicates desilication and chemical control of feed-water for boiler units, since only ion-dispersed silicic acid responds to ion-exchange treatment processes.

1·5 (8) Oxidability. Oxidability to some extent is a measure of the contamination of water by organic substances. It is generally expressed in milligrams of oxygen required to oxidize, under given conditions, the organic substances present in 1 kg of water. However, it should be noted that oxidability does not represent the total content of organic substances in water, since under the given experimental conditions used for the determination of oxidability, complete oxidation of the organic substances does not take place. Oxidability can also be expressed by the amount of $KMnO_4$ (mg/kg) spent to oxidize the organic substances.

1·6. Methods of Treatment of Water for Domestic and Industrial Purposes

Municipal water supply has been one of the most challenging problems of water technology. Water supplied by municipalities for domestic purposes must be free from pathogenic bacteria. It should be clear, colourless and pleasant to taste. It should be free from excessive dissolved salts, suspended impurities and harmful microorganisms.

Water for domestic purposes should be obtained from such a source which is least contaminated by animal and vegetable matter as well as industrial effluents. Rivers, lakes and wells are the most common sources of water used by municipalities. Generally the treatment of these waters involves removal of suspended impurities and removal of colloidal impurities if any, followed by sterilization. If the water is very hard, certain amount of softening may be needed, which is very rare. Sedimentation, coagulation, filtration and sterilization are the treatment techniques usually employed depending on the requirements of the situation.

1.6.1. Sedimentation. Sedimentation is a process of removing relatively large particles (suspended solids) into large reservoirs of settlement tanks in which it is left for a few days or even weeks, where the suspended impurities partially sink to the bottom. The principle involved is to slow down the flow of water so that substances held up by the turbulence of fast moving water can fall gravitationally to the bottom of the tank when water flow is stilled. Periodically the accumulations of the debris are to be scraped away. In order to remove floating impurities, screens of various kinds (*e.g.,* Bar screen, Band and drum screens and microstrainers) are employed. These screens also must be continuously cleaned.

The rate of settling in still water at 10°C is known as the hydraulic settling value of a particle and is generally expressed in millimeters/second.

During sedimentation, solid particles settle by gravity on the bottom of a settling tank in which the water being clarified is at rest or in slow horizontal or upward motion. The velocity with which a particle in water will fall under the action of gravity depends upon (*i*) the horizontal flow velocity of the water, (*ii*) the size of the particle, (*iii*) the specific gravity of the particles, (*iv*) the shape of the particle and (*v*) the temperature of the water. Accordingly, several formulae have been given to calculate the velocity of falling spherical particles in slowly moving water on the basis of which several types of sedimentation tanks have been designed. The sedimentation tanks commonly used are of horizontal flow rectangular type and circular shaped upward flow type.

Sedimentation takes a long time, requires large-capacity settling tanks and cannot ensure complete removal of coarse-dispersed impurities from water. Plain sedimentation usually removes only 70 to 75% of the suspended matter.

1.6.2. Coagulation. Finely divided silica, clay and organic matter does not settle down easily and hence cannot be removed by simple sedimentation. Most of these are in colloidal form (*e.g.,* sols, gels or emulsions) and are generally negatively charged and hence do not coalesce because of mutual repulsion. Such impurities are generally removed by chemically assisted sedimentation, in which certain chemicals are added which produce ions of right electrical charge that neutralize the oppositely charged colloidal particles and brings about their coalescence. This process is called coagulation. This permits the particles to aggregate together until a denser particle is formed which falls through still water at a reasonable rate and is called flocculation.

Aluminium sulphate is the most common coagulating agent used for removing clay particles and is generally called filter alum. Other coagulants which also find application in water treatment include ferric sulphate, ferrous sulphate (copperas), chlorinated copperas, alum, ammonia alum or potash alum and sodium aluminate.

Aluminium sulphate, when added to natural waters, hydrolyses to form colloidal aluminium hydroxide and an equivalent quantity of sulphuric acid as follows:

$$Al_2(SO_4)_3 + 6H_2O \longrightarrow 2Al(OH)_3 + 3H_2SO_4$$

The $Al(OH)_3$ so formed acts as a floc or coagulant, which has an enormous surface area per unit volume and removes the finely divided and colloidal impurities by neutralizing the charge on them as well as by other mechanisms like adsorption and mechanical entrainment. Thus the smaller particles join together to form denser particles which settle down to the bottom. Some bacteria and colour associated with these particles also get removed simulataneously. In order to render the $Al(OH)_3$ filterable and also in order to neutralize the H_2SO_4 liberated to permit the hydrolysis reaction to completion, some alkali will have to be added if the water is not sufficiently alkaline. The reaction taking place in alkaline waters may be represented as follows:

$$Al_2(SO_4)_3 + 3Ca(HCO_3)_2 \longrightarrow 2Al(OH)_3 + 3CaSO_4 + 6CO_2$$

With water having a little or no natural alkalinity (*e.g.*, moorland waters) an alkali such as calcium hydroxide or sodium carbonate is added; the latter is more commonly used as it does not increase the hardness of the water.

$$Al_2(SO_4)_3 + 3Na_2CO_3 + 3H_2O \longrightarrow 2Al(OH)_3 + 3Na_2SO_4 + 3CO_2$$

For treatment of acidic waters, sodium aluminate can be used as a source of $Al(OH)_3$ and it is often used in conjunction with aluminium sulphate. The reactions taking place may be represented as under:

$$NaAlO_2 + 2H_2O \longrightarrow NaOH + Al(OH)_3$$
$$Al_2(SO_4)_3 + 6NaAlO_2 + 12H_2O \longrightarrow 8Al(OH)_3 + 3Na_2SO_4$$

Thus the alkalinity due to aluminate is neutralized by the acidity of the sulphate and $Al(OH)_3$ is thus precipitated from both the reagents.

for treatment of alkaline water $FeSO_4$ can be used

$$FeSO_4 + H_2O \longrightarrow Fe(OH)_2 + H_2SO_4$$
$$FeOH)_2 + \tfrac{1}{2}O_2 \longrightarrow Fe(OH)_3 \downarrow$$

In order to increase the efficiency of the coagulation process, coagulant aids such as lime, Fuller's earth, bentonite clay and polyelectrolytes are also added.

The coagulants are generally added in solution form to the water with the help of mechanical flocculators provided with slow-moving rotating baffles or stationary baffles turning the flow of water; thus ensuring a gentle contact with the water and the reagents.

After any sedimentation process, especially after that utilizing a chemical flocculent, there will be a substantial reduction in the bacterial count in the water. In addition, many coagulants release oxygen to the water as the chemical transformations take place. This oxygen helps in destroying some bacteria, in breaking up some organic compounds and also in partial removal of colour and taste producing organisms present in water. Thus the process of coagulation, flocculation and sedimentation has many beneficial results and is one of the most important processes of purification of water.

Coagulation and settling equipment should be so designed that quick and thorough mixing of the coagulant and raw water should be achieved so that the coagulant dosage required is minimised. For this purpose, mechanical type mixers and baffled mixing troughs are widely used. Modern types of coagulation and settling equipment include the floc-former type and the sludge-blanket type.

The floc-former type coagulation and settling equipment (Fig. 1.3) consists of (*i*) a flash mixer where the coagulant is quickly and efficiently mixed with the raw water with the help of a rotating stirrer (*ii*) rolling mix chambers where only a gentle mixing of the gelatinous precipitate formed to aid in the even distribution and formation of a large-particled, easily settled-floc, and (*iii*) settling basins where the floc is allowed to settle and the sludge thus formed is removed continuously or intermittently.

The sludge-blanket type of water-treatment equipment came into extensive use for coagulation and settling as well as for water softening by cold lime-soda process. It majorly differs from the floc-fomer type of equipment in that it filters all of the coagulated water upwardly through a suspended sludge blanket instead of merely dropping it to the bottom by gravity. This prolonged intimate contact with the floc helps in

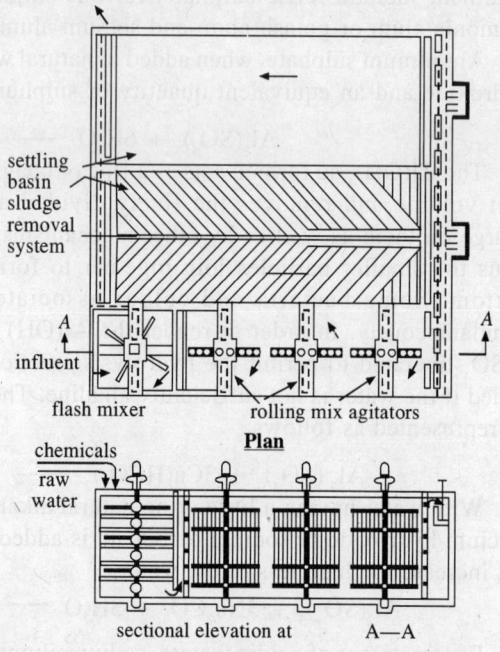

Fig. 1.3. Floc-fomer type coagulation and settling equipment.

fully utilizing its absorbing capacity which effects savings in the amount of coagulant or other adsorbents (such as activated carbon used for removing colours and odours) required. The equipment is made in both vertical and horizontal designs.

For certain industrial uses where turbidities upto 10 ppm are permissible and where the raw water is turbid but not highly coloured, the effluent from the sludge-blanket type of coagulation and settling equipment may be used directly. However, if colour removal and a greater degree of clarity is required, filtration is also necessary.

1.6.3. Filtration. Filtration is a process of clarification of water by passing the water through a porous material, which is capable of retaining coarse impurities on its surface and in the pores. The porous material used is called the filtering medium and the equipment used for filtration is known as a filter.

Filtration of water takes place due to the difference between the pressures at the of the bed of the filtering material and that underneath it. The difference in the two pressures, Δh is called the pressure drop through the filtering medium, but also the drop in pressure through the filter proper (in the pipelines, distribution devices, etc.). This pressure drop through the filtering bed depends on the rate of filtration, height of the filtering bed, diameter (size) of the grains of the filtering material and the extent of its contamination by the trapped impurities. Greater the numerical values of these factors (excepting grain size), greater will be the pressure drop.

WATER TREATMENT

When contamination of a filter cannot be tolerated any longer, it is taken out of service for subsequent washing, in the course of which the impurities trapped in the filtering bed are washed out. Then the filter can be pressed into service again. The common materials used for the filtering medium are quartz sand (grain size 0.5 – 1.0 mm) containing not more than 96% SiO_2, crushed anthracite (piece size 0.8 – 1.5 mm) and porous clay (piece size 0.8 – 1.5 mm). It should be remembered that when quartz sand filters are used with alkaline water, the filtrate (filtered water) gets enriched with silicic acid (due to the solubility of quartz sand in alkaline water).

Slow sand filtration

Water for domestic use may be filtered through large area of finely graded sand beds at a slow rate (about 2 gal/sq. ft./hour). The rate of filtration slowly diminishes due to the accumulation of sediment in the capaillaries of the filter bed, and finally the rate becomes so slow that the bed must be cleaned. Cleaning is usually done by scarping the surface of the sand bed, or excavating, washing, and then relaying the entire filtering medium,

Slow gravity filters are not capable of removing colloidal impurities. These can be filtered out only when the water has been chemically treated (such as alum treatment) and the slow sand filters cannot deal with the gelatinous type of precipitates which are produced in such treatments. A typical equipment for Municipal water treatment with coagulation, settling and filtration tanks is shown in Fig. 1.4.

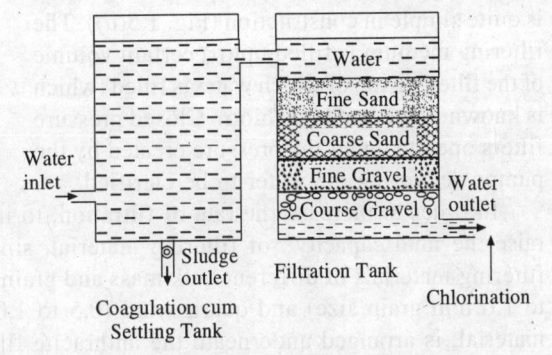

Fig. 1.4. Municipal water treatment with coagulation, settling and filtration tanks

Rapid-gravity filtration

Rapid gravity filters are capable of producing potable waters at flow-rates as high as 100 gal/sq. ft./hour. This is achieved by using carefully graded quartz sand and collecting the filtrate as evenly as possible over the entire bottom area of the sand bed which avoids undesirable channeling. The sand bed may be cleaned either by agitation with compressed air or in small units mechanically. This is followed by a flush back with clean water to wash away the accumulated impurities.

Rapid Gravity filtration has the following advantages:
(1) The filter bed and the quanlity of the filtrate can be easily inspected (2) The filter is unaffected by pressure variations on either the inflow or draw-off sides, and (3) Large reinforced concrete filters can be constructed at relatively low cost.

Rapid pressure filtration

Pressure filters are much more widely used than gravity filters, particularly in industrial installations. Rapid pressure filters are used preferentially where filtration is to be effected in a rising main without breaking the hydraulic head, or where water from an elevated source can be passed through the filter and delivered in storage. The operation and cleaning of these filters are more or less similar to gravity sand filters. Filtration rates are of the order of 80 to > 200 gallons/sq. ft./hour.

Pressure filters are manufactured in vertical and horizontal types. These filters consist of cylindrical steel shells fitted with dashed heads, containing a layer of a granular filter medium (sand or anthrafilt) supported by graded gravel (or anthrafilt), and is equipped with the required accessories *e.g.,* piping, underdrains, valves, etc., for carrying out the cycle of operations, viz., filtration, backwashing and filtering to waste. Filter installations may consist of one or more units depending upon the requirements.

Several types and designs of mechanical filters are employed (Fig. 1.5). The most commonly used clarifying filter is the single-flow, single-bed, closed pressure filter, which is quite simple in construction (Fig. 1.5(*a*)). The filtering medium is filled upto a certain volume of the filter, on top of which water is filled (which is known as the water cushion). Closed pressure filters operate under the pressure created by the pumps delivering the water to be clarified.

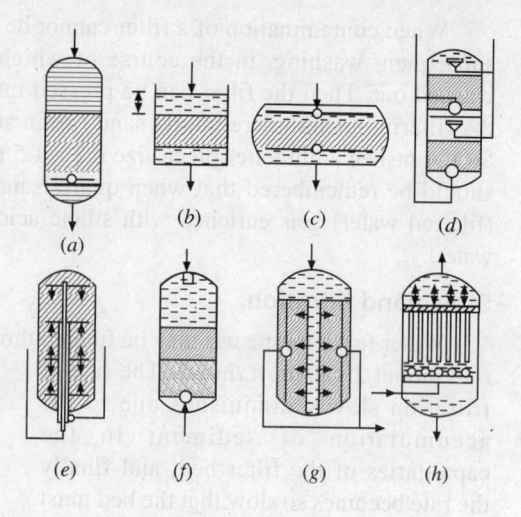

Fig. 1.5. Mechanical filters.

In order to increase the rate of filtration, to improve the quality of clarified water and to raise the mud capacity* of filtering material, single-flow filters are often charged with two filtering materials of different bulk mass and grain size, as for instance, crushed anthracite (0.8 to 1.6 mm grain size) and quartz sand (0.5 to 1.0 mm grain size). The sand, being a denser material, is arranged underneath the anthracite filtering bed. Two-bed filters of this kind are often used in practice (Fig. 1.5(*f*)).

In open-type mechanical filters (Fig. 1.5(*b*)) which are widely used in purification and treatment of drinking water; filtration takes place due to the pressure exerted by the column of water, *h* in the filter. However, the relatively small head, *h* restricts the use of these filters at increased filtering rates. Horizontal closed pressure filters (Fig. 1.5(*c*)) are provided with larger filtering surface and hence have larger throughout per filter. However, they require more floor space when installed indoor and cumbersome to operate. Multiflow mechanical filters (Fig. 1.5(*e*)) are free from such limitations. These are filled completely with filtering material, inside which are fitted the draining and distributing devices. They permit the water to be passed in several flows (three in the illustration given) and ensuring increased output of a filter as many times. In radial flow filter (Fig. 1.5(*g*)) the filtering material is charged into the annuus formed between the filter shell (wall) and an internal tube. The water treated is delivered into the central tube which distributes it through the filtering material heightwise. Water is passed in radial flows through the filtering material.

The clarifying filters, with a granular filtering bed discussed above, contain a high bed of coarse-grained filtering material and a small filtering surface. In contrast to these filters, tubular-element precoat type filters (Fig. 1.5(*h*)) contain a large filtering surface, thin filtering bed (5 to 10 mm thick), and mainly use powders as filtering media (*e.g.,* powdery or fibrous cellulose, ionite and perlite powders). Precoat-type ionite-exchanger filters are capable of operating at high temperatures (100 – 110°C), and hence can be used to purify hot condensate and water. These filters are considered promising for service in high-capacity steam-generator-turbine units operated at nuclear power plants. A precoat type filter with powdery perlite (porous SiO_2)

---

* The mud capacity of a filter, M is defined as the quantity of impurities entrapped, in the filter during the filtering period, related to 1 m^3 of the filtering material and expressed in kilograms

$$M = \frac{Q.C_{ss}}{10^3 hf}$$, where h = height of the filtering bed, m:

f = filter cross-sectional areas, m^2; Q = amount of water filtered, m^3; and C_{ss} = concentration of suspended solids in the filtered water. The greater the mud capacity of a filter at a given C_{ss}, the longer the duration of the filter run.

as filtering material is suitable for fine clarification of water. A mixture of powdery perlite and activated carbon is used for removing oil from condensate.

Clarification for domestic and industrial purposes is often achieved by a combination of two to four of the different processes as listed below:

(a) Sedimentation (b) Filtration (c) Coagulation + filtration (d) Coagulation + settling (e) Coagulation + settling + filtration (f) Sedimentation + coagulation + settling + filtration (g) Chlorination and (h) Special filters (e.g., activated carbon for removing taste and odour; manganese zeolite for removing Fe and Mn; and Neutralizing filters such as graded calcite for removing CO_2).

1.6.4. Sterilization of water. Clarification of water by sedimentation, filtration and storage removes suspended solids and also reduces the number of bacteria in the water. Total elimination of bacteria can be achieved only by sterlization.

The organisms which should be eliminated by disinfection are divergent in character. They include (a) the enteric bacteria belonging to Salmonella, Shigella and Vibrio groups (b) the intestinal protozoa such as Entamoeba histolytica (c) some types of worms such as Schistosomes (d) Viruses such as those of infectious hepatitis and (e) Coliform organisms which indicate water pollution, though not pathogentic. Coliforms and enteric bacteria can be easily destroyed while viruses and cysts of E. histolytica are resistant to disinfection.

Chlorine is the most common sterilizing agent in water treatment. It is capable of removing B. Coli and substantially reducing other bacteria. Chlorine may be added in the form of bleaching powder, or directly as a gas or in the form of concentrated solution in water. Whatever may be the method employed, the treatment should give accurate dosage, good distribution and sufficient time of contact (~ 30 minutes) so as to ensure effective sterilization.

$$CaOCl_2 + H_2O \longrightarrow Ca(OH)_2 + Cl_2$$

$$Cl_2 + H_2O \longrightarrow HOCl + HCl$$

$$\underset{\substack{\text{Hypochlorous}\\\text{acid}}}{HOCl} \longrightarrow \underset{\substack{\text{Nascent}\\\text{oxygen}}}{[O]} + HCl$$

The nascent oxygen so liberated destroys the germs and bacteria by oxidation. The chlorine itself, the hypochlorous acid and other chlorine compounds are also believed to have powerful germicidal properties. The OCl^- ions are capable of rupturing the cell membranes of the disease producing microbes.

Bleaching powder has the limitations of being unstable during storage and also it increases the calcium content of water rendering it more hard. Chlorine as a sterilizing agent has the advantages of being economical, efficient, limited space requirement and convenience. Further, it does not introduce any other impurities in the water. Both chlorine as well as bleaching power when used in excess produce disagreeable odour in the water. Too much excess may cause irritation to the mucous membranes. The unpleasant taste of this excess chlorine can be removed by treatment with ammonia, which reacts with chlorine to form the tasteless compound chloramine (NH_2Cl). Thus, the ammonia-chlorine treatment (chloramine process) is particularly useful where traces of impurities are present (e.g., phenols) which produce unpleasant tastes when chlorine alone is used. Further, chloramine provides a more lasting effect than that of chlorine.

$$Cl_2 + NH_3 \longrightarrow \underset{\text{Monochloramine}}{NH_2Cl} + HCl$$

$$2Cl_2 + NH_3 \longrightarrow \underset{\text{Dichloramine}}{NHCl_2} + 2HCl$$

$$NH_2Cl + Cl_2 \longrightarrow NHCl_2 + HCl$$

$$NH_2Cl + H_2O \longrightarrow HOCl + NH_3$$
$$HOCl \longrightarrow HCl + [O]$$
nascent oxygen

The chloramine process consists of adding ammonia to water in the form of gas together with chlorine resulting in the formation of dichloramine and monochloramine as shown above. Dichloramine is a relatively stable compound and is perhaps not a sterilizing agent by itself. However, it slowly decomposes with evolution of chlorine. Thus, the addition of ammonia stabilizes the chlorine to provide a prolonged effect. This is particularly useful when the water is passed into storage after treatment.

Chlorine, ammonia-chlorine, or sodium hypochlorite are used widely for the sterilization of swimming-pool water.

A disadvantage of chlorination is the potential formation of trihalomethanes (such as chloroform, bromodichloromethane, dibromochloromethane and bromoform) which are carcinogenic. Trihalomethanes (THMs) may be formed when chlorine combines with natural organic substances, such as decaying vegetation, etc., that may be present in water itself. One approach to tackle this problem is to remove the organics completely before subjecting it to water chlorination.

In future, the actual removal of THMs from the treated water, perhaps by aeration or adsorption on activated carbon, may become necessary.

Superchlorination

In **superchlorination,** a large excess of chlorine is added to the water, thereby destroying not only the micro-organisms but also the other organic impurities present. This process ensures rapid and complete sterilization and successfully used for waters derived from wells and rivers. This process is usually followed by dechlorination by NH_3 or SO_2.

Break-point chlorination is a more precisely controlled process in which just sufficient chlorine is added to oxidize all the organic matter, destroy bacteria and react with any ammonia, leaving a slight excess of free chlorine. **The break-through point,** *i.e.,* the appearance of free chlorine, in the water, must be determined experimentally as follows:

If chlorine is added to a sample of water and after a few minutes, the residual chlorine available in the water is estimated, it will be found that the residual chlorine in water is less than the amount added initially. This is due to the fact that some of the chlorine added initially is consumed by oxidising bacteria and other organic matter. Now, if we take a few more aliquots of the same volume of the water sample and add increasing doses of chlorine to different samples and analyse the residual chlorine after the same interval of time (a few minutes), a curve of the type shown below is obtained (Fig. 1.6).

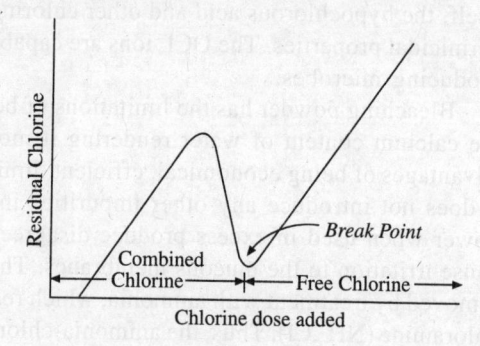

Fig. 1.6. Break-point chlorination.

It can be seen from Fig. 1.6 that the quantity of residual chlorine increases with increasing dose of chlorine added giving a straight line until at a definite chlorine dose, a sudden decrease in the residual chlorine is noticed. This is known as the break-point, after which the residual chlorine appearing more or less agrees with the chlorine dose added. The reason for such a behaviour is due to the fact that some organic compounds which defy oxidation at lower chlorine concentrations, get oxidized when the break-point chlorine concentration is reached. Since it is these organic compounds which are generally responsible for bad tastes and odours in water, it is obvious that break-point chlorination eliminates bad tastes and odours.

WATER TREATMENT

Determination of free chlorine in a water sample

The principle involved in the estimation of free chlorine in water is that when a measured quantity of water is treated with excess of potassium iodide, the free chlorine present in the water oxidizes the corresponding amount of potassium iodide to iodine. The liberated iodine is estimated by titrating against standard sodium thiosulphate solution using starch as indicator

$$Cl_2 + 2KI \longrightarrow 2KCl + I_2$$
$$I_2 + 2Na_2S_2O_3 \longrightarrow Na_2S_4O_6 + 2NaI$$

Dechlorination

The water treated by the process of break-point chlorination may be filtered through activated carbon, in order to remove the decomposition products formed and the excess chlorine remaining. Other methods of dechlorination include treatment with SO_2 or Na_2SO_3.

$$SO_2 + Cl_2 + 2H_2O \longrightarrow H_2SO_4 + 2HCl$$
$$Na_2SO_3 + Cl_2 + H_2O \longrightarrow Na_2SO_4 + 2HCl$$

Sterilization by Ozone

Ozone is a powerful disinfectant and is readily absorbed by water. Ozone being unstable decomposes as follows giving nascent oxygen which is capable of destroying the bacteria.

$$O_3 \longrightarrow O_2 + [O]$$

However, this process is relatively expensive but has the advantage of removing bacteria, colour, odour and taste without leaving any harmful residual effects in the water being treated.

Sterilization by ultraviolet radiation

Ultraviolet radiation (190 - 380 nm) emanating from electric mercury vapour lamp is capable of sterilizing water. This process is particularly useful for sterilizing swimming pool waters. However, this process cannot be economical for water works. Irradiation of water by ultraviolet light is commonly used for disinfection in food industries.

Water for domestic purposes on a smaller scale may be sterilized by boiling the filtered water for about 20 minutes. Chemicals like potassium permanganate and tincture iodine are also used occasionally but chlorine tablets and bleaching powder are more commonly employed.

1·7. Removal of Dissolved Salts: Softening of Water

The concentration of dissolved impurities mostly determines the hardness of water. The process of removing the hardness causing salts from water is called softening of water. It has been pointed out under the section on "hardness" that soft water is essential for many industries such as a in textiles, laundries, paper, rayon, ice, brewing, distilleries, canning etc. Water used for generation of steam should be perfectly soft to minimize troubles like scale formation in the boilers, which leads to loss of efficiency and even ultimate failure of the boiler tubes.

The following methods are generally used for softening of water. Temporary hardness can be removed by prolonged heating of the water. The dissolved gases e.g., CO_2 and O_2 are also removed simultaneously.

$$\left.\begin{matrix} Ca \\ Mg \end{matrix}\right\} (HCO_3)_2 \xrightarrow{\text{boiling}} \left.\begin{matrix} Ca \\ Mg \end{matrix}\right\} CO_3 + H_2O + CO_2$$

1.7.1. Lime-Soda process. This is the most important method of chemical water-softening. The principle involved in this process is to chemically convert all the soluble hardness-causing impurities into insoluble precipitates which may be removed by settling and filtration.

In the lime-soda process, a suspension of milk of lime is added in requisite amount to the water, together with a calculated amount of sodium carbonate (soda) solution. Generally about 10% excess of chemicals are added for quick completion of the reactions. The reactions taking place may be summarized as follows:

(i) lime removes the temporary hardness:

$$Ca(HCO_3)_2 + Ca(OH)_2 \longrightarrow 2CaCO_3\downarrow + 2H_2O$$
$$Mg(HCO_3)_2 + 2Ca(OH)_2 \longrightarrow Mg(OH)_2\downarrow + 2CaCO_3 + 2H_2O$$

(ii) lime removes the permanent magnesium hardness:

$$MgCl_2 + Ca(OH)_2 \longrightarrow Mg(OH)_2\downarrow + CaCl_2$$
$$MgSO_4 + Ca(OH)_2 \longrightarrow Mg(OH)_2\downarrow + CaSO_4$$

(iii) lime removes dissolved iron and aluminium salts:

$$FeSO_4 + Ca(OH)_2 \longrightarrow Fe(OH)_2\downarrow + CaSO_4$$
$$2Fe(OH)_2 + 2H_2O + O \longrightarrow 2Fe(OH)_3\downarrow$$
$$Al_2(SO_4)_3 + 3Ca(OH)_2 \longrightarrow 2Al(OH)_3\downarrow + 3CaSO_4$$

(iv) lime removes free mineral acids:

$$2HCl + Ca(OH)_2 \longrightarrow CaCl_2 + 2H_2O$$
$$H_2SO_4 + Ca(OH)_2 \longrightarrow CaSO_4 + 2H_2O$$

(v) lime removes dissolved CO_2 and H_2S:

$$Ca(OH)_2 + CO_2 \longrightarrow CaCO_3\downarrow + H_2O$$
$$Ca(OH)_2 + H_2S \longrightarrow CaS\downarrow + 2H_2O$$

(vi) Soda removes all the soluble calcium permanent hardness (*i.e.*, that which is originally present as well as that which is introduced during the removal of Mg^{2+}, Fe^{2+}, Al^{3+}, HCl, H_2SO_4 etc., by lime).

$$CaCl_2 + Na_2CO_3 \longrightarrow CaCO_3\downarrow + 2NaCl$$
$$CaSO_4 + Na_2CO_3 \longrightarrow CaCO_3\downarrow + Na_2SO_4$$

Since natural waters generally contain a large proportion of temporary hardness, it is often convenient and economical to remove temporary hardness by lime treatment. Lime is rather cheap and it removes temporary hardness efficiently without introducing any soluble salts in the water. Magnesium hydroxide produced in the above reactions precipitates as an insoluble sludge. The reaction of soda with the permanent calcium hardness produces insoluble $CaCO_3$. Addition of a coagulant such as sodium aluminate or alum helps in accelerating the coagulation of the carbonate sludge, which is subsequently removed by filtration. Water softened by this process contains appreciable concentrations of soluble salts, such as sodium sulphate, and cannot be used in high-pressure boiler installations.

Types of cold lime-soda softeners

There are four basic types of cold lime soda softeners:
(1) The intermittent type (batch process)
(2) The conventional type
(3) The catalyst or spiractor type ⎫ continuous processes
(4) The sludge blanket type ⎭

(1) Intermittent or batch process

The intermittent type of cold lime-soda softener consists of a set of two tanks which are used in turn for softening of water. Each tank is provided with inlets for raw water and chemicals, outlets for softened water and sludge, and a mechanical stirrer (Fig. 1.7). Raw water and calculated

WATER TREATMENT

quantities of the chemicals are slowly sent into the tank simultaneously under agitation with the help of the stirrer. Some sludge from a previous operation is also added which forms nucleus for fresh precipitation and thus accelerates the process. Thus by the time the tank is full, the reaction is more or less complete. Stirring is stopped and the sludge formed is allowed to settle. The clear softened

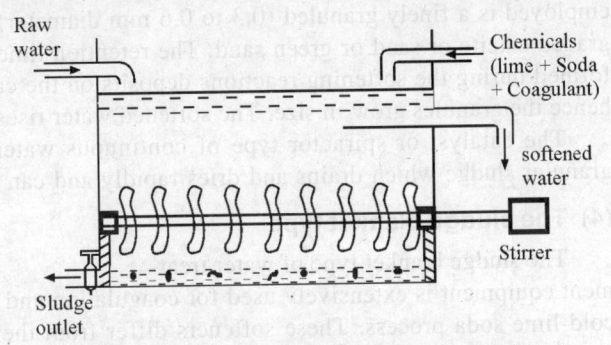

Fig. 1.7. Intermittent cold lime-soda softener.

water is collected through a float pipe and sent to the filtering unit. The sludge formed in the tank is removed through the sludge outlet. By employing a set of tanks planned for alternate cycles of reaction and settling, continuous supply of softened water may be ensured.

(2) Conventional type

In this process, the raw water and the chemicals in calculated quantities are continuously fed from the top into an inner chamber of vertical circular tank provided with a paddle stirrer (Fig. 1.8). The raw water and the chemicals flowing down the chamber come into close contact because of the continuous stirring and the softening reactions take place. The sludge formed settles down to the bottom of the outer chamber from where it is periodically removed through the sludge outlet. The softened water rising up passes through the fibre filter where traces of sludge are removed and filtered soft water passes through the outlet provided.

Water treated by the cold lime-soda process generally produces softened water containing about 50 – 60 ppm of residual hardness.

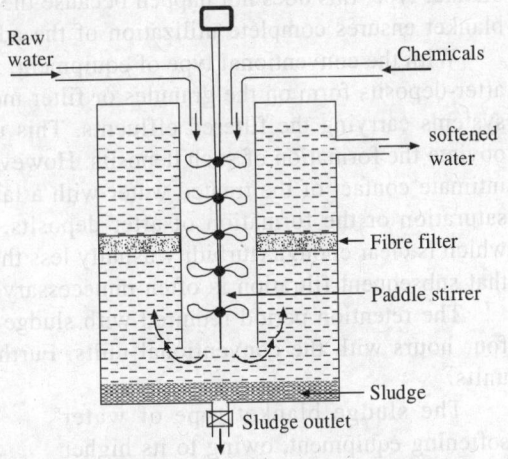

Fig. 1.8. Conventional type of lime-soda softener.

(3) Catalyst or spiractor type

The spiractor consists of a conical tank which is about two-thirds filled with finely divided granular catalyst (Fig. 1.9). The tank used may be either open (for gravity operation) or closed (for operation under pressure). In both the cases the raw water and the calculated quantities of chemicals enter the tank tangentially near the bottom of the cone and spiral upwards through the suspended catalyst bed. The catalyst

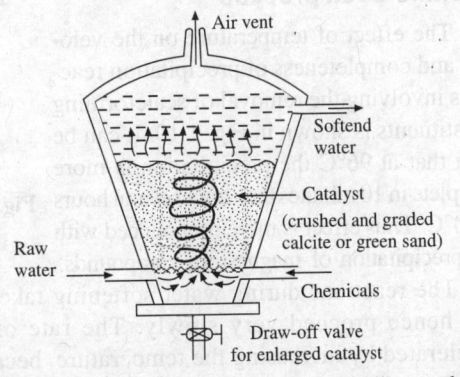

Fig. 1.9. Catalyst or spiractor type cold lime-soda water softener.

employed is a finely granuled (0.3 to 0.6 mm diameter) insoluble mineral substance such as graded calcite or sand or green sand. The retention time is about 8 to 12 minutes. The sludge formed during the softening reactions deposits on the catalyst grains in an adherent form and hence the granules grow in size. The softened water rises to the top from where it is drawn off.

The catalyst or spiractor type of continuous water softener is of interest as it gives a granular sludge which drains and dries rapidly and can be handled easily.

(4) The sludge blanket type

The sludge blanket type of water treatment equipment is extensively used for coagulation and settling as well as water softening by cold lime soda process. These softeners differ from the conventional type in that the treated water is filtered upwardly through a suspended sludge blanket composed of previously formed precipitates. Thus in a single unit, all the three processes namely mixing, softening and clarification take place.

In the conventional type of equipment, some of the added lime suspension is carried down in the sludge formed by the precipitates, before it has time to dissolve and react with the hardness causing impurities of the raw water and thus some of the lime is wasted. In the sludge blanket type, this does not happen because the upward filtration through the suspended sludge blanket ensures complete utilization of the added lime.

With the conventional type of equipment, it is generally observed that after-precipitates or after-deposits form on the granules or filter media employed, and in pipe lines or distribution systems carrying the filtered effluents. This usually necessitates recarbonation with CO_2 to obviate the formation of such deposits. However, in the sludge blanket type of equipment, the intimate contact of the treated water with a large mass of solid phase mostly prevents supersaturation or the formation of after deposits. This results in the production of the effluent which is clear enough (turbidity usually less than 10 mg/l) for many industrial applications, so that subsequent filtration is often unnecessary.

The retention period required with sludge blanket type equipment is one hour as against four hours with the conventional units. Further, silica is removed better in sludge blanket units.

The sludge blanket type of water softening equipment, owing to its higher efficiency, shorter detention period and smaller space requirements, is rapidly displacing the conventional type.

Hot lime-soda process

The effect of temperature on the velocity and completeness of precipitation reactions involving the removal of scaleforming constituents is shown in Fig. 1.10. It can be seen that at 96°C the precipitation is more complete in 10 minutes than after several hours at 10°C. Thus effect is more pronounced with the precipitation of magnesium compounds.

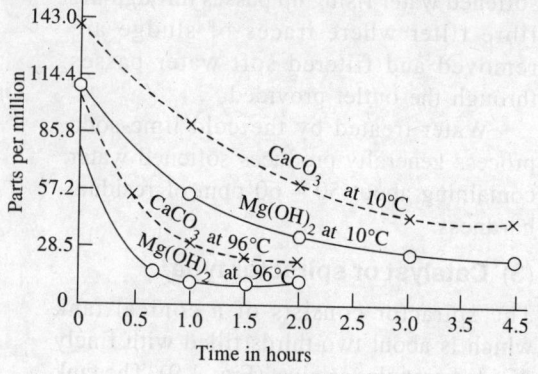

Fig. 1.10. Effect of temperature on rate of precipitation of $CaCO_3$ and $Mg(OH)_2$ from $CaSO_4$ and $MgSO_4$ solutions respectively.

The reactions during water softening take place in very dilute solutions (about 0.001 M) and hence proceed very slowly. The rate of these precipitation reactions can be greatly accelerated by increasing the temperature, because, this not only increases the rate of the ionic reactions themselves, but also the rate at which particles of measurable size are formed.

WATER TREATMENT

Hot lime-soda plants carry out softening at 94° – 100°C which has several advantages. For efficient softening, cold lime-soda softening plants must be of considerable area and water-storage capacity, whereas hot lime-soda softeners are much more rapid in operation and therefore for a given through-put, much more compact. Elevated temperatures not only accelerate the actual chemical reactions but also reduce the viscosity of the water and increase the rate of aggregation of the particles. Thus, both the settling rates and filtration rates are increased. Thus the softening capacity of the hot lime-soda process will be several times higher than the cold process. Since the sludge formed settles down rapidly, there is no need of adding any coagulants. A smaller excess of chemicals is needed than with the cold process. Further, dissolved gases are driven out of the solution to some extent at the high temperature. The hot lime-soda process yields softened water having relatively lower residual hardness (about 17 to 34 ppm) as against the cold process (about 50 – 60 ppm). A typical hot lime-soda water softening unit is shown in Fig. 1.11, which includes a reaction cum settling tank and a filter. If the water is alkaline, filtration through sand and gravel beds might contaminate the water with dissolved silica, particularly if the quartz used is of inferior quality. Other filtering media used are anthracite coal, calcite and magnetite. If the precipitation is incomplete in the softening tank, "after-precipitation" occurs in pipes, storage tanks and even in boiler itself. If slight excess of chemicals are used over that theoretically required, more rapid and more complete removal of hardness will result. But if larger excess of chemicals are used, naturally they will appear in the softened water. Lime-soda plants do not produce water of zero hardness.

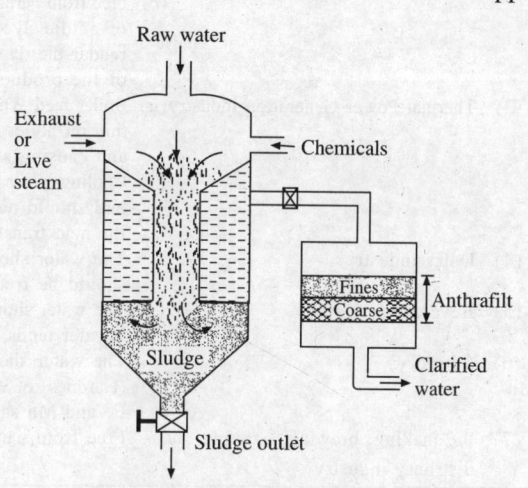

Fig. 1.11. Hot lime-soda water softener.

1.7.2. Barium carbonate-lime softening (cold lime-barium process). This modification of the lime-soda process is used for treating boiler feed waters high in calcium sulphate or magnesium sulphate in order to obtain both softening as well as reduction in the total dissolved solids. In this process, temporary hardness and magnesium salts are removed by the lime in the usual way. Then, further interaction of calcium sulphate and barium carbonate results in the precipitation of calcium carbonate and barium sulphate.

$$BaCO_3 + CaSO_4 \longrightarrow CaCO_3\downarrow + BaSO_4\downarrow$$
$$MgSO_4 + Ca(OH)_2 + BaCO_3 \longrightarrow BaSO_4\downarrow + CaCO_3\downarrow + Mg(OH)_2\downarrow$$

This process has a limited application. The cost of $BaCO_3$ per ton is about four times as high as that of soda as equivalent because of its higher molecular weight and higher cost per ton. In using $BaCO_3$, it should also be kept in mind that it is very poisonous.

NOTE :

(1) Although $BaCO_3$ will react with sodium sulphate, an equivalent amount of Na_2CO_3 will be left in the effluent so that this process is of value only in removing $CaSO_4$ and $MgSO_4$.

(2) If Na_2CO_3 is used instead of $BaCO_3$, the Na_2SO_4 equivalent to the $CaSO_4$ removed passes into solution.

Specifications of Water used for Different Industries and other purposes

Purpose	Specifications and Remarks
(1) Paper Industry	(a) Free from alkalinity (alkaline water consumes more alum, thereby increasing the cost of production).
	(b) Free from hardness: (Calcium and magnesium salts increase the ash content of the paper produced).
	(c) Free from colour, turbidity and salts of Fe and Mn : (colour and brightness of the paper are affected by the above impurities).
	(d) Free from Silica : (Silica causes cracks in the paper).
(2) Textile industry	(a) Free from turbidity : (turbidity causes uneven dyeing).
	(b) Free from colour, and salts of Fe and Mn : (these impurities cause stains on the fabric).
	(c) Free from hardness and organic matter : (Hard water reduces the solubility of acidic dyes and causes precipitation of basic dyes. They also render the dyeing non-uniform. Organic matter may cause foul smell of the product).
(3) Thermal Power Generation industry	(a) Boiler feed Water : Free from hardness : (hard water causes scaleformation on boiler metal surface, thereby reducing heat transfer efficiency and causing shut-down or even accidents).
	(b) Cooling water : The water should be non-scale forming, non-corrosive, and should not permit the growth of algae. Scale and algae reduce the heat transfer efficiency and interfere with free flow of water.
(4) Dairy industry	— The water should be colourless, odourless, and tasteless. It should be free from pathogenic organisms.
(5) Beverage industry	— The water should be pure. It should not be alkaline, because alkalinity inwater tends to neutralise the fruit acids and distorts the taste.
(6) Laundry	— The water should be free from colour, hardness and salts of Fe and Mn : (Hardness of water increases the consumption of soaps and detergents. Fe and Mn salts impart undesirable colour to the fabric.
(7) Ice making, brewing, canning and distillery industry	— Free from hardness and bacteria.

Inter-Relationship Between Hardness and Alkalinity of Water

The Inter-relationship between hardness and alkalinity of water can be illustrated by the following examples.

	Analytical Results (as ppm of $CaCO_3$)				Interpretation (as ppm of $CaCO_3$)				
Sample Number	Total Hardness	Calcium Hardness	Magnesium Hardness	Total Alkalinity	Calcium Alkalinity	Magnesium Alkalinity	Sodium Alkalinity	Magnesium Non-carbonate Hardness	Calcium Non-carbonate Hardness
1.	385	290	100	250	250	NIL	NIL	100	40
2.	300	180	220	350	180	170	NIL	50	NIL
3.	420	120	280	460	120	280	40	NIL	NIL

Notes: Equivalent amounts of cations and anions are paired as follows:

(a) The HCO_3 alkalinity is frist paired with Ca^{2+} and then, if any alkalinity is left over, is paired with Mg^{2+}. If any HCO_3^- alkalinity is still remaining, it is paired with Na^+.

(b) Non-carbonate hardness will be present only if the total hardness is greater than carbonate hardness or total alkalinity.

(c) Non-carbonate hardness is first paired with Mg^{2+} and then the remaining non-carbonate hardness is paired with Ca^{2+}.

1.7.3. Problems on water-treatment by Lime-soda process. On the basis of the various reactions taking place in lime-soda process given earlier, the following deductions can be made:

(i) One equivalent of calcium temporary hardness requires one equivalent of lime

$$Ca(HCO_3)_2 + Ca(OH)_2 \longrightarrow 2CaCO_3\downarrow + 2H_2O$$

(ii) One equivalent of magnesium temporary hardness requires two equivalents of lime

WATER TREATMENT

$$Mg(HCO_3)_2 + 2Ca(OH)_2 \longrightarrow Mg(OH)_2\downarrow + 2CaCO_3\downarrow + 2H_2O$$

(iii) One equivalent of calcium permanent hardness requires one equivalent of soda,

$$CaSO_4 + Na_2CO_3 \longrightarrow CaCO_3\downarrow + Na_2SO_4$$
$$CaCl_2 + Na_2CO_3 \longrightarrow CaCO_3\downarrow + 2NaCl$$

(iv) One equivalent of magnesium permanent hardness requires one equivalent of lime and one equivalent of soda.

$$MgSO_4 + Ca(OH)_2 + Na_2CO_3 \longrightarrow Mg(OH)_2\downarrow + CaCO_3\downarrow + Na_2SO_4$$
$$MgCl_2 + Ca(OH)_2 + Na_2CO_3 \longrightarrow Mg(OH)_2\downarrow + CaCO_3\downarrow + 2NaCl$$

(v) Lime reacts with HCl, H_2SO_4, CO_2, H_2S, salts of iron, aluminium, etc. Accordingly, their respective equivalents must be considered for calculating the lime requirement.

$$Ca(OH)_2 + 2HCl \longrightarrow CaCl_2 + 2H_2O$$
$$Ca(OH)_2 + H_2SO_4 \longrightarrow CaSO_4 + 2H_2O$$
$$Ca(OH)_2 + CO_2 \longrightarrow CaCO_3\downarrow + H_2O$$
$$Ca(OH)_2 + H_2S \longrightarrow CaS\downarrow + 2H_2O$$
$$Ca(OH)_2 + FeSO_4 \longrightarrow CaSO_4 + Fe(OH)_2$$
$$2Fe(OH)_2 + H_2O + \frac{1}{2}O_2 \longrightarrow 2Fe(OH)_3$$
$$2Ca(OH)_2 + Al_2(SO_4)_3 \longrightarrow 2Al(OH)_3 + 3CaSO_4$$

(vi) Lime, while reacting with HCl, H_2SO_4, $MgSO_4$, $MgCl_2$, $Mg(NO_3)_2$, salts of Fe, Al etc., generates the corresponding quantities of calcium permanent hardness. Accordingly, these constituents also should be considered while calculating the soda requirement.

(vii) Two equivalents of HCO_3^- reacts with two equivalents of lime as follows:

$$\underset{\text{(2 equivalents)}}{2\ HCO_3^-} + \underset{\text{(2 equivalents)}}{Ca(OH)_2} \longrightarrow CaCO_3 + 2H_2O + \underset{\text{(2 equivalents)}}{CO_3^{2-}}$$

It is evident that in the above reaction, 2 equivalents of CO_3^{2-} are generated. Thus for every one equivalent of HCO_3^- present, the corresponding reduction in the dose of soda has to be made in the calculations for soda requirement.

For solving numerical problems on lime-soda requirements for softening of hard water, the following steps may be followed:

1. The units in which the impurities analysed are expressed i.e., ppm (or mg/l), grains per gallon (or degrees Clark), etc., are to be noted.

2. Substances which do not contribute towards hardness (e.g., KCl, NaCl, SiO_2, Na_2SO_4, Fe_2O_3, K_2SO_4, etc.) should be ignored while calculating lime and soda requirements. This fact should be explicitly stated.

3. All the substances causing hardness should be converted into their respective $CaCO_3$ equivalent, as a matter of convention and convenience.

$$\text{CaCO}_3 \text{ equivalent of a hardness causing impurity} = \frac{\text{Wt. of the impurity}}{\text{Chemical equivalent wt. of the impurity}} \times 50$$

(since chemical equivalent weight of $CaCO_3$ = 50).

For instance, 136 parts by weight of CaSO$_4$ would contain the same amount of Ca as that of 100 parts by weight of CaCO$_3$. Hence, in order to convert the weight of CaSO$_4$ as its CaCO$_3$ equivalent, the weight of CaSO$_4$ should be multiplied by a factor of $\dfrac{100}{136}$ or $\dfrac{50}{68}$.

Conversion factors for some of the impurities in water which commonly come across are given in Table-3.

Table 3

Salt	Multiplication factor to convert into its CaCO$_3$ equivalent
(1)	(2)
Ca(HCO$_3$)$_2$	100/162
Mg(HCO$_3$)$_2$	100/146
CaSO$_4$	100/136
CaCl$_2$	100/111
MgSO$_4$	100/120
MgCl$_2$	100/95
Mg(NO$_3$)$_2$	100/148
Ca^{2+}	100/40
Mg^{2+}	100/24
HCO$_3^-$	$\dfrac{100}{61 \times 2}$
HCl	$\dfrac{100}{36.5 \times 2}$
H$_2$SO$_4$	100/98
CO$_2$	100/44
Al$_2$(SO$_4$)$_2$	$\dfrac{100}{342/3}$
FeSO$_4$. 7H$_2$O	100/278
CaCO$_3$	100/100
MgCO$_3$	100/84
NaAlO$_2$	$\dfrac{100}{82 \times 2}$

Notes :

(*a*) If the impurities are given as CaCO$_3$ or MgCO$_3$, these should be considered to be due to Ca(HCO$_3$)$_2$ or Mg(HCO$_3$)$_2$ respectively and they are only expressed in terms of CaCO$_3$ and MgCO$_3$.

(*b*) The amount expressed as CaCO$_3$ does not require any further convertion. However, the amount expressed as MgCO$_3$ should be converted into its CaCO$_3$ equivalent by multiplying with 100/84.

4. Calculate the lime and soda requirements as follows:

(A) $\begin{Bmatrix} \text{Lime required} \\ \text{for softening} \end{Bmatrix} = \dfrac{74}{100} \times \begin{Bmatrix} \text{Temporary calcium hardness} + (2 \times \text{Temporary magnesium} \\ \text{hardness}) + \text{Perm. Mg harndess} + CO_2 + HCl + H_2SO_4 \\ + HCO_3^- + \text{salts of } Fe^{2+}, Al^{3+} \text{ etc.}, - Na\,AlO_2 \text{; all expressed} \\ \text{in terms of their } CaCO_3 \text{ equivalents.} \end{Bmatrix}$

WATER TREATMENT

(B) $\left.\begin{array}{l}\text{Soda required}\\\text{for softening}\end{array}\right\} = \dfrac{106}{100} \times \left\{\begin{array}{l}\text{Permanent Ca--hardness + permanent Mg hardness + Salts}\\\text{of } Fe^{2+}, Al^{3+}, \text{etc.}, + HCl + H_2SO_4 - HCO_3^- - Na\,AlO_2;\\\text{all expressed in terms of their } CaCO_3 \text{ equivalents}\end{array}\right\}$

(C) If the analytical report shows the quantities of Ca^{2+} and Mg^{2+}, then; 1 eq. of soda is required for Ca^{2+}; whereas one eq. of lime and 1 eq. of soda is required for Mg^{2+}.

5. If the lime and soda used are impure and if the % purity given, then the actual requirements of the chemicals should be calculated accordingly. Thus, if lime is 90% pure, then the value obtained under 4-A above should be multiplied by $\dfrac{100}{90}$ to get the actual requirement of lime. Similarly, if soda is 95% pure, then the value obtained under 4-B above should be multiplied by $\dfrac{100}{95}$ to get the actual soda requirement.

Example 1. *Calculate the amount of lime (84% pure) and soda (92% pure) required for treatment of 20,000 litres of water whose analysis is as follows:*

$$\begin{array}{ll}
Ca(HCO_3)_2 & - \quad 40.5\ ppm\\
Mg(HCO_3)_2 & - \quad 36.5\ ppm\\
MgSO_4 & - \quad 30.0\ ppm\\
CaSO_4 & - \quad 34.0\ ppm\\
CaCl_2 & - \quad 27.75\ ppm\\
NaCl & - \quad 10.0\ ppm
\end{array}$$

Also, calculate the temporary and permanent hardness of the water sample.

Solution.

Salt		$CaCO_3$ equivalent
$Ca(HCO_3)_2$	—	$40.5 \times \dfrac{100}{162} = 25$ ppm or mg/l
$Mg(HCO_3)_2$	—	$36.5 \times \dfrac{100}{146} = 25$ ppm or mg/l
$MgSO_4$	—	$30.0 \times \dfrac{100}{120} = 25$ ppm or mg/l
$CaSO_4$	—	$34.0 \times \dfrac{100}{136} = 25$ ppm or mg/l
$CaCl_2$	—	$27.75 \times \dfrac{100}{111} = 25$ ppm or mg/l
$NaCl$	—	Ignored as it does not contribute to hardness.

Temporary hardness = $[Ca(HCO_3)_2] + [Mg(HCO_3)_2]$ = 25 + 25 = 50 ppm
Permanent hardness = $[MgSO_4] + [CaSO_4] + [CaCl_2]$ = 25 + 25 + 25 = 75 ppm

Lime is required for $Ca(HCO_3)_2$, $Mg(HCO_3)_2$ and $MgSO_4$. Soda is required for $CaSO_4$, $CaCl_2$ and the $CaSO_4$ generated as a result of reaction of lime with $MgSO_4$.

Hence, 84% pure lime required for treating 20,000 litres of water

$= \dfrac{74}{100}\ [\underset{Ca(HCO_3)_2}{25} + \underset{Mg(HCO_3)_2}{(2 \times 25)} + \underset{MgSO_4}{25}] \times \underset{\substack{\text{Purity}\\\text{factor}}}{\dfrac{100}{84}} \times \underset{\substack{\text{grams for}\\\text{2,000 litres}}}{\dfrac{20,000}{1,000}}$

= 1761.905 g
= **1.7619 kg**

92% pure soda required for softening 20,000 litres of the water sample

$$= \frac{106}{100} [\underset{CaSO_4}{25} + \underset{CaCl_2}{25} + \underset{MgSO_4}{25}] \times \underset{\substack{\text{Purity} \\ \text{factor}}}{\frac{100}{92}} \times \underset{\substack{\text{grams for} \\ \text{2,000 litres}}}{\frac{20,000}{1,000}}$$

= 1728.26 g   = **1.72826 kg**

Example 2. *A water sample has the analytical report as under:*

$MgCO_3$	—	84 mg/l
$CaCO_3$	—	40 mg/l
$CaCl_2$	—	55.5 mg/l
$Mg(NO_3)_2$	—	37 mg/l
KCl	—	20 mg/l

Calculate the amount of lime (86% pure) and soda (83% pure) needed for the treatment of 80,000 litres of water.

Solution.

$CaCO_3$ and $MgCO_3$ should be regarded as being present in the form of their bicarbonates and only their weights have been expressed in terms of $CaCO_3$ and $MgCO_3$. KCl does not react with lime or soda and also it does not contribute to hardness.

On converting the weights of each constituent in terms of their $CaCO_3$ equivalent, we get the following:

Salt		$CaCO_3$ equivalent
$MgCO_3$	—	$84 \times \frac{100}{84} = 100$ mg/l
$MgCO_3$	—	$40 \times \frac{100}{100} = 40$ mg/l
$CaCl_2$	—	$55.5 \times \frac{100}{111} = 50$ mg/l
$Mg(NO_3)_2$	—	$37 \times \frac{100}{148} = 25$ mg/l
KCl	—	Ignored

86% pure lime required for softening 80,000 litres of the water

$$= \frac{74}{100} [\underset{MgCO_3}{(2 \times 100)} + \underset{CaCO_3}{40} + \underset{Mg(NO_3)_2}{25}] \times \underset{\substack{\text{Purity} \\ \text{factor}}}{\frac{100}{86}} \times \underset{\substack{\text{grams for} \\ \text{80,000 litres}}}{\frac{80,000}{1,000}}$$

= 18241.86 g. = **18.24186 kg**

83% pure soda required for softening 80,000 litres of the water

$$= \frac{106}{100} [\underset{CaCl_2}{50} + \underset{\substack{Ca(NO_3)_2 \\ \text{generated} \\ \text{from reaction} \\ \text{of lime with} \\ Mg(NO_3)_2}}{25}] \times \underset{\substack{\text{Purity} \\ \text{factor}}}{\frac{100}{83}} \times \underset{\substack{\text{grams for} \\ \text{80,000} \\ \text{litres}}}{\frac{80,000}{1,000}}$$

= 7662.65 g. = **7.6625 kg.**

WATER TREATMENT

Example 3. *A water sample, on analysis, gave the following data :*

$MgCl_2$	—	95 ppm
$CaSO_4$	—	272 ppm
$MgSO_4$	—	120 ppm
H_2SO_4	—	49 ppm
SiO_2	—	4 ppm

Calculate the amount of lime (95% pure) and soda (97% pure) needed for treating 1 million litres of water. If the costs of lime and soda are Rs. 1000 and Rs. 2000 per 100 kg each respectively, calculate the total cost of chemicals used for treating 1 million litres of the water.

Solution.

Impurity	$CaCO_3$ equivalent
$MgCl_2$	$95 \times \dfrac{100}{95} = 100$ mg/l
$CaSO_4$	$272 \times \dfrac{100}{136} = 200$ mg/l
$MgSO_4$	$120 \times \dfrac{100}{120} = 100$ mg/l
H_2SO_4	$49 \times \dfrac{100}{98} = 50$ mg/l
SiO_2	Ignored

Amount of 95% pure lime required for softening 1 million litres of the water

$$= \frac{74}{100} \underbrace{[100}_{MgCl_2} + \underbrace{100}_{MgSO_4} + \underbrace{50]}_{H_2SO_4} \times \underbrace{\frac{100}{95}}_{\substack{\text{purity}\\\text{factor}}} \times \underbrace{\frac{10^6}{10^6}}_{\substack{\text{kg for}\\10^6 \text{ litres}}}$$

$= \mathbf{194.74}$ **kg**

∴ Cost of the lime required at the rate of Rs. 1000 per 100 kg

$$= \frac{194.74 \times 1000}{100} = \textbf{Rs. 1947.4}$$

Similarly, amount of 97% pure soda needed for 1 million litres of the water

$$= \frac{106}{100} [\underbrace{100}_{\substack{CaCl_2\\\text{derived}\\\text{from}\\MgCl_2\\\text{during}\\\text{lime}\\\text{treatment}}} + \underbrace{200}_{CaSO_4} + \underbrace{100}_{\substack{CaSO_4\\\text{derived}\\\text{from}\\MgSO_4\\\text{during}\\\text{lime}\\\text{treatment}}} + \underbrace{50}_{\substack{CaSO_4\\\text{derived}\\\text{from}\\H_2SO_4\\\text{during}\\\text{lime}\\\text{treatment}}}] \times \underbrace{\frac{100}{97}}_{\substack{\text{purity}\\\text{factor}}} \times \underbrace{\frac{10^6}{10^6}}_{\substack{\text{kg for}\\10^6 \text{ litres}}}$$

$= \mathbf{491.75257}$ **kg**

∴ Cost of the soda required at the rate of Rs. 2000 per 100 kg

$$= \frac{491.75257 \times 2000}{100} = \textbf{Rs. 9835.05}$$

∴ Total cost of the chemicals needed for softening 1 million litres of the water
= Rs. (1947.4 + Rs. 9835.05)
= **Rs. 11782.45**

Example 4. *A water sample, on analysis, gave the following constitutents in grains per gallon*

$MgCl_2$	—	9.5
$CaSO_4$	—	3.4
$CaCO_3$	—	5.0
$Mg(HCO_3)_2$	—	7.3
$MgSO_4$	—	6.0
SiO_2	—	2.4

Calculate the cost of the chemicals required for softening 20,000 gallons of water if the purities of lime and soda are 95% and 90% respectively. The costs per 100 pounds each of lime and soda are Rs. 490 and Rs. 960 respectively.

Solution.

Salt	$CaCO_3$ equivalent
$MgCl_2$	$9.5 \times \dfrac{100}{95} = 10$ gpg
$CaSO_4$	$3.4 \times \dfrac{100}{136} = 2.5$ gpg
$CaCO_3$	$5.0 \times \dfrac{100}{100} = 5.0$ gpg
$Mg(HCO_3)_2$	$7.3 \times \dfrac{100}{146} = 5.0$ gpg
$MgSO_4$	$6.0 \times \dfrac{100}{120} = 5.0$ gpg
SiO_2	Ignored

Lime is required for [Temporary calcium hardness
+ (2 × Temporary magnesium hardness)
+ permanent magnesium hardness]
= 5 + (2 × 5) + 10 + 5 = 30 gpg

∴ Cost of 95% pure lime required for softening 20,000 gallons at the rate of Rs. 490 per 100 lbs

$$= \underset{\substack{\text{lime required}\\\text{per gallon}}}{\dfrac{74}{100} \times 30} \times \underset{\substack{\text{gallons}\\\text{treated}}}{20{,}000} \times \underset{\substack{\text{1 lb = 7000}\\\text{grains}}}{\dfrac{1}{7000}} \times \underset{\substack{\text{purity}\\\text{factor}}}{\dfrac{100}{95}} \times \underset{\substack{\text{cost per}\\\text{100 lb}}}{\dfrac{490}{100}}$$

= **Rs. 327.15**

Similarly, soda is required for [Ca permanent hardness + Mg permanent hardness which generated equivalent quantity of Ca permanent hardness during lime treatment]
= 2.5 + (10 + 5) = 17.5 gpg

∴ Cost of 90% soda required for treating 20,000 gallons at the rate of Rs. 960 per 100 lbs

$$= \underset{\substack{\text{Soda required}\\\text{per gallon}}}{\dfrac{106}{100} \times 17.5} \times \underset{\substack{\text{gallons}\\\text{treated}}}{20{,}000} \times \underset{\substack{\text{1 lb = 7000}\\\text{grains}}}{\dfrac{1}{7000}} \times \underset{\substack{\text{purity}\\\text{factor}}}{\dfrac{100}{90}} \times \underset{\substack{\text{cost per}\\\text{100 lb}}}{\dfrac{960}{100}}$$

= **Rs. 565.3**

WATER TREATMENT

Example 5. *A water sample, using $FeSO_4 \cdot 7H_2O$ as a coagulant at the rate of 139 ppm, gave the following results on analysis:*

Ca^{2+} — 160 ppm; Mg^{2+} — 72 ppm

CO_2 — 88 ppm; HCO_3^- — 488 ppm

Calculate the lime and soda required to soften 1,00,000 litres of water.

Solution.

Impurity		Equivalents of $CaCO_3$
Ca^{2+}	$160 \times \dfrac{100}{40}$	= 400 ppm
Mg^{2+}	$72 \times \dfrac{100}{24}$	= 300 ppm
CO_2	$88 \times \dfrac{100}{44}$	= 200 ppm
HCO_3^-	$488 \times \dfrac{100}{61 \times 2}$	= 400 ppm
$FeSO_4 \cdot 7H_2O$	$139 \times \dfrac{100}{278}$	= 50 ppm

Lime required $= \dfrac{74}{100} \times [Mg^{2+} + CO_2 + HCO_3^- + FeSO_4 \cdot 7H_2O]$

$= \dfrac{74}{100} \times [300 + 200 + 400 + 50]$

$= \dfrac{74}{100} \times 950 = 703$ ppm

$= 703$ mg/l

Lime required for softening 1,00,000 litres of water

$= 703 \times \dfrac{10^5}{10^5} =$ **70.3 kg**

Similarly, soda required

$= \dfrac{106}{100} \times [Ca^{2+} + Mg^{2+} + FeSO_4 \cdot 7H_2O - HCO_3^-]$

$= \dfrac{106}{100} \times [400 + 300 + 50 - 400]$

$= \dfrac{106}{100} \times 350 = 371$ ppm

$= 371$ mg/l

∴ Soda required for softening 10^5 litres of water

$= 371 \times \dfrac{10^6}{10^6} =$ **37.1 kg**

Note 1 $Mg^{2+} + Ca(OH)_2 \longrightarrow Mg(OH)_2 + Ca^{2+}$

Note 2 $2HCO_3^- + Ca(OH)_2 \longrightarrow CO_3^- + CaCO_3 + H_2O$
(Two equivalents) (Two equivalents) (Two equivalents)

Example 6. *Calculate the quantities of lime and soda required for cold softening of 2,00,000 litres of water using 16.4 ppm of sodium aluminate as a coagulant. The results of the analysis of raw water and softened water are as follows:*

	Raw water		Softened water
Ca^{2+}	— 160 ppm	CO_3^{2-}	— 30 ppm
Mg^{2+}	— 72 ppm	OH^-	— 17 ppm
HCO_3^-	— 732 ppm		
Dissolved CO_2	— 44 ppm		

Solution.
Converting each of the constituents into their respective equivalents of $CaCO_3$, we have

Raw Water

Ion or salt	ppm	$CaCO_3$ equivalent (ppm)	Lime required, (ppm)	Soda required (ppm)
Ca^{2+}	160	$160 \times \dfrac{100}{40} = 400$	—	400
Mg^{2+}	72	$72 \times \dfrac{100}{24} = 300$	300	300
HCO_3^-	732	$732 \times \dfrac{100}{61 \times 2} = 600$	600	− 600
CO_2	44	$44 \times \dfrac{100}{44} = 100$	100	—
$NaAlO_2$	16.4	$16.4 \times \dfrac{100}{82 \times 2} = 10$	− 10	− 10

Softened Water

CO_3^{2-}	30	$30 \times \dfrac{100}{60} = 50$	—	+ 50
OH^-	17	$17 \times \dfrac{100}{17 \times 2} = 50$	50	+ 50
			1040	**190**

Lime required = $\dfrac{74}{100} \times 1040$ mg/l

Lime required for 2,00,000 litres of water

$$= \dfrac{74}{100} \times 1040 \times \dfrac{2,00,000}{10^6} \text{ kg}$$

$$= \mathbf{153.93 \text{ kg}}$$

Similarly, soda required

$$= \dfrac{106}{100} \times 190 \text{ mg/l}$$

∴ Soda required for 2,00,000 litres of water

$$= \dfrac{106}{100} \times 190 \times \dfrac{2,00,000}{10^6} \text{ kg}$$

$$= \mathbf{40.28 \text{ kg}.}$$

WATER TREATMENT

Notes:

1. 1 equivl. of Ca^{2+} requires 1 equiv. of soda.
$$Ca^{2+} + Na_2CO_3 \longrightarrow CaCO_3 + 2Na^+$$

2. 1 equiv. of Mg^{2+} requires 1 equiv. of lime and 1 equiv. of soda.
$$Mg^{2+} + Ca(OH)_2 \longrightarrow Mg(OH)_2 + 2Ca^{2+}$$
$$Ca^{2+} + Na_2CO_3 \longrightarrow CaCO_3 + 2Na^+$$

3. 1 equiv. of HCO_3^- requires 1 equiv. of lime which simultaneously produces 1 equiv. of CO_3^{2-}, which may be considered to be equivalent to 1 equiv. of soda.
$$\underset{2 \text{ equiv.}}{Ca(OH)_2} + \underset{2 \text{ equiv.}}{2HCO_3^-} \longrightarrow CaCO_3 + \underset{2 \text{ equiv.}}{CO_3^{2-}} + H_2O$$

4. 1 equiv. of CO_2 requires 1 equiv. of lime.

5. 1 equiv. of $NaAlO_2$ requires neither lime nor soda. But, however, $NaAlO_2$ produces 1 equiv. of OH^-, which may be imagined to be 1 equiv. of lime
$$NaAlO_2 + 2H_2O \longrightarrow NaOH + Al(OH)_3$$
Hence the corresponding quantity of $NaAlO_2$ in equivalents should be deducted both from lime as well as soda requirements.

6. Since the treated water is shown to contain OH^- and CO_3^{2-}, the required amount of OH^- should have been supplied by its equivalent amount of $Ca(OH)_2$. But, however, the corresponding amount of Ca^{2+} so incorporated should have been removed by adding equivalent amount of soda. The CO_3^{2-} required to be present in the treated water must have been supplied by its equivalent amount of Na_2CO_3.

Example 7. *The analytical report of a raw water sample is as follows:*

Mg Cl$_2$ — 47.5 mg/l
Ca Cl$_2$ — 55.5 mg/l
Ca SO$_4$ — 4.06 mg/l
Turbidity — 120 mg/l

10 mg/l alum dose was found to be sufficient to remove the entire turbidity of the water sample. Calculate the total weight of the dry sediment in a lime-soda softening plant for 20,000 litres of water. Alum contains 7% Al.

Solution.

1. $\underset{95}{MgCl_2} + Ca(OH)_2 \longrightarrow \underset{58}{Mg(OH)_2\downarrow} + \underset{111}{CaCl_2}$

   $\underset{111}{CaCl_2} + Na_2CO_3 \longrightarrow \underset{100}{CaCO_3\downarrow} + 2NaCl$

   95 mg of $MgCl_2$ yields 58 mg of $Mg(OH)_2$

   ∴ 47.5 mg of $MgCl_2$ yields $\dfrac{58 \times 47.5}{95}$ mg

   $= 29$ mg of $Mg(OH)_2$ ...(1 a)

   Again,
   95 mg of $MgCl_2$ will produce 111 mg of $CaCl_2$ which gives 100 mg of $CaCO_3$ on treatment with soda,

   ∴ 29 mg of $MgCl_2$ will produce $29 \times \dfrac{100}{95}$ mg of $CaCO_3$

   $= 30.526$ mg of $CaCO_3$ ...(1 b)

2. $\underset{111}{CaCl_2} + Na_2CO_2 \longrightarrow \underset{100}{CaCO_3\downarrow} + 2NaCl$

   111 mg of $CaCl_2$ yields 100 mg $CaCO_3$

$\therefore$ 55.5 mg of CaCl$_2$ yields $\dfrac{100 \times 55.5}{111}$ mg of CaCO$_3$

$\qquad = 50$ mg of CaCO$_3$ ...(2)

3. $\underset{136}{CaSO_4} + Na_2CO_2 \longrightarrow \underset{100}{CaCO_3\downarrow} + Na_2SO_4$

136 mg of CaSO$_4$ gives 100 mg of CaCO$_3$

$\therefore$ 4.08 mg of CaSO$_4$ gives

$\dfrac{4.08 \times 100}{136} = \dfrac{408}{136} = 3$ mg of CaCO$_3$ ...(3)

4. Turbidity 120 mg/l ...(4)

5. $\underset{\underset{27}{\text{(From alum)}}}{Al^{3+}} \xrightarrow{\text{hydrolysed}} \underset{78}{Al(OH)_3\downarrow}$

100 mg of alum contains 7 mg of Al^{3+}

$\therefore$ 10 mg of alum contains $\dfrac{7 \times 10}{100} = 0.7$ mg of Al^{3+}

(Since it is given that alum contains 7% Al and the alum dose used is 10 mg/l)

Now, 27 mg of Al^{3+} (from alum) gives 78 mg of Al(OH)$_3$

$\therefore$ 0.7 mg of Al^{3+} gives

$= \dfrac{78 \times 0.7}{27} = 2.222$ mg of Al(OH)$_3$ ...(5)

$\therefore$ Dry sediment obtained from 1 litre of the water sample

$\qquad = (1a + 1b) + (2) + (3) + (4) + (5)$
$\qquad = (29 + 30.526 + 50 + 3 + 120 + 2.222)$ mg
$\qquad = 234.748$ mg

$\therefore$ Total weight of dry sediment obtained from softening 20,000 litres of the water sample
$\qquad = 234.748 \times 20,000$ mg $= \mathbf{4.69496}$ **kg.**

Example 8. *A water sample gave the following analytical results*:
Total alkalinity   = 290 ppm as CaCO$_3$
Calcium hardness   = 242 ppm as CaCO$_3$
Magnesium hardness = 63 ppm as CaCO$_3$
Calculate the lime and soda required to soften 1 million litres of such a water sample.

Solution.
Alkalinity present in water is first bound to Ca^{2+}, then to Mg^{2+}, and only then to Na$^+$.
Further, in the present case,
Ca-alkalinity   = (Ca-hardness or Total alkalinity, whichever is small)
$\qquad = 242$ ppm as CaCO$_3$
$\therefore$ Ca-Temporary hardness = 242 ppm as CaCO$_3$
Mg-alkalinity = (Total alkalinity – Ca-hardness) since, the total alkalinity is greater than
$\qquad\qquad$ Ca hardness but lesser than total hardness
$\qquad = (290 - 242) = 48$ ppm as CaCO$_3$
$\therefore$ Mg-temporary hardness = 48 ppm as CaCO$_3$
Mg-permanent hardness = (Mg hardness – Mg alkalinity)
$\qquad = (63 - 48) = 15$ ppm as CaCO$_3$
Ca-permanent hardness = (Ca-hardness – Ca-alkalinity)
$\qquad = (242 - 242) = 0$
$\therefore$ Lime required for softening 10^6 litres of water sample

WATER TREATMENT

$$= \frac{74}{100} \times [\text{Ca Temp} + (2 \times \text{Mg Temp}) + \text{Mg Perm}] \times 10^6 \text{ mg}$$

$$= \frac{74}{100} \times [242 + (2 \times 48) + 15] \times 10^6 \text{ mg}$$

$$= \frac{74}{100} \times 353 \times 10^6 \text{ mg}$$

$$= \frac{74}{100} \times 353 \text{ kg}$$

$$= 261.22 \text{ kg.}$$

Soda reqd. for softening 10^6 litres of the water sample

$$= \frac{106}{100} \times [\text{Ca perm} + \text{Mg perm}] \times 10^6 \text{ mg}$$

$$= \frac{106}{100} \times [0 + 15] \times 10^6 \text{ mg}$$

$$= \frac{106}{100} \times 15 \times 10^6 \times \frac{1}{10^6} \text{ kg} = \mathbf{15.9 \text{ kg.}}$$

1.7.4. Ion-exchange process of water softening. As applied to water treatment, an ion-exchange process may be defined as a reversible exchange of ions between a liquid phase and a solid phase. Materials capable of exchange of cations are called cation exchangers and those which are capable of exchanging anions are called anion exchangers. Both cation and anion exchangers are widely used in industry, including preparation of boiler feed and process waters.

Ion-exchangers commonly used in water treatment include:
  (*i*) Natural and synthetic zeolites
  (*ii*) Carbonaceous ion-exchangers
  (*iii*) Synthetic resins.

1.7.4. (*i*) Natural and synthetic Zeolites. The name Zeolite is derived from two Greek words (Zein + lithos) which mean "boiling stone". The name was first used by Cronstedt, a Swedish geologist, in 1736, to a certain group of natural minerals, which released their water of hydration (or combination) in the form of steam. These natural Zeolites include:

  (*a*) Thomsonite $(Na_2O, CaO) \cdot Al_2O_3 \cdot 2SiO_2 \cdot 2\frac{1}{2} H_2O$
  (*b*) Natrolite $Na_2O, Al_2O_3 \cdot 3SiO_2 \cdot 2H_2O$
  (*c*) Laumontite $CaO \cdot Al_2O_3 \cdot 4SiO_2 \cdot 4H_2O$
  (*d*) Harmotome $(BaO, K_2O) \cdot Al_2O_3 \cdot 5SiO_2 \cdot 5H_2O$
  (*e*) Stilbite $(Na_2O \cdot CaO), Al_2O_3 \cdot 6SiO_2 \cdot 6H_2O$
  (*f*) Brewsterite $(BaO, SrO, CaO) \cdot Al_2O_3 \cdot 6SiO_2 \cdot 5H_2O$
  (*g*) Ptilolite $(CaO, K_2O, Na_2O), Al_2O_3 \cdot 10SiO_2 \cdot 5H_2O$

The chemical structure of Sodium Zeolite may be represented as $Na_2O, Al_2O_3, X_{2-10} \cdot SiO_2, Y_{2-6} \cdot H_2O$, so that they may be regarded as hydrated sodium alumino-silicates which are capable of exchanging their sodium ions for multivalent ions of alkaline earth group and for the divalent ions of some of the metals in water. Thus, Zeolites find application in softening of water for domestic and industrial purposes. Two types of Zeolites are in common use:

  (*a*) Natural Zeolites *e.g.,* non-porous green sands, and
  (*b*) Synthetic Zeolites which are porous and possess a gel structure.

The natural zeolites are derived from green sands by washing, heating and treatment with NaOH. The synthetic zeolites are prepared from solutions of sodium silicate and aluminium hydroxide. They may also be prepared by heating together (*a*) China clay, felspar and soda ash and granulating the resultant mass after cooling (b) solutions of sodium silicate, $Al_2(SO_4)_3$ and $NaAlO_2$ (c) solutions of sodium silicate and $Al_2(SO_4)_3$ (d) solutions of sodium silicate and $NaAlO_2$. Natural zeolites are more durable, whereas synthetic zeolites have higher exchange capacity per unit weight. Zeolites are now being replaced slowly by high capacity resins having greater stability.

Zeolites are also known as permutits. Sodium zeolites are used in water softening and are simplistically represented as Na_2Z, where Z stands far the insoluble zeolite radical framework. Since these are capable of exchanging basic radicals, these are generally known as base exchangers.

When hard water passes through a bed of active granular sodium zeolite, the Ca^{2+} and Mg^{2+} ions are taken up by the zeolite and simultaneously releasing the equivalent sodium ions in exchange for them. The various reactions taking place may be indicated as follows:

$$Ca(HCO_3)_2 + Na_2Z \longrightarrow CaZ + 2NaHCO_3$$
$$Mg(HCO_3)_2 + Na_2Z \longrightarrow MgZ + 2NaHCO_3$$
$$CaSO_4 + Na_2Z \longrightarrow CaZ + Na_2SO_4$$
$$MgSO_4 + Na_2Z \longrightarrow MgZ + Na_2SO_4$$
$$CaCl_2 + Na_2Z \longrightarrow CaZ + 2NaCl$$
$$MgCl_2 + Na_2Z \longrightarrow MgZ + 2NaCl$$

Relatively small quantities of iron and manganese, present as the divalent bicarbonates, may also get removed simultaneously.

$$Fe(HCO_3)_2 + Na_2Z \longrightarrow FeZ + 2NaHCO_3$$
$$Mn(HCO_3)_2 + Na_2Z \longrightarrow MnZ + 2NaHCO_3$$

(**Note.** Soluble iron and manganese are always present in the divalent form in waters containing bicarbonate alkalinity).

Regeneration

When the Zeolite bed is exhausted (i.e., saturated with Ca^{2+} and Mg^{2+}), it can be regenerated and reused. The chemicals used for regeneration are concentrated sodium chloride solution (brine), sodium nitrate, sodium sulphate, potassium chloride or ptassium nitrate. Brine is most widely used on account of its cheapness, relatively low molecular weight and also because the products formed by the regeneration reactions are chiefly $CaCl_2$ and $MgCl_2$ which are highly soluble and can be readily rinsed out from the zeolite bed.

$$CaZ + 2NaCl \longrightarrow Na_2Z + CaCl_2$$
$$MgZ + 2NaCl \longrightarrow Na_2Z + MgCl_2$$

Zeolite water softeners

Zeolite water softeners are made in both pressure type and gravity type and both the types are available in automatic, semi-automatic and manually operated designs. All these units operate on the same principles, involving alternate cycles of softening run and the regeneration run. During the softening run, the water is softened by passing it through the sodium zeolite bed, where Ca^{2+} and Mg^{2+} are removed from the water and held by the Zeolite and simultaneously releasing equivalent amount of Na^+ in exchange. The regeneration step comprises of (a) backwashing (b) salting (or brining) and (c) rinsing before reuse.

A simple zeolite softener is depicted in Fig. 1.12 and the scheme for softening and regeneration processes are represented in Fig. 1.13.

Advantages of Zeolite process

1. Hardness is completely removed.
2. Equipment used is compact and occupies less place.

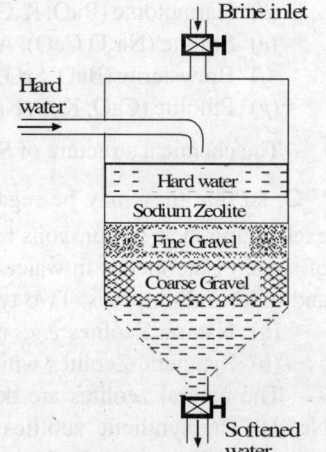

Fig. 1.12. Zeolite softener.

WATER TREATMENT

3. It automatically adjusts itself to waters of different hardness.
4. It can work under pressure. Hence the plant can be installed in the water supply line itself, avoiding double pumping.
5. In this process, the hardness causing ions are simply exchanged with sodium ions. As the process does not involve any precipitation, there is no problem of sludge formation and after-precipitation in the softened water at later stages.

Limitations

1. Water, having turbidity and suspended matter should not be directly fed to the zeolite softener because the pores of the zoelite bed will be clogged and the rate of flow will be unduly decreased. Therefore, the raw water turbidity and suspended matter should be removed (*e.g.*, by sedimentation, coagulation and filtration) before subjecting it to the Zeolite treatment.

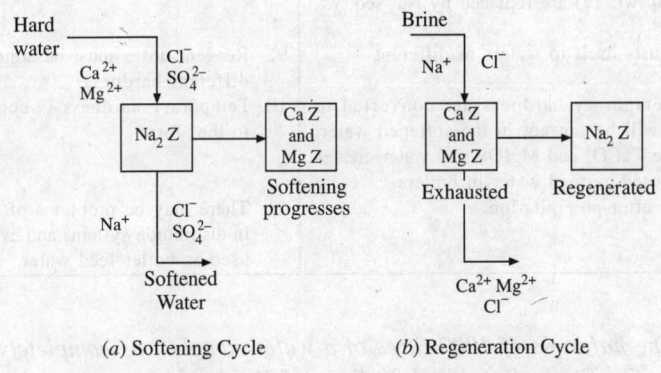

Fig. 1.13. Schemes for softening of water by Zeolite process.

2. Water containing excess of acidity or alkalinity may attack the Zeolite. It is preferable to have the pH of the water passing through the Zeolite softener around 7.
3. Water containing large quantities of Fe^{2+} and Mn^{2+} when passed through the Zeolite bed are converted into their respective Zeolites which cannot be easily regenerated.
4. Hot water should not be used as the zeolite tends to dissolve in it.
5. Acid radicals are not removed by this process. Hence, the temporary hardness present in the raw water is converted to $NaHCO_3$ which goes into soft water effluent. If such a water is used in the boilers, it dissociates as follows under the boiler conditions.

$$NaHCO_3 \longrightarrow NaOH + CO_2$$

This NaOH may cause caustic embrittlement of boiler metal. Further, CO_2 goes along with the steam and renders the condensed water acidic and corrosive. Hence, raw waters containing large amounts of temporary hardness should be subjected to prolonged boiling or lime treatment to remove temporary hardness, before subjecting to the Zeolite process.
6. Water treated by the Zeolite process contains about 25% more dissolved solids than that treated by lime-soda process.

Comparison between the Zeolite process and the lime-soda process

The salient features of Zeolite process and lime-soda process are summarized in the following table.

Zeolite process	Lime-soda process
1. This process produces water of almost zero hardness	1. This process produces water having hardness of 15 to 60 ppm depending on whether it is hot process or cold process.
2. The cost of the plant and the Zeolite are higher. Hence the capital cost is higher.	2. The capital cost is lower.

3. The exhausted Zeolite bed can be regenerated with brine which is very cheap. Hence the operating cost is lesser	3. The chemicals needed viz., lime, soda, and coagulant are consumed in the process. Hence the operating cost is higher.
4. The plant is compact and occupies less space. The size of the plant depends on the hardness of water being treated.	4. The plant occupies more space. The size of the plant depends on the amount of water being handled
5. Cannot be used for hot water, acidic waters, and waters having turbidity and suspended impurities.	5. The process is free from such limitations.
6. This process can operate under pressure and can be designed for fully automatic operation.	6. This process cannot be operated under pressure.
7. This process does not involve cumbersome operations such as settling, coagulation, filtration.	7. This process involves all the problems associated with settling, coagulation and filtration.
8. Water treated with the Zeolite process contains larger amounts of sodium salts and greater percentage of dissolved salts than the raw water since Ca^{2+} (eq. wt. 20) and Mg^{2+} (eq. wt. 12) are replaced by Na^+ (eq. wt. 23).	8. Treated water contains lesser percentage of dissolved solids and lesser quantities of sodium salts.
9. This process adjusts itself to waters of different hardness.	9. Reagent doses must be adjusted for waters of different hardness.
10. Salts causing temporary hardness are converted to $NaHCO_3$ which will be present in the softened water. Such of insoluble $CaCO_3$ and $Mg(OH)_2$. a water creates problems when used as feed water in boilers.	10. Temporary hardness is completely removed in the form
11. No problems of after-precipitation.	11. There may be problems of after-precipitation in distribuion systems and even in boilers when used as boiler feed water.

Numericals

Example 1. *The hardness of 1000 litres of a water sample was completely removed by a Zeolite softener. The Zeolite had required 30 litres of NaCl solution, containing 1,500 mg/l of NaCl for regeneration. Calculate the hardness of the water sample.*

Solution.
30 litres of NaCl solution contains
$$= 1.5 \times 30 \text{ g} = 45 \text{ g of NaCl.}$$
$$\equiv 45 \times \frac{50}{58.5} \text{ g of } CaCO_3 \text{ equivalent hardness.}$$

This much hardness may be deemed to be present in 1000 litres of water sample.
∴ Hardness of the water sample
$$= 45 \times \frac{50}{58.5} \times \frac{1,000}{1,000} \text{ mg/l}$$
$$= \textbf{38.46 ppm.}$$

Example 2. *An exhausted Zeolite softener was regenerated by passing 150 litres of NaCl solution, having a strength of 150 g/l of NaCl. How many litres of hard water sample, having hardness of 600 ppm can be softened, using this softener?*

Solution.
150 litres of NaCl solution contains
$$150 \times 150 \text{ g} = 22,500 \text{ g Na Cl}$$
$$\equiv 22,500 \times \frac{50}{58.5} \text{ g of } CaCO_3 \text{ equivalent hardness.}$$

Given that 1 litre of hard water contains 600 ppm hardness ≡ 600 mg of $CaCO_3$ = 0.6 g of $CaCO_3$
∴ The amount of hard water that can be softened by this softener
$$= \frac{22,500 \times 50}{0.6 \times 58.5} = \textbf{32,051 litres.}$$

WATER TREATMENT

Lime required for 8,00,000 litres of water

$$= \frac{74}{100} \times \frac{100}{80} \times \frac{8,00,000}{100} \text{ Kg}$$

$$= 490.82 \text{ Kg}$$

Similarly, Soda required

$$= \frac{106}{100}\left[Ca^{++} + Mg^{++} + OH^- + CO_3^{--} - HCO_3^- - NaAlO_2 \right]$$

$$= \frac{106}{100}[375 + 300 + 132.35 + 41.66 - 150 - 21.34] \times \frac{100}{90}$$

$$= \frac{106}{100}[677] \times \frac{100}{90} \times \frac{800000}{10^6} \text{ Kg}$$

$$= 638.17 \text{ Kg}$$

ii) Hardness of water $= [Ca^{++}] + [Mg^{++}]$
$= 300 + 375$
$= 675$ ppm

15000 Liters of water is softered
Hardness in 15000 litres of water $= 15000 \times 675$
$= 10125000$ mg/15000 litres

let NaCl required for regneration is χ mg

$$\therefore \text{NaCl required} = \frac{\chi \times 50}{58.5}$$

Hardness in 15000 liters of Water $=$ NaCl required

$$10125000 = \frac{\chi \times 50}{58.5}$$

$$\therefore \chi = 11.846 \times 10^6 \text{ gm}$$
$$= 11.84 \text{ Kg.}$$

iii) Let volume of 0.01 M EDTA required for 100 ml of water is γ ml.

1 mole of EDTA $=$ 1 mole of Ca^{++}
1 ml of 1 M EDTA $=$ 100 mg of $CaCO_3$

Total hardness present in 100 ml of the water sample
γ ml of 0.01 M EDTA $= 100 \times \gamma \times 0.01$ mg/10ml

The hardness present in 1000 ml of water sample $\therefore 675$

$$\therefore 675 = 100 \times \gamma \times 0.01 \times \frac{1000}{100}$$

$$\therefore \gamma = 67.5 \text{ ml.}$$

Example 5. Water on analysis gave the following results.

$Ca(HCO_3)_2 = 40.5$ ppm $MgCl_2 = 71.25$ ppm

$Mg(HCO_3)_2 = 58.4$ ppm $MgSO_4 = 168$ ppm

$CaCl_2 = 44.4$ ppm   Turbidity $= 11$ ppm

Calculate

i) Amount of lime and soda required to soften 1 million liters of water using 24.6 ppm sodium aluminate as coagulent

ii) Calculate the total weight of the dry sediments in a lime soda softening plant for 1 million liters of water.

Solution :

i) Converting each constituents into their respective equivalent of $CaCO_3$.

Salt	$CaCO_3$ equivalent
$Ca(HCO_3)_2$	$\dfrac{40.5 \times 100}{162} = 25$ ppm
$Mg(HCO_3)_2$	$\dfrac{58.4 \times 100}{146} = 40$ ppm
$CaCl_2$	$\dfrac{44.4 \times 100}{111} = 40$
$MgCl_2$	$\dfrac{71.25 \times 100}{95} = 75$
$MgSO_4$	$\dfrac{168 \times 100}{120} = 140$
Turbidity	100
$NaAlO_2$	$\dfrac{24.6 \times 100}{2 \times 82} = 15$

$$\text{Lime required} = \frac{74}{100}\left[2 \times Mg(HCO_3)_2 + Ca(HCO_3)_2 + MgCl_2 + [MgSO_4 + NaAlO_2\right]$$

$$= \frac{74}{100}[25 \times 2 + 40 + 75 + 140 - 15] \text{ ppm}$$

$$= \frac{74}{100}[305] \times \frac{10^6}{10^6} \text{ Kg}$$

$$= 225.7 \text{ Kg}$$

WATER TREATMENT

$$\text{Soda required} = \frac{106}{100}\left[CaCl_2 + MgCl_2 + MgSO_4 - NaAlO_2\right]$$

$$= \frac{106}{100}[40+75+140-15] \text{ ppm}$$

$$= \frac{106}{100}[240] \times \frac{10^6}{10^6} \text{ Kg}$$

$$= 254.4 \text{ Kg}$$

ii) Sediments formation $Ca(HCO_3) + Ca(OH)_2 \rightarrow 2CaCO_3 \downarrow + CO_2 + H_2O$

162   2 x 100 mg
40.5 mg   ?

$$\frac{2 \times 100 \times 40.5}{162} = 50$$

$Mg(HCO_3)_2 + 2Ca(OH)_2 \rightarrow Mg(OH)_2 \downarrow + 2CaCO_3 \downarrow + CO_2 + H_2O$

146 mg   58 mg   200 mg
58.4 mg

$$CaCO_3 \text{ formed from } Mg(HCO_3)_2 = \frac{58.4 \times 58}{146} = 23.2$$

$$Mg(OH)_2 \text{ formed from } Mg(HCO_3)_2 = \frac{58.4 \times 200}{146} = 80$$

$CaCl_2 + Na_2CO_3 \rightarrow CaCO_3 \downarrow + 2NaCl$
111   100
44.4

$$CaCO_3 \text{ formed from } CaCl_2 = \frac{44.4 \times 100}{111} = 40$$

$MgCl_2 + Ca(OH)_2 \rightarrow Mg(OH)_2 \downarrow + CaCl_2 \rightarrow CaCO_3$
95mg   58   100 mg
71.25

$$\frac{71.25 \times 58}{95} = 43.5; \quad \frac{71.25 \times 100}{95} = 75$$

$MgSO_4 + Ca(OH)_2 \rightarrow Mg(OH)_2 \downarrow + CaSO_4 + CaCO_3 \downarrow$
120   58   100
168   ?   ?

$$CaCO_3 \text{ formed from } MgSO_4 = \frac{168 \times 58}{120} = 81.2$$

$$Mg(OH)_2 \text{ formed from } MgSO_4 = \frac{168 \times 100}{120} = 140$$

Sediments = $CaCO_3$ + MgOH + Turbidity
= 385 + 147.9 + 100
= 642.9 mg/l

Weight of dry sediments for 1 million liters of water
= 642.9 x 10^6
= 642.9 Kg

Lime Zeolite Process

Water treated by lime-soda process is frequently supersaturated with respect to $CaCO_3$ and $Mg(OH)_2$ (sometimes free lime also) which leads to after-precipitation. In order to obviate this difficulty of delayed precipitation, the following process may be employed: the water is first treated with a slight excess of lime to precipitate most of the bicarbonate and magnesium. Now, this water containing precipitates of $CaCO_3$ (~ 40 ppm) and $Mg(OH)_2$ (~30 ppm) and also the permanent hardness is fed to a pressure filter containing a bed of carbonaceous Zeolite in the calcium form. This bed acts as a filter for suspended matter, completes the lime-bicarbonate reaction, retains the precipitated carbonates, hydroxides etc., and absorbs the excess lime. The residual permanent hardness may now be removed by conventional Zeotile softening. The stabilizing carbonaceous Zeolite may be regenerated with CO_2 gas or with dilute HCl or dilute H_2SO_4.

1.7.4. (ii) Carbonaceous Ion-Exchangers. If certain coals are granulated and treated with fuming sulphuric acid, they develop ion-exchange properties. The products obtained are called sulphonated coals which are physically and chemically stable towards acid and to some extent towards alkalis (upto pH 9.5). The most widely employed carbonaneous cation exchanger is made by sulphonating a special grade of bituminous coal with either fuming sulphuric acid or sulphur trioxide. This introduces — SO_3H groups into the complex organic compounds present in the coal which are responsible for the cation exchange properties. After sulphonation is complete, the excess of acid is washed out and the last traces are removed by treating with sodium carbonate, which also converts the material into the sodium form. The material is then dried, screened to size and packed. When used on the sodium cycle, the carbonaceous cation exchangers have several advantages over the siliceous Zeolites such as that (ah there is no possibility of silica pick-up and (b) $CaCO_3$ coating (which) may occur when the carbonaceous exchangers are used after cold lime soda treatment) and iron deposits formed on the exchanger can be safely and readily cleaned by acid treatment.

A lot of work has been done on sulphonation of coal, peat, lignite and wood with fuming sulphuric acid, for the preparation of carbonaceous ion-exchangers. Some of the commercial products available include Zeokarb, Catex 27 and Catex 55. These exchangers can be operated both on sodium and hydrogen cycles and are not deteriorated by waters of even very low pH or by iron bearing waters. However, these carbonaceous exchangers are vulnerable to oxidation by chlorine.

1.7.4. (iii) Synthetic ion-exchange resins. In an ion-exchange process, a reversible exchange of ions takes place between the stationary ion-exchange phase and the external liquid mobile phase. An ion-exchange resin consists of an insoluble polymeric matrix that is permeable. Fixed charge groups and mobile counter ions of opposite charge are incorporated in the matrix. These counter ions can be exchanged for ions in the external liquid phase.

The purely synthetic organic exchangers are made by:

1. Polycondensation *e.g.*, between a substituted phenol and formaldehyde, with some unsubstituted phenol to get the desired cross-linking between the linear chains.

2. Polymerisation *e.g.*, of stryrene in presence of divinyl benzene, to give the desired cross-linking.

The functional groups are then introduced into the cross-linked resin network, either by subsequent treatment of the resin or by introducing the functional groups into the starting material itself. It is

WATER TREATMENT

these functional groups which decide the nature of the exchanger (*i.e.*, cationic or anionic). Based on the acidity or basicity of the functional group, exchangers are further classified as follows:

Types of ion-exchanger	Functional groups
Strongly acidic cation-exchangers	—SO$_3$H
Moderately strong cation-exchangers	—PO(OH)$_2$
Weakly acidic cation-exchangers	—COOH or —OH
Strongly basic anion-exchanger	—NR$_3^+$; ≡ P$^+$—CH$_3$; etc.
Weakly basic anion-exchanger	—NH$_2$; —(C$_2$H$_4$)$_x$(NH)$_y^-$

Hypothetical formulations of a typical cationic and a typical anionic resin based on styrene and divinyl benzene network are shown in Figs. 1.14 and 1.15.

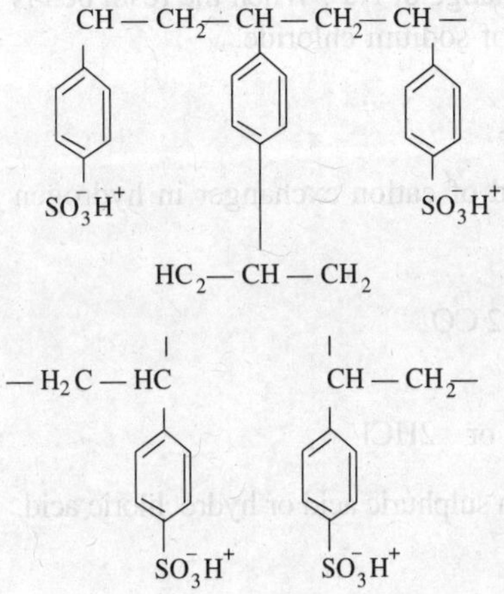

Fig. 1.14. Hypothetical structure of a typical cation exchanger.

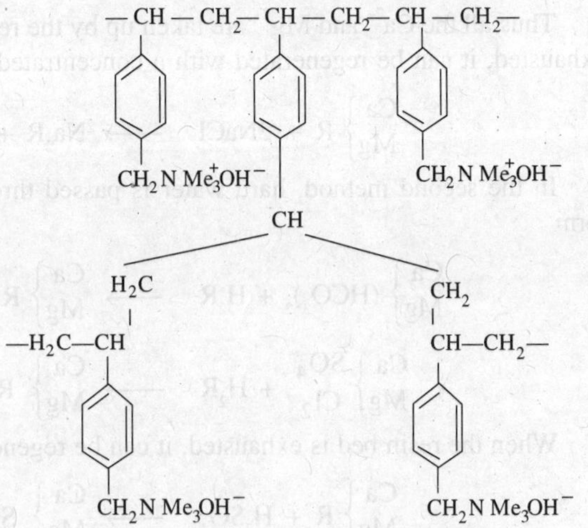

Fig. 1.15. Hypothetical structure of a typical anion exchanger.

Variation in polymer type and cross-linking affects the insolubility and life of the resin and the diffusibility of ions in an exchange process.

Resins available for water treatment include cation-exchange resins of the strongly and weakly acidic types, anion-exchange resins of the strongly and weakly basic types and highly porous modifications of strong acid cation exchangers and strong base anion exchangers. These resins are available in granular or bead-like form and may be obtained in an effective size and uniformity coefficient.

For effective water treatment, ion exchangers should possess the following properties:
1. They should be non-toxic.
2. They should not discolour the water being treated.
3. They should possess a high ion-exchange capacity. (It depends upon the total number of ion active groups per unit weight of the exchanger and is expressed as mill-equivalents per gram of the exchanger).
4. They should be physically durable.
5. They should be resistant to chemical attack.
6. They should be cheap and commonly available.
7. They must be capable of being regenerated and back-washed easily and economically.
8. They should have a large surface area since ion-exchange is a surface phenomenon. At the same time, their resistance to flow must be compatible with hydraulic requirements.

National Chemical Laboratory, Pune, developed polystyrene based cation-exchangers and melamine based anion exchangers which are available in the market with different trade names.

1.7.5. Softening of water by ion-exchange. The ion-exchange resins used in water-softening may be used on either the sodium cycle or the hydrogen cycle (*i.e.*, the active exchanging cationic group may be in the sodium form or hydrogen form). In the first method, hard water is passed through a bed of a cation-exchanger in sodium form:

$$Ca \begin{cases} (HCO_3)_2 \\ SO_4 \\ Cl_2 \end{cases} + Na_2R \longrightarrow \begin{matrix} Ca \\ Mg \end{matrix} \} R + Na_2 \begin{cases} (HCO_3)_2 \\ SO_4 \\ Cl_2 \end{cases}$$

Thus all the Ca^{2+} and Mg^{2+} are taken up by the resin in exchange of Na^+. When the resin bed is exhausted, it can be regenerated with a concentrated solution of sodium chloride.

$$\begin{matrix} Ca \\ Mg \end{matrix} \} R + 2NaCl \longrightarrow Na_2R + \begin{matrix} Ca \\ Mg \end{matrix} \} Cl_2$$

In the second method, hard water is passed through a bed of cation exchanger in hydrogen form

$$\begin{matrix} Ca \\ Mg \end{matrix} \} (HCO_3)_2 + H_2R \longrightarrow \begin{matrix} Ca \\ Mg \end{matrix} \} R + 2H_2O + 2CO_2$$

$$\begin{matrix} Ca \\ Mg \end{matrix} \} \begin{matrix} SO_4 \\ Cl_2 \end{matrix} + H_2R \longrightarrow \begin{matrix} Ca \\ Mg \end{matrix} \} R + H_2SO_4 \text{ or } 2HCl$$

When the resin bed is exhausted, it can be regenerated with sulphuric acid or hydrochloric acid.

$$\begin{matrix} Ca \\ Mg \end{matrix} \} R + H_2SO_4 \longrightarrow \begin{matrix} Ca \\ Mg \end{matrix} \} SO_4 + H_2R$$

For many industrial water supplies, chemical softening (*e.g.*, lime soda process) is used first to remove most of the hardness followed by a cation exchange process for completing the softening process. For softening water, usually sodium cycle is preferred.

1.7.6. Demineralisation or deionisation. For many years, distillation was the only method available for complete removal of soluble salts from water. However, with the advent of the development of commercial cation-exchange and anion exchange resins, it has been possible to produce de-ionised water of highest purity by a two-step ion exchange process.

First, the hard water is passed through a cation-exchange resin in the hydrogen form

$$H_2R + \begin{matrix} Ca^{2+} \\ Mg^{2+} \\ 2Na^+ \end{matrix} \} \begin{matrix} SO_4 \\ Cl_2 \end{matrix} \longrightarrow R \begin{cases} Ca \\ Mg \\ Na_2 \end{cases} + \begin{matrix} H_2SO_4 \\ \text{or} \\ 2HCl \end{matrix}$$

The effluent from this step is then passed through an anion exchanger bed.

$$R'(OH)_2 + \begin{matrix} H_2SO_4 \\ \text{or} \\ 2HCl \end{matrix} \longrightarrow R' \begin{cases} SO_4 \\ \text{or} \\ Cl_2 \end{cases} + 2H_2O$$

$$R'(OH)_2 + H_2CO_3 \longrightarrow R'CO_3 + 2H_2O$$

Thus, all the ions present in water are removed and demineralised or deionised water is produced.

Strongly basic anion-exchanger is used when silica also is to be removed. Otherwise, weakly basic anion exchangers are used generally. If CO_2 removal is also desired, forced draft or vacuum degasifier unit is provided. The anion exchanger may be regenerated by treating it with dilute NaOH solution. The cation exchanger may be regenerated by passing a solution of HCl or H_2SO_4.

WATER TREATMENT

$$R'\begin{Bmatrix} SO_4 \\ Cl_2 \end{Bmatrix} + 2NaOH \rightarrow R'(OH)_2 + Na_2 \begin{Bmatrix} SO_4 \\ Cl_2 \end{Bmatrix}$$

$$R\begin{Bmatrix} Ca \\ Mg \end{Bmatrix} + 2HCl \longrightarrow RH_2 + \begin{Bmatrix} Ca \\ Mg \end{Bmatrix} Cl_2$$

A typical demineralisation unit is shown in Fig. 1.16.

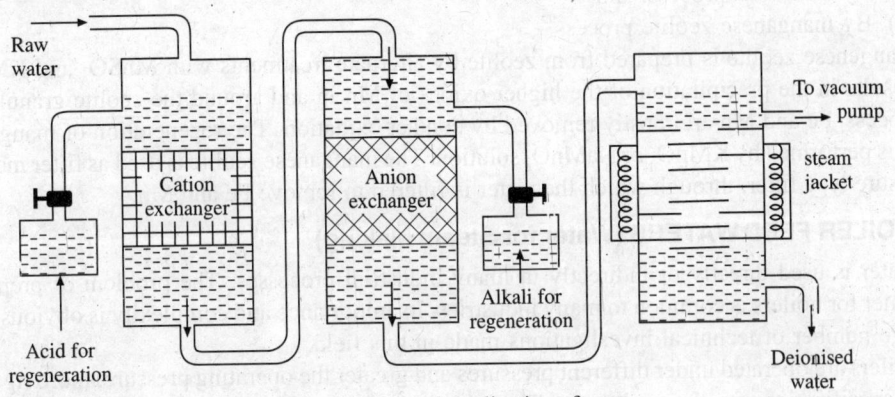

Fig. 1.16. Demineralization of water.

Demineralisation of water can be better achieved by employing a "mixed bed" of cationic and anionic resins. This method produces an effluent which is far superior to that produced by the two bed operation. When water is passed through a "mixed bed" consisting of anion and cation exchange resins, each pair of contrasting resin particles functions as a stage in the treatment and thus the total effect is that of a multiple-cycle-deionisation. Thus the process is highly efficient.

When the resins are exhausted the bed is backwashed when the two resins are separated in different layers due to difference in their densities. Then the resins are separately regenerated, washed and mixed again by injecting air and reused for a fresh cycle.

The deionisation process can also be used to purify highly acidic or alkaline waters. All ionizable impurities are removed and many commercial plants produce water having as low as 3 ppm dissolved solids and 0.1 ppm silica. However, the process is costly. The equipment and the resins as well as the regenerants are expensive as compared to those used in zeolite process. Water having turbidity reduces the efficiency of the process. Hence it is advisable to treat it first by lime-soda process and then removing the residual hardness by this process. This process is recommended for very high pressure boilers, where the specifications are more exacting.

Demineralized water is also used in the manufacture of photographic materials, fine chemicals, synthetic rubber, pharmaceuticals, ceramics, explosives, plastics, paper products such as dielectrics, soluble coffee, cosmetics, liquid soaps, storage batteries, catalysts, ice, cellulose products, televison tubes, for cooling purposes in broadcasting stations and diesel locomotives, power plants, oil refineries, electroplating plants and in laboratories.

Distillation

This method of removing hardness and scale forming constituents is used in large power stations and on steamships, where the amount of make-up water is small compared with the total amount of water evaporated in the boilers. However, until cheaper sources of energy are available, water softening by distillation is going to be an expensive proposition.

Removal of iron and manganese from water

Iron and manganese may be present in water as bicarbonates, sulphates, hydroxides or as chelates. The presence of iron and manganese in water even in such low concentrations as 0.1 to 0.3 ppm may be objectionable in industrial processes such as textile dyeing and finishing, paper making, rayon processing, beverage manufacture, laundering and tanning.

Iron and manganese in water may be removed by the following processes:
   (i) By aeration followed by settling and filtration, the metals are removed in the form of Fe$(OH)_3$ and $Mn(OH)_3$.
   (ii) By ion exchange process using zeolites, carbonaceous or synthetic cation exchangers.
   (iii) By lime-soda process, and
   (iv) By manganese zeolite process

Mangenese zeolite is prepared from zeolite by alternate treatments with $MnSO_4$ and $KMnO_4$. This results in the precipitation of the higher oxides of Mn in and around the zeolite granules. In this process, Fe and Mn are mostly removed by contact oxidation. The regeneration of manganese zeolite is performed by $KMnO_4$ or $NaMnO_4$ solution. The manganese zeolite is used as filter medium in pressure type filter, through which the water is filtered to remove Fe and Mn.

1·8. BOILER FEED WATERS (Water for Steam Making)

Water is used directly or indirectly in many industrial processes. The problem of preparing feed water for boilers is common to many industries. Its importance and complexity is obvious from the large number of technical investigations made in this field.

Boilers are operated under different pressures and greater the operating pressure, more rigorous are the specifications for the feed water. The operating pressures for different types of boilers are given below:

	Operating pressures
Low-pressure boilers	Upto 200 lb/sq. in
Intermediate pressure boilers	200 to 500 lb/sq. in
High-pressure boilers	500 to 2,000 lb/sq. in
Very high-pressure boilers	2,000 to 3,209 lb/sq. in
Supercritical boilers	Above 3,209 lb/sq. in

(1 lb/sq. in = 0.0703 kg/cm^2; 1 kg/cm^2 = 14.223 lb/sq. in.)

There are a large number of different types of boilers in use which may be divided into two main classes, viz., shell boilers and water tube boilers. A shell boiler essentially consists of a cylinder in which the water is contained and heated by the passage of hot gases through flues or tubes immersed in the water. The shell boilers in turn, can be further subdivided into (a) Horizontal flue-type boilers (e.g., the Cornish boilers) and (b) Horizontal fire-tube boilers (e.g., the Locomotive and the Economic boilers). In the water-tube boilers, of which many types exist, it is the water which passes through tubes, either by forced or by natural circulation.

An important difference in the two main classes of boilers mentioned above is that the simple shell boilers are not usually provided with the ancillary equipment with which the water tube boilers are fitted. Thus water tube boilers are provided with economizers and feed-water heaters (pre-boiler equipment) and with super-heaters. While designing a water-treatment process for water-tube boilers, therefore, it is essential to take into consideration the conditions existing in such equipment.

Any natural source of water does not supply a perfectly suitable boiler feed water. It is generally believed that a boiler feed water should satisfy the following requirements:

Hardness	—	< 0.2 ppm
Caustic alkalinity	—	0.15 – 0.45 ppm
Soda alkalinity	—	0.15 – 1 ppm
Excess soda ash	—	0.3 – 0.55 ppm

WATER TREATMENT

External treatment of boiler feed-water make-up

Steam produced in a boiler is condensed after use and returned as pure feed water to the boiler. There are inevitable losses, however, the "make-up" water must be supplied from time to time. The term "boiler feed-water make-up" means the treated water which is required to make up the water losses from the boiler caused by use of open steam in process work or steam otherwise lost plus the water lost in the boiler blow-down. In certain cases, there are no condensate returns or none fit for re-use in the boiler, so that the feed water is 100% make-up water. In some other cases, particularly in large power plants, surface condensers employed are capable of recovering almost all of the water evaporated and make-up may comprise of < 2% of the water fed to the boiler. All other cases lie between the above two extremes.

The treatment of boiler feed waters may be broadly divided into two categories:

1. External treatment. In this, the scale-forming and corrosive impurities are removed from the water before it enters the boiler.

2. Internal treatment. In this, dosages of various materials are added in the boiler to react with the non-carbonate hardness, to reduce the sticking tendency of scale-forming materials and to neutralise or overcome corrosive tendencies.

In many cases, both the above forms of treatment are employed.

In the external treatment of boiler feed water, one or more of the following processes are usually employed:

1. Removal of "hardness" or scale forming salts
   (a) Lime-soda process
   (b) Zeolite process
   (c) Demineralization
   (d) Distillation.
2. Removal of silica:
   Magnesia process, ferric sulphate process, anion-exchange process, Fluosilicate process.
3. Removal of suspended matter:
   (a) pre-treatment: sedimentation, coagulation, settling and filtration.
   (b) softening processes.
4. Removal of dissolved gases: CO_2 and O_2
   (a) Decarbonation by aeration
   (b) Deaeration
   (c) Chemical treatment.
5. Removal of oil
   (a) Mechanical oil-separators
   (b) Sodium aluminate treatment.

Most of these processes have been discussed earlier and others will be discussed later, along with the methods for internal treatment.

1·9. BOILER TROUBLES

The problem of making up feed water for boilers is common to almost all industries and has been a subject of detailed study by several workers. In treatment of boiler feed water, total elimination of all the impurities is generally not attempted. Only those impurities which give rise to certain operational troubles are either eliminated or maintained within tolerable limits. The major boiler troubles caused by the use of unsuitable water are (1) carry over: priming and foaming (2) scale formation (3) corrosion and (4) caustic embrittlement. All these troubles increase with increasing operating pressure of the boilers. The limits of tolerance for boiler feed waters, suggested by the committee on water quality tolerances for industrial users of the New England Water Works Association are, given under Table-4. These tolerance limits take into account that (a) reduction in turbidity colour, oxygen consumed value and total solids decrease carry over (b) reduction in hardness, silica and alumina decrease scale formation (c) elimination of oxygen, reduction of bicarbonate ions and increase in pH help in suppressing corrosion and (d) maintaining a high sulphate to carbonate ratio checks caustic embrittlement.

Table 4. Suggested limits of Tolerances for Boiler Feed Waters

Characteristic	\multicolumn{4}{c}{Pressure, MN/m^2 (Mega-newton/meter square)}
	0 – 1.0	1.0 – 1.7	1.7 – 2.8	over 2.8
1. Turbidity (silica scale units)	20	10	5	1
2. Colour (Hazen units)	80	40	5	1
3. Oxygen consumed	15	10	4	3
4. Dissolved oxygen	2	0.2	0	0
5. Hydrogen sulphide	5	3	0	0
6. Total hardness ($CaCO_3$)	80	40	10	2
7. Sulphate to carbonate ratio	1:1	2:1	3:1	3:1
8. Aluminium oxide	5	0.5	0.05	0.01
9. Silica	40	20	5	1
10. Bicarbonate	200	100	40	20
11. Hydroxide	50	40	30	15
12. Total solids	3000-1500	2000-1500	1500-100	50
13. pH Value (min)	8.0	8.4	9.0	9.6

(a) Units for items 3 to 12 : mg/l
(b) 1 MN/m^2 = 145 psi

1.9(1) Carry over. As steam rises from the surface of the boiling water in the boiler, it may be associated with small droplets of water. Such steam, containing liquid water, is called wet steam. These droplets of water naturally carry with them some suspended and dissolved impurities present in the boiler water. This phenomenon of carrying of water by steam along with the impurities is called "carry over". This is mainly due to priming and foaming.

The solid and liquid contaminations in the steam are expressed by steam purity and steam quality respectively. The steam purity is expressed in parts per million of the impurity present and accordingly if steam contains 2 parts of solids contamination by weight per million parts by weight of steam, it is called 2 ppm steam. Similarly the liquid contamination in the steam is expressed in percentage by weight, of the steam in the mixture. Thus if the steam contains 0.2% moisture, its steam quality is reported as 99.8%.

Steam used for power production is usually superheated for achieving greater efficiencies in the turbine or engine in which steam is used. Wet steam causes corrosion in the inlet ends of the superheaters. If the steam contains high percentage of moisture, the extent of superheating will decrease with the consequent reduction in efficiency of the turbine or engine. Further, the water carried over with the steam contains salts and sludges, these are carried into the superheater where they may deposit as the water evaporates. This will seriously restrict the flow of steam. Moreover, due to the insulating effect of these deposits, the superheater tubes also may burn out. A part of the dried salts may be carried along with the steam farther and deposit on the high-pressure turbine blades or in engine valves. Even a small amount of deposit on the turbine blades decreases its efficiency considerably. In order to eliminate the bad effects of moisture, mechanical steam purifiers are often installed in the steam drums of the boiler or between the boiler and the superheater. These devices force the steam to take curved paths whereby due to centrifugal action, the moisture is thrown out of the steam.

Priming is a very rapid boiling of water occuring in the boiler in such a way that some water particles are carried away along with steam in the form of spray into the steam outlet. Priming is mainly attributed to the presence of suspended impurities and to some extent to dissolved impurities in the water. Thus it may be caused by imperfect filtration or precipitation of insoluble salts from the improperly softened or unsoftened feed water. Algae and vegetable growth in tanks can also

WATER TREATMENT

lead to priming. Feed water containing even a small quantity of scale forming salts may cause priming particularly if the total solids in the feed exceed 300 ppm. Priming may also be caused by: (1) steam velocities high enough to carry droplets of water into the steam pipe (2) very high water level in the boiler (3) presence of excessive foam on the surface of the water which substantially fills the foam space (4) sudden steam demands leading to sudden drop of pressure in the steam line followed by ebullition, which causes a large mass of water to be interspersed with fine bubbles and (5) faulty boiler design.

Priming can be minimised by (1) good boiler design providing for proper evaporation of water maintaining uniform heat distribution and adequate heating surfaces and also providing with antipriming pipes and dash-plates, etc. (2) maintaining low water levels (3) avoiding rapid changes in the steaming rate caused by sudden steam demands (4) maintaining as low a concentration of boiler water (with respect to dissolved impurities) as is consistent with scale and corrosion prevention and (5) minimizing foaming.

Foaming is the formation of small but presistent bubbles at the water surface. These bubbles are carried along with steam leading to excessive priming.

According to Bancroft, foams are formed when there is a difference in concentration of solute or suspended matter between the surface film and the bulk of the liquid. Substances which increase the viscosity of the film favour production of foam. The bubbles may also be protected by finely divided solids forming a protective "shell" around each of them. Any material which lowers the surface tension of the water will collect at the interface and thus increase the foaming tendency of the liquid.

In steam raising, the bubbles of steam may be stabilized because of the accumulation of soluble salts in water. Clay or organic matter in raw water, oil or grease in condensed make-up water, and finely divided particles of sludge may also cause foaming.

If the steam bubble does not collapse on reaching the surface of the water, the foam may be drawn into the preheater or steam lines. The accompanying liquid film may carry along with it dissolved salts, suspended solids or other stabilizing materials which will be deposited on the cylinder walls, turbine blades, or in steam lines. If the stability of the bubble is such that it breaks near the steam outlet, tiny droplets of liquid from the collapsing film may be swept along into the steam lines giving "wet steam".

Foaming (and the consequent priming) can best be prevented by removal of the foaming and stabilizing agents from the water. Clay and other suspended solids as well as droplets of oil and grease can be removed by treatment of the feed water with clarifying agents such as hydrous silicic acid and aluminium hydroxide. The concentration of salts and sludge in the boiler can be controlled by intermittent or continuous blow-down. Spiral or "cyclone baffles" or a series of baffle plates near the steam outlets help to prevent water droplets from entering the steam lines.

Foaming can also be controlled by adding antifoaming agents. Some of these act by counteracting the reduction in surface tension while others reduce foaming by simple mechanical action. For instance, castor oil spreads on the surface of water and prevents foaming. This is used only for low pressure boilers. Addition of a small concentration of polyamide antifoamer alters the surface tension and leads to the formation of only large unstable bubbles at the same heat input. Antifoaming agents may also act by reducing the charge on the protective film or its components. A foam may be destroyed by the addition of another good foaming agent. The foam is finally destroyed by mutual antagonistic effect of the difference in charge (positive and negative) on the colloidal particles of the two foams. This can be illustrated by the fact that the foam obtained by a solution of an anionic detergent (*e.g.,* Aerosol OT) and that obtained from a cationic detergent (Ethyl Cetab) will destroy each other on mixing the two solutions.

For avoiding priming and foaming, the following limits on total dissolved solids (in the boiler water) are usually suggested.

Type of boiler	Total dissolved solids, ppm
Water tube	2000 to 5000
Lancashire	1000 to 1500
Vertical	≤ 3500

In order to keep down the feed water concentration within safe limits, it is customary to remove a portion of the concentrated boiler water by blow-down and replace it with fresh feed water. Blow-down may be done periodically or continuously. However, blow-down will result in loss of heat.

Removal of oil

Oil present in boiler water may lead to (a) the formation of heat insulating oil films, and (b) carry over (if the oil is fatty). Natural waters generally do not contain oil. Small amount of oil may come from the oil used for lubrication of the pumps. A large quantity of boiler water is obtained from the condensation of exhaust steam which usually gets contaminated with lubricating oil in the steam cylinder. In order to remove oil, the exhaust steam is passed through mechanical separators which contain a number of metal plates. As the steam passes over the metal plates, the oil droplets are retained on the metal surface.

Oil is also removed by coagulation with sodium aluminate or alum and soda ash. The floc produced by the $Al(OH)_3$ and $Mg(OH)_2$ precipitates enmesh the oil. The floc is then removed by filtering through anthrafilt.

Oil can also be removed from water by cataphoresis.

Removal of Silica

Silica may cause hard deposits on turbine blades due to carry over. Silica also forms thermally resistant boiler scales of magnesium or calcium silicate.

Silica is removed from the boiler feed water by the following methods:

(1) By the addition of magnesia (MgO) to the water after the removal of the temporary hardness.
(2) By the addition of magnesia to the water in the boiler when magnesia silica sludge is deposited.
(3) By using magnesia or dolomitic lime in hot lime soda process of water softening. Magnesia acts as $Mg(OH)_2$. In all these processes, silica is removed by magnesia by adsorption or by the formation of magnesium silicate.
(4) By the use of ferrous sulphate or sodium aluminate as coagulants. They act as $Fe(OH)_2$ or $Mg(OH)_2$ and $Al(OH)_3$, which enmesh finely suspended and colloidal impurities including oil and silica.
(5) By passing the demineralised water through a strongly basic special anion-exchanger which removes silica as H_2SiO_3 or $HSiO_3^-$ or in some other form. The silica content is usually reduced to a fraction of a ppm.
(6) By converting the silica into fluosilicate by the addition of Hydrofluoric acid and then removing it by passing through an anion exchange resin.*

$$H_2SiF_6 + 2R_3N \longrightarrow (R_3N)_2 \cdot H_2SiF_6$$

(7) By distillation.

1.9. (2) Scale formation. In a boiler, water is continuously converted into steam. This results in the concentration of the dissolved impurities until the water becomes saturated. Then the salts start separating out from the solution in order of their solubility, the least soluble one separating out first. Some of the solids separate in the body of the liquid in the form of soft and muddy deposits or in the form of suspension which can be flushed out easily. Such deposits are known as *sludges*. On the contrary, some of the solids deposit on a solid surface to form a sticky and coherent *scale*. The

---

* Fluosilicic acid is strongly ionized (unlike silicic acid, H_2SiO_3) and hence can be removed by passing through a weakly basic anion exchanger. Silica in water can be converted into fluosilicic acid by adding sodium fluoride and passing through a cation exchanger:

$$SiO_2 + 6NaF + 3H_2R \longrightarrow 3Na_2R + H_2SiF_6 + 2H_2O.$$

WATER TREATMENT

formation of sludges, unless excessive, has not much deleterious effects. But, however, the deposition of solid scale on the metal surfaces of the boiler has a serious effect on the efficient operation of boilers.

The commonest solids that separate from boiler water are the sparingly soluble calcium salts e.g., $CaSO_4$, $Ca(OH)_2$, $CaCO_3$, $Ca_3(PO_4)$, $Ca(OH)_2$, and the magnesium compounds e.g., $Mg(OH)_2$; and calcium and magnesium silicates. Some of these are deposited as sludges and others as coherent scales.

The suspended matter may be coagulated owing to the increase in temperature in the boiler, or it may be entrapped by the crystallising soluble salts. For the soluble materials to be crystallized, their respective solubility products must be exceeded. This may be caused by (1) the normal increase in concentration that occurs due to continuous evaporation of water in the boiler. (In a high pressure boiler, the total quantity of water, equal to the capacity of the boiler, is converted into steam within 15 minutes) (2) the decrease of solubility of some dissolved substances e.g., $CaSO_4$ or $CaCO_3$ at high temperature or (3) by the reactions that produce insoluble salts, such as by the decomposition of soluble bicarbonates e.g., $Ca(HCO_3)_2$, forming insoluble carbonates (e.g., $CaCO_3$).

$$2HCO_3^- = CO_3^{--} + CO_2\uparrow + H_2O$$

Calcium sulphate and calcium carbonate both have negative temperature solubility curves, and may, therefore, deposit scale. $CaCO_3$ may in certain circumstances be deposited as a scale due to decomposition of calcium bicarbonate. Calcium sulphate, calcium silicate and magnesium silicate form the most troublesome scales. Calcium and magnesium hydroxides and calcium carbonate together with iron and aluminium oxides are often entrapped in the sulphate and silicate scales during their formation. In very high pressure boilers, sodium aluminium silicate scales resembling the zeolite minerals are also noticed, particularly in regions of excessive local evaporation. Several other salts and colloidal and suspended impurities also may get deposited or entrapped in the precipitates of salts forming the scales.

Disadvantages of scale formation

1. Scale is a poor conductor of heat. Its effect is like that of an insulator coating on the metal surface. This results in the reduced rate of heat transfer, and thus the evaporative capacity of the boiler will be reduced. Thus, scale formation will result in wastage of fuel and reduction in boiler efficiency.

2. Scale formation on the boiler tubes or other heated surfaces insulates the metal so well that it becomes overheated. The metal becomes soft and weak thus making the boiler unsafe particularly at high pressures. The overheating also causes burning out of the metal plates and tubes and breakdown of the expanded joints.

3. In addition to the loss of strength due to overheating, rapid reaction between water and iron occurs at high temperatures, causing additional thinning of the tube wall.

$$3Fe + 4H_2O \longrightarrow Fe_3O_4(s) + 4H_2(g)$$

4. Since the scale acts as heat insulator, the metal of the boiler is overheated. Under the high pressure of steam existing in the boiler, the metal expands until the scale on it cracks. Sudden entry of the water through these cracks to the very hot metal causes sudden cooling of the boiler metal with the simultaneous conversion of water into steam. The sudden increase in pressure due to this large quantity of steam thus formed may lead to explosion.

5. Excessive scaling may cause clogging of tubes. Considerable quantities of sludges may also be entrapped in the scale which may reduce the water circulation and impair the efficiency of the boiler.

Prevention of scale formation

Scale formation can be prevented by the following methods:

(a) External treatment

This involves removal of hardness causing impurities (such as calcium and magnesium salts) and silica from the water before entering the boiler. The various methods of external treatment of water are already discussed earlier.

(b) Internal treatment

Internal treatment consists of adding chemicals directly to the water in the boilers for removing dangerous scale forming salts which were not completely removed in the external treatment for water softening. This is mainly used as a corrective treatment to remove the slight residual hardness and also sometimes to remove the corrosive tendencies in water. This treatment is not usually applied to raw waters, except for small boilers, but it is usually practised in larger power-stations. In modern heavy-duty high pressure boilers, water of zero hardness is required, since even an egg-shell thickness of scale may be extremely detrimental.

1. Carbonate conditioning

For a salt to be precipitated, sufficient amount of the ions forming the salt must be present so that the product of their concentrations (*i.e.*, ionic product) exceed a limiting value known as the solubility product. Thus, for a salt like $CaCO_3$ to be precipitated, the product of the concentration of Ca^{2+} and CO_3^{2-} must exceed the solubility product of $CaCO_3$, represented as K_{CaCO_3}.

The formation of $CaCO_3$ may be represented by

$$Ca^{++} + CO_3^{--} \rightleftharpoons \underset{\text{solution}}{CaCO_3} \rightleftharpoons \underset{\text{solid}}{CaCO_3} \qquad ...(1)$$

When the equilibrium is established,

$$\underset{\substack{\text{(Concentration} \\ \text{of } Ca^{2+})}}{[Ca^{++}]} \quad \underset{\substack{\text{(Concentration} \\ \text{of } CO_3^{2-})}}{[CO_3^{--}]} = \underset{\substack{\text{(Solubility product} \\ \text{of } CaCO_3)}}{K_{CaCO_3}}$$

at any given temperature.

If under these conditions, some Na_2CO_3 is added, the CO_3^{2-} increases. Then, in order to maintain the K_{CaCO_3} constant, some Ca^{2+} will get precipitated.

Similarly in a saturated $CaSO_4$ solution,

$$Ca^{++} + SO_4^{--} \rightleftharpoons \underset{\text{solution}}{CaSO_4} \rightleftharpoons \underset{\text{solid}}{CaSO_4}$$

and

$$\underset{\substack{\text{(Concentration} \\ \text{of } Ca^{2+})}}{[Ca^{++}]} \quad \underset{\substack{\text{(Concentration} \\ \text{of } SO_4^{2-})}}{[SO_4^{--}]} = \underset{\substack{\text{(Solubility product} \\ \text{of } CaSO_4)}}{K_{CaSO_4}} \qquad ...(2)$$

Thus, if a solution is saturated with $CaCO_3$ and $CaSO_4$ (as occurring in boiler water conditions), both the relations (1) and (2) apply, and by dividing (1) by (2), we get

$$\frac{[CO_3^{--}]}{[SO_4^{--}]} > = \frac{K_{CaCO_3}}{K_{CaSO_4}} \qquad ...(3)$$

From this, it is clear that if

$$\frac{[CO_3^{--}]}{[SO_4^{--}]} > \frac{K_{CaCO_3}}{K_{CaSO_4}} \text{ or } K';$$

or in other words, if $[CO_3^{--}] > K'[SO_4^{--}]$, then only $CaCO_3$ will be precipitated in preference to $CaSO_4$ (because the solubility product of $CaSO_4$ cannot be attained under these conditions).

These principles are used in the carbonate conditioning. When sodium carbonate solution is added to boiler water, the $[CO_3^{--}]$ increases and when it becomes greater than $K' \times [SO_4^{--}]$, only $CaCO_3$ gets precipitated and $CaSO_4$ remains in solution. Thus the deposition of scale-forming $CaSO_4$ is prevented.

$$Na_2CO_3 + CaSO_4 \longrightarrow CaCO_3\downarrow + Na_2SO_4$$

Carbonate conditioning is used only for low pressure boilers. In high pressure boilers the excess Na_2CO_3 might be converted into NaOH due to hydrolysis as follows:

$$Na_2CO_3 + H_2O \rightleftharpoons 2\,NaOH + H_2CO_3$$

WATER TREATMENT

$$H_2CO_3 \rightleftharpoons H_2O + CO_2\uparrow$$

NaOH causes caustic embrittlement in high pressure boilers.

2. Phosphate conditioning

Just as in the case of carbonate conditioning, if we consider a solution saturated with both $Ca_3(PO_4)_2$ and $CaSO_4$, the solubility product equations for both salts must be satisfied.

$$Ca_3(PO_4)_2 \rightleftharpoons 3\,Ca^{++} + 2PO_4^{---}$$

$$\therefore \quad [Ca^{++}]^3\,[PO_4^{---}]^2 = K_{Ca_3(PO_4)_2} \text{ solubility product of } Ca_3(PO_4)_2 \qquad ...(4)$$

$$\therefore \quad [Ca^{++}]\,[PO_4^{---}]^{2/3} = [K_{Ca_3(PO_4)_2}]^{1/3} \qquad ...(5)$$

Similarly, for $CaSO_4$

$$CaSO_4 = Ca^{++} + SO_4^{--}$$

$$\therefore \quad [Ca^{++}]\,[SO_4^{--}] = K_{CaSO_4} \text{ solubility product of } CaSO_4 \qquad ...(6)$$

By dividing (5) by (6) we have

$$\frac{[PO_4^{---}]^{2/3}}{[SO_4^{--}]} = \frac{[K_{Ca_3(PO_4)_2}]^{1/3}}{[K_{CaSO_4}]} \text{ or } K' \qquad ...(7)$$

or $\quad [PO_4^{---}]^{2/3} = K'\,[SO_4^{--}] \qquad ...(8)$

Therefore, as long as $[PO_4^{---}]^{2/3} > K'\,[SO_4^{--}]$, only $Ca_3(PO_4)_2$ will be precipitated in preference to $CaSO_4$ (because the solubility product of $CaSO_4$ cannot be exceeded under these conditions) and hence $CaSO_4$ scale will not form.

These principles are made use in the phosphate conditioning. In this, an excess of a soluble phosphate is added to the boiler water to precipitate the residual calcium ions in the form of a non-adherent precipitate of calcium phosphate and thus the scale formation is prevented.

The three sodium orthophosphates *viz.*, Na_3PO_4; Na_2HPO_4 and NaH_2PO_4 have been used for phosphate conditioning as also sodium pyrophosphate ($Na_4P_2O_7$) and sodium metaphosphate ($NaPO_3$). The typical reactions of the various phosphates with the hardness represented as $CaCO_3$, may be summarized as follows:

$$2\,Na_3PO_4 + 3\,CaCO_3 \longrightarrow Ca_3(PO_4)_2 + 3\,Na_2CO_3$$
$$2\,Na_2HPO_4 + 3\,CaCO_3 \longrightarrow Ca_3(PO_4)_2 + 2\,Na_2CO_3 + CO_2 + H_2O$$
$$2\,NaH_2PO_4 + 3\,CaCO_3 \longrightarrow Ca_3(PO_4)_2 + Na_2CO_3 + 2\,CO_2 + 2\,H_2O$$
$$2\,NaPO_3 + 3\,CaCO_3 \longrightarrow Ca_3(PO_4)_2 + Na_2CO_3 + 2\,CO_2$$

The quality of the feed water dictates the choice of a particular phosphate to be used. For instance, if the feed water tends to produce an acidic condition in the boiler, the alkaline Na_3PO_4 should be chosen. This treatment could be supplemented with NaOH if the required alkalinity could not be maintained with Na_3PO_4 alone. If the feed water produces almost the right alkalinity desired in the boiler, it is preferable to use Na_2HPO_4 which is practically neutral. If the boiler water becomes too alkaline, the acidic NaH_2PO_4 would be selected.

Both sodium pyrophosphate and metaphosphate are rapidly hydrolysed under boiler water temperatures to orthophosphate.

$$NaPO_3 + H_2O = NaH_2PO_4$$
$$Na_4P_2O_7 + H_2O = 2\,Na_2HPO_4$$

Thus, their behaviour within the boiler is identical with that of orthophosphates mentioned above. However, $NaPO_3$ solutions are practically neutral, whereas NaH_2PO_4 solutions are acidic. Hence the former would be preferred if the use of NaH_2PO_4 causes feedline corrosion.

The use of internal treatment combined with suitable blow-down to remove sludge has contributed largely to the operation of the modern high-pressure steam boilers without the formation of hard scales. However, precaution should be taken to inspect them at least once in six months and remove the scale and sludge accumulations.

3. Colloidal conditioning

Scale formation can also be minimised by introducing into the boiler some colloidal conditioning agents such as glue, agar agar, tannins, starches and sea-weed extract. These substances act as protective colloids. They function by surrounding the minute particles of $CaCO_3$ and $CaSO_4$ and prevent their coalescence and coagulation. Thus, the precipitated scale-forming salts are maintained in loose suspended form which can easily be removed by blow-down operation. Thus the scale formation is prevented.

4. Calgon conditioning

Another approach for preventing scale formation is to convert the scale forming salts into highly soluble complexes which are not easily precipitated under the boiler conditions. In order to achieve this, sodium hexameta phosphate $(Na\,PO_3)_6$ or $Na_2[Na_4P_6O_{18}]$ (its trade name is calgon) is generally employed. This substance interacts with the residual calcium ions forming highly soluble calcium hexametaphosphate and thus prevents the precipitation of scale-forming salts.

$$Na_2[Na_4P_6O_{18}] \rightleftharpoons 2\,Na^+ + [Na_4P_6O_{18}]^{2-}$$
$$2\,Ca^{2+} + [Na_4P_6O_{18}]^{2-} \longrightarrow 4\,Na^+ + [Ca_2P_6O_{18}]^{2-}$$

5. Conditioning with EDTA

Phosphate treatment fails to prevent the formation of iron oxide and cuprous depositions. Sometimes, the phosphate chemicals themselves become a source of deposit formation (*e.g.,* iron phosphate). In distinction to phosphate treatment, the conditioning of boiler water with complexing agents can secure scale-free and sludge free operation of boiler units and sometimes, also corrosion-free operation under certain conditions. These reagents are known as complexones or complexing agents such as ethylenediamine tetraacetic acid (EDTA) and its disodium salt (Na_2EDTA) known as Trylon-B, or even its tetrasodium salt (Na_4EDTA).

6. Boiler compounds

Once, during the early times of Watt's engine, it is said that workmen, after cleaning the boiler and refilling with water, hung a bag containing potatoes in the boiler to cook them. They forgot about them and put the boiler into operation after closing it. Then, when the boiler was shut down again for manual removing of the scale, it was found that the scale formation was much less, much of it having come down in the form of loose sludge. The story seems to be true because, for many years, engineers used to throw some potatoes in the boiler after every cleaning operation. Perhaps, this is the forunner of internal treatment of boiler salines to reduce the adherent scale formation by rendering it in the form of softer sludge which can be easily removed by blow-off. Slowly, potatoes were replaced by starches and later by tannins and other organic and inorganic materials such as soda ash and caustic soda, phosphates, coagulants such as sodium aluminate, and deoxidizers like sodium sulphite. Natural wood and plant extracts and several synthetic products have also been used. U.S. Navy was reportedly using a boiler compound consisting of 47% anhydrous disodium phosphate, 44% soda ash and 9% starch. Castor oil compounds and various synthetic antifoam compounds were also used in locomotive boilers to reduce foaming tendency. These boiler compounds do help in reducing scale formation when properly used under proper situations.

WATER TREATMENT

(c) Blow-off

In spite of external and internal treatment, the concentration of impurities in the boiler water goes on increasing continuously as the generation of steam continues. In order to keep the concentration of boiler water below a set figure, a part of the boiler water is blown off to waste and fresh feed water is filled in to replace it. The blow-down (or blow-off) may be done periodically or, preferably, continuously. During the blow-down process, a part of concentrated water containing dissolved impurities and suspended sludge is removed and the remaining salines get diluted with the fresh feed water make-up.

1.9. (3) Corrosion. Corrosion of boiler tubes, plates, economisers and pipe lines can be majorly attributed to one or more of the following factors:

(a) Presence of free acids in water.
(b) Acids generated as a result of hydrolysis of some salts in water.
(c) Acids formed by the hydrolysis of fatty lubricating oils.
(d) Presence of dissolved gases such as O_2, CO_2, H_2S etc., in water.
(e) Presence of salts like $MgCl_2$ which directly attack the boiler metal:

$$MgCl_2 + Fe + 2H_2O \longrightarrow Mg(OH)_2 + FeCl_2 + H_2$$

(f) Presence of salts like MnS_2 which may generate H_2SO_4 due to oxidation and hydrolysis.
(g) Formation of galvanic cells.

Some of these factors are discussed below:

(i) Dissolved oxygen

Dissolved oxygen is the chief source of corrosion in boilers and ancillary equipment. It is highly desirable to remove the dissolved oxygen or air from softened boiler waters by physical or by chemical means. The concentration of oxygen in boiler waters should be below 0.05 ppm for low pressure boilers and less than 0.01 ppm for high pressure boilers (~ 500 psi).

Oxygen enters the boiler through raw make up water and also through infiltration of air into the condensate system. When the water containing dissolved oxygen is heated in the boiler, the free gas is evolved which corrodes the metal parts under the conditions obtaining in the boiler.

The solubility of a gas is directly proportional to pressure and inversely proportional to temperature (Dalton's law and Henry's law). These two principles are made use of in the design of mechanical deaerators.

In mechanical deaerators, dissolved oxygen is removed by injecting hot feed-water, as a fine spray, into a vacuum chamber heated externally by steam. Equipments with various designs are available.

For complete removal of dissolved oxygen, chemical methods of treatment are adopted.

Sodium sulfite is suitable for boilers operating below 650 psi. The sulphite reacts with the dissolved oxygen to form sulphate.

$$Na_2SO_3 + \tfrac{1}{2} O_2 \longrightarrow Na_2SO_4$$

This method is very effective for removing oxygen in low pressure boilers but cannot be used in high pressure boilers because (i) increasing dissolved salts' concentrations produce foaming and priming and (ii) sodium sulphate decomposes and liberates SO_2 and/or H_2S.

Ferrous sulphate is also used sometimes. It reacts with dissolved oxygen giving a precipitate of $Fe(OH)_2$ which is oxidized to $Fe(OH)_3$.

$$FeSO_4 + 2NaOH \longrightarrow Fe(OH)_2 + Na_2SO_4$$
$$2Fe(OH)_2 + H_2O + O \longrightarrow 2Fe(OH)_3$$

Hydrazine, N_2H_4 is now extensively used to remove dissolved oxygen in high pressure boilers. The pure compound is an explosive inflammable liquid (B.Pt. 113.5°C) but the 40% aqueous solution used for water treatment is quite safe to handle. However, rubber gloves must be worn to guard against the danger of dermatitis in some cases. One of the important advantage of hydrazine treatment is that combination with oxygen does not produce any salts. Nitrogen and water are the only reaction products obtained.

$$N_2H_4 + O_2 \longrightarrow N_2 + 2H_2O$$

The molecular weight of hydrazine and oxygen are the same (*viz.* 32), so that complete elimination of 1 ppm of oxygen would require only 1 ppm of hydrazine (which is 1/8 of the necessary concentration of sodium sulphite). The residual hydrazine can be measured easily by a colorimeter.

The amount of hydrazine added must be closely controlled. Any excess reagent decomposes in the boiler, liberating ammonia:

$$3N_2H_4 = 4NH_3 + N_2$$

Dissolved ammonia can bring about corrosion of some alloys, *e.g.*, copper alloy condenser tubes.

(*ii*) Dissolved mineral acids

Most of the natural waters are alkaline excepting waters from mining areas or those polluted from acidic industrial wastes or those in which wet oxidation of sulphide minerals occur.

Some inorganic salts like magnesium chloride and calcium chloride are also corrosive agents. $MgCl_2$ hydrolyzes completely at 200°C producing hydrochloric acid as follows:

$$MgCl_2 + 2H_2O \longrightarrow Mg(OH)_2 + 2HCl$$

$CaCl_2$ also undergoes hydrolysis but to a lesser extent. At 600°C, its hydrolysis is about 25%. Silicic acid catalyzes the reaction so that in water containing silica, appreciable quantities of HCl may be formed at lower temperatures. The HCl thus produced reacts with iron as follows:

$$Fe + 2HCl = FeCl_2 + H_2$$
$$FeCl_2 + 2H_2O = Fe(OH)_2 + 2HCl$$

Hence, even a small amount of $MgCl_2$ can cause considerable corrosion of the metal. If the amount of HCl formed is small, it might get neutralized by the alkalinity present in the water, but otherwise it should be neutralized by the addition of alkali.

(*iii*) Dissolved carbon dioxide

Natural waters contain CO_2. Water containing bicarbonates release CO_2 on heating. If CO_2 is released inside the boiler, it will go along with the steam and as the steam condenses, the CO_2 is dissolved in water forming carbonic acid. This produces intense local corrosion called pitting.

CO_2, along with oxygen in water, can be removed by mechanical deaeration.

CO_2 in water can be removed by lime treatment. Another method of removing CO_2 from water is to filter it through lime stone:

$$CaCO_3 + CO_2 + H_2O \longrightarrow Ca(HCO_3)_2$$

but this reaction produces temporary hardness in the water.

CO_2 can be converted into ammonium carbonate by the addition of ammonia.

$$CO_2 + 2NH_3 + H_2O \longrightarrow (NH_4)_2CO_3$$

However, the excess ammonia added to the boiler feed water may go along with steam to the condenser. If some O_2 is also present in the condensate, the ammonia may attack the condenser tubes made of copper. Hence, a safe limit of 10 mg of NH_3/litre of the condensate is prescribed.

(*iv*) Formation of galvanic cells

Corrosion can also occur because of galvanic cell formation between iron and other metals present in the alloys used in boiler fittings. This may also lead to pitting corrosion. This can be prevented by suspending zinc plates which act as sacrificial anodes.

1.9. (4) Caustic embrittlement. Caustic embrittlement is a form of corrosion caused by a high concentration of sodium hydroxide in the boiler water. It is characterised by the formation of irregular intergranular cracks on the boiler metal, particularly at places of high local stress, such as

WATER TREATMENT

riveted seams, bends and joints. It is caused by the high concentration of NaOH which is capable of reacting with steels stressed beyond their yield point. It is most likely to occur in boilers operating at higher pressures, where NaOH is produced in the boiler by the hydrolysis of Na_2CO_3 as follows:

$$Na_2CO_3 + H_2O \longrightarrow 2NaOH + CO_2\uparrow$$

The extent of the hydrolysis increases with temperature and may reach even 90% of the carbonate present. The rate and extent of corrosion by caustic embrittlement increases with the concentration of NaOH and temperature and hence with increasing operating pressure.

NaOH reacts with iron forming magnetic oxide and hydrogen.

$$2NaOH + Fe \longrightarrow Na_2FeO_2 + H_2\uparrow$$
$$3(Na_2FeO_2) + 4H_2O \longrightarrow 6NaOH + Fe_3O_4 + 4H_2\uparrow$$
$$3Fe + 4OH^- \longrightarrow Fe_3O_4 + 4H$$

Under normal conditions in unstressed metal a fairly continuous film of oxide is produced. When the metal is stressed beyond its yield point, the oxide coating cracks and chemical attack continues into the metal mainly along grain boundaries. This may be due partly to the energy stored there and partly to the increased E.M.F. produced as a result of the stress. The products of the reaction *viz.*, Fe_3O_4 and hydrogen also tend to favour penetration along grain boundaries. The attack is considerable under these conditions because of the large area of the metal exposed.

It is observed that boiler waters containing sodium sulphate or sodium phosphate inhibit the caustic embrittlement either by crystallizing out and plugging the capillaries and crevises with the solid salts (and prevent the infiltration of NaOH) before a dangerously high concentration of NaOH has been produced or that these salts act as buffer solutions and lower the E.M.F. to such an extent that the eorrosion cannot occur.

Generally, the concentrations of Na_2SO_4: NaOH are maintained at 1 : 1, 2 : 1 or 3 : 1 for operating pressures of 10, 20 and > 20 atmospheres respectively to check caustic embrittlement in boilers. However, the requirements for water conditioning and prevention of caustic embrittlement may clash with each other, because, a very high concentration of Na_2SO_4 may lead to the formation of $CaSO_4$ scales. Under these conditions, Na_3PO_4 should be used. Na_3PO_4 is an effective conditioning agent and also is over 300 times as effective as Na_2SO_4 in suppressing embrittlement.

Caustic embrittlement may also be prevented by adding lignins or tannins which help in blocking the infiltration of NaOH through the hair-cracks. Addition of $NaNO_3$, neutralization of excess alkali are also used to prevent embrittlement.

1.10. Cooling waters: Requisites and Treatment

A large proportion of industrial waters is used for cooling purposes. Cooling water for industrial use may be drawn from various sources such as surface water, ground water or sea water. The type of treatment required and the specification for cooling waters depend upon the source of water, the type of the cooling system employed and the nature of application. The general requisites of cooling waters used for industrial purposes are:

(*i*) The water should not form heat-insulating scales and clogging deposits in the cooling system employed, and

(*ii*) It should not be unduly corrosive under the specified conditions of use or reuse.

Hence, the cooling waters have to be treated to prevent scale formation, slime formation and corrosion in heat exchangers, condensers, cooling towers and other cooling surfaces.

In treating cooling waters, the carbonate hardness is removed by lime treatment. Alum is used as coagulant. Sometimes, sodium hexametaphosphate treatment is used for preventing scale formation. Demineralization by ion-exchange process is employed only in closed-recirculation cooling systems discussed later. Turbidity due to coarse or colloidal suspensions may form clogging deposits and hence has to be removed by coagulation, settling and filtration. Slime and algae growths in heat exchangers and condensers cause serious operational problems in cooling water systems. They are prevented by adding chemicals like chlorine, chloramine or copper sulphate.

Corrosion of metals in cooling water systems are prevented (*a*) by control of pH value of water, (*b*) by deposition of protective layer of $CaCO_3$, (*c*) by deoxygenation by mechanical or chemical means, and (*d*) by use of corrosion inhibitors, *e.g.*, calgon, sodium chromate, sodium nitrite, sodium benzoate, amines and tannins.

The nature and extent of treatment mostly depend upon the type of cooling water system that is employed.

The various cooling water systems that are commonly employed fall under the following three categories:

(*i*) **Once-through systems:** These are used where a plenty of water supply is available. In such systems, the water after use, is usually discharged to the source from which it was originally withdrawn. Since the quantity of water passing through such systems is very large, treatment by even small dosage of chemicals becomes very expensive and hence prohibitive. If at all any treatment is contemplated, it should be very cheap. Further, if the water supplied through a once-through system is to be used subsequently either for process purposes or as boiler feed-water, contamination by the substance being cooled or the addition of chemicals must be avoided. The only form of treatment commonly used in such systems is the cheap process of intermittent chlorination to prevent slime formation. Apart from this, prevention of corrosion and scale formation in these systems should be mainly achieved by proper design and correct choice of materials of construction. Physical or mechanical deaeration, possibly in conjunction with addition of small amounts of Na_2SO_3 was used in USA for corrosion prevention. The use of quaternary ammonium compounds as corrosion inhibitors at moderate cost was also suggested in some cases. However, while deciding upon any treatment, the possible pollution of the body of water to which the cooling water is returned must be duly considered.

Sea-water is used as a coolant only in "once-through systems" and in such a case, the corrosion problems are counteracted by addition of inhibitors such as chromates and dichromates.

(*ii*) **Open-recirculating systems.** In these systems, the cooling water is used repeatedly. The heat picked up by the water during its passage through the coolers is removed by direct contact with atmospheric air in some form of evaporative cooling device, such as a water cooling tower, cooling pond or spray pond.

The pH of water used in open recirculating cooling systems should be maintained in the range 7.7 to 8.2 and the total alkalinity in the range 100 to 400 ppm (as $CaCO_3$). If necessary, an alkali such as Na_2CO_3 (for hard waters), lime (for soft waters, where $CaCO_3$ protection is used) or chalk (for waters into which there is much ingress of acid) should be added to the water to maintain the desired pH value and total alkalinity. Addition of ammonia is usually undesirable.

If the total hardness of the circulating water is ≤ 100 ppm (as $CaCO_3$), it may be corrosive. Then, $CaCO_3$ protection may be applied either by increasing the concentration factor* or by adding lime. Alternatively, calgon (sodium hexametaphosphate) may be used as an inhibitor. If the recirculating water has an alkaline hardness of ≥ 100 ppm (as $CaCO_3$), corrosion becomes less likely, but treatment

---

* **Concentration Factor.** The efficiency of evaporative coolers depends on the evaporation of a part of the circulating water. Evaporation of 1% of the circulating water corresponds roughly to a fall of 7°C in the water temperature. Besides this loss by evaporation, there is also a loss of water due to carry-over, called windage loss which accounts for about 10% of the evaporative losses. Hence, certain make-up water is almost always required in open-recirculating systems. However, it is usual to purge deliberately some water out to reduce the dissolved solids content. This increases the make-up water requirement correspondingly. The degree to which the concentrations of dissolved substances are increased in the system is called the *"concentration Factor"* which is equal to,

$$\frac{\text{(Water lost by carry over and purge + water lost by evaporation)}}{\text{(water lost by carry over and purge)}}$$

The value of Concentration Factor can be varied by changing the amount of purge.

WATER TREATMENT

may then be required to prevent scale formation.

(*iii*) **Closed-recirculating systems.** In these systems also, the water is used repeatedly but the heat picked by the cooling water is removed in some kind of heat exchanger in which there is no direct contact between the water and the coolant. Typical examples of such systems are the cooling systems used in diesel and other internal combustion engines or jacket cooling of compressors.

Since evaporative coolers are not used in closed-recirculating systems, there is very little loss of water from such systems, and it is easy to maintain, at reasonable cost, the necessary concentrations of treatment chemicals in the water to prevent corrosion. Scaling is rarely a problem because of the small amounts of make-up needed, and in any case, water of high purity, such as condensate, can often be used, both for filling the system and for make-up. The water in a closed recirculating system will not be saturated with air, but will tend to be slightly alkaline. A particular problem with closed recirculating systems is that they are often made up of dissimilar metals in contact, so that the risk of corrosion is increased. Further, in practice, the most often used coolant used is ethylene glycol/water antifreeze mixture; and any treatment chemicals used should be such that they must not react with it. In aircraft engine cooling systems, corrosion may occur rapdily because of the high value of the ratio of heat-transfer area to coolant volume, while in the high temperature zones of the system, the ethylene glycol may be oxidized to give acidic decomposition products.

The use of inhibitors is the most satisfactory method of preventing corrosion in these systems. Na_2CrO_4 is preferred in the absence of organic anti-freeze compounds, otherwise, sodium benzoate is most suitable. For aircraft engine cooling systems, triethanolamine phosphate in conjunction with sodium mercaptobenzothiozole, has been used successfully as corrosion inhibitor.

Langelier Index (L.I.)

Water used for cooling purposes should be non-scaling and non-corrosive. Water is considered to be stable when it neither deposits scales of $CaCO_3$ nor dissolves the small protective coating of $CaCO_3$ that guards the metal pipes against corrosion. The scaling or corrosive behaviour of natural waters can be reckoned on the basis of "$CaCO_3$ saturation index" or "Langelier index." The Langelier index (L.I.) is defined as the algebraic difference between the actual pH and the saturation pH, denoted as $(pH)_S$:

$$L.I. = pH - (pH)_S$$

The saturation pH, in turn, is given by the following formula:

$$(pH)_S = \log \frac{K_{SP}}{K_2} - \log [Ca^{2+}] - \log [HCO_3^-]$$

where K_{SP} is the solubility product of $CaCO_3$

$$K_{SP} = [Ca^{2+}][CO_3^{2-}]$$

and K_2 is the dissociation constant for the reaction

$$HCO_3^- \rightleftharpoons H^+ + CO_3^{2-}$$

$[Ca^{2+}]$ is the calcium hardness expressed as $CaCO_3$ in mg l^{-1} and $[HCO_3^-]$ is the bicarbonate alkalinity.

Alternatively the saturation pH may also be expressed as follows:

$$(pH)_S = (PK_2 - PK_S) + P[Ca^{2+}] + P[Alk]$$

where the operator, 'P' represents the logarithm of reciprocal (*i.e.*, $px = \log_{10} x^{-1} = -\log_{10} x$); K_2 and K_S are the dissociation constants of the bicarbonate ion and of $CaCO_3$ respectively, both expressed in concentration units and [Alk] is the titratable equivalents of base per litre.

If L.I. = 0, the water is said to be stable, if L.I. is positive, it indicates scale forming tendency; and if L.I. is negative it indicates the corrosive tendency of the water under consideration. In practice, the L.I. is usually adjusted between 0.6 to 1.0.

The term "Saturation pH" has often been the subject of confusion. It should be noted that $(pH)_S$

is not that value of pH to which the water would come if $CaCO_3$ were stirred with it until equilibrium, were reached, because this would permit the Ca-and the total alkalinity values to change. The saturation pH is that pH value which the water would attain if the various forms of alkalinity were adjusted, keeping the total alkalinity constant, until the water was just saturated with $CaCO_3$. Thus a hard bicarbonate water which is of low pH value and unsaturated with $CaCO_3$ could be brought to the saturation pH in Langelier's sense by abstracting a certain amount of bicarbonate alkalinity. The pH value at which the solution would then be just saturated with $CaCO_3$ is the saturation pH, *i.e.*, $(pH)_S$.

The significance of the term "saturation pH" has to be clearly understood. The "saturation pH" represents the pH value which the water would have if its Ca-content and total alkalinity were unchanged but it was just saturated with $CaCO_3$. The saturation index (or Langelier Index) is thus a criterion of the degree of saturation of the water with $CaCO_3$ and, therefore of its tendency either to form scale on one hand or, on the other, to dissolve any scale already formed and thus to be corrosive towards iron. The index does not give any information on the rate at which these two processes take place. A negative value of the index indicates that the water is unsaturated with $CaCO_3$ and it is then said to be aggressive. Conversely, a positive value of the index shows that the water is supersaturated and is non-aggressive and scale forming. In a cooling water system, the saturation index may vary from one part to another since it is a function of temperature.

For $CaCO_3$-protection to be effective, the saturation index should be appreciably positive; a value of atleast 0.4 is aimed at. In view of the uncertainties of the fundamental solubility data on which the application of Langelier's index equation depends, this value of 0.4 is quite reasonable.

Control of Langelier index in practice entails the adjustment of one or more of the following: pH value, alkalinity and calcium hardness. The index may be increased by adding Na_2CO_3 or lime; or reduced by lowering the concentration factor (by increasing the purge) or by adding the acid or by softening the water.

1.11 DESALINATION OF WATER

Water with high levels of dissolved salt is not suitable for domestic, industrial or irrigation uses. Sea water contain 3.5% dissolved salt. Sea water can be desalinated to separate into fresh water by the process called desalination. The process of removing salt from sea water to make it suitable for agricultural purposes or for drinking is called desalination. The people are trying to develop the desalination process in an economical way for providing fresh water for domestic use in regions where the availability of water is limited. Desalination can be performed by two process. membrane process and thermal process

Reverse osmosis (RO) and electrodialysis are two membrane processes for desalination. Reverse osmosis is a separation process forcing a solvent from a region of high solute concentration through a membrane to a region of low solute concentration by applying a pressure in excess of the osmotic pressure. The flow of water (or other solvent) through a semipermeable membrane called osmosis. Normal osmosis is a process were natural movement of solvent flow from low to high concentration through a membrane. A membrane which is permeable to solvent and not to solute is called a semipermeable membrane.

When a solution is separated from pure water by a semipermeable membrane, osmosis of water occurs from water to solution. This osmosis can be stopped by applying pressure equal to osmotic pressure, on the solution. If pressure greater than osmotic pressure is applied, osmosis is made to proceed in the reverse direction to normal osmosis i.e. from solution to water.

The membranes used for reverse osmosis have a dense barrier layer in the polymer matrix

WATER TREATMENT

where most separation occurs. In most cases the membrane is designed to allow only water to pass through the dense layer while preventing the passage of solutes. This process requires a high pressure to exerted on the high concentration side of a membrane usually 30-250 psi for fresh and brackish water and 600 - 1000 psi for sea water. This process is well known for its use in desalination, it has also been used to purify fresh water for medical, industrial and domestic applications since 1970s.

Electrodialysis :

Electrodialysis is used to transport salt ions from one solution through ion-exchange membranes to anather solution under the influence of an applied electric potential difference. The cell consist of a feed (dilute) compartment and a concentrate (brine) compartment formed by an anion exchange membrane and a cation exchange membrane placed between two electrodes. Cation exchange membranes allow only cations to pass through where as onion exchange membranes allow passage of only anions through them.

Multiple electrodialysis cells consists of alternating anion and cation exchange membranes and electrode. Under the influence of an electrical potential difference the negatively charged ions (Cl') migrate toward the positively charged anode. These ions pass through the positively charged anion exchange membrane, but are prevented from further migration toward the anode by the negatively charged cation exchange membrane, and stay in the C steam which becomes concentrated with the anions. The positively charged species e.g. Na^+ in the D steam migrate toward the negatively charged cathod and pass through, the negatively charged cation exchange membrane, these cations also stay in the C steam, prevented from further migration towards the cathode by the positively charged anion exchange membrane. As a result of the anion and cation migration, electric current flows between the cathode and anode, equal number of anion and cation charge equivalents are transferred from the D stream into the C steam and so the charge balance is maintained in each stream.

The result of the electrodialysis process is an ion concentration increase in the concentration stream with a depletion of ions in the diluate solution feed steam. E steam is the electrode stream may consist of the same composition as the feed stream (ex. NaCl or Na_2SO_4) depending on the stack configuration anions and cations from the electrode stream may be transported into the C stream or anions and cations from the D stream may be transported into the E-stream.

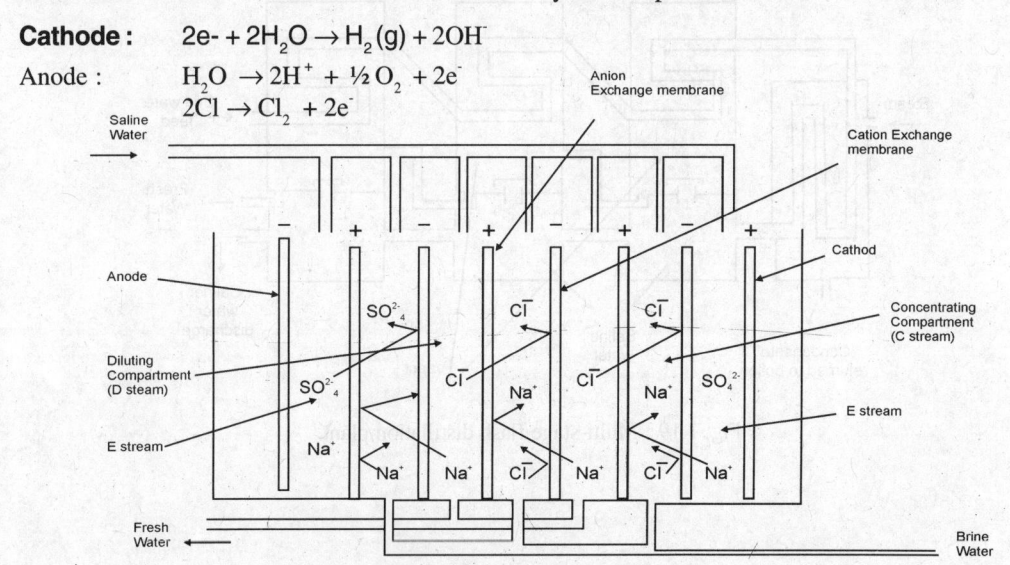

Cathode : $2e^- + 2H_2O \rightarrow H_2(g) + 2OH^-$

Anode : $H_2O \rightarrow 2H^+ + \frac{1}{2}O_2 + 2e^-$

$2Cl^- \rightarrow Cl_2 + 2e^-$

Fig. 1.17 : Removal of ions by electrodialysis

Electrodialysis systems can be operated as continuous production or batch production process. Electrodialysis is applied to deionization of aqueous solutions. The major application of electrodialysis has been the desalination of sea water as an alternative to RO for portable water production and sea water concentration for salt production.

RO is more cost-effective when total dissolved solids (TDS) are 3000 ppm or grater, while electrodialysis is more cost-effective for TDS feed concentration less than 3000 ppm.

Multi-Stage Flash Distillation

Multi-stage flash distillation is a thermal process in which the fresh water is separated from sea water by heating. evaporation and condensation. Multi-stage flash distillation technique is used for production of large quantities of desalted water. The process is carried out in a series of closed containers (stages) set at progressively lower pressures.

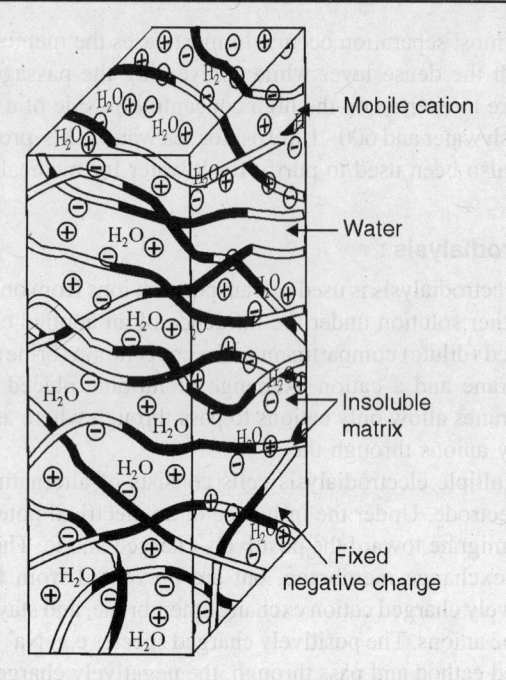

Fig. 1.18 Cut section of cation exchange membrane

Sea water is heated using steam, when preheated saline water enters a container at low pressure it get boil into vapour. The generated vapour get condensed into fresh water on heat exchanger tube and than collected in trays under the tube. The leftout saline water flow into next container set at even low pressure, where some of it also condensed and the process is continued. Some plant have up to 40 containers.

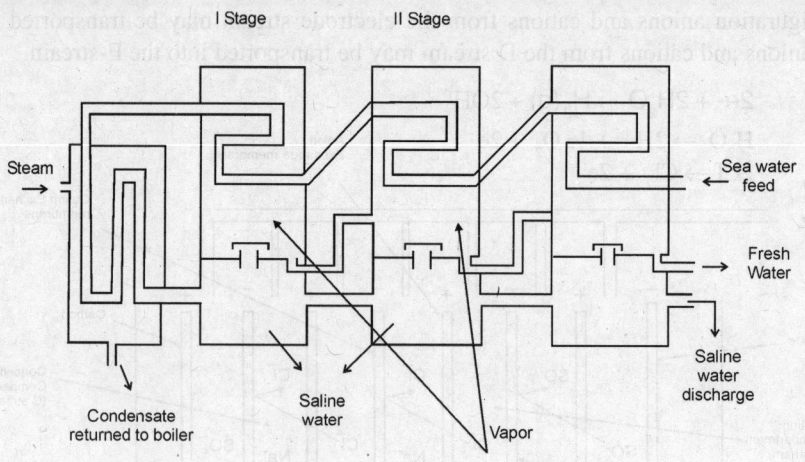

Fig. 1.19 : Multi-stage flash distillation plant.

WATER TREATMENT 75

QUESTIONS

1. Answer the following as directed:
   (a) "If hard water contains sufficient magnesium hardness, silica can be removed effectively in lime soda softening process". State whether the statement is true or false. Justify your answer.
   (b) "Phosphate conditioning is suitable at all-operating pressures". Give reasons to support this statement. (Nagpur Univ. W-2005, S-2007)
   (c) Mention any two points of contrast in lime soda and zeolite water softening processes.
   (d) Explain any two major drawbacks of scale formation.
   (e) A water sample contains Ca^{++} – 20 mg/l, Mg^{++} – 12 mg/l. What is the total hardness of the sample?
   (f) "Coagulents are not required in hot lime - soda process" Jusify the statmeent (Nagpur University W - 2005)
2. What are boiler troubles and what are their consequences? How can be boiler troubles be minimized?
3. (a) What do you mean by hardness of water? How is it classified?
   (b) What process of water softening would you recommend for obtaining feed water for the modern high pressure boilers, and why? (Nagpur Univ. S-2006)
4. Compare the hot lime soda process and the zeolite process of water softening with respect to the principles involved, advantages and limitations. (Nagpur Univ. W- 2003)
5. Compare carbonate conditioning and phosphate conditioning of boiler water with respect to the principles involved, advantages and limitations.
6. (a) Give principal of zeolite process? Write advantages, disadvantages and limitations of zeolite process. (Nag. Univ. S-2007).
   (b) Explain with chemical equations the demineralization of hard water and regeneration of ion exchange resine. (Nag. Univ. S-2007).
   (c) What are scales? what are its disadvantages and causes, (Nag. Univ. S-2003, S-2006, W-2006).
7. (a) Explain the various steps involved in the purification of water for municipal supply. (Nag. Univ. S-2005).
   (b) Give the chemical reaction involved during heating of hard water (Nag. Univ. S-2005).
8. (a) Establish the relation between ppm and mg/l.
   (b) Calculate the lime and soda required to soften 10,000 litres of the water sample having the following analysis:
   Calcium hardness — 250 ppm as $CaCO_3$
   Magnesium hardness — 100 ppm as $CaCO_3$
   Total alkalinity present — 300 ppm as $CaCO_3$
9. Explain how boiler corrosion takes place due to presence of (i) dissolved O_2 (ii) dissolved CO_2 and (iii) Mg salts in boiler feed water. Give the removal methods for each. (Nag. Univ. W-2005).
10. A zeolite softener was 75% exhausted by removing the hardness completely when 7000 litres of a water sample are passed through it. The zeolite bed required 180 litres of 2.5% NaCl solution for complete regeneration. Calculate the harndess of the water sample.
11. A water sample on analysis gave the following data. (Nag. Univ. S-2004)
    $MgSO_4$— 60 ppm, $CaCl_2$ — 111 ppm
    $Mg(HCO_3)_2$ — 73 ppm, NaCl — 58 ppm
    $CaSO_4$— 68 ppm, SiO_2 — 2.5 ppm
    (a) Calculate the amount of lime (90% pure) and soda (95.% pure) required to soften 80,000 litres of water.
    (b) If 80,000 litres of the same water is passed through a zeolite bed, how much NaCl will be required for its regeneration?
    (c) Also give the chemical reactions involved in lime soda softening process for this water.
12. A water sample on analysis gave the following data:
    Ca^{2+} — 20 ppm, Mg^{2+} — 24 ppm
    CO_2 — 30 ppm, HCO_3^- — 150 ppm
    K^+ — 10 ppm

Calculate the lime (87% pure) and soda (91% pure) required to soften 1 million litres of the water sample.

13. Write short notes on the following:
    (a) Sludge blanket type of lime soda softener
    (b) Cold lime-soda process
    (c) Sedimentation
    (d) Coagulation - (Nagpur Univ. S -2007)
    (e) Filtration of water - (Nagpur Univ. S -2007)
    (f) Sterilization. - (Nagpur Univ. S -2007)

14. Write short notes on the following:
    (a) Caustic embrittlement - (Nagpur Univ. S -2004)
    (b) Internal treatment
    (c) Break-point chlorination (Nagpur Univ. S - 2003, S-2004, S-2007)
    (d) Impurities in water and their effects
    (e) Slow sand filtration
    (f) Rapid pressure filtration.

15. Write information notes on the following:
    (a) Advantage and disadvantages of chlorination of drinking water.
    (b) Effect of fluoride in drinking water. (Nagpur Univ. W -2004)
    (c) Effect of excessive nitrates in drinking water.
    (d) Effect of arsenic in drinking water. (Nagpur Univ. S -2004)

2

Fuels and Combustion

"For many years to come, the greatest portion of the world's power will come from the combustion of fuels."

2.1. INTRODUCTION

Efficiency in the use and production of heat and power is one of the major engineering problem of the present day. The worlds energy demands are constantly increasing and to meet these expanding requirement we look to various kinds of fuels. Material which possess chemical energy is known as fuel. The important fuels are carbon compounds and whole industrial society is based upon the reaction.

Fuel + O_2 ⟶ Combuction Product + Heat

$C(s) + O_2(g) \longrightarrow CO_2(g) + \Delta H = -94.1$ Kcal / mole

$H_2 + \frac{1}{2} O_2 \longrightarrow H_2O + \Delta H = -67.5$ Kcal / mole

When C and H atom from coal and oil reacts with oxygen atoms chemical energy is relased, such reaction involves a rearrangement of the outer electron only at the atoms and the atomic nuclei is unaffected.

A fuel may be defined as any combustible substance which is obtainable in bulk, which may be burnt in atmospheric air in such a manner that the heat evolved is capable of being economically used for domestic and industrial purposes for heating and generation of power.

Fuels in the broad sense, include the stored fuels that are available in the earth's crust, viz., "fossil fuels" or those derived from them by an industrial process.

2·2. CLASSIFICATION

Fuels may be divided into two types:

(i) primary fuels which occur in nature as such, and

(ii) secondary fuels which are derived from the primary fuels.

Fuels may also be classified into three groups:

(a) Solid fuels,

(b) Liquid fuels, and

(c) Gaseous fuels.

The examples of each of these main classes of fuels are summarised in Table-1. This classification is of practical significance because the equipment used for handling and burning of each class of fuels are usually different for these three types of fuels.

Table-1: Classification of Fuels

Solid Fuels	Liquid Fuels	Gaseous Fuels
Primary Wood, peat, lignite, brown coal, bituminous coal, anthracite, oil shales, tar sands and bitumen. **Secondary** Semicoke, coke, charcoal briquettes, petroleum coke, pulverised coal and colloidal fuels. Solid rocket fuels such as thiokol, hydrazine, nitrocellulose, etc.	**Primary** Crude oil or petroleum. **Secondary** Gasoline or motor spirit, diesel oil, kerosene, fuel, oils, coal tar and its fractions, alcohols and synthetic sprits.	**Primary** Natural gas. **Secondary** Coal gas, coke oven gas, water gas, producer gas, carburetted water gas, oil gas, blast furance gas, refinery oil gas, synthesis gas, acetylene and liquid petroleum gas (LPG).

2·3. CALORIFIC VALUE

One of the important properties of a fuel on which its efficiency is judged is its *calorific value*. The calorific value of a fuel is defined as the amount of heat obtainable by the complete combustion of a unit mass of the fuel.

Units of heat

The units of heat generally employed are calories, kilogram calories, British thermal units and centigrade heat units.

(a) Calorie (cal) or gram calorie (g cal)

For all practical purposes, the calorie or gram calorie may be defined as the amount of heat required to raise the temperature of 1 g of water through 1°C (more precisely from 15°C to 16°C)*.

1 calorie = 4.185 Joules = 4.185×10^7 ergs.

(b) Kilocalorie or kilogram calorie or kilogram centigrade unit (Kcal or Kg cal or K.C.U.)

This is equal to 1000 calories and is, thus, the amount of heat required to raise the temperature of 1 kg of water through 1°C (more precisely from 15°C to 16°C).

1 K cal = 1000 Cal

(c) British thermal unit (B.Th.U. or B.T.U.)

A British thermal unit is the amount of heat required to rise the temperature of 1 lb of water through 1°F (more precisely from 60°F to 61°F).

1 B.Th.U. = 1,054.6 Joules = $1,054.6 \times 10^7$ ergs.

(d) Centigrade heat unit (C.H.U.)

The centigrade heat unit is the amount of heat required to raise the temperature of 1 lb of water through 1°C.

Interconversion of the various units of heat

The various heat units described above can be easily interconverted on the basis that 1 kg = 2.2 lb and 1°C = 1.8°F. Accordingly,

1 K cal = 1000 Cals = 3.968 B.Th.U. = 2.2 C.H.U.
1 B.Th.U. = 252 Cals
100,000 B.Th.U. = 1 Therm.

---

* *Thermal capacity or specific heat*. It is the quantity of heat required to produce unit change of temperature in unit mass of a substance. Alternatively, specific heat of a substance is the ratio of the thermal capacity of the substance to that of water at 15.5°C (or 60°F), since the thermal capacity of water at 15.5°C (or 60°F) = 1.000. Units of specific heat are B.Th.U./lb/°F and Cal/g/°C.

FUELS AND COMBUSTION

Units of calorific value

Calorific values of solid and liquid fuels are usually expressed in calories per gram (Cals/g) or Kilocalories per kilogram (K cals/Kg) or British Thermal Units per pound (B.Th.U./lb.); whereas the calorific values of gases are expressed as Kilocalories per cubic metre (K cals/m^3) or British thermal units per cubic foot (B.Th.U./ft^3) or C.H.U./lb or C.H.U./ft^3.

These units can be inter-converted as follows:

1 cal/g = 1 K cal/Kg = 1.8 B.Th.U./lb
1 K cal/m^3 = 0.1077 B.Th.U./ft^3
1 B.Th.U./ft^3 = 9.3 Kcals/m^3

Gross calorific value and net calorific value

The Gross Calorific Value or Higher Calorific Value is the total heat generated when a unit quantity of fuel is completely burnt and the products of combustion are cooled down to 60°F or 15°C (room temperature).

When a fuel containing hydrogen is burnt, the hydrogen present undergoes combustion and will be converted into steam. As the products of combustion are cooled to room temperature, the steam gets condensed into water and the latent heat is evolved. Thus the latent heat of condensation of steam so liberated is included in the gross calorific value.

The calorific value determination by Bomb calorimeter gives the Gross or Higher Calorific Value.

The Net Calorific Value or Low Calorific Value is the net heat produced when a unit quantity of fuel is completely burnt and the products of combustion are allowed to escape. Thus,

Net Calorific Value = Gross Calorific Value — Latent heat of Condensation of the water vapour produced.

= Gross Calorific Value — (Mass of Hydrogen per unit weight of the fuel burnt × 9 × latent heat of vapourization of water).

1 part by weight of hydrogen gives 9 parts by weight of water as follows:

$$H_2 + O \longrightarrow H_2O$$
$$2g \quad 16g \quad \quad 18g$$
$$1g \quad 8g \quad \quad 9g$$

The latent heat of steam is 587 Cal/g (or Kcal/Kg) or 1060 B.Th.U./lb of water vapour produced.

Net C.V. = Gross C.V. $- 9 \times \dfrac{H}{100} \times 587$

= Gross C.V. $- 0.09 \times H \times 587$

where H = % of hydrogen in the fuel.

In actual practical use of a fuel, it is rarely feasible to cool the combustion products to the room temperature to allow the condensation of water vapour formed and utilise that latent heat; hence the water vapour formed also is allowed to escape along with the hot combustion gases.

Calorific value at constant pressure

The calorific value at constant pressure, $Q_{C.P.}$ can be calculated on the basis of the following:

$$Q_{C.P.} = Q_{C.V.} - (\Delta_n).R.T.$$

where

$Q_{C.V.}$ = Calorific value at constant volume (as determined in a Bomb Calorimeter).
Δ_n = *increase* in number of gaseous molecules after reaction.
R = gas constant
T = absolute temperature.

If there is a decrease in the number of gaseous molecules formed after the reaction, then Δ_n will have a negative value and consequently, $Q_{C.P.}$ will be higher than $Q_{C.V.}$.

2·4. CALORIFIC INTENSITY AND FLAME TEMPERATURE

When the fuel burns without appreciable flame, the whole of the heat liberated because of combustion is concentrated over a relatively smaller area. Under such conditions, a solid fuel burning rapidly results in a high local fuel-bed temperature. Then it is said that the fuel is burning with a high **calorific intensity**.

During the combustion of fuels (solid, liquid or gaseous) having appreciable amounts of volatile matter, a flame is produced almost invariably. Then, the total heat produced by the fuel is liberated over the entire area of the burning mass and the larger the area of the flame the lesser the concentration of heat. Hence, in general, it can be said that flaming fuels burn with lower calorific intensities than the flameless fuels.

The flame temperature represents the maximum temperature to which an object can be heated by the flame. Moreover, the rate of heating of an object increases as the difference in temperature between the flame and the object increases. The temperature of a flame mainly depends on the calorific value of the gas and on the quantities of the total gaseous products formed and their respective specific heats. However, flame temperatures calculated solely from the above considerations are invariably higher due to the following reasons:

1. the combustion does not take place instantly and completely.
2. some of the heat liberated is lost to the surroundings by radiation.
3. some heat is lost as latent heat of steam produced.
4. a little heat may be absorbed by dissociating gas molecules, and
5. the specific heats of gases increase with temperature and the data regarding the specific heats at very high temperatures is not fully available.

However, the actual flame temperatures of different fuel gases will be in the same order as their respective theoretical flame temperatures and can therefore be used for the comparison of different fuels. Theoretical flame temperature can be calculated by the following formula:

Theoretical flame temperature

$$= \frac{\text{(Heat of combustion + Sensible heat in fuel and air)}}{\text{(total quantity of the combustion products)} \times \text{(their mean specific heats)}}$$

2·5. FLEXIBILITY AND CONTROL

Although calorific value is the major factor in determining the cost of a fuel, some other factors also need to be considered. A fuel which can be easily ignited, which can be burnt at varied rates so that the heat generation is slow or rapid as desired, which can be easily handled and easily controlled or, to sum up, which has a greater flexibility, proves itself to be more preferable and worthy than a fuel which has higher calorific value but lower flexibility. Flexibility may be considered as the rate of response in heat liberation with the variation in operating conditions such as fuel or air supply.

2·6. DETERMINATION OF CALORIFIC VALUE OF SOLID AND NON-VOLATILE LIQUID FUELS

Bomb calorimeter

The calorific value of a solid or non-volatile liquid fuel can be satisfactorily determined with a high-pressure oxygen bomb calorimeter (Fig. 2.1). It consists of :

1. a strong cylindrical bomb made of stabilized austenitic steel which is corrosion resistant and which is capable of withstanding a pressure of at least 50 atmospheres. The bomb is provided with a gas-tight screw-cap to which a couple of stainless steel electrodes and a release valve are fitted. One of these electrodes is tubular which can therefore act also as an oxygen inlet. A small ring is attached to this electrode which acts as a support for the crucible.
2. a copper calorimeter vessel in which the bomb stands during the experiment.
3. an outer water-jacketed enclosure.

FUELS AND COMBUSTION

4. a stirrer provided for stirring the water in the calorimeter at a uniform rate.
5. a thermometer graduated in one hundredths of a degree. Fixed zero or Beckman thermometers are generally used.
6. other necessary accessories for compressing the coal into a pellet, filling the bomb with oxygen, etc.

Procedure

1. About 0.5 to 1 g of finely ground air-dired coal (preferably compressed into a pellet) is accurately weighed into the crucible of the calorimeter.
2. A piece of fine platinum wire (0.0075 cm thick) is tighly stretched across the pole pieces of the bomb and one end of a piece of a sewing cotton thread is tied round the wire. The crucible is placed in position and the lose end of the cotton is arranged so as to be in contact with the coal. Alternatively, a longer platinum wire is used, and bent into a loop so as to touch the coal pellet.
3. About 10 ml of distilled water are introduced into the bomb to absorb vapours of sulphuric and nitric acids formed during the combustion and the lid of the bomb is screwed.

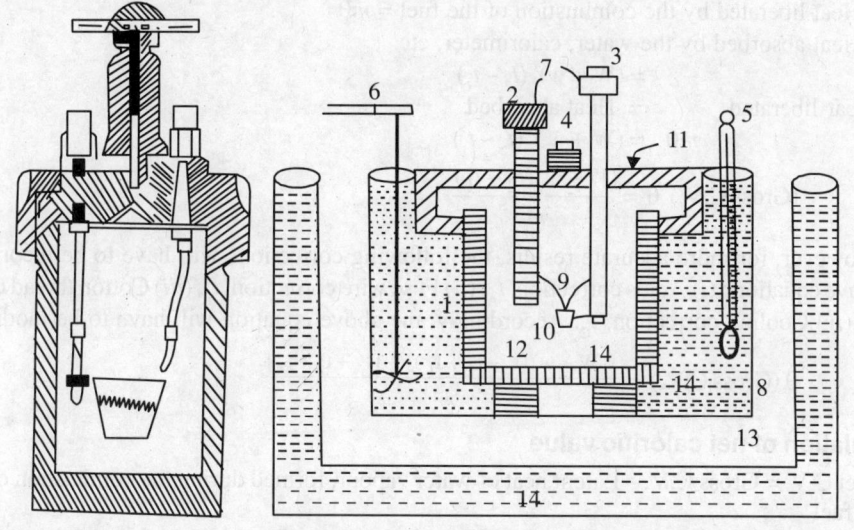

Fig. 2.1(A). Section of the Bomb. (1) Strong cylindrical bomb made of stabilized austenitic steel, (2) Stainless steel rod electrode, (3) Stainless steel tube electrode cum oxygen inlet, (4) Release valve, (5) Beckman thermometer, (6) Electrically driven stirrer, (7) Firing terminals, (8) Calorimeter, (9) Fuse wire,
(10) Crucible, (11) Screw cap, (12) Fuel pellet, (13) Air jacket, (14) Water.

Fig. 2.1(B). Bomb-calorimeter assembly.

4. The bomb is filled with oxygen upto 25 atmospheres pressure and the firing wires are attached to the terminals.
5. The calorimeter vessel is weighed, sufficient water is weighed in completely to submerge the cover of the bomb. The calorimeter vessel is then kept in the outer jacket on the insulating feet provided. The bomb is then lowered in the calorimeter. The stirrer and the lid of the calorimeter vessel are placed in position and the Beckman thermometer is adjusted.
6. The stirrer is started.
7. After 5 minutes, the temperature of the water is noted to the nearest 0.002°C and 5 more readings are taken at one minute intervals.

8. At the end of the fifth minute, the electrodes are connected to a 6 to 12-volt battery to ignite the charge and readings are continued at one-minute intervals. After the maximum temperature is attained, readings are still continued until the rate of fall is uniform.
9. The stirrer is stopped and the bomb removed from the calorimeter. After allowing about half an hour for the settlement of the acid mist within the bomb, the contents of the bomb are washed into a beaker and the amounts of H_2SO_4 and HNO_3 present in this solution are determined.

Calculations

Let

Weight of the fuel sample taken = m grams
Higher or Gross calorific value of the fuel = θ cals/gram
Weight of the water taken in the calorimeter = W grams
Water equivalent of the calorimeter, bomb, thermometer, stirrer, etc. } = w grams
Initial temperature = $t_1 °C$
Final temperature = $t_2 °C$
Heat liberated by the combustion of the fuel = $m\theta$
Heat absorbed by the water, calorimeter, etc.

$$= (W + w)(t_2 - t_1)$$

Heat liberated = Heat absorbed

$$\therefore m\theta = (W + w)(t_2 - t_1)$$

$$\therefore \text{(Gross C.V.)} \; \theta = \frac{(W+w)(t_2 - t_1)}{m}$$

However, for more accurate results, the following corrections will have to be incorporated in the above equation : (i) Acids correction, t_A, (ii) Fuse wire correction, t_F, (iii) Cotton thread correction, t_T and (iv) Cooling correction, t_C. Accordingly, the above equation will have to be modified as:

$$\theta \text{ (Gross C.V.)} = \frac{[(W+w)(t_2 - t_1 + t_C)] - [t_A + t_F + t_T]}{m}$$

Calculation of net calorific value

Net C.V. = Gross C.V. − Latent heat of water vapour, formed during the combustion of m grams of the fuel

= Gross C.V. − 0.09 × H × 587

(where H is the percentage of hydrogen present in the fuel and latent heat of steam is 587 cals/g).

Corrections

(i) **Acids correction (t_A):**

The sulfur present in the coal is converted into H_2SO_4 in the bomb.

$$S + O_2 \longrightarrow SO_2$$

$$2SO_2 + O_2 + 2H_2O \longrightarrow 2H_2SO_4; \; \Delta H = -144{,}000 \text{ calories}$$
(4 × 49)
(Eq. Wt. of H_2SO_4 = 49)

Similarly the nitrogen present in the coal and part of that in the air in the bomb are converted into HNO_3.

$$2N_2 + 5O_2 + 2H_2O \longrightarrow 4HNO_3; \; \Delta H = -57{,}160 \text{ calories}$$
(4 × 63)
(Eq. Wt. of HNO_3 = 63)

Since the above two reactions are exothermic and since the heat thus liberated is not obtainable in practical use of coal (because SO_2 and NO_2 pass off into the atmosphere) correction must be made

FUELS AND COMBUSTION

for the heat liberated in the bomb by the formation of H_2SO_4 and HNO_3, as follows:
   (a) 3.6 calories should be subtracted for each ml of N/10, H_2SO_4 formed.
   (b) 1.43 calories must be deducted for each ml of N/10 HNO_3 formed
              (as per the equations given above)

(ii) **Fuse wire correction, (t_F):**
Correction has to be made for the amount of heat equivalent in calories derived from the amount of fuse wire burnt as per the instructions furnished by the supplier of the fuse wire.

(iii) **Cotton thread correction, (t_T):**
The correction for the cotton thread used for firing the charge is calculated form the weight of the dry cotton thread actually used and on the basis that the calorific value of cellulose is 4140 cals per gram.

(iv) **Cooling correction, (t_C):**
If the time taken for the water in the calorimeter to cool from the maximum temperature attained to the room temperature is x minutes and the rate of cooling is $dt°$/minute, then the cooling correction $= x \times dt$. This should be added to the observed raise in temperature.

For more accurate results, the cooling correction by Regnault and Pfaundler's formula should be used.

Cooling correction, (t_C)

$$= nv + \frac{v'-v}{t'-t} \left\{ \sum_1^{n-1}(t) + \frac{1}{2}(t_0+t_n) - nt \right\}$$

$$= nv + KP$$

where
n = number of minutes between the time of firing and the first reading after the temperature begins to fall from the maximum.
v = the rate of fall of temperature per minute during the period before firing.
v' = the rate of fall of temperature per minute after the maximum temperature.
t and t' = the average temperatures during the prefiring and final periods respectively.
$\sum_1^{n-1}(t)$ = the sum of the readings during the period between firing and the start of cooling.
$\frac{1}{2}(t_0+t_n)$ = mean of the temperatures at the moment of firing and the first temperature after which the rate of change of temperature is constant, and
$K = \frac{v'-v}{t'-t}$ which is known as the "cooling constant" of the calorimet.

The colling correction so obtained should be added to the observed rise of temperature.

Example 1. *During the determination of calorific value of a coal sample, the following time-temperature data have been recorded with the help of an accurate Beckman thermometer. Determine the cooling correction and find out the corrected rise of temperature.*

Solution.

Initial Period or Prefiring Period		Combustion Period or Heating Period		Final Period or Cooling Period	
Time (mts)	Temp°	Time (mts)	Temp°	Time (mts)	Temp°
0	1.628	t_1 6	3.37	t_n 10	3.774
1	1.629	t_2 7	3.77	11	3.772
2	1.631	t_3 8	3.777	12	3.770
3	1.633	t_4 9	3.779	13	3.768
4	1.635			14	3.766
t_e 5	1.637			15	3.764

$$t = \frac{9.793}{6} = 1.632$$

$$v = \frac{(1.628-1.637)}{5} = -\frac{0.009}{5} = -0.0018$$

$$n = (10-5) = 5$$

$$\therefore nv = -0.009$$

$$\overset{4}{\underset{1}{t}} = 14.696$$

$$\frac{1}{2}(t_0 + t_n) = \frac{(1.637 + 3.774)}{2} = \frac{5.411}{2} = 2.706$$

$$-nt = (-1.63 \times 5) = -8.150$$

$$P = \overset{4}{\underset{1}{t}} = \frac{1}{2}(t_0 + t_a) - nt$$
$$= 14.696 + 2.706 - 8.15$$
$$= 17.402 - 8.150$$
$$= 9.252$$

$$t' = \frac{22.614}{6} = 3.769$$

$$v' = \frac{(3.774 - 3.764)}{6} = \frac{0.011}{6} = 0.002$$

$$K = \frac{v' - v}{t' - t} = 0.0018$$

$\therefore$ Cooling correction $= nv + \text{KP}$
$= -0.009 + (0.0018 \times 9.252)$
$= -0.009 + 0.0166$
$= 0.0076°$

Uncorrected rise of temperature
$= t_n - t_0 = 2.137°$

Corrected rise of temperature
$=$ (uncorrected rise of temperature + cooling correction)
$= 2.1370 + 0.0076$
$= \mathbf{2.1446°}.$

DETERMINATION OF WATER EQUIVALENT OF THE APPARATUS

The water equivalent of the apparatus is best determined by burning a known weight (preferably about 1.2g) of pure and dry benzoic acid (in a pellet form) in the bomb under identical conditions as described above. The rise in temperature is noted. The standard calorific value of benzoic acid is taken as 6324 calories per gram. Since all the other values in the formula are known, the water equivalent of the apparatus can be calculated. Other substances suitable as standards are:

		Calorific Value
Naphthalene	...	9622 cals/g
Salicylic acid	...	5269 cals/g
Camphor	...	9292 cals/g

Example 2. *Determine the water equivalent of the bomb calorimeter apparatus which gave the following data in an experiment:*

Weight of benzoic acid taken	=	*1.346 g*
Weight of the calorimeter can	=	*1025 g*
Weight of the calorimeter can + water	=	*3025 g*
Initial temperature	=	*11.872°C*
Final temperature	=	*14.625°C*
Cooling correction	=	*0.015°C*
Heat from fuses	=	*22 cals.*

The washings of the bomb on analysis indicated the presence of 3.5 ml of N/10 HNO^3 and there was no H_2SO_4 formed.

FUELS AND COMBUSTION

Solution.
Weight of benzoic acid taken = 1.364 g
Weight of water taken in the calorimeter
    can = (3025 g − 1025 g) = 2000 g
Corrected rise in temperature
    = (14.625 − 11.872) + 0.015
    = 2.753 + 0.015 = **2.768°C**
Heat evolved due to benzoic acid =
(Wt. of benzoic acid × calorific value)
    = 1.364 × 6324 = 8626 cals ...(1)
Heat from fuses = 22 cals ...(2)
Heat from HNO_3 = 3.5 × 1.43 = 5 cals ...(3)
Total heat given to the calorimeter
    = (1) + (2) + (3) = 8626 + 22 + 5 cals
    = **8653 cals**

$$\text{Total water equivalent} = \frac{\text{(Heat given to the calorimeter)}}{\text{(Corrected rise of temperature)}}$$

$$= \frac{8653}{2.768} = 3126 \text{ g}.$$

Water taken in the calorimeter = 2000 g
∴ Water equivalent of the calorimeter and auxiliaries
    = (3126 − 2000)g = **1126g.**

Example 3. *The following data is obtained in a bomb calorimeter experiment:*
Weight of the crucible = 3.649 g
Weight of the crucible + fuel = 4.678 g
Water equivalent of the calorimeter = 570 g
Water taken in the calorimeter = 2200 g
Observed rise in temperature = 2.3°C
Cooling correction = 0.047°C
Acids correction = 62.6 calories
Fuse wire correction = 3.8 calories
Cotton thread correction = 1.6 calories.
Calculate the gross calorific value of the fuel sample. If the fuel contains 6.5% hydrogen, determine the net calorific value.

Solution.

$$\text{High C.V.} = \frac{(W+w)(t_2 - t_1 + t_c) - (T_A + T_F + T_t)}{m}$$

$$= \frac{(2200+570)(2.3+0.047)-(62.6+3.8+1.6)}{(4.678-3.649)}$$

$$= \frac{(2770 \times 2.347)-68}{1.029}$$

$$= \frac{6510-68}{1.029}$$

    = 6261 cals/g

Low C.V. = (H.C.V. − 0.09 × H × 587) cals/g
    = 6261 − 0.09 × 6.5 × 587
    = 6261 − 343.4
    = **5917.6 cals/g**

2·7. DETERMINATION OF CALORIFIC VALUE OF GASES AND VOLATILE LIQUID FUELS

Boy's calorimeter

The calorific value of gaseous and volatile liquid fuels is usually measured by Boy's calorimeter or Junker's calorimeter. The principle involved in this method is to burn the gas at a known constant rate in a vessel under such conditions that the entire amount of heat produced is absorbed by water which is also flowing at a constant rate. From the volume of the gas burnt, the volume of water collected and the mean rise of temperature of the water, the calorific value of the gaseous fuel is calculated.

Description of the apparatus

Several types of gas calorimeters are available which differ in the arrangements for burning the gas and the heat transmitting system. The Boy's non-recording gas calorimeter (Fig. 2.2) consists of two flat-flame burners 'B' situated in a chimney 'C' which forms the centre of the annual vessel 'V'. The lower portion of 'V' is provided with a trough where water condensed from the products of combustion is collected and can be removed through the side tube 'T' for measurement of its volume.

The products of combustion pass up to the chimney and are deflected downwards by the watercooled head 'H' over a spiral of copper tubing through which cold water flows. The gas is again deflected upwards by an insulating baffle over another spiral of similar copper tubing and finally passes out through a number of holes in the lid of the calorimeter. The inlet water passes through the outer coils downwards to return upwards through the inner coils. Finally, it flows around suitable channels on the exterior of the metal casting immediately above the chimney and passes into a mixer. The temperatures of the inlet and outlet water are measured with different thermometers. The gas flows into the burners through an accurate gas-meter.

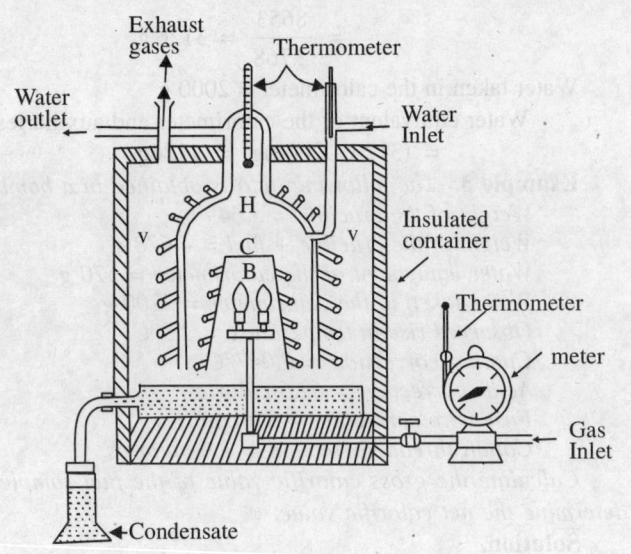

Fig. 2.2. Boy's calorimeter.

Procedure

The apparatus is assembled as shown in the diagram. After ensuring that the assembly is leak-proof, the gas is turned on and lighted. The water is turned on and the rate of flow is so adjusted that the rise in temperature of the water in passing through the calorimeter is as nearly as possible to 20°C. After the conditions are allowed to stabilize for about 45 minutes, the following readings are noted: (a) the volume of the gas burnt at a given temperature and pressure during a certain time interval (b) the amount of water passed through the cooling coils during the above time interval (c) the steady difference in temperature of the outlet and inlet water and (d) the amount of the water condensed from the products of combustion during the experiment. Also, the atmospheric pressure and the gas pressure are recorded from the barometer and manometer respectively. Further, the temperatures of the effluent gas, the ambient air and the gas meter are also noted to the nearest degree.

FUELS AND COMBUSTION

Then the high calorific value of the gaseous fuel can be calculated from the following relation:

Gross calorific value, $\theta = \dfrac{w(t_2 - t_1)}{V}$

where
w = wt. of the cooling water passed in time t.
V = volume of gas burnt at S.T.P. in time t.
t_1 = temperature of the incoming water.
t_2 = temperature of the outgoing water.

Further, the amount of water condensed from the steam produced by burning 1 m³ of gas = m/V; and the latent heat of steam per 1 m³ of gas at 15°C = 587 Kcal)

$\therefore$ Net calorific value = $\left(\text{Gross calorific value} - \dfrac{m}{V} \times 587\right)$

where m = weight of steam condensed in time, t.

Since the calorific value of gases is expressed on volumetric basis, it is essential to define the conditions of temperature and pressure for the volume being referred.

Example 1. *During the determination of calorific value of a gaseous fuel by Boy's calorimeter, the following results were recorded:*

Volume of the gaseous fuel burnt at normal temperature and pressure (N.T.P. or S.T.P.)	=	0.093 m³
Weight of the water used for cooling the combustion products	=	30.5 kg
Weight of steam condensed	=	0.031 kg
Temperature of inlet water	=	26.1°C
Temperature of outlet water	=	36.5°C

Determine the gross and net calorific values of the gaseous fuel per cubic meter at NTP, provided that the heat liberated in condensation of water vapour and cooling the condensate is 587 Kcal/kg.

Solution.

Gross calorific value, $\theta = \dfrac{w(t_2 - t_1)}{V}$

$= \dfrac{30.5(36.5 - 26.1)}{0.093}$

$= \mathbf{3410\ Kcal/m^3}$

Net calorific value $= \text{Gross C.V.} - \left(\dfrac{m}{V} \times 587\right)$

$= 3410 - \dfrac{0.031}{0.093} \times 587$

$= 3410 - 195.7$

$= \mathbf{3214.3\ Kcal/m^3}.$

2·8. CRITERIA FOR SELECTING A FUEL

The following characteristics are taken into consideration for the selection of a fuel for a particular purpose:

1. The fuel selected should be most suitable for the process. For instance, coke made out of bituminous coal is most suitable for blast furnace and also as a foundry fuel.
2. The fuel should posses a high calorific value.
3. The fuel should be cheap and readily available.

4. It should possess a moderate ignition temperature. Too high ignition temperatures cause difficulty in kindling while too low ignition temperatures may create safety problems during storage, transport and use of the fuel.
5. The supply position of the fuel should be reliable.
6. The velocity of combustion should be moderate.
7. The fuel should have reasonable flexibility and control.
8. The fuel should be such that a safe and clean operation is ensured. Too much smoke and obnoxious odours are not desirable.
9. It should be safe, convenient and economical for storage and transport.
10. It should have low moisture content.
11. In case of a solid fuel, the ash content should be less and the size should be more or less uniform.

2·9. SOLID FUELS

The main solid fuels include wood, peat, lignite, coal, charcoal and briquetted fuels. In addition to these, certain agricultural and industrial wastes such as bagasse, spent tan, rice husk, coconut and nut shells are also employed as fuels.

2.9.1. Wood. Wood has been the main source of fuel until recent times on account of its relatively rapid growth and production and ease of obtaining the supplies. However, large scale deforestation and the increasingly large demands of energy by the industries led to the more extensive use of other types of fuels. Freshly cut wood possesses greater water content (25 to 50%) than dry wood (15%) and its heating value is directly proportional to the water content. Wood mainly consists of cellulose, ligno-cellulose as well as some cell sap associated with traces of mineral ash. On the dry and ash-free basis, the average composition of wood is 50% C, 43% O, 6% H and 1% nitrogenous and resinous material. As a rule, dry wood is very combustible, easily kindled and burns with a long non-smoky flame. It gives maximum heat intensity very quickly. However, the calorific value of dry wood is only 19.7 to 21.3 MJ/kg (4710 to 5085 Kcal/kg). The ash content of wood is very low and lies in the range 0.3 and 0.6%. Wood is largely used as domestic fuel. It is rarely used in industry except for special purposes where dirt and smoke are undesirable. Due to its high flame emissivity, it is preferentially used for space heating.

Wood charcoal is obtained by the destructive distillation of wood. The carbonisation is performed usually in closed retorts. The charcoal is not pure carbon because even when the carbonisation is conducted at high temperatures, it rapidly absorbs some gas and moisture. It also contains some inorganic residues derived from the wood. Charcoal was widely used for metallurgical operations formerly, but it has now been replaced by coke excepting for some special applications. The major use of wood charcoal today is for producing activated carbon which finds extensive application for decolorisation (*e.g.*, in sugar industry), adsorption of gases and vapours and recovery of solvents from gases and air. Charcoal is also used in the production of CaC_2, ferro-alloys, and special quality pig iron in small furnaces.

2.9.2. Peat. Peat is generally considered as the first stage in the conversion of vegetable debris to coal and is produced under water-logged conditions by the action of fungi and anaerobic bacteria. Peat is generally found in high altitudes.

Three main types of peat are usually distinguished: (1) upland type consisting mainly of decomposed heaths and mosses (2) lowland type derived from sedges, grasses and willows and (3) Forest peat formed from accumulations of leaves, twigs, etc. and is mostly found in tropical countries. The bog peats derived from the smaller forms of vegetation are mainly found in USSR, Ireland, U.K., Canada, Finland, Poland, Siberia, France, Germany and Italy. About 42% of the world's peat deposits occurs in the USSR.

Freshly won peat from a well-drained bog may contain even 90% of water, which can be brought down to the level of 15 to 20% by air-drying. Most of the commercial peat blocks contain

FUELS AND COMBUSTION

about 20-25% moisture. Thus the utilisation of peat depends on its economical drying. On the dry basis its calorific value is around 5450 Kcal/Kg. Peat as such is used only as a local fuel. It is not considered as an economic fuel on account of the cost of drying and handling, cost of transportation as it is voluminous, relatively low calorific value and also because of its property of getting powdered during burning. It is mostly used, after briquetting with other substances, as a domestic fuel. Carbonisation of peat at low temperature (500-600°C) produces a char, oils and light spirit. Carbonisation of peat under proper conditions may yield coke and gas as well. Peat is largely used as domestic fuel in Europe but its use as such in India is limited. Dried peat is used as a fuel for the domestic appliances (*e.g.,* cookers and space heaters), for steam raising, for thermal insulation, packing, gas purification and soil conditioning. Due to the shortage of coal, peat is used for generation of electricity in Russia. In India, peat deposits are found in Nilgiri Hills (Tamil Nadu), Sundarbans (West Bengal), Kashmir and Calcutta.

2.9.3. Lignite (brown coal). These belong to the intermediate stage between peat and black coals. Their moisture content lies in the range 35 to 50%. Their carbon content, on the "dry, ash-free basis", ranges from 60 to 75% while the oxygen is over 20%. Lignites have high volatile matter content, usually upto 48 to 50% and an ash content of 4% or more. They generally occur at shallow depths and won by open cast mining techniques. Owing to their high volatile matter, they burn with long smoky flame. When dried, they tend to disintegrate into small pieces and powder and hence much of it is lost in transport and storage in open. Their calorific value is only in the range 24.3 to 29.3 MJ/kg (5800 – 7000 Kcals/kg) on dry ash free basis. Lignites absorb oxygen readily on exposure to air and gets ignited spontaneously. Lignites are not considered as a good fuel as such but are better than peat. Lignites are commonly made into briquettes after dehydration and marketed as such or after carbonisation. They are used as fuels in power plants and in the form of briquettes for domestic use. They are also used for production of producer gas. On further carbonisation, they give tar and ammonium sulfate. The tar on further hydrogenation gives motor spirit. Ammonium sulfate is used as fertilizer.

Lignites are classified into the following types on the basis of their maturity and external characteristics: (1) earthy brown coals '*e.g.,* Australian Morewel brown coal (2) wood brown coals *e.g.,* Italian Valderno brown coal and (3) laminated lignites which are rather black. The brown varieties generally tend to darken on exposure to air. Large amounts of lignites are consumed in Germany and Russia for steam raising. Lignite deposits occur in Assam, Kashmir, Rajasthan, Tamil Nadu, Travancore, Malabar coast and in M.P. Important lignite deposits of our country occur at Neyveli (South Arcot district of Tamil Nadu) which spread over 250 sq. km. and the deposits are estimated to be 200 million tonnes. The Neyveli project under the Lignite Corporation of India produces millions of tonnes of lignite briquettes for domestic and industrial use, and also urea.

2.9.4. Bituminous coal. These coals burn with smoky yellow flame like bitumen and their product of distillation is coal tar which is bituminous in nature.

Sub-bituminous coals form a group between lignite and bituminous coal. They are harder and denser than lignite. They are black in colour and possess a dull waxy lustre. Its moisture content is 12 to 25% and calorific value is about 7000 Kcals/kg. Their carbon content is 75 to 83%. Their oxygen content varies from 10 to 20%.

The bituminous coals are black in colour with a banded appearance, laminated structure and cubical fracture. Their carbon content is 78 to 90% and volatile matter 20 to 45%. They vary widely in their properties. Their calorific value is around 8000 to 8500 Kcals/kg. They may possess the property of caking (strong, medium or weak) or they may be non-caking. They are easy to handle and good in heating qualities. They are the most widely used coals in the world. They are used for domestic and industrial purposes. They are used for steam raising, coke and gas production and by-product manufacture.

The semi-bituminous coals form a group between bituminous coal and anthracite. They are characterised by low volatile matter (9 to 20%). Their carbon content is 90-93% and calorific value is around 8600 Kcals/kg. They are used for the manufacture of coke.

The Gondwana coals of India are bituminous in nature and the Raniganj bituminous coals are characterised by high volatile matter. The bituminous coals of Bihar, Bengal, M.P. and Orissa are the most important deposits of our country.

2.9.5. Anthracite. Anthracites are considered to be the highest rank of coal and contain the maximum percentage of carbon (92 to 96%). They are black, hard and lustrous. They have a conchoidal fracture. They have a very low percentage of volatile matter (about 6%) and hence produce a very little flame which is short, non-smoky and blue. Their calorific value is about 8600 Kcals/kg. They burn with intense local heating. Owing to their smokeless combustion, they are used for domestic heating in Canada and in the Continent. Anthracites are used in metallurgical operations, naval purposes, slow combustion stoves, central heating furnaces, generation of producer gas, for drying malt and hops (because of its low arsenic content), and curing rubber. They are also used for special applications such as manufacture of cathodes for aluminium industry and as a filter medium of water treatment. In India anthracites are found in Jammu, Darjeeling and Rajhara.

The variation in average composition from wood to anthracite are summarised in the following Table -2.

Table 2. Variation in average composition from wood to anthracite during different stages of coalification.

Fuel	Moisture at 40°C and 60% relatives humidity	Volatile matter	C	H	N	O	Calorific value MJ/kg (1 MJ = 238.8 Kcal/kg)
Wood	25	75	50	6	0.5	43.5	20.9
Peat	25	65	57	5.7	2.0	35.3	23.0
Lignite	18	50 to 56	67	5.0	1.5	26.5	27.2
Sub-bituminous coal	11	45 to 50	77	5.0	1.8	16.2	30.2
Bituminous coal	4	20 to 45	83	5.0	2.0	10.0	36.0
Sem-bituminous coal	1	9 to 20	90	4.5	1.5	4.0	36.4
Anthracite	1.5	5.6	93	3.0	0.7	3.0	35.6

2.9.6. Rank of Coal. The different stages of coalification are called as peat-anthracite series. Each stage in the above series is considered to belong to higher rank or maturity than their respective preceding member in the series. As the coalification progresses, the percentage of carbon and hardness increases, while the percentage of hydrogen, oxygen, moisture and volatile matter generally decrease. The calorific value gradually increases from peat to semi-bituminous coal. There is a slight fall in the calorific value of anthracite because the percentage of hydrogen decreases.

2·10. COAL

2.10.1. Origin of coal formation. Coal is regarded as a fossil fuel produced from large accumulations of vegetable debris due to partial decay and alteration by the action of heat and pressure over millions of years.

The formation of coal is explained by the following two theories:

(1) "In situ" theory states that coal seams are formed in the same area where vegetation grew and accumulated. The great purity of many coal seams holds testimony to this theory.

(2) The "drift" or "transportation" theory contends that the coal seams are not formed where the vegetation grew and accumulated originally. These materials were drifted or transported by rivers to lakes or estuaries and got deposited there. The great thickness of coal seams support this theory. Thus evidences are available in support of both the above theories.

The various agencies responsible for the conversion of plant tissues to coal include (i) Bacteria

FUELS AND COMBUSTION

(under water), (ii) Time (millions of years), (iii) Temperature (> 300°C) and (iv) Pressure.

The time required for the formation of young brown coals is of the order of 10^7 years while that for the most mature coals is 3×10^8 years.

The vegetable matter fallen on the ground undergoes microbial degradation in presence of air and eventually gets converted to carbon dioxide and water without leaving any organic matter remaining. However, the course of decay when it is buried under water is different. The transformation of the vegetable debris to coal takes place in two stages; (i) the biochemical or peat stage and (ii) the metamorphic stage during which peat is transformed into coal.

The effect of temperature and pressure caused by the depth of burial on the rank of a coal is brought out by Hilt's law which states that in any vertical section the rank of the seams increases with depth.

The formation of coal from decaying plant debris to bituminous stage is explained by two alternative theories.

(i) **Serial evolution.** This is the commonly accepted theory according to which the evolution of coals occurs through geo-chemical metamorphism of peat to anthracite as follows:

(peat ⟶ lignite ⟶ bituminous coal ⟶ anthracite)

(ii) **Parallel evolution.** This theory is based on the concept of entirely **biochemical** origin of coals of various ranks. According to this theory, lignites, bituminous coals and anthracites may not form a continuous series but may be the end-products resulting from the differences in the extent of the aerobic decomposition of peat, the subsequent composition of the overlying strata and the depth of the burial. This theory may be represented as follows:

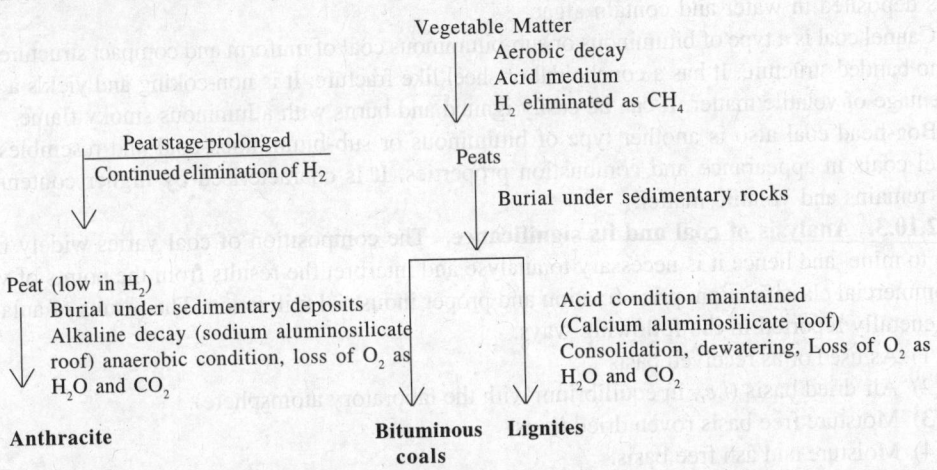

```
                    Vegetable Matter
                    | Aerobic decay
                    | Acid medium
                    ↓ H₂ eliminated as CH₄

  Peat stage prolonged            Peats
  | Continued elimination of H₂    | Burial under sedimentary rocks
  ↓                                ↓

Peat (low in H₂)                  Acid condition maintained
| Burial under sedimentary         (Calcium aluminosilicate roof)
|  deposits                        Consolidation, dewatering, Loss of O₂ as
| Alkaline decay (sodium           H₂O and CO₂
|  aluminosilicate roof)
|  anaerobic condition, loss of O₂ as
↓ H₂O and CO₂

Anthracite              Bituminous      Lignites
                          coals
```

2.10.2. Composition of Coal. A mined sample of coal contains the coal substance intermixed with mineral constituents such as Kaolin, shale, chlorides, sulfides, etc. These mineral constituents contribute to the ash content of the coal which is the residual mass obtained on combustion of coal. The coal also contains considerable amounts of free and hygroscopic moisture. The determination of moisture, volatile matter, ash and fixed carbon is known as proximate analysis which is of significance in commercial classification and industrial utilization of coal.

Carbon, hydrogen and oxygen constitute the true coal substance and the properties of coal mostly depend upon these constituents. Small quantities of sulfur, arsenic and phosphorous also exist in many coal samples. If these substances are present in appreciable amounts, the coal may be unsuitable for metallurgical applications. The ultimate analysis of coal consists of determination of C,H,S,N and O. The details regarding proximate and ultimate analyses and their significance are discussed later.

On the basis of difference in proportions of plant ingredients, the following types of coal are usually distinguished:

Banded coals. Bituminous coals usually possess banded structure consisting of alternate layers of very bright, laminated and dull layers. These bands are identified as Vitrain, Clarain, Durain and Fusain.

Vitrain has a black brilliant, glassy lustre. It breaks with a conchoidal fracture. It is very low in ash content and the ash has creamy or purplish colour.

Clarain has a black and silky lustre, duller than vitrain and has a tendency to break irregularly. Its ash content is more than that of vitrain and is usually reddish in colour.

Durain has a dull earthy lustre and is hard and tough. It breaks irregularly. The ash content is higher than that of bright coal. The ash is generally grey or white and is rather infusible.

Fusain is a soft black powdery material with satiny lustre similar to charcoal in appearance. It occurs in thin layers. This is derived from woody tissue and is porous and soft. It is a minor constituent of most coal seams. The volatile matter is low and ash content is high which is often fusible.

Splint coal. It is a type of bituminous or sub-bituminous coal having dull lustre and greyish black colour. It is hard, tough and has compact structure. It burns freely without swelling. It breaks with an irregular fracture.

Cannel coals and boghead coals

Cannel is a further type of coal found in large lenticular masses in many coal seams. This has a greasy lustre and a white clayey ash. Channels and bogheads may be regarded as fossil vegetable muds deposited in water and contain algae.

Cannel coal is a type of bituminous or sub-bituminous coal of uniform and compact structure and has no banded structure. It has a conchoidal or sheel like fracture. It is non-coking and yields a high percentage of volatile matter. It can be easily ignited and burns with a luminous smoky flame.

Bog-head coal also is another type of bituminous or sub-bituminous coal and resembles the cannel coals in appearance and combustion properties. It is characterised by higher contents of algal remains and volatile matter.

2.10.3. Analysis of coal and its significance. The composition of coal varies widely from mine to mine and hence it is necessary to analyse and interpret the results from the points of view of commercial classification, price fixation and proper industrial utilization. The results of anlaysis are generally reported in the following ways:

(1) As used or as received basis
(2) Air dried basis (*i.e.,* in equilibrium with the laboratory atomsphere).
(3) Moisture free basis (oven dried basis)
(4) Moisture and ash free basis.

The quality of a coal is ascertained by the following two types of analysis.

(1) The proximate analysis, which includes the determination of moisture, volatile matter, ash and fixed carbon. This gives quick and valuable information regarding commercial classification and determination of suitability for a particular industrial use.

(2) The ultimate analysis, which includes the estimation of ash, carbon, hydrogen, sulfur, nitrogen and oxygen. The ultimate analysis is essential for calculating heat balances in any process for which coal is employed as a fuel.

Significance of proximate analysis

The proximate analysis is an assay rather than true analysis, since the results have no absolute significance. However, if the assay is carried out in accordance with standard specifications, reproducible results can be obtained, thus enabling the coal to be classified and opinion formed regarding its cost, quality and probable use in a particular industry. Each constituent determined

FUELS AND COMBUSTION

under proximate analysis has its own implication and importance in the assessment of the coal sample.

Moisture. Moisture increases the transport costs. Excessive surface moisture may cause difficulties in handling the coal. Moisture reduces the calorific value. A considerable amount of heat is wasted in evaporating the moisture avaliable in coal during combustion. Hence high percentage of moisture is undesirable.

Volatile matter. The volatile matter is not a constituent of coal, but consists of a complex mixture of gaseous and liquid products resulting from the thermal decomposition of the coal substance. The amount of decomposition and the yield of volatile matter depends on the conditions of heating particularly the temperature. Hence, stipulated conditions should be strictly followed during its determination.

The volatile matter content of a coal is related to the length of the flame, smoke forming tendency and the ignition characteristics. High volatile matter coals give long flames, high smoke and relatively low heating values. Coal with low volatile content burns with a shorter flame. Thus, the higher the volatile matter content the larger is the combustion space required. Hence, the volatile matter content of a coal influences the furnace design. Further, the % of volatile matter in a coal denotes the proportion of the coal which will be converted into gas and tar products by heat. Hence, high volatile matter content is preferable in coal gas manufacture and in carbonization plants, particularly when the main objective is the byproduct recovery. For the manufacture of metallurgical coke, a coal with low volatile matter and high fixed carbon is preferred.

The volatile matter content is more in bituminous coals than in anthracite coals. The volatile matter percentage gives some idea about coking property of the coal.

Ash. The ash which is intimately interspersed within the mass of the coal is called fixed ash or inherent ash whereas the ash which occurs in different layers of the coal is known as free ash or extraneous ash. Only the extraneous or free ash can be removed by washing. Many Indian coals have high ash content.

The nature of the ash and its amount in a coal and the softening temperature are very vital in determining the quality of a coal. Ash reduces the heating value of coal.

Ash usually consists of silica, alumina, iron oxide and small quantities of lime, magnesia, etc. Its composition is of considerable importance in metallurgical operations as it affects the slag and metal composition and consequently is a prime consideration in selecting the flux.

When coal is used in a boiler, the fusion temperature of the ash is of particular significance. Fusion temperature of coal ash is generally between 1000° – 1700°C. Ash with fusion temperatures below 1200°C is called fusible ash and that above 1430°C is called refractory ash. If the ash fuses at the working temperatures when coal is burnt on grates, it leads to the formation of clinkers (lumps of ash) which reduces the primary air supply and the efficiency of production and distribution of heat are advsersely affected. Clinkers cause uneven temperature on the grates, which may contribute to further clinker formation. Some coal particles also get embedded in the fused ash thereby causing loss of fuel. Fused particles of the ash may stick to the boiler tubes and reduce heat transfer. Removal of clinkers from grates is difficult and laborious. Ash with low melting point forms molten slag which is absorbed in the pores of the refractory lining of the boiler furnace. On account of differences in the coefficients of expansion and contraction of the refractory material and ash, the life of the refractory material might be reduced due to spalling. In view of all the above considerations, coals used in boilers should have high ash fusion temperatures.

Fixed carbon. It is reported as the difference between 100 and the sum of percentages of moisture, volatile matter and ash content of a coal. The fixed carbon content increases from low ranking coals such as lignite to high ranking coals such as anthracite. It is the Fixed Carbon which burns in the solid state. Hence, information regarding the percentage of fixed carbon helps in designing of the furnace and the fire box.

Procedure for proximate analysis

1. Moisture. Moisture is generally determined by heating a known quantity of air-dried coal to 105°C to 110°C for one hour and calculating the loss in weight as percentage.

2. Volatile Matter. Volatile matter is determined by heating 1 g of air-dried coal exactly for 7 minutes in a translucent silica crucible of specified dimensions at a steady temperature of 925°C in a muffle furnace. The loss in weight calculated as percentage minus the % moisture gives the % volatile matter.

3. Ash. Ash is determined by heating at 400°C a known quantity of the powdered sample until most of the carbonaceous matter is burnt off and then heating for 1 hour at 750°C to complete the combustion. The weight of the residue remaining in the crucible corresponds to the ash content of the coal, which is reported on percentage basis.

4. Fixed Carbon. The sum total of the percentages of volatile matter, moisture and ash subtracted from 100 gives the percentage of fixed carbon.

Ultimate Analysis

Methods of determination

Carbon and hydrogen are determined by burning a known weight of a sample in a stream of pure oxygen in a combustion apparatus similar to that used for the analysis of organic compounds. C and H present in the sample are converted into CO_2 and H_2O respectively, which are absorbed separately in suitable absorption tubes. The % of C and H are calculated from the increase in weight of the respective absorption tubes.

Nitrogen is determined by digesting 1 g of the coal sample in a Kjeldahl flask with *Con.* H_2SO_4, K_2SO_4 and $HgSO_4$ when the nitrogen present is converted to ammonium salts. The sample is then made alkaline with NaOH and the liberated ammonia is distilled into a measured amount of standard acid. The residual acid is determined by back titration with NaOH. From the amount of the standard acid neutralised by the liberated ammonia, the nitrogen present in the sample is calculated.

Sulfur is determined conveniently from the bomb washings obtained from the combustion of a known mass of coal in the bomb calorimeter experiment for the determination of calorific value. The washings contain sulfur in the form of sulfate from which it is precipitated as $BaSO_4$. The precipitate is filtered, ignited and weighed. From the weight of $BaSO_4$ obtained the sulfur present in the coal is calculated.

Sulfur can also be determined by heating the coal with Eschka mixture (2 parts of MgO : 1 part of Na_2CO_3) and estimating the sulfates produced as $BaSO_4$.

Ash is determined as described under proximate analysis.

Oxygen is determined by difference as follows:

% Oxygen = 100 – % of (C + H + S + N + Ash)

Significance

Carbon and Hydrogen in coal directly contribute towards the calorific value of the coal. Higher the percentages of C and H, better is the quality of the coal and higher is its calorific value. Hydrogen is mostly associated with the volatile matter of the coal and thus influences the use of coal for the byproduct manufacture or otherwise. Nitrogen in the coal does not contribute any useful value to the coal and since it is generally present only in small quantities (~ 1%), its presence is not of much significance. Sulfur present in coal contributes towards the heating value of the coal but its combustion products (SO_2 and SO_3) have corrosive effects on the equipments, particularly in presence of moisture. Further, the oxides of sulfur are undesirable from the atmospheric pollution point of view. Sulfur containing coal is not suitable for the preparation or metallurgical coke as it adversely affects the properties of the metal. Oxygen content of coal is generally associated with moisture. The lower the oxygen content, the more is the maturity of the coal and greater is its calorific value. As the oxygen content increases, the capacity of the coal to hold moisture increases and the caking power decreases.

FUELS AND COMBUSTION

Use of proximate and ultimate analysis in the theoretical determination of the calorific value of coal:

The calorific value is determined by burning 1 g of coal sample in an oxygen bomb calorimeter equipment and measuring the rise of temperature thus produced in the water content of the calorimeter. However, quite often, an engineer may have to estimate the thermal efficiency of a process when the calorific value of the fuel has not been determined. In such circumstances, formulae for the calculation of calorific value from ultimate and proximate analysis are very helpful.

Formulae Based on Ultimate Analysis

1. Dulong's Formula

Calorific value in B.Th.U./lb

$$= 14{,}544\,C + 62{,}028\left(H - \frac{O}{8}\right) + 4{,}050\,S$$

where C, H, O and S represent the respective percentages of carbon, hydrogen, oxygen and sulfur. Several modifications to it have been proposed to this formula and one of them is as follows:

$$\text{Gross C.V.} = \frac{1}{100}\left[8{,}080\,C + 34{,}500\left(H - \frac{O}{8}\right) + 2240\,S\right] \text{ K cals/kg}$$

2. Davies Formula

$$\left.\begin{array}{l}\text{Calorific value in}\\ \text{B.Th.U. per lb}\end{array}\right\} = (6.543\,H + 403)\left(\frac{C}{3} + H - \frac{O-S}{8}\right)$$

where C, H, O and S are their respective percentages in the coal.

3. Seyler's Formula

$$\left.\begin{array}{l}\text{Calorific value in}\\ \text{B.Th.U. per lb}\end{array}\right\} = 223.1\,C + 698.6\,H - 7684 + 0.45\,O$$

where C, H and O are their respective percentages in the coal.

2.10.4. Formula Based on Proximate Analysis

1. Gouthal's Formula

$$\left.\begin{array}{l}\text{Calorific value in}\\ \text{B.Th.U. per lb}\end{array}\right\} = 147.6\,C + aV$$

where C is the % of carbon, V is the % of volatile matter and 'a' is a constant depending on V. The relation between V and a is as follows:

V	1–4	10	15	20	25	30	35	40
a		270	261	210.6	196.2	185.4	176.4	171	144

2. Nakamura's Formula

$$\left.\begin{array}{l}\text{Calorific value in}\\ \text{B.Th.U. per lb}\end{array}\right\} = a\left(V - \frac{\%\,\text{Ash}}{10}\right) + 140.4\,C$$

where 'a' depends on the % volatiles and caking propensity as shown in Table below. The deduction of $\dfrac{\text{Ash}}{10}$ from the volatiles is made to allow for that part of the volatiles as determined which is derived from the combined water and pyritic sulfur in the mineral matter and not from the coal substance.

Caking quality	Natural moisture	V 20–25	25–35	35–40	40–45
Non-caking	15			108	115
Non-caking	15			122	126
Caking (slight contraction)				137	142
Caking (slight swelling)			151	151	153
Caking (strong swelling)		225–219	219–198	173	171

The main use of all these formulae is that they provide a means for calculating the calorific values of coals approximately when their compositions are known but of which the samples are not available.

2·11. INDIAN COALS AND THEIR PROPERTIES

On the basis of geological division, Indian coals can be classified into the following four types:

1. Gondwana coals. In the geological terminology, Gondwana coals refer to a particular type of rock system which is about 200 million years old. This type of coal fields are present in India at Raniganj, Jharia, Bokaro, Ramgarh, Rewa, Bilaspur, Ballarpur, Chanda, Chhindwara, Yeotmal, Rampur, Raigarh, Talcher, Kothagudem, Singareni, etc. The Gondwana coals account for about 98% of our coal reserves.

Some coal seams of Jharia, Giridh and East Bokaro are the only deposits of prime coking coals in India. Their volatile matter is 28–32% and moisture about 2%. Their calorific value is about 7500 K cal/kg. They are suitable for steam raising and manufacture of metallurgical coke. The ash of these coals have a greater tendency to clinker formation.

Some coals in Raniganj, Central India and Assam give weak cokes due to their high volatile matter (about 35%). They burn with long flame. The moisture content is about 5% and calorific value is about 7000 Kcals/kg. They are good for making gas. However the ash from these coals does not have much tendency to form clinkers.

2. Tertiary coals. These coals are next in importance to Gondwana coals and are only 60 million years old. These contribute about 2% of the Indian coals. These being relatively younger are lower in rank (*e.g.,* lignites) excepting for a few coal fields of Assam and Jammu. They contain 3 to 8% of sulfur and are liable to spontaneous combustion. Important tertiary coal fields of India are located in Assam, Kashmir and Rajasthan. Most of these coals are obviously unsuitable for metallurgical purposes.

3. Jurassic coals. Coal seams of Cutch and Lameta Ghat in Narbada valley belong to this type.

4. Cretaceous coals. This type of coals are situated in Garo Hills and some coal fields of Khasi and Jaintia Hills in Assam.

Indian coals are generally dull in appearance and interbanded. The moisture content of coking coals is about 2% but that of weakly caking and non-caking coals is 3 to 12%. The ash content is generally high (15 to 30%) and it is generally refractory. Sulfur in Indian coals is less than 1% excepting for Assam coals which have 2 to 8%.

Coking coals are scarce in India and insufficient to meet the demands of metallurgical industry. However, there are extensive deposits of inferior variety of coals.

2·12. CLASSIFICATION OF COALS

The various parameters used for classification of coal include the rank of the coal, the proximate analysis, the ultimate analysis, calorific value and caking properties. Several systems of classification have been made on the basis of the above parameters.

2·13. GRADING OF COALS

Grading of coal is done with the objective of commercial classification and price fixation. In India, the Coal Controller's grading system is based on moisture content for low moisture containing coals while for high moisture containing coals, it is based on the total of ash and moisture contents. The details of the grading system are summarized in Table-3.

Table – 3. Coal Commissioner's Grading Scheme for Indian Coals.

I. Coal from Bihar and Bengal Coal fields

Grade	Non-coking coals		Coking coals	
	Ash + moisture % for coals having 2% moisture	Ash % for coals having 2% moisture	Grade	Ash %
Selected A	< 17.5	< 15	A	<13
Selected B	17.5 – 19	15–17	B	13–14
Grade I	19–24	17–20	C	14–15
Grade II	24–28	20–24	D	15–16
			E	16–17
Grade III A	—	24–28	F	17–18
Grade III B	—	28–35	G	18–19
			H	19–20
			HH	20–24

II. Coal from seams other than Bihar, Bengal, Andhra Pradesh and Assam

Grade	(Ash + moisture) %
Selected grade	<19
Grade I	19–24
Grade II	24–28
Grade III	28–35

III. Coals from Assam Coalfields

No grading. Different prices are fixed for different collieries.

IV. Coals from Andhra Pradesh Coalfields

No grading. Statutory fixation of prices is done on the size of coal.

2·14. CHARACTERISTICS OF COAL

Apart from proximate and ultimate analysis, the assessment of a coal is based on the following characteristics:

1. Colour. Lignites are brown or brownish black and the colour darkens with increasing rank of the coal.

2. Texture. Lignites are earthy and fibrous in structure. With increasing maturity, the coals tend to be more tough, hard and brittle.

3. Specific gravity. The specific gravity depends upon the type of the coal and its ash content. The specific gravity increases from lignite (1.2) to anthracite (1.5).

4. Heat of combustion (calorific value). The calorific value increases with increasing rank of the coal (except in case of anthracites whose calorific value may be lesser than semi-bituminous coals because of lesser percentage of hydrogen). High volatile coals having long flame have less heating value than those of low volatile short-flame coals. Coals with 20% volatiles generally have the best heating value.

5. Grindability. This shows the ease with which a coal can be ground and is generally expressed as grindability index. It is a measure of the power required for grinding a coal and is of special significance for pulverized coals. The coals which can be easily pulverised have grindability index of about 100.

6. Friability. This is the tendency of coal to break to pieces on handling and is tested by drop shatter test. Non-friability is essential for coals used with stoker firing. Splint and cannel coal are less friable than others.

7. Caking and coking properties. Caking is the ability of coal to form a coherent cake on carbonisation. This is an important property used to assess the value of a coal and is tested by various tests such as Swelling Number, Gray-King Assay, Roga Index and Audibert Arnu test. If the residue formed on carbonisation is strong and porous, then the coal is called coking coals, which is used for preparing metallurgical coke.

8. Weathering or slaking index. This is a measure of the tendency of a coal to break on exposure to weather or alternate cycles of dry and wet climate.

9. Bulk density. This is an important characteristic of coal on which the design of bunkers and containers for storage of coal and for manufacturing coke depends.

2·15. SELECTION OF COAL

The selection of coal for different applications is mainly made on the basis of the following factors:

1. Calorific value. Higher the calorific value, more preferred is the coal.

2. Moisture content. A coal with low moisture content is more economical and hence is more preferred.

3. Ash content. A coal with low ash content is preferred. Further, the ash should have high fusion temperature so that there is no danger of clinker formation. The significance of the quality, quantity, fusibility and composition of the ash has already been discussed under proximate analysis.

4. Calorific intensity and flexibility. A coal having a high calorific intensity and flexibility is always preferred. These factors have already been discussed earlier.

5. Uniformity. A coal having uniformity in size is better from the points of view of storage, handling and efficiency in operation.

6. Sulfur and Phosphorus content. A coal having very low sulfur and phosphorous contents is essential for metallurgical purposes because these impurities may contaminate the metal and adversely affect its properties. P and S tend to make the metal brittle.

7. Coking quality. Coking coals are selected for preparation of strong and porous coke suitable for metallurigcal purposes.

2·16. COMMERCIAL TYPES OF COAL

1. Steam coals. High volatile lignitous coals and low volatile carbonaceous coals are more suitable for steam raising. They are either non-caking or slightly caking. The high volatile coals have somewhat low calorific value and require a large combustion space for efficient and smokeless combustion. These coals burn freely and are quite flexible. On the other hand, the low volatile steam coals have high calorific values. They burn freely liberating bulk of the heat in the fuel bed.

2. Gas coals. These coals must be strongly caking. Coals with high volatile matter are preferable if the gas manufacture is the primary objective. If manufacture of hard metallurgical coke is required, coals having low volatile matter are more satisfactory.

3. House coals. These coals should burn freely and should not be strongly caking.

2·17. COAL TECHNOLOGY

Preparation of coal:

After mining, the coal is usually subjected to some sort of preparation or physical treatment after which it is despatched to the consumers. The preparation consists of steps such as (*i*) Sorting, (*ii*) Sizing (by crushing and screening), (*iii*) Washing followed by dewatering and (*iv*) Blending. In India, non-coking coals are usually subjected to sizing before use while coking coals are beneficiated by washing.

FUELS AND COMBUSTION

Coal beneficiation

The impurities present in the coal as mined are generally removed by "wet washing" or "cleaning". These methods are based on the difference in density between the clean coal and its impurities. The relative density of coal is 1.25–1.50 whereas the relative densities of its common impurities, *viz.*, shale, sandstone, calcite, siderite and pyrites are 2.6, 2.45, 2.7, 3.8 and 4.95 respectively. The "wet washing" is affected with the help of water or a dense medium as the sorting liquid. The equipment employed for coal washing in India include heavy media separators, heavy medium cyclones, water cyclones and jig washers. For heavy media separation in the laboratory, chemicals like bromoform, carbon tetrachloride, sulfuric acid and zinc chloride solution are used whereas in industrial practice, fine grains of sand, barytes, magnetite and shale in water are usually employed. After the washing is complete, the dewatering of the coal is affected by methods such as centrifugation and oscillating screening for coarse fractions and filtration, thickening and flocculation for finer fractions. Methods like froth flatation and selective agglomeration are also used for beneficiation of coal.

Cleaning of coal has the following advantages:
(a) It reduces the ash content and clinker formation.
(b) As it reduces the impurities in coal, savings are affected in the transportation, storage and handling costs.
(c) Some of the objectionable impurities such as sulfur and phosphorous are partially removed.
(d) The efficiency of the coal is increased.

Blending of coal

If a coal processing unit or a coal consuming plant receives different types of coals having different characteristics, they have to be intimately intermixed or blended for efficient operation. Blending is also done when there is a shortage of good quality coal.

In view of the shortage of good quality coking coals and also in view of the increasing demands for such coals by the metallurgical industry, it is essential to upgrade the inferior grades of coking coal and also in utilizing the middlings so produced for steam raising and domestic consumption as follows:

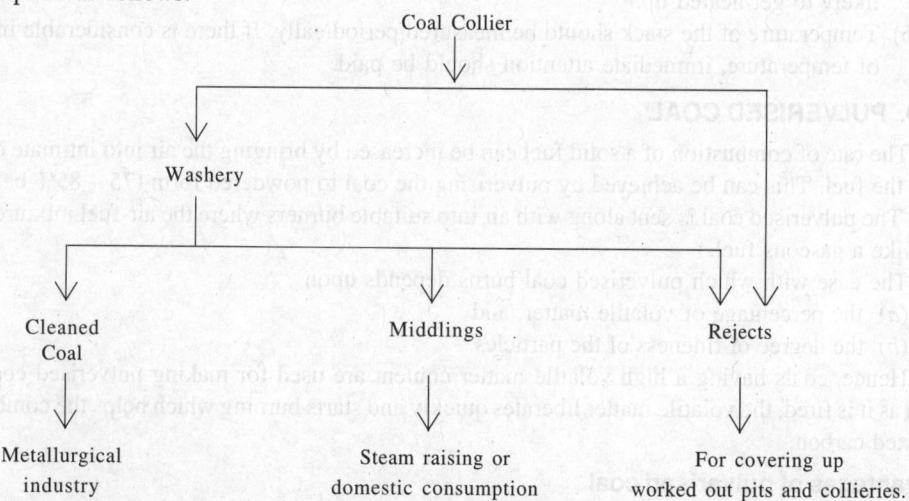

```
                        Coal Collier
                             ↓
                         Washery
         ┌───────────────────┼───────────────────┐
         ↓                   ↓                   ↓
      Cleaned             Middlings            Rejects
       Coal
         ↓                   ↓                   ↓
   Metallurgical        Steam raising or      For covering up
     industry         domestic consumption   worked out pits and collieries.
```

2·18. STORAGE OF COAL

Coal is stored in order to ensure uninterrupted supply to the consumers. The storage is done in underground, ground as well as over head systems. However, coal should be stored only for a minimum period required because it involves not only the locking up of capital but also slow oxidation leading to deterioration of the coal and possible risk of spontaneous combustion.

If coal is stored improperly for prolonged periods, the following difficulties may arise:
(1) Decrease in the calorific value because of oxidation.
(2) Deterioration in caking power and coking properties.
(3) Increase in the proportion of fines.
(4) Hazard of spontaneous combustion.

The basic cause for spontaneous combustion of coal is oxidation. When a coal is freshly mined, it has a greater tendency of oxidation due to the presence of some hydrocarbons on the surface. This is the reason why coal is not usually transported immediately after mining. Oxidation of coal is an exothermic process and in case the heat is not removed by proper circulation of air, the temperature increases which further enhances the rate of oxidation (For every 10°C rise in temperature, the rate of oxidation is doubled). Thus this process continues until it reaches a particular temperature when the coal starts burning by itself. This phenomenon is called spontaneous combustion. Generally, the chances of spontaneous combustion decrease with increasing maturity of the coal. Thus lignite is more susceptible to spontaneous combustion during storage than anthracite.

Factors favourable for spontaneous combustion are:
(1) Any condition which tends to conserve the heat such as mixing lumps of coal with fine powder when the gaps are filled and cooling air does not circulate freely.
(2) High temperatures which increase the rate of oxidation, and
(3) Presence of sulfides which tend to get oxidised leading to expansion in volume. This increase of surface area promotes further oxidation.

The following safety measures are to be employed during storage of coal to avoid the risk of spontaneous combustion:
(1) Coal of uniform size only should be stacked without the presence of fines.
(2) Coal should be stored in heaps not higher than 1.8 metres.
(3) Coal may be stored under water to exclude air. If complete submergence of the stack with water is not possible, it is better to avoid the use of water.
(4) Care should be taken to avoid storing of coal against walls of boiler house, etc., which are likely to get heated up.
(5) Temperature of the stack should be measured periodically. If there is considerable increase of temperature, immediate attention should be paid.

2·19. PULVERISED COAL

The rate of combustion of a solid fuel can be increased by bringing the air into intimate contact with the fuel. This can be achieved by pulverising the coal to powdered form (75 – 85% below 74 μm). The pulverised coal is sent along with air into suitable burners where the air-fuel mixture burns just like a gaseous fuel.

The ease with which pulverised coal burns depends upon
(a) the percentage of volatile matter, and
(b) the degree of fineness of the particles.

Hence, coals having a high volatile matter content are used for making pulverised coals. As soon as it is fired, the volatile matter liberates quickly and starts burning which helps the combustion of fixed carbon.

Advantages of pulverised coal

(1) Its handling and transportation are easier. The transportation can be done by forcing a stream of air or by screw conveyors.
(2) The rate of combustion can be easily controlled as in the case of a liquid or gaseous fuel. Heating the furnace can be readily started or stopped as required without undue wastage of

FUELS AND COMBUSTION

fuel. The temperature of the finance can be easily increased or decreased by controlling the fuel supply.

(3) Pulverised coal can be intimately mixed with air so that the combustion is uniform and complete.

(4) The excess air needed for complete combustion is less than that needed for solid fuels. This helps in minimizing heat losses and achieving higher thermal efficiency.

(5) Low-grade coals with high percentage of ash can be used for preparing pulverised coal. This is of particular significance for the Indian coals as most of them have high ash content.

(6) Problems of clinker formation are not there.

(7) Both oxidizing and reducing atmosphere can be maintained in the furnace as required. This is of particular importance in metallurgical operations.

(8) Responds well to automatic control.

(9) A high temperature flame at the correct position in the furnace can be obtained.

Disadvantages of pulverised coal

(1) Additional cost is involved for pulverising and sieving the fuel.

(2) The ash produced by burning pulverised coal is in a finely divided state and about 75% of it is deposited in the surrounding places. This is known as "Fly ash".

(3) Special type of turbulent burners must be used for achieving good mixing of the air and the fuel.

The disadvantages could be overcome to some extent by using dust catchers and modern pulverising techniques and advanced methods of firing the fuel.

Since pulverised coal produces a long flame, large combustion space is required. In order to obviate this difficulty, turbulent burners are generally employed. In such burners, pulverised coal along with primary air is given rotary motion in the burner and then secondary air is pushed in spirally to complete the combustion process. Two typical turbulent burners are shown in Fig. 2.3.

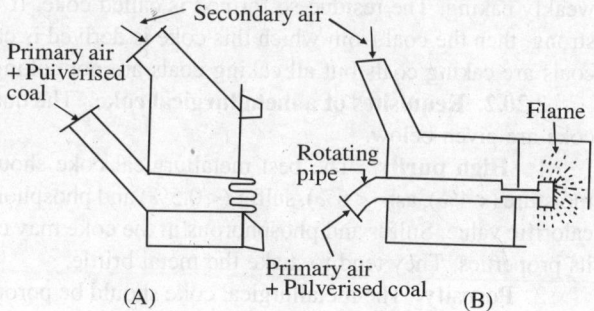

Fig. 2.3. Burners for pulverised coal.

Several types of firing techniques are employed with pulverised fuel. These include (i) Front wall firing, (ii) Side wall firing, (iii) U-flame (long flame) firing and (iv) Corner firing. In shell boilers, spiral burners are used to produce the maximum swirl and thus keep the flame short, at the same time maintaining efficient combustion and minimizing smoke formation. A typical tangential firing system from corners of the furnace walls is shown in Fig. 2.4.

Pulverised fuel furnaces may be distinguished also by the method of removal of ash, such as (i) Dry-bottom (ii) Slag-tap and (iii) Cyclone burner furnaces. As mentioned earlier, a disadvantage in the use of powdered coal is the production of a fine ash which is carried away by the exhaust gases and

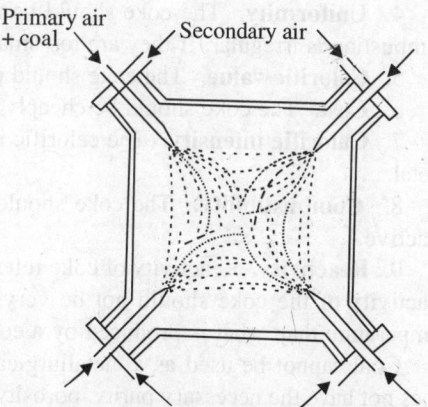

Fig. 2.4. Tangential firing system from the corners of the furance using pulverised coal.

deposits in the surrounding places. This "fly ash" can be reduced if the pulverised coal is burnt in a "slag-tap" furnace in which the flame is directed towards and sweeps over a bed of molten slag where the particles of ash are caught.

Pulverised coal is mostly used in the generation of steam and in some metallurgical furnances.

2·20. SECONDARY SOLID FUELS

Important secondary solid fuels include charcoal, briquettes and coke. Pulverised coal may also be considered as a secondary fuel. A charcoal is the residue from destructive distillation of wood. It has limited application as a fuel in industrial practice. Briquettes are made by compressing low grade fuels with or without a binder. Briquetting provides a method of utilizing small-sized waste coal produced in mining and hence is likely to be an important method as the best coal seems get depleted. In briquetting, the fine coal is mixed with 5 to 8% of a suitable binder (*e.g.*, asphalt pitch, coal tar, molasses, starch, gilsonite and sulfite liquor from paper industry) and compressed into briquettes under pressure. The briquettes so produced may be baked to remove some volatile matter.

Coke is an important secondary fuel of industrial importance and is produced by strongly heating the coal out of contact with air. This process is known as carbonization.

2.20.1. Caking coals and coking coals. Coals that soften on heating producing a "pasty" or "plastic" mass which fuse together yielding coherent masses impervious to air are called *caking coals*. Coals from which very little of such plastic material is formed are either non-caking or weakly caking. The residue so formed is called coke. If the coke so produced is hard, porous and strong, then the coal from which this coke is derived is called a *coking coal*. Obviously, all coking coals are caking coals but all caking coals are not coking coals.

2.20.2. Requisites of a metallurgical coke. The quality requirements of a good metallurgical coke are given below.

1. **High purity.** The best metallurgical coke should contain lowest possible percentage of moisture (< 4%), ash (< 6%), sulfur (< 0.5%) and phosphorous (< 0.1%). Moisture and ash reduce the calorific value. Sulfur and phosphorous in the coke may contaminate the metal and adversely affect its properties. They tend to make the metal brittle.

2. **Porosity.** The metallurgical coke should be porous to provide intimate contact between the carbon and oxygen and to ensure efficient combustion of the fuel in the furnace.

3. **Strength.** The coke should be strong enough to withstand the abrasion and over-burden of the ore, flux and the fuel itself in the furnace. If the coke breaks into fine particles during charging of the furnace, they may hinder the flow of gases and choke the air passages.

4. **Uniformity.** The coke should be uniform and medium in size. If the lumps are too big, combustion is irregular. If they are too small, choking may result.

5. **Calorific value.** The coke should possess a high calorific value.

6. **Cost.** The coke should be cheaply available near the plant site.

7. **Calorific intensity.** The calorific intentisty of the fuel should be high enough to melt the metal.

8. **Combustibility.** The coke should burn easily but at the same time should not be very reactive.

9. **Reactivity.** Reactivity of coke refers to its ability to react with CO_2, steam, air and O_2. The reactivity of the coke should not be very high. Coke of low reactivity gives a higher fuel bed temperature than what is produced by a coke of high reactivity.

Coal cannot be used as a metallurgical fuel (excepting in reverberatory furnaces) because it does not have the necessary purity, porosity and strength. During the process of carbonisation from coking coals, much of the volatile matter and sulfur compounds are removed and a strong and porous coke is produced.

2.20.3. Process of Carbonisation. When bituminous coal is slowly heated, moisture and

FUELS AND COMBUSTION

occluded gases are driven out first. When the temperature reaches about 270°C, some H_2S and olefine gases are evolved. At about 350°C, active decomposition of the coal substance take place with brisk evolution of gas and tarry vapours. Below 450°C, the liberated gases mainly comprise of hydrocarbons. At about 700°C, amount of hydrogen liberated increases and above 800°C, gas is the main product.

If the coal is a strongly caking one, along with decomposition between 300 to 450°C, it also melts and becomes plastic. Gases liberating during this plastic stage blow the molten coal into a sort of froth, but with the progress of decomposition, it becomes more and more viscous and finally rigid. If the coal is free to expand during this plastic stage, a highly swollen open-textured coke is obtained from strongly coking coals. If the expansion is suppressed by the retaining walls, a dense coke having a fine pore structure is produced.

Coal is a poor conductor of heat and when thick layers of coal are heated, the various stages of carbonisation take place in well-defined zones which gradually progress from the heated surface inwards (towards the interior layers). Particularly, the thin plastic zone formed with caking coals acts as a heat barrier which results in a considerable temperature gradient. There may be a drop of about 250°C across half an inch thickness, and with a carbonisation temperature of 1000°C, the rate of coking is only about 0.8 inch per hour.

Non-caking coals do not produce plastic layer on heating and hence have much higher rates of carbonisation. However, they do not produce good coherent coke. Hence, by *blending* coking coals with non-caking coals, an increased rate of heat penetration and increased rate of carbonisation can be achieved. The caking coal acts as a binder under these conditions. Thus, blending of coking coals with non-caking coals for carbonisation gives the duel benefits, namely the increase of rate of carbonisation on one hand and overcoming the shortage of good coking coals on the other. Further, when small amounts of non-caking coal, coke-breeze, low temperature coke or fusain dust are used as "fillers" for strongly coking fusible coals, in addition to obtaining the improved rate of heat penetration, expansion can also be controlled and shrinkage cracks reduced resulting in the production of stronger coke.

The properties of the coke produced mainly depend on the type of coal used, the temperature of carbonisation and the rate of carbonisation. These factors control the porosity, reactivity and the amount of volatile matter retained in the coke.

Carbonisation is done in special plants called retorts or coke-ovens.

2.20.4. Types of Carbonisation. Depending on the temperature of carbonisation, the following three types of carbonisation are usually performed:

1. Low temperature carbonisation (500 – 750°C)

The low temperature carbonisation (LTC) is carried out mainly for the manufacture of domestic fuel. The coke produced is called low temperature coke and the yield is about 75% to 80%. The coke thus produced contains 5 to 15% volatile matter and is not sufficiently strong to be used as a metallurgical fuel. Since it is easily ignited, it is a valuable, smokeless, domestic fuel. The by-product gas produced during the low-temperature carbonisation process is richer than that produced in the high-temperature carbonisation process. Its calorific value is about 6300 to 9300 Kcals/m^3 depending on the coal and the process used and the yield is about 150 to 330 m^3/tonne of the coal carbonised. Hence it is a more valuable gaseous fuel.

The low-temperature coking process is carried out both with coking and non-coking coals. When the coke is too friable to use, it may be briquetted or pulverised. The old method of open stack burning of coal to produce "soft coke" is still in practice in India. However, this process has several disadvantages as follows: (1) produces heavy air pollution, (2) poor quality of the soft coke produced, (3) loss of liquid and gaseous by products formed.

An important practical difficulty for low temperature carbonisation of coals is the poor transmission of heat through the coal charge. In order to obviate this difficulty modern commercial plants employ either externally heated metal or refractory retorts of narrow width or internally heated stationary or rotary retorts or a combination of the two. Some such processes developed to commercial or prototype scale include

1. Krupp-Lurgi and the Brennstoff technique
2. Koppers and Didier Werke
3. Fuel Research Board (UK) Vertical retorts.
4. CFRI (India) vertical retorts with gas circulation
5. Rexco, and
6. Disco process.

2. Medium temperature carbonisation (MTC) (700 – 900°C)

In medium temperature carbonization, the yields of gas will be lesser and the yields of NH_3 are same as that of high temperature carbonisation. However, the yields of tar are more and its properties are different than the tar obtained from high temperature carbonzation. Further, the yields of total light oils and phenols (particularly cresol) are more in medium temperature carbonisation and the light oil contains more of paraffinic and diolefinic hydrocarbons. Thus medium temperature carbonisation is done to obtain greater yields of particular types of byproducts

3. High Temperature Carbonisation (HTC) (> 900°C)

High temperature carbonisation (HTC) is used for the production of pure, hard, strong and porous metallurgical coke. The yield of the coke is 65 to 75% and the volatile matter is 1 to 3%. The byproduct gas produced is about 370 to 480 m^3/tonne but its calorific value is lower (5000 to 6000 Kcals/m^3) than that produced in LTC. Hence the process has been used to produce town gas and gas coke at the gas works of cities.

Manufacture of Metallurgical Coke

There are two methods of manufacturing metallurgical coke which are named according to the character of the oven in which the distillation is carried out:

(i) Beehive ovens method.
(ii) Chamber ovens or Otto-Hoffman ovens or Byproduct ovens method.

(i) Beehives ovens method

Almost upto the end of the 19th century, practically the entire requirement of the metallurgical coke of the world was manufactured in "beehive ovens". The beehive oven, as illustrated in Fig. 2.5, is a dome shaped structure made of bricks, which is about 4 metres in diameter and about 2 to 2.4 metres high. It has two doors, one at the top for charging the coal and the other at the side through which the coke formed is removed. The side door also acts as an inlet for air as and when desired.

The oven is charged with coal to give a layer of about 0.75 metre thick. The coal is ignited and just sufficient air is admitted through the side

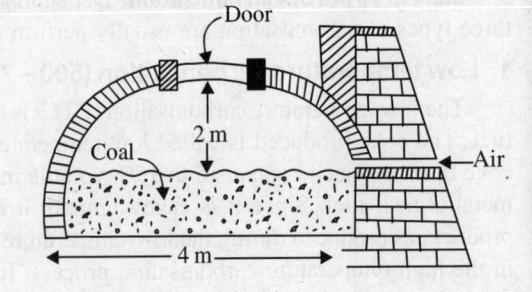

Fig. 2.5. Beehive ovens.

door to maintain the desired temperature. The heat furnished by this partial combustion melts or fuses the coal and volatile matter that is evolved burns at the door. Carbonisation takes place slowly from the top layer to the bottom layer and it completes within about 3 days. Then the oven is allowed to cool and the hard metallurgical coke formed is quenched with water and taken out through the discharge door. The yield of the coke is about 60% of the coal charged and averages about 5 tons per oven.

FUELS AND COMBUSTION

The process can be made economical by operating a series of ovens in such a way that the hot gases escaping from an oven is utilized to ignite the charge in the adjacent oven.

This is the cheapest and earliest process for the manufacture of coke although it is time-taking. However, the disadvantage of this method is that most of the volatile matter, which contains a treasure of chemicals, is allowed to eascape out into the atmosphere as waste. This also creates air pollution problems. Nevertheless, beehive ovens are still in use in India and elsewhere as "emergency" units to meet high demands for hard coke.

Fast coking beehive ovens have been designed in India which complete the coking process within 1/3rd of the time required in the conventional ovens and produce better quality coke in greater yields. A wide range of coals and blends can be utilized for the production of metallurgical coke. A plant having coal carbonising capacity of 0.4 million tons/year is in operation by TISCO at Sijua.

The beehive-ovens have been mostly replaced by the chamber ovens which operate on the regenerative principle of heat economy. This process is economical and clean because the valuable byproducts are recovered.

(ii) Chamber ovens or Otto Hoffman's ovens or By-product ovens

These ovens consist of narrow rectangular chambers made of silica bricks having length, height and width as 12 m, 4 m and 0.5 m respectively (Fig. 22). They are tightly closed so that no air is admitted. The heat for coking is furnished by burning gas (*e.g.*, coke oven gas, producer gas or blast furnace gas) in flues contained in the side walls of the oven. The chambers are also fitted with charging doors having 3 to 4 openings at the top and discharging doors at the bases. Each oven is separated from the neighbouring one by a vertical flue in which the fuel gas burns. Thus the ovens get heated from both sides. These ovens work on the regenerative principle of heat economy. Generally, a battery of such ovens (consisting about 25 to 1000 ovens) are used which are placed over the regenerators having a chequer brick-work which helps in the utilization of the heat of the flue gases. The hot gases leaving the flues are allowed to pass through the chequer brick-work in the regenerator of the next oven while air and fuel gas pass through other chambers which have been already heated. After some time, the directions are reversed so that the waste hot flue gases preheat other chambers through which the air and fuel gas were passing.

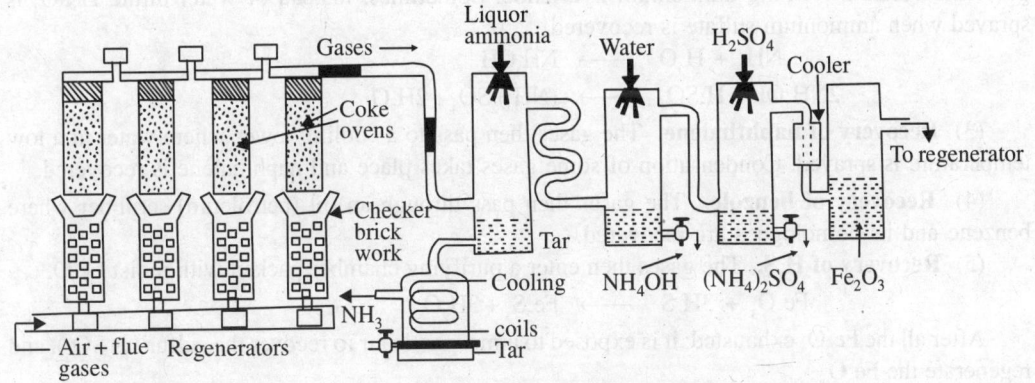

Fig. 2.6. Chamber ovens with byproduct recovery system.

As the coal is heated in the coke oven, moisture is first expelled, then the decomposition of the coal substance takes place at 300 to 450°C. At about 500°C, the coal passes through a plastic state but at about 550° to 600°C, the plasticity ceases and semicoke is produced. This is black in colour and is low in strength. As the temperature raises further above 600°C, the semicoke decomposes with loss in volatile matter and gets transformed into steel grey hard coke. The process of carbonisation takes place layer by layer in the coal charge starting from the two side walls of the oven and moving towards the centre.

Each oven holds about 20 tonnes of coal charge and the time taken for carbonisation is about 12 to 20 hours. The temperature goes around 1100°C and the yield of the coke is about 70% of the coal charged. Nearly 40% of the coke oven gas generated is sufficient to heat the ovens and the rest is available for other uses in the steel plant or is sold out. This gas has high calorific value and can be transported to distant places.

After the carbonisation is complete, the discharging doors are lifted by a crane and the red hot coke is pushed out mechanically into a coke car. The car carries it to quenching station where the coke comes into contact with a spray of cooling water. The excess water on the coke is allowed to get evaporated and the coke is screened to different sizes. Then it is supplied for different uses in the plant *e.g.*, foundry and blast furnace and also for domestic purposes.

Quenching the hot coke with water has the following disadvantages:
(a) The sensible heat of the coke is wasted.
(b) Large quantity of steam, formed during quenching is contaminated with acid gases which are corrosive to the nearby metal structures in the plant.
(c) Large quantity of coke breeze is formed which is a nuisance.
(d) Large quantity of dirty water is produced which can be disposed off only after clarification in settling ponds.

In view of these difficulties, sometimes dry cooling is practised in which the sensible heat of the coal is also utilized by transferring the hot coke from the oven to a brick-lined chamber and air is forced through the coke. The oxygen present in the air is consumed and the Nitrogen and inert gases cool the coke bed. The hot gases (after passing through a dust catcher) are used to preheat the boiler-feed water etc. and then recirculated for cooling a fresh charge of coke.

The gas coming out of the coke ovens contains ammonia, sulfur (as H_2S) volatile hydrocarbons, tar, etc. Important byproducts such as high calorific value gas, light oil, tar, ammonium sulfate, finely divided sulfur and ammonium thiocyanate can be recovered from the byproduct-oven gases.

(1) **Recovery of tar.** The gas from the coke ovens is passed through a tower where liquor ammonia trickles from the top. Tar and dust are removed into a tar tank. Tar and ammonia are recovered. Ammonia liquor is again sent to the top of the trickling tower.

(2) **Recovery of ammonia.** The gases now enter another tower when water is sprayed. Ammonia goes into solution forming ammonium hydroxide. Sometimes, instead of water dilute H_2SO_4 is sprayed when ammonium sulfate is recovered.

$$NH_3 + H_2O \longrightarrow NH_4OH$$
$$2NH_4OH + H_2SO_4 \longrightarrow (NH_4)_2SO_4 + 2H_2O$$

(3) **Recovery of naphthalene.** The gases then pass to a cooling tower where water at a low temperature is sprayed. Condensation of some gases takes place and naphthalene is recovered.

(4) **Recovery of benzole.** The gases then pass through an oil (petroleum) scrubber where benzene and its homologues are recovered.

(5) **Recovery of H_2S.** The gases then enter a purifying chamber packed with moist Fe_2O_3

$$Fe_2O_3 + 3H_2S \longrightarrow Fe_2S_3 + 3H_2O$$

After all the Fe_2O_3 exhausted, it is exposed to atmospheric air to recover the sulfur (as SO_2), and regenerate the Fe_2O_3

$$Fe_2S_3 + 4O_2 \longrightarrow 2FeO + 3SO_2$$
$$4FeO + O_2 \longrightarrow 2Fe_2O_3$$

(6) **Recovery of gas.** The gas after passing through the various scrubbers and condensers to remove the various byproducts mentioned above is finally collected in a gas holder. The gas has the calorific value of about 5000 Kcal/m^3

FUELS AND COMBUSTION

2·21. COMBUSTION OF COAL

Coal is converted into energy by combustion. 75% of the coal in the world is used in this manner for generation of steam for producing electricity. The C, H and S present in the coal undergo combustion in presence of air liberating their heats of combustion. The flue gases contain CO_2, water vapour, SO_2 and N_2 from air.

If the elementary composition of a carbonaceous fuel is known, the theoretical amount of air required for the complete combustion of the fuel can be calculated. Maximum efficiency would be obtained if the fuel is burnt completely with the theoretical quantity of air. However, in actual day to day practice, considerable amount of potential heat is lost when only the theoretical quantity of air is used and smoke formation would result in many cases. About 20% excess of air is generally used to achieve the best possible firing conditions. However, if excess air is supplied, the loss of sensible heat in the stack gases increases, due to the greater quantity of stack gases and also due to their higher temperature. The optimum amount of excess air required depends upon the design of the furnace, the fuel used and the equipment used to burn it. For a given furnace, the optimum excess air is that which keeps the sum of the potential heat loss and the sensible heat in the stack gases at a minimum.

During the firing of coal on grates, air is supplied in two or more streams. The **primary air** sent through the fuel initiate the combustion reactions. The **secondary air** sent over the fuel bed helps to burn the volatile products formed by these reactions. The "tertiary air" (heated) may be supplied at the end of the grate to ensure more complete combustion of any smoke that might have been formed.

The primary air sent through the fuel bed must be evenly distributed and channels should not be allowed to form through which air escapes without reaction. Mechanical stokers are employed to achieve this objectives.

Both hand firing and mechanical firing methods are commonly used. In mechanical firing, stokers such as "overfeed", "underfeed", "crossfeed", "chain grate" and "spreader" are employed in practice. Where combustion of coal is carried out without grates, pulverised fuel in suspension or crushed coal in cyclone furnaces with slagging of ash are employed. Burning of coal in a fluidised bed is the latest development in this field.

2·22. LIQUID FUELS

The important liquid fuels include petroleum, petroleum products, tar, alcohols and colloidal fuels. Liquid fuels are also obtained synthetically from the hydrogenation of coal. Low boiling fractions of petroleum are used in petrol engines and higher boiling fractions in diesel engines and oil fired furnaces. Kerosene is used for heating, lighting and cooking. Liquid fuels find extensive use in domestic and industrial fields.

2.22.1. Merits and demerits of liquid fuels

Merits:
(1) Liquid fuels are easy to handle, store and transport. They can be transported cheaply to distant places through pipe lines. They require less volume for storage as compared to solid and gaseous fuels. Fuel oils weigh 30% less and occupy 50% less space than coal of equal heating value.
(2) After burning, they do not leave any appreciable amount of ash. Hence the problem of ash disposal is not there as in the case of solid fuels. Moreover, the problem of clinker formation is totally absent in case of liquid fuels.
(3) Liquid fuels can be easily kindled. The combustion can be started or stopped at once. Maximum temperature can be attained soon unlike in the case of solid fuels. Further, the rate of combustion can be easily controlled as desired. Hence the process is under better control and uniform combustion rate can be maintained.

(4) There is no need for maintaining "banked fires" in case of liquid fuels unlike solid fuels. Thus better fuel economy is possible in case of liquid fuels.//
(5) Liquid fuels are handled through pipes and one man can regulate the fuel supply to many furnaces at a time. Continuous manual feeding as in the case of solid fuels is unnecessary. Hence these are more economical.//
(6) Less excess air is needed in case of liquid fuels than in the case of solid fuels.//
(7) The furnace space required is lesser than in the case of solid fuels.//
(8) No danger of spontaneous combustion. If not volatile, they do not deteriorate on storage.//
(9) Operation is cleaner than in the case of solid fuels.

Demerits:

(1) Liquid fuels give unpleasant odours particularly when the combustion is incomplete.//
(2) They are more expensive than solid fuels.//
(3) Possibility of losses due to evaporation during storage and leakage in containers.//
(4) Risk of the hazards is greater particularly in case of inflammable and volatile liquid fuels.//
(5) Special type of burners and sprayers are needed for efficient combustion. Careful supervision is necessary to avoid difficulties like leaking and choking which may lead to accidents.

2.22.2. Petroleum. Petroleum or crude oil is the main source of liquid fuels. 60% of the world reserves of petroleum are in the Middle East while 15% are in the western hemisphere.

The word petroleum is derived from the two Latin words, petra (= rock) and oleum (= oil). It is also known as crude oil or rock oil or mineral oil. It is a dark coloured viscous liquid found deep in the earth's crust. It is a complex mixture of hydrocarbons (e.g., paraffinic, olefinic and aromatic) with small quantities of optically active organic compounds containing O, N and S, and traces of metallic constituents.

The average composition of crude oil is as follows:

$$
\begin{array}{lcl}
C - 80 & \text{to} & 87.1\% \\
H - 11.1 & \text{to} & 15.0\% \\
S - 0.1 & \text{to} & 3.5\% \\
O - 0.1 & \text{to} & 0.9\% \\
N - 0.4 & \text{to} & 0.9\%
\end{array}
$$

In some grades of petroleum the contents of O, S and N may reach 2.1, 8 and 0.93 per cent respectively. Besides these, some heavier grades of petroleum may contain compounds of heavy metals e.g., Fe, Ni, V etc.

2.22.2.1. Origin of Petroleum. There are two theories proposed on the origin of petroleum:

(1) **Inorganic theory (or carbide theory or Mendeef's theory),** which is mainly based on laboratory experiments, postulates that metal carbides are formed under the earth's crust due to the interaction of metals and carbon under high temperature and pressure. These carbides react with steam or water giving lower hydrocarbons, which further undergo hydrogenation and polymerization and eventually generate a complex mixture of paraffinic, olefinic and aromatic hydrocarbons,

$$Ca + 2C \xrightarrow{\text{Under high temperature}} \underset{\text{Calcium carbide}}{CaC_2}$$

$$4Al + 3C \xrightarrow{\text{and pressure}} \underset{\text{Aluminium carbide}}{Al_4C_3}$$

$$CaC_2 + 2H_2O \longrightarrow Ca(OH)_2 + \underset{\text{Acetylene}}{C_2H_2}$$

$$Al_4C_3 + 12H_2O \longrightarrow 4Al(OH)_3 + \underset{\text{methane}}{3CH_4}$$

FUELS AND COMBUSTION

$$C_2H_2 \xrightarrow{H_2} C_2H_4 \xrightarrow{H_2} C_2H_6$$
$$\text{Acetylene} \qquad \text{Ethylene} \qquad \text{Ethane}$$

$$3C_2H_2 \xrightarrow{\text{Polymerisation}} C_6H_6$$
$$\text{Acetylene} \qquad\qquad \text{Benzene}$$

$$3C_2H_4 \xrightarrow{\text{Polymerisation}} C_6H_{12}$$
$$\text{Ethylene} \qquad\qquad \text{Hexane}$$

This theory has been further reinforced by the claim of Moison that by the action of water on uranium carbide, a liquid similar to petroleum has been obtained. But, however, this theory fails to explain the presence of S and N, chlorophyll, haemen, prophyrin and optically active compounds in the petroleum.

2. **Organic theory (or Engler's theory),** which suggests that petroleum is generated from animal and vegetable debris accumulating in sea basins or estuaries and buried by sand and silt. The debris may have been decomposed by anaerobic bacteria under reducing conditions, so that most of the oxygen has been removed, or oil may have been distilled from the partially decayed debris by heat generated by earth movements or by depth of burial ultimately resulting in the production of black viscous liquid called petroleum. The role of high temperature and pressure metamorphism and radioactive emanations have also been predicted.

The organic theory has been largely supported by the geologists because of the presence of S and N, optically active organic compounds, fossils and brine in petroleum deposits. Further, the fact that a petroleum like liquid is obtained by distillation of fish lends further support to this theory.

However, no unanimously acceptable mechanism is yet available on the origin of petroleum.

2.22.2.2. Classification. The proportions in which the paraffinic, olefinic and aromatic hydrocarbons are present mainly define the character of the petroleum. There are three main types of crude oil generally distinguished.

1. **Paraffin base oils.** Which mainly consist of saturated paraffinic hydrocarbons (from methane to solid waxes) together with smaller amounts of naphthenes and aromatic hydrocarbons. These oils on distillation leave a residue of solid paraffin wax. (It is worth mentioning here that a lubricating oil obtained from paraffin-base oils usually has lower specific gravity and higher viscosity index than that obtained from asphaltic base oils. Also, more solid waxes separate from the paraffin base lubricating oils on cooling).

2. **Asphaltic base oils,** which principally consist of non-paraffinic hydrocarbons such as aromatic and naphthenic hydrocarbons. These oils on distillation leave behind asphalt or bitumen.

3. **Mixed base oils,** which are intermediate between the paraffin and asphaltic base oils and hence contain varying proportions of asphaltic, naphthenic, and aromatic hydrocarbons.

2.22.2.3. Petroleum resources of India. The Indian resources of petroleum are estimated to be around 300 million tonnes as against the world reserves of about 30,000 million tonnes. The major oil fields in India are in Assam (Naharkatiya) and Gujarat (Ankaleswar and Cambay). The Ankaleswar and Cambay crudes are predominantly of paraffin base type whereas the Naharkatiya crudes are of mixed base type. Extensive exploration for oil in all prospective sedimentary basins of India is in progress. The latest findings of oil deposits are in the Godavari deltas of Andhra Pradesh. The Oil and Natural Gas Commission (ONGC) of India with foreign collaboration has undertaken successfully the exploration of petroleum in Persian Gulf and Gulf of Kutch. One such exploration station is "Sagar Samrat". The production of petroleum from Bombay High is in progress. Further drillings are also in progress.

India produces only about one-third of its total petroleum requirements within the country. The rest is imported from the Middle East (Iran, Iraq, Saudi Arabia, Kuwait, etc.) and from Indonesia in the East. The Government of India has a share in Rostam crude from its joint venture in the Parsian Gulf.

The petroleum refineries of India are located in Trombay (near Bombay), Visakhapatnam (A.P.), Digboi, Barauni (Bihar), Gauhati (Assam), Koyali (Gujarat), Cochin (Kerala), Manali, Haldia (West Bengal), Bongaigaon (Assam) and Mathura (U.P.).

2.22.2.4. Detection of Oil Deposits. Oil deposits are detected by visual, geological and geophysical methods.

1. **Visual methods.** These are based on observations of oil seepages at the surface and fossils occurring in the strata.

2. **Geological methods.** These methods comprise of mapping the age of rocks, their nature and the types of formation present. These are sometimes assisted by aerial photography.

3. **Geophysical methods.** These include :

(a) **Gravimetric methods,** in which the variations in density of the earth's crust are measured with sensitive instruments.

(b) **Seismic methods,** in which the reflectance of the shock waves passed through the earth's crust are measured. This gives the depth of hard reflecting layers such as lime stone.

(c) **Magnetic methods,** in which local variations in the intensity and direction of the earth's magnetic field show the distribution of the various rocks in the earth's crust.

The field studies are supported by to elaborate analytical studies using sophisticated gadgets such as electron microscope, mass spectrograph, X-ray chromatograph, nuclear magnetic resonance (NMR), U.V. and I.R. spectroscopy, differential thermal analysis (DTA). On the basis of all these studies, the most suitable location for drilling is selected.

4. **Drilling**

This provides the final proof for the occurrence of oil-bearing strata. Cores for the drill are examined for fossil formation and evidence of porous or non-porous rocks. When the existence of oil deposit is proved, drilling is continued until the depth of the oil or oil-bearing strata is ascertained.

The drilling equipment consists of a tall huge tower called 'derrick' anchored to the ground, engines, mud pumps, water tanks draw-works and many other modules.

The success of drilling mostly depends on the quality of the specially prepared slurry of water, called the "mud", used during the drilling operation. Oil-based mud or water-based are used. The mud consists of various chemicals, adhesives and special additives which serve important functions such as

(i) removing the drill cuttings to the surface
(ii) cooling the drill bit which gets heated due to friction
(iii) lubricating the drill bit
(iv) providing buoyancy to the drill string to reduce the hook load
(v) retaining the sidewall of the well from caving in
(vi) facilitating perusal of the hole by lowering various instruments, and
(vii) blancing the formation pressure that prevents the formation fluids from running into the well.

The various chemicals used in water-based mud and the purpose they serve are given in the following Table 4.

FUELS AND COMBUSTION

CH_4 — 25 to 30%
H_2 — 50 to 55%
CO — 11%
CO_2 — 3%

The calorific value of oil gas is about 6700 Kcal/m^3. It is used as a laboratory gas. It is also used to improve the calorific value of water gas and the mixture of the two gases is called carburetted water gas.

Onia Geigi process

In this process, a light oil or heavy oil is gasified with steam in presence of a catalyst. The gas produced can be used as a town gas for domestic consumption and as a synthesis gas for industrial purposes.

2.23.7. Producer gas.

Producer gas is essentially a mixture of carbon monoxide and nitrogen. It is prepared by passing a mixture of air and steam over a bed of incandescent coke, anthracite coal, high volatile non-coking coal or lignite. Several types of "gas producers" are in use and one typical producer is shown in Fig. 2.20. The producer is made up of cylindrical steel furnace lined with fire-bricks. The fuel is fed from the hopper at the top. A mixture of air and steam (about 0.35 kg per kg of coal) is passed over the red hot coal bed through the inlet at the bottom. The producer is also provided with an ash-pit at the bottom to remove the ash formed during the combustion reactions.

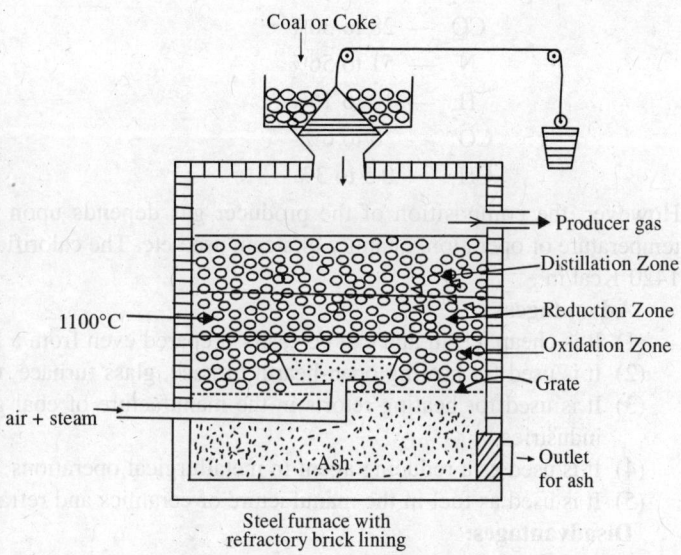

Fig. 2.20. Manufacture of producer gas.

When a mixture of air and steam is passed over red hot coke or coal bed at about 1000° to 1100°C in the producer, the following reactions take place in different zones of the fuel bed.

(1) Oxidation Zone

This is the lowest part of the coal bed. Here, the carbon of the coal burns in presence of excess of air to give carbon dioxide

$$C + (O_2 + 3.76\,N_2)_{\text{air}} \rightleftharpoons CO_2 + 3.76\,N_2 + 96.7 \text{ Kcal (oxidation)}$$

(2) Reduction Zone

This is the middle part of the fuel bed. Here, the CO_2 produced in the oxidation zone and the steam move up through the red hot fuel bed and the following reactions take place:

$$C + CO_2 \rightleftharpoons 2CO - 38.7 \text{ Kcal (Reduction)}$$
$$C + H_2O \text{ (steam)} \rightleftharpoons CO + H_2 - 28.7 \text{ Kcal (water gas)}$$
$$C + 2H_2O \text{ (steam)} \rightleftharpoons CO_2 + 2H_2 - 18.6 \text{ Kcal}$$

All these reactions in the reduction zone are endothermic and the temperature falls. In order to maintain the temperature of the bottom of the coal bed at about 1100°C (which is essential for the oxidation reaction) the quantity of the steam admitted along with air can not be increased beyond a certain limit. Also, it is not desirable to increase the H_2 content in the producer gas. Moreover, undecomposed steam in the producer gas carried away sensible and latent heat from the system.

(3) **Pre-heating and Distillation Zone**

This is the upper part of the fuel bed. In this zone some of the CO formed reacts with steam as follows:

$$CO + H_2O \rightleftharpoons CO_2 + H_2 + 10.1 \text{ Kcal}$$

Further, all the gases formed namely, CO, CO_2, H_2, N_2 etc. pass out through the gas outlet.

The average composition of the producer gas is as follows:

CO — 28 to 30%
N_2 — 51 to 56%
H_2 — 10 to 15%
CO_2 — 3 to 6%
CH_4 — 0.5 to 3%

However, the composition of the producer gas depends upon the nature of the fuel used, the temperature of operation, quantity of steam used etc. The calorific value of producer gas is 1160 to 1420 Kcal/m^3.

Advantages:
(1) It is cheap, clean and can be easily prepared even from a low grade coal.
(2) It is used for heating open-hearth furnace, glass furnace, muffle furnace, etc.
(3) It is used for heating retorts in the manufacture of coal gas and ovens in the carbonizing industries.
(4) It is used as a reducing agent in metallurgical operations.
(5) It is used as fuel in the manufacture of ceramics and refractories.

Disadvantages:
(1) It has a very low calorific value.
(2) If coal is used as fuel, the gas should be purified to remove the tar. Hence, most of the producer gas plants use coke as fuel.

Advantages of using steam along with air:
(1) The reaction between C and oxygen is exothermic and the temperature of the fuel bed of an air-blown producer tends to rise. This leads to clinker and slag formation. If a mixture of steam and air are used simultaneously, the heat evolved by the carbon-oxygen reactions is balanced by the strongly endothermic steam-carbon reactions. Thus, the temperature of the oxidation zone is reduced and clinker formation is prevented. The fuel bed temperature is maintained at 1000° to 1100°C by maintaining the required quantities of steam and air.
(2) The steam-carbon reactions convert some of the sensible heat into potential heat and thus increase the calorific value. The coal efficiency of the plant is also increased.
(3) The % of N_2 in the gas is reduced.

Note : If all the carbon were converted into CO, it should be theoretically possible to make a producer gas containing 34.7% of CO and 65.3% N_2, but in practice, the CO percentage is always lower than this and some CO_2 and H_2 are always present

2.23.8. Water gas (or Blue gas). Water gas is essentially a mixutre of CO (40 to 42%) and H_2 (48.51%). It is prepared by passing steam over red hot coke. The water gas is also known as blue gas because it burns with a blue flame due to the combustion of carbon monoxide.

Water gas is preapred by reacting steam over red hot coke in a cyclic process comprising of the following two main steps :

FUELS AND COMBUSTION

(1) Air is passed through the coke bed when carbon burns to CO_2 and the temperature of the fuel bed increases to over 1000°C.

$$C + (O_2 + 3.76\ N_2) \longrightarrow CO_2 + 3.76\ N_2 + 96.7\ \text{Kcal}$$
air   (exothermic)

The outgoing gases containing CO_2 and N_2 are used to preheat the incoming air and steam.

(2) Steam is now blown through the red hot coke when CO and H_2 are produced:

At higher temperatures:
$$C + H_2O \rightleftharpoons CO + H_2 - 28.7\ \text{Kcal (endothermic)}$$

At lower temperatures:
$$C + 2H_2O \rightleftharpoons CO_2 + 2H_2 - 18.6\ \text{Kcal (endothermic)}$$

These reactions being endothermic, the temperature of the fuel bed falls below 1000°C. At this stage, steam supply is cut off and air is again blown to raise the temperature to 1000° – 1100°C. Then the steam is blown in and the process is repeated.

In order to maintain the temperature of the fuel bed above 1000°C, the air and steam supply are regulated alternately to achieve the desired effects. The period for which air is passed is called "hot blow" and that for which steam is passed is known as "cold blow". From the technical and economic points of view, the "hot blow" is carried on for 10 minutes followed by the "cold blow" for about 1 minute and so on. Several types of water gas producers have been designed and one such is shown in Fig. 2.21.

The calorific value of water gas is about 2580 to 2670 $Kcal/m^3$. The normal composition of water gas is as follows:

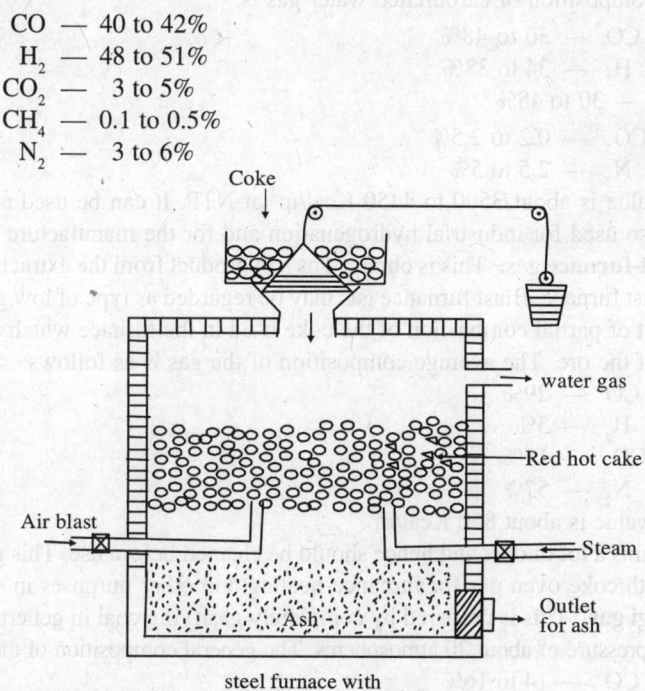

CO — 40 to 42%
H_2 — 48 to 51%
CO_2 — 3 to 5%
CH_4 — 0.1 to 0.5%
N_2 — 3 to 6%

Fig. 2.21. Water gas generator.

Water gas burns with a blue flame. It gives a short but hot flame. It can be used for heating purposes in industrial operations. It cannot be used as an illuminant. As it contains large proportion of CO, it should be used with caution. Further, its calorific value is low, and it cannot be transported to distant places. Hence it is not used as town gas.

Water gas is used for the production of hydrogen and in synthesis of ammonia. It is used as a fuel in many industrial processes such as enamelling, annealing etc., in glass and ceramic industries.

2.23.9. Carburetted water gas. The calorific value of water gas can be conveniently increased by enriching it with cracked oil vapours (oil gas), which can be made in an auxiliary plant called a carburettor (Fig. 2.22). This enriched gas is called the carburetted water gas. This can be obtained by heating a checker work of fire-bricks by burning a portion of the water gas. Then, oil is sprayed over it. The oil gets vaporized and cracked and at this stage, the water gas is blown through it. The resulting carburetted water gas is led through a hydraulic main and a purifier followed by scrubbers to remove the impurities.

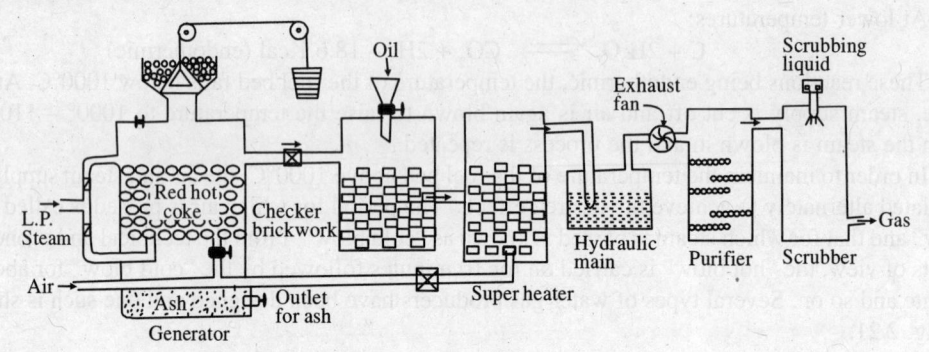

Fig. 2.22. Carburetted water gas plant.

The average composition of carburetted water gas is:

CO — 30 to 48%
H_2 — 34 to 38%
Hydrocarbons — 30 to 48%
CO_2 — 0.2 to 2.5%
N_2 — 2.5 to 5%

Its calorific value is about 3500 to 4450 Kcal/m^3 at NTP. It can be used as a fuel and as an illuminant. It is also used for industrial hydrogenation and for the manufacture of alcohol.

2.23.10. Blast-furnace gas. This is obtained as a byproduct from the extraction of metals from their ores using blast furnace. Blast furnance gas may be regarded as type of low grade producer gas obtained as a result of partial combustion of the coke used in the furnace which is modified by the partial reduction of the ore. The average composition of the gas is as follows:

CO — 29%
H_2 — 3%
CO_2 — 11%
N_2 — 57%

The calorific value is about 850 Kcal/m^3.

This gas contains a lot of dust and hence should be cleaned before use. This gas is used as such or after mixing with coke oven gas for furnance-heating and other purposes in steel plants.

2.23.11. Lurgi gas. This is prepared by completely gasifying coal in generators using oxygen and steam under a pressure of about 20 atmospheres. The general composition of the gas is as under:

CO — 14 to 16%
H_2 — 35 to 42%
CO_2 — 30%
CH_4 — 10 to 15%

The calorific value is about 3000 Kcal/m^3. The gas is used as town gas and also for the production of synthesis gas. Lurgi process has been proved to be a versatile process as the generator used can deal with a variety of fuels including low-grade ones and also because it can produce town gas as well as gas useful for synthesis of liquid fuels.

FUELS AND COMBUSTION

2.23.12. Kopper's gas. This is prepared by gasification of pulverised coal in presence of oxygen and super-heated steam in the Kopper's Totzek entrained bed gasifier at atmospheric pressure. High ash content low-grade coals can also be used for producing this synthesis gas. The composition of this gas is as follows:

CO — 51.1%
CO_2 — 12.6%
H_2 — 34.0%
N_2 — 1.9%
O_2 — less than 0.1%
CH_4 — less than 0.1%

The calorific value is around 2500 Kcal/m^3. The gas after suitable scrubbing, shift conversion and purification may be used for synthesis of ammonia.

2.23.13. Winkler's gas. Synthesis gas is prepared in the Winkler's process by gasification of crushed coal with steam and O_2 (or air) in a fluidised bed. The temperature is maintained below the ash fusion temperature. The gas is used as fuel as well as a synthesis gas for manufacturing urea, etc. Such process is used at the fertiliser plant at Neyveli (Tamil Nadu–India) where synthesis gas for the manufacture of urea is prepared from gasification of lignite. The typical composition of Winkler's gas is given below:

CO — 46%
H_2 — 36%
CO_2 — 14%
N_2 — 1.1%
CH_4 — 2.4%

Its calorific value is about 2500 Kcal/m^3.

2.23.14. Synthesis gas. This is so called due to the fact that it is useful for synthetic manufacture of chemicals and fertilizers. It provides N_2 and H_2 for the synthesis of ammonia, CO and H_2 for the manufacture of methanol, and CO and H_2 for generating liquid fuels. Water gas, coke oven gas, Kopper's gas, Winkler's gas, etc were used as sources of synthesis gas. Modern synthesis gas plants are based on natural gas or petroleum.

Synthesis Gas (Syngas)

Modern processes have been developed to produce CO - H_2 mixture from coal gas and steam more efficiently than the old water-gas and producer-gas plants. The gases produced are of low heat content (3.7 to 7.5 MJ/m^3) if steam and O_2 are used. The principal difference between the low-heat gas and the medium-heat gas is N_2, which enters the system as part of the air. The low-heat gas is mainly used as an on-site industrial fuel and as an intermediate in the production of formaldehyde and ammonia; while the medium heat gas can be economically transported by pipeline for distances of 160 to 320 Km. Much of the gas is methanated to produce substitute natural gas (SNG).

The types of gasification systems are classified according to gasifier bed type : fixed bed, fluidised bed, and entrained bed. The oldest examples of these are the Lurgi process (fixed bed), Winkler's process (fluidized bed), and Kopper-Totzek process (entrained flow). All of these processes have been commercially used worldwide, although none have been commercially operated in the U.S. Efforts to create improved reactors for very large scale SNG production (daily throughputs of 5000 to 10,000 t of coal) have led to the development of several so-called second-generation reactors, which are usually modifications of these older processes.

Composition of typical producer gases (dry basis) mole %

Gas constituent	Lurgi	Kopper-Totzek	Winkler
H_2	38.0	36.7	41.8
CO	20.2	55.8	33.3
CO_2	28.6	6.2	20.5
CH_4	11.4	0.0	3.0
C_2H_6	1.0	0.0	0.0
H_2S or COS*	0.5	0.3	0.4
N_2	0.3	1.0	1.0

*COS - compounds of sulphur.

2.23.15. Purification of gaseous fuels. Fuel gases, as manufactured, are hot and contain dust, grit, water vapour, tar, sulfur compounds, etc. Sometimes, useful by products such as ammonia, tar and benzol may be there as in coke oven gas. The gases should be purified by removing the impurities by cooling, scrubbing (by water or oil) and purification (using purifiers containing thylox, ferricyanide, Fe (OH)$_3$, etc to remove H_2S, CS_2, thiophene and mercaptons).

2.24 Biodiesel

There has been an increase in efforts to reduce the reliance on petrolium fuel for energy generation and transportation. Among the alternative fuels, biodiesel and diesohol have receive the much attention for diesel engines due to their advantages as the renewable, domestically produced energy resources and they are environmentally friendly because there is substantial reduction of unburned hydrocarbon, CO and particulate matter when it is used in conventional diesel engine.

Biodiesel can be produced from vegitable oil via transestrification process. Biodiesel has been used not only as an alternative to the fossil fuels but also as an additive for diesohol. The presence of biodiesel or vegitable oil can improve the lubricating properties of fuel, however there are some concerns regarding reduction of flash point and startability problem with ethanol-gasoline blend and ethanol-diesel blend.

The term 'biodiesel' refers to a mixtures of alkyl esters of long-chain fatty acids. Since it is biodegradable spills cause little or no environmental threat. A series of animal and vegitable oil feedstocks can be used to produce biodiesel. Biodiesel can be produced by base - catalysed transestrification. The reaction scheme is as follows.

1) Transestrification (Main reaction) : In transestrification ethanol is mostly used as it is a renewable resource and does not raise toxicity (as in case of methanol).

$$\text{Rapeseed oil} + \text{Alcohol} \underset{\text{(NaOH, KOH)}}{\overset{\text{alkali}}{\rightleftharpoons}} \text{Mixed ester} + \text{Glycerol}$$

$$\begin{array}{l} CH_2COOR_1 \\ CHCOOR_2 \\ CH_2COOR_3 \end{array} + R'OH \underset{}{\overset{60-70°C}{\rightleftharpoons}} \begin{array}{l} R_1COOR' \\ R_2COOR' \\ R_3COOR' \end{array} \begin{array}{l} CH_2OH \\ +CHOH \\ CH_2OH \end{array}$$

Triglyceride   Alcohol   Ethylester   Glycerol

FUELS AND COMBUSTION

$C_{17}H_{35}COOCH_3$ - Stearic ester
$C_{17}H_{33}COOCH_3$ - Oleic ester
$C_{15}H_{31}COOCH_3$ - Palmitic ester

$$\begin{array}{l} CH_2COO(CH_2)_{16}CH_3 \\ CHCOO(CH_2)_{16}CH_3 \\ CH_2COO(CH_2)_{16}CH_3 \end{array} + 3CH_3OH \rightarrow 3CH_3(CH_2)_{16}COOCH_3 + \begin{array}{l} CH_2OH \\ CHOH \\ CH_2OH \end{array}$$

Vegetable oil   Methanol   Biodiesel   Glycerol

ii) Neutralisation : unavailable side reaction in presence of free fatty acids presence of free fatty acids

Fatty acid + Alkali $\rightleftharpoons$ Soap + Water
            water

iii) Saponification Undesirable side reaction

Rapeseed oil + Alkali $\rightleftharpoons$ Soap + Glycerol

Mostly the biodiesel is produced by base - catalysed transestrification process. The blocks diagram is as follows

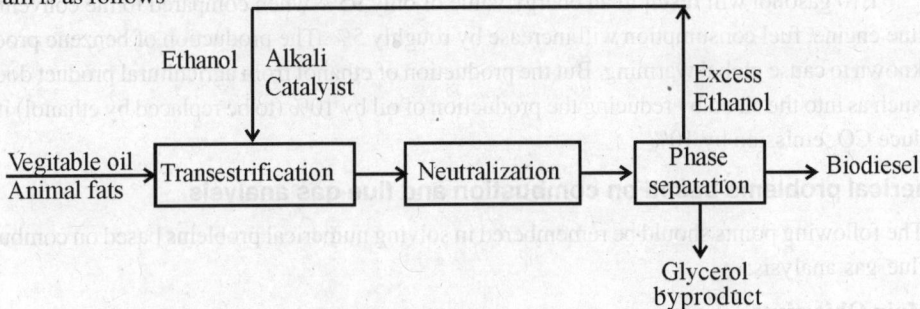

```
                Ethanol  Alkali                           Excess
                         Catalyst                         Ethanol
                            │                                │
Vegetable oil    ┌──────────▼──────┐   ┌──────────────┐   ┌──▼──────────┐
─────────────►  │ Transestrification├──►│ Neutralization├──►│   Phase     ├──► Biodiesel
Animal fats     └─────────────────┘   └──────────────┘   │  sepatation │
                                                          └──────┬──────┘
                                                                 │
                                                                 ▼
                                                             Glycerol
                                                             byproduct
```

Transestrification process gives byproduct glycerine and oilcake. The oilcake is a good source of organic manure which contains about 30% protein. Biodiesel contains no petroleum, it can be blended at any level with petrolium diesel. A pure 100% biodiesel fuel is referred as B 100. Biodiesel blends are referred as BXX, where XX indicates the amount of biodiesel in the blend ex. B20 blend is 20% by volume biodiesel and 80% by volume petrodiesel.

In USA, biodiesel is made from soybean oil. It can also be made from tallow. In Europe, the biodiesel is made from rapeseed oil. In India it is produced from seeds of plant Jatropa curcas and Karanja (Non edible oil).

In India Jatropa plant is grown in marginal and poor soil with minimum cultural practices or in waste lands with low fertility rockiness of soils. Biodiesel has higher cetane number, this oil reduces emission of CO_2 by 78% CO by 44%, Sulphate 100%, polycyclic aromatic hydrocarbon (PAA) which is carsiogenic compound by 75-85% and particulate matter.

Diesohol

Diesohol is an emulsion of hydrated ethanol in diesel fuel. A diesohol blend consist of 84.5% diesel, 15% hydrated ethanol and 0.5% emulsifier. Diesohol has been known as one of the preference to decrease the use of regular diesel fuel. The use of ethanol can decrease the amount of petroleum fuel imports. India was one of the earliest countries to recognize the merits of burning ethanol in diesel engines. Diesohol reduces CO_2, CO, NOx emission and smoke.

Ethanol has lower heating value than the regular diesel and use of pure ethanol as an alternative can cause drawbacks to the effectiveness of engines. To avoid such problem one of the approache is to blend the ethanol with regular diesel fuel and use the blend known as diesohol as the power source for diesel engine. However ethanol and diesel fuel are immiscible because of their difference in chemical structure and characteristics. These liquids can be emulsified by mechanical blending with suitable emulsifiers. The emulsifier would reduce the interfacial tension force and increase the affinity between the two liquid phase.

Know a days diesohol is become so important because it has the potential to create a large market for ethanol.

Gasohol :

Gasohol is a mixture of gasoline and alcohol (mostly ethanol since methanol is toxic). Its primary intention is to reduce the consumption (import) of gasoline.

E10 called gasohol a fuel, mixture of 10% ethanol and 90% gasoline that can be used in the internal combustion engines. E10 blends are rated as 2 to 3 octane higher than regular gasoline. For ethanol-based gasohol the ethanol is made by fermentation of agricultural crops or crap wastes i.e. corn, suger cane or cellustic sources such as paper waste blend. Methanol-based gasohol is expensive to produce and is toxic and corrosive and the emission produce cancer causing form aldehyele. Use of E10 can reduce CO emission by 20-30% and CO_2 by 2%.

E10 gasohol will result in an energy value of only 95% when compared to the conventional gasoline engine, fuel consumption will increase by roughly 5%. The production of benzene produces CO_2 known to cause global warming. But the production of ethanol from agricultural product does not emit such as into the air so by reducing the production of oil by 10% (to be replaced by ethanol) it help to reduce CO_2 emission by 10%

Numerical problems based on combustion and flue gas analysis

The following points should be remembered in solving numerical problems based on combustion and flue gas analysis:

(1) Main Objectives

The problems are generally based on calculating:
(a) The weight or volume of air theoretically required or used for the combustion of 1 kg of fuel.
(b) The percentage of excess air used per kg. of fuel burnt,
(c) The percentage composition of flue gas obtained (by weight or by volume) by burning a known quantity of fuel, and
(d) The percentage composition of the fuel burnt.

(2) Concept of mole

A mole of a substance is that quantity whose weight is numerically equal to its molecular weight.

$$\text{Number of moles, } n = \frac{\text{Weight of the substance, W}}{\text{Molecular weight, M}}$$

The "mole" (or "mol") is a general unit; when expressed in grams, it is called "gram mole"; when expressed in kilograms, it is called "kilogram mole" or "kilo mole"; and when expressed in pounds, it is called "pound mole". Hence, the term "mole" should be interpreted as "gram mole", "kilogram mole" or "pound mole" depending upon whether the basis of calculation is in grams, kilograms or pounds respectively. Also, "mol" is a volume unit and hence composition in terms of m^3) can be taken as the same as composition by volume (or composition in terms of m^3) i.e., volume % = mole %. Further, 1 gram mole (g mol) of a gas at NTP occupties 22.4 litres (or dm^3); 1 kilogram mole occupies 22.4 m^3 and 1 pound mole occupies 359 ft^3.

(3) Composition and mean molecular weight of air

The composition of air is taken as 21% of O_2 and 79% of N_2 (by volume); and 23% of O_2 and

FUELS AND COMBUSTION

77% of N_2 (by weight). The mean molecular weight of air is taken as 28.95. It may be noted that air consists of 21.00 mols of O_2, 78.06 mols of N_2 and 0.94 mols of Ar (argon). Since argon is inert, it is considered together with nitrogen for combustion calculations. That is why the % N_2 in air by volume is taken as 79%. However, the mean molecular weight of air is taken as 28.95 on the basis of the following:

$$\left[\frac{(21\times32)+(78.06\times28)+(0.94\times39.34)}{100}\right] = 28.9522$$

Further, 1 m³ of O_2 is supplied by $\frac{1\times100}{21}$ = 4.76 m³ of air; and 1 kg. of O_2 is supplied by $\frac{1\times100}{23}$ = 4.35 kg. of air.

(4) Density of air

The density of air at N.T.P. is 1.290 kg. m⁻³ or 1.290 × 10⁻³ g cm⁻³.

(5) Minimum Oxygen Required

The minimum O_2 required) = (Theoretical O_2 required — O_2 present in the fuel)

The minimum O_2 required should be calculated on the basis that complete combustion is taking place according to theoretical and stoichiometric combustion reactions.

In case of partial combustion, the combustion products contain CO.

In case of irregular combustion, the combustion products contain both CO and O_2. In such a case, the excess O_2 is calculated after subtracting the amount of O_2 required to burn CO to CO_2.

(6) Combustion of Carbon

The combustion of carbon in air may be represented as:

C	+	(O_2	+	N_2)	→	$CO_2\uparrow$	+	N_2
1 mol		1 mol		3.76 mols		1 mol		3.76 mols
or		or		or		or		or
12 kg.		32 kg.		107.2 kg.		44 kg.		107.2 kg.

Thus, combustion calculation can be done on mol basis or by using the stoichiometric weight relationships between the reactants and products. However, the mol method is considered to be more convenient and simpler.

(7) Combustion of hydrogen

The hydrogen in coal is present as

(a) Combined hydrogen in the form of H_2O

$2H_2$	=	O_2	=	$2H_2O$
4 kg.		32 kg.		2 × 18 = 36 kg.
1 kg.		8 kg.		9 kg.

(b) Available hydrogen:

Combined hydrogen in coal present as moisture does not undergo combustion. It is only the available hydrogen which is equivalent to (H—O/8) that takes part in combustion.

(8) Weight of theoretical amount of air required

For complete combustion of 1 kg. of solid or liquid fuel, the theoretical amount of air required

$$= \frac{100}{23}\left[\frac{32}{12}\times C + 8\left(H - \frac{O}{8}\right) + S\right] \text{kg.}$$

where C, H, O and S are the respective weights of carbon, hydrogen oxygen and sulfur present in 1 kg. of the fuel.

(9) Calorific Value

If the ultimate analysis of coal is available, its calorific value may be calculated by Dulong's formula as follows:

Calorific value (Kcals/kg).

$$= \left[8080\,C + 34460\left(H - \frac{O}{8}\right) + 2250\,S \right]$$

where C, H, O and S represent the respective weights of carbon, hydrogen, oxygen and sulfur per kg. of coal.

(10) % Excess Air

$$\% \text{ Excess Air} = \frac{(\text{Actual air used} - \text{Theoretical air})}{\text{Theoretical air}} \times 100$$

(10) The mass of dry flue gases formed should be calculated by balancing the carbon in the fuel and the carbon present in the flue gases.

(11) The composition of a solid or liquid fuel is usually expressed on weight basis whereas the composition of a gaseous fuel is expressed on volume basis, unless otherwise stated.

(12) The composition of flue gases is usually given on dry basis and on volume basis, unless otherwise stated.

2·25. Efficiency of combustion and flue gas analysis

In order to achieve efficient combustion of coal in a boiler furnace, it is essential that the coal and its distillation products are brought into intimate contact with sufficient quantity of air to burn all the combustible matter under appropriate conditions. The correct conditions are: (a) intimate mixing of air with the combustible matter, and (b) sufficient time to allow the combustion process to be completed. If these factors are inappropriate, inefficient combustion occurs. The factors of good mixing and duration in the furnace are largely determined by the design of the equipment and the combustion is controlled by the operator by varying the rate of feed of the fuel and the amount of air admitted. The rate of feed of the fuel is varied mechanically and is measured by meters or by the known characteristics of the feeding mechanism, e.g., depth of fuel bed and speed of grate in case of chain grate stokers. The control of the air is by means of dampers, in the case of natural draught, or by the forced and/or induced draught fans when these are fitted; but the usual and the only practical way of determining the amount of air used in the furnace is to analyse the flue gases for the constituents CO_2, CO and O_2.

Under normal conditions of operation, in which air in excess of the quantity theoretically required for complete combustion is admitted to the furnace, it is usual for control purposes to measure only the proportion of CO_2, CO and O_2 in the flue gases, which gives the correct idea about the combustion of the fuel). Sometimes N_2, H_2 and hydrocarbons such as CH_4, C_2H_6 are also determined. If the analysis shows the presence of free CO, it means that the combustion is incomplete which calls for an immediate increase in the amount of air used for combustion or a diminution in the rate of supply of the coal. If a large amount of free oxygen is shown in the analysis, it means that the air supply is very much in excess. In ordinary furnaces, 50 to 100% excess air is generally supplied. But too much excess of air results in the loss of heat. In such cases, the supply of air to the hearth is cut short or the rate of fuel supply is increased. However, if the presence of appreciable amounts of both CO and O_2 together is indicated in the flue gas analysis, it shows that the combustion is irregular and non-uniform. That means, in some parts of the furnace, there is excess of air and in some other parts, the supply of air is insufficient, and obviously, such a situation needs immediate attention.

FUELS AND COMBUSTION

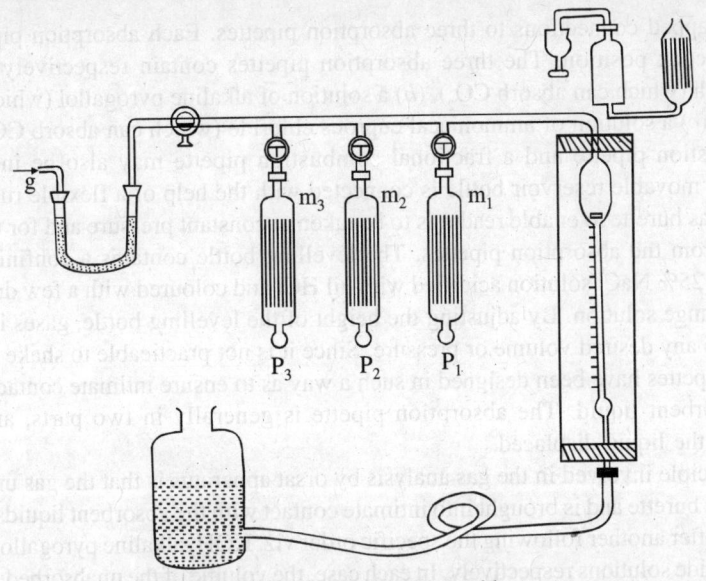

Fig. 2.23. Orsat's apparatus assembly.

The flue gas analysis is generally carried out by Orsat's apparatus (Fig. 2.23). Two types of orsat apparatus are commercially available.

(1) A portable model, known as "short orsat" which is used where the number of components to be determined is relatively limited and where moderate accuracy is sufficient.

(2) A much larger apparatus known as a "long orsat" or a precision model orsat, which can be used for the analysis of quite complicate mixtures of gases and which gives very accurate results in competent hands. The precision orsat assemblies are mounted on substantial stands, and are provided with compensators for gas burettes. Three gas absorption pipettes, a fractional combustion unit and a slow combustion unit are also usually supplied.

The fractional combustion pipette consists of a pyrex U-tube charged with cupric oxide (preferably in wire form) which can be heated in a small, thermostatically controlled electric furnace at 280–295°C. At this temperature, hydrogen is oxidised to water, and CO (if not previously removed by absorption) is oxidised to CO_2. The diminution in volume of the gas sample indicates the volume of hydrogen. The CO_2 formed may be determined by absorption in potassium hydroxide pipette and its volume is equal to that of the CO originally present in the gas mixture. ($2CO + O_2 = 2CO_2$). It may be noted that saturated hydrocarbons, such as methane and ethane, are not oxidised by copper oxide at 283-295°C. The slow combustion pipette is provided with a platinum wire spiral which can be heated electrically and its temperature controlled by a rheostat. The confining liquid is mercury. It is generally used for the determination of saturated hydrocarbons (CH_4, C_2H_6 etc.). These are mixed under appropriate conditions, with a small excess (10 to 15%) of O_2 and the mixture is carefully burned in contact with the heated platinum wire. Heat is generated during the combustion and precautions against explosion (due to accumulation of un-brunt gases) must be taken. The contraction in volume and measurement of the volume of CO_2 produced by absorption in potassium hydroxide solution, enable the calculation of the percentage of saturated hydrocarbons (provided their nature is known).

Other types of gas analysis apparatus are also available commercially. Hempel's apparatus, Bunte's apparatus, Ambler's apparatus, Haldane's air analysis apparatus and Bone and Wheeler's apparatus are the examples.

A simple portable orsat apparatus essentially consists of a water-jacketed and graduated burette of 100 ml capacity, which is connected at the top to a stop-cock manifold constructed of capillary tubing and tapped connections to three absorption pipettes. Each absorption pipette is connected at each stopcock position. The three absorption pipettes contain respectively (a) a solution of caustic potash (which can absorb CO_2), (b) a solution of alkaline pyrogallol (which can absorb CO_2 and O_2) and (c) a solution of ammoniacal cuprous chloride (which can absorb CO_2, O_2 and CO). A slow combustion pipette and a fractional combustion pipette may also be incorporated in the apparatus. A movable reservoir bottle is connected with the help of a flexible rubber tubing to the base of the gas burette to enable readings to be taken at constant pressure and for use in transferring gas to and from the absorption pipettes. The levelling bottle contains a confining liquid such as mercury or a 25% NaCl solution acidified with dil HCl and coloured with a few drops of methyl red or methyl orange solution. By adjusting the height of the levelling bottle, gases in the burette may be brought to any desired volume or pressure. Since it is not practicable to shake the apparatus, the absorption pipettes have been designed in such a way as to ensure intimate contact between the gas and the absorbent liquid. The absorption pipette is generally in two parts, an absorber and a reservoir for the liquid displaced.

The principle involved in the gas analysis by orsat apparatus is that the gas under investigation is taken in the burette and is brought into intimate contact with the absorbent liquids in the absorption pipettes one after another following the specific order viz. KOH, alkaline pyrogallol and ammoniacal cuprous chloride solutions respectively. In each case, the volume of the unabsorbed gases is measured separately at atmospheric pressure. The reduction in volume in each case corresponds respectively to the amounts of CO_2, O_2 and CO present in the gas under test. The result is usually expressed as the percentage composition of the gas by volume.

2·26. Combustion Calculations

In all combustion reactions, definite relationships exist between the masses of the fuel fired, the air used and the flue gases formed as well as the amounts of heat evolved or absorbed during the reactions, all of which are governed by certain well-known laws. These are:
1. The law of conservation of mass
2. The law of definite proportion
3. The gas laws (Boyle's law and Charle's law)
4. The law of conservation of energy.

As the composition of air used is taken as uniform, the relationship existing between the above mentioned quantities can be accurately calculated. Thus, if the values of any two of the following are given the third can be found out: (1) the composition of the fuel, (2) the composition of the flue gas and (3) the nature of combustion. Nature of combustion indicates whether the combustion is complete or not and whether any excess air was used.

The General Gas Equation:

All perfect gases obey the equation of state:

$$PV = nRT$$

where
- P = absolute pressure
- V = Volume
- n = number of moles of gas
- R = The "Gas constant", and
- T = absolute temperature.

Gram-mol and pound-mol

The composition of a solid or liquid fuel is usually expressed by weight whereas the composition of a gaseous fuel is given by volume. However, the composition of the flue gases is generally given by volume.

FUELS AND COMBUSTION

The masses of gaseous substances can be calculated from their volumetric compositions and vice versa in accordance with the gas laws and Avogadro's Hypothesis. From the corollary of the Avogadro's law, it follows that gram molecular volume of all gases at N.T.P. (*i.e.*, Normal temperature and pressure which are 0°C and 760 mm abs, or 32°F and 14.7 lbs/sq. in abs) is 22.4 litres. The corresponding volume in English system is 359 cu. ft., which is known as pound molecular volume. Thus, a gram mol of carbon is 12 grams; a kilogram mol of carbon is 12 kilograms, and a pound mol of carbon is 12 lbs. Similarly, a kilogram mol of CH_4 is 16 kgs and a pound mol of CH_4 is 16 lbs and a gram mol of CH_4 is 16 g. Each gram mol, kilogram mol and pound mol occupy 22.4 litres, 22.4 m^3 and 359 cu. ft. at N.T.P. respectively.

The Avogadro's hypothesis is true not only for gases but also for mixtures of gases. A mixture of gases behaves as if it were single with a molecular weight equal to the average molecular weight of its constituents. Thus, $\frac{1}{3}$ mol of N_2 + $\frac{1}{3}$ mol of H_2 + $\frac{1}{3}$ mol of O_2 will have an average molecular weight of 20.67 $\left(\frac{1}{3} \times 28 + \frac{1}{3} \times 2 + \frac{1}{3} \times 32 = \frac{62}{3} = 20.67\right)$. Hence, 20.67 grams of this gaseous mixture at 0°C, and 760 mm pressure occupies 22.4 litres. Similarly, 20.67 lbs of the gaseous mixture at 32°F and at 1 atmosphere pressure (*i.e.*, 14.7 lbs/sq. in) occupies 359 cu. ft. Thus, we can easily calculate the weight per cubic foot (c. ft.) or cubic centimeter (c.c.) or cubic meter (m^3) of any gas or any gaseous mixture of known composition.

The molecular weights of some common gases are as follows:

Hydrogen, H_2	... 2;	Oxygen O_2	... 32
Carbon monoxide, CO	... 28;	Carbon dioxide, CO_2	... 44
Methane, CH_4	... 16;	Ethylene C_2H_4	... 28
Nitrogen, N_2	... 28;	Water vapours, H_2O	... 18

$$\text{Number of moles} = \frac{\text{Weight of the substance}}{\text{Molecular weight}}$$

Composition of air

In all combustion calculations, the composition of air is taken as 21% O_2 and 79% N_2 (by volume); or as 23% O_2 and 77% N_2 (by weight). The average molecular weight of air is taken as 28.952.*

Calculation of volume of a gas at a given temperature and pressure

Boyle's and Charle's Laws can be used for reducing the volume of a gas at a given temperature and pressure to the corresponding volume at any other specified conditions of temperature and pressure with the help of the equation

$$\frac{PV}{T} = \frac{P_1 V_1}{T_1}$$

where T and T_1 are absolute temperatures of the gases and P and P_1 and V and V_1 denote the initial and final values of pressures and volumes expressed in identical units. If the temperature is i°C, then the absolute temperature, T = 273 + i°. Similarly, if the temperature is i°F, then the corresponding value in absolute scale is 460 + i°.

---

* O_2 = 21 mols = (21.0 × 22) = 672
N_2 = 78.06 mols = (78.06 × 28) = 2135.68
A_r = 0.94 mols = 0.94 × 39.94 = 37.54

Average Mol. wt. = $\frac{2895.2}{100}$ = 28.952.

Combustion reactions

The reactions most commonly encountered in combustion calculations are given below:

$$C + O_2 \longrightarrow CO_2 \quad \text{...(1)}$$
$$\underset{1\text{ mol}}{C} + \underset{1\text{ mol}}{O_2} \longrightarrow \underset{1\text{ mol}}{CO_2}$$

$$2C + O_2 \longrightarrow 2CO \quad \text{...(2)}$$

$$2CO + O_2 \longrightarrow 2CO_2 \quad \text{...(3)}$$

$$2H_2 + O_2 \longrightarrow 2H_2O \quad \text{...(4)}$$

$$\underset{1\text{ mol}}{CH_4} + \underset{2\text{ mol}}{2O_2} \longrightarrow \underset{1\text{ mol}}{CO_2} + \underset{2\text{ mol}}{2H_2O} \quad \text{...(5)}$$

$$C_2H_4 + 3O_2 \longrightarrow 2CO_2 + 2H_2O \quad \text{...(6)}$$

$$C_2H_2 + 2\frac{1}{2}O_2 \longrightarrow 2CO_2 + H_2O \quad \text{...(7)}$$

$$C_2H_6 + 3\frac{1}{2}O_2 \longrightarrow 2CO_2 + 3H_2O \quad \text{...(8)}$$

Note: In air, the volume of O_2 and N_2 are in the ratio of 21 : 79 = 1 : 3.76.

Equation (1) shows that 1 mol of carbon (12 lbs) combines with 1 mol of oxygen (32 lbs or 359 cu. ft. at N.T.P.) to form 1 mol of CO_2 (44 lbs or 359 cu. ft. at N.T.P.).

Similarly, Equation (5) shows that 1 mol of CH_4 reacts with 2 mols of O_2 to give 1 mol of CO_2 and 2 mols of H_2O. From this equation, we can calculate the air required for the complete combustion of 1 mol of methane. Since 21 mols of oxygen come from 100 mols of air and also since 2 mols of oxygen are required to burn completely 1 mol of CH_4, the air required = $2 \times \frac{100}{21}$ = 9.52 mols.

This also means that 1 cft of CH_4 required 9.52 cft of air (since volume % = mol %).

Analyses

Analyses of solids are always reported on weight basis. In order to convert them into mol basis, it is necessary to divide each constituent by its molecular weight. Thus a coal containing 72% C, 4% H_2 and 6% O_2 would contain $\frac{72}{12}$ = 6 mols of C; $\frac{4}{2}$ = 2 mols of H_2; and $\frac{6}{32}$ = 0.187 mol of O_2 per 100 kg of the coal sample.

Analysis of gases is always reported on volume basis and hence directly gives the number of mols of each constituent present per 100 mols of the mixture. Thus, the analyses of gases obtained by the Orsat apparatus are always expressed as % by volume and also on dry basis. Hence, this gas analysis gives molar composition directly since the mol is a volume unit. (A pound mol corresponds to 359 cft at N.T.P.). For example, 100 mols of air (100 × 359 = 35900 cft) contains 21 mols of O_2 (21 × 359 cft) and 79 mols of N_2 (79 × 359 cft).

Excess air

Combustion seldom takes place efficiently with the theoretically minimum quantity of air. Invariably, an excess of air is used in the furnace. Excess air is the amount of air used over and above that required for complete combustion.

Sample problem

Let us consider the following data obtained with methane, CH_4 as a fuel:

Fuel gas		Flue gas		
CH_4 = 100%	Constituent	%	Mols of C	Mols of O_2
	CO_2	5.5	5.5	5.5
	O_2	11.1	—	11.1
	N_2	83.4	—	—
	Total	100	5.5	16.6

FUELS AND COMBUSTION

(A) Ratio of Flue gas: Fuel gas

If we take 100 mols of dry flue gas as the basis for calculation, there are 5.5 mols of C, 16.6 mols of O_2 and 83.4 mols of N_2. It is obvious that the entire amount of carbon present in the fuel gas must have been present in the flue gas. Since all the carbon came from the fuel gas, it is clear that 5.5 mols of CH_4 (containing 5.5 mols of C) were used to form 100 mols of dry flue gas. Therefore,

$$\frac{\text{Dry flue gas}}{\text{Fuel gas}} = \frac{100}{5.5} = 18.2$$

i.e., 18.2 mols (or cft) of dry flue gas per mol (or cft) of fuel gas.

(B) Dry flue gas: dry air

We know that N_2 in air is equal to N_2 in flue gas. In the above example, there are 83.4 mols of N_2 per 100 mols of dry flue gas. In order to obtain 83.4 mols of N_2, the air used must have been $83.4 \times \frac{100}{79}$ mols.

$$\therefore \frac{\text{Dry flue gas}}{\text{Dry air}} = \frac{100}{83.4 \times \frac{100}{79}}$$

$$= 0.948$$

i.e., 0.948 mols of dry flue gas per mol of air or 0.948 cft. of dry flue gas per cft. of air.

(C) Air: Fuel gas ratio

From (A) and (B) above, we have

$$\frac{\text{Air}}{\text{Fuel gas}} = \frac{\text{Flue gas}}{\text{Fuel gas}} \div \frac{\text{Flue gas}}{\text{Air}} = \frac{18.2}{0.948} = 19.2$$

i.e., 19.2 mol or cft of air per mole or cft of fuel (CH_4).

(D) Excess air

If means the % of air in excess of that is theoretically required for complete combustion. It can be determined by the following methods:

(i) $\qquad CH_4 + 2O_2 \longrightarrow CO_2 + 2H_2O$

i.e., 1 mol of methane requires 2 mols of O_2 which can be obtained by $2 \times \frac{100}{21} = 9.52$ mols of air.

From (C) above, we calculated that the air actually used was 19.2 mols (or cft.) per mol (or cft.) of fuel gas.

$$\therefore \text{\% Excess air} = \frac{(19.2 - 9.52)}{9.52} \times 100 = \mathbf{101.7\%}$$

(ii) The N_2 in the gas analysis indicates the total air used and the free oxygen indicates the excess air (provided that the fuel does not contain N_2 and O_2 respectively. In case they are present in fuel, the corresponding quantities will have to be accounted for in the calculation). Thus in the present example, 83.4 mols of N_2 are present which should have come from $\frac{83.4}{79} \times 100$ mols of air, which contains $83.4 \times \frac{21}{79}$ mols of O_2 = 22.17 mols of O_2. However, the free oxygen (O_2) in the flue gas is 11.10 mols. Hence the O_2 required = (22.17 – 11.10) = 11.07 mols.

$$\therefore \text{Excess air} = \frac{(22.17 - 11.07)}{11.07} \times 100 = \mathbf{100.3\%}$$

Notes:

1. The difference in the value of excess air in the above two methods of calculation might be due to the slight errors in the gas analysis.
2. When carbon monoxide is present in the flue gas, the amount of O_2 necessary to burn it must be deducted from the free O_2 in the gas before determining excess air:

$2CO + O_2 \rightarrow 2CO_2$. Hence for every mole of CO, $\frac{1}{2}$ mol of O_2 needed.

For example, if the gas from combustion of a nitrogen free fuel has 12% CO_2, 5% O_2, 2% CO and 81% N_2, then the total oxygen (O_2) corresponding to N_2 is $81 \times \frac{21}{79} = 21.5$ mols and the excess O_2 would be $\left(5 - \frac{2}{2}\right) = (5-1) = 4$ mols.

Hence, the % excess air $= \frac{4}{(21.5 - 4.0)} \times 100 = 22.85\%$.

(E) Net hydrogen

The net hydrogen in fuel gas, or the hydrogen in excess of that which can combine with the O_2 present in the fuel can also be calculated from the flue gas analysis. H_2 burns with O_2 in air to form H_2O, which does not appear in the gas analysis. Hence it has to be calculated from the oxygen balance. For example, in the above sample problem on methane, while the O_2 appearing in the flue gases is only 16.6 mols, the O_2 corresponding to 83.4 mols of N_2 is $83.4 \times \frac{21}{79} = 22.17$ mol. Hence, $(22.17 - 16.6) = 5.57$ mols of O_2 must have reacted with $(2 \times 5.57) = 11.14$ mols of H_2 forming 11.14 mols of H_2O from each 100 mols of flue gas. Further, $\frac{\text{Net hydrogen}}{\text{Carbon}} = \frac{11.14}{5.5} = 2$, i.e., 2 mols of H_2 per mol of C.

(F) Moisture

The moisture in the flue gas is the sum of the water from free and combined moisture in the fuel, that from combustion of H_2 and that from moisture in the air used. The last one is so small that it may be neglected. Thus if the fuel analysis is known, the moisture in the flue gas (moisture in the fuel + H_2O from combustion of H_2) can be calculated. Thus, if the fuel used is dry methane, the volume of water vapour would be 2 mols per mol of fuel gas (since $CH_4 + 2O_2 \rightarrow CO_2 + 2H_2O$); and $\frac{2}{18.2} \times 100 = 11.1$ mols per 100 mols of dry flue gas i.e., for every 100 cft of dry flue gas, there would be 11.1 cft of wet flue gas.

Example 1. *A coal sample has the following percentage composition; C = 84.0%; H_2 = 3.5%, O_2 = 3.0%, S = 0.5%, Moisture = 3.5%, N_2 = 0.5% and ash = 5.0%. Calculate (a) the theoretical weight of air required for the complete combustion of 1 kg of coal (b) its volume in m^3 at NTP, and (c) percentage composition by weight and volume of the dry products of combustion.*

Solution

The composition of solid fuels is expressed by weight.

Method-1 (Mole Method)

Let the basis of calculation be 100 kg. of coal.

Constituent	% By wt.	Mol. wt.	No. of Kmols	Kmols of O_2 reqd.	Kmols of dry products
C	84.0	12	$\frac{84}{12} = 7.0$	7.0	CO_2 – 7.0
H_2	3.5	2	3.5/2 = 1.75	0.875	—

FUELS AND COMBUSTION

O_2	3.0	32	$\frac{3.0}{32} = 0.094$	-0.094	—
S	0.5	32	$\frac{0.5}{32} = 0.016$	0.016	SO_2 — 0.016
Moisture	3.5	18	—	—	—
N_2	0.5	28	$\frac{0.5}{28} = 0.0178$	—	N_2 – 0.0178
Ash	5.0	—	—	—	—
				7.797 k mols	

$\left.\begin{array}{l}\text{K mols of } O_2 \text{ required}\\ \text{per 100 kg of coal}\end{array}\right\} = \textbf{7.797 Kmols.}$

(a) Hence, the theoretical quantity of O_2 required for the combustion of 1 kg of coal
$$= 7.797 \times 10^{-2} \text{ Kmols}$$
$$= 7.797 \times 10^{-2} \times 32 \text{ kg.}$$

$\therefore \left.\begin{array}{l}\text{Weight of air required for}\\ \text{the complete combustion}\\ \text{of 1 kg of coal}\end{array}\right\} = 7.797 \times 10^{-2} \times 32 \times \frac{100}{23} \text{ kg.}$

$$= \textbf{10.848 kg.}$$

(b) Theoretical quantity of O_2 required per kg. of coal = 7.797×10^{-2} K mols.
$\therefore$ Theoretical quantity of air required per kg. of coal

$$= 7.797 \times 10^{-2} \times \frac{100}{21} = 0.3713 \text{ K mols}$$

$\therefore$ Volume of air supplied at NTP for 1 kg of coal
$$= 0.3713 \times 22.4 \text{ m}^3 = 8.3171 \text{ m}^3$$

(c) **Composition of dry products of combustion by volume**

The dry products formed by the combustion of 1 kg of coal are:
$CO_2 = 0.07$ K mols
$SO_2 = 0.00016$ K mols

$$N_2 = \underbrace{0.000178}_{\text{from fuel}} + \underbrace{\left(0.3713 \times \frac{79}{100}\right)}_{\text{from air}} = 0.2935 \text{ K mols}$$

$\therefore$ Total volume of the dry products of combustion = 0.3637 K mols
$\therefore$ Volumetric composition of the products of combustion is:

$$CO_2 = \frac{0.07 \times 100}{0.3637} = 19.25\%$$

$$SO_2 = \frac{0.00016 \times 100}{0.3637} = 0.04\%$$

$$N_2 = \frac{0.2935 \times 100}{0.3637} = 80.7\%$$

By weight

The dry products of combustion formed from 1 kg of coal are:
$$CO_2 = 0.07 \times 44 = 3.08 \text{ kg}$$
$$SO_2 = 0.00016 \times 64 = 0.01024 \text{ kg}$$
$$N_2 = 0.2935 \times 28 = 8.218 \text{ kg}$$
Total weight = 11.3082 kg.

∴ Gravimetric composition of the dry products of combustion is:

$$CO_2 = \frac{3.08 \times 100}{11.3082} = 27.237\%$$

$$SO_2 = \frac{0.01024 \times 100}{11.3082} = 0.091\%$$

$$N_2 = \frac{8.218 \times 100}{11.3082} = 72.673\%$$

Method 2: (Stoichiometric method)

$$\begin{array}{c} C + O_2 \longrightarrow CO_2 \\ 12\text{ kg} \quad 32\text{ kg} \quad 44\text{ kg} \end{array}$$

$$\begin{array}{c} H_2 + \frac{1}{2} O_2 \longrightarrow H_2O \\ 2\text{ kg} \quad 16\text{ kg} \quad 18\text{ kg} \end{array}$$

$$\begin{array}{c} S + O_2 \longrightarrow SO_2 \\ 32\text{ kg} \quad 32\text{ kg} \quad 64\text{ kg} \end{array}$$

Let 100 kg of the coal be the basis of calculation:

Constituent	Weight in kg	Weight of O_2 required, kg.	Weight of dry products, kg
C	84.0	84 × 32/12 = 224	$CO_2 = \frac{84 \times 44}{12} = 308$
H_2	3.5	$3.5 \times \frac{16}{2} = 28.0$	—
O_2	3.0	– 3.0	—
S	0.5	$0.5 \times \frac{32}{32} = 0.5$	$SO_2 = 0.5 \times \frac{64}{32} = 1.0$
Moisture	3.5	—	—
N_2	0.5	—	$N_2 = 0.5$
Ash	5.0	—	—

Weight of O_2 required for the combustion of 100 kg coal $\bigg\} = 249.5$ kg.

(a) Hence, the theoretical quantity of O_2 required for the combustion of 1 kg of coal = 2.495 kg.

∴ The theoretical quantity of air required for the combustion of 1 kg. of coal = $2.495 \times \frac{100}{23}$ kg

= **10.848 kg.**

(b) 32 kg. of O_2 occupies 22.4 m³ at N.T.P.

∴ 2.495 kg. of O_2 occupies $\frac{2.495 \times 22.4}{32} = 1.7465$ m³

Now, 21 m³ of O_2 are present in 100 m³ of air at N.T.P.

∴ 1.7465 m³ are present in $\frac{1.7465 \times 100}{21} = 8.317$ m³ of air.

∴ Volume of the air at N.T.P. = 8.317 m³.

(c) Composition of dry products of combustion.

FUELS AND COMBUSTION

By weight

The dry products of combustion formed from 1 kg of coal are:
$$CO_2 = 3.08 \text{ kg}$$
$$SO_2 = 0.01 \text{ kg}.$$
$$N_2 = [\underset{\text{from fuel}}{0.005} + \underset{\text{from air}}{(10.848 - 2.495)}] \text{ kg.} = 8.358 \text{ kg.}$$

Total weight = 11.448 kg.

∴ Gravimetric composition of the dry products of combustion is

$$CO_2 = \frac{3.08 \times 100}{11.448} = 26.90\%$$

$$SO_2 = \frac{0.01 \times 100}{11.448} = 0.087\%$$

$$N_2 = \frac{8.358 \times 100}{11.448} = 73.01\%$$

By Volume

$$CO_2 = \frac{3.08}{44} = 0.07 \text{ Kmol}$$

$$SO_2 = \frac{0.01}{64} = 0.0001563 \text{ Kmol}$$

$$N_2 = \frac{8.358}{28} = 0.2985 \text{ Kmols}$$

Total volume = 0.3687 Kmols

∴ Volumetric composition of the dry products of combustion is:

$$CO_2 = \frac{0.07 \times 100}{0.3687} = 18.986\%$$

$$SO_2 = \frac{0.001563 \times 100}{0.3687} = 0.0424\%$$

$$N_2 = \frac{0.2985 \times 100}{0.3687} = 80.96\%$$

Example 2. *Find the volume of air required for complete combustion of 1 m³ of acetylene and the weight of air necessary for the combustion of 1 kg of fuel.*

Solution

$$2C_2H_2 + 5O_2 = 4CO_2 + 2H_2O$$
2 vols   5 vols   4 vols   2 vols

O_2 required per m³ of C_2H_2 = 2.5 m³

∴ Air required per m³ of C_2H_2 = $2.5 \times \frac{100}{21}$ = **11.9 m³**

Further,

2 mols of C_2H_2 require 5 mols of O_2

∴ (2 × 26) kg of C_2H_2 requires (5 × 32) kg of O_2

(Since mol. wt. of C_2H_2 = 26 and Mol. wt. of O_2 = 32)

O_2 to be supplied per kg of fuel (C_2H_2) = $\frac{160}{52}$ kg

∴ Air to be supplied per kg of C_2H_2 = $\dfrac{160}{52} \times \dfrac{100}{23}$ = **13.378 kg**.

Example 3. *The % analysis by volume of producer gas is H_2 – 18.3%, CH_4 – 3.4%, CO – 25.4%, CO_2 – 5.1%, N_2 – 47.8%. Calculate the volume of air required m^3 of the gas.*

Solution.
100 mols of the producer gas contains:

Constituents	% and Mols	Mols of O_2 required
H_2	18.3	9.15
CH_4	3.4	6.80
CO	25.4	12.70
CO_2	5.1	—
N_2	47.8	—
	Total O_2 required ...	28.65

∴ Air required for 100 mols of fuel gas = $28.65 \times \dfrac{100}{21}$ = 136.43 mols

∴ Air required per m^3 of gas = $\dfrac{136.43}{100}$ = **1.3643 m^3**.

Example 4. *A gas has the following composition by volume: H – 22%, CH_4 – 4%, CO – 20%, CO_2 – 6%, O_2 – 3% and N_2 – 45%. If 25% excess air is used, find the weight of air actually supplied per m^3 of this gas.*

Solution.
100 mols of the fuel gas contains

Constituents	% and Mols	Mols of O_2 required
H_2	22	11
CH_4	4	8
CO	20	10
CO_2	6	—
O_2	3	– 3
N_2	45	—
	Total O_2 required	26

∴ Air required for 100 mols of fuel gas = $26 \times \dfrac{100}{21}$ = 123.8 mols

But air supplied actually (25% excess) for 100 mols of gas = $123.8 \times \dfrac{125}{100}$ = 154.75 mols.

∴ Air supplied for 1 mol of gas = $\dfrac{154.75}{100}$ = 1.5475 mols

∴ Air supplied for 1 m^3 of gas = 1.5475 m^3 = $\dfrac{1.5475 \times 28.97}{22.4}$ = **2 kg**.

(Since 22.4 m^3 of air *i.e.* 1 kg mol at N.T.P. weighs 28.95 kg).

Example 5. *A sample of coal contained: C – 81%, H_2 – 4%, O_2 – 2% and N_2 – 1%. Estimate the minimum quantity of air required for complete combustion of 1 kg of the sample. Find the composition of the dry flue gas by volume if 40% excess air is supplied.*

FUELS AND COMBUSTION

Solution.
Let the basis of calculation be 100 kg of coal.

Element	Kg	K Mols	K mols of O_2 required	Product
C	81	6.7500	6.7500	CO_2 – 6.750
H	4	2.0000	1.0000	H_2O – 2.000
O	2	0.06250	–0.0625	—
N	1	0.0360	Nil	N_2 – 0.036
		Total	7.6875	

Oxygen required for 100 kg of coal = 7.6875 k mols.

$\therefore$ Air required for 100 kg of coal = $7.6875 \times \dfrac{100}{21}$ K mols

= 36.6 k mols
= 36.6 × 28.95 kg
= 10.59.6 kg.

$\therefore$ Air required per kg of coal = $\dfrac{10596}{100}$ = **10.596 kg.**

Air actually used is 40% excess, and hence it is equal to

$36.6 \times \dfrac{140}{100} = 51.24$ k mols .

$\therefore$ Excess air used = (Actual air used – theoretical quantity of air required) = (51.24 – 36.60) k mols.

= 14.64 k mols

Nitrogen in the total air used = $31.24 \times \dfrac{79}{100}$ = 40.48 k mols.

Oxygen in the excess air used which remains in free state in the flue gas

= $14.64 \times \dfrac{21}{100}$ = 3.0744 k mols

The dry flue gases thus contain

CO_2 = 6.75 k mols
N_2 = (40.48 + 0.036) = 40.51 k mols
      from   from
      air   fuel
O_2 = 3.0744 k mols

Total – 50.3344 k mols

Percentage composition of the dry flue gases would be

$CO_2 = \dfrac{6.75}{50.3344} \times 100 = 13.4\%$

$N_2 = \dfrac{40.51}{50.3344} \times 100 = 80.5\%$

$O_2 = \dfrac{3.0744}{50.3344} \times 100 = 6.1\%$

Notes
1. Air contains 21% by volume of O_2 and 79% by volume of N_2.
2. Average mol. wt. of air is 28.95.
3. H_2O in flue gases is neglected for dry flue gas analysis.

Example 6. *A gas fired engine uses a fuel gas having the following composition:*
$$CO_2 - 9\%,\ CO - 42\%,\ H_2 - 33\%,\ N_2 - 16\%$$
The combustion takes places with the theoretical quantity of air. Calculate the volume of air used per cubic meter of fuel gas burnt. Determine the dry flue gas analysis.

Solution.
100 mols of fuel gas contains

	%	Mols	Mols of O_2 required	Products and Mols
CO_2	9	9	Nil	$CO_2 - 9$
CO	42	42	21	$CO_2 - 42$
H_2	33	33	16.5	$H_2O - 33$
N_2	16	16	Nil	$N_2 - 16$
			Total 37.5 mols	

Mols of O_2 required per 100 mols of fuel gas = 37.5

∴ Mols of air required per 100 mols of fuel gas

$$= 37.5 \times \frac{100}{21} = 178.5$$

∴ Mols or cubic meters (m³) of air required per 1 mol or 1 cubic meter of the fuel gas = $\frac{178.5}{100}$ = **1.785**.

The dry product of combustion:
$$CO_2 \longrightarrow 9 + 42 = 51\ \text{mols}$$

$$N_2 \longrightarrow 16 + \left(37.5 \times \frac{79}{21}\right) = 157\ \text{mols}$$
$$\quad\quad\quad\quad \text{(from fuel)}\quad \text{(from air)}$$
$$\quad\quad\quad\quad\quad\quad\quad \text{Total}\quad 208\ \text{mols}$$

∴ The dry exhaust gas contains

$$CO_2 - \frac{51 \times 100}{208} = 24.52\%$$

$$N_2 - \frac{157 \times 100}{208} = 75.48\%$$

Example 7. *A furnace fired with a hydrocarbon fuel oil has a stack gas analysing as follows: $CO_2 - 10.2\%,\ O_2 - 8.3\%,\ N_2 - 81.5\%$. Calculate (a) the composition of the original fuel oil, (b) % excess of air used and (c) volume of air supplied per kg of oil.*

Solution.
Basis for calculation: 100 k mols of dry stack gas

	% and mols	K Mols of C	K mols of O_2	K mols of N_2
CO_2	10.2	10.2	10.2	—
O_2	8.3	—	8.3	—
N_2	81.5	—	—	81.5
	Total	10.2	18.5	81.5

(a) 100 mols of dry flue gas contains 18.5 k mols of O_2 and 81.5 k mols of N_2 which must have come from the air supplied.

Hence, the air supplied = $81.5 \times \frac{100}{79}$ = 103.17 k mols;

and the O_2 accompanying it = $103.17 \times \frac{21}{100}$ = 21.69 k mols.

The difference in quantity of O_2 supplied and that used in the formation of water which is not appearing in the dry flue gas analysis will be

FUELS AND COMBUSTION

$$= 21.69 - 18.5 = 3.19 \text{ k mols}$$

∴ Water vapour formed = 2 × 3.19 = 6.38 k mols

$$\left(H_2 + \frac{1}{2}O_2 \rightarrow H_2O\right)$$

∴ Hydrogen present in the fuel oil
$$= 6.38 \text{ k mols}$$

The same quantity of fuel oil contains
$$= 10.2 \text{ k mols of C}$$

∴ The fuel oil contains 10.2 k mols of C or (10.2 × 12) = 122.4 kg of C and (6.38 × 2) = 12.76 kg of H.

That means, the oil weighing (122.4 + 12.76) = 135.16 kg was burnt and it should be composed of 122.4 kg of C and 12.76 kg of H.

Hence the composition of the fuel oil is:

$$C = \frac{122.4}{135.16} \times 100 = 90.5\%$$

$$H = \frac{12.76}{135.16} \times 100 = 9.5\%$$

(b) Total O_2 supplied = 21.69 k mols
O_2 needed for C = 10.20 k mols
O_2 needed for H = 3.19 k mols

O_2 actually needed for C and H present in 100 mols of fuel oil $\Big\} = 13.39$ k mols

∴ Excess O_2 = (21.69 – 13.39) = 8.30 k mols

∴ Excess O_2 or air used $= \dfrac{8.30}{13.39} \times 100 = 61.99\%$

(c) Total oil burnt = 135.16 kg

Total air supplied $= 21.69 \times \dfrac{100}{21} = 103.29$ k mols

$= 103.29 \times 22.4 = 2313.7 \text{ m}^3$ at NTP

(Avogadro's law)

∴ Air used per kg of fuel oil $= \dfrac{2313.7}{135.16} =$ **17.12 m³**.

Example 8. *A coal containing 62.4% C, 4.1% H, 6.9% O, 1.2% N, 0.8% S and 15.1 moisture and 9.7% ash was burnt in such a way that the dry flue gases contained 12.9% CO_2, 0.2% CO, 6.1% O and 80.8% N. Calculate (a) the weight of air theoretically required per kg of coal, (b) the weight of air actually used and (c) the weight of dry flue gas produced per kg of coal.*

Solution.
(a) **Basis of calculation.** 100 kg of coal, which contains

Analysis	Kg	K mols	K mols of O_2 required	K mols of product formed
C	62.4	5.2	5.2	CO_2 — 5.200
H	4.1	2.05	1.025	H_2O — 2.050
O	6.9	0.217	– 0.217	Nil
N	1.2	0.043	Nil	N_2 — 0.043
S	0.8	0.025	0.025	SO_2 — 0.025
H_2O	15.1	0.840	Nil	H_2O — 0.840
Ash	9.7	—	Nil	Nil
			Total 6.033 k mols	

Thus, O_2 theoretically required for complete combustion of 100 kg of coal = 6.033 mols.

∴ Air theoretically required for complete combustion

$$= 6.033 \times \frac{100}{21} = 28.73 \text{ mols}$$

$$= (28.73 \times 28.95) = \textbf{831.7 kg}$$

Wt. of air required per kg of coal $= \frac{831.7}{100} = \textbf{8.317 kg}$

(b) **Actual air used**:

Basis : 100 k mols of dry flue gas, which contains

Constituent	%	K Mols	K Mols C	K Mols of C	K Mols of N
CO_2	12.9	12.9	12.9	12.9	—
CO	0.2	0.2	0.2	0.1	—
O	6.1	6.1	—	6.1	—
N	80.8	80.8	—	—	80.8
			Total 13.1	19.1	80.8

13.1 k mols of C must have been supplied by $\frac{13.1}{5.2} \times 100$ kg of coal = 252 kg of coal containing

$0.043 \times \frac{252}{100} = 0.108$ k mols of N.

(Since 100 kg of coal contains 5.21 mols of C and 0.043 k mols of N).

∴ K Mols of N from air in 100 k mols of flue gas

$$= (80.800 - 0.108) = 80.692 \text{ k mols}$$

∴ K Mols of air actually supplied to obtain 100 k mols of dry flue gas or to burn 252 kg of coal would therefore be

$$= 80.692 \times \frac{100}{79} = 102.15 \text{ k mols}$$

∴ Wt. of air supplied $= (102.15 \times 28.97)$ kg

$$= 2959.3 \text{ kg}$$

∴ Weight of air supplied per kg of coal

$$= \frac{2959.3}{252} = \textbf{11.74 kg}$$

(c) **Wt. of dry flue gas obtained per kg of coal**:

100 kg mols of dry flue gas contains:

Constituent	% K Mols	Weight in kg
CO_2	12.9	$12.9 \times 44 = 567.6$
CO	0.2	$0.2 \times 28 = 5.6$
O_2	6.1	$6.1 \times 32 = 195.2$
N_2	80.8	$80.8 \times 28 = 2262.4$
		Total = 3030.8 kg

This much flue gas is obtained by burning 252 kg of coal.

∴ Wt. of dry flue gas obtained per kg of coal burnt

$$= \frac{3030.8}{252} = \textbf{12.03 kg.}$$

Example 9. *A hydrocarbon fuel, on burning, gave a flue gas having the following volumetric analysis: $O_2 = 3.56\%$, $CO_2 = 13.53\%$; $N_2 = 82.91\%$. What is the composition of the fuel by weight and the percentage of excess air used? What is the volume of air supplied per kg of the fuel oil?*

FUELS AND COMBUSTION

Solution.
Basis of calculation: 100 k mols of dry flue gas.

Constituent	% and K Mols	Mols of C	K Mols of O_2	K Mols of N_2
CO_2	13.53	13.53	13.53	—
O_2	3.56	—	3.56	—
N_2	82.91	—	—	82.91
	100.00	13.53	17.09	82.91

100 k mols of dry flue gas contions 17.09 k mols of O_2 and 82.91 k mols of N_2 which must have come out from the air supplied, *i.e.*, $82.91 \times \dfrac{100}{79}$ = 104.95 k mols; and the O_2 accompanying this air

$$= 104.95 \times \dfrac{21}{100} = 22.04 \text{ k mols.}$$

Hence, the difference in quantity of O_2 supplied and used in the formation of water vapour, which is not appearing in dry flue gas analysis, will be (22.04 – 17.09) = 4.95, k mols.

∴ The water vapour formed = 2 × 4.95 = 9.90 k mols.
∴ Mols of H_2 present in the oil should also be = 9.90 k mols.
$(2H_2 + O_2 \rightarrow 2H_2O)$
The same quantity of the fuel oil contains 13.53 k mols, of C, $(C + O_2 \rightarrow CO_2)$.
∴ The fuel oil contains 13.53 mols of C ≡ 13.53 × 12 kg
 = **162.36 kg of C** and 9.9 × 2 = **19.8 kg of H_2.**
∴ The composition by weight of the hydrocarbon fuel is as follows:

$$C = \dfrac{162.36 \times 100}{182.16} = 89.13\%$$

$$H = \dfrac{19.8 \times 100}{182.16} = 10.87\%$$

Now,
 Total O_2 supplied = 22.04 k mols,
 O_2 needed for C = 13.53 k mols,
 O_2 needed for H = 4.95 k mols.
∴ Total O_2 actually needed} = (13.53 + 4.95) = 18.48 k mols
∴ Excess O_2 = (22.04 – 18.48) mols = 3.56 k mols.
∴ % of Excess O_2 (or air) used

$$= \dfrac{3.56 \times 100}{18.48} = \mathbf{19.264\%}$$

Further,
 Total fuel oil burnt = 182.16 kg

 Total air supplied = $22.04 \times \dfrac{100}{21}$ kg mols

 = $22.04 \times \dfrac{100}{21} \times 22.4 \text{ m}^3$

 = 2319.4476 m³
∴ Air used per kg of fuel oil

$$= \dfrac{2319.4476}{182.16} = \mathbf{12.733 \text{ m}^3.}$$

---

* For other worked out examples on combustion calculations, the students may refer to the "Text Book on Experiments and Calculation in Engineering Chemistry" By S.S. DARA.

Example 10. *A example gave the following analysis: C – 84%, H – 4%, O – 4%, ash – 8%. The composition of the dry flue gas obtained by using the above coal is as follows:*
   CO_2 – 9%, CO – 1%, O_2 – x% and N_2 – (90 – x)%.
What will be the value of x?

Solution.
The problem can be solved by using the principle of oxygen balance.
100 kg of the coal sample contains:

Constituent	Kg	k Mols
C	84	$\frac{84}{12} = 7$
H	4	$\frac{4}{2} = 2$
O	4	$\frac{4}{32} = 0.125$

100 mols of the flue gas contains

Constituent	k Mols	k Mols of C	k Mols of O	k Mols of N
CO_2	9	9	9	—
CO	1	1	0.5	—
O_2	x	—	x	—
N_2	(90 – x)	—	—	(90 – x)
Total	100	10	(9.5 + x)	(90 – x)

The coal burnt to produce 100 k mols of the flue gas

$$= \frac{100}{7} \times 10 = 142.8571 \text{ kg}$$

k mols of H present in 142.857 kg of coal

$$= \frac{2}{100} \times 142.8571 \text{ k mols} = 2.8571 \text{ k mols}$$

As the fuel does not contain any N_2, it may be assumed that the entire N_2 present in the flue gas is obtained from air only.

Mols of O_2 supplied for the combustion of 142.8571 kg of coal

$$= (90 - x) \times \frac{21}{79} = 0.2658 (90 - x) \text{ k mols}$$

Further, the k mols of O_2 supplied = k mols of O_2 present in dry flue gas + k mols of O_2 that combined with H_2 to form water vapour.

i.e., $\quad 0.2658 (90 - x) = (9.5 + x) + \left(2.8571 \times \frac{1}{2}\right)$

∴ $\quad 23.922 - 0.2658\, x = 9.5 + x + 1.4285$
∴ $\quad 1.2658\, x = 23.922 - 9.5 + 1.4285$
∴ $\quad x = \mathbf{10.265}.$

Example 11. *A petroleum gas has the following composition. Ethane – 5%, Propane – 10%, Butane – 40%, Butene – 10%, Isobutane – 30%, Propene – 5%.*

Calculate the volume of air required for complete combustion of 100 M^3 of the gas and the percentage composition of the dry flue gases if 35% excess air is supplied.

FUELS AND COMBUSTION

Solution.

The combustion reactions of the constituents of the fuel gas are:

$$C_2H_6 + 3.5\,O_2 = 2CO_2 + 3H_2O$$
$$\text{1 mol} \quad \text{3.5 mols} \quad \text{2 mols} \quad \text{3 mols}$$

$$C_3H_8 + 5O_2 = 3\,CO_2 + 4H_2O$$
Propane

$$C_4H_{10} + 6.5\,O_2 = 4\,CO_2 + 5H_2O$$
Butane
or
Isobutane

$$C_4H_8 + 6O_2 = 4CO_2 + 4H_2O$$
Butene

$$C_3H_6 + 4.5O_2 = 3CO_2 + 3H_2O$$
Propene

Let 100 mols of the petroleum gas be the basis for calculation:

Constituent	Mols	Mols O_2 reqd.	Products and Mols
C_2H_6	5	5 × 3.5 = 17.5	CO_2 = 10; H_2O = 15
C_3H_8	10	10 × 5 = 50	CO_2 = 30; H_2O = 40
C_4H_{10}	40 (Butane)	40 × 6.5 = 260	CO_2 = 160; H_2O = 200
	30 (Isobutane)	30 × 6.5 = 195	CO_2 = 120; H_2O = 150
C_4H_8	10	10 × 6 = 60	CO_2 = 40; H_2O = 40
C_3H_6	5	5 × 4.5 = 22.5	CO_2 = 15; H_2O = 15
	Total	605 mols	375 mols

Now,

O_2 required for complete combustion of 100 mols of the petroleum gas = 605 mols.

∴ Air required for the complete combustion of 100 mols of the petroleum gas = $605 \times \dfrac{100}{21}$ mols = 2880.9524 mols.

∴ Air required for the complete combustion of 100 m³ of the petroleum gas = 2880.9524 m³.

Dry flue gas analysis with 35% excess air

If 35% excess air is used, the actual air supplied would be = $2880.9524 \times \dfrac{135}{100}$ mols = 3889.2857 mols.

The N_2 contributed by the above quantity of air

$$= 3889.2857 \times \dfrac{79}{100} = 3072.5357 \text{ mols.}$$

Excess O_2 supplied during the combustion of 100 mols of the petroleum gas = $605 \times \dfrac{35}{100}$ = 211.75 mols.

∴ The flue gases produced by the combustion of 100 mols of the petroleum contain

CO_2 = 375 mols
N_2 = 3072.5357 mols
O_2 = 211.75 mols.
Total = 3659.2857 mols

∴ The dry flue gas analysis would be

$$CO_2 = \dfrac{375}{3659.2857} \times 100 = 10.25\%$$

$$N_2 = \frac{3072.5375 \times 100}{3659.2857} = 83.96\%$$

$$O_2 = \frac{211.75 \times 100}{3659.2857} = 5.79\%.$$

Example 12. *The following data for flue gas analysis has been recorded in a Lancashire boiler with an economiser.*

Constituent	Percentage Analysis of Flue gas Entering the Economiser	Leaving the Economiser
CO_2	9.3	7.2
CO	0.4	0.3
O_2	10.2	12.6
N_2	80.1	79.9

Calculate the air leakage into the economiser per kg of the fuel stoked if the carbon in the fuel was 0.735 kg/kg of the fuel.

Solution.
Let 100 ml of dry flue gas be entering the economiser. If x mols of air leaked into the economiser during a certain time, the volume of the gases leaving the economiser will be $(100 + x)$ mols.

Hence, the % CO_2 in the flue gas leaving the economiser $= \dfrac{9.3 \times 100}{(100+x)}$ which should be equal to 7.2 mols as per the data.

$$\therefore \quad \frac{9.3 \times 100}{(100+x)} = 7.2$$

$\therefore x = 29.16$ mols $= (29.16 \times 28.97)$ kg $= 846$ kg.

Hence 846 kg of air leaked into 100 mols of dry flue gas entering the economiser.

Fuel burnt to produce 100 mols of flue gas entering the economiser $= \dfrac{\text{Mols of C in flue gas}}{\text{Moles of C in 100 kg of fuel}} \times 100$

$$= \frac{(9.3+0.4)}{6.125} \times 100 = 158 \text{ kg}$$

(Since 100 kg fuel contains 73.5 kg carbon $\equiv \dfrac{73.5}{21}$ mols $= 6.125$ mols of C)

$\therefore$ Air leakage into the economiser per kg of fuel

$$= \frac{846}{168} \text{ kg} = \mathbf{5.42 \text{ kg}}.$$

Example 13. *A coal sample of the following analysis by weight was used for a boiler trial:*
$$C = 82\%, \; H = 4.9\%, \; O = 4.8\%$$
The flue gas analysis by Orsat's apparatus gave the following data:
$CO_2 = 10\%, \; O_2 = 8\%, \; CO = 1.5\%$ and $N_2 = 80.5\%$.
The temperature of ambient air in the boiler house is 18°C and that of the flue gases is 320°C. Calculate the heat carried away by the excess of air per kg of coal. The average specific heat of air is 0.238.

Solution.

Vol. of constituents per m^3 of flue gas	Proportional weight
$CO_2 = 10/100 = 0.1 \text{ m}^3$	$0.1 \times 44 = 4.40$ kg
$O_2 = 8/100 = 0.08 \text{ m}^3$	$0.08 \times 32 = 2.56$ kg
$CO = 1.5/100 = 0.015 \text{ m}^3$	$0.015 \times 28 = 0.42$ kg
$N_2 = 80.5/100 = 0.805 \text{ m}^3$	$0.805 \times 28 = 22.54$ kg
	Total = 29.92 kg

FUELS AND COMBUSTION 173

Weight of CO_2/kg of flue gas = $\dfrac{4.4}{29.92}$ = 0.147 kg

Weight of O_2/kg of flue gas = $\dfrac{2.56}{29.92}$ = 0.085 kg

Weight of CO/kg of flue gas = $\dfrac{0.42}{29.92}$ = 0.014 kg

Weight of N_2/kg of flue gas = $\dfrac{22.54}{29.92}$ = 0.747 kg.

Total weight of carbon/kg of flue gas = (Wt. of C in CO_2 + Wt. of C in CO/kg of flue gas

$$= \left(\dfrac{12 \times 0.147}{44} + \dfrac{12 \times 0.014}{28}\right)$$

$$= (0.04 + 0.006) \text{ kg} = 0.046 \text{ kg}.$$

(Since 12 kg of C is present in 44 kg of CO_2 and in 28 kg of CO respectively).
But, weight of C per kg of fuel = 82/100 = 0.82 kg.
∴ Wt. of dry flue gas/kg of fuel burnt = 0.82/0.046 = 17.82 kg

Weight of O_2 required for the combustion of CO per kg of flue gas = $\dfrac{16 \times 0.014}{28}$ = 0.008 kg

But, wt of O_2/kg of flue gas = 0.085 kg as shown above.
∴ Excess O_2/kg of flue gas = (0.085 – 0.008) = 0.077 kg
∴ Excess O_2/kg of fuel burnt = 0.077 × 17.82 = 1.37 kg
∴ Excess air/kg of fuel burnt = 1.37 × $\dfrac{100}{23}$ = 6.0 kg

∴ Heat carried away by the excess air
= M × S × $(t_2 - t_1)$ Kcals
  Mass Sp. heat rise in temp.
= 6 × 0.238 × (320 – 18) Kcals
= 431.256 Kcals.

Example 14. *A gaseous fuel has the following volumetric composition : CH_4 = 40%; C_2H_2 = 6%; C_2H_6 = 24%; C_4H_8 = 16%; O_2 – 3.0%; CO = 5% and N_2 = 6%. If 40% excess air was used for the combustion, calculate the air; fuel ratio and the analysis of dry flue gas.*

Solution.
The combustion reactions of the constituent gases in the fuel are:

CH_4 + $2O_2$ ⟶ CO_2 + $2H_2O$
1 mol   2 mole   1 mol   2 mols

C_2H_4 + $3O_2$ ⟶ $2CO_2$ + $2H_2O$

C_2H_6 + 3.5 O_2 ⟶ $2CO_2$ + $3H_2O$

C_4H_8 + $6O_2$ ⟶ $4CO_2$ + $4H_2O$

CO + $\dfrac{1}{2}$ O_2 ⟶ CO_2

Let 1 mol of the fuel gas be the basis for calculation:

Constituent & mols	Mols of O_2 reqd.	Mols of Products formed
CH_4 — 0.4	0.4 × 2 = 0.80	CO_2 — 0.4; H_2O — 0.8
C_2H_4 — 0.06	0.06 × 3 = 0.18	CO_2 — 0.12; H_2O — 0.12
C_2H_6 — 0.24	0.24 × 3.5 = 0.84	CO_2 — 0.48; H_2O — 0.72
C_4H_8 — 0.16	0.16 × 6 = 0.96	CO_2 — 0.64; H_2O — 0.64

O_2 — 0.03	= − 0.03	—
CO — 0.05	0.05 × 0.5 = 0.025	CO_2 — 0.05
N_2 — 0.06	—	—
Total	= 2.775 mols	CO_2 — 1.69 mols.

∴ O_2 required/mol of the fuel gas = 2.775 mols
or O_2 required/m³ of the fuel gas = 2.775 m³
(Since mol% = m³% = Vol%)
But 40% excess air was actually used.
∴ O_2 actually supplied/m³ of fuel gas

$$= 2.775 \times \frac{140}{100} = 3.885 \text{ m}^3$$

∴ Air actually supplied/m³ of fuel gas

$$= 3.885 \times \frac{100}{21} = 18.5 \text{ m}^3$$

∴ Air : fuel ratio = **18.5 m³**.

Analysis of dry products of combustion/m³ of fuel gas

CO_2 = 1.69 m³

$O_2 = 18.5 \times \frac{40}{100} \times \frac{21}{100}$ m³ = 1.554 m³

[from excess air]

$N_2 = \left[\left(18.5 \times \frac{79}{100}\right) + 0.06\right]$ m³ = 14.675 m³

(from air + fuel)

Total volume of combustion products = 17.919 m³
Whence the volumetric composition of the flue gas is

$$CO_2 = \frac{1.69 \times 100}{17.919} = 9.43\%$$

$$O_2 = \frac{1.554 \times 100}{17.919} = 8.67\%$$

$$N_2 = \frac{14.675 \times 100}{17.919} = 81.9\%.$$

QUESTIONS

1. How is calorific value of a solid fuel determined using a bomb calorimeter?
   The following data is obtained in a bomb calorimeter experiment:
   Mass of the fuel pellet = 0.85 g
   Mass of water taken in the calorimeter = 2000 g
   Water equivalent of the calorimeter = 540 g
   Difference in final and initial temperature = 1.9°C
   Cooling correction = 0.041°C
   Fuse wire correction = 3.8 calories
   Acids correction = 48.8 calories
   Calculate the net calorific value of the fuel if it contains 3.6% hydrogen and 1.2% oxygen.
2. Distinguish clearly between the following:
   (a) Gross and net calorific values of a fuel (Nag. Univ. S - 2006)
   (b) proximate and ultimate analysis (Nag. Univ. S - 2004)
   (c) High and Low temperature corbonization (Nag. Univ. S - 2005, S-2005)
   (d) coking coals and caking coals (Nag. Univ. S - 2006)

FUELS AND COMBUSTION 175

    (e) octane number and cetane number
    (f) Thermal and catalytic cracking.(Nagpur Univ. S-2005)
3. How is the proximate analysis of a coal conducted and what is its 'significance in determining' the utility of a coal for a particular purpose?
4. Write short notes on the following:
    (a) Fractional distillation of crude petrolium oil (Nagpur Univ. S-2007)
    (b) Regnault and pfaundler's formula
    (c) Water equivalent of a Bomb calorimeter.
    (d) Catalytic converters.
    (e) Commercial types of coal
    (f) Blending of coal
    (g) Storage and spontaneous combustion of coal (Nagpur Univ. S-2003, S-2007)
5. What are the difficulties arising if coal is stored improperly for prolonged periods? What are the factors favourable for spontaneous combustion? What safety measures do you suggest for storage of coal?
6. What is coke? why is it preferred as a metallurgical fuel over coal? Give the properties of metallurgical coke (Nagpur Univ. S-2004).
7. What are the types of carbonization and for what purposes they are employed? (Nagpur Univ. S-2003)
8. Describe the Otto. Hoffman's process for preparing coke? What are its advantages over the earlier methods? (Nagpur Univ. S-2003, S-2005)
9. Write informative notes on the following:
    (a) Advantages and limitations of liquid fuels
    (b) Detection of oil deposits
    (c) Refining of petroleum
    (d) Bubble tower and Top-flasher
    (e) Thermal and catalytic processes
    (f) Reforming.
10. What is cracking and what for it is used? What are the types of cracking? Describe the working of fluid bed catalytic cracking unit. (Nagpur Univ. S-2006)
11. How is synthetic gasoline obtained from non-petroleum sources? (Nagpur Univ. S-2006)
12. What are the requisites of an ideal gasoline? What is meant by blending and doping?
13. What is meant by knocking in a petrol engine and what is it due to? What is octane number of a potrol? How is knocking related to chemical structure of the constituents of petrol and how can it be reduced? (Nagpur Univ. S-2006)
14. How do you explain knocking in a diesel engine? How can it be controlled? What is cetane number?
15. Write informative notes on the following:
    (a) Rocket fuels
    (b) Power alcohol
    (c) Liquid petroleum gas (LPG)
    (d) Coal gas and coke oven gas
    (e) Kopper's gas
    (f) Winkler's gas
    (g) Synthesis gas.
16. How is producer gas manufactured? What are its disadvantages and advantages?
17. Justify the following statements :
    (i) Carbon determined in ultimate analysis is more than that determined in proximate analysis. (Nagpur Univ. S - 2004)
    (ii) Gasoline containing tetraethyl lead is used in internal combustion engines (Nagpur Univ. S - 2004)
    (iii) Good compression ignition engine fuels are poorer spark ignition engine fuels and vice-versa (Nagpur Univ. S - 2004)
18. What is the common apparatus used for flue gas analysis? How it is useful to determine the efficiency of combustion?
19. A producer gas has the following composition by volume: CH_4 = 3.5%: CO = 25%, H_2 = 10%; CO_2 = 10.8%, N_2 = 50.7%. Calculate the theoretical quantity of air required per cubic meter of the gas. If 22% excess air is used, find the percentage composition of the dry products of combustion.
20. A liquid hydrocarbon fuel contains 89.4% C and 10.6% H by weight. If 100 kg of the fuel is burnt, calculate (a) the theoretical amount of air required and (b) the volumetric composition of the products of combustion if 20% excess air was used,

(**Ans.** (a) 148.1 K moll (b) $CO_2 = 12.34\%$, $H_2O = 8.78\%$; $N_2 = 75.53\%$ and $O_2 = 3.35\%$)

21. A coal sample gave the following analysis by weight: C = 88%, H = 4%, S = 1%, O = 5%, ash and moisture 1% each. The coal sample on combustion in a boiler gave a flue gas having the following composition by volume:
   $CO_2 = 11\%$; $O_2 = 7\%$, $CO = 1.5\%$, $SO_2 = 0.5\%$ and $N_2 = 80\%$. Calculate (a) the weight of air required for complete combustion of 1 kg of the coal (b) weight of the wet flue gases per kg of coal, (c) percentage of excess air used and (d) percentage of carbon burnt to CO.
   (**Ans.** (a) 11.42 kg/kg of coal; (b) 18.08 kg/kg of coal; (c) 49.4%(d) 12% C burnt to CO)

22. How does knocking occur in I.C. engines. How can it be prevented?

23. Establish the relationship between knocking in I.C. engines and the nature and molecular structure of the constituents in petrol and diesels fuel. (Nagpur Univ. S - 2005)

24. How are water gas and producer gas manufactured? Discuss the theoretical principles involved.

25. How is coal gas manufactured? How is coal gas purified? What types of coal is preferred for gas-making?

26. Define gross and net calorific values. Calculate the approximate calorific value (by Dulong's formula) of a coal sample having the following ultimate analysis:
   C = 80%; H = 3.5%; S = 2.8%; O = 5.0; N = 1.5% and ash = 7.2%.

27. A hydrocarbon fuel on combustion gave the flue gas having the following composition by volume: $CO_2 = 13\%$; $O_2 = 6.5\%$; and $N_2 = 80.5\%$. Calculate (a) the composition of fuel by weight and (b) percentage of excess air used.

28. A producer gas has the following composition by volume; CO = 30%; $H_2 = 12\%$; $CO_2 = 4\%$; $CH_4 = 2\%$ and $N_2 = 52\%$. When 100 m³ of the gas is burnt with 50% excess air used, what will be the composition of the dry fiue gases obtained.

29. A coal sample has the following composition; C = 90%; H = 3.5%; O = 3%; S = 0.5; and $N_2 = 1\%$; the remaining being ash. Calculate the theoretical volume of air required at 27°C and 1 atm pressure when 100 kg of the coal is burnt. (**Ans.** 1100 m³)

30. The composition by weight of a coal sample is: C = 81%; H = 5%; O = 8.5%; S = 1%; N = 1.5%; Ash = 3%. (a) Calculate the amount of air required for the complete combustion of 1 kg of the coal, (b) Calculate the gross and net calorific values of the coal sample. Given that the calorific values of ? C, H and S are 8060 Kcals/kg; 34000 Kcals/kg and 2200 Kcal/kg respectively.
   (**Ans.** 10.8 kg; 7.910 kcal, 7,667 kcal/kg)

31. A coal gas has the following volumetric composition: $CO_2 = 2.5\%$; $C_2H_4 = 3.5\%$; CO = 8%; $H_2 = 50\%$; $CH_4 = 34\%$ and $N_2 = 2\%$. Calculate the gross calorific value of the gas at N.T.P. Given that 1 kg of C while burning to CO develops 2440 kcals; while burning to CO_2 develops 8060 Kcals of heat; and 1 kg of H_2 while burning to H_2O develops 34,400 Kcals, 1 kg mol of gas at N.T.P. occupies 22.4 m³. (**Ans.** 5,848 Kcals/m³)

32. A coke-oven gas of the following volumetric composition was used for firing gas retorts:
   $CO_2 = 1.4\%$; CO = 5.1%; $O_2 = 0.5\%$; $H_2 = 57.4\%$; $CH_4 = 28.5\%$; $C_2H_4 = 2.9\%$ and $N_2 = 4.2\%$. The dry flue gases on analysis by Orsat's apparatus gave the following results; $CO_2 = 15\%$; CO = 0%; $O_2 = 2.5\%$; $N_2 = 82.5\%$. Calculate the excess air supplied for burning 100 m³ of the gas. (**Ans.** 32.4 m³)

33. A coal containing 62.4% C; 4.1% H; 6.9% O; 1.2% N; 0.8% S; 15.1% moisture and 9.5% ash was burnt in such a way that the dry flue gases contained 12.9% CO_2; 0.2% CO; 6.1% O_2 and 80.8% N_2. Calculate (a) the weight of air theoretically required per kg of coal, (b) % of excess air used and (c) the weight of the dry flue gas produced per kg of coal. (**Ans.** 8.3246 kg; 41.0984% 12.0306 kg)

34. A gasoline sample contains 86% C and 14% H_2. If the air supplied is only 95% of the theoretically required value, calculate the dry flue gas analysis. Assume that all the H_2 underwent complete combustion and no carbon is left free.
   [Hint. For every 0.5 mol of O_2 supplied less, 1 mol of C is converted into CO].
   (**Ans.** $CO_2 = 13.47\%$; CO = 2.36% and $N_2 = 84.17\%$)

35. Explain the following:
   (a) The knocking tendency of petrol or diesel oil can be predicted on the basis of the nature and molecular structure of its constituent compounds. (Nagpur Univ. S - 2007)
   (b) All coking coals are caking but all caking coals are not coking. (Nagpur Univ. S - 2005)

36. A furnace utilises a mixture of blast furnace gas and coke oven gas as a fuel. The gas mixture is fired along with air to obtain the required temperature. The analysis of the individual gases used and the flue gas evolved due to the combustion of the above gas mixture are given on dry basis and in volume % as below:

Constituent	Coke oven gas	Blast furnace gas	Air	Flue gas
CO_2	2.4	13.4	—	13.5
O_2	2.0	0.2	21.0	6.9

FUELS AND COMBUSTION

$$CH_4 \text{ — 25 to 30\%}$$
$$H_2 \text{ — 50 to 55\%}$$
$$CO \text{ — 11\%}$$
$$CO_2 \text{ — 3\%}$$

The calorific value of oil gas is about 6700 Kcal/m^3. It is used as a laboratory gas. It is also used to improve the calorific value of water gas and the mixture of the two gases is called carburetted water gas.

Onia Geigi process

In this process, a light oil or heavy oil is gasified with steam in presence of a catalyst. The gas produced can be used as a town gas for domestic consumption and as a synthesis gas for industrial purposes.

2.23.7. Producer gas.

Producer gas is essentially a mixture of carbon monoxide and nitrogen. It is prepared by passing a mixture of air and steam over a bed of incandescent coke, anthracite coal, high volatile non-coking coal or lignite. Several types of "gas producers" are in use and one typical producer is shown in Fig. 2.20. The producer is made up of cylindrical steel furnace lined with fire-bricks. The fuel is fed from the hopper at the top. A mixture of air and steam (about 0.35 kg per kg of coal) is passed over the red hot coal bed through the inlet at the bottom. The producer is also provided with an ash-pit at the bottom to remove the ash formed during the combustion reactions.

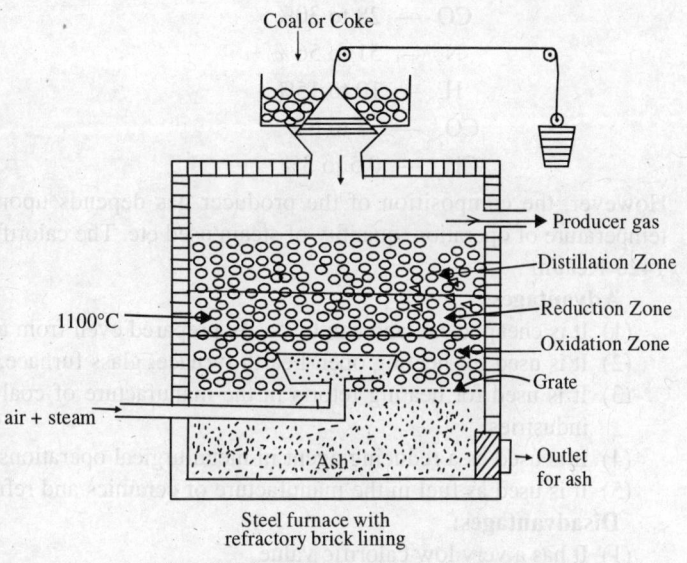

Fig. 2.20. Manufacture of producer gas.

When a mixture of air and steam is passed over red hot coke or coal bed at about 1000° to 1100°C in the producer, the following reactions take place in different zones of the fuel bed.

(1) Oxidation Zone

This is the lowest part of the coal bed. Here, the carbon of the coal burns in presence of excess of air to give carbon dioxide

$$C + (O_2 + 3.76 N_2) \rightleftharpoons CO_2 + 3.76 N_2 + 96.7 \text{ Kcal (oxidation)}$$
$$\text{air}$$

(2) Reduction Zone

This is the middle part of the fuel bed. Here, the CO$_2$ produced in the oxidation zone and the steam move up through the red hot fuel bed and the following reactions take place:

$$C + CO_2 \rightleftharpoons 2CO - 38.7 \text{ Kcal (Reduction)}$$
$$C + H_2O \rightleftharpoons CO + H_2 - 28.7 \text{ Kcal (water gas)}$$
$$\text{(steam)}$$
$$C + 2H_2O \rightleftharpoons CO_2 + 2H_2 - 18.6 \text{ Kcal}$$
$$\text{(steam)}$$

All these reactions in the reduction zone are endothermic and the temperature falls. In order to maintain the temperature of the bottom of the coal bed at about 1100°C (which is essential for the oxidation reaction) the quantity of the steam admitted along with air can not be increased beyond a certain limit. Also, it is not desirable to increase the H_2 content in the producer gas. Moreover, undecomposed steam in the producer gas carried away sensible and latent heat from the system.

(3) **Pre-heating and Distillation Zone**

This is the upper part of the fuel bed. In this zone some of the CO formed reacts with steam as follows:
$$CO + H_2O \rightleftharpoons CO_2 + H_2 + 10.1 \text{ Kcal}$$
Further, all the gases formed namely, CO, CO_2, H_2, N_2 etc. pass out through the gas outlet.

The average composition of the producer gas is as follows:

- CO — 28 to 30%
- N_2 — 51 to 56%
- H_2 — 10 to 15%
- CO_2 — 3 to 6%
- CH_4 — 0.5 to 3%

However, the composition of the producer gas depends upon the nature of the fuel used, the temperature of operation, quantity of steam used etc. The calorific value of producer gas is 1160 to 1420 Kcal/m^3.

Advantages:
(1) It is cheap, clean and can be easily prepared even from a low grade coal.
(2) It is used for heating open-hearth furnace, glass furnace, muffle furnace, etc.
(3) It is used for heating retorts in the manufacture of coal gas and ovens in the carbonizing industries.
(4) It is used as a reducing agent in metallurgical operations.
(5) It is used as fuel in the manufacture of ceramics and refractories.

Disadvantages:
(1) It has a very low calorific value.
(2) If coal is used as fuel, the gas should be purified to remove the tar. Hence, most of the producer gas plants use coke as fuel.

Advantages of using steam along with air:
(1) The reaction between C and oxygen is exothermic and the temperature of the fuel bed of an air-blown producer tends to rise. This leads to clinker and slag formation. If a mixture of steam and air are used simultaneously, the heat evolved by the carbon-oxygen reactions is balanced by the strongly endothermic steam-carbon reactions. Thus, the temperature of the oxidation zone is reduced and clinker formation is prevented. The fuel bed temperature is maintained at 1000° to 1100°C by maintaining the required quantities of steam and air.
(2) The steam-carbon reactions convert some of the sensible heat into potential heat and thus increase the calorific value. The coal efficiency of the plant is also increased.
(3) The % of N_2 in the gas is reduced.

Note : If all the carbon were converted into CO, it should be theoretically possible to make a producer gas containing 34.7% of CO and 65.3% N_2, but in practice, the CO percentage is always lower than this and some CO_2 and H_2 are always present

2.23.8. Water gas (or Blue gas). Water gas is essentially a mixutre of CO (40 to 42%) and H_2 (48.51%). It is prepared by passing steam over red hot coke. The water gas is also known as blue gas because it burns with a blue flame due to the combustion of carbon monoxide.

Water gas is preaprd by reacting steam over red hot coke in a cyclic process comprising of the following two main steps :

FUELS AND COMBUSTION

(1) Air is passed through the coke bed when carbon burns to CO_2 and the temperature of the fuel bed increases to over 1000°C.

$$C + (O_2 + 3.76 N_2) \longrightarrow CO_2 + 3.76 N_2 + 96.7 \text{ Kcal}$$
air   (exothermic)

The outgoing gases containing CO_2 and N_2 are used to preheat the incoming air and steam.

(2) Steam is now blown through the red hot coke when CO and H_2 are produced:

At higher temperatures:

$$C + H_2O \rightleftharpoons CO + H_2 - 28.7 \text{ Kcal (endothermic)}$$

At lower temperatures:

$$C + 2H_2O \rightleftharpoons CO_2 + 2H_2 - 18.6 \text{ Kcal (endothermic)}$$

These reactions being endothermic, the temperature of the fuel bed falls below 1000°C. At this stage, steam supply is cut off and air is again blown to raise the temperature to 1000° – 1100°C. Then the steam is blown in and the process is repeated.

In order to maintain the temperature of the fuel bed above 1000°C, the air and steam supply are regulated alternately to achieve the desired effects. The period for which air is passed is called "hot blow" and that for which steam is passed is known as "cold blow". From the technical and economic points of view, the "hot blow" is carried on for 10 minutes followed by the "cold blow" for about 1 minute and so on. Several types of water gas producers have been designed and one such is shown in Fig. 2.21.

The calorific value of water gas is about 2580 to 2670 $Kcal/m^3$. The normal composition of water gas is as follows:

CO — 40 to 42%
H_2 — 48 to 51%
CO_2 — 3 to 5%
CH_4 — 0.1 to 0.5%
N_2 — 3 to 6%

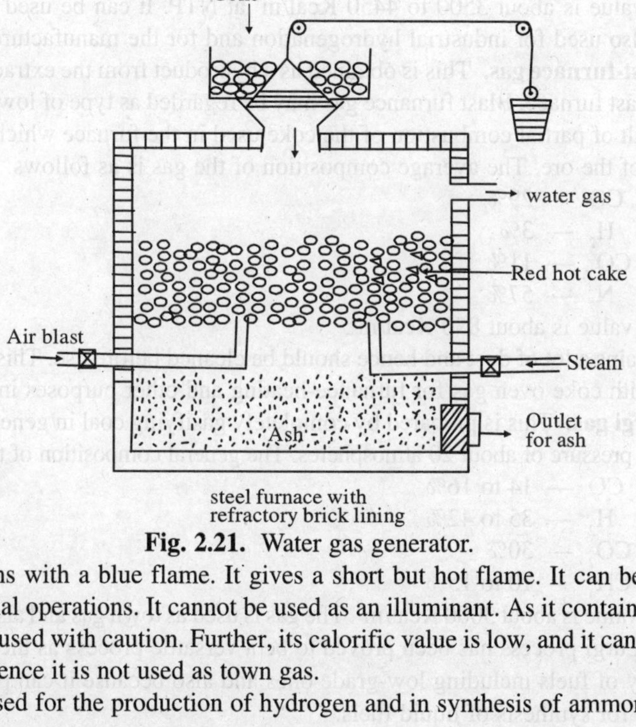

Fig. 2.21. Water gas generator.

Water gas burns with a blue flame. It gives a short but hot flame. It can be used for heating purposes in industrial operations. It cannot be used as an illuminant. As it contains large proportion of CO, it should be used with caution. Further, its calorific value is low, and it cannot be transported to distant places. Hence it is not used as town gas.

Water gas is used for the production of hydrogen and in synthesis of ammonia. It is used as a fuel in many industrial processes such as enamelling, annealing etc., in glass and ceramic industries.

2.23.9. Carburetted water gas. The calorific value of water gas can be conveniently increased by enriching it with cracked oil vapours (oil gas), which can be made in an auxiliary plant called a carburettor (Fig. 2.22). This enriched gas is called the carburetted water gas. This can be obtained by heating a checker work of fire-bricks by burning a portion of the water gas. Then, oil is sprayed over it. The oil gets vaporized and cracked and at this stage, the water gas is blown through it. The resulting carburetted water gas is led through a hydraulic main and a purifier followed by scrubbers to remove the impurities.

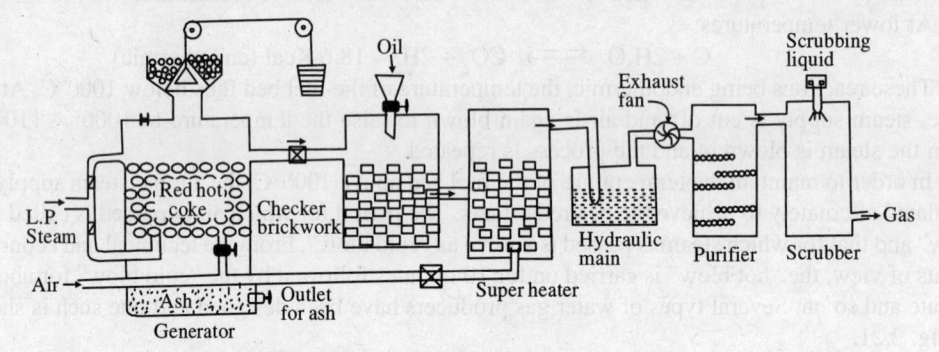

Fig. 2.22. Carburetted water gas plant.

The average composition of carburetted water gas is:

CO — 30 to 48%
H_2 — 34 to 38%
Hydrocarbons — 30 to 48%
CO_2 — 0.2 to 2.5%
N_2 — 2.5 to 5%

Its calorific value is about 3500 to 4450 Kcal/m^3 at NTP. It can be used as a fuel and as an illuminant. It is also used for industrial hydrogenation and for the manufacture of alcohol.

2.23.10. Blast-furnace gas. This is obtained as a byproduct from the extraction of metals from their ores using blast furnace. Blast furnance gas may be regarded as type of low grade producer gas obtained as a result of partial combustion of the coke used in the furnace which is modified by the partial reduction of the ore. The average composition of the gas is as follows:

CO — 29%
H_2 — 3%
CO_2 — 11%
N_2 — 57%

The calorific value is about 850 Kcal/m^3.

This gas contains a lot of dust and hence should be cleaned before use. This gas is used as such or after mixing with coke oven gas for furnace-heating and other purposes in steel plants.

2.23.11. Lurgi gas. This is prepared by completely gasifying coal in generators using oxygen and steam under a pressure of about 20 atmospheres. The general composition of the gas is as under:

CO — 14 to 16%
H_2 — 35 to 42%
CO_2 — 30%
CH_4 — 10 to 15%

The calorific value is about 3000 Kcal/m^3. The gas is used as town gas and also for the production of synthesis gas. Lurgi process has been proved to be a versatile process as the generator used can deal with a variety of fuels including low-grade ones and also because it can produce town gas as well as gas useful for synthesis of liquid fuels.

FUELS AND COMBUSTION

2.23.12. Kopper's gas. This is prepared by gasification of pulverised coal in presence of oxygen and super-heated steam in the Kopper's Totzek entrained bed gasifier at atmospheric pressure. High ash content low-grade coals can also be used for producing this synthesis gas. The composition of this gas is as follows:

CO — 51.1%
CO_2 — 12.6%
H_2 — 34.0%
N_2 — 1.9%
O_2 — less than 0.1%
CH_4 — less than 0.1%

The calorific value is around 2500 Kcal/m^3. The gas after suitable scrubbing, shift conversion and purification may be used for synthesis of ammonia.

2.23.13. Winkler's gas. Synthesis gas is prepared in the Winkler's process by gasification of crushed coal with steam and O_2 (or air) in a fluidised bed. The temperature is maintained below the ash fusion temperature. The gas is used as fuel as well as a synthesis gas for manufacturing urea, etc. Such process is used at the fertiliser plant at Neyveli (Tamil Nadu–India) where synthesis gas for the manufacture of urea is prepared from gasification of lignite. The typical composition of Winkler's gas is given below:

CO — 46%
H_2 — 36%
CO_2 — 14%
N_2 — 1.1%
CH_4 — 2.4%

Its calorific value is about 2500 Kcal/m^3.

2.23.14. Synthesis gas. This is so called due to the fact that it is useful for synthetic manufacture of chemicals and fertilizers. It provides N_2 and H_2 for the synthesis of ammonia, CO and H_2 for the manufacture of methanol, and CO and H_2 for generating liquid fuels. Water gas, coke oven gas, Kopper's gas, Winkler's gas, etc were used as sources of synthesis gas. Modern synthesis gas plants are based on natural gas or petroleum.

Synthesis Gas (Syngas)

Modern processes have been developed to produce CO - H_2 mixture from coal gas and steam more efficiently than the old water-gas and producer-gas plants. The gases produced are of low heat content (3.7 to 7.5 MJ/m^3) if steam and O_2 are used. The principal difference between the low-heat gas and the medium-heat gas is N_2, which enters the system as part of the air. The low-heat gas is mainly used as an on-site industrial fuel and as an intermediate in the production of formaldehyde and ammonia; while the medium heat gas can be economically transported by pipeline for distances of 160 to 320 Km. Much of the gas is methanated to produce substitute natural gas (SNG).

The types of gasification systems are classified according to gasifier bed type : fixed bed, fluidised bed, and entrained bed. The oldest examples of these are the Lurgi process (fixed bed), Winkler's process (fluidized bed), and Kopper-Totzek process (entrained flow). All of these processes have been commercially used worldwide, although none have been commercially operated in the U.S. Efforts to create improved reactors for very large scale SNG production (daily throughputs of 5000 to 10,000 t of coal) have led to the development of several so-called second-generation reactors, which are usually modifications of these older processes.

Composition of typical producer gases (dry basis) mole %

Gas constituent	Lurgi	Kopper-Totzek	Winkler
H_2	38.0	36.7	41.8
CO	20.2	55.8	33.3
CO_2	28.6	6.2	20.5
CH_4	11.4	0.0	3.0
C_2H_6	1.0	0.0	0.0
H_2S or CO S*	0.5	0.3	0.4
N_2	0.3	1.0	1.0

*COS - compounds of sulphur.

2.23.15. Purification of gaseous fuels. Fuel gases, as manufactured, are hot and contain dust, grit, water vapour, tar, sulfur compounds, etc. Sometimes, useful by products such as ammonia, tar and benzol may be there as in coke oven gas. The gases should be purified by removing the impurities by cooling, scrubbing (by water or oil) and purification (using purifiers containing thylox, ferricyanide, Fe $(OH)_3$, etc to remove H_2S, CS_2, thiophene and mercaptons).

2.24 Biodiesel

There has been an increase in efforts to reduce the reliance on petrolium fuel for energy generation and transportation. Among the alternative fuels, biodiesel and diesohol have receive the much attention for diesel engines due to their advantages as the renewable, domestically produced energy resources and they are environmentally friendly because there is substantial reduction of unburned hydrocarbon, CO and particulate matter when it is used in conventional diesel engine.

Biodiesel can be produced from vegitable oil via transestrification process. Biodiesel has been used not only as an alternative to the fossil fuels but also as an additive for diesohol. The presence of biodiesel or vegitable oil can improve the lubricating properties of fuel, however there are some concerns regarding reduction of flash point and startability problem with ethanol-gasoline blend and ethanol-diesel blend.

The term 'biodiesel' refers to a mixtures of alkyl esters of long-chain fatty acids. Since it is biodegradable spills cause little or no environmental threat. A series of animal and vegetable oil feedstocks can be used to produce biodiesel. Biodiesel can be produced by base - catalysed transestrification. The reaction scheme is as follows.

1) Transestrification (Main reaction) : In transestrification ethanol is mostly used as it is a renewable resource and does not raise toxicity (as in case of methanol).

$$\text{Rapeseed oil + Alcohol} \underset{\text{alkali (NaOH, KOH)}}{\rightleftharpoons} \text{Mixed ester + Glycerol}$$

$$\begin{array}{l} CH_2COOR_1 \\ CHCOOR_2 \\ CH_2COOR_3 \end{array} + R'OH \underset{}{\overset{60\text{-}70°C}{\rightleftharpoons}} \begin{array}{l} R_1COOR' \\ R_2COOR' \\ R_3COOR' \end{array} + \begin{array}{l} CH_2OH \\ CHOH \\ CH_2OH \end{array}$$

Triglyceride   Alcohol   Ethylester   Glycerol

FUELS AND COMBUSTION

$C_{17}H_{35}COOCH_3$ - Stearic ester
$C_{17}H_{33}COOCH_3$ - Oleic ester
$C_{15}H_{31}COOCH_3$ - Palmitic ester

$$\begin{matrix} CH_2COO(CH_2)_{16}CH_3 \\ CHCOO(CH_2)_{16}CH_3 \\ CH_2COO(CH_2)_{16}CH_3 \end{matrix} + 3CH_3OH \rightarrow 3CH_3(CH_2)_{16}COOCH_3 + \begin{matrix} CH_2OH \\ CHOH \\ CH_2OH \end{matrix}$$

Vegetable oil   Methanol   Biodiesel   Glycerol

ii) Neutralisation : unavailable side reaction in presence of free fatty acids presence of free fatty acids

Fatty acid + Alkali $\rightleftharpoons$ Soap + Water
                          water

iii) Saponification Undesirable side reaction

Rapeseed oil + Alkali $\rightleftharpoons$ Soap + Glycerol

Mostly the biodiesel is produced by base - catalysed transestrification process. The blocks diagram is as follows

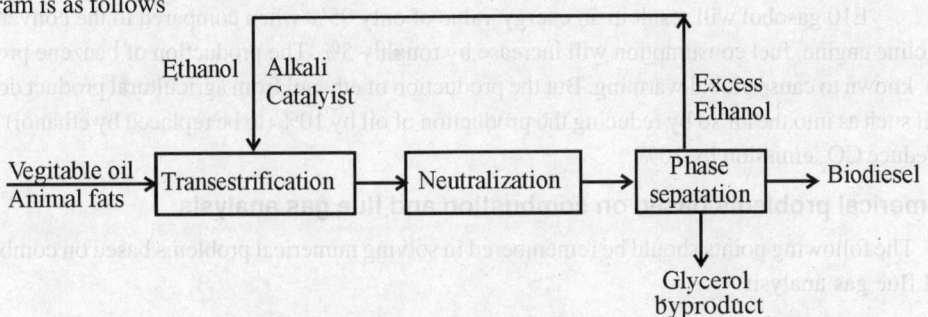

```
                Ethanol   Alkali                              Excess
                          Catalyst                            Ethanol
                             ↓                                   ↓
Vegetable oil    ┌──────────────────┐    ┌──────────────┐    ┌───────────┐
──────────────→  │ Transestrification│ → │ Neutralization│ → │   Phase   │ → Biodiesel
Animal fats      └──────────────────┘    └──────────────┘    │ sepatation│
                                                              └───────────┘
                                                                   ↓
                                                               Glycerol
                                                               byproduct
```

Transestrification process gives byproduct glycerine and oilcake. The oilcake is a good source of organic manure which contains about 30% protein. Biodiesel contains no petroleum, it can be blended at any level with petrolium diesel. A pure 100% biodiesel fuel is referred as B 100. Biodiesel blends are referred as BXX, where XX indicates the amount of biodiesel in the blend ex. B20 blend is 20% by volume biodiesel and 80% by volume petrodiesel.

In USA, biodiesel is made from soybean oil. It can also be made from tallow. In Europe, the biodiesel is made from rapeseed oil. In India it is produced from seeds of plant Jatropa curcas and Karanja (Non edible oil).

In India Jatropa plant is grown in marginal and poor soil with minimum cultural practices or in waste lands with low fertility rockiness of soils. Biodiesel has higher cetane number, this oil reduces emission of CO_2 by 78% CO by 44%, Sulphate 100%, polycyclic aromatic hydrocarbon (PAA) which is carsiogenic compound by 75-85% and particulate matter.

Diesohol

Diesohol is an emulsion of hydrated ethanol in diesel fuel. A diesohol blend consist of 84.5% diesel, 15% hydrated ethanol and 0.5% emulsifier. Diesohol has been known as one of the preference to decrease the use of regular diesel fuel. The use of ethanol can decrease the amount of petroleum fuel imports. India was one of the earliest countries to recognize the merits of burning ethanol in diesel engines. Diesohol reduces CO_2, CO, NOx emission and smoke.

Ethanol has lower heating value than the regular diesel and use of pure ethanol as an alternative can cause drawbacks to the effectiveness of engines. To avoid such problem one of the approache is to blend the ethanol with regular diesel fuel and use the blend known as diesohol as the power source for diesel engine. However ethanol and diesel fuel are immiscible because of their difference in chemical structure and characteristics. These liquids can be emulsified by mechanical blending with suitable emulsifiers. The emulsifier would reduce the interfacial tension force and increase the affinity between the two liquid phase.

Know a days diesohol is become so important because it has the potential to create a large market for ethanol.

Gasohol :

Gasohol is a mixture of gasoline and alcohol (mostly ethanol since methanol is toxic). Its primary intention is to reduce the consumption (import) of gasoline.

E10 called gasohol a fuel, mixture of 10% ethanol and 90% gasoline that can be used in the internal combustion engines. E10 blends are rated as 2 to 3 octane higher than regular gasoline. For ethanol-based gasohol the ethanol is made by fermentation of agricultural crops or crap wastes i.e. corn, suger cane or cellustic sources such as paper waste blend. Methanol-based gasohol is expensive to produce and is toxic and corrosive and the emission produce cancer causing form aldehyele. Use of E10 can reduce CO emission by 20-30% and CO_2 by 2%.

E10 gasohol will result in an energy value of only 95% when compared to the conventional gasoline engine, fuel consumption will increase by roughly 5%. The production of benzene produces CO_2 known to cause global warming. But the production of ethanol from agricultural product does not emit such as into the air so by reducing the production of oil by 10% (to be replaced by ethanol) it help to reduce CO_2 emission by 10%

Numerical problems based on combustion and flue gas analysis

The following points should be remembered in solving numerical problems based on combustion and flue gas analysis:

(1) **Main Objectives**

The problems are generally based on calculating:
 (a) The weight or volume of air theoretically required or used for the combustion of 1 kg of fuel.
 (b) The percentage of excess air used per kg. of fuel burnt,
 (c) The percentage composition of flue gas obtained (by weight or by volume) by burning a known quantity of fuel, and
 (d) The percentage composition of the fuel burnt.

(2) **Concept of mole**

A mole of a substance is that quantity whose weight is numerically equal to its molecular weight.

Number of moles, $n = \dfrac{\text{Weight of the substance, W}}{\text{Molecular weight, M}}$

The "mole" (or "mol") is a general unit; when expressed in grams, it is called "gram mole"; when expressed in kilograms, it is called "kilogram mole" or "kilo mole"; and when expressed in pounds, it is called "pound mole". Hence, the term "mole" should be interpreted as "gram mole", "kilogram mole" or "pound mole" depending upon whether the basis of calculation is in grams, kilograms or pounds respectively. Also, "mol" is a volume unit and hence composition in terms of mols can be taken as the same as composition by volume (or composition in terms of m^3) i.e., volume % = mole %. Further, 1 gram mole (g mol) of a gas at NTP occupties 22.4 litres (or dm^3); 1 kilogram mole occupies 22.4 m^3 and 1 pound mole occupies 359 ft^3.

(3) **Composition and mean molecular weight of air**

The composition of air is taken as 21% of O_2 and 79% of N_2 (by volume); and 23% of O_2 and

FUELS AND COMBUSTION

77% of N_2 (by weight). The mean molecular weight of air is taken as 28.95. It may be noted that air consists of 21.00 mols of O_2, 78.06 mols of N_2 and 0.94 mols of Ar (argon). Since argon is inert, it is considered together with nitrogen for combustion calculations. That is why, the % N_2 in air by volume is taken as 79%. However, the mean molecular weight of air is taken as 28.95 on the basis of the following:

$$\left[\frac{(21\times32)+(78.06\times28)+(0.94\times39.34)}{100}\right] = 28.9522$$

Further, 1 m³ of O_2 is supplied by $\frac{1\times100}{21}$ = 4.76 m³ of air; and 1 kg. of O_2 is supplied by $\frac{1\times100}{23}$ = 4.35 kg. of air.

(4) Density of air

The density of air at N.T.P.
is 1.290 kg. m⁻³ or 1.290×10^{-3} g cm⁻³.

(5) Minimum Oxygen Required

The minimum O_2 required) = (Theoretical O_2 required — O_2 present in the fuel)

The minimum O_2 required should be calculated on the basis that complete combustion is taking place according to theoretical and stoichiometric combustion reactions.

In case of partial combustion, the combustion products contain CO.

In case of irregular combustion, the combustion products contain both CO and O_2. In such a case, the excess O_2 is calculated after subtracting the amount of O_2 required to burn CO to CO_2.

(6) Combustion of Carbon

The combustion of carbon in air may be represented as:

$$C + (O_2 + N_2) \longrightarrow CO_2\uparrow + N_2$$

1 mol	1 mol	3.76 mols	1 mol	3.76 mols
or	or	or	or	or
12 kg.	32 kg.	107.2 kg.	44 kg.	107.2 kg.

Thus, combustion calculation can be done on mol basis or by using the stoichiometric weight relationships between the reactants and products. However, the mol method is considered to be more convenient and simpler.

(7) Combustion of hydrogen

The hydrogen in coal is present as

(a) Combined hydrogen in the form of H_2O

$2H_2$	O_2	$2H_2O$
4 kg.	32 kg.	$2 \times 18 = 36$ kg.
1 kg.	8 kg.	9 kg.

(b) Available hydrogen:

Combined hydrogen in coal present as moisture does not undergo combustion. It is only the available hydrogen which is equivalent to (H—0/8) that takes part in combustion.

(8) Weight of theoretical amount of air required

For complete combustion of 1 kg. of solid or liquid fuel, the theoretical amount of air required

$$= \frac{100}{23}\left[\frac{32}{12}\times C + 8\left(H - \frac{0}{8}\right) + S\right] \text{ kg.}$$

where C, H, O and S are the respective weights of carbon, hydrogen oxygen and sulfur present in 1 kg. of the fuel.

(9) Calorific Value

If the ultimate analysis of coal is available, its calorific value may be calculated by Dulong's formula as follows:

Calorific value (Kcals/kg).

$$= \left[8080\,C + 34460 \left(H - \frac{O}{8} \right) + 2250\,S \right]$$

where C, H, O and S represent the respective weights of carbon, hydrogen, oxygen and sulfur per kg. of coal.

(10) % Excess Air

$$\% \text{ Excess Air} = \frac{(\text{Actual air used} - \text{Theoretical air})}{\text{Theoretical air}} \times 100$$

(10) The mass of dry flue gases formed should be calculated by balancing the carbon in the fuel and the carbon present in the flue gases.

(11) The composition of a solid or liquid fuel is usually expressed on weight basis whereas the composition of a gaseous fuel is expressed on volume basis, unless otherwise stated.

(12) The composition of flue gases is usually given on dry basis and on volume basis, unless otherwise stated.

2·25. Efficiency of combustion and flue gas analysis

In order to achieve efficient combustion of coal in a boiler furnace, it is essential that the coal and its distillation products are brought into intimate contact with sufficient quantity of air to burn all the combustible matter under appropriate conditions. The correct conditions are: (a) intimate mixing of air with the combustible matter, and (b) sufficient time to allow the combustion process to be completed. If these factors are inappropriate, inefficient combustion occurs. The factors of good mixing and duration in the furnace are largely determined by the design of the equipment and the combustion is controlled by the operator by varying the rate of feed of the fuel and the amount of air admitted. The rate of feed of the fuel is varied mechanically and is measured by meters or by the known characteristics of the feeding mechanism, e.g., depth of fuel bed and speed of grate in case of chain grate stokers. The control of the air is by means of dampers, in the case of natural draught, or by the forced and/or induced draught fans when these are fitted; but the usual and the only practical way of determining the amount of air used in the furnace is to analyse the flue gases for the constituents CO_2, CO and O_2.

Under normal conditions of operation, in which air in excess of the quantity theoretically required for complete combustion is admitted to the furnace, it is usual for control purposes to measure only the proportion of CO_2, CO and O_2 in the flue gases, which gives the correct idea about the combustion of the fuel). Sometimes N_2, H_2 and hydrocarbons such as CH_4, C_2H_6 are also determined. If the analysis shows the presence of free CO, it means that the combustion is incomplete which calls for an immediate increase in the amount of air used for combustion or a dimunition in the rate of supply of the coal. If a large amount of free oxygen is shown in the analysis, it means that the air supply is very much in excess. In ordinary furnaces, 50 to 100% excess air is generally supplied. But too much excess of air results in the loss of heat. In such cases, the supply of air to the hearth is cut short or the rate of fuel supply is increased. However, if the presence of appreciable amounts of both CO and O_2 together is indicated in the flue gas analysis, it shows that the combustion is irregular and non-uniform. That means, in some parts of the furnace, there is excess of air and in some other parts, the supply of air is insufficient, and obviously, such a situation needs immediate attention.

FUELS AND COMBUSTION

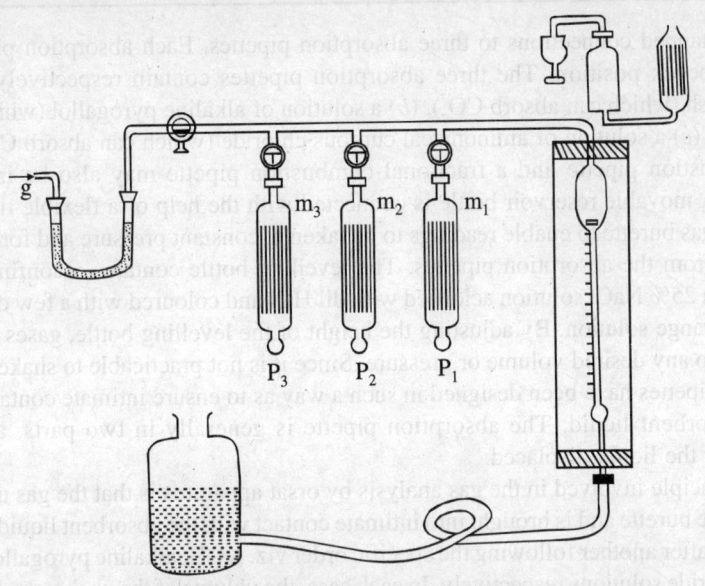

Fig. 2.23. Orsat's apparatus assembly.

The flue gas analysis is generally carried out by Orsat's apparatus (Fig. 2.23). Two types of orsat apparatus are commercially available.

(1) A portable model, known as "short orsat" which is used where the number of components to be determined is relatively limited and where moderate accuracy is sufficient.

(2) A much larger apparatus known as a "long orsat" or a precision model orsat, which can be used for the analysis of quite complicate mixtures of gases and which gives very accurate results in competent hands. The precision orsat assemblies are mounted on substantial stands, and are provided with compensators for gas burettes. Three gas absorption pipettes, a fractional combustion unit and a slow combustion unit are also usually supplied.

The fractional combustion pipette consists of a pyrex U-tube charged with cupric oxide (preferably in wire form) which can be heated in a small, thermostatically controlled electric furnace at 280–295°C. At this temperature, hydrogen is oxidised to water, and CO (if not previously removed by absorption) is oxidised to CO_2. The diminution in volume of the gas sample indicates the volume of hydrogen. The CO_2 formed may be determined by absorption in potassium hydroxide pipette and its volume is equal to that of the CO originally present in the gas mixture. ($2CO + O_2 = 2CO_2$). It may be noted that saturated hydrocarbons, such as methane and ethane, are not oxidised by copper oxide at 283-295°C. The slow combustion pipette is provided with a platinum wire spiral which can be heated electrically and its temperature controlled by a rheostat. The confining liquid is mercury. It is generally used for the determination of saturated hydrocarbons (CH_4, C_2H_6 etc.). These are mixed under appropriate conditions, with a small excess (10 to 15%) of O_2 and the mixture is carefully burned in contact with the heated platinum wire. Heat is generated during the combustion and precautions against explosion (due to accumulation of un-brunt gases) must be taken. The contraction in volume and measurement of the volume of CO_2 produced by absorption in potassium hydroxide solution, enable the calculation of the percentage of saturated hydrocarbons (provided their nature is known).

Other types of gas analysis apparatus are also available commercially. Hempel's apparatus, Bunte's apparatus, Ambler's apparatus, Haldane's air analysis apparatus and Bone and Wheeler's apparatus are the examples.

A simple portable orsat apparatus essentially consists of a water-jacketed and graduated burette of 100 ml capacity, which is connected at the top to a stop-cock manifold constructed of capillary tubing and tapped connections to three absorption pipettes. Each absorption pipette is connected at each stopcock position. The three absorption pipettes contain respectively (a) a solution of caustic potash (which can absorb CO_2), (b) a solution of alkaline pyrogallol (which can absorb CO_2 and O_2) and (c) a solution of ammoniacal cuprous chloride (which can absorb CO_2, O_2 and CO). A slow combustion pipette and a fractional combustion pipette may also be incorporated in the apparatus. A movable reservoir bottle is connected with the help of a flexible rubber tubing to the base of the gas burette to enable readings to be taken at constant pressure and for use in transferring gas to and from the absorption pipettes. The levelling bottle contains a confining liquid such as mercury or a 25% NaCl solution acidified with dil HCl and coloured with a few drops of methyl red or methyl orange solution. By adjusting the height of the levelling bottle, gases in the burette may be brought to any desired volume or pressure. Since it is not practicable to shake the apparatus, the absorption pipettes have been designed in such a way as to ensure intimate contact between the gas and the absorbent liquid. The absorption pipette is generally in two parts, an absorber and a reservoir for the liquid displaced.

The principle involved in the gas analysis by orsat apparatus is that the gas under investigation is taken in the burette and is brought into intimate contact with the absorbent liquids in the absorption pipettes one after another following the specific order viz. KOH, alkaline pyrogallol and ammoniacal cuprous chloride solutions respectively. In each case, the volume of the unabsorbed gases is measured separately at atmospheric pressure. The reduction in volume in each case corresponds respectively to the amounts of CO_2, O_2 and CO present in the gas under test. The result is usually expressed as the percentage composition of the gas by volume.

2·26. Combustion Calculations

In all combustion reactions, definite relationships exist between the masses of the fuel fired, the air used and the flue gases formed as well as the amounts of heat evolved or absorbed during the reactions, all of which are governed by certain well-known laws. These are:
1. The law of conservation of mass
2. The law of definite proportion
3. The gas laws (Boyle's law and Charle's law)
4. The law of conservation of energy.

As the composition of air used is taken as uniform, the relationship existing between the above mentioned quantities can be accurately calculated. Thus, if the values of any two of the following are given the third can be found out: (1) the composition of the fuel, (2) the composition of the flue gas and (3) the nature of combustion. Nature of combustion indicates whether the combustion is complete or not and whether any excess air was used.

The General Gas Equation:

All perfect gases obey the equation of state:
$$PV = nRT$$
where
- P = absolute pressure
- V = Volume
- n = number of moles of gas
- R = The "Gas constant", and
- T = absolute temperature.

Gram-mol and pound-mol

The composition of a solid or liquid fuel is usually expressed by weight whereas the composition of a gaseous fuel is given by volume. However, the composition of the flue gases is generally given by volume.

FUELS AND COMBUSTION

The masses of gaseous substances can be calculated from their volumetric compositions and vice versa in accordance with the gas laws and Avogadro's Hypothesis. From the corollary of the Avogadro's law, it follows that gram molecular volume of all gases at N.T.P. (*i.e.,* Normal temperature and pressure which are 0°C and 760 mm abs, or 32°F and 14.7 lbs/sq. in abs) is 22.4 litres. The corresponding volume in English system is 359 cu. ft., which is known as pound molecular volume. Thus, a gram mol of carbon is 12 grams; a kilogram mol of carbon is 12 kilograms, and a pound mol of carbon is 12 lbs. Similarly, a kilogram mol of CH_4 is 16 kgs and a pound mol of CH_4 is 16 lbs and a gram mol of CH_4 is 16 g. Each gram mol, kilogram mol and pound mol occupy 22.4 litres, 22.4 m^3 and 359 cu. ft. at N.T.P. respectively.

The Avogadro's hypothesis is true not only for gases but also for mixtures of gases. A mixture of gases behaves as if it were single with a molecular weight equal to the average molecular weight of its constituents. Thus, $\frac{1}{3}$ mol of N_2 + $\frac{1}{3}$ mol of H_2 + $\frac{1}{3}$ mol of O_2 will have an average molecular weight of 20.67 $\left(\frac{1}{3} \times 28 + \frac{1}{3} \times 2 + \frac{1}{3} \times 32 = \frac{62}{3} = 20.67\right)$. Hence, 20.67 grams of this gaseous mixture at 0°C, and 760 mm pressure occupies 22.4 litres. Similarly, 20.67 lbs of the gaseous mixture at 32°F and at 1 atmosphere pressure (*i.e.,* 14.7 lbs/sq. in) occupies 359 cu. ft. Thus, we can easily calculate the weight per cubic foot (c. ft.) or cubic centimeter (c.c.) or cubic meter (m^3) of any gas or any gaseous mixture of known composition.

The molecular weights of some common gases are as follows:

Hydrogen, H_2	... 2;	Oxygen O_2	...	32
Carbon monoxide, CO	... 28;	Carbon dioxide, CO_2	...	44
Methane, CH_4	... 16;	Ethylene C_2H_4	...	28
Nitrogen, N_2	... 28;	Water vapours, H_2O	...	18

$$\text{Number of moles} = \frac{\text{Weight of the substance}}{\text{Molecular weight}}$$

Composition of air

In all combustion calculations, the composition of air is taken as 21% O_2 and 79% N_2 (by volume); or as 23% O_2 and 77% N_2 (by weight). The average molecular weight of air is taken as 28.952.*

Calculation of volume of a gas at a given temperature and pressure

Boyle's and Charle's Laws can be used for reducing the volume of a gas at a given temperature and pressure to the corresponding volume at any other specified conditions of temperature and pressure with the help of the equation

$$\frac{PV}{T} = \frac{P_1 V_1}{T_1}$$

where T and T_1 are absolute temperatures of the gases and P and P_1 and V and V_1 denote the initial and final values of pressures and volumes expressed in identical units. If the temperature is i°C, then the absolute temperature, T = 273 + i°. Similarly, if the temperature is i°F, then the corresponding value in absolute scale is 460 + i°.

---

* O_2 = 21 mols = (21.0 × 22) = 672
N_2 = 78.06 mols = (78.06 × 28) = 2135.68
A_r = 0.94 mols = 0.94 × 39.94 = 37.54

Average Mol. wt. = $\frac{2895.2}{100}$ = 28.952.

Combustion reactions

The reactions most commonly encountered in combustion calculations are given below:

$$C + O_2 \longrightarrow CO_2 \qquad \text{...(1)}$$
$$\text{1 mol} \quad \text{1 mol} \qquad \text{1 mol}$$
$$2C + O_2 \longrightarrow 2CO \qquad \text{...(2)}$$
$$2CO + O_2 \longrightarrow 2CO_2 \qquad \text{...(3)}$$
$$2H_2 + O_2 \longrightarrow 2H_2O \qquad \text{...(4)}$$
$$CH_4 + 2O_2 \longrightarrow CO_2 + 2H_2O \qquad \text{...(5)}$$
$$\text{1 mol} \quad \text{2 mol} \qquad \text{1 mol} \quad \text{2 mol}$$
$$C_2H_4 + 3O_2 \longrightarrow 2CO_2 + 2H_2O \qquad \text{...(6)}$$
$$C_2H_2 + 2\tfrac{1}{2} O_2 \longrightarrow 2CO_2 + H_2O \qquad \text{...(7)}$$
$$C_2H_6 + 3\tfrac{1}{2} O_2 \longrightarrow 2CO_2 + 3H_2O \qquad \text{...(8)}$$

Note: In air, the volume of O_2 and N_2 are in the ratio of 21 : 79 = 1 : 3.76.

Equation (1) shows that 1 mol of carbon (12 lbs) combines with 1 mol of oxygen (32 lbs or 359 cu. ft. at N.T.P.) to form 1 mol of CO_2 (44 lbs or 359 cu. ft. at N.T.P.).

Similarly, Equation (5) shows that 1 mol of CH_4 reacts with 2 mols of O_2 to give 1 mol of CO_2 and 2 mols of H_2O. From this equation, we can calculate the air required for the complete combustion of 1 mol of methane. Since 21 mols of oxygen come from 100 mols of air and also since 2 mols of oxygen are required to burn completely 1 mol of CH_4, the air required = $2 \times \dfrac{100}{21}$ = 9.52 mols.

This also means that 1 cft of CH_4 required 9.52 cft of air (since volume % = mol %).

Analyses

Analyses of solids are always reported on weight basis. In order to convert them into mol basis, it is necessary to divide each constituent by its molecular weight. Thus a coal containing 72% C, 4% H_2 and 6% O_2 would contain $\dfrac{72}{12}$ = 6 mols of C; $\dfrac{4}{2}$ = 2 mols of H_2; and $\dfrac{6}{32}$ = 0.187 mol of O_2 per 100 kg of the coal sample.

Analysis of gases is always reported on volume basis and hence directly gives the number of mols of each constituent present per 100 mols of the mixture. Thus, the analyses of gases obtained by the Orsat apparatus are always expressed as % by volume and also on dry basis. Hence, this gas analysis gives molar composition directly since the mol is a volume unit. (A pound mol corresponds to 359 cft at N.T.P.). For example, 100 mols of air (100 × 359 = 35900 cft) contains 21 mols of O_2 (21 × 359 cft) and 79 mols of N_2 (79 × 359 cft).

Excess air

Combustion seldom takes place efficiently with the theoretically minimum quantity of air. Invariably, an excess of air is used in the furnace. Excess air is the amount of air used over and above that required for complete combustion.

Sample problem

Let us consider the following data obtained with methane, CH_4 as a fuel:

Fuel gas			Flue gas		
CH_4 = 100%	Constituent	%		Mols of C	Mols of O_2
	CO_2	5.5		5.5	5.5
	O_2	11.1		—	11.1
	N_2	83.4		—	—
	Total	100		5.5	16.6

FUELS AND COMBUSTION

(A) Ratio of Flue gas: Fuel gas

If we take 100 mols of dry flue gas as the basis for calculation, there are 5.5 mols of C, 16.6 mols of O_2 and 83.4 mols of N_2. It is obvious that the entire amount of carbon present in the fuel gas must have been present in the flue gas. Since all the carbon came from the fuel gas, it is clear that 5.5 mols of CH_4 (containing 5.5 mols of C) were used to form 100 mols of dry flue gas. Therefore,

$$\frac{\text{Dry flue gas}}{\text{Fuel gas}} = \frac{100}{5.5} = 18.2$$

i.e., 18.2 mols (or cft) of dry flue gas per mol (or cft) of fuel gas.

(B) Dry flue gas: dry air

We know that N_2 in air is equal to N_2 in flue gas. In the above example, there are 83.4 mols of N_2 per 100 mols of dry flue gas. In order to obtain 83.4 mols of N_2, the air used must have been $83.4 \times \frac{100}{79}$ mols.

$$\therefore \frac{\text{Dry flue gas}}{\text{Dry air}} = \frac{100}{83.4 \times \frac{100}{79}} = 0.948$$

i.e., 0.948 mols of dry flue gas per mol of air or 0.948 cft. of dry flue gas per cft. of air.

(C) Air: Fuel gas ratio

From (A) and (B) above, we have

$$\frac{\text{Air}}{\text{Fuel gas}} = \frac{\text{Flue gas}}{\text{Fuel gas}} \div \frac{\text{Flue gas}}{\text{Air}} = \frac{18.2}{0.948} = 19.2$$

i.e., 19.2 mol or cft of air per mole or cft of fuel (CH_4).

(D) Excess air

If means the % of air in excess of that is theoretically required for complete combustion. It can be determined by the following methods:

(i) $\quad CH_4 + 2O_2 \longrightarrow CO_2 + 2H_2O$

i.e., 1 mol of methane requires 2 mols of O_2 which can be obtained by $2 \times \frac{100}{21} = 9.52$ mols of air.

From (C) above, we calculated that the air actually used was 19.2 mols (or cft.) per mol (or cft.) of fuel gas.

$$\therefore \quad \text{\% Excess air} = \frac{(19.2 - 9.52)}{9.52} \times 100 = \mathbf{101.7\%}$$

(ii) The N_2 in the gas analysis indicates the total air used and the free oxygen indicates the excess air (provided that the fuel does not contain N_2 and O_2 respectively. In case they are present in fuel, the corresponding quantities will have to be accounted for in the calculation). Thus in the present example, 83.4 mols of N_2 are present which should have come from $\frac{83.4}{79} \times 100$ mols of air, which contains $83.4 \times \frac{21}{79}$ mols of O_2 = 22.17 mols of O_2. However, the free oxygen (O_2) in the flue gas is 11.10 mols. Hence the O_2 required = (22.17 – 11.10) = 11.07 mols.

$$\therefore \quad \text{Excess air} = \frac{(22.17 - 11.07)}{11.07} \times 100 = \mathbf{100.3\%}$$

Notes:

1. The difference in the value of excess air in the above two methods of calculation might be due to the slight errors in the gas analysis.

2. When carbon monoxide is present in the flue gas, the amount of O_2 necessary to burn it must be deducted from the free O_2 in the gas before determining excess air:

$2CO + O_2 \rightarrow 2CO_2$. Hence for every mole of CO, $\frac{1}{2}$ mol of O_2 needed.

For example, if the gas from combustion of a nitrogen free fuel has 12% CO_2, 5% O_2, 2% CO and 81% N_2, then the total oxygen (O_2) corresponding to N_2 is $81 \times \frac{21}{79} = 21.5$ mols and the excess O_2 would be $\left(5 - \frac{2}{2}\right) = (5-1) = 4$ mols.

Hence, the % excess air = $\frac{4}{(21.5-4.0)} \times 100 = 22.85\%$.

(E) Net hydrogen

The net hydrogen in fuel gas, or the hydrogen in excess of that which can combine with the O_2 present in the fuel can also be calculated from the flue gas analysis. H_2 burns with O_2 in air to form H_2O, which does not appear in the gas analysis. Hence it has to be calculated from the oxygen balance. For example, in the above sample problem on methane, while the O_2 appearing in the flue gases is only 16.6 mols, the O_2 corresponding to 83.4 mols of N_2 is $83.4 \times \frac{21}{79} = 22.17$ mol. Hence, $(22.17 - 16.6) = 5.57$ mols of O_2 must have reacted with $(2 \times 5.57) = 11.14$ mols of H_2 forming 11.14 mols of H_2O from each 100 mols of flue gas. Further, $\frac{\text{Net hydrogen}}{\text{Carbon}} = \frac{11.14}{5.5} = 2$, i.e., 2 mols of H_2 per mol of C.

(F) Moisture

The moisture in the flue gas is the sum of the water from free and combined mositure in the fuel, that from combustion of H_2 and that from moisture in the air used. The last one is so small that it may be neglected. Thus if the fuel analysis is known, the moisture in the flue gas (moisture in the fuel + H_2O from combustion of H_2) can be calculated. Thus, if the fuel used is dry methane, the volume of water vapour would be 2 mols per mol of fuel gas (since $CH_4 + 2O_2 \rightarrow CO_2 + 2H_2O$); and $\frac{2}{18.2} \times 100 = 11.1$ mols per 100 mols of dry flue gas i.e., for every 100 cft of dry flue gas, there would be 11.1 cft of wet flue gas.

Example 1. *A coal sample has the following percentage composition; C = 84.0%; H_2 = 3.5%, O_2 = 3.0%, S = 0.5%, Moisture = 3.5%, N_2 = 0.5% and ash = 5.0%. Calculate (a) the theoretical weight of air required for the complete combustion of 1 kg of coal (b) its volume in m^3 at NTP, and (c) percentage composition by weight and volume of the dry products of combustion.*

Solution

The composition of solid fuels is expressed by weight.

Method-1 (Mole Method)

Let the basis of calculation be 100 kg. of coal.

Constituent	% By wt.	Mol. wt.	No. of Kmols	Kmols of O_2 reqd.	Kmols of dry products
C	84.0	12	$\frac{84}{12} = 7.0$	7.0	$CO_2 - 7.0$
H_2	3.5	2	3.5/2 = 1.75	0.875	—

FUELS AND COMBUSTION

O$_2$	3.0	32	$\frac{3.0}{32}$ = 0.094	– 0.094	—
S	0.5	32	$\frac{0.5}{32}$ = 0.016	0.016	SO$_2$ — 0.016
Moisture	3.5	18	—	—	—
N$_2$	0.5	28	$\frac{0.5}{28}$ = 0.0178	—	N$_2$ – 0.0178
Ash	5.0	—	—	—	—
					7.797 k mols

$\left.\begin{array}{l}\text{K mols of O}_2 \text{ required}\\ \text{per 100 kg of coal}\end{array}\right\}$ = **7.797 Kmols.**

(a) Hence, the theoretical quantity of O$_2$ required for the combustion of 1 kg of coal
$$= 7.797 \times 10^{-2} \text{ Kmols}$$
$$= 7.797 \times 10^{-2} \times 32 \text{ kg.}$$

$\therefore \left.\begin{array}{l}\text{Weight of air required for}\\ \text{the complete combustion}\\ \text{of 1 kg of coal}\end{array}\right\}$ = $7.797 \times 10^{-2} \times 32 \times \dfrac{100}{23}$ kg.

$$= \textbf{10.848 kg.}$$

(b) Theoretical quantity of O$_2$ required per kg. of coal = 7.797×10^{-2} K mols.

$\therefore$ Theoretical quantity of air required per kg. of coal
$$= 7.797 \times 10^{-2} \times \frac{100}{21} = 0.3713 \text{ K mols}$$

$\therefore$ Volume of air supplied at NTP for 1 kg of coal
$$= 0.3713 \times 22.4 \text{ m}^3 = 8.3171 \text{ m}^3$$

(c) **Composition of dry products of combustion by volume**

The dry products formed by the combustion of 1 kg of coal are:
CO$_2$ = 0.07 K mols
SO$_2$ = 0.00016 K mols

$$N_2 = 0.000178 + \left(0.3713 \times \frac{79}{100}\right) = 0.2935 \text{ K mols}$$
 from fuel from air

$\therefore$ Total volume of the dry products of combustion = 0.3637 K mols
$\therefore$ Volumetric composition of the products of combustion is:

$$CO_2 = \frac{0.07 \times 100}{0.3637} = 19.25\%$$

$$SO_2 = \frac{0.00016 \times 100}{0.3637} = 0.04\%$$

$$N_2 = \frac{0.2935 \times 100}{0.3637} = 80.7\%$$

By weight

The dry products of combustion formed from 1 kg of coal are;
CO$_2$ = 0.07 × 44 = 3.08 kg
SO$_2$ = 0.00016 × 64 = 0.01024 kg
N$_2$ = 0.2935 × 28 = 8.218 kg
Total weight = 11.3082 kg.

∴ Gravimetric composition of the dry products of combustion is:

$$CO_2 = \frac{3.08 \times 100}{11.3082} = 27.237\%$$

$$SO_2 = \frac{0.01024 \times 100}{11.3082} = 0.091\%$$

$$N_2 = \frac{8.218 \times 100}{11.3082} = 72.673\%$$

Method 2: (Stoichiometric method)

$$\underset{12 \text{ kg}}{C} + \underset{32 \text{ kg}}{O_2} \longrightarrow \underset{44 \text{ kg}}{CO_2}$$

$$\underset{2 \text{ kg}}{H_2} + \underset{16 \text{ kg}}{\frac{1}{2} O_2} \longrightarrow \underset{18 \text{ kg}}{H_2O}$$

$$\underset{32 \text{ kg}}{S} + \underset{32 \text{ kg}}{O_2} \longrightarrow \underset{64 \text{ kg}}{SO_2}$$

Let 100 kg of the coal be the basis of calculation:

Constituent	Weight in kg	Weight of O_2 required, kg.	Weight of dry products, kg
C	84.0	84 × 32/12 = 224	$CO_2 = \frac{84 \times 44}{12} = 308$
H_2	3.5	$3.5 \times \frac{16}{2} = 28.0$	—
O_2	3.0	– 3.0	—
S	0.5	$0.5 \times \frac{32}{32} = 0.5$	$SO_2 = 0.5 \times \frac{64}{32} = 1.0$
Moisture	3.5	—	—
N_2	0.5	—	$N_2 = 0.5$
Ash	5.0	—	—

Weight of O_2 required for the combustion of 100 kg coal $\Big\} = 249.5$ kg.

(a) Hence, the theoretical quantity of O_2 required for the combustion of 1 kg of coal = 2.495 kg.

∴ The theoretical quantity of air required for the combustion of 1 kg. of coal = $2.495 \times \frac{100}{23}$ kg

$$= 10.848 \text{ kg.}$$

(b) 32 kg. of O_2 occupies 22.4 m³ at N.T.P.

∴ 2.495 kg. of O_2 occupies $\frac{2.495 \times 22.4}{32} = 1.7465$ m³

Now, 21 m³ of O_2 are present in 100 m³ of air at N.T.P.

∴ 1.7465 m³ are present in $\frac{1.7465 \times 100}{21} = 8.317$ m³ of air.

∴ Volume of the air at N.T.P. = 8.317 m³.

(c) Composition of dry products of combustion.

FUELS AND COMBUSTION

By weight

The dry products of combustion formed from 1 kg of coal are:
$$CO_2 = 3.08 \text{ kg}$$
$$SO_2 = 0.01 \text{ kg}.$$
$$N_2 = [\underset{\text{from fuel}}{0.005} + \underset{\text{from air}}{(10.848 - 2.495)}] \text{ kg} = 8.358 \text{ kg}.$$

Total weight = 11.448 kg.

∴ Gravimetric composition of the dry products of combustion is

$$CO_2 = \frac{3.08 \times 100}{11.448} = 26.90\%$$

$$SO_2 = \frac{0.01 \times 100}{11.448} = 0.087\%$$

$$N_2 = \frac{8.358 \times 100}{11.448} = 73.01\%$$

By Volume

$$CO_2 = \frac{3.08}{44} = 0.07 \text{ Kmol}$$

$$SO_2 = \frac{0.01}{64} = 0.0001563 \text{ Kmol}$$

$$N_2 = \frac{8.358}{28} = 0.2985 \text{ Kmols}$$

Total volume = 0.3687 Kmols

∴ Volumetric composition of the dry products of combustion is:

$$CO_2 = \frac{0.07 \times 100}{0.3687} = 18.986\%$$

$$SO_2 = \frac{0.001563 \times 100}{0.3687} = 0.0424\%$$

$$N_2 = \frac{0.2985 \times 100}{0.3687} = 80.96\%$$

Example 2. *Find the volume of air required for complete combustion of 1 m³ of acetylene and the weight of air necessary for the combustion of 1 kg of fuel.*

Solution

$$2C_2H_2 + 5O_2 = 4CO_2 + 2H_2O$$
2 vols   5 vols   4 vols   2 vols

O_2 required per m³ of C_2H_2 = 2.5 m³

∴ Air required per m³ of C_2H_2 = $2.5 \times \dfrac{100}{21}$ = **11.9 m³**

Further,

2 mols of C_2H_2 require 5 mols of O_2

∴ (2 × 26) kg of C_2H_2 requires (5 × 32) kg of O_2

(Since mol. wt. of C_2H_2 = 26 and Mol. wt. of O_2 = 32)

O_2 to be supplied per kg of fuel (C_2H_2) = $\dfrac{160}{52}$ kg

∴ Air to be supplied per kg of C_2H_2 = $\dfrac{160}{52} \times \dfrac{100}{23}$ = **13.378 kg.**

Example 3. *The % analysis by volume of producer gas is H_2 – 18.3%, CH_4 – 3.4%, CO – 25.4%, CO_2 – 5.1%, N_2 – 47.8%. Calculate the volume of air required m^3 of the gas.*

Solution.
100 mols of the producer gas contains:

Constituents	% and Mols	Mols of O_2 required
H_2	18.3	9.15
CH_4	3.4	6.80
CO	25.4	12.70
CO_2	5.1	—
N_2	47.8	—
	Total O_2 required ...	28.65

∴ Air required for 100 mols of fuel gas = $28.65 \times \dfrac{100}{21}$ = 136.43 mols

∴ Air required per m^3 of gas = $\dfrac{136.43}{100}$ = **1.3643 m^3.**

Example 4. *A gas has the following composition by volume: H – 22%, CH_4 – 4%, CO – 20%, CO_2 – 6%, O_2 – 3% and N_2 – 45%. If 25% excess air is used, find the weight of air actually supplied per m^3 of this gas.*

Solution.
100 mols of the fuel gas contains

Constituents	% and Mols	Mols of O_2 required
H_2	22	11
CH_4	4	8
CO	20	10
CO_2	6	—
O_2	3	– 3
N_2	45	—
	Total O_2 required	26

∴ Air required for 100 mols of fuel gas = $26 \times \dfrac{100}{21}$ = 123.8 mols

But air supplied actually (25% excess) for 100 mols of gas = $123.8 \times \dfrac{125}{100}$ = 154.75 mols.

∴ Air supplied for 1 mol of gas = $\dfrac{154.75}{100}$ = 1.5475 mols

∴ Air supplied for 1 m^3 of gas = 1.5475 m^3 = $\dfrac{1.5475 \times 28.97}{22.4}$ = **2 kg.**

(Since 22.4 m^3 of air *i.e.* 1 kg mol at N.T.P. weighs 28.95 kg).

Example 5. *A sample of coal contained: C – 81%, H_2 – 4%, O_2 – 2% and N_2 – 1%. Estimate the minimum quantity of air required for complete combustion of 1 kg of the sample. Find the composition of the dry flue gas by volume if 40% excess air is supplied.*

FUELS AND COMBUSTION

Solution.
Let the basis of calculation be 100 kg of coal.

Element	Kg	K Mols	K mols of O_2 required	Product
C	81	6.7500	6.7500	CO_2 – 6.750
H	4	2.0000	1.0000	H_2O – 2.000
O	2	0.06250	–0.0625	—
N	1	0.0360	Nil	N_2 – 0.036
			Total 7.6875	

Oxygen required for 100 kg of coal = 7.6875 k mols.

$\therefore$ Air required for 100 kg of coal = $7.6875 \times \dfrac{100}{21}$ K mols

$= 36.6$ k mols
$= 36.6 \times 28.95$ kg
$= 10.59.6$ kg.

$\therefore$ Air required per kg of coal = $\dfrac{10596}{100}$ = **10.596 kg.**

Air actually used is 40% excess, and hence it is equal to

$36.6 \times \dfrac{140}{100} = 51.24$ k mols

$\therefore$ Excess air used = (Actual air used – theoretical quantity of air required) = (51.24 – 36.60) k mols.

$= 14.64$ k mols

Nitrogen in the total air used = $31.24 \times \dfrac{79}{100} = 40.48$ k mols.

Oxygen in the excess air used which remains in free state in the flue gas

$= 14.64 \times \dfrac{21}{100} = 3.0744$ k mols

The dry flue gases thus contain

$CO_2 = 6.75$ k mols
$N_2 = (40.48 + 0.036) = 40.51$ k mols
     from   from
     air   fuel
$O_2 = 3.0744$ k mols

Total – 50.3344 k mols

Percentage composition of the dry flue gases would be

$CO_2 = \dfrac{6.75}{50.3344} \times 100 = 13.4\%$

$N_2 = \dfrac{40.51}{50.3344} \times 100 = 80.5\%$

$O_2 = \dfrac{3.0744}{50.3344} \times 100 = 6.1\%$

Notes

1. Air contains 21% by volume of O_2 and 79% by volume of N_2.
2. Average mol. wt. of air is 28.95.
3. H_2O in flue gases is neglected for dry flue gas analysis.

Example 6. *A gas fired engine uses a fuel gas having the following composition:*
$CO_2 - 9\%$, $CO - 42\%$, $H_2 - 33\%$, $N_2 - 16\%$
The combustion takes places with the theoretical quantity of air. Calculate the volume of air used per cubic meter of fuel gas burnt. Determine the dry flue gas analysis.

Solution.
100 mols of fuel gas contains

	%	Mols	Mols of O_2 required	Products and Mols
CO_2	9	9	Nil	$CO_2 - 9$
CO	42	42	21	$CO_2 - 42$
H_2	33	33	16.5	$H_2O - 33$
N_2	16	16	Nil	$N_2 - 16$
			Total 37.5 mols	

Mols of O_2 required per 100 mols of fuel gas = 37.5

∴ Mols of air required per 100 mols of fuel gas

$$= 37.5 \times \frac{100}{21} = 178.5$$

∴ Mols or cubic meters (m³) of air required per 1 mol or 1 cubic meter of the fuel gas $= \frac{178.5}{100}$
= **1.785**.

The dry product of combustion:

$CO_2 \longrightarrow 9 + 42 = 51$ mols

$N_2 \longrightarrow 16 + \left(37.5 \times \frac{79}{21}\right) = 157$ mols
    (from fuel)   (from air)
         Total   208 mols

∴ The dry exhaust gas contains

$CO_2 - \dfrac{51 \times 100}{208} = 24.52\%$

$N_2 - \dfrac{157 \times 100}{208} = 75.48\%$

Example 7. *A furnance fired with a hydrocarbon fuel oil has a stack gas analysing as follows:*
$CO_2 - 10.2\%$, $O_2 - 8.3\%$, $N_2 - 81.5\%$. *Calculate (a) the composition of the original fuel oil, (b) % excess of air used and (c) volume of air supplied per kg of oil.*

Solution.
Basis for calculation: 100 k mols of dry stack gas

	% and mols	K Mols of C	K mols of O_2	K mols of N_2
CO_2	10.2	10.2	10.2	—
O_2	8.3	—	8.3	—
N_2	81.5	—	—	81.5
Total		10.2	18.5	81.5

(a) 100 mols of dry flue gas contains 18.5 k mols of O_2 and 81.5 k mols of N_2 which must have come from the air supplied.

Hence, the air supplied = $81.5 \times \dfrac{100}{79} = 103.17$ k mols;

and the O_2 accompanying it = $103.17 \times \dfrac{21}{100} = 21.69$ k mols.

The difference in quantity of O_2 supplied and that used in the formation of water which is not appearing in the dry flue gas analysis will be

FUELS AND COMBUSTION

$= 21.69 - 18.5 = 3.19$ k mols

∴ Water vapour formed $= 2 \times 3.19 = 6.38$ k mols

$$\left(H_2 + \frac{1}{2}O_2 \rightarrow H_2O\right)$$

∴ Hydrogen present in the fuel oil
$= 6.38$ k mols

The same quantity of fuel oil contains
$= 10.2$ k mols of C

∴ The fuel oil contains 10.2 k mols of C or $(10.2 \times 12) = 122.4$ kg of C and $(6.38 \times 2) = 12.76$ kg of H.

That means, the oil weighing $(122.4 + 12.76) = 135.16$ kg was burnt and it should be composed of 122.4 kg of C and 12.76 kg of H.

Hence the composition of the fuel oil is:

$$C = \frac{122.4}{135.16} \times 100 = 90.5\%$$

$$H = \frac{12.76}{135.16} \times 100 = 9.5\%$$

(b) Total O_2 supplied $= 21.69$ k mols
O_2 needed for   C $= 10.20$ k mols
O_2 needed for   H $= 3.19$ k mols

$\left.\begin{array}{l}O_2 \text{ actually needed for C and H}\\ \text{present in 100 mols of fuel oil}\end{array}\right\} = 13.39$ k mols

∴ Excess $O_2 = (21.69 - 13.39) = 8.30$ k mols

∴ Excess O_2 or air used $= \dfrac{8.30}{13.39} \times 100 = 61.99\%$

(c) Total oil burnt $= 135.16$ kg

Total air supplied $= 21.69 \times \dfrac{100}{21} = 103.29$ k mols
$= 103.29 \times 22.4 = 2313.7$ m^3 at NTP
(Avogadro's law)

∴ Air used per kg of fuel oil $= \dfrac{2313.7}{135.16} = \mathbf{17.12\ m^3}$.

Example 8. *A coal containing 62.4% C, 4.1% H, 6.9% O, 1.2% N, 0.8% S and 15.1 moisture and 9.7% ash was burnt in such a way that the dry flue gases contained 12.9% CO_2, 0.2% CO, 6.1% O and 80.8% N. Calculate (a) the weight of air theoretically required per kg of coal, (b) the weight of air actually used and (c) the weight of dry flue gas produced per kg of coal.*

Solution.

(a) **Basis of calculation.** 100 kg of coal, which contains

Analysis	Kg	K mols	K mols of O_2 required	K mols of product formed
C	62.4	5.2	5.2	CO_2 — 5.200
H	4.1	2.05	1.025	H_2O — 2.050
O	6.9	0.217	– 0.217	Nil
N	1.2	0.043	Nil	N_2 — 0.043
S	0.8	0.025	0.025	SO_2 — 0.025
H_2O	15.1	0.840	Nil	H_2O — 0.840
Ash	9.7	—	Nil	Nil
			Total	6.033 k mols

Thus, O_2 theoretically required for complete combustion of 100 kg of coal = 6.033 mols.

∴ Air theoretically required for complete combustion

$$= 6.033 \times \frac{100}{21} = 28.73 \text{ mols}$$
$$= (28.73 \times 28.95) = \mathbf{831.7 \text{ kg}}$$

Wt. of air required per kg of coal $= \dfrac{831.7}{100} = \mathbf{8.317 \text{ kg}}$

(b) **Actual air used**:

Basis : 100 k mols of dry flue gas, which contains

Constituent	%	K Mols	K Mols C	K Mols of C	K Mols of N
CO_2	12.9	12.9	12.9	12.9	—
CO	0.2	0.2	0.2	0.1	—
O	6.1	6.1	—	6.1	—
N	80.8	80.8	—	—	80.8
			Total 13.1	19.1	80.8

13.1 k mols of C must have been supplied by $\dfrac{13.1}{5.2} \times 100$ kg of coal = 252 kg of coal containing

$0.043 \times \dfrac{252}{100} = 0.108$ k mols of N.

(Since 100 kg of coal contains 5.21 mols of C and 0.043 k mols of N).

∴ K Mols of N from air in 100 k mols of flue gas
$$= (80.800 - 0.108) = 80.692 \text{ k mols}$$

∴ K Mols of air actually supplied to obtain 100 k mols of dry flue gas or to burn 252 kg of coal would therefore be

$$= 80.692 \times \frac{100}{79} = 102.15 \text{ k mols}$$

∴ Wt. of air supplied $= (102.15 \times 28.97)$ kg
$$= 2959.3 \text{ kg}$$

∴ Weight of air supplied per kg of coal

$$= \frac{2959.3}{252} = \mathbf{11.74 \text{ kg}}$$

(c) **Wt. of dry flue gas obtained per kg of coal:**

100 kg mols of dry flue gas contains:

Constituent	% K Mols	Weight in kg
CO_2	12.9	$12.9 \times 44 = 567.6$
CO	0.2	$0.2 \times 28 = 5.6$
O_2	6.1	$6.1 \times 32 = 195.2$
N_2	80.8	$80.8 \times 28 = 2262.4$
		Total = 3030.8 kg

This much flue gas is obtained by burning 252 kg of coal.

∴ Wt. of dry flue gas obtained per kg of coal burnt

$$= \frac{3030.8}{252} = \mathbf{12.03 \text{ kg.}}$$

Example 9. *A hydrocarbon fuel, on burning, gave a flue gas having the following volumetric analysis: $O_2 = 3.56\%$, $CO_2 = 13.53\%$; $N_2 = 82.91\%$. What is the composition of the fuel by weight and the percentage of excess air used? What is the volume of air supplied per kg of the fuel oil?*

FUELS AND COMBUSTION

Solution.
Basis of calculation: 100 k mols of dry flue gas.

Constituent	% and K Mols	Mols of C	K Mols of O_2	K Mols of N_2
CO_2	13.53	13.53	13.53	—
O_2	3.56	—	3.56	—
N_2	82.91	—	—	82.91
	100.00	13.53	17.09	82.91

100 k mols of dry flue gas contions 17.09 k mols of O_2 and 82.91 k mols of N_2 which must have come out from the air supplied, *i.e.*, $82.91 \times \dfrac{100}{79} = 104.95$ k mols; and the O_2 accompanying this air

$$= 104.95 \times \dfrac{21}{100} = 22.04 \text{ k mols.}$$

Hence, the difference in quantity of O_2 supplied and used in the formation of water vapour, which is not appearing in dry flue gas analysis, will be $(22.04 - 17.09) = 4.95$, k mols.

∴ The water vapour formed $= 2 \times 4.95 = 9.90$ k mols.
∴ Mols of H_2 present in the oil should also be $= 9.90$ k mols.
$(2H_2 + O_2 \rightarrow 2H_2O)$
The same quantity of the fuel oil contains 13.53 k mols, of C, $(C + O_2 \rightarrow CO_2)$.
∴ The fuel oil contains 13.53 mols of $C \equiv 13.53 \times 12$ kg
$= 162.36$ kg of C and $9.9 \times 2 = $ **19.8 kg of H_2.**
∴ The composition by weight of the hydrocarbon fuel is as follows:

$$C = \dfrac{162.36 \times 100}{182.16} = 89.13\%$$

$$H = \dfrac{19.8 \times 100}{182.16} = 10.87\%$$

Now,
Total O_2 supplied $= 22.04$ k mols,
O_2 needed for C $= 13.53$ k mols,
O_2 needed for H $= 4.95$ k mols.
∴ Total O_2 actually needed$\} = (13.53 + 4.95) = 18.48$ k mols
∴ Excess $O_2 = (22.04 - 18.48)$ mols $= 3.56$ k mols.
∴ % of Excess O_2 (or air) used

$$= \dfrac{3.56 \times 100}{18.48} = 19.264\%$$

Further,
Total fuel oil burnt $= 182.16$ kg
Total air supplied $= 22.04 \times \dfrac{100}{21}$ kg mols

$$= 22.04 \times \dfrac{100}{21} \times 22.4 \text{ m}^3$$

$$= 2319.4476 \text{ m}^3$$

∴ Air used per kg of fuel oil

$$= \dfrac{2319.4476}{182.16} = \textbf{12.733 m}^3.$$

---

* For other worked out examples on combustion calculations, the students may refer the "Text Book on Experiments and Calculation in Engineering Chemistry" By S.S. DARA.

Example 10. *A example gave the following analysis: C – 84%, H – 4%, O – 4%, ash – 8%. The composition of the dry flue gas obtained by using the above coal is as follows:*
CO_2 – 9%, CO – 1%, O_2 – x% and N_2 – (90 – x)%.
What will be the value of x?

Solution.

The problem can be solved by using the principle of oxygen balance.
100 kg of the coal sample contains:

Constituent	Kg	k Mols
C	84	$\frac{84}{12} = 7$
H	4	$\frac{4}{2} = 2$
O	4	$\frac{4}{32} = 0.125$

100 mols of the flue gas contains

Constituent	k Mols	k Mols of C	k Mols of O	k Mols of N
CO_2	9	9	9	—
CO	1	1	0.5	—
O_2	x	—	x	—
N_2	(90 – x)	—	—	(90 – x)
Total	100	10	(9.5 + x)	(90 – x)

The coal burnt to produce 100 k mols of the flue gas

$$= \frac{100}{7} \times 10 = 142.8571 \text{ kg}$$

k mols of H present in 142.857 kg of coal

$$= \frac{2}{100} \times 142.8571 \text{ k mols} = 2.8571 \text{ k mols}$$

As the fuel does not contain any N_2, it may be assumed that the entire N_2 present in the flue gas is obtained from air only.

Mols of O_2 supplied for the combustion of 142.8571 kg of coal

$$= (90 - x) \times \frac{21}{79} = 0.2658 (90 - x) \text{ k mols}$$

Further, the k mols of O_2 supplied = k mols of O_2 present in dry flue gas + k mols of O_2 that combined with H_2 to form water vapour.

i.e., $0.2658 (90 - x) = (9.5 + x) + \left(2.8571 \times \frac{1}{2}\right)$

∴ $23.922 - 0.2658 x = 9.5 + x + 1.4285$
∴ $1.2658 x = 23.922 - 9.5 + 1.4285$
∴ $x = \mathbf{10.265}$.

Example 11. *A petroleum gas has the following composition. Ethane – 5%, Propane – 10%, Butane – 40%, Butene – 10%, Isobutane – 30%, Propene – 5%.*

Calculate the volume of air required for complete combustion of 100 M^3 of the gas and the percentage composition of the dry flue gases if 35% excess air is supplied.

FUELS AND COMBUSTION

Solution.

The combustion reactions of the constituents of the fuel gas are:

$$C_2H_6 + 3.5\,O_2 = 2CO_2 + 3H_2O$$
$$\text{1 mol} \quad \text{3.5 mols} \quad \text{2 mols} \quad \text{3 mols}$$

$$C_3H_8 + 5O_2 = 3\,CO_2 + 4H_2O$$
Propane

$$C_4H_{10} + 6.5\,O_2 = 4\,CO_2 + 5H_2O$$
Butane
or
Isobutane

$$C_4H_8 + 6O_2 = 4CO_2 + 4H_2O$$
Butene

$$C_3H_6 + 4.5O_2 = 3CO_2 + 3H_2O$$
Propene

Let 100 mols of the petroleum gas be the basis for calculation:

Constituent	Mols	Mols O_2 reqd.	Products and Mols
C_2H_6	5	5 × 3.5 = 17.5	CO_2 = 10; H_2O = 15
C_3H_8	10	10 × 5 = 50	CO_2 = 30; H_2O = 40
C_4H_{10}	40 (Butane)	40 × 6.5 = 260	CO_2 = 160; H_2O = 200
	30 (Isobutane)	30 × 6.5 = 195	CO_2 = 120; H_2O = 150
C_4H_8	10	10 × 6 = 60	CO_2 = 40; H_2O = 40
C_3H_6	5	5 × 4.5 = 22.5	CO_2 = 15; H_2O = 15
	Total	605 mols	375 mols

Now,

O_2 required for complete combustion of 100 mols of the petroleum gas = 605 mols.

∴ Air required for the complete combustion of 100 mols of the petroleum gas = $605 \times \dfrac{100}{21}$ mols = 2880.9524 mols.

∴ Air required for the complete combustion of 100 m³ of the petroleum gas = 2880.9524 m³.

Dry flue gas analysis with 35% excess air

If 35% excess air is used, the actual air supplied would be = $2880.9524 \times \dfrac{135}{100}$ mols = 3889.2857 mols.

The N_2 contributed by the above quantity of air

$$= 3889.2857 \times \frac{79}{100} = 3072.5357 \text{ mols.}$$

Excess O_2 supplied during the combustion of 100 mols of the petroleum gas = $605 \times \dfrac{35}{100}$ = 211.75 mols.

∴ The flue gases produced by the combustion of 100 mols of the petroleum contain

$$CO_2 = 375 \text{ mols}$$
$$N_2 = 3072.5357 \text{ mols}$$
$$O_2 = 211.75 \text{ mols.}$$
$$\overline{\text{Total} = 3659.2857 \text{ mols}}$$

∴ The dry flue gas analysis would be

$$CO_2 = \frac{375}{3659.2857} \times 100 = 10.25\%$$

172 A TEXTBOOK OF ENGINEERING CHEMISTRY

$$N_2 = \frac{3072.5375 \times 100}{3659.2857} = 83.96\%$$

$$O_2 = \frac{211.75 \times 100}{3659.2857} = 5.79\%.$$

Example 12. *The following data for flue gas analysis has been recorded in a Lancashire boiler with an economiser.*

Constituent	Percentage Analysis of Flue gas Entering the Economiser	Leaving the Economiser
CO_2	9.3	7.2
CO	0.4	0.3
O_2	10.2	12.6
N_2	80.1	79.9

Calculate the air leakage into the economiser per kg of the fuel stoked if the carbon in the fuel was 0.735 kg/kg of the fuel.

Solution.

Let 100 ml of dry flue gas be entering the economiser. If x mols of air leaked into the economiser during a certain time, the volume of the gases leaving the economiser will be $(100 + x)$ mols.

Hence, the % CO_2 in the flue gas leaving the economiser $= \dfrac{9.3 \times 100}{(100+x)}$ which should be equal to 7.2 mols as per the data.

$$\therefore \quad \frac{9.3 \times 100}{(100+x)} = 7.2$$

$\therefore x = 29.16$ mols $= (29.16 \times 28.97)$ kg $= 846$ kg.

Hence 846 kg of air leaked into 100 mols of dry flue gas entering the economiser.

Fuel burnt to produce 100 mols of flue gas entering the economiser $= \dfrac{\text{Mols of C in flue gas}}{\text{Moles of C in 100 kg of fuel}} \times 100$

$$= \frac{(9.3+0.4)}{6.125} \times 100 = 158 \text{ kg}$$

(Since 100 kg fuel contains 73.5 kg carbon $\equiv \dfrac{73.5}{21}$ mols $= 6.125$ mols of C)

$\therefore$ Air leakage into the economiser per kg of fuel

$$= \frac{846}{168} \text{ kg} = \mathbf{5.42 \text{ kg}}.$$

Example 13. *A coal sample of the following analysis by weight was used for a boiler trial:*

$C = 82\%, \ H = 4.9\%, \ O = 4.8\%$

The flue gas analysis by Orsat's apparatus gave the following data:

$CO_2 = 10\%, \ O_2 = 8\%, \ CO = 1.5\% \text{ and } N_2 = 80.5\%.$

The temperature of ambient air in the boiler house is 18°C and that of the flue gases is 320°C. Calculate the heat carried away by the excess of air per kg of coal. The average specific heat of air is 0.238.

Solution.

Vol. of constituents per m^3 of flue gas	Proportional weight
$CO_2 = 10/100 = 0.1 \text{ m}^3$	$0.1 \times 44 = 4.40$ kg
$O_2 = 8/100 = 0.08 \text{ m}^3$	$0.08 \times 32 = 2.56$ kg
$CO = 1.5/100 = 0.015 \text{ m}^3$	$0.015 \times 28 = 0.42$ kg
$N_2 = 80.5/100 = 0.805 \text{ m}^3$	$0.805 \times 28 = 22.54$ kg
	Total $= 29.92$ kg

FUELS AND COMBUSTION

Weight of CO_2/kg of flue gas = $\dfrac{4.4}{29.92}$ = 0.147 kg

Weight of O_2/kg of flue gas = $\dfrac{2.56}{29.92}$ = 0.085 kg

Weight of CO/kg of flue gas = $\dfrac{0.42}{29.92}$ = 0.014 kg

Weight of N_2/kg of flue gas = $\dfrac{22.54}{29.92}$ = 0.747 kg.

Total weight of carbon/kg of flue gas = (Wt. of C in CO_2 + Wt. of C in CO/kg of flue gas

$$= \left(\dfrac{12 \times 0.147}{44} + \dfrac{12 \times 0.014}{28}\right)$$

= (0.04 + 0.006) kg = 0.046 kg.

(Since 12 kg of C is present in 44 kg of CO_2 and in 28 kg of CO respectively).

But, weight of C per kg of fuel = 82/100 = 0.82 kg.

∴ Wt. of dry flue gas/kg of fuel burnt = 0.82/0.046 = 17.82 kg

Weight of O_2 required for the combustion of CO per kg of flue gas = $\dfrac{16 \times 0.014}{28}$ = 0.008 kg

But, wt of O_2/kg of flue gas = 0.085 kg as shown above.

∴ Excess O_2/kg of flue gas = (0.085 – 0.008) = 0.077 kg
∴ Excess O_2/kg of fuel burnt = 0.077 × 17.82 = 1.37 kg

∴ Excess air/kg of fuel burnt = 1.37 × $\dfrac{100}{23}$ = 6.0 kg

∴ Heat carried away by the excess air
= M × S × $(t_2 - t_1)$ Kcals
   Mass Sp. heat rise in temp.
= 6 × 0.238 × (320 – 18) Kcals
= **431.256 Kcals.**

Example 14. *A gaseous fuel has the following volumetric composition : CH_4 = 40%; C_2H_2 = 6%; C_2H_6 = 24%; C_4H_8 = 16%; O_2 – 3.0%; CO = 5% and N_2 = 6%. If 40% excess air was used for the combustion, calculate the air; fuel ratio and the analysis of dry flue gas.*

Solution.

The combustion reactions of the constituent gases in the fuel are:

$$CH_4 + 2O_2 \longrightarrow CO_2 + 2H_2O$$
  1 mol 2 mole   1 mol 2 mols

$$C_2H_4 + 3O_2 \longrightarrow 2CO_2 + 2H_2O$$
$$C_2H_6 + 3.5 O_2 \longrightarrow 2CO_2 + 3H_2O$$
$$C_4H_8 + 6O_2 \longrightarrow 4CO_2 + 4H_2O$$

$$CO + \dfrac{1}{2} O_2 \longrightarrow CO_2$$

Let 1 mol of the fuel gas be the basis for calculation:

Constituent & mols	Mols of O_2 reqd.	Mols of Products formed
CH_4 — 0.4	0.4 × 2 = 0.80	CO_2 — 0.4; H_2O — 0.8
C_2H_4 — 0.06	0.06 × 3 = 0.18	CO_2 — 0.12; H_2O — 0.12
C_2H_6 — 0.24	0.24 × 3.5 = 0.84	CO_2 — 0.48; H_2O — 0.72
C_4H_8 — 0.16	0.16 × 6 = 0.96	CO_2 — 0.64; H_2O — 0.64

O_2 — 0.03	= – 0.03	—
CO — 0.05	$0.05 \times 0.5 = 0.025$	CO_2 — 0.05
N_2 — 0.06	—	—
Total	= 2.775 mols	CO_2 — 1.69 mols.

∴ O_2 required/mol of the fuel gas = 2.775 mols
or O_2 required/m³ of the fuel gas = 2.775 m³
(Since mol% = m³% = Vol%)
But 40% excess air was actually used.
∴ O_2 actually supplied/m³ of fuel gas

$$= 2.775 \times \frac{140}{100} = 3.885 \text{ m}^3$$

∴ Air actually supplied/m³ of fuel gas

$$= 3.885 \times \frac{100}{21} = 18.5 \text{ m}^3$$

∴ Air : fuel ratio = **18.5 m³**.

Analysis of dry products of combustion/m³ of fuel gas

$$CO_2 = 1.69 \text{ m}^3$$

$$O_2 = 18.5 \times \frac{40}{100} \times \frac{21}{100} \text{ m}^3 = 1.554 \text{ m}^3$$

[from excess air]

$$N_2 = \left[\left(18.5 \times \frac{79}{100}\right) + 0.06\right] \text{ m}^3 = 14.675 \text{ m}^3$$

(from air + fuel)
Total volume of combustion products = 17.919 m³
Whence the volumetric composition of the flue gas is

$$CO_2 = \frac{1.69 \times 100}{17.919} = 9.43\%$$

$$O_2 = \frac{1.554 \times 100}{17.919} = 8.67\%$$

$$N_2 = \frac{14.675 \times 100}{17.919} = 81.9\%.$$

QUESTIONS

1. How is calorific value of a solid fuel determined using a bomb calorimeter?
   The following data is obtained in a bomb calorimeter experiment:
   Mass of the fuel pellet = 0.85 g
   Mass of water taken in the calorimeter = 2000 g
   Water equivalent of the calorimeter = 540 g
   Difference in final and initial temperature = 1.9°C
   Cooling correction = 0.041°C
   Fuse wire correction = 3.8 calories
   Acids correction = 48.8 calories
   Calculate the net calorific value of the fuel if it contains 3.6% hydrogen and 1.2% oxygen.
2. Distinguish clearly between the following:
   (a) Gross and net calorific values of a fuel (Nag. Univ. S - 2006)
   (b) proximate and ultimate analysis (Nag. Univ. S - 2004)
   (c) High and Low temperature corbonization (Nag. Univ. S - 2005, S-2005)
   (d) coking coals and caking coals (Nag. Univ. S - 2006)

FUELS AND COMBUSTION

    (e) octane number and cetane number
    (f) Thermal and catalytic cracking.(Nagpur Univ. S-2005)
3. How is the proximate analysis of a coal conducted and what is its 'significance in determining' the utility of a coal for a particular purpose?
4. Write short notes on the following:
    (a) Fractional distillation of crude petrolium oil (Nagpur Univ. S-2007)
    (b) Regnault and pfaundler's formula
    (c) Water equivalent of a Bomb calorimeter.
    (d) Catalytic converters.
    (e) Commercial types of coal
    (f) Blending of coal
    (g) Storage and spontaneous combustion of coal (Nagpur Univ. S-2003, S-2007)
5. What are the difficulties arising if coal is stored improperly for prolonged periods? What are the factors favourable for spontaneous combustion? What safety measures do you suggest for storage of coal?
6. What is coke? why is it preferred as a metallurgical fuel over coal? Give the properties of metallurgical coke (Nagpur Univ. S-2004).
7. What are the types of carbonization and for what purposes they are employed? (Nagpur Univ. S-2003)
8. Describe the Otto. Hoffman's process for preparing coke? What are its advantages over the earlier methods? (Nagpur Univ. S-2003, S-2005)
9. Write informative notes on the following:
    (a) Advantages and limitations of liquid fuels
    (b) Detection of oil deposits
    (c) Refining of petroleum
    (d) Bubble tower and Top-flasher
    (e) Thermal and catalytic processes
    (f) Reforming.
10. What is cracking and what for it is used? What are the types of cracking? Describe the working of fluid bed catalytic cracking unit. (Nagpur Univ. S-2006)
11. How is synthetic gasoline obtained from non-petroleum sources? (Nagpur Univ. S-2006)
12. What are the requisites of an ideal gasoline? What is meant by blending and doping?
13. What is meant by knocking in a petrol engine and what is it due to? What is octane number of a potrol? How is knocking related to chemical structure of the constituents of petrol and how can it be reduced? (Nagpur Univ. S-2006)
14. How do you explain knocking in a diesel engine? How can it be controlled? What is cetane number?
15. Write informative notes on the following:
    (a) Rocket fuels
    (b) Power alcohol
    (c) Liquid petroleum gas (LPG)
    (d) Coal gas and coke oven gas
    (e) Kopper's gas
    (f) Winkler's gas
    (g) Synthesis gas.
16. How is producer gas manufactured? What are its disadvantages and advantages?
17. Justify the following statements :
    (i) Carbon determined in ultimate analysis is more than that determined in proximate analysis. (Nagpur Univ. S - 2004)
    (ii) Gasoline containing tetraethyl lead is used in internal combustion engines (Nagpur Univ. S - 2004)
    (iii) Good compression ignition engine fuels are poorer spark ignition engine fuels and vice-versa (Nagpur Univ. S - 2004)
18. What is the common apparatus used for flue gas analysis? How it is useful to determine the efficiency of combustion?
19. A producer gas has the following composition by volume: CH_4 = 3.5%: CO = 25%, H_2 = 10%; CO_2 = 10.8%, N_2 = 50.7%. Calculate the theoretical quantity of air required per cubic meter of the gas. If 22% excess air is used, find the percentage composition of the dry products of combustion.
20. A liquid hydrocarbon fuel contains 89.4% C and 10.6% H by weight. If 100 kg of the fuel is burnt, calculate (a) the theoretical amount of air required and (b) the volumetric composition of the products of combustion if 20% excess air was used,

(Ans. (a) 148.1 K moll (b) CO_2 = 12.34%, H_2O = 8.78%; N_2 =75.53% and O_2 = 3.35%)

21. A coal sample gave the following analysis by weight: C = 88%, H = 4%, S = 1%, O = 5%, ash and moisture 1% each. The coal sample on combustion in a boiler gave a flue gas having the following composition by volume:
CO_2 = 11%; O_2 = 7%, CO =1.5%, SO_2 = 0.5% and N_2 = 80%. Calculate (a) the weight of air required for complete combustion of 1 kg of the coal (b) weight of the wet flue gases per kg of coal, (c) percentage of excess air used and (d) percentage of carbon burnt to CO.
(Ans. (a) 11.42 kg/kg of coal; (b) 18.08 kg/kg of coal; (c) 49.4%(d) 12% C burnt to CO)
22. How does knocking occur in I.C. engines. How can it be prevented?
23. Establish the relationship between knocking in I.C. engines and the nature and molecular structure of the constituents in petrol and diesels fuel. (Nagpur Univ. S - 2005)
24. How are water gas and producer gas manufactured? Discuss the theoretical principles involved.
25. How is coal gas manufactured? How is coal gas purified? What types of coal is preferred for gas-making?
26. Define gross and net calorific values. Calculate the approximate calorific value (by Dulong's formula) of a coal sample having the following ultimate analysis:
C = 80%; H = 3.5%; S = 2.8%; O = 5.0; N = 1.5% and ash = 7.2%.
27. A hydrocarbon fuel on combustion gave the flue gas having the following composition by volume: CO_2 = 13%; O_2 = 6.5%; and N_2 = 80.5%. Calculate (a) the composition of fuel by weight and (b) percentage of excess air used.
28. A producer gas has the following composition by volume; CO = 30%; H_2 = 12%; CO_2 = 4%; CH_4 = 2% and N_2 = 52%. When 100 m^3 of the gas is burnt with 50% excess air used, what will be the composition of the dry fiue gases obtained.
29. A coal sample has the following composition; C = 90%; H = 3.5%; O = 3%; S = 0.5; and N_2 = 1%; the remaining being ash. Calculate the theoretical volume of air required at 27°C and 1 atm pressure when 100 kg of the coal is burnt. **(Ans.** 1100 m^3)
30. The composition by weight of a coal sample is: C = 81%; H = 5%; O = 8.5%; S = 1%; N = 1.5%; Ash = 3%. (a) Calculate the amount of air required for the complete combustion of 1 kg of the coal, (b) Calculate the gross and net calorific values of the coal sample. Given that the calorific values of ? C, H and S are 8060 Kcals/kg; 34000 Kcals/kg and 2200 Kcal/kg respectively.
(Ans. 10.8 kg; 7.910 kcal, 7,667 kcal/kg)
31. A coal gas has the following volumetric composition: CO_2 = 2.5%; C_2H_4 = 3.5%; CO = 8%; H_2 = 50%; CH_4 = 34% and N_2 = 2%. Calculate the gross calorific value of the gas at N.T.P. Given that 1 kg of C while burning to CO develops 2440 kcals; while burning to CO_2 develops 8060 Kcals of heat; and 1 kg of H_2 while burning to H_2O develops 34,400 Kcals, 1 kg mol of gas at N.T.P. occupies 22.4 m^3. **(Ans.** 5,848 Kcals/m^3)
32. A coke-oven gas of the following volumetric composition was used for firing gas retorts:
CO_2 = 1.4%; CO = 5.1%; O_2 = 0.5%; H_2 = 57.4%; CH_4 = 28.5%; C_2H_4 = 2.9% and N_2 = 4.2%. The dry flue gases on analysis by Orsat's apparatus gave the following results; CO_2 = 15%; CO = 0%; O_2 = 2.5%; N_2 = 82.5%. Calculate the excess air supplied for burning 100 m^3 of the gas. **(Ans.** 32.4 m^3)
33. A coal containing 62.4% C; 4.1% H; 6.9% O; 1.2% N; 0.8% S; 15.1% moisture and 9.5% ash was burnt in such a way that the dry flue gases contained 12.9% CO_2; 0.2% CO; 6.1% O_2 and 80.8% N_2. Calculate (a) the weight of air theoretically required per kg of coal, (b) % of excess air used and (c) the weight of the dry flue gas produced per kg of coal. **(Ans.** 8.3246 kg; 41.0984% 12.0306 kg)
34. A gasoline sample contains 86% C and 14% H_2. If the air supplied is only 95% of the theoretically required value, calculate the dry flue gas analysis. Assume that all the H_2 underwent complete combustion and no carbon is left free.
[**Hint.** For every 0.5 mol of O_2 supplied less, 1 mol of C is converted into CO].
(Ans. CO_2 = 13.47%; CO = 2.36% and N_2 = 84.17%)
35. Explain the following:
    (a) The knocking tendency of petrol or diesel oil can be predicted on the basis of the nature and molecular structure of its constituent compounds. (Nagpur Univ. S - 2007)
    (b) All coking coals are caking but all caking coals are not coking. (Nagpur Univ. S - 2005)
36. A furnace utilises a mixture of blast furnace gas and coke oven gas as a fuel. The gas mixture is fired along with air to obtain the required temperature. The analysis of the individual gases used and the flue gas evolved due to the combustion of the above gas mixture are given on dry basis and in volume % as below:

Constituent	Coke oven gas	Blast furnace gas	Air	Flue gas
CO_2	2.4	13.4	—	13.5
O_2	2.0	0.2	21.0	6.9

FUELS AND COMBUSTION

CO	7.2	24.2	—	0.4
CH_4	30.8	—	—	—
H_2	51.3	3.0	—	—
H_2	6.3	59.2	79.0	79.2

Calculate (a) the ratio of blast furnace gas to coke oven gas used, and (b) the % excess of air used.

(Ans. 5.38; 74.7%)

37. A steam raising plant of a refinery uses 500 kg per hour of tar as a fuel. The analytical report on the tar indicates that it contains

C – 8.8%
S – 0.5%

Combustible Hydrogen, moisture and ash – 11.5%. The ash content of the tar is found to be four times its moisture content.

The flue gas orsat analysis for a given test run gave the following data:

CO_2 – 14.5%; CO – 1.5%; O_2 – 2.5% and N_2 – 81.5%.

Dry air inlet condition : 30°C and 1.1 bar.

Calculate the following:
(a) The combustible hydrogen, moisture and ash content of the tar.
(b) The kg of water produced per kg of dry flue gas produced.
(c) The kg of Sulfuric acid that could have been produced per hour presuming that all the Sulfur in the tar were converted into H_2SO_4.

(**Note:** In the flue gas orsat analysis, SO_2 is detected as CO_2).

[**Answer:**
(a) H – 6.98 g, Ash – 3.7 kg, Moisture – 0.92 g;
(b) 0.445 kg per kg of dry flue gas
(c) 7.65 kg of H_2SO_4]

3

Nuclear Fuels and Nuclear Power Generation

> *"In spite of the negative impact of the issues concerning nuclear safety, radioactive waste disposal, weapons proliferation potential, etc., nuclear power generation, whether one likes it or not, is going to be a primary, if not main, source of energy."*

The energy released during a nuclear reaction is called nuclear energy. The materials which make such energy available are called nuclear fuels. The energy released during induced nuclear fission and fusion reactions is several orders of magnitude greater than that obtained from chemical fuels because the forces involved in binding the nucleons in the nucleus are very high. Since the fossil fuel resources on the earth are finite and are likely to be exhausted sooner or later, attention has been focussed majorly on nuclear energy which seems to offer an infinite source of energy.

Future dependence on nuclear power in a big way seems to be an inevitable imperative because of strategic and economic considerations. The strategic considerations lie basically in the fact that a single large nuclear power plant can save over 50,000 barrels of oil per day. Such a power plant can recover its capital cost in a few years. The unit costs per kilowatt hour for nuclear energy are now comparable to the unit costs for coal in many parts of the world. Other advantages of nuclear power include the lack of environmental problems that are associated with coal-or oil-fired power plants and the near absence of problems related to mine safety, labour and transportation. Although natural gas is a relatively clean fuel, it is not available in many parts of the world and even if it is available, it has to be conserved for domestic and other industrial uses. The hydroelectric power generation though having many advantages is limited only to a few potential and feasible sources which has reached near saturation in many parts of the world. Solar power useful in domestic space, outer space and for water heating is at-themost considered as a supplementary energy source. Thus, in spite of the negative impact of the issues concerning nuclear safety, radioactive waste disposal, weapons proliferation potential, etc., nuclear power generation, whether one likes it or not, is going to be a primary, if not the main, source of energy.

3.1. NUCLEAR BINDING ENERGY

The masses of the various atomic nuclei were invariably found to be lesser than the sums of the masses of their corresponding constituent nucleons. This mass difference is referred to as "Mass Defect" of the corresponding nucleus. This mass is converted into energy which can be calculated from Einstein's mass-energy equation, $E = mc^2$. This energy which is released when the required number of protons and neutrons coalesce to form the nucleus is known as

NUCLEAR FUELS AND NUCLEAR POWER GENERATION

the Binding Energy which also indicates the energy needed to be supplied to disrupt the nucleus into the constituent protons and neutrons. The Binding Energy is usually expressed in terms of million electron volts (MeV), where 1 MeV = 1.602×10^{-6} ergs.

1 amu (Mass Defect) = 931 MeV (Nuclear Energy)

The Binding energies of nuclides increase progressively as their mass numbers increase and the rate of increase is somewhat less at higher mass numbers. Binding energy divided by the number of nucleons gives the Average Binding Energy per nucleon. Fig. 3.1 shows that the Average Binding Energy is relatively small for light elements. It rises to a maximum of about 9 MeV with isotopes of intermediate mass number such as $^{59}_{27}Co$ and then falls gradually to about 7.57 MeV for the heavy elements such as uranium.

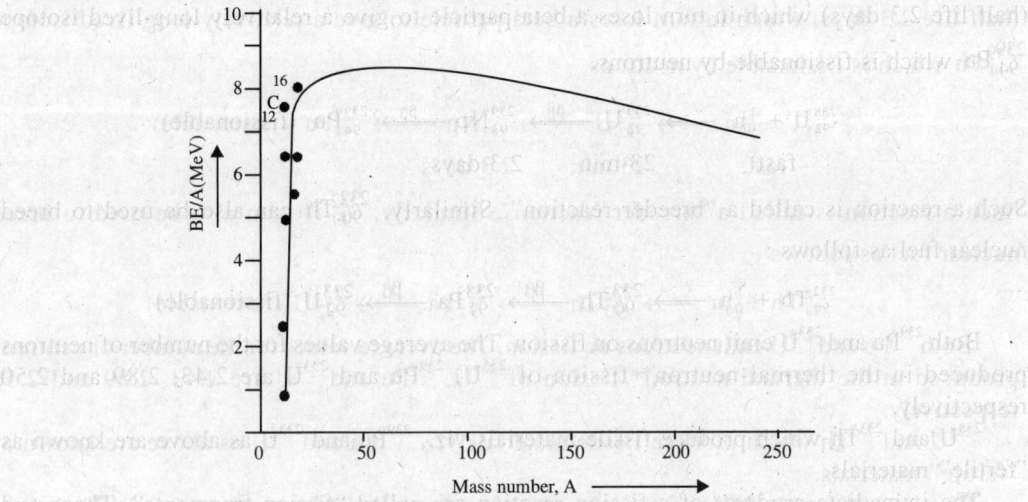

Fig. 3.1. Graph for Binding Energy/nucleon versus Mass number

Thus, if Binding Energy per nucleon is taken as a measure of the relative stabilities of nuclei, it is clear that elements of intermediate mass numbers such as $^{59}_{27}Co$ are more stable as compared to both the lighter elements like Hydrogen and heavier elements like $^{238}_{92}U$.

The process of spontaneous decomposition of a heavy nucleus into fragments of lighter nuclei is called *Nuclear Fission*.

Ex : $^{235}_{92}U + ^{1}_{0}n \longrightarrow [^{236}U] \longrightarrow ^{141}_{56}Ba + ^{92}_{36}Kr$

$+ 2$ to $3\ ^{1}_{0}n + 200$ MeV

The process in which two or more light nuclei fuse to give a heavier nucleus is called Nuclear Fusion.

Ex : $^{2}_{1}H + ^{3}_{1}H \longrightarrow ^{4}_{2}He + ^{1}_{0}n + 17.6$ MeV

$^{2}_{1}H + ^{2}_{1}n \longrightarrow ^{3}_{1}H + ^{1}_{1}H + 4.0$ MeV

Thus, it is obvious that tremendous amounts of energy will be liberated during Nuclear Fission and Nuclear Fusion reactions, which can be harnessed for power generation.

3.2. NUCLEAR FISSION

Nuclear fission can be brought about by bombarding the positively charged nuclei of certain isotopes of high mass numbers such as ^{235}U, ^{233}U or ^{239}Pu by neutrons when the heavy nucleus splits into two lighter elements called "fission fragments" accompanied by liberation of 2 or 3 fresh neutrons and energy.

Ex : $\quad ^{235}_{92}U + ^{1}_{0}n \longrightarrow ^{137}_{56}Ba + ^{97}_{36}Kr + 2^{1}_{0}n + 193.6$ Mev

Such a fission can be brought about by neutrons of high, moderate or low speeds without being repulsed.

Although nuclear fission can be brought about by other high speed particles, neutrons are the only practical projectiles that result in a sustained chain reaction. Only isotopes such as ^{235}U, ^{239}Pu and ^{233}U are fissionable by neutrons of all energies whereas isotopes like ^{238}U, ^{232}Th and ^{240}Pu are fissionable only by high-energy neutrons only. Although $^{238}_{92}U$ by itself is not a nuclear fuel, it can absorb fast neutrons to form $^{239}_{92}U$ which is an unstable radioactive isotope with a half-life of 23 minutes. This emits a beta particle resulting in the formation of $^{239}_{93}Np$ (half life 2.3 days) which in turn loses a beta particle to give a relatively long-lived isotope $^{239}_{94}Pu$ which is fissionable by neutrons.

$$^{238}_{92}U + ^{1}_{0}n \xrightarrow{\text{fast}} ^{239}_{92}U \xrightarrow[\text{23 min}]{\beta} ^{239}_{93}Np \xrightarrow[\text{2.3 days}]{\beta} ^{239}_{94}Pu \text{ (fissionable)}$$

Such a reaction is called a "breeder reaction". Similarly, $^{232}_{90}Th$ can also be used to breed nuclear fuel as follows :

$$^{232}_{90}Th + ^{1}_{0}n \longrightarrow ^{233}_{90}Th \xrightarrow{\beta} ^{233}_{91}Pa \xrightarrow{\beta} ^{233}_{92}U \text{ (fissionable)}$$

Both ^{239}Pu and ^{233}U emit neutrons on fission. The average values for the number of neutrons produced in the thermal-neutron* fission of ^{235}U, ^{239}Pu and ^{233}U are 2.43, 2.89 and 2.50 respectively.

^{238}U and ^{233}Th which produce fissile materials viz., ^{239}Pu and ^{233}U as above are known as "fertile" materials.

The immediate products of a fission reaction are called "fission fragments". These and their decay products together are called "fission products". The fission of a nucleus usually results in the production of radio-active fission fragments, charged particles, γ-rays and neutrons. Since a heavy nucleus having a low binding energy per nucleon is undergoing fission into lighter fission fragments each of which having a greater binding energy per nucleon, the fission reactions are accompanied by the liberation of energy.

A schematic representation of fission of ^{235}U and capture of neutrons by ^{238}U in a nuclear reactor is shown in Fig. 3.2.

The probability of occurrence of a given nuclear reaction (e.g., fission, scattering or absorption) depends on the cross-section of the target nucleus, which depends upon the property of the nucleus, type of the reaction and energy associated with the incident neutrons. *Nuclear cross-sections* (σ) are generally in the range of 10^{-22} cm^2 to 10^{-26} cm^2 and are expressed in barn (*b*) which corresponds to a value of 10^{-24} cm^2.

When a beam of projectile particles strikes a target nucleus, the different reactions taking place may involve different types of mechanisms. According to the Bohr's mechanism of *nuclear reaction*, the projectile particle enters the nucleus to form a *compound nucleus*. The compound nucleus possesses a high excitation energy (of the order of several MeV). In case a particular particle acquires enough kinetic energy to escape out, it would be emitted. Otherwise the excess energy is emitted in the form of γ-rays. The mode of disintegration of a compound nucleus depends upon factors such as its excitation energy, the heights of the potential barriers for

---

* Neutrons associated with energies in the range of about 0.025 eV which is the energy of the atoms and molecules of the medium at ordinary temperatures are called "thermal neutrons". Such neutrons which possess energies in the range of a few million electron volts are called "fast neutrons".

NUCLEAR FUELS AND NUCLEAR POWER GENERATION

different reactions and the stabilities of the residual nuclei but not on its mode of formation. In fact, a compound nucleus may be formed in different ways and may break down also in several different ways.

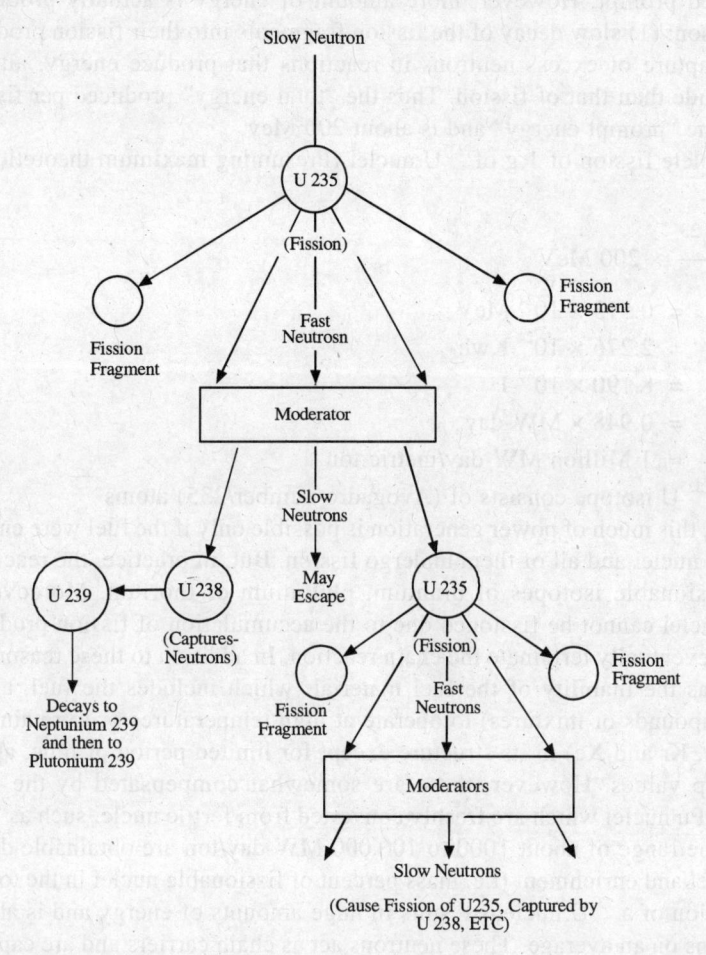

Fig. 3.2. A schematic representation of fission of ^{235}U and capture of neutrons by ^{238}U in a nuclear reactor.

Natural uranium is essentially a mixture of two isotopes, viz., ^{238}U (99.27%) + ^{235}U (0.72%), with traces of ^{234}U. The various isotopes of uranium can be separated by converting them to UF_6 followed by gas centrifugation or other separation techniques. Among these isotopes, only ^{235}U can undergo fission by slow or fast neutrons, the probability of fission being greater in case of slow neutrons. ^{238}U is fissionable by only fast neutrons and the fission probability in this case is low. Slow neutrons are absorbed by ^{238}U resulting only in non-fission (n, γ) reactions. When a neutron is absorbed by ^{235}U nucleus the latter undergoes fission producing two fission fragments and an average of 2.5 neutrons. A portion of the mass is converted to energy. The fission fragments formed are not always the same. For instance, ^{235}U nucleus can undergo 40 different modes of fission reactions by thermal neutrons. The following types of binary fission are most common :

$$^{235}_{92}U + ^{1}_{0}n \longrightarrow \left[^{236}_{92}U\right] \longrightarrow A^{85-105} + B^{130-150} +$$

compound nucleus   1 to 3 neutrons + energy

On the average, the fission of ^{235}U nucleus yields about 193 MeV. The fission of ^{233}U or ^{239}Pu also yields energy of the same order. This much of energy is released at the time of fission and hence is called prompt. However, more amount of energy is actually produced due to the following reason: (1) slow decay of the fission fragments into their fission products and (2) the non-fission capture of excess neutrons in reactions that produce energy, although of much lesser magnitude than that of fission. Thus the "total energy" produced per fission reaction is greater than the "prompt energy" and is about 200 Mev.

The complete fission of 1 g of ^{235}U nuclei (Presuming maximum theoretical burn-up) can thus produce :

$$\frac{6.023 \times 10^{23}}{235} \times 200 \text{ MeV}$$

$$= 0.513 \times 10^{24} \text{ MeV}$$
$$= 2.276 \times 10^{24} \text{ Kwh}$$
$$= 8.190 \times 10^{10} \text{ J}$$
$$= 0.948 \times \text{MW-day}$$
$$\approx 1 \text{ Million MW-day/metric ton}$$

Note : 1 g of ^{235}U isotope consists of (Avogadro number/235) atoms

However, this much of power generation is possible only if the fuel were entirely composed of fissionable nuclei and all of them undergo fission. But, in practice, the reactor fuel contains other non-fissionable isotopes of uranium, plutonium or thorium. Moreover, even all the fissionable nuclei cannot be fissioned due to the accumulation of fission products that absorb neutrons and eventually terminate the chain reaction. In addition to these reasons, metallurgical factors such as the inability of the fuel material (which includes the fuel, alloying or other chemical compounds or mixtures) to operate at high temperatures or to retain gaseous fission products (e.g., Kr and Xe) in its structure except for limited period of time, also contribute to lower burn-up values. However, these are somewhat compensated by the fission of some fissionable ^{239}Pu nuclei which are freshly converted from fertile nuclei such as ^{238}U. In practice, burn-ups in the range of about 1000 to 100,000 MW-day/ton are obtainable depending on the type of the fuel and enrichment (i.e. mass percent of fissionable nuclei in the total reactor fuel).

Each fission of a ^{235}U nucleus results in huge amounts of energy and is also accompanied by 2.5 neutrons on an average. These neutrons act as chain carriers and are capable of bringing about a sustaining chain reaction. If this fission chain reaction proceeds uninterrupted, a tremendous amount of energy will be released in a fraction of a second, which can lead to a nuclear explosion. If this chain reaction is controlled, it can be used to produce power in a device called a "nuclear reactor".

3.3. Conditions for maintaining a sustaining chain reaction

The development of a sustaining nuclear chain reaction will be hindered if the neutrons produced in the fission reaction are lost by the following:

(a) escape of neutrons from the lump of the active material without causing fission, and

(b) absorption of neutrons by non-fissionable atoms (e.g., 238U atoms) or by other structural materials employed.

These two factors can be taken care of by (a) having sufficient concentration of ^{235}U or ^{239}Pu to provide a good neutron flux and (b) bringing together a sufficient quantity of the enriched fuel so that only a small portion of the neutrons escape and a multiplying chain reaction occurs to release tremendous amount of energy instantaneously. This quantity of fissionable material required to initiate the chain reaction is called "critical mass". The critical mass increases as the percentage of the fissile material in the fuel decreases.

NUCLEAR FUELS AND NUCLEAR POWER GENERATION

Maintaining a sustaining chain reaction and at the same time managing it without going out of control is the main factor in nuclear power generation.

3.4. Nuclear Power Reactors

For producing power from nuclear fission, the rate of the chain reaction must be controlled in such a way that on an average, one neutron created in a fission process is capable of activating another nucleus to cause fission and so on. This condition is defined by the multiplication factor, K :

$$K = \frac{\text{Number of neutrons in second generation}}{\text{Number of neutrons in first generation}}$$

If K > 1, the number of fission events increases for each successive neutron generation with a resultant increase in the power production of the reactor. If K = 1, the number of fissions per unit time (and thus the energy production) is constant. Hence, a straight but continuous energy chain sustains. If K < 1, the chain reaction cannot be maintained.

A reactor operating at K = 1 is said to be critical; that operating at K < 1 is said to be subcritical and that operating at K > 1 is called super critical. In supercritical operation, multiple branching energy chains will be generated as in the case of an atom-bomb. In a nuclear reactor, the multiplication factor, K has to be carefully regulated by inserting control rods made up of Cd, Hf or B which absorb neutrons. If the control rods are withdrawn, the neutron flux increases. Dropping of the entire set of control rods into their tubes in the reactor causes shut down of the reactor.

3.5. Reactor Concepts

Nuclear reactors are used for production of heat, mechanical and electrical power, radioactive isotopes, weapons material and nuclear research. Accordingly, the design of the nuclear reactor depends on the purpose for which it is going to be used. If the heat energy generated in the fission is utilized for generation of electricity, the reactor is known as a "power reactor". There is enough flexibility in choices of design for a controlled nuclear chain reaction and each concept has its own merits and drawbacks. The design considerations for a Reactor System may also be influenced by the cost and availability of the raw materials and skilled manpower, safety considerations and the risk for proliferation of reactor materials for weapons. Several types of full-scale or prototype nuclear power reactor concepts have been developed and operated.

There are three basic principles for reactor design :

(1) according to neutron energy (*e.g.*, thermal or fast reactors), (2) according to core configuration (homogeneous or heterogeneous reactors), and (3) according to fuel utilization (*e.g.*, burner, converter or breeder type reactors).

In a thermal reactor, the fission neutrons are slowed down by a "moderator" whereas in a fast reactor, the fission is caused by fast neutrons and there is no need to bring their energy down by a moderator. In the homogeneous reactors, the core may comprise of a molten metal, molten salt, an aqueous or an organic solution. In heterogeneous reactors, the fuel is mostly in the form of rods filled with metal oxide. The fuel material may be almost any combination of fissile and fertile atoms in a mixture or separated as in the core (fissile) and blanket (fertile) concept. The moderator can be H_2O, D_2O, graphite, beryllium or an organic solvent. The coolant can be molten metal, molten salt, liquid H_2O, D_2O or organic solvent as well as gaseous CO_2, helium or even steam.

The following are the most common reactor types :

(1) AGR — Advanced gas-cooled graphite moderated reactor
(2) BHWR — Boiling heavy-water-cooled and moderated reactor
(3) BWR — Boiling light-water-cooled and-moderated reactor

(4) FBR — Fast breeder reactor
(5) GCFBR — Gas-cooled fast breeder reactor
(6) GCR — Gas-cooled graphite-moderated reactor
(7) HTGR — High temperature gas-cooled graphite-moderated reactor.
(8) HWGCR — Heavy water moderated gas-cooled reactor.
(9) LWGR — Light-water-cooled graphite-moderated reactor
(10) HWLWR — Heavy-water-moderated light-water cooled reactor
(11) LMFBR — Liquid metal-cooled fast breeder reactor
(12) OMR — Organic-moderated and-cooled reactor.
(13) PHWR — Pressurized heavy-water-moderated and-cooled reactor.
(14) PWR — Pressurized light-water-moderated and-cooled reactor.
(15) SGR — Sodium-cooled graphite-moderated reactor.

In any nuclear reactor, the heat generated by the controlled fission chain reaction of the nuclear fuel confined in the core of the reactor is transferred to a coolant and this heat energy is used to generate high pressure steam that turns a turbine connected to a generator, thus producing electric power.

Sustained (critical) chain reaction is not possible in a mass of natural uranium or in a homogeneous mixture of natural uranium and moderator. In fact, one can store natural uranium plates in contact with each other to any desired height without fear of a critical reaction. In order to obtain criticality or steady power, the following three methods are often used.

(i) building a heterogeneous reactor
(ii) enriching the fuel by artificially increasing the percentage of ^{235}U or other fissionable material in it, and
(iii) both the above.

3.6. Components of a Nuclear Power Reactor

A typical assembly of a nuclear power reactor indicating the various important components is represented in Fig. 3.3.

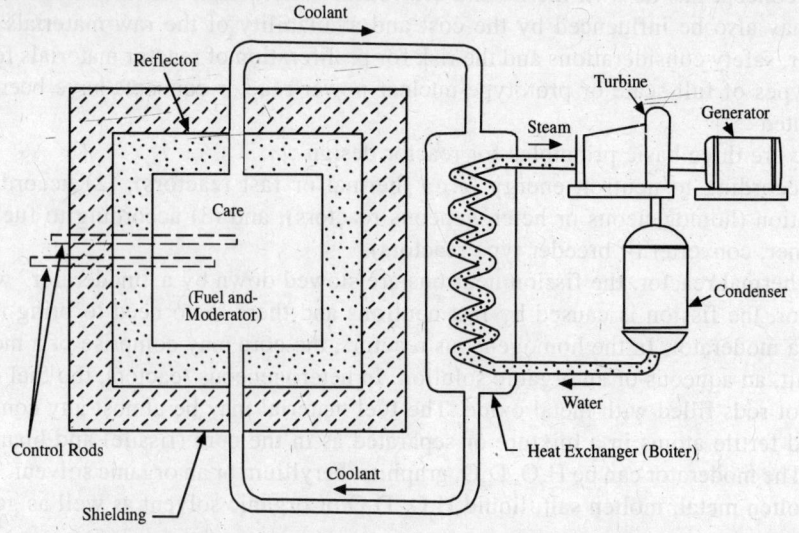

Fig. 3.3. Schematic diagram of a nuclear reactor system for generating electric power.

1. Reactor Core

This is the principal component of any nuclear reactor. It is in this part where the fissionable

NUCLEAR FUELS AND NUCLEAR POWER GENERATION

fuel material is located and where the heat energy is liberated due to the fission chain reaction. The reactor core comprises of an assembly of fuel elements, moderator, coolant and control rods.

(a) Fuel elements

The material containing the fissile isotope is called the reactor fuel or nuclear fuel. Its composition can vary from natural uranium to material highly enriched in ^{235}U, ^{239}Pu or ^{233}U. The basic source materials for nuclear fuels are uranium and thorium. ^{235}U present to the extent of about 0.71 percent by weight in the natural uranium is fissile and this is the only natural material from which nuclear energy can be directly produced.

The transformation of ^{238}U into ^{239}Pu; and the conversion of ^{232}Th into fissile ^{233}U has already been described earlier. Although ^{238}U and ^{232}Th will undergo fission by fast neutrons (> 1 MeV), it is not possible to sustain a chain reaction in these isotopes. The only substances that can be used for this purpose are the fissile materials ^{235}U, ^{233}U and ^{239}Pu since they can suffer fission due to capture of neutrons of all energies. Since ^{238}U and ^{232}Th can be respectively converted into the fissile species ^{239}Pu and ^{233}U, they are called fertile materials.

In all commercial reactors, the nuclear fuel used is one of the following:
(a) Natural uranium which contains 0.71% of fissile ^{235}U isotope.
(b) enriched uranium in which ^{235}U has been increased to a few percent.
(c) ^{232}Th in which fissile ^{233}U is bred.
(d) ^{239}Pu mixed with ^{238}U as a supplement to the ^{235}U (mixed fuel).

The fuel element comprises of the fissile and fertile material enclosed in the cladding and is usually in the form of a thin rod. The fuel must be charged in the form of regular sized elements arranged in the coolant channels in a particular geometry in order to achieve (a) proper utilisation of the fissionable material, (b) efficient removal of heat generated during the fission process and (c) uniform and dependable reactor performance.

The nuclear fuel material should satisfy the following criteria :
(i) It should be free from impurities, particularly those having high neutron absorption cross-section.
(ii) It should be resistant to radiation damage.
(iii) It should possess suitable physical and mechanical properties to enable easy and economic fabrication.
(iv) It should possess high melting point and high thermal conductivity.
(v) It should have chemical stability, particularly to the coolant, to ensure minimum interaction in case of cladding failure.

In most reactors, solid fuel elements made up of a uranium metal or an alloy, UO_2, UC or UC_2 are used. The fuel material is usually clad with a metal jacket which serves the following purposes :
(i) To protect the fuel material from chemical reaction with air, water, or other material used as coolant.
(ii) Preventing the escape of fission products and protecting the coolant from fission product contamination.
(iii) Ensuring geometrical integrity of the fuel element.

The cladding material should possess good mechanical and corrosion resistance properties, high temperature stability and low neutron absorption cross-section. Only a few materials available at reasonable cost satisfy these requirements. These are Al, Mg, Zr (or their alloys), and stainless steel. Al or Mg is preferred as cladding material for moderate temperatures whereas stainless steel or Zircaloy (a corrosion resistant alloy of Zr and Al) are used for high temperatures existing in power reactors. Zircaloy seems to be the best for thermal reactors using ordinary or heavy water as moderator or coolant. However in fast reactors, stainless steel is suitable as cladding material, since parasitic neutron capture is not important in that case. In UK, an alloy of magnesium, called Magnox has been widely used as cladding in large thermal reactors using CO_2 as coolant.

(b) Moderators

The function of moderator in a thermal reactor is to rapidly reduce (within a fraction of a second) the high energy of the fission neutrons. A good moderator should satisfy the following requirements :

(1) It should have a high slow-down-power(SDP), *i.e.*, it should reduce the energy of the fast neutrons from 1 or 2 MeV to about 0.025 eV. Such a decrease in neutron energy per collision is possible only with low mass number-elements.
(2) It should possess a low neutron-absorption-cross-section and large neutron-scattering cross-section.
(3) It should be cheap and abundantly available in pure form.
(4) It should have good chemical, thermal and radiation stability.
(5) It should be corrosion resistant, if it is a solid.
(6) It should have high melting point, if it is a solid.

The commonly used moderators are water, heavy water and graphite while beryllium and some organic compounds have also been considered. Owing to its extremely low neutron-absorption cross-section, heavy water is considered to be the best moderator. It is used in many reactors, particularly those employing natural uranium as fuel.

Natural water contains about 0.015% of heavy water (D_2O). The various methods used for the production of heavy water include (*i*) distillation of natural water, (*ii*) distillation of hydrogen, (*iii*) electrolysis of water and (*iv*) chemical exchange between H_2S-H_2O or H_2-H_2O or H_2-NH_3.

Graphite is considered as a good moderator because of its cheapness, availability in pure state, good thermal conductivity, high temperature resistance, non-toxic properties and easy machinability into desired shapes. However, the nuclear properties of graphite are inferior to those of heavy water or beryllium. It has a relatively low impact strength. It reacts with O_2 and CO_2 at high temperatures. The porous nature of graphite, reaction with some metals to form their carbides and dimensional instability under radiation are some of its limitations under normal operating conditions.

Beryllium is also considered as a good moderator and reflector due to its high temperature stability and stability to radiation and chemical action. However, its brittleness and susceptibility to air and water at high temperatures, its high cost of production and its toxicity restrict its use only in special cases.

However, on considerations of price and other properties, H_2O finds favour in many commercial reactors.

(c) Coolants

The heat generated in the core of the reactor is removed by a coolant. In case of a power reactor, the heat from the coolant is exchanged to a working fluid to produce steam or a hot gas which is then used in a conventional turbo-generator system. In some reactors, water is boiled within the core of the reactor to directly generate steam. An ideal reactor coolant should possess the following properties :

(*i*) High specific heat capacity
(*ii*) High thermal conductivity
(*iii*) Stability to heat and radiation
(*iv*) Low neutron-absorption-cross-section
(*v*) Low cost and low power requirement for pumping

(vi) Non-corrosive and non-toxic
(vii) High boiling point and low melting point in case of liquid coolants.

The materials used as coolants may be divided into four broad categories:

(i) ordinary water and heavy water (ii) liquid metals
(iii) organic liquids (iv) gases

Both ordinary and heavy water are considered to be good coolants for thermal reactors because of their excellent heat exchanging properties. They have also the advantage of serving as both coolants as well as moderators. However, due to their low boiling point, high pressures are required to prevent their boiling in the reactors operating at high temperatures. Further, since water is corrosive at high temperatures, special claddings such as zircaloy or stainless steel are needed for the fuel elements.

Liquid metals such as Na and K are recommended as coolants for use at high temperatures in fast reactors. Liquid sodium has excellent heat removing properties. It does not require pressurization even at high temperatures and its rate of attack on zircaloy and stainless steel claddings is believed to be small and tolerable. However, Na is very reactive with oxygen (in air) and water and the liquid metal solidifies at well above ordinary temperatures. Moreover when passed through the reactor, it is converted into radioactive ^{24}Na (due to the capture of neutrons) with a half-life of 15 hrs. It decays with the emission of β particles and γ-rays which poses a potential hazard for the operating personnel and hence calls for a special protecting shield for the sodium system outside the reactor.

As a compromise between water (which requires the use of high pressures) and liquid sodium (which is chemically reactive and potentially hazardous), organic compounds such as benzene and polyphenyls (e.g. diphenyl and terphenyl) were considered. Although their heat removing properties are inferior to those of water and liquid sodium, they have the following advantages :

(i) They do not require pressurization
(ii) They are not corrosive
(iii) They are fairly stable to heat and radiation.

Gaseous coolants such as air, CO_2, CH_4, H_2 and He were also considered. Several earlier research reactors and low-power natural uranium-graphite reactors used air as coolant. For high-power reactors where high temperatures are confronted with, air is not satisfactory as coolant because of its chemical reactivity with the reactor materials such as Zr, Mg, Al, Be and graphite. CO_2 is the most widely used gaseous coolant in large power reactors. It is relatively inexpensive, does not attack metals even at fairly high temperatures (upto about 500°C) and the neutron absorption cross-sections of its constituent atoms viz., C and O are small. For operating temperatures above 600°C , H_2 would be the best gaseous coolant from theoretical considerations. However, since it forms an explosive mixture with air and also since it renders many metals brittle, gaseous H_2 is considered as hazardous. The next choice is He gas, since it is stable, non-reactive and has negligibly small neutron-capture-cross-section. However, He is very expensive and special precautions are to be taken to minimize losses due to leakage. He is used as a coolant in some experimental power reactors. A gaseous coolant must be circulated under a pressure of many atmospheres for efficient heat removal and for lowering the pumping costs.

(d) Control rods

The control of power level in a reactor can be achieved by controlling the neutron flux which in turn is achieved by altering the neutron leakage rate or the quantity of fuel in the reactor core or by incorporating a suitable neutron-absorbing material. Control in thermal reactors is invariably achieved by means of a neutron absorbing material which does not become radioactive as a result of neutron capture (otherwise, the radiations can be hazardous during repair and maintenance works). In earlier reactors cadmium was extensively used as a control material due to its availability and ease of fabrication. However, due to its relatively low melting point, cadmium could be used only in reactors operating at low temperatures. Further, the neutron absorption cross-section of Cd is not large in the energy region above 5 eV, where the neutron density is high in water-moderated reactors. In such reactors, an alloy made of (80% Ag + 5% Cd + 15% In) has been used because of its relatively higher melting point and larger neutron-capture cross-section over wider energy range than that of Cd. Control material made up of Hf would have been preferred for water-moderated reactor but for its high cost.

Boron is the most common material used for reactor control due to its very high melting point and very large neutron-absorption cross-section over a wide energy range.

It is commonly employed in the form of boron steel (boron incorporated in stainless steel) or Boral (boron carbide, B_4C dispersed in Al) which have good corrosion-resistance.

The control elements of a reactor are often referred to as control rods (which may be in the form of rods or plates). These are commonly located in the core. However, in some reactors, it was considered to be more convenient to have them in the reflector close to the core.

The control rods are meant to serve the following purposes :
   (i) For increasing the neutron flux (i.e. to increase the fission rate) in a thermal reactor, the control rods are lifted up; whereas they are pushed in to decrease the neutron flux i.e., to decrease the fission rate.
   (ii) For start-up or for shutting off the reactor under routine or emergency conditions.
   (iii) For maintaining the reactor at a steady state of power production.
   (iv) For maintaining the nuclear chain reaction under perfect control to avoid explosion or damage to the reactor.

Generally, a large number of control rods are used in such a way that they are disposed throughout the reactor core so as to maintain a uniform distribution of the neutron flux.

Neutron-absorbing materials are not much useful for the control of fast reactors because of their small cross-sections. In such cases, control is achieved by adjusting the escape of neutrons by moving either part of the reflector or blanket or some of the fuel elements. Owing to the escape of fast neutrons, the fission rate will be decreased.

2. Reflector

The reactor core is usually surrounded by a neutron reflector which reflects back many of the neutrons that leak out from the core. The requisites of a good reflector obviously are low neutron absorption cross-section, high neutron scattering cross-section, high radiation stability and high oxidation resistance. The reflector helps to decrease the critical mass (i.e. the minimum quantity of the fissile material required to maintain the chain reaction after it has been initiated) and to increase the average power output for a given quantity of the fuel. In thermal reactors, the same material used as moderator (viz., D_2O, graphite, Be or H_2O) can also be used in the reflector. However, in a fast reactor, the use of moderator is omitted and a reflector comprising of a dense high-mass numbered element is employed.

3. Pressure Vessel

The core and the reflector are enclosed in a pressure vessel which is designed to withstand pressures of the order of 200 kg/cm^2.

4. Structural Materials

Important structural components of the reactor include supports for fuel and moderator, pipes, valves, fittings, reactor shell, baffle plates and sleeves for control rods. A good structural material should be strong and eminable for easy fabrication, thermal and radiation stability, high corrosion resistance and low neutron absorption cross-section. The commonly used structural materials are Al, Mg, Zircaloy and stainless steel.

NUCLEAR FUELS AND NUCLEAR POWER GENERATION

5. Shielding

It is an important component of a reactor installation although it has nothing to do with the operation of the reactor. Shielding is provided for attenuation of the gamma rays and neutrons emerging from the reactor which helps to protect the operating personnel from radiation hazard. In high pressure reactors, two shields are usually provided.

(*i*) *the thermal shield* is fairly close to the core and comprises of a few inches of iron or steel. It absorbs much of the gamma radiation and protects the biological shield from possible damage due to overheating. It is located in such a way that it can be readily cooled by circulation of water.

(*ii*) *the biological shield* comprises of a layer of concrete (several feet thick) which surrounds the reactor core, reflector, and the thermal shield. It absorbs both gamma rays and neutrons.

As a precaution against possible leakage of radioactive materials in the unlikely event of an accidental damage to the reactor core, the entire reactor system including the shield and heat exchanger is usually enclosed in a steel containment.

3.7. BREEDER REACTORS

Nuclear reactors can be designed in such a way that they can generate their own fuel by converting fertile material into fissile material. Such reactors are called regenerative reactors or converters. Plutonium production reactors belong to this category. In these reactors, U-235 helps in maintaining the fission chain to generate energy but some of the neutrons are captured by the fertile U-238 with the ultimate formation of Pu-239. Thus, the U-235 consumed by fission is replaced partially by another fissile material, Pu-239.

Since U-235 is the only fissile material in nature, this will have to be used in the nuclear reactors for some more time to come. Since there is no known method to regenerate U-235, it may be assumed that this material will be eventually exhausted. At this stage, it may be assumed that an approximately equivalent quantity of Pu-239 (or U-233) might have been generated by conversion of fertile material. Since natural uranium contains about 140 times more U-238 than U-235, considerable quantities of U-238 will still remain. Considerable quantities of Th-232 will also still remain. Therefore, the long-term future of nuclear energy depends on the efficiency of converter reactors in which Pu-239 (or U-233) is consumed as fuel and, at the same time, is also regenerated from U-238 (or Th-232). The following three general cases may be considered in this context :

(*i*) If for every fissile nucleus consumed, less than one neutron is available for capture by the fertile material, the quantity of the fissile material regenerated will be less than what is consumed. Then the stockpile of the fissile material will steadily decrease until exhaustion. Then there will not be a long-term future for fission energy as a source of power.

(*ii*) If one fissile nucleus is regenerated for every one consumed in the reactor, the stockpile of the fissile material will then remain constant. In that case, the maximum rate of energy production will depend on the available fissile material and no increase will be possible. Thus, although the entire available fertile material would be converted into fissile species, no further expansion of the nuclear power industry will be possible.

(*iii*) If for every Pu-239 nucleus undergoing fission, more than one neutron on the average were captured by U-238 to regenerate Pu-239, then it is possible to design a reactor in which more fissile material (*i.e.* Pu-239) was formed than what was used up in the operation. Such reactors are called *breeder reactors* and the regeneration process is called *breeding*. Similarly, it is possible to conceive design of a reactor with U-233 as the fissile material in which for every one of these nuclei undergoing fission, more than one nuclei would be regenerated by the capture of neutrons in Th-232.

With the help of breeding, the stockpile of fissile material could be steadily increased and so also the rate of power production. However, this would not go on indefinitely because a time would come in the ultimate future when all the fertile material was consumed.

The prospects for achieving breeder reactors naturally depend upon the number of fission neutrons liberated for each neutron absorbed by the fission material. One neutron is needed to perpetuate the fission chain and in addition, more than one on an average is required to make breeding possible. Hence, apart from the neutrons lost by leakage or parasitic capture, somewhat more than two fission neutrons must be available for each neutron absorbed. In other words, breeding is possible only if the fission factor, n > 2. On the basis of known facts, breeding of Pu-239 is feasible only in fast reactors whereas breeding of U-233 from Th-232 is favoured in thermal reactors. Since the accessible resources of U-238 are more abundant than those of Th-232, the fast breeder reactors employing Pu-239 as fuel and U-238 as fertile materials are expected to play a vital role in the future development of nuclear power.

The fertile material may be located in different ways with respect to the core of the breeder reactor. As a general rule 75 to 80% of the fertile material is included in the fuel element. In addition, the reactor core is surrounded by a blanket of the fertile material in order to capture the neutrons escaping from the core. Such a blanket also serves the purpose of a reflector for a fast reactor.

In the thermal breeder reactor, ^{233}U is produced from ^{232}Th. The reactor may be designed either with the reactor containing a mixture of ^{232}Th and ^{233}U, or with a central core of ^{233}U surrounded by an outer blanket of ^{232}Th. Several other modifications have also been experimented.

$$^{232}\text{Th}(n, \gamma) \, ^{233}\text{Th} \xrightarrow{\beta^-} \, ^{233}\text{Pa} \xrightarrow{\beta^-} \, ^{233}\text{U} \quad (t^{1/2} = 1.62 \times 10^5 \text{ y})$$

In the fast breeder reactor, fission occurs in central core of plutonium which is surrounded by an outer blanket of ^{238}U where the neutrons are captured to form new ^{239}Pu. This blanket is surrounded by a reflector, usually of iron.

$$^{238}\text{U}(n, \gamma) \, ^{239}\text{Th} \xrightarrow{\beta^-} \, ^{239}\text{Np} \xrightarrow{\beta^-} \, ^{239}\text{Pu}$$

$$(t^{1/2} = 23 \text{ min}) \quad (t^{1/2} = 2.3 \text{ d}) \quad (t^{1/2} = 24,400 \text{ y})$$

Some of the breeder reactors under investigation include (1) Light-water breeder reactor (LWBR), (2) Liquid metal fast breeder reactor (LMFBR), (3) Gas cooled fast breeder reactor (GCFBR) and (4) Molten salt breeder reactor (MSBR).

The essential features of a gas-cooled fast breeder reactor includes the reactor core, helium circulators, steam turbine and generator. The fuel used is (UO_2 + PuO_2) and the coolant is helium gas at 8.6 MN/m^2. The coolant enters the reactor core at 315°C and exits at 650°C. After generating steam in steam generators, it is recirculated back by axial flow compressors.

Fast breeder reactor concept extends the fission energy resources by about a factor of 100. Hence uranium may be considered as the largest energy resource presently available on earth.

3.8. NUCLEAR POWER STATIONS IN INDIA

(1) Nuclear Power Station at Tarapur (Maharashtra) which started production of electricity in 1969. Its installed capacity is 420 MW. It contains two boiling water reactors which use enriched uranium oxide (in zircaloy cladding) as fuel and light water as coolant as well as moderator.

(2) Rajasthan Atomic Power Station at Ranapratap Sagar in Kotah district (Rajasthan) which has two reactors of 220 MW each which were installed in 1972 and 1980 respectively. These reactors use natural uranium in the form of UO_2 (in zircaloy cladding) as fuel and heavy water as moderator as well as coolant.

NUCLEAR FUELS AND NUCLEAR POWER GENERATION 191

(3) Madras Atomic Power Station at Kalpakkam near Madras (Tamil Nadu). It has two reactors of 235 MW each which started production in 1984. These reactors use natural uranium (in the form of UO_2) as fuel and heavy water as moderator as well as coolant.

(4) Narora Atomic Power Project at Narora (U.P.) which has two reactors of 235 MW each and are similar in design to those at Kalpakkam.

(5) Kakrapara Atomic Power Project at Kakrapara (Gujrat) which has two reactors of 235 MW each and are of similar design as those at Kalpakkam.

(6) Koodamkulam Nuclear Project (Tamilnadu) with two reactors of 1000 MW each with Russian Collaboration.

3.9. ENVIRONMENTAL ASPECTS OF NUCLEAR POWER GENERATION

Besides thermal pollution which is common with almost all types of power generation, the environmental effects of nuclear power generation mainly arise from:

(*i*) the nuclear fuel cycle,
(*ii*) low-level dose radiations from nuclear-power-plant effluents,
(*iii*) low-and high-level dose radiations from wastes.

The nuclear fuel cycle include mining of uranium ore, milling and refining the ore to produce uranium concentrates (U_3O_8), processing to produce UF_6, isotopic enrichment of UF_6 by gaseous diffusion process, fabrication of the reactor fuel elements, power generation in the reactor giving rise to irradiated or spent fuel, short term storage of the spent fuel, reprocessing of the spent fuel and waste management.

Low-level dose radiations from nuclear-power plant-effluents are mainly gases and liquids. Environmental concerns about nuclear power plants are mainly due to the effects of these radiations on the people living near the plants. Tritium, (^3T or $_1^3H$) is the major radioactive gaseous effluent. The major liquid effluent is the coolant contaminated with various radioactive isotopes.

High-level radioactive wastes result from the spent fuel. Owing to their intense exothermic activity, the high level wastes generate too much of heat. Hence they have to be cooled by air circulation or other means, possibly for decades, before they can be permanently stored. Alternatively, the spent fuel can be reprocessed for recovering the unused uranium, converted plutonium and other radio-isotopes for use in a variety of applications such as isotopic generators, agriculture, medicine, research and industry. During this reprocessing of the spent fuel, radioactive gaseous wastes (*e.g.*, ^3T, ^{14}C, ^{85}Kr and ^{129}I), liquid wastes (high, medium and low-level), solid wastes (high, medium and low-level) and organic liquid wastes are generated.

Dilute liquid wastes may be treated by processes such as chemical precipitation, ion-exchange and evaporation to obtain a concentrated stream for storage or confinement and a dilute stream for dispersal into the environment.

High-level liquid wastes are stored at site, in liquid form, in underground tanks of suitable material (specially designed high integrity stainless steels for high-level acidic wastes and specially designed carbon steel tanks for the alkaline intermediate level wastes) fitted with cooling coils to remove the decay heat. The tanks are built in concrete vaults and are provided with a secondary confinement. Systems for condensing vapours, ventilation and critical surveillance are also installed. Alternatively, the liquid wastes are also converted into solid form by processes such as calcination and vitrification to achieve volume reduction, chemical stability and low leaching rate of the material.

The current approaches for management of high-level and intermediate-level radioactive wastes are (*i*) immobilization and (*ii*) isolation. "Immobilization" comprises of "fixing" the radioactive pollutants in the wastes on solid matrices of proven stability. This retards the migration of the radioactivity into the sub-soil environment surrounding the spot where the waste is buried. High-level acidic waste concentrates are usually immobilised in borosilicate glass matrix or polycrystalline "synroc"; whereas intermediate level alkaline waste concentrates are immobilized in bitumen and concrete.

"Isolation" involves retaining the solidified wastes by engineered barriers. The bitumen solidified waste products, collected in steel drums, are disposed into tile holes which are shallow land burial sites. However, for the disposal of the waste canisters containing the vitrified wastes, a two stage approach is favoured : (1) Retrievable interim storage in sub-terranean caverns, and (2) ultimate disposal in selected geological formations, called "repositories".

Although all possible efforts are made to isolate high level wastes from the biological system, it is essential to study their deleterious effects on the body cells in the vicinity of the irradiated region. The effects are classified as "somatic" and "genetic". "Somatic effects" are caused on the exposed individuals and the cell damage caused may manifest in some form of malignancy, such as leukaemia or cancer.

Genetic effects are transmitted to the descendants of exposed individuals and thus can effect unexposed generations. The radiation-induced changes in the genes may manifest themselves in (a) gene mutations, (b) chromosome aberrations and (c) changes in the number of chromosomes. Such changes can result in the offspring abnormalities ranging from mild to lethal.

3.10. ENERGY FROM NUCLEAR FUSION

As stated earlier, the binding energy curve indicates that the binding energy per nucleon in case of lightest nuclei and heaviest nuclei is lesser than that for nuclei of intermediate mass number. Therefore, the combination of two or more of the lightest nuclei by a process of fusion should, like fission, result in liberation of energy.

Ex :
$$^2_1H + ^3_1H \longrightarrow ^4_2He + ^1_0n + 17.6 \text{ MeV}$$
$$^2_1H + ^2_1H \longrightarrow ^3_1H + ^1_1H + 4 \text{ MeV}$$

Since the percentage efficiency accompanying the process of nuclear fusion is greater than that in the case of nuclear fission, a greater percentage of mass is converted into energy in the case of fusion. For instance, the energy obtainable per unit mass of the fusion of deuterium and tritium nuclei is about four times than that from the fission of U-235 and about 10 million times than that from petrol. However, a fusion reaction can occur only when the two lighter nuclei participating in the fusion are brought near each other within a distance of one fermi when the nuclear attractive forces can overcome the coulombic repulsive forces between the positively charged nuclei. Lighter nuclei such as ^1_1H, ^2_1H, ^3_1H and ^4_2He are preferred for fusion reactions because the electrostatic repulsion increases with the mass number of the participating nuclei. The kinetic energy of the interacting nuclei must be increased to very high value to bring about nuclear fusion. This can be done either by thermal means or by particle acceleration. The thermal methods are generally preferred. For bringing about a fusion reaction, temperatures of the order of 10^6 °C will be required. For example, the temperature required for fusion of ^2_1d and ^3_1T is about 50×10^6 °C. Such reactions taking place at very high temperatures are called "**thermonuclear reactions**".

H.A. Bethe of United States suggested in the year 1938 that most of the energy of the sun and the stars is due to thermonuclear reactions involving isotopes of hydrogen. To explain this, he suggested two sets of thermonuclear reaction sequences. The first is known as the "**carbon cycle**" in which carbon acts as catalyst to facilitate the combination of four protons to form a helium nucleus.

$$4^1_1H \longrightarrow ^4_2He + 2\,^0_{+1}e + \text{energy}$$

The second is known as the proton-proton chain in which two protons combine to form a deuterium nucleus which then combines with another proton to yield helium-3 as follows :

$$^1_1H + ^1_1H \longrightarrow ^2_1D + ^0_{+1}e + \text{energy}$$

$$^2_1D + ^1_1H \longrightarrow ^3_2He + \text{energy}$$

Two helium-3 nuclei then interact as follows :

$$^3_2He + ^3_2He \longrightarrow ^4_2He + 2^1_1H + \text{energy}$$

The net result in both the above reaction sequences is the formation of a helium-4 nucleus and the energy released also is the same, namely 26.7 MeV for each helium atom produced. Although the above reaction sequences seem to be the main sources of energy in stars, other nuclear reactions accompanying the release of energy might also be involved particularly in older or hotter stars.

The thermonuclear reactions of practical use on the earth seems to be those involving the lightest nuclei, viz., isotopes of hydrogen. The following thermonuclear reactions appear to offer prospects of success :

$$^2_1D + ^2_1D \longrightarrow ^3_2He + ^1_0n + 3.2 \text{ MeV}$$

$$^2_1D + ^2_1D \longrightarrow ^3_1T + ^1_1H + 4.0 \text{ MeV}$$

$$^3_1T + ^2_1D \longrightarrow ^4_2He + ^1_0n + 17.6 \text{ MeV}$$

If suitable conditions could be designed for achieving these reactions, the deuterium present in all natural waters on this planet (one atom of deuterium is present with about 6500 atoms of hydrogen) would represent an almost inexhaustible source of energy. The energy equivalent of the deuterium in one gallon of water is calculated to be the same as that obtainable from the combustion of 300 gallons of gasoline. Thus, with 10^{20} gallons of estimated water in the oceans, thermonuclear fusion of the available deuterium nuclei could meet the power requirements on the earth for several million years.

3.11. CONTROLLED THERMONUCLEAR REACTORS

Since controlled thermonuclear reactors (CTR) have a potential to meet the energy demands of the world for millions of years to come, considerable research effort has been directed during the last four decades to achieve this objective. Controlled fusion reactions involving atoms of deuterium and tritium (^2_1D and ^3_1T) have been receiving greater attention because these fusion reactions are relatively easier to achieve due to the low coulomb barrier and favourable wave mechanical transmission factor. The D-T reactions and D-D reactions are believed to be most important in this context.

Physical Principles

For a controlled fusion reaction to provide useful energy, the process must be self-sustaining. That means, when once the required quantity of heat has been supplied to raise the temperature of deuterium (or a mixture of deuterium and tritium) to the point at which nuclear fusion can occur at an appreciable rate, the energy released must be sufficient at least to maintain that temperature. In calculating the minimum temperature at which a particular thermonuclear reaction becomes self-sustaining, a balance must be struck between the energy released in the fusion reactions and the energy lost, mostly due to escaping radiation. Both these quantities increase with increase of temperature, but the release of energy increases more rapidly than the loss of energy due to radiation; hence at a certain temperature, known as critical or ideal ignition temperatue, the fusion reaction should be self-sustaining. Calculations have shown that the critical energy for deuterium-tritium system is about 5 KeV and about 40 KeV for deuterium-deuterium system which correspond to temperatures of the order of 10^7 K and 10^8 K respectively. At the temperatures of thermonuclear interest, the various gases are in the form of plasma consisting only of nuclei and electrons. Confinement of such a hot plasma poses yet another problem.

For achieving a self-sustaining thermonuclear fusion reaction, the following two prerequisites are to be satisfied :

(1) The particle temperature should be $> 10^8$ K to achieve the necessary thermonuclear reaction energy.

(2) The Lawson limit according to which the product n τ must be >1020 particles S$^-$ m^{-3}, where n is the particle density (*i.e.* the number of reacting nuclei per cubic centimeter) and τ the confinement time in seconds during which thermonuclear reaction takes place, (*i.e.* the time during which the high temperature plasma can be confined). The *Lawson criterion* is based on the requirement that, in operation of a practical reactor for producing power from nuclear fusion, the total useful recoverable energy shall be atleast sufficient to maintain the temperature of the reacting species. The calculated minimum values for n τ are 6×10^{13} for the deuterium-tritium system (D-T) and 2×10^{15} for a deuterium system (D-D). Concentrated research is being directed to achieve the above conditions in either *steady state or pulsed operation*.

On the basis of heat transfer and other considerations, the steady state reactor is limited in power density to about $n = 10^{20} - 10^{21}$. The fusion power density varies as the square of the particle density because each collision involves two particles. At 1 Pa (*i.e.* 3×10^{20} particles m^{-3}), the power density would be tens of MW per m^3; at atmosphere densities, it would be 10^{10} times larger. This leads to a required confinement time of about 0.1–1 second.

A number of machines based on *"magnetic confinement"* of hydrogen ions have been built. However, when higher temperature (~108 K) were achieved, the confinement times were very short. On the other hand, longer confinement times (~1 second) were achieved when n is about 10^{17} m^{-3}. The best results seem to have been achieved with Tokamak machine (Moscow) with T~10^7 K, n~5×10^{19}, and τ 0.02 simultaneously. The principle of magnetic confinement in the Tokamak machine is shown in Fig. 3.4.

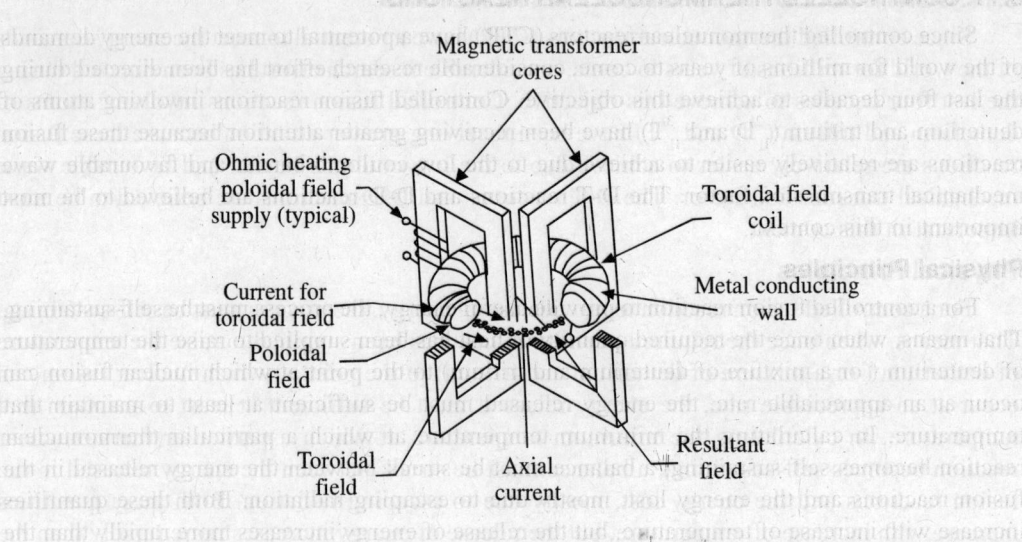

Fig. 3.4. The Tokamak Plasma Confinement Scheme

Studies are in progress in USA and USSR on pulsed operation using *"inertia confinement"*. In these operations, small pellets of solid D_2 and/or T_3 are dropped into the middle of a chamber where they are irradiated by intense beams of photons from lasers or from electrons from accelerators located around the chamber. The surface of the pellet vaporizes at once, resulting in the production of a jet-stream of particles away from the pellet and an impulse (i.e.

NUCLEAR FUELS AND NUCLEAR POWER GENERATION

temperature-pressure wave) travelling into the pellet and increasing the central temperature to $\geq 10^8$K. This causes a small explosion and the energy thus released can be collected. Since the particle density is high, the pulse time will be very short but still meet the Lawson criterion. Considerable progress with electron beams has been reported from USSR and USA in which temperatures of $\sim 10^9$K could be reached and fusion neutrons have been produced. Larger machines of this type are being developed. For photon beams, the lasers must be of higher energy ($\sim 10^5 - 10^6$ J pulse^{-1}), which are not available at present. An experimental CO_2 laser with 100 kJ pulse^{-1} is reported to be in operation in USA. With the expected energy output of $10^7 - 10^8$ J pulse^{-1} and with a repetition frequency of 100 pellets per second a power output of 1–10 GW can be achieved.

3.12. ENVIRONMENTAL ASPECTS OF THERMONUCLEAR POWER GENERATION

Since the present design of power stations based on nuclear fusion is more or less the same as that of the nuclear fission reactors, the thermal efficiency is expected to be of the same order. However, the necessarily larger power output will lead to a larger amount of the heat energy removed by the coolant. Hence, the thermal pollution from fusion reactors is expected to be higher than that from the present fission power reactors. Large scale development of magneto hydrodynamic systems may minimize this problem.

A fusion reactor would contain several kilograms of tritium, ^{3}T. Every 10 kg of ^{3}T roughly correspond to an activity of 10^8 Ci. ^{3}T is more difficult to contain and it can pass through many metals like Nb and more so at high temperatures. Further, ^{3}T can exchange with 1H atoms in water and thus can pose ingestion hazard. The permissible leak out of ^{3}T is less than 10^{-7}. Thus, prevention of tritium leakage, which is already a problem with fission reactors, poses a much greater technological problem with fusion reactors.

Similar to the fast fission breeder reactors, the fusion reactors also would contain large quantities of liquid metal. Very high activities would be induced in the structural materials (the preferred materials as of today are V and Nb) which poses a maintenance hazard.

Huge magnetic energy ($\sim 10^{11}$ J) would be stored in a fusion reactor, the consequences of which are yet to be clearly understood.

Other health and environmental problems of the fusion reactors are more or less similar to those of the fission reactors discussed earlier.

QUESTIONS

1. Discuss the theoretical principles involved in the generation of power by nuclear fission and nuclear fusion.
2. Describe the various components of a nuclear power reactor and their functions.
3. Discuss the environmental aspects of nuclear power generation.
4. Discuss the problems involved in the development of thermonuclear reactors.
5. Write informative notes on :
    (*a*) Breeder reactors.
    (*b*) Energy from nuclear fusion.

4

Corrosion

"The secret of effective engineering lies in controlling rather than preventing corrosion, because it is impossible to eliminate corrosion."
— *Michael Henthorne*

4.1. INTRODUCTION

Corrosion is the destructive attack of a metal by its environment and it is a general chemical or electrochemical phenomenon that is observed in day-to-day life. Common examples of corrosion the rusting of iron with the formation of corrossion products consisting of hydrous ferric oxides. In industry corrosion is serious it can lead to weakening of metal structures fexilure of plant.

Metals occur in nature most commonly as oxide or sulphide are in which they are in a higher oxidation state than that of the free metal. Extraction of the metal from its are involves reduction of the oxidised form to free metal, resulting in an increase in internal free energy. Consequently the metal will try to lose its excess energy by oxidation again.

This oxidising tendency of metal is the driving force for corrosion and it is found in virtually all metals except in nobel metals such as gold and platinum.

	Extraction		Corrosion	
Ore	$\xrightarrow{\text{Reduction}}$	Metal	$\xrightarrow{\text{Oxidation}}$	Corrosion product
Stable		Metastable		Stable

4·2. NERNST THEORY

According to Nernst's theory, no metal was truly insoluble, and all metals have a tendency to pass into solution. Thus, if a piece of copper is immersed in water, some copper atoms will shed their valence electrons and the positively charged copper ions, Cu^{++} go into solution. This leaves an excess of negative charge on the metal electrode and thus a potential difference exists between the metal and the solution. The positively charged Cu^{++} are held near the negatively charged metal piece because of electrostatic attraction. The process continues until an equilibrium is established when the number of Cu^{++} ions returning to the copper electrode and regaining two electrons forming Cu atoms is equal to the number of copper atoms going into solution as Cu^{++};

$$Cu^0 - 2e \rightleftharpoons Cu^{++}$$

At this stage, the process stops unless this equilibrium is disturbed.

CORROSION

If the copper electrode is put into a concentrated solution of a copper salt, there will be a greater tendency for Cu^{++} to deposit on the electrode as copper atoms, than for the copper atoms of the electrode to go into solution as ions. At a particular concentration of the Cu^{++} in solution, the potential difference between the metal and the solution will be zero.

Zinc shows a much greater tendency than copper to go into solution in water as Zn^{++} and, for the equilibrium to be established, a much greater concentration of Zn^{++} in solution is required. In case of sodium, the tendency to go into solution as Na^+ is still greater. This tendency of a metal to go into solution as ions when dipped in water is reckoned in terms of its electrode potential.

The relation between the electrode potential and the concentration (or more precisely the "activity") of the metal ion is given by Nernst's equation;

$$E = \frac{0.0591}{n} \log \frac{C}{K}$$

where C = concentration (more precisely "activity") of the metal ions in gram ions/litre, n = the charge or valence of the ions and K = equilibrium constant. Obviously, if C = K, the potential is zero. Nernst suggested that the numerical values of K for each ion gave a quantitative estimate of the tendency of the metal to form an electrolytic solution and hence is known as "Electrolytic solution-tension constant."

4·3. STANDARD ELECTRODE POTENTIALS

To determine the potential difference between an electrode and a solution, it is necessary to have another electrode and a solution whose potential difference is accurately known. The two half cells can then be combined to form a voltaic cell, the e.m.f. of which can be directly measured. Since the e.m.f. of the cell is the arithmetical sum or difference of the 2 electrode potentials, the value of the unknown potential can be calculated. The primary reference electrode is the normal or standard hydrogen electrode. This can be prepared by bubbling hydrogen gas at 1 atmosphere pressure over platinised platinum, immersed in a solution of hydrochloric acid containing hydrogen ions at unit activity. By convention, the potential of the standard hydrogen electrode is taken as zero volts at all temperatures. By connecting the standard hydrogen electrode with a metal electrode in contact with a solution of its ions of unit activity, the standard electrode potentials of various metals are determined. These values are summarized in Table – 1. When metals are arranged in the order of their standard electrode potentials, it is called "electrochemical series" of the metals.

Thus, when C = 1, the Nernst equation becomes:

$$E_0 = \frac{0.0591}{n} \log \frac{1}{K}$$

where E_0 is the "Standard Electrode potential".

The standard electrode potentials of the metals are also called as standard oxidation potentials, because the conversion of a neutral atom to a positive ion is called "oxidation."

Table – 1 provides a quantitative estimate to the electrochemical series. The greater the negative value of the potential, the greater is the tendency of the metal to pass into solution in the ionic state. Thus, potassium having greater propensity to go into solution has $E_0 = -2.92$, whereas gold with least tendency to go into solution has $E_0 = +1.42$.

Table –1. Standard Electrode Potentials in aqueous solutions (at 25°C)

Metal ion electrode system	Electrode Reaction (Acid Solutions)	Standard Electrode Potential (E_0) in volts	
Li/Li^+	$Li^+ + e^- = Li$	– 3.045	(Reactive metals)
K/K^+	$K^+ + e^- = K$	– 2.925	↑
Ca/Ca^{++}	$Ca^{2+} + 2e^- = Ca$	– 2.866	
Na/Na^+	$Na^+ + e^- = Na$	– 2.714	
Mg/Mg^{++}	$Mg^2 + 2e^- = Mg$	– 2.363	
Al/Al^{+++}	$Al^{3+} + 3e^- = Al$	– 1.662	
Mn/Mn^{++}	$Mn^2 + 2e^- = Mn$	– 1.180	
Zn/Zn^{++}	$Zn^{2+} + 2e^- = Zn$	– 0.763	
Cr/Cr^{+++}	$Cr^3 + 3e^- = Cr$	– 0.744	
Fe/Fe^{++}	$Fe^{2+} + 2e^- = Fe$	– 0.441	
Cd/Cd^{++}	$Cd^{2+} + 2e^- = Cd$	– 0.441	
Co/Co^{++}	$Co^{2+} + 2e^- = Co$	– 0.277	
Ni/Ni^{++}	$Ni^{2+} + 2e^- = Ni$	– 0.25	
Sn/Sn^{++}	$Sn^{2+} + 2e^- = Sn$	– 0.136	
Pb/Pb^{++}	$Pb^{2+} + 2e^- = Pb$	– 0.126	
$H_2/2H^+$	$2H^+ + 2e^- = H_2$	0.000	(Reference)
Cu/Cu^{++}	$Cu^{2+} + 2e^- = Cu$	0.337	
Cu/Cu^+	$Cu^+ + e^- = Hg$	0.522	
Hg/Hg^+ or $Hg_2/2H^+$	$Hg^+ + e^- = Hg$	0.799	
Ag/Ag^+	$Ag^+ + e^- = Ag$	0-800	
Hg/Hg^{++}	$Hg^{2+} + 2e^- = Hg$	0.854	
Pd/Pd^{++}	$Pd^{2+} + 2e^- = Pd$	0.987	
Pt/Pt^{++}	$Pt^{2+} + 2e^- = Pt$	1.2	↓
Au/Au^{+++}	$Au^{3+} + 3e^- = Au$	1.42	(Noble metals)

Further, a metal will normally displace any other metal below it in the series from solutions of its salts and precipitate it in the metallic form. Thus, Mg, Al, Zn or Fe will displace Cu from solutions of its salts. Similarly, Pb will displace Cu, Hg or Ag; while Cu will displace Ag.

The standard electrode potential is a quantitative measure of the readiness of the element to act as a reducing agent in aqueous solution; the more negative the potential of the element, the more powerful is its action as a reducing agent.

It should be emphasized that standard electrode potential values relate to an equilibrium condition between the metal electrode and the solution of its ions of unit activity. Potentials determined under such conditions as referred as "reversible electrode potentials".

4·4. GALVANIC SERIES

Although electrochemical series gives very useful information regarding chemical reactivity of metals, it may not be able to provide sufficient information in predicting the corrosion behaviour in a particular set of environmental conditions. However, in many practical situations, several side reactions may take place which influence the corrosion reactions. In view of this, oxidation potential measurements of various metals and alloys in common use have been made using standard calomel electrode as the reference electrode and immersing the metals and alloys in sea water. These are arranged in Table 2, in decreasing order of activity and this series is known as "galvanic series". This gives a more practical information on the relative corrosion tendencies of the metals and alloys. In general, the speed and severity of corrosion depends upon the difference in potential between the anodic and cathodic metals in contact. However, the exact position of a metal or alloy in the galvanic series in relation to its respective corrosion tendency may be still influenced by other interfering factors.

CORROSION

Table –2. Galvanic series (on the basis of relative oxidation potentials in sea water)

More Anodic or Active	(or Corroded end)
Magnesium	
Magnesium alloys	
Zinc	
Aluminum	
Aluminium alloys	
Low carbon steel	↑
Cast iron	Active
Stainless steel (active)	(anodic)
Lead-tin alloys	
Lead	
Tin	
Brass	
Copper	
Bronze	
Copper-nickel alloys	
Inconel	
Silver	Noble
Stainless steel (passive)	(cathodic)
Monel	↓
Graphite	
Titanium	
Gold	
Platinum	
More cathodic or inactive	

4·5. Galvanic or electric cells

The cells used for electrolysis are called electrolytic cells while those used for generation of electrical energy from chemical reactions are called galvanic or voltaic cells. A galvanic or electric cell (chemical battery) can be obtained by combining any two half cells having different electrode potentials. Thus, if a standard zinc-half cell is coupled with a standard copper half cell by wire and a salt bridge as shown in Fig. 4.1, the resulting cell will have a voltage of

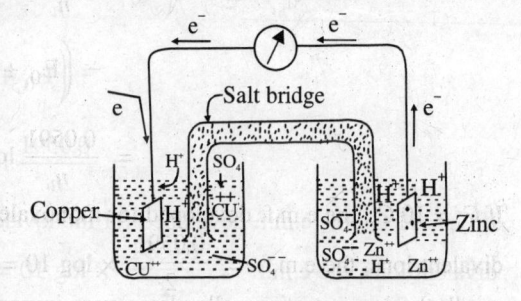

Fig. 4.1. Daniell cell.

$E_{0_{Cu}} - E_{0_{Zn}} = 0.345 - (-0.762) = 1.107$ volts

In the Daniell cell so formed, the electrons flow from the zinc electrode to the copper electrode through the connecting metal wire, while the Cu^{++} ions in the solution move to the copper cathodes where they reduce their electron "pressure" and get deposited. The electron "pressure" is maintained by the generation of the corresponding stream of zinc ions from the dissolution (or "corrosion") of the zinc anode:

$$Zn^° \longrightarrow Zn^{++} + 2e \quad \text{(oxidation)}$$
$$Cu^{++} + 2e^- \longrightarrow Cu^° \quad \text{(reduction)}$$

For concentrations other than 1 gm ion per litre, the e.m.f. of the cell may be calculated from the Nernst equation:

$$E_C = \frac{0.059}{n} \log \frac{C}{K}$$

$$= \frac{0.059}{n} \log \frac{1}{K} + \frac{0.059}{n} \log C$$

$$= E_0 + \frac{0.059}{n} \log C$$

For example, if the concentration of Cu^{++} and Zn^{++} are 0.01 M and 0.1 M respectively, then the voltage of the resultant Daniell cell would be

$$E_{Cu} - E_{Zn} = \left\{ E_{0_{Cu}} + \frac{0.059}{2} \log [Cu^{++}] \right\} - \left\{ E_{0_{Zn}} + \frac{0.059}{2} \log [Zn^{++}] \right\}$$

$$= \{0.345 + 0.0295 \log 10^{-2}\} - \{-0.762 + 0.0295 \log 10^{-1}\}$$

$$= 0.345 - 0.059 + 0.762 + 0.0295$$

$$= 1.0775 \text{ volts.}$$

4·6. CONCENTRATION CELLS

An electrode potential varies with the concentration of the ions in the solution. Hence it follows from the Nernst equation that two electrodes of the same metal, but immersed in solutions containing different concentrations of its ions, may form a cell (when connected with a wire and salt bridge). Such a cell is known as a concentration cell (Fig. 4.2)

The e.m.f. of the cell will be the algebraic difference of the two potentials. Thus,

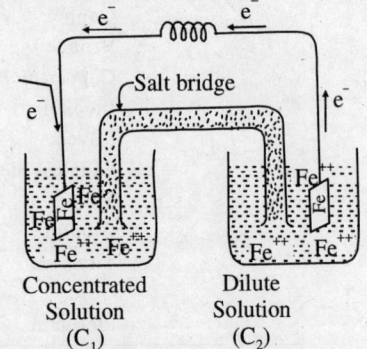

$$E = E_{C_1} - E_{C_2} = \left(E_0 + \frac{0.0591}{n} \log C_1 \right)$$

$$- \left(E_0 + \frac{0.0591}{n} \log C_2 \right)$$

$$= \frac{0.0591}{n} \log \frac{C_1}{C_2}, \text{ where } C_1 > C_2$$

Fig. 4.2. Concentration cell.

If $C_1 = 10 C_2$, the e.m.f. developed for monovalent ions = $0.0591 \times \log 10$ = 0.0591 volt; while for divalent ions, the e.m.f. = $\frac{0.0591}{2} \times \log 10$ = 0.0295 volt. Although these potentials seem to be small, the concentration cells play a very important role in corrosion.

In a concentration cell, the metal immersed in the dilute solution has a tendency to go into solution (*i.e.*, it acts as anode) while the metal immersed in the concentrated solution behaves as cathode and metal ions in the solution around the cathode get discharged and deposited on it.

4·7. REVERSIBLE CELLS

If, to the Daniell cell described earlier, a small e.m.f. from an external source is applied in the reverse direction, the rate of zinc going into solution and of copper depositing on the cathode are reduced. If this applied e.m.f. in the reverse direction is increased further, the rates of zinc dissolution and copper deposition will be decreased still further until the applied e.m.f. balances the cell e.m.f. and no current flows. At this stage, there will be no corrosion of Zn and no deposition of Cu. This principle is of vital importance because it provides a method for preventing corrosion. If the

CORROSION

applied e.m.f. is still further increased (in reverse direction) so that this is greater than the cell e.m.f., then the reverse reactions, namely, corrosion of copper and deposition of zinc, take place.

The reactions taking place in the Daniell cell are also slowed down by a different mechanism. As the current is withdrawn, the concentration of zinc ions around the zinc anode in the cell increases and the concentration of copper ions around the copper cathode decreases. This reduces the tendency of further corrosion of zinc and deposition of copper. Thus the e.m.f. generated falls down and the cell is said to be run out or *polarised*.

4·8. POLARIZATION

When a cell is under use, the concentration of the ions surrounding the electrodes differ from that in the bulk of the electrolyte. These concentration gradients at the electrodes set up a back e.m.f., and the cell potential drops. This is known as *concentration polarisation*, which increases with time and current density. In the commonly used dry cells, this polarization is counteracted by the NH_4Cl present which removes zinc ions in the form of a complex $Zn(NH_3)_4^{++}$.

$$Zn^{++} + NH_4Cl \rightleftharpoons Zn(NH_3)_4^{++} + 4HCl$$

There is another mechanism by which a second type of polarisation may occur in a cell. This is caused by the formation of a resistant film of adhering atoms or molecules of gas on the electrode. This is known as *gas polarisation* which may occur when O_2 or Cl_2 is liberated at anodes or when hydrogen ions are discharged at the cathode liberating H_2.

(It is interesting to note that copper is not easily displaced from its solution by aluminium although aluminium is much more electropositive. This is due to the presence of a very stable oxide film on aluminium, which considerably reduces the electrode potential. However, chloride ions are found to be particularly effective in destroying the oxide films on aluminium. Hence, Al exhibits its true reactivity in a copper salt solution containing added chloride ions; copper is then deposited and aluminium goes into solution).

In a dry cell cathodic polarization occurs at the carbon electrode. The depolarizing agent employed is MnO_2 which reacts with the H_2 to produce water and lower oxides of manganese.

$$H_2 + MnO_2 \longrightarrow H_2O + MnO; \quad H_2 + 2MnO_2 \longrightarrow H_2O + Mn_2O_3$$

Films formed on the cathode or the anode surface separate the cathodic or the anodic areas from the bulk of the electrolyte and increase the cathodic or anodic polarization mainly due to the following reasons:

(a) they cause an increase in the resistance of the path between the electrodes, and

(b) they decrease the rate of diffusion of reactants and products of the electrode reactions towards and away from the electrode surfaces. This results in the increase of concentration polarization to much greater extent than what would prevail in the absence of the films.

The more adherent and non-porous the film formed, the more it contributes to the increase of polarization at an electrode. However, even porous and loosely adhering films may also exert a considerable effect. These films, whether naturally formed on metal surfaces during manufacture or artificially formed by the application of special coatings or whether they are formed due to the products of corrosion, are of considerable importance in determining the rate of corrosion of a metal or an alloy and also for the protection from corrosion.

4·9. DECOMPOSITION POTENTIAL

If a small voltage (about 0.5 V) is applied to two smooth platinum electrodes immersed in 1 M-sulfuric acid solution, then an ammeter placed in the circuit initially shows that an appreciable amount of current is flowing. However, its strength quickly decreases and falls practically to zero within a short time. If the applied voltage is gradually increased, there is a slight increase in the current until, when the applied voltage reaches a certain value, the current suddenly increases rapidly with increase in the e.m.f. In general, it is observed that at the point at which there is a sudden increase in current, copious evolution of gas at the electrodes commences. The voltage at this point is called the *decomposition voltage or decomposition potential*. Thus, the *decomposition potential* of an electrolyte may be defined as the minimum external voltage that must be applied in order to bring about continuous electrolysis. (Fig. 4.3)

If the circuit is broken after the e.m.f. has been applied, it can be observed that the voltmeter reading which is fairly steady at first, rapidly falls to zero. That means, the cell is behaving as a source of current, and is said to exert a back e.m.f. (or counter e.m.f. or polarisation e.m.f.) because it acts in a direction opposite to that of the applied e.m.f. This back e.m.f. arises due to the accumulation of O_2 and H_2 at the anode and cathode respectively, which consequently act as gas electrodes. The potential difference between them opposes the applied e.m.f. When the primary current from the battery is cut off, the cell produces a moderately steady current until the gases at the electrodes are either used up or have diffused away, and then the voltage falls to zero. This back e.m.f. is present even when the current from the battery passes through the cell. The minimum value of the back e.m.f. can be calculated because it is the algebraic difference of the electrode potentials which exist at the anode and the cathode respectively.

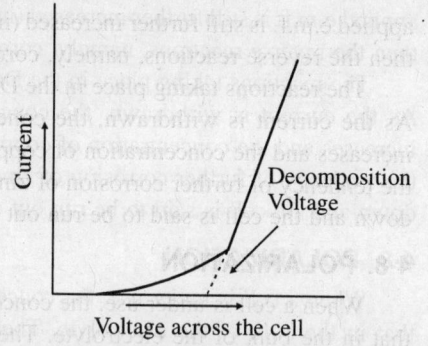

Fig. 4.3. Decomposition votlage

4·10. OVERVOLTAGE OR OVER-POTENTIAL

It has been experimentally found that the decomposition voltage of an electrolyte varies with the nature of the electrodes employed for the electrolysis and is, in many cases, considerably higher than that calculated from the difference of the reversible electrode potentials. This excess voltage over the calculated back e.m.f. is known as *overvoltage or overpotential*. Overvoltage may occur both at the anode and at the cathode. Hence, the decomposition voltage, E_D is given by the following relation:

$$E_D = (E_{cathode} + E_{o.c.}) - (E_{anode} + E_{o.a.})$$

where $E_{o.c.}$ and $E_{o.a}$ are the overvoltages at the cathode and anode respectively.

For instance, the e.m.f. of a standard chlorine-hydrogen cell is 1.36 V. But unless one uses a platinized platinum electrode, the decomposition voltage required to generate hydrogen and chlorine by electrolysis of the electrolyte is considerably greater than 1.36 V. This extra voltage above the cell voltage is due to the over-potential or overvoltage.

Although the possibility of concentration polarisation at the electrodes and film resistances are eliminated, still some polarisation effects persist which are attributed to overvoltage. Overvoltage may be considerably high in electrode reactions involving gases, such as the discharge of hydrogen or the reduction of oxygen at cathode or the discharge of O_2, Cl_2 etc. at the anodes. By and large, overvoltages in electrode reactions involving the metals and their ions are quite small. The overvoltages at the anode or cathode depends upon the following factors:

1. *The nature and the physical state of the metal employed for the electrodes.* For the electrolysis of an electrolyte solution (e.g. of 1N H_2SO_4) using platinised platinum (platinum black) electrodes, the decomposition potential is 1.23V. This is the same as the e.m.f. of the standard oxygen-hydrogen cell. However, when the platinised platinum black electrodes are replaced by smooth platinum electrodes, the decomposition voltage is 1.7 V. That means, there is an overvoltage of (1.70 – 1.23) = 0.47 V.

The fact that reactions involving gas evolution generally require much less overvoltage at platinised rather than polished platinum electrode is due to the much larger surface area of the platinised electrode and the smaller current density at a given electrolysis current.

The magnitude of the gas overvoltages depends upon the nature of the electrode. For instance, hydrogen overvoltages are low on Pt, Au and Ag; intermediate on Fe, Ni, Co, C and Cu; and high on Pb, Cd and Zn. Further, gas overvoltages are lower on rough, abraded metal surfaces than on smooth and polished surfaces.

2. *Current density.* The overvoltage increases with the current density (amperes per unit area of the electrode surface). Overvoltage increases rapidly at first with current density (upto 0.01 ampere cm^{-2}), but increases less rapidly afterwards until if finally reaches a maximum value at high current densities.

3. *Temperature.* Overvoltage decreases, often considerably, with increase in temperature.

4. *The physical state of the substance deposited.* The over-potential is greater for gases such as O_2 or H_2 than for metals.

5. *The change in concentration, or the concentration gradient existing in the immediate vicinity of the electrodes.* If the concentration increases the overpotential increases. The concentration gradient in turn depends upon the current density, temperature and the rate of stirring of the solution.

6. *Presence of inhibitors.* The overvoltages may be increased by the addition of certain substances called "corrosion inhibitors". For example, hydrogen overvoltages are increased by the addition of substances like gelation, glue and other inhibitors.

7. Overvoltage increases by the formation of passive films on the metal surface.

Hydrogen overvoltage is of particular significance in many electrolytic reactions and especially in electroplating and corrosion. The hydrogen overvoltages on different electrodes at 25°C and at current densities of 1 and 10 milliamperes/cm^2 are shown in Table 3.

The importance of hydrogen overvoltage in electroplating of metals can be illustrated by the case of Zn. If a solution of 1N $ZnSO_4$ in 1N H_2SO_4 is electrolyzed using platinum electrodes, hydrogen is liberated rather than Zn being plated out. This is because of the fact that the electrode potential of Zn is – 0.76 V, while that of hydrogen is zero and hence Zn shows a tendency to stay in the form of ions. However, if Hg is used as cathode, the hydrogen overvoltage is so great (1.04 V at a current density of 10 milliamp cm^2) that it is now easier by 0.28 V (1.04 V – 0.76 V) to plate out Zn than to drive off H_2. The application of overvoltage in corrosion will be discussed later.

Table – 3. Hydrogen overvoltages at 25°C

Electrode	Hydrogen overvoltage in volts at current density	
	1 milliamp/cm^2	10 milliomp/cm^2
Pt, black	0.01	0.01
Pt, smooth	0.02	0.03
Au	0.24	0.39
Fe	0.40	0.56
Ag	0.48	0.76
Cu	0.48	0.58
Pb	0.52	1.09
Ni	0.56	—
Al	0.57	0.83
C (graphite)	0.60	—
Zn	0.72	0.75
Bi	0.78	—
Sn	0.86	1.08
Hg	0.88	1.04
Cd	0.98	1.13

4·11. CORROSION

4.11.1. Introduction. Corrosion may be defined as the gradual eating away or disintegration or deterioration of a metal by chemical or electrochemical reaction with its environment.

Corrosion is an important factor in any chemical process plant. It makes all the difference between a trouble free operation and a costly shut-down.

All metals and alloys are susceptible to corrosion. No single material may be suitable for all applications. For instance, gold has excellent resistance to corrosion under atmospheric conditions, but it gets readily corroded when exposed to mercury at ambient temperature. On the other hand, iron readily gets rusted in the atmosphere, but it does not corrode in mercury. Thus, several metals and alloys perform excellently in a given environment. Further, there are several methods to effectively control and minimize corrosion. "The secret of effective engineering lies in controlling rather than preventing corrosion, because it is impracticable to eliminate corrosion."

4.11.2. Consequences of corrosion. The economic and social consequences of corrosion include, (1) plant shutdown due to failure (2) replacement of corroded equipment, (3) preventive maintenance (such as painting), (4) necessity for overdesign to allow for corrosion, (5) loss of efficiency, (6) contamination or loss of the product (*e.g.*, from a corroded container), (7) Safety (*e.g.*, from a fire hazard or explosion or release of a toxic product or a collapse of construction because of a sudden failure), and (8) Health (*e.g.*, from pollution due to a corrosion product or due to an escaping chemical from a corroded equipment).,

4.11.3. Cause of corrosion. It is generally easier to understand why corrosion occurs than how. Most metals exist in nature (in ores and minerals) as compounds such as oxides, sulfides, sulfates, etc. because these compounds represent their thermodynamically stable state. The metals are extracted from these ores after expending lot of energy. Hence, unless the nature of the metal is susbtantially changed (*e.g.*, by alloying), the metal will have a natural tendency to revert back to its natural thermodynamically stable state. This is the basic reason for metallic corrosion. However, metals that exist in elemental state in nature (*e.g.*, gold) naturally have excellent corrosion resistance in natural environment.

4.11.4. Classification. Corrosion processes are usually classified broadly on the basis of one of the following factors:

(*a*) **Nature of the corrodent.** Corrosion can be classified as "dry" or "wet". *Dry corrosion* usually involves reaction with gases at high temperature. *Wet corrosion* occurs in presence of water or a conducting liquid.

(*b*) **Mechanism of Corrosion.** This may involve direct chemical attack or indirect electrochemical attack.

(*c*) **Appearance of the corroded metal.** Corrosion may be either uniform and the metal gets corroded at the same rate over its entire surface; or it is localized and only small areas are attacked. Further distinction of macroscopically localised (*e.g.*, pitting) and microscopically localised corrosion may also be made.

However, several different forms of corrosion are often distinguished as follows: (1) Galvanic corrosion, (2) Erosion, (3) Crevice corrosion, (4) Pitting corrosion, (5) Exfoliation, (6) Selective leaching, (7) Intergranular corrosion, (8) Stress corrosion cracking, (9) Water-line corrosion, (10) Micro-biological corrosion etc.

4.11.5. Theories of corrosion

(1) *The acid theory.* This theory suggests that the presence of acids (such as carbonic acid) is essential for corrosion.

This theory is particularly applicable to rusting of iron in the atmosphere. According to this theory, rusting of iron is due to the continued action of oxygen, carbon dioxide and moisture,

CORROSION

converting the metal into a soluble ferrous bicarbonate which is further oxidised to basic ferric carbonate and finally to hydrated ferric oxide.

$$Fe + O + 2CO_2 + H_2O \longrightarrow Fe(HCO_3)_2$$
$$2Fe(HCO_3)_2 + H_2O + O \longrightarrow 2Fe(OH)CO_3 + 2CO_2 + 2H_2O$$
$$2Fe(OH)CO_3 + 2H_2O \longrightarrow 2Fe(OH)_3 + 2CO_2$$

This theory is supported by the facts that (1) rust analysis generally shows the presence of ferrous and ferric carbonates along with hydrated ferric oxide and (2) retardation of rusting in presence of added lime or NaOH to the water in which iron is immersed.

(2) *Direct chemical attack.* This theory explains the so-called chemical or dry corrosion. Direct chemical attack by dry gases on a metal at atmospheric temperature is rather uncommon. However, whenever corrosion takes place by direct chemical attack, a solid film of the corrosion product is usually formed on the surface of the metal which protects the metal from further corrosion. However, if a soluble or liquid corrosion product is formed, then the metal is exposed to further attack. For instance, chlorine and iodine attack silver generating a protective film of silver halide. Similarly, during detinning of tinned low-carbon steel cans using chlorine gas at high temperatures (over 120°C) Sn is converted into volatile $SnCl_4$ and hence all the Sn is readily removed from the metal surface. However, under these conditions, dry Cl_2 attacks the base metal iron only superficially because the $FeCl_3$ formed on the surface is a non-volatile solid which protects the rest of the metal.

One of the most common ways in which metals are attacked by direct action is by interaction with oxygen. Alkali and alkaline earth metals (*e.g.*, Na, Ca, Mg, etc.) suffer extensive oxidation even at low temperatures whereas at high temperatures, practically all metals excepting Ag, Au and Pt are oxidised. Alkali and alkaline earth metals on oxidation produce oxide deposits of smaller volume than the respective metals from which they were formed. This results in a the formation of a porous layer through which oxygen can diffuse to bring about further attack of the metal. On the other hand heavy metals and Al form oxide layers of greater volume than the metal from which they were produced. These non-porous continuous oxide films prevent the diffusion of oxygen and hence the rate of further attack decreases with increase in the thickness of the oxide film (pilling-Bedworth's Rule). However, as the films grow thicker, their tendency to crack and scale out becomes greater and particularly at high temperatures, intermittent attack of metal may take place at the exposed parts of the metal (due to cracking of the oxide film).

Practically all metals on exposure to air get covered with a film of oxide a few Angstrom units thick (1 A = 10^{-8} cm). The thickness of the oxide film varies with the metal and the temperature. At high temperatures, the scale formation proceeds in a two-way process: One is by the diffusion of O_2 to the metal through the intervening oxide layers already formed and the other is by diffusion of the metal (*e.g.*, Fe, Al, and Zn) outward through the oxide layers. According to an ionic theory of oxidation, the thickness of the oxide films on metals increases because of the diffusion of metal ions outwards through the oxide layer via interstitial positions in the oxide lattice and eventually oxidises by interaction with adsorbed oxygen at the free surface.

Aluminium and chromium form extremely protective oxide layers and hence these metals are usually employed for alloy formation with the other more readily attacked metals to provide increased resistance to oxidation.

Example 1 : Which of the following coatings will not be protective and why?

Oxide	Oxide density gm/cc.	Atomic weight of metal	Metal density gm/cc.
MgO	3.65	24.31	1.74
CaO	3.40	40.08	1.55
Al_2O_3	3.70	27.00	2.70

Solution :

As per pillig Bedworth's Rule, if the oxide deposit volume is more than the metal volume the oxide deposited is non-porous and film is continous oxide and rate of further corrosion decreases with the thickness of oxide film In other words of if ratio of Metal Oxide Volume/ Metal Volume > 1 Protective film form and if the ratio is < 1 Non-protective film formed.

Oxide	Volume of metal V_1	Volume of metal oxide V_2	V_2/V_1
MgO	24.31/174=13.97	40.3/3.65=11.04	11.04/13.97 = 0.79 Non Protective
CaO	40.08/1.55 = 25.85	56.07/3.40 = 16.49	16.49 / 25.85=0.637 Non Protective
Al_2O_3	24/2.70 = 10	101.93/3.7 = 27.54	27.54/10 = 2.754 Protective

(3) *The electrochemical theory*: This theory explains the indirect or wet corrosion. The modern electrochemical theory is based on Nernst theory according to which all metals have a tendency to pass into solution. If a Zn electrode is dipped in a solution of zinc sulfate, the positive zinc ions in the metal electrode are in a continuous vibration and occasionally an ion receives sufficient energy to escape from the metal and pass into surrounding solution. Then the electrode acquires a negative charge; thus each Zn^{++} passing from metal to solution leaves the electrode depleted of two electrons. Zinc ions in solution are attracted to the negatively charged zinc electrode and some of the ions may be redeposited (or "plated out") on the solid electrode. Thus, an equilibrium between positive ions in solution and the metal is rapidly established. In the case of copper, the tendency for Cu^{++} to go into solution is lesser and their tendency to "plate out" on the copper electrode is greater.

The tendency of a metal to pass into solution when immersed in a solution of its salt is measured in terms of its electrode potential. As has been already described under III.1.1. and III.1.2, the standard electrode potentials of the various metals have been determined in comparison with the standard electrode potential of hydrogen taken as zero.

If a metal having a higher electrode potential comes into contact with another metal having a lower electrode potential, a "galvanic cell" is set up and the metal having higher electrode potential becomes anodic and goes into solution to a measurable extent. If the surrounding liquid is sufficiently acidic, H_2 gas will be evolved at the cathodic metal while the anodic metal dissolves. If the acidity of the surrounding liquid falls below a certain value, the rate of dissolution of the anodic metal slows down and is controlled by the rate at which oxygen can diffuse to the cathode and depolarise the corrosion cell.

The extent of galvanic corrosion depends upon the difference in the electrode potentials of the two electrodes and their respective areas. The greater the difference in the potentials of the cathode and anode, the greater will be the corrosion. Also, the smaller the area of the anode as compared to the cathode, the more severe will be the attack.

Corrosion reactions

It has already been emphasized that two metals having different electrode potentials form a galvanic cell when they are immersed in a conducting solution. The e.m.f. of the cell on "open circuit" (*i.e.,* when no current is flowing) is given by the difference between the electrode potentials. When the electrodes are joined by a wire, electrons flow from the electropositive "anode" to the electronegative "cathode". The various reactions that might be taking place at the electrodes are as follows:

CORROSION

Anode reactions

At the anode, the metal atoms lose their electrons to the environment and pass into solution in the form of positive ions (oxidation). Ex: $Fe \rightarrow Fe^{++} + 2e^{-1}$. This will continue as long as the electrons and ions are removed from the environment. If they are not removed, the corrosion will not proceed further. Thus, the extent of corrosion of the metal anode depends upon the reactions at the cathode which "mop up" electrons flowing from the anode and convert the metal ions formed at the anode into insoluble corrosion products. Ex.:

$$Fe^{++} + 2OH^- \longrightarrow Fe(OH)_2$$

Cathode reactions

The electrons released at the anodes are conducted to the cathodes and are responsible for the various cathodic resections:

(1) *Electroplating*: The metal ions at the cathode collect the electrons and "plate out" (deposit) on the cathode surface. Ex.:

$$Cu^{++} + 2e^{-1} \longrightarrow Cu$$

(2) *Liberation of hydrogen*. In acid solution (in the absence of oxygen) hydrogen ions acquire electrons and H_2 gas is formed.

$$2H_3O^+ + 2e^{-1} \longrightarrow H_2\uparrow + 2H_2O$$

In neutral and alkaline media and in the absence of oxygen, the reaction taking place will be as follows:

$$2H_2O + 2e^{-1} \longrightarrow H_2\uparrow + 2OH^-$$

Corrosion processes in which H_2 is evolved are called "hydrogen type".

(3) *Formation of hydroxyl ions*: (a) In presence of dissolved oxygen and in neutral or alkaline media:

$$2H_2O + O_2 + 4e^{-1} \longrightarrow 4OH^-$$

(b) In presence of dissolved oxygen and in acid media:

$$4H^+ + O_2 + 4e^{-1} \longrightarrow 2H_2O$$

Corrosion processes involving O_2 are called "oxygen type".

The above mentioned reactions only represent the predominant reactions. In many cases, many of these reactions may take place simultaneously.

Thus, from the foregoing discussion, it is clear that the essential requirements of electrochemical corrosion are (a) formation of anodic and cathodic areas, (b) electrical contact between the cathodic and anodic parts to enable the conduction of electrons, and (c) an electrolyte through which the ions can diffuse or migrate. This is usually provided by moisture.

This theory has successfully explained the corrosion of metals which are in contact with a more "noble" metal or alloy; and also of metals containing impurities that are capable of acting as cathodic areas in the bulk of anodic base metal. However this theory was unable to explain why concentrated attack such as pitting takes place so often and also why corrosion often occurs at places which are not easily accessible to oxygen. In order to explain these phenomena, U.R. Evans and others suggested a modified electrochemical theory which states that corrosion takes place in many cases due to "differential aeration". Anodic and cathodic areas may be generated even in a perfectly homogeneous and pure metal due to different amounts of oxygen reaching different parts of the metal which form oxygen concentration cells. In such circumstances, those areas which are exposed to greater amount of air (or oxygen) become cathodic while the areas which are little exposed or not exposed to air (or oxygen) become anodic and suffer corrosion.

Some typical examples of corrosion due to differential aeration are shown in figures given below. Fig. 4.4 -A shows the part of a metal surface covered with dirt which is less accessible to air than the rest of the metal. Hence, the area covered with dirt becomes anodic and suffers corrosion. The most common reaction taking place at the cathode is $H_2O + \frac{1}{2}O_2 + 2e^{-1} = 2\,OH^-$ which consumes electrons with the help of oxygen while at the anode the reactions taking place will be such as : $Fe = Fe^{++} + 2e^{-1}$; $Fe^{++} + 2OH^- = Fe(OH)_2$, which is oxidised to $Fe(OH)_3$ in presence of an oxidizing environment. As the corrosion proceeds further, more rust is formed and the area becomes still more inaccessible to air. This promotes further corrosion producing deep cavities in the metal. This sort of intense local corrosion is called "pitting". Fig. 4.4-B shows a wire fence in which the areas where the wires cross, are less accessible to air than the rest of the fence and hence corrosion takes place at the wire crossings which are anodic. Similarly, a crack in a metal is less accessible to air and hence forms an anodic area where corrosion takes place.

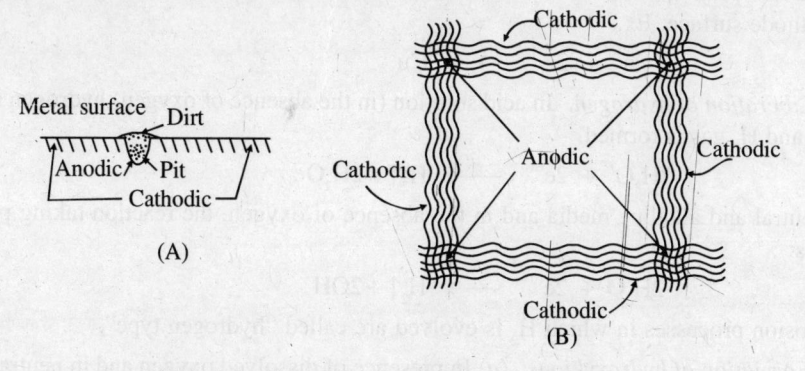

Fig. 4.4. Corrosion by differential aeration.

Fig. 4.5, depicts an iron pipe which is **incompletely** covered by an oxide surface layer. In presence of moisture, a corrosion cell is set up in which the oxide surface becomes cathodic with respect to the metal part which is directly exposed to the liquid film at a crack in the oxide layer. Although the electronic conductivity of the oxide layer is small and the difference in electrode potentials is less, the flow of current is measurable. The metal exposed gets corroded and Fe^{++} pass into solution. The cathode reaction is usually the same as described earlier, namely the formation of OH^- from dissolved oxygen: $H_2O + \frac{1}{2}O_2 + 2e^{-1} = 2OH^-$. As the corrosion proceeds, Fe^{++} and OH^- accumulate in solution and combine to form insoluble

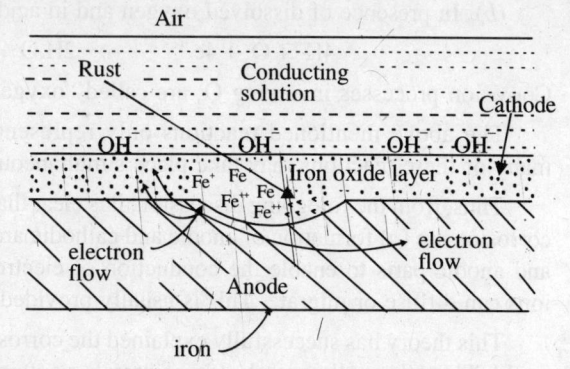

Fig. 4.5. Corrosion of an iron tube at a crack on the oxide surface layer.

$Fe(OH)_2$ which eventually oxidises to hydrous ferric oxide, $Fe_2O_3\,XH_2O$. If the cathode has a very large surface area e.g., the surface of a long pipeline, the cathode reaction may be considerable. This puts forth a higher demand of electrons on the relatively small anode and hence heavy corrosion at the anode takes place, leading to even perforation of the pipe where the oxide film is defective. In general, rust may form in the solution some distance away from the corroding metal. However, if the insoluble metal hydroxide formed can seal up the cracks in the oxide layer, corrosion may be inhibited.

CORROSION

It has already been pointed out that electrode potential of a metal electrode depends on the nature and concentration of the solution surrounding the electrode. It has also been made clear that the electrode potential of a metal depends upon the concentration of dissolved oxygen in the solution; the metal in oxygen-free water becomes more electropositive than the metal in water containing dissolved oxygen. The formation of an oxygen concentration cell can be interestingly illustrated from Fig. 4.6 which represents a drop of water resting on an iron or steel surface. The thin layer of water at the perimeter of the drop has a greater dissolved oxygen concentration than the water in contact with the metal at the centre of the drop because it is away from the external atmosphere which is a source of oxygen. Hence, the metal at the drop-centre becomes anodic and suffers corrosion, Fe^{++} ions pass into solution and OH^- ions are formed in the water drop by the cathodic reaction taking place at the perimeter of the drop as described earlier. In this case also, $Fe(OH)_2$ and, subsequently $Fe(OH)_3$ (rust) will form in the solution and a circle of corrosion products may eventually encompass the pitted metal. The rust so formed around the anode further prevents the entry of oxygen from the surfaces of the drop thereby accelerating the rate of corrosion. The various reactions taking place under the drop can followed with the help of ferroxyl indicator (phenolphthalein + potassium ferricyanide). The reactions can be accelerated by adding a drop of NaCl solution. The solution at the centre turns blue indicating the formation of ferrous ions at the anodic part which react with potassium ferricyanide giving blue colour. The indicator turns pink at the periphery of the drop because of the formation of the OH^- at the cathode which reacts with the phenolphthalein in the indicator giving pink colour.

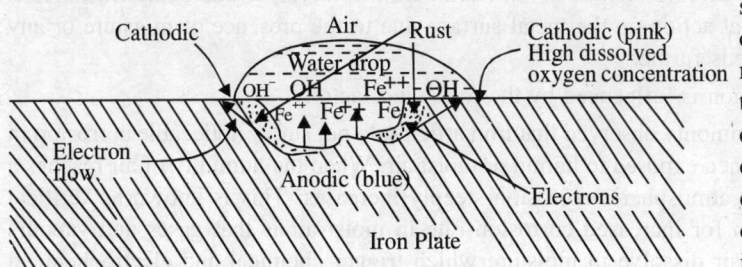

Fig. 4.6. Formation of an oxygen concentration cell on a metal plate on which a drop of water is put.

Anodic and cathodic areas are formed on metal structures due to several other factors which lead to the formation of corrosion couples. For instance, anodic areas are formed at (*a*) strained metal, (*b*) areas having relatively low concentration of dissolved oxygen, (*c*) areas at which protective surface film has been cracked, (*d*) external or internal metallic impurities which are in electrical contact with the metal, etc. Cathodic areas may be formed at (*a*) relatively unstrained metal, (*b*) areas at which the oxygen concentration is relatively high, (*c*) impurities of the metal upon which there is a low oxygen or low hydrogen overvoltage, (*d*) external or internal impurities in contact with the metal which are cathodic to the metal, etc.

The various factors affecting corrosion and different types of corrosion are discussed in the subsequent sections.

The electrochemical theory effectively explains the corrosion occurring under cracks, pits, scales and accumulated debris. But, however, it could not explain by itself as to why aerated metals do not corrode. The work of **pilling, Bedworth and Tanman** provided an answer for this. According to their work, a metal under suitable conditions undergoes direct oxidation producing very thin films of oxide on the surface and these films protect the metal from further oxidation. At low temperatures, the diffusion of gases and ions through the films is extremely small and thus the oxide or hydroxide films formed on the metal surface imparts the protecting or ennobling effect. At high temperatures also, even an invisible oxide film can afford a considerable protection.

Atmospheric Corrosion

When a metal is exposed to the atmosphere, an oxide film is formed on its surface due to interaction with atmospheric air. As long as this protective oxide film is maintained on the entire metal surface, further attack of corrosion does not occur. But, however, if this oxide film breaks down due to electro-chemical action on the metal surface due to the presence of moisture or any electrolyte, corrosion proceeds further.

The atmospheric corrosion is influenced by the following factors:

(1) **Humidity.** It is commonly observed that iron articles do not suffer noticeable corrosion in dry air but they are rusted when exposed to humid or moist air. When the humidity of air reaches a particular value, the rate of atmospheric corrosion steeply increases. This is known as "critical humidity". The main reason for increased corrosion rate in moist air is that gases and vapours present in the atmospheric air dissolve in moisture which trigger chemical and electrochemical reactions on the metallic surface leading to corrosion. The extent of attack depends upon the properties of the metal. Metals like Cr which provide a highly protective oxide layer are the most resistant. However, if the oxide film develops cracks or discontinuities, then local electrochemical cells are formed due to the action of humid atmospheric air and corrosion of the metal takes place at the exposed areas of the metal which become anodic. The oxide film may also be disturbed due to the impact of rain water which may lead to enhanced atmospheric corrosion.

(2) **Impurities present in the atmospheric air.** In marine atmosphere, the presence of electrolytes such as NaCl in the humid atmospheric air leads to increase in corrosion of the metals. Similarly, the atmospheric air near industrial areas is contaminated with acid fumes and gases such as H_2S, SO_2, SO_3 and CO_2. If the water in which these gases or fumes are dissolved come into contact with a metal surface, its electrical conductivity increases. This naturally increases the corrosion current flowing in the local and tiny electrochemical cells on the exposed areas on the metal surface. Thus, enhanced corrosion rates are usually observed in marine atmospheres and industrial areas.

Further, the presence of suspended particles in the atmospheric air also influences corrosion. Chemically inert particles such as coal, dust or charcoal particles may absorb moisture or gases and hence lead to increased corrosion of the metal exposed. Similarly, particles of chemically active salts like NaCl, $(NH_4)_2 SO_4$ etc. can absorb moisture and provide electrolytes necessary for electrochemical corrosion attack.

(3) **Nature of corrosion products formed.** Metals like Ni, Cr, Al and alloys like stainless steel exhibit good resistance for atmospheric corrosion due to their tendency to form thin and firm protective oxide film. However, metals like Cu and Pb, though not capable of forming highly protective films, develop atmospheric corrosion resistance due to the formation of secondary layers of corrosion products. These layers offer corrosion resistance by forming physical barriers between the metal and the environment. The extent of protection naturally depends upon coherence of the protective layer formed and its critical humidity. Copper forms a coherent and adherent layer of basic copper carbonate or basic copper sulphate which offers resistance to atmospheric corrosion. However, in case of iron, the primary corrosion product, ferrous hydroxide is oxidised to basic ferric carbonate and then to hydrated ferric oxide (rust), which is non-adherent and incoherent. One of the reasons for this behaviour is that the oxidation of $Fe(OH)_2$ to $Fe(OH)_3$ is accompanied by decrease in volume. The rate of atmospheric corrosion of iron depends upon the extent of humidity present in the atmosphere, the degree of pollution of the atmospheric air and the frequency of exposure to rain.

4.10.6. Factors influencing corrosion. The rate and extent of corrosion depend majorly on:

(1) the nature of the metal, and (2) the nature of the environment. It will be very useful to discuss these factors in detail.

1. Nature of the metal

(a) **Oxidation potential.** The extent of corrosion depends upon the position of the metal in the electrochemical series (Table -1, Section 4.2) and galvanic series (Table-2 Section 4.3). When two metals are in electrical contact in presence of an electrolyte, the metal higher up in the galvanic series becomes anodic and suffers corrosion. Further, the more the two metals are apart in the galvanic series, the greater will be the difference in their oxidation potential and hence the faster will be the corrosion of the anodic metal.

(b) **Overvoltage.** If pure Zn is placed in 1N H_2SO_4, it undergoes corrosion forming a film and generating bubbles of hydrogen gas on the immersed metal surface. Despite the high position of the metal as compared to hydrogen in the electrochemical series (Table-1), the initial rate of reaction is very slow because of the high overvoltage (about 0.70V), which reduces the electrode potential or the driving force for corrosion to a small fraction of a volt. However, the corrosion of Zn can be accelerated by adding a drop of $CuSO_4$ because some copper plates out on the zinc forming minute cathodes at which the hydrogen overvoltage is only 0.33 V. If a drop of platinic chloride is added, the corrosion is still faster because the overvoltage on the platinised spots is still less (0.2V). Thus, the reduction in overvoltage plays an important part in accelerating corrosion.

(c) **Relative areas of the anode and cathode.** When two steel plates of the same area are separately connected, one to a copper plate (cathode) having the same area and the other to a copper plate having a much larger area, it can be observed that the latter couple produces greater amount of current and the corrosion of the steel anode in this case will be much greater than the first one. It may be broadly concluded that when anodic polarization is negligible and conductance remaining more or less constant, corrosion of the anode is directly proportional to the area of the cathode.

Although the corrosion current is same at both cathode and anode, the current density at the smaller anode will be much greater. The great demand for electrons by the larger cathodic area will have to be met by the smaller steel anode by forming more Fe^{++} ions and hence the attack will be more severe.

(d) **Purity of the metal.** The impurities present in a metal form minute galvanic cells with the metal under appropriate environment and the anodic part gets corroded. For instance, impurities such as Pb, Fe or C in zinc lead to the formation of tiny electrochemical cells at the exposed part of the impurity and the corrosion of zinc around the impurity takes place due to "local action". The rate of corrosion increases with the increasing exposure of the impurities. The effect of even traces of impurities on the rate of corrosion of zinc can be seen from the following data:

Metal	% purity	Corrosion rate
Zinc	99.999	1
Zinc	99.99	2,650
Zinc	99.95	5,000

However, impurities which form solid solution in an alloy are homogeneous and hence do not form local action cells. But alloys having grain structure in which the electrode potentials of the crystals and the matrix are different may undergo appreciable corrosion. Dezincing of brass provides an example of this kind.

(e) **Physical state of the metal.** The rate of corrosion is influenced by the physical state of a metal. For instance, the smaller the grain size of the metal or alloy, the greater will be the solubility (as compared to the macroscopic crystals).

The corrosion rate may also be influenced by the orientation of the crystals at the metal surface. For example, the corrosion rate of copper ions was found to be different on different faces of a pure copper crystal.

Further, even in a pure metal, the areas under stress tend to be anodic and corrosion takes place. Stress corrosion cracking has been observed in several alloys such as alpha brass and iron and Al alloys in appropriate environments. Electron microscopic studies on stainless steels have also proved the formation of chrominum oxide platelets at the points under stress while they were not detected at the points where there was no stress.

The effect of stress on corrosion can be demonstrated by dropping a nail in ferroxyl indicator. It will be observed that a blue colour will be seen at the head and point of the nail (which were placed under stress during its manufacturing process) indicating the anodic areas while the region along the nail turns red showing the cathodic area. Similar behaviour can be observed with a piece of iron wire hammered at the middle which shows the corrosion at that point under stress.

Residual stresses in a metal could be relieved by annealing at suitable temperatures.

(f) Nature of the oxide film. This aspect has already been discussed under direct chemical attack theory. In aerated atmosphere, practically all metals get covered with an oxide film having a thickness of a few angstroms ($1A° = 10^{-10}$m). The film may contain one or more forms of the metal oxide and its thickness depends upon the nature of the metal and the temperature. Metals such as Mg, Ca, Ba, Li, Na and K form oxides whose specific volume is lesser than that of the metal atom. Hence the oxide film formed will be porous through which oxygen can diffuse and bring about further corrosion. On the other hand, Al and heavy metals form oxides whose specific volume is greater than that of the metal atoms and the impervious oxide film so formed will protect the metal from further oxidation, unless a crack or fissure in the film develops (pilling - Bedworth's Rule).

The ratio of the volumes of the metal oxide to the metal is known as specific volume ratio.. The specific volume ratios for W, Cr and Ni are 3.6, 2.0 and 1.6 respectively, which indicates that the rate of oxidation at elevated temperatures is least for tungsten.

The oxide films may be rendered thicker by prolonged heating, by chemical oxidizing agents and by anodizing.

A passive film can be formed on iron by dipping in strong oxidizing agents like HNO_3. However, this passive film is rather fragile and the protective quality is lost when the film is fractured. On the contrary, passive films formed on stainless steel are quite stable and offer adequate protection, unless the oxide film is destroyed by reduction.

Thus, corrosion depends upon the specific volume ratio of the oxide and the metal under consideration, the thickness and conductance of the oxide film and its reaction with the environment.

(g) Solubility of the products of corrosion. Solubility of the corrosion products formed is an important factor in electro-chemical corrosion. If the corrosion product is soluble, corrosion of the metal will proceed faster. On the other hand, if the corrosion product formed is insoluble (*e.g.*, many metal oxides) or if it forms another insoluble product by interaction with the medium (*e.g.*, $PbSO_4$ in the case of Pb in H_2SO_4 medium), then the protective film formed tends to suppress corrosion.

(2) Nature of the environment

(a) Temperature. The rate of chemical reactions, and the rate of diffusion increase with temperature and polarisation decreases. Hence, in general, it may be said that corrosion increases with temperature. But, however, solubility of gases such as O_2 which affect corrosion decreases with temperature. A passive metal may become active at a higher temperature. Intergranular corrosion such as caustic embrittlement takes place only at high temperatures in high-pressure boilers.

(b) Presence of moisture. Atmospheric corrosion of iron is rather slow in dry air but increases rapidly in the presence of moisture. This is mostly due to the fact that moisture acts as the solvent for oxygen in the air, other gases or salts to furnish the electrolyte essential for setting up a corrosion cell. Moisture also reacts with the metal or the oxide in some cases. Metals like Mg, Al, Zn, Cr, Mn and Fe may be corroded in presence of water even in the absence of oxygen.

Rusting of iron increases appreciably when the relative humidity of air reaches 60 to 80% and the rusted spots act as corrosion centre. Similarly, particles of dust, soot, fly ash and charcoal in presence of moisture may act as corrosion centres.

(c) Effect of pH. The hydrogen ion concentration of the medium is another important factor in corrosion reactions as well as corrosion control. Acidic media are generally more corrosive than alkaline and neutral media. Amphoteric metals like Al, Zn and Pb form complex ions in alkaline solutions and go into solution. The corrosion of iron is slow in oxygen-free water until the pH falls below 5. The corrosion rate is much higher in presence of oxygen. At pH 4, corrosion is stimulated by the oxidation of Fe^{++} to Fe^{+++} by dissolved oxygen and the subsequent reduction of the Fe^{+++} to Fe^{++} at the cathode. In less acidic solutions, the excess of OH^- ions in the area combine with Fe^{++} to form $Fe(OH)_2$ which undergoes further oxidation to form rust.

Many materials are resistant to alkalis although they are readily attacked by acids. In such cases, their corrosivity can be reduced by increasing the pH of the solution. pH can also have a major role to play on the resistance to stress-corrosion cracking and pitting.

The possibility of corrosion with respect to pH of the solution and the oxidation potential of the metal can be correlated with the help of Pourbaix diagram. It deals with thermodynamic equilibria and gives valuable information regarding the pH-potential conditions where corrosion can be expected and how the corrosion could be minimized. For example, let us consider the Pourbaix diagram for iron in water, depicted in Fig. 4.7-A and B.

Fig. 4.7(A) clearly shows the zones of corrosion, immunity from corrosion and passivity and hence provides a pictorial idea of corrosion control methods. Fig. 4.7(B) shows that for the system of iron in neutral water (pH = 7) with a corrosion potential of –0.4 V, represented by the point X, corrosion is possible on the basis of thermodynamic considerations and this has been proved in practice also (that iron rusts in water under these conditions). From the diagram it can be seen that iron would be immune to corrosion if the potential is changed to about –0.8V. This can be achieved by the application of an external circuit as in the case of "cathodic protection" discussed later. On the other hand, the corrosion rate can be greatly reduced by moving into the passive range, *i.e.*, to potentials of zero volt or higher. This also can be achieved with an external circuit and is called "anodic protection" discussed later. Yet another alternative is to increase the pH of the originally neutral water by adding an alkali. However, both the methods namely anodic protection and increasing the pH should be done under carefully controlled conditions because, a border line situation can be even more dangerously corrosive than that of the original situation. Pourbaix diagrams are available for many other metals also.

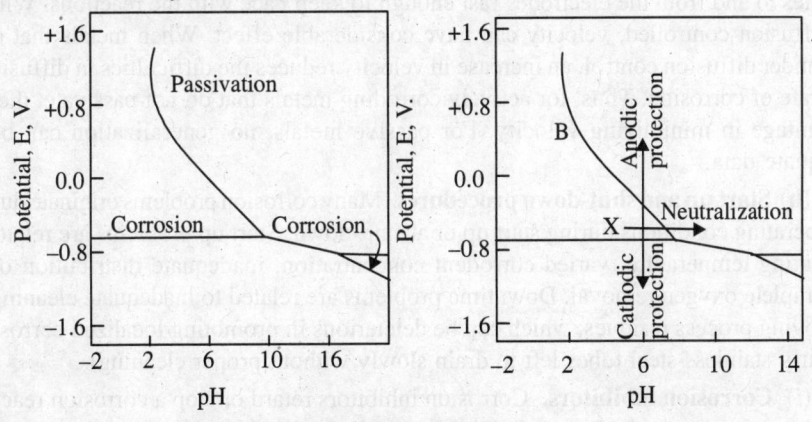

Fig. 4.7 A & B. Pourbaix diagram

Zinc is rapidly corroded even in weakly acidic solutions such as carbonic acid. Even fermenting organic matter in a galvanised container can corrode the zinc from the contrainer. Zn suffers minimum corrosion at pH 11 but at higher alkalinities, it goes into solution as complex. Al has minimum corrosion rate at about pH 5.5 while Sn is rapidly corroded at pH greater than 8.5.

(d) Nature of anions and cations present. Chloride ions present in the medium destroy the passive film and corrode many metals and alloys. On the contrary, some anions like silicate may form an insoluble reaction product (*e.g.,* silica gel) which inhibits corrosion.

Traces of Cu or more noble metals accelerate the corrosion of the iron pipes carrying mine waters. Many metals including iron corrode more rapidly in ammonium salts than in sodium salts of identical concentrations.

(e) Conductance of the medium. The conductance of the medium is of profound importance in the corrosion of underground or submerged structures because the corrosion current depends on this factor. Conductance of clay and mineralized soils is higher than that of dry sandy soils. Hence, stray currents from power leakages will damage the metal structures buried under soils of higher conductance to a greater extent than those under dry sandy soils having higher resistance.

(f) Concentration of oxygen and formation of oxygen concentration cells. The rate of corrosion increases with increasing supply of oxygen. Hydrogen depolarization also increases under these conditions. Further, differential aeration will also promote corrosion by the formation of concentration cells. The region where oxygen concentration is lesser becomes anodic (with respect to the regions which are more exposed to oxygen) and suffers corrosion. The mechanism of corrosion by differential aeration has already been discussed under the electrochemical theory of corrosion.

Corrosion often takes place under metal washers where oxygen cannot diffuse readily. Similarly, buried pipe-lines and cables passing from one type of soil to another suffer corrosion due to differential aeration. Lead-pipeline passing through clay and then through cinders may undergo corrosion because the pipeline under cinders is more aerated which gives rise to potential difference.

Water-line corrosion and crevice corrosion are also due to differential aeration leading to the formation of oxygen concentration cells.

(g) Flow velocity of process streams. In order to understand the effects of velocity, we must remember that corrosion is controlled by polarization (*i.e.,* slowing down of the reactions) at the anodes and the cathodes. There are two main types of polarization. Activation polarization results from slow step in a chemical reaction occurring at the anode or cathode, *e.g.,* slow transfer of electrons. Concentration polarization is a slowing down of reaction due to difficulty in diffusing species to and from the electrodes fast enough to keep pace with the reactions. When the corrosion is diffusion controlled, velocity can have considerable effect. When metals that do not passivate are under diffusion control, an increase in velocity reduces the difficulties in diffusion and increases the rate of corrosion. Thus, for actively corroding metals that do not passivate, there may be some advantage in minimizing velocity. For passive metals, no generalization can be made without adequate data.

(h) Start up and shut-down procedures. Many corrosion problems originate due to irregularities in operating conditions during start up or at shut-down. Start up problems are related to too high an operating temperature, varied corrodent concentration, inadequate distribution of inhibitors and incomplete oxygen removal. Downtime problems are related to inadequate cleaning procedures for removing process residues, which can be deleterious in promoting localized corrosion (*e.g.,* pitting of dirty stainless steel tubes left to drain slowly without proper cleaning).

(i) **Corrosion inhibitors.** Corrosion inhibitors retard or stop a corrosion reactions. They may be inorganic or organic. Inorganic inhibitors such as silicates, chromates, borates etc., suppress the rate of corrosion by acting on the anode. Alkaline sodium nitrite alone or in combination with other inhibitors such as phosphate has been used to control corrosion of tankers and pipe lines. Sodium benzoate has been used as an inhibitor for mild steel and in preventing corrosion in cooling

systems such as automobile radiators. A mixture containing 0.1% $NaNO_2$ and 1.5% sodium benzoate has been used as inhibitor in antifreeze solutions. Lime acts as a cathodic inhibitor by precipitating $CaCO_3$ in water containing temporary harndess or dissolved CO_2. Colloidal particles of $CaCO_3$ having positive charge are attracted to the cathodic areas and get deposited there, thus reducing the corrosion current.

Organic inhibitors act by different mechanisms. For instance, organic colloidal inhibitors form protective layers on the metal surface by adsorption. Surface active reagents (surfactants) containing polar groups promote spreading and oriented adhesion to the surface thus forming a protective film. Organic bases *e.g.,* amines, pyridine, quinoline and their derivatives contain hydrophobic groups or radicals. These positively charged cationic groups attach themselves to the cathodic areas through the nitrogen and provides inhibition. The efficiency of inhibition depends upon the size and number of the alkyl groups. For example, primary amyl amine ($C_5H_{11}NH_2$) has been found to be more effective as inhibitor than primary ethyl amine ($C_2H_5NH_2$).

High molecular weight amines derived from rosin have been used as corrosion inhibitors. The product obtained by rosin amine and pentachlorophenol was used in castings for underground pipelines to protect them from corrosion by soil bacteria. Organic inhibitors (such as soluble salts of rosin *e.g.,* stearate or naphthenate) have been used as inhibitors in the various metal cleaning operations like acid "pickling" of metals, cleaning of boilers, condensers, heat exchangers, chemical equipment, pipe lines, etc., to prevent undue and excess corrosion of the metal while removing rust, scales and deposits.

It should be emphasized here that inhibitors should be used with caution. Improper or insufficient use of an inhibitor may even accelerate corrosion. For instance, addition of chromate as inhibitor in presence of hydrogen polarization may have depolarization effect and corrosion may be aggravated. Similarly, insufficient use of an inhibitor may not provide the protective film over the entire anode and the small areas left exposed, suffer intense corrosion leading to pitting.

Vapour phase inhibitors (VPI) are such organic inhibitors which readily vaporize and form a protective layer of the inhibitor on the metal surface. These are conveniently used to avoid corrosion in enclosed spaces and also during storage, packing, shipping etc. Vapour phase inhibitors are used to protect steam lines in the presence of CO_2. They are also used to protect iron and steel parts in presence of moisture and corroding gases like SO_2. Small metal parts can be protected from atmospheric corrosion by keeping them in envelopes made from paper impregnated with a suitable vapour phase inhibitor. For storing spare aircraft engines, crystals of VPI are blown into each cylinder through the spark-plug hole and then the entire engine is wrapped in plastic-coated low-porosity paper containing some VPI. Dicyclohexylammonium nitrite and cyclohexyl amine carbonate are amongst the most successful vapour phase indicators widely used. Octadecyl amine, hexadecyl amine and dioctadecyl amine do not come under the VPI but they provide protective action by forming impervious non-wettable surface films. These are used in radiator compounds and are also added to boiler waters. Amine salts, morpholine and sodium benzoate are particularly useful for impregnation of packaging materials.

4.11.7. Passivity. A metal is said to be passive in a certain environment if it exhibits much lower corrosion rate than what is expected thermodynamically or from its position in the electro-chemical series. A passivated metal can be rendered active by a change in the environmental conditions.

Metals which exhibit passivity include iron, chromium, nickel, titanium and alloys of these metals. Passivation is usually associated with oxidizing environment and are characterized by the formation of a very thin protective oxide films. However, if should be emphasized that a metal when is passive in one environment may not necessarily be so in a different environment, because passivity results from a continuing reaction between a metal and its environment. When a metal is passive, it corrodes at a very slow rate. Hence, passivity offers an excellent method of corrosion control.

The passivity of iron in concentrated solutions of nitric acid has been studied for over a century because of its potential to develop methods for corrosion suppression. Passivation treatment used for stainless steel comprises of immersing it in an oxidizing solution *e.g.*, HNO_3 for about half an hour. The prime purpose of such a treatment is not (as is generally believed) to form passive film, but rather to clean the steel by removing surface inclusions, iron particles, etc., which might act as nucleation sites for possible attack in future service. The passive film conditions that exist on the surface of the steel when it is dipped in HNO_3, do not persist when it is transferred to another environment. The passivation treatment is only helpful in that it creates surface conditions that may make the steel more amenable to passivation in future service. A clean stainless-steel surface will passivate itself in clean air irrespective of whether it has received a passivation treatment.

4.11.8. Types of Corrosion. Different forms of corrosion are diagramatically shown in Fig. 4.8

(1) *Uniform and galvanic corrosion.* This is the most common form of corrosion and can be dry or wet, chemical or electrochemical. Galvanic corrosion may take place when two different metals in contact (or connected by an electrical conductor) are exposed to a conductive solution. The difference in electrical potential between the two metals provides the driving force to pass current through the corrodent and results in corrosion of the anodic metal close to the junction of the two metals (vide Fig. 4.8). The larger the potential difference between the two metals, the greater may be the galvanic corrosion. The relative areas of the cathodic and anodic metals are also important. A much larger area of the more noble metal compared to the active metal, accelerates the attack. The various methods for protection from corrosion will be discussed later.

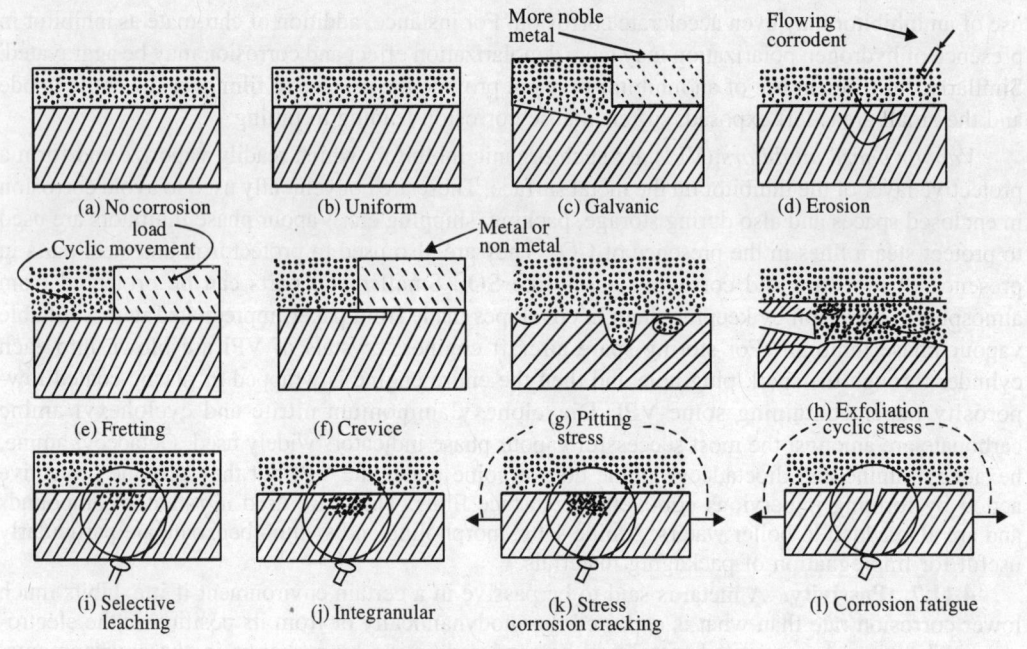

Fig. 4.8. Different types of corrosion.

(a) No corrosion (b) Uniform (c) Galvanic (d) Erosion
(e) Fretting (f) Crevice (g) Pitting stress (h) Exfoliation cyclic stress
(i) Selective leaching (j) Integranular (k) Stress corrosion cracking (l) Corrosion fatigue

(2) *Erosion corrosion.* When movement of a corrodent over a metal surface increases the rate of attack due to both mechanical wear and corrosion, the type of attack so produced is known as erosion corrosion. The major cause for erosion is the removal of protective surface films, *e.g.*, air-formed protective-oxide films or adherent corrosion products. Erosion corrosion generally occurs during high-velocity conditions, turbulence, impingement, etc. and is frequently observed on pump impellers, agitators and piping at bends and elbows. Erosion corrosion generally appears as smooth bottomed shallow pits (vide Fig. 4.8). Cavitation and fretting corrosion are special forms of

erosion corrosion. Cavitation is caused by the formation and collapse of vapour bubbles at the metal surface and the high pressure so produced may deform the underlying metal or destroy the protective film on it. Fretting corrosion occurs during the sliding of metals over each other which causes mechanical damage as well as the increased corrosion due to the continuous mechanical removal of protective films and increased heat of friction, both of which accelerating the oxidation of the metal. Erosion corrosion can be minimized by using harder metals and design changes to avoid excess of friction and using proper lubrication.

(3) *Crevice corrosion.* This is a type of local corrosion and is usually created by dirt deposits, corrosion products, crack in paint coatings, etc. This can be commonly observed near the gaskets, bolts, rivets, lap joints, etc. Crevice corrosion is usually attributed to changes in acidity in the crevice, lack of O_2 in the crevice and concentration of a detrimental ionic species in the crevice. Selection of resistant materials, proper design to minimize crevices and maintaining clean surfaces are the measures taken to control crevice corrosion.

(4) *Pitting corrosion.* Pitting corrosion is a localised attack resulting in the formation of holes in an otherwise relatively unattacked surface. The shape of the pit is often responsible for its continued growth for the same reasons mentioned under crevice corrosion. A pit may be considered to be a self-formed crevice.

Pitting is generally due to heterogeneity in the metallic surface in presence of a corrording environment, particularly chloride solutions in presence of a depolarizer. Cracking of a protective surface film on a metal at some points in the presence of an appropriate environment leads to concentrated attack at those points (because of very small anodic area as compared to very large cathodic area) forming holes or pits (vide Fig. 4.8).

A pure and homogeneous metal with a highly polished surface will be much more resistant to pitting than the one with many inclusions, defects and a rough surface. Surface cleanliness and selection of proper materials known to be resistant to pitting in the given environment are the usual methods to combat this problem.

(5) *Exfoliation and selective leaching.* Exfoliation is a subsurface corrosion which starts on a clean surface and spreads below it. It differs from pitting in that the attack has a rather laminated appearance (vide Fig. 4.8) and whole layers of material are corroded. The attack is usually recognized from a flaky or blistered surface. This type of corrosion is known to take place in aluminium alloys and can be avoided by heat treatment and proper alloying.

Selective leaching (parting) is the removal of one of the elements in an alloy. Dezincification from copper-zinc alloys is the most common example. This type of corrosion is highly undesirable as it yields a porous metal with poor mechanical properties (vide Fig. 4.8). This can be remedied by the use of nonsusceptible alloys.

Selective leaching of zinc is noticed in the condenser tubes (made of brass) in marine boilers, where sea water is used as condenser water. Although both Zn and Cu go into solution, only copper (being more noble) is redeposited as a spongy layer while dezincing proceeds continuously.

(6) *Inter-granular corrosion.* When a molten metal is cast, the solidification starts at many randomly distributed nuclei. Each of them grow in a regular atomic array to form what are called grains. The arrangement of atoms and the spacing between the layers of atoms are identical in all the grains of a given metal. However, because of random nucleation, the planes of atoms in neighbouring grains do not match up. Such areas of mismatch between the grains is called grain boundaries. A one inch long line drawn on an alloy surface would cross as many as thousand grain boundaries.

Grain boundaries are generally more susceptible for attack by a corrodent because of segregation of specific elements or the formation of a compound in the boundary (because of the atomic mismatch existing there). Corrosion occurs because of preferential attack by the corrodent near the grain boundaries or zones adjacent of them which might have lost an element necessary for adequate corrosion resistance (due to the precipitation of certain compounds at the grain boundaries thus leaving the solid solution adjacent to them depleted in one of the constituent elements). In extreme cases of grain-boundary corrosion the affected grains are totally dislodged due to complete deterioration of their boundaries (vide Fig. 4.8).

Alloys are generally more susceptible to intergranular corrosion. During welding of staninless steel, chrominum carbide gets precipitated at the grain boundaries thus depleting the adjacent region of chromium, which therefore becomes more anodic to the alloy composition in the grain centre and also to the precipitated chromium carbide. This renders the regions near the grain boundaries more susceptible for corrosion. (The regions low in chromium have lesser tendency to become passive as compared to the main portion of the grain). The grain boundary corrosion usually is a microscopic attack and proceeds along the grain boundary until the affected grain is totally dislodged. Thus it leads to sudden failure without any apparent indication of a severe attack.

Susceptibility to intergranular attack usually results from a heat treatment (such as welding or stress-relieving operations) and it can be remedied by another heat treatment or by using a modified alloy.

(7) *Stress-corrosion cracking.* The combined action of a tensile stress and a corrodent sometimes brings about cracking of a metal or alloy. Many alloys are susceptible to this type of cracking. Stresses that cause cracking result from residual cold working and quenching, welding, thermal treatment or due to applied loads during service. In such cases, the metal under stress become more anodic and tend to increase the rate of corrosion.

Stress corrosion cracking can be combated by (1) suitable treatment to relieve internal stresses, (2) adjusting the composition and eliminating certain impurities to prevent the discrepancies in composition between the grains and the grain-boundary material, (3) suitable heat treatment to prevent the above mentioned heterogeneity, (4) removing the critical environmental species and (5) selecting a more resistant material.

Stress corrosion depends on the metallurgical condition of the alloy. It requires a threshold tensile stress below which cracking does not occur. Stress corrosion occurs only in some specific environments for a given alloy. It may take a long time before cracks become visible but then the cracks propagate quite rapidly resulting in an unexpected failure.

Failure often observed in cold-drawn brass articles such as cartridge cases when exposed to environments containing ammonia are due to the so called "season cracking". Such a tendency can be eliminated by a low temperature annealing to remove internal stresses without much loss of hardness.

Caustic embrittlement of boiler plates is another type of stress corrosion resulting in intergranular attack and failure.

Corrosion fatigue is a special type of stress corrosion cracking. In this, failures are caused by repeated cyclic stresses (such as shaking, vibration, tapping, "shuttering" and flexing), in presence of a corrosive environment although the stress is well below the normal fatigue limit. Alloy steels are often more susceptible to this type of corrosion. The stress alterations "work-harden" small regions of metal around internal cavities or surface flaws, thereby distorting the metal structure and making it less elastic. Such a work-hardened material becomes anodic with respect to the bulk metal in a moist atmosphere. Corrosion of the anode occurs and any surface crack or internal flaw spreads through the metal.

(8) *Water-line corrosion.* This type of corrosion results from differential aeration leading to the formation of oxygen concentration cells. When water is stagnant in a steel tank for long time, it is observed that corrosion takes place just below the water level (Fig. 4.9). The concentration

of dissolved oxygen at the water surface is greater than that under the surface. This generates an oxygen concentration cell in which the metal at the water level is cathodic with respect to the metal below the water level. Owing to the poor conductivity of water, the ions just below the water level are more readily available for the reactions and hence the metal just below the water level gets corroded. The intensity of attack may not be considerable under normal conditions but in some practical circumstances, the attack may be intense. If the liquid is in motion and contains chlorides and anodic inhibitors such as hydroxide, carbonate, phosphate, chromate or silicate are inadequate to stifle the corrosion reactions completely, intense attack at the water line may take place. Changing position of the water line also favours an intense attack because, a porous layer of the corrosion product becomes more adherent when it is above the water level; and then, when it is submerged again, the part of the metal under the deposit receives much less oxygen and becomes anodic. Water line corrosion is a serious problem for ocean-going ships because the attack is further accelerated due to the marine plants attaching to the sides of the ship. This can be combated by using special anti-fouling paints.

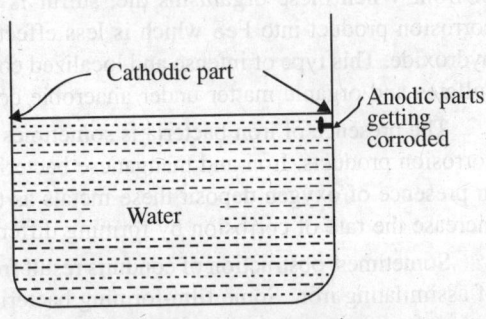

Fig. 4.9. Water-line corrosion.

(9) *Soil corrosion.* The enormous amount of money invested on underground structures and pipe lines (for oil, gas and water) encouraged considerable research on the constitution of soils and their corrosive properties. The various factors that are responsible for soil corrosion include (*a*) acidity of the soil, (*b*) moisture content, (*c*) content of electrolytes, (*d*) micro-organisms present, (*e*) content of organic matter and (*f*) physical properties of the soil.

In non-acid soils, the conductivity (which depends on moisture and electrolyte content) is the major factor governing the corrosive character. In such soils, corrosion takes place due to the formation of differential aeration couples and the rate of corrosion majorly depends upon the resistance between the anodic and cathodic areas and the rate of arrival of oxygen to the cathodic area. Thus when a cable or pipe passes under a paving, the portion under the paving has less access to oxygen than the one lying under unpaved soil. Hence the portion under the paving becomes anodic and suffers corrosion.

In highly acidic soils, the corrosion taking place is of hydrogen type with anodic and cathodic areas being very close to each other. The conductivity of the soil is not of much importance in this case.

In soils which contain larger amounts of organic matter, the formation of soluble metal complexes and the peptization of corrosion products may accelerate the corrosion process as compared to what is expected in the soils relatively free from organic matter.

Soils rich in gravel and sand are more porous and hence are generally more aerated and the corrosion of a metal pipe under such soils depends mostly on the moisture and salt content. Water-logged soils may generate anaerobic bacteria which may create conditions for microbiological corrosion.

Buried pipelines or cables passing from one type of soil to another say, from clay (less aerated) to cinders (more aerated) may get corroded due to differential aeration. Other corrosion factors in this case are the differences in pH and the presence of unburnt carbon in the cinders.

Air-pockets in the soil, if present, may also cause corrosion due to differential aeration.

(10) *Micro-biological corrosion.* Some types of bacteria exert some influence on the corrosion process. Oxygen-consuming bacteria present in water or soils decrease the concentration of oxygen in the medium in contact with a metal structure. If the metal structure is totally exposed to the

medium in which oxygen has been depleted by bacteria, then it may reduce corrosion. But, if the structure is only partially exposed to the oxygen-depleted environment, then the bacterial action may intensify the corrosion due to differential aeration.

Anaerobic bacteria such as "microspira or vibrio desulfuricnas" reduce sulfates to sulfur which is used to prepare their protoplasm. During this process, it is supposed that these bacteria convert the O_2 from the sulfates into such a form which brings about a depolarization effect on the corrosion of iron. When these organisms die, sulfur is liberated as H_2S which converts a portion of the corrosion product into FeS which is less effective in suppressing the corrosion than the oxide or hydroxide. This type of intense and localized corrosion is come across with cast iron in presence of sulfates and organic matter under anaerobic conditions.

The presence of iron bacteria is sometimes connected with pitting or blockage of pipes due to corrosion products. Iron and manganese bacteria which thrive on iron and manganese compounds in presence of oxygen deposit these metals as their insoluble hydrated oxides. These may further increase the rate of corrosion by forming differential aeration couples.

Sometimes, obstruction of conduits result from the growth of low-form algae which are capable of assimilating iron. Some film-forming bacteria are also known to induce corrosion.

Desulfovibrio and Clostridria are known as corrosive bacteria. Slime forming bacteria *e.g.,* Flavobacterium, mucoids, Aerobacter and Pseudomonos lead to corrosion in cooling tower systems.

(11) *Stray current corrosion.* Metal structures such as water pipes, gas pipes and cable sheaths, adjacent to D.C. circuits may get corroded due to leakages from the main circuit.

Whenever a material is selected for a particular job, special attention must be paid regarding the possibility of localized corrosion, particularly, pitting and stress corrosion cracking.

4.11.9. Testing and Measurement of Corrosion.
Measurement of corrosion in a process plant is usually done to achieve the following objectives:

(1) Monitoring the corrosion process taking place in the plant.
(2) Evaluation of the quality of a specific lot of material being used (such as whether correct heat treatment has been given or not).
(3) Evaluation of materials and environmental effects to serve as guidelines for future use.
(4) Studying the mechanism of corrosion.

Corrosion tests which are performed to achieve these objectives may be classified into the following three groups:

1. Service tests in actual process streams or natural environments.

2. Simulated-service tests in which the anticipated service conditions are simulated in the laboratory. These may vary from beaker tests to pilot plant operations.

3. Accelerated tests in which the test media may or may not be related to the intended service conditions. These are further subdivided as follows:

(a) *Quality control tests*: These are the accelerated tests in which the test media may not be related much to the intended service conditions but, at the same time, they are capable of quick identification of a metallurgical or surface condition which may be deleterious under actual service performance.

(b) *Screening tests*: These are used for new alloys or deciding suitable materials for a particular job. In such cases, the materials are ranked by testing in an environment which is similar but not identical with the expected or intended service conditions. For instance, materials used for a strongly oxidizing process stream is screened on the basis of performance in a nitric acid solution. Similarly materials used for automobile trim are screened on the basis of high-humidity or salt-spray tests.

CORROSION

(c) *Accelerated simulated service tests.* These are used to study corrosion mechanisms. In these tests, one or more amongst the intended service conditions (such as temperature) are changed to obtain in a short time, the same type and extent of corrosion that would occur in actual service.

The measurement of corrosion is usually done by the following methods:

1. Weight-loss method. Weight-loss tests are most commonly used for measurement of corrosion. In this, a clean metal coupon or a standard test piece is measured, weighed and exposed to the corrodent for a known time. The piece is then taken out, cleaned to remove the corrosion products and reweighed. The rate of corrosion of the metal, R is then calculated from the equation:

$$R = \frac{KW}{ATD}$$

where K is a constant, T is time of exposure to the nearest 100th of an hour, W is the weight loss to the nearest milligram, A is the area to the nearest sq. cm., and D is density, g/cm^3. The results are expressed in many different units such as $mg/dm^2/day$ or $oz/ft^2/day$ or month or inches per year (IPY) or millimeters per year (MPY). Assuming that the surface corrosion was uniform, these units may be converted to depth of corrosion or penetration.

2. Microscopic Examination. The comparison of the numerical results obtained by the above method should be reinforced by microscopic examination of the test specimen to check the nature of the corroded surface, the number of pits formed and their depth, and the presence of intergranular corrosion.

3. Measurement of corrosion potentials. The corrosion tendency of a material can be estimated by coupling the test plate to a standard half cell and determining the corrosion current.

4. Measurement of electrical resistance. If the test sample is in the form of a thin wire or strip, its electrical resistance increases as corrosion decreases its cross-section. Hence, periodic or continuous measurement of the resistance between the ends of the specimen can be used to monitor the corrosion. The electrical resistance measurement has nothing to do with the electrochemistry of the corrosion reaction. Here, we are only measuring a bulk property which depends upon the cross-sectional area of the material.

5. Corrosometer. Electrical instruments such as corrosometers measure the change in resistance of a standard probe as corrosion converts the metal to a corrosion product. A second reference probe covered with a highly corrosion-resistance coating is connected to a bridge arrangement. The ratio of the resistance of the corroding test piece to the resistance of its non-corroding counterpart is directly related to the extent of corrosion.

In this method, corrosion rates can be measured, without removing out the test piece and without interrupting the process. Changes of corrosion can be detected early and remedial measures can be initiated.

6. Electrochemical tests. These are mostly used in laboratory work, particularly in studies of mechanism and for new alloys. Commercial instruments are available and are based on the electrochemical nature of corrosion. The amount of externally applied current needed to change the corrosion potential of a freely corroding specimen by a few millivolts is measured. This current is related to the corrosion current and hence the corrosion rate of the sample. These instruments are useful for in-plant measurements also.

4.11.10. Protection from corrosion or corrosion control

(A) Design and material selection:

1. Selection of the right type of material is the main factor for corrosion control. The choice of the metal should be made not only on its cost and structure but also on its chemical properties and its environment.

2. Noble metals are most immune to corrosion but they cannot be used for general purposes for economical reasons. The next choice is to use the purest possible metal. Even minute amounts of some impurities may lead to severe corrosion. For example, minute quantities of iron in magnesium or lead in zinc die-casting alloys may be highly detrimental.

3. Both corrosion resistance and strength of many metals can be improved by alloying. Several corrosion-resistant alloys have been developed for specific purposes and environments. Ex:

(i) Stainless steels containing chromium produce an exceptionally coherent oxide film which protects the steel from further attack.

(ii) Highly stressed 'Nimonic" alloys (Ni-Cr-Mo alloys) used in gas turbines are very resistant to hot gases.

(iii) Cupro-nickel (70% Cu + 30% Ni) alloys containing 0.2% Fe are now used extensively for condenser tubes and for bubble-trays used in fractionating columns in oil refineries.

4. Heat treatment like annealing helps to reduce internal stresses and reduce corrosion.

5. If an active metal is used, it should be insulated from more cathodic metals.

6. If two metals have to be in contact, they should be so selected that their oxidation potentials are as near as possible. Further, the area of the more noble or inactive metal should be smaller than that of the anodic or active metal. A protective coating over both the metals will reduce the chance of pitting. Special care should be taken to coat the anode completely because any scratch or crack on the anode coating can lead to intense attack at the exposed areas.

7. Moisture should be excluded wherever practicable (Stored metal parts sealed in a low permeability plastic in presence of activated silica or alumina gel can be protected for years together). If moisture or electrolyte solution is present, suitable inhibitors should be employed.

8. Each metal shows a minimum corrosion at a specific pH. Therefore, corrosion can be controlled by suitably adjusting the acidity or alkalinity of the environment. When control of pH is not practicable, corrosion can be reduced by using inert coating and inactive metals.

9. When contact of dissimilar metals is unavoidable, suitable insulators should be inserted between them to reduce current flow and attack on the anode.

10. Impingement attack can be reduced by careful filtration of suspended solids from the liquid stream and by preventing turbulent flow.

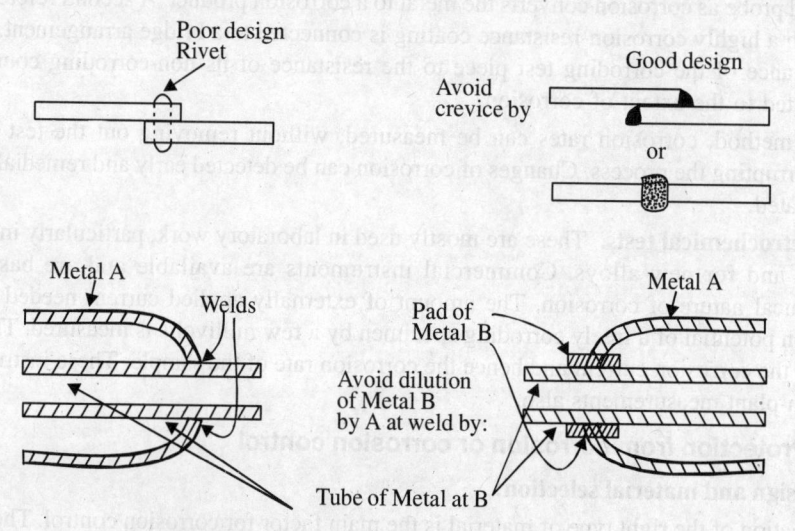

Fig. 4.10. Proper welding reduces corrosion.

CORROSION

11. When a structure consists of two dissimilar metals, it is beneficial to use a more active third metal in contact so that the structure will be saved from corrosion at the expense of the third metal.

12. The equipment should be so designed as to avoid localised stresses. Further, equipment design should avoid sharp bends, baffles and lap joints, which may produce stagnant areas with scales or sediments which contribute to corrosion by the formation of concentration cells. Tanks and pipelines should be free from obstructions or crevices (*e.g.,* badly riveted seams) in which water may stagnate, giving rise to corrosion because of differential aeration.

As far as possible, crevices should be avoided between adjacent parts of a structure to avoid the formation of concentration cells. Bolts and rivets should be replaced by proper welding. As far as possible, metal washers should be replaced by rubber or plastic washers as they do not absorb water and do not generate galvanic couples. Further, they act as insulators. A good design should take care of avoiding accumulation of dirt, etc., stagnation of water and allow for free circulation of air. On the basis of these general guidelines, some poor and good engineering designs are given in the following figures (Figs. 4.10-14).

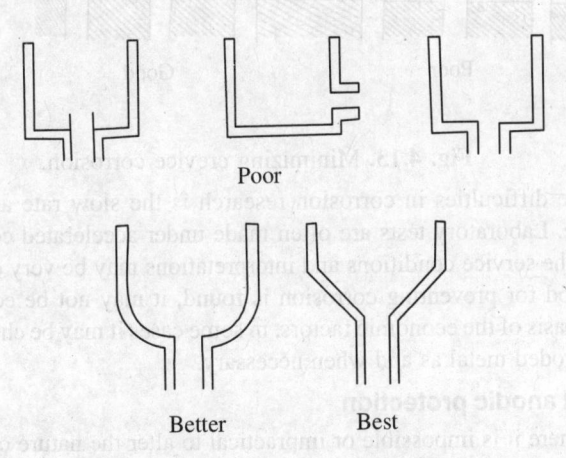

Fig. 4.11. Drainage affects corrosion.

Special Remarks

(1) The electrochemical series (or redox potentials) is not always reliable in predicting galvanic corrosion because conditions in corrosion cells are usually very different from those in which standard potentials are measured due to the involvement of other factors such as exchange current densities and polarization effects. Hence, indiscriminate application of standard electrode potentials may be dangerous. Many metal couples which are widely separated in a galvanic series may not fail by galvanic corrosion. Similarly, materials relatively close in the galvanic series may show considerable galvanic corrosion if their exchange current-densities (*i.e., reaction rates*) for their anode or cathode reactions are widely different. For instance, copper-base alloys and stainless steels are often used together in marine applications without problems. Thus, it is wise to measure potential differences between the metals and alloys in actual use rather than to estimate the effects from the electromotive series.

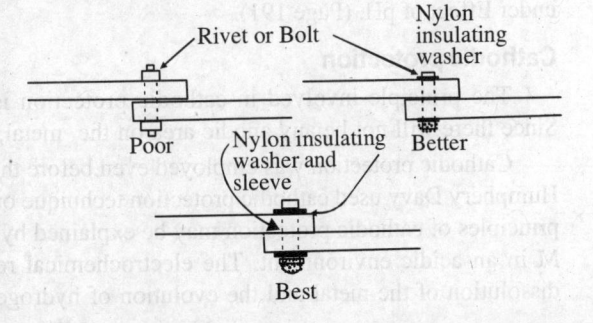

Fig. 4.12. Insulation avoids galvanic corrosion.

(2) Each corrosion problem should be tackled as a special case on the basis of the general guidelines and actual experimental data.

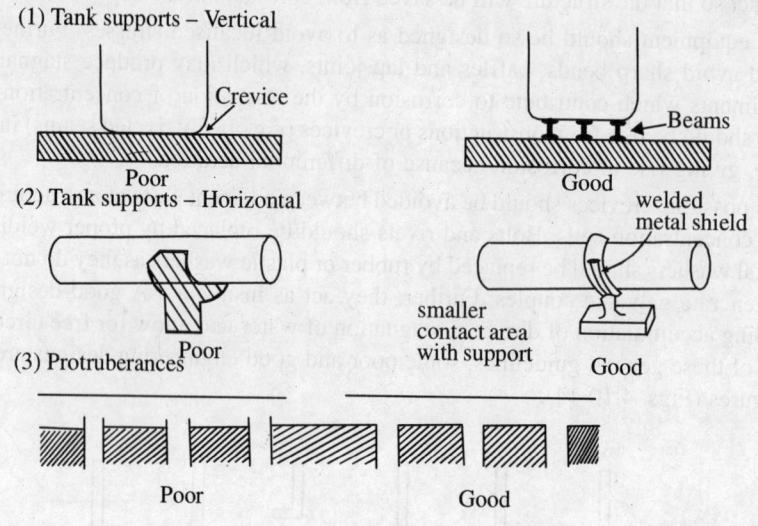

Fig. 4.13. Minimizing crevice corrosion.

(3) One of the difficulties in corrosion research is the slow rate at which many corrosion reactions take place. Laboratory tests are often made under accelerated conditions which may not truly represent the service conditions and interpretations may be very difficult. Further, even if a satisfactory method for preventing corrosion is found, it may not be economical to use it on a large scale. On the basis of the economic factors, in some cases it may be cheaper to accept corrosion and replace the corroded metal as and when necessary.

(B) Cathodic and anodic protection

In situations where it is impossible or impractical to alter the nature of the corrosion medium, corrosion control may be achieved by cathodic protection or by anodic protection.

When electrical current flows between the anodic and cathodic areas and on a corroding metal surface, the higher the current the greater and faster will be the corrosion at the anode. The rate of corrosion can be controlled by imposing additional current on the metal using an external circuit. If an opposing current is applied to nullify corrosion, it is called cathodic protection. If the potential of the metal is so adjusted that the corrosion is very much suppressed because the metal is rendered passive, then it is called anodic protection. Both the techniques are of practical importance in corrosion control and can be envisaged with reference to the Pourbaix diagram discussed earlier under Effect of pH. (Page 191).

Cathodic protection

The principle involved in cathodic protection is to force the metal behave like a cathode. Since there will not be any anodic area on the metal, corrosion does not occur.

Cathodic protection was employed even before the science of electrochemistry was developed. Humphery Davy used cathodic protection technique on British Naval ships as early as in 1824. The principles of cathodic protection may be explained by considering the corrosion of a typical metal, M in an acidic environment. The electrochemical reactions occurring in such a system are the dissolution of the metal and the evolution of hydrogen gas as per the following equations:

$$M \longrightarrow M^{+n} + ne$$
$$2H^+ + 2e \longrightarrow H_2$$

Cathodic protection can be achieved by supplying electrons to the metal structure to be protected. By examining the above two equations, it is obvious that the addition of electrons to the structure will tend to suppress metal dissolution and increase the rate of hydrogen evolution. If current is considered to flow from (+) to (−), as in conventional electrical theory, then a structure is projected if current enters it from the electrolyte. Conversely, accelerated corrosion occurs if current passes from the metal to the electrolyte. This current convention has been adopted in cathodic protection technology.

There are two types of cathodic protection:
(1) By using galvanic or sacrificial anode (galvanic protection).
(2) By using impressed current (Impressed current cathodic protection).

Galvanic protection

In galvanic protection, a more-active metal is connected to the metal structure to be protected so that all the corrosion is concentrated at the more-active metal and thus saving the metal structure from the corrosion. The more-active metal so used is known as "sacrificial anode." Metals commonly used as sacrificial anodes are Mg, Zn, Al and their alloys. Magnesium has the most negative potential and can provide highest current output and hence is widely used in high resistivity electrolytes such as soils. Zinc is generally used as sacrificial anode in good electrolytes such as sea water. Aluminium anodes are also used successfully but the more noble oxide films formed on them may create problems in some cases. While using sacrificial anodes other than the most active ones, it should be once again emphasised that electromotive series is not the only consideration in determining the galvanic effects because polarization behaviour and exchange-current densities may also play important role.

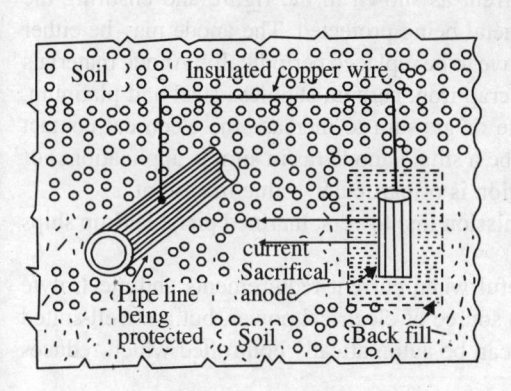

Fig. 4.14. Galvanic protection using sacrificial anode.

Important applications of cathodic protection include (a) protection from soil corrosion of underground cables and pipelines, and (b) protection from marine corrosion of cables, ship hulls, piers etc. Cathodic protection is frequently used in conjunction with coatings to reduce the cost and current capacity of the system. Magnesium bars are bolted along the sides of ships near the bilge keel for protecting the hulls. Megnesium rods are inserted into domestic water boilers or tanks to prevent the formation of rusty water. Calcium metal slugs are used to suppress engine corrosion. Galvanic protection of an underground pipeline is represented in Fig. 4.14.

Galvanic protection is often preferred to impressed current techniques when the current required is low and the resistivity of the electrolyte is relatively low (less than about 10,000 ohm-cm). Obviously, this is the method of choice where there is no source of electricity and when a completely underground system is desirable. This is the most economical method particularly for short-term protection because the capital investment is low.

Impressed current cathodic protection

In this method, an impressed current is applied to convert the corroding metal from an anode to a cathode. This can be accomplished by applying sufficient amount of Direct Current from a D-C source to an anode (e.g., graphite or high silican iron) buried in the soil or immersed in the corrosion medium, and connected to the corroding metal structure which is to be protected, as in Fig. 4.15.

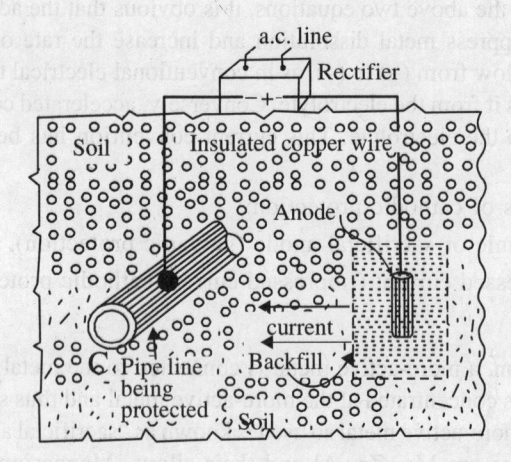

Fig. 4.15. Impressed current cathodic protection

This is usually done by rectifying an A–C current as shown in the figure and ensuring the electrical path between the new electrode and the metal being protected. The anode may be either an inert material or one which deteriorates and will have to be replaced periodically. Anode materials commonly used include graphite, high silica iron, scrap iron, carbon, stainless steel and platinum. The anode is buried in backfill such as coke breeze or gypsum to increase the electrical contact between itself and the surrounding soil. There may be a single anode (as in simple applications) or many anodes (as in pipe lines). This type of protection is of particular value in case of buried structures such as tanks and pipelines, transmission line towers, marined piers, laid-up ships etc.

Impressed current systems are particularly useful when current requirements and electrolyte resistivity are high. They, however, require a cheap source of electrical power, but are well-suited for large structures and long-term operation. They can be automatically controlled which reduces maintenance and operating costs.

Problems and limitations with cathodic protection

(1) A cathodic protection system which is efficiently protecting a pipeline may increase the corrosion of an adjacent pipeline or an adjacent metal structure because of stray currents. This may lead to unexpected legal and technical problems.

(2) Capital investment and maintenance costs.

(3) Chemical reactions occurring at the surface of the protected structure (which is the cathode in the structure) may cause problems. If the cathodic reaction produces hydrogen, it may have a deliterious effect (*e.g.,* blistering) on the metal itself. If the metal is coated, the coating may get peeled off.

(4) Special care should be taken to see that the structure is not overprotected, that is, we should avoid the use of potentials much higher than the open-circuit voltage for the metal/metal-ion couple or use of sacrificial anodes that are too active relative to the metal being protected. Otherwise, the problems associated with cathodic reaction such as evolution of H_2 or formation and accumulation of OH^- ions will be more.

(5) Possibility of soil and microbiological corrosion effects should also be considered in case of underground structures.

In spite of these problems and limitations, cathodic protection has been widely used with success, when suitable preautions are taken.

CORROSION

Anodic protection

This is another important form of corrosion control in which metal is passivated by applying current in a direction that renders it more anodic. The technique is only applicable to such metals and alloys which exhibit active-passive behaviour (vide pourbaix diagram in Fig. 4.7 given earlier). It has been applied mostly in case of steel and stainless steel and to some extent in case of Fe, Ti, Al and Cr. Like cathodic protection, this is also applicable when the corrodent is an electrolyte.

As stated above, anodic protection is based on the formation of a protective film on metals by externally applied anodic currents. Considering the two equations:

$$M \longrightarrow M^{+n} + ne$$
$$2H^+ + 2e \longrightarrow H_2$$

It appears that the application of anodic current to a metal structure should tend to increase the dissolution rate of a metal and decrease the rate of hydrogen evolution. This normally does occur except for those metals with active-passive transitions, such as iron, nickel, chromium, titanium and their alloys. If carefully controlled anodic currents are applied to these materials, they are passivated and the rate of metal dissolution is decreased. To protect a structure anodically, a device called a "potentiostat" is used. A potentiostat is an electronic device which maintains a metal at a constant potential with respect to a reference electrode. A potentionstat usually has three terminals; one is connected to the tank or structure that is to be protected, the other is connected to an auxiliary cathode (a platinum or a platinum clad electrode) and the third to a reference electrode (e.g., a calomel electrode). In actual operation, the potentiostate maintains a constant potential between the tank (or the structure being protected) and the reference electrode. The optimum potential for protection is determined by electro-chemical measurements. Anodic protection can decrease corrosion rate substantially. The primary advantages of cathodic protection are its low current requirements and its applicability in extremely corrosive environments.

The important parameters in anodic protection are:

(a) Potential range over which the metal is passive should be wider. A range of 50 mV is reasonably satisfactory for commercial use.
(b) The current density needed to start the protection should be as low as possible.
(c) The lower the passive current needed to maintain protection the lesser will be the operating costs.

An anodic protection system for tanks is represented in Fig. 4.16.

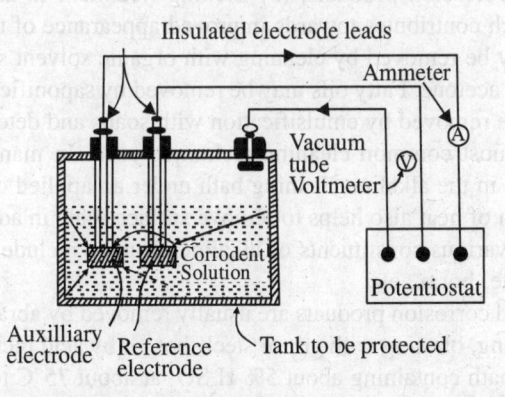

Fig. 4.16. Anodic protection system for tanks.

The cathode material used must be such that it does not suffer much corrosion in the environment. Platinum clad metals or some corrosion resistant alloys are often used. Anodic protection system has a large throwing power and hence quite complex structures can be protected with proper cathode placement. A major disadvantage of anodic protection is the high current needed to induce passivity. Further, ability to accurately maintain the desired potential is very critical in anodic protection. Advantages and limitations of anodic protection are summarized below:

Anodic Protection

Advantages	Limitations
1. Low operating costs.	1. Suitable for only these metal-corrodent systems which show active-passive behaviour.
2. Applicability to wide range of severe corrodents.	2. High installation cost requireing a potentiostat, reference electrode and auxiliary electrode.
3. Ability to protect complex structures.	3. High starting current required.
4. Needs few auxiliary electrodes.	4. If the system goes out of control, high corrosion rates may occur.
5. Feasibility of the process can be predicted by laboratory experiments.	
6. Protection current gives an indication of corrosion rate.	

Cathodic protection Vs. anodic protection

Cathodic protection	Anodic protection
1. Applicable to all metals.	1. Applicable to only those metals which show active passive behaviour.
2. Used where there is no source of power by employing sacrificial anodes.	2. More aggressive corrodents can be handled.
3. Lower installation costs.	3. Operating costs are lower although installation costs are higher.
4. Standard and well-established method.	4. Better throwing power and hence fewer electrodes are needed
	5. Feasibility can be predicted in laboratory and design is easier.

C. Protective coatings

Another important method for protecting a metal from corrosion is to apply a protective coating. The protective coatings may be metallic, inorganic non-metallic or organic substances. For applying any type of coating, it is very essential to prepare the metal surface properly. This usually involves three steps: (1) Removal of grease and other surface contamination, (2) Removal of oxide scale rust and corrosion products, (3) Etching treatment to aid in proper adhesion or buffing or polishing which contributes towards improved appearance of the coating to be applied.

Oils and greases may be removed by cleaning with organic solvent such as trichloroethylene, toluene, xylene, CCl_4 and acetone. Fatty oils may be removed by saponification with strong alkalies while mineral oils may be removed by emulsification with soaps and detergents. Immersion in hot alkaline solution is the most common cleaning technique used. In many cases, the metal to be cleaned is made cathodic in the alkaline cleaning bath under an applied current of about 10 amp./sq. ft. The rapid evolution of heat also helps to remove surface scale in addition to accelerating the degreasing process. The various constituents of alkaline cleaners include NaOH, Na_2CO_3, sodium phosphate, sodium silicate, borax, etc.

Oxide scales, rust and corrosion products are usually removed by abrasion such as by grinding, wire brushing, sand blasting, blasting with grit or steel shot, or by acid pickling. Iron and steel parts are dipped in a pickling bath containing about 5% H_2SO_4 at about 75°C to remove the mill scales. Iron and steel parts which are to be painted are dipped in H_3PO_4 or in H_2SO_4 followed by a dip in 2 to 10% H_3PO_4. Brass and bronze articles are pickled by immersion in a mixture of HCl and H_2SO_4 to which a little HNO_3 is added. Pickling inhibitors are sometimes used to reduce the undue attack on the metal. In case of sand castings, a preliminary dip in HF is sometimes used.

A final etching or abrasion of the metal surface often helps better adhesion and improved appearance of the coating.

Metallic coatings

Metallic coatings are mostly applied on iron and steel because they are the cheap and commonly used construction materials and are also the most susceptible ones for corrosion. The metallic coatings often used are of Zn, Sn, Ni, Cu, Cr, Al and Pb.

Metalic coatings are usually imparted by the following methods:

1. Hot dipping. In this process, the metal to be coated is dipped in the molten bath of the coating metal for sufficient time and then removed out along with the adhering film. This method is widely used for applying coating of low-melting metals and alloys such as Zn, Sn, Pb, solder and terneplate (Pb – Sn alloy). The process of providing a zinc coating is called galvanising and the one providing a tin coating is called tinning.

2. Electroplating. In this method, the freshly cleaned base metal is made cathodic in a bath of suitable composition containing, along with other constituents, a solution of a compound of the metal being deposited. The nature of the deposit depends upon the current density, the bath composition, the temperature and the presence of other additives. In many cases, anodes of the metal being deposited are used and the conditions are adjusted in such a way that the rate of dissolution of the anode is same as the rate of metal deposited at the cathode. However, insoluble anodes (e.g., Pb or Sb-Pb alloy) are used in chromium plating and the bath composition is maintained by adding chromic acid periodically. This is an important method for imparting protective coatings of Cu, Zn, Sn, Ni, Cr, Cd, Pb, Ag, Au and various alloys of controlled composition such as bronze, brass and coronite (Zn–Ni alloy).

3. Metal Spraying. In this method, the molten metal is sprayed on the cleaned base metal with the help of a spraying gun or "pistol" which can be held in hand to direct the molten metal stream as required. In many cases the metal is fed through a central barrel in the form of a wire. A gaseous mixture (e.g., oxyacetylene or H_2 or town gas) passing through a tube around the wire barrel burns at the orifice to melt the wire as it protrudes out. Compressed air is admitted through an outer tube surrounding the gas inlet which atomizes the molten metal and projects it against the surface to be coated. In another type of the pistol, the metal is supplied in the form of powder or in the molten form, which is converted into a cloud of globules by the blower and adsorbed on the base metal surface.

The advantages of the method include: (1) coating can be applied to the finished structure of the base metal in place, (2) coating can be applied to any desired spot, (3) thickness of the coating is controllable, and (4) uniform coating is obtainable.

However, sprayed coatings are generally more porous, less adherent and harder as compared to the coatings applied by other methods. Careful cleaning and roughening of the base metal surface improves adherence. This method is widely used for applying coating of Cu, Pb, Ni, Sn, Al, Zn, brass, monel metal etc.

4. Metal cladding. Many processes for cladding a base metal with another metal or alloy have been developed recently to impart corrosion and wear resistance. In one of the methods. a duplex ingot is cast with the coating material on the outside and subsequently the inget is rolled into a plate, sheet or bar or drawn into a wire form. Steel sheets clad with stainless steels, copper-covered steel articles and tin-cladded lead foils are prepared by this method. Other methods of cladding include, (a) applying the clean sheets or plates of the two materials together, (b) applying the coating sheet by spot welding or resistance welding, (c) fusing the cladding material over the surface of the base metal.

5. Cementation. In this method the base metal articles are packed in the powdered coating metal, or a mixture of the powdered metal and a filler, and are heated to a temperature just below the melting point of the more fusible metal. Generally an inert or reducing atmosphere is usually maintained during the process. An alloy of the two metals is formed by diffusion of the coating metal into the base metal. This method is used for producing alloy layer on iron and steel surfaces with Zn (sherardizing), Al (calorizing), Cr, (chromizing), Si (siliconizing), etc. Steel can be case-hardened by cementation with carbonaceous materials in the pack carburizing process.

6. Miscellaneous methods. Cathode sputtering consists of applying a high voltage (500 to 2000 V) between two electrodes in a partial vacuum (0.01 to 0.1 mm) inducing a glow discharge when the cathode metal disintegrates into vapour which is deposited as a thin film on the base metal parts to be coated which are placed nearby. If the chamber is suitably designed and the object suitably located, uniform films of the cathode metal on the object can be obtained. This method is particularly useful for depositing thin metallic fabrics on non-conducting materials such as fabrics and phonographic recording waxes.

Metallic coatings can also be obtained by the *condensation of the metal vapours.* The coating metal is evaporated in an evacuated system and the vapours produced are allowed to condense on the surface of a base metal article placed in the system properly.

Metal films can also be produced by *decomposition* of gaseous compounds on the metal *e.g.,* deposition of Ni by the decomposition of $Ni(CO)_4$.

Important metallic coatings

Zinc coatings are generally obtained by hot-dipping of the base metal in a molten zinc bath and the process is called galvanizing. Iron and steel articles are protected from corrosion usually by galvanizing because of the low cost of zinc, easy application of the coating and efficient anodic protection afforded. Even if there is a scratch on the coating, the base metal is still protected from corrosion because of the electrochemical "sacrificial" protection offered by the zinc.

Tin coatings are obtained by hot dipping, or electro-deposition from acid stannous sulfate or from alkaline sodium stannate baths, or by spraying. Tin forms a thin, fairly uniform and resistant coating over iron, copper and other metals. However, tin is cathodic to iron in most environments and hence cannot offer galvanic protection of the base metal as is offered by zinc.

Nickel coatings are usually applied on iron, steel and brass articles mostly by electrodeposition from plating baths containing $NiCl_2$, $NiSO_4$ and boric acid at a temperature of 50 to 60°C. Application of nickel coatings by spraying and cladding is also gaining importance. Nickel affords decorative, protective and wear-resistant coatings on the base metals. Nickel coatings are widely used on automobile parts, turbine blades, electrotypes, propellers, paper machine rolls etc. A composite coating composed of Ni followed by Cu and an outer nickel coating again is supposed to have low porosity and offers excellent protection.

Chromium coatings are non-tarnishing, lustrous and decorative and are widely used for automobiles, hardware, radiator-grilles, metal furnitures, etc. The hardness and wear-resistance of chromium coatings also make them useful for machine parts, printing plates, gauges etc. Chromium coatings are mostly applied by electrodeposition from plating baths containing chromic acid and H_2SO_4 using insoluble anodes *e.g.,* steel, Pb-Sb alloys. Chromium surface on exposure to air gets coated with a thin, invisible film of oxide which protects the metal from further attack. The passivating oxide film in most aerated aqueous environments is permanent and "self-healing" but in the absence of O_2 and particularly in presence of chlorides, the film may be damaged. However, since commercial coatings are generally porous and contain cracks, the protection afforded may not be effective. Hence, chromium coatings are better made on a bright nickel or copper-nickel under coat, which gives both effective protection as well as brightness.

Aluminium coatings on steel are generally applied by spraying hot dipping, or cemetation and are useful in preventing scaling oxidation of ferrous materials even at high temperatures. Aluminium alloys *e.g.,* Duralumin may be protected from corrosion by cladding them with a pure aluminium coating (Alclad). Aluminium coatings do not afford cathodic protection to ferrous metals as expected from its high position in the electromotive series. However, Al readily forms an adherent oxide film in many environments which offers protection.

Copper coatings and *lead coatings* are also used for some limited applications.

Inorganic non-metallic coatings

The inorganic non-metallic protective coatings include surface conversion or chemical-dip coatings, anodized oxide coatings and vitreous enamel coatings.

Surface conversion or chemical dip coatings

These coatings are produced by covering the surface of a metal or alloy by chemical or electrochemical methods. The metal is immersed in a solution of a suitable chemical which reacts with the metal surface producing an adherent coating. These coatings afford good protection of the base metal from corrosion in some environments and sometimes are of decorative value. Further, many of these coatings are particularly useful to serve as excellent bases (under-coats) for the application of paints, enamels and other organic protective coatings. The most commonly used surface conversion coatings are chromate coatings, phosphate coatings and oxide coatings. Chromate coatings are applied on Al, Zn, Mg etc. by immersing first in acid chromate bath followed by that of neutral chromate. Coloured shades may be obtained by adjusting the bath composition. Oxide coatings can be obtained by treating the metal with alkaline oxidizing agent. Protective film of magnetic oxide on iron can be obtained by passing steam over the heated metal. Corrosion resistant black coating on stainless steel can be obtained by immerising it in fused sodium dichromate at about 280°C. Phosphate coatings may be obtained by reacting the base metal with phosphoric acid and phosphates of Zn, Fe or Mn in presence of a copper salt which acts as an accelerator. Corrosion resistance of Zn may be increased by immersion in $Na_2Cr_2O_7$ solution acidified with dilute H_2SO_4 or by the application of phosphate coating. Mg metal and its alloys can be better protected by first treating with $Na_2Cr_2O_7$ and con. HNO_3 or $Na_2Cr_2O_7$ and Na_3PO_4, followed by painting. A smooth grey coating may be obtained on Al by immersing it in a boiling solution of $K_2Cr_2O_7$ and Na_2CO_3. Selenium films can be formed by immersing the base metal in acidified sodium selenate.

Anodized oxide coatings

Protective oxide films are produced on Al and its alloys in air spontaneously. A more protective, thicker and stronger oxide film can be produced by making Al as the anode in an electrolytic bath cotaining chromic acid or oxalic acid or sulfuric acid. After anodizing, the oxide coating is "sealed" by immersing in boiling water. This treatment decreases the porosity and increases corrosion resistance of the film. Anodized coatings, when properly prepared, give high electrical insulation to the base metal and imparts good corrosion resistance and resistance to abrasion and stain. The films may be coloured with organic dyes and inorganic pigments to give decorative effects. If coloured coatings are required, anodizing is performed at higher temperatures and in higher acid coucentrations so as to produce a porous coating and then immersing the article in a dye solution or precipitating an inorganic pigment in the film before the "sealing" operation.

Anodized coatings on Zn can be obtained by making Zn as anode in an electrolytic bath containing chromic acid solution or chromates.

Vitreous enamel coatings

The coatings are widely applied for ferrous materials used for equipment in the pharmaceutical, chemical, dairy, food and beverage industries. Enameled steel is used for refrigerators, stoves, table tops, kitchen utensils, etc. Enamelled cast iron is widely used for bath-tubs, sanitary ware and pans, basins etc used in hospitals.

Vitreous or procelain enamels are modified glass-like materials having different compositions which are usually applied on steel and cast iron equipment. The metal part to be enamelled is first cleaned carefully to remove grease and oxide scale. The vitreous material for the enamel, called "frit" is prepared by fusing together refractory acidic substances (*e.g.,* quartz and feldspar) with basic fluxes (*e.g.,* borax, cryolite, fluorspar, soda ash, sodium nitrate and litharge). The "frit" is applied to the metal by any of the following methods:

In the wet process, the "frit" is ground in a pebble mill with water, floating agent (*e.g.,* clay), opacifier (*e.g.,* Sn, Zn, or Sb_2O_3) or coluoring agent. The mixture is called the "slip" to which electrolytes such as soda ash or borax may be added before application for proper "set up" on the metal surface. The metal part which is to be enamlled is dipped in the slip. The slip may also be sprayed on the metal part. The coating thus formed is dried and fused or "burnt" in an enameling furnace. Two or three coats may be given in this manner. The ground coat usually contains cobalt oxide which promotes adherence.

In the dry promotes, the "frit" is ground in a dry mill together with opacifiers or coloring agents. The finely ground mixture is dusted on the heated surface which has an undercoat of a moist adhesive or a ground coat by the wet process. The dusted metal is then fused in the enameling furnace.

High proportions of silica and feldspar and lower proportions of fluxes give acid-resistant enamels.

Cermic protective coatings made from high refractory oxides such as Cr_2O_3 afford protection of the metal from corrosion even at high temperatures.

The ceramic coatings have high chemical resistance, corrosion resistance, high temperature stability, wear resistance and attractive appearance. The enamelled articles cannot be bent without deformation of the coating. The enamelled articles tend to crack when subjected to thermal or mechanical shock. Even microscopic cracks in the coating lead to intense corrosion.

Organic coatings

Protection of a metal surface from corrosion by using organic protective coatings is an established practice. Important organic protective coatings include paints, varnishes, enamels and lacquers. When applied on cleaned metal surfaces, they act as effective inert barriers which not only protect the metal from corrosion but also afford decorative and aesthetic appeal.

Paints

Paint is a viscous suspension of finely divided solid pigment in a fluid medium which on drying, yields an impermeable film having considerable hiding or obliterating power. The drying of oil paints is mainly due to oxidation and polymerisation of the oil "vehicle" whereas the drying of water and cellulose paints is mainly by evaporation of the solvents.

A good paint should satisfy the following requirements:
(1) It should form a good, impervious and uniform film on the metal surface so that effective protection from corrosion is achieved.
(2) It should have a high hiding power.
(3) The film should not crack on drying.
(4) It should have good resistance to the atmospheric conditions in which it is used.
(5) It should have the required consistency for the required purpose so that it can be spread on the metal surface easily.
(6) It should give a glossy film.
(7) The film produced should be washable.
(8) It should give a stable and decent colour on the metal surface.

CORROSION

All the required properties in a paint can be achieved by a proper selection of pigments, vehicles and extenders and their propertions. Mostly this has to be done by trial and error. An empirical property called the *pigment volume concentration* (P.V.C.) provides a good guideline for this purpose. It is generally defined as the volumetric concentration of the pigments expressed as percentage of the total volume of the non-volatile constituents in the paint.

$$P.V.C. = \frac{\text{Volume of pigment in paint}}{\text{Total volume of non-volatile constituents in the paint}}$$

$$= \frac{\text{Volume of pigment in paint}}{(\text{Volume of pigment in paint} + \text{Volume of non-volatile vehicle constituents in the paint})}$$

P.V.C. is very important in controlling important properties of the paint such as durability, washability, consistency, gloss and adhesion. Hence, P.V.C. should be maintained within proper limits.

Significance of P.V.C.

(1) If the P.V.C. is increased (*i.e.*, when the volume of pigment is increased relative to the non-volatile vehicle), gloss tends to decrease until the paint becomes flat.

(2) If P.V.C. is unduly increased the washability, durability and adhesion of the paint decreases.

(3) Addition of extenders to the paint increase the P.V.C. and hence decrease the gloss. In case of expensive pigments having good hiding power, it is economical to substitute a part of the pigment with a suitable extender, without undue sacrifice of efficiency.

The recommended ranges of pigment volume concentration of some types of paints are given below:

	P.V.C.
Gloss paints ...	25 to 35
Semi-gloss paints ...	35 to 45
Flat paints ...	50 to 75
Exterior house paints ...	28 to 36

Constituents of paints and their functions

The important constituents of a paint are (1) Pigment, (2) Vehicle or medium (drying oils), (3) Thinner, (4) Drier, (5) Fillers or extenders and (6) Plasticizers.

(1) **Pigments.** The important properties of pigments in a paint are colour hiding power, oil absorption, rheology (flow characteristics) and chemical behaviour. The chemical composition of the pigments, their size distribution, shape of the particles, refractive index and the proportion of the pigment to vehicle, all affect the properties of the paint.

The following types of pigments are generally used on paints :

(a) Natural or mineral pigment *e.g.*, clay, chalk, mica, talc, diatomaceous earth, iron ores, barytes, etc.

(b) Synthetic chemical pigments *e.g.*, white lead $2PbCO_3 \cdot Pb(OH)_2$, sublimed white lead, zinc oxide, lithophone (about 71% $BaSO_4$ + 28% ZnS + less than 1% ZnO), Titanium dioxide TiO_2, Barium sulfate, and several organic and inorganic colouring substances.

(c) Reactive pigments which react with drying oils or their fatty acids to form soaps, *e.g.*, red lead and ZnO.

The most commonly used pigments in paints are white pigments (*e.g.*, white lead, ZnO, lithophone, $BaSO_4$ and TiO_2), Blue pigments (*e.g.*, prussian blue and ultramarine blue), Black pigments (*e.g.*, graphite, carbon black, lamp black), Red pigments (*e.g.*, red lead, Fe_2O_3, basic lead chromate and cadmium reds). Green pigments (*e.g.*, chromium oxide Cr_2O_3), Brown pigments (*e.g.*, iron containing clays such as burnt sienna, burnt ochre and burnt umber) and yellow pigments (*e.g.,* ochre, chrome yellow, litharge PbO and zinc yellow, $4\,ZnO.K_2O.4\,CrO_3.3H_2O$).

Functions of a good pigment in a paint are:
(1) It gives opacity and colour to the film.
(2) It provides an aesthetic appeal to the film.
(3) It gives strength to the film.
(4) It protects the film by reflecting u.v. radiations which are deleterious to the film as they catalyze destructive oxidation of the paint film.

Criteria of a good pigment in a paint are, (1) It should be chemically inert, (2) It should be non-toxic, (3) It should be opaque and possess high covering power, (4) It should be cheap and easily available, (5) It should mix freely with film-forming constituents of the oil.

(2) **Vehicle or Medium.** This is the film-forming constituent of a paint. In oil paints, the vehicle or medium comprises of drying oils (*e.g.,* linseed oil, tung oil, dehydrated castor oil, Perilla oil, etc. (and semidrying oils such as soyabean oil, fish oil, rosin oil etc. The semi-drying oils are slow-drying and hence are used only in admixture with drying oils. The important properties of drying oil are (*a*) Acid value, (*b*) Saponification value and (*c*) Iodine value. The drying oils are usually refined before use by acid refining (with con. H_2SO_4) or by alkali refining (with 10% NaOH solution).

The important functions of vehicle or medium are:
1. they hold the pigment on the metal surface.
2. they form the protective film by evaporation or by oxidation and polymerisation of the unsaturated constituents of the oil.
3. they impart water-repellency, durability and toughness to the film.
4. they improve the adhesion of the film.

(3) **Thinners.** Thinners are added to the paints to reduce the consistency (or viscosity) of the paints so that they can be easily applied to the metal surface. The thinners are volatile substances which evaporate easily after application of the paint. The thinners commonly used are dipentine, turpentine, petroleum spirits and aromatic hydrocarbons such as toluol, xylol and methylated naphthalene. The following are the important functions of a thinner:
1. They reduce the visocity of the paint to render it easy to handle and apply to the metal surface.
2. They dissolve the film-forming material and also the other desirable additives in the vehicle.
3. They suspend the pigments in the paint.
4. The thinners evaporate rapidly and help the drying of the film.
5. They increase the elasticity of the film.
6. They increase the penetration of the vehicle.

(4) **Driers.** Driers are used to accelerate or catalyze the drying of the oil film by oxidation, polymerisation and condensation. The commonly used driers are naphthenates, linoleates, borates, resinates and tungstates of heavy metals (*e.g.,* Pb, Zn, Co and Mn). Too much of a drier tends to produce hard and brittle paint films.

The important function of driers is that they act as oxygen carrier catalysts which help the absorption of oxygen and catalyze the drying of the oil film by oxidation, polymerisation and condensation.

(5) **Fillers or extends.** These are the inert materials which improve the properties of the paint although they have a low opacity themselves. The commonly used extenders include gypsum, chalk, silica, diatomaceous earth, talc, clay, barytes, $CaCO_3$, $CaSO_4$ magnesium silicate etc.

The important functions of an extender are:
1. They serve to fill the voids in the film.
2. They increase random arrangement of the primary pigment particles.
3. They act as carriers for the pigment colour.
4. They reduce cracking of the paints after drying and improve the durability of the film.
5. They reduce the cost of the paint. Expensive pigments like $ZnSO_4$ and TiO_2 which have excellent hiding power can be used in admixture with cheap extenders which reduce the cost without reducing the efficiency.

(6) **Plasticizers.** Plasticizers are sometimes used in paints to give elasticity to the film and to prevent cracking of the film. Plasticizers in common use are triphenyl phosphate, dibutyl tartarate, tributyl phthalate, tricresyl phosphate, diamyl phthalate, etc.

(7) **Antiskinning agents.** These are sometimes added to some paints to prevent gelling and skinning *e.g.*:—polyhydroxy phenols.

The manufacture of paint essentially involves thorough grinding of the required amounts of the constituents in roller mills or pebble mills.

The common methods of application of paints are hand brushing, spraying, dipping, tumbling and roller coating. Industrial spray methods make use of electrophoresis principle because paint spray is a typical aerosol which consist of negatively charged droplets. The article to be painted is suspended from the positively charged monorail which moves between the two negatively charged electrodes where it meets a fixed spray gun, Potentials of about 100,000 V are used. The paint droplets converge on the article giving a uniform coat.

A paint film may fail because of chalking, cracking, flaking or blistering.

Varnishes

Varnishes are solutions of natural or synthetic gums or resins in a vehicle. The vehicle used may be a drying oil or a volatile solvent (or "spirit") and accordingly they are called oil varnishes (or oleoresinous varnishes) and spirit varnishes. Varnishes differ from paints in two respects, (1) they do not contain pigments and (2) a part or whole of the oil is replaced by resin. Varnishes also give protective and decorative coatings on the surfaces on which they are coated. Varnish also dries by evaporation, oxidation and polymerisation giving a transparent, durable, lustrous and glossy film on the surface. Important constituents of varnish are:

1. Drying oils (*e.g.*, linseed oil, tung oil, castor oil, tall oil, fish oil, coconut oil, soya been oil, etc.).
2. Resins (natural resins such as rosin, shellac, kauri, copal etc., and synthetic resins such as phenol formaldehyde, urea and melamine-formaldehydes, alkyd resins, chlorinated rubbers, vinyl resins, etc.)
3. Driers (*e.g.*, naphthenates, linoleates, rosinates of Co, Mn, Zn and Pb.
4. Thinners or solvents such as turpentine, naphthas, kerosene, xylol, toluol, alcohol, etc.)
5. Antiskinning agents such as tertiary amyl phenol, guaiacol, etc.

The functions of these constituents are similar to those used in case of paints.

The properties of a varnish depends upon the nature of the resin and the oil used. Varnishes are used for protection from corrosion. They are mainly used for wooden furniture to improve their brightness.

Enamels

Enamels are nothing but pigmented varnishes and pigmented lacquers. They contain pigments, oil or resin vehicle, driers and thinners. Enamels give the combined advantages of a paint and a varnish and produce harder, smoother, more lustrous and more glossy finish than ordinary paints.

Lacquers

Lacquers are compounded from the following substances:
1. A cellulose derivative e.g., cellulose acetate, cellulose nitrate, ethyl cellulose, etc., to give hardness and durability to the film.
2. A solvent e.g., methyl ethyl ketone, ethyl acetate, amyl acetate, acetone, cellosolve, dioxane etc. to dissolve the cellulose derivative and the resin.
3. A resin such as phenol formaldehyde, alkyd, copal etc. to give thickness, gloss and adhesion to the film.
4. Plasticizers e.g., dibutyl phthalate, dibutyl phosphate, tricresyl phosphate, castor oil and some polyesters, which are added to give a smooth and flexible film. They act as internal lubricants to the resin.
5. Diluents or extenders or non-solvents which are low cost liquids which decrease the viscosity of the medium by acting as thinners. Examples are petroleum naphtha and toluol.

Lacquers are used for protection from corrosion and for interior decoration. They give tough, durable and glossy coating. They are widely used for protecting the interior of tank cars and for automobiles. Methacrylate lacquers used in automobile industry give high gloss, moisture resistance and heat resistance.

Some lacquers and enamels are "cured" by heating or "baking" the applied film. Thermosetting resins are generally used in such cases which give harder and more stable film due to the formation of cross-linkages.

Emulsion paints

There are dispersions of synthetic resin latex in water containing oil, resin, pigment and other constituents. The continuous phase consists of water and hence these can be readily diluted with water. The latex or rubber like materials used are styrene butadiene copolymers, polyvinyl acetate, polystyrene etc. Water dispersable oils such as linseed oil and resins such as alkyds and ester gums are generally used. Other constituents include pigments and extenders (e.g., TiO_2, ZnS, magnesium silicate, etc.), emulsifying agents (e.g., tetra sodium pyrophosphate), stabilizers (e.g., caesin, bentonite, methyl cellulose, etc.) and the usual driers and preservatives. The advantages of emulsion paints include that they are quick drying, easily washable, readily diluted with water and easily applicable over the existing coating many times.

Special paints

Apart from the organic coatings described above, several special paints such as heat resistant paints, fire retardant paints, temperature indicating paints, antifouling and fungicidal paints, phosphorescent and fluorescent paints are also available. Antifouling and fungicidal paints are of particular interest to engineers dealing with marine constructions. These paints contain mercuric oxide, cuprous oxide, phenyl mercury naphthenate, pentachlorophenol etc., which retard the fouling of ships and piers by barnacles, marine worms, fungi, etc. and help in checking corrosion.

QUESTIONS

1. State whether the following statements are true or false. If they are false, rewrite the statements by incorporating the necessary corrections.
    (a) The relation between the electrode potential (E) and the concentration of the metal ion (C) is given by Nernst's equation as follows:

    $$E = \frac{0.0591}{C} \log \frac{n}{K}$$

    (b) The standard electrode potential is a quantitative measure of the readiness of the element to act as a reducing agent in aqueous solution; the more positive the potential of the element, the more powerful

is its action as a reducing agent.
- (c) In general, it may be said that the more the two metals are apart in the galvanic series, the lesser will be the rate of corrosion when they are in electrical contact in presence of an electrolyte.
- (d) In a concentration cell, the metal immersion in the more concentrated solution acts as an anode and suffers corrosion.

  (**Answer:** All the above statements are wrong. Appropriate corrections may be made by going through the chapter carefully).

2. Write informative notes on the following:
   - (a) Concentration cells
   - (b) Galvanic series
   - (c) Polarization
   - (d) Overvoltage
   - (e) Electrochemical theory of corrosion (Nag. Univ. S-2004)
   - (f) Anodic protection
   - (g) Cathodic protection (S-2006) (Nag. Univ. S-2007)
   - (h) Importance of design and material selection in minimizing corrosion (Nag. Univ. S-2004)
   - (i) Corrosion inhibitors
   - (j) Passivity
3. Discuss the various factors which influence corrosion.
4. Discuss the relative merits and demerits of cathodic and anodic protection methods for controlling corrosion.
5. Discuss the electrochemical theory of corrosion (Nag. Univ. S-2007)
6. Write informative notes on the following:
   - (a) Testing and measurement of corrosion
   - (b) Microbiological corrosion. (Nag. Univ. W - 2003)
   - (c) Intergranular and stress corrosion (Nag. Univ. W - 2005)
   - (d) Water-line corrosion (Nag. Univ. S-2007)
   - (e) Organic coating for corrosion control
   - (f) Galvanizing vs. tinning
   - (g) Pitting corrosion (Nag. Univ. S. 2007)
   - (h) Anodic and cathodic inhibitors (Nag. Univ. 2004)
   - (i) Differential aeration and its significance in corrosion. (Nag. Univ. S. - 2006)
   - (j) Pourbaix diagram.
7. Discuss the importance of design and material selection in controlling corrosion.
8. Discuss the various types of protective coatings used for controlling corrosion.
9. Corrosion is reverse of extractive metallurgy explain (Nag. Univ. W. 2003)

5

Lubricants

"Modern machinery could not have been developed to meet the increasing demand for greater precision and higher speeds, if lubricants technology had been unable to keep pace with the expanding engineering sciences."

5.1. HISTORICAL

Archaeological investigations reveal that the usefulness of lubrication was known to the chariot drivers of 1400 B.C. Leonardo da vinci (1452–1519) discovered the fundamental principles of friction and lubrication and described the effects of lubrication on the coefficient of friction between two rubbing surfaces. The importance of lubrication was widely recognized with the advent of industrial era and systematic investigation into the nature of lubrication afforded by fluids was initiated in 1886 by Osborn Reynold which was followed up by many workers. Large scale consumption of lubricants started since 1947 in various fields such as automotive applications, industrial machinery and aviation.

5·2. INTRODUCTION

The ward "lubricate" cames from the latin lubricus which means slippery. The early concept of lubrication was slipperiness and a lubricant was regarded as a substance which promoted the sliding of one body again anather.

Lubricants is defines as a substance that will when interposed between moving parts of machinery make the surface slippery and reduce friction, eliminate asperites and prevent cohesion.

Lubrication results in the reduction of friction and wear. Friction and wear arise from the relative motion of solid surfaces in contact. Friction is a force of resistance to the relative motion of two contacting surfaces. Wear results when this resistance is overcome by applied forces. Lubrication may be defined as the reduction of friction between two relatively moving surfaces by the interposition of some other substance between the surfaces. A study of what happens when two solid surfaces contact each other must obviously start with the nature and phenomena peculiar to surfaces themselves.

5·3. SURFACE TENSION AND SURFACE ENERGY

A surface is generally considered as an interface between two media or phases. It has been suggested that the surface of a liquid behaves as if it were a stretched membrane or film, always tending to contract. This gave rise to the ideas of surface tension of a liquid or the energy of the surface film. Support for this idea can be produced by the fact that a greased pin can float on water. If the water is frozen it assumes the solid state. The surface forces remain but it is no longer feasible to liken the surface to a membrane. The concept of surface tension cannot be applied to a solid but a solid can have surface energy which exists in the form of strain energy. This strain energy may be reduced when a layer of liquid covers a solid surface.

As the surface layer can be likened to a stretched membrane, there is an analogy with a stretched spring in which energy is stored up since the stretching of the string requires work to be done. The

LUBRICANTS

stored up energy is given out when the spring is unloaded. This gives the alternative concept of surface energy resulting from the molecular forces exerted on the surface layer. The concept of surface energy may be applied to both solids and liquids. Surface energy varies with the constitution of the material forming the surface. The surface energy of metals increases as the temperature falls, while in air metals become coated with oxide which reduced the surface energy. The effect of lubricant applied to the surface is to considerably reduce the surface energy.

5·4. ADSORPTION

If, due to the forces of attraction at an interface between two phases, a layer of different composition from that of the two phases is formed, the new layer is said to be adsorbed and the process of attraction is that of adsorption. An adsorbed layer will have characteristics entirely its own. In the case of liquid/gas or liquid/liquid interfaces, the degree of adsorption can be assessed by the change of surface tension produced but this is not possible in the case of adsorption at a solid surface though adsorption does reduce surface energy. The substance adsorbed is referred to as the adsorbate, the phase responsible for adsorption being the adsorbent. Adsorption may be of two kinds, purely physical adsorption or chemisorption which involves more powerful forces. Adsorbed layers of the lubricant play a very important part in cases of boundary lubrication.

5·5. SURFACE ROUGHNESS

Even the best polished metal surface, when examined under a microscope giving high magnification (*e.g.*, electron microscope), is seen to be more or less rough, having peaks and valleys of different heights and depths as shown in Fig. (5.1a). This is known as surface roughness. The highest peaks are called "asperities". When two flat surfaces are placed over one another, the asperities of the upper surface rest on those of the lower surface. Thus, the surfaces make contact at these points only while over most of the area they are separated. Thus the real area of contact is very much smaller than the apparent area of contact (Fig. 5.1b).

5·6. SURFACE ATTRACTION

Surface forces are effective over only very small distances. When two solid surfaces approach each other, they are ultimately held apart by contact of their asperities. The smaller the asperities, the closer the contact, that is the finer the surface finish of the contacting surfaces, the more intimate the degree of contact. Clean, highly polished metal surfaces have relatively high surface energies because of which such surfaces when pressed together, adhere very strongly. This explains why friction and wear are very high when such surfaces are in relative motion. In the absence of any adsorbed films, it is possible to weld two finely finished surfaces together by pressure alone since the cohesive forces come into play fully. This is the basis of the "cold welding" (Fig. 5.1c).

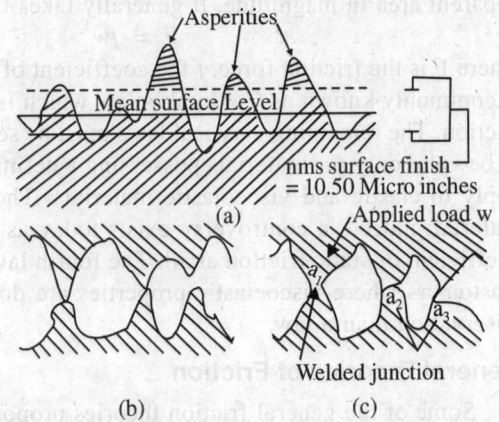

Fig. 5.1. Surface roughness and intimacy of contact.

Mechanism of surface attraction

Interaction between atoms or molecules in close proximity may create momentary dipoles even where permanent dipoles do not exist. These may synchronise and in turn induce further moments. The resultant state can be extremely complex involving oscillation frequencies of the atoms and molecules, but from the theory of wave mechanics the energy of such forces can be calculated. The effect is known as the dispersion effect and the force as the dispersion force. Further, the attractive or bonding forces between molecules in a solid can be very complicated in character and depend on the structure of the solid itself. In general, the principal forces are termed van der Waals forces and these consist of forces due to permanent dipoles, induced dipoles and dispersion forces.

5·7. CLASSICAL LAWS OF FRICTION

Leonardo da vinci (1452–1519) was the first to develop the basic concepts of friction, which were further developed by the work of Amonton and Coulomb. The classic laws of friction as evolved from these earlier studies can be summarized as follows:
1. Friction force is proportional to load.
2. Coefficient of friction is independent of apparent area of contact.
3. Static coefficient is greater than kinetic coefficient.
4. Coefficient of friction is independent of sliding speed.
5. The coefficient of friction is material dependent.

The classic laws have survived the years without significant amendment until recent times. However, in the light of recent advancements, most of the laws have been found to be incorrect and need to be properly qualified. At the same time, considering the limitation of tools available to the early workers, the classic laws reflect remarkable insight into the mechanism of dry friction, which served as the basis for later and more precise investigations.

The first law is correct except at high pressure when the actual contact area approaches the apparent area in magnitude. It generally takes the form

$$F = fw$$

where F is the friction force, f the coefficient of friction, and w the normal load. This equation is commonly known as Coulomb's law which is regarded as a definition for the coefficient of friction. The remaining classical laws must be severely qualified. Thus, the second law appears to be valid only for materials possessing a definite yield point (such as metals), and it does not apply to elastic and viscoelastic materials. The third law is not obeyed by any viscoelastic material; indeed, a controversy exists today as to whether viscoelastic materials possess any coefficient of static friction at all. The fourth law is nearly valid for metals but not at all so for elastomers where viscoelastic properties are dominant. The fifth law is regarded more as an observation than a law.

General Theories of Friction

Some of the general friction theories proposed to explain the nature of dry friction are as follows:

(a) *Mechanical Interlocking.* Amontons and de la Hire in 1699 proposed that metallic friction can be attributed to the mechanical interlocking of surface roughness elements. This mechanism gives an explanation for the existence of a static coefficient of friction, and it also explains dynamic friction as the force required to lift the asperities of the upper surface over those of the lower surface.

(b) *Molecular attration.* Tomlinson in 1929 and Hardy in 1936 attributed friction forces to energy dissipation when the atoms of one material are "plucked" out of the attractive range of their counterparts on the mating surface. Later work attributed adhesional friction to a molecular-kinetic bond rupture process in which energy is dissipated by the stretch, break and relaxation cycle of surface and subsurface molecules.

(c) *Electrostatic forces.* This theory is presented as recently as in 1961 according to which stick-slip phenomena between rubbing metal surfaces can be explained by the initiation of a net flow of electrons, which produces clusters of charges of opposite polarity at the interface. These charges are assumed to hold the surfaces together by electrostatic attraction.

(d) *Welding theory.* This theory is based on the work of Bowden and co-workers at Cambridge. Even the most highly finished surfaces, when examined under the electron microscope, show a hill and valley formation–even the smallest asperities are large compared with the largest molecules. When two such dry, clean, surfaces are placed on each other, contact takes place between a relatively small number of asperities on each so that the actual contact area will be very small compared with the nominal surface area. Consequently, an applied load will cause local pressures sufficiently high to bring about plastic yielding of the metal thus distributing the load over a greater area of contact. Ultimately a state of equilibrium will be reached when plastic deformation will cease. By then cold welding will have taken place between the junctions.

If now, one surface is made to slide over the other, these welded junctions will ultimately fail by shearing. However, as one junction is broken the released asperity on the one surface will collide with one on the other surface so that as junctions are broken others are formed by collision. Shearing will not take place along the plane of the weld but within the bulk of the metal. In the case of two dissimilar metals, shearing will almost always take place within the bulk of the softer metal and wear particles may result. Also, hard surface asperities may plough out grooves in a softer metal so that the frictional resistance in general will consist of a shearing force and a ploughing force though in most cases the ploughing will be relatively small.

$$F = S + P = AS + A'P'$$

where F = friction force, P = the ploughing force, A = the real area of contact, S = the force per unit area to shear junctions, A' = the cross-sectional area of grooved track, P' = the mean pressure per unit area required to displace the metal in the surface.

The theory presupposes clean contacting surfaces, but while some freshly machined or chemically cleaned surfaces may be temporarily free from contaminants, most surfaces are contaminated by films, of moisture, adsorbed gases, liquids, or oxide films, and their effect will be to restrict the amount of cold welding between contacting surfaces.

This view is, however, not shared by Kragelshi and Russian co-workers, who draw attention to the fact that "when a penetrating asperity slides over a surface a bulge is formed in front of it whose dimensions increase with the strength of the adhesive bonds." The material is raised thus leading to appreciable deformation of the surface layer. Repeated deformation will result in work hardening of the material and ultimate destruction of the embrittled layer. Points of stress concentration result in surface damage and the removal of small particles. Kragelski and co-workers opined that the primary factor determining friction is the bulk deformation of the solid material which was not taken into account by Bowden and Tobor.

5·8. WEAR

Wear can be defined as the progressive loss of substance from the surface of the body brought about by mechanical action. This definition includes such processes as abrasion, pitting, scuffing and corrosion galling. In any particular case such mechanisms may operate singly or together *e.g.*, a hard metal then act as a fine abrasive and wear both surfaces. It has been suggested that 'wear' is a composite term, which includes:

    (1) adhesive or galling wear
    (2) abrasive and cutting wear
    (3) corrosive wear
    (4) surface fatigue
    (5) miscellaneous factors

It is generally believed that two or three types of wear occur together and that this combined wear rate was dependent on velocity and load.

Theories of wear

Archard's theory rests on the fact that the true area of contact between surfaces is generally small compared with the apparent area, so that during sliding there is a series of encounters as localised regions of the surfaces come temporarily into true contact. At each encounter a wear particle either forms or does not, the magnitude of the probability of formation being represented by a constant, K, and that the wear particle is a lump of material with dimensions comparable with those of the local contact area. If the wear particle is assumed to be a hemisphere with a radius equal to that of the contact region, assumed circular, the wear rate is given by $W = KPS/P_m$ where S is the sliding distance, P is the applied load and P_m is the flow pressure of the softer material. Thus according to this theory, the wear should be directly proportional to the sliding distance and the load, and independent of the apparent area of contact. With the qualifications made earlier, this is in coformity with experimental evidence.

The theory does not give any direct indication of the expected variation of wear rate with hardness because the probability constant, K, may also be expected to depend upon the material. When different metals are examined under similar states of wear, it is observed experimentally that K tends to decrease with hardness and also that the value of K is so small (ranging from 10^{-2} to 10^{-7}) that any individual localized area in the surface must be loaded many times before suffering damage by wear.

Archards treatment requires modification to allow for the fact that the wear process may involve several stages. Rightmire put forth his theory recently on the basis of Kerridge's experiment in which the wear of a soft tool-steel pin loaded against a hardened tool-steel ring was studied. The wear process comprised of three stages namely, transfer from pin to ring, oxidation of the transferred material and rubbing-off of the oxide to form a loose wear product. Rightmire was able to formulate by statistical methods an expression for the rate of removal of the layer of oxide, assuming that this occurred by a succession of small stages.

However, a comprehensive theory which can account for an absolute magnitude of the wear rate, is yet to be developed.

5·9. LUBRICATION

When a surface slides over another there are three basic physical factors which can affect the overall wear between them:

(*a*) the distance between the surfaces
(*b*) the force acting on the surfaces
(*c*) the surface texture.

Lubrication may be defined as the reduction of friction and wear between two relatively moving surfaces by the interposition of some other substance between the surfaces. The substance thus introduced is known as a lubricant.

The lubricant is forced into the clearances and prevents the surfaces from meeting and engaging thus decreasing the frictional forces and minimizing the wear and tear. Thus, the presence of a lubricating medium decreases the waste of energy, surface wear, damage and deformation and brings about reduction in operational costs.

There are two basic mechanisms by which the reduction of friction may be achieved.

(*i*) *Solid lubrication* in which the two surfaces are coated with a substance, such as graphite, which lowers the coefficient of friction between the two surfaces as they slide over each other.

(*ii*) *Fluid lubrication* in which a fluid film maintained between the two surfaces keeps them from contact with each other so that the only resistance to motion is that due to the "stickiness" or "viscosity" of the fluid. This method is most widely used in practice.

The presence of a lubricant between the moving surfaces might serve several other functions apart from reducing the friction and wear:

1. It acts as a coolant by dissipating the frictional heat generated because of the rubbing surfaces. Thus the expansion of metal by local frictional heat and the resultant

LUBRICANTS

deformation and damage will be reduced.
2. In internal combustion engines, the lubricant also acts as fuel gasket between the piston and cylinder wall at the compression rings and prevents the leakage of gases at high pressure in the combustion chamber, thus minimizing the power loss.
3. It prevents the entry of moisture, dust and dirt between the moving parts.
4. It acts as a cleaning agent and as a scavenger to wash off and transport solid particles produced in combustion or wear.
5. In aircraft, the lubricating oil may be used as a hydraulic fluid to change the pitch of the propeller or to operate other mechanisms. The viscosity index of the oil is the important property in such cases.

5·10. MECHANISM OF LUBRICATION

When two metal surfaces are placed in such a way that one rests on the other, contact is established between their asperities, the real area of contact often being only about 1/10,000 th of the apparent area of contact. When the high pressures developed at these points of contact exceed the elastic limit of the metals, plastic, flow will occur and the two mating surfaces get welded together. When one metal surface slides over the other, the motion will not be smooth and continuous but there is a sequence of "slips" and "sticks", since successive welded junctions are broken and re-made. Friction between the surfaces depends on the ease with which the welds break, which, in turn, depends on the load and on the hardness of the materials concerned. The coefficient of friction for a hard metal sliding over a soft metal is usually small: welded junctions gouge out grooves in the softer material when sliding starts. Particularly strong welds may form between two hard metals in contact, and as these welds break, considerable physical damage may be done to both surfaces. Formation of welded junctions becomes more extensive as the speed of sliding increases since the heat evolved in sliding softens the metal surfaces.

Since friction usually results from the welded junctions between moving surfaces, the main action of a lubricant will be to hinder the formation of these junctions, or to replace them with new junctions which shear more easily, thus reducing the frictional resistance and minimizing the wear and damage. The following three types of lubrication mechanisms are generally distinguished; and the factors determining the choice of a suitable lubricant differ in each case accordingly.

1. Fluid or hydrodynamic lubrication. In fluid or hydrodynamic lubrication, moving surfaces are separated from each other by a fluid film at least 1,000 Å thick, so that the surface-to-surface contact and welding rarely occur. Under hydrodynamic conditions of lubrication, the coefficient of friction will be as low as 0.001 to 0.03 (for unlubricated surfaces, the coefficient of friction ranges from 0.5 to 1.5). The coefficient of friction, $f = F/W$, where F is the force required to cause motion and W is the applied load.

Hydrodynamic lubrication occurs in the case of a shaft running at a fair speed in a well-lubricated bearing, with not too high a load. An exaggerated account of a shaft rotating in a journal bearing is illustrated in Fig. 5.2. If the centre line of the shaft is displaced from the journal axis, a wedge-shaped lubricant film can be drawn in, and under these circumstances, hydrodynamic theory predicts development of a sufficient pressure to keep shaft and journal apart; the shaft "floats" in the lubricant. The efficiency of lubrication by this mechanism depends on the design of the bearing, the loading, the rate of rotation of the shaft and on the viscosity of the lubricant. For a given load and rate of

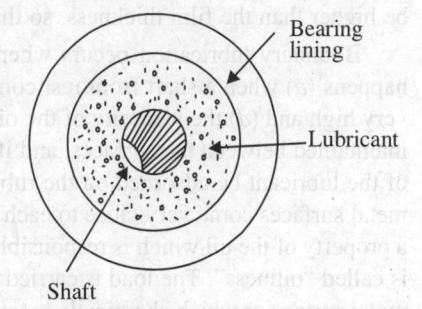

Fig.5.2. Journal bearing hydrodynamic lubrication.

rotation, the greater the lubricant viscosity, the greater the hydrodynamic pressure developed. However, there is a practical limit to the viscosity which can be used, since a large amount of energy would be needed to circulate and maintain a very viscous lubricant film.

In hydrodynamic lubrication (Fig. 5.2), a film of the lubricating oil covers the shaft as well as the bearing surfaces. The oil film is sufficiently thick to cover the irregularities of the surfaces and the metal surfaces do not come into contact with each other. Thus practically there is no wear. The resistance to movement is only due to the resistance between the particles of the lubricant moving over each other. Thus, in hydrodynamic lubrication, the lubricant chosen must have enough viscosity so that the bearing force due to the rotation of the journal, will be sufficient to drag enough oil between the journal and the bearing and at the same time, it should not be too much viscous to offer resistance for the free motion of the lubricant particles over each other.

Hydrodynamic lubrication is maintained in case of delicate mechanical systems such as watches, sewing machines and scientific instruments.

The selection of a suitable fluid lubricant is complicated by changes of viscosity with temperature. The viscosity of a typical hydrocarbon oil decreases as the temperature rises, so that an oil which is satisfactory when an engine is cold may become too 'thin' to maintain an adequate lubricant film at normal running temperatures. In order to maintain suitable viscosity of the oil for adequate lubrication in all seasons of a year, ordinary hydrocarbon lubricants are usually blended with selected long chain polymers.

Hydrocarbon oils are considered to be satisfactory lubricants. Their viscosity increases with increasing molecular weight. Suitable blends of appropriate fractions from petroleum refining plants can be selected for different applications. However, these fractions generally contain small quantities of unsaturated compounds which will oxidize under operating conditions, forming gums or lacquers. Antioxidants, such as aminophenols, must therefore be blended with these oils. Further, these oils may also undergo some decomposition in practice with the formation of solid carbon partices. In order to keep these carbon particles in suspension in the lubricating oil, organometallic "detergent" compounds are generally added.

Thin film or Boundary Lubrication. Fluid or hydrodynamic lubrication is effective only if the lubricant film is at least 1000 Å thick. Film thickness, however, is often very much smaller than this value under conditions of high loading and slow rate of rotation, and lubrication must be maintained by a 'boundary' film, whose thickness may not exceed one or two molecular layers. If the maximum pressure is so great that the oil film is only two or three molecules thick hydrodynamic or fluid lubrication is replaced by boundary lubrication. Some of the hills or asperities in the surfaces may be higher than the film thickness, so that some wear or erosion will take place.

Boundary lubrication occurs whenever a continuous fluid film cannot be maintained which happens (*a*) when a shaft from rest comes into operation, (*b*) the speed is very low, (*c*) the load is very high and (*d*) the viscosity of the oil is very low. Under such conditions, a fluid film cannot be maintained between the surfaces, and if lubrication has to be maintained, it is essential that a layer of the lubricant be adsorbed on the rubbing surfaces by physical and/or chemical forces. Then the metal surfaces come very close to each other but yet are separated by a layer of the lubricant. Such a property of the oil which is responsible for maintaining a "boundary" film of the oil by adsorption is called "oiliness". The load is carried by layers of the lubricant which have been adsorbed on the metal surface or which chemically reacts with metal surface forming a thin film of metal soap which acts as a lubricant. This kind of lubrication is called boundary lubrication. The coefficient of friction in such cases is usually between 0.05 to 0.15.

When fluid film or hydrodynamic lubrication prevails, the oil film separating the journal from the bearing surface is of such thickness as to ensure that the asperities of each surface are kept well apart and the motion takes place by the shearing of successive layers of the lubricant so that the coefficient of friction in a given bearing is dependent upon viscosity of the lubricant. Under boundary lubrication the distance apart of the mating surfaces is very much less, being of the order of the height of the surface asperities. If, when the oil film is squeezed out and, metal to metal contact took place, the load would be taken on the high spots of the journal and the bearing, appreciable heat would be generated and the two surfaces would tend to become welded together. When the two surfaces adhere together and so prevent motion, "seizure" is said to take place. If motion proceeds with the removal of some metal from one of the surfaces the result is known as "scuffing".

In practice, seizure or scuffing are delayed by the fact that metals tend to form films on their surfaces and these temporarily prevent metal to metal contact. Such films may result from oxidation or other chemical action at the surface of the metal or by absorption of lubricant, that is, thin films of a lubricant becomes intimately attached to the surface.

In the case of absorbed films the reduced resistance to motion results from the separation of the surfaces by these very thin films, and the reduced resistance is ascribed to "oiliness". Different lubricants of the same viscosity have different oiliness. Oiliness was defined by Herschel as "the property which causes a difference in the friction when two lubricants of the same viscosity at the temperature of the film are used under identical conditions." This definition is not completely satisfactory and conceptions of oiliness are still somewhat vague. There is no recognized method of assessing it though it is well known that the fixed oils, *i.e.*, animal and vegetable oils, have a higher oiliness than petroleum oils. This has been ascribed to the chemical constitution of these oils, particularly their fatty acid content. Mineral oils have a symmetrical molecule with a CH_3 group at each end but the fatty acids have a CO_2H group at one end and this has been offered as an explanation of the improved oiliness of the fixed oils. Hardy suggested that through its much greater affinity for metal surfaces, this CO_2H group would become attached to a metal surface and the CH_3 group would be repelled. This presents a picture, then, of two mating surfaces being separated by molecular chains attached to each surface by the CO_2H end groups (Fig. 5.3). He also suggested that friction between surfaces was due to the mutual action of surface fields of force, so that the greater the separation of the surface, the lesser the interaction. This very simply explains the known fact that as the length of the molecular chain of fatty acids is increased, the friction is diminished. Bowden and Tabor have however clearly shown that chemical action is of primary importance and that the increased oiliness of the fatty oils is bound up with the lubricating properties of the metallic soaps resulting from the action of the fatty acid constituent upon the bearing surfaces.

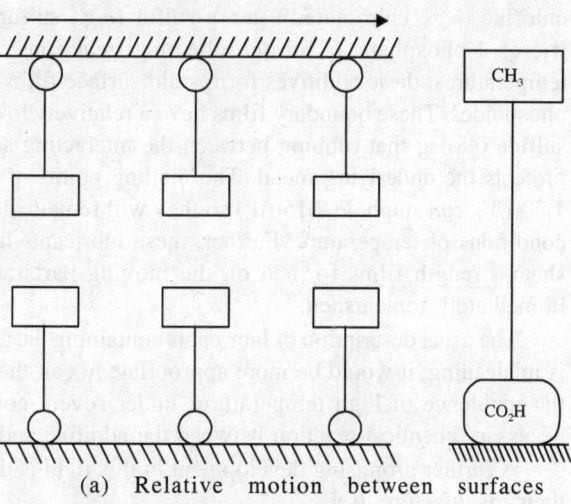

(a) Relative motion between surfaces
(b) Molecular orientation
Fig. 5.3. Oiliness of fatty acids

Boundary lubrication conditions are not ideal but they can prevail at starting and stopping of machinery and also under conditions of high pressure and low speed such as occur at the end of the stroke of an engine piston, so it is advantageous, when other things being equal, to have a lubricant with a high oiliness value. Although the fatty oils have greater oiliness than mineral oils, they break down at high temperatures and are consequently not suitable for use in the cylinder of an internal combustion engine. Early attempts to improve the oiliness of mineral oils, while retaining their great advantage of thermal stability, consisted of adding small amounts of fatty oils, to mineral oils to give "compounded oils". Later, fatty acids themselves were added and became known as oiliness improvers or additives.

The effectiveness of boundary lubrication depends, even more than it does in fluid lubrication, on the structure and chemical properties of the oil. A lubricant for high speed gears may be subjected to pressures of 135,000 psi. The thin molecular film must withstand displacement by these high pressures. For boundary lubrication, the molecules should have, (1) long hydrocarbon chains, (2) lateral attraction between the chains, (3) polar groups to promote wetting or spreading and orientation over the surface, and for high pressures, (4) active atoms or groups to form chemical bonds with the metal or other surface.

High viscosity index, resistance to heat and oxidation, detergent qualities, adherence, low pour point, "oiliness" and other properties necessary for special uses, are all determined by the chemistry of the lubricant.

3. Extreme pressure lubrication. If the rubbing surfaces are subjected to very high pressure and speed, excessive frictional heat will be generated. The high local temperatures thus produced at the surfaces render the commonly used liquid lubricants ineffective due to decompositon or evaporation. In order to provide effective lubrication under these high local pressures and temperatures and onerous conditions, special additives called "extreme pressure additives" are used along with the lubricants. The active chemicals that are in general use are compounds of chlorine (*e.g.,* chlroinated esters), sulfur (*e.g.,* sulfurized fats and oils), and phosphorous (*e.g.,* tricresol phosphate). Through chemical reactions with the metal surfaces, at the prevailing temperatures, these additives form solid surface films of metallic chlorides, sulfides and probably phosphides. These boundary films have a relatively low shear strength (*e.g.,* iron chloride, 0.2, iron sulfide 0.5) so that rubbing between the interacting surfaces occurs in the additive film and thus protects the underlying metal. The melting points of the E.P. layers are high (*e.g.,* iron chloride 1,200°F; iron sulphide 2,150°F) so they will remain attached to the base metal even under extreme conditions of temperature. Further, these lubricants have an additional advantage that if the low shear strength films formed on the moving parts are broken by the rubbing action, they are immediately replenished.

The usual description of lubricants containing the above additives as extreme pressure lubricants is misleading; it would be more appropriate to call them extreme temperature lubricants since it is the incidence of high temperatures under severe conditions of service which brings about the necessary chemical reaction between the additive and the bearing surface.

A further promising development in this field is the formation of films on mating parts during their manufacture itself.

There are certain mechanisms in which the loading between moving metal surfaces is so high that ordinary fluid or boundary film lubricants are unable to prevent metal-to-metal junctions. This is the case, for example, in the 'hypoid' gears used in the rear axle drive of cars; in these spiral level gears there is a longitudinal sliding motion in addition to the normal rolling movement, and very high temperature are developed. The breakdown of ordinary lubricants is, in fact, a consequence of these high temperatures, rather than the direct influence of "extreme pressures". Extreme pressure lubricants are substances which combine with metal surfaces at high temperatures, giving surface layers which shear much more easily than the junction between the underlying metals. One such compound is tricresyl phosphate, which reacts with steel at high temperatures to form an iron phosphide surface layer, with a melting point much lower than that of steel itself. Organic compounds containing reactive chlorine or sulfur atoms have also been used; they react to give metal chloride or sulphide surface layers.

These films are formed only at high temperatures, and extreme pressure lubricants are not generally good lubricants under conditions of high loading, where the temperatures developed are comparatively small. Lubricants are therefore usually made by blending an extreme pressure additive with a hydrocarbon lubricant, to which some organic acid has been added; adequate lubrication at lower temperatures is maintained by the fluid and boundary film lubricants, while the extreme pressure additive takes over at high temperatures where the boundary film melts.

Extreme pressure additives are also needed in the "cutting fluids" used as lubricants in machining of tough metals, a continuous stream of the fluid, which may contain a hydrocarbon oil, a small amount of fatty acid as a boundary lubricant, and an organic chloride or sulfide additive, is fed to the cutting surface. In light cutting operations, a simple oil-water emulsion may be adequate.

Extreme pressure additives are also used as lubricants in wire drawing. Whereas mild steel, and even tungsten, can be drawn through suitable dies if graphite is used as a lubricant, other metals, titanium for example, cannot be drawn under these conditions. Titanium, in fact, can only be drawn into wire in the presence of a chlorine-containing additive which reacts with the stable oxide film of the metal surface.

5.11. LUBRICANTS FOR EXTREME AMBIENT CONDITIONS AND FOR SPECIAL APPLICATIONS

1. Gas lubrication: Gas lubrication of bearings has received special consideration due to its resistance to radiation, resistance for high speeds and for extreme temperatures. The gases that were successfully used for bearing lubrication include air, hydrogen, argon, helium, nitrogen, oxygen, CO_2, and uranium hexafluoride. A useful property of gases is that their viscosity and hence their capacity to generate hydrodynamic pressure, increases with temperature. The viscosity of gases is usually independent of pressure upto about 1 MPa.

When gases are employed as lubricants at high speeds, start-stop wear can be minimised by using a wear-resistant coating material (*e.g.*, tungsten carbide, chromium oxide) for journal and bearing, or by employing solid lubricants (*e.g.*, MoS_2, teflon).

Applications: Gas bearings have been used in precision spindles, motor-driven or turbine-driven circulators, fans, gyroscopes, compressors, environmental simulation tables, memory drums, turbomachinery, etc.

Limitations: Because of very low viscosity of gases, a limiting load of 15-30 KPa only are generally used in case of self-acting (hydrodynamic) gas bearings and upto 70 KPa for operation with external gas-lifting pressure in hydrostatic operation. Further, in view of the small bearing-clearances in gas bearings, care should be taken to minimise the entry of dust particles, moisture and wear debris.

2. Liquid metal lubricants : Liquid metal lubricants (*e.g.*, low melting point metals, like Cs, Ga, Hg, K, Na, Rb, and Na-K) have been used in situations when operating temperatures go beyond 250°C where many organic fluids decompose and water exerts high vapour pressure.

Liquid metals having lower melting points only are used as lubricants:

Metal	Melting Point, °C	Boiling Point, °C
Mercury	–39	360
Sodium-potassium eutectic	–11	780
Sodium	98	880
Potassium	62	760
Gallium	30	1980
Cesium	28	670
Rubidium	39	700

The viscosity of liquid metals is similar to that of water at room temperature but it approaches the viscosity of gases at high temperatures. Hydrodynamic load capacity with both liquid metals and water in a bearing is about 0.1 of that with oils.

The Na-K eutectic is commercially available for use as a liquid over a wide temperature range. However, due to its excessive oxidizing tendency in air, its handling and disposal is hazardous. It is for this reason that Na-K eutectic should be used as lubricant only in closed vacuum or in an inert gas atmosphere of helium, argon or nitrogen. Further, selection of a suitable bearing material is critical for liquid metal bearings. Tungsten carbide cermet with 10-20 wt % cobalt binder gives excellent performance when running against molybdenum under heavy loads at low speeds at temperatures upto 815°C.

A low-melting Ga-In-Sn alloy was found to give superior performance of spiral-groove bearings in vacuum for X-ray tubes at speeds upto 7000 rpm.

3. Cryogenic bearing lubrication

Cryogenic fluids (*e.g.,* liquid oxygen, liquid nitrogen and liquid hydrogen) are used as fuels in liquid rocket-propulsion systems, in pumps used to transfer large quantities of liquefied gases and for turbine expanders in liquefaction and refrigeration. Bearings operating in cryogenic fluids are amply cooled for dissipating the frictional heat but due to the low viscosity of fluids, only marginal lubrication is obtained.

Coatings of Teflon (PTFE) or molybdenum disulfide (MoS_2) are generally used for wear resistance and low friction levels.

Since cryogenic liquids have low viscosities, rolling element bearings seem to be more satisfactory than hydrodynamic bearings for turbo pumps. Stainless steel (AISI 440C) balls and rings are generally preferred due to their superior corrosion-resistance.

4. Lubrication in nuclear reactor systems

Lubricants are required in control-rod drives, compressors or coolant circulating pumps, motor-operated valves and fuel-handling valves of nuclear reactor systems.

The extent of damage suffered by the lubricant in nuclear reactor system depends on the total radioactive energy absorbed due to Y-radiation or neutron bombardment. The unit for radioactive energy absorbed is gray (Gy), which is equal to 1×10^{-5} J absorbed per gram of material, or 0.01 Gy = 1 rad. The first changes observed with petroleum based lubricating oils at a dosage of about 10^4 Gy is the evolution of hydrogen and light hydrocarbon gas as fragments from the original molecules. The resulting unsaturation leads to decreased oxidation-stability and in cross-linking, polymerisation, or scission.

Several petroleum oils exhibit a trend for increasing viscosity with increased radiation dose. In general, a dose which gives a 25% increase in 40° C-viscosity is taken as a tolerance limit for many lubricant applications. Higher dosage results in rapid thickening, sludging and operating problems which interfere with the performance of the lubricating oil.

Greases consisting of petroleum oils thickened with lithium soaps, calcium soaps or sodium soaps suffer significant breakdown of the soap gel structure at doses above 10^5 Gy followed by softening of the greases to become fluid. At still higher doses, polymerisation of the oil takes place leading to overall grease hardening.

Some greases with radiation-resistant components, e.g., polyphenylether oil and non-soap thickeners, maintain satisfactory consistency for lubrication purposes even upto a dosage of 10^7 Gy.

5. Lubrication using glass

Soft glass types such as fused silica, 96% silica-soda-lime, borosilicates, and alumino-silicates are used as lubricants for extrusion, forming and other hot-working processes with steel and nickel-based alloys up to about 1000°C. They are also used for extrusion and forming of titanium and zirconium alloys and less frequently for extruding copper alloys. To serve as an ideal hydrodynamic lubricant, the composition of the glass is so selected that it gives proper viscosity (10 – 100 Pa) at the mean temperature of the die and workpiece. Glass lubricants may be applied as fibers or powder to the die or hot workpiece, or as a slurry with a polymeric bonding agent to the workpiece before

LUBRICANTS

heating. Alternately, heated steel billets are rolled across glass sheets, where the glass then wraps around the billet before passing to the die extrusion chamber.

Glass lubricants were also used for "forming"* ceramic materials at high temperatures.

Environmental and health factors

Regulations in several countries prohibit the disposal of lubricants in streams, chemical dumps, and other environmental channels. About 50% of the disposed lubricants are burnt as fuel, usually mixed with virgin residual and distillate fuels. Waste aqueous metal working fluids may be successfully treated by conventional means for removal of tramp oil, surfactants, etc. to achieve the desired effluent water quality.

Polycyclic aromatics in the base oil and the additives are the most toxic and hazardous constituents in lubricating oils. Lubricant additives such as phenyl-2-naphthyl amine, chlorinated naphthalenes, sodium nitrite plus amines, tricresylphosphate high in the ortho-cresol isomer, lead compounds and some sulfur compounds are among the hazardous constituents. Used motor oils exhibit increased carcinogenic activity as compared to fresh oils. Contact with lubricants, metal working oils, and quench oils which were highly degraded or were in service at extremely high temperatures or are contaminated with toxic metals or bacteria, should be avoided.

Lubricants used in food processing

Food grade lubricants must be formulated from highly purified petroleum products that were fully refined either by acid treatment or by hydrogenation to eliminate all the unsaturated compounds, aromatics and colouring materials.

For exacting requirements where lubricants come into contact with food materials, food grade lubricants such as the following are used:

(a) Food-grade white mineral oils or food-grade petrolatums only are used for lubrication purposes e.g., release agents in bakery products, confectioneries, dehydrated vegetables or fruits, and egg whites.

(b) Technical white oils of special grades are used in processing aluminium foil for food packaging, in manufacturing animal feed and fiber bags, and for machinery used in food processing.

5.12. Biodegradable lubricants

Lubricating agents are used on a very large scale. Every mechanically operated device requires some form of lubrication. Part of these lubricants are emitted to the environment. The aquatic pollution caused by the use of mineral oil based lubricants is unacceptable. Lubricant emission to the surface water have the largest hazard potential due to diffuse emission of lubricants to the surface water. Small amount of mineral oil have injurious effects on the quality of drinking water. Spilled oil can cantaminate steams kill vegetation, harm wild life and lead to costly remediation.

Biodegradable lubricants are those which 60% of the corban of base stock biodegrades with in 28 days and degradation continus thereafter. Base stock of biodegradable lubricants include polyalkylene glycol, esters and vegitable oil. Polyethylene glycols and polypropylene glycols known as polyglycols manufactured by polymerizing ethylene oxide and polyethylene oxide Polyethylene glycol - $H(OCH_2CH_2)_nOH$

Polypropylene glycol - $H(OCH_2CH_2CH_2)_nOH$

One of the important requirement of biolubricants should be that they are easily biodegradable. Secondly biolubricants should not be toxic to the environment they must be ecofriendly. Biolubricants are derived primarily from plant material renewable resource, relatively non-toxic, biodegradable and more easily extracted and processed than petrolubricants. The composition of biolubricants trends to consist of higher molecular weight

---

*"Forming" is a generic term used in powder metallurgy describing the first step in changing a loose powder into a solid specific configuration.

/ lower vapour pressure components. Biolubricants form thicker film, lower shear strength. But it has low oxidation stability, ex. Soybean oil high viscosity index.

Biolubricants have many uses in industrial and automotive applications, especially in hydrallic fluids. Biolubricants must be used in on and around water such as ships, mobile equipments, hydroelectric power plants etc. Chemists Terry A Isbell and Steven C. Carmak made environmentally friendly effective lubricants, containing estolides which are fatty acids from oleic oilseeds, such as high oleic sunflower. These lubricants are favourable to Soybean and Canda oil which have used to produce biodegradable lubricants. Environment friendly hydrolic, gear, engine oil has been developed.

Polygiycols are ethers and contain oxygen in their structure and hence easy for biodegration. Ethers - organic compounds containing the group -O- in their structure molecule.

ex. dimethylether CH_3OCH_3
diethylether $C_2H_5OC_2H_5$ (ethoxy ethane)

30-40% oil released into environment through, spills, leaks, evaporation and indirect means. Biolubricant derived from plant materials. It has low vapour pressure It form thick film., It obtained from renewable resource. VOC emission is less, ex Soybean oil.

5·13. CLASSIFICATION OF LUBRICANTS

Lubricants may be broadly classified as follows:
(1) Solid lubricants:
   e.g.,: Soap stone, graphite, talc, chalk, mica, teflon, molybdenum disulfide, etc.
(2) Semi-solid lubricants:
   e.g.,: Greases, vaselines, etc.
(3) Liquid lubricants:
   (i) Vegetable oils: *e.g.,*:–olive oil, palm oil, castor oil, etc.
   (ii) Animal oils: *e.g.,*:–Whale oil, Lard oil, Tallow oil, etc.
   (iii) Mineral oils: *e.g.,*:–Petroleum fractions.
   (iv) Blended oils or compounded oils: *e.g.,*: Mineral oils with various additives to induce desired properties.
   (v) Synthetic oils: *e.g.,*: Silicones, Fluolubes, etc.
(4) Emulsions:
   (a) Oil-in-water type:
      e.g.,: cutting emulsions
   (b) Water-in-oil type:
      Ex.:–cooling liquids

5·14. SOLID LUBRICANTS

Solid lubricants are used in situations such as:
1. Heavy machinery working on a crude job at very high loads and slow speeds.
2. Where a liquid or a semi-solid lubricant film cannot be maintained or their presence is undesirable as in the case of commutator blades of electric motors and generators.
3. Where parts to be lubricated are not easily accessible, and
4. Where the operating temperatures and pressures are too high to use the easily combustible liquid lubricants.

Many solid lubricants contain grains of particles which may damage delicate parts of the

---

Compatibility : A lubricants ability to be mixed with another lubricants without detriment to either lubricant. Also the abililty to come into contact with other components or materials without deterimental effects

LUBRICANTS

machinery. Hence, they are used only in special cases similar to those mentioned above.

Classification of solid lubricants

A solid lubricant is a material that separates two moving surfaces under boundary conditions and decrease the amount of wear.

The various types of solid lubricants may be conveniently divided into various classes as under:

1. Structural lubricants. These include materials like graphite, molybdenum disulfide, talc, mica, vermiculite etc. whose lubricating properties are due to their layer lattice structure. These function by cleaving within themselves and fixing themselves on or into the bearing surface.

2. Mechanical lubricants. These include metals and plastics and are characterised by their sacrificial wear. They form a continuous adherent film on the rubbing surfaces and reduce the wear.

3. Soaps. They function both as solid lubricants in their own right and also by formation of compounds 'in situ' in the metal surface by the interaction of fatty acids and the metal.

4. Chemically active lubricants. These include extreme pressure additives and other chemicals which interact with the metal surface to produce a lubricating layer. Examples are phosphates, chlorides and oxidizing agents.

5. Refractories, ceramics and glass. These are used in defence programmes and rocketry. Combinations of refractory materials work satisfactorily as lubricants for short periods at high temperatures. Glass functions by softening at the operating temperature and assists in hydrodynamic lubrication.

This classification, though arbitrary, is very useful.

Theory of action as lubricants

When two moving surfaces are in contact, the real area of contact (A) is much smaller than the apparent area and may be represented as follows:

$$A = \frac{W}{P_m} \quad ...(1)$$

where W = load and P_m = Yield point.

For metal-metal contact,

$$F = A S_m \quad ...(2)$$

where S_m = shear strength of metal junctions. Hence,

$$\frac{F}{W} = f = \frac{A S_m}{A P_m} = \frac{S_m}{P_m} \quad ...(3)$$

which shows that the coefficient of friction (f) is independent of the real area of contact and can be reduced by lowering the shear strength or increasing the yield point. This can be achieved by using solid lubricants which have a lower shear strength than the metal substance and which are bonded firmly on to the metal surface.

If the surfaces are covered with a solid lubricant film so that the amount of metal-to-metal contact is greatly reduced, then the above eqn. (2) can be expanded as:

$$F = [a S_m + (1-a) S_f] A \quad ...(4)$$

where a = ratio of apparent area to real area of contact and S_f = shearing strength of solid lubricant.

$$\therefore \quad f = \frac{F}{W} = \frac{a S_m}{P_m} + \frac{(1-a)}{P_m} S_f \quad ...(5)$$

(from eqns. (1) and (4))

With adequate coverage of the surfaces by the solid lubricant and 'a' tending to zero,

$$f = \frac{S_f}{P_m}$$

That means, the frictional force is determined by the shearing strength of the solid lubricant and the yield pressure.

The commonly used solid lubricants include graphite, molybdenum disulfide, talc, mica, french chalk, boron nitride, etc. Amongst these, the most widely used solid lubricants are graphite (in colloidal form) and molybdenum disulfide (MoS_2), both having laminar structure. Graphite consists of a meshwork of hexagonal carbon rings separated from the upper layer of the unit crystal cell by a distance of about 6.79 Å. Molybdenum sulfide has a sandwich like structure in which a layer of molybdenum atoms lie between two layers of sulfur atoms, which are 6.26Å apart (Fig. 5.4).

When graphite is incorporated as a lubricant between uneven surfaces, it fills into the valleys on the surface, thus rendering the surface more even, and further the particles slide over each other as the surfaces of the machinery are in motion. Further, the carbon atoms in graphite are arranged in regular hexagons in flat parallel layers and each atom is linked by covalent bonds to three other atoms but its distance from the fourth one is more than double. Hence, this fourth valency bond is not fixed but moves about and hence there is no strong bonding between different layers (Fig. 5.4). This is the reason for the softness and lubricating property of graphite.

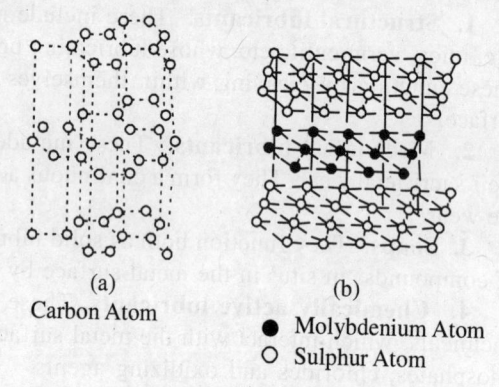

(a) Carbon Atom
(b) ● Molybdenium Atom ○ Sulphur Atom

Fig. 5.4. Structures of graphite and molybdemum sulfide.

Molybdenum disulfide has a higher specific gravity than graphite but is slightly softer. Both these substances can be used as a dry powder, or as an aerosol from a freon pressurised container, or as a paste, grease or liquid dispersion. A dispersion of graphite in water is called "aquadag" and that in oil is called "oil dag". Molybdenum disulfide preparations are commercially available as "molykotes". "Oil dag" has been found to be useful in internal combustion engines as it forms a film between the cylinder and the piston rings giving a tight fit to increase the compression of the air-ful mixture. "Aquadag" has been found to be particularly useful in situations where a lubricant free from oil is desirable. It has been used in air compressors, lathes and other machine shop operations, and equipments used for processing foods. Graphite and molybedenum disulfide are also used in some types of greases.

Graphite and molybdenum disulfide are particularly valuable as lubricants at high temperatures and extreme pressures, as they give a low coefficient of friction as long as their decomposition temperatures are not reached.

Oil-less bearings are prepared by sintering metal in which graphite or molybdenum sulfide are used to fill the pores or impregnating plastic washers. A solid-film lubricating surface useful for space vehicles is made from (70% MoS_2 + 7% graphite) bonded with 23% silicates, which can withstand extreme temperatures, low pressures and nuclear radiations.

Teflon has a very low coefficient of friction and has been used as lubricant for periscopes, brass cartridges, gasoline gear pumps, oxygen valves and underwater mechanisms.

Metals themselves act as lubricants if they melt during sliding. Bearing alloys made from low melting point metals such as Sn and Pb, have long been used. More recently, thin films of Indium (M.Pt. 155°C) have been used to lubricate steel bearings, which reduced the coefficient of friction from 0.8 (on clean steel) to 0.05 (on steel coated with Indium).

5·15. SEMI-SOLID LUBRICANTS

The most important semi-solid lubricants are greases and vaselines.

Lubricating greases are employed in the following situations:

1. When a machine is worked at slow speeds and high pressures.
2. In situations where spilling or spurting oil from the bearings is deterimental to the product being manufactured as in the case of textile mills, paper and food product manufacture, etc. Greases are ideal in such cases because they do not spill or splash as they are designed to "stay put".
3. In situations where oil cannot be maintained in position due to bad seal or intermittent operation.
4. In situations where the bearing has to be sealed against entry of dirt, water, dust and grit.

A lubricating grease is generally defined as "a semisolid or solid combination of a petroleum product and a soap or a mixture of soaps, with or without fillers, suitable for certain types of lubrication." Greases are essentially thixotropic gels in which the fibrils or structural elements are metallic soaps, and the liquid entrained is the lubricating oil.

Greases are generally prepared by dispersing a gelling agent in a lubricating oil which is generally a petroleum fraction. However, in case of special greases, synthetic oils such as aliphatic diesters, siloxanes and fluorocarbons are used. The gelling agents commonly employed are soaps of Ca, Na, Li, Al, Ba, Pb, Mg and K, as well as fatty acids such as oleic acid, stearic acid, and carboxylic acids obtained from tallow, etc.

Important functions of soap in a grease are (*a*) it acts as a thickener, (*b*) it enables the grease to stick to the metal surface firmly, (*c*) the nature of the soap determines the temperature upto which the grease can be used, its consistency, water resistance, oxidation resistance, resistance to break down on continued use and its ability to stay in place.

The properties of a grease depend upon the nature and amount of the thickener used, the additives used (if any), the characteristics of the base oil used and the way in which the grease is prepared.

Most of the common greases are made with soap thickeners. The soap content is generally between 7 to 18% but in special cases it may be as high as 50% or as low as 3%.

Non-soap thickeners such as carbon black, asphaltenes, bentonite treated with quaternary ammonium compounds, etc. have also been used for manufacturing greases. The term "grease" has also been used for semi-solid lubricants prepared from graphite, chalk, talc, mica in admixture with mineral oils, and also for some combinations of fats, waxes and petroleum residues.

Some important aspects about the use of greases are as follows:

1. Greases show higher coefficients of friction than oils because of the greater amount of work that must be done in shearing the lubricating film. Therefore, wherever possible, it is better to use an oil instead of grease, barring the special situations listed above.
2. Greases cannot effectively dissipate heat from the bearing as a result of which the grease-lubricated bearing works at relatively higher temperature as compared to the oil-lubricated bearing.
3. Greases on storage tend to separate into oil and soap.
4. On constant use, oil in the grease volatalizes off.
5. Greases do not require as much attention as oils and are thus more convenient in use.
6. Greases are capable of supporting greater load at lower speed due to their high shear resistance.

Soda-based and lime-based soap-oil greases are most commonly used. These greases are generally manufactured by saponification of a solution of fatty oil in mineral oil with NaOH or Ca(OH)$_2$ followed by agitation with sufficient oil to obtain the desired consistency. "Boiled greases" are prepared with soaps produced by heating a fatty oil with alkali and mixing the soap formed with spindle oil or thicker lubricating oil of the correct viscosity suitable for the particular application.

Lime or calcium-soap-base greases, known as *"cupgreases"*, are the cheapest and widely used. They have good resistance to displacements by water and are suitable for lubricating water pumps, tractors, caterpillar treads, etc. They can be prepared in a wide range of consistency, from soft paste to hard, smooth solid by varying the amount of lime soap from 10 to 30%. However, these greases cannot be used above 65°C as they deteriorate losing combined water. Cup-greases are prepared from fats, lubricating oil and slaked lime.

Sodium-soap greases have good high temperature properties as they can hold water more firmly due to their high melting point and fibrous structure and hence can be used upto 175°C. However, as the sodium soaps are soluble in water, these greases are not suitable for bearings exposed to wet conditions.

Aluminium soap greases have excellent stringiness, attractive appearance and good clarity and are used in situations where adhesiveness is of prime importance. They usually contain about 5% more mineral oil than lime-base greases of same consistency. Due to lower soap content, they are relatively water-proof and have slightly higher dropping points than the calcium-base greases. These greases cannot be used beyond 90°C as they tend to become hard and coarse.

Lithium-soap greases have combined advantages of both calcium-base and soda-base greases, namely, good high-temperature properties and good water resistance. Hence, they are excellent multipurpose greases for a wide range of applications. They have high mechanical stability, low oxidation and are stable in storage. They have a high melting point (about 150°C). Properly formulated lithium-base lubricants can be used for aircraft applications where temperatures as low as –55°C may exist at extreme heights and the lubricant used should permit functioning of the controls under such conditions. Due to their high cost, they are used for special applications only.

Barium-base greases are characterised by extreme resistance to removal from bearings by water, good cohesiveness and adhesiveness, high melting point and maintenance of consistency during service. They are widely used as a multipurpose lubricant for automative and farm equipment and for several industrial applications.

Rosin soap grease is prepared from rosin oil which contains several saponifiable acids such as abietic acid. The rosin oil is dissolved in the lubricating oil and allowed to react at 58°C with a slurry of slaked lime, emulsified oil and water called "sett". The resulting grease is known as "cold set grease". It is mainly used as axle grease for farm wagons and low-speed machinery. This is the cheapest of the greases.

In order to prevent the grease from being squeezed out under heavy loads, fillers like talc, mica, etc. are added to these "set" greases. Rosin greases are also used on any heavy, slow-moving bearings as gear greases and for lubricating the curves in street railway tracks.

Fire-cooked greases are prepared by driving off water at about 260°C from sodium soaps of lard oil, tallow oil or rosin oil mixed with spindle or cylinder oils. They have high melting points and fibrous structures. "Sponge or fibre" greases are stirred to remove volatiles while "block" greases are prepared in the same way but are unstirred. The "block" greases have melting points upto 540°C and are employed for the lubrication of very heavy bearings. "Cold" greases are prepared at relatively low temperatures, by neutralizing the fatty acids obtained from fish oil or vegetable oil and mixing the soap thus produced with spindle oil. Lime-base grease made in this way from rosin oil is called "sett grease" which has been already discussed.

Non-soap greases are usually prepared from non-soap thickeners such as carbon black, silica gel, modified clays, organic dyes, etc. Many of them are suitable for high temperature applications.

LUBRICANTS

The properties of the soap-greases described above may be modified by incorporating additives such as other soaps to yield mixed soap base, special fatty acids, stabilizing agents, inorganic thickeners and solid lubricants to enhance their range of usefulness. For instance, copper phthalate decreases the oxidation tendency of the grease and helps maintaining its consistency up to 150°C. Silica gel, modified clays and acetylene black are used as thickeners.

Silicone greases made from synthetic lubricating oils have high viscosity index, oxidation resistance and water resistance. Greases made from fluorinated diesters have good high temperature stability.

For selecting greases for practical applications, the following tests are carried out : consistency or yield value or penetration test, drop-point test, ash, water stability and neutralization number.

Consistency test

The consistency or yield value is defined as the "distance in tenths of a millimeter that a standard cone penetrates vertically into the sample under the standard conditions of load, temperature and time". The standard values selected for the test are as follows:

Load	...	150 g
Temperature	...	25°C
Time	...	5 seconds

The consistency of a grease is a very important property from practical point of view and it depends upon the structure and properties of the gelling agent and the oil (*e.g.*, viscosity) used in the preparation of the grease. The apparatus used for the determination of consistency of a grease is called "penetrometer" (Fig. 5.5). It consists of a heavy base made from a cast iron alloy and is provided with levelling screws, spirit level and a plain "table" for keeping the box

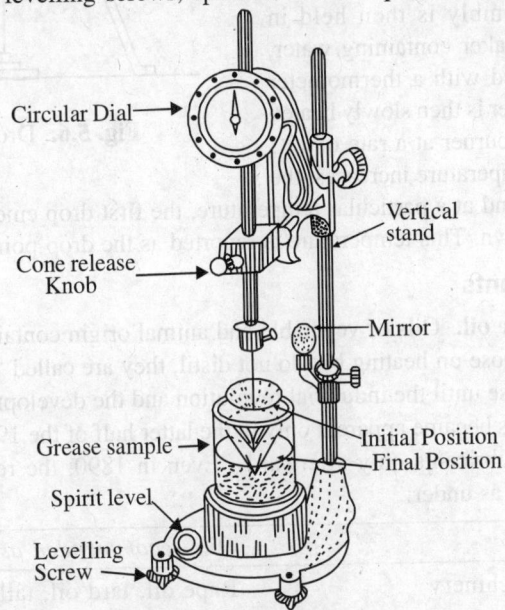

Fig. 5.5. Penetrometer.

containing the grease being tested. The base is fitted with an iron rod (called vertical support) provided with a movable holder and screw with the help of which it can be firmly fixed at any of the slotted marks on the rod. The holder is fitted with a circular dial (15 cm diameter) graduated in millimeters. A moving dial rod is attached behind the dial. The dial rod is provided with a clutch arrangement with the help of which it can be connected or disconnected to the circular dial. The standard cone (150 g) is fitted at the lower end of the moving dial rod. An adjustable mirror attached to the vertical support aids in proper positioning of the cone before starting the experiment without parallax error.

To start with the test, the apparatus is levelled with the help of the levelling screws and the spirt level. The standard cone is thoroughly cleaned. The box containing the grease sample is placed below the cone on the "table" on the base. With the help of the mirror, the position of the cone is so adjusted that the tip of the cone just touches the surface of the grease sample. The initial reading on the dial is noted and the cone is released exactly for 5 seconds by pressing the button. After exactly 5 seconds measured with a stop-watch, the button is released and the final reading on the dial is noted. The difference between the final and initial dial readings is reported as the yield value.

Drop-point test

Drop-point is defined as the temperature at which the grease passes from semi-solid state to the liquid state. This determines the upper limit of temperature upto which a grease can function as a satisfactory lubricant. The apparatus used for the determination of drop-point is shown in Fig. 5.6.

The grease sample is taken in a metal cup having an opening of a standard dimension at the bottom. The cup is fitted in a glass case with a tightly fitting lid. A thermometer is inserted into the cup in such a way that the bulb of the thermometer is just above the surface of the grease sample under test. The entire assembly is then held in position in a glass beaker containing water, which is also provided with a thermometer and a stirrer. The beaker is then slowly heated electrically or with a burner at a rate of 1°C per minute. As the temperature increases, the grease sample melts and at a particular temperature, the first drop emerging out of the opening of the metal cup falls down. This temperature is reported as the drop-point of the grease sample.

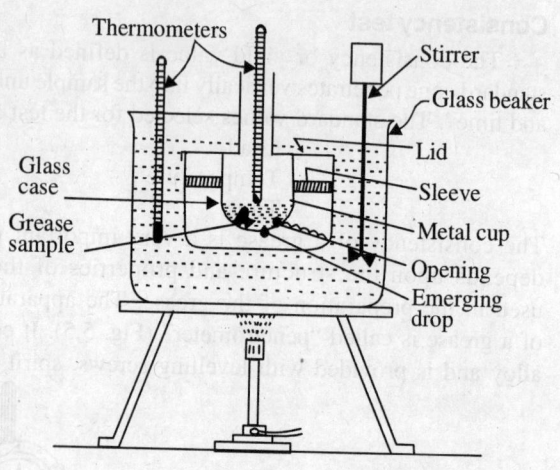

Fig. 5.6. Drop-point test apparatus.

5·16. Liquid lubricants

5.16.1. Vegetable oil. Oils of vegetable and animal origin contain glycerides of higher fatty acids. As they decompose on heating but do not distil, they are called "fixed oils". These were the lubricants in general use until the industrial revolution and the development of petroleum industry. In fact, their limitations became apparent only in the latter half of the 19th century and were largely replaced by mineral lubricants only afterwards. Even in 1890, the recommended lubricants for various purposes were as under:

Sr. No.	Purpose	Oils recommended as lubricants
1.	Ordinary machinery	Rape oil, lard oil, tallow-oil, medium mineral oils.
2.	Steam cylinders	Heavy mineral oils, tallow, lard, rape oil, etc.
3.	Watches, clocks, etc.	Light mineral oils, clarified sperm, neatsfoot, olive, purpoise.
4.	For great pressures with low speed	Tallow, lard oil, palm oil, grease, etc.
5.	For high pressures and high speeds	Sperm oil, rape oil, castor oil, medium mineral oils.
6.	For light pressures and high speeds	Sperm, refined petroleum, cottonseed, rape, olive and mineral oils.

It was also stated in the above recommendations that a mineral oil should always be used along with these oils wherever possible, particularly in case steam cylinders, because the fatty oils may decompose at the high temperatures producing fatty acids which attack, Cu, Fe, etc. forming the respective metallic soaps which cause damage to the fittings. The significance of boundary lubrication was not appreciated at that time. Relative cost was an important factor which influenced the replacement of vegetable and animal oils by mineral oils later.

Fixed oils have a property called "oiliness" by virtue of which they are adsorbed on the metal surface tenaciously and offer lower coefficient of friction and higher load carrying capacity than the petroleum oils. Hence they are often added to the mineral oils to improve the oiliness of the latter and such mixtures are called blended oils. Fixed oils are also used in the preparation of greases and cutting emulsions.

Fixed oils are more expensive and scarce. They hydrolyze in presence of water. They have a greater tendency for oxidation and decomposition at higher temperatures forming acidic and gummy products which are deliterious to the metal parts. Because of all the above limitations, they are generally not used now as such but are used for blending with mineral oils to incorporate desired properties like "oiliness".

Vegetable oils such as castor oil, rape oil and cotton seed oil are obtained by crushing the seeds. The old mills were operated by manual or animal power. At the end of the 18th century, hydraulic presses replaced the earlier wind and water-driven stamper mills. These have been further replaced by the continuous processes of expelling and solvent extraction. At present, it is a common practice to use high pressure expellers or a low pressure expeller followed by solvent extraction. Solvent extraction process consists of washing the compressed seeds in petrol to extract the residual oil and distilling it to recover the oil and petrol. Solvent extraction is more expensive than high pressure expelling but the oil recovery is more.

Some of the important vegetable oils having potential use as lubricants are given below:

(*a*) **Olive oil (or sweet oil).** This is obtained by compressing the fruits of the olive tree or by extracting the fruits with carbon disulfide. It is greenish yellow in colour with a specific gravity of 0.914 to 0.919. It is a good lubricant for machine parts working under low pressures and high speed. It is expensive. It tends to decompose on storage and congeals at 0°C.

(*b*) **Castor oil.** It is obtained by crushing the seeds. It has yellow to pale brown colour. Its specific gravity is 0.950 to 0.974. Its viscosity is 1370 seconds Saybolt universal at 38°C. It has a high content of ricinoleic acid, a high viscosity, a low coefficient of friction and is the only one among common fatty oils which is soluble in alcohol but not in petroleum. It has demand in medical, printing and plastic industry. It is a very good lubricant for bearings and machinery operating at low pressure and high speeds. It has been used in the past in racing cars and early aero engines of rotary type. Its major drawback is that when used in heavy duty engines, it oxidizes into gummy and acidic substances.

(*c*) **Rape oil or Rape seed oil or Colza oil.** This is the most extensively used fatty oil in the lubricating oil industry. At one time it was extensively used as burning oil. It is a good lubricant for ordinary machinery, for machinery operating under high speed and high pressure and for steam cylinders. It is used for compounding with mineral oils. When thickened by blowing, it can be used to increase the viscosity of other oils.

(*d*) **Palm oil and Palm Kernel oil.** Palm oil is obtained by boiling and pulping the fruit of the palm tree, while palm kernel oil is obtained by crushing the kernels. The specific gravity of the palm oil is 0.921 to 0.925 and that of the kernel oil is 0.930. These are used for lubricating delicate instruments such as scientific equipment, watches etc.

(*e*) **Cotton-seed oil.** This is a by-product of the cotton plant which is primarily grown for the cotton fibre. The oil obtained by crushing the seed after separating the fibre is in demand as an edible oil. Its viscosity can be increased by blowing and hence it finds application as a thickener for lubricating oils. Its specific gravity is 0.913 to 0.930.

5.16. 2. Animal fats and oils. Animal fats and oils are extracted from the crude fat by a process called "rendering" in which the enclosing tissue is broken by treatment with steam or with the combined action of steam and water.

(*a*) **Lard and lard oil.** Lard is a "rendered" hog fat. Inedible lard with a low free fatty acid content is sometimes sold as a grease which is designated white or yellow grease, according to the part of the animal from which it is derived. When lard is allowed to crystallize at a carefully controlled temperature, the lard stearine crystallizes out and lard oil can be extracted by pressure. Lard oil is generally used as a cutting oil and for lubricating ordinary machine parts. Its specific gravity is 0.915 and its Saybolt universal viscosity at 38°C is 206 seconds. The freezing point of the oil is 27°C.

Tallow and tallow oil

Tallow is obtained by "rendering" the fat of cattle and sheep. At one time, this was used as a cylinder lubricant in slow-speed steam engines. Tallow oil is obtained by melting tallow and allowing it to stand at 26° to 32°C. The stearine so separated is used for manufacturing candles. Tallow oil has a specific gravity of 0.94 and a Saybolt universal viscosity at 38°C is 230 seconds. It can be used as lubricant as such or blended with mineral oils.

Neatsfoot oil

This is obtained from the hoofs of slaughtered cattle and sheep by extraction in steam-heated kettles and skimming off the fats. The fat is subjected to such a temperature where the stearine crystallizes out and is separated from the oil by pressing. The oil is graded according to its cloud point, which in turn depends on the degree of processing of the oil. It has a pale yellow colour with specific gravity of 0.916 and Saybolt universal viscosity of 215 seconds at 38°C. It is used as a dressing for leather and for lubricating instruments such as clocks and sewing machines.

Fish oils
Sperm oil

This is the lightest of the fatty oils, having a specific gravity of 0.880 to 0.884. The two main sources for this oil are the Arctic Sperm or bottlenose whale and the Cacholot whale. The oil is generally obtained by boiling the blubber (or the head cavities in case of cacholot whale) and when cooled, it deposits spermacetti which can be extracted by refrigeration and filtration under pressure Sperm oil is pale yellow in colour having a tendency to become gummy but is not corrosive. It is a glyceride like other animal and vegetable oil but has the constitution of a wax. Sperm oil is a good spindle oil which is particularly suitable for light machinery.

Whale oil. It is obtained from a number of other types of whales such as Greenland whales. The oil is extracted from the blubber by boiling. Its specific gravity lies in the range of 0.917 to 0.922, and its colour pale yellow to reddish brown. It is a relatively cheap oil which can be used by itself or blended with mineral oil.

Seal oil. Seals are hunted for their skins and oil. The oil is extracted from the blubber just like whale oil. Its specific gravity is 0.924 to 0.933 and colour varies from straw to brown with fishy odour.

Refining of fatty oils. The animal and vegetable oils require further treatment before use. The oil is cooled until stearine separates out. The oil is then filtered through animal carbon or treated with fuller's earth for removing the colour and brightening the oil. Free fatty acids are neutralized by adding calculated quantity of caustic soda.

LUBRICANTS

Chemical refining consists of treatment with H_2SO_4 followed by removal of traces of acid by washing. H_2SO_4 removes suspended impurities by carbonizing and causing them to coagulate and settle out. The oil is finally filtered.

5.16.3. Mineral oils. Mineral oils are prepared from the heavier fractions obtained from fractionation of crude oil at atmospheric pressure. The heavy residual fraction is subjected to vacuum distillation (below 40 mm pressure) in a bubble tower and the lubricating oil fraction is collected. By employing modern refinery techniques, a wide range of good lubricating oils could be produced from almost any type of crude oil but the greater the degree of refining needed, higher will be the cost of the finished product.

A good lubricating oil should have reasonably good viscosity, high viscosity index, good stability, low pour point and oxidation resistance.

In general, paraffin base oils have good viscosity index, but they contain wax which settles out at low temperatures and interferes with the free flow of the oil. Naphthenic base oils contain asphalt which, at high temperatures, tends to gum formation. Aromatic constituents also readily give deterioration products which are most undesirable and have poor viscosity indices.

Asphalt base oils give only distillate lubricating oil fractions (since asphalt is left as residue on distillation) but paraffinic crudes may have both distillate and residual (heavy) lubricating oils. However, the majority of crudes are of the mixed type and may give distillate or residual lubricant yields. Distillates from naphthenic base crudes have to be treated with caustic soda to remove the naphthenic acids and are then redistilled.

The main operations in refining a lubricating oil (suitable for cylinder engines, etc.) are the removal of wax, asphaltic matter and aromatic constituents.

Solid waxes crystallize at low temperature and interfere with the flow properties of the lubricating oil in service. Easily oxidizable impurities cause sludge formation during the operating conditions. Asphaltic, naphthenic and resinous impurities decompose at higher temperatures leading to the deposition of carbon and sludges. Hence, the impurities are generally removed by the following methods of refining:

(1) *Dewaxing*. In the older conventional method, the lubricating oil is slowly chilled as such and the wax thus crystallized is removed by a filter pressing operation.

For removing amorphous wax, the oil is diluted with twice the volume of naphtha and chilled or refrigerated. The precipitated wax is separated from the oil naphtha solution by high-speed centrifugation. The naphtha is removed by distillation.

The more recent method is to mix the oil with proper volume of a suitable solvent (*e.g.,* propane; mixture of benzene and acetone, trichloroethylene; a mixture of benzene and ethylene dichloride) and refrigerating the mixture to the required temperature when the wax in the oil precipitates out. The oil and wax are separated using continuous filters or centrifuges. The oil is separated from the solvent by distillation.

(2) *Acid-refining*. The dewaxed lubricating oil fractions still contain naphthenic, asphaltic and other undesirable constituents, which must be removed to produce a finished lubricating oil. This is done by thoroughly agitating the oil with the required amount of concentrated sulfuric acid, which acts as a solvent for some of the constituents and react chemically with others to form a tarry sludge. The oil is separated and neutralized with ammonia or caustic soda or by being brought in contact with finely divided fuller's earth at 108 to 230°C. The latter treatment also helps to decolourize and stabilize the oil. The sludge can be used as a fuel. Percolation filtration through fuller's earth is the final refining step in many cases.

(3) *Solvent refining*. In this process, a suitable solvent is selected in which the solubility of the undesirable impurities such as naphthenic, asphaltic, and resinous constituents is more than that in

the oil. The important solvents in use today include furfural, phenol, pp'- dichloro-diethyl ether (chlorex) and nitrobenzene. Mixtures of SO_2 and benzene; propane (which dissolves the desired paraffinic fraction and decreases the solubility of undesirable constituents in this fraction) and a mixture of phenol and cresol (which is immiscible with the oil-propane solution is a good solvent for the naphthenic, asphaltic and resinous constituents) are also used.

The general procedure for the solvent refining method is to bring the oil into intimate contact with a suitable solvent (which is immescrible with the oil) in a counter-current continuous operation. The oil fraction recovered consists of the refined lubricating oil containing mostly paraffinic hydrocarbons and a little solvent which can be separated by distillation. The solvent fraction containing the various impurities extracted from the oil is distilled to recover the solvent which can be reused. The residue remaining after distillation can be used as a fuel or as a source of asphalt, or as cracking stock blending material.

Solvent refining is more economical than acid treatment. Further, solvent refining produces higher yields of finished oil which has lower carbon residue and higher viscosity index. But, however solvent refined oils are generally inferior in oxidation resistance and oiliness. They tend to thicken on use and form films of lacquer-like deposits on the heated parts of the engines. The natural inhibitors to oxidation present in the raw lubricating oil fractions seem to be almost completely removed in the solvent refining process unlike in the acid refining process. One important advantage with solvent refining process is that it enabled the production of lubricating oils at reasonable cost from crude oils which were not considered to be a possible source of lubricants when only the older acid-refining process was available.

Mineral oils have largely replaced animal and vegetable oils because the former are cheaper, available in bulk quantities, stable under service conditions and can be re-refined after use.

Recognition of the drawbacks of solvent treated oils mentioned above led to the successful search for substances that could be added in small quantities to these oils to impart or improve the desirable properties such as oiliness, oxidation resistance, resistance to thickening during use, etc.

5.16.4. Blended or Compounded oils. The main objective of the refining methods described above is to improve the desirable characteristics of the lubricating oils by removing the undesirable constituents. In recent years, commendable progress has been made in further improvement of refined oils by adding small quantities of various additives. The oils thus prepared are known as blended oils or compounded oils.

The main purpose of the various additive systems and the broad range of chemicals used in their manufacture for this purpose are summarized in the following Table-1.

Table–1. Additives and their functions

Sr. No.	Name of the additive	Chemical used	Functions
(1)	(2)	(3)	(4)
1.	Detergents and deflocculents	Normal or basic calcium and barium salts of Phosphonates and sulfonates, some salts of phenol, etc.	Reduce or prevent deposits in engines operated at high temperature.
2.	Dispersants	Polymers such as nitrogen containing Polymethacrylates, alkyl succinimides and high molecular weight amines and amides.	Prevent or retard sludge formation and deposition under low temperature operating conditions.

LUBRICANTS

Sr. No. (1)	Name of the additive (2)	Chemical used (3)	Functions (4)
3.	Anti-oxidants	Phenols, amines, organic sulfides, organic phosphides, etc.	Retard the oxdiation of oils, Minimize the formation of resins, varnish, acids, sludges and polymers.
4.	Corrosion inhibitors	Organo metallic compounds, zinc dithiophosphates, sulfurized terpenes, phosphorized terpenes.	Protect bearings and other metal surfaces from corrosion.
5.	Rust inhibitors	Amine phosphates, Alkyl succinic acids, fatty acids, sodium and calcium petroleum sulfonates.	Protects ferrous metals from rusting.
6.	Anti-wear additives	Zinc dialkyl dithio-phosphate, Tricresyl phosphate, alkyl earth phenolates.	Reduce rapid wear in steel-on-steel applications.
7.	Metal deactivators	Triaryl phosphites, sulfur compounds, diamines.	Stop the catalytic effect of metals on oxidation and corrosion.
8.	Extreme pressure additives	Sulfurized fats, chlorinated hydrocarbons, lead salts of organic acids, organic phosphorous compounds, metallic soaps such as lead naphthenates.	Adsorbed on the metal surface or react chemically with the metal forming a surface layer of low shear strength which prevent tearing up, seizure and welding of the metals.
9.	Oiliness	Fatty acids, fats and fatty amines, vegetable oils	Increase the strength of the oil film and prevent the rupture of the oil film.
10.	Polymeric thickeners and viscosity index improvers.	Long chain polymers such as polyisobutylene, polystyrene or alkyl styrene polymers, long chain alkyl acrylates and polyesters, "Plexol", "Paratone" and "Exanol" are the trade names of such products commercially available.	Reduce the rate of change of viscosity with temperature.
11.	Pour-point depressants	Wax alkylated naphthalene, wax alkylated pheols, polymethacrylates. "Para flow" and "Santapour" are the patented commercial products available.	Lower the pour-point of the oil.
12.	Antifoam additives	Silicone polymers, oil insoluble liquids like glycols and glycerols	Prevent formation of stable foams.

Sr. No. (1)	Name of the additive (2)	Chemical used (3)	Functions (4)
13.	Emulsifiers	Sodium salts of sulfonic acids, sodium salts of organic acis, fatty amine salts, nonionic emulsifiers such as monoesters of polyhydric alcohols.	Make mineral oil miscible with water or help the formation of emulsions of lubricating oils with water.
14.	Tackiness improvers	Soaps, polyisobutylene and polyacrylate polymers	Provide the oil with greater cohesion.

The types and the quantities of the various additives used for a lubricating oil for a particular purpose is a complex and specialized job and should be decided only after careful and judicious balancing of the effects of the individual additives.

5.16.5. Synthetic lubricating oils. Synthetic lubricants are oily liquids which are not found naturally or not produced directly during the normal manufacturing and refining processes of the petroleum industry. Synthetic lubricants are designed for special jobs.

In case of machinery operating at severe or extreme conditions, petroleum lubricants even with suitable additives have been found to be unsatisfactory. Such situations include:

1. In metal forming processes such as die casting in which fire resistant hydraulic fluids are particularly desirable.

2. Hot running bearings and hot rolling mills, where extreme temperatures and pressures are encountered.

3. Air-craft turbines where the lubricant must have a sufficiently low viscosity at starting to permit adequate oil flow to the pump, it must remain stable for prolonged periods at a very high operating temperature and finally must effectively lubricate heavily loaded high speed gears. For modern jet turbines, the temperature range is extremely wide, from as low as $-55°C$, for which a viscosity of 3,000 to 10,000 centistokes is required, to as high as $+260°C$ at which temperature bearings must be lubricated without incurring excessive evaporation loss or formation of objectionable deposits.

4. **In reactive environments.** In order to meet the lubricating requirements under such peculiar operating conditions, viscous fluids have been prepared from various organic and inorganic substances, which are called synthetic lubricants. Many of these were previously known to industries other than the lubrication industry. The most important specific chemical classes of compounds which have been found to possess useful lubricating functions are:

(1) **Dibasic acid esters.** They are finding widespread use as bases for low-volatility greases and lubricants for gas turbines. Their outstanding advantages are excellent viscosity temperature characteristics, very low volatility and high thermal stability. Further, they are non-corrosive to metals, non-toxic, and stable to hydrolysis. Their pronounced solvent action on rubbers is a disadvantage. The most widely used diester is di-2-ethyl hexyl sebacate, which has satisfactory performance from $-50°C$ to $230°C$ for lubrication in turbojets.

(2) **Organo-phosphate esters.** Phosphate esters have been used for some years as additives in petroleum lubricants for improving boundary lubrication properties. Those phosphate esters having inherent chemical stability are used as the main components of synthetic lubricants, while those which are chemically active find application as extreme pressure additives. Most of the phosphate esters are stable upto about $150°C$ but rapidly deteriorate at higher temperatures. Tricresyl phosphate

is widely used as additive for petroleum lubricants. Organo-phosphorous compounds are widely used in air-craft hydraulic oils as they possess good fire-resistant properties.

(3) **Polyalkylene glycols and their derivatives.** These are available in plenty at reasonable cost and are very widely used. The compounds belonging to this class include polyethylene glycol, polypropylene glycol and their ethers and esters, higher polyalkane oxides, polyglycidyl ethers, polythioglycols, etc. The first two are water soluble and hence used in water diluted lubricants for use in rubber bearings and joints. They have also the advantage of being easily removed by water flushing, when used for metal surfaces. Higher polyalkylene-oxides and polyglycidyl ethers are not water soluble but can absorb lot of water. They have high viscosity index, low viscosity at sub-zero temperatures and low freezing points. They decompose at high temperatures into volatile parts which undergo oxidation subsequently. Hence they are useful as residue-free high temperature lubricants also, *e.g.,* in roller bearings of sheet glass machinery.

Polypropylene glycol is an excellent winter grade crank-case oil which is characterszed by zero carbonization, low ash, low starting temperature and from 60 to 350% greater mileage use. Polyalkylene glycols may be oil soluble or water soluble, liquid or solid. The aqueous lubricating solutions are also used as hydraulic fluids.

Polyalkylene glycols or polyethers are also widely used for combustion engines, gears, compressors, pumps and also for formulation as fire-resistant water based hydraulic fluids. They are attractive as aircraft turbine lubricants because of their thermal stability, lack of corrosive action on metals and resistance to breakdown at high rates of mechanical shear.

(4) **Chlorinated and flourinated hydrocarbons.** The main useful features of chlorinated hydrocarbons (*e.g.,* chlorinated diphenyl compounds) are low-inflammability and good extreme pressure lubricating properties. Unfortunately, the high activity of the chlorine atoms which may be released under conditions of severe loading and high temperatures, results in the creation of highly corrosive or toxic end products.

Fluorinated hydrocarbons (and also completely fluorinated tertiary amines, ethers, esters, etc.) show a greater degree of chemical stability, high-degree of non-inflammability, thermal, chemical and oxidation resistance and high viscosity index. This chemical inertness means that they are inherently poor boundary lubricants. These compounds are very expensive.

The commercially available fluolubes have high chemical and thermal stability and they are less susceptible to oxidation and cracking.

Fluolubes have an interesting application in submarines.

(5) **Silicate esters.** These are characterized by high viscosity index (150 to 200), exceptionally low volatility, poor oxidation resistance at elevated temperatures, non-corrosive to metals, compatibility with plastics and rubbers (they tend to harden rubber only after extended contact at high temperatures, similar to silicones), tendency to hydrolyse and incompatibility to moisture in many cases. Their industrial use has been mostly limited to heat-transfer fluids.

(6) **Silicones.** Silicones are the products obtained by hydrolysis and polymerization of organochlorosilanes *e.g.,* $(CH_3)_2 SiCl_2$ and $(CH_3) SiCl_3$. For lubrication at elevated temperatures, several types of silicone fluids are available. Their general formula may be represented as

$$R_3 SiO \begin{bmatrix} R \\ | \\ Si-O \\ | \\ R \end{bmatrix}_n - SiR_3$$

where R is an organic radical.

One of the most important properties of silicones is their remarkably high viscosity index. They have good oxidation resistance at moderate temperatures (say, upto 200°C), but they are prone to oxidize quickly at higher temperatures forming gels. Hence they are not suitable for use as aircraft turbine lubricants. They are chemically more inert than other synthetics, and they do not attack rubber, plastics or paints. Their physical properties are highly desirable under hydrodynamic conditions but they exhibit very poor boundary properties towards bearing metals, particularly ferrous material. Their lack of adsorption on steel renders them inadequate as rust protectors. Silicones give good result only when one of the rubbing surfaces is non-ferrous. They are very useful for low temperature lubrication of small parts.

Improvement in the properties of silicones as boundary lubricants has been made by leaving some unreacted – OH groups and by variation in type of substituted group. The incorporation in the silicone of phenyl groups containing halogen substituents increases their load carrying capacity. The most commonly used silicones at present are the dimethyl and methylphenyl-silicone polymers.

Various types of silicone oils having wide range of viscosities are available for specific uses. Silicone-200 fluid series are recommended for rubber and plastic surfaces, including moving picture film, slide rules, gears, bushings, bearings, and as moisture-repellent, dielectric lubricants for clocks, timers and other electronic devices. Silicone-510 fluids are suitable for low-temperature lubrication while Silicone-710 fluid is a thermal-resistant lubricant for oven hinges, oven timers, automatic toasters, conveyor and dolly wheels or rollers exposed to high temperatures, high humidity or weathering.

Important characteristics of synthetic lubricants

The most compelling reason for using synthetic lubricants is the demand for lubricants which can perform satisfactorily over a wide range of temperature, say from – 50°C upto + 260°C. A straight petroleum product having sufficiently low viscosity to be mobile at –50°C will be susceptible to unacceptable losses at high temperatures. Conversely, some high molecular weight hydrocarbons have been stabilized for use upto 175°C but they refuse to pour below –18°C. Most synthetic lubricants exhibit appreciably lower viscosity for a given volatility than any equivalent petroleum product. For instance, a synthetic lubricant like di-2-ethyl-hexyl sebacate loses less than 0.05% of its weight after heating for 168 hours at 65°C while a petroleum oil selected for comparable viscosity may lose 10% of its weight under identical conditions.

Other superior characteristics of synthetic lubricants as compared to the conventional petroleum-based lubricants are their high thermal stability, oxidation resistance, resistance to hydrolysis, flash-point, high fire resistance, rust preventive qualities and especially, much better viscosity-temperature characteristics (*i.e.,* they possess very high viscosity index values). For instance, a very good pennsylvanian oil has a viscosity index of 104 whereas the viscosity index values of di-2-ethyl hexyl sebacate and di-methyl silicone are 152 and > 200 respectively.

Furthermore, many synthetic oils are excellent solvents for important additives *e.g.,* viscosity-index improvers, oxidation inhibitors and rust inhibitors. Certain synthetic lubricants also possess a high degree of natural detergency which is a valuable property in situations where the lubricant has to act as a scavenger for dirt and fuel deposits, without itself getting deteriorated rapidly.

However, many synthetic lubricants are expensive. Some of them are toxic. Some are corrosive particularly at high temperature. Their solvent action on sealing and packing materials also should be studied before use.

5·17. LUBRICATING EMULSIONS

In various machining operations such as milling, threading, turning and boring, the tool employed gets heated to a very high temperature, particularly at the cutting edge. In a cutting process, the pressure at the knife-edge may sometimes reach as high as 100,000 psi and a lot of heat is generated. This may lead to oxidation and rusting of the metal under work. In order to prevent overheating in such cases and the consequent injury to the tool, efficient cooling and lubrication have to be provided. This is usually done by employing emulsions of oil droplets in water, which are called cutting oils or cutting fluids or cutting emulsions. Oil has a poor specific heat, but it has good lubricating properties. On the other hand, water is a poor lubricant but is an excellent cooling medium because of its high specific heat and a high heat of vaporization. Hence, the combination of the two in the form of an emulsion can provide both lubrication and cooling effects. The corrosive action of water on the tools, the machines and the work piece are objectionable and is therefore checked by the addition of soaps or other inhibitive alkaline substances. Even then, the use of water is generally limited to simple operations such as grinding and rough turning.

The important criteria of a cutting emulsion include (*a*) to get itself drawn between the chip and the face of the tool and to provide efficient lubrication, (*b*) to conduct off the heat so as to prevent wear and damage of the metal, (*c*) to wash away the fragments of the metal, (*d*) to give a stable emulsion with water, (*e*) it should not cause rusting of the metal, (*f*) it should be antiseptic so that in case the worker gets injured, the wound should be rendered aseptic.

A good cutting oil increases the accuracy of the cuts and reduces the cost of the work, (*a*) by making possible to achieve higher cutting speeds, (*b*) by prolonging the life of the cutting tool, and (*c*) by reducing the power demand and the number of rejects.

Straight chain petroleum oils are very poor cutting lubricants. Fatty oils such as lard oil and sperm oil are very good for cutting oil although blended oils, pine oil, turpentine and rosin oil also find some use. In a good cutting oil, a sulfonated additive is present. For low speeds and light cuts, a chlorinated lubricant may be used. For machining brass, the emulsified oil would be a paraffin oil containing copper oleate or free fatty acid because a sulfurized oil may discolour the work piece.

Two types of emulsions are used for lubricating jobs:

1. Oil-in-water type emulsions or cutting emulsions are prepared by mixing together an oil containing about 3 to 20% of a water soluble emulsifying agent (*e.g.*, water soluble soap, alkyl or aryl sulfonate, alkyl sulfates, etc) and a suitable quantity of water. Chemicals like glycols, glycerols and triethanol amine are also added sometimes. Oil-in-water type emulsions are used as coolant cum lubricant for cutting tools and in diesel motor pistons and large internal combustion engines.

2. Water-in-oil type emulsions or cooling liquids which are prepared by mixing together water and oil containing 1 to 10% of water insoluble emulsifiers (*e.g.*, alkaline earth metal soaps).

Emulsions containing 50% lube oil and water are used for the lubrication of steam cylinders, giving cooler walls and lesser oil consumption. Such emulsions have also been successfully used in lubricating compressors handling fuel gases.

5·18. PROPERTIES OF LUBRICANTS AND TESTS

To make an intelligent choice of a lubricant for a particular application, a lubrication engineer should be well informed about some important properties of lubricating oils.

1. Colour. The colours of lubricating oils vary from almost complete transparency to pitch black with all intermediate shades of yellow, red and brown. Some mineral oils exhibit green or blue fluorescence in reflected light.

The colour of an oil, to some extent, indicates its origin. Paraffin base oils show a green bloom while napthenic base oils have rather a bluish appearance. In general, the higher the boiling point of a petroleum fraction, the darker it will be.

Colour can be totally removed by treatment. In general, colour cannot be taken as an indication of viscosity or performance of a lubricant. Light coloured oils of high viscosity still find difficulty in marketing because of traditional prejudices associated with colours. However, in textile, paper and food industries etc., where the possibility of staining the finished product exists, the colour of the lubricating oil is naturally important.

Sometimes, the degree of deterioration or contamination of a lubricating oil is reckoned by comparing with the colour of the unused oil.

2. Specific gravity and A.P.I. gravity. The specific gravity of an oil virtually conveys no information regarding its lubricating properties, but since oil is sold by volume, this information on weight to volume ratio may be useful.

Specific gravity is a dimensionless quantity which expresses the ratio of the density of the oil to the density of water at a specified temperature. In the petroleum industry, the specific gravity of oils is usually determined at 60°F (15.55°C). Most lubricating oils have specific gravity values between 0.85 to 0.9 at 60°F/60°F.

In the U.S.A. specific gravity is sometimes replaced by A.P.I. (American Petroleum Institute) gravity, to provide a simple scale eliminating decimal points. In this, pure water has degrees A.P.I of 10 and for the zero a specific gravity of 1.076 has been adopted. Then,

$$\text{A.P.I.}° = \frac{141.5}{\text{Sp. Gr. } 60°\text{F}/60°\text{F}} - 131.5$$

3. Specific heat. The specific heats of most lubricating oils lie in the range 0.44 to 0.49. Information on specific heat is required in heat transfer problems such as those pertaining to the design of plain bearings where the lubricating oil functions both as a lubricant and also as a coolant.

4. Neutralization number. The acidity or alkalinity of lubricating oil is determined in terms of neutralization number. Determination of acidity is more common and is expressed as acid value or acid number. It is defined as the number of milligrams of potassium hydroxide required to neutralize all the free acid present in 1 gram of the oil.

Even the most carefully refined oil may have a slight acidity. This is due to the presence of minute amounts of organic constituents not completely neutralized during the refining treatment or to traces of residues from the refining process. This small intrinsic acidity may not be harmful in itself but the degree to which it increases in used oil is usually taken as a measure of the deterioration of the oil due to oxidation or contamination. In fact, acid number greater than 0.1 is usually taken as an indication of oxidation of the oil.

5. Saponification number. The saponification value of an oil is defined as the number of milligrams of potassium hydroxide required to saponify one gram of the oil. This is usually determined by refluxing a known quantity of the oil with a known excess of standard KOH solution and determining the alkali consumed by titrating the unreacted alkali.

Animal and vegetable oils undergo saponification but mineral oils do not. Further, most of the animal and vegetable oils possess their own characteristic saponification values. Hence, the determination of saponification value helps to ascertain the presence of animal and vegetable oils (*i.e.*, fixed oils) in a lubricant. Conversely since each of the fixed oil has got its own specific saponification number, any deviation from this value in a given sample indicates the probability and extent of adulteration.

6. Oxidation. Oxidation of straight mineral oils proceeds slowly even at room temperature but is greatly accelerated at higher temperatures (particularly above 200°C). Oxidation is also accelerated by the presence of moisture in the environment as well as by the presence of oxidation catalysts like Fe, Al and especially Cu, particularly when they are in finely divided state (*e.g.*, as wear products).

The resistance of various oils to oxidation depends largely on the nature of the crude oil and the method of refining. In most commercial oils, the rate of oxidation is retarded by adding sacrificial

LUBRICANTS 267

oxidation inhibitors such as phenyl-β-naphthylamine.

Oxidation products are undesirable because: (1) the insoluble oxidation products or sludge may clog oil holes, oil pipe lines, filters and other parts of the lubricating system, (2) The soluble oxidation products circulating with the oil have an acidic tendency and may corrode or pit bearing surfaces or may form harmful and tenacious varnish-like deposits and gums.

Several tests have been suggested for testing oxidation resistance of the oil but none is universally accepted. The only reliable test is the one in which nearly all the service conditions are simulated.

7. Corrosion. Lubricating oils are frequently employed in contact with systems containing Cu and brass. Hence the corrosive properties of lubricating oils are ascertained by copper corrosion test and steel corrosion test.

Corrosive substances like sulfur, H_2S and polysulfides are found in petroleum and are removed or converted into relatively harmless organic sulfides by the refining processes. Refinery chemists usually test for sulfur both during refining and in the final products.

The copper corrosion test is a valuable criterion for products like cutting oils used for machining of non-ferrous metals and for lubricants used in rolling contact bearings which have non-ferrous cages. The so-called "copper strip test" comprises of keeping a polished copper strip in the lubricating oil at a specified temperature for a specified time and then examining the strip after taking it out. Any tarnishing of the strip indicates the presence of corrosive substances in the oil.

The steel corrosion test for oils is designed to determine the ability of the oil to prevent corrosion of ferrous parts in the presence of water.

Corrosion inhibitors like zinc dithiophosphate and organometallic compounds are usually added to the lubricating oils.

8. Emulsification. When pure oil is mixed with pure water, the liquids separate out into layers fairly quickly. But, if the oil is contaminated by finely divided dust, dirt, metal particles or acids, alkalis or soaps, the rate of separation is decreased and an emulsion of either oil in water or water in oil may be formed. Emulsions tend to collect impurities which may cause abrasion, or to form sludges which clog the oil lines, etc. Hence, in a large number of situations, it is essential that the lubricating oil should form such an emulsion with water which breaks off rapidly. This particular property of the lubricant is called "demulsification number" and is determined by noting the time required in seconds for a given volume of oil to separate out in distinct layer from an equal volume of condensed steam under standard conditions. It is also called "steam emulsion number". The lower the steam emulsion number, the quicker the oil separates out from the emulsion formed and the better the lubricating oil for most purposes.

Steam turbine oils are often contaminated with steam or condensate and it is essential that the oil should separate out rapidly, otherwise sludges may be formed which may clog oil lines and pumps. Hence, in all such cases where the oil may come into contact with water during service, it should have a low demulsification number. On the other hand in certain industrial lubricating oils such as cutting oils persistent emulsion formation is required because the emulsion acts as a coolant as well as a lubricant.

Many additives used for oxidation inhibition and corrosion inhibition of industrial lubricating oils have remarkable property of reducing the demulsification number of the oils.

9. Aniline point. Aniline point of an oil gives an indication of the possible tendency of deterioration of an oil when it comes into contact with packing, rubber sealing, etc. Generally, aromatic hydrocarbons have a tendency to dissolve natural and certain types of synthetic rubbers. Hence, the aromatic hydrocarbon content of the oil has much significance from this point of view. This is usually determined on the basis of "Aniline point" of an oil which is defined as "the minimum equilibrium solution temperature for equal volumes of aniline and oil sample". A higher aniline point means lower percentage of aromatic hydrocarbons. A higher aniline point is therefore desirable. Aniline point is determined by thoroughly mixing (mechanically) equal volumes of aniline and the oil sample in a tube and heating the mixture until a homogenous solution is obtained. Then it is allowed to cool at a specified rate until the two phases just separate out. The temperature corresponding to this particular observation is reported as the "Aniline Point".

10. Ash Content. The ash content of a perfectly refined mineral oil is very low. In the case of used oils the 'ash' will include metal particles and attempts have been made to assess the rate of cylinder wear from the iron content of the ash from used cylinder oils.

11. Decomposition stability. The stability of the oil towards hydrolysis and pyrolysis reactions is also important because the products from these reactions are detrimental to the machine parts.

12. Precipitation number. This shows the percentage of asphalt present in an oil. This is determined by dissolving a known weight of the oil in petroleum ether and separating the asphalt precipitated by centrifugation. The asphalt is dried and weighed and reported as weight percentage of the oil taken.

13. Oiliness. Oiliness is the property of the lubricant by virtue of which a lubricating oil can stick on to the surface of the machine parts operating under high pressures. Oiliness of an oil is the most important property of a lubricant under boundary or thin film lubrication conditions. Mineral oils have very poor oiliness whereas animal and vegetable oils have good oiliness. Hence, oiliness of mineral oils is generally improved by adding small quantities of high molecular weight fatty acids like oleic acid, stearic acid, chlorinated esters of these acids, etc. There is no perfect method for the determination of absolute oiliness of on oil. Only relative oiliness is considered while selecting a lubricating oil for a particular job.

14. Volatility. If the lubricating oil is exposed to high temperatures as in heavy machinery, some of it may volatalize off. Apart from loss of the volatalized lubricant, the residual oil left behind may have different properties (such as high viscosity and different viscosity index) than the original oil. A good lubricant naturally should have a low volatility. The volatility of a lubricating oil is usually determined by an apparatus called vaporimeter shown in Fig. 5.7. It consists of a furnace heated by a gaseous fuel in the middle of which a coiled copper tubing is placed. Air can be passed through the copper tubing. A known weight of oil sample is taken in a platinum crucible and it is introduced at the centre of the copper tube as shown in the diagram. Now, dry air at a rate of 2 litres per minute is passed through the copper tube for an hour. Then the tray is withdrawn and it is cooled and weighed. The loss in weight of the oil, calculated as percentage weight of the oil taken, is reported as volatility of the oil.

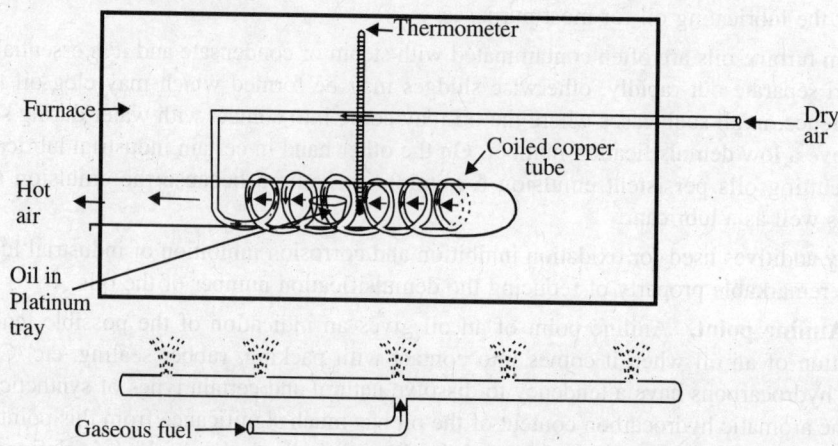

Fig. 5.7. Vaporimeter for determining volatility of a lubricating oil.

LUBRICANTS

15. Carbon residue test. The carbon forming tendency of a lubricating oil on combustion is significant particularly for internal combustion engines. Oils which deposit minimum amount of carbon are naturally preferable. Carbon deposition in an internal combustion engine results both from incomplete combustion of the fuel as well as the carbonizing of the lubricating oil carried up past the piston rings into the combustion chamber. Excessive build-up of carbon deposits in the combustion chamber results in decreased volume of the charge at the end of the compression stroke giving increased compression ratio which eventually leads to detonation. Deposition of carbon residues by the lubricant may be objectionable in other situations also.

Two tests are commonly used for the determination of carbon residue: (1) Conradson test and (2) Ramsbottom test.

The apparatus for the Conradson test is shown in Fig. 5.8. It consists of an insulator such as asbestors block. An outer iron crucible is fixed in the insulating box. The crucible is fitted with a lid and rests on a triangular support. It contains a layer of sand on which rests an inner crucible fitted with a ventilated cover. This inner crucible in turn contains a glazed porcelain crucible having a capacity of 29 to 31 ml and a rim diameter of 46–49 mm. The hood is circular in form with a diameter of 4¾–5¼ inches. The chimney has a diameter of 2–2¼ inches and a height of 2–2½ inches, the cone-shaped piece bringing the height of the complete hood to 5-1/8 inches. The bridge is made of 1/8 inch iron or nichrome wire and is 2 inches above the top of the chimney.

The weight of the oil sample which is taken in the crucible should be 10 g. Heat is applied to the outer crucible by a Meker burner such that the pre-ignition period is 10 ± 1.5 minutes. When smoke is seen at the chimney the flame is played on the side of the crucible in order to ignite the

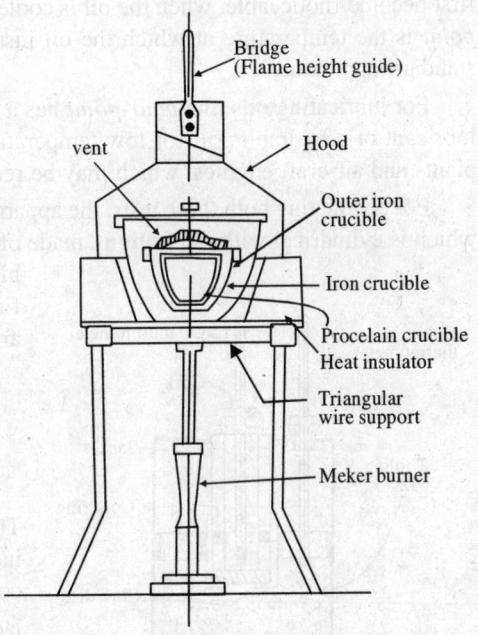

Fig. 5.8. Conradson's apparatus for the determination of carbon residue.

vapours which should burn uniformly with the flame above the chimney, but not above the bridge, in 13 ± 1 minutes, after which the crucible bottom is heated to redness for an additional 7 minutes so that the total heating time is 30 minutes. After this, the porcelain crucible is removed, allowed to cool and weighed. The result is reported as % of the residue in the weight of the oil taken.

In the Ramsbottom test, a hard glass coking bulb (Fig. 5.9) was originally used, but now the alternative of a stainless steel bulb is acceptable. 1 to 4 g of the oil sample (depending upon the expected percentage of carbon residue as follows; Below 2% –4 g, 2 to 4% – 2 g, Above 4% – 1g) is taken in the bulb and the weight of the oil taken is determined. The bulb is then placed in a sheath consisting of an iron tube approximately 76 mm

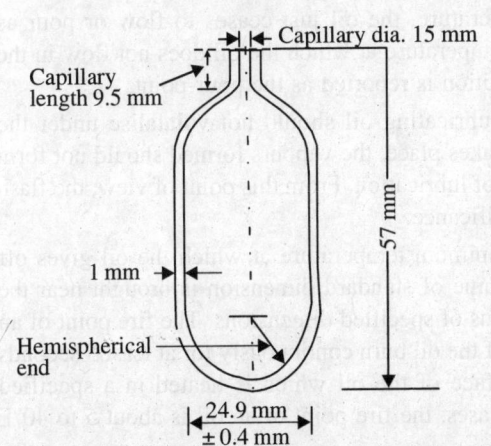

Fig. 5.9. Coking bulb used in Ramsbottom carbon residue test.

long and 25.5 mm internal diameter, having a flat closed end. The sheath is immersed in bath of molten lead to a depth of not less than 2–7/8 inch, then the lead bath is heated to 550°C for 20 minutes. The bulb is then taken out, cooled and weighed. Each test should be made in duplicate and care should be taken to see that there is no loss of the oil due to frothing. The heating may also be done in an electrically heated furnace at 500–550°C for 20 minutes. The result is reported as % carbon residue of the weight of the oil taken for the experiment.

16. Cloud point and pour point. Petroleum oils are complex mixtures of chemical compounds and do not show a fixed freezing point. When they are sufficiently cooled, they become plastic solids due to the formation of solid crystals or due to congealing of the hydrocarbons present. The *cloud-point* is the temperature at which this crystallization of solids in the form of a cloud or haze first becomes noticeable, when the oil is cooled in a standard apparatus at a standard rate. The *pour point* is the temperature at which the oil just ceases to flow when cooled at a standard rate in a standard apparatus.

For lubricating oils, the *pour-point* has a greater significance. It determines the suitability of a lubricant or a hydraulic oil for low temperature installations. Important examples are refrigerator plants and air-craft engines, which may be required to start and operate at sub-zero temperatures.

For conducting both these tests, the apparatus shown in Fig. 5.10 is used. It consists of a test jar, which is cylindrical with flat bottom, made of clear glass and is about 3 cm in diameter and 12 cm high. It is enclosed in a glass or metal jacket which is firmly fixed in a cooling bath. The cooling baths used are as follows:

Upto 10°C — Ice and water
Upto — 12°C — Crushed ice and salt
Upto — 26°C — Ice and $CaCl_2$
Upto — 57°C — Solid CO_2 and petrol

The oil is poured into the test jar to a height of 2 to 2¼ inch. Thermometers are introduced in the oil and the cooling bath. As the cooling takes place via the air-jacket, the temperature of the oil falls. At every degree fall of temperature of the oil, the test jar is momentarily withdrawn for examination and replaced immediately.

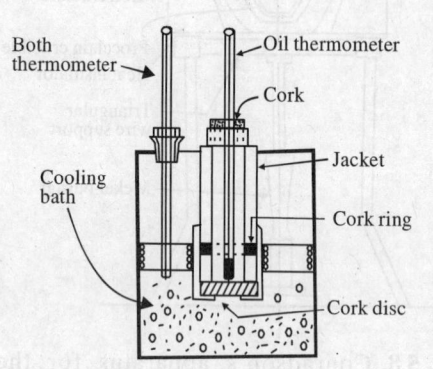

Fig. 5.10. Pour-point apparatus.

The temperature at which cloudiness or hazyness is fist noticed represents the cloud-point. As the cooling is further continued, at a particular temperature, the oil just ceases to flow or pour as observed from tilting the test jar. This particular temperature at which the oil does not flow in the test jar for 5 seconds on tilting it to horizontal position is reported as the pour-point.

17. Flash-point and Fire-point. A good lubricating oil should not volatalise under the working temperatures. Even if some volataliation takes place, the vapours formed should not form inflammable mixture with air under the conditions of lubrication. From this point of view, the flash point and fire point of a lubricating oil are of significance.

The flash-point of an oil is defined as the minimum temperature at which the oil gives off sufficient vapour to ignite momentarily when a flame of standard dimension is brought near the surface of the oil at a prescribed rate in an apparatus of specified dimensions. The fire point of an oil is the lowest temperature at which the vapours of the oil burn continuously for at least 5 seconds when the standard flame is brought near the surface of the oil which is heated in a specified apparatus at a specified rate. In a majority of the cases, the fire point of an oil is about 5 to 40°F higher than its flash point.

LUBRICANTS

A lubricating oil selected for a job should have a flash-point which is reasonably above its working temperature. This ensures safety against fire hazards during the storage, transport and use of the lubricating oil. In addition, the flash point of an oil is often used as a means of identification and also for detection of contamination of the lubricating oils.

Flash point of an oil is determined by either open-cup or closed-cup apparatus. In the open-cup apparatus, the oil is heated with its upper surface exposed to the atmosphere. The open-cup apparatus commonly employed is Cleveland's apparatus. The closed-cup apparatus in common use are Abel's apparatus and Pensky-Martens apparatus. The closed-cup apparatus gives more reproducible results. The flash-point obtained with an open-cup apparatus is generally about 10 to 30°F higher than that obtained with a closed-cup apparatus.

Cleveland's open-cup apparatus is generally used for determination of flash-point of fuel oils and other oils having flash-point below 175°F. The Abel's closed-cup apparatus is best used for oils having flash point below 120°F, while the Pensky Marten's apparatus is used for oils with flash points above 120°F.

The flash point of lubricating oils is usually determined by Pensky-Marten's apparatus.

This is the most commonly used apparatus for determination of flash-points of oils having flash points between 50°C to 370°C. The essential features of the apparatus are shown in Fig. 5.11. It consists of a brass cup which is 5 cm in diameter and 5.5 cm in depth. The level upto which oil is to be filled in the cup is marked at about 1 cm below the top of the cup. The cup is supported by its flange over a heating vessel in such a way that there is clearance between the cup and the heating vessel. The cover for the cup is provided with four openings of standard dimensions, which are meant for a special type of stirrer, a standard thermometer, an air inlet and a device for introducing the standard flame. The shutter provided at the top of the cup has a lever mechanism. When the shutter is turned, openings for the test flame and air are opened and the flame exposure device dips into the opening over the surface of the oil. The test flame gets extinguished when it is introduced into the opening for the test, but as soon as it returns to its original position on closing the shutter, the flame is automatically lighted again by the poilt burner.

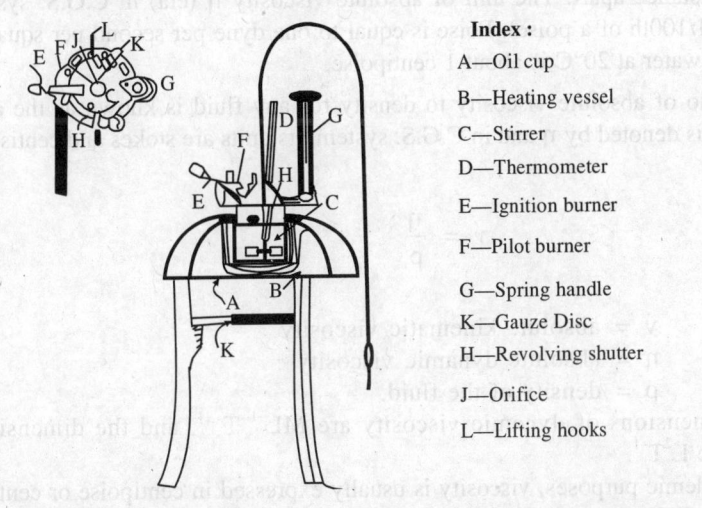

Index :
A—Oil cup
B—Heating vessel
C—Stirrer
D—Thermometer
E—Ignition burner
F—Pilot burner
G—Spring handle
K—Gauze Disc
H—Revolving shutter
J—Orifice
L—Lifting hooks

Fig. 5.11. Flash-point determination by Pensky-Marten's apparatus.

The oil sample under test is poured into the oil cup upto the mark. The cover incorporating the stirring device, the thermometer and the flame exposure device is fixed on the top. The test flame is lighted and adjusted until it is the size of a bead approximately 4 mm in diameter. The apparatus is heated so that the oil temperature increases by about to 6°C per minute while the stirrer is rotated at approximately 60 revolutions per minute. When the temperature rises to within upto 15°C of the anticipated flashpoint, the test flame is dipped into the oil vapour for about 2 seconds at every degree rise of temperature. This is done by twisting the knob which lowers the test flame and simultaneously opens the shutter. These spring back to their original positions as the knob is released. The flash-point is taken as that minimum temperature at which, on introducing the test flame into the oil cup, a distinct flash is observed.

Oils containing minute quantities of volatile organic substances are liable to flash below the true flash-point of the oil. Although a small flash may be observed in such cases, it should not be confused with the true flash, since its intensity does not increase with increased temperature, as occurs when the true flash-point is reached.

The minimum closed cup flash-point required for turbine oils is 165°C and that for insulating oils is 146°C.

18. Viscosity and Viscosity index. Viscosity is one of the most important properties of a lubricating oil. The formation of a fluid film of a lubricant between the friction surfaces and the generation of frictional heat under particular conditions of load, bearing spreed and lubricant supply mostly depend upon the viscosity of the lubricant and to some extent on its oiliness.

If the viscosity of the oil is too low, the fluid lubricant film cannot be maintained between the moving surfaces as a result of which excessive wear may take place. On the other hand, if the viscosity of lubricating oil is too high, excessive friction due to the shearing of oil itself would result. Hence in hydrodynamic lubrication, the lubricant selected must possess a sufficiently high viscosity to adhere to the bearing and prevent its being squeezed out due to high pressure and yet fluid enough so that the resistance to the shear is not too high. Hence, it is essential to have a knowledge of the viscosity of a lubricating oil.

Viscosity is a measure of the internal resistance to motion of a fluid and is mainly due to the forces of cohesion between the fluid molecules. Absolute viscosity may be defined as the tangential force per unit area required to maintain a unit velocity gradient between two parallel planes in the fluid unit distance apart. The unit of absolute viscosity η (eta) in C.G.S. system are poise and centipoise (1/100th of a poise). Poise is equal to one dyne per second per square centimeter. The viscosity of water at 20°C is about 1 centipoise.

The ratio of absolute viscosity to density for any fluid is known as the absolute kinematic viscosity. It is denoted by η and in C.G.S. system, its units are stokes and centistokes (1/100th of a stoke).

$$v = \frac{\eta}{\rho}$$

where
v = absolute kinematic viscosity
η = absolute dynamic viscosity
ρ = density of the fluid.

The dimensions of dynamic viscosity are $ML^{-1}T^{-1}$, and the dimensions of kinematic viscosity are L^2T^{-1}.

For academic purposes, viscosity is usually expressed in centipoise or centistoke, but a more common practical measure of the viscosity of an oil is the time in seconds for a given quantity of the oil to flow through a standard orifice under the specified set of conditions. Thus, viscosities are usually determined with Redwood Viscometer in the Commonwealth countries, with Engler's Viscometer in the Europe and with Seybolt's viscometer in the U.S.A. In these commercial viscometers, a fixed volume of the liquid is allowed to flow through a capillary tube of specified

LUBRICANTS

dimensions under given set of conditions and the time of flow is measured at a particular temperature. The results are usually expressed in terms of the time (as seconds) taken by the oil to flow through the standard orifice of the particular standard apparatus used, *e.g.,* viscosity of the oil is 156 Redwood (No. 1) seconds at 25°C. The viscosity of the oil so determined in the time units is sometimes called as relative viscosity. Since the instruments used are of standard dimensions, kinematic viscosity of the oil in centristokes can be calculated from the time taken by the oil to flow through the standard orifice of the instrument, with the help of the following equations:

$$\mu = Ct \text{ (for fluids whose kinematic viscosity is more than 10 centistokes)}$$
and
$$\mu = Ct - \beta/t \text{ (for fluids having kinematic viscosities lesser than or equal to 10 centistokes)}$$

where

μ = kinematic viscosity in centistokes
t = time of flow in seconds
C = viscometer constant
β = coefficient of kinetic energy which may be determined experimentally or eliminated by choosing long flow-times.

For routine purposes, the test viscometer may be calibrated and the constant C determined, by using solutions of known viscosity. The primary standard used is freshly distilled water whose kinematic viscosity is 1.0008 centistokes. Other standards usually employed are:

40% sucrose solution:
$$\nu = 4.390 \, cs \text{ at } 25°C; \; \rho = 1.17395$$

60% sucrose solution:
$$= 33.66 \, cs \text{ at } 25°C; \; \rho = 1.28335$$

For Redwood Viscometer No. 1 the values for the constants are as below:

Time of flow, t	β	C
40 to 85 seconds	190	0.264
85 to 2000 seconds	65	0.247

These constants are based on the results of the work carried out at the National Physical Laboratory at a temperature of 70°F (21.11°C), and with the ranges of viscosity, at that temperature the results are accurate to ± 1%.

Note. For calibration purposes, the kinematic viscosity in absolute units is determined in U-tube viscometers of standard dimensions:

Standard U-tube Viscometer	Range covered	Standard used
No. 0	0.5 to 2.0	distilled water
No. 1	1.5 to 6.0	distilled water
No. 2	5.4 to 43	40% sucrose soln.
No. 3	32 to 260	60% sucrose soln.
No. 4	190 to 1500	No primary standard available*

Redwood No. 2 Viscometer is used for very viscous liquids and gives 1/10th the value of Redwood No. 1 Viscometer.

Viscosity index

The viscosity of an oil decreases with increase of temperature as a result of decrease in intermolecular attraction due to expansion. Hence it is always necessary to state the temperature at which the viscosity was determined.

In many applications, a lubricating oil will have to function in a machinery over a considerably wide range of operating temperatures. If this is due to seasonal variations in atmospheric temperatures, adjustments can be affected by selecting different oils of appropriate viscosity for different seasons. However, in case of internal combustion engines, aeroplanes etc., the lubricant used must function both at low starting temperatures as well as at very high operating temperatures. Since the viscosity of lubricating oils decreases with temperature, it is impossible to select an oil having same viscosity over such a wide range of operating temperatures. However, one can select an oil whose variation in viscosity with temperature is minimum. This variation can either be indicated by viscosity-temperature curves or by means of the viscosity index. Viscosity index is the numerical expression of the average slope of the viscosity temperature curve of a lubricating oil between 100°F to 210°F. The oil under examiantion is compared with two standard oils having the same viscosity at 210°F as the oil under test. Oils of the Pennsylvanian type crudes thin down the least with increase of temperature; whereas oils of Gulf coast origin thin down the most as the temperature is increased. Hence the viscosity index of Pennsylvanian (Paraffinic) oil is taken as 100 and that of the Gulf (Napthenic) oil as zero. Then the viscosity of the oil under investigation is deduced as follows:

$$\text{Viscosity index of the under} = \frac{V_L - V_X}{V_L - V_H} \times 100$$

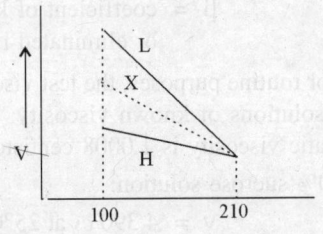

where

V_L = Viscosity at 100°F of Gulf oil standard which has the same viscosity at 210°F as that of the oil under test

V_X = Viscosity of the oil under test

V_H = Viscosity at 100°F of Pennsylvanian standard oil which has the same viscosity at 210°F as that of the oil under test.

Thus, the higher the viscosity index the lower the rate at which its viscosity decreases with increase of temperature. Hence, oils of high viscosity index *i.e.,* those having flat viscosity temperature curves are demanded for air-cooled internal combustion engines and aircrafts engines. In general, oils of high specific gravity have steeper viscosity-temperature curves. However, all oils tend to attain the same viscosity above 300°C.

By and large, light oils of low viscosity are used in plain bearings for high-speed equipment such as turbines, spindles and centrifuges whereas high viscosity oils are used with plain bearings of low speed equipment.

The Redwood Viscometer

The Redwood Viscometer is made in two sizes. The Redwood–1 Viscometer is commonly used for determination of viscosities of lubricating oils and has an efflux time of 2,000 seconds or less. The Redwood 2 viscometer is similar to the 1 type but the jet for the outflow of the oil is of a larger diameter and hence gives an efflux time of approximately 1/10th of that obtained with 1 instrument under otherwise identical experimental conditions. Redwood 2 instrument is therefore used for the

---

* To calibrate No. 4 Viscometer a viscous liquid such as castor oil must be used; its viscosity must first be determined by means of a No. 3 viscometer whose constant is already known and the time of flow in the No. 4 instrument must then be determined. If the respective times of flow are t_3 and t_4 seconds, then, since $v = k_3 t_3, = k_4 t_4$, where v is the viscosity of the liquid $k_4 = k_3 t_3 / t_4$. This process of "stepping up" may be used for any pair of viscometers, provided the time of flow of the liquid in either is less than 100 seconds.

LUBRICANTS

oils having higher viscosities, such as the fuel oils.

The Redwood Viscometer does not give a direct measure of viscosity in absolute units but it enables the viscosities of oils to be compared by measuring the time of efflux of 50 ml of oil through the standard orifice of the instrument under standard conditions. The results given by these two viscometers are reported as "Redwood 1 Viscosity" or "Redwood 2 Viscosity" followed by the efflux time in seconds at the experimental temperature.

Description

The Redwood 1 Viscometer shown in Fig. 5.12 essentially consists of a standard cylindrical oil cup made up of brass and silvered from inside and has 90 mm height and 46.5 mm in diameter. The cup is open at the upper end. It is fitted with an agate jet in the base*. The diameter of the orifice is 1.62 mm and the internal length is 10 mm. The upper surface of the agate is ground to concave depression into which a small silver plated brass ball attached to a stout wire can be placed in such a way that the channel is totally closed and no leakage of the oil from the cup through the orifice can take place. The cup is provided with a pointer which indicates the level upto which the oil should be filled in the cup. The lid of the cup is provided with an arrangement to fix a thermometer to indicate the oil temperature. The oil cup is surrounded by a cylindrical copper vessel containing water which serves as a water-bath used for maintaining the desired oil temperature with the help of electrical heating coils or by means of a gas burner as the case may be. A thermometer is provided to measure the temperature of water. A stirrer with four blades is provided in the water-bath to maintain uniform temperature in the bath and hence enabling uniform heating of the oil. The stirrer contains a broad curved flange at the top to act as a shield for preventing any water splashing into the oil cylinder. The entire apparatus rests on a sort of tripod stand provided with levelling screws at the bottom of the three legs. The water bath is provided with an outlet for removing water as and when needed. A spirit level is used for levelling the apparatus and a 50 ml flask for receiving the oil from the jet outlet are also provided.

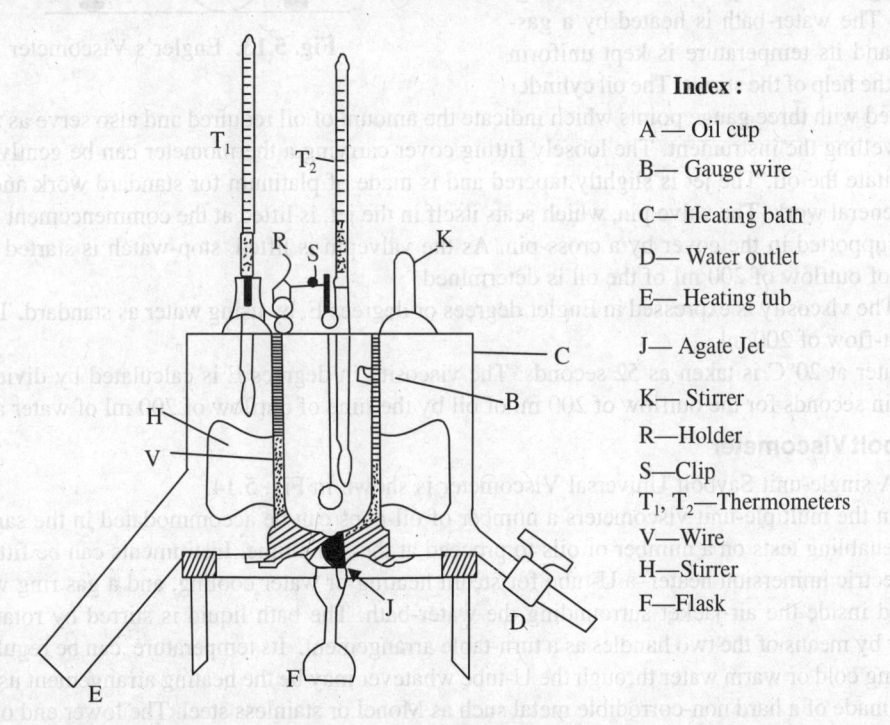

Index :

A— Oil cup
B— Gauge wire
C— Heating bath
D— Water outlet
E— Heating tub
J— Agate Jet
K— Stirrer
R—Holder
S—Clip
T_1, T_2—Thermometers
V—Wire
H—Stirrer
F—Flask

Fig. 5.12. The Redwood Viscometer.

---

* For Redwood No. 2. Viscometer, the diameter of the orifice is 3.8 mm and internal length is 50 mm.

Working. The instrument is levelled with the help of the levelling screws on the tripod. The water-bath is filled with water to the height corresponding to the tip of the indicator upto which the oil is to be filled in the cylindrical cup. The orifice is sealed by keeping the brass ball in position. Then the oil under test is carefully poured into the oil cup upto the tip of the indicator. The 50 ml flask is placed in position below the jet. The oil and water are kept well stirred and their respective temperatures are noted. The ball is raised and suspended from the thermometer bracket. Simultaneously, a stop-watch is started. When the level of the oil dropping into the flask just reaches the 50 ml mark, the stop-watch is stopped and the time is noted in seconds. The ball valve is replaced in the original position to prevent the overflow of the oil. The experiment is repeated and the mean value of time of flow for 50 ml of the oil is reported as t seconds, Redwood-1 at T°C". The usual test temperatures stipulated are 21.11°C (70°F), 60°C (140°F) and 93.33°C (200°F).

During the test, the measuring flask should be shielded from draughts with the help of metal shields usually supplied with the instrument.

Engler Viscometer. This instrument is diagrammatically represented in Fig. 5.13. The water-bath is heated by a gas-ring and its temperature is kept uniform with the help of the stirrer. The oil cylinder is fitted with three gauge points which indicate the amount of oil required and also serve as a means of levelling the instrument. The loosely fitting cover carrying a thermometer can be gently rotated to agitate the oil. The jet is slightly tapered and is made of platinum for standard work and nickel for general work. The valve pin, which seats itself in the jet, is lifted at the commencement of a test and supported in the cover by a cross-pin. As the valvepin is lifted, stop-watch is started and the time of outflow of 200 ml of the oil is determined.

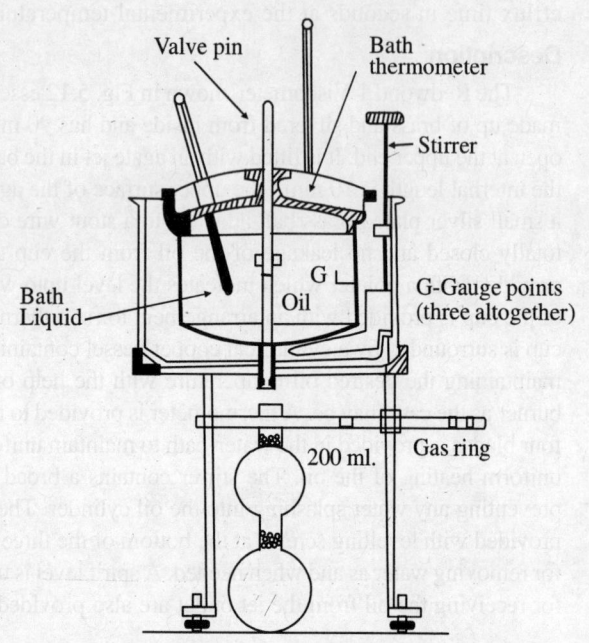

Fig. 5.13. Engler's Viscometer.

The viscosity is expressed in Engler degrees or degrees E, by using water as standard. The time of out-flow of 200 ml of water at 20°C is taken as 52 seconds. The viscosity in degrees E is calculated by dividing the time in seconds for the outflow of 200 ml of oil by the time of outflow of 200 ml of water at 20°C.

Saybolt Viscometer

A single-unit Saybolt Universal Viscometer is shown in Fig. 5.14.

In the multiple-unit viscometers a number of oil-cups can be accommodated in the same bath thus enabling tests on a number of oils to proceed at the same time. Instruments can be fitted with an electric immersion heater–a U-tube for steam heating or water cooling, and a gas ring which is placed inside the air-jacket surrounding the water-bath. The bath liquid is stirred by rotating the cover by means of the two handles as a turn-table arrangement. Its temperature can be regulated by running cold or warm water through the U-tube whatever may be the heating arrangement used. The jet is made of a hard non-corrodible metal such as Monel or stainless steel. The lower end of the jet opens into a larger tube. This tube, when stoppered by a cork, becomes a closed air chamber preventing the oil flowing out.

LUBRICANTS

To start with the test, the bath is brought to the test temperature and the oil is heated to the same temperature in a separate vessel. The oil is then poured into the oil cylinder and stirred with the oil thermometer, any excess oil flowing over into the surrounding gallery. When the oil and the bath are at the same temperature, the oil thermometer is removed, the excess oil drawn off from the gallery with a pipette, the cork is withdrawn and the stop-watch started. The collecting flask is so arranged that the oil stream will strike its neck and thus avoiding the formation of foam. The time of outflow of 60 ml of the oil is the viscosity in seconds Saybolt Universal at the test temperature.

For very viscous fuels, a viscometer with a larger jet known as the Saybolt Furol Viscometer is used. The Saybolt Universal viscometer can be used for oils having flow times of more than 32 seconds. There is no maximum limit, but in general, for liquids having flow times over 1000 seconds, Saybolt Furol Viscometer is better.

Example. *An oil sample under test has a Saybolt universal viscosity of 64 seconds at 210°F and 564 seconds at 100°F. The low viscosity standard (Gulf oil) possesses a Saybolt universal viscosity of 64 seconds at 210°F and 774 seconds at 100°F. The high viscosity standard (Pennsylvanian oil) gave the Saybolt Universal viscosity values of 64 seconds at 210°F and 414 seconds at 100°F. Calculate the viscosity index of the oil sample under test.*

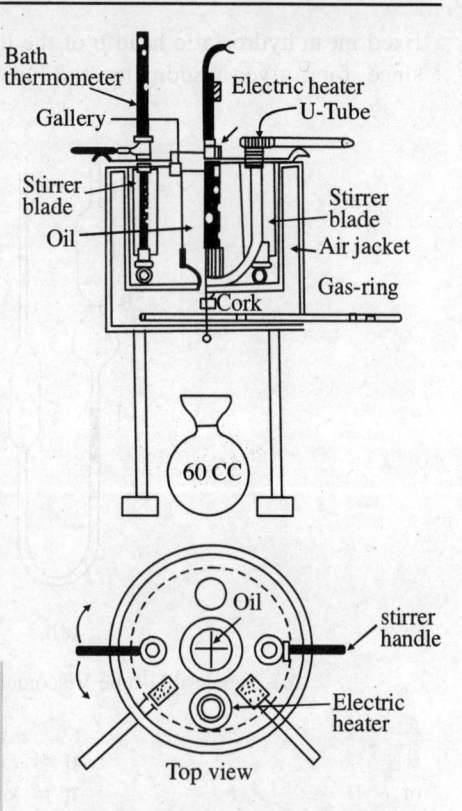

Fig. 5.14. Saybolt Viscometer.

Solution.

$$\text{Viscosity index of the oil under test} = \frac{V_L - V_X}{V_L - V_H} \times 100$$

$$= \frac{(774-564)}{(774-414)} = \frac{210}{360} \times 100$$

$$= 58.33.$$

U-tube Viscometer

The apparatus used today for the determination of the absolute viscosity of lubricating oils is the Standard U-tube Viscometer, which is an improved form of the Ostwald Viscometer. The modern form of the U-tube Viscometer is shown in Fig. 5.15a, which is made in 8 sizes.

The determination of absolute viscosity of lubricating oils by the U-tube Viscometer is based on poiseuille's law:

$$v = \frac{p \pi r^4 t}{8 l \eta}$$

where v is the volume of the liquid flowing through a capillary tube of length 'l' (cm) of uniform radius 'r' (cm) in a time 't' (seconds) and 'η' (poise) is the coefficient of viscosity of the liquid at the particular temperature. The relation is valid only if the flow is non-turbulent and slow enough for the kinetic energy to be negligible.

The determination of absolute viscosity by the U-tube Viscometer essentially consists of measurement of the time of passage through the capillary of a fixed volume of liquid under a fixed mean hydrostatic head p of the liquid. If the density of the liquid is 'd', then $p \propto d$, and since, for a given viscometer, the Poiseuille's equation shows that

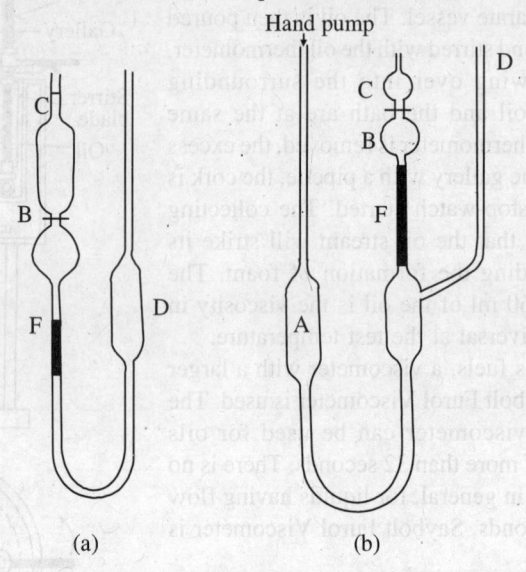

(a) Standard U-tube Viscometer (b) Ubbelohde Suspended Level Viscometer.

Fig. 5.15.

$$\eta \propto t\,d$$
or
$$\eta = k\,t\,d$$

where k is the constant of proportionality and has to be determined for each viscometer from its known dimensions or by calibration with a liquid (such as water) whose viscosity coefficient is known.

Absolute kinematic viscosity, v (nu), in centistokes, can be determined by dividing the absolute dynamic viscosity η by ρ, the density of the liquid. Errors may arise due to Standard U-tube Viscometer being set out of the vertical. Such error can be minimized by modifying the viscometer so that the discharge bulb is vertically over the receiving bulb as in the Cannon-Fenske Viscometer. This involves inclining the capillary tube accordingly. Another draw back of the Standard U-tube Viscometer is that an exact volume of liquid must be used in it. When readings are required over a range of temperatures, the volume of liquid has to be adjusted at each change of temperature. This tedious adjustment becomes unnecessary if the level of the liquid at the discharge end of the capillary can be kept constant. Several instruments have been designed with this object in mind, an example being the Suspended Level Viscometer due to Ubbelohde (Fig. 5.15b). In this, the liquid leaving the capillary enters the discharge bulb via a large bell-mouth which is ventilated to the atmosphere by a tube. The liquid leaving the capillary is thus induced to trickle over the inner surface of the bell-mouthed end so keeping the discharge level constant at the point of exit from the capillary.

Reverse flow viscometers are used to determine the viscosity of opaque liquids. The time of flow is the time required for the liquid level to rise between lower and the upper timing marks.

Viscosity of transparent liquids is also determined by Standard Falling Sphere Viscometer.

Conversion of Redwood, Engler and Saybolt viscosities into absolute units

Redwood, Engler and Saybolt viscosities can be converted to absolute units (centistokes). However, since these instruments are not the ideal methods of determining the absolute viscosities, the conversion values are only considered as good approximations and that too only when taken at the same temperature. For instance, Redwood viscosities at 30°C cannot be converted into absolute

LUBRICANTS

units at say 40°C, because, different fluids have different viscosity temperature relationships.

The conversion of the above relative viscosities to absolute viscosities is done with the help of the following equation:

$$v = Ct - \beta/t$$

where v = kinematic viscosity in centistokes, t = time of flow in seconds, and C and β are constants. The following values are taken for the constants:

Instrument	Value of C	Value of β
Redwood No. 1	0.25	172
Redwood No. 2	2.72	1120
Saybolt Universal	0.22	180
Engler	0.147	374

Notes:—
1. t = (Degrees Engler) × 52
2. Redwood Seconds No. 1 = 0.88 Saybolt Seconds Universal
3. Degrees Engler = 0.0328 Redwood Seconds No. 1
4. For accurate results, conversion constants should be determined experimentally at the temperature under consideration.

19. Mechanical Tests. Several mechanical tests have been devised to test the performance of a lubricating oil under a given set of conditions of temperature, load etc; one among them is the "four ball extreme pressure lubricant test." The working portion of such a machine consists three steel balls held in a ring and an upper fourth ball in contact with them and this ball is held at the end of a vertical shaft which is rotated at a fixed speed by an electric motor (Fig. 5.16). The three stationary balls are pressed upwards against the fourth ball by a lever carrying an adjustable load. The torque thus transmitted to the three fixed balls can be measured and if required, the coefficient of friction can be continuously recorded throughout a test. The points of the balls are lubricated by the test lubricant contained in a cup surrounding the ball assembly. In a simple lubricant test, the highest load that a ball can stand for 1 minute without squeezing can be taken as a measure of lubricant quality.

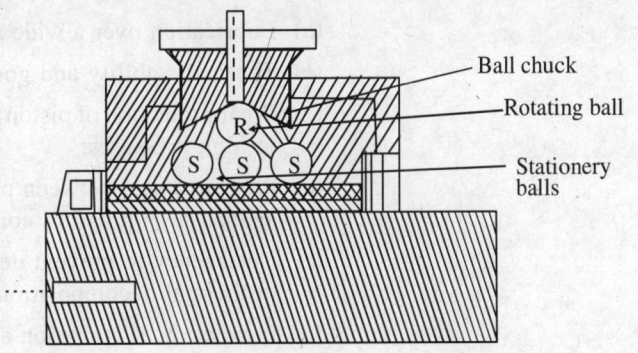

Fig. 5.16. Four-ball extreme pressure testing machine.

5·19. SELECTION OF LUBRICANTS FOR DIFFERENT PURPOSES

Industrial oils can be broadly classified into one of the following types:
1. Machine and Engine oils.
2. Spindle oils
3. Refrigeration oils
4. Circulating oils
5. Gear oils
6. Steam cylinder oils.

Obviously, the properties required for each of the above class of lubricating oils are different. The selection of industrial lubricants in any mechanised industry involves a consideration of the requirements of the equipments, available methods for handling and application of the lubricant itself and environmental conditions. In general, it is wiser to use the lubricants as recommended by the manufacturers of the equipment being used or by the standard oil companies in order to ensure maximum life of the operating equipment.

In selecting a lubricant for a given application, it is essential to consider the various properties of the lubricant required in relation to the service conditions.

The main machine elements that require the use of lubricants in any equipment are bearings, gears and cylinders. These simple elements work under a variety of operating conditions in different machines and equipment, and lubricants have to be specially designed to provide adequate protection to them. Some of the main factors to be taken into account are the effects of load, temperature, and speed at which these elements operate and also the contaminants which may affect the performance of the lubricants in use. The properties of lubricants required for different types of machinery are summarized in the following Table-2.

Table-2. Properties of lubricants required for different types of machinery

	Type of Machinery	Functions and properties of the lubricating oil required
1.	Automotive Engine Oils (Internal Combustion Engines)	Automative engines are the most difficult piece of equipment from the point of view of lubrication. The properties of the foil required are: (a) Lubrication over a wide range of temperatures. (b) Thermal stability and good heat transfer properties (c) Wear protection of piston rings and cylinder liners subjected to high pressures. (d) Provide a seal between piston rings and cylinder walls against high pressure combustion gases. (e) Detergency to prevent deposits and lacquer formation due to thermal decomposition of the oil at high temperatures. (f) Prevention of corrosion and rusting of internal parts of the engine. (g) Prevention of contaminants from precipitating to form sludge deposits (Dispersancy).
2.	Spindle oils	For the lubrication of lightly loaded spindles at very high speeds, thin oils are essential. It is desirable to add oxidation and rust inhibitors. Oils with viscosities ranging from 30 to 105 SUS (Saybolt Universal, Seconds) at 100°F are suitable for many applications.
3.	Refrigeration oils	Oils with low pour–, cloud–, and flow-points are needed in refrigeration systems. Naphthenic base oils only have such characteristics. Many manufacturers also stipulate a minimum dielectric strength so as to ensure that the oil is dry and free from free or dissolved moisture. ISI specification IS 4578–1968 for, refrigeration oils cover 4 grades of viscosities of 85, 160, 200 and 325 SUS at 100°F. The pour-point requirements are – 40°F Max. for the lightest grade and –13°F for the heaviest grade.
4.	Circulating oils (a) Turbines	Oils with viscosities around 150, 220 and 320 SUS at 100°F are commonly used for direct driven and geared turbines. Oils

LUBRICANTS

for marine applications have viscosities of 400 SUS at 100°F. Lubrication conditions in steam turbines are stringent. Very high oxidation stability and chemical stability are needed. Anti-oxidation, anti-rusting and anti-foaming additives are also required.

(b) **Hydraulic systems** — The important characteristics required are proper viscosity, high viscosity index, good demulsibility characteristics, good oxdiation stability, rust preventive properties, anti-wear characteristics and sufficiently low pour-point. Viscosities of oils required in hydraulic systems are 150, 210, 310 and 400 SUS at 100°F for the highest grade, medium, medium heavy and heavy grades respectively.

5. **Gear oils** — Various types of gears like spur, bevel, helical, herring, bone hypoid and worm are used in enclosed gearboxes for the transmission of power. The oils should have good oxidation resistance, proper viscosity, high viscosity index, extreme pressure characteristics, water separation properties and good foam resistance.

6. **Steam cylinder oils** — Premium quality high viscosity index oils are required. Viscosities of 165, 220 and 300 SUS at 210°F are adequate to cover the requirements of most cylinder oils. Straight mineral oils are used for super-heated steam while compounded oils have to be used for wet or saturated steam. Compounded oils contain additives of fixed oils and emulsifiers which help to form an inverted emulsion with water.

7. **Cutting oils** — Good lubricating property, low viscosity to enable it to fill the cracks formed on the work piece, high thermal conductivity, chemical stability, and anti-corrosive and anti-septic properties are required. Good lubrication and cooling are essential.

8. **Transformer oils** — The lubricating oils used in electrical transformers must possess high dielectric properties to insulate the windings. It should have low viscosity, optimum oxidation resistance, and good chemical stability under the operating conditions. Highly refined oils, without even traces of moisture and dirt, and possessing good dielectric properties, optimum oxidation resistance and good chemical stability are used.

9. **Lubricants used for machinery running at extreme pressures and low speeds e.g., tractor rollers, lathes, concrete mixers, and railway track joints where a film of lubricating oil or grease cannot be maintained.** — Solid lubricants such as graphite are used either as dry powder or in the form of emulsion *e.g.,* aquadag or oil dag.

10. **Machines operating at high pressures and low speeds, as in wire ropes** — Thick blended oils or greases are used.

and rail axle boxes	
11. Delicate equipment such as watches, clocks, scientific equipment and sewing machines.	Fixed Oils (animal or vegetable oils) such as clarified sperm oil, neatsfoot oil, olive oil, palm oil, hazelnut oil are used. Apart from other conventional types of solid, liquid or semi-solid lubricants, gases are also used as lubricants. The wear problems during starting or shut down of machinery in this case can be minimized by choosing wear resistant compounds or by coating machinery with molybdenum disulfide or Teflon.

5·20. METHODS OF LUBRICATION

Intermittent lubrication is used for low speeds whereas in other cases, continuous lubrication is employed. Various methods of lubrication such as gravity feed methods, force feed methods and mechanical feed methods are available to suit different purposes.

5·21. RE-REFINING OF USED LUBRICATING OIL

When lubricants put into use lubricants results into solid contaminate the water, with metals like lead, cromium, metal complexes and bound elements of S, P, Cl. etc. various contaminants in used lubricating oils are as follows

Contamination in used Lubricants :
Production of deterioration
Sludge
Oil soluble product
Oil insoluble product 5 - 15%

Extraneous
Surrounding impurities (dust, dirt, moisture) - 3.5%
metallic particles (wear & dust)

Acid clay process is used for re-refining. The flow diagram is as follows

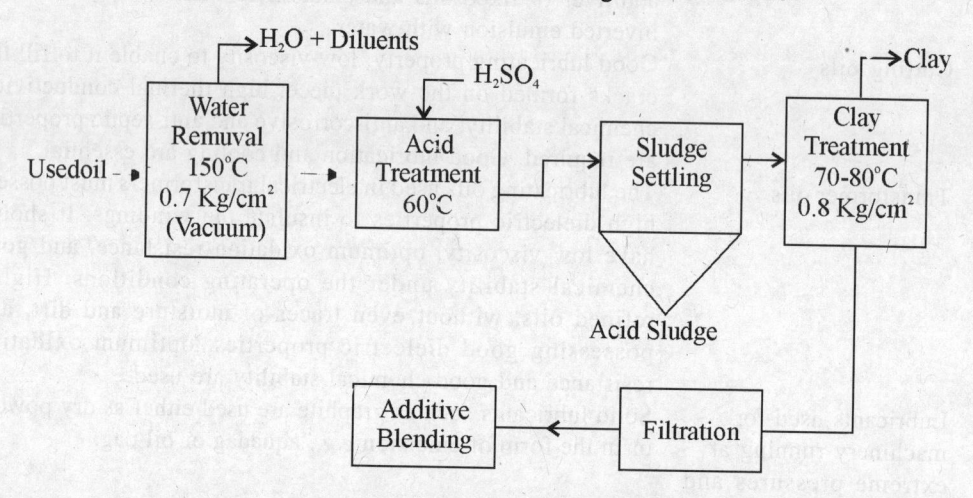

```
                    →H₂O + Diluents                                          →Clay
                    ↑                  H₂SO₄                                  ↑
          ┌─────────────┐         ┌───────────┐    ┌──────────┐    ┌─────────────┐
          │   Water     │         │           │    │          │    │    Clay     │
          │  Removal    │         │   Acid    │    │  Sludge  │    │  Treatment  │
Usedoil → │   150°C     │    →    │ Treatment │ →  │ Settling │ →  │   70-80°C   │
          │ 0.7 Kg/cm²  │         │   60°C    │    │          │    │ 0.8 Kg/cm²  │
          │  (Vacuum)   │         │           │    │          │    │             │
          └─────────────┘         └───────────┘    └────┬─────┘    └──────┬──────┘
                                                       ↓                   │
                                                  Acid Sludge              │
                                  ┌───────────┐    ┌──────────┐            │
                                  │ Additive  │ ←  │Filtration│ ←──────────┘
                                  │ Blending  │    │          │
                                  └───────────┘    └──────────┘
```

The re-refining technologies used by several oil refineries require that used oil to vacuum distilled, to remove the remaining chemicals and contaminants from the base oil to restore it to its original condition. Once that process is complete, the highest quality additives are blended into the base stock to fortify and bring the oil to the desired performance standards. These process are similar to the processes applied to virgin crude oil.

Re-refining oil offers both economic and environmental advantages. Used oil can be re-refined and reused indefinietly, Thus eliminating the need to purchase vergin oil. Hence preserving a non-renewable resource - oil (extends the life of non-renewable natural resource).

Used oil use to manufacture re-refined oil is not discarded into the waste stream, the recycting re-refining process serve to eliminate air pollution from oil incineration and potential water pollution caused by improper dumping. Helping to protect the environment.

Re-refining is an energy-efficient and environmentally beneficial method of managing the used oil. Less energy is required to produce agallon of re-refined base stock than a gallon of crude oil. 2.5 quart of re-refined lubricating oil can be produced from 1 gallan of used oil, while 42 gallons of crude oil are necessary to produce the same amount.

5·22. MARKETING AND MANUFACTURE OF LUBRICANTS IN INDIA

Lubricants are marketed in India mainly by (1) Bharat Petroleum, (2) Hindustan Petroleum, (3) Indian Oil Corporation, (4) Indrol Ltd., Bombay, (5) Tide Water India Ltd., Bombay, (6) Industrial Products Pvt. Co. Ltd. (IPCOL), Mumbai, (7) National Oil Co., Calcutta and (8) Asian Oil Co. Kolkata.

Lube base stocks are obtained by vacuum distillation of crude oil and blended to get the required properties. Lube base stocks may be obtained from paraffinic and naphthenic oils as well. Paraffinic base stocks give high and medium, viscosity index oils, whereas naphthenic base stocks give medium and low viscosity index oils having low pour points. The selective refineries which produce lube base oils are (1) Lube India Ltd., Mumbai (managed by HPCL); (2) M.R. Ltd., Chennai, (3) Haldia Refineries (W.B.); (4) Digboi Refineries (Assam); and (5) Barouni Refineries (Bihar). Turbine base stocks are available indigenously as well as imported; but naphthenic basic stocks are totally imported. These are used for refrigeration applications because of their low pour points. In India, no refinery is processing naphthenic crudes.

Consumption of lubricants in India at present is about 7 lakh tons per annum of which 4 lakh tons go for automotive applications.

Indian standard methods of testing petroleum and its products are given in IS : 1448.

5·23. CONCLUSION

"Modern machinery could not have been developed to meet the increasing demand for greater precision and higher speeds if lubricants technology had been unable to keep pace with expanding engineering sciences". A lubricant should not be developed in isolation. It should be an integral part of the overall engineering design which should be made by the combined efforts of designer, metallurgist and chemist to achieve a more efficient and less costly solution. It should be clearly recognized that a penny saved on capital expenditure can be pound foolish in maintenance.

5.24 ELECTRICAL INSULATING OILS

Depending upon the specific circumstances, an electrical insulating oil has to satisfy the following requirements :
  (i)   It should have high electrical resistance.
  (ii) It should not lead to substantial dielectric-losses in an alternating field.
  (iii) It should be chemically stable in presence of atmospheric oxygen during service.
  (iv) It should not suffer breakdown easily with the passage of a spark or arc under an applied voltage gradient.
  (v)   Under high electric stresses, it should not liberate gaseous hydrogen that might result in electrical breakdown or cause mechanical problems.
  (vi) It should have a low pour point.

(a) Transformer oil

Transformer units are sub-merged in a refined low viscosity lube oil, called transformer oil, having good insulating properties. Apart from insulation, the fluid should dissipate the heat developed in the core and windings and also to convect the heat to the external cooling system. The life of the transformer and its performance largely depend on the type and quality of the transformer oil used. The requirements for transformer and switch gear oils as per IS : 325 - 1972 are given in Table 3.

A transformer oil can function effectively over its long service period if it possesses the following characteristics :
(1) Good chemical stability
(2) Adequate low viscosity
(3) Good compatibility with the material of construction
(4) Necessary electrical properties eg., dielectric strength, resistivity and power factor.
(5) Resistance to gas formation under electrical stresses.

Transformer oils made from naphthenic crudes have excellent stability and electrical properties. However, there is scarcity of naphthenic crudes in the world. Many countries therefore are forced to use alternative paraffinic crudes. In India, the two refineries viz., Hindustan Petroleum Corporation Limited (HPCL), Bombay and Madras Refinery Limited (MRL), Madras are producing transformer oil feedstock (TOFS). The composition of the crude processed is of vital importance because the presence of polar comounds, sulphur and nitrogen impairs the electrical properties (e.g., resistivity) of the transformer oil.

Table 3 : IS : 335 - 1972 Specifications for Transformer and Switch Gear Oils

S.No.	Characteristics	Requirement
1.	Density at 27°C, g/cm^2, max.	0.89
2.	Kinematic viscosity at 27°C, cst, max.	27
3.	Interfacial tension at 27°C, N/m, min.	0.04
4.	Flash point, Pensky-Martens (closed), °C, min.	140
5.	Pour point, °C max.	-10
6.	Neutralization value, mg KOH/g, max.	0.03
7.	Corrosive sulphur	None
8.	Electric Strength (Breakdown voltage), KV	
	(a) New untreated oil	30
	(b) After treatment	50
9.	Dielectric dissipation factor at 90°C, max	0.005
10.	Specific resistance (resistivity)	
	(a) At 90°C, min	13×10^{12}
	(b) At 27°C, min.	500×10^{12}
11.	Oxidation stability	
	(a) Neutralization value after oxidation, mg KOH / g, max.	0.4
	(b) Total sludge after oxidation, wt. %	0.1
12.	Presence of oxidation inhibitor	None
13.	Water content, ppm, max.	50

Transformer oil is used in the system for a long period (5 to 10 years) in contact with metals (eg., copper), air and high temperature which are conducive for oxidation of the oil. The oxidation of oil leads to the formation of acids and sludge which in turn affect the electrical properties of the oil. This is the reason why a transformer oil must possess excellent oxidation stability. The oxidation stability is improved by treatment with about 10 to 15 wt. % of oleum or sulphuric acid, separating the sludge formed and neutralizing the oil by washing with alcoholic alkali followed by removal of oil soluble sulphonates.

(b) Switch oil

Large electric switches are filled with an oil to quench the arc formed when the contacts are

LUBRICANTS

opened or closed. Such an oil is called switch oil. It should have a good chemical stability and high breakdown voltage. In practice, the requirements of the switch oil and transformer oil are more or less similar.

(c) Cable oil

Mineral oils play vital role in insulation of high tension electrical cables. Two types of cables are generally used : (1) solid cables (2) oil filled cables.

In the solid cables the conducting core is surrounded by various strata of insulating material, including winding of oil-impregnated paper. The oil used for impregnation is rather thick, having viscosity of the order of 100 CSt at 60°C and is generally compounded with rosin.

Hollow cables contain one or more internal passages filled with oil to keep the insulating material fully saturated. The cables are kept full of oil by gravity flow from tanks at different points along their length. A thin oil of about the same viscosity as transformer oil is employed. Oil-filled cables are designed to carry electricity at very high voltages without breakdown. Hence the oil must have good dielectric properties and excellent resistance to oxidation because the oxidation products impair the dielectric properties. Further, cable oils should not liberate hydrogen when subjected to high electric stresses. It is for this reason that non-gassing oils containing some proportion of unsaturated compounds are specially prepared for this purpose.

(d) Capacitor oil

Electric capacitors / condensers are generally impregnated with oils having properties such as low dielectric losses, high resistance to breakdown, low tendency to liberate gases, and oxidation stability. The viscosity of these oils may be similar to that used in hollow-core cable oils.

5.25 WHITE OILS

These are made from petroleum distillates by treating with sulphuric acid. The important by-products from the manufacture of white oils are petroleum sulphonic acids. Petroleum sulphonates find extensive use in the formulation of huge quantities of important products such as heavy duty motor oils, cutting oils, textile oils, leather oils, spray oils, rust preventives and crude oil emulsion breakers (demulsifiers). Large quantities calcium and barium salts prepared from sulphonates are used in the manufacture of motor oils as detergents generally used along with oxidation and corrosion inhibitors.

White oils can be produced from any of the three types of crude, namely paraffinic, intermediate or naphthenic. However, naphthenic crudes are used to produce white oil having high viscosity and high specific gravity.

The various steps involved in the manufacture of white oil and sulphonate are as follows :
  (i)   Acid treatment (with oleum)
  (ii) Neutralization with caustic alkali or Na_2CO_3 solution and extraction of sulphonates
  (iii) Solvent removal and steaming to remove light ends.
  (iv) Removal of colour constituents by treatment with clay.
  (v)   Recovery of solvent and purification of sulphonates.
  (vi) Disposal of sludge and recovery of acid.

Uses of white oils : The medicinal grade white oils are used for the treatment of chronic constipation. Large quantities of medicinal grade oil is sold in the form of oil-in-water type emulsion using agar-agar, accacia or gelatin as emulsifying agents. The lighter white oils are used as a vehicle in ointments and as a base for nasal sprays. White oil mixed with appropriate quantity of germicides is widely used as baby oil to prevent rash and chafing.

White oils, being inert and oxidation resistant, are extensively used in process industries and in technical applications. White oils are extensively used in cosmetic industry as an indispensable ingredient in cold creams, vanishing creams, cleansing creams, etc. The lighter white oils are used to prevent caking and drying in tooth pastes and shaving creams. A lot of white oil is used in the manufacture of petroleum jelly.

QUESTIONS

1. Explain the following properties of lubricants and discuss their significance:
   (a) Viscosity and viscosity index
   (b) Flash point and Fire point (Nagpur Univ. S - 2007)
   (c) Aniline point (Nagpur Univ. S - 2005)
   (d) Saponification value. (Nagpur Univ. S - 2005)
   (e) Cloud and pour point (Nag. Univ. S - 2007)
2. What are different mechanisms of lubrication? Explain boundary film lubrication. (Nagpur Univ. S-2006)
3. Suggest suitable lubricants for each of the following systems with proper justification:
   (a) Internal combustion engines
   (b) Steam engine cylinders (Nagpur Univ. S - 2004)
   (c) Cutting tools   (d) Steam turbines   (e) Gears
4. Distinguish between fluid film and boundary lubrication. (Nagpur Univ. W - 2003)
5. A lubricating oil has the same viscosity as standard naphthenic and paraffinic type oils at 210°F. Their viscosities at 100°F are 320 S.U.S. 430 S.U.S. and 260 S.U.S. respectively. Find the viscosity index of the oil. (Nagpur Univ. W - 2003)
6. What do you mean by viscosity index of a lubricating oil?

    A lubricating oil has a S.U.S. of 58 seconds at 210°F and of 600 seconds at 100°F. The high viscosity index standard (*i.e.,* Pennsylvanian) oil has S.U.V. of 58 seconds at 210°F and 400 seconds at 100°F. The low viscosity index standard (*i.e.,* Gulf) oil has a S.U.V. of 58 seconds at 210°F and 800 seconds at 100°F. Calculate the viscosity index of the oil.
7. Write an essay on solid lubricants with emphasis on their classification, mechanism of action, examples and applications.
8. How are semi-solid lubricants prepared? In what situations a semisolid lubricant is preferred? Mention some important tests for evaluating semisolid lubrications. (Nagpur Univ. S - 2005, S - 2006, S-2007)
9. What do you mean by blended or compounded oils? What are the various additives used to induce or improve the necessary properties of a lubricating oil?
10. What are the various types of synthetic lubricants available? Discuss their merits and demerits.
11. Discuss the use of lubricating emulsions.
12. Write informative notes on the following:
    (a) Graphite
    (b) Fixed oils
    (c) Neutralization number
    (d) Extreme pressure lubrication
        (Nagpur Univ. S-2003)
    (e) Oiliness
    (f) Carbon residue test
    (g) Pensky Marten's apparatus
    (i) Gas lubrication
    (j) Biodegradable lubricants
    (k) Cryogenic bearing lubricants
    (h) Redwood Viscometer.
    (l) Lubricants for nuclear reactor systems
    (j) Lubricants for food processing
    (k) Environmental and health factors in the use of lubricants
13. Justify the following statements:
    (a) Flash point determination by the closed cup apparatus gives a lower value than that determined by an open cup apparatus.
    (b) Closed cup apparatus gives a more reliable value than the open cup apparatus for the determination of flash point.
    (c) The relative viscosity determined by Saybolt viscometer or Bedwood viscometer can be converted into absoute kinmatic viscosity by calculations.
14. Write an essay on silicone lubricants with particular emphasis on their preparation, properties, uses and limitations.

6
Portland Cement

> "Portland cement is the essential bonding material in concrete, which is the most widely used non-metallic material of construction, in our industrial age."

6.1. INTRODUCTION

Concrete is the most widely used non-metallic material of construction in our industrial age. Concrete is used for the construction of bridges and highways, run-ways for the aircraft, dams, buildings and structures and monuments of architectural beauty. The essential bonding material in concrete is portland cement, which is mixed with rock, sand and water.

Common lime and hydraulic lime are the earlier cementing materials having historical records. In prehistoric days, perhaps even before man learnt how to kindle a fire, clay was used as a cementous material to make sun-dried bricks. The same bricks were laid and bound in clay mortar for construction of huts for living.

Hydraulic cements are those cements which are capable of setting and hardening under water by virtue of interaction of the water with the constituents of the cement. Romans were the first to use such cements in antiquity and the structures built with them are still standing in excellent state of preservation. These cements were formerly made from low burnt lime stone and volcanic tuff. John Smeaton, an English engineer in 1760 was the first to conduct a systematic study to develop the best composition for a hydraulic cement. He found that burning of impure lime stone gave better hydraulic cements rather than the pure stone formerly preferred. He also found that the quality of the resulting cement has a relationship to the clay content of the rock used.

While burning the lime stone, lumps of sintered material were sometimes found in regions subjected to the highest heat. These lumps were commonly discarded because they were difficult to grind. The superior value of these rejected sintered lumps was adequately recognized by Joseph Aspidin, a brick-layer of Leeds, who patented his findings in 1824. He found that the hard clinkers when ground and mixed with water, produced a highly superior hydraulic cement. The resemblance of the colour of the set product (made of cement, sand and water) to a natural stone quarried at Portland, England, prompted Aspidin to name this product as Portlant Cement.

This material slowly attracted the attention and preference over the low-burnt types of cement. For about three-quarters of a century it continued to be manufactured by crude and hand-labour methods. The earlier kilns for burning the limestone were small and intermittent ones. The first rotary kiln 25 ft long and 5 ft in diameter erected at Coplay in Pennsylvania in United States by the Keystone Company in the year 1889 revolutionized the cement industry. Now, Portland cement is being manufactured in millions of tonnes every year. In India, about 20 million tonnes per year of Portland Cement were produced as early as 1978.

The first cement factory in India was started in 1904 in Chennai, by the South India Industrials Ltd. which existed for a short period. The manufacture of Portland cement on a large scale was started by the Indian Cement Co. Ltd. at Porbandar (Kathiawar) in 1914. The First World War gave an impetus to the cement industry and by year 1924, there were 10 cement factories in India with a total production of 5.81 lakh tons per year. By the time of Second World War, the cement industry in India was well established with 12 factories of A.S.C., 5 factories of Dalmia group and four others. The total capacity of these factories by the end of the war was 26.15 lakh tons. Consequent to the partition of India in August 1947, 5 factories went to Pakistan and 18 factores remained in India with total installed capacity of 21.15 lakh tons. Since then several new cement factories were started in India.

The production pattern of Portland cement in India during 1961 to 2005 is as follows:

Year	Cement produced in Megatons (Mt) (i.e. 10^6 tons)
1961	8.32
1965	10.59
1970	13.95
1975	16.21
1985	43.42
1995	75
2005	115 (projected)

The size of the kilns used before 1940 were suitable to produce about 300 tons/day whereas today the trend is to install much larger single kilns. The largest cement factory of India in terms of output capacity is in Madhya Pradesh, which is rated to produce 1.1 Megaton of Portland blast furnace slag cement per annum.

Since Portland cement is an important ingredient of cement-concrete which is the most widely used building material throughout the world, the quantity of cement produced may be largely regarded as an index of industrial and economic development of a country. The following table gives the cement produced in different countries in 1969:

Country	Cement produced (Mt)	Per capita production (kg)
India	13.6	26
UK	17.6	305
USA	70.5	342
Russia	89.4	343
Japan	48.1	365

6·2. RAW MATERIALS

The raw materials used in the manufacture of Portland cement are:

(a) **Calcareous materials** (which supply lime) e.g., lime stone, cement rock (a soft argillaceous lime stone), chalk, marl or marine shells, and waste calcium carbonate from industrial processes.

A lime stone high in magnesia cannot be used unless its magnesia content is reduced by some means, as by flotation, or dilution with low-magnesia rock so that the product will not contain more than 5% MgO. Similarly, chalk containing flint has to be freed from that impurity, and seams of gypsum or other materials such as pyrite may require selective handling before use.

(b) **Argillaceous materials** (which supply silica, alumina and iron oxide) e.g., clay, shale, blast-furnace slag, ashes and cement rock. Clay or shale are most commonly used. Cement rock was sometimes used as such without any further addition, since it contains both limestone and clay minerals. The modern demands, however, have made necessary more precise control of composition and hence, such a simple procedure is rarely employed today.

The demands of modern cement industry are so exacting with reference to the permissible clinker composition that small variations in the established ratios of the principal components of the ground rock mixture might considerably change the burning characteristics of the mix or the properties of the final product. If the CaO content is too low, the tricalcium silicate content will be inadequate to produce satisfactory early strengths. On the other hand, if the CaO content is too high, free CaO will be left in the clinker producing an "unsound" cement. The term "unsoundness" refers to an excessive expansion of the hardened cement paste upon storage in water or moist air. The alumina and iron oxide also are the essential constituents in the cement manufacture because they are the chief flux-forming components of the mix. If some liquids were not formed during burning, the reactions would be much slower, require much higher temperatures and perhaps may be incomplete. Tricalcium aluminate is undesirable above limiting values in some special cements. Hence an optimum balance must be sought, with the amount of flux just adequate for rapid clinkering reactions, but at the same time low enough to avoid undesirable effects in the cement paste. The introduction of iron oxide will markedly affect the formation of the tricalcium aluminate. Therefore, rigid limitations on certain oxide ratios in the raw mixture are usually imposed to take care of all above mentioned restrictions in composition.

6·3. IMPORTANT PROCESS PARAMETERS FOR MANUFACTURING A GOOD CEMENT CLINKER:

(1) The lime saturation factor

$$\frac{CaO}{2.8\,SiO_2 + 1.2\,Al_2O_3 + 0.65\,Fe_2O_3}$$

should be in the range 0.66 to 1.02. This will ensure the formation of C_3S, C_2S and C_3A, which are responsible for giving strength, in desired proportions.

(2) Silica modulus, $\dfrac{SiO_2}{Al_2O_3 + Fe_2O_3}$ should be 2.2 to 3.5.

(3) Fine grinding of the raw materials which helps the kinetics of reaction.

(4) Maintaining the MgO content below the specified limits which ensures that the cement is "sound".

(5) Maintaining alkali chlorides within the specified limits.

6·4. Methods of Manufacturing Cement

Portland cement is generally manufactured by the following three methods:

(1) **Wet process.** This process was in predominant use in India and Europe until recently. In this process, the raw materials are finely ground and blended in the desired proportion and the mix is brought to the condition of a free flowing slurry containing 30 – 40% water. The slurry is thoroughly homogenised with the help of compressed air and introduced into a rotary kiln. The charge slowly moves down the kiln due to the rotary motion while a blast of burning coal dust is blown up from the other end of the kiln. In the upper part of the kiln (drying zone) where the temperature is 400°C, the slurry loses all its water. When the charge enters the middle portion of the kiln (calcining zone) where the temperature is about 900–1000°C, lime stone decomposes to form CaO and CO_2. Then the charge enters the lower portion of the rotary kiln (the burning zone), where the temperature ranges from 1400–1600°C, lime and clay combines to form calcium silicates and aluminates.

$$[CaCO_3 + Al_2O_3 + Fe_2O_3 + SiO_2 + H_2O]$$
$$(1400 - 1600°C) \quad -H_2O$$
$$-CO_2$$

$2CaO \cdot SiO_2$	Dicalcium Silicate (C_2S)
$3CaO \cdot SiO_2$	Tricalcium Silicate (C_3S)
$3CaO \cdot Al_2O_3$	Tricalcium Aluminate (C_3A)
$4CaO \cdot Al_2O_3 \cdot Fe_2O_3$	Tetracalcium Alumino ferrite (C_4AF)

The resulting product is known as cement clinker which is in the form of small balls or pellets of varying size. The clinkers are cooled in a rotary cooler and pulverised together with 2–3% gypsum in grinding machines. The resulting powder is called Portland cement which is filled in air tight bags to exclude moisture.

(2) **Dry process.** In this process, the calcareous and argillaceous materials are crushed in gyratory crushers to small pieces, dried and mixed in proper proportion, pulverised in tube mills and homogenised in a mixing mill with the help of compressed air. This "raw meal" is introduced into the upper end of the rotary kiln while a blast of burning coal dust is blown from the other end. The reactions taking place and the rest of the process is same as described under wet process.

(3) **Semi-dry process.** In this process, the raw materials are initially ground dry, but instead of feeding as a powder the 'raw meal' is nodulised with 10–14% water in a pan or drum type noduliser. The nodules are fed on a travelling grate where they get dried and preheated before entering a short rotary kiln where they are burnt to form cement clinker.

6·5. DRY PROCESS VS. WET PROCESS

The cost of grinding is more in dry process than that in wet process. In the wet process, a longer kiln is needed and the consumption of fuel is more in order to drive off the excess water. In the dry process a shorter kiln is sufficient and the fuel consumption is also lesser. More accurate control of composition can be attained in the wet process than that obtainable in dry process.

The choice between the two processes is governed by factors such as physical conditions of the available raw materials, climatic conditions where the factory is located and price of fuel. The wet process is preferred when the raw materials are soft, the climate is fairly humid and cost of the fuel is low. The dry process is employed when limestone and shale are hard and the cost of the fuel is high. If the principal raw material has an inherent moisture content of 15% or more, obviously the wet process would have to be adopted, as it would be uneconomical to drive away initially this excessive quantity of moisture and thereafter process the material further to enable its blending with other dry raw materials by homogensiation. Similarly where the calcareous material is obtained from chemical precipitation (*i.e.,* precipitated $CaCO_3$) or where low grade limestone is beneficated by froth flotation, the process to be employed is necessarily the wet process. If the raw materials contain high percentage of alkali, chlorine and sulfur, employment of the efficient Humboldt type dry process of manufacture is not recommended because of the possibility of formation of incrustations inside the cyclones.

The main consideration for adopting dry process and semidry process is economy in fuel consumption. In dry process, the heat energy required per kg of clinker produced is 3000–4000 KJ/kg while equivalent figure for the wet process is 5500–6000 KJ/Kg. As against this, the dry process plant may consume about 10–15% extra electrical energy for grinding. The dry process, particularly that with the suspension preheater system, is superceding the wet process except where the wet process has to be prefered for special reasons. More heat is required in wet process due to evaporation of water. In fact, in most parts of the world including India, the shift is towards the dry process to achieve economy of fuel and also to achieve higher output of cement clinker from the same kiln.

PORTLAND CEMENT

6.6. SEQUENCE OF OPERATIONS

1. Selecting of raw materials. The primary calcareous material used in India is lime stone and the argillaceous materials used are clay or any other residual clay like bauxite, laterite, slate, moorum, etc. Lime stone can be mined by closed pit or open pit quarry method. Low grade lime stones can be concentrated by froth flotation process. This process is used only when high grade lime stone is not available near the factory. The process is costly and only four factories in India are employing it. Gypsum and water (for wet process) are the other raw materials needed for cement manufacture.

Coal, furnace oil and natural gas can be used as fuel. Natural gas is scarce in India. Oil is expensive and is generally used only for the manufacture of white cement and special quality cements. In most of the cases, only Grade 1 coal (22–30% ash) is allotted for cement manufacture in India.

2. Crushing and grinding. The raw materials are crushed in large jaw crushers or coarse gyratory crushers. Then they are ground in the raw mills. About 80% of the total electrical energy consumed in the manufacture of cement is used up in crushing, grinding and blending of materials. (The energy required for crushing and grinding of the raw mix is about 23–35 Kwh per tonne of raw mix or about 40–55 Kwh per tonne of equivalent cement. Wet grinding consumes about 25% less power than dry grinding, and closed circuit operation of the mill saves around 20% or less in the power demand over the open mill).

In dry milling, the raw materials are dried with hot air and pulverised. In wet milling, the raw materials are ground with 30–40% of water in tube mills. The finer the particles the better is the cement produced.

3. Storage of the slurry. The ground raw materials are stored in concrete silos where they are kept agitated with compressed air. From there, the raw materials are sent in fixed proportion to correction vats. When the vat is full a sample is drawn and analysed. If the slurry is not of the required composition, it is corrected by adding the requisite quantity of the deficient raw material. Then, the slurry from the correction vats goes to the kiln feed basins where the slurry is kept in agitation until it is fed to the rotary kiln.

4. Burning the ground mix to clinker in a rotary kiln. The rotary kiln is a long horizontal steel cylinder lined with refractory bricks and rotating at a speed of 0.5 to 2 rotations per minute. The process and the quantity of production determines the length (30 to 50 metres) and diameter (2 to 5 metres) of the kiln. Both dry and wet process rotary kilns are set at a slight inclination (of about 3°–6°), to allow the material fed at one end to travel slowly to the firing and discharge end. The kiln is supported by several tiers which run on rollers and the kiln is driven by an AC commutator motor.

As against a mere 50, tonnes/day kiln capacity, some 60 years back, the capacity of the kilns today has reached as high as 4000 to 5000 tonnes/day in some countries.

A typical rotary kiln is illustrated in Fig. 6.1.

The slurry of the raw materials enters from the upper end of the rotary kiln while the burning fuel (pulverised coal, oil or natural gas) and air are induced from the lower end of the kiln. The slurry gradually descends in the kiln into different zones of increasing temperature:

1. The upper part of the kiln is known as **drying zone** where the temperature is about 400°C. In this zone, most of the water is driven out of the slurry because of the hot gases.

2. The upper central part having a temperature of about 400–700°C is known as **pre-heating zone.** In this zone, clay and magnesium carbonate decompose.

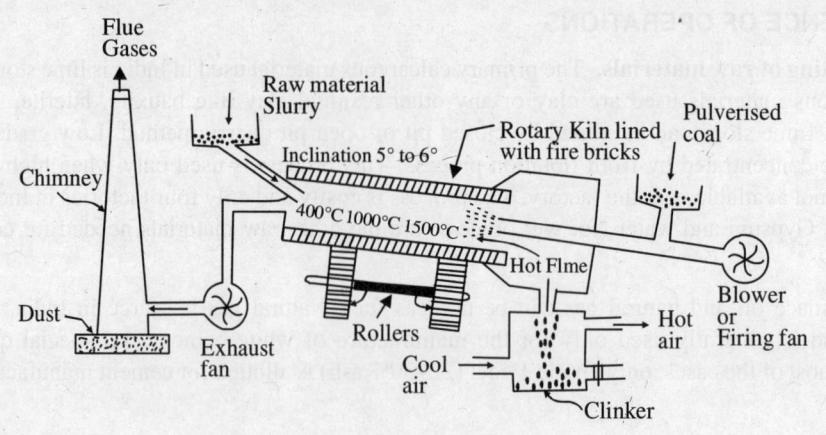

Fig. 6.1. Rotary Kiln for Cement manufacture.

3. The lower central part of the kiln is known as **calcining or decarbonating zone** where the temperature ranges from 7000 to 1000°C. Here, lime stone is decomposed to give CaO and CO_2.

$$CaCO_3 \rightleftharpoons CaO + CO_2$$

4. The material then enters the hottest zone (1350 to 1500°C) known as **burning and clinkering zone**, where lime and clay react with each other forming aluminates and silicates:

$$2\,CaO + SiO_2 \longrightarrow 2\,CaO \cdot SiO_2 \quad (C_2S)$$
$$\text{dicalcium silicate}$$

$$3\,CaO + SiO_2 \longrightarrow 3\,CaO \cdot SiO_2 \quad (C_3S)$$
$$\text{tricalcium silicate}$$

$$3\,CaO + Al_2O_3 \longrightarrow 3\,CaO \cdot Al_2O_3 \quad (C_3A)$$
$$\text{tricalcium aluminate}$$

$$4\,CaO + Al_2O_3 + Fe_2O_3 \longrightarrow 4\,CaO \cdot Al_2O_3 \cdot Fe_2O_3 \quad (C_4AF)$$
$$\text{tetracalcium alumino ferrite}$$

The compounds then combine together to form small, hard, greyish pellets called cement clinkers. Clinkers of size 3-20 mm in diameter are formed.

The composition of the clinker depends upon the ratio $\dfrac{C_3A + C_4AF}{C_3S}$, which is known as the *burnability* index. This is usually kept in the range 0.45 to 0.85. Too much of flux leads to balling and too little, to bad coating. The clinker formation is an exothermic reaction.

5. **Cooling of hot clinker.** The hot clinker emerging from the kiln is cooled by various systems such as rotary coolers, planetary coolers or air quench type coolers. In the coolers, the clinker is cooled with atmospheric air. The hot air so produced is used for drying the coal before pulverization.

The quality of cement produced also depends upon the rate of cooling. Cooling of the clinker should be controlled to produce a definite degree of crystallization of the molten clinker.

6. **Grinding the clinker with gypsum.** The cooled clinker is then finely pulverised together with 2 to 6% gypsum (which acts as a setting time retarder of cement water paste) in long tube mills. The finer the cement, the greater is the strength of the concrete made from it.

7. **Storage and packing.** The cement coming out of the grinding mills is stored in concrete storage silos. Moisture-free compressed air is used to agitate the cement and to keep it free from compaction by its own weight. In India, cement is usually packed in jute bags each holding 50 kg nett of cement.

Portland Cement Flow Sheet

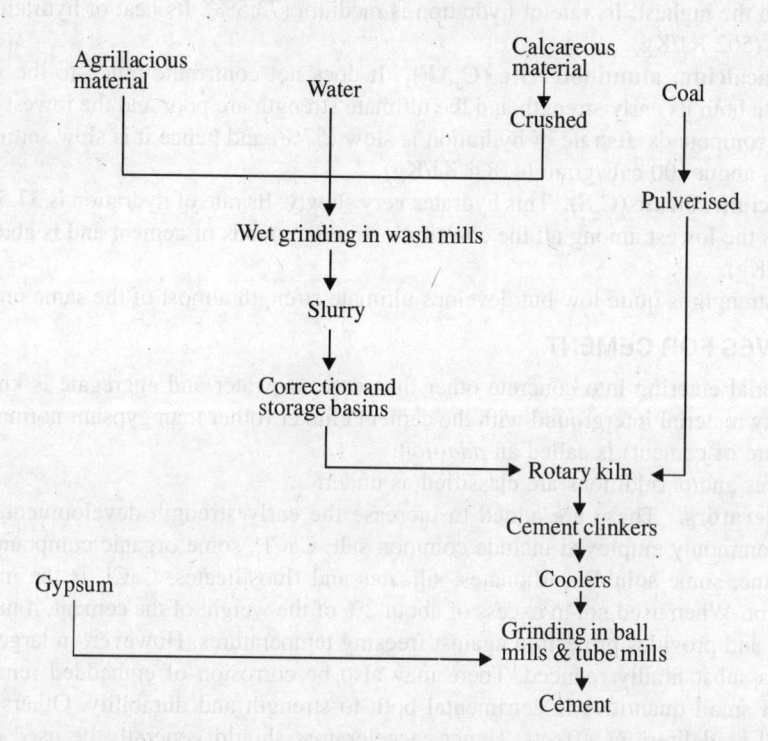

```
Agrillacious              Calcareous
material      Water       material      Coal
    |           |            |           |
    |           |         Crushed        |
    |           |            |           |
    |           |            |       Pulverised
    v           v            v           |
  Wet grinding in wash mills             |
            |                            |
            v                            |
          Slurry                         |
            |                            |
            v                            |
     Correction and                      |
     storage basins                      |
            |                            |
            +-------> Rotary kiln <------+
                         |
                         v
                   Cement clinkers
                         |
                         v
                      Coolers
  Gypsum                 |
     \                   v
      \----> Grinding in ball
             mills & tube mills
                         |
                         v
                      Cement
```

6·7. CHARACTERISTICS OF THE CONSTITUTIONAL COMPOUNDS IN CEMENT

The properties of cement depend upon the relative proportions of the constitutional compounds present and each of them has different characteristic properties. These constitutional compounds are also called microscopic coostituents.

(1) Tricalcium aluminate (C_3A). The strength developed by different constitutional compounds in cement with time is represented in Fig. 6.2. Tricalcium aluminate undergoes hydration at a very fast rate. It is responsible for the initial set or flash set. Its early strength is good but the ultimate strength is quite low as shown in Fig. 6.2. Its heat of hydration is about 210 cals/gram* (879 KJ/Kg), which is the highest amongst all the constitutional compounds of cement.** Its rate of hydration is 82.5%, as followed by X-ray diffraction studies.

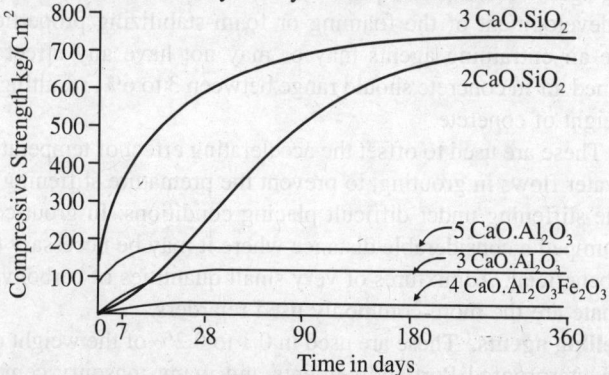

Fig. 6.2. Strength developed by different constitutional compounds in cement.

Curves labeled: $3\,CaO.SiO_2$, $2\,CaO.SiO_2$, $5\,CaO.Al_2O_3$, $3\,CaO.Al_2O_3$, $4\,CaO.Al_2O_3Fe_2O_3$. Y-axis: Compressive Strength kg/Cm². X-axis: Time in days (0, 7, 28, 90, 180, 360).

---
* KJ/Kg = Kilo joules per kilogram. This is the unit of heat in MKS system. 1 cal/g = 4.185 Joules/g = 4.184 KJ/Kg.
** The rate on hydration is the percentage of hydration over 7 days using Type I cement at water to cement ratio of 0.4. It is followed by X-ray diffraction studies.

(2) **Tricalcium silicate (C_3S).** It develops very high strength quite early and the ultimate strength is also the highest. Its rate of hydration is medium (73.5%). Its heat of hydration is about 120 cals/gram (502 KJ/Kg).

(3) **Tetracalcium aluminoferrite (C_4AF).** It does not contribute much to the strength of cement because both its early strength and the ultimate strength are poor and the lowest among the constitutional compounds. Its rate of hydration is slow (57%) and hence it is slow setting. Its heat of hydration is about 100 cals/gram (418.4 KJ/Kg).

(4) **Dicalcium silicate (C_2S).** This hydrates very slowly. Its rate of hydration is 37.5%. Its heat of hydration is the lowest among all the constitutional compounds of cement and is about 60 cals/gram (251 KJ/Kg).

Its early strength is quite low but develops ultimate strength almost of the same order as C_3S.

6·8. ADDITIVES FOR CEMENT

Any material entering into concrete other than cement, water and aggregate is known as an *admixture*. Any material interground with the cement clinker (other than gypsum normally used in the manufacture of cement) is called an *addition*.

Admixtures and/or additions are classified as under:

1. Accelerators. These are added to increase the early strength development. Chemical accelerators commonly employed include common salt, $CaCl_2$, some organic compounds such as triethanol amine, some soluble carbonates, silicates and fluosilicates. $CaCl_2$ is the most widely used accelerator. When used not in excess of about 2% of the weight of the cement, it increases the early strength and provides protection against freezing temperatures. However, in larger amounts, the strength is substantially reduced. There may also be corrosion of embedded reinforcement. NaCl, even in small quantities is detrimental both to strength and durability. Others have both merits as well as deleterious effects. Hence, accelerators should generally be used only under competent technical advice or with personal experience.

2. Air-entrainment agents. These have assumed great importance since 1940, primarily from the standpoint of pavement durability against alternating cycles of severe cold weather and the injurious action of salts used in snow removal. The action of air-entraining agents is similar to that of a foam or froth stabilizer. On account of favourable effects of air entrainment on workability and texture of the concrete, air entrainment is being frequently extended from pavement concrete to other constructions where durability is not a serious problem. Vinsol resin and Darex are the commonly used commercial air-entrainment agents which are introduced as "additions" during grinding of the clinker. Other organic compounds such as natural resins, fats and oils are not in themselves foaming agents but depend upon a saponification reaction with the alkali constituents of the cement for development of the foaming or foam-stabilizing properties when interground with cement. These air-entraining agents may or may not have any effect on the hydration of cements. The entrained air in concrete should range between 3 to 6%, resulting in a reduction of 4.5 to 9 lb/cft in the weight of concrete.

3. Retarders. These are used to offset the accelerating effect of temperature from hot weather concreting or hot-water flows in grouting, to prevent the premature stiffening of some cements, or to actually delay the stiffening under difficult placing conditions. In grout-cement slurries where the grout must be pumped a considerable distance where it may be necessary to redrill grout holes, retarders are often beneficial. Admixtures of very small quantities of carbohydrate derivatives and calcium lignosulfonate are the more commonly used retarders.

4. Water-repelling agents. These are used in 0.1 to 0.2% of the weight of the cement and are usually present in waterproofed Portland cements and many masonry cements. The commonly used water-repelling agents include soaps or other fatty acid compounds such as calcium–, ammonium–, aluminium–, or sodium stearates or oleates and petroleum oils or waxes.

The water-soluble soaps are not water-repellent as added to the concrete mixture but can become so by reaction with the hydrated lime evolved during the hydration of the cement to form water-insoluble calcium soaps. Concretes containing water-repellent materials are used in floors and walls in contact with soil where the concrete is subject to capillary flow of moisture but not to the flow of water under a high pressure head. Water repellancy may also be a useful property in masonry mortars to decrease capillary raise or travel of the moisture that produce discoloration and efflorescence upon evaporation from exposed surfaces. When used in permissible amounts, (0.1 to 0.2% of the cement), water-repelling agents increase the workability of the fresh concrete and decrease the absorption by capillarity of the hardened concrete, but they may increase its permeability under high head and decrease the strength. Greater amounts increase the effects noted.

5. Workability agents. These are usually employed to offset deficiencies in grading that tend to produce harshness or segregation and jeopardize successful placement under inaccessible difficult conditions. Examples are bentonite clay and diatomaceous earth which are used upto 3 to 5% by weight of cement. Other examples are fly ash, clay, finely divided silica, fine sand, hydrated lime, talc and pulverized stone, some of which are added even upto 20% by weight. Some of the commonly used air-entraining agents also increase workability.

6. Gas forming agents. Aluminium powder is the widely used gas-forming agent. It reacts with the hydrating hydroxide in concrete to permeate the mass with minute hydrogen bubbles. Amounts added are of the order 0.005 to 0.02% by weight of the cement. However, larger quantities are used to produce the light-weight, low-strength, sound-or heat-insulation filler concrete known as Acrocrete.

7. Pozzolanic materials. *Finely divided siliceous and aluminous substances *e.g.,* fly ash (fine flue dust, which is obtained as a by-product of thermal power-plants), volcanic ash, heat treated diatomaceous earths, heat treated raw clays and shales which are not cementitious in themselves, combine with hydrated lime and water to form stable compounds of cementitious value. These are generally used upto 10 to 35% of the cement (as cement substitutes) in large hydraulic structures (mass concrete works) to lower the heat of hydration and to instill greater resistance to sea water, sulfate bearing soils, or natural acid waters. The specific gravity is lower than that of cement and substitution by equal weights increases the relative bulk of fine material thereby improving workability and reducing bleeding and segregation. The rate of gaining strength is slower but under favourable curing conditions, the later strengths are higher with most of the pozzolanic admixtures.

8. Natural cementing materials. These are the natural cementing materials such as hydraulic lime, water-quenched blast furnace slag and lime. These are used upto 10 to 25% by weight of Portland cement. These may increase workability, decrease the bleeding and segregation, decrease the heat of hydration and usually decrease the strength when used in larger quantities. Some of them may contribute to the strength of the concrete through their own chemical activity. Natural cementing materials generally require a longer curing period for the development of their potential strength.

9. Miscellaneous admixtures. These include colouring pigments, integral floor hardeners, pore fillers and additives for resistance to wear and decrease of dusting.

6·9. PROPERTIES OF CEMENT

(1) Setting and hardening of cement. When water is mixed with cement powder, heat is liberated and generally some initial setting known as 'flash set' takes place. The setting and hardening of cement are mainly due to hydration and hydrolysis reactions taking place when the different constitutional compounds present in the cement intimately interact with water. Anhydrous compounds react with water and undergo hydration. The solubility of these hydrated compounds

---

* The term "Pozzolanic" comes from "Pouzzoles", a city near Naples where volcanic silico-aluminate calcium ash is found.

is much less and hence are precipitated as insoluble gels or crystals. These have the ability to surround the inert materials like sand or crushed stones (in mortars or concretes) and bind them very strongly.

According to the colloidal theory of Michaelis, hardening of cement is due to the hardening of the silicate gels formed during hydration. According to the crystalline theory of Le Chatlier, the hardening of cement is due to the interlocking of the crystalline products formed during hydration of the constitutional compounds. It is generally agreed that setting and hardening of cement are essentially due to the formation of interlocking crystals reinforced by the rigid gels formed by the hydration and hydrolysis of the constitutional compounds. It is generally believed that the setting times of C_2S, C_3S, C_3A and C_4AF are 28 days, 7 days, 1 day and 1 day respectively.

When cement is mixed with water, the paste becomes quite rigid within a short time which is known as initial set or flash set. This is due to C_3A which hydrates rapidly as follows:

$$3\,CaO \cdot Al_2O_3 + 6\,H_2O \longrightarrow 3\,CaO \cdot Al_2O_3 \cdot 6\,H_2O$$
$$\text{(crystals)}$$

However, these crystals prevent the hydration reactions of other constitutional compounds forming barrier over them. In order to retard this flash set, gypsum $CaSO_4 \cdot 2H_2O$ or plaster of paris $CaSO_4 \cdot \tfrac{1}{2}H_2O$ is added during the pulverisation of cement clinkers. Gypsum retards the dissolution of C_3A by interacting with it forming insoluble complex sulfo aluminate which does not have quick hydrating property.

$$3\,CaO \cdot Al_2O_3 + x\,H_2O + y\,CaSO_4 \cdot 2H_2O \longrightarrow 3CaO \cdot Al_2O_3 \cdot y\,CaSO_4 \cdot z\,H_2O$$
$$(1-3) \qquad\qquad (10-33)$$

The tetracalcium aluminoferrite (C_4AF) then reacts with water forming both gels and crystalline compounds as follow:

$$4\,CaO \cdot Al_2O_3 \cdot Fe_2O_3 + 7\,H_2O \longrightarrow \underset{\text{gels}}{3CaO \cdot Al_2O_3 \cdot 6H_2O} + \underset{\text{crystals}}{CaO \cdot Fe_2O_3 \cdot H_2O}$$

These gels shrink with passage of time and leave some capillaries for the water to come in contact with C_3S and C_2S to undergo further hydration and hydrolysis reactions enabling the development of greater strength over a length of time.

$$3\,CaO \cdot SiO_2 + x\,H_2O \longrightarrow \underset{\text{gels}}{2\,CaO \cdot SiO_2\,(x-1)\,H_2O} + \underset{\text{crystals}}{Ca(OH)_2}$$

$$2\,CaO \cdot SiO_2 + x\,H_2O \longrightarrow \underset{\text{gels}}{2\,CaO \cdot SiO_2 \cdot x\,H_2O}$$

The setting and hardening of cement may be summarized dia-grammatically as follows:

```
                    Unhydrated cement
                       Hydration
                   ↙              ↘
           Metastable Gel      Crystalline hydration products
                 ↓
            Stable Gel    →    Crystalline products
```

An abnormal type of set is sometimes encountered where the cement paste stiffens quickly, but without the evolution of considerable heat and may again be rendered fluid by remixing. This condition is called *false set*. This is due to (1) the dehydration of the gypsum during grinding process brought about by excessive temperature in the mills. The resulting cement then contains anhydrite, $CaSO_4$ which quickly sets by hydrating to gypsum, $CaSO_4, 2H_2O$, and (2) presence of alkali carbonates in the cement, which may form during storage of alkali containing cements. The

PORTLAND CEMENT

alkali carbonate could then react with $Ca(OH)_2$ produced by the rapid hydrolysis of C_3S, thus generating $CaCO_3$ which brings about some rigidity of set.

2. Heat of hydration. When water is mixed with Portland cement, some amount of heat is liberated due to hydration and hydrolysis reactions leading to setting and hardening of cement. On the average, the quantity of heat evolved during complete hydration of cement is of the order of 500 KJ/Kg. As described earlier, the heats of hydration of the different constitutional compounds are in the following order:

$$C_3A > C_3S > C_4AF > C_2S$$
$$\text{KJ/Kg} \quad 878 \quad 502 \quad 418 \quad 251$$

Therefore, wherever large masses of concrete are poured into positions (such as construction of dams), it is necessary to dissipate the heat generated during hydration as quickly as possible to avoid the formation of shrinkage cracks on setting and hardening.

3. Soundness. Refers to the ability of the cement paste to retain its volume after setting and is related to the presence of the excessive amount of free lime or magnesia in the cement. MgO should not be more than 6%. If a cement on hydration produces only very small volume changes and that such volume changes are well within tolerance limits laid down in the specifications, the cement is said to be "sound". Presence of excessive quantities of crystalline magnesia contributes to delayed expansion or unsoundness. The soundness is determined by Le Chatlier's test in which the expansion of a test piece in boiling water for 3 to 5 hours is measured. Recently this test is replaced by the "Autoclave test".

4. Fineness : Finer cement react faster with a corresponding increase in early strength development during the first 7 days. Fineness also influence workability, since the finer the material, the greater the surface area.

Fineness of hydraulic cement 150 μm (No.-100) sieves
75 μm (No.-200) sieves
45 μm (No.-325) sieves

Most portland cement particle are less than 45 μm (0.045 mm) (No. 325 sieve) in diameter.

6.10. TESTING OF CEMENT

In order to maintain the quality of cement, various tests are conducted from raw material stage right upto the cement in packing stage, at every half an hour to one hour intervals. The final product cement is tested for various physical and chemical characteristics. Different types of cements have to satisfy their relevant specifications. Some important specifications for ordinary portland cement, as per Indian Standard : 269 – 1967, are given below.

Chemical requirements

1. *Lime saturation factor.*

$$\frac{CaO - 0.7 SO_3}{2.8 SiO_2 + 1.2 Al_2O_3 + 0.65 Fe_2O_3} = 0.66 \text{ to } 1.02 \text{ applied to cement}$$

$$\frac{CaO}{2.8 SiO_2 + 1.2 Al_2O_3 + 0.65 Fe_2O_3} = 0.92 - 0.98 \text{ applied to clinkers}$$

2. Silica ratio = $SiO_2 / (Al_2O_3 + Fe_2O_3) = 2.0$-$3.0$

3. Alumina ratio $\dfrac{Al_2O_3}{Fe_2O_3}$ Not lesser than 0.66 not apply to special type such as white cement or sulphate resisting cement

4. *Insoluble residue* : Not more than 2%

5. *MgO* : Not more than 6%.

6. SO_3 : Not more than 2.75%
7. *Loss on Ignition*: Not more than 4%

Physical requirements
1. *Setting time*:
   Initial : Not less than 30 minutes
   Final: Not more than 600 minutes
2. *Compressive strength*:
   (1 : 3 cement mortar cubes cement and blended Ennore sand)
   3 days   Not less than 1.6 Kgf/mm^2
   7 days   Not less than 2.2 Kgf/mm^2
   (Kgf = Kilogram force = 9.807 Newtons).

$$0.92 - 0.98 \leftarrow \frac{CaO}{2.8 SiO_2 + 1.2 Al_2O_3 + 0.65 Fe_2O_3} \text{ applies to cliners}$$

3. *Soundness*:
   By Autoclave method : Expansion not more than 0.8%
   By Le Chatiler method : Unaerated cement : max 10 mm
                          Aerated cement : max 5 mm
4. *Fineness*:
   As specific surface by ⎫ Not less than
   Blain permeability method ⎭ 215 m^2/Kg

6·11. GENERAL COMPOSITION OF ORDINARY PORTLAND CEMENT:

Average Chemical Composition	*Average Composition with respect to Constitutional Compounds*
CaO : 60 to 66%	C_3S – 48%
SiO_2 : 17 to 25%	C_2S – 27%
Fe_2O_3 : 0.5 to 6%	C_3A – 10%
Al_2O_3 : 3 to 8%	C_4AF – 8%
MgO : 0.1 to 5.5%	Free CaO – 0.9%
Na_2O & K_2O : 0.5 to 1.5%	MgO – 2.5%
SO_3 : 1 to 3%	$CaSO_4$ – 2.8%

6.12. REACTIONS TAKING PLACE DURING BURNING OF THE RAW MATERIALS IN THE ROTARY KILN:

< 800°C — Formation of CA, C_2F and C_2S starts

800 – 900°C — Formation of $C_{12}A_7$ starts

900 – 1100°C — Formation and decomposition of C_2AS; Formation of C_3A and C_4AF starts; Full decomposition of $CaCO_3$ takes place and formation of free CaO reaches a maximum.

1100 – 1200°C — Maximum formation of the constitutional compounds viz., C_3A, C_4AF, and C_2S

1200 – 1450°C — Formation of C_3S with gradual disappearance of free CaO takes place.

PORTLAND CEMENT

6.13. THERMOCHEMICAL CHANGES TAKING PLAC DURING CEMENT FORMATION:

Temperature range	Reactions occurring	Nature of the heat-change
> 100°C	Evaporation of free water	Endothermic
≥ 500°C	Evolution of combined water from clay	Endothermic
≥ 900°C	Amorphous dehydration products of clay start crystallizing	Exothermic
≥ 900°C	Evolution of CO_2 from $CaCO_3$	Endothermic
900 – 1200°C	Reaction between clay and lime	Exothermic
1250 – 1280°C	Liquid formation starts	Exothermic
> 1280°C	Further liquid formation and completion of the formation of constitutional compounds	The net heat-change may be endothermic

6.14. ACTION OF SOME CHEMICALS ON CONCRETE:

Chemical	Reactions on concrete
Chlorine	Continuous exposure to water containing 5 to 10 ppm can cause surface etching of concrete.
Bleaching Powder	Acidic solutions of bleaching powder can attack concrete
Ink	The acid types of ink, containing free organic acids and H_2SO_4, attack concrete.
Calcium bisulphite	Solution of calcium bisulphite can attack concrete. High alumina cement is more resistant for attack.
Sodium sulphide	Solutions of moderate concentration can attake Portland cement, forming sulphides. High alumina cement is more resistant for attack.
Sodium sulphite and bisulphite	Solutions can attack concrete. High alumina cement is more resistant for attack.
Sodium thiosulphate (hypo)	Leakage of photographic wastes containing hypo as well as pure hypo solutions, can attack and disintegrate concretes, brickwork motors and renderings.
Borax	Slight attack may be there.
Tri-sodium phosphate	No appreciable effect upto 5% solution.
Tan liquors	Destructive action on concrete.
Formaldehyde, Formic acid, or Acetic acid	Aqueous solutions are destructive to concrete
Detergents	Acid detergents, such as those containing phosphoric acid, may attack Portland cement concrete slowly. High alumina cement is less resistant to detergents containing free alkali hydroxides and more resistant to acid detergents.

6.15. COMPUTATION OF THE AMOUNTS OF CONSTITUENT COMPOUNDS IN CEMENTS:

Microscopic and other investigations have shown that Porland cement essentially consists of the following constituents : C_2S, C_3A, C_2S, a ferrite similar to C_4AF, MgO and small quantities of free CaO. It also consists of gypsum, added during grinding to control the setting time of the cement and traces of TiO_2, Mn_2O_3, Na_2O, K_2O. These oxides together may not be more than 2% and are usually ignored in the calculations of the potential compound composition of cement. Hence, if the analytical composition of a cement and the compounds formed in it during burning in the rotary kiln are known, the composition of the cement can be approximately calculated as follows:

(1) The SO_3 content is considered first and the equivalent amount of CaO required to form $CaSO_4$ is calculated.

(2) The sum of the CaO calculated as above and the free-lime content of the cement estimated are subtracted from the total lime-content reported in the analysis.

(3) The remaining lime, together with the content of SiO_2, Al_2O_3 and Fe_2O_3, are now apportioned to give the constitutional compounds C_4AF, C_3A, C_3S and C_2S. This can be done by calculating first the amounts of CaO and Al_2O_3, required to combine with the Fe_2O_3 to form C_4AF and thus derive the content of this compound.

(4) The remaining alumina is now calculated to C_3A.

(5) The CaO remaining after subtraction of the amounts used up in forming C_4AF and C_3A, as stated above, is finally allotted to SiO_2 and the contents of C_3S and C_2S are calculated.

(6) The molecular proportions of the various oxides present in different constitutional compounds and their respective ratios are given below, which are very helpful in the above calculations.

Chemical equations

$$CaSO_4 \longrightarrow CaO + SO_3$$
$$136 \qquad\qquad 56 \quad\; 80$$

$$4\,CaO.Al_2O_3.Fe_2O_3 \longrightarrow 4\,CaO + Al_2O_3 + Fe_2O_3$$
$$485 \qquad\qquad\qquad 224 \quad\; 102 \quad\; 159$$

$$3CaO.Al_2O_3 \longrightarrow 3CaO + Al_2O_3$$
$$270 \qquad\qquad 168 \quad\; 102$$

$$2CaO.SiO_2 \longrightarrow 2CaO + SiO_2$$
$$172 \qquad\qquad 112 \quad\; 60$$

$$3CaO.SiO_2 \longrightarrow 3CaO + SiO_2$$
$$228 \qquad\qquad 168 \quad\; 60$$

$$2CaO.SiO_2 + CaO \longrightarrow 3CaO.SiO_2$$
$$172 \qquad\quad 56 \qquad\qquad 228$$

Ratios of oxides

$$\frac{CaO}{SO_2} = \frac{56}{80} = 0.7$$

$$\frac{3CaO}{Al_2O_3} = \frac{168}{102} = 1.65$$

$$\frac{Al_2O_3}{Fe_2O_3} = \frac{102}{159} = 0.64$$

$$\frac{4CaO}{Fe_2O_3} = \frac{224}{159} = 1.4$$

PORTLAND CEMENT

$$\frac{3CaO.Al_2O_3}{Al_2O_3} = \frac{270}{102} = 2.65$$

$$\frac{4CaO.Al_2O_3.Fe_2O_3}{Fe_2O_3} = \frac{485}{159} = 3.04$$

$$\frac{2CaO}{SiO_2} = \frac{112}{60} = 1.87$$

$$\frac{3CaO.SiO_2}{CaO} = \frac{228}{56} = 4.07$$

$$\frac{2CaO.SiO_2}{CaO} = 3.07$$

The method of calculation is illustrated with the help of an example:

Analytical data :

CaO	—	63.80%
SiO_2	—	20.70%
Al_2O_3	—	5.60%
Fe_2O_3	—	2.40%
MgO	—	3.70%
TiO_2	—	0.23%
Na_2O	—	0.21%
K_2O	—	0.51%
SO_3	—	1.60%
Loss by difference	—	0.25%
Total		100.00%

Free CaO — 0.4%

Calculations:

Free CaO — 0.4%

1.6% SO_3 = 1.6 × 0.70% CaO = 1.12% CaO
Free CaO = 0.4%
Total = 1.52%

Remaining CaO = 63.80 – 1.52 = 62.28%
2.4% Fe_2O_3 = 2.4 × 0.64% Al_2O_3 = 1.436% Al_2O_3
= 2.4 × 1.4% CaO = 3.36% CaO
= 2.4 × 3.04% C_4AF = 7.296% C_4AF

Remaining Al_2O_3 = 5.600 – 1.436 = 4.064%
4.064% Al_2O_3 = 4.064 × 1.650% CaO = 6.7056% CaO
= 4.064 × 2.65% C_3A = 10.77% C_3A

Remaining CaO = 63.8 – (3.36 + 6.706)
= (63.8 – 10.066) = 53.734%

Now, the contents of C_3S and C_2S can be calculated as per the following formulae:

%C_3S = 4.07x – 7.60 Y
%C_2S = 8.60Y – 3.07 X

where X is the % CaO content remaining for silicate formation and Y is the % of SiO_2 content.

Accordingly, in the present example:

$$\% C_3S = (4.07 \times 53.734) - 7.60 \times 20.70)$$
$$= 61.38\%$$

and
$$\% C_2S = (8.60 \times 20.70) - (3.07 \times 53.734)$$
$$= 13.06\%$$

Thus the compound composition is

$C_4AF \longrightarrow 7.3\%$
$C_3A \longrightarrow 10.8\%$
$C_3S \longrightarrow 61.4\%$
$C_2S \longrightarrow 13.1\%$

In addition, it contains free CaO, $CaSO_4$, MgO, Na_2O, K_2O, TiO_2 and moisture and CO_2, which are lost on ignition.

6·16. TYPES OF PORTLAND CEMENT AND ITS DERIVATIVES

The properties and uses of various types of Portland Cement and its derivatives are summarised in Table 1.

Table-1. Types of portland cement and its derivatives

Sr. No. (1)	Type (2)	Characteristics (3)	Uses and remarks (4)
	True portland cements		
1.	Orindary cement (Type I)	As discussed earlier, C_3S – 48%, C_2S – 27%, C_3A – 10%, C_4AF – 8% CaO – 0.9%, MgO – 2.5%, $CaSO_4$ – 2.8%, Loss on ignition – 0.8%	An all purpose construction material most widely used.
2.	Rapid hardening cement (Type III)	Manufactured in the same manner as that of ordinary Portland cement excepting that the lime saturation factor is maintained relatively higher and the final product is ground to more fineness. Contains greater proportion of C_3S than ordinary cement so that a more rapid gain of strength is achieved for the mortar or concrete. More expensive to produce composition: C_3S – 54%, C_2S – 18%, C_3A – 12%, C_4AF – 8%, CaO – 1.2%, MgO – 2.5%, $CaSO_4$ – 3%	For emergency constructions for high early strength and for use in prestressed concrete constructions. Used for urgent constructions as in the constuction of border roads during emergency, where high early strength is desired. This is achieved by having greater amount of C_3S (by introducing high percentage of lime in the raw material mix) and carrying out grinding to a greater fineness.
3.	Low heat cement (Type IV)	Heat of hydration is low and within specified limits at specified ages. High percentages of C_4AF and C_2S and low percentages of C_3S and C_3A	For mass concrete work where low liberation of heat is desirable such as in dams and other monolithic works. However, in recent times,

PORTLAND CEMENT

(1)	(2)	(3)	(4)
		than those of ordinary type. The heat of hydration is only about half of that of the ordinary cement, so that the shrinkage cracks are reduced to a minimum. The raw materials are to be selected in such a way as to maintain C_3A to minimum. The specficiation for low heat Portland cement is that the heat of hydration measured in an adiabetic calorimeter should be less than 65 and 75 cals/g (272 and 314 KJ/kg) for 7 and 28 days respectively Composition : C_3S – 20%, C_2S – 53%, C_3A – 5%, C_4AF – 16%, CaO – 0.4%, MgO – 1.8%, $CaSO_4$ – 3%.	blended cements are preferred to low heat cements for mass concrete works. For example, in Bhakra Dam, low heat cement was not used. Instead of that, both pozzolanic cement and chilling of concrete were employed in the mas concrete work.
4.	Oil well cement	Very expensive. Proportion of of C_3A should be absolute minimum which is achieved by having a high iron content in the raw materials which helps in producing C_4AF in preference to C_3A. Special retarders like sugars, cellulose derivatives and organic acids are added in controlled quantities for lengthening the setting time.	For cementing steel castings of oil and gas wells which go to depths of the order 1000 m where temperature and pressure are very high.
5.	Hydrophobic or water-proof cements	This is nothing but ordinary portland cement to which a water-repellent agent (*e.g.,* calcium stearate or rosin, oleic, lauric and stearic acids or pentachlorophenol which are hydrophobic in nature) is added during grinding.	For rendering concrete more impermeable to water and for imparting better storage properties under high humidity conditions.
6.	White cement	The raw materials used (limestone and clay) should be free from iron.	For making decorative pastel shades for cement paints and for making coloured cements.
7.	Sulfate-resisting cement (Type V)	Composition so adjusted to have higher C_4AF and lower C_3A than ordinary cements.	Resistant to sulfate-or chloride-bearing waters.
8.	Expansive cements	Ordinary portland cement has the drawback of certain	This is mostly used in France and USA. Used where

(1)	(2)	(3)	(4)
		amount of shrinkage after setting and hardening, which produces small cracks. In case of concrete pavements, some gaps are inevitably formed which have to be filled with special compounds. In order to overcome these drawbacks, expansive shrinkage compensating cements are prepared. The chief ingredient for this purpose is calcium aluminate (CA & C_5A_3) prepared by heating a mixture of high purity lime stone, bauxite and gypsm (25 : 25 : 50) to control the rate of formation of calcium sulfo-aluminate during hardening.	shrinkage characteristics of the cements are undesirable.
9.	Moderate heat cement (Type II)	Contains lesser quantity of C_3A which is more susceptible for sulfate action. Composition : $C_3S - 44\%$, $C_2S - 30\%$, $C_3A - 7\%$, $C_4AF - 12\%$, $CaO - 0.7\%$, $MgO - 2.9\%$, $CaSO_4 - 2.8\%$	Used where the construction should resist moderate sulfate action and also where a lower heat of hydration is desired.
	Blended Cements:		
1.	Portland Pozzolana cement	Pozzolana is a matrial which does not have any hydraulic property in itself but is activated into a hydraulic material. Portland pozzolana cement is obtained by grinding together Portland cement clinker and burnt clay or pulverised fly ash (from power plants) or bricks or burnt shale or any other pozzolanic material in the proportion of 3: 1 together with 6% gypsum. Such a product has lower heat of hydration, lower porosity and better sulfate resistance than ordinary Portland cement.	Particularly suited for mass concrete works such as construction of dams and piers. Because of lower porosity, this type of cement is useful for lining of canals. Offers resistance to sulfate attack. Improves workability and reduces liberation of heat. About 7 million tonnes of fly ash are currently produced annually. It has latent hydraulic properties. Fly ash having more than 6 – 7% unburnt carbon may cause difficulties in the final product.
2.	Portland blast furnace slag cement	Most important type of blended cements manufactured in India	This product meets the twin objectives of increased

(1)	(2)	(3)	(4)
		and in other countries also. As per the I.S.I. specifications, the granulated blast furnace slag should form 25 – 65% of the blend. In actual practice, the cement and granulated slag are in the ratio 1 : 1. When molten blast furnace slag at a temperature of about 1400°C is brought into contact with a jet of water under high pressure, a granulated slag is produced because of the sudden quenching and exfoliation. The product containing 7 to 22% moisture is dried and transported to cement factories. When this finely granulated slag powder is ground with portland cement or hydraulic lime, it is activated into a cementitious material so that the entire blend is as good as portland cement itself. According to the ISI specifications, this cement should have the same strength, soundness and setting time as that of ordinary portland cement. Further, it is claimed to have some special characteristics such as low heat of hydration and greater resistance to sea and sulfatic waters.	production of a good cementing material as well as utilizing a largely produced industrial waste material. It is an all purpose cement with low heat of hydration and volume stability concrete. Hence this is useful in mass concrete works and in grouting oil well casings. However, this type of blended cement has to be ground to greater fineness than the ordinary portland cement, to get identical strength. Thus the cost of grinding is higher.
3.	Super-sulfated cement.	Prepared by intergrinding a mixture containing 80–85% granulated slag, 10–12% anydride and 5 – 6% portland cement. This product is highly resistant to sulfatic and marine waters. When exposed to a temperature of ≥ 40°C over long periods, this type of cement deteriorates in strength.	Suitable in situations where the concrete is exposed to sea water and sulfate-bearing soils. Owing to its low heat liberation, it may be used for mass concrete jobs.

(1)	(2)	(3)	(4)
4.	Masonry cement	Produced by intergrinding portland cement with ground limestone or an inert filler along with air-entraining and/or plasticiser additives. This improves plasticity and water retaining power and reduces shrinkage.	For producing mortars having better plasticity than that of ordinary portland cement.

6·17. OTHER TYPES OF CEMENTS

Some other types of cements are also used in special types of construction work, and these are not portland cements or their derivatives.

(1) High alumina cement. It is manufactured by heating, until molten, a mixture of limestone and bauxite to produce essentially calcium aluminates (CA and $C_{12}A_7$) and quickly cooling the product. The finely ground product is quite dark in colour. It gains strength very quickly and is particularly useful under extremely low temperature conditions where ordinary Portland cement does not gain strength. This kind of cement is known as "Ciment Fondu" in western countries. Another remarkable feature about this cement is that it can be used as refractory material at 1000°C and is used for making castable refractories and for making 'in situ' castings.

(2) Magnesium oxychloride cement. This is also called Sorel cement. It is prepared by the reaction of magnesia and a solution of $MgCl_2$. The reaction is exothermic. The composition of the product formed is roughly $3\ Mg\ O.Mg\ Cl_2.11\ H_2O$. This kind of cement is generally prepared 'in situ' or in moulds. It can be used as flooring material with coloured pigments and other inert fillers producing very pleasing surfaces. However, it is rapidly eroded by water and hence is limited in its use.

(3) Strontium and Barium Cements. These cements are extensively used in concrete shielding for atomic piles where resistance to penetration of radioactive emanations is essential. These are manufactured by employing strontium and barium salts instead of calcium salts during the cement manufacture.

6.18 USE OF FLY ASH AS CEMENTING MATERIAL

Fly Ash

Fly ash is a fine glass like powder generated in the combustion of a coal by coal fired electric power generation. Fly ash solidifies while suspended in the exhaust gases and is collected by electrostatic precipitators. Fly ash particales are generally spherical in shape and range in size from 0.5-100 mm. They consist mostly of silicon dioxide SiO_2, aluminium oxide Al_2O_3 and iron oxide Fe_2O_3.

ASTM C618 classify two types of ash class F fly ash and class C fly ash. Class F fly ash produced by burning of bituminous and anthracite coal, which contains less than 10% CaO. It required additive (cementing agent) such as portland cement, quick line with the presence of water in order to react and produce cementious compounds. Class C fly ash produced by burning of younger lignite or subbituminous coal. It contains more than 20% CaO (Lime) and due to its high concentration of CaO it has self cementing properties. In presence of water class C fly ash will harden and gain strength over time.

Uses of fly ash

Portland cement : Earlier the fly ash produced from coal combustion was simply entrained in flue gases and dispersed into the atmosphere. This created environmental and health concerns that promoted laws

– PORTLAND CEMENT

which have reduced fly ash emission less than 1% of produced ash. Due to its pozzolanic properties fly ash is used in portland cement it improves strength, durability, chemical resistance, segregation and ease of pumping of the concrete. It can replace up to 30% by mass of portland cement. Due to spherical shape of particles it reduces internal friction thereby increases the concrets consistency and mobility it also increase the workability of cement (less water is needed resulting in less segregation of the mixture).

The replacement of portland cement with fly ash also reduces the green house gasses. As production of one ton of portland cement produces nearly one ton of CO_2, replacement of 30% by fly ash could reduce global carbon emissions. Ash used for cement must have fineness of 45 mm or less and loss on igniton less than 4%. 65% of fly ash produced from coal power station is disposed of in landfills.

Soil stabilization - Soil stabilization involves the addition of fly ash to improve the engineering performance of a soil. This is used for a soft, clayey subrade beneath a road, that will experience frequent loadings. Improvement can be done with both class C and class F fly ash. Use of class F-fly ash cementing agent such as lime or cement is needed.

Bricks - Fly ash is used in ingredient in brick, block and structural fills. When bricks come into contact with moisture it expands due to the chemical reaction. Class C fly ash can be easily used due to its self cementing property. Fly ash bricks can save 20% manufacturing cast compared to traditional clay brick.

6·19. MORTARS AND CONCRETES

Cement alone cannot be used for construction activities as it is sensitive to moisture and also the internal stresses developed in cement lead to cracking and reduction in strength. In order to obviate these difficulties, cement is generally used in admixture with sand and crushed stones. Aggregates are those inert materials, which when bound together by cement, form the substance called concrete.

Mortar is a mixture of cement and sand in water. Sometimes, other fine aggregates (< 0.63 cm) may also be used. Mortars are utilized for bonding in masonry works and also for surface covering. The fine aggregates commonly used are natural sand or rock screenings.

Concrete is a mixture of cement, sand and coarse aggregates (> 0.63 cm) in the proportion 1 : (2 to 3.5) : (4 to 7), with water. This has high compressive strength but relatively low tensile strength. The proportions of the different ingredients used vary with the purpose to which the concrete is used. The coarse (or large) aggregates commonly used are crushed rock, natural gravel, crushed brick, cinders or blast furnace slag.

Reinforced concrete is the ordinary concrete reinforced with steel rods or heavy wire mesh. The concrete on setting bonds very strongly with the reinforcements giving high compressive and tensile strengths.

Even if cracks develop in the concrete, considerable reinforcement is maintained.

Steel serves as a suitable reinforcement because its coefficient of thermal expansion is nearly the same as that of concrete. Further, steel is not rapidly corroded in the cement environment, and a relatively strong adhesive bond is formed between it and the cured concrete. This adhesion may be enhanced by the incorporation of contours in the surface of the steel members which permits a greater degree of mechanical interlocking.

Portland cement concrete may also be reinforced by mixing into the fresh concrete fibers of a high-modulus material such as glass, nylon, steel and polyethylene. However, in such a case, care must be exercised regarding possible deterioration of such materials in the cement environment.

6.20. PRESTRESSED CONCRETE

Another reinforcement technique for strengthening concrete involves the introduction of residual compressive stresses into the structural member. Such a material is called *Prestressed Concrete*. A prestressed concrete member fractures only when the applied tensile stress exceeds the magnitude of the precompressive stress. A common prestressing technique comprises of positioning high-strength steel wires inside the empty molds and stretching with a high tensile force, which is maintained constant. Concrete is then placed in the mold and allowed to harden. Then the tension on the steel-wires is released. As the wires undergo contraction, they put the structure in a state of compression because the stress is transmitted to the concrete via the concrete-wire bond that is formed.

Concrete that is prestressed should be of a high quality, having a low shrinkage and a low creep rate. Prestressed concretes, usually prefabricated, are commonly used for highway and railway bridges.

6.21. POST-TENSIONING

This is a technique in which stresses are applied after the concrete hardens. Sheet metal or rubber tubes are situated inside and pass through the concrete forms, around which concrete is cast. After the cement is hardened, steel wires are fed through the resulting holes, and tension is applied to the wires by means of Jacks attached and abutted to the faces of the structure. Again, a compressive stress is imposed on the concrete piece, this time by the Jacks. Finally, the empty spaces inside the tubing are filled with a grout to protect the wire from corrosion.

6.22. CURING

Curing is a process which facilitates the hydration reactions of cement constituents to enable the cement to develop the desired properties to sufficient extent as to meet the requirements. Curing is usually performed by maintaining the required moisture content and a favourable temperature in the concrete during and after its placement in order to provide ideal conditions for the quicker completion of the hydration of cement constitutions and development of the desired properties.

6.23 OVERALL SCENARIO OF CEMENT INDUSTRY

Production and consumption pattern

While the global cement production crossed 1000 million tons per year, the Indian cement industry is marching to cross 100 million tons.

The per capita consumption of cement in India in 1989 was 43 Kg as compared with world average of about 200 Kg.

Technology

The following three technologies are used for the manufacture of cement :
(i) The largely outmoded wet process technology.
(ii) The more modern dry process which utilizes only 19% of coal as against 30% for the wet process.
(iii) The latest pre-calcinator technology which enables optimum utilization of power. In this process, the raw materials are partly or completely calcined before they enter the rotary kiln. This reduces the thermal load on the rotary kiln. Precalcinators can be fitted on to the existing kilns used in the dry process. This technology not only saves power but also enables increase in installed capacity by 30 to 50%. The modern 3000 ton per day plants being set up in our country are using this technology.

Dry process plants predominate now because of savings in heat and possibility of accurate control with the present advances in technology. Conversion of wet process to dry process and

PORTLAND CEMENT

modernization by adding precalciners to the kilns used in the dry process should be taken up immediately. R & D programmes should be encouraged to remove technological obsolescence and to develop appropriate technologies to suit our conditions. High grade and beneficiated circuit rock may be supplemented by blending with burnt clays and blast furnace slag for conserving native limestone deposits.

Quantitative requirements

Basis : 1 ton of Type-I portland cement

Requirements :
- Lime stone - 1.2 to 1.3 tons
- Clay - 0.1 to 0.3 ton
- Gypsum - 0.03 to 0.05 ton
- Coal - 0.25 to 0.40 ton
- Water - 3 tons
- Electricity - 80 KWH

Plant capacities : 200 to 1200 tons / day.

Characteristics of different types of portland cement

Type	Characteristics
Type I (Regular)	C_3S - 40 to 60%, C_2S - 10 to 30%, C_3A - 7 to 13%. Hardens to full strength in 28 days.
Type II (Modified)	Higher C_2S / C_3S to resist sulphate attack.
Type III (High early strength)	High C_3S and C_3A. Attains good strength within 3 days. Heat rate is high and hence unsuitable for mass concrete structures.
Type IV (Low heat)	Low C_3S and C_3A, whose heats of hydration are high. Used for mass concrete structures.
Type V (Sulphate resistant)	$C_3A < 4\%$. Good for structures with sea water contact.

Relative compressive strength characteristics

Portland Cement Type	Compressive strength, Kg / cm²		
	1 day	3 days	28 days
Type I (Regular)	37	120	340
Type II (Modified)	28	83	260
Type III (High early strength)	103	240	440
Type IV (Low heat)	20	49	177
Type V (Sulphate resistant)	28	88	214

Major problems

(A) Engineering problems

(i) **Raw material preparation :** The first step in cement production is cement rock beneficiation. Many sources of lime-stone contain unacceptably high content of silica and iron. The undesirable constituents in the rock are usually removed by ore dressing or beneficiation methods based on fluid mechanics and adsorption. The beneficiation process involves grinding, classification, flotation and thickening. The rock is ground in wet condition, fed to a hydroseparator where the overflow goes to the final thickener if the composition is satisfactory. Otherwise it is subjected to flotation separation to remove silica, talc and mica. The choice of flotation agent used is very vital. In older days, oleic acid was used but nowadays, new types of chemicals are used to achieve better selectivity and economy.

Further, it is important to optimize the particle size range with the power input for grinding as discussed in the next section.

The quantitative requirements for beneficiation of 1 ton of low grade limestone are as follows:

Water	-	2 - 3 tons
Chemicals	-	50 - 200 g
Electricity	-	2.5 KWH

Plant capacity is of the order of 300 - 1000 tons / day.

(ii) Grinding : Most of the new cement plants prefer dry grinding rather than wet grinding. This operation has to be properly designed to ensure optimum use of power. This is particularly important because about 80% of the total power consumed in a cement plant is used in the crushing, grinding and blending operations.

(iii) Kiln design : Calcining and firing take place in the kiln. During the calcining process, $CaCO_3$ is decomposed to CaO. This is followed by firing at 1400 - 1500°C to facilitate the formation of the constitutional compounds (microscopic constituents) of cement. Heat duty is also required for evaporation of water, oxidation of organic material, partial volatalization of sulphates, chlorides and alkalies. In wet process, the length of the kilns is 90 - 170 m and diameter of 2.5 - 6 m. The speed of rotation varies in the range 2 to ½ rpm. The kilns used in dry process may be as short as 50 m. Kiln design is an important aspect of a cement plant.

(iv) Heat economy : The theoretical requirement of heat is 430 K cal / Kg of portland cement clinker. However, in practice, the heat requirement is in the range of 700 - 1000 K cal / Kg for processes using dry grinding, and 1300 - 1800 K cal / Kg for processes using wet grinding.

(v) Quality control : The quality of cement and its performance is very much dependent on the composition of the raw material blend, particle size and degree of calcining. Dependable instrumentation and automatic control of the calcining kiln are of paramount importance. Indeed, investments in this proved to be rewarding for the Indian cement industry in the world market.

(B) Other problems faced by the Indian cement industry

(i) Cement industries in India are mainly located in western and southern regions, which produce 71% of the total output as against their consumption of 57%. On the other hand, Northern and Eastern regions account for 29% of the total output as against their consumption of 43%. This is resulting in lot of transportation cost.

(ii) Cement industry is capital intensive. Entrepreneurs are facing problems to raise the necessary capital in the backdrop of declining profits.

(iii) India's contribution in the global cement trade is extremely low.

(iv) Cement industry is power intensive. Frequent power cuts too affect production. Cement produced using captive power has been expensive.

(v) Coal shortage leads to idle capacity and affects production. Further, low quality coal affects the quality of cement produced.

(vi) Non-availability of transport wagons, pilferage and damage by rain in open railway wagons, primitive, inefficient and uneconomical bagging and transportation.

(vii) Technological obsolescence.

(viii) Non-utilization of full capacity of some of the large plants due to various constraints.

Mini cement plants

In places where only limited quantities of lime stone are available, establishment of mini

PORTLAND CEMENT

cement plants is an attractive proposition because of the following reasons :
- (a) Lesser infrastructural requirements
- (b) Lesser gestation period for commissioning
- (c) Lower cost per installed ton capacity
- (d) Lower cost for providing captive power
- (e) Better quality control
- (f) Effective distribution system
- (g) Considerable relief in excise duty and other incentives by the Government.

QUESTIONS

1. Match the following and write appropriate statements in each case:

   Group A
   - (i) Tricalcium silicate
   - (ii) Dicalcium silicate
   - (iii) Tricalcium aluminate
   - (iv) Tetracalcium aluminaferrite
   - (v) Gypsum
   - (vi) Argillaceous material
   - (vii) Setting of cement
   - (viii) Calcareous material
   - (ix) Mortar
   - (x) Concrete
   - (xi) High alumina cement
   - (xii) Barium and strontium cement

   Group B
   - (a) Flash set
   - (b) High ultimate strength
   - (c) Lowest heat of hydration
   - (d) Retarder
   - (e) Lowest ultimate strength
   - (f) Chalk
   - (g) Shale
   - (h) Hydration and hydrolysis reactions
   - (i) Cement + sand + fine aggregates
   - (j) Cement + Sand + coarse aggregates
   - (k) Superior chemical resistance to sea water
   - (l) Concrete shield for atomic piles

   Answers : 1b, 2c, 3a, 4e, 5d, 6g, 7h, 8f, 9i, 10j, 11k, 12l.

2. Draw a labelled diagram of a rotary kiln used for the manufacture of Portland cement by wet process and discuss the various reactions taking place in the furnace. (Nag. Univ. S - 2003, S. 2006)

3. What are the microscopic constituents (or constitutional compound) present in Portland cement? How do they contribute towards the properties of the cement? (Nag. Univ. S - 2007)

4. What do you mean by setting and hardening of cement? Discuss the various reactions involved with the help of equations. (Nag Univ. S - 2006, S-2007)

5. "The properties of Portland cement depend upon the relative proportions of its constitutional compounds". Justify this statement. (Nag. Univ. W-2003)

6. Explain setting and hardening of portland cement with major chemical reactions involved. (Nag. Univ. W-2005).

7. Write informative notes on the following:
   - (a) Important process parameters for the manufacture of good cement clinker. (Nag. Univ. S-2005)
   - (b) Reactions taking place in the rotary kiln
   - (c) Constitutional compounds in cement and their properties
   - (d) Additives for cement (Nag Univ. S-2005)
   - (e) Important properties of cement

8 a. Differentiate between Dry and Wet process of manufacturing of portland cement (Nag. Univ. S - 2007).
   b. Discuss the different types of Portland cement and its derivatives.

9. Write short notes on the following:
   (a) Setting and hardening of cement
   (b) Soundness of cement (Nag. Univ. W - 2003, S-2006)
   (c) Cement additives (Nag. Univ. W-2003, S-2006, S-2007)
   (d) Mortar and concrete
   (e) Curing
   (f) High alumina cement (Nag. Univ. W-2005)
   (g) Portland Pozzolonic cement (Nag. Univ. W-2005)
   (h) Rapid hardening cement (Nag. Univ. S-2003)
   (i) White cement (Nag. Univ. S-2003)
10. Discuss the significance of the following with respect to the manufacture and properties of Portland cement: (a) lime saturation factor, (b) Silica modulus, (c) Cooling of clinker, (d) Heat of Hydration, (e) Soundness.

7

Phase Rule

> *"With the help of phase rule, it has been possible to predict qualitatively by means of a diagram the effect of changing temperature, pressure and concentration on a heterogeneous system in equilibrium".*

7.1. INTRODUCTION

When two or more different phases exist in equilibrium with each other, then it is called a polyphase or heterogeneous equilibrium. Such a system can be conveniently studied with the help of a generalization called *Phase Rule*. This rule was deduced on the basis of thermodynamic principles by J. Willard Gibbs (1876) and thoroughly studied further by H.W.B. Roozeboom (1884). With the help of phase rule, it has been possible to predict qualitatively, by means of a diagram, the effect of changing temperature, pressure and concentration on a heterogeneous system in equilibrium.

7·2. GIBB'S PHASE RULE

Gibb's phase rule states that in every heterogeneous system in equilibrium, the sum of the number of phases and degrees of freedom is greater than the number of components by 2. This is also expressed in the form :

$$P + F = C + 2$$
or
$$F = C - P + 2$$

where P is the number of phases present in equilibrium, C is the number of components for the system and F is the number of degrees of freedom for the equilibrium.

This is valid for any system at equilibrium at a definite temperature and pressure provided the equilibrium between any number of phases is not influenced by gravity, by electrical or magnetic forces or by surface action; and is only influenced by temperature, pressure and concentration.

For an accurate and effective interpretation of the phase rule, a clear understanding of the terms involved is absolutely essential.

Phase. A phase is defined as any homogeneous and physically distinct part of a heterogeneous system which is separated from other parts of the system by definite bounding surfaces.

A homogeneous system is one which is uniform throughout in physical and chemical properties. *e.g.*, : A solution of NaCl in water.

A heterogeneous system, on the other hand, comprises of two or more different parts, each of which is homogeneous in itself and is separated from others by bounding surfaces. These homogeneous, physically distinct and mechanically separable parts of a heterogeneous system existing in equilibrium are called *"phases"*.

Ex:
(1) In a freezing water system, ice, water and water vapour are the three phases, each of which is physically distinct and homogeneous, and there are definite boundaries between ice and water, between ice and vapour and between liquid water and vapour.
(2) Each crystalline form of ice constitutes a separate phase since it is clearly distinguished and marked off from the other forms. These various forms of ice exist at high pressures.
(3) Every solid in a system is an individual phase and each is separated from others by a definite bounding surface. Hence, a heterogeneous mixture of solids consists of as many phases as the number of solid substances present in the mixture. A mixture of CaO and $CaCO_3$ consists of two phases.
(4) A solid solution, being perfectly homogeneous, is considered as a single phase irrespective of the number of chemical compounds it may contain.
(5) A liquid layer is considered as one phase as long as it is homogeneous and irrespective of whether it consists of a pure substance or a mixture. Thus two perfectly miscible liquids form a single phase while two perfectly immiscible liquids in contact form two different phases.
(6) A mixture of two or more gases is always homogeneous because of the intimate mixing of the molecules. Hence, such a, system constitutes only one phase.

Component. The components of a system do not represent the number of the constituents or chemical individuals present in the system. As given in the statement of phase rule, the number of *components* of a system at equilibrium is the smallest number of independently variable constituents by means of which the composition of each phase present can be expressed, either directly or in the form of a chemical equation. While expressing the composition of a phase in terms of the components, zero and negative quantities of the components are permissible.

Ex: (1) In water system, we have three phases *viz.*, ice, water and vapour in equilibrium. Each of these phases is the different physical form of the same chemical substance, H_2O. Hence, this may be regarded as a one-component system. Although the molecular complexity of H_2O is different in the three phases, the number of components is not affected.

Even though the H_2O is made up of hydrogen and oxygen, they cannot be considered as components because :
(i) they are combined in definite proportions (2 : 1) to form water and their quantities cannot be varied independently, and
(ii) they do not take part in the equilibrium. Hence, the water system is regarded as a one-component system.

Further, strictly speaking, each isotopic form of water should be considered as a separate component, so that ordinary water has, theoretically, two components, namely H_2O and D_2O. But, however, since the ratio of these isotopes remains constant for all practical purposes, it may be assumed to consist of one component only. The form HDO is, in any case, not a separate component, because it can be expressed in terms of H_2O and D_2O by a chemical equation.

(2) The acetic acid system is a one-component system because the composition of each phase can be expressed in terms of CH_3COOH (or $C_2H_4O_2$), although acetic acid exists in the form of double molecules entirely in the solid state, to a great extent in the liquid state and only to a small extent in the vapour state. There may be equilibria such as

$$2CH_3COOH \rightleftharpoons (CH_3COOH)_2$$

in the solid and liquid phases due to association, but the only chemically independent species in all the three phases is CH_3COOH.

PHASE RULE

(3) In the decomposition of calcium carbonate, the following equilibrium exists.
$$CaCO_3(S) \rightleftharpoons CaO(s) + CO_2(g)$$

The system consists of three phases, namely; solid $CaCO_3$, solid CaO and gaseous CO_2. Although this system has three different constituents, it is considered as a two-component system because the composition of each of the above phases can be expressed in terms of any two of the three constituents present, as follows :

(a) When CaO and CO_2 are considered as components,

Phase		Components
$CaCO_3$	=	$CaO + CO_2$
CaO	=	$CaO + OCO_2$
CO_2	=	$OCaO + CO_2$

(b) When $CaCO_3$ and CaO are selected as components,

Phase		Components
$CaCO_3$	=	$CaCO_3 + OCaO$
CaO	=	$OCaCO_3 + CaO$
CO_2	=	$CaCO_3 - CaO$

(c) When $CaCO_3$ and CO_2 are taken as components,

Phase		Components
$CaCO_3$	=	$CaCO_3 + OCO_2$
CaO	=	$CaCO_3 - CO_2$
CO_2	=	$OCaCO_3 + CO_2$

(4) In the following salt hydrate equilibrium,
$$CuSO_4.5H_2O(s) \rightleftharpoons CuSO_4.3H_2O(s) + 2H_2O(g)$$
the composition of each phase can be expressed with the help of the two components, $CuSO_4$ and H_2O.

(5) In the equilibrium
$$Fe(s) + H_2O(g) \rightleftharpoons FeO(s) + H_2(g)$$
the composition of each phase can be expressed in terms of the components, Fe, O and H_2. Hence it is a three-component system.

(6) In the dissociation of ammonium chloride in vacuum, the following equilibrium occurs :
$$NH_4Cl \rightleftharpoons NH_3 + HCl$$
$$\text{(Solid)} \quad \text{(Gas)} \quad \text{(Gas)}$$

The system consists of two phases, viz., solid NH_4Cl and the homogeneous gaseous mixture consisting NH_3 and HCl present in the same proportion in which they are present in the solid NH_4Cl. Since the phase rule does not distinguish between a chemical compound and its constituents present in the same proportion in a homogeneous mixture, this system is regarded as a one-component system. The composition of both the phases can be represented by the same chemical individual, viz., NH_4Cl.

However, if an excess of NH_3 or HCl is introduced into the system, it becomes a two-component system, because the vapour no longer has the same ultimate composition as the solid.

(7) Let us consider the following equilibrium :
$$PCl_5 \rightleftharpoons PCl_3 + Cl_2$$
$$\text{(Solid)} \quad \text{(Liquid)} \quad \text{(Gas)}$$

This system has three phases and three species but the number of components is only 2. The concentration (number of moles) of any two of these three species can be altered arbitrarily. The alteration in the concentration of the third is then automatically fixed in accordance with the following relation :

$$K_e = \frac{[PCl_3][Cl_2]}{[PCl_5]},$$

where K_e = equilibrium constant; and $[PCl_3], [Cl_2]$ and $[PCl_5]$ are their respective concentrations. Thus any two of the above three species are chemically independent and the third is not. Which two are chosen is immaterial. Hence this is a two-component system at low temperatures. However, at higher temperatures, all the three constituents viz., PCl_5, PCl_3 and Cl_2 will be in gaseous state and under these conditions, it will be a one component system.

Degrees of freedom (or variance)

The number of degrees of freedom of a system is the number of variable factors, such as temperature, pressure and concentration, which must be fixed in order that the condition of the system at equilibrium may be completely defined. A system having one, two, three or zero degrees of freedom are usually called univariant, bivariant, trivariant and invariant respectively.

Ex:

(1) If we consider a sample of a pure gas, it should satisfy the gas equation, PV = RT. Hence, if temperature and pressure are fixed, then its volume automatically gets fixed at a definite value. Thus, in such a case, only two factors need to be fixed in order to describe the system completely. Hence this system has two degrees of freedom (bivariant).

(2) If a gaseous system having two or more gases is considered, it can be completely defined if the temperature, pressure and composition of the phases are given. For instance, a mixture containing 71% N_2 and 29% O_2 at 25°C and 1 atmosphere pressure is perfectly defined and any sample of the mixture of the two gases under the said conditions will behave in the same manner. Hence, the number of degrees of freedom of such a system is three (Trivariant).

(3) A system containing a, saturated solution of sodium chloride in contact with the solid and vapour is completely defined if the temperature is specified. Hence this system has only one degree of freedom (univariant).

(4) For the system containing ice-water-vapour in equilibrium no specification of conditions is necessary because the three phases can occur in equilibrium only at a particular temperature and pressure. This situation exists only at triple point end hence this system does not have any degree of freedom (invariant).

Derivation or the Phase Rule

Let us consider a heterogeneous system in equilibrium consisting of 'C' components distributed in 'P' phases. As has been already defined, the number of degrees of freedom of a system in equilibrium is that number of variable factors (such as temperature, pressure and composition) that must be arbitrarily fixed to define the system completely. Obviously, the number of such variables is given by the total number of variables of the system minus the number of variables which are defined automatically by virtue of the system being in equilibrium.

When a system is in equilibrium, there can be only one temperature and one pressure : hence, the total of these variables is two only. However, the number of concentration (or composition) variables can be more. In order to define the composition of each phase, it is necessary to specify only (C—1) composition variables because the composition of the remaining components can be obtained by difference. Since there are P phases, the total number of composition or concentration variables will be P(C—I). On adding the temperature and pressure variables, the total number of variables of the system are P(C—1) + 2.

PHASE RULE

On the basis of thermodynamic considerations, when a heterogeneous system is in equilibrium, at a constant temperature and pressure, the chemical potential, μ of a given component must be same in every phase. Thus, if there is one component in three phases (say α, β, and γ) and one of these (say, α) is referred to as standard phase, then, this fact may be expressed in the form of the following two equations.

$$\mu_\alpha = \mu_\beta$$
$$\mu_\alpha = \mu_\gamma$$

Thus, for each component in equilibrium in 3 phases, 2 equations are possible. Hence, for each component in 'P' phase, (P—I) equations can be written. If there are 'C' components, the number of equations or variables possible, from the conditions of equilibrium are C(P—I). Since chemical potential is a function of temperature, pressure and concentration, each equation must represent on variable. Hence, the number of possible variables or degrees of freedom, F = {P(C — I) + 2} — C(P — I).

$$\therefore F = C - P + 2.$$

7·3. APPLICATION OF PHASE RULE TO ONE-COMPONENT SYSTEM

The equilibrium conditions in a one-component system may be conveniently represented with the help of diagrams taking pressure and temperature as the two axes. These are known as pressure-temperature (P—t) diagrams. In these diagrams, "*areas*" represent *bivariant* systems because to define the system completely at any point in the area, both temperature and pressure should be fixed. Any "*line*" in the diagram represents a "*monovariant*" system because the equilibrium conditions at any point on the line could be completely defined by just fixing temperature or pressure, the other variable gets fixed up automatically. A "*point*" on the diagram where all the three phases are in contact with each other represents an invariant system because it is completely defined by itself and any further statement regarding pressure or temperature is unnecessary.

7.3.1. The Water System. In this system, there is only one component namely H_2O. The three phases taking part in the system are ice, water and vapour. However, the actual number of these phases existing in equilibrium at a particular instance depends upon the conditions of temperature and pressure. On the basis of actual experimentation at different conditions of temperature and pressure, phase diagram (pressure-temperature diagram) for the water system has been prepared, which is represented in Fig. 7.1.

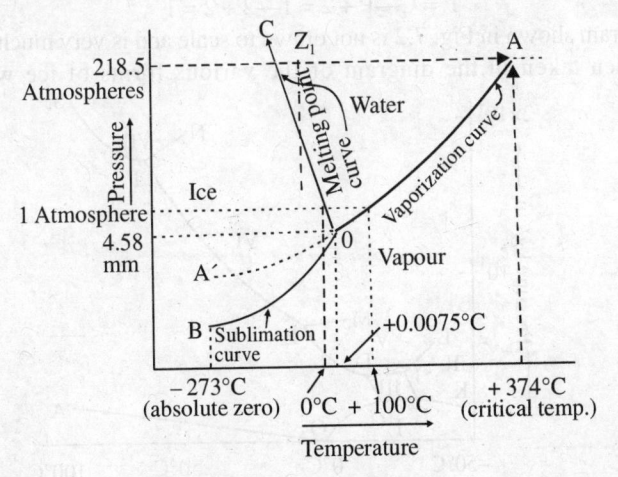

Fig. 7.1. Water System

Areas. The areas in the diagram *viz.*, BOC, COA and AOB represent the conditions of temperature and pressure under which only one of the three phases, namely ice, water or vapour is capable of stable existence. In order to define the system completely at any point in the areas, it is essential to spell out both the temperature and pressure (*i.e.*, abscissa and ordinate). Thus, the areas in the diagram representing ice, water or vapour have two degrees of freedom and hence these are called *bivariant systems*. We will arrive at the same conclusion by the application of the phase rule equation also :

$$F = C - P + 2 = 1 - 1 + 2 = 2$$

Curves

Curve OC (Melting point curve) :

Curve OC represents the melting point curve or the freezing point curve or the fusion curve. This curve shows the equilibrium between water and ice and is expressed by the effect of pressure on the freezing point of ice. The slope of the curve OC shows that ice melts with decrease of volume. This follows from Le Chatelier's principle, which states that whenever a constraint is applied to a system in equilibrium, a change takes place within the system so as to nullify the effect of the constraint and to restore its original equilibrium. Thus, if the temperature is kept constant in the above system in equilibrium and the pressure is increased, such a change will take place which is accompanied by the decrease in volume. The vertical line XYZ represents change of pressure at constant temperature. If the pressure is increased when the system is at point Y, the temperature being maintained constant, the new conditions are represented by Z, *i.e.*, liquid has been formed. Hence, the melting of ice is accompanied by decrease in volume.

The inclination of the line OC towards the pressure axis also indicates that the melting point of ice is slightly lowered by increase of pressure. This is also a consequence of Le Chatelier's principle, according to which an increase in pressure on the ice-water equilibrium will cause a shift in such a direction that there is a decrease in volume. That means that some ice melts into water taking latent heat of ice from the system itself. Thus the melting point of ice will be lowered by an increase of pressure.

Along the curve OC, there are two phases in equilibrium, *i.e.*, ice and water. In order to describe the system at any time, it is enough to state either temperature or pressure, so that the other is automatically fixed. For instance, at atmospheric pressure, ice and water can be in equilibrium only at one temperature, *i.e.*, the freezing point of water. Thus ice-water system has only one degree of freedom (univariant). The same conclusion can be drawn by the application of phase rule equation also :

$$F = C - P + 2 = 1 - 2 + 2 = 1.$$

The phase diagram shown in Fig. 7.2 is not drawn to scale and is very much simplified. In fact, no account has been taken in the diagram of the various forms of ice which exist at high

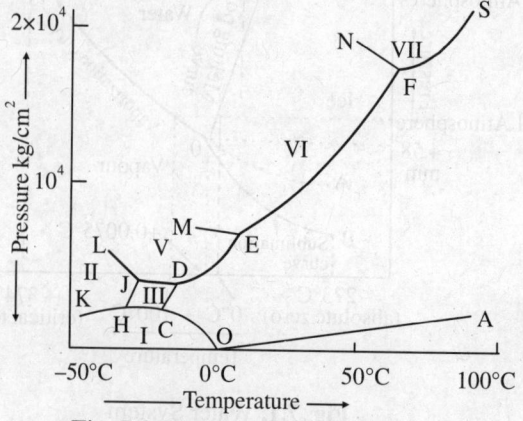

Fig. 7.2. Phase Diagram of Ice

PHASE RULE
319

pressures. G. Tamman (1900) and P.W. Bridgman (1912) showed that the upper limit of the curved OC is marked by the appearance of five other forms of ordinary ice at high pressures (Fig. 7.2).

It can be seen that at increasing pressures, the solids separating from the liquid water are ice I (ordinary ice) III, V, VI and VII ; whereas ice II can only be obtained from other solid forms, *e.g.,* I. III or V by suitable alteration of temperature or pressure.* At triple points C, D, E and F, the three phases in equilibrium are two forms of ice and liquid water. However, at H and J, three solids can coexist in each case. The curves CH, DJ, HK, JL, EM and FN represent two solid phases in equilibrium. In other words they show the effect of pressure on the transition points.

Curve OA (Vaporization Curve)

This curve, also called the vapour pressure curve, shows various temperatures and pressures at which water and its vapour are in equilibrium in a closed vessel from which all the air has been removed. For describing the system completely at any point on the curves, it is necessary to state either temperature or pressure, the other gets fixed up automatically. For instance, at atmospheric pressure, water and vapour can exist in equilibrium only at one temperature, *i.e.,* the boiling point of water. Thus, the system water-vapour has only one degree of freedom (univariant). This fact follows from the phase rule equation also :
$$F = C - P + 2 = 1 - 2 + 2 = 1.$$

The curve OA ends abruptly at the point, A which corresponds to the critical temperature (374°C) and pressure (218·5 atm) of water, beyond which the liquid and vapour phases merge into each other resulting in a homogeneous phase. However, at the lower end, the curve would normally terminate at the point O, where water freezes to form ice. But by careful elimination of solid particles which induce crystallization, water may be cooled far below its freezing point without the crystallization of ice. Thus by preventing water from freezing at the point O, it is possible to follow the vapour pressure curve even below the normal freezing point. In fact, the dotted curve OA has been obtained in this fashion and represents the vapour pressure of "supercooled" water. It represents a metastable equilibrium and is situated above the sublimation curve, OB in the phase diagram because the vapour pressure of the metastable phase at any point is greater than that of the stable phase. The super-cooled water will at once change into solid ice even with the slightest disturbance.

Curve OB (Sublimation Curve)

This curve represents the conditions when ice and vapour are in equilibrium. It can be seen from the phase diagram that this curve is not the prolongation of the curve OA and it falls more steeply. To describe the system corresponding to any point on the curve, only one of the two variables, namely, temperature or pressure has to be specified. In other words, for any temperature arbitrarily fixed, the vapour pressure can have only one value and vice versa. Hence the ice-vapour system represented by curve OB has only one degree of freedom and hence it is a univariant system. This follows from the phase rule equation also :
$$F = C - P + 2 = 1 - 2 + 2 = 1.$$

At the lower limit, the curve OB terminates at absolute zero where no vapour can be present and only ice exists.

Point 'O' :

It has been found experimentally that the curves OA, OB and OC meet in a point 'O'. This is called the "triple point", and is the point where ice, water and vapour can coexist. Since this is a "point", only one set of conditions is possible for the three phases to exist together, as predicted by the phase rule. The triple point corresponds to a temperature of 0·0075°C and a pressure of 4·58 mm. It represents a non-variant system (degree of freedom = 0) and it is unnecessary to specify either

---

* Ice-iv is absent among these because the existence of what was earlier considered to be Ice-iv could not be confirmed by later work.

temperature or pressure under these conditions because only one possible temperature and one possible pressure permit all the 3 phases to exist together in equilibrium. If we vary either temperature or pressure from the conditions of triple point, one of the three phases will vanish. If the temperature is raised at constant pressure, the ice will melt and the liquid will vaporize, leaving only vapour. If the temperature is lowered, there will be only ice. If the pressure on the vapour is increased at constant temperature, the vapour will condense. If the pressure is lowered, the liquid and ice will evaporate.

The salient features of the phase diagram of water system are summarized in Table 1.

Table 1. Water System

Number of components = 1

Name of the system as represented in the phase diagram	Phases in equilibrium	Degrees of freedom or variance $F = C - P + 2$
Areas		
(1) BOC (ice)	Ice	Two
(2) COA (water)	Water	(Bivariant)
(3) AOB (vapour)	Vapour	$F = 1 - 1 + 2 = 2$
Curve		
(1) OC (Melting point curve)	Ice & water	One
(2) OA (Vaporization curve)	Water & Vapour	(univariant)
(3) OB (Sublimation curve)	Ice & Vapour	$F = 1 - 2 + 2 = 1$
Point		
(1) O (Triple point)	Ice, water & vapour	Zero (Invariant) $F = 1 - 3 + 2 = 0$

7·3·2. The Sulfur System. The sulfur system is another one component system and the only chemical individual representing the different phases is sulfur itself. This is a classical example of a one-component system where polymorphism and solid-solid transformation are exhibited. Polymorphism is the phenomenon in which the same chemical substance exists in more than one different crystalline forms and in the case of elements, the phenomenon is called allotropy.

According to the phase rule, the maximum number of phases that can co-exist in this system are 3 :

$$F = C - P + 2 \quad \text{or} \quad P = C - F + 2$$

If $F = 0$ and $C = 1$, then

$$P = 1 - 0 + 2 = 3.$$

The phase diagram (P-T diagram) of this system is represented in Fig. 7.3.

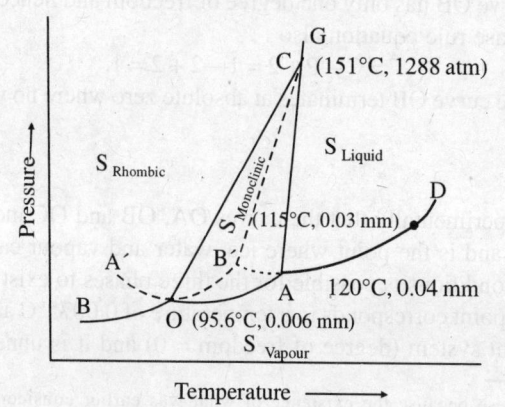

Fig. 7.3. Sulfur System.

PHASE RULE

In the sulfur system, the following four phases exist in equilibrium :

(1) Rhombic Sulfur (S_R)
    (or α-sulfur)
(2) Monoclinic sulfur (S_M)
    (or β-sulfur)
(3) Liquid sulfur (S_L)
(4) Sulfur vapour (S_V)

There are two solid phases $(S_R$ and $S_M)$, a liquid phase (S_L) and a vapour phase (S_V). There are other solid phases capable of existence but they are metastable. For a one-component system, it is not possible to have all the four phases in equilibrium because F cannot have a negative value.

$$F = C - P + 2 = 1 - 4 + 3 = -1,$$

which is impossible. Hence, as pointed out earlier, the maximum number of phases which can exist in equilibrium in the sulfur system is only 3.

The phase diagram shows that the following systems are possible :

I. Areas : (Bivariant systems) :

1. Area, below BOAD is the region where only sulfur vapour (S_V) exists.
2. Area on the right side of GCAD is the region where only liquid sulfur (S_L) exists.
3. Area on the left side of GCOB is the region where only rhombic sulfur (S_R) exists.
4. Area enclosed by COA, is the region where only monoclinic sulfur (S_M) exists.

Thus it is clear that in each area given above, only one phase exists and in order to define the conditions of the system at any point in the area, both temperature and pressure should be specified. Hence, they are bivariant systems.

II. Curves : (Univariant systems) :

Stable Curves

(1) Curve BO is the "sublimation curve" or vapour pressure curve of S_R and it separates the regions of S_R and S_V. Along this curve S_R and S_V are in equilibrium and the degree of freedom is one. The point B corresponds to the lower limit upto which the vapour pressure of rhombic sulfur can be measured.

Point 'O' is the "transition" point which corresponds, to 95·6°C. At this temperature, the two crystalline forms of sulfur are in equilibrium with each other : $S_R \rightleftharpoons S_M$. At the transition temperature, both S_R and S_M are stable. Below this temperature only S_R is stable and S_M gradually changes into S_R. If the heating is rapid the transition from one crystalline form to the other does not occur and the vapour pressure curve of S_R extends even beyond 'O' and upto B′, which is the melting point of the metastable S_R. S_R is metastable above the transition temperature and the dotted curve OB′ is the vapour pressure curve of the metastable S_R.

(2) Curve OA is the sublimation curve or vapour pressure curve of S_M and it separates the regions of S_M and S_V. S_M and S_V exist in equilibrium along this line and the degree of freedom is one (univariant). This curve meets the vapour pressure curve S_R at 'O'. Point 'O' is common to both the vapour pressure curves of S_R and S_M and hence this is a triple point where S_R, S_M and S_V coexist. The other end of the curve is terminated at A (120°C, 0·04 mm) which is the melting point of S_M.

The dotted curve OA' is the prolongation of this curve AO and represents the vapour pressure curve of metastable S_M. It can be seen here also that the metastable phase has the higher pressure than the stable phase at the same temperature.

(3) The curve AD is the vapour pressure curve of S_L and separates S_L and S_V, which exist in equilibrium along this curve. The curve AD meets the curve OA at the point 'A' which is a triple point. At this point 'A'; S_L, S_M and S_V are in equilibrium and degree of freedom is zero. At the other end, the curve AD terminates at D which is the critical temperature and beyond which, only S_V could exist.

The dotted curve AB' is a continuation of DA and represents the metastable vaporization curve S_L. The metastable curves AB' and OB' meet at B' which is the melting point of metastable S_R. Here, metastable S_R, metastable S_B and vapour coexist and this is a metastable triple point.

(4) The transition curve, OC shows the effect of pressure on the transition point 'O'. This curve is inclined away from the pressure axis, indicating that the transition point is raised by the application of pressure (This can be explained as follows: The change from S_R to S_M is accompanied by an absorption of heat and by an increase in volume. Hence, according to Le Chatelier's principle, the increase of pressure will shift the equilibrium in such direction that accompanies with decrease of volume. Thus, with increase of pressure, the equilibrium, $S_R + Q$ (heat) $\rightleftharpoons S_M$ will be displaced towards the left forming more and more of S_R. Since the formation of S_R is accompanied by liberation of heat, the transition temperature will be raised by the increase of pressure. It also indicates that S_R is heavier than S_M, which in turn is heavier than S_L. The transition curve OC terminates at the point C beyond which S_M disappears.

(5) The curve AC is the melting point curve or fusion curve of S_M. This curve also is slightly inclined away from the pressure axis indicating that the melting point of S_M is raised by the increase of pressure. This is due to the fact that S_M melts taking its heat of fusion, which is accompanied by increase in volume : $S_M + Q$ (heat) $\rightleftharpoons S_L$. According to Le Chatelier's principle, the increase in pressure in such an equilibrium will shift the reaction towards left (backward reaction) and since this results in the evolution of heat, the melting point would be raised.

The curves AC and OC meet at the point C, which corresponds to a temperature of 151°C and 1288 atmospheres pressure. The point 'C' is a triple point where S_M, S_R and S_L are in equilibrium.

(6) Curve CG is the melting point or freezing point curve of S_R and it separates the regions of S_R and S_L.

Metastable Curves

(1) Curve OB' represents the metastable sublimation curve of S_R. It separates the regions of S_R and S_V. Along the curve OB', S_R and S_V are in metastable equilibrium. This curve is the continuation of curve BO.

(2) Curve B'C represents the metastable melting point or freezing curve of S_R. This is continuation of GC. Along this line, S_R and S_L are in equilibrium.

(3) Curve B' A is the metastable vaporization curve of S_L. Along this line, S_L and S_V occur in metastable equilibrium. It is the continuation of DA.

(4) Curve OA' is the continuation of AO. It represents the metastable sublimation curve of S_M. S_M and S_V exist in equilibrium along this curve.

III. Triple Points

(1) Triple point 'O' occurs at a temperature of 95·6°C and pressure 0·006 mm. The 3 phases in equilibrium are S_R, S_M and S_V.

(2) Triple point "A" represents the conditions where S_M, S_L and S_V are in equilibrium. It

occurs at temperature of 120°C and 0·04 mm pressure. It is the melting point of S_M.

(3) Triple point 'C' represents a temperature of 151°C and 1288 atmospheres pressure where S_M, S_L and S_R exist in equilibrium. Although the curves OC and AC are both inclined away from the pressure axis, their slopes are such that they meet at the point 'C'. This restricts the region of stability of S_M which cannot exist beyond the conditions of temperature and pressure lying outside the triangle AOC.

(4) Metastable triple point, B' which is sometimes reached when S_R is rapidly heated. The 3 phases existing in equilibrium at this point are S_R, S_L, and S_V. It represents the melting point of S_R under the metastable conditions. It occurs at the temperature of 115°C and 0·03 mm pressure.

The salient features of the phase diagram of the sulfur system are summarised in Table 2.

Table 2. The Sulfur System
Number of Components = 1
Metastable systems are shown in brackets

Name of the system	Phases in equilibrium	Degree of freedom (Variance) $F = C-P + 2$
I. Areas	One	
(a) Below BOAD	S_V	
(b) Towards the right of GCAD	S_L	$F = 1-1 + 2 = 2$
(c) Towards the left of GCOB	S_R	(Bivariant) Two
(d) AOC	S_M	
II. Curves (or lines)	Two	
(a) BO, (OB')— Sublimation curve of S_R	S_R and S_V	
(b) OA, (OA')— Sublimation curve of S_M	S_M and S_V	
(c) AD, (B'A)— vaporisation curve of S_L	S_L and S_V	$F = 1-2 + 2 = 1$ (Univarient) One
(d) AC—melting point curve of S_M	S_M and S_L	
(e) CG, (B'C)—Melting or freezing point curve of S_R	S_R and S_L	
(f) OC—Transition curve of S_R	S_R and S_M	
III. Triple Points		
(a) 'O'—the transition temperature	S_R, S_M and S_V	
(b) 'A'—Melting point of S_M	S_M, S_L and S_V	$F = 1-3 + 2 = 0$
(c) 'C'—Melting point of S_R	S_M, S_L and S_R	(Invariant)
(d) (B')—Melting point of S_R under metastable conditions	S_R, S_L and S_V	(Zero)

7·4. TWO-COMPONENT SYSTEMS
General characteristics of two-component systems

(1) The maximum number of phases in a two-component system will be four.
$$P = C-F + 2 = 2-0 + 2 = 4$$
(Maximum number of phases exist when degrees of freedom = 0. Negative degree of freedom cannot exist).

(2) The maximum number of degrees of freedom in a two-component system will be three (*i.e.*, when the system exists as a single phase).
$$F = C - P + 2 = 2 - 1 + 2 = 3$$
(3) The system will have three variables namely, temperature, pressure and concentration.
(4) The composition of all the individual phases of the system can be expressed by means of not less than two components.
(5) For constructing a phase diagram of a two-component system, a three dimensional space model is required using the three variables (*viz.*, temperature, pressure and concentration) as its coordinates.

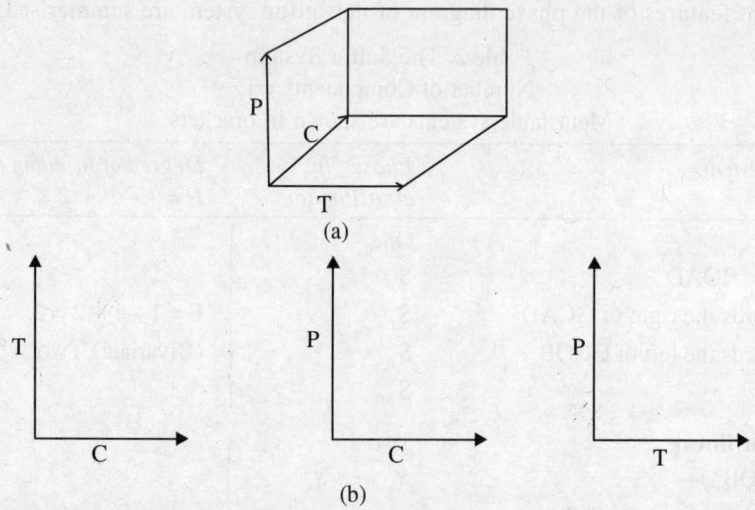

Fig.7.4. (*a*) Three dimensional space model of a two component system (*b*) Two dimensional split up models.

It is possible to split such a three-dimensional diagram into three two-dimensional diagrams by keeping the third variable in each case as constant.

In many cases, it is convenient to prepare temperature-composition (t-c) diagrams (keeping the pressure constant at atmospheric value) and such diagrams are called *isobaric*. Similarly, P-C diagrams at constant temperature are called *isothermal* and P-T diagrams at constant composition are called *isoplethal*. Any such restriction in the phase rule equation regarding the constancy of one of the variables reduces the phase rule equation to the following form :

$$F = C - P + 1$$
or
$$P + F = C + 1$$

This is known as the *reduced phase rule equation*.

Solid-liquid systems of two-components. The liquid phase generally comprises of a solution of components which could be either partially or completely miscible in the liquid state. A homogenous liquid phase when cooled may deposit one or two solid phases which could be (1) pure components, (2) solid solutions or (3) inter-component compounds.

7·4·1. Silver-lead system. This is an example of a simple eutectic system in which pure components alone separate out as solid phases. This occurs when the components are almost immiscible in the solid state and do not form intercomponent compounds.

Molten silver and molten lead mix together in all proportions resulting in a homogeneous solution. They do not react chemically to form inter-component compounds. The Pb-Ag system

PHASE RULE

consists of the following four possible phases taking part in the equilibrium : (1) solid lead, (2) solid silver, (3) solution of silver and lead, and (4) vapour.

But, however, if we study the system at constant pressure such as the atmospheric pressure, then the vapour phase can be ignored and the condensed phase rule equation can be made applicable :

$$P + F = C + 1$$

In this equation, the maximum number of phases which call exist in equilibrium is 3 and this is possible only when $F = 0$.

$$P = C - F + 1 = 2 - 0 + 1 = 3$$

Taking Ag as component A and Pb as component B, the phase diagram [temperature-composition (t—c) diagram] is shown in Fig. 7.5. In this diagram, point A represents the melting point of pure silver and point B that of pure lead. It can be seen from the diagram that addition of Pb to pure Ag lowers the melting point of Ag. Similarly, addition of Ag to pure Pb lowers the melting point of Pb. All the points on the curves, AO and BO represent the melting points of various mixtures of Ag and Pb. AO represents the melting point curve or freezing point curve of silver and shows the effect of addition of lead on the melting point of pure silver. Similarly

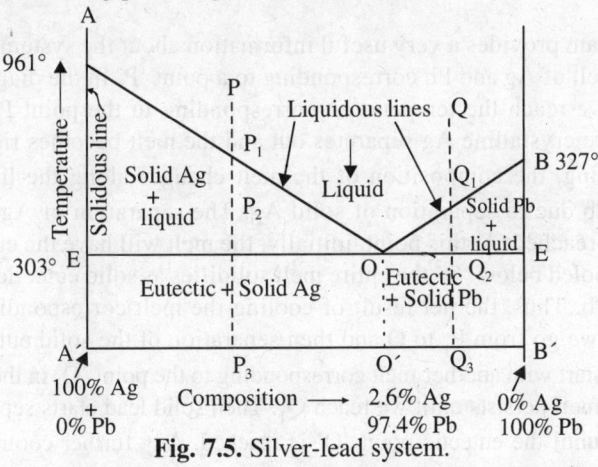

Fig. 7.5. Silver-lead system.

curve BO represents the melting point curve or freezing point curve of lead and shows the effect of addition of silver on the melting point of lead. All along AO, solid Ag and liquid are in equilibrium while along BO, solid lead and liquid are in equilibrium. Hence, AO and BO both represent univariant systems.

$$F = C - P + 1$$
$$= 2 - 2 + 1$$
$$= 1$$

The curves AO and BO intersect at a point 'O' at a temperature of 303°C. The point 'O' is called the eutectic temperature or the eutectic point and it corresponds to the composition (2·6% Ag + 97·4% Pb). This is called the eutectic composition and its melting point (*i.e.*, 303°C) is lower than any alloy of Ag and Pb. The name eutectic comes from the Greek word "eutectos" meaning "easy melting". The eutectic mixture melts mostly at the eutectic temperature (303°C) to form a liquid of the same composition whereas other mixtures melt over a range of temperatures. Because of this sharp melting point, the eutectic mixture was originally thought to be a compound. But later, on examination under microscopes having high magnification revealed that eutectic is a mixture and not a compound. The eutectic point 'O' has no degrees of freedom (non-variant). At this point, solid Ag, solid Pb and liquid coexist in equilibrium.

$$F = C - P + 1$$
$$= 2 - 3 + 1$$
$$= 0$$

The phase diagram shows five distinct regions as follows :

1. *The region above AOB*. Any composition and temperature in this region form a single, homogeneous stable liquid phase consisting of the liquid alloy or melt of Ag and Pb. In this area, the degrees of freedom are two and both temperature and composition of the liquid phase can be independently varied. Since the liquid solution phase in this area is not in contact with any solid, these solutions may be considered to be "unsaturated". If any liquid composition above AOB is cooled to a temperature where a solid phase separates out, the liquid phase then becomes saturated with respect to this solid phase.

2. *The region enclosed by the curve AOE*. In this region solid crystalline silver plus the liquid alloy are stable.

3. *The region enclosed by the curve BOE'*. In this alloy, solid crystalline lead plus the liquid alloy are stable.

4. *The region enclosed by EOO'A'*. In this region, we find Ag crystals and solid eutectic crystals are stable.

5. *The, region enclosed by E'OO'B'*. In this region, we find Pb crystals and solid eutectic crystals are stable.

The phase diagram provides a very useful information about the system. For instance, if we start with a liquid melt of Ag and Pb corresponding to a point 'P' in the diagram and cool it, the melt persists until we reach the temperature corresponding to the point P'. At this stage, a small amount of pure crystalline Ag separates out and the melt becomes richer with respect to Pb. On further cooling, the composition of the melt changes along the line P_1O as the melt becomes richer in Pb due to separation of solid Ag. The separation of Ag continues until the eutectic point 'O' is reached. At this point, initially, the melt will have the eutectic composition, but as soon as it is cooled below 'O' the entire melt solidifies as solid eutectic mixture containing 2·6% Ag + 97·4% Pb. Thus, the net result of cooling the melt corresponding to point P is the separation of Ag as we go from P_1 to O and then separation of the solid eutectic below 'O'.

Similarly, if we start with another melt corresponding to the point 'Q' in the diagram, and allow it to cool, the liquid melt persists until we reach Q_1. Then solid lead starts separating out as we go along the line Q_1O until the eutectic point 'O' is reached. Any further cooling to just below the eutectic point 'O' results in the solidification of the entire melt to form the eutectic mixture (2·6% Ag + 97·4% Pb). This behaviour of lead-silver system on cooling has been utilized in the Pattinson's process for enriching argentiferous lead with respect to silver. The enrichment of Ag—Pb solutions having less than 2·6% Ag can be affected by cooling the liquid and removing Pb as it separates out until the eutectic composition has been obtained. Then it is subjected to cupellation.

Systems giving rise to eutectic points are known as eutectic systems. The eutectic temperatures and compositions of some of the eutectic systems are given in Table 3.

Table 3. Properties of some eutectic systems

Components and their melting points		Eutectic composition and eutectic temperature	
Ag (961°C);	Cu (1083°C)	(71·8% Ag + 29·2% Cu)	778°C
Ag (961°C);	Pb (327°C)	(2·6% Ag + 97·4% Pb)	303°C
Zn (419°C);	Cd (323°C)	(67% Zn + 33% Cd)	270°C
Zn (419°C);	Al (658·7°C)	(95·64% Zn + 4·36% Al)	380·5°C
Bi (273°C);	Cd (323°C)	(60% Bi + 40% Cd)	140°C
KCl (773°C);	H_2O (0°C)	(20% KCl + 80% H_2O)	−11°C
KI (682°C);	H_2O (0°C)	(52% KI + 48% H_2O)	−22°C

Characteristics of eutectic point

1. Eutectic point represents the lowest or limiting temperature at which a liquid phase can exist in the system.

2. No other mixture containing the two components will have a melting point lower than the eutectic temperature.

3. Eutectic point has precise values of temperature and composition and it represents an invariant system.

4. If the liquid is cooled to just below the eutectic point, both the components of the eutectic simultaneously solidify without any change in the composition or temperature of the liquid phase.

5. An eutectic system can maintain its temperature constant over long periods.

6. When the liquid is cooled below the eutectic point, the components solidify in the form of small crystals intimately mixed with each other which fill in the spaces between the larger crystals of the pure components which are already separated out. Eutectic mixtures appreciably contribute towards the strength of the solid structures in case of alloys.

7. Recent electron microscopic studies show that eutectics are mixtures of the components but not their compounds.

Uses of eutectic systems

1. Eutectic mixtures are used in preparing "solders" used for joining two metal pieces together. Ex: Pb—Sn solders.

2. They are used in "safety fuses" used in buildings to protect them against fire hazards. Ex: Woods metal, which is an alloy containing (50% Bi + 25% Pb + 12·5% Sn + 12·5% Cd). This alloy melts at 65°C, and is used for plugging water sprayers in buildings. In case of accidental fire in the building, the plug made of the eutectic melts and the water forces out as a spray to put off the fire.

7·4·2. Cadmiun-Bismuth System. The equilibrium diagram for the Cd-Bi system is shown in Fig. 7.6. Cadmium has a melting point at 321°C and bismuth melts at 271°C. They are completely miscible in the liquid state. It can be seen from the phase diagram that addition of cadmium to pure liquid bismuth lowers the freezing point of bismuth to temperatures below 271°C.

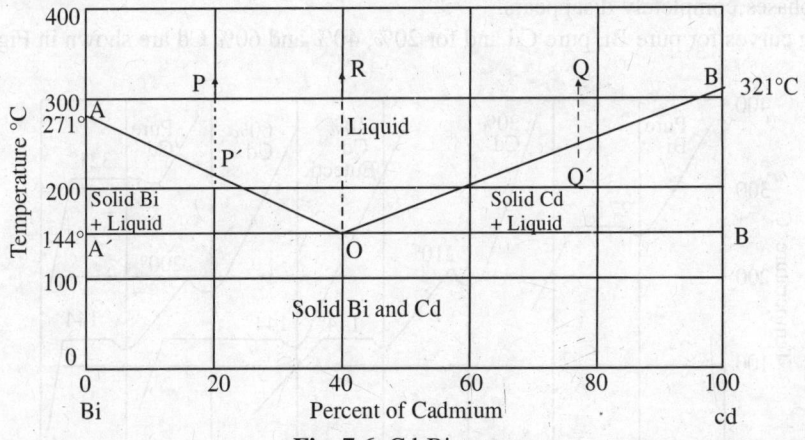

Fig. 7·6. Cd-Bi system.

Likewise, addition of bismuth to liquid cadmium progressively lowers the freezing point to temperatures below 321°C. The freezing point curve of bismuth, AO and the freezing point curve of cadmium BO, intersect at 'O' which is the eutectic point. It corresponds to a temperature of 144°C and the composition (40% Cd + 60% Bi). The eutectic point represents an invariant system and the three phases existing in equilibrium at this point are solid Bi, solid Cd and the liquid. Even on a slight lowering of temperature, the entire liquid solidifies as a mixture of minute crystals of Bi and Cd with a characteristic laminated eutectic structure.

If a liquid, having temperature and composition corresponding to the point 'P' in the diagram, is slowly cooled, the liquid phase persists until it reaches P'. At this stage, a small amount of solid Bi starts separating out and hence the liquid becomes richer with respect to Cd. On further cooling, the composition of the liquid changes along the line P'O as the melt becomes richer and richer with respect to Cd. The solidification of Bi is accompanied by evolution of heat and the cooling rate thus decreases. The separation of solid Bi continues until the eutectic point 'O' is reached. At this point, initially the liquid will have the eutectic composition and eutectic temperature (144°C) but the moment it is further cooled, solid cadmium and solid bismuth separate out simultaneously and the temperature remains constant until the entire mass has solidified.

Likewise, if a liquid of the eutectic composition (*i.e.* 40% Cd) at a temperature corresponding to point 'R' in the diagram is cooled, no separation of solid takes place until the eutectic temperature (*i.e.* 144°C) reaches, but the moment it is further cooled, complete solidification of the mass occurs at constant temperature and the solid alloy will show eutectic composition and eutectic structure throughout.

On the other hand, if a liquid having the composition and temperature corresponding to point 'Q' in the diagram is cooled, no separation of solid occurs until the point Q, beyond which solid Cd starts separating out. The solidification is accompanied by evolution of heat and the cooling rate thus decreases. At the same time, the liquid phase becomes richer and richer in bismuth and the composition of the liquid changes along the line Q'O. When the eutectic point 'O' is reached, complete solidification of the liquid takes place at constant temperature, as described earlier.

Cooling curves. A plot of temperature of a hot substance with time is known as its cooling curve. The rate of cooling of a substance depends on the temperature of the substance and its surroundings. However, if a phase change is involved during the cooling process, it occurs with an evolution of heat, which naturally interferes with its normal rate of cooling. Thus, occurrence of such phase changes are indicated by an abrupt break in the cooling curve and the slow rate of cooling also appears in the curve. Further, if during the process of phase change, the number of existing phases is such that it renders the system invariant, the temperature remains steady with time, which is shown from the horizontal part of the cooling curve. This situation continues until one of the phases completely disappears.

Cooling curves for pure Bi, pure Cd and for 20%, 40% and 60% Cd are shown in Fig. 7.7.

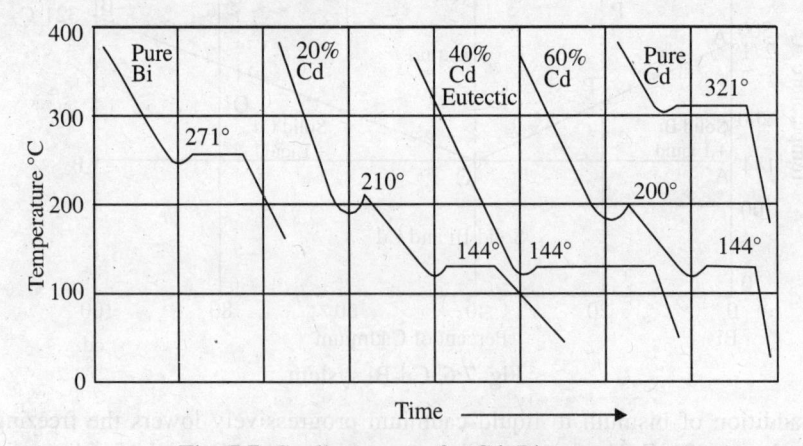

Fig. 7·7. Cooling curves for Cd, Bi system.

For preparing the cooling curves, the temperature of the system at different intervals of time are recorded with the help of a thermocouple. The first breaks in the cooling curves for 20% and 60% Cd indicate the location of the freezing point curve for these compositions. It is also evident from the curves that some super-cooling takes place before the first separation of the respective solids. The

PHASE RULE

cooling then continues at a slower rate due to the evolution of heat accompanying the separation of solids. This continues right upto the eutectic point when there is a constancy of temperature during the final solidification of the entire mass. The cooling curve for the eutectic composition is similar to those of pure substances. It is possible to construct the complete phase diagram for the system on the basis of a large number of cooling curves at various compositions.

7·4·3. Aluminium-Copper Systems. Aluminium and copper are used in binary casting alloys. The essential features of the equilibrium diagram of the aluminium-copper system (upto 60%

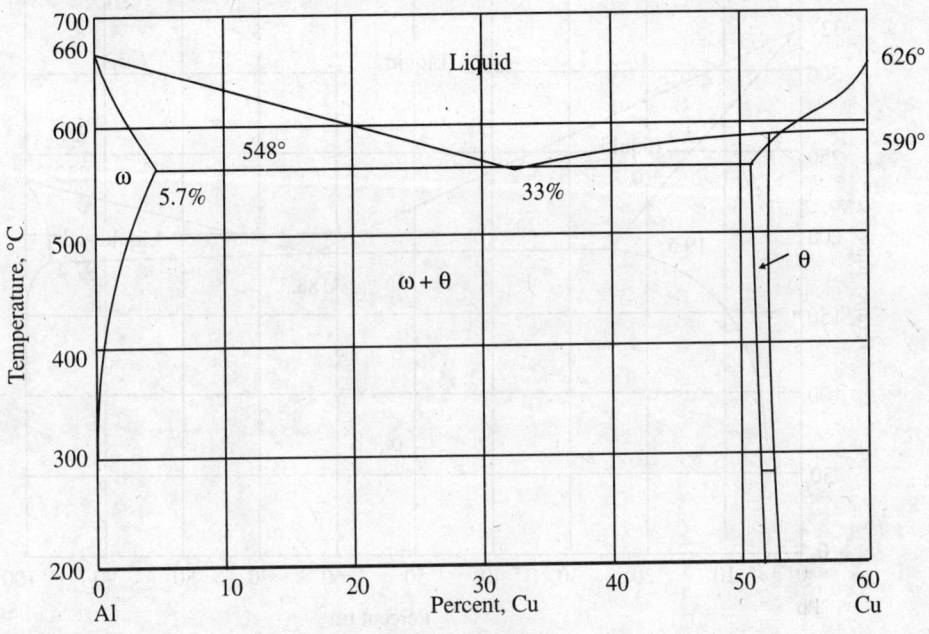

Fig. 7·8. Al-Cu System.

Cu) are shown in Fig. 7·8. It can be seen that at lower concentrations of copper, an ω solid solution of copper in aluminium is formed. At higher concentrations of Cu, an eutectic between this solid solution and an intermediate θ solid solution occurs at a temperature of 548°C. The ω solid solution formed in the eutectic solidification contains 5·7% Cu whereas the eutectic liquid contains 33% Cu, as can be seen from the phase diagram. The commonly employed copper aluminium alloy, containing 8% Cu and 92% Al, solidifies over a temperature range of about 100°C. The solid alloy contains primary ω, separated above 548°C and the eutectic of ω and θ solid solutions. This alloy accounts for about 50% of the aluminium foundry castings, which may occasionally contain small amounts of Si, Fe and Zn. These alloys have reasonably good mechanical properties and are widely used in automobile industry for crank-cases, transmission housings and also for vacuum cleaners, washing machines, etc.

Casting alloys containing 4% Cu are also in common use. It can be expected from the phase diagram that this alloy should solidify as a single-phase ω solid solution. However, at lower temperatures, θ solid solution begins to precipitate out due to decrease in solid solubility. This precipitation would result in a hardening of the alloy. These alloys generally contain Si and Fe also.

7·4·4. Lead-Tin System. Lead-tin systems are often used in soldering. A solder is a readily fusible alloy used to join two metal pieces together. The capacity of solder to join the metal parts together depends upon the formation of a surface alloy between the solder and the parts being soldered. The selection of a solder alloy depends upon the melting point desired and the metals to be joined. An alloy of Sn and Pb, known as "soft or tinner's solder" is most widely used.

The important features of the equilibrium diagram for Pb-Sn system are shown in Fig. 7·9. It can be seen from the diagram that alloys containing not more than 19·5% Sn on one side and those containing not more than 2·6% Pb on the other, solidify as solid solutions. All other alloys of these two metals solidify with the formation of the eutectic containing about 38% Pb. Most of the commercially used Pb-Sn alloys contain 25 to 75% Sn and they comprise of primary crystals of α- or β-solid solutions (depending on the proportions of the metals used) held in a matrix of tiny crystals of the eutectic.

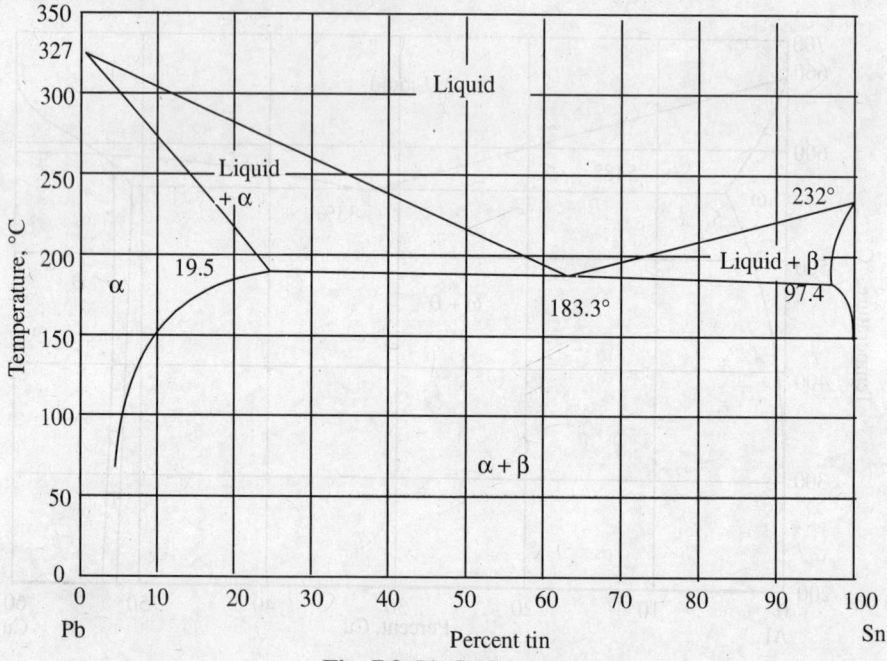

Fig. 7.9. Pb-Sn System.

The solid phase at the left-end of the diagram is the lead-rich α, which dissolves only a limited amount of tin. This solubility decreases with decreasing temperature. This limit of the solid solubility is indicated by the phase boundary between α and α + β, called "solvus". The solid phase at the right end is the tin-rich β, with only a very small quantity of lead dissolved in it. The phase boundaries on this phase diagram are given below :

Solidus I	:	Boundary between regions of α and (Liquid + α)
Solidus II	:	Boundary between regions of β and (Liquid + β)
Liquidus I	:	Boundary between regions of Liquid and (Liquid + α)
Liquidus II	:	Boundary between regions of Liquid and (Liquid + β)
Solvus I	:	Boundary between α and (α + β) regions
Solvus II	:	Boundary between β and (α + β) regions.

(The phase boundary between the liquid and the two-phase region is called "liquidus", whereas the phase boundary between the solid and the two-phase-region is called the "solidus").

The three two-phase regions, namely (L + α), (L + β) and (α + β), are separated by horizontal (isothermal) line corresponding to the "eutectic temperature" (T_e = 183.3°C). Below the eutectic temperature, the entire material is solidified for all compositions. The composition which remains fully liquid upto the eutectic temperature during cooling is called the "eutectic composition".

$$(C_e = 62\% \text{ Sn} + 38\% \text{ Pb})$$

At the eutectic temperature, the eutectic reaction taking place is as follows :

$$\text{Liquid} \underset{\text{heating}}{\overset{\text{cooling}}{\rightleftharpoons}} (\alpha + \beta)$$

In this context, "cooling" refers to the heat that is extracted from the system at the eutectic temperature whereas "heating" refers to the heat that is added to the system. As there is some heat evolution or absorption during the reaction, it is possible to add or subtract heat at constant temperature. The eutectic horizontal line is used as a "tie-line" the ends of which give the compositions of the α-phase ($C\alpha_e$ = 19.5% Sn + 80.5% Pb) and β-phase ($C\beta_e$ = 97.4% Sn and 2.6% Pb) at this temperature.

The alloy containing 67% Pb and 33% Sn is called "Plumber's solder". On cooling through its solidification range, this alloy acquires a plastic consistency similar to that of a baker's dough. In this condition, it may be moulded into the required shape in the so-called "wiping of joints" in plumbing. Further cooling makes it hard. Solders containing 67% Pb assume plastic state at about 235°C while those containing 67% Pb assume plastic state at about 243°C. Both of them finally solidify at the eutectic point (183°C). Thus, they are plastic over a temperature range of about 52 to 60°C.

The alloys containing 55 to 60% Pb melt in the range of 215 to 230°C and are quite fusible and free-flowing for ordinary soldering, such as soldering joints in electric wiring. However, the most favourite solder alloy is that containing 50% each of Pb and Sn (commonly known as half and half), which melts rapidly, flows freely and provides a bright surface finish after soldering. Despite many good properties of this alloy, it is expensive and such a high content of Sn is not essential for many applications.

The solder containing 72.5% Pb yields the greatest tensile strength but unfortunately this alloy is not fusible enough for general purpose soldering. They may be used for coating iron or steel sheets for roofing, for filling hollow castings and similar applications.

7·4·5. Iron-Carbon System. Iron has three allotropic forms as below :

Sl. No.	Name	Lattice pattern	Range of temperature in which the phase exists in equilibrium
1.	α-iron	Body-centred cubic	Upto 890°C
2.	γ-iron	Face-centred cubic	890 to 1400°C
3.	δ-iron	Body-centred cubic	1400 to 1535°C

Iron melts at 1535°C and the liquid iron boils at about 3000°C.

The solid solutions of carbon in the various forms of iron are interstitial solutions. They have the same crystal form as the iron; but the carbon atoms fit between the iron atoms in the lattice instead of replacing them. The various important micro-constituents are as follows :

Austenite is the solid solution of carbon in γ-iron. The maximum solubility is 1.7% carbon. Austenite does not exist in equilibrium below 723°C, as it decomposes on cooling.

Cementite is the name given to the compound Fe_3C. It exists at all temperatures studied from 0 to 1800°C. It undergoes only a magnetic change at 200°C. It is hard, brittle and wear-resistant.

Ferrite is the solution of carbon in α-iron. The solubility of carbon is very small, maximum being 0.035% at 723°C and the solubility at room temperature is only 0.007%. Ferrite is often considered to be practically pure α-iron but even by this minute amount of dissolved carbon, some of its properties (particularly its magnetism) are strongly affected.

Pearlite is the eutectoid of ferrite and cementite formed by the decomposition of austenite. It contains 0.8% carbon, it is a fine-grained mixture of ferrite and cementite and normally has a lamellar structure.

Martensite is the transformed form of austenite. It consists of a supersaturated solution of carbon (or Fe_3C) in highly stressed α-iron. It is magnetic. It has needle-like structure. Next to cementite, it is the hardest constituent of cast iron and steel.

Trootsite is formed when the martensite is tempered in the range 230 to 400°C. It consists of finely dispersed aggregate or superfine particles of cementite and ferrite. It is softer but tougher than martensite.

Bainite is the transformed form of austenite in the temperature range of 260 to 540°C. It has a ferrite matrix in which cementite is embedded. It is magnetic, moderately ductile and harder than pearlite and tougher than martensite.

Sorbite is formed when martensite is tempered at 400 to 600°C. It is an important constituent in some special steels such as rails. It consists of ferrite and globules of iron carbide. Its properties are intermediate between those of pearlite and troutsite.

A simplified iron-iron carbide phase diagram is depicted in Fig. 7·10.

Any point in the diagram represents a definite alloy at a definite temperature, the carbon content being shown on the horizontal axis perpendicular to the point while the temperature

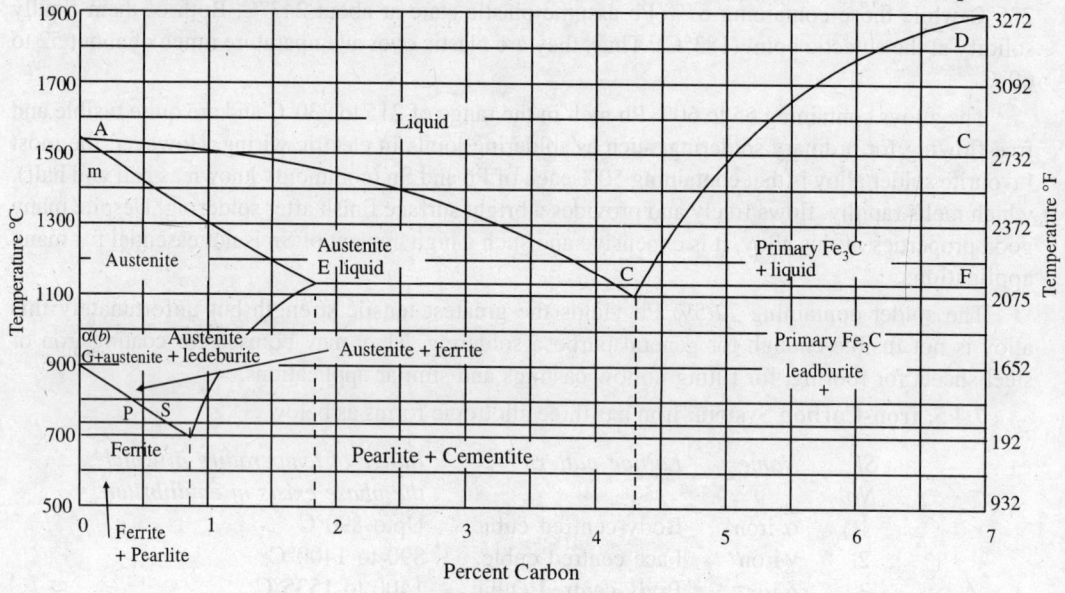

Fig. 7.10. (*a*) Fe-C System.

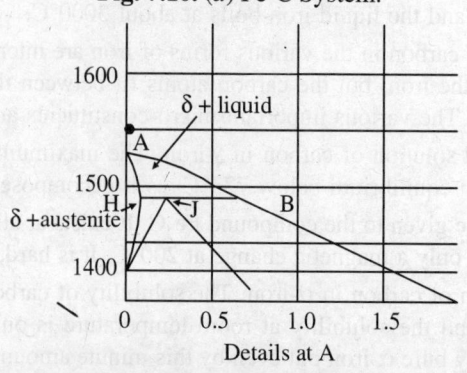

Fig. 7.10. (*b*) Delta Region.

being shown on the vertical axis perpendicular to the point. Further, whenever an alloy is heated or cooled in such a way that a line on the diagram is crossed, a phase change occurs. All alloys corresponding to any point in the region above ACD in the diagram exist in the liquid phase only. When any liquid alloy containing below 4.3% of carbon is cooled to temperatures on the curve AC, solid crystals begin to form. Hence AC is called the "liquidus" for these alloys. If the liquid alloys containing 0 to 0.55% of carbon is cooled, the first crystals formed will be of the solid solution of carbon in δ-iron. If the liquid alloys containing 0.55 to 4.3% of carbon are cooled, the first crystals formed are of "austenite", which is the solid solution of carbon in γ-iron. In the region enclosed by AEC, the liquid melt and austenite exist in equilibrium. When an alloy containing upto 4.3% carbon is cooled to the conditions represented by the lines AE or EC, it is completely converted into solid. These lines represent the "solidus" for alloys in this composition range.

When liquid alloys containing greater than 4.3% carbon are cooled to temperatures represented on the curve CD, solid crystals of "cementite" (Fe_3C) start separating out and hence the curve CD is called the "liquidus" for these alloy compositions. When an alloy containing greater than 4.3% of carbon is cooled to the temperatures on the curve represented by CF, it is completely solidified. Hence CF represents the "solidus" for these alloy compositions. The alloy containing 4.3% carbon is the eutectic alloy and it will completely solidify at C which corresponds to the temperature 1130°C, with the simultaneous formation of the solid eutectic mixture containing austenite and cementite.

Iron-carbon alloys containing upto 1.7% of carbon are classified as steels while those containing over 1.7% carbon are called cast irons.

With the help of the iron-carbon phase diagram, the course of events taking place and the various phase transformations occurring, while cooling liquid alloys having different carbon contents can be clearly understood.

7.5. USES OF PHASE RULE

1. It applies to physical as well as chemical phase reactions.
2. It provides a convenient basis for classification of equilibrium states of systems with the help of phases, components and degree of freedom.
3. It applies to macroscopic systems and hence information about molecular structures is not essential.
4. Phase rule does not take any cognizance of the nature or the amounts of substances present in the system.
5. It indicates that different systems having the same degrees of freedom behave in a similar fashion. Further, it helps in predicting the behaviour of a system under different conditions of the governing variables.
6. It helps in deciding whether the given number of substances together would exist in equilibrium under a given set of conditions or whether some of them will have to be inter-converted or eliminated.

7.6. Limitations of Phase Rule.

1. Phase rule can be applied only for systems in equilibrium. It is not of much help in case of systems which attain the equilibrium state very slowly.
2. All the phases of the system must be present under the same conditions of temperature, pressure and gravitational force.
3. It applies only to a single equilibrium state. It does not indicate the other possible equilibria in the system.
4. Phase rule considers only the number of phases but not their quantities. Even a minute quantity of the phase, when present, accounts towards the number of phases. Hence much care has to be taken in deciding the number of phases existing in the equilibrium state.
5. The solid and liquid phases should not be so finely sub-divided as to bring about deviation from their normal values of vapour pressure.

Example 1. *What metal will separate out when a liquid alloy of copper and aluminium containing 25% copper is cooled, if the eutectic includes 33.5% Cu. How many grams of that metal can be separated from 200 g of the alloy ?*

Solution

Alloy. The alloy contains 25% Cu and, *ipso facto*, 75% of Al.

∴ 200 g of alloy will contain 50 g of Cu and 150 g of Al.

Eutectic. Let the eutectic mass formed = x g

It is given that the eutectic formed contains 32.5% Cu and hence 67.5% Al.

Since the percentage of Cu in the alloy is less than that in the eutectic, all the copper present in the alloy will be included in the eutectic. Therefore, Al separates out when the alloy is cooled.

Further, if 100 g of eutectic is formed then the Cu in it is 32.5 g. In other words, if 32.5 g of Cu are present in the alloy, 100 g of eutectic is formed. Therefore if 500 g of Cu is present, the eutectic formed = $\dfrac{50 \times 100}{32.5}$ = 153.85 g.

∴ The eutectic formed, x = 153.85 g.

∴ The weight of Al separated out } = Total wt. of the alloy — wt. of the eutectic formed

= (200 – 153.85) g

= **46.15 g.**

QUESTIONS

1. (a) State phase rule and explain the terms involved with the help of suitable examples.
   (b) Identify the number of phases and components involved in each of the following systems.
      (i) Decomposition of $CaCO_3$
      (ii) Decomposition of NH_4Cl in vacuum
      (iii) Decomposition of PCl_5
      (iv) Acetic acid at room temperature
         $2CH_3COOH \rightleftharpoons (CH_3COOH)_2$
2. Discuss the application of phase rule to water system. State the significance of the triple point.
3. What are the possible phases existing in sulfur system? What do you mean by transition temperature? Discuss the application of phase rule to the sulfur system.
4. State and explain the Phase rule. Discuss the application of phase rule to a two-component system of industrial importance.
5. What is the significance of Gibb's Phase rule. Discuss its applications and limitations.
6. Discuss the application of Phase rule to Ag-Pb system. State the characteristics of eutectic point and uses of eutectic systems.
7. Write informative notes on the following :
   (a) Cd-Bi system, (b) Cooling curves, (c) Pb-Sn system, (d) Cu-Al system, (e) Eutectic point, (f) Transition temperature, (g) Triple point, and (h) Uses and limitations of Phase rule.
8. Discuss the salient features of Fe-C system.
9. When various mixtures of metal A and metal B in the form of melt were cooled, the following freezing points were observed :

%A	100	90	80	70	60	50	40	30	20	10	0
Freezing point, °C	240	225	210.5	195	180	166	150	162	175	188	200

   (a) Draw the phase diagram. What is the eutectic temperature and what is the eutectic composition? Label all the phases in the diagram.
   (b) If 200 g of this alloy containing 80% A is cooled from 280°C to 190°C, what will be the weight of the solid deposited and that of the remaining melt?
   (c) If the alloy is cooled below the eutectic temperature, what will be the weight of the solid deposited and that of the remaining melt?

8

Chemical Bonding

> "One of the most exciting endeavours in man's investigation of the world in which he lives has been his attempts to understand the basic units of matters that makeup the material world. For the chemist, the basic units are the molecules and the atoms of which they are composed. Even when these particles have been identified, the question of why and how atoms are held together into molecules remain. Some of the long-sought-for answers to these questions can now be given, and in this respect the description of the nature of the chemical bond represents the culmination on one aspect of man's efforts to unravel the secrets of matter.
> Gordon M. Barrow

A chemical bond represents a strong force of attraction between two atoms.

According to the *Lewis' octet rule,* atoms of all elements have a tendency to acquire an electronic configuration similar to that of inert gases because it represents the most stable electronic configuration. All atoms having an unstable or incomplete outer shell have a propensity to gain or lose electrons so as to acquire an electronic configuration of the nearest inert gas in the periodic table. It is this tendency of atoms to complete and hence stabilize their outermost orbit of electrons which is mainly responsible for chemical combination between the atoms.

According to a more general rule of *electron pairing,* atoms of elements have a tendency to get their electrons paired in all the occupied orbitals and thus attain greater stability.

The requirements of electron pairing are mainly accomplished by the following modes:
(1) By complete transfer of one or more electrons from one atom to another. This results in the formation of ions and hence is called *ionic bonding.*
(2) By sharing of electrons between the two combining atoms, which results in the formation of a *covalent bonding.*

8.1. IONIC OR ELECTROVALENT BOND

This type of bond is formed by the transference of electrons from one atom to the other. The two participating, atoms must be dissimilar in character, one with a tendency to lose electrons and the other having a tendency to gain electrons. The atom which loses electrons acquires positive charge and contracts while the atom which gains electrons acquires negative charge and increases in size. The cation and anion thus produced are held together by electrostatic lines of force and hence the linkage is called electrovalent bond and the compound so produced is called an electrovalent compound. These are also called polar compounds since their molecules acquire polarity due to the formation of ions by electron-transfer. In the formation of the electrovalent bond, the electrons gained or lost are always from the outermost shell and the ions produced acquire the nearest inert gas configuration (with a few exceptions).

Ex. :

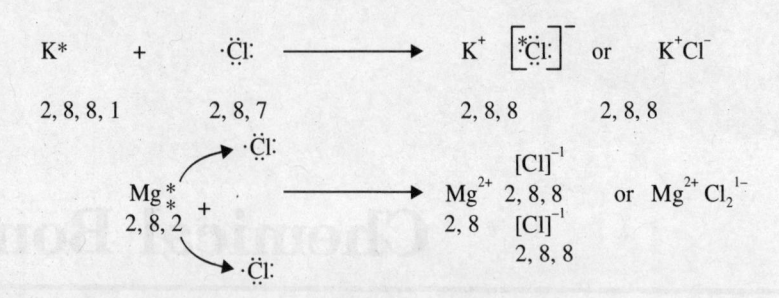

$$K^* + \cdot\ddot{C}l\!: \longrightarrow K^+ \left[\!\!\begin{array}{c}\!\!\ddot{\cdot\!\!C}l\!:\!\!\end{array}\!\!\right]^{\!-} \text{ or } K^+Cl^-$$

2, 8, 8, 1   2, 8, 7   2, 8, 8   2, 8, 8

$$Mg^*_{**} + \begin{array}{c}\nearrow \cdot\ddot{C}l\!:\\ \\ \searrow \cdot\ddot{C}l\!:\end{array} \longrightarrow Mg^{2+}\begin{array}{c}[Cl]^{-1}\\ 2, 8, 8\\ [Cl]^{-1}\\ 2, 8, 8\end{array} \text{ or } Mg^{2+}Cl_2^{1-}$$

2, 8, 2   2, 8

Fig. 8·1. Electrovalent compounds

The formation of an ionic compound is dependent upon the ease of formation of anions and cations from the neutral atoms, which in turn depends upon the following major factors.

(1) *Electronic structure:* Ions are more easily formed when the electronic structure of the ion is more stable.

(2) *Ionic charge:* The tendency of the ion formation is more when the charge on the ion is smaller.

(3) *Size:* Ion formation tendency is greater when the atom forming the ion is smaller for anion formation and larger for the cation formation.

(4) *Electron affinity:* The higher the electron affinity of an atom, the greater is the ease of formation of the anion.

(5) *Ionization potential:* The lower the ionization potential of an atom, the greater is the ease of formation of the cation.

(6) *Lattice energy:* The higher the lattice energy of the resultant ionic compound, the greater is the ease of its formation.

The electrostatic field of an ion extends equally in all directions, and hence a crystal of an ionic compound is a sort of cluster of ions in which a positive ion is surrounded spatially by a number of negative ions and a negative ion is similarly surrounded by a number of positive ions. As the process of clustering of cations and anions continues, the potential energy of the system decreases and the system becomes more and more stable. However, the geometry of the crystal sets a limit on its size. The lattice energy of a crystal of an ionic compound is defined as the energy released when the requisite number of infinitely separated gaseous cations and anions are brought together and condensed into the ionic crystal to form one mole of the compound.

Energetics of formation of an ionic substance :

The experimental determination of the crystal lattice energy is quite difficult in many cases. However, it can be calculated from the available thermochemical data by a cyclic process known as the Born-Haber cycle. The calculations involved in the Born-Haber cycle result from the principle that the sum of the energy changes occurring in a closed cycle from the same initial and final states is zero. This principle obviously is based on the First law of thermodynamics.

The formation of an ionic compound from its constituent elements may be considered to proceed in a number of simple steps, as represented by the Born-Haber cycle given below for the formation of 1 mole of NaCl from sodium and chlorine.

CHEMICAL BONDING

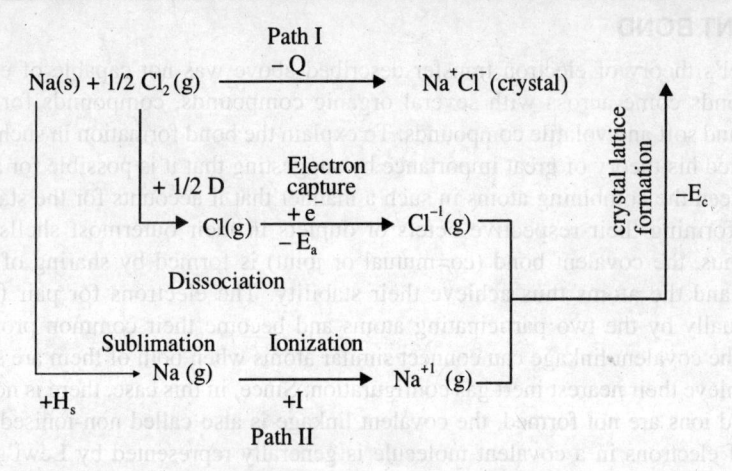

```
                        Path I
                         – Q
Na(s) + 1/2 Cl₂(g) ──────────────────→ Na⁺Cl⁻ (crystal)

           + 1/2 D        Electron
                          capture
                            + e⁻                              –E_c
              Cl(g) ─────────────────→ Cl⁻¹(g)       crystal lattice
                            – E_a                       formation
           Dissociation
    Sublimation      Ionization
           Na(g) ─────────────────→ Na⁺¹(g)
   +H_s              +I
                        Path II
```

Fig. 8·2. Born-Haber Cycle

If the amount of heat liberated in the overall reaction be represented by —Q (the negative sign indicating the liberation of heat), then it is equal to the heat of formation of NaCl as above (Hess's Law of constant heat summation). Hence, $-Q = H_s + I + \frac{1}{2}D - E_a - E_c$ where

H_s = Heat of sublimation of sodium metal
I   = Ionization potential
D   = Heat of dissociation of molecular chlorine
E_a = Electron affinity of chlorine
E_c = Lattice energy of NaCl

Amongst these energy terms, I, E_a and E_c are considerably greater than H_s and D and hence exert greater influence on the value of the heat of formation. It is obvious that the larger the negative value of heat of formation, the greater will be the stability of the ionic compound formed. Thus, from the above equation, it can be inferred that the formation of an ionic compound will be favoured by the low ionization potential (I) of the metal, high electron affinity (E_a) of the other element and high lattice energy of the ionic compound being formed (E_c).

Characteristics of ionic compounds :

(1) They consist of oppositely charged ions held together by electrostatic forces of attraction.

(2) The linkage between the ions is non-rigid and non-directional. Hence, the ionic compounds do not exhibit any type of space isomerism.

(3) They are generally soluble in water and insoluble in organic solvents.

(4) Owing to the powerful electrostatic force existing between the ions in the crystal, considerable amount of energy is required to separate the molecules from each other. Hence electrovalent compounds have got high melting points and high boiling points.

(5) When dissolved in water or in the molten state, they conduct electricity because, then the binding forces in the crystal lattice are weakened.

(6) X-ray diffraction studies indicate that the component parts of the ionic crystals are ions and not molecules.

(7) Chemical reactions of electrovalent compounds in solution are nothing but the reactions of their constituent ions. Hence, the reactions involving them take place instantaneously.

(8) They are herd crystalline compounds.

8·2. COVALENT BOND

The Kossel's theory of electron transfer described above was not capable of explaining the non-ionised bonds come across with several organic compounds, compounds formed by non-metals, gases, and soft and volatile compounds. To explain the bond formation in such cases, Lewis in 1919 advanced his theory of great importance by suggesting that it is possible for an electron to be shared between the combining atoms in such a manner that it accounts for the stability of both the atoms by forming their respective octets or duplets in their outermost shells (with a few exceptions). Thus, the covalent bond (co=mutual or joint) is formed by sharing of one or more electron pairs and the atoms thus achieve their stability. The electrons for pair formation are contributed equally by the two participating atoms and become their common property. Unlike ionic linkage, the covalent linkage can connect similar atoms when both of them are short of a few electrons to achieve their nearest inert gas configuration. Since, in this case, there is no transference of electrons and ions are not formed, the covalent linkage is also called non-ionised linkage. The arrangement of electrons in a covalent molecule is generally represented by Lewi's structure, in which the valency shells only are depicted.

Examples

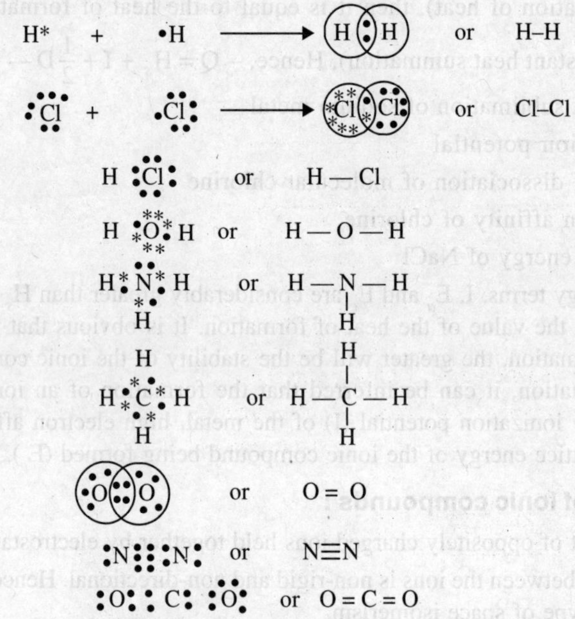

$$H^* + \cdot H \longrightarrow H(:)H \quad \text{or} \quad H-H$$

$$:\overset{..}{\underset{..}{Cl}}\cdot + \cdot \overset{..}{\underset{..}{Cl}}: \longrightarrow :\overset{**}{\underset{**}{Cl}}(\overset{*}{\underset{.}{\cdot}})\overset{..}{\underset{..}{Cl}}: \quad \text{or} \quad Cl-Cl$$

$$H : \overset{..}{\underset{..}{Cl}} : \quad \text{or} \quad H-Cl$$

$$H : \overset{**}{\underset{**}{O}} : H \quad \text{or} \quad H-O-H$$

$$H * \overset{*}{\underset{H}{N}} * H \quad \text{or} \quad H-\underset{\underset{H}{|}}{N}-H$$

$$H * \overset{H}{\underset{H}{\overset{*}{C}}} * H \quad \text{or} \quad H-\overset{\overset{H}{|}}{\underset{\underset{H}{|}}{C}}-H$$

$$:\overset{\cdot\cdot}{\underset{\cdot\cdot}{O}}(::)\overset{\cdot\cdot}{\underset{\cdot\cdot}{O}}: \quad \text{or} \quad O=O$$

$$:N:::N: \quad \text{or} \quad N\equiv N$$

$$:\overset{..}{\underset{..}{O}}::C::\overset{..}{\underset{..}{O}}: \quad \text{or} \quad O=C=O$$

Fig. 8·3. Covalent Compounds.

Polar and non-polar covalent bonds

It has been pointed out already that a covalent bond is formed by sharing of electrons between the two participating atoms. However depending upon whether the electron pairs are shared equally between them or not, a covalent bond may be non-polar or polar. In homo-diatomic molecules like H_2, Cl_2, etc., and between identical atoms with identical neighbours (as in the case of the two carbon atoms in the ethane $H_3C : CH_3$), the sharing of the electron pair is equal. Such covalent bonds are called *non-polar*. But, if the two bonded atoms are dissimilar, as in HF or HCl, or identical but in different molecular surroundings as the two carbon atoms in $H_2C : CCl_3$, the sharing of the electron pair is unequal because one of the two atoms is likely to attract the electrons more strongly than the other. The atom that attracts electrons more strongly develops some negative charge; and the other atom develops some positive charge. These fractional charges are designated as δ^- and $\delta+$; so that they are not confused with unit charges. For instance, a chlorine atom is more electron-attracting than a H atom and hence, hydrogen chloride may be depicted as follows :

$$\overline{H^{\delta+} - Cl^{\delta-}}$$

Such covalent bonds are said to be *polar;* and the bond is said to have partial ionic character. The non-polar covalent bond and the ionic bond are the two observed extremes for the distribution of electron pairs between two nuclei. Between these two extremes, several intermediate conditions of charge distribution can exist.

The relative tendency of a bonded atom in a molecule to attract electrons is denoted by its *electronegativity*. This word does not mean the actual content of negative charge, but it only means the tendency to acquire it. Thus F is highly electronegative but F^- is not.

Characteristics of Covalent Compounds

(1) Covalent linkage is rigid and directional because the participating atoms are held together by the shared pair (or pairs) of electrons. Hence, different spatial arrangements are possible and the covalent compounds exhibit stereo-isomerism.

(2) The atoms in a covalent molecule are firmly held together by the shared pair (or pairs) of electrons and hence covalent compounds do not dissociate or conduct electricity when put in water or melted. However, substances like graphite which consist of separate layers conduct electricity due to the passage of electrons between the two flat layers.

(3) Covalent compounds are generally insoluble in water and soluble in organic solvents.

(4) Normal covalent compounds are liquids or gases at room temperature. However, covalent compounds with high molecular weights exist as solids.

(5) A covalent molecule is non-polar and hence do not have external field of force. Because of this, covalent compounds generally have low melting and boiling points.

(6) Covalent compounds in solution react more slowly as compared to ionic compounds.

(7) Covalent compounds exist in the following three crystalline types :

   (*i*) Crystals of sulfur, iodine, organic compounds, etc., which are volatile, soft and easily fusible and in which the molecules are small and held together only by weak van-der Waal's forces.

   (*ii*) Those crystals in which each atom is linked with the others by covalent bonds giving rise to the formation of giant molecules. For example, in diamond, each carbon atom is linked with four carbon atoms held at the corners of a regular tetrahedron at the centre of which the carbon atom under consideration exists. Since the carbon-carbon bonds are very strong and pervade through the entire crystal, diamond is a very hard solid with a high melting point. SiC and AlN afford other examples of this type.

   (*iii*) Those crystals which exist in separate layers *e.g.,* graphite. X-ray studies revealed that graphite exists in a series of separate layers of hexagonal crystals. The carbon atoms constituting a crystal are held very closely together (1·415 Å apart), but the distance between the layers is far greater (about 3·35 Å). Hence the layers can easily slide away from each other which accounts for the soft and greasy touch as well as lubricity of graphite. Graphite is a conductor of electricity which is due to the fact that on account of all the carbon bonds being not satisfied, some of the electrons are free to move through the crystal.

8·3. EXCEPTIONS TO THE OCTET RULE

1. Atoms with less than an octet of electrons

When an atom having lesser than four valence electrons shares them to form covalent bonds, it may sometimes have less than an octet of electrons. Examples of this are afforded by ˟B˟ and Be˟˟. Boron combines with fluorine to form the covalent compound boron trifluoride, BF_3.

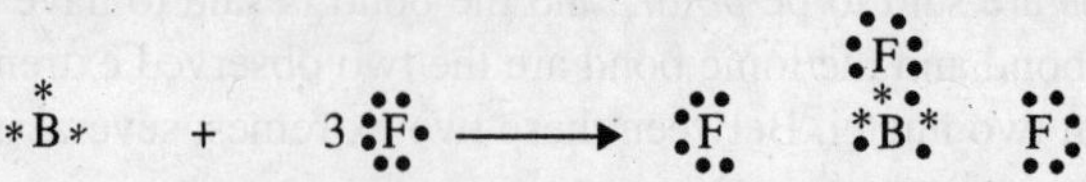

B + 3 :F· ⟶ :F: *B* :F:

Fig. 8·4. Boron-trifluoride.

In BF_3, boron has only six electrons. However, the tendency to acquire an octet predisposes BF_3 to further reaction.

When beryllium reacts with fluorine, it forms an ionic compound BeF_2; but with the less electronegative chlorine, it forms a covalent compound, $BeCl_2$, in which Be has only four electrons.

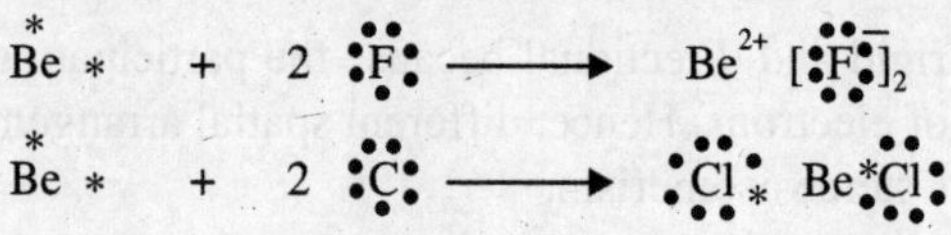

Be* + 2 :F: ⟶ Be^{2+} [:F:]$_2$

Be* + 2 :Cl: ⟶ :Cl: Be :Cl:

Fig. 8·5. Beryllium difluoride and dichloride.

2. Free radicals

Molecules in which one or more electrons are unpaired are known as free radicals. Substances consisting of free radicals are mainly characterized by paramagnetic behaviour and colour.

In free radicals, the octet rule may not be followed. In compounds like nitric oxide, nitrogen peroxide and chlorine dioxide, the N or O may not have satisfied their octets.

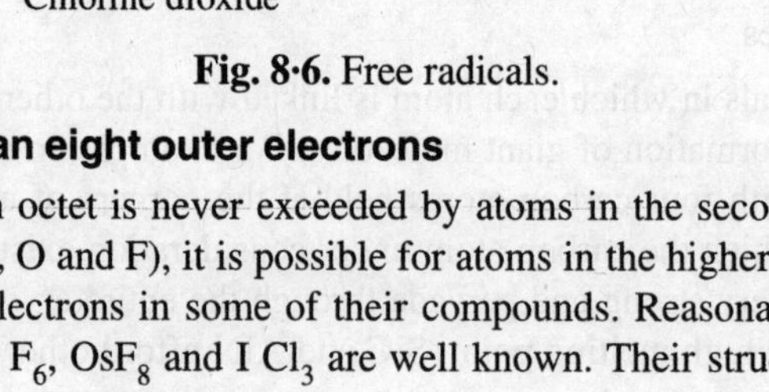

:N: :O: Colourless in gaseous state
Nitric Oxide   and blue in liquid state

:O: :N: :O: Brown gas
Nitrogen peroxide

:O: Cl :O: Yellow gas
Chlorine dioxide

Fig. 8·6. Free radicals.

3. Atoms with more than eight outer electrons

Although the electron octet is never exceeded by atoms in the second period of the periodic table (viz., Li, Be, B, C, N, O and F), it is possible for atoms in the higher periods to be surrounded by more than 8 valence electrons in some of their compounds. Reasonably stable molecules and ions such as PCl_5, SF_6, SiF_6, OsF_8 and ICl_3 are well known. Their structural formulae are given below.

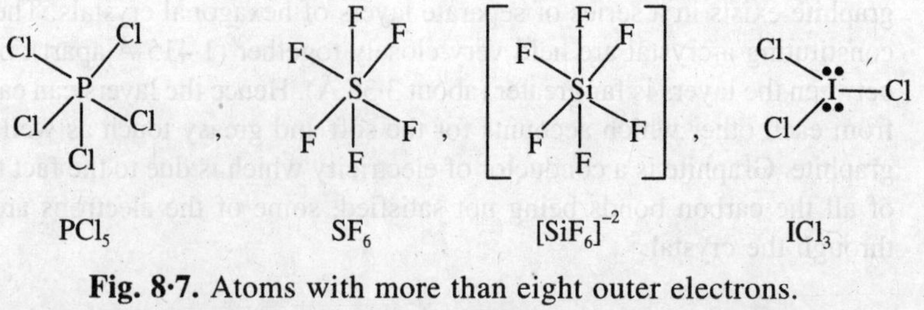

PCl_5 SF_6 $[SiF_6]^{-2}$ ICl_3

Fig. 8·7. Atoms with more than eight outer electrons.

CHEMICAL BONDING

Sidgwick believed that in compounds like PCl$_5$, SF$_6$, and OsF$_8$, the octets might have exceeded on the central P ; S; and Os atoms respectively. All these are gaseous compounds and are connected with their respective central atoms by 5, 6 and 8 covalent links respectively. However, Sugden accounted for the stability of these compounds by assuming that some or all the halogen atoms in these compounds might have joined to the central atom by linkages involving the sharing of a single electron, which he called as singlet bonds or singlets. Thus the formation of PCl$_5$, SF$_6$ and OsF$_8$ can be explained by the formation of 2, 4 and 8 singlet linkages respectively.

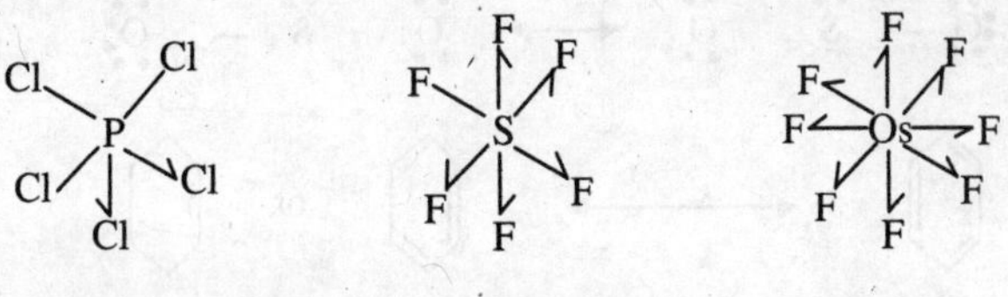

PCl$_3$ SF$_6$ [OsF$_6$

Fig. 8·8. Singlet linkages.

4. Ions of transition elements and of certain representative elements

The ions of many transition elements and heavier representative elements are not isoelectronic with noble gases; and hence do not necessarily achieve an octet of valence electrons. Examples include Fe^{2+} $(3s^2\ 3p^6\ 3d^5)$ and Pb^{2+} $(6s^2)$.

8·4. Resonance

Nitrous oxide (N$_2$O) can be represented by the following two reasonable Lewis formulae, both of which fit into the octet rule and possess the required skeleton :

$$:N=\overset{+}{N}=\overset{..}{\underset{..}{O}}: \quad \text{or} \quad :N\equiv \overset{+}{N}-\overset{..}{\underset{..}{O}}:^{-}$$

Bond lengths, Å :

Calculated 1·20   1·15   1·10   1·36
Observed   1·12   1·19   1·12   1·19

(Formal charges are shown).

The bond lengths are calculated from the single, double and triple bond covalent radii of the individual atoms. Neither set of the predicted bond lengths matches the observed values. The consequent assignment of electronic formula is resolved by the concept of resonance.

According to the concept of resonance, the structure of a molecule such as N$_2$O cannot be accurately depicted by a single Lewis formula. Instead, the molecule is depicted by two or more formulae, such as those shown above for N$_2$O, which taken together, serve as a better description than any single one. The separate formulae are called **contributing or resonance forms** and the actual molecule is said to be a **resonance hybrid** of the resonance forms. A double headed arrow, ↔, is written between the contributing forms to indicate resonance. The resonance formulae should satisfy the following two conditions :

1. The relative positions of all atoms must be the same in each formula—only the positions of the electrons may differ.

2. There should be the same number of pairs of electrons in all the contributing formulae. For instance.

:Ö::Ö: and :Ö·Ö: are not the contributing forms of O$_2$, because the number of electron pairs is different.

Other examples of resonance structures are of CO_2, SO_2 and benzene, as shown below :

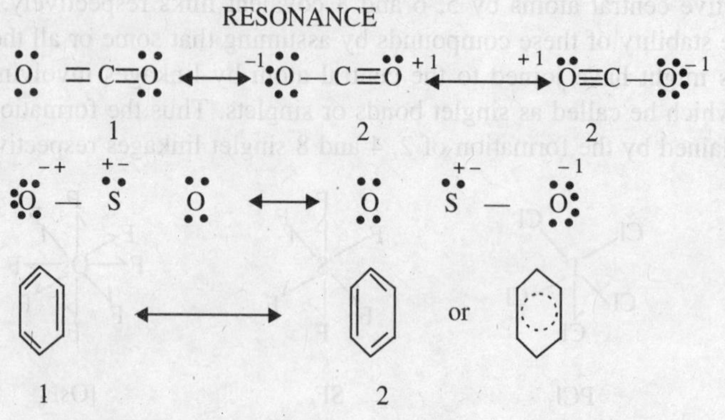

RESONANCE

[Resonance structures of CO_2 shown: structure 1 with O=C=O, structure 2 with $^{-1}$:Ö—C≡Ö:$^{+1}$, and $^{+1}$Ö≡C—Ö:$^{-1}$]

[Resonance structures of SO_2 shown as structures 1 and 2]

[Resonance structures of benzene (Kekulé structures 1 and 2, or delocalized representation)]

Fig. 8·9. Resonance structures of CO_2, SO_2 and C_6H_6.

8·5. VARIABLE VALENCY

Some elements exhibit two or more valencies during their chemical combinations with other elements. This may be due to any of the following reasons :

1. Atoms of some elements during chemical combination may lose not only the electrons from the outermost shell but also from the next inner shell. This situation exists with some transitional elements which exhibit variable valency. Let us examine the electronic configurations of Fe, Co, Ni and Cu:

Fe	2,	8,	14,	2
Co	2,	8,	15,	2
Ni	2,	8,	16,	2
Cu	2,	8,	18,	1

In case of atoms of some of these elements, since the next inner orbit is unstable, one or more electrons can be drawn out from it for valency purposes. Thus, in the case of iron, two electrons from the N shell can be removed to form Fe^{2+}. Under suitable conditions, one more electron can be drawn out from the M-shell to form Fe^{3+}. In the case of Co, it is more difficult to draw the third electron from the M—shell against the stronger attraction of the nucleus whose positive charge is one unit greater than that of iron. Hence, trivalent cobaltic ion, Co^{3+} is rather difficult to form. In case of Ni, the tendency to form Ni^{3+} is not at all there due to the further increase in the nuclear charge. In case of Cu, Cu^{1+} is formed by the loss of the only electron in the N shell. As this is unstable, one more electron may be drawn from the M shell to form Cu^{2+} having electronic configuration of 2, 8, 17.

2. In the formation of ionic compounds, the atoms of an element may lose electrons in two stages. This usually happens with elements like Sn and Pb (+2 and +4) and As and Sb (+3 and +5) which have inert electron pairs.

3. During the formation of covalent bonds, after the unpaired electrons of an atom are paired, the lone pairs of electrons may also be shared with electron deficient atoms. The variable valency exhibited by the halogens can be explained on this basis.

8·6. COORDINATE OR DATIVE BOND

This is a special type of covalent linkage in which both the shared electrons are contributed by one atom only and not one by each atom as in the case of a covalent linkage. The atom

CHEMICAL BONDING

which donates the electron pair is called a **donor** and the other atom which does not contribute any electron towards the shared electron pair and also which generally tries to pull it more towards itself, is called the **acceptor**. The pair of the valency electrons possessed by the donor is called a **lone pair**. The bond formed as above is called the **coordinate** or coordinate **covalent** or **dative bond**. Further, since such a bond always renders the molecule polar, it is also called **semi-polar bond**. Compounds containing one or more of such coordinate bonds in their molecules are termed coordination compounds. A coordinate bond is usually represented by an arrow pointing from the donor to the acceptor atom. Some examples of coordinate bond formation are given below.

COORDINATION BOND FORMATION

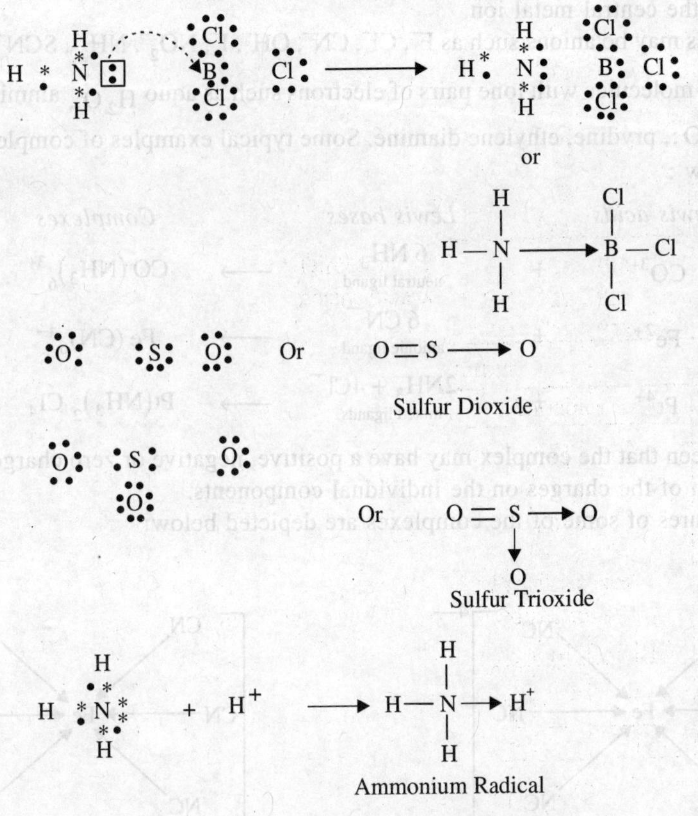

Sulfur Dioxide

Sulfur Trioxide

Ammonium Radical

Fig. 8.10. Examples of co-ordinate bond.

Characteristics of compounds containing coordinate linkages

1. The coordinate bond is rigid and directional and hence provides opportunity for exhibiting sterio-isomerism of molecules.

2. These compounds are generally insoluble in water and soluble in organic solvents.

3. The compounds containing coordinate linkages have their melting and boiling points relatively higher than those of covalent compounds and lower than those of electrovalent compounds.

4. The coordinate linkage being partly covalent in nature, the atoms are firmly held by electrons and hence these compounds do not dissociate when placed in water or melted.

5. The coordinate compounds are stable like covalent compounds. Exceptions to this are the so called molecular compounds like $BF_3.NH_3$, in which both the donor and the acceptor are capable of independent existence. In such cases, the coordinate linkage is broken readily.

8·7. COMPLEX IONS

A complex (molecule or ion) is a species comprising of several parts, each of which has some independent existence in solution. It usually consists of a positive metal ion and a number of electron-rich **ligands**. (The species that binds directly to the central metal ion is called a coordinating group or more usually, a **ligand**. The cation is a Lewis acid and the ligand is a Lewis base. The ligands may be ions or neutral molecules. They may be bound at only one position to the central ion or they may be linked at two, three, or even up to six positions, and accordingly the respective ligands are called mono-, bi-, tri- and so on upto hexadentate. Ligands that can bind at more than one position (*i.e.,* polydentate) are also called **chelates.** In Latin, chela means crab's claw. In all cases, the ligands possess unshared electrons which can be donated to the central metal ion.

The ligands may be anions such as F^-, Cl^-, CN^-, OH^-, I^-, NO_2^-, NH_2^-, SCN^-, $C_2O_4^{2-}$, CO_3^{2-} and the neutral molecules with lone pairs of electrons such as aquo $H_2\ddot{O}:$, ammine : NH_3, : PH_3, carbonyl : $C\equiv O:$, prydine, ethylene diamine. Some typical examples of complex compositions are given below :

Lewis acids		Lewis bases		Complexes
CO^{3+}	+	6 NH_3 neutral ligand	$\longrightarrow$	$CO(NH_3)_6^{3+}$
Fe^{2+}	+	6 CN^- anionic ligand	$\longrightarrow$	$Fe(CN)_6^{4-}$
Pt^{4+}	+	$2NH_3 + 4Cl^-$ mixed ligands	$\longrightarrow$	$Pt(NH_3)_2 Cl_4^\circ$

It can be seen that the complex may have a positive, negative or zero charge. The charge is always the sum of the charges on the individual components.

The structures of some of the complexes are depicted below:

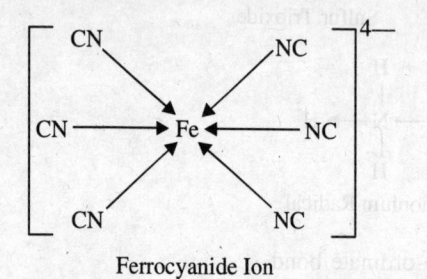

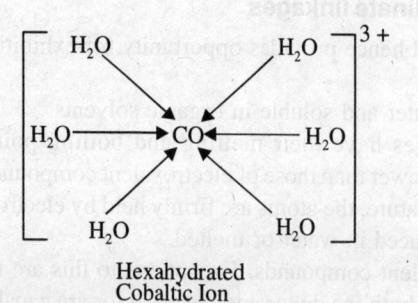

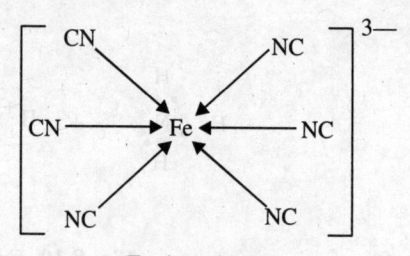

Ferrocyanide Ion

Ferricyanide Ion

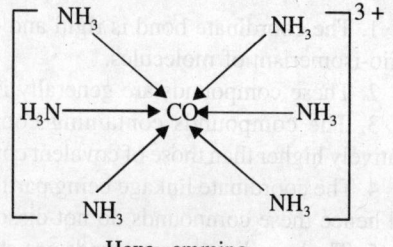

Hexahydrated Cobaltic Ion

Hexa-ammino Cobaltic Ion

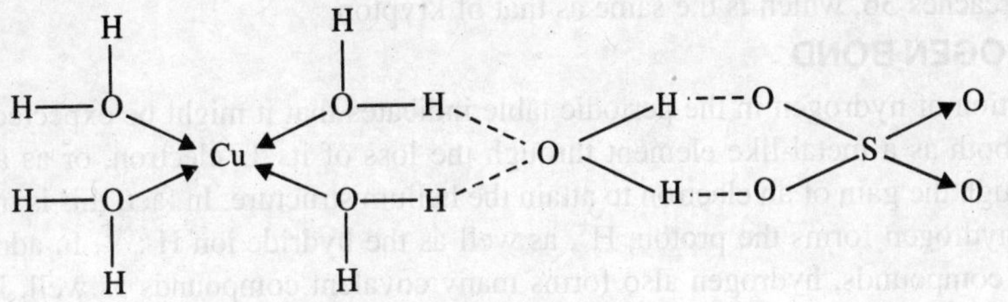

CHEMICAL BONDING

[Figure showing Copper(2) Sulfate Pentahydrate structure and Cupri-Ethylene Diamine Complex (Chelate) formation]

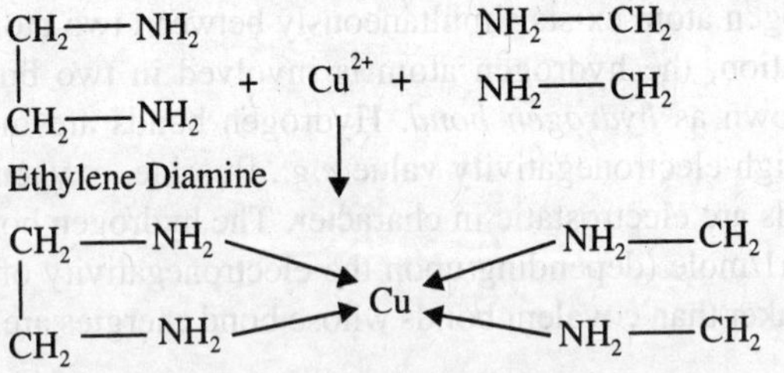

Fig. 8·11.

8·8. COORDINATION NUMBER OR LIGANCY

The number of atoms attached to the central atom is called the **coordination number** or **ligancy**. Ligancy is influenced by the charge on the cation, the charge on the ligand, the relative sizes of the cation and the ligand and the repulsion among ligands. However, the interplay of these factors cannot be quantitatively evaluated and hence reliable predictions cannot be made. The numerical value of coordination number increases as the size of the central atom increases or the size of the surrounding units decreases. It is also determined by the number of available empty orbitals or hybridized orbitals on the central atom. Some typical coordination numbers are 2 for Ag^+, 4 for Cu^{2+}, 6 for CO^{3+}, 4 for Ni^{2+} and 6 for Fe^{3+}. The same cation can have more than one ligancy, as in FeF_6^{3-} and $FeCl_4^-$.

8·9. WERNER'S COORDINATION THEORY

Alfred Werner, in 1893, proposed a theory to explain the structure of complexes. He suggested that in addition to common valence (also called primary or principal valence) certain metals possess "auxiliary" (or secondary or subsidiary) valencies. The primary valency is ionizable but the secondary type of valency is non-ionizable. Cr (III) has a valency of three in the compound $Cr_2(SO_4)_3$ and has an auxiliary valency of 6 in the complex $(CrCl_6)^{3-}$. The common valences are satisfied only by anions, but auxiliary valences are satisfied by either anions or molecules. The coordination complexes studied by Wernor behave in solution as stable units. For example, an aqueous solution of $PtCl_4(NH_3)_2$ gives no precipitate of AgCl with $AgNO_3$, and does not neutralize with H_2SO_4. Werner concluded that Cl^- and NH_3 are not present in an uncombined state but are intimately associated with Pt (IV) in a stable unit.

On the basis of the modern electronic theory of valency, the primary valency is identified as the electrovalency (ionizable) whereas the auxiliary valency as the coordinate valency (non-ionizable) given by the coordination number. The coordination compounds are formed due to the tendency of the central atom to attain the number of electrons equal or as nearly equal as possible to that of a noble gas. Hence, it behaves like an acceptor and attaches other atoms or groups or molecules by coordinate linkages as dictated by its coordination number. Thus in the formation of $Co(NH_3)_6^{3+}$, the Co^{3+} ion (2, 8, 14) combines with six NH_3 molecules by coordinate linkages, thereby gaining 12 electrons (that is, 2 electrons from each NH_3 molecule) so that the total number of electrons around the Co^{3+} ion reaches 36, which is the same as that of krypton.

8·10. HYDROGEN BOND

The position of hydrogen in the periodic table indicates that it might be expected to behave in its reactions both as a metal-like element through the loss of its $1s$ electron, or as a halogen-like element, through the gain of an electron to attain the helium structure. In fact, this is true because of the fact that hydrogen forms the proton, H^+, as well as the hydride ion $H:^{-1}$. In addition to these electrovalent compounds, hydrogen also forms many covalent compounds as well. Hydrogen can also form another type of bond known as *"hydrogen bond"*.

In many compounds, the hydrogen atom exists simultaneously between two atoms, acting as a bridge between them. In this situation, the hydrogen atom is involved in two bonds : one is a covalent bond and the other is known as *hydrogen bond*. Hydrogen bonds are most commonly found to form with atoms having high electronegativity value *e.g.*, fluorine, oxygen and nitrogen. It is now recognized that these bonds are electrostatic in character. The hydrogen bonds have bond energies varying from 3 to 10 K Cal/mole (depending upon the electronegativity of the atom; $F > O > N \approx Cl$) and hence are much weaker than covalent bonds whose bond energies are of the order of 80 to 100 K Cal/mole.

The formation of a hydrogen bond accounts for some anomalous properties observed in many compounds. Thus, anomalous boiling points and molecular weights in solution can be explained on the basis of this phenomenon. Hydrogen bonds exist in compounds such as water, HF, formic acid, NH_3, etc.

Formic acid forms a dimer in non-polar solvents which results in a molecular weight twice that of the expected value. The dimer is formed through hydrogen bonding as shown below :

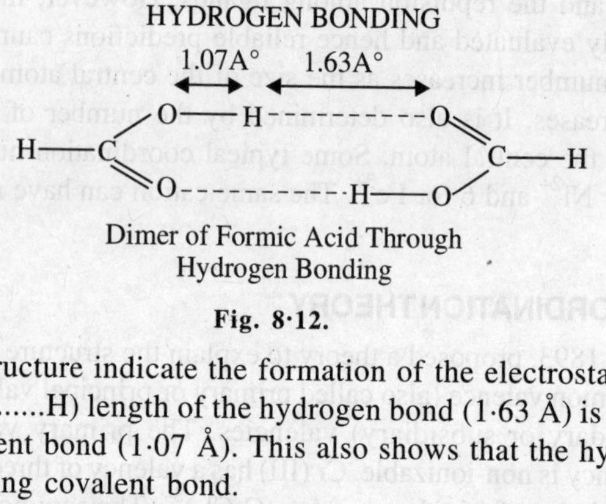

HYDROGEN BONDING

Dimer of Formic Acid Through
Hydrogen Bonding

Fig. 8·12.

The dotted lines in this structure indicate the formation of the electrostatic hydrogen bond. It may be noted that the (O.........H) length of the hydrogen bond (1·63 Å) is much greater than the (O–H) length of the covalent bond (1·07 Å). This also shows that the hydrogen bond is much weaker than a corresponding covalent bond.

One of the strongest hydrogen bonds is observed in hydrogen fluoride, whose formula should be written as $(HF)_x$, since it exists as linear or cyclic aggregates as shown below :

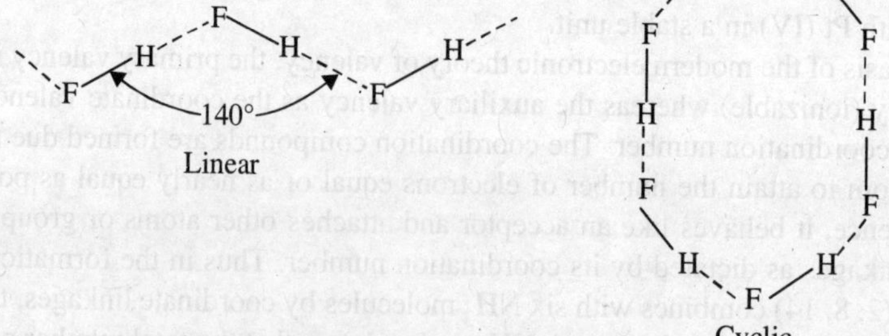

Fig. 8·13. Hydrogen Bonding in HF

CHEMICAL BONDING

The hydrogen difluoride ion, HF_2^-, possesses the strongest known hydrogen bond, in which the H atom exists amidst the two F atoms. The ion thus consists of two fluoride ions shielded from each other by a proton as shown below :

$$[F......H......F]^- \text{ or } [F^- H^+ F^-]^-$$

Although the hydrogen bond in $(HF)_x$ is stronger than the hydrogen bond in water, how can we account for the higher boiling point of water (100°C; Mol. wt. 18) as compared to the boiling point of HF (19.4°C; Mol. wt. 20)? How to account for the anomalous decrease in density of water when it freezes? How to account for the relatively high melting and boiling points of water as compared with the other hydrides of VI group elements (Table-1)? The answers for all the above questions can be derived from the geometry of the hydrogen bonded system in water. In case of HF, any one F atom can be surrounded by only two hydrogen atoms and hence can participate in only one hydrogen bond. This limitation leads to only linear or cyclic arrays, as shown Fig. 8·13. On the other hand, in ice, the oxygen atom is surrounded tetrahedrally by 4 hydrogen atoms, and so participates in two hydrogen bonds. Thus, the presence of two hydrogen bonds per H_2O molecule increases the attraction between the individual H_2O molecules. The unusually high melting point of 0°C for a molecule with a molecular weight of 18 g/mole is accounted for by the cross-linked nature of the hydrogen bonds. Since cross-linked hydrogen bonds persist to some extent even in the liquid state, water also has an inordinately high boiling point. As ice melts, the hydrogen bonding becomes more random and the molecules at first are able to move closer together. Thus, the density of water increases and reaches maximum between 0° to 4°C. Above 4°C, the increase in kinetic energy of the molecules is sufficient to cause the molecules to begin to disperse, and density steadily decreases with increasing temperature. The hydrogen bonding in ice and water are represented in Fig. 8.14.

The relatively high melting and boiling points of HF, H_2O and NH_3 can also be explained on the basis of association due to hydrogen bonding. From Table 1, we can see that in the

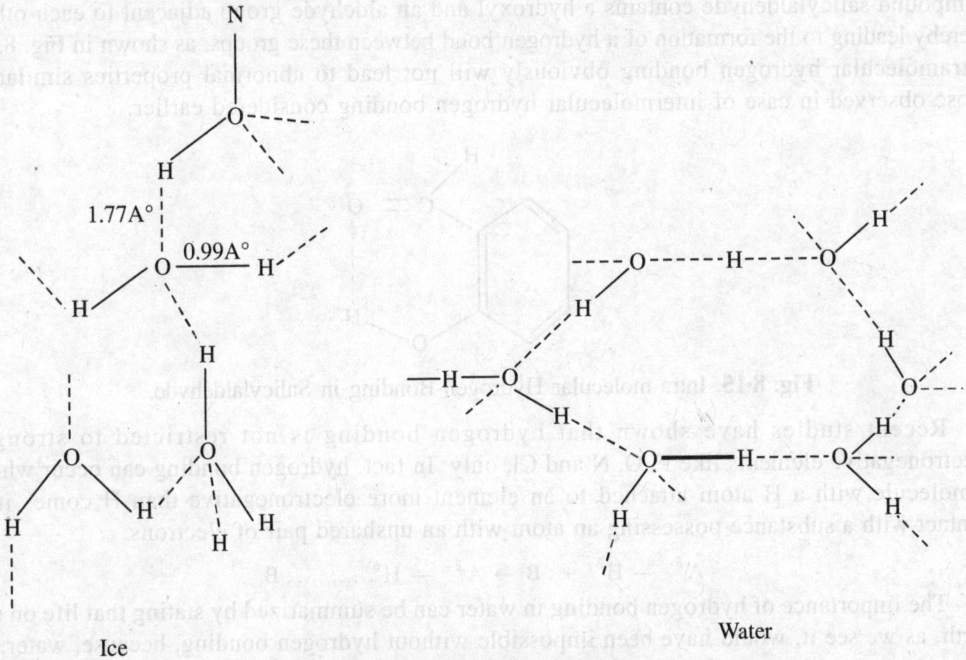

Ice   Water

Hydrogen Bonding in Ice And Water

Fig. 8·14.

Group IV series of hydrides, CH_4, SiH_4, GeH_4 and SnH_4, the melting and boiling points increase progressively with the increase of molecular weights. This is because there is no hydrogen bonding in these compounds. On the other hand, it can be seen that the first members of the hydrides of Groups V, VI and VII (*i.e.*, NH_3, H_2O and HF) show abnormal melting and boiling points relative to their molecular weights. This is due to the hydrogen bonding existing in these compounds. More energy is required to these solids or liquids in order to rupture these bonds and set the molecules free to melt or vaporize.

Table-1. Melting and Boiling points of hydrides of Group IV to VII.

	Group IV			*Group V*	
	M. Pt.	*B. Pt.*		*M. Pt.*	*B. Pt.*
CH_4	—184	—161.5	NH_3	—77.3	—33.4
SiH_4	—185	—111.8	PH_3	—135	—88
GeH_4	—165	—90	AsH_3	—113.5	—55
SnH_4	—150	—52	SbH_3	—88	—17
	Group VI			*Group VII*	
	M. Pt.	*B. Pt.*		*M. Pt.*	*B. Pt.*
H_2O	0.0	100	HF	—92.3	19.4
H_2S	82.9	—61.8	HCl	—112	—83.7
H_2Se	—64	—42	HBr	—88.5	—67
H_2Te	—51	—2	HI	—50.8	—36

So far, we have considered hydrogen bonding between different molecules (Intermolecular hydrogen bonding). However, cases of hydrogen bonding between groups within the same molecule are also known. This is called *Intramolecular hydrogen bonding*. For, instance, the compound salicylaldehyde contains a hydroxyl and an aldehyde group adjacent to each other, thereby leading to the formation of a hydrogen bond between these groups, as shown in Fig. 8.15. Intramolecular hydrogen bonding obviously will not lead to abnormal properties similar to those observed in case of intermolecular hydrogen bonding considered earlier.

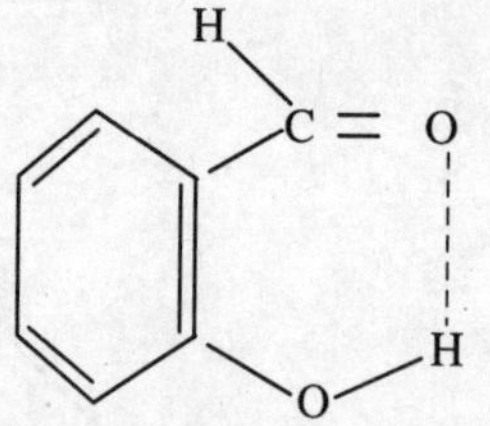

Fig. 8·15. Intra-molecular Hydrogen-Bonding in Salicylaldehyde.

Recent studies have shown that hydrogen bonding is not restricted to strongly electronegative elements like F, O, N and Cl, only. In fact, hydrogen bonding can occur where a molecule with a H atom attached to an element more electronegative than H comes into contact with a substance possessing an atom with an unshared pair of electrons.

$$A^{\delta-} - H^{\delta+} + :B \rightarrow A^{\delta-} - H^{\delta+} \dots\dots\dots :B$$

The importance of hydrogen bonding in water can be summarized by stating that life on the earth, as we see it, would have been impossible without hydrogen bonding, because, water, in that case, would have boiled at about –100°C.

CHEMICAL BONDING

8·11. VALENCE BOND THEORY

It was seen that the octet rule was inadequate to explain the various exceptions to the rule discussed above. In fact, the octet rule is nothing more than a guide for writing structures but it is not capable of answering the fundamental questions concerning the forces of interaction and the energy of the bond, nor does it give any clue as to the geometry of the molecule.

A more comprehensive theory called the valence bond theory was developed in 1927 by Heitler and London to explain the nature of the covalent bond. This theory was further modified by Pauling and Slater to take into account the directional orientation of the bond in space, which led to an accurate picture of the geometry of the molecule.

According to the valence bond theory, when two atoms come closer to each other, energy changes are produced due to the mutual rearrangements (or overlapping) of their electron cloud. As a result of this overlapping, the maximum electron density occurs somewhere between the two atoms. In case of atoms of the same element (such as 2 hydrogen atoms forming a hydrogen molecule), it would be just in the middle. The formation of a stable bond requires (1) the greater overlapping of the electron clouds and (2) the electrons should have opposite spins.

Linus Pauling and J.C. Slater extended this theory further as follows :

(1) The strength of a bond is proportional to the extent of overlapping of the electron wave functions. The strongest bond will be formed between the orbitals of two atoms that overlap to the maximum extent.

(2) The direction of the bond formed will be in that direction in which the orbitals are concentrated.

(3) Among two orbitals of identical stability (or energy), the one which is more directionally concentrated would form a stronger bond.

(4) A spherically symmetrical orbital, like S-orbital does not show preference to any direction. However, non-spherical orbitals like P– or d– orbitals prefer to form a bond in the direction of maximum electron density with the orbital.

(5) The overlapping of the orbitals of only those electrons would occur which take part in the bond formation but not with the electrons of other atoms.

These assumptions help in predicting the strength and the direction of the bonds.

Types of overlapping and orbital diagrams

(1) *S-S overlapping:*

Ex. In the formation of a hydrogen molecule, the spherical 1–s orbital of each of the hydrogen atoms overlaps with one another forming a single covalent bond.

(2) *S-P overlapping :*

Ex. Formation of a HF molecule. The electronic configuration of fluorine atom is as follows:

$$1s^2, 2s^2, 2p_x^2, 2p_y^2, 2p_z^1$$

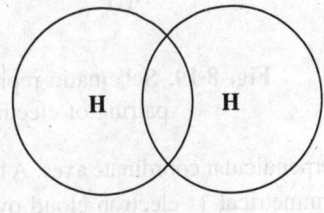

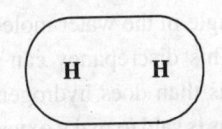

Overlap Of Orbitals
Fig. 8·16. S-S Overlapping.

It is clear that one of the electrons $(2p_z)$ is unpaired. Hence, HF molecule is obtained by the overlapping of 1-s orbital of hydrogen with the $2p_z$ orbitals of fluorine.

(3) *P-P Overlapping:*

Ex. Formation of a chlorine molecule. The electronic configuration of chlorine is as under:

$$1s^2, 2s^2, 2p^6, 3s^2, 3p_x^2, 3p_y^2, 3p_z^1$$

In this, the electron in the $3p_z$ orbital is unpaired. Hence, chlorine molecule would be formed by the overlapping of the $3p_z$ orbitals of two chlorine atoms.

On the basis of these concepts of overlapping of orbitals, let us now consider the structures of some simple molecules like H_2O, NH_3, BeF_2, BF_3, CH_4 and C_2H_4.

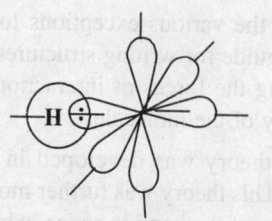

Molecules of HF produced as a result of s - p Overlapping

Fig. 8·17. s-p Overlapping.

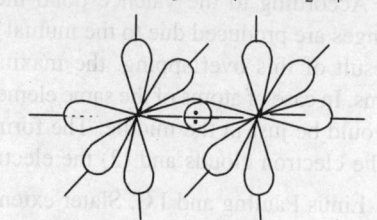

A water molecule is formed from two hydrogen atoms with 1s electrons and an oxygen atom with two unpaired electrons. The formation of a stable bond due to the overlap of electron clouds can be pictured schematically in terms of the pairing of electron spins from each atomic orbital involved in the bond formation for H_2O (Fig. 8·19). The 1s charge distribution of hydrogen is spherically symmetrical, whereas the p-orbitals of oxygen have maximum electron density along the mutually

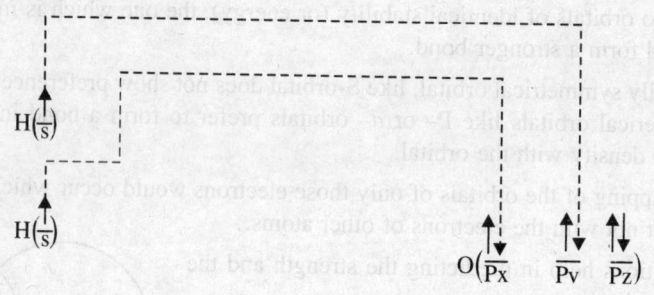

Formation of Cl – Cl molecule by the overlapping of their $3P_z$ Orbitals

Fig. 8·18. p–p Overlapping.

Fig. 8·19. Schematic representation of bond formation in H_2O, in terms of the pairing of electron spins for each bond formation

perpendicular coordinate axes. A bond is formed between hydrogen and oxygen when the spherically symmetrical 1s electron cloud overlaps the 2p orbital of oxygen, as is shown at the right in Fig. 8·20. Maximum overlap occurs along the mutually perpendicular axes, and therefore the H—O—H bond angle in the water molecule should be 90°. However, this angle is found experimentally to be 105°. This discrepancy can be explained on the basis of that oxygen has a greater affinity for electrons than does hydrogen, due to its higher effective nuclear charge. This greater affinity for electrons is said to make oxygen more electronegative, which leads to a greater relative concentration of negative charge around the oxygen atom. The net result is that the oxygen atom has a partial negative charge (δ^-) and the two hydrogens have a partial positive charge (δ^+). These positive charges lead to repulsion, causing the hydrogen atoms to move away from each other and thereby enlarging the bond angle. An alternate explanation for the bond angle in the water molecule will be given later.

CHEMICAL BONDING

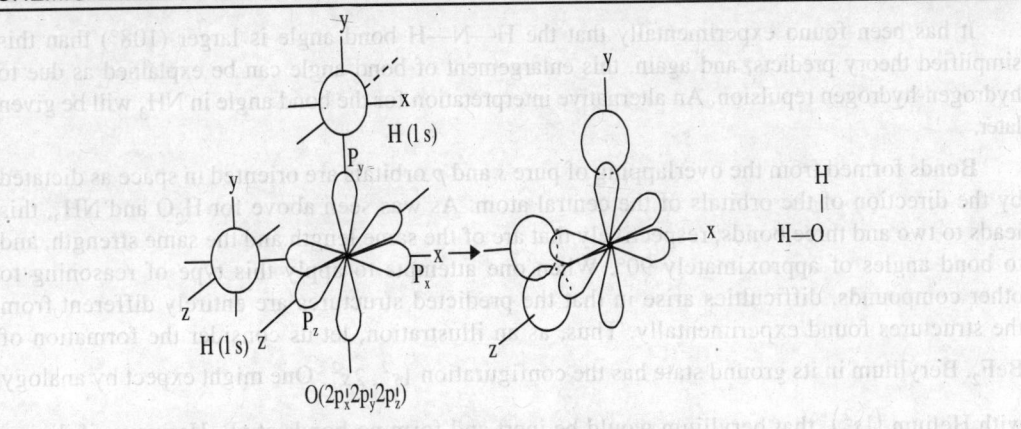

Fig. 8·20. Bond formation in H₂O, using pure s and pure p orbitals.

An analogous situation occurs in the bonding in the ammonia molecule, NH₃. Nitrogen $(1s^2, 2s^2, 2'p_x, 2'p_y, 2'p_z)$ can form three equivalent, mutually perpendicular bonds by overlap of its p-orbitals with three spherically symmetrical 1s orbitals of hydrogen. This is shown in Fig. 8·21. Again, the formation of a stable bond due to the overlap of electron clouds can be illustrated schematically in terms of the pairing of the electron spins from each atomic orbital involved in the bonding as shown below for NH₃. (Fig. 8·22)

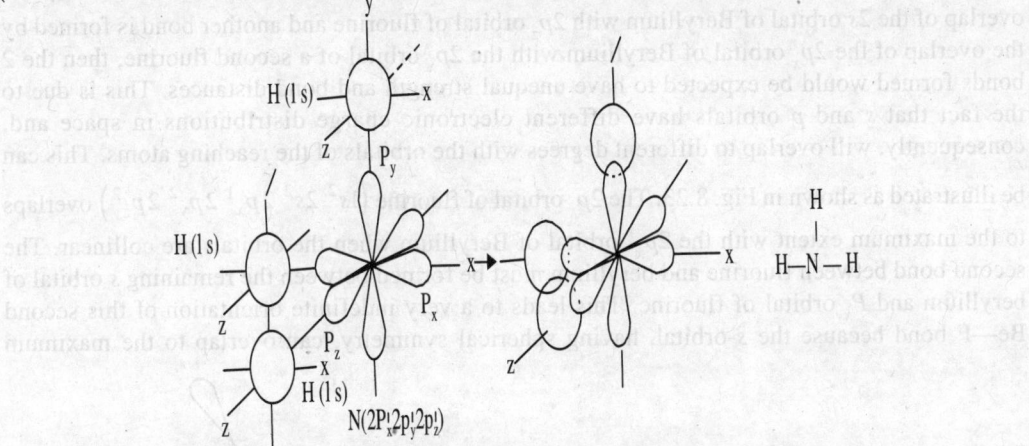

Fig. 8·21. Bond formation in NH₃ using pure s and pure p orbitals.

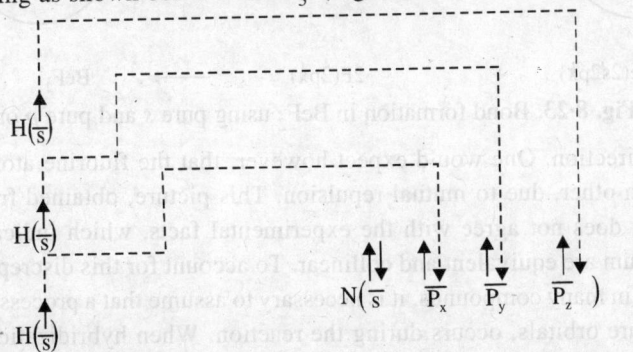

Fig. 8·22. Schematic representation of stable bond formation in NH₃ due to overlap of electron clouds, in terms of pairing of electron spins from each atomic orbital involved in the bonding.

It has been found experimentally that the H—N—H bond angle is larger (108°) than this simplified theory predicts, and again, this enlargement of bond angle can be explained as due to hydrogen-hydrogen repulsion. An alternative interpretation for the bond angle in NH_3 will be given later.

Bonds formed from the overlapping of pure s and p orbitals are oriented in space as dictated by the direction of the orbitals of the central atom. As was seen above for H_2O and NH_3, this leads to two and three bonds, respectively that are of the same length and the same strength, and to bond angles of approximately 90°. When one attempts to apply this type of reasoning to other compounds, difficulties arise in that the predicted structures are entirely different from the structures found experimentally. Thus, as an illustration, let us consider the formation of BeF_2. Beryllium in its ground state has the configuration $1s^2, 2s^2$. One might expect by analogy with Helium $(1s^2)$, that beryllium would be inert and form no bonds at all. However, if during the chemical reaction enough energy were available to promote one of the 2s electrons, an excited state of the atom would result having 2 unpaired electrons, a pure s and a pure p type, i.e., the configuration would be $1s^2 \, 2s' \, 2p'_z$. Now if one electron-pair bond is formed by the overlap of the 2s orbital of Beryllium with $2p_z$ orbital of fluorine and another bond is formed by the overlap of the $2p_x$ orbital of Beryllium with the $2p_x$ orbital of a second fluorine, then the 2 bonds formed would be expected to have unequal strength and bond distances. This is due to the fact that s and p orbitals have different electronic charge distributions in space and, consequently, will overlap to different degrees with the orbitals of the reaching atoms. This can be illustrated as shown in Fig. 8.23. The $2p_x$ orbital of fluorine $(1s^2 \, 2s^2 \, 2p_x^1 \, 2p_y^2 \, 2p_z^2)$ overlaps to the maximum extent with the $2p_x'$ orbital of Beryllium when the orbitals are collinear. The second bond between fluorine and beryllium must be formed between the remaining s orbital of beryllium and P_x orbital of fluorine. This leads to a very indefinite orientation of this second Be—F bond because the s-orbital, having spherical symmetry, can overlap to the maximum

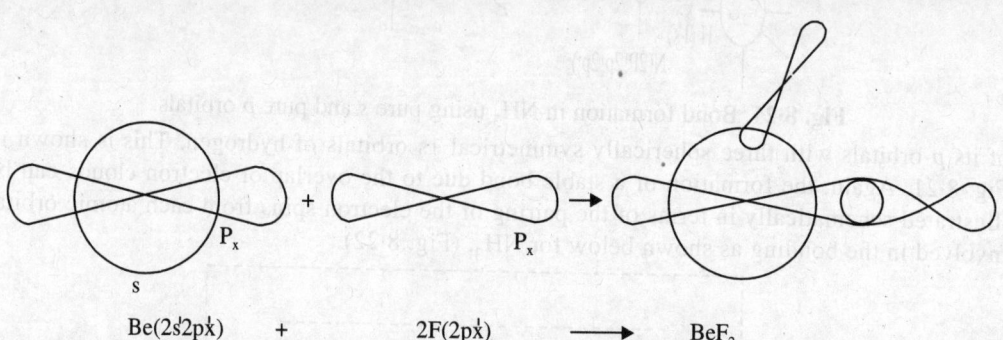

Be(2s2px)   +   2F(2px)   →   BeF$_2$

Fig. 8·23. Bond formation in BeF_2 using pure s and pure p orbitals.

extent in any direction. One would expect however, that the fluorine atoms would stay as far away from each other, due to mutual repulsion. This picture, obtained from using pure s and pure p orbitals, does not agree with the experimental facts, which indicate that the bonds in bivalent beryllium are equivalent and collinear. To account for this discrepancy between theory and experiment in many compounds, it is necessary to assume that a process called hybridization, or mixing of pure orbitals, occurs during the reaction. When hybridization occurs, one pure s and one pure p, orbital combine to form two equivalent orbitals called sp hybrid orbitals. This is shown in Fig. 8·24 below.

CHEMICAL BONDING

Pures orbital   Pure porbital   Hybrid sp orbital

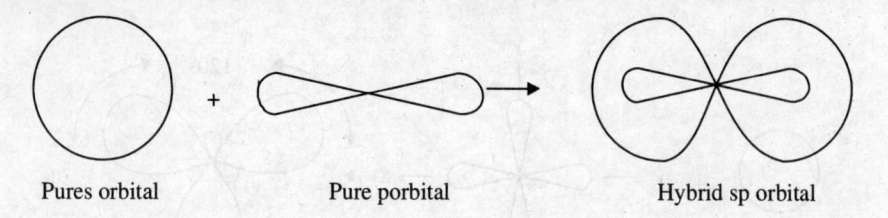

Fig. 8·24. Formation of hybrid *sp* orbitals from the combination of one *s* and one *p* orbital.

It can be seen that the hybrid *sp* orbital consists of two lobes, one having much greater extension in space than the other, and that the lobes are 180° from each other. This extension along an axis of the *sp* orbital as compared to a pure *p* orbital leads to more effective overlap and, consequently, to a stronger bond. Now, one can visualize the formation of Be—F bonds in beryllium difluoride as due to the overlap of the $2p_x$ orbitals of fluorine with the two equivalent collinear *sp* orbitals of beryllium. This is illustrated in Fig. 8·25 below.

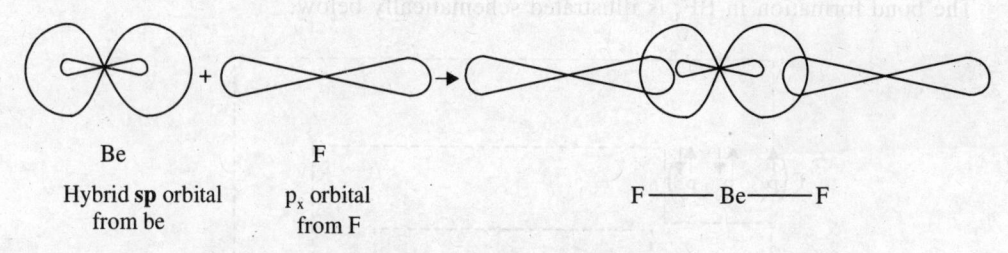

Be — Hybrid **sp** orbital from be
F — p_x orbital from F
F——Be——F

Fig. 8·25. Bond formation in BeF_2, using *sp* hybrid orbital.

The important steps in hybridization can be summarized as follows:

(1) The first step is the formation of an excited state which involves uncoupling of electrons, followed by promotion of the electron to an orbital of higher energy (from 2*s* to 2*p* in the case of Be). In the second row elements, this promotion occurs between orbitals with the same principal quantum number *n*. However, it will be shown below that promotion and hybridization can occur between orbitals with the different *n* values, since the energy difference between the orbitals is not as large as for the second-row elements.

(2) Second, the pure orbitals in the excited atom are mixed or hybridized to form equivalent orbitals which have definite orientations in space.

The first step requires energy to achieve the excited state. However, this expenditure of energy is more than gotten back, since the hybrid orbitals, because of their greater extension in space, can overlap more effectively than the pure orbitals, and this greater overlap results in a more stable bond and a resultant compound which is lower in energy.

Other types of hybrid orbitals can be formed using combination of pure *s*, *p* and *d* orbitals. In boron $\left(1s^2\ 2s^2\ 2p_x^{\ 1}\ 2p_y^{\ 1}\ 2p_z\right)$, one 2*s* electron is promoted to give the excited state configuration $1s^2\ 2s^1\ 2p_x{'}\ 2p_y{'}\ 2p_z$. The 2*s* and two 2*p* orbitals hybridize to give three equivalent sp^2 orbitals which are planar and result in a bond angle of 120°. Compounds formed from overlap with sp^2 orbitals, then, should be planar and have bond angles of 120°. Therefore, the BF_3 molecule can be thought of as being formed by the overlap of the sp^2 hybrid orbitals of boron with the pure $2p_x$ orbitals of three fluorine atoms. This is illustrated in Fig. 8·26 below.

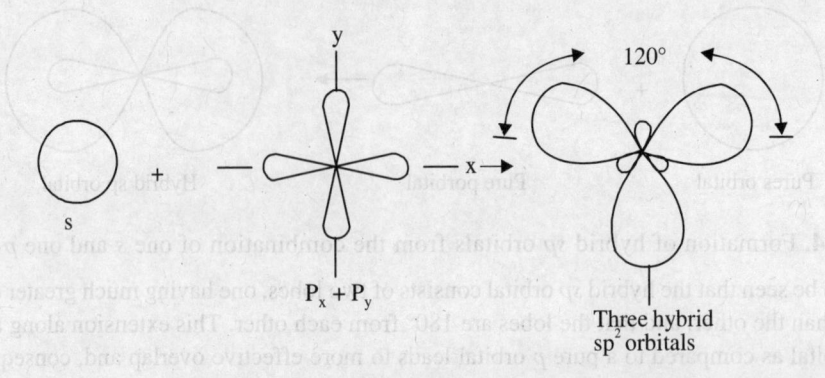

Fig. 8·26. Formation of sp^2 hybrid orbital from the combination of the s and two p orbitals.

The bond formation in BF_3 is illustrated schematically below:

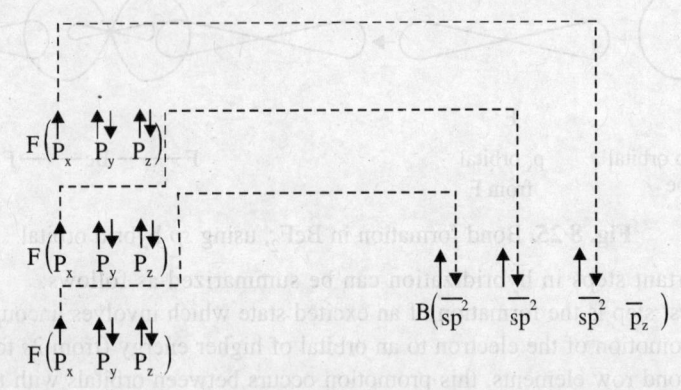

Fig. 8·27. Schematic representation of the bond formation in BF_3.

In the case of carbon $(1s^2\ 2s^2\ 2p'_x\ 2p'_y\ 2p_z)$, an excited state can be formed, $(1s^2\ 2s'\ 2p'_x\ 2p'_y\ 2p'_z)$, which hybridizes to give four equivalent sp^3 orbitals that are tetrahedrally arranged in space. The formation of the methane molecule, then can be, pictured as being formed by the overlap of sp^2 hybrid orbitals of carbon with the pure s orbitals of four hydrogen atoms as is illustrated below :

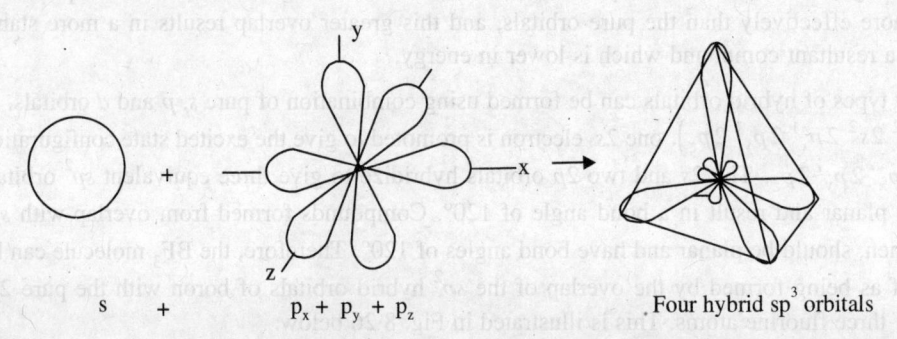

Fig. 8·28. Formation of hybrid sp^3 from the combination one s and three p orbitals.

CHEMICAL BONDING

(The smaller lobe of each hybridized orbital is not involved in bonding, and therefore, will be omitted in all the following illustrations). A schematic illustration for the bond formation in CH_4 is given below :

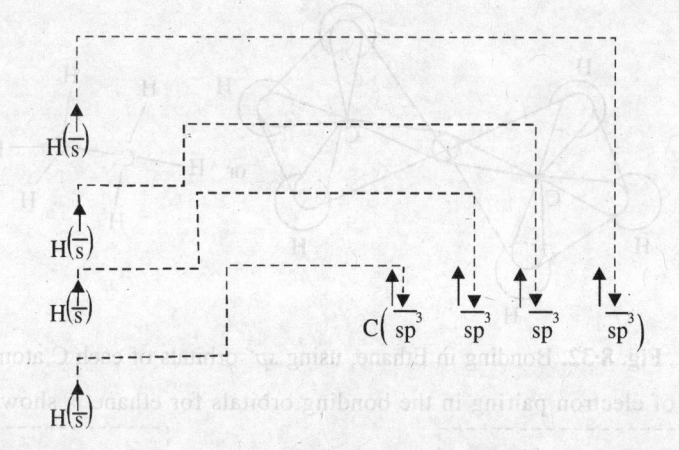

Fig. 8·29. Schematic illustration for the bond formation in CH_4.

The angle between sp^2 orbitals is about 109·5°, which is the tetrahedral angle. Therefore, one would expect that compounds formed from overlap of these orbitals with the orbitals of other atoms would have tetrahedral bond angles of 109·5°.

One could explain the bonding in water and ammonia on the basis of sp^3 hybridization. With a light expenditure of energy, an s and $3p$ orbital in oxygen can hybridize to four equivalent sp^3 orbitals. Two of these orbital contain unshared pairs of electrons. The other two overlap with the $1s$ electrons of hydrogen atoms to give a structure that has a bond angle of 109·5°. In this structure, the 2 hydrogens and the 2 unshared pairs of electrons are located at alternate apices of a tetrahedron. This is shown in Fig. 8·30 below. In a similar manner, the nitrogen atom, in the ammonia, molecule can be thought of as having four sp^3 orbitals, three of which are bonded to the $1s$ orbital of three hydrogen atoms and the fourth sp^3 orbital contains one shared electron pair which is also located at an apex of the tetrahedron. This is shown in Fig. 8·31.

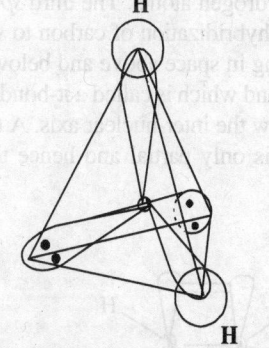

Fig. 8·30. Bonding in H_2O, using hybrid sp^3 orbitals of Nitrogen.

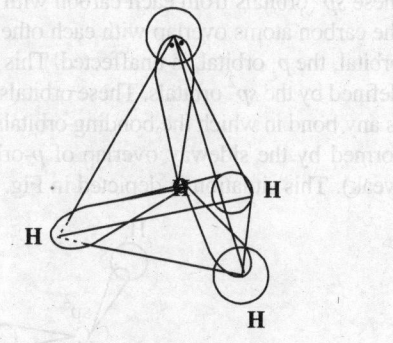

Fig. 8·31. Bonding in NH_3 using hybrid sp^3 orbital of oxygen.

The bonds in ethane are formed from the overlapping of three of the four sp^2 orbitals from each carbon atom, with the $1s$ orbitals of three hydrogen atoms and the fourth orbital of each carbon overlapping each other. This leads to a single, or σ, bond between the carbons. A σ bond is any bond in which the bonding orbitals overlap along the inter-nuclear axis. Sigma (σ)

bonds are strong bonds formed as a result of maximum overlapping of *s-s*, *s-p* or *p-p* orbitals along their internuclear axis. The structure can be depicted as shown in Fig. 8.32. The schematic

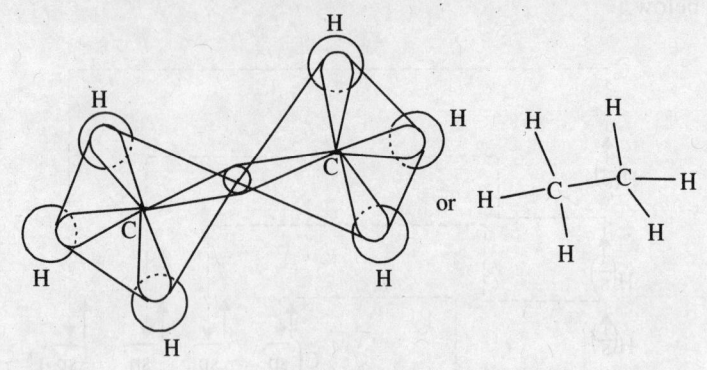

Fig. 8.32. Bonding in Ethane, using sp^3 orbitals of each C atom.

representation of electron pairing in the bonding orbitals for ethane is shown in Fig. 8.33.

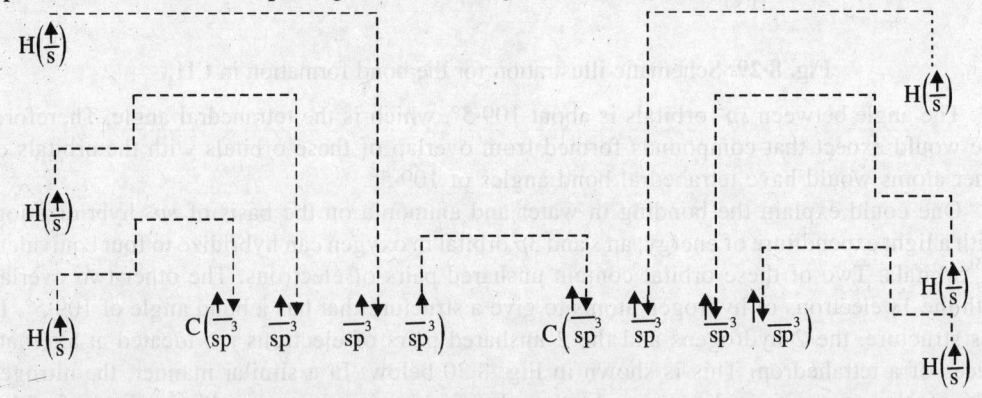

Fig. 8.33. Schematic representation of the electron pairing in the bonding orbitals for ethane.

In ethylene, the carbon atoms hybridize to form three equivalent sp^2 orbitals which are planar and make an angle of 120° with each other. The carbon-hydrogen bonds are formed by overlap of two of these sp^2 orbitals from each carbon with the 1s orbitals of two hydrogen atoms. The third sp^2 orbital of the carbon atoms overlap with each other to form a σ bond. The hybridization of carbon to sp^2 left one orbital, the p_z orbital, is unaffected. This p_z orbital is left projecting in space above and below the plane defined by the sp^2 orbitals. These orbitals can overlap to form a bond which is called a π-bond. A π-bond is any bond in which the bonding orbitals overlap above and below the inter-nuclear axis. A (π)-bond is formed by the sideway overlap of *p*-orbitals. The overlapping is only partial and hence the bond is weak). This situation is depicted in Fig. 8.34.

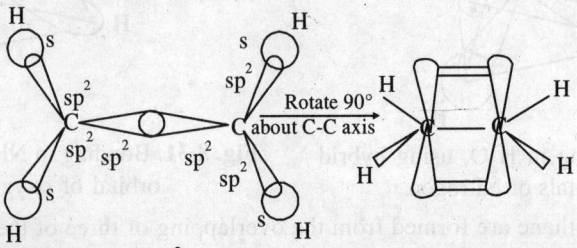

Formation of σ bond from overlap of sp^2 with sp^2 and sp^2 with s orbitals

Pure p_z orbitals overlap to from π bonds

Fig. 8.34. Bonding in ethylene, using three hybrid Sp^2 orbitals and one pure *p* orbital of each C atom.

CHEMICAL BONDING

The electron pairing in ethylene occurs as under : (Fig. 8·35).

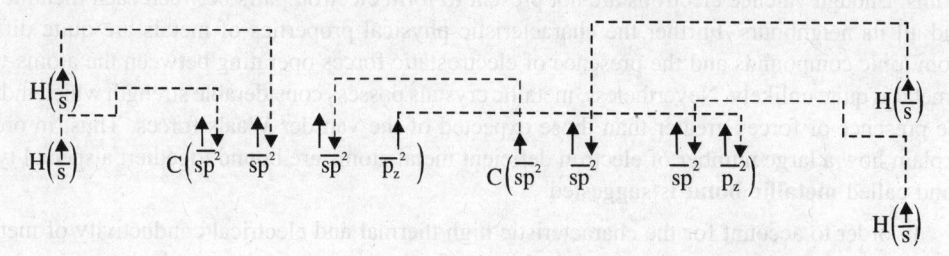

Fig. 8·35. Schematic representation of electron pairing in ethylene.

The π bond, due to its less effective overlap is weaker than σ bonds, and this explains the greater reactivity of alkenes as compared to alkanes. The double bonds represented by the two electron-pair bond in the Lewis structures consist of a σ and a π bond.

Hybridization involving d-orbitals

Elements in the second period of the periodic table and beyond, where d-orbitals begin to become involved in bonding, may form new types of hybrid orbitals. These hybrids involve mixing of pure d orbitals with s and p orbitals.

One such hybrid orbital is that formed from one d-orbital, one s orbital and two p orbitals. The four equivalent orbitals so formed are called $d\,sp^2$ hybrid orbitals, and they are coplanar and at right angles to one another. Ex:— $Ni(CN)_4^{-2}$.

Another type of hybridization involving d orbitals is the one formed from the combination of two d orbitals, one s orbital and three p orbitals. This is known as d^2sp^3 hybrid orbitals and leads to six equivalent orbitals which are directed towards the apices of a regular octahedron. Ex:—$[PtCl_6]^{-2}$.

The use of dsp^2 and d^2sp^3 hybrid orbitals in bonding is found in many complex or coordination compounds.

8·12. METALLIC BOND

Nearly three-fourths of the elementary substances are metals. Metals are characterized by properties such as good thermal and electrical conductivity, a bright appearance called metallic lustre, high malleability, ductility and tensile strength. In order to account for the various characteristics of the metallic state two theories have been advanced :

1. The quantum-mechanical theory of metals developed by Sommerfeld, Bloch and others which is comparable with the molecular orbital approach to molecular structure, in as much as the electrons in a metallic crystal are regarded as belonging to the crystal as a whole rather than to individual atoms.

2. Pauling's theory of metallic bonding which involves resonance between a number of structures with one-electron and electron-pair bonds. According to Pauling, the metallic bond is essentially of a covalent character. This approach is similar to the Heitler-London concept of electron-pair bonds in molecular structures.

Metals possess a unique structural arrangement of atoms which is widely different from that in non-metals. Most metals crystallize with an atomic arrangement in which the atoms acquire as many neighbouring atoms as is geometrically possible. X-ray investigation of metals indicate that the closest packing arrangement of atoms may produce the following three forms of crystals :

1. Face-centred cubic lattice
2. Close-packed hexagonal lattice
3. Body-centred cubic lattice (occasionally).

The close-packing structure prevailing in metals results in a high coordination number, usually 8 or 12. These high coordination numbers and the fact that atoms of metals are relatively poor in valence electrons, rule out any possibility of the existence of normal covalent linkages between metal atoms. Enough valence electrons are not present to form electron pairs between each metallic atom and all its neighbours. Further the characteristic physical properties of metals are quite different from ionic compounds and the presence of electrostatic forces operating between the atoms within a metal is quite unlikely. Nevertheless, metallic crystals possess considerable strength which indicates the presence of forces greater than those expected of the van der Waals forces. Thus, in order to explain how a large number of electron deficient metal atoms are bound together, a special type of bond called **metallic bond** is suggested.

In order to account for the characteristic high thermal and electrical conductivity of metals, it was proposed that metal atoms are packed in the form of ions in a close-packed structure and that the electrons acting as a gas are diffused throughout the crystal lattice. Pauli suggested that the free electrons are restricted to a set of continuous or partially continuous energy levels in such a manner that they are identified not with any particular atoms but rather with the crystal as a whole. Thus the metallic solids may be visualized as a sort of collection of positive atomic cores embedded in a "sea of electrons" which serves as an "electrostatic glue" for holding the positively charged atoms in the metal crystal.

On the basis of this assumption almost all the general physical properties of metals such as metallic lustre, malleability, ductility, high thermal and electrical conductivity, high melting and boiling points can be explained. The mobile sea of electrons count towards the electrical conductivity and lustre of metals. A beam of light striking the metal surface excites electrons, which are free to oscillate. When they revert back to their normal positions, energy is emitted in the form of light which appears to be reflecting from the surface thus counting towards the bright lustre of the metals. When a shearing stress is applied, the metal crystals change their shape and this distortion results in a shift in the metal lattice along a plane. The metal lattice is readily deformed without any visible change in the relative disposition of metal cores. This explains the malleability of metals.

Bloch's theory

Bloch and others further extended this theory, according to which the nucleus of each atom in a metallic crystal is surrounded by a series of concentric zones, each representing some definite potential. Such a configuration is manifested due to the wave nature of the electron, which restricts the electron energies to certain "permitted" bands called Brillouin zones, between which some empty bands of "forbidden" energies exist. A large amount of energy is needed to cross these empty bands. The inner electrons do not participate in interatomic bonding. The outermost zones of highest potential may overlap to some extent due to the close packing of metal atoms. The high-energy electrons which exist in the overlapping zones are considered to be mobile and are no longer identified with any particular atoms but belong to the entire crystal as a whole. It is these mobile electrons which bring about the metallic bond. The overlapping zones known as molecular energy levels are governed by the Pauli's exclusion principle. Accordingly, not more than two electrons can enter the same molecular orbital and, that too, only if they have opposite spins.

Pauling's theory

Pauling's approach to the metallic bonding is similar to the valence bond theory for covalent bonding. According to this, the structure of metals and alloys may be described in terms of covalent bonds that resonate amongst the alternate interatomic positions in the metals.

Even though the number of nearest neighbours of an atom exceeds its valence electrons, the resonance permits the existence of normal covalencies resonating between all the possible equivalent pairs of atoms. This theory has been proved to be valuable in explaining more satisfactorily the properties such as lattice energies, bond distances and magnetic behaviour and also in correlating them with the valence theory.

CHEMICAL BONDING

Delocalized orbitals

The representations of the formation of σ and π bonds in ethylene (vide Fig. 8·36) indicate that the *p*-electrons are fixed in the region between the carbon nuclei. Hence, they are called **localized** electrons. However, if we consider a molecule like butadiene, $H_2C=CH—CH=CH_2$, which has a conjugated double bond system (*i.e.* alternating single and double-bond arrangement), we find a different situation. The butadiene molecule (depicted in Fig. 8·36a) may be considered

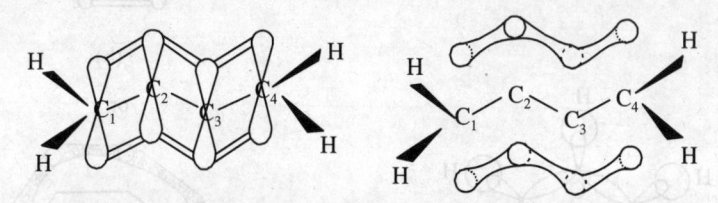

Fig. 8·36. A molecule of butadiene $H_2C=CH—CH=CH_2$
(a) Overlap of the pure *p*-orbital of each carbon atom
(b) Delocalization of electrons in the π-orbitals.

to have been formed from the overlap of sp^2 planar orbitals of carbon with the 1s orbitals of hydrogen and sp^2 orbitals of the other carbon atoms. This gives rise to the planar zig zag carbon chain indicated by solid lines in the figure. The *p*-orbitals have lobes extending above and below this zig zag plane, just as in the case of ethylene discussed earlier. However, an important difference arises in the manner in which these *p*-orbitals can overlap as compared with ethylene. Owing to the fact that there are four *p*-orbitals on the four consecutive carbon atoms, electrons are not only capable of overlapping between carbon atoms 1 and 2, and between carbon atoms 3 and 4, but also between carbon atoms 2 and 3, and 1 and 4. Consequently, the *p*-electrons can be no longer considered localized between any particular two carbon atoms, but that a truer picture of the overlap is obtained by considering the electrons free to move in the entire region between carbons 1 and 4. The electrons then are said to be **delocalized** and are pictured as a region of charge density above and below the molecular plane extending along the entire length of the chain, as shown in Fig. 8.36(*b*). This delocalization of the *p*-electrons leads to a lowering of the energy and therefore, to an increased stability of the butadiene molecule. This increased stability is measured by the delocalization energy and is exactly equivalent to the resonance energy discussed earlier. In fact, the description for the two butadiene structures in terms of *p*-orbital overlap corresponds to the following two resonance forms, and the true state of the molecule can be better described as the delocalized form, as depicted in Fig. 8·36(*b*).

$$H_2C=C-C=CH_2 \leftrightarrow \overset{+1}{H_2C}-\overset{-1}{C}=\overset{-1}{C}-\overset{..}{CH_2} \leftrightarrow \overset{..}{H_2C}-\overset{-1}{C}=C-\overset{+1}{CH_2}$$

Benzene is another example in which a conjugated double bond system exists, but in this case, the conjugation is around the ring. The plane of the benzene molecule is formed by overlap with sp^2 orbitals of the six carbon atoms. This leads to the formation of six carbon-carbon bonds in the form of a hexagon, Overlap of the remaining sp^2 orbitals of the carbons with the 1s electron of the six hydrogen atoms form six carbon-hydrogen σ bonds, also in the plane of the molecule, as shown in Fig. 8·37(*a*). The remaining pure *p*-orbitals from the carbon atoms overlap to from π-bonds, above and below the plane [Fig. 8·37(*b*)]. Just as in the case of

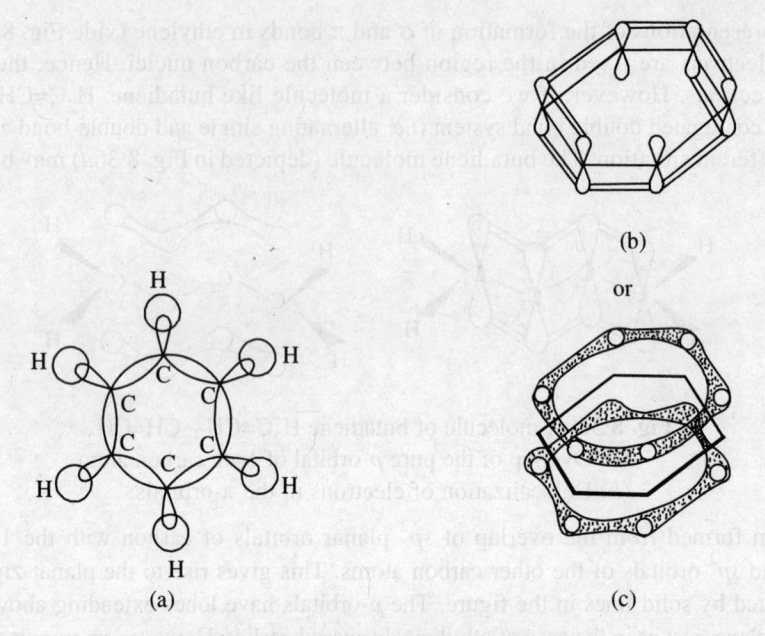

Fig. 8·37. Benzene molecule illustrating :
(a) overlap of the sp^2 orbitals of the coplanar carbon atoms and with s-orbitals of coplanar hydrogen atoms.
(b) the overlap of pure p-orbitals of the carbon atoms, which are perpendicular to the plane containing carbon atoms.
(c) the delocalization of electrons in the π-orbitals.

butadiene, these p-orbital electrons become delocalized and are free to move around the ring, with the result that the charge density above and below the molecular plane has the shape of hexagonal streamers [Fig. 8·37(c)]. Thus, the localization of the electron density between the carbon atoms leads to the two resonance forms, which are of higher energy than the structure

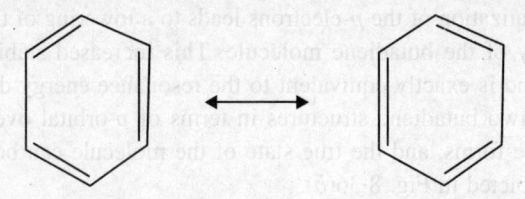

Fig. 8·38. Resonance forms of benzene molecule.

with the delocalized orbital. This difference in energy is called the delocalization or resonance energy, which contributes towards greater stability of the benzene molecule. This true structure, obtained from the delocalized orbital treatment just described, agrees much better with experimental results for the bond energies and bond distances, than those predicted with the structures having alternate single and double-bonds.

Molecular orbital theory

Hund and Mullikan developed an alternate theory to describe the formation of the chemical bond and this is known as the molecular orbital theory. In valence bond theory, a bond is supposed to have been formed when, for instance, two hydrogen atoms are brought close enough together to achieve maximum overlap of the two 1s atomic orbitals, which then leads to a stable hydrogen molecule with a given internuclear distance. Thus, in valence bond theory, the inner atomic orbitals from each atom forming the bond are undisturbed, *i.e.,* each atom retains its own identity. When they form a bond, only an electron from each bonded atom loses its identity and moves in the outer atomic orbitals of both bonded atoms. However, molecular orbital theory begins with the nuclei of the bonded atoms, stripped of all their electrons, already at their equilibrium internuclear distance. This structure consists of quantized molecular orbitals of varying energy levels, surrounding both nuclei. The molecular orbitals are formed when the atoms to be bonded come together so that the individual atomic orbitals coalesce into molecular orbitals. Then according to molecular orbital theory, the orbitals of the bonded atoms lose their individual identity. Electrons (which are equal to the sum of the electrons originally present in the atoms) are placed one at a time into the molecular orbitals to obtain the electronic configuration of the molecule. Similar to the case of electrons in atoms, the electrons in molecules fill the lowest lying molecular orbital first. The number of electrons in a molecular orbital is restricted by the Pauli'a Exclusion principle to two electrons with opposite spins. As the molecular orbitals are occupied, series of closed electron groups are built up in a similar fashion as that of the building up of electrons in atoms.

In the molecular orbital system, the s, p and d orbitals in the isolated atoms, which correspond to angular momentum quantum number l of 0, 1, 2 are replaced by the σ, π, δ molecular orbitals, respectively. These orbitals correspond to the component of angular momentum about the molecular axis given in the units of $\lambda h/2\pi$, where $\lambda = 0$, 1, 2 respectively. When $\lambda = 0$, this corresponds to the charge distribution being symmetrically disposed about the molecular bond (Compare with the case in the free atom when $\lambda=0$, which gives $1s$ spherical charge distribution about the nucleus). The bond formed from the use of these orbitals then is symmetrical about the bond axis. When $\lambda=1$, one obtains a π orbital, in which the charge distribution is concentrated above and below the inter-nuclear axis.

Two atomic orbitals, one from each bonded atom, whose energies are comparable in value and possess a large amount of overlap, coalesce to form two molecular orbitals. As represented in Fig. 8·39, one of these molecular orbitals is lower in energy than either of the atomic orbital from which it was formed and hence, gives rise to an attractive state. The other molecular orbital is

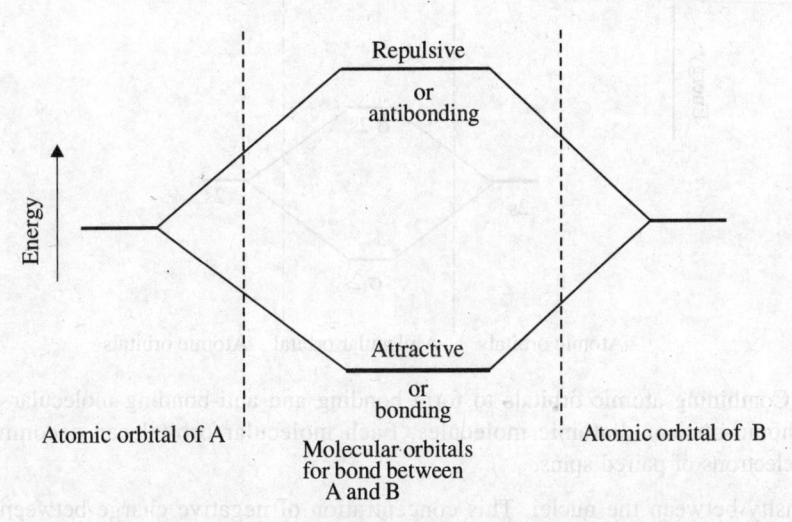

Fig. 8·39. Formation of two molecular orbitals from the coalescence of two atomic orbitals in atoms of A and B.

higher in energy and, therefore, gives rise to a repulsive state. This higher energy molecular orbital is called the anti-bonding orbital, because electrons placed in this kind of orbital decrease the stability of the bond. The anti-bonding orbitals are represented by superscript asterisks. The lower energy molecular orbital is called the bonding orbital, since electrons placed in this kind of orbital increase the stability of the bond.

The molecular orbitals for homo-nuclear diatomic molecules formed from given atomic orbitals are designated as per the following scheme.

Molecular orbital	Formed from
$\sigma\, 1s, \sigma^*\, 1s$	$1s$ atomic orbitals
$\sigma\, 2s, \sigma^*\, 2s$	$2s$ atomic orbitals
$\sigma\, 2p, \sigma^*\, 2p$	$2p_x$ atomic orbitals
$\pi_y\, 2p, \pi_y^*\, 2p$	$2p_y$ atomic orbitals
$\pi_z\, 2p, \pi_z^*\, 2p$	$2p_z$ atomic orbitals

These molecular orbitals are schematically represented in an energy level diagram shown in Fig. 8·40. The electronic charge distribution for these molecular orbitals is depicted in Fig. 8·41 from which it can be seen that the bonding orbitals are responsible for the build-up of electron

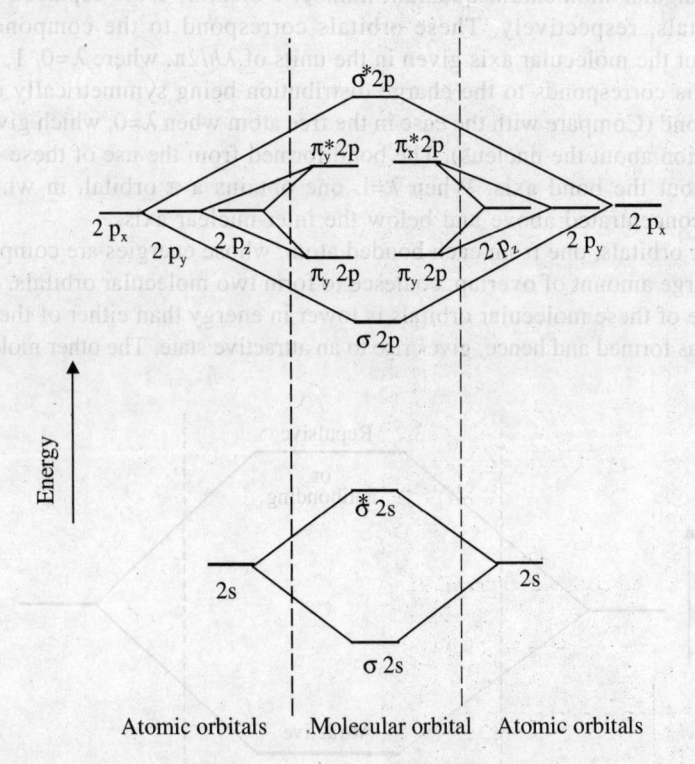

Fig. 8·40. Combining atomic orbitals to form bonding and anti-bonding molecular orbitals for homo-nuclear diatomic molecules. Each molecular orbital can accommodate two electrons of paired spins.

charge density between the nuclei. This concentration of negative charge between the nuclei lowers the repulsion between the positive nuclei. This increased charge density gives rise to a bonding or attractive state. The anti-bonding orbitals do not give rise to bonding, since the charge density between the nuclei is small. Hence, there is a net repulsion between the atoms because of the small amount of shielding between the positive nuclei.

CHEMICAL BONDING

The molecular orbitals are filled up according to the order of increasing energy as follows:

$\sigma 1s < \overset{*}{\sigma} 1s < \sigma 2s < \overset{*}{\sigma} 2s < \sigma 2p < \pi_y 2p = \pi_z 2p < \overset{*}{\pi}_y 2p = \overset{*}{\pi}_z 2p < \overset{*}{\sigma} 2p$.

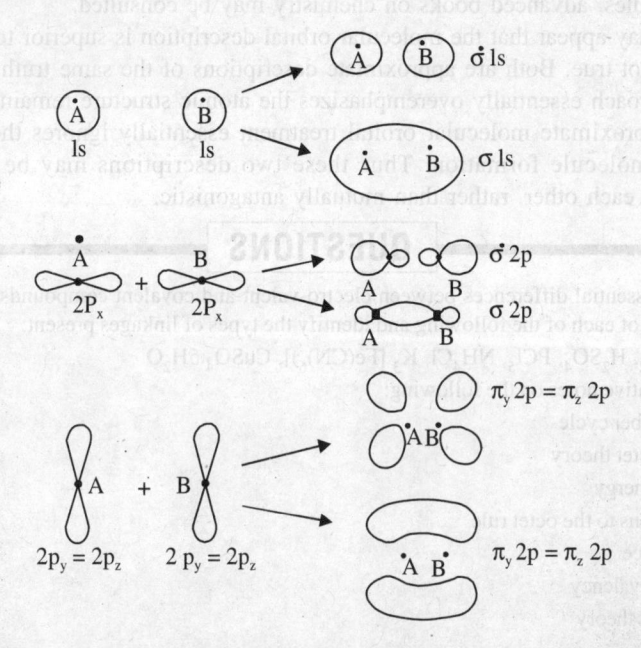

Fig. 8·41. Illustration of the electron charge density of molecular orbitals, formed from the combination of atomic orbitals.

It may be noted that in the case of p_y and p_z atomic orbitals, the $\pi_y 2p$ and $\pi_z 2p$ bonding and antibonding molecular orbitals formed are of the same energy, that is, they are degenerate. Molecular orbitals having two nodal planes are called δ orbitals which are come across with certain transition metal compounds. The δ molecular orbitals cannot be formed with s and p atomic orbitals but the overlap of suitable atomic d-orbitals, such as, two d_{xy}, or two $d_{x^2-y^2}$ orbitals, will give rise to the formation of a δ molecular orbital.

Let us now apply the above principles to the build-up of electron configuration of some homo-nuclear diatomic molecules.

Hydrogen molecule is formed when the two electrons from each atom enter the lowest-lying σ 1s orbital with paired spins. Its molecular orbital configuration may be represented as $[(\sigma 1s)^2]$, the superscript indicating the number of electrons in the molecular orbital. Since the two electrons are in a bonding orbital, a single bond is formed in the H_2 molecule. According to the "aufbau" principle, the next electron must enter into the anti-bonding $\overset{*}{\sigma}$ 1s orbital to give the configuration $[(\sigma 1s)^2 (\overset{*}{\sigma} 1s)^1]$ which is that of the helium molecule ion, He_2^{+1}. Since there are more bonding than anti-bonding electrons, the bond is stable. This prediction was confirmed, since He_2^{+1} is known to exist under certain conditions. However, in the case of He_2, the four electrons, two from each helium atom will have to occupy the σ 1s bonding and $\overset{*}{\sigma}$ 1s anti-bonding orbitals giving the configuration $[(\sigma 1s)^2 (\overset{*}{\sigma} 1s)^2]$. This molecule does not exist, because the bonding energy of the 1s orbital is more than cancelled by the antibonding $\overset{*}{\sigma}$ 1s orbital. The molecule of Li_2, which can be detected in the vapour state, has 2 more electrons than He_2, and these must occupy the next lowest-lying molecular orbital, namely the σ 2s bonding orbital. The configuration for Li_2 is represented as $[K K(\sigma 2s)^2]$. The symbol KK denotes the closed K shell structure $(\sigma 1s)^2 (\overset{*}{\sigma} 1s)^2$. It is believed that these closed shells do not enter into bonding, and hence, are known as non-bonding electrons. Thus it is seen that Li_2 bond is a σ type bond. The diatomic molecules formed from the alkali metals have analogous configurations. For instance, the configuration of Na_2 is represented as [KK LL(σ $3s)^2$].

For the application of molecular orbital theory to hetero-nuclear diatomic molecules and polyatomic molecules, advanced books on chemistry may be consulted.

Although it may appear that the molecular orbital description is superior to the valence bond description, it is not true. Both are approximate descriptions of the same truth. The approximate valence bond approach essentially overemphasizes the atomic structure remaining after molecule formation; the approximate molecular orbital treatment essentially ignores the atomic structure remaining after molecule formation. Thus these two descriptions may be considered to be complementary to each other, rather than mutually antagonistic.

QUESTIONS

1. Discuss the essential differences between electro-valent and covalent compounds. Give the electronic configuration of each of the following and identify the types of linkages present:
   CH_4, CO_2, N_2, H_2SO_4, PCl_5, NH_4Cl, K_3 [Fe(CN)$_6$)], $CuSO_4.5H_2O$.
2. Write informative notes on the following:
   (a) Born-Haber cycle
   (b) Lewis octet theory
   (c) Lattice energy
   (d) Exceptions to the octet rule
   (e) Resonance
   (f) Variable valency
   (g) Werner's theory
   (h) Chelates
   (i) Hydrogen bonding
   (j) Metallic bond
   (k) Delocalized orbitals.
3. Discuss the salient features of the valence bond theory. Give an example of sp^3 hybridization.
4. Match the following:

Orbital of central atom used in bonding	Spatial arrangement	Examples
p^2	Angular	H_2O, NO_2^{-1}
sp	Linear	C_2H_2, N_2O
p^3	Trigonal pyramid	NH_3, SO_3^{-2}
sp^2	Trigonal plane	BF_3, NO_3^{-1}
sp^3	Tetrahedral	SiF_4, CCl_4
dsp^2	Square plane	$[Cu(NH_3)_4]^{+2}$
d^2sp^3	Octahedron	$[Co(CN)_6]^{-3}$

   Ans. The order given is totally correct.
5. Discuss the essential features of the valence bond theory and the molecular orbital theory.
6. State whether the following statements are correct and justify your answers :
   (a) Diamond is a very hard solid with high melting point.
   (b) Graphite is a good conductor of electricity.
   (c) AlF_3 is predominantly ionic while $AlCl_3$ is rather covalent.
   (d) H_2O is a liquid at room temperature while a comparable compound H_2S is a gas.
   (e) HF is more polar than HI.
   (f) Melting points of ionic compounds are higher than those of covalent compounds.
   (g) Van der Waal's forces of attraction are responsible for the condensation of inert gases.

9

Polymers

"In view of the macro-molecular versatility, plastics can be tailor-made for specific uses by selecting the monomer, by combining several materials and by the introduction of suitable substituents."

9.1 INTRODUCTION

The word "Polymer" is derived from two Greek words, Poly (Many) and meros (Parts or units). A polymer is a large molecule formed by combining small molecules. The individual small molecules from which the polymer is formed are known as "Monomers" and the process by which the monomer molecule are linked to form big polymer molecule is called "Polymerization"

$$n\,CH_2=CH_2 \xrightarrow{\text{Polymerization}} -(CH_2-CH_2)_n-$$
Ethylene → Polyethylene
(Monomer) (Polymer)

The length of the polymer chain is specified by the number of repeat unit in the chain, the average number of repeat units in the chain is called "degree of polymerization" (n). It is obtained by dividing the average molecular weight of polymer by the molecular weight of the monomer. The polymer with a high degree of polymerization are called "high polymers" and those of with low degree of polymerization is called "Oligopolymer" or "Oligomers" (less than 10 repeating unit). High polymers have very high molecular weight (10^4 - 10^6) and hence are called as macromolecules.

The polymers are generally called as "plastics". The term plastics derived from the Greek word "Plastikos" which means 'fit for moulding'. Hence plastics can be defined as organic, inorganic, natural or synthetic substance which can be moulded. Hence it must be noted that all plastics are polymers while all polymers are not necessarily plastics.

Polymerization is a process which allow monomer to combine and form polymer. In order to farm polymer, monomer should have a capability to react at least with two other molecules of same or other compound. The number of reactive site (bonding) available in a molecule for a particular reaction is called "functionality" i.e. monomer should have a functionality of at least two. Monomer assumes functionality because of presence of reactive functional group like -OH, -COOH, -NH$_2$, -SH etc. Some compounds do not contain any reactive functional group but the presence of double or triple bonds in the molecule give functionality. Ethylene has functionality of two, acetylene has functionality of four. There are some compands in which presence of easily replaceable hydrogen atom imparts functionality ex. Phenol when replace one H it is monofunctional when it is replace 3H it became trifunctional. Bifunctional monomer give linear polymer e.g. styrene. Trifunctional monomer give non-linear polymer.

9.1.1 Classification of Polymers
Polymeric material can be classified into different ways.

i) Classification based on structure : Polymer have different molecular arrangement. The term "linear Polymer" is used for polymer in which monomeric units are joined together in a continuous long, chains (linearly) eg. High density polyethylene HDPE, Polystyren etc. The molecular chains of linear polymer is linked by physical force, rather than chemical. In these polymers, monomers are closely packed which posses high metling point high density and tensile strength. A polymer with a site chain extending from the main backbone are known as "branched polymers". ex. Low density polyethylene.

A polymer in which adjacent linear molecular chain are joined at various positions by covalent bonds called "crosslinked polymer". They are characteristics property of thermosetting polymer. Cross linking inhibit close packing of polymer chain, preventing the formation of crystalline regions. eg. cross linked poly(isoprene), cross linking in volcanization. When the cross linked polymer is stretched, the cross-links prevent the individual chains from sliding past each other. Polymer structures are given below fig. 9.1

Linear polymer 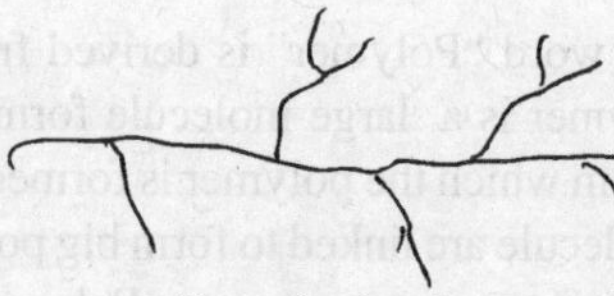   Branched polymer

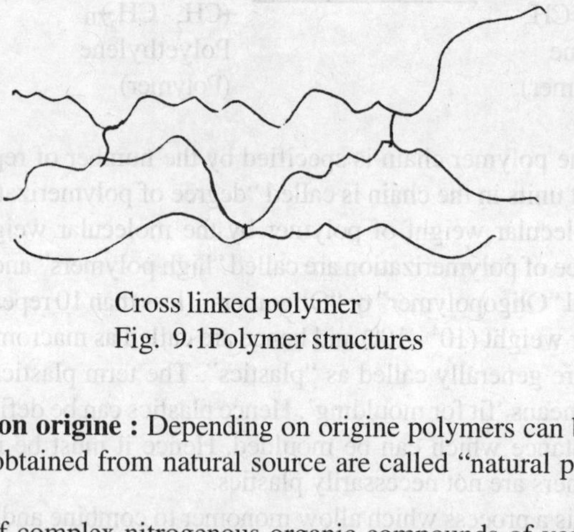

Cross linked polymer
Fig. 9.1 Polymer structures

ii) Classification based on origine : Depending on origine polymers can be classified as natural or synthetic. The polymer obtained from natural source are called "natural polymers". Proetins - silk, collagen, keratin, wool.

Proteins : It is a class of complex nitrogenous organic compounds of very high molecular weight (6000 - several millions) and are of great importance to all living matter. Protein molecules consist of hundreds or thousands of amino acids arranged in a linear chain and joined together by the peptide bonds between carboxyl and amino groups of adjacent amino acid residues. Some twenty different amino acids occure in proteins and each protein molecule is likely to contain all of them, arranged in a variety of sequence. It is the sequence of the different amino acids which give individual proteins their specific properties. The formation of a peptide linkage between two amino acid (glycine) is shown below. fig .

POLYMERS

$$H_2N-CH_2-\overset{\overset{O}{\|}}{C}-OH + H-\overset{\overset{H}{|}}{N}-CH_2-\overset{\overset{O}{\|}}{C}-OH$$
(Glycine)

$$\downarrow -H_2O$$

$$H_2N-CH_2\underset{}{\overline{\underline{-\overset{\overset{O}{\|}}{C}-\overset{\overset{H}{|}}{N}-}}}CH_2-\overset{\overset{O}{\|}}{C}-OH$$
$\downarrow$
(Peptide bond)
Glycyl glycine

Silk : It is a polypeptide formed from amino acids.

Wool : It is also a polypetide formed from nearly 20 different kinds of ∝ - amino acids. Natural rubber: Carbohydrates (hydrates of carbon) or saccharides are simple large group of organic compounds with general formula $C_x(H_2O)_y$, that are aldehydes or ketones with many hydroxyl groups. It include monosaccharides (glucose, fructose), disaccharides (lactose, sucrose) and polysaccharides (starch and cellulose) Carbohydrates play numerous roles in the metabolism of all living organisms such as storage and transport of energy (starch, glycogen) and structural components (cellulose in plant and chitin in animals)

Natural rubber : Natural rubber is a polymer containing isoprene units. The degree of polymerization is around 500 and average molecular weight is about 3,40,000. Natural rubber is an elastic hydrocarbon polymer that naturally occurs as a milky colloidal suspension in the sap of rubber tree (Hevea brasiliensis).

Synthetic polymer : polymer synthesysed from low molecular weight compound called synthetic polymer. eg. Polyethylene, Polystyrene, Polyvinyl chloride, Nylone, Polyester etc.

Based on number of monomer

Homopolymer : Polymer is made from all identical monomer molecules eg. Polyethylene, Polystyrene, Polyvinylchloride. etc.

Copolymers : If varing characteristics are needed for a polymer different monemers are polymerised together. The polymerization of two monomer are called copolymerization. Polymer made from two different type of monomer units eg. styrene butadiene rubber, styrene acrylonitrile rubber etc.

Based on the arrangement of monomers within the chain : copolymers are classified into four type : the alternating copolymer, random copolymer, block copolymer, graft copolymers. Random copolymer contains a random arrangement of the monomers. Alternating copolymer contains a alternate arrangement of the monomer. eg. Nylon is an alternating copolymer with two monomers, a six carbon diacid and six carbon diamine. Block copolymer contains blocks of monomers of the same type. Block copolymer is linear molecule. Graft copolymer contains a main chain polymer consisting of one type of monomer with branches made up of other monomers. Graft copolymer has branched structure arising from the attachment of side chains of one type of monomer unit onto a backbone of other type.

Terpolymer : Polymer made from three different types of monomer units eg. Acrylonitrile butadiene styrene ABS.

ABABABABABABABABABAB   Alternating copolymer

ABAABABBBABABBBAABBAAA   Randarm copolymer

AAAAAAABBBBBBBAAAAA   Block copolymer

```
                    B
                    B
                    B
                    B
AAAAAAAAAAAAAAAAAAAAAAAAA        Graft copolymer
    B
    B
    B
    B
```

ABCBCBABCBCACBABCBAC   Randam terpolymer\

Fig. 9.2 The structures of basic copolymers

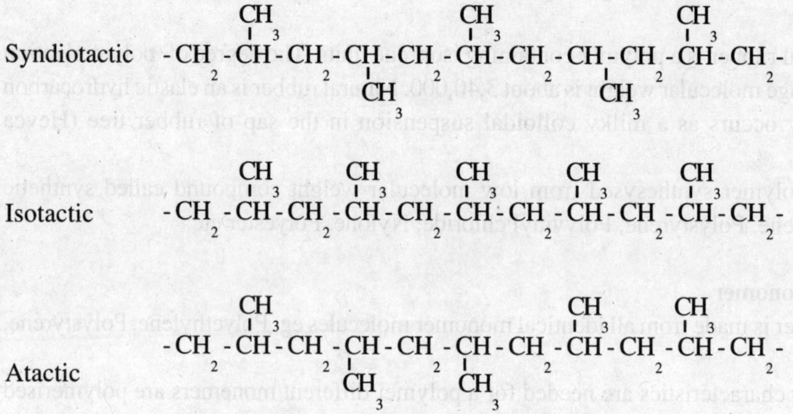

Syndiotactic $-CH_2-\underset{CH_3}{CH}-CH_2-CH-CH_2-\underset{CH_3}{CH}-CH_2-CH-CH_2-\underset{CH_3}{CH}-CH_2-$

Isotactic $-CH_2-\underset{CH_3}{CH}-CH_2-\underset{CH_3}{CH}-CH_2-\underset{CH_3}{CH}-CH_2-\underset{CH_3}{CH}-CH_2-\underset{CH_3}{CH}-CH_2-$

Atactic $-CH_2-\underset{CH_3}{CH}-CH_2-CH-CH_2-\underset{CH_3}{CH}-CH_2-CH-CH_2-\underset{CH_3}{CH}-CH_2-$

Fig. 9.3 Syndiotactic, isotactic and atactic forms of polypropylene

Based on configuration : Sterioregularity is the term used to describe the configuration of polymer chain. Isotactic is an arrangement were all substituents are on the same side of the polymer chain. Because of more symmetric arrangement of the chain, the molecules fit better into a cystal lattice and the polymers generally are highly crystalline. Syndiotactic polymer chain is composed of alternating groups and atactic is a random combination of the group. They do not conform to a certain shape and they do not fit easily into a crystalline pattern.

Based on mode of polymerization. : Polymers are classified by the characteristics of the reactions by which they are formed. If all the atoms in the monomers are incorporated into the polymer, the polymer is called an addition polymer eg. polyethylene, polystyrene. If some of the atoms of the monomers are released into small molecules such as water, the polymer is called a condensation polymer. Usually two different monomer combine with the loss of a small molecule eg. Polyesters and polyamides (nylon).

POLYMERS

Addition polymers (Chain Polymerization) : Majority of the addition polymers are formed through the polymerization of simple olefinic monomers such as ethylene, propylene, styrene and vinyl chloride. Addition polymerization many proceed through a free-radical, ionic (anionic, cationic), coordination mechanism, depending on the nature of the catalyst used. Polyethlene is a addition polymer formed through a free radical chain reaction involving the addition of ethylene units.

$$n\,CH_2 = CH_2 \xrightarrow{\text{Peroxide}} \text{\textlbrackdbl} CH_2 - CH_2 \text{\textrbrackdbl}_n$$
$$\text{Ethylene} \qquad\qquad\qquad \text{Polyethylene}$$

$$n\,CH_2 = CH\text{-}(C_6H_5) \rightarrow \text{\textlbrackdbl} CH_2 = CH\text{-}(C_6H_5) \text{\textrbrackdbl}_n$$
$$\text{Styrene} \qquad\qquad \text{Polystyrene}$$

Condensation polymer : Polyamides are prepared from the reaction of dicarboxylic acid with a diamine. Nylon (66) or polyhexamethylene adipamide is one of the best known polyamides. This polymer is prepared by the condensation of adipic acid with hexamethylene-diamine. Nylon is a polymer with high tensile strength and can be converted into fibres or moulded into articles.

Polyesters : Polyethylene terephthalate is one of the important polyesters that is commercially manufactured. Fibres such as dercon are terylene are made from this polymer. Polyethylene terephthalate is manufactured by an ester-exchange reaction between dimethyl terephthalate and ethylene glycol. This reaction is facilitated by the use of calcium acetate or other salts of carboxylic acids (these are weak basic catalysts)

$$H_3C-\overset{O}{\underset{\|}{C}}-C_6H_4-\overset{O}{\underset{\|}{C}}-OCH_3 + 2\,HO-CH_2-CH_2-OH \xrightarrow[200°C]{\text{Base}}$$

$$HO-CH_2-CH_2-O-\overset{O}{\underset{\|}{C}}-C_6H_4-\overset{O}{\underset{\|}{C}}-O-CH_2-CH_2-OH$$

$$+\,2CH_3-OH \xrightarrow{280°C} \left[=O-CH_2-CH_2-O-\overset{O}{\underset{\|}{C}}-C_6H_4-\overset{O}{\underset{\|}{C}}= \right]_n^-$$
$$\text{Polyethylene terephthalate}$$

$$HOOC-(CH_2)_4-COOH + H_2N-(CH_2)_6-NH_2 \xrightarrow{\Delta}$$
$$\text{Adipic acid} \qquad\qquad \text{hexamethylene diamine}$$

$$\left[\overset{O}{\underset{\|}{C}}\text{-}(CH_2)_4\text{-}\overset{O}{\underset{\|}{C}}\text{-}\overset{H}{\underset{|}{N}}\text{-}(CH_2)_6\text{-}\overset{H}{\underset{|}{N}} \right]_n + H_2O$$
$$\text{Nylon}$$

Chain polymerization	Condensation polymerization
1) Reaction generall requires initiators initiatiors or catalysts	1) Reaction can be proceeds without catalysts also.
2) Only active species (macroradicals) can add further mehomer molecules in the propagation process	2) Both monomer and polymer molecules with suitable functional end groups can reacts.
3) Monomer concentration decreases with reaction time	3) Monomer Molecues disappear quickly, more than 99% monomer molecules have already reacted when the degree of polymerization is 10.
4) Macromolecules are formed from the begining of the reaction	4) Monomer molecules first give oligomers high polymer is formed only twoards the end of reaction.
5) Average Molecular weight of the polymer changes little with reaction time	5) The average molecular weight increases linearly with reaction time. Long reaction time are needed to produce high molecular weight

Based on their behaviour when heated :

Plastics are classified into two categories according to what happens to them when they are heated to high temperature. Thermoplastics keep their plastic properties, they melt when heated than harden again when cooled. ex. Polyethylene, Polystyrene. Thermoplastics has primary strong covalent bonds within the chains and weaker secondary vander waals forcess between the chains. Usually secondary forces can be easily broken when object is heated, making thermoplastics mouldable at high temperature. Few common applications of thermoplastics are bottles, cable insulators, mixer bowls, medical syringes, mugs, packing etc. Thermoplastic polymers reforms when the plastic object is cooled. Hence thermoplastics are more popular because heating and cooling may be repeated.

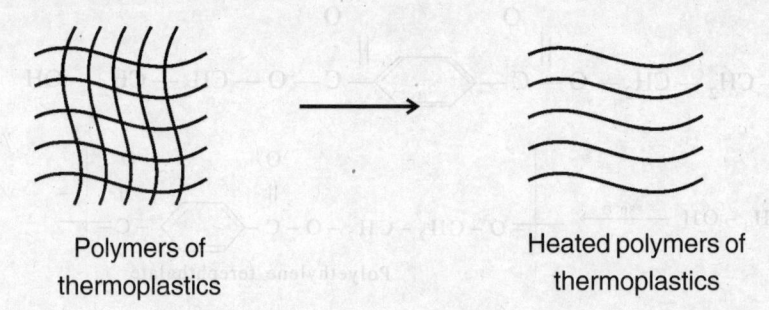

Polymers of thermoplastics → Heated polymers of thermoplastics

Thermosets are permanently set once they are initially formed and cant be melted. If they are exposed to enough heat they will crack or decompose or become charred. In thermoset, polymers are crosslinked strongly chemically bonded, this prevents a thermoset objects from being melted and reformed.

POLYMERS

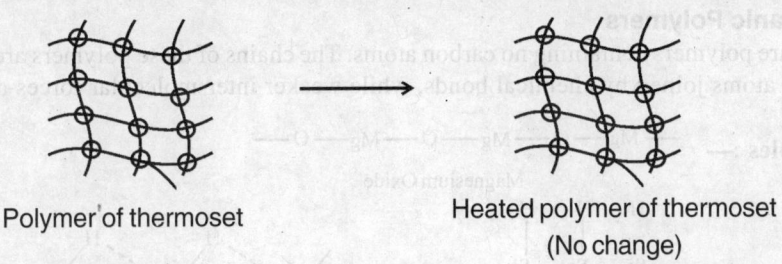

Polymer of thermoset → Heated polymer of thermoset (No change)

ex. Bakelite, Polyurathines which is used in handles of pots and pans, dishes, electric outlets.

Based on chemical composition

(1) Organic polymers

These include-compounds containing, apart from carbon atoms, hydrogen, oxygen, nitrogen, sulfur and halogen atoms, even if the oxygen, nitrogen, or sulfur is in the back-bone (or main) chain. Organic polymers also include polymeric substances containing other elements in their molecules provided the atoms of these elements are not in the main chain and are not connected directly to carbon atoms.

Examples :—

Polyethylene	...$-CH_2-CH_2-$...

Polyvinyl alcohol	...$-CH_2-CH-$...
	                                  $\|$
	                                OH

PVC	...$-CH_2-CHCl^-$...
Epoxy polymers	...$-O-R-O-CH_2-CH-CH_2-$... OH

Polyurethane	$-C-NH-(CH_2)_x-NH-C-O-(CH_2)_y-O-$
	   $\|\|$ $\|\|$
	   O O

Polysulfides	...$-R-(S)_z-R'-(S)_z-$...

starch, cellulose, etc.

(2) Elemento-organic or hetero-organic polymers

These include :

(a) Compounds whose chains are composed of carbon atoms and hetero-atoms (excepting N, S and O).

(b) Compounds with organic chains if they contain side groups with C atoms connected directly to the chain.

(c) Compounds whose main chains consist of carbon atoms and whose side groups contain hetero-atoms (excepting N. S, O and halogen atoms) connected directly to the C atoms in the chain.

Examples. :—Polysiloxanes

...—Si—O—Si—O—...
      | |
      R   R

      R   R
      | |
...—Ti—O—Ti—O—...
      | |
      R   R

(3) Inorganic Polymers

These are polymers containing no carbon atoms. The chains of these polymers are composed of different atoms joined by chemical bonds, while weaker inter-molecular forces act between the chains.

Examples :—

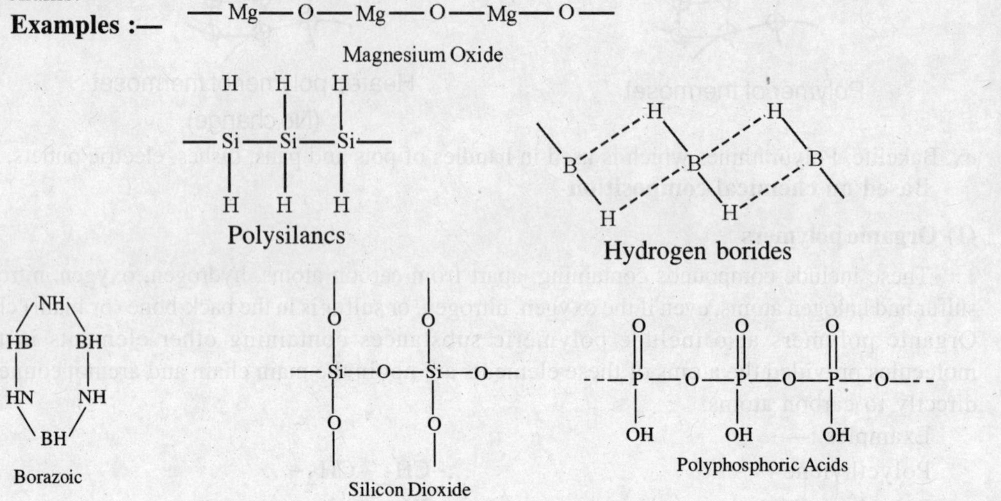

—Mg—O—Mg—O—Mg—O—
Magnesium Oxide

Polysilancs

Hydrogen borides

Borazoic

Silicon Dioxide

Polyphosphoric Acids

Fig. 9·4. Inorganic polymers.

9·1·2. Types of Polymerization : Two types of polymerization are generally distinguished:

(1) Addition or chain polymerization : In addition polymerization, the polymer is formed from the monomer, without the loss of any material, and the product is an exact multiple of the original monomeric molecule.

Application of energy in the form of heat, light, pressure, ionizing radiation or the presence of a catalyst, is usually necessary for initiating the chain polymerization.

In general, addition ploymerization proceeds by the initial formation of some reactive species, such as free radicals or ions and by the addition of the reactive species to another molecule, with the regeneration of the reactive feature. Some examples of free-radical addition polymerization is given below. The curved arrows represent electron shifts, occurring during the reaction :–

Examples of Addition Polymerization :–

$CH_2=CH_2$ ⟶ $-CH_2-CH_2-$ ⟶ Polymerization ⟶ $(-CH_2-CH_2-)_n$

Ethylene Monomer — Molecular rearrangement — Polythylene (Polymer)

Styrene (Monomer) ⟶ (Molecular rearrangement) ⟶ Polymerization ⟶ Polystyrene (Polymer)

HO• + Acrylonitrile ⟶ Free redical adduct addition product

Hydroxyl radical from decomposition of H_2O_2

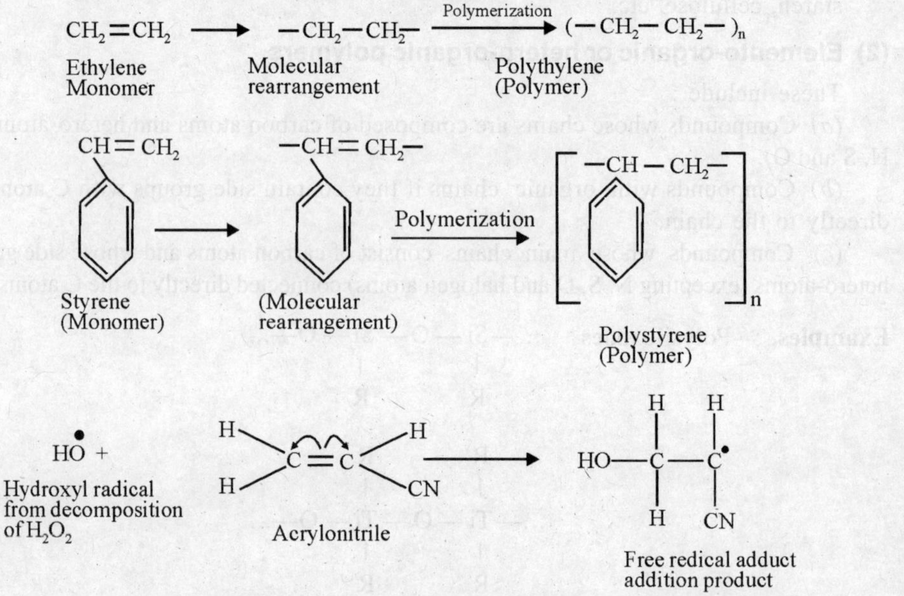

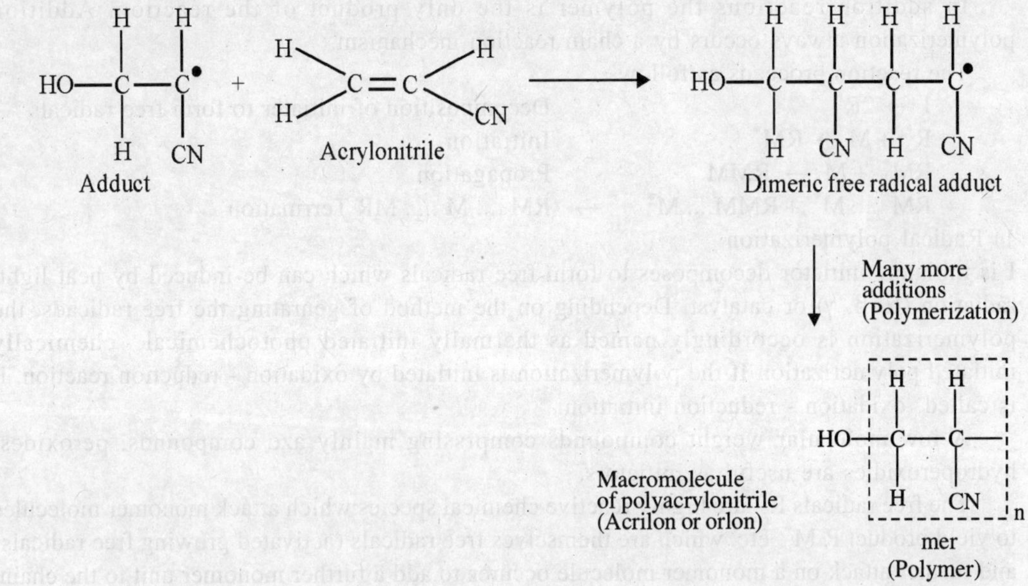

POLYMERS

Fig. 9·5. Addition or chain polymerization

(2) Condensation Polymerization : In condensation polymerization, the chain growth is accompanied by elimination of small molecules, such as H_2O, CH_3OH, etc.

Polymides :

(1) $\underset{\text{Adipic acid}}{HOOC\,(CH_2)_4\,COOH} + H_2N\,(CH_2)_6\,NH_2$

$\downarrow$ Condensation polymerization with elimination of H_2O molecule.

$$H-\left[\begin{array}{c} H \\ | \\ -N-(CH_2)_6- \end{array} \begin{array}{c} H \quad O \\ | \quad \| \\ N-C-(CH_2)_4-\overset{O}{\overset{\|}{C}}- \end{array}\right]_n -OH + H_2O$$

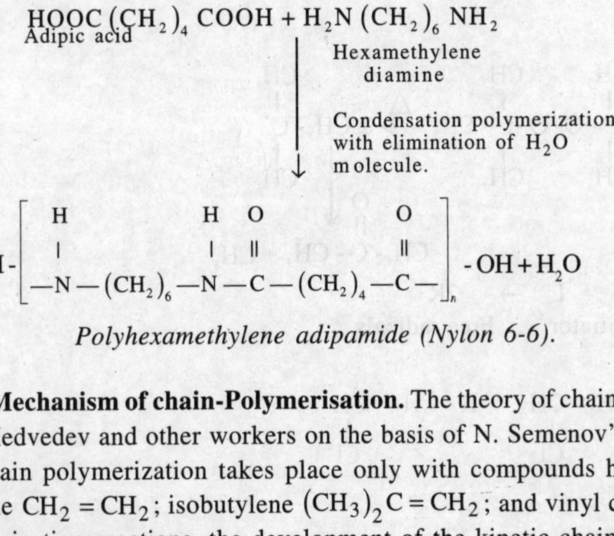

Polyhexamethylene adipamide (Nylon 6-6).

9.1.3. Mechanism of chain-Polymerisation. The theory of chain polymerization was developed by S.Medvedev and other workers on the basis of N. Semenov's theory of chain reactions.

Chain polymerization takes place only with compounds having multiple bonds such as ethylene $CH_2 = CH_2$; isobutylene $(CH_3)_2 C = CH_2$; and vinyl chloride $CH_2 = CHCl$. In chain polymerization reactions, the development of the kinetic chain is accompanied by the growth of molecular chain.

The chain polymerization reaction comprises of the following three major steps.

Initiation, propagation and termination

Addition Polymerization is written as

$$nM \rightarrow M_n$$

In addition reactions the polymer is the only product of the reaction. Addition polymerization always occurs by a chain reaction mechanism.

The reaction proceeds as follow

$I \rightarrow 2R^\bullet$	Decomposition of initiater to form free radicals
$R^\bullet + M \rightarrow RM^\bullet$	Initiation
$RM^\bullet + M \rightarrow RMM$	Propagation
$RM.....M^\bullet + RMM....M^\bullet \rightarrow RM....M.....MR$	Terrination

1) Radical polymerization

I is initiator initiator decomposes to form free radicals which can be induced by heat light, radiation (α, β, γ) or catalyst. Depending on the method of genrating the free radicads. the polymerization is occordingly named as thermally initiated photochemical, chemically initiated polymerization If the polymerization is initiated by oxidation - reduction reaction, it is called oxidation - reduction initiation.

A low molecular weight compounds comprssing mainly azo compounds, peroxides, hydroperoxidies are useful as initiators.

The free radicals $R^\bullet$ are highly reactive chemical species which attack monomer molecules to yield product $R M^\bullet$ etc. which are themselves free radicals (activated growing free radicals) and further attack on a monomer molecule occures to add a further monomer unit to the chain. This process continues called propagation. The process come to an end only if the supply of monomer is exhausted or if free radicals combine together to terminate the chain raction.

The mechanism of a free rodical chain raction can be shown by the polymerisation of vinyl chlaride. Tertiary butyl peroxide initiator can be used.

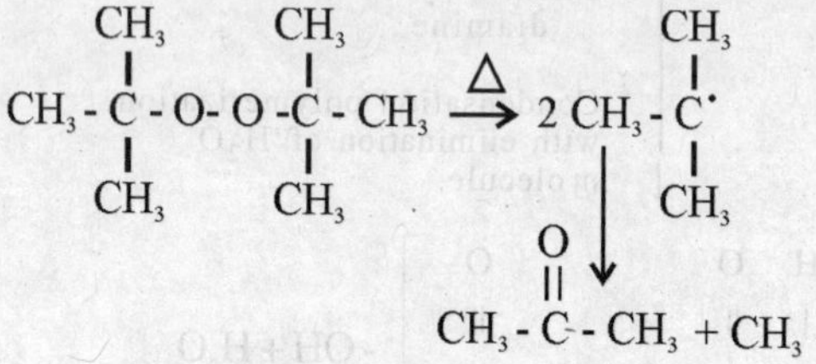

i) Initiation

$$\underset{\underset{CH_3}{|}}{\overset{\overset{CH_3}{|}}{CH_3-C-O}}-\underset{\underset{CH_3}{|}}{\overset{\overset{CH_3}{|}}{O-C-CH_3}} \xrightarrow{\Delta} 2\,\underset{\underset{CH_3}{|}}{\overset{\overset{CH_3}{|}}{CH_3-C^\bullet}}$$

$$\downarrow$$

$$CH_3-\overset{\overset{O}{\|}}{C}-CH_3 + CH_3$$

$$I \rightarrow 2R^\bullet$$
Initiator   Free radicals

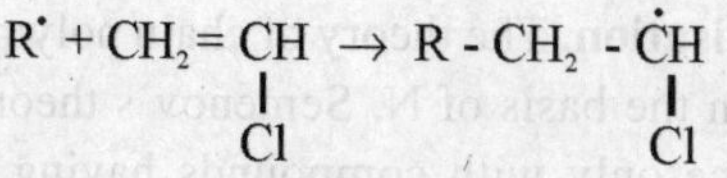

ii) Propagation

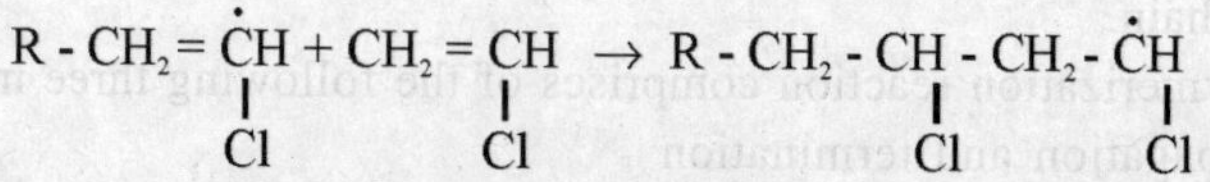

$$R^\bullet + CH_2 = \underset{\underset{Cl}{|}}{CH} \rightarrow R-CH_2-\underset{\underset{Cl}{|}}{\overset{\bullet}{C}H}$$

$$R-CH_2=\underset{\underset{Cl}{|}}{\overset{\bullet}{C}H} + CH_2 = \underset{\underset{Cl}{|}}{CH} \rightarrow R-CH_2-\underset{\underset{Cl}{|}}{CH}-CH_2-\underset{\underset{Cl}{|}}{\overset{\bullet}{C}H}$$

iii) Termination

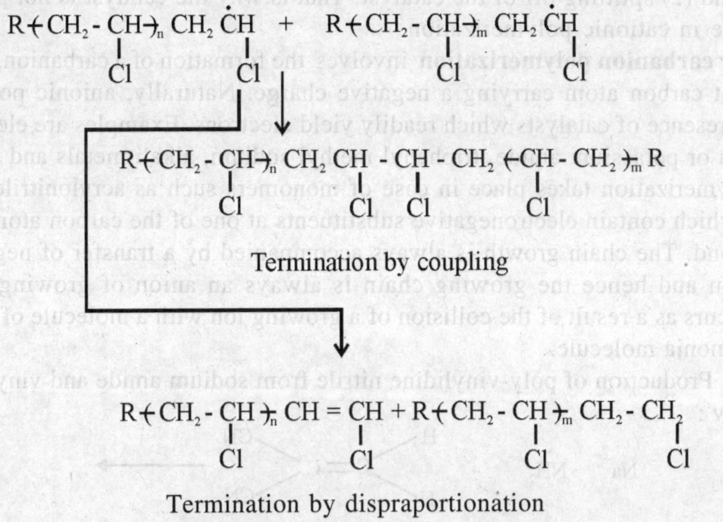

$$R\text{-}(CH_2\text{-}CH)_n\text{-}CH_2\overset{\cdot}{C}H + R\text{-}(CH_2\text{-}CH)_m\text{-}CH_2\overset{\cdot}{C}H$$
$$\qquad\quad\;|\qquad\qquad\quad|\qquad\qquad\quad\;|\qquad\qquad\quad|$$
$$\qquad\quad Cl\qquad\qquad Cl\qquad\qquad\;\; Cl\qquad\qquad Cl$$

$$R\text{-}(CH_2\text{-}CH)_n\text{-}CH_2\text{-}CH\text{-}CH\text{-}CH_2\text{-}(CH\text{-}CH_2)_m\text{-}R$$
$$\qquad\quad\;|\qquad\qquad\quad|\quad\;\;|\qquad\qquad\quad|$$
$$\qquad\quad Cl\qquad\qquad Cl\;\; Cl\qquad\qquad Cl$$

Termination by coupling

$$R\text{-}(CH_2\text{-}CH)_n\text{-}CH=CH + R\text{-}(CH_2\text{-}CH)_m\text{-}CH_2\text{-}CH_2$$
$$\qquad\quad\;|\qquad\qquad\;|\qquad\qquad\quad|\qquad\qquad\quad|$$
$$\qquad\quad Cl\qquad\quad\; Cl\qquad\qquad\; Cl\qquad\qquad Cl$$

Termination by dispraportionation
Fig. 9·6. Radical polymerization

(2) Ionic polymerization

In ionic polymerization, the active centres initiating the chain reaction are "ions". Ionic polymerization proceeds due to the presence of catalysts and hence it is called catalytic polymerization. Depending upon the charge of the ion formed, the polymerization may be cationic or anionic.

Cationic polymerization (or carbonium polymerization) takes place with the formation of a carbonium ion, which is a polar compound with a tri-covalent carbon atom having a positive charge :

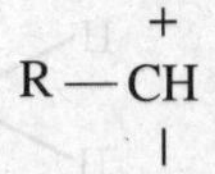

$$R - \overset{+}{\underset{\underset{R'}{|}}{CH}}$$

The catalysts used in carbonium polymerization (or cationic polymerization) are compounds with pronounced electron acceptor properties *e.g.*, $AlCl_3$, $SnCl_2$, $TiCl_4$ and boron fluoride. The polymerizing monomer is an electron donor such as styrene in the presence of $SnCl_4$.

A carbonium ion interacts with a monomeric molecule, and the reaction of the chain growth is accompanied with the communication of a positive charge along the chain. Thus, the growing chain itself is a cation and the molecular mass increases in the course of polymerization

Examples :

During the polymerization of styrene in presence of $SnCl_4$ leading to the formation of polystyrene, the pattern of the growing chain can be represented as follows:

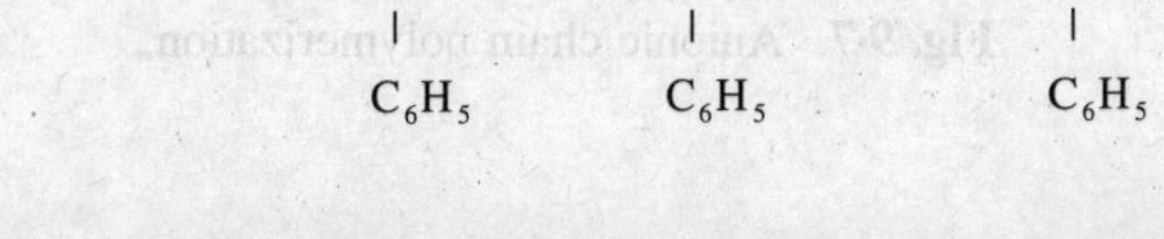

$$[SnCl_4]^- \; CH_2 - \underset{\underset{C_6H_5}{|}}{CH} - CH_2 - \underset{\underset{C_6H_5}{|}}{CH} - ... - CH_2 - \underset{\underset{C_6H_5}{|}}{\overset{+}{C}H}$$

Chain termination takes place as a result of (1) the mutual collision of the ends of a growing ion, end (2) splitting off of the catalyst. That is why the catalyst is not present in the macro-molecule in cationic polymerization.

Anionic or carbanion polymerization involves the formation of a carbanion, a compound with a trivalent carbon atom carrying a negative charge. Naturally, anionic polymerization occurs in the presence of catalysts which readily yield electrons. Examples are electron donors such as sodium or potassium amide, triphenyl methyl sodium, alkali metals and alkyl alkalis. Carbanion polymerization takes place in case of monomers such as acrylonitrile and methyl methacrylate which contain electronegative substituents at one of the carbon atoms connected by a double bond. The chain growth is always accompanied by a transfer of negative charge along the chain and hence the growing chain is always an anion of growing size. Chain termination occurs as a result of the collision of a growing ion with a molecule of the medium, such as an ammonia molecule.

Examples. Production of poly-vinylidine nitrile from sodium amide and vinylidine nitrile as shown below :

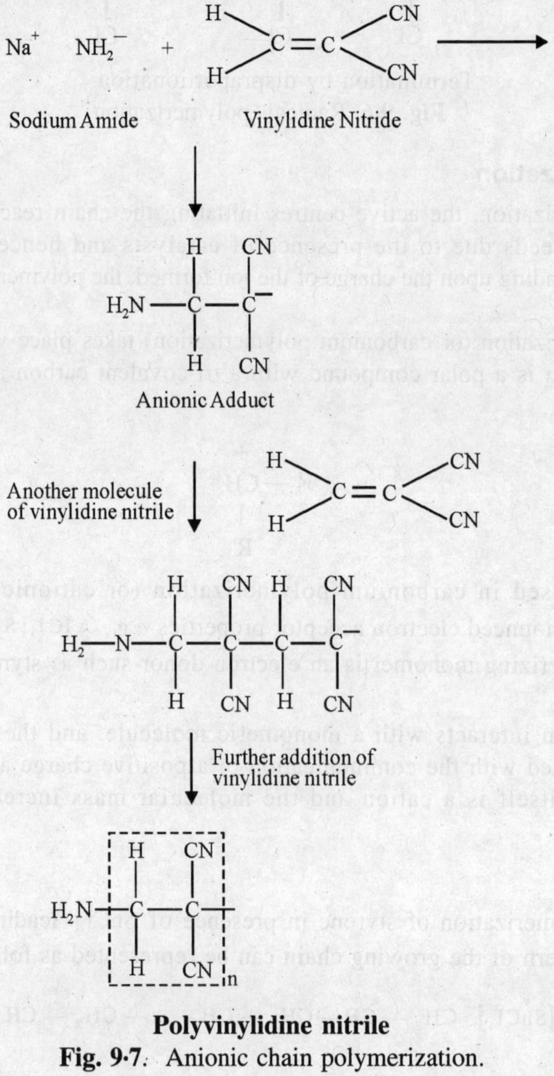

$$Na^+ \quad NH_2^- \ + \ \underset{H}{\overset{H}{>}}C=C\underset{CN}{\overset{CN}{<}} \longrightarrow$$

Sodium Amide   Vinylidine Nitrile

$$H_2N-\underset{\underset{H}{|}}{\overset{\overset{H}{|}}{C}}-\underset{\underset{CN}{|}}{\overset{\overset{CN}{|}}{C^-}}$$

Anionic Adduct

$$\downarrow \text{Another molecule of vinylidine nitrile} \quad \underset{H}{\overset{H}{>}}C=C\underset{CN}{\overset{CN}{<}}$$

$$H_2N-\underset{\underset{H}{|}}{\overset{\overset{H}{|}}{C}}-\underset{\underset{CN}{|}}{\overset{\overset{CN}{|}}{C}}-\underset{\underset{H}{|}}{\overset{\overset{H}{|}}{C}}-\underset{\underset{CN}{|}}{\overset{\overset{CN}{|}}{C^-}}$$

Further addition of vinylidine nitrile

$$H_2N {\left[\begin{array}{c} \overset{H}{|} \ \ \overset{CN}{|} \\ -C-C- \\ \underset{H}{|} \ \ \underset{CN}{|} \end{array}\right]}_n$$

Polyvinylidine nitrile

Fig. 9·7. Anionic chain polymerization.

POLYMERS

Anionic polymerization is a modern method of preparing high molecular weight monomolecular polymers. It has great advantage since spontaneous chain termination does not occur. In anionic polymerization, the end group of a growing macro-molecule possesses high activity and great stability. Hence, polymers produced by this method retain active centres at the end of the chain. Such polymers are known as "living polymers" because the active centres are capable of initiating the polymerisation later when a fresh batch of monomer is added to it later. Such a technique is used to produce block copolymers.

9·1·4. Sterio-specific polymerization. Sterio-specific polymerization is one which results in the production of sterio-regular polymers. Sterio-regular polymers are those in which all units and substituents are arranged in space in some definite order. It may proceed by an ionic or a radical mechanism.

A polymer lacking a regular order of its units and substituents is called sterio-irregular. These differences in spatial arrangement of units or substituents give rise to configurational isomerism (*e.g.*, cis-trans and D—L isomerism) of polymers.

The term coordination polymerisation is sometimes used to the processes that yield polymers with ordered structure (sterio-polymers) because the coordination complex between the monomer and the organometallic compound regulates the polymer structure through coordination.

Ionic copolymerization

In ionic copolymerisation, two or more monomers may be copolymerized by ionic mechanism, similar to free-radical copolymerization.

Ex: (1) Commercial elastomer of polyisobutylene (butyl rubber) is obtained by cationic copolymerisation of isobutylene with 0·5- 2% isoprene which acts as co-monomer and provides unsaturated sites for vulcanisation.

(2) Triblock copolymer SBS, having a central block of butadiene with styrene blocks at each end of the chain, is a thermoplastic elastomer that is elastic at ambient temperature but can be molded at higher temperatures. It is synthesized by adding styrene monomer to an active butadiene chain having anionic sites at both ends (i.e., a butadiene dianion).

Coordination polymerization

Polyethylene was first synthesized in 1939 by a high-pressure, free radical polymerization process developed at Imperial chemical Industries (ICI) in England. This had a backbone of $-CH_2-CH_2-$ having some short and long alkane branches. It possesses a moderate crystallinity and such thermal and mechanical properties that are amenable for film and bottle applications. This grade of polyethylene is now called as low-density polyethylene (LDPE).

Karl Triegler in Germany developed a steriochemical process to polymerize ethylene at much lower temperature and pressure than what were required for free-radical polymerization by using Zeigler-Natta catalyst. The product thus formed had fewer branches and possess a higher degree of crystallinity than that of LDPE and hence is called high density polyethylene (HDPE). Zeigler and Natta were awarded the Nobel prize in 1963 for this work. A Ziegler-Natta catalyst, in general, comprises of a metal-organic complex of a metal cation from I to III Group of the periodic table [e.g. triethyl aluminium, $Al(C_2H_5)_3$] and a transition metal compound from IV to VIII Group [e.g., titanium tetrachloride, $TiCl_4$].

For example, HDPE can be produced by bubbling ethylene into a suspension of $Al(C_2H_5)_3$ and $TiCl_4$, in hexane at ambient temperature. Similarly, polypropylene of about 90% isotacticity can be prepared by polymerising propylene in presence of $TiCl_3$ and diethylaluminium chloride,

Al $(C_2H_5)_2$ Cl, at 50°C. It is postulated that the growing polymer-chain is bound to the metal atom of the catalyst and that monomer insertion involves a coordination of the monomer with the atom. Thus, it is this coordination of the monomer which is responsible for steriospecificity of the polymerization. Coordination polymerisation processes can be terminated by the introduction of hydrogen, water, metals, such as Zinc and aromatic alcohols.

9·1·5. Step polymerization. Step polymerization involves the combination of several molecules

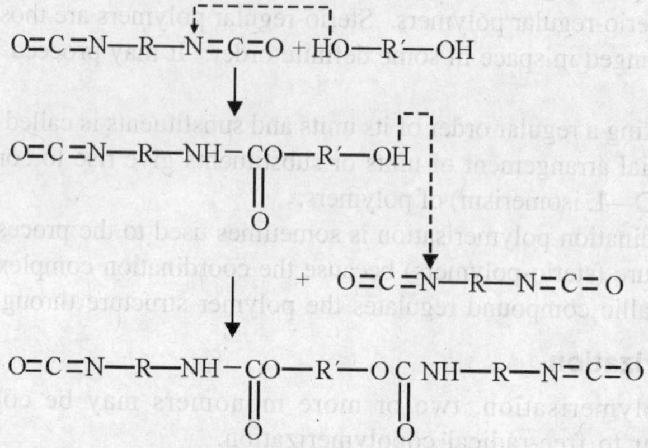

```
         O=C=N—R—N=C=O + HO—R'—OH
                        │
                        ↓
         O=C=N—R—NH-CO—R'—OH
                    ‖
                    O
                         +  O=C=N—R—N=C=O
                        │
                        ↓
         O=C=N—R—NH-CO—R'—OC NH—R—N=C=O
                    ‖           ‖
                    O           O
```

Fig. 9·8. Step polymerization.

to one another as result of the migration of some mobile atom (generally a hydrogen atom) from one molecule to another. An example of step polymerization is provided by the polymerization of di-isocyanates and dihydric alcohols into linear polyurethanes, as shown in Fig. 9.5.

9.1.6. Polymerizability of a monomer. The polymerizability of a monomer depends majorly on the polarizability of the double bond (*i.e.*, the ease of displacement of electron cloud). In symmetrically arranged molecules (e.g., ethylene. symmetrical butylene, etc.) the double bond is not polarized easily and hence their polymerization is difficult as compared to propylene and isobutylene. The degree of polarization increases upon an introduction of polar substituents (*e.g.*, $CH_2 = CH—, — C \equiv N, CH_2 = CH — COOH$, etc.). The double bond in a styrene molecule is also polarized and hence it is more easily polymerized than ethylene.

Non-polar molecules which are not amenable for polarization by themselves may do so under the influence of the polar molecules of another monomer. Thus, a monomer which does not polymerize by itself may copolymerize with another monomer.

However, the polarizability of a monomer influences only the initiation stage. The subsequent stage of growth depends on the activity of a free radical that is formed.

9.1.7. Thermodynamics of a polymerization process. According to the general thermodynamic considerations, a process can take place spontaneously at constant temperature and pressure only if it is accompanied by a reduction of free energy (G). That is, the free energy of the polymer should be less than the free energy of the monomer. Let us examine this possibility by first taking the example of polymerization of monomers having multiple double bonds. The process of the formation of long polymer molecules from randomly arranged monomers always results in their ordering and hence the entropy of a polymer is less than the entropy of monomers. Thus, the process of polymerization is accompanied by a decrease in entropy. Therefore, as per the equation

$$\Delta G = \Delta H — T. \Delta S, \text{ (where H = enthalpy and S = entropy)}$$

POLYMERS

this situation does not promote the polymerization process. If the process has to take place, it must be accompanied with a decrease in enthalpy and ΔH should be greater than $(T.\Delta S)$.

During the polymerization of a monomer having a double bond, the double bond breaks up resulting in the formation of two ordinary single bonds. The energy of the double bond $C = C$ is 145.5 Kcal/mole, whereas the energy of the two single C—C bonds is $2 \times 84 = 168$ Kcal/mole. Thus, 145.5 Kcal/mole are consumed while 168 Kcal/mole are saved. Hence, the difference (168—145.5), i.e., 22.5 Kcal/mole is the enthalpy of the polymerization. It follows that polymerization is an exothermal reaction. However, during polymerization of most monomers, less than this 22.5 Kcal/mole are lost which is ascribed to the losses of conjugation energy, the interaction of side groups, etc. Nevertheless, the polymerization process takes place if the absolute value of ΔH is greater than that of $T.\Delta S$, which is usually 7—10 Kcal/mole. Otherwise, the monomer will not polymerize.

Similar thermodynamic considerations on the polymerization of ring compounds show that the process of polymerization of the non-strained rings is thermodynamically disadvantageous. (The lower rings are known to be strained and the strain energy decreases as the number of carbon atoms in the ring decreases. Six-membered rings are virtually not strained and strain grows with a further increase in the number of carban atoms); Three and four-membered rings are easily polymerized.

Effect of temperature on polymerization. The rate of reaction in radical polymerization usually increases with increase of temperature. Temperature also influences the size, structure and degree of branching of a polymer. Polymers of high molecular weight are formed generally at lower reacting temperatures in case of many monomers. Increase of temperature increases the degree of branching of a polymer.

Effect of pressure on polymerization. Increase of pressure increases the number of collisions between active centres and monomers and hence increases the rate of polymerization. Further, raising pressure enables a lower temperature of polymerization and hence polymers with high molecular weight can be produced.

9.1.8 Practical methods of polymerization. Depending upon the experimental conditions involved in the preparation, the following techniques of polymerisation are usually employed.

a) Bulk polymerization. This technique is simple and produces the polymer of highest purity, high molecular weight and high rates of polymerization. It requires only a monomer, a monomer-soluble initiator and a chain-transfer agent to control the molecular weight of the polymer. This technique provides a high yield per reactor volume, easy recovery of the polymer and the option for casting the polymerization mixture into a find product form. The disadvantage of the technique is the difficulty in the removal of unreacted monomer and the problem of dissipating the heat of polymerization. Since the thermal conductivity of monomers and the problem of dissipating the heat of polymerization, since the thermal conductivity of monomers and polymers is low, and as the viscosity builds up, the ability for heat transfer by convection decreases. If the heat energy cannot be dissipated, temperature rises, and at higher temperature the rate of polymerization will increase and again more heat generate.

Bulk polymerization technique can be used for free-radical polymerization of polystyrene, polymethyl methacrylate and high pressure polymerization of polyethylene.

b) Solution Polymerization : Here the disadvantage of heat control in bulk polymerization is overcome if the monomers are dissolved in suitable solvent and the polymerization carried out in solution. The heat created during the reaction is dissipated over the whole solvent, solute system. In this process the monomer and the initiators are dissolved in a non monomeric liquid solvent at the begining of the polymerization reaction. Both solutions were mixed and agitated to start the polymerization. After specific time, the non solvent is added to precipitate the polymer.

Solvent removes the heat of polymerization. Solvent has to select carefully so that they do not undergo chain transfer reaction with the polymer. Many free readical and ionic polymerization can be performed in solution in an organic solvent (ex. Polystyrene, Polymethyl methcrylate, Polyvinyl acetate) or in water polyacrylic acid, polyvinyal alcohol.

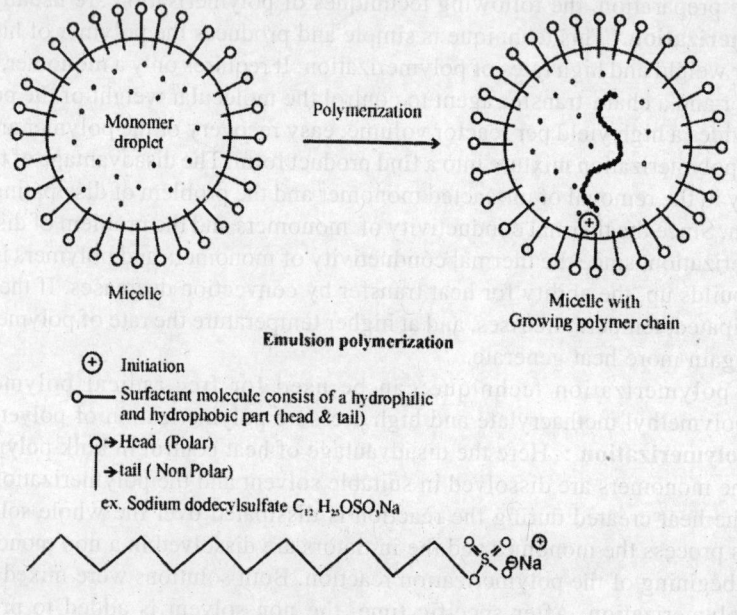

The advantage of solution polymerization over bulk polymerization is better heat control. The disadvantage of solution polymerization is that the reaction is slow and low average molecule weight product is obtained. The removal of the solvent from the polymer requires a distillation and that cost an appreciable money.

c) Suspension polymerization :

Suspension polymerization is a polymerization process that uses mechanical agitation to mix the monomer in a water, polymerizing the monomer droplets while they are dispersed by continuous agitation. The average particle size of the dispersed phase is 10-100 mm. The monomer (water insoluble) is dispersed as droplets in water by vigorous stirring. Initiator (monomer soluble) dibenzoyl peroxide is added to initiate the polymerization. The dispersion of the monomer in water can be assisted by the addition of suspension stabilizer (Calcium phosphate, Magnesium phosphate).

The limitations of suspension polymerization are reaction is highly agitation sensitive which is difficult to control reactor, capital costs are high since large reactor volume is taken up by water and low polymer purity because of the presence of suspending and stabilizing additives which are difficult to remove completely. Free radical suspension polymerization method can be used for producing syrenic ion-exchange resins, extrusion and injection molding grade of PVC, Poly(styrene-co-acrylonitrile) etc.

The advantages are better heat control of the reaction and separation is much easier than in solution polymerization the tendency to absorb impurities due to large surface area of small particals is less due to large droplets. The disadvantage is that few monomers are water soluble.

d) Emulsion polymerization :

It is limited to addition polymerization. The basic principal is to finely (d = 1mm) disperse the water insoluble monomer in water. The dispersion of the monomer taken place in the presence of

Fig. 9.9. Emulsion polymerization.

POLYMERS

surfactant (emulsifier) (sodium, potassium salt of fatty acid, soap) that form micelles. The dispersed monomer particles are stopped from coagulating with each other since each particles is surrounded by the surfactant.

Water soluble initiator (Potassium peroxodisulphate) enters the micelle and polymerization starts. The monomer consumed in the micellers is replaced by diffusion from the monomer droplets through the aqueous phase, by this method high molecular weight polymer is obtained at faster polymerization rate. Since each polymerization sites are isolated. The continuous water phase is an excellent conductor of heat which removes the heat from system. The dispersion (polymer) resulting from emulsion polymerization is called a latex. These emulsions has applications in adhesives, paints, papers and textile coatings.

Advantage of method is that it has better heat control; the size of the emulsion polymer is usually 0.05 to 5 microns. Emulsion polymerization is used for large scale production of polymers of high average molecular weight, However due to pressence of emulsifiers (which is difficult to remove) in the reaction mixture make polymer unsuitable in applications involving high optical clarity and outstanding electrical insulating properties. The large surface area of the small particles of polymer provides large surface area for impurities to be absorbed.

(*e*) **Solid-state polymerisation :** Some monomers in their crystalline state can be polymerized to yield extended chain polymers oriented along crystallographic directions. Polymer single crystals with interesting optical properties can be produced by this technique.

(*f*) **Gas-phase polymerization :** This technique is used for polymerization of olefins such as for producing HDPE from ethylene. Gaseous ethylene and a solid catalyst (e.g. chromium or other complexes) are made to react in a continuous fluidized-bed reactor. Since the polymerization reaction is highly exothermic, efficient heat transfer management is very critical to prevent agglomeration of the particles and termination of the process.

This technique can be used for the polymerization of propylene and co-polymerization of propylene and ethylene.

(*g*) **Plasma polymerization :** This technique can be used to produce graft copolymers or depositing a thin corrosion-resistant polymer coating on a metal or a thin photo resistant polymer film on a silicon wafer. The polymerization is performed in plasma environment produced from a low-pressure glow-discharge of positively charged species, electrons, excited and neutral species.

9.1.9 Molecular weight of polymers

Synthetic polymers generally comprise of chains with a wide distribution of chain lengths. The breadth of the molecular weight distribution depends upon the conditions of polymerization. Therefore, any individual polymer sample is characterized on the basis of average molecular weight, $\overline{M}$ which is defined as follows :

$$\overline{M} = \frac{\sum_i N_i M_i^{\alpha}}{\sum_i N_i M_i^{\alpha-1}}$$

where, N_i = number of moles of molecules having a molecular weight M_i

α = weighing factor which defines particular average of the molecular weight distribution

then, $W_i = N_i M_i$

where W_i is the weight of molecules having a molecular weight, M_i

The following three types of molecular-weight averages are generally distinguished for determining the polymer properties :

$\overline{M}_n$ The number average molecular weight ($\alpha = 1$),

$\overline{M}_w$ the weight average molecular weight ($\alpha = 2$), and

$\overline{M}_z$ the Z average molecular weight ($\alpha = 3$).

The molecular weight distribution of commercial polymers is normally a continuous function. The molecular weight averages can therefore be determined by integration if the mathematical form of the molecular weight distribution is known (In other words if N as a function of M is known). For instance, the number average molecular weights for discrete and continuous distributions are given as follows :

$$\overline{M}_n \text{ (for discrete distribution)} = \frac{\sum_{i=1}^{n} N_i M_i}{\sum_{i=1}^{n} N_i}, \text{ and}$$

$$\overline{M}_n \text{ (for continuous distribution)} = \frac{\int_o^M N M \, dM}{\int_o^M N \, dM},$$

where, N = total number of polymer species in the distribution.

The corresponding relationships for the weight-average molecular weight are given as follows :

$$\overline{M}_w = \frac{\sum_{i=1}^{N} N_i M_i^2}{\sum_{i=1}^{N} N_i M_i}, \text{ and}$$

$$\overline{M}_w = \frac{\int_o^M NM^2 \, dM}{\int_o^M NM \, dM}$$

The number average molecular weight of high-molecular weight polymers can be directly determined by osmometry, whereas their weight-average molecular weight is determined by light-scattering and other similar techniques.

The ratio of the weight-average-molecular-weight to that of the number-average-molecular-weight is known as the polydispersity index (PDI), which gives a measure of the breadth of the molecular-weight distribution :

$$\text{PDI} = \frac{\overline{M}_w}{\overline{M}_n}$$

A method widely used for routine molecular-weight determination is based on the determination of the intrinsic viscosity, [η], of a polymer in solution through measurements of solution viscosity. Molecular weight is related to [η] by the Mark-Houwink-Sakurada equation

POLYMERS

given below:

$$[\eta] = K \overline{M}_v^a, \text{ where}$$

$\overline{M}_v$ is the viscosity-average molecular weight defined for a discrete distribution of molecular weights as

$$\overline{M}_v = \frac{\sum_{i=1}^{N} N_i M_i^{a+1}}{\sum_{i=1}^{N} N_i M_i}$$

Similarly the corresponding relationship for the weight-average molecular weight are given as follows:

$$\overline{M}_w = \frac{\sum_{i=1}^{N} N_i M_i^2}{\sum_{i=1}^{N} N_i M_i}, \text{ and}$$
(for discrete distribution)

$$\overline{M}_w = \frac{\int_0^M N M^2 \, dM}{\int_0^M N M \, dM}$$
(for continuous distribution)

In case of high-molecular-weight polymers, $\overline{M}_n$ is directly determined by membrane osmometry and vapour-pressure osmometry, whereas the $\overline{M}_w$ is determined by light scattering and other methods.

A ratio of $\overline{M}_w / \overline{M}_n$ is known as polydispersity index (PDI) and this provides a measure of the breadth of the molecular-weight distribution.

Routine determination of molecular weight of commercial polymers is made by the intrinsic viscosity $[\eta]$ of a polymer solution using Mark-Houwink-Sakurada equation given below:

$$[\eta] = K \overline{M}_v^\alpha, \text{ where}$$

$\overline{M}_v$ = viscosity average molecular weight.

Another most widely used method for the routine determination of molecular weight is by gel-permeation chromatography using the principle of size-exclusion chromatography. 'K' and 'a' are empirical Mark-Houwink constants that are specific for a given polymer, solvent and temperature. The value of exponent 'a' normally lies between 0.5 to 1.0. Mark-Houwink constants are available for most commercially important polymers.

9.1.10 ENGINEERING AND SPECIALITY POLYMERS

Engineering and speciality thermo-plastics offer some remarkable properties (i.e., high thermal stability, high flexural, tensile and impact strength, low creep compliance and good chemical resistance) as compared to commodity plastics like polystyrene and polyolefins, and that too with reasonable price tags. Hence, these polymers offer in many cases a viable alternative for metals, especially where high strength-to-weight ratio is a prime consideration, as in aerospace and automotive applications.

The distinction between engineering polymers and speciality polymers is rather arbitrary. *Engineering polymers* generally include those polymers which are used in the manufacture of premium plastic products that exhibit high impact strength, high temperature resistance, high chemical resistance and such special properties. **Ex:-** Aliphatic polyamides (e.g. Nylon-6, Nylon-66) acetal, polysulfones, polycarbonates, Acrylonitrile butadiene and styrene (ABS) resin, poly (phenylene oxide) resins, poly (p-phenylene sulfide), and fluroplastics (*e.g.*, Teflon TM) and engineering polyesters. These engineering plastics are generally priced 1 to 5 times the cost of the commodity plastics.

Speciality polymers : Speciality polymers are usually priced at 10 time more than the commodity plastics. **Ex:-** Polyetherimide, poly (amide-imide), polybismaleimides, ionic polymers, polyphosphazines, poly (aryl ether ketones), polyarylates and related aromatic polyesters, and ultra-high- molecular-weight polyethylene.

These polymers exhibit exclusively remarkable performance in certain special, limited but critical applications such as in aerospace composites, membranes for gas liquid separations, fire-retardant textile fabrics used by fire-fighters and race-car drivers, and as sutures and surgical implants. Polymides represent the most important group of speciality polymers.

9.1.11 Electrically Conducting Polymers

Generally polymers including plastics, elastomers and fibers are regarded as insulators ($\sigma = 10^{-7} - 10^{-18}$ S Cm^{-1}) because of the intrinsic property of carbon-carbon covalent bonds and were used as insulating materials in electric wires. Polyethylene, PVC, rubber used as insulating sheath. Under certain circumstancess they can be made to behave like a metal ($10^3 - 10^7$ S Cm^{-1}). For this discovery Alan T. Heegar, Alan G. MacDiarmid and Hideki Shirakawa receive the Nobel prize in Chemistry 2000.

Conductivity generally increases with decreasing bandgap which is defined as the amount of energy needed to promote an electron from the highest occupied energy level or valence band to the empty band (conduction band). Metals have zero band gaps, whereas insulators like many polymers have large bandgap (1.5 - 4 eV) which hinder electron flow. By careful design of the chemical structure of the polymer backbone, it may be possible to obtain bandgaps as low as 0.5 to 1 eV.

In becaming electrically conductive its electron need to be free to move. The polymer should have alternate single and double bond called conjugated double bond and the polymer has to be disturbed either by removing electron from (Oxidation) or inserting them into (Reduction) the polymer. The process is known as doping .

Some important electrically conducting polymers are polysulphur nitrile PSN, polyacetylene PA, Poly-p-phenylene PPP, Polypyrrole PPY, Polythiophene PTh, Polyaniline PANI etc.

Polysulphur nitride : Polysulphur nitride was discovered by V. V. Walatka in 1973. It give conductivity 5 x 10^3 S.Cm^{-1} and is due to their nonbonding electrons of sulphur and nitrogen. It behaves as a superconductor at -272°C. Polysulphur nitrite could not find wider use due to its extreme reactivity.

The reduction of S_2Cl_2 with ammonia forms cyclic tetrasulfur tetranitride (A) which is orange yellow crystalline solid which can explode if stored as a solid. The cyclic tetramer is converted to cyclic dimer (B) by heating at 85°C in vacuum. The vapours are passed through heated silver wool at 200 - 300°C. The cyclic dimer is condensed as white solid on a cold finger cooled in liquid N_2. The formation of polymeric sulphur nitride (C) occure by solid state polymerization of cyclic dimer at 25°C for 3 days followed by heating in vacuum of 75°C for 2h.

$$\begin{matrix} N=S-N \\ | \quad \quad || \\ S \quad \quad S \\ || \quad \quad | \\ N-S=N \end{matrix} \longrightarrow \begin{matrix} S=N \\ | \quad | \\ N=S \end{matrix} \longrightarrow \{S=N\}_n$$

A → B → C

Polyacetylene In 1970 H. Shirakawa reported high molecular weight polyacetylene film by passing dry acetylene gas over the Ziegler-Natta catalyst of titanium tetrabutoxide triethyl aluminium in toluene at -78°C.

$$n\,CH=CH \xrightarrow{\dfrac{Ti(OBu)_4}{Al(C_2H_5)_3}} \{CH=CH\}_n$$

The polyacetylene film was obtained on the surface of the catalyst solution. The cis/trans ratio of polyacetylene obtained depends on the nature of the catalyst and the polymerization conditions.

catalyst system	Polymerization condition	Cis / trans
$Ti(O\text{-}n\text{-}C_4H_9): 4\,Al\,(C_2H_5)_3$	Toluene, -78°C	95/5
$Co(NO_3)_2 : NaBH_4$	C_2H_5OH, -78°C	50/50
Ni(acetylacetonate)$_2$	THF, dark 100°C	5/95

The conductivity of cis and trans form of polyacetylene is as follows

Trans - PA $\{CH{=}CH{-}CH{=}CH\}$ 4×10^{-5} S. cm^{-1}

Cis - PA $\{CH{=}CH{-}CH{=}CH\}$ 2×10^{-9} S. cm^{-1}

The conductivity of polyacetylene varies with the nature of dopant. Plyacetylene exists as two geometric isomers. viz Cis PA and trans PA

Dopant ion	σ, S Cm^{-1}
I_3^-	550
AsF_6^-	1100
BF_4^-	100

Poly(p-pherylene) Poly(p-pherylene) PPP, is prepared by the step polymerization process involving the reaction of 1, 4 - dibromobenzene with magnisium in ether in presence of nickel chloridebipyridyl catalyst.

$$n\,Br\text{-}\langle O \rangle\text{-}Br + Mg \xrightarrow[Ni(Py)_2]{Ether} \{\langle O \rangle\}_n$$

PPP is black in colour. It show conductivity 10^{-11} S. Cm^{-1} and conductivity increases when the polymer is doped with I_2, AsF_5, BF_4^-.

Dopant	Conductivity, σ SCm^{-1}
I_3	$< 10^{-5}$
A_5F_5	$\sim 10^2$
BF_4^-	~ 70

Poly (phenylene sulfide) : Polypherylene sulphide PPS, is synthesized by reacting p-dichlorobenzene with unhydrous sodium sulphide in methyl pyrrolidone at 260°C under normal pressure. It precipitate as white powder.

$$nCl-\langle O \rangle-Cl + Na_2S \longrightarrow (\langle O \rangle-S)_n$$

Without dopant the conductivity of PPS is 10^{-16} S.Cm^{-1}. While on doping with arsenic pentafluoride (AsF_5) is increases to 5 S.Cm^{-1} and with Iodine it is 10^{-5} S.Cm^{-1}.

Polypyrrole : Polypyrrole PPY was synthesized by electrochemical technique. Pyrrole (0.06 M) in aqueous acetonitrile solution was electrolyzed by using platinum electrode and tetraethyl ammonium boron tetrafluoride as the supporting electrolyte (0.1M). A blue block film of polymer precipitated on the anode.

The conductivity of PPY with different dopent is as followes :

Dopant	Conductivity σ SCm^{-1}
I_2	600
AsF_5	100
BF_3	100

Polythiophene : Polythiophene can be synthesize chemically as well as electrochemically. Thiophene polymerizes in acetonitrile and tetrabutyl ammonium tetrafluoroborate at 1.7 Vs SCE (saturated calomel electrode).

Poly(3-alkylthiophene) can be synthesized from 3-alkylthiophene Via oxidative polymerization with $FeCl_3$.

Polyaniline : PANI is the oxidative product of aniline under acidic conditions and is known in 1862 as aniline black. PANI is prepared either by chemical or electrochemical oxidation of aniline under aqueous acidic media. Whenever thin films and better ordered polymer is required an electrochemical method is preferred. The quality of prepared polymer is affected by a number of factors. These includes electrode material, current density, solvent, temperature and concentration of oxidant.

Chemical synthesis : An aqueous solution of ammonium persulphate is added slowly to a solution of aniline dissolved in aqueous HCl at lower temperature (0-5°C) under agitation. After 1h the precipitate formed was removed by filtration. The polymer thus obtained is emeraldine hydrochloride. The emeraldine hyelrochleride was converted into the emeraldine base by stirring with 0.1 M solution of ammonium hydroxide for several hours.

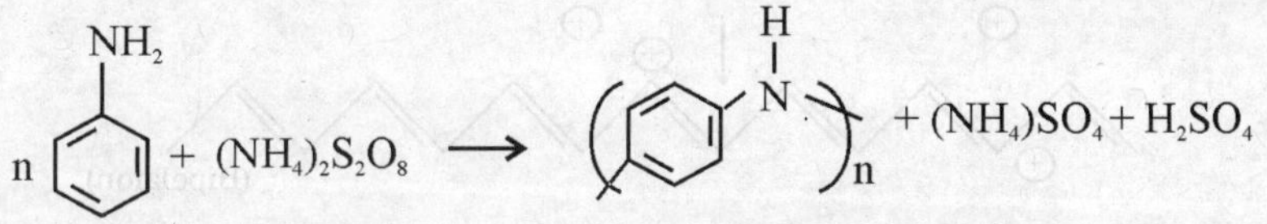

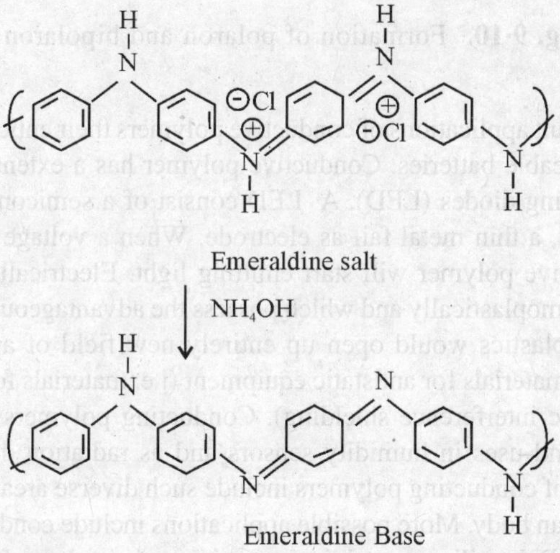

$$n\ C_6H_5NH_2 + (NH_4)_2S_2O_8 \longrightarrow (C_6H_4-NH)_n + (NH_4)SO_4 + H_2SO_4$$

Emeraldine salt

$$\downarrow NH_4OH$$

Emeraldine Base

Electrochemical synthesis : Electrochemical polymerization has technological potential, electrochemical synthesis is achieved by using potentiostate or galvaoistate. The electrochemical cell consists of three electrodes dipped in an electrolyte solution containing monomer in aqueous acid. When potential 0.7 - 1.1 V is used it will lead to the deposition of PANI film on the surface of platinum foil.

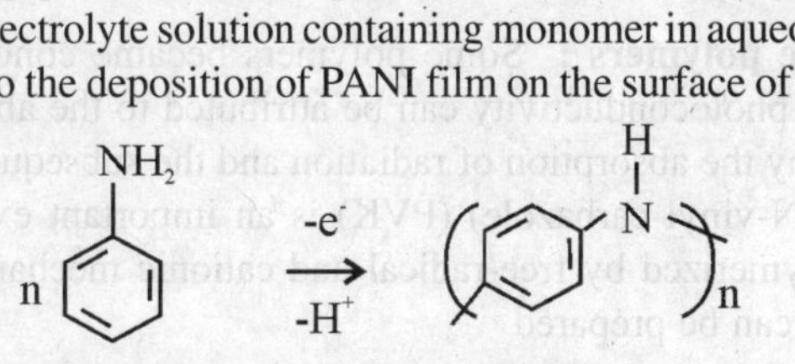

$$n\ C_6H_5NH_2 \xrightarrow{-e^-, -H^+} (C_6H_4-NH)_n$$

Doping for better molecule performance : Doped polyacetylene is comparable to good conductor such as cupper and silver, whereas in its original form is semiconductor. Polymer molecule consists of σ bond and π- bonds. The σ - bond are fixed and immobile. They form the covalent bonds between the carbon atoms. π electrons in a conjugated double bond are relatively localized. One or more electrons have to be removed or inserted. If an electrical field is applied, the electrons constituting the π bonds can move rapidly along the molecular chain. On oxidation polyacetylene get electron deficient called radical cation or polaron. The lonely electron of the double bond from which an electron was removed can move easily. As a consequence the double bond moves along the molecule. The positive charge is fixed by electrostatic attraction to the chloride ion. If the polyacetylene chain is havily oxidised polarons condense pair-wise into so called solitons. These solitons are then responsible for the transport of charge along the polymer chains.

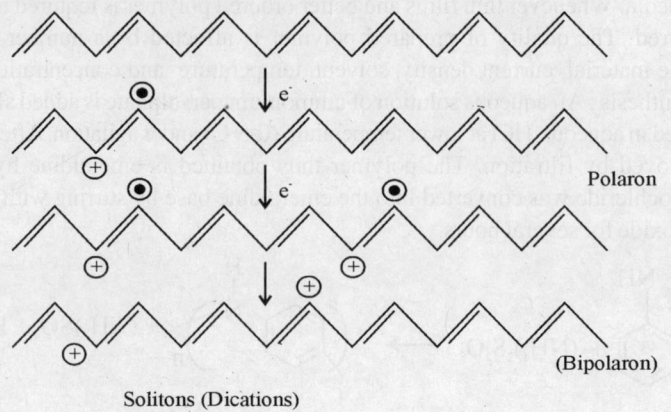

Fig. 9·10. Formation of polaron and bipolaron

Applications:

One of the most important applications of conductive polymers their enticipated use as electrodes in light weight and rechargeable batteries. Conductive polymer has a extensive commerical use in photodiodes and light emitting diodes (LED). A LED consist of a semiconductive polymer in the middle and at the other end, a thin metal fail as electrode. When a voltage is applied between the electrodes, the semiconductive polymer will start emitting light. Electrically conducting polymers which can be processed thermoplastically and which possess the advantageous mechanical behaviour and corrosion stability of plastics would open up entirely new field of applications. Cables for conducting electricity, new materials for antistatic equipment (i.e. materials for avoiding electrostatic charges and electromagnetic interference shielding). Conducting polymers such as polyfuran and polythiophene have profound uses in humidity sensors and as radiation detectors as well as gas sensors. Other applications of conducting polymers include such diverse areas as solar cells and drug delivery system for the human body. More possible applications include conductive paints, toners for reprographics and printing. Polyaniline containing organic coatings have found to offer corrosion protection of steel in acid and saline media.

9.1.12 Photoconductive polymers : Some polymers became conductive when illuminated. In general, the property of photoconductivity can be attributed to the ability to generate free-charge-carriers (electron hole) by the absorption of radiation and the subsequent transport of these carriers to the electrode. Poly (N-vinyl carbazole) (PVK) is an important example of a photoconductive polymer. It can be polymerized by free-radical and cationic mechanisms. Both amorphous (free radical) and tactic PVK can be prepared

Poly (N-vinyl Carbazole)

In the dark, PVK behaves as an insulator, but however, it becomes conductive when exposed to UV radiation. Incorporating sensitising dyes or electron acceptors extends the photoconductive response to the visible and IR region.

Photoconductive polymers are of great importance for xerox-copying, laser printing and duplicator industries.

Ionic polymers : Ionic sites can be introduced into polymers by copolymerization or chemical modification. When the ionic polymers derived from synthetic organic polymers contain upto 10 - 15 mole % ionic content, they are called "Ionomers". Those polymers which contain ionic content much higher than 15 mol % are called "polyelectrolytes". These are insoluble in common organic solvents but are soluble in water.

Ionomers are used in ion-exchange resins, superacid catalysts, membranes for liquid and facilitated gas separations, and particularly as separators in chloralkaline electrolytes.

Ionic sites that are generally incorporated are carboxylic or sulfonic acid groups that are partially or completely neutralized to form the polymeric salt. Counter ions used include Na, Zn, ammonium or a halide. Typical non-ionic polymer backbones include polyethylene, polystyrene, and copolymers of fluorocarbons such as tetrafluoroethylene.

Speciality polyolefins : Ultra high-molecular weight-polyethylene (UHMWPE) having molecular weight in the range 1 to 5 million, exhibits exceptional impact and tensile strength, tear-resistance, puncture-resistance, high abrasion-resistance, chemical inertness, low coefficient of friction and good fatigue resistance. UHMWPE is used in orthopedic implants, grocery sacks, battery separators, and as additives for improving the sliding and wear behaviour of other thermoplastics, UHMWPE may be processed mainly by compression sintering and ram extrusion into sheets and rods. Fibers with high tensile strength and low density can also be obtained by gel spinning.

Polyaryletherketones (PEEK) : These are new group of important engineering thermoplastics of which polyetheretherketone (PEEK) is the most important. It is partially crystalline (20 to 30%) but can be melt processed at elevated temperatures. It has good abrasion resistance, low flamability and low emission of smoke and toxic gases, low water absorption, and resistance to hydrolysis, wear, radiation and super-heated steam. Its high solvent-resistance, good impact strength and good thermal stability make it extremely useful as a thermoplastic matrix for graphite composites. It is also used as solvent-resistant tubing for chromatography, wire and cable insulation for hostile environments, and as magnet-wire coating.

Liquid-crystal polymers (LCP) : Low-molecular weight liquid crystals were known as early as 1888 when it was reported that some cholesteryl esters which are colored and opaque became clear on heating. Liquid-crystal behaviour for a polymeric system was first observed in 1974 in a concentrated solution of poly (benzyl-L-glutamate), which forms a rod-like α-helical conformation in a variety of organic solvents.

Liquid-crystal polyesters and copolyesters have highly aromatic structures.

Ex:- Vectra (Hochst celanese)

Vectra

The LCP^s, by virtue of their morphology, possess remarkable properties such as exceptional oxygen and barrier properties (100 times that of PETP), good thermal stability, good dimensional stability, high modulus and strength and high chemical and solvent resistance.

LCP^s, are used for speciality applications such as in electronic components (e.g. computer memory modules), housings of light-wave conductors, a variety of aerospace applications and filaments to compete with aramid fibers.

Other new entrants in the group of speciality polymers include inorganic polymers (such as poly(organophosphazenes) and poly(silastyrene), polybenzimidazoles, poly(p-phenylene benzobisthiazole) and polybenzoazole used for high-temperature composites and high performance fibers, and polyphthalamide (Amodel), poly(aryl ether ketones) etc. which exhibit superior creep resistance, excellent chemical resistance and dimensional stability, high rigidity and tensile strength and good thermal properties for special applications.

Application for polymers in separations : Biomedical devices polymeric membranes have great versatility as compared to metals, ceramics, and microporous carbon. Production of potable water by reverse osmosis and the separation of industrial gases are the major applications of polymeric membranes. Membranes can also be used for several other important applications such as filtration of particulate matter from liquid suspensions, air, and industrial flue-gases and the separation of liquid mixtures such as the dehydration of ethanol azeotropes. Polymeric membranes are also used for specialized applications such as ion separation in electrochemical processes, membrane dialysis of blood and urine, artificial lungs and skin, the controlled release of therapeutic drugs, the affinity separation of biological molecules, membrane-based sensors for gas and ion detection, and membrane reactors.

Intelligent polymers Intelligent polymers are among the novel "smart materials". These materials combine sensors, actuators, information processing and energy storage/conversion functions into the single material or composite material system. The smart material is capable of detecting a change in its environment (e.g., the onset of corrosion) and actuates an appropriate response (e.g., the release of a corrosion inhibitor) autonomously and be self-powered. Several multi-functional integrated intelligent polymer systems are being developed nowadays for a variety of applications e.g., electronic nose, fabric strain gauges, artificial muscles, etc.

Crystallinity of polymers, and inter-relationship between molecular make-up polymers and their properties and uses are described under Chapter-13 (Structure of Solids)

9.1.13 Structure - Property Relationships in Polymers

Polymeric substance exhibit wide varieties of physical and chemical properties, because of the many possible kinds of macromolecular composition and arrangement. The chemical nature of the monomeric unit is the primary factor which determines the properties of the polymer. Difference in the thermal stability and mechanical strength of different polymers are related to differences in bonding and structure of the monomer chemical reactivity of a polymer is mostly due to the reactivity of its molecular components. Polymers have two types of bonds chemical and intermolecular which differ in energy and length. The atoms in the polymeric chain are jointed each other by strong chemical bond about 1 to 1.5 A° in length and intermolecular forces between the chain of about 3 to 4 A°.

Hence the propertys of polymeric material depends on constituent monomer unit, the way they are arranged in polymer chain and the nature of intermolecular interaction that hold them together

Chemical resistances of polymers: The solubility of a polymer in solvent is inverse to its molecular weight. Chemical resistance of polymer depends upon the chemical nature of monomer. Polar polymers such as PVA, PVC polyamide etc. dissolved in polar solvent such as water, alcohol, phenol, formic acid etc. While non-polar polymer PE, PP, PS can dissolve in non polar solvents ex. benzene, toluene, xylene etc. High molecular weight polymer on dissolution gives a solution of high viscosity. Solubility decreases with the cross linking of polymer. Polymer containing double bond are vulenerable to oxidation and scission ex. polybutadiene, polyisoprene compared to PE, PVC. Polymers containing ester, amide and corbonate groups are susceptible to hydrolysis. The presence of a benzene ring adjacent to such groups offers some protection against hydrolysis.

Crystallinity and amorphousness : Crystallinity is favoured by symmetrical chain structure that allow close packing of the polymer units to take greatr advantage of secondary forces. Crystallinity is also favoured by high interchain (Secondary forces) interactions. Thus linear HDPE and PTFE both with high symmertrical chain structure have a high tendency to form crystalline solids. Increased of hydrogen bonding increase Tg and Tm. HDPF is highly symmertrical has a low Tm (135°C), where as nylon 66, both symmertrical and possessing hydrogen bonding show a Tm of about 270°C.

Tacticity (Sterioregularity) and geometrical isomerism affect the tendency toward crystallization. The tendency increase as the tacticity is increased and when the geometrical isomer is predominantly trens. Thus isotactic PS is crystalline, whereas atactic PS is amorphous and Cis-polyisoprene is amorphous, whereas the more easily packed trans isomer is crystalline. Most unstretched elastomers and many plastics are amorphous most fibers are highly crystalline.

The density of a polymer is directly related and the transparency is incersely related to the degree of crystallinity. Atactic PS, PMMA and polycarbonate is amorphous HDPE and nylon 66 is highly crystalline. The density of a polymer is directly related and the transparency is inversely related to the degree of crystalllinity. Atactic PS PMMA and polycarbonate is amorphous HDPE and nylon 66 is highly crystalline.

Thermal properties : The glass transition temperature Tg increases and the intermolecular force in the polymer and the regularity of the polymer structure increase. Thus PVC has a higher Tg than linear polyethylene (HDPE) because of the presence of dipole-dipole interactions between the chains in PVC similarly the Tg of isotactic PP is greater than the Tg of the less regular atactic PP. The Tg of cis-polyisoprene (178 K) is lower than that of trans-polyisoprene (190 K).

Polyesters produced by the reaction of a glycol with phthalic acid have less regularity, are less crystalline and have lower T values than those produced from terephthalic acid. The T_m (538°C) values of aromatic polyester (Ekonol) are higher than those of the corresponding aliphatic esters, polyethylene adipate (65 °C). The T_m of aromatic polyester can be reduced by the insertion of methylene group in the polymer chain. Similarly T_m values of aromatic polyamides.

Similarly the T_m values of aromatic polyamide (aramides) are higher than those of the corresponding aliphatic polyamides nylons - 66. The T_m of nylon 8 is 200°C and the T_m of nylon 9 is 209°C. There are eight methylene groups in nylon 9 and they fit more tightly than the seven methylene groups in nylon 8.

Regularly arranged bulky pendant groups such as phenyl groups increase Tg Thus. Polystyrene has a Tg of 100°C while that of PE is 120°C.

Isotactic PP crystal have T_m of 171°C and syndiotactic has a T_m of 138°C whereas atactic PP is amorphous and has a Tg of 255 K.

Plastic Deformation : Plastic deformation is found in thermoplastic polymers structure is deformed on application of heat and / or pressure. Since theramoplastic polymer has covalent bond within the chain and weaker secondary Vander woal bonds between the chain and secondary forces, can be easily broken when object is heated. At high temperature the vander waal forces acting between different molecule become more and more weak. In heated state it takes the shape of mold on cooling polymer become hard. Hence polymers like PE, PVC, PS, PMMA etc. undergo physical deformation. However in thermosetting polymers only chemical bond (Covalent bond) are present in the polymer structure. Hence deformation does not occure on heating or on pressure ex. Backelite, polysilicons.

Mechanical Properties : Various factors that affect mechanical properties of polymers are molecular weight, extent and distribution of crystallinity and composition of polymer. Mechanical properties of polymers improve rapidly as the molecular weight increases up to the threshold molecular weight.

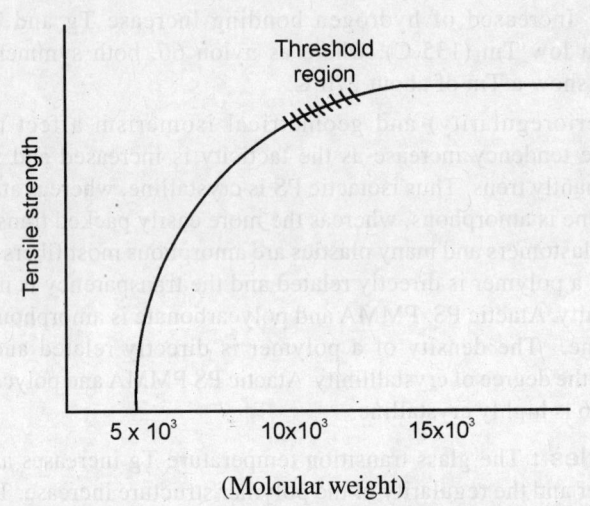

Fig. 9·11. Effect of molecular weight on tensile strength of PVC.

Polar group increases the intermolecular force of attraction and hence the strength of polymer increases. Polyester, Polyamide has higher strength than the PE, PP. Fibres are characterized by high tensile strength. These properties are associated with molecular symmetry and high cohesive energies between chains both of which are associated with high degree of crystallinity. Fibers are linear and oriented in one direction and thus has high tensile strength. Branching and cross linking in fibers are undesirable since they disrupt crystal formation.

9.2. Resins and Plastics

Plastics are high molecular weight organic materials, which can be moulded or formed into stable shapes by the application of heat and pressure. "Plastics derive their name from their existence, at some stage of their formation or fabrication, as plastic masses capable of being shaped by flow induced by heat or pressure or both".

Plastics are relatively new materials of *construction* which rapidly found extensive industrial applications. Plastics having a variety of properties are available at present. They have low specific gravities, ease of fabrication, resistance to solvent action and low thermal and electrical conductivities. Some of them are completely transparent and have a high refractive index. Many plastics can take wide range of colour to enable them useful for decorative purposes. Their light weight, low fabrication *expenses,* low thermal expansion coefficient, chemical and corrosion resistance have conferred several engineering advantages on plastics as compared to other materials of construction. Furthermore, plastics used in conjunction with metals, helped in overcoming many design problems. Many plastics have greater strength than metals on unit weight basis. Some plastics have wear resistance which enabled them useful for making bearings, gears, etc.

Plastic parts of complicated shapes and accurate dimensions can be cheaply produced by moulding. Yet another advantage of plastics is that they are capable of minimizing vibration and noise in machines.

Plastics are widely used in making electrical instruments, telephones, panelling for walls, instrument boards, automobile parts, lamps, goggles, optical instruments, household appliances, etc. Plastics have been helpful in making non-metallic bearings which last much longer than metallic bearings in some applications. These bearings havs been made from laminated phenolics with canvas or paper base and also from moulded phenolics with wood fluor, fabric or inorganic fillings. These bearings can be lubricated by grease or water. Water lubricated plastic bearings are extensively used in food industry (where oil-lubricated metallic bearings may cause contamination) and in rolling mills for steel.

Many plastics are electrical insulators and hence find extensive use in electrical industry. Their optical clarity combined with good strength enabled them useful for preparation of windscreens for automobiles and aircraft.

Many plastics cannot be used at elevated temperatures. Many of them are inflammable but some plastics are fireproof. Thermosetting polymers have a tendency to shrink.

Plastics are supplied to the manufacturer usually in the form of powder, flakes or granules. From these materials, the required articles can be produced by pressure moulding or by forcing the softened plastic through a die to form rods, sheets, tubes and other shapes. The substance mary be used in liquid form to impregnate paper, cloth, glass matte, wood etc., for preparing laminates. The impregnated sheets of the desired shape can be bonded by the application of heat and pressure. Laminated plastics are used to prepare a variety of articles such as automobile bodies, insulation panels, luggage, table tops, etc. Some types of plastics are nowadays produced in the form of solid foams useful for insulation, cushioning, packaging or to produce buoyancy.

9.2.1 Constituents of plastics. Plastics may contain a, number of constituents such as (1) Binders (2) Fillers (3) Dyes and pigments (4) Plasticizers (5) Lubricants (6) Solvents and (7) Catalysts.

(1) Binders. A plastic is usuelly classified depending on the type of binder used for its manufacture. The binder used also determines the type of treatment needed to mould the articles from the plastic material. The main purpose of a binder, as the name itself implies, is to hold the other constituents of the plastic together. A binder may compose of 30 to 100% of the plastic. The binders used may be natural or synthetic resins or cellulose derivatives, both of which are polymeric materials having large molecular weights. Thus, *resins* are the basic binding materials in the plastic, which account for a major part of the plastic and which determines the type of treatment needed in the moulding operations.

Two types of resins are usually distinguished from engineering point of view :

(*i*) Thermoplastic resins, and

(*ii*) Thermosetting resins.

On the basis of the type of the resin used in its preparation, the plastic itself is called thermoplastic or thermosetting plastic.

(*i*) **Thermo-plastic resins.** Thermoplastic resins soften on heating and become plastic so that it can be converted to any shape by moulding. On cooling, they become hard and rigid. On re-heating, they soften again and the material can be remoulded to any desired shape. Thus, their softness and hardness are temporary phenomena which are attainable by heating or cooling. The resins that are formed by addition polymerization are thermoplastic and have linear long chain polymeric structure without any cross-linkings. In thermoplastic resins, the chemical structure or the molecular weight are not changed during the heating or moulding operations. Only the secondary bonds between the individual molecular chains are broken on heating which results in their softening and flow properties. On cooling, these secondary bonds are re-established as a result of which they become hard again. Obviously, thermoplastic resins are weaker, softer and less brittle as compared to thermosetting resins. The molecular weights of thermoplastic resins are smaller than those of thermosetting resins. Also their inter-molecular forces are weaker than in thermosetting plastics. That is why thermoplastics swell or dissolve in some solvents.

Examples : Polyethylene, polystyrene, polyvinyls, (such as nylons), acrylics and cellulose derivatives.

(*ii*) **Thermo-setting resins.** Thermosetting resins are those which set upon heating and cannot be reformed when once they are set. In general, those resins which are formed by condensation are thermosetting. The thermosetting resins have three-dimensional network structure and have very high molecular weights. These resins have predominant covalent cross links between the long chain molecules which are responsible for the three-dimensional network structure. When these materials are moulded, additional cross-linkings are formed between the long chains leading to further increase in molecular weight. When cross-linkings are formed, the thermo-setting resins acquire some of their characteristic properties such as hardness, toughness, non-swelling and non-softening properties, brittleness, etc. The strength of these bonds are retained even on heating and hence they cannot be softened or remoulded or reclaimed when once they are cured as can be done in the case of thermoplastic resins.

Examples. Phenol formaldehyde (Bakelite), urea and melamine formaldehydes, alkyds, polyesters, silicones, etc,

(2) Fillers or extenders. Fillers serve two important functions :

(*i*) To reduce the cost of the plastic per unit weight.

(*ii*) To impart certain specific properties to the finished product. For instance, barium salts are used to render the plastic impervious to X-rays; quartz and mica are used to improve hardness; inorganic fillers like asbestos are added to improve heat resistance and corrosive resistance; shredded textiles are used to increase impact strength and so on. In addition to those mentioned above, other materials in common use as fillers include cotton, corn husks, graphite, carbon black, clay, paper pulp, wood flour, pumice, metallic oxides (*e.g.,* ZnO, PbO), saw dust, carborundum and metal powders (such as Fe, Cu, Pb and Al). The proportion of the filler added can reach as high as 50% of the plastic.

3. Dyes and Pigments. These are meant for providing decorative colours to the plastic.

4. Plasticizers. The important functions of a plasticizer in plastics is to improve plasticity and flexibility, so as to reduce the temperature and pressure required for moulding. Platicizers also play a vital part in determining the properties of the finished product and hence are chosen accordingly. Their plasticizing properties are believed to be mainly due to the partial neutralization of the intermolecular forces of attraction in the resin molecules. The porportion of plasticizers can be even upto 60% of the plastic. Plasticizers are often used with thermosetting plastics, particularly with cellulose derivatives. For instance, cellulose acetate has a tendency to discolour when moulded unless the moulding temperature is reduced by the addition of a suitable plasticizer. A variety of organic materials are used as plasticizers. These include non-drying vegetable oils, tributyl phosphate, triphenyl phosphate, triacetin, camphor, esters of oleic. stearic or phthalic acids, etc. When used with cellulose acetate, camphor increases surface hardness; tributyl phosphate and triphenyl phosphate imparts flame-proofness, triacetin and tributy phosphate improves toughness, and so on. However, a plasticizer may tend to reduce the tensile strength and chemical resistance of the plastic.

5. Lubricants. The lubricants in common use include oils, waxes and soaps. They help in easy moulding and elegant finish to the final product.

6. Catalysts. These are used in the case of thermosetting plastics to accelerate the condensation polymerization to form the cross-linked product.

9·2·2. Fabrication of plastic articles. The commonly used methods for fabrication of plastic articles include casting, blowing, extrusion, lamination and moulding.

1. Casting. This is a simple method of forming used for both thermoplastics and thermosetting plastics. It consists of pouring the molten resin in a suitable mould (usually made of lead) and curing at about 70°C for several hours at atmospheric pressure. For phenolic resins, the curing time is about 50 hours. Cast phenolics may be transparent or translucent, colour well and are finding wide application. The products formed are free from internal stresses end can take a high polish.

2. Blowing. In this process, the softened thermoplastic resin is blown by air or steam into a closed mould, just like industrial glass-blowing.

3. Extrusion. This method is generally used for the, manufacture of articles, like sheets, tubes, rods, etc., having uniform cross-section. The method is used only for thermoplastics. In this method, the material of the required composition is forced by a screw conveyer into a heated chamber where it softens, and then is forced through a die having the required shape. The finished product that extrudes out is cooled by atmospheric exposure or by blowing air or by spraying water as it is carried away by a long conveyer.

4. Lamination. Laminated plastics are generally prepared from sheets of cloth, wood or paper by impregnating them with a resin solvent solution. Phenolic and urea type resins are most commonly used. These are then piled one over the other until the desired thickness is obtained, heated to remove the excess of the solvent, and pressed together between two highly polished steel surfaces to get the laminated product. Laminated platics have high tensile strengths and impact resistances.

5. Moulding. Moulding is an important method of fabrication of plastics and is done in several ways as given below. Generally, the moulding of the plastic is done around a metal insert so that the finished product has a metal part firmly bonded to the plastic.

(*a*) **Cold moulding**. In this, the object is formed in a mould by the application of high pressure in the cold. It is then taken out of the mould, heated (cured) to remove the excess solvent and to promote further condensation and hardening (as in the case of phenolic resin compositions).

(*b*) **Compression moulding**. This is the most widely used method particularly for thermosetting plastics. In this method, the raw materials of the right composition are placed in a mould and the mould is carefully closed under low pressure. Then, high pressures of the order of 100 to 500 kg/cm^2 and temperatures of the order of 100 to 200°C are applied to complete the curing process. Nowadays, automatic compression moulding presses are available for conducting the process rapidly.

(c) **Injection moulding**. This method is generally used for thermoplastics. The moulding composition is heated in a suitable chamber connected by a duct leading to the mould. The hot softened plastic is then forced under high pressure into the relatively cool mould where it is set by cooling and the moulded object is then ejected. The temperatures used are 90° to 260°C.

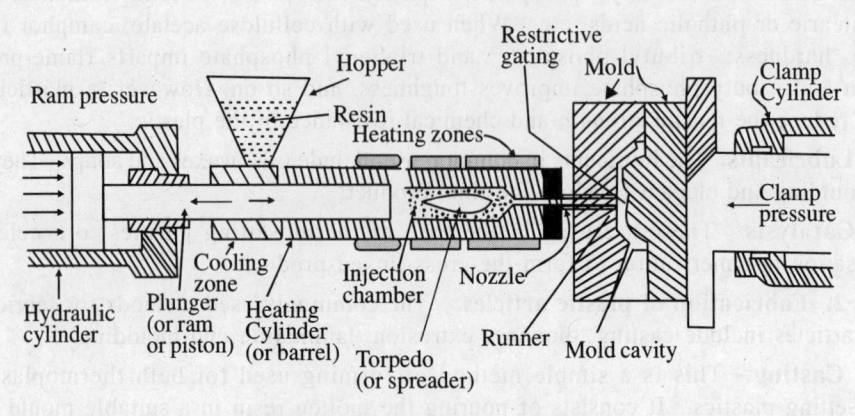

Fig. 9·12. Injection moulding.

(d) **Transfer moulding**. This is a modification of injection moulding suitable for use with thermosetting plastics. The moulding composition is first plasticized by applying minimum heat and pressure in a chamber outside of the mould and then it is injected into the mould where curing takes place under the influence of heat and pressure. This method is used for more complicated shapes and permits the use of delicate inserts.

9·2·3. Thermoplastic Resins

(1) **Natural Resins**. Naturally occurring materials such as shellac, rosin or calophony, copal, amber, kauri, mastic, elemi, dammar, sanderac and some asphaltic and bituminous materials can act as resins and can be used as binders in plastics.

Excepting shellac, all the natural resins orignate from the exudations from plants and trees. When wounded accidentally or intentionally, these plants and trees exude water-insoluble resinous materials from their resin duct as a protection mechanism against the wound. These resins as they are exuded are rather fluid and are known as balsams. These are readily soluble in many solvents. When the balsams are exposed, their volatile constituents are lost the residue undergoes oxidation and other complex reaction as a result of which the resins become insoluble in the usual solvents, chemically inert and their softening temperatures also increase.

Rosin or calophony is the residue left after turpentine has been distilled from the balsam obtained from certain varieties of pine. Rosin majorly consists of abietic acid, but other constituents which restrain crystallization are also present. Rosin is graded according to its colour. 'A' is the darkest grade while 'WG' (window glass) and 'WW' (water white) are the lighter grades. Its importance to the paint and varnish industries lies in the fact that it can be hydrogenated, oxidized, esterified and converted into other film-forming materials. Chief among them is "Easter gum" which is made by esterification of the rosin acid with glycerin. Copal, amber and kauri are fossil resins obtained from trees and they are widely used in varnishes as well. Shellac is processed from the exudation of parasitic female insects called "coccus" on some types of trees found mainly in India. Shellac is soluble in alcohol and this is best used in making spirit varnishes. Natural resins are generally hard and possess low thermal conductivity and low dieletric constant. They can be easily moulded. These resins find use as binders for phonographic discs, grading wheels, as electrical insulators and for several other applications.

(2) **Cellulose derivatives**. Cellulose is a naturally occurring polymeric material containing thousands of glucose-like rings each of which contain three alcoholic OH-groups. Its general formula is represented as $(C_6H_{10}O_5)_n$. The OH-groups present in cellulose can be esterified or etherified. The most important aellulose derivatives are the esters *e.g.,* cellulose nitrate, cellulose acetate and cellulose acetobutyrate, and the ethers, methyl and ethyl cellulose. These substances possess most of the film-forming characteristics of the resins but their films are not adherent and have a tendency to wrinkle. Hence suitable resins are incorporated in lacquers to promote film-adherence and plasticizers are added to improve their flow properties, toughness, elasticity and smoothness of the film.

(i) **Cellulose nitrate**. This is also known as nitro-cellulose. This is prepared by reacting cellulose with nitric acid in presence of sulphuric acid which acts as a dehydrating agent. The partially nitrated cellulose mixed with camphor gives the so called celluloid, which can be easily softened and moulded. The function of camphor is to act as a plasticizer to enhance the moulding properties.

$$n[C_6H_7O_2.(OH)_3] + 2n.HNO_3 \xrightarrow{H_2SO_4} n[C_6H_7O_2(OH)(NO_3)_2]$$
cellulose dinitrate
(contains 11·1% nitrogen)
$+2n\,H_2O$

$$n[C_6H_7O_2.(OH)_3] + 3n.HNO_3 \xrightarrow{H_2SO_4} n[C_6H_7O_2(NO_3)_3]$$
cellulose trinitrate
(contains 14·1% nitrogen)
$+3n\,H_2O$

The properties of the final product depend upon the degree of nitration, which in turn depends upon the experimental conditions. For instance, the solubility of the product in organic solvents increases to a maximum at about 11·5% nitrogen.

Cellucose nitrate plastics are transparent, flexible, strong and touch. The specific gravity is 1.35 to 1.40. Its dimensional stability is poor as it slowly shrinks due to slow loss of plasticizer and hence is not used for precision parts. It is inflammable. It is resistant to water but is attacked by strong acids and alkalis. It tends to discolour with age. It becomes brittle at low temperatures.

Nitrocellulose containing 10.5 to 11.5% N is used for moulded extruded articles; that containing 11.5 to 12.1% N is used in lacquers, and that with 12.1 to 13.8% N is used for explosive gun-cotton. Cellulose nitrate plastics are used for tool handles, spectacle frames, drawing instruments, pens, toothbrush handles, table tennis balls, radio dials, picture films and many fancy articles. The first "artificial leather" was made by coating fabric with nitrocellulose suitably covered and embossed, which is still used in coverings for books, luggage and upholstery.

(*ii*) **Cellulose acetate**. This is obtained by treating cellulose with concentrated acetic acid or acetic anhydride in presence of a catalyst such as H_2SO_4

$$n[C_6H_7O_2.(OH)_3] + 3n.CH_3COOH \xrightarrow{H_2SO_4}$$
cellulose

$$n[C_6H_7O_2.(OCOCH_3)_3] + 3n.H_2O$$
cellulose triacetate

The resulting cellulose triacetate is partially hydrolysed to render it soluble in organic solvents such as acetone. It has good clarity, stability, toughness, impact strength and resistance to U.V. radiation. It has a high dielectric strength, high tensile strength and resistance to mineral acids. It has good film and plastic strength. It is used for preparing fibres for textiles on one hand and for preparing transparent sheets on the other. Thick sheets are made from blocks and polishing. Thinner films can be cast from solution on to polished metal surfaces. The films are also widely used in photography, for wrapping, and for making small envelopes, bags and boxes for packaging. It is also used for combs, goggles, handles, radio-appliances, steering wheels, etc. Since cellulose acetate is much less inflammable than cellulose nitrate it has replaced the latter in the manufacture of photographic films.

(*iii*) **Cellulose acetate butyrate**. It is a copolymer having lower water absorption that the cellulose acetate alone. It is prepared by treating cellulose with a mixture of acetic acid and butyric acid in presence of H_2SO_4 (catalyst). Its thermal stability and other properties are similar to that of cellulose acetate plastics but has better chemical and moisture stability, better dimensional stability and impact strength. It is used for toothbrush handles, combs, buttons, pens automobile hardware, etc.

(*iv*) **Ethyl cellulose**. It is prepared by treating cellulose first with caustic soda to produce the caustic cellulose and then allowing it to react with ethyl chloride to give an average of 2.5 ethoxyl groups per 1/2 ethyl cellulose monomer.

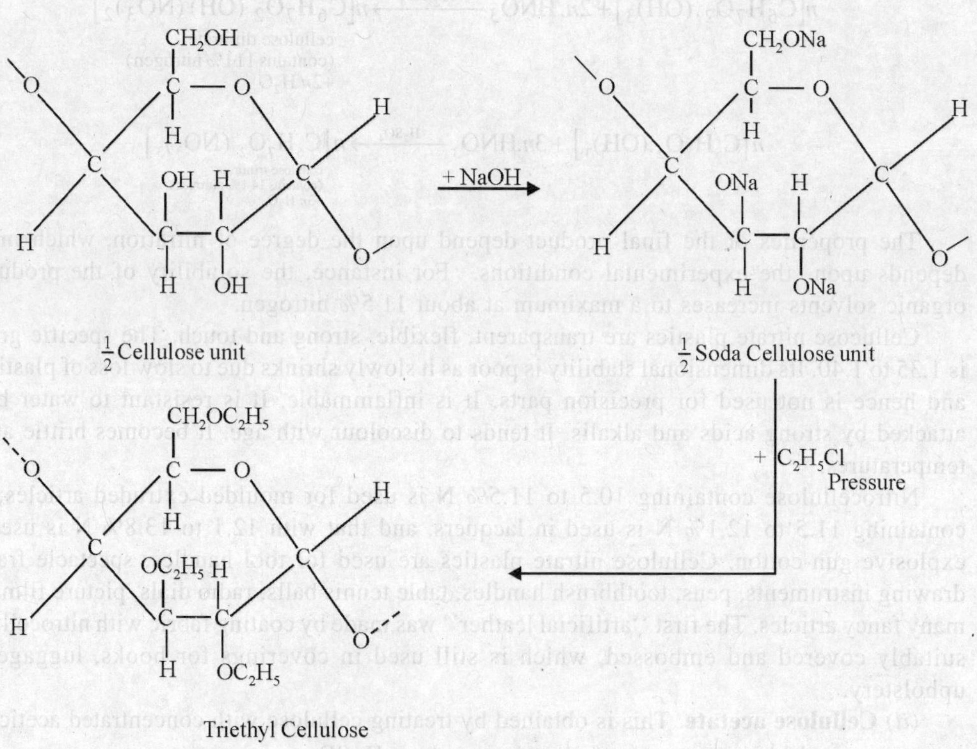

Fig. 9·13. Ethyl cellulose.

Ethyl cellulose is the toughest of all the cellulose plastics. Chisel handles, mallet heads and other objects which have to withstand impact have been made from this material. It has excellent solubility in many cheap solvents and hence foils and films with good strength and toughness can be prepared by casting its solutions. Its toughness and flexibility are retained through wide range of temperature. It has good chemical resistance and stable to heat and light. It has good electrical properties and hence films, foils and sheets made from it are used for electrical insulation purposes. Special compositions have been used to produce raincoats, surgical tapes, laminated glass sheets and straw like extrusions for weaving furniture seats, etc.

(3) Polyethylene. Polyethylene or polythene is a versatile plastic made from ethylene gas. Ethylene can be obtained (*a*) as a byproduct from the cracking of hydrocarbons in oil refining (*b*) by the dehydrogenation of ethane, and (*c*) by the dehydration of ethyl alcohol. During polymerization, the ethylene molecules are activated by the breaking of one of the coordinate double bonds, leaving in active electron on each of the two carbon atoms. Under suitable conditions, these activated molecules join together to form a very long chain polymeric hydrocarbon molecule.

$$n(CH_2 = CH_2) \rightarrow n(-CH_2 - CH_2 -)$$
molecular rearrangement to produce
a bifunctional molecule

$$\xrightarrow{polymerization} (-CH_2 - CH_2 -)_n$$
polyethylene

Polyethylene was first produced in the laboratories of Imperial chemical Industries Ltd. (England) at 1400 atmospheres of pressure and 70°C. Traces of oxygen caused the polymerization to take place.

High polymers of ethylene are made commercially at pressures of 1000 to 3000 atmospheres and at temperatures as high as 250°C.

Traces of oxygen, peroxides (e.g., bonzoyl. diethyl), hydroperoxides, and azo compounds have been used as initiators. Rapid exothermic reactions and even violent explosions may take place and hence proper precautions should be taken.

The original high-pressure process gave some branched polymers and the product is called low-density polyethylene. It is a white, waxy, translucent material having a specific gravity of 0.91 to 0.93. It is chemically inert and has good resistance to acids and alkalis. It is used as container for acids (including HF) and alkalis. It does not dissolve in any solvent at room temperature, but is slightly swollen in liquids like benzene and carbon tetrachloride in which it is soluble at high temperatures. It has excellent electrical properties. It has good toughness. It is used as films and sheets for packaging applications, including bags and wrappings, drapes, table cloths; etc. It is also used for production of house-wares, wire and cable insulations, for coating paper for preparing bottles, industrial containers, pipes for water supply, etc.

High density polyethylene

Polyethylene formed at low pressures it has higher melting point, higher density (0.941 to 0.965 g/cm^3), and higher tensile strength. This is a near crystalline polymer. Commercial production of high-density linear polyethylenes is being done now by coordination polymerization of ethylene (in presence of a coordination catalyst made from aluminium alkyl and TiCl$_4$ in a solvent like heptane) or by polymerization of ethylene with supported metal oxide catalysts.

Apart from the applications described above, high density polyethylene is used for housewares, toys, detergent bottles, pipes, textiles, drapes, upholstery, industrial clothes and filters.

(4) Polytetrafluoroethylene (Teflon). Polytetrafluoroethylene (Teflon) is prepared by polymerization of tetrafluoroethylene $(CF_2 = CF_2)$ at elevated pressures in presence of water using initiators (catalysts) such as benzoyl peroxide, persulfates, H$_2$O$_2$, etc.

$$n . \underset{\underset{F}{|}}{\overset{\overset{F}{|}}{C}} = \underset{\underset{F}{|}}{\overset{\overset{F}{|}}{C}} \longrightarrow -\underset{\underset{F}{|}}{\overset{\overset{F}{|}}{C}} - \underset{\underset{F}{|}}{\overset{\overset{F}{|}}{C}} -$$

Tetrafluoro-ethylene   PTFE (Teflon)

Polymerization can lead to reactions of explosive violence. Since the heat of polymerization is high, precautions must be taken to prevent local overheating leading to explosive disproportionation the monomer into carbon and carbon tetrafluoride.

It is a highly crystalline and orientable polymer. It has many remarkable properties.

(1) It is extremely resistant to attack by corrosive reagents and solvents. Only alkali metals either molten or dissolved in liquid ammonia attack the polymer, probably by removing fluorine from the chain. Fluorine itself can degrade the polymer under pressure on prolonged contact.

(2) It is totally unaffected by water.

(3) It has an excellent thermal stability. Even at 250°C. Its mechanical and electrical properties do not change for months together. It decomposes at still higher temperatures.

(4) Its melting point is so high that techniques similar to those employed in powder metallurgy have to be used in forming sheets and moulding articles.

(5) Moulded polytetrafluoroethylene articles have high impact strength but are easily strained beyond the point of elastic recovery.

(6) Its density is unusually high (2.1 to 2.3 g/cm^3).

(7) Its refractive index is unusually low (1.375).

(8) It is not hard, but is slippery and waxy to the touch.

(9) It has a very low coefficient of friction.

(10) It has extremely good electrical properties. Its dielectric constant is low (2.0), and its loss factor for all frequencies tested, including radar and T.V., is one of the lowest known for solids.

(11) Its mechanical properties including resistance to wear and deformation under load, stiffness and compressive strength can enhanced by the use of additives or fillers such as asbestos, glass fibre, graphite, bronze powder, zirconium oxide etc.

(12) It can be machined, punched and drilled.

Applications

Teflon is widely used because of its excellent toughness, chemical and heat resistance, electrical properties and low frictional coefficient.

(1) Its electrical applications include wire and cable insulation, insulation for motors, generators, coils transformers and capacitors and high frequency electronic uses.

(2) Its low-friction and antistick applications include its use a solid lubricant film on ammunition, gun mechanisms and as a bearing surface for light loads. It is also used in non-lubricated bearings, linings for trays in bakeries, etc. It is used in mould release devices, package machines, etc. It does not stick to other materials and is hence called an *abhesive*. It is accordingly used for making non-sticking stopcocks for burettes and as a thin tape for severing pipe threads. Low molecular-weight PTFE can be dispersed as aerosols for effective dry lubrication.

(3) It is used in chemical equipment *e.g.*, gaskets, pumps, valve packing, and pump and valve parts.

(4) Its use as fibre include gasketing, belting, pump and valve picking, filter cloths and for various industrial functions where absolute chemical and thermal resistance upto 250°C is required.

(5) **Polystyrene**. Polystyrene is prepared by polymerization of styrene usually in presence of a peroxide (*e.g.*, benzoyl peroxide) as initiator. The monomer, styrene is prepared from benzene and ethylene under pressure at 90°C in presence of a catalyst $(AlCl_3)$. The resulting ethyl benzene is dehydrogenated to styrene by passing over an iron oxide or magnesium oxide

or aluminium oxide catalyst about 600°C. The styrene is then refined by distillation.

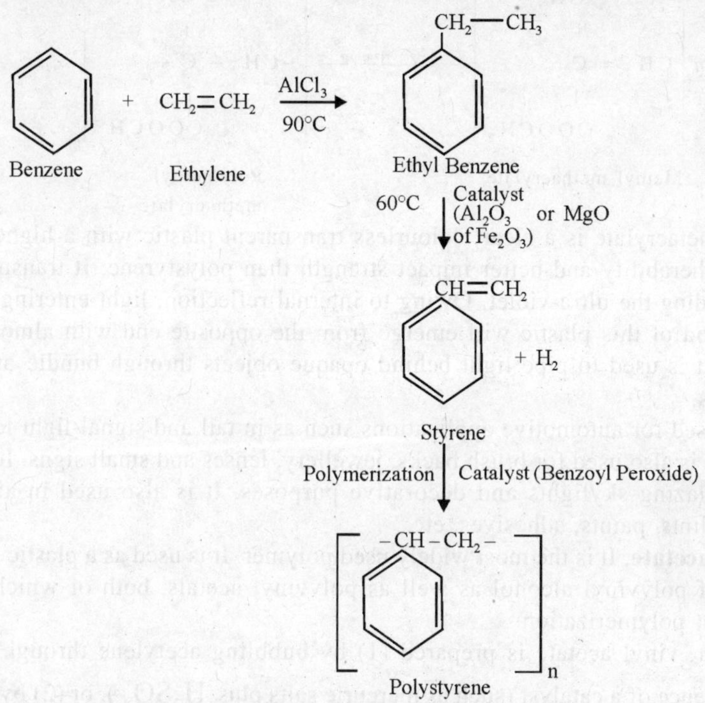

Fig. 9·14. Polystyrene.

Polystyrene is a linear polymer which is relatively inert chemically. It is quite resistant to alkalis, halide acids and oxidizing an reducing agents. It can be nitrated by fuming HNO_3 and sulfonated by con. H_2SO_4 at 100°C to a water-soluble resin.

Polystyrene is transparent and its high refractive index (1.60) renders it useful for plastic optical components. It is a good electrical insulator and has a low dielectric loss factor at moderate freequencies. Its tensile strength reaches about 8000 psi.

On the other hand, polystyrene is readily attacked by a number of solvents including dry-cleaning agents. It has a poor stability for outdoor weathering. It turns yellow and crazes on exposure. Its two major mechanical defects are its brittleness and low heat distortion temperature (softening temperature) which is only about 82—100°C. Hence polystyrene articles cannot be sterilized.

Polystyrene is used for injection moulding of articles such a combs, buttons, toys, buckles, high-frequency electric insulators, lenses, radio, T.V. and refrigerator parts, indoor lighting panels, etc. High impact strength polystyrene made by compounding polystyrene with styrene butadiene synthetic rubber is used as an injection moulding material for producing all the above mentioned articles.

6. Polymethyl methacrylate. Polymethyl methacrylate (or lucite or plexiglass) is the best known plastic from the acrylates because of its outstanding optical properties. It is obtained by the polymerization of methyl methacrylate in presence of acetyl peroxide.

$$n \begin{bmatrix} \text{CH}_2 = \underset{\underset{\text{COOCH}_3}{|}}{\overset{\overset{\text{CH}_3}{|}}{\text{C}}} \end{bmatrix} \xrightarrow{\text{Polymerization}} \begin{bmatrix} -\text{CH}_2 - \underset{\underset{\text{COOCH}_3}{|}}{\overset{\overset{\text{CH}_3}{|}}{\text{C}}} - \end{bmatrix}_n$$

Methyl methacrylate   Polymethyl methacrylate

Polymethyl metacrylate is a clear, colourless transparent plastic with a higher softening point, better weatherability and better impact strength than polystyrene. It transmits 98% of the sunlight including the ultra-violet. Owing to internal reflection, light entering one end of a spiral or bent rod of this plastic will emerge from the opposite end with almost the same intensity. Hence it is used to pipe light behind opaque objects through bundle of fibers and appropriate lenses.

It is mostly used for automotive applications such as in tail and signal-light lenses, dials, medallions, etc. It is also used for brush backs, jewellery, lenses and small signs. Its sheets are used for signs, glazing skylights and decorative purposes. It is also used in aircraft light fixtures, bone splints, paints, adhesives, etc.

7. Polyvinyl acetate. It is the most widely used polymer. It is used as a plastic and also for the preparation of polyvinyl alcohol as well as polyvinyl acetals, both of which cannot be prepared by direct polymerization.

The monomer, vinyl acetate is prepared (1) by bubbling acetylene through hot glacial acetic acid in presence of a catalyst (such as mercuric salts plus H_2SO_4), or (2) by passing the mixed gases of acetylene and acetic acid at about 200°C over a catalyst containing Zn or Cd salts on charcoal (vapour phase synthesis).

$$\underset{\text{acetylene}}{CH \equiv CH} + \underset{\substack{\text{acetic} \\ \text{acid}}}{CH_3COOH} \longrightarrow \underset{\text{Vinyl acetate}}{CH_3COOCH = CH_2}$$

Polyvinyl acetate is prepared from vinyl acetate by heating in presence of a small quantity of benzoyl peroxide or acetyl chloride as catalyst. Emulsion or suspension-polymerization is most commonly used although bulk or solution polymerization can also be used

$$n. \text{CH}_2 = \underset{\underset{\text{OCOCH}_3}{|}}{\text{CH}} \xrightarrow{\text{Polymerization}} \begin{bmatrix} \text{CH}_2 - \underset{\underset{\text{OCOCH}_3}{|}}{\text{CH}} - \end{bmatrix}_n$$

This polymer is atactic and hence amorphous. Its glass transition temperature is only 28°C as a result of which the polymer although tough and stable at room temperature (below 28°C), becomes sticky and undergoes severe cold flow at slightly higher temperatures. Hence the articles formed from polyvinylacetate are distorted even at room temperature under the influence of compressive and tensile forces. Lower molecular weight polymers are brittle but become gum-like when masticated, and hence are used in chewing gums. Polyvinyl acetate is a clear, colourless and transparent material. It has got water and heat resistance. It is fairly soluble in organic solvents.

It is used for the manufacture of polyvinyl alcohol. Its major use is in the production of water-based emulsion paints, lacquers and adhesives. It is used for making chewing gums, surgical dressings, for coating on wrapping paper and cardboard for packing purposes, for the finishing of textile and other fabrics and for bonding papers, textiles, leather, metals, etc.

8. Polyvinyl chloride (PVC). It is prepared from the monomer, vinyl chloride. The vinyl chloride is a gas, boiling at —14°C and is commercially produced by the catalytic addition of dry hydrogen chloride to acetylene. The two gases are passed together at 100 to 250°C over charcoal catalysts containing mercuric or other heavy metal salts. An alternative source is vapour phase cracking.

Polyvinyl chloride is prepared by the polymerisation of vinyl chloride by heating its water emulsion in presence of a small quantity of benzoyl peroxide or H_2O_2 in an autoclave under pressure.

$$CH \equiv CH + HCl \longrightarrow CH_2 = CH\text{—}Cl$$
$$\text{Acetylene} \qquad\qquad \text{vinyl chloride}$$

$$n.\ CH_2 = CH\text{—}Cl \longrightarrow [-CH_2-CH(Cl)-]$$
$$\text{Vinyl chloride} \qquad\qquad \text{Polyvinyl chloride}$$

Owing to the high intermolecular attraction forces present, commercial polyvinyl chloride (Exon, Geon, Velon, Marvinol, Vinylite, Tygon) is a hard and stiff amorphous plastic. It has glass transition temperature of 81°C and is soluble in cyclohexanone and tetrahydrofuran. PVC is characterized by excellent flame resistance and low-cost. Unplasticized PVC can be extruded, calendered or press laminated. It has outstanding strength, lightness and chemical resistance. However, for many industrial uses, PVC is plasticized.

Flexible films may be obtained by the addition of plasticizers such as dioctyl phthalate, dibutyl phthalate and tricresyl phosphate. The plasticized polymers are used for wire coating, electrical insulations like coverings of electric cables for distribution and wiring, upholstery, film and tubing. The unplasticized rigid polymer is used for the production of pipe, sheet and moulded parts. Rigid PVC sheets are used for tanks, linings, display, light fittings, safety helmets, refrigerator components, trays, and cycle and motor cycle mudguards. PVC compounds with low plasticizer contents are used for injection moulding of articles such as toys, tool handles. Bobbins and radio components while PVC compounds with higher plasticizer content are used for injection moulding of flexible articles.

9. Polyamides. *Nylon* is a generic term for synthetic polyamides capable of forming fibres. Nylons are described by a numbering system which indicates the number of carbon atoms in the monomer chains. Aminoacid polymers are designated by a single number, as *Nylon : 6* for poly(w-amino-caproic acid) or *polycaprolactum*. Nylons from diamines and dibasic acids are designated by two numbers, the first representing the diamine, as Nylon 6 : 6 for the polymer of hexamethylene diamine and adipic acid and Nylon 6 : 10 for that of hexamethylene diamine and sebacic acid. Polyamides of commercial importance are Nylon 6; 6 : 6, 6 : 10 and 11 (Poly w-amino-undecanoic acid), and their copolymers.

Nylon 6 : 6 is prepared by the condensation of adipic acid and hexamethylene diamine in the absence of air. The amine and carboxyl groups condense to form the amide linkage with the evolution of water.

$$HOOC(CH_2)_4 COOH + H_2N(CH_2)_6 NH_2$$
$$\text{Adipic acid} \qquad\qquad \text{Hexamethylenediamine}$$
$$\downarrow \text{Polymerization}$$

$$\left[NH_3(CH_2)_6 NH_3OOC(CH_2)_4 COO\right]_n$$
$$\text{Nylon 6 : 6 salt}$$

$$-H_2O \downarrow \text{ Further heating}$$

$$H-\left[-\overset{H}{\underset{|}{N}}-(CH_2)_6-\overset{H}{\underset{|}{N}}-\overset{O}{\overset{\|}{C}}-(CH_2)_4-\overset{O}{\overset{\|}{C}}-\right]_n-OH + H_2O$$
Nylon 6:6 polymer

Both as a plastic as well as a fiber, Nylon 6:6 is characterized by a combination of high strength, elasticity, toughness and abrasion resistance. Its solvent resistance is good, only formic acid, phenols and cresols dissolve the polymer at room temperature. Strong acids degrade it somewhat. Its good mechanical properties are retained well upto 125°C. Its outdoor weathering properties are only fair unless it is specially stabilized or pigmented with carbon black. Its melting point is 264°C and specific gravity is 1.14 and is fairly resistant to moisture. Owing to the polar groups present, its electrical uses are restricted to low frequencies only.

Its most important fiber applications include automobile tire cords, ropes, threads, cords having high tenacity and good elasticity, belting and filter cloths resistant to abrasion and chemical attack. It is also used in ladies' hose, undergarments, dresses and in carpets.

Its most important plastic applications include its use as an engineering material as a substitute for metal in bearings, gears, cams, etc., because of its high tensile and impact strength, high temperature stability, good abrasion resistance and self-lubricating bearing properties. It is also used in making rollers, slides and door latches, bearings and thread guides in textile machinery.

Nylon 6 (polycaprolactum) is produced by the self condensation of ω-amino caproic acid.

$$H_2N(CH_2)_5-COOH \longrightarrow \left[-\overset{H}{\underset{|}{N}}-(CH_2)_5-\overset{O}{\overset{\|}{C}}-\right]_n$$
Nylon — 6

Its properties are similar to those of Nylon 6:6 but its melting point is 225°C and is somewhat softer and less stiff than Nylon 6:6. It is majorly used in tire cords.

Nylon 6:10 (polyhexamethylene sebacamide) is prepared from reaction between hexamethylene diamine and sebacic acid to produce hexamethylene diammonium sebacate (Nylon 6 : 10 salt) followed by polymerization further. The polymer formed is extruded in the form of ribbon which is out into chips. This is not used as a fiber, but it has several properties making it suitable for monofilaments (brushes, bristles, sports equipment, etc.)

Nylon 11 (poly w-aminoundecanoic acid) is produced by the self-condensation of w-aminoundecanoic acid.

$$NH_2(CH_2)_{10}COOH \longrightarrow \left[-\overset{H}{\underset{|}{N}}-(CH_2)_{10}-\overset{O}{\overset{\|}{C}}-\right]_n$$
Aminoundecanoic acid   Nylon — 11

The polymer is less water sensitive than the other nylons because of its greater hydrocarbon character. It is used as textile fibre.

In addition to the above, there are several other types of Nylon such as Nylon 7, Nylon 9, Nylon 13, etc.

10. Alkyd Resins (or Glyptal Resins)

Alkyd or Glyptal resins are synthesised from polybasic acids or anhydrides with polyhydric alcohols (such as glycerol) by condensation polymerisation at elevated temperatures in presence of a catalyst.

Since the above polymer is synthesized from glycerol and phthalic anhydride, the alkyd resin thus prepared is called *"glyptal resin"*.

Glyptal resins are hard, dimensionally stable, infusible, insoluble, and resistant to chemicals and corrosion. They are used in paints, lacquers, varnishes, enamels, switches, gears, circuit breaker insulation, as binder in asbestos, cement and automobile parts.

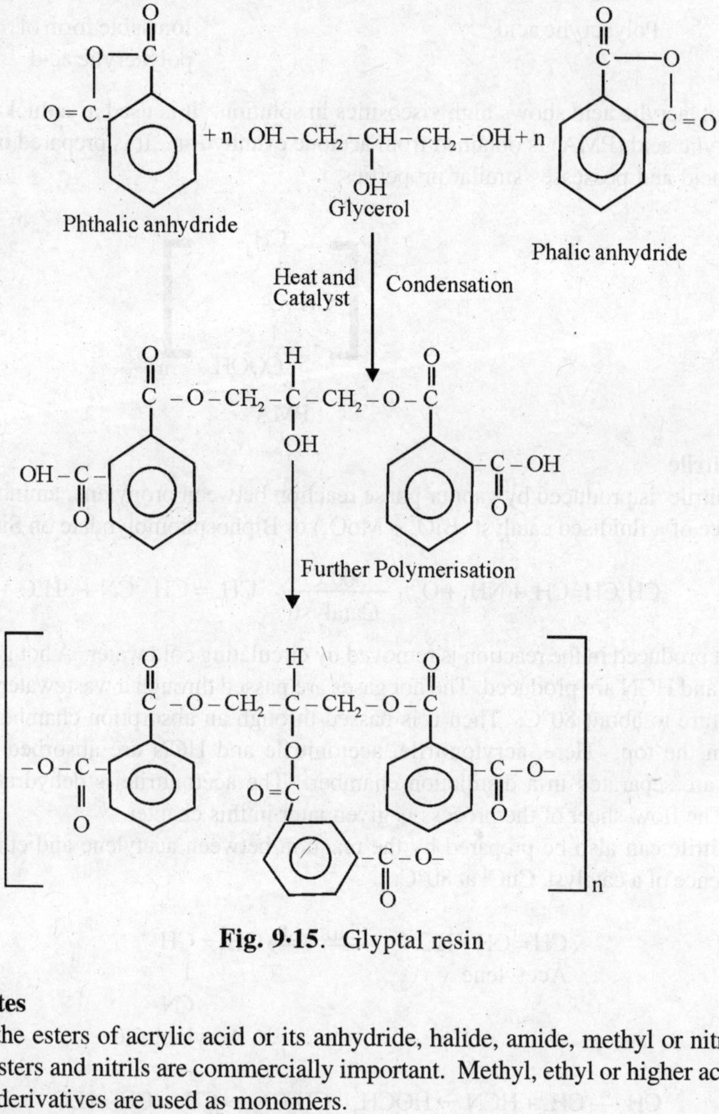

$$\text{Phthalic anhydride} + n \cdot OH-CH_2-CH(OH)-CH_2-OH + n \text{ Phthalic anhydride}$$

$$\xrightarrow[\text{Catalyst}]{\text{Heat and Condensation}}$$

(intermediate structure with) $-C(=O)-O-CH_2-CH(OH)-CH_2-O-C(=O)-$

↓ Further Polymerisation

$$\left[-O-C(=O)-\text{Ar}-C(=O)-O-CH_2-CH(-O-C(=O)-\text{Ar}-C(=O)-O-)-CH_2-O-C(=O)-\text{Ar}-C(=O)-O- \right]_n$$

Fig. 9·15. Glyptal resin

11. Polyacrylates

These are the esters of acrylic acid or its anhydride, halide, amide, methyl or nitril derivatives. Among them, esters and nitrils are commercially important. Methyl, ethyl or higher acrylates as well as methacrylic derivatives are used as monomers.

Polyacrylic acid and polymethacrylic acid are obtained from acrylic acid and polymerisation of methyl acrylate gives polymethacrylate. Polymerization of acrylic monomers can yield polymers in the form of casting, moulding powders, emulsions or solution polymers depending on the method of polymerization.

Acrylic acid monomer is obtained when a mixture of ethylene cyanhydrin and sulfuric acid blown with steam, when 50% acrylic acid solution distils out. The sodium salt of this acrylicacid, on polymerization and subsequent acidification yields polyacrylic acid. This is insoluble in its monomer but highly soluble in water. Polyacrylic acid behaves like a polyelectrolyte because it contains insoluble groups in its repeating units :

$$\left[\begin{array}{c} -CH=CH- \\ | \\ COOH \end{array} \right]_n \xrightarrow{\text{when dissolved in water}} \left[\begin{array}{c} -CH=CH- \\ | \\ COO^-H^+ \end{array} \right]_n$$

Polyacrylic acid   Ionisable form of polyacrylic acid

Thus, polyacrylic acid shows high viscosities in solution. It is used as a thickener in adhesives. Polymethacrylic acid (PMA) is obtained from acetone cyanhydrin. It is prepared in a similar way as polyacrylic acid and possesses similar properties.

$$\left[\begin{array}{c} CH_3 \\ | \\ -CH-C- \\ | \\ COOH \end{array} \right]_n$$

PMA

Polyacryloritrile

Acrylonitrile is produced by vapour phase reaction between propylene, ammonia and excess of air in presence of a fluidised catalyst ($BiO_3 + MoO_3$) or Biphosphomolybdate on SiO_2.

$$CH_3CH=CH_2 + NH_3 + O_2 \xrightarrow[\text{Catalyst}]{500°C} CH_2 = CH - CN + 3H_2O + \text{heat}$$

The heat produced in the reaction is removed by circulating cold water. A hot gaseous mixture of acrylonitrile and HCN are produced. The hot gases are passed through a wastewater boiler to cool the gaseous mixture to about 80°C. Then it is passed through an absorption chamber where H_2SO_4 is sprayed from the top. Here, acrylonitrile, acetonitrile and HCN are absorbed in water. These components are separated in a distillation chamber. The acetonitrile is dehydrated by azeotropic distillation. The flow-sheet of the process is given later in this chapter.

Acrylonitrile can also be prepared by the reaction between acetylene and ethylene oxide with HCN in presence of a catalyst, CuCl at 80°C.

$$CH \equiv CH + HCN \xrightarrow{\text{Catalyst}} \begin{array}{c} CH_2 = CH \\ | \\ CN \end{array}$$

Acetylene   Acrylonitrile

$$\underset{\text{Ethylene oxide}}{CH_2 - CH_2 \atop \diagdown O \diagup} + HCN \rightarrow \underset{\text{Cyanohydrin}}{HOCH_2 - CH_2CN} \rightarrow \underset{\text{Acrylonitrile}}{\begin{array}{c} CH_2=CH + H_2O \\ | \\ CN \end{array}}$$

Polyacrylonitrile PAN is produced from acrylonitrile by radical polymerization using peroxide initiators. The polyacrylonitrile, which is commonly known as orlan or acrilon, is soluble in dimethylsulfoxide, adiponitrile, dimethylformamide, dimethylacetamide, nitrophenols etc. It is chemical resistant and is also resistant to weathering. It is hard and has a high melting point. It has remarkable resistance to heat upto about 220°C and possesses very good mechanical properties. Polyacrylonitrile (PAN) is also known as polyvinyl cyanide and is used to produce fibers.

9.2.4 Thermoset resins

1. Phenolic resins. These are the condensation products of phenol or phenolic derivatives (*e.g.*, resorcinol) and aldehydes (*e.g.*, formaldehyde and furfural). These resins are also called phenoplasts.

Bakelite was the earliest thermosetting resin named after the Belgian-American chemist, Bakeland, who patented it in 1909. It was prepared by condensation of phenol and formaldehyde. Today, there are several types of phenol-formaldehyde resins which are extensively used in a variety of applications such as water-soluble adhesives, laminating adhesives, varnish and lacquer resins and thermosetting moulding powders. The nature of the product formed depends upon (1) the proportion of the reactants and (2) the nature of the catalyst, i.e., acidic or basic. If the mole ratio of phenol to formaldehyde (P/F) is greater than 1 (that is, if phenol is in excess), the reaction proceeds in an almost linear fashion. On the other hand, if the mole ratio of phenol to formaldehyde (P/F) is lesser than 1 (*i.e.*, if the formaldehyde is in excess) and if an alkaline catalyst is present, three dimensional network structure would be produced.

(*a*) **Methylolation.** The first step in the reaction between phenol and formaldehyde is the formation of addition compounds known as methylol derivatives, the reactions taking place in ortho and para-positions. These products may be considered as the monomers for subsequent polymerization. These are formed most satisfactorily under neutral or alkaline conditions.

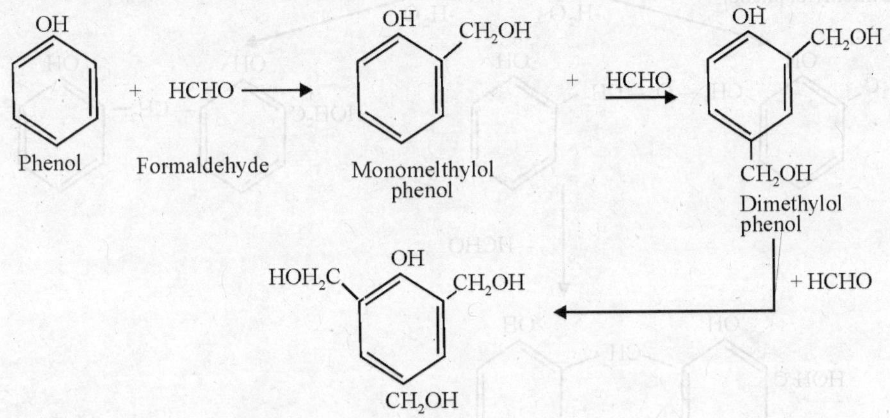

Fig. 9·16. (*a*) Trimethylol phenol.

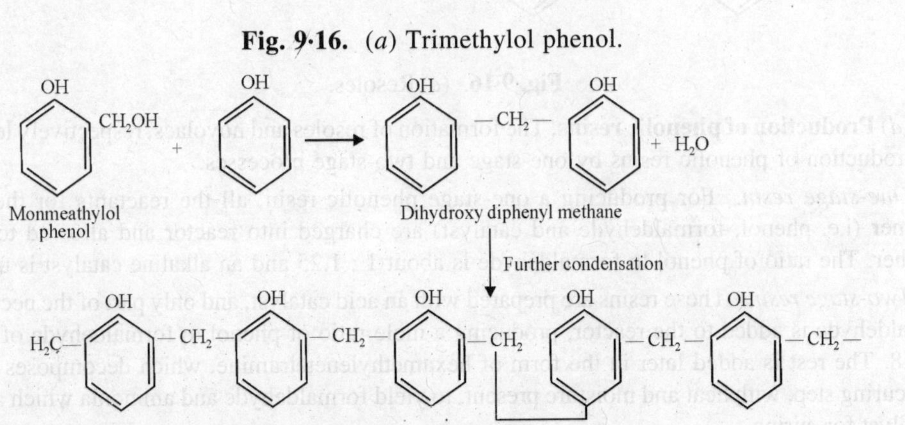

Fig. 9·16. (*b*) Novolacs.

(b) **Novolac formation.** In the presence of acid catalysts and with the mole ratio of phenol to formaldehyde greater than 1, the methylol derivatives condense with phenol to first form dihydroxy-diphenyl methane. On further condensation and methylene bridge formation, fusible and soluble linear low polymers called "Novolacs" are formed, which have the structure where ortho and para links occur at random.

Molecular weights may range as high as 1000, corresponding to about 10 phenyl residues. These materials by themselves do not react further to give cross-linked resins, unless the mole ratio of phenol to formaldehyde (P/F) is lowered to less than 1.

(c) **Resole formation.** In the presence of alkaline catalysts and with more formaldehyde (i.e. P/F less than 1), the methylol phenols can condense either through methylene linkages or through other linkages. In the latter case, subsequent loss of formaldehyde may occur with methylene bridge formation.

Products of this type which are soluble and fusible but containing alcohol groups are called resoles. If the reactions leading to their formation are carried further, large numbers of phenolic nuclei can condense to give network structure.

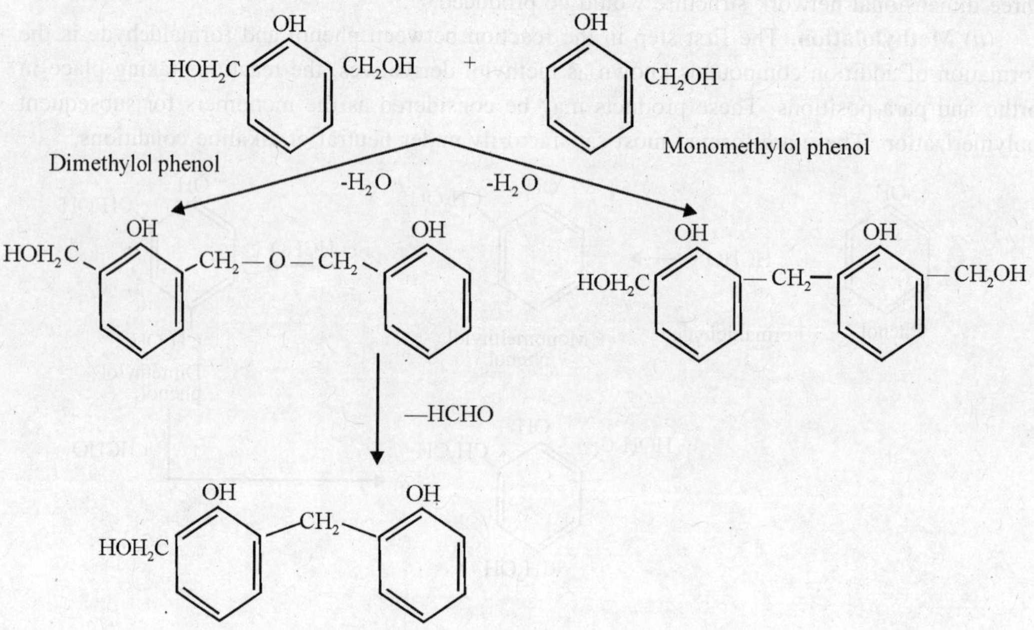

Fig. 9·16. (c) Resoles.

(d) **Production of phenolic resins.** The formation of resoles and novolacs, respectively leads to the production of phenolic resins by one-stage and two-stage processes.

One-stage resin. For producing a one-stage phenolic resin, all the reactants for the final polymer (i.e. phenol, formaldehyde and catalyst) are charged into reactor and allowed to react together. The ratio of phenol to formaldehyde is about 1 : 1.25 and an alkaline catalyst is used.

Two-stage resins. These resins are prepared with an acid catalyst, and only part of the necessary formaldehyde is added to the reactor, producing a mole ratio of phenol to formaldehyde of about 1 : 0.8. The rest is added later in the form of hexamethylenetetramine, which decomposes in the final curing step, with heat and moisture present, to yield formaldehyde and ammonia which acts as a catalyst for curing.

Resin formation. The procedures for one-and two-stage resins are similar and the same equipment is used for both. The reaction is exothermic and cooling is necessary. The formation of a resole or a novolac is evidenced by an increase in viscosity. Water is then driven off under vacuum and a thermoplastic A-stage resin, soluble in organic solvents, remains. This is taken out of the reactor and ground to fine powder. At this stage, fillers, colorants, lubricants (and enough hexamethylene tetramine in case of two-stage resin) are added and the mixture is rolled on heated mixing rolls, where the reactions are carried out further, to give B-stage resin. This is nearly insoluble in organic solvents but still fusible under heat and pressure. The resin is then cooled and cut into final form. The C-stage, the final stage of infusible cross-linked polymer, is reached on subsequent fabrication, e.g. by moulding.

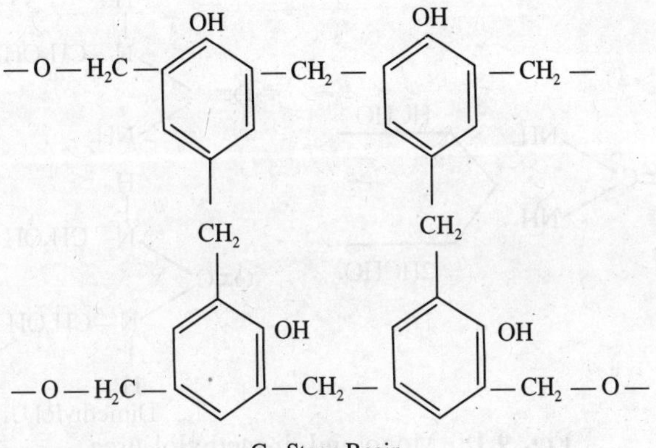

Fig. 9·17. Phenol formaldehyde C-stage resin.

Properties and uses

Phenol formaldehyde moulding resins have outstanding heat resistance, dimensional stability and resistance to cold flow. They are widely used, because of their good dielectric properties, in electrical, automotive, radio and TV parts. Fillers are usually added in phenolic moulding applications, both to reduce the cost and improve the properties. The commonly used fillers are wood fluor, asbestos, chopped rags, etc.

Phenolic resins withstand very high temperatures and hence are used in missile nose cones. Phenolic resins are used for impregnating paper, wood and other fillers, for producing decorative laminates and wall coverings, and industrial laminates for electrical parts including printed circuits.

Phenolic resins have excellent adhesive properties and bonding strength. Hence they are used for producing brake linings, abrasive wheels and sand paper, and sand filled foundry moulds.

Phenolic resins are widely used in varnishes, electrical insulation and protective coatings.

Phenolics are also widely used in the production of ion-exchange resins with a variety of functional groups such as amine, sulfonic acid, hydroxyl, phosphoric acid, etc.

2. Amino resins. These are the condensation products of urea and melamine with formaldehyde. Resins formed by the condensation of formaldehyde with other nitrogenous compounds such as aniline and amides also belong to this group but these are of limited use.

Urea and melamine formaldehyde resins are considered together here because of the similarity in their production and applications. By and large, the melamine resins have somewhat better properties but are more expensive.

Melamine is a trimer of cyanamide. Both melamine and urea react with formaldehyde,

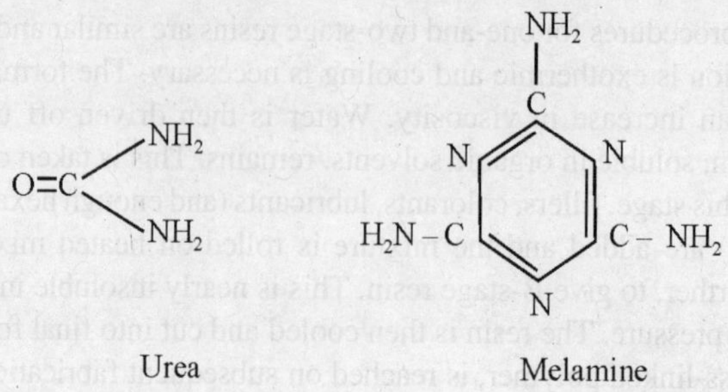

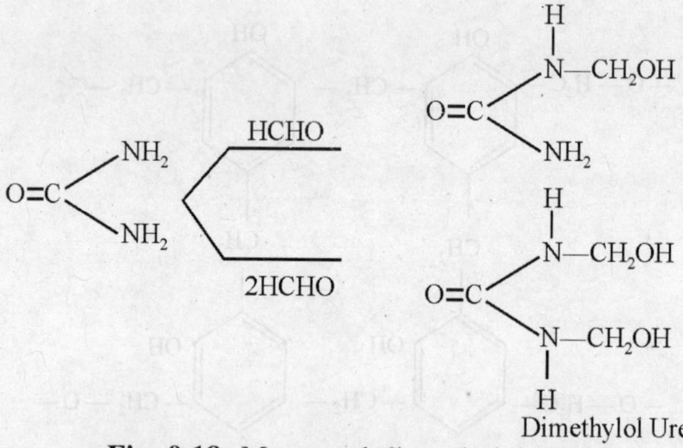

410   A TEXTBOOK OF ENGINEERING CHEMISTRY

Urea (structure)   Melamine (structure)

Mono-methylol urea and Di-methylol urea structures with HCHO and 2HCHO

Fig. 9.18. Mono- and di-methylol urea.

first by addition to form methylol compounds and then by condensation in reactions much

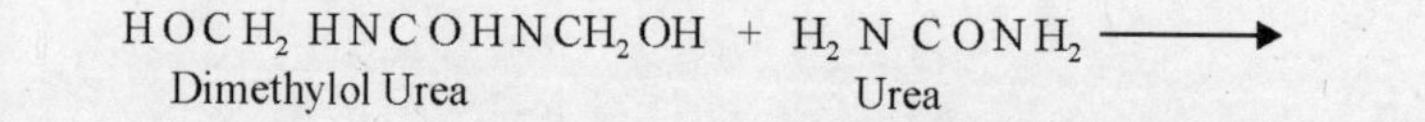

$$HOCH_2\,HNCOHNCH_2OH + H_2NCONH_2 \longrightarrow$$
Dimethylol Urea   Urea

$$HOCH_2\,HNCOHNCH_2OH$$
$$|$$
$$HNCOHN_2$$
A Trimer

Condensation / Polymerization
H_2O Molecules

Urea-formaldehyde network structure

Fig. 9.19. Urea-formaldehyde

similar to those between phenol and formaldehyde. The mono and di-methylol urea may then react with more urea with the elimination of water. But if the formaldehyde is in excess, both the hydrogens on some amine groups may react forming linkages to give a thermosetting resin. Melamine also reacts with formaldehyde in the same fashion forming bonded networks of melamine rings.

The production of the amino-resins is similar to that of phenolic resins. A-stage resin, being water-soluble, is only partially dehydrated and the water solution is used for impregnating the filler. The moulding resins are almost always filled with cellulose. Impregnation is carried out in a vacuum mixer, and the subsequent drying step carries the resin to the B-stage. It is then ground to the desired particle size in ball mills.

The amino-resins have a distinct advantage over the phenolics in that the former are very clear and colourless, so as to enable the production of light or pastel coloured objects. The hardness and tensile strength of the amino-resins are better than those of phenolics but their impact strength and the heat and moisture-resistance are lower. The melamine resins possess better hardness, heat resistance and moisture resistance than the urea formaldehyde resins.

Amino-resins are also used for moulding and as a laminating adhesive. They are also used to impregnate wood to prevent cracking. Melamine resins (called Melamac) are used for dishes, for increasing the wet strength of paper, for shrink-proofing wool and for adhesives.

Because of their colourability, solvent and grease resistance, surface hardness the urea resins are widely used for cosmetic container closures, appliance housings and stove hardware. The production of high-quality dinner-wares from cellulose-filled compounds is the largest single use for the melamine resins.

Amino-resins are used as adhesives for plywood and furniture. The melamine resins give excellent, boil-resistant bonds. They are usually blended with the urea resins for economy.

Melamine resins are extensively used for the production of decorative laminates. Amino-resins are used to modify textiles such as cotton and rayon by imparting crease resistance, stiffness, shrinkage control, water-repellancy and fire-retardation. They also improve the wet strength, rub resistance and bursting strength of paper. The urea-based enamels are used for refrigerator and kitchen appliances and melamine formulations are used in automotive finishes.

3. Epoxy resins. Epoxy resins are basically polyethers, but they are given this name due to the fact that their starting material is epichlorohydrin and the polymer before cross-linking contains epoxide groups.

The epoxy resin is prepared by the condensation of epichlorohydrin with bis-phenol A (diphenylol propane). An excess of epichlorohydrin is generally used to leave epoxy groups on each end of the low-molecular weight (900 to 3000) polymer. Depending on molecular weight, the polymer is a viscous liquid or a brittle solid having a high melting point, (about 150°C).

Other hydroxyl-containing compounds such as resorcinol, glycols, glycerol and hydroquinone may be used instead of bisphenol-A. However, epoxides other than epichlorohydrin are expensive.

Epoxy resins show outstanding toughness, chemical inertness, flexibility and adhesion in the finished products in this three dimensional cross-linked state only.

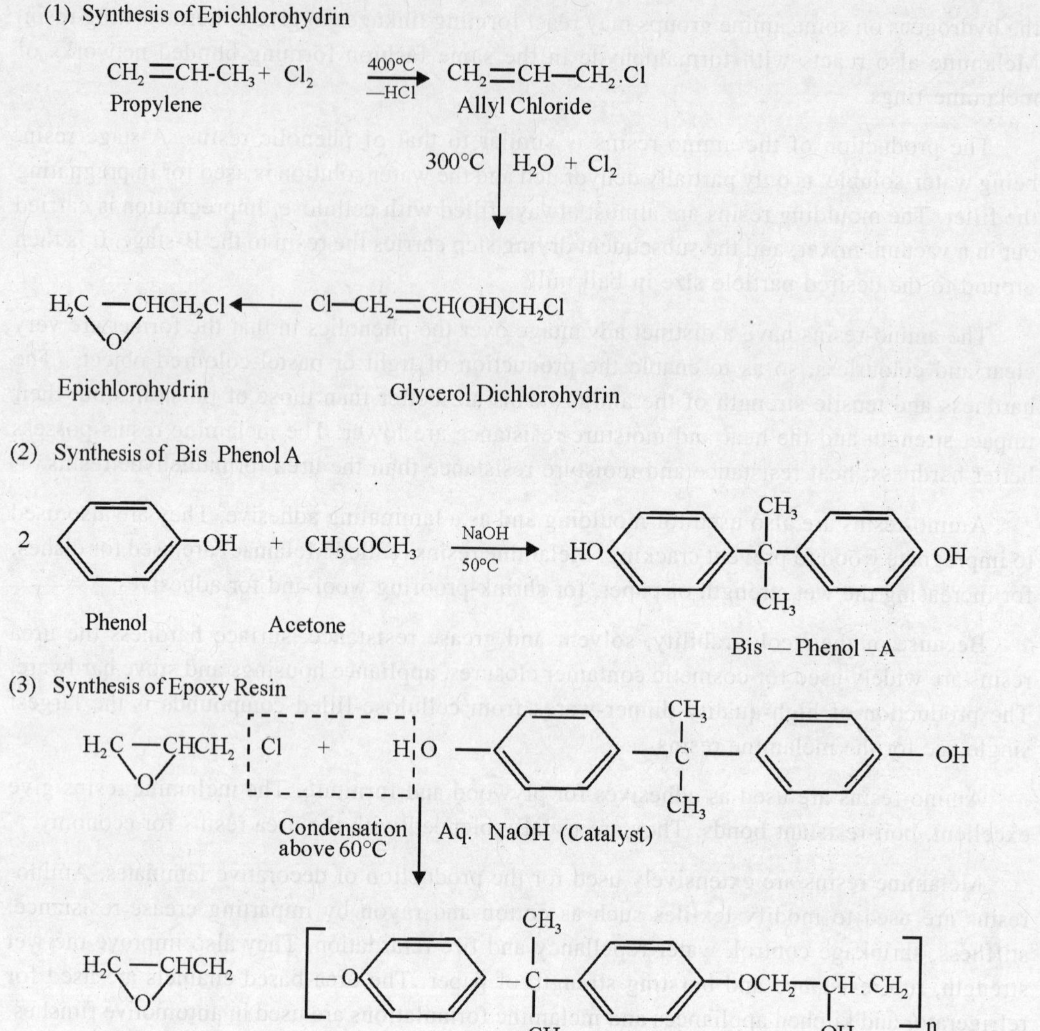

(1) Synthesis of Epichlorohydrin

$$CH_2=CH-CH_3 + Cl_2 \xrightarrow[-HCl]{400°C} CH_2=CH-CH_2.Cl$$
Propylene → Allyl Chloride

$$\xrightarrow{300°C \;|\; H_2O + Cl_2}$$

H₂C—CHCH₂Cl ← Cl—CH₂=CH(OH)CH₂Cl
   \\O/
Epichlorohydrin   Glycerol Dichlorohydrin

(2) Synthesis of Bis Phenol A

$$2 \; C_6H_5-OH + CH_3COCH_3 \xrightarrow{NaOH, 50°C} HO-C_6H_4-C(CH_3)_2-C_6H_4-OH$$

Phenol + Acetone → Bis—Phenol—A

(3) Synthesis of Epoxy Resin

H₂C—CHCH₂ ⌈ Cl + H O—⟨C₆H₄⟩—C(CH₃)₂—⟨C₆H₄⟩—OH ⌉
   \\O/

Condensation above 60°C | Aq. NaOH (Catalyst)

H₂C—CHCH₂ [—O—⟨C₆H₄⟩—C(CH₃)₂—⟨C₆H₄⟩—OCH₂—CH.CH₂—]ₙ
   \\O/   OH

Epoxy Resin

Fig. 9·20. Epoxy Resin

Epoxy resins can be used in both moulding and laminating techniques to prepare glass fibre-reinforced articles with good mechanical strength; chemical resistance and electrical insulating properties. Casting, potting, encapsulation and embedment are widely practised with the epoxy resins in the electrical and tooling industries. Liquid resins are frequently used, while hot melt solids also have occasional application.

Epoxy resins are also used in industrial floorings, adhesives and solders, foams, highway surfacing and patching materials, glass-fiber boards; as stabilizers for vinyl resins, as laminating materials in electrical equipment and to impart crease-resistance and shrinkage control on cotton and rayon fabrics.

4. Polyesters. Among this class of compounds, poly (ethylene terephthalate) or (Decaron or Terylene) is the most important one from the commercial point of view. The raw materials

POLYMERS

required for its preparation are terephthalic acid, methyl alcohol and ethylene glycol. Terephthalic acid is made by the oxidation of *p-xylene*. The terephthalic acid is first converted to the dimethyl ester by treating with methyl alcohol. This is purified by distillation or crystallization. Then it is allowed to react with ethylene glycol by ester interchange. The reaction takes place in two steps. First, a low molecular weight polyester is made with an excess of glycol to ensure hydroxyl end groups. Then the temperature is raised and pressure lowered to effect condensation of these molecules by ester interchange with the lots of the glycol.

The glass transition point of this resin is about 80°C and its crystalline melting point is 265°C. Its chemical and solvent resistance is good and similar to that of nylon. It retains its good mechanical properties up to 150—175°C.

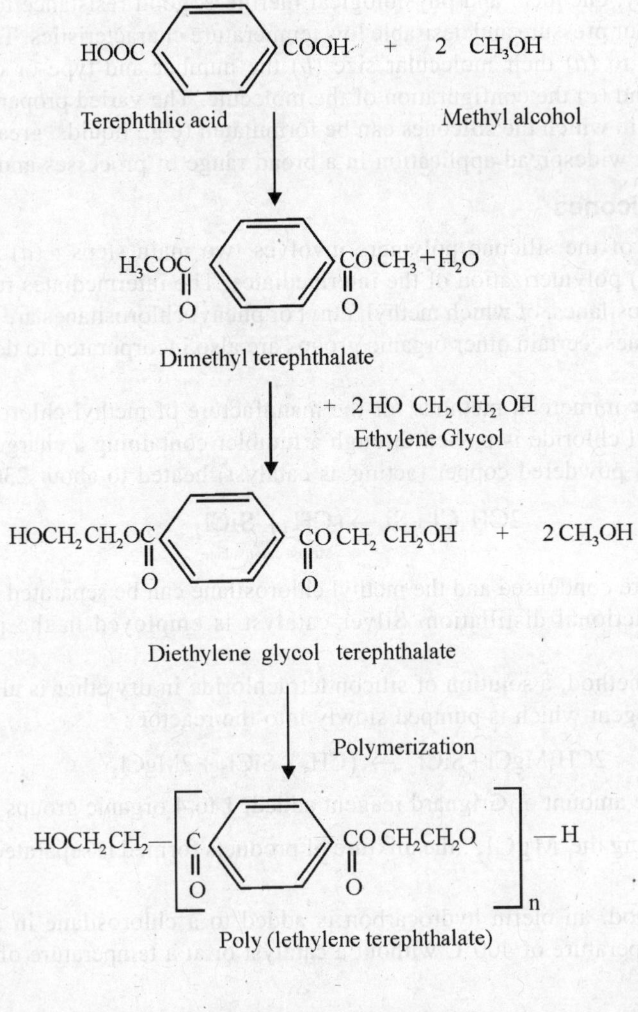

HOOC—⟨⟩—COOH + 2 CH$_3$OH

Terephthlic acid   Methyl alcohol

↓

H$_3$COC(=O)—⟨⟩—CO CH$_3$ + H$_2$O
(O)

Dimethyl terephthalate

+ 2 HO CH$_2$CH$_2$OH
Ethylene Glycol

↓

HOCH$_2$CH$_2$OC(=O)—⟨⟩—CO CH$_2$CH$_2$OH + 2 CH$_3$OH
(O)

Diethylene glycol terephthalate

↓ Polymerization

HOCH$_2$CH$_2$—[C(=O)—⟨⟩—CO CH$_2$CH$_2$O]$_n$—H

Poly (lethylene terephthalate)

Fig. 9·21. Decoran or Terylene.

The important properties of poly (ethylene terephthalate) fiber are its outstanding crease resistance and work recovery and its low moisture absorption. These properties arise from the stiff polymer chain and the resulting high modulus and the fact that the inter-chain bonds are not susceptible to moisture. Hence, garments made from the polyester fibres are resistant to wrinkling and can be repeatedly washed without the need for subsequent ironing. The polyester fibres are generally blended with cotton or wool to make summer and medium-weight suiting and other goods.

The tensile strength, impact strength and tear strength of this polymer are quite high and this toughness has the major advantage in its typical applications such as in magnetic recording tape.

5. Silicones. The development of these commercial polymeric compounds owes its beginning mostly to : (a) synthesis of tetraethyl silicon (tetraethyl silane) around 1865, and (b) the fundamental contributions of F.S. Kipping concerning the analogous characteristics of the silicon and carbon compounds. The silicone products possess certain unusual but very useful properties such as remarkable stability over wide range of temperature (from —70° to 250°C), good water repellency, chemical and physiological inertness, good resistance to the effects of weathering, low vapour pressure and desirable low temperature characteristics. These properties of siloxanes are due to (a) their molecular size (b) the number and type or organic groups attached to silicon, and (c) the configuration of the molecule. The varied properties, combined with the many forms in which the silicones can be formulated (e.g., liquids, greases, resins and rubbers) lead to their widespread application in a broad range of processes and industries.

Manufacture of Silicones

The preparation of the silicone polymers involves two main steps : (a) preparation of intermediates, and (b) polymerization of the intermediates. The intermediates required are the various organic chlorosilanes, of which methyl, ethyl or phenyl chlorosilanes are of commercial significance. Sometimes, certain other organic groups are also incorporated to develop specific properties.

There exit three commercial methods for the manufacture of methyl chlorosilanes. In the direct method, methyl chloride is passed through a tumbler containing a charge consisting of powdered silicon and powdered copper (acting as catalyst) heated to about 230–290°C.

$$2CH_3Cl + Si \rightarrow \underset{\text{Methyl chlorosilane}}{(CH_3)_2 SiCl_2}$$

The exist gases are condensed and the methyl chlorosilane can be separated from the other components by a fractional distillation. Silver catalyst is employed in the preparation of phenyl chlorosilane.

In the Grignard method, a solution of silicon tetrachloride in dry ether is allowed to react with the Grignard reagent which is pumped slowly into the reactor :

$$2CH_3MgCl + SiCl_4 \rightarrow (CH_3)_2 SiCl_2 + 2MgCl_2$$

Depending on the amount of Grignard reagent added, 1 to 4 organic groups may be added to silicon. After filtering the $MgCl_2$, the mixture of products formed is separated by fractional distillation.

In the third method, an olefin hydrocarbon is added to a chlorosilane in a bomb under pressure and at a temperature of 400°C without a catalyst or at a temperature of 45°C using a peroxide catalyst :

$$CH_2 = CH_2 + HSiCl_3 \rightarrow CH_3CH_2SiCl_3$$

POLYMERS

Preparation of Silicon

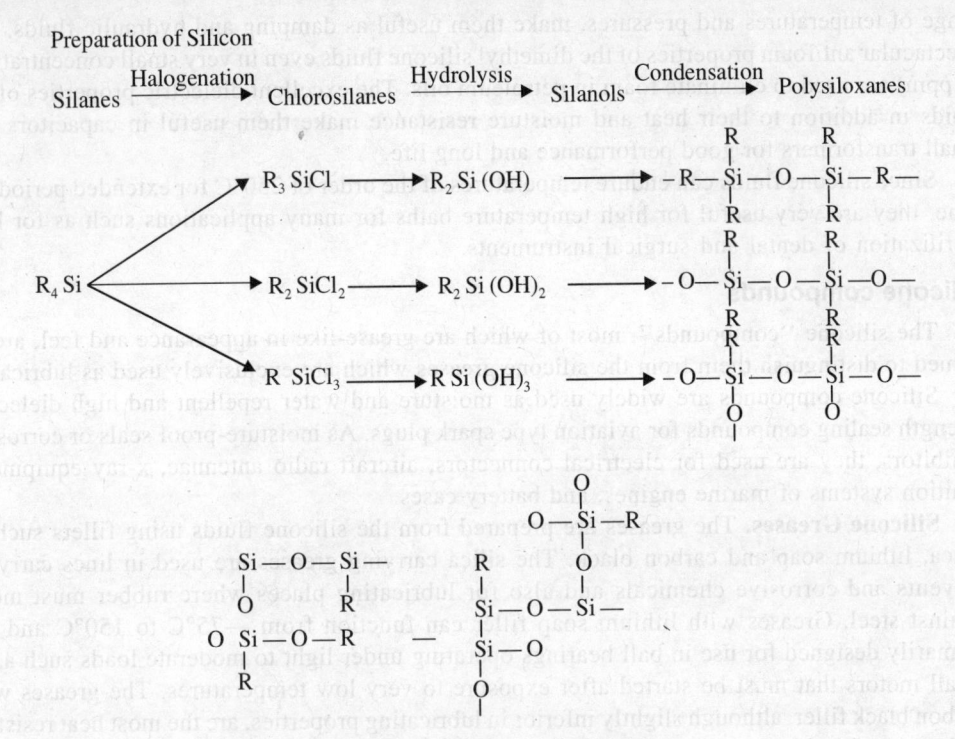

```
                    Halogenation              Hydrolysis              Condensation
           Silanes ─────────────► Chlorosilanes ──────────► Silanols ─────────────► Polysiloxanes
```

```
                    ┌──► R₃SiCl  ───► R₂Si(OH)          R   R
                    │                                   │   │
                    │                              ─R─Si─O─Si─R─
                    │                                   │   │
                    │                                   R   R
                    │                                   │   │
          R₄Si ─────┼──► R₂SiCl₂ ───► R₂Si(OH)₂    ─O─Si─O─Si─O─
                    │                                   │   │
                    │                                   R   R
                    │                                   │   │
                    │                                   R   R
                    │                                   │   │
                    └──► RSiCl₃  ───► RSi(OH)₃     ─O─Si─O─Si─O─
                                                        │   │
                                                        O   O
                                                        │   │
                                                        O   O
                                                        │   │
                                                   O─Si─R
                                                        │   R
                                                        O   │
                                              Si─O─Si─   Si─O─Si─
                                              │   │      │   │
                                              O   R      R   │
                                              │          │   
                                          O─Si─O─R      Si─O
                                              │          │
                                              R          O
```

High Molecular Weight Cross-Linked Polymer

Fig. 9.22. Manufacture of silicones.

The silicone compositions are generally classified on the basis of their physical state and appearance at room temperature, *e.g.*, fluids, greases, resins and rubbers.

Silicone fluids

Dimethyl, and the methyl and phenyl fluids are the two types of silicone fluids which are commonly available. Dimethyl fluids are prepared by pumping very pure dimethyl dichlorosilane into strongly acidified water. When low-molecular weight (or low viscosity) products are required, a suitable amount of trimethyl chlorosilane is added to serve as an end-blocker and also to stabilize the viscosity against further polymerization upon standing.

$$x(CH_3)_2 SiCl_2 + xH_2O \rightarrow (CH_3)_2 SiO + 2x\ HCL$$

The fluids containing both methyl and phenyl groups can be prepared in the similar fashion.

By virtue of their unusual properties the silicone fluids have got a large number of applications. The high degree of water repellency coupled with the spreading efficiency due to low surface tension allows deposition of silicone fluid films over plastics, metals, glass and ceramic ware. Silicone fluids are widely used for treating the interiors of the vials for aqueous suspensions such as penicillin. The "caking" of organic or inorganic powders can be avoided by coating the particles with a film of silicone fluid.

Since dust and dirt come off easily from surface coated with silicone fluids, they are used with advantage in polishes for cars and furniture. The availability of silicone fluids in a wide range of viscosities, coupled with their high viscosity index and ability to operate over wide range of temperatures and pressures, make them useful as damping and hydraulic fluids. The spectacular antifoam properties of the dimethyl silicone fluids even in very small concentrations (1 ppm) are used to eliminate foam in petroleum oils. The excellent dielectric properties of the fluids in addition to their heat and moisture resistance make them useful in capacitors and small transformers for good performance and long life.

Since silicone fluids can endure temperatures of the order of 250°C for extended periods of time, they are very useful for high temperature baths for many applications such as for heat sterilization of dental and surgical instruments.

Silicone compounds

The silicone "compounds", most of which are grease-like in appearance and feel, are so named to distinguish them from the silicone greases which are exclusively used as lubricants.

Silicone compounds are widely used as moisture and water repellent and high dielectric strength sealing compounds for aviation type spark plugs. As moisture-proof seals or corrosion inhibitors, they are used for electrical connectors, aircraft radio antennae, x-ray equipment, ignition systems of marine engines, and battery cases.

Silicone Greases. The greases are prepared from the silicone fluids using fillers such as silica, lithium soap and carbon black. The silica carrying greases are used in lines carrying solvents and corrosive chemicals and also for lubricating places where rubber must move against steel. Greases with lithium soap filler can function from —75°C to 150°C and are primarily designed for use in ball bearings operating under light to moderate loads such as in small motors that must be started after exposure to very low temperatures. The greases with carbon black filler, although slightly inferior in lubricating properties, are the most heat resistant and are used in high temperature conveyors, oven doors, etc. Silicone greases thickened with aryl urea compounds can function even at 260°C and are reported to be superior to the other greases with respect to thixotropy (property of flowing when shaken) at high temperatures and evaporation.

Resins Silicone. The resins are prepared by hydrolyzing and condensing (or polymerizing) mixtures of bifunctional or trifunctional alkyl chlorosilanes under controlled conditions. The properties of the resins can be varied by controlling the relative amounts of bifunctional and trifunctional units, the type of organic units attached to silicon, the polymer size or configuration, and the addition of other organic resins. It is usually necessary to cure the silicone resin at a temperature somewhat higher than that at which it is likely to be used.

The silicone resins can be broadly classified into six categories, although there is overlapping because some can be applied in more than one way.

Coating resins. These have got distinctive properties of heat resistance, water repellency and resistance to most of the aqueous chemicals and corrosive gases. The pure silicone (unmodified) films are useful on hot stocks, exhausts, electric motors and turbines, stoves, boilers, steam pipes and other places where conditions are too severe for the usual type of protective coatings. Silicone coating resin formulations with aluminium flakes or zinc dust and with or without the addition of organic resin can withstand temperatures of the order of 500°C and can be applied on high temperature processing equipments.

Silicone modified alkyd resins have got the additional advantages of giving finishes with greater flexibility and hardness, stability at elevated temperatures, better colour retention and toughness. They are recommended for application to refrigerators, washing machines, stoves and similar household equipments.

Laminating resins. They possess high temperature stability and good electrical properties and can be prepared in the form of sheets or tubes. Resin laminated structures are widely used in electrical industry in the form of insulated board, coil-former, rods, tubes and slotwedges in motors operating at high temperature. They can be used in aircraft instruments such as radomes.

Release resins. The release characteristics of the silicone fluids and compounds are incorporated into the resins. The release resins are widely applied to bread pans, waffle irons, cooky sheets candy pans as well as to "smoke sticks" used in smoking of meats.

Water repellent resins. These resins applied as dilute solution coat the pores (but do not fill them) in masonry and concrete. They do not allow water to enter the pores because of their water repellent character and as a result keep the masonry dry.

Electrical resins. Silicone electrical insulations possess greater heat resistance, water repellency and resistance to combustion. They can be used with advantage where conditions are too severe for conventional insulators. Impregnating silicone resin compositions are also used in the treatment of capacitor insulations, transformer coils, motor windings and similar equipment demanding excellent chemical, mechanical and dielectric stability.

Foamed resins. Silicone resins may be foamed in situ or may be foamed and then moulded to the desired shape by the use of blowing agents. They can be advantageously used for thermal and electrical insulation, firewall structures, buoyancy applications and high temperature vibration dampers.

Silicone rubbers. Rubbers are prepared by milling together a dimethyl silicone polymer, an inorganic filler (*e.g.*, TiO_2, ZnO, iron oxide, silica, etc.) and a vulcanizer (*e.g.*, peroxides such as benzoyl peroxide). Curing is accompanied by heating in a mould in the absence of air to about 150°C. This apparently results in the abstraction of hydrogen atoms from methyl groups with a subsequent cross-linking of the polymer at these points. A subsequent cure at 250°C for at least 4 hrs in a circulating air oven is required to develop the best properties.

The outstanding characteristic of all silicone rubbers is the remarkable stability at elevated temperatures. Their life is practically unlimited at 150°C and they are flexible even at as low a temperature as —90°C. Silicone rubbers are highly water repellent and have exceptional resistance to corona discharge, while most organic rubbers get quickly hardened under the same conditions. However, their tensile strength and elongation are inferior to organic rubbers at normal temperatures.

Silicone rubber is available in a variety of forms varying from thin sheets to heavy mouldable stocks. It is also prepared as smooth or heavy pastes. Either the paste or the stocks may be dispersed in organic solvents to give free-flowing dispersion which can be applied to surfaces by brushing, spraying or dipping.

Belting, having a coating of silicone rubber, is useful for handling frozen food which do not stick because of the release characteristics of the rubbers.

The pastes are used for coating organic and inorganic fabrics, for preparing reinforced gasketting or compressible pads, or for instrument mountings for use at high and low temperatures. Hoses for carrying hot gas or liquid are prepared from the coated fabrics. The pastes adhere well to ceramic, glass and metalware and can provide an electrically insulating coating.

In Pharmacy and Medicine

The total absence of physiological response to the silicones has encouraged investigations regarding their use in pharmacy and medicine. Hand lotions containing silicone fluids are reported to be useful for treating skin affections where contact with water is to be avoided. A dental polyantibiotic silicone paste containing penicillin, bacotracin and streptomycin has been developed for root canal therapy. Silicone antifoam has been reported to be useful in treating extreme cases of pneumonia.

The properties of organic acids and amines change by the presence of silicon in the molecule. Thus by superimposing silicone chemistry on organic chemistry, materials with entirely new properties can be created. Knowledge of these new products may be helpful in working out better machine design and manufacturing process.

9.2.5 Low Dielectric Constant Polymers

A low dielectric constant (dielectric constant = k) is an insulating polymer that exhibits weak polarization when subjected to an externally applied field. There are many guildlines employed to design low - k polymers. One is to choose a nonpolar dielectric system. Polarity is weak in a polymer with no polar chemical groups and with symmetry, to cancel the dipoles of chemical bonds between dissimilar atoms. Dielectric constant of air ≈ 1. Dielectrics can be lower by incorporation of some porosity into the chemical structure. Anather approach is to minimize the moisture content in the dielectric or design a dielectric with minimum hydrophilicity, since dielectric contents of water ≈ 80, a low k - dielectric needs to absorb only very small traces of water.

The currently available dielectric materials are undoped plasma. SiO_2 (k = 4 - 4.5), fluorinated SiO2 (k = 3.5), spin - on glasses (silsesquio xanes k = 2.2 - 3.0) and organic polymers such as polyarylene ethers, polyimides, parylene - F etc. Poly(arylene ether) polymer has repeat unit of the following structure.

$(O-Ar_1-O-Ar_2)_m (O-Ar_3O-Ar_4)_n$

Where, Ar_1, Ar_2, Ar_3 and Ar_4, are identical or different aryl radicals. m is 0-1, n is 1-m and at least one of the aryl radicals is grafted to the backbone of the polymer.

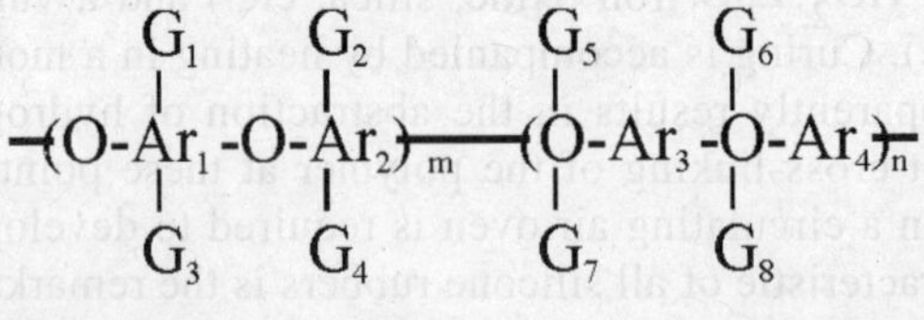

$$-(O-\underset{G_3}{\overset{G_1}{Ar_1}}-O-\underset{G_4}{\overset{G_2}{Ar_2}})_m-(O-\underset{G_7}{\overset{G_5}{Ar_3}}-O-\underset{G_8}{\overset{G_6}{Ar_4}})_n-$$

Where G_{1-8} are individually H, alkyl, alkylene or functicnalized alkylene.

Polyarylene ethers consist of aromatic groups (Ar) connected by ether oxygen linkages in a linear fashion compared to other functional groups (eg carboxylic acids, aldehydes, esters, and ketones), ether linkages provide strong c-o bonds for improved thermal stability, while introducing relatively weak for low dielectric constant. Schumacher PAE - 2 (k = 2.8) is a variety of polyarylene ether

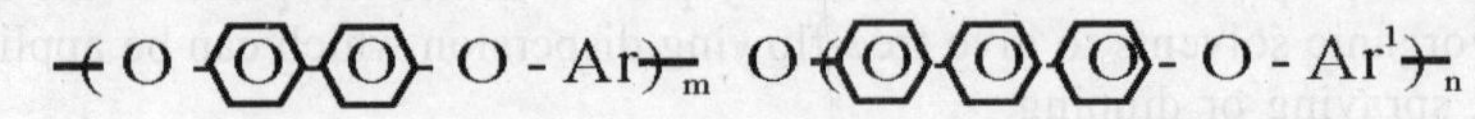

$$-(O-\bigcirc\bigcirc-O-Ar)_m-O-(\bigcirc\bigcirc\bigcirc-O-Ar^1)_n-$$

Structural formula of polyarylene ether (schumacher PAE - 2) Ar, Ar' are proprietary aromatic groups. Polyarylene ethers possess many desirable properties of interlevel dielectrics integration but require extensive crossliking for good thermal stability and solvent resistance. Curing the polymer in oxygen improves crosslinking, but high temperature exposure to oxygen may not be suitable for integration. Florinated polyarylene ether, exhibit lower k (k = 2.5 - 2.6) than nonfluorinated counterparts. However, the instability of fluorine in this polymer resulted in severe metal corrosion, thus rendering integration to be unfeasible.

Polyimides : Dupoint (kepton) offered the first commercial polyimide in 1960s, made by condensation between pyromellitic dianhydride and 4-4 diamino diphenyl ether. Polyimide (commercial name kepton) chemical name is poly (4, 4 - oxydiphenylene -pyromellitimide). According to NASA internal report, space shuttle wires were coated with an insulator known as kepton. The polymer has an exceptionally high heat distortion temperature (357^0C) and thermo - oxidative resistance. It can continuously used at about 270^0 C. Kepton insulated wiring has been widely used in civil and military avionics (electrical wiring for aircraft) because of its very light weight compased to other insulators as well as good insulating and temperature characteristics. Vespel is the trademark of a durable high-performance polyimide - based polymer manufactured by Dupoint. Polyimide is aften used in the electronic industry for flexible cables as an insulating film on magnet wire and medical tubing e.g.. in a laptop

POLYMERS

computer, the cable that connects the main logic board to the display (which must flex every time the laptop is opened as closed) is often a polyimide base with copper conductors. The semiconductor industry uses polyimide as a high temperature adhesive.

Dupoint offer insulating wire enamel, varnishes and laminating resins. The varnish and film are mostly used for exceptionally high temperature electrical insulation application.

Kepton

Table No.
Properties of unfilled polyimides

Properties	Vespel	Torlon	Ultem
Tensile Strength (M pa) 150°c	67	105	46
Modulas (G Pa) 150°c	2.7	3.6	2.5
Heat distortion temperature	357	282	200
Limiting oxygen index	35	42	47

Fluorinated polyimide exhibits lower k, lower moisture uptake and better isotropy than conventional counterports.

Parylene - F

Poly (tetrafluoro - p - xylylene) or parylene - F is a vapor-deposited crystalline polymer with k = 2.3 - 2.4

The process involves the vaporization of the di-tetrafluoro - p- xylylene dimer at about 150°c. The dimer is led into a reactor where it is cracked at 650°C to form two reactive tetrafluoso - p - xylylene monomers. The monomers are led to a vacuum chamber, condense on the wafer surface which is maintained at - 15°C, diffuse into the bulk of the parylene - F film and then polymerize by reacting with the ends of the free-redical polymer chains. After deposition, the film must undergo a vacuum anneal at 350°C to stabilize its properties.

9.2.6 Biopolymers

Synthetic polymers are resistive to chemical and physical degradation and they are durable, they are relatively cheap compared to other products and hence they are widely accepted and hence polymers replace glass, paper and cardboard. These convenience enhance the quality and comfort of ife in modern society.

One major drawback is their disposal. After use it is discarded into the environment and it create different environment problems. Since conventional plastics are non-degradable and environmentally unfriendly in the public perception. It block the domastic pipeline and sewage lines. During burning it lead to the emission of CO_2 and toxic fumes which causes global warming and affects human health. One way to solve these problem is by replacing conventional bioresistant synthetic polymers, which are in use today with biopolymers.

Biopolymers are polymers that are biodegradable. The initial materials for the production of these polymers may be either renewable or synthetic. A degradable material in which degradation result from the action of microorganisms and the material are converted to water, carbon dioxide and/or methane. Biodegradation is a process by which organic substance are broken down by the enzymes produced by living organisms.

There are several different types of degradation that can occure in the environment these include, biodegradation, photodegradation, oxidation and hydrolysis. From the study the following conclusion that has been drawn.

i) Naturally occuring polymers are biodegradable.
ii) Synthetic addition polymer with carbon as the only atom in the backbone do not biodegrade at molecular weight above 500.
iii) If an addition polymer contains atom other than carbon in the backbone it may biodegrade depending on any attached functional groups.
iv) Synthetic condensation polymers are generally biodegradable to different extent depending on chain coupling (ester > ether > amide > urethane).
v) Morphology (amorphous > crystalline)
vi) Molecular weight (lower > higher)
vii) Hydrophilic polymers degrade faster than hydrophobic

However, if a polymers is water soluble that does not necessarily mean that it is biodegradable. The biodegradation of these polymers proceeds by hydrolysis and oxidation. The presence of hydrolysable and/or oxidisable linkages in the polymer main chain, the presence of suitable substitutents, correct stereo configuration, balance of hydrophobicity and hydrophilicity contribute to the biodegradability of the polymers.

Biodegradable polymers can be divided into three classes
i) Natural polymers originating from plant or animals resources e.g. cellulose, starch, protein, collagen etc.
ii) Biosynthetic polymers produced by fermentation process by microorganisums e.g. polyhydroxy alkanoates.
iii) Certain synthetic polymers possesing the biodegradable properties e.g. polycaprolactone and polylactic acid.

Polylactic acid or polylactide (PLA)

PLA is a biodegradable thermoplastic aliphatic polyester derived from renewable resources such as corn starch or sugar canes. Lactic acid is produced by bacterial fermentation of sugarcane or from the conversion of starch from corn. The lactic acid is oligomerized and then catalytically dimerized to make the lactide monomer. High molecular weight PLA is produced from the lactide monomer by ring opening polymerization using a stannous octoate catalyst.

Polymerization of a racemic mixture of L and D-Lactides leads to the synthesis of Poly-DL-lactide. PDLLA which is not crystalline but amorphous. Poly-L-lactide PLLA is the product resulting from polymerization of L, L-lactide. PLLA has a crystallinity of around 37%, a glass transition temperature 50-80°C and T_m 173 - 178°C. PLLA can be processed like most thermoplastics into fiber (using conventional melt spinning) and film.

Biodegradation of PDLA is slower than for PLLA due to the higher crystallinity of PDLA. PLA is used in a number of biomedical applications, such as sutures, stents, dialysis media and drug delivery devices. Because of its biodegradability it can be used in preparation of bioplastics useful for food packaging and disposable table ware.

Polycaprolactone (PCL)

PCL is a biodegradable polyester with a low melting point of 60°C and T_g of -60°C. PCL is derived by chemical synthesis from crude oil. It can be prepared by ring opening polymerization of ε-caprolactone using stannous octoate catalyst. PCL has good water, oil, solvent and chlorine resistance.

PCL is used as an additive for resins to improve their processing characteristics and their end use properties (i.e. impact resistance). PCL is degraded by hydrolysis of its ester linkages in physiological conditions (in human body) and hence use as an implantable biomaterial. PCL is used in a drug delivery device and suture. A variety of drugs have been encapsulated within PCL beads for controlled release and targeted drug delivery. In Dentistry (as composite) is used in root canal filling.

Polyhydroxyalkanotes

Two types of polyhydroxy alkanoates PHA are known polyhydroxybutyrate PHB and polyhydroxyvalerates PHV. These are based on fermented sugars (sucrose, glucose and lactose) with different starch crops as starting materials. PHB is produced by micro-organisms (like Alcaligenes entrophus or Bacillus megaterium) in response to conditions of physiological stress.

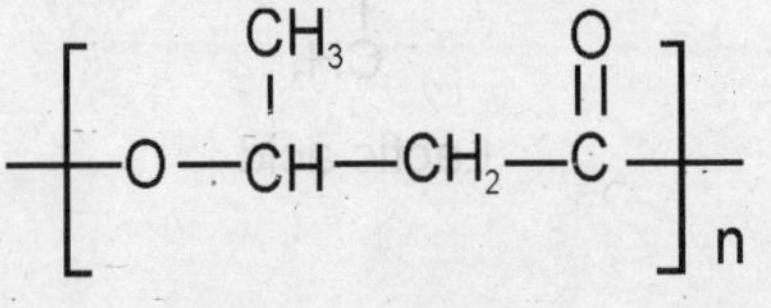

PHB is the most common type of polyhydroxyalkanoate other polymers of this class produced by a variety of organisms are polyhydroxyvalerate (PHV), polyhydroxy hexanoate (PHH) and their copolymer.

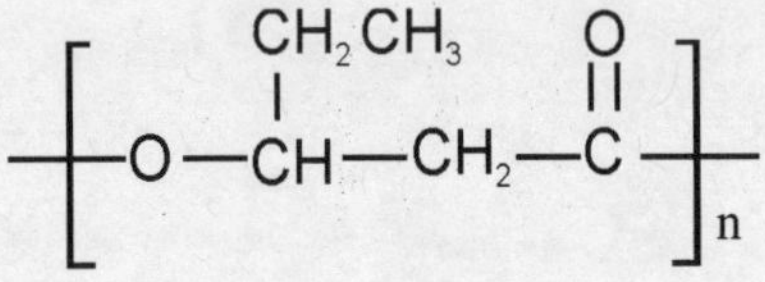

Polyhydroxybutyrate

Polyhydroxyvalerate

PHB has attracted commercial interest because of its physical properties similar to polypropylene. It shows high degree of crystallinity, high melting point 180°C, T_g - 15°C and is rapidly biodegradable. If PHB becames as cheap as plastics produced from petrochemicals, then it will probably become widely used for packing products like bottles, bages, wrapping film and disposable nappies. PHB is also used for tissue engineering scaffolds and for controlled drug-release carriers owing to its biodegradability, optical activity and isotacticity. PHB is water insoluble and relatively resistant to hydrolytic degradation. soluble in chloroform and other chlorinated hydrocarbons, good ultra-violet resistance but poor resistance to acid and base.

Polygycolide or Polyglycolic acid (PGA)

PGA is a biodegradable thermoplastics, linear aliphatic polyester. It can be prepared from glycolic acid by polycondensation or ring-opening polymerization. Glycolic acid is heated at atmospheric pressure and a temperature of about 175-185°C is maintained until water ceases to distill. Subsequantly pressure is reduced to 150 mm Hg, keeping the temperature constant for two hours the low molecular weight polygycolide is obtained. Ring opening polymerization of glycolide can be catalyzed using different catalysts such as antimony trioxide or antimony trihalides. Stannous octoate is the most commonly used initiator. Initiator is added to glycolide under a nitrogen atmosphere at 195°C. The reaction is proceed for two hours, then temperature is raised to 230°C for half an hour. After solidification the high molecular weight polymers can be obtained.

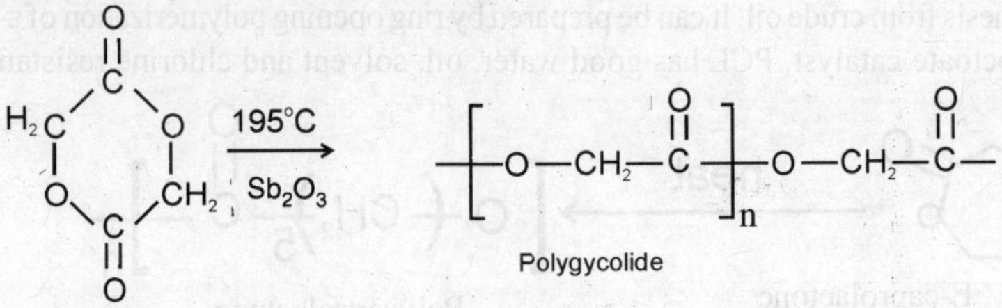

Glycolide

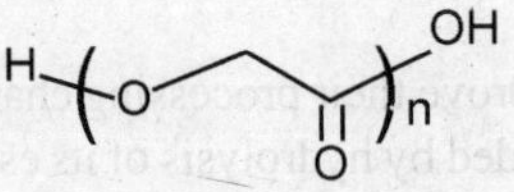

Polygycolide

Polygycolide

Properties : PGA has a T_g 35-40°C and T_m 225-230°C also show 45-55% degree of crystallinity. It is insoluble in water. High molecular weight form is insoluble in all common organic solvents acetone, dichloromethane, chloroform, ethyl acetate, THF, while low molecular weight oligomers differ in their physical properties to be more soluble. PGA is soluble in highly fluorinated solvents like hexafluoroisopropanol (HFIP) and hexafluoroacetone sesquihydrate that can be used to prepare solutions of the high moecular weight for melt spinning and film preparation.

Degradation : PGA is characterized by hydrolytic instability owing to the presence of the ester linkage in its backbone. First water diffuses into the amorphous (non-crystalline) regions of the polymer matrix, cleaving the ester bonds, the second step starts after the amorphous regions have been eroded, leaving the crystalline portion of the polymer susceptible to hydrolytic attack. Upon collapse of the crystalline regions the polymer chain dissolves. When exposed to physiological conditions PGA is degraded by random hydrolysis and it is also broken down by certain enzymes. The degradation product of glycolic acid is non toxic and it can enter the tricarboxylic acid cycle after which it is excreted as water and carbon dioxide. A part of the glycolic acid is also excreted by urine. PGA made sutures loses half of its strength after two week and 100% after four weeks. The polymer is completely resorbed by the organism in four to six months.

Uses : PGA was marked under the trade name of Dexon was used as absorbable suture in surgical applications. Since PGA give strong fibers and degrade into water soluble monomers and there is no need for further medical attention to remove them. Implantable medical devices have been produced from PGA such as anastomosis rings, pins, rods, plates and screws.

Applications of biodegradable plastics : Biodegradable plastics have captured tremendous public attention in recent years. Some of their applications are given below :

(a) Applications in agriculture : Time-controlled biodegradable polyolefins are used in agriculture for mulching, netting, twine, etc. They have yielded considerable economic dividends in increasing crop yields and reducing crop management costs. Controlled release of fertilizers, pesticides, etc. are the other important applications.

(b) Biomedical applications : Synthetic resorbable polyesters are used in various biomedical applications such as in absorbable surgical implants and sutures, controlled release of drugs, absorbable skin grafts and bone places to support the body recovery systems.

(c) Waste management applications : Polymer wastes account for about 15% of the total wastes. Polymer waste management requires sound complementary practices of conservation, recycling, incineration and biodegradation–bioconversion. Since biodegradation is potentially the most environmentally friendly of all these practices, biodegradable polymers are receiving greater attention particularly for use as packaging materials. e.g. Blends of starch derivatives with biodegradable polymers, Copolymers of poly-R-3-hydroxy-butyrate (PHB), etc.

On the other hand, plastics used at present cannot be substituted with degradable resin because of requirements related to specific mechanical, optical, electrical durability and other properties required for a specific applications. Moreover, degradable plastics have higher prices than commodity resins.

Degradation cannot offer a magic solution for the plastic waste disposal problem. Further, degradation in certain conditions is not as fast as expected. Future solutions to plastic wastes will probably establish a balance among recycling, degradation and incineration.

At present, the main applications of degradable plastics are in agriculture (including mulching films), medical and pharmaceutical fields and in packaging, where discarded in a public place is a matter of concern and collection for recycling is difficult (e.g. in beaches and roadsides).

Future Scenario of Biodegradable plastics

1. Polymers with induced biodegradability (initiated by light and/or heat) in contact with the soil or in municipal composters are likely to become increasingly important in the future.

2. Polymers with programmed "closed loop" recyclability (e.g. in automobiles and in some major packaging applications) will be recovered by recycling and will not be made degradable.

3. The most suitable disposable systems for polymers in collected mixed plastics waste are pyrolysis with recovery of useful chemicals or incineration with recovery of heat energy. Degradable plastics will not interfere with these processes.

4. Potential uses for photodegradable polymers, such as E/CO, will exist wherever littering of plastics occurs. **Ex :-** Food packaging films, carriers for beverage canes, marine packaging, fishing gear etc.

5. Solutions for solid waste disposal problems include recycling, incineration and land-fill wherever feasible. However, in cases where collection is prohibitively expensive degradable plastics is a viable solution.

9.2.7 Liquid crystal polymers

Liquid crystal polymers LCPs are a unique class of partially crystalline aromatic polyesters that provide previously unavailable high performance properties. They are thermoplastic based on p-hydroxybenzdic acid and related monomers. LCPs are manufactured in a stepwise polycondensation by either a batch or continuous process.

LCPs are capable of forming regions of highly ordered structure in the molten state, however the degree of order is somewhat less than that of a regular solid crystal. LCPs give morphology results in verygood properties in the direction of flow. Due to its highly crystalline structure it has outstanding dimentional stability, high strength, stiffness, chemical and solvent resistance, highly resistant to fire and ease moldability. Processing of LCPs from liquid crystal phases ((or mesophases) give to fibers and injected materials having high mechanical properties.

Specific gravity	1.38 to 1.95 g/cc
Modulus	8530 to 17200 Mpa
Tensile strength	52.8 to 185 Mpa

The example of lytropic LCPs is the commercial aramid known as Kevlar. Chemical structure of this aramid consists of linenarrly substituted aromatic rings linked by amide groups.

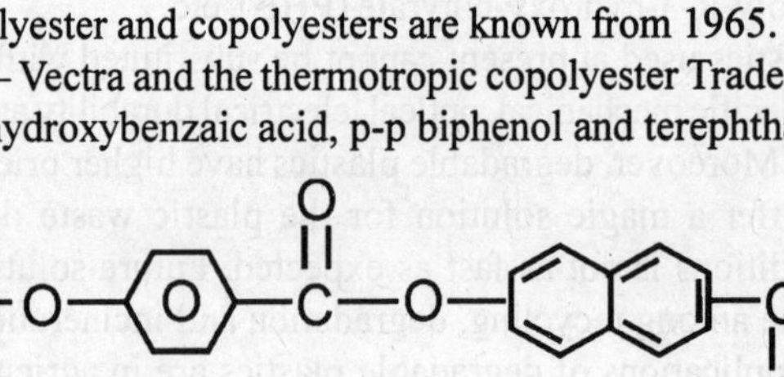

Linked by amide groups

Kevelar

The liquid crystal polyester and copolyesters are known from 1965. LPCs have highly aromatic structure eg. Trade name – Vectra and the thermotropic copolyester Trade name Xydar made from the polycondensation of p-hydroxybenzaic acid, p-p biphenol and terephthalic acid.

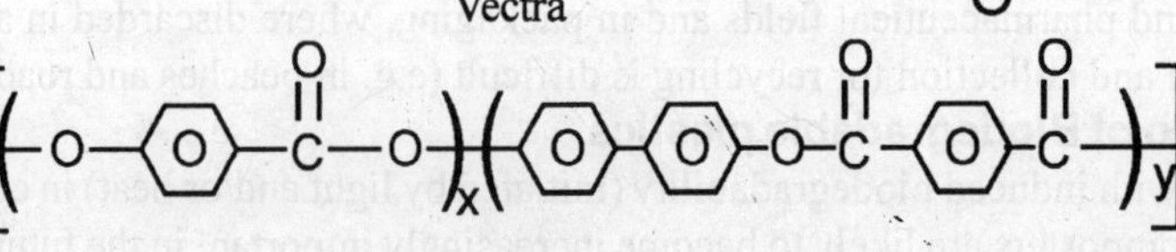

Vectra

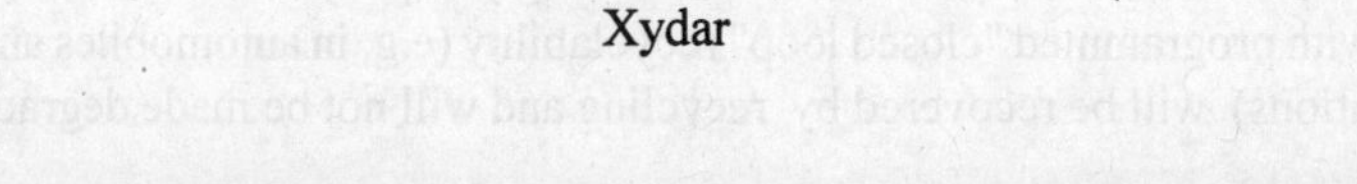

Xydar

POLYMERS

LCPs are use for speciality applications such as electrical / electronic application (e.g. computer memory modules, optical application, Automotive application, connectors, surgical devices and a variety of aerospace application.

LCPs exhibits a highly ordered structure in both the melt and solid states, due to this LCPs can replace material such as ceramics, metals, composites and other plastics because of its outstanding strength at extreme temperature and resistance to all chemicals, weathering, radiation and burning.

9.3 Rubbers (Elastomers)

9.3.1. Natural Rubber.
The main source of natural rubber is the species of tree known as "Hevea Brasiliensis", although it is found in several other plants, shrubs and vines. The sole source of rubber before the full development of rubber industry was the wild rubber trees from Brazil. Today, more than 95% of the rubber is obtained from the rubber trees. "Hevea Brasiliensis", grown on plantation mostly in Ceylon and the Malay peninsula. Small quantities of the rubber are produced in Brazil from the uncultivated wild rubber trees and from the "guayule" shrub in Mexico and southwestern United States. Natural rubber is a polymerized form of isoprene (2 methyl-1, 3-butadiene) :

$$CH_2 = \underset{\underset{Isoprene}{}}{\overset{\overset{CH_3}{|}}{C}} - CH = CH_2 \longrightarrow \underset{Polyisoprene}{+ (CH_2 - \overset{\overset{CH_3}{|}}{C} = CH-CH_2)_n}$$

Rubber latex is a milky colloidal emulsion containing about 25 to 45% of rubber, while the remainder is made up of mainly water and small quantities of protein and resinous material. The latex exudes from the bark of the tree when it is cut or wounded. A cut is made halfway through the bark, extending about 2/3rds around the tree and the liquid oozing out is collected in containers. The flow of latex from the cut diminishes with time, necessitating the removal of another thin layer of bark. This process, called tapping, is continued at intervals throughout the life of the tree. A plantation grown tree continues to yield for as long as 40 years and gives latex to the extent that 3 to 6 lb of rubber can be obtained every year.

Latex is treated in two ways to obtain rubber goods : (1) the crude rubber is coagulated from it by acids or heat, and then processed (2) the latex itself is mixed with appropriate compounding materials and then precipitated directly from solution in the shape to be used, *e.g.*, rubber gloves. This is a new and important technique.

In the recovery of rubber from "guayule", the entire plant, containing about 20% by weight of rubber, is harvested for 4 years, then ground up and soaked in water. The latex coming to the surface is skimmed off. The resin content in "guayule" rubber is about 18 to 20% as compared to 4% in "hevea" rubber. The resin can be extracted by solvents for recovery.

As mentioned earlier, the latex of the rubber is an aqueous dispersion of rubber containing 25 to 40% rubber hydrocarbon, stabilized by a small quantity of protein material (about 2%. The rubber droplets are quite small, having a diameter of 0·5 to 3·0 microns. The emulsion may be broken by the action of enzymes on the protective colloid of the natural latex or by the addition of coagulating agents.

Generally, a small amount of ammonia is added as a preservative to the latex collected. The latex is then coagulated by the addition of 5% solution of acetic acid or formic acid of 90% strength. Ammonium or potassium alum are also used as coagulants. The coagulum is washed and dried. Then it is subjected to any of the following processes :

1. Crepe rubber is prepared by adding a small amount of sodium bisulfite to bleach the rubber and the coagulum is then rolled out into sheets of about 1 mm thickness and dried in air at about 50°C.

2. Smoked rubber is made by eliminating the bleaching with sodium sulfite and rolling the coagulum into somewhat thicker sheets. These are then dried in smokehouses at about 50°C in the smoke from burning wood or coconut shells.

Crude rubber does not have all the desirable properties. It becomes soft and sticky in hot summer, while in cold weather it becomes hard and brittle. Both these properties are detrimental for many common applications. However, its properties can be substantially improved by the addition of suitable materials and further heat treatment, etc.

Mastication. (or plasticization). In 1824, Hancock discovered that rubber becomes a soft and gummy mass when subjected to severe mechanical working. This process is called mastication. This process greatly facilitated the addition of compounding agents to rubber, which is usually carried out on roll mills or in internal mixers or Banbury mixer or plasticators which are similar to extruders. Mastication is accompanied by a marked decrease in the molecular weight of the rubber. Oxidative degradation is an important factor in mastication, since the decrease in viscosity and the other changes in properties do not take place if the rubber is masticated in the absence of oxygen. After mastication is complete, compounding ingredients are added and the rubber mix is prepared for vulcanization process, which will be discussed later.

Compounding of rubber

The masticated rubber is mixed with other substances by the rolls to incorporate the desired properties. All the ingredients may be worked into the rubber thoroughly in a mill or Banbury mixer.

1. Vulcanizers. Vulcanization process was discovered by Charles Goodyear in 1839. He patented the process in 1884. He observed that when rubber is heated with sulfur, its tensile strength, elasticity and resistance to swelling are increased tremendously. This process is named as **vulcanization**. The sulfur combines chemically at the double bonds in the rubber molecule bringing about excellent changes in its properties *e.g.*, resistance to changes in temperature, increased elasticity and tensile strength, durability and chemical resistance.

Vulcanisation is an irreversible process, similar to other termoset this irreversible cure reaction defines vulcanized rubber compand as thermoset, which do not melt on heating and placed them away from the class of thermoplastic material. This is a fundamental difference between rubber and thermosplastic.

Vulcanization brings about a stiffening of the rubber by anchoring and restricting the intermolecular movement of the rubber springs. This is due to the chemical combination of the sulfur at the double bonds of different rubber springs and providing cross-linking between the chains.

The vulcanization (or the curing as it is some times called) can be carried out in several ways:
1. The articles to be vulcanized are heated with steam under pressure.
2. The article is immersed in hot water under pressure.

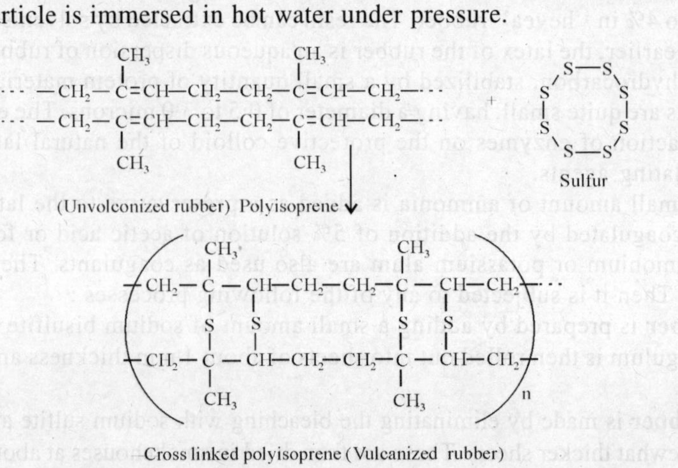

Fig. 9·22. Polyisoprere and cross linked polyisoprene

3. By heating the article in air or in carbon dioxide.
4. By passing steam directly into the article such as fire hose.
5. By vulcanizing the article in the mould in which it is shaped.

The temperature used in 110 to 140°C. The curing time may vary from a few minutes to 3 hours; overcured stock decreases stretch and tensile strength whereas undercured stock is too soft with excessive stretch but lower tensile strength. The amount of sulfur used for ordinary soft vulcanized rubber is 1 to 5% whereas for hard rubber, it is 40 to 45% of the rubber.

Vulcanization of very thin sheets of rubber can be accomplished by either dipping the articles in S_2Cl_2 or exposing them to vapours of S_2Cl_2. Other vulcanizing agents used include Se, Te, ZnO, benzoyl chloride, trinitrobenzene, alkyl phenol sulfides, H_2S, MgO, benzoyl peroxide etc.

2. Accelerators. These are meant for catalyzing the vulcanization process thus reducing the time required for vulcanization. The inorganic accelerators include lime, magnesia, litharge and white lead, whereas the organic accelerators are complex organic compounds such as aldehyde amines, thiocarbamates, Guanidines, zinc alkyl xanthate and 2-mercaptol benzothiozole, which are more useful and more commonly used. Sometimes, accelerator activators like ZnO are also added. Generally 0.5 to 1% of the accelerator is used.

3. Antioxidants. These substances when used in small quantities (about 1%) retard the deterioration of rubber by light and air. These are complex organic amines like phenyl naphthyl amine, phenolic substances and phosphites.

4. Reinforcing agents. These are usually added to give strength, rigidity and toughness to the rubber and may form as much as 35% of the rubber compound. The commonly used reinforcing agents include carbon black (for automobile tyres, etc.), ZnO, $MgCO_3$, $BaSO_4$, $CaCO_3$ and some clays.

5. Inert fillers. The main function of the inert fillers are to alter the physical properties of the mix to achieve simplification of the subsequent manufacturing operations, or to lower the cost of the product.

6. Plasticizers or softeners. These are added to impart greater tenacity and adhesion to the rubber. The most commonly used plasticizers are vegetable oils, waxes, stearic acid, rosin, etc.

7. **Colouring agents.** These are added to impart to the rubber the desired colour, as follows:

TiO_2, zinc sulfide or lithophone	...	White
Lead chromate	...	Yellow
Ferric oxide	...	Red
Antimony sulfide	...	Crimson
Ultramarine	...	Blue
Chromium trioxide	...	Green

8. Miscellaneous agents. These include baking soda for sponge rubber, abrasives (*e.g.*, silica and pumice), stiffening agents to stiffen the stock until vulcanization, etc.

Calendering. The rubber compound is passed through a calendering machine to convert it into sheets. In this, the material is passed between rolls which press it into thin sheets (0.003 to 0.1 inch thickness). If thicker sheets are required, several thin sheets are rolled together which eliminates air pockets.

Properties and Uses of rubber
Properties

Vulcanized rubber has got the ideal elastomeric properties such as rapid extensibility to great elongations, high stiffness and strength when stretched and rapid and complete retraction on release of the external stress. Its chief properties which render it valuable are its resiliency, low thermal and electrical conductivity, low permanent set, resistance to abrasion, resistance to water bases and dilute acids. Concentrated acids react with rubber and destroy it. Rubber is attacked by oxygen in air and by sunlight and the rate of such attack is reduced by anti-oxidants. Rubber is soluble in various organic solvents such as petrol, petroleum ether, benzene, turpentine, CS_2, CCl_4 etc. Solutions of rubber in benzene are used as adhesives. When rubber is treated with a solvent, the solvent is first absorbed by the rubber, which swells and forms a jelly-like mass and when enough of the solvent has been absorbed, the mass assumes the liquid state. Thus, in a sense the solvent may be said to have dissolved in the rubber.

The specific heat of raw rubber at room temperature is 0.502 and its specific gravity at 0°C is 0.950 and that at 20°C is 0.934. Its coefficient of cubical thermal expansion is 670×10^{-6}. When rubber is extended, heat is evolved and this is called Joule's effect. When it is stretched to 82%, 680 calories of heat/g are liberated. When rubber is cooled to low temperatures, it becomes stiff. When it is frozen it attains a fibrous structure.

When rubber is heated with about 1% organic sulfonyl chloride or an organic sulfonic acid at about 130°C, it is converted into a tough, thermoplastic resin which resembles Gutta-percha. Such products are known as thermoprenes.

Rubber reacts with chlorine giving chlorinated rubber. With HCl, rubber forms rubber hydrochloride. Rubber is oxidized by oxidizing agents such as HNO_3, peroxybenzoic acid, peroxide and $KMnO_4$. The oxidation reactions are catalyzed by Cu and Mn. During the normal oxidation of rubber, an unstable peroxide of rubber is first formed followed by transformation into a stable oxide.

Uses

1. Rubber is used for the manufacture of gaskets used for sealing refrigerator cabinet doors, etc.

2. Rubber is mainly used for the manufacture of tyres.

3. It is used for preparing V-belts for the power transmission and conveyor belts for conveying several types of materials. These products are compact, non-slipping, clean and shock-absorbing.

4. Rubber lined tanks (steel, Al, etc.) are widely used in chemical plants where protection from corrosive chemicals is required.

5. Rubber mountings are prepared from sandwiching the rubber between two metal plates. They reduce machine vibrations and prolong the life of the machines besides reducing noise.

6. Rubber is used for manufacturing hoses.

7. Rubber threads and sponge rubber have good shock absorbing and thermal insulation properties. Rubber threads are used in shock absorber cords, heat bands for goggles and helmets, golf balls, etc.

8. Rubber is used for various related products like chlorinated rubber, oxidized rubber, rubber hydrochloride, cyclized rubber and ebonites. All these substances have many industrial uses.

9. Foam rubber is used in the manufacture of cushions, mattresses, paddings, etc.

10. Rubber is also used for manufacturing toys, sports items, etc.

11. Rubbers when blended with plastics give improved strength, hardness, flexibility and chemical and thermal-resistance.

9.3.1. Natural rubber

1. Chlorinated rubber. Chlorinated rubber was traditionally prepared by allowing chlorine gas to react with a solution of masticated rubber in a chlorinated solvent. The newer methods of preparing chlorinated rubber involve the direct chlorination of the latex or passing of Cl_2 over thin sheets of rubber swollen with a solvent like CCl_4.

The mechanism of chlorination involves substitution, addition as well as cyclization.

Chlorinated rubber is mainly used for preparing thermal and chemical-resistant paints, varnishes and lacquers. Films, impregnating solutions and adhesives also can be prepared from chlorinated rubber.

2. Oxidized rubber. This is prepared by controlled oxidation of rubber by mastication in air in presence of a catalyst. This is used for impregnating paper and card-board and for protective coatings. Varnishes prepared with oxidized rubber have outstanding electrical insulating properties.

3. Rubber hydrochloride. When HCl gas is passed into a solution of previously milled rubber in benzene, rubber hydrochloride is produced, due to the addition of HCl to the double bonds of rubber as follows :

$$-CH_2 - \underset{\underset{Cl}{|}}{\overset{\overset{CH_3}{|}}{C}} - CH_2 - CH_2 -$$

The presence of chlorine atom on the tertiary carbon has been confirmed by X-ray studies and is also in accordance with Markownikoff's rule.

Stretched and plasticized films of rubber hydrochloride have good mechanical properties, including high tear resistance. Rubber hydrochloride is highly resistant to chemical attack but it is susceptible to thermal and photochemical decomposition, stabilizers are helpful in retarding this decomposition.

Rubber hydrochloride is extensively used for wrapping precision machines, machine parts, food materials, etc.

4. Cyclized rubber. Cyclized rubbers are commercially prepared by treating rubber with either H_2SO_4 or various sulfonic acids or sulfonyl chlorides, or chlorostannic acid. The products are non-elastic and are primarily used as compounding ingredients in shoe soles and heels and for rubber-to-metal bonding adhesives.

Guayule Rubber

It is obtained from Guayule shrub and is a source of natural rubber in North America. Rubber latex from Hevea tree exists in a canal system but the rubber Guayule is enclosed in the cells. Rubber obtained from the Guayule bush is recovered by cutting the shrub (after removing the leaves) into small pieces and then milling them in pebble mills in presence of water. The material is then sent to flotation tanks. The rubber floats to the top which is collected. This rubber material contains 70% rubber hydrocarbon, 20% resin and 10% cellulose, lignin and other insolubles.

Hevea rubber and Guayule rubber are chemically identical and both exist in the form of cis-polymer of isoprene. However, the molecular weight of Guayule rubber is lower than that of Hevea rubber.

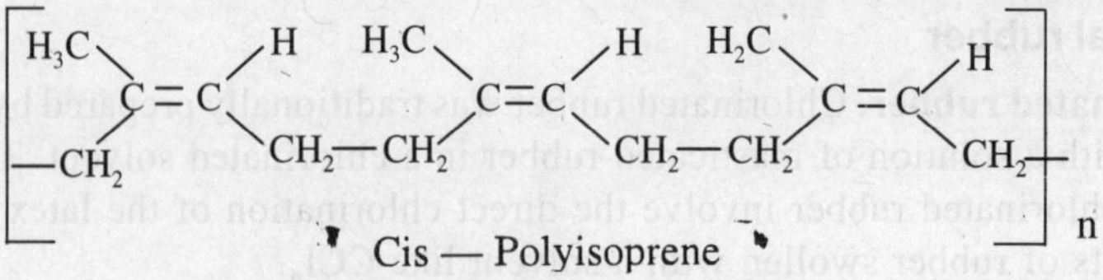

$$\left[\begin{array}{c} H_3C \\ \diagdown \\ -CH_2 \end{array} C = C \begin{array}{c} H \\ \diagup \\ CH_2- \end{array} \begin{array}{c} H_3C \\ \diagdown \\ CH_2- \end{array} C = C \begin{array}{c} H \\ \diagup \\ CH_2- \end{array} \begin{array}{c} H_2C \\ \diagdown \\ CH_2- \end{array} C = C \begin{array}{c} H \\ \diagup \\ CH_2 \end{array} \right]_n$$

Cis — Polyisoprene
(Hevea and Guayule Rubber)

Fig. 9·23. Hevea rubber and Guayule rubber (Cis-polyisoprene)

Gutta Percha and Balata

This is obtained from the mature leaves of the trees known as Dichopsis gutta and palaqium gutta belonging to Sapotaceae family, which are found mostly in Sumatra, Borneo and Malaya. The mature leaves are carefully ground in mills and treated with water at 70°C for about half an hour and then dropped into cold water. Gutta-percha floating on the surface is collected. Very pure Gutta-percha can be recovered by solvent extraction, so that insoluble gums and resins are separated.

Gutta percha is tough and horny at room temperature but turns soft and tacky at about 100°C. It is soluble in chlorinated and aromatic hydrocarbons but not in aliphatic hydrocarbons. Gutta percha is used in the manufacture of submarine cables, golf ball covers, tissue for adhesive and surgical purposes.

Balata is obtained from wild trees in Central and South America and the processing and uses are to Gutta percha. Both are the trans-polymers of isoprene.

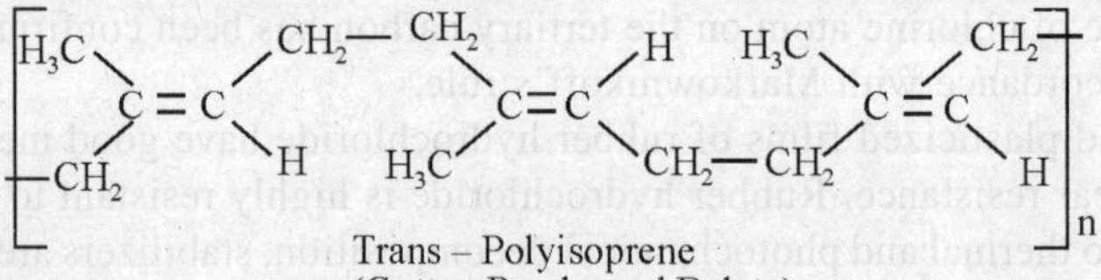

$$\left[\begin{array}{c} H_3C \\ \diagdown \\ -CH_2 \end{array} C = C \begin{array}{c} CH_2-CH_2 \\ \diagup \\ H \end{array} \begin{array}{c} \diagdown \\ H_3C \end{array} C = C \begin{array}{c} H \\ \diagup \\ CH_2-CH_2 \end{array} \begin{array}{c} H_3C \\ \diagdown \\ \end{array} C = C \begin{array}{c} CH_2- \\ \diagup \\ H \end{array} \right]_n$$

Trans – Polyisoprene
(Gutta - Percha and Balate)

Fig. 9·24. Gutta percha and Balata (Trans-polyisoprene)

Reclaimed Rubber

Rubber can be reclaimed in a usable form from wornout rubber articles and rubber waste from the factories. This can be done by various methods but the most widely used one is the "alkali process". In this process, the used rubber is separated from metal (by electromagnetic separation) and fiber and heated at about 200°C under pressure for about 8 to 15 hours with an aqueous solution of an alkali (e.g., NaOH) in a closed iron vessel (or an autoclave). This treatment removes the remaining fiber and converts the free sulfur present into alkaline sulfide. The fibre gets hydrolysed and the rubber becomes devulcanized. The resulting material is carefully washed and dried. This may now be mixed with small quantities of reinforcing and processing agents such as clay, carbon black, softeners, etc.

Reclaimed, rubber, known as "rubber shoddy", generally has much lower elasticity, tensile strength and wearing quality than fresh rubber. Nevertheless, it may be superior to some of the poorer grades of crude rubber. The properties of the reclaimed rubber mostly depend on the degree to which the plasticity is regenerated by the reclaiming process.

Reclaimed rubber is cheaper than fresh rubber. Reclaimed rubber is quicker and easier to process, faster in curing and has better aging properties as compared with fresh rubber.

Reclaimed rubber is used for the manufacture of tyres, tubes, belting, hoses, automobile

POLYMERS

floor mats, hard rubber containers for batteries, soles, heels, steering wheels, mountings, couplings, etc.

Manufacture of rubber articles directly from latex

Manufacture of rubber goods directly from latex is a recent technique and has several advantages :

(1) There is no need of expensive machinery.

(2) Mastication of rubber is not required in this process and hence higher tensile strength can be obtained.

(3) Time required for vulcanization is much less.

In this process, the compounding ingredients are first emulsified in water and then added to the latex. Then, the finished articles are made in any of the following ways :

(1) Insulated wires may be made by passing the wire through the latex compound.

(2) Cords and fabrics can be impregnated by dipping.

(3) Sponge rubbers can be prepared by forcing air into the latex compound mechanically, followed by adding a coagulant.

(4) Rubber thread can be prepared by extruding the latex into a coagulating bath.

Several other techniques are also available and in all these cases, the forming process should invariably be followed by vulcanization.

Sponge rubber

It can be produced by milling the elastomer to the desired extent and mixing with all the necessary ingredients (as in the compounded rubbers) in a mill or in Banbury along with chemical blowing agents (such as Na_2CO_3) and the modifiers (such as fatty acids) which react with the blowing agents to produce gas under vulcanizing conditions. The mixed stock is then moulded and cured.

Chemically blown sponge may have an open (inter-connecting) or closed cell structure depending on the process conditions. Closed cell sponge can be produced by adding nitrogen producing organic materials (*e.g.*, diazoamino-benzene and benzenesulphonyl hydrazide) in place of CO_2 producing chemicals (*e.g.*, Na_2CO_3 with a fatty acid).

Foam rubber

This can be prepared by bubbling a gas into a compounded liquid latex. The product is gelled in a mould with the help of gelling agents like sodium or potassium silicofluoride to give it the desired shape, cured, washed and dried. Latex foam has an open cell structure.

Foam rubber is generally produced by the following two processes.

(1) In the Talalay process, the foam is blown in the latex by adding chemical blowing agents which release oxygen (such as H_2O_2) and a modifier (such as yeast). The foam is quickly frozen in a mould and coagulated with carbon dioxide.

(2) In the Dunlop process, the foam is formed by whipping air into the latex.

Gelling agent like sodium silicofluoride is added. The latex foam is vulcanized or cured before it is dried.

9.3.2 Synthetic Rubbers. Greville Williams (in 1860) was the first to isolate isoprene as a decomposition product by heating rubber to a high temperature. The monomer-polymer structural correlation between isoprene and rubber gave a clue that rubber can be synthesized by the polymerization of monomers like isoprene. It was confirmed by the end of the 19th century that rubber-like products could be made from isoprene by treating it with hydrogen chloride or allowing it to polymerize spontaneously on storage. The materials could be vulcanized with sulfur rendering them more elastic, tougher and more heat-resistant.

Around 1900, it was discovered that other dienes like butadiene and 2, 3—dimethyl butadiene could be polymerized spontaneously into rubber like materials in presence of alkali metals or by free radicals. Production of hard rubber continued, making use of the above knowledge, during first World War. After the war, the research on synthesis of rubber-like products took a new turn with the ready availability of butadiene (which was produced at that time from a acetaldehyde via aldol condensation). Since the alkali metals were used as initiators for the polymerization of butadiene, the products were called "*buna* rubbers" (from the first letters of butadiene and the symbol, *Na* for sodium). However, the product prepared at that time was poor in tensile strength, aging characteristics, etc., and the polymerization was slow and difficult to control. Improvements in processing and properties were made (1) by developing copolymers of butadiene with vinyl monomers, particularly styrene and (2) by adopting emulsion polymerization. By the time of the Second World War, acceptable polymers (such as buna—S) were produced in Germany. This contained 68—70% butadiene and 30—32% styrene. A persulfate initiator was used. Production of styrene-butadiene rubber, then known as GR—S; started in USA during the Second World War.

(1) SBR (GR—S or Buna—S or Ameripol or Styrene-butadiene rubber). This is a copolymer of about 75% butadiene and 25% styrene. The two components are allowed to react in a mixing vessel containing an aqueous solution of an emulsifying agent,. Initiators like cumene hydroperoxide and *p*-methane hydroperoxide were used in presence of anti-freeze components to produce "cold" SBR or "cold rubber."

The mastication and vulcanization process are more or less similar to that with natural rubber. A reinforcing filler (*e.g.*, carbon black) is essential to achieve good physical properties. Less sulfur is required for vulcanization. The tensile strength and flexibility of these rubbers are inferior to those of natural rubber.

It is used for lighter-duty tires, hoses, belts, molded goods, unvulcanized sheet, gum, floorings, rubber shoe soles and for electrical insulations.

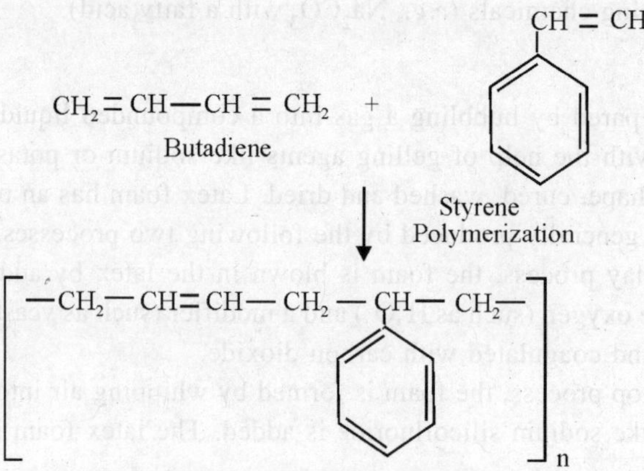

$$CH_2 = CH—CH = CH_2 \; + \; C_6H_5—CH=CH_2$$
Butadiene Styrene

Polymerization

$$\left[—CH_2—CH=CH—CH_2—CH(C_6H_5)—CH_2— \right]_n$$

Buna S or SBR or GR—S

Fig. 9.25

(2) Nitrile rubbers. Nitrile rubbers are polymers of butadiene and acrylonitrile. These are also prepared in emulsion systems. They are noted for their oil resistance but are not suitable for tyres. Compounding and vulcanization are similar to those of natural rubber.

These rubbers have low swelling, low solubility, good tensile strength and abrasion resistance even after immersion in gasoline or oils. The rubbers have good heat resistance. They are inherently less resilient than natural rubber. Nitrile rubbers are extensively used for fuel tanks, gasoline hoses, creamery equipment, etc. They are also used in adhesives, and in the form of latex, for impregnating paper, leather and textiles.

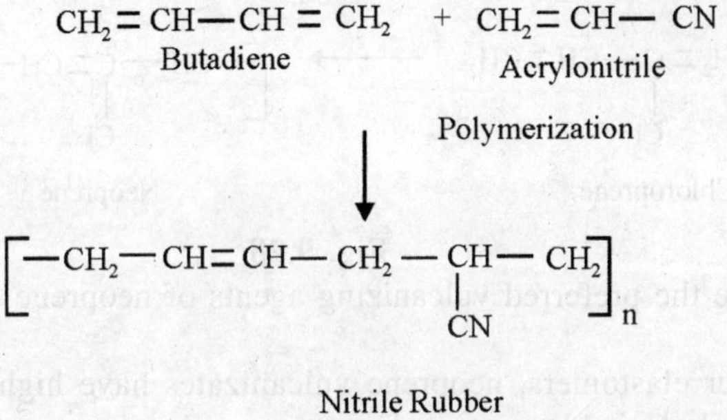

$$CH_2=CH-CH=CH_2 \quad + \quad CH_2=CH-CN$$
$$\text{Butadiene} \qquad\qquad\qquad \text{Acrylonitrile}$$

$$\downarrow \text{Polymerization}$$

$$\left[-CH_2-CH=CH-CH_2-\underset{\underset{CN}{|}}{CH}-CH_2\right]_n$$

Nitrile Rubber

Fig. 9·26.

(3) Polyisobutylene and Butyl rubber. Butyl rubbers are copolymers of isobutylene with a small amount of isoprene added in order to render them vulcanizable. It is manufactured by mixing isobutylene with 1·5 to 4·5% isoprene and methyl chloride as solvent. The mixture is fed to stirred reactors cooled to —95°C by liquid ethylene anhydrous $AlCl_3$ is used as catalyst in methyl chloride. The polymer forms was suspended in the reaction mixture.

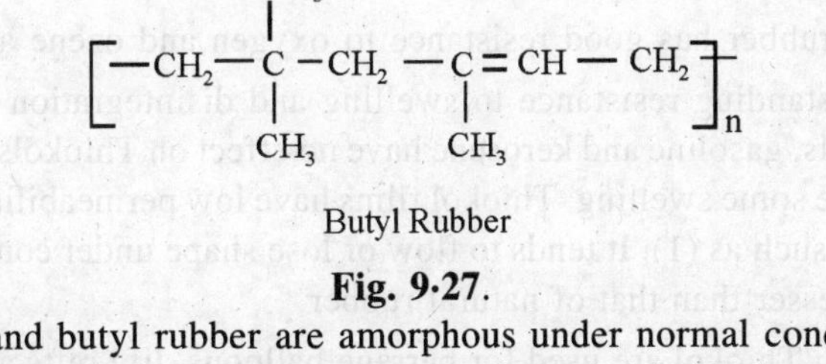

$$CH_3-\underset{\underset{CH_2}{||}}{C}-CH_3 \quad + \quad CH_2=\underset{\underset{CH_3}{|}}{C}-CH=CH_2$$
$$\text{Isobutylene} \qquad\qquad\qquad \text{Isoprene}$$

$$\downarrow$$

$$\left[-CH_2-\underset{\underset{CH_3}{|}}{\overset{\overset{CH_3}{|}}{C}}-CH_2-\underset{\underset{CH_3}{|}}{C}=CH-CH_2\right]_n$$

Butyl Rubber

Fig. 9·27.

Polyisobutylene and butyl rubber are amorphous under normal conditions but crystallize on stretching. Unstabilized polyisobutylenes are degraded by light or heat to sticky low-molecular-weight products. The usual rubber antioxidants or retarders of free radical reactions stabilize the polymer well.

Butyl rubber is used as inner tubes because of its superior impermeability to gases. It is used for wire and cable insulation. It is used in the production of tyres.

(4) Neoprene (Polychloroprene). Neoprene represents all rubber-like polymers and copolymers of chloroprene (2-chloro-1, 3-butadiene). They are primarily known for their oil resistance, but they are good general-purpose rubbers which can replace natural rubber in many of its uses.

Chloroprene is prepared by the catalytic addition of hydrogen chloride to vinylacetylene, which in turn is made by the catalytic dimerization of acetylene. The neoprenes are produced by emulsion polymerization. Some types are polymerized in the presence of sulfur, which introduces some cross-linking in the polymer. In such cases, the latex is allowed to age in the presence of an emulsion of tetraethylthiuram disulfide, which restores the plasticity of the ploymer. The latex is then coagulated by acidification followed by freezing.

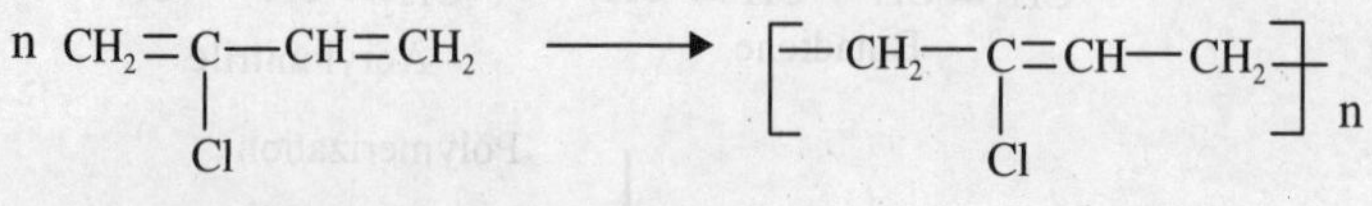

$$n\ CH_2=\underset{Cl}{C}-CH=CH_2 \longrightarrow \left[-CH_2-\underset{Cl}{C}=CH-CH_2-\right]_n$$

Chloroprene → Neoprene

Fig. 9.28

ZnO and MgO are the preferred vulcanizing agents of neoprene as sulphur vulcanizes it very slowly.

Unlike many other elastomers, neoprene vulcanizates have high tensile strength in the absence of carbon black. Suitably protected neoprene vulcanizates are extremely resistant to oxidative degradation. Weathering resistance and ozone resistance are also good. Oil resistance is inferior to nitrile rubber but superior to natural rubber.

Neoprenes are excellent for tyres but are more expensive than those prepared from other elastomers. It is mainly used in wire and cable coatings, industrial hoses and belts, shoe heels and solid tyres. Gloves and coated fabrics can be prepared from neoprene latex.

(5) Polysulphide rubbers : The polysulphide rubber commonly known in the trade name as "Thiokols" J. G. Patrik, an American chemist prepared polysulphide rubber by heating sodium polysulphide with ethylene dichloride

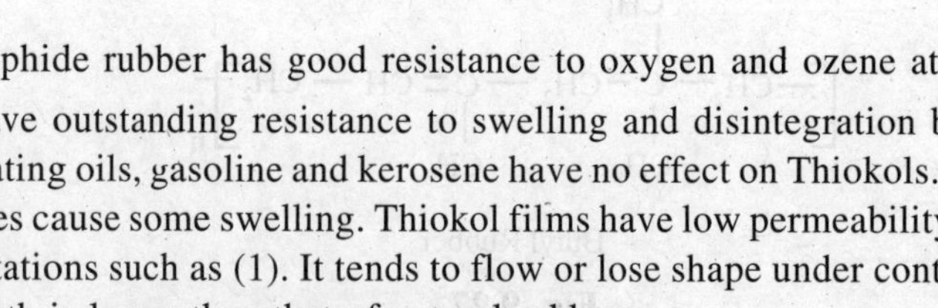

$$n\ Cl\ CH_2\ CH_2\ Cl\ +\ n\ Na_2S\ x$$
Ethylene dichloride   Sodium Polysulphide

$$\downarrow$$

$$(CH_2-CH_2-S-\underset{\underset{S}{\|}}{S}-)\ n + 2n\ NaCl$$

Polysulphide rubber

The polysulphide rubber has good resistance to oxygen and ozene attack.

Thiokols have outstanding resistance to swelling and disintegration by organic solvents. Fuel oils, lubricating oils, gasoline and kerosene have no effect on Thiokols. However, benezene and its derivatives cause some swelling. Thiokol films have low permeability to gases. However, it has some limitations such as (1). It tends to flow or lose shape under continuous pressure (2). Its tensile strength is lesser than that of natural rubber.

Fabrics coated with Thiokol are used for barrage balloons, life rafts and jackets which are inflated by CO_2. Thiokols are also used for lining hoses for conveying gasoline and oil, in paints, for gaskets, diaphragms and seals in contact with solvents and for printing rolls.

(6) Silicone rubbers. The preparation, properties and uses of silicone rubbers have been described in the previous section under "silicones." Careful condensation of pure silicon diols such as dimethyl silicon diol produces long elastic chains upto a molecular weight of 5×10^5. The properties of these elastomers (also known as silastics) can be modified by suitable selection of the alkyl radicals attached to the silicon, by regulating the chain length and selection of fillers in compounding of the elastomer.

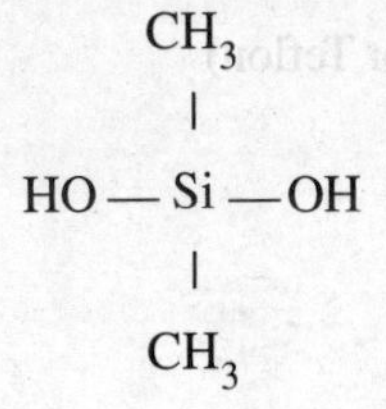

$$\text{HO—Si(CH}_3\text{)}_2\text{—OH}$$

Silicates are resistant to extremes of temperatures, —50°C to 300°C. They have a good thermal stability and good dielectric properties. Their tensile strength is lower than natural rubber. Silicon rubbers find extensive uses as seals, gaskets, diaphragms and rollers, whenever resilience at very low or moderately high temperatures is required, such as in aircraft or domestic oven doors.

(7) Polyurethanes. These are polymers containing the group typically formed by reaction between diisocyanates, $R(CNO_2)_2$

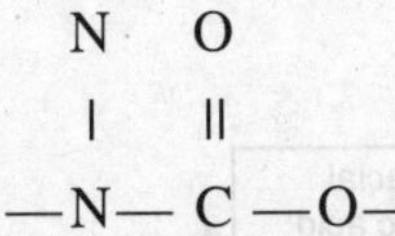

$$-N(N)-C(=O)-O-$$

(*e.g.*, ethylene diisocyanate), and polyalcohols (*e.g.*, ethylene glycol).

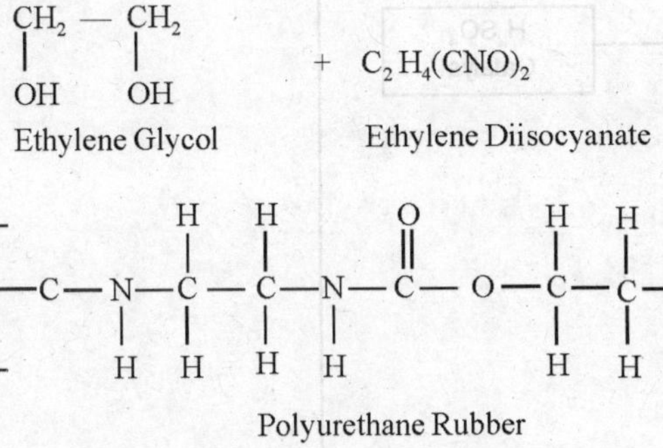

$$\underset{\text{Ethylene Glycol}}{\text{HOCH}_2\text{—CH}_2\text{OH}} + \underset{\text{Ethylene Diisocyanate}}{C_2H_4(CNO)_2}$$

$$\left[-\underset{H}{\overset{}{C}}-\underset{H}{\overset{H}{N}}-\underset{H}{\overset{H}{C}}-\underset{H}{\overset{H}{C}}-\underset{}{\overset{H}{N}}-\underset{}{\overset{O}{\overset{\|}{C}}}-O-\underset{H}{\overset{H}{C}}-\underset{H}{\overset{H}{C}}-O- \right]_n$$

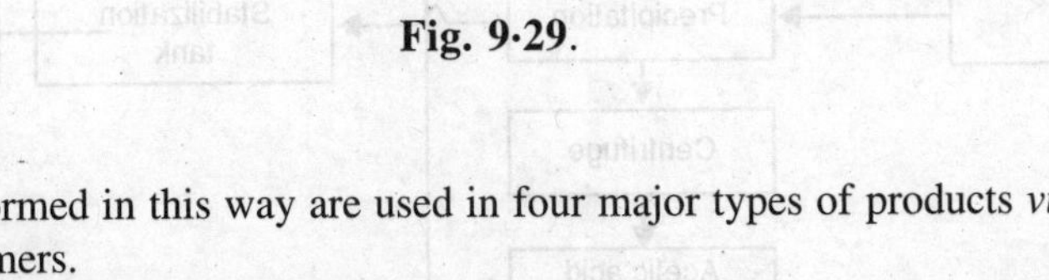

Polyurethane Rubber

Fig. 9·29.

The polymers formed in this way are used in four major types of products *viz.*, fibers, foams, coatings, and elastomers.

Polyurethane elastomers have outstanding abrasion resistance and hardness, combined with good elasticity and resistance to oils, greases and solvents. They enhance the life of tyre treads. They are used in applications where extreme abrasion resistance is required such as in heel lifts and small industrial wheels.

9.4. Flow-sheets for producing some important polymers

The flow-sheets for producing the following polymers are given in the following pages

(1) Cellulose acetate
(2) Polytetrafluoroethylene (PTFE or Teflon)
(3) Vinyl polymer
(4) Vinyl chloride from acetylene
(5) Nylon 6:6
(6) Nylon-6
(7) Acrylonitrile
(8) Phenol formaldehyde resin products
(9) Silicones via direct monomer process
(10) Silicones by Grignard's method
(11) Polyolefins by Ziegler process
(12) Butadiene - Styrene (SBR) rubber process

Flow-Sheet diagrams

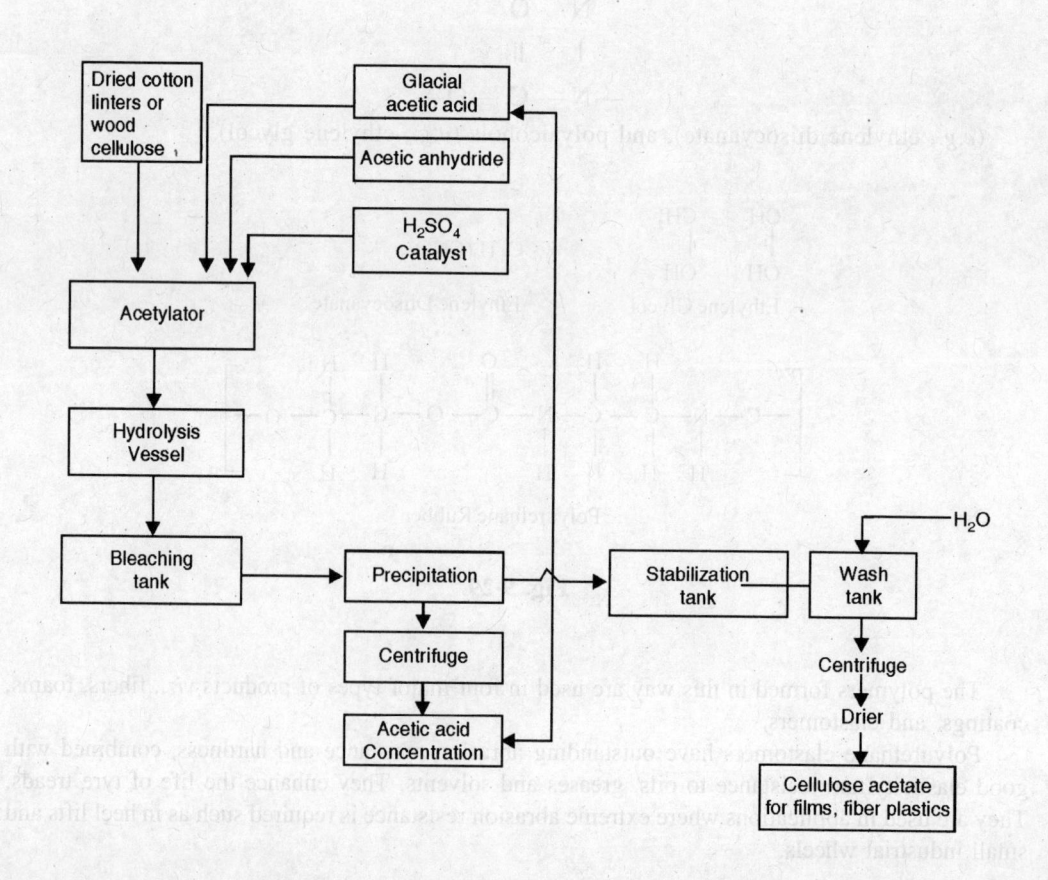

1. Flow-sheet for producing cellulose acetate

POLYMERS

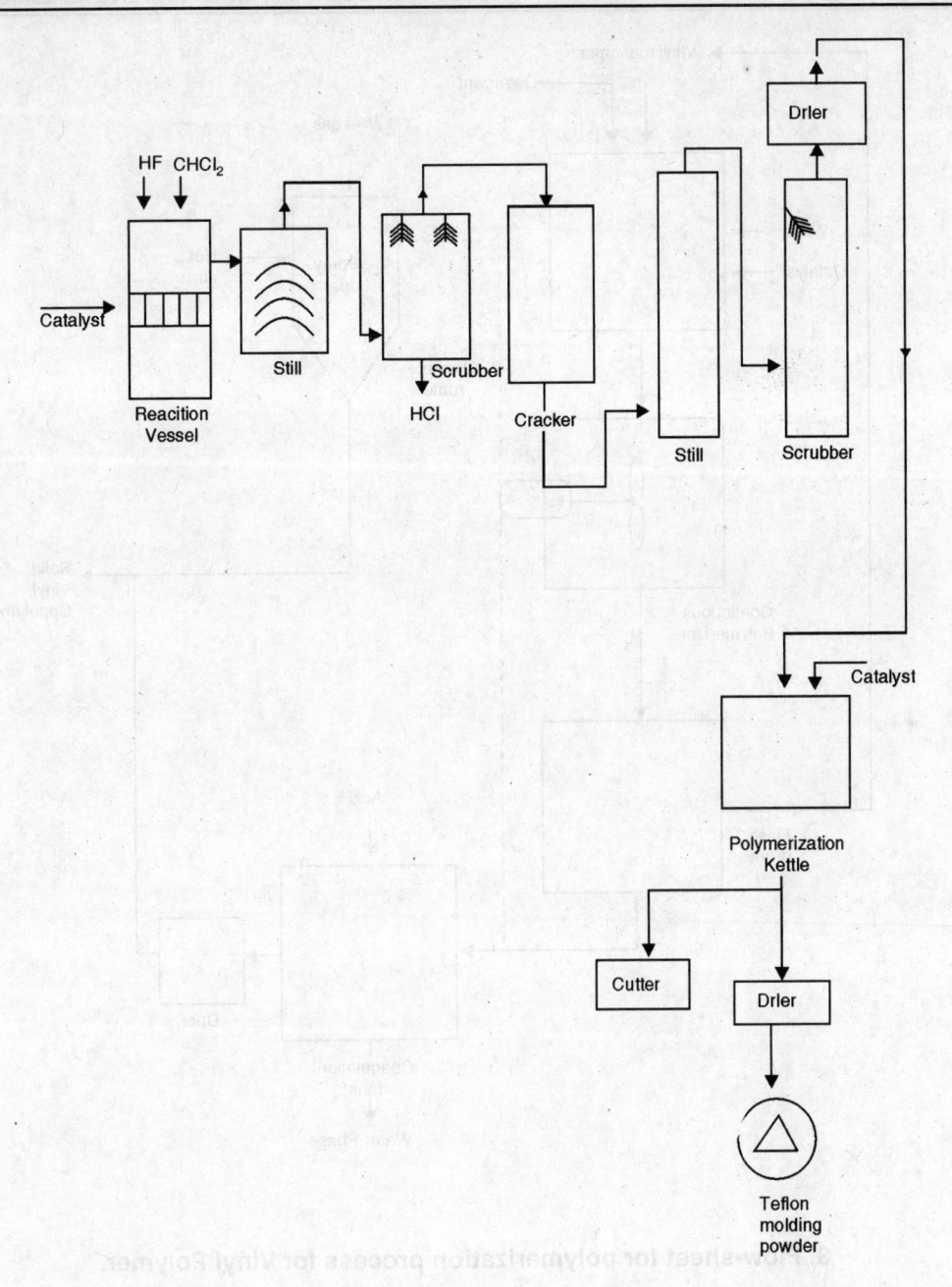

2. Flow-sheet for producing polytetrafluorethylene (Teflon)

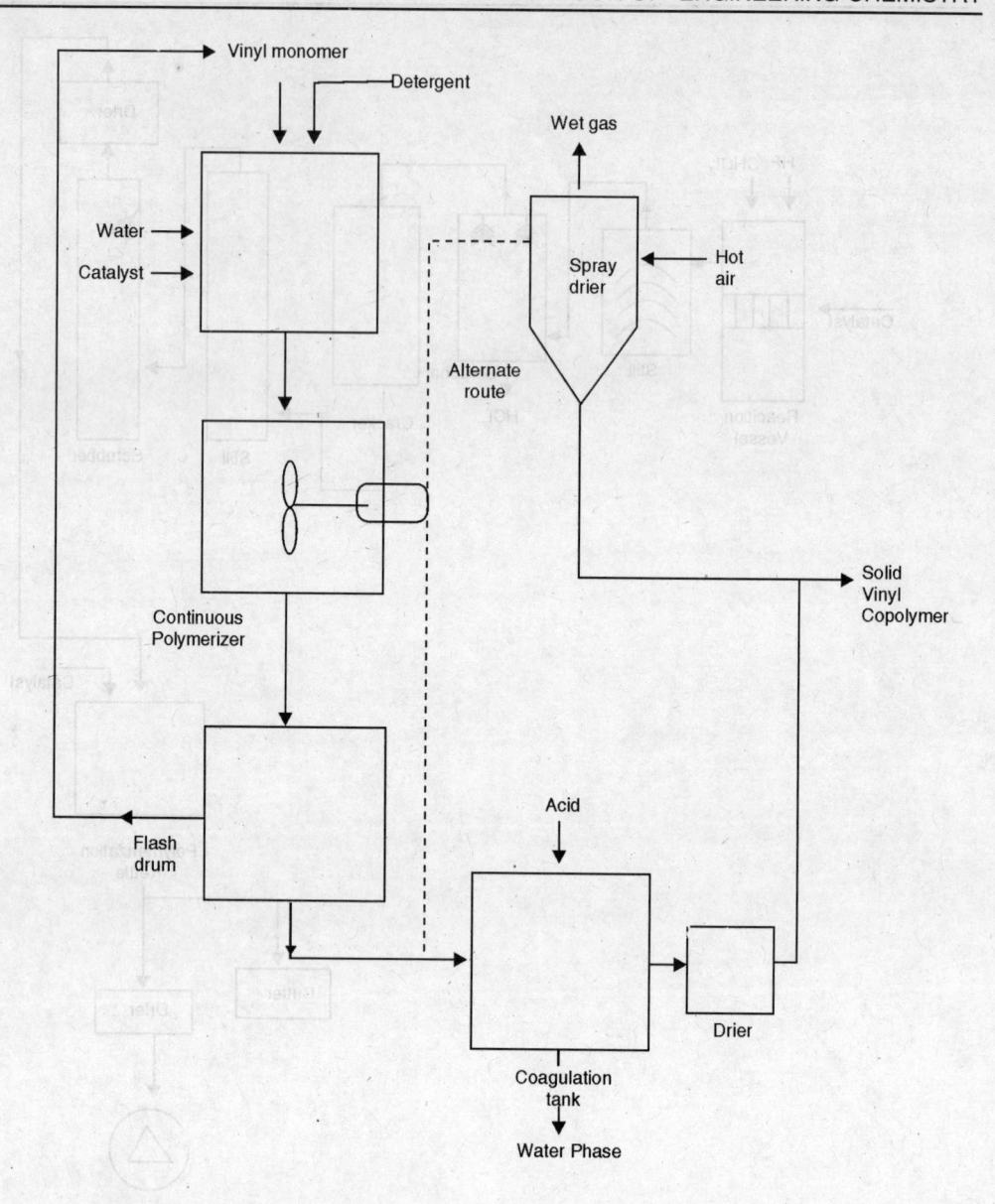

3. Flow-sheet for polymerization process for Vinyl Polymer.

POLYMERS

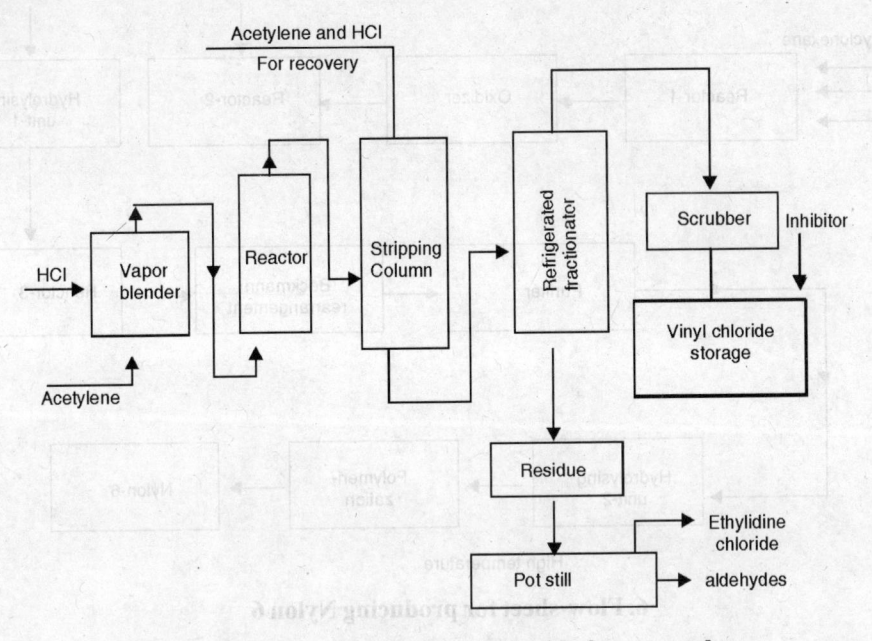

4. Flow-sheet for producing vinyl chloride from acetylene.

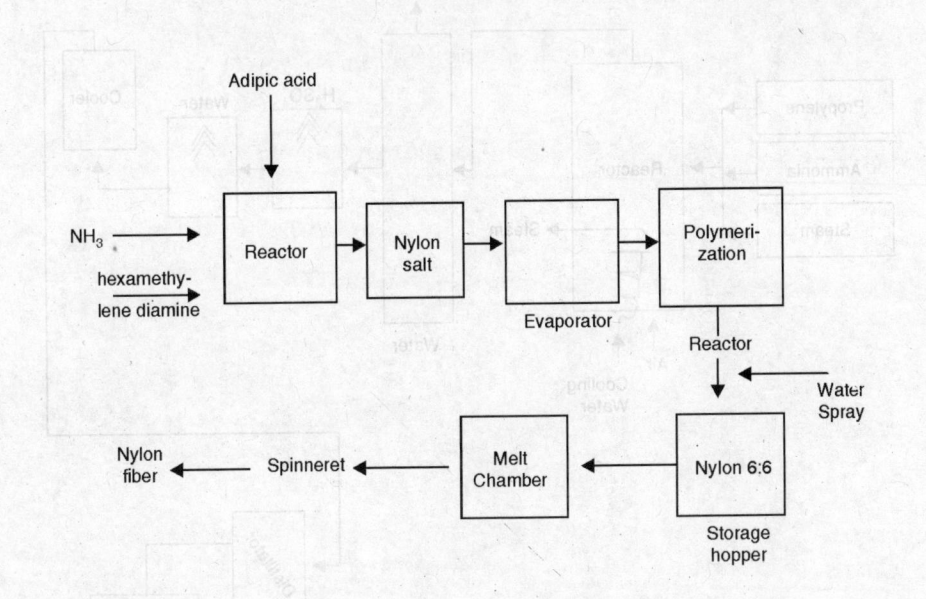

5. Flow-sheet for producing Nylon 6 : 6

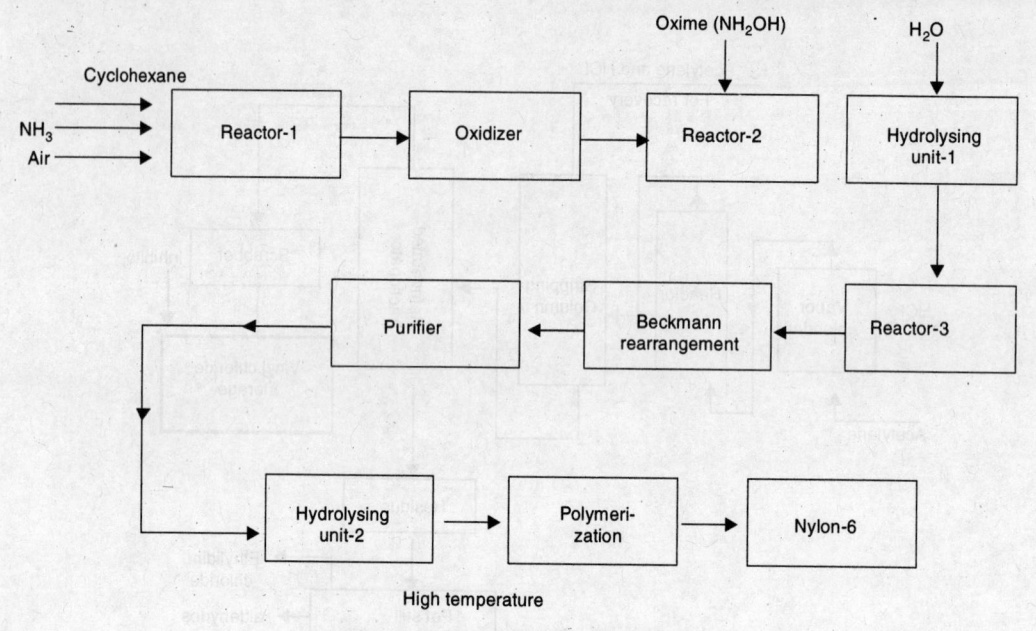

6. Flow-sheet for producing Nylon 6

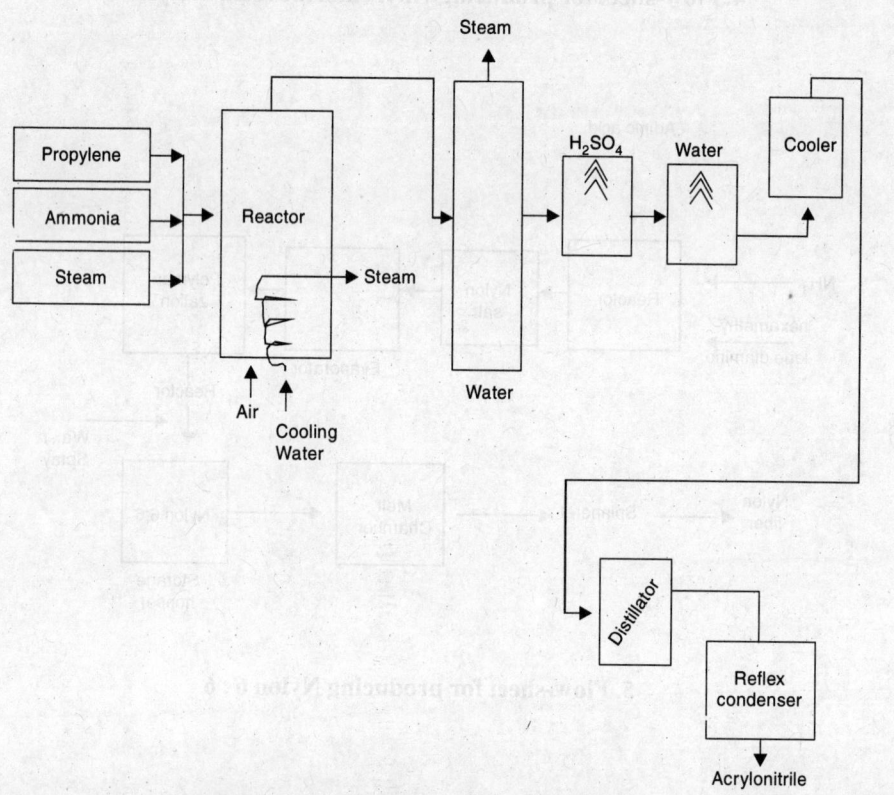

7. Flow-sheet for producing acrylonitrile

POLYMERS

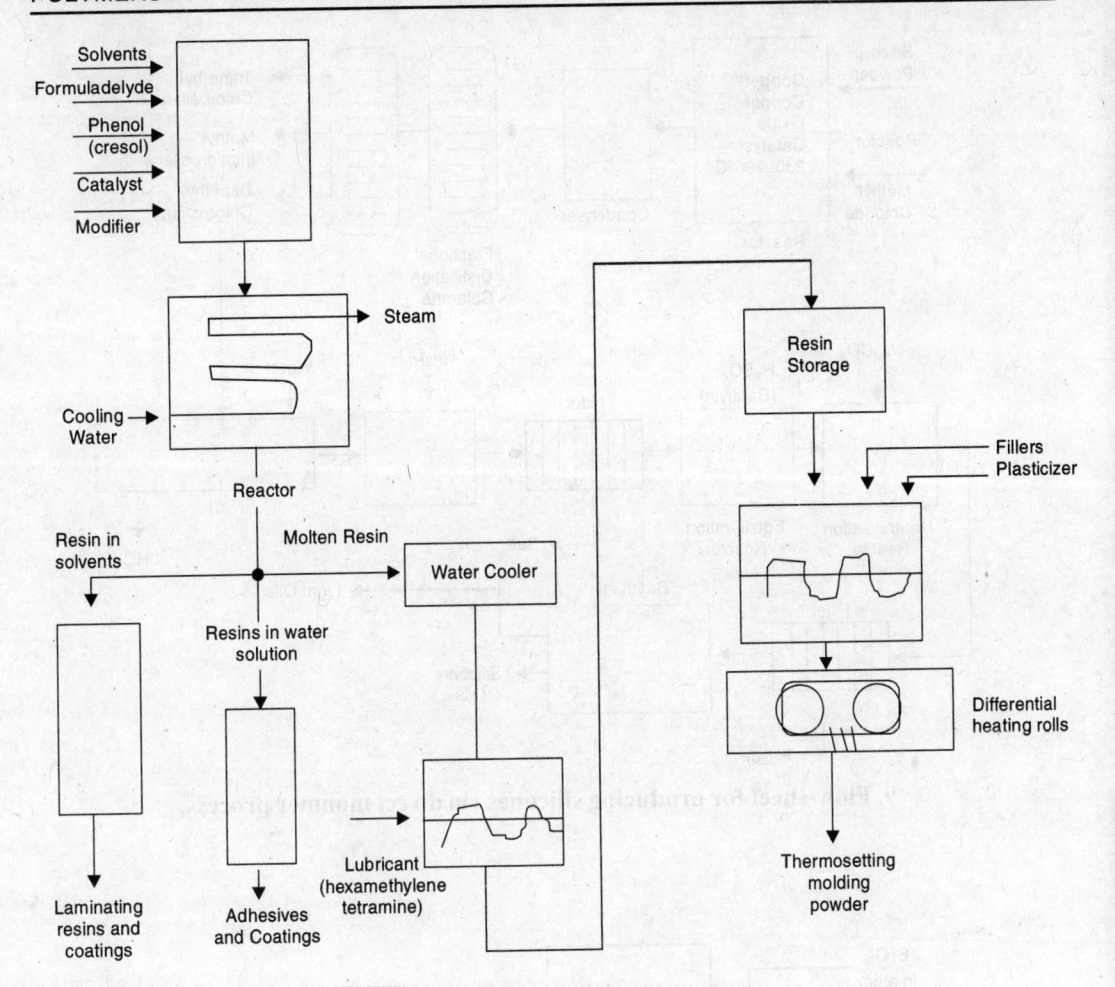

Production of Phenol-Formaldehyde Resin Products

Phenol formaldehyde C-Stage resin

8. Flow-sheet for producing phenol formaldehyde resin products.

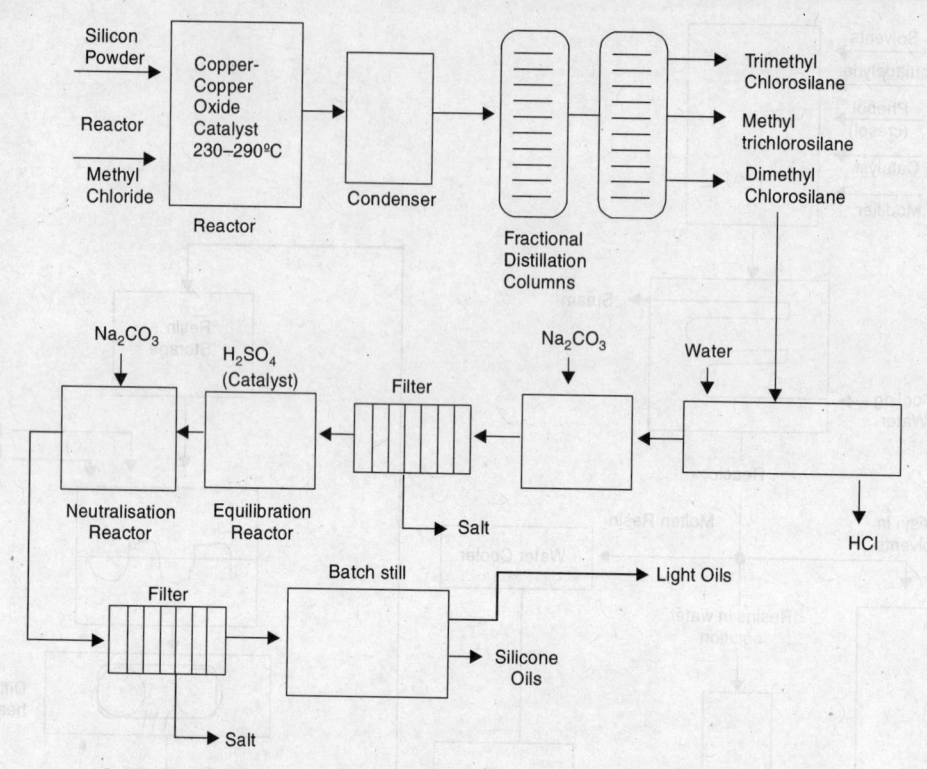

9. Flow-sheet for producing silicones via direct monmer process.

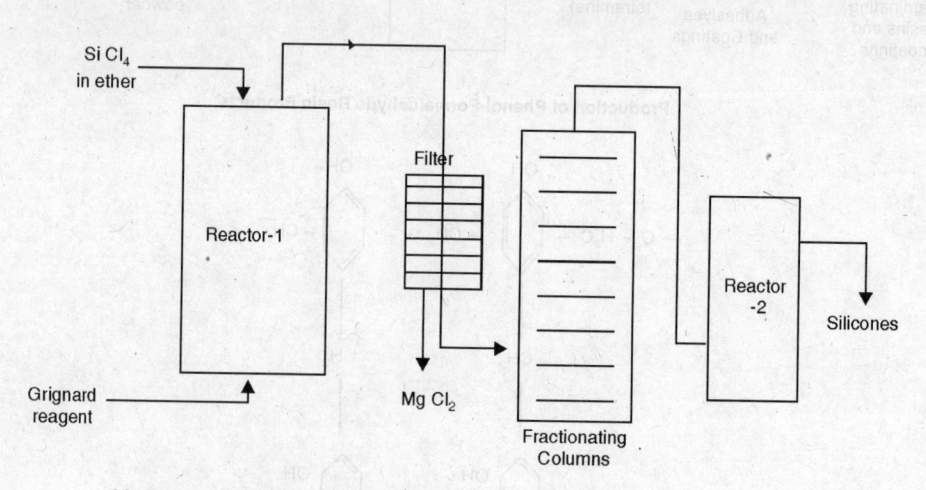

10. Flow-sheet for producing silicones by Grignard's method.

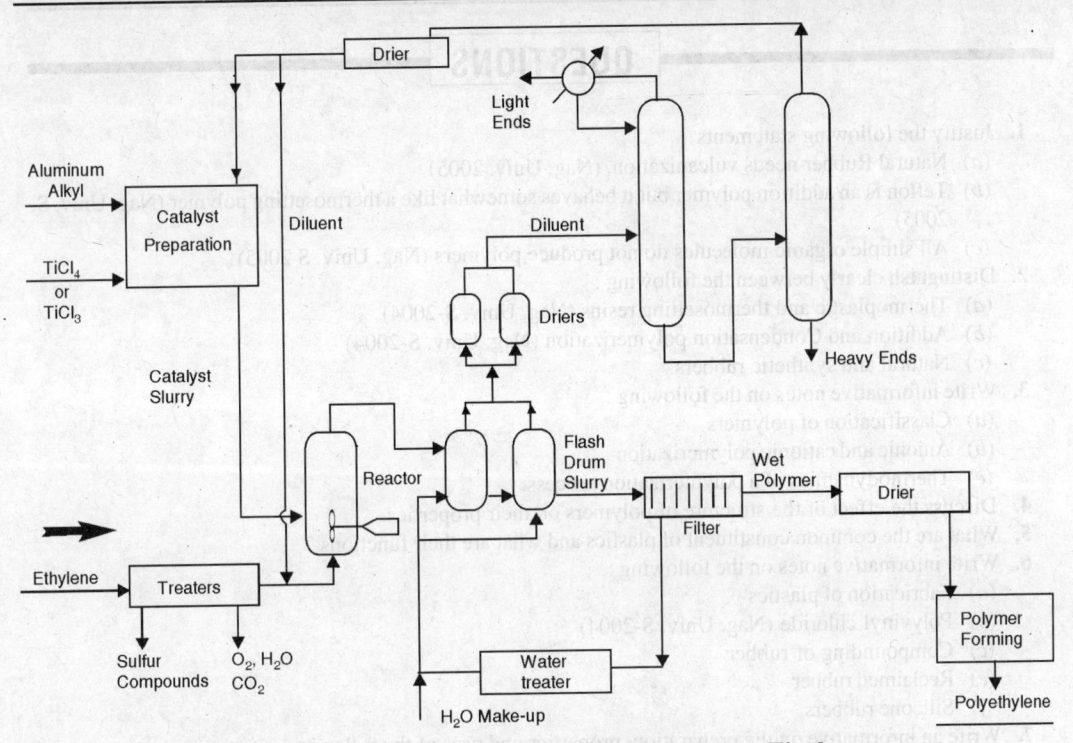

POLYMERS

11. Flow-sheet for producing Polylefins by low pressure Ziegler process.

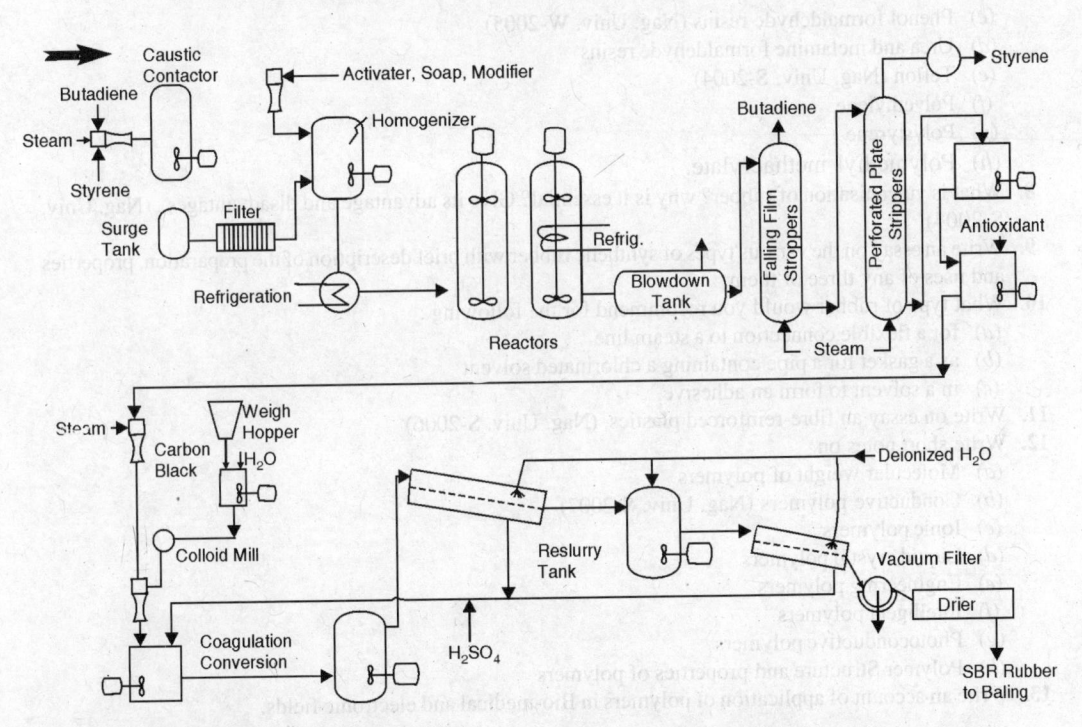

12. Flow-sheet for the manufacture of Butadiene styrene (SBR) rubber process.

QUESTIONS

1. Justify the following statements
   (a) Natural Rubber needs vulcanization. (Nag. Univ. 2005)
   (b) Teflon is an addition polymer but it behaves somewhat like a thermosetting polymer (Nag. Univ. S. 2005)
   (c) All simple organic molecules do not produce polymers (Nag. Univ. S 2005)
2. Distinguish clearly between the following :
   (a) Thermoplastic and thermosetting resins (Nag. Univ. S-2004)
   (b) Addition and Condensation polymerization (Nag. Univ. S-2004)
   (c) Natural and synthetic rubbers.
3. Write informative notes on the following :
   (a) Classification of polymers
   (b) Anionic and cationic polymerization
   (c) Thermodynamics of a polymerization process.
4. Discuss the effect of the structure of polymers on their properties.
5. What are the common constituent of plastics and what are their functions ?
6. Write informative notes on the following :
   (a) Fabrication of plastics
   (b) Polyvinyl chloride (Nag. Univ. S-2004)
   (c) Compounding of rubber
   (e) Reclaimed rubber
   (f) Silicone rubbers
7. Write an informative on the preparation, properties and uses of the following :
   (a) Polyvinyl acetate
   (b) Cellulose acetate
   (c) Phenol formaldehyde resins (Nag. Univ. W-2005)
   (d) Urea and melamine formaldehyde resins
   (e) Teflon (Nag. Univ. S-2004)
   (f) Polyethylene
   (g) Polystyrene
   (h) Polymethyl methacrylate.
8. What is vulcanisation of rubber? why is it essential? Give its advantage and disadvantages . (Nag. Univ. S-2004)
9. Write an essay on the various types of synthetic rubber with brief description of the preparation, properties and uses of any three of them.
10. What type of rubber would you recommend for the following :
    (a) for a flexible connection to a steam line
    (b) as a gasket for a pipe containing a chlorinated solvent
    (c) in a solvent to form an adhesive.
11. Write on essay an fibre-reinforced plastics. (Nag. Univ. S-2006)
12. Write short notes on.
    (a) Molecular weight of polymers
    (b) Conductive polymers (Nag. Univ. S-2007)
    (c) Ionic polymers
    (d) Liquid crystal polymers
    (e) Engineering polymers
    (f) Intelligent polymers
    (g) Photoconductive polymers
    (h) Polymer Structure and properties of polymers
13. Write an account of application of polymers in Bio-medical and electronic fields.

10

Composite Materials

> "The Voyager aircraft made a 25,000 mile non-stop flight around the world without mid-air refuelling. High-strength, low-density structural members of the Voyager are constructed from a series of crossplies consisting of graphite fibers that are aligned and embedded within an epoxy matrix."
>
> "William D. Callister, Jr."

10.1. INTRODUCTION

Our modern technologies, demand materials with unusual and extraordinary combinations of properties that cannot be provided by the conventional metal alloys, ceramics and polymeric materials. Aircraft engineers are in search of structural materials having low density, stiffness, high strength, abrasion-resistance, impact-resistance and corrosion-resistance. Such a combination of properties are rather difficult to achieve in the conventional materials because strength is usually associated with high-density materials and also increase in strength and stiffness usually results in decrease in impact strength. The ever increasing demand for such special materials having unusual combination of properties for use in modern technologies is responsible for the development of composite materials.

A composite is multiphase material that exhibits a significant proportion of the properties of both the constituent materials. Scientists and engineers have designed various composite materials by the combination of metals, ceramics and polymers to produce new generation of extraordinary materials having combination of superior mechanical characteristics such as toughness, stiffness and high-temperature strength.

A composite, is an artificially prepared multiphase material in which the chemically dissimilar phases are separated by a distinct interface. Thus, most of the ceramic materials, metallic alloys (*i.e.*, pearlitic steels), and natural materials like wood (which consists of strong and flexible cellulose fibers surrounded and held together by a stiffer material called Lignin) and bone (which is a composite of the strong, yet soft, protein collagen and the hard, brittle mineral apartite) are not considered as composite materials as per this difinition.

10·2 CONSTITUTION

Composite materials comprise of two phases, one is called the **matrix** which is continuous and surrounds the other phase called the **dispersed phase** (reinforcement). The final properties of the composites are determined by the properties of the constituent phases, their relative amounts and the geometry of the dispersed phase namely the shape and size of the particles, their distribution and orientation.

Role of matrix phase

The matrix phase is required to peform several functions

1. It binds the fibers together and acts as the medium by which an externally applied stress is transmitted and distributed to the fibers. Only a very small portion of the applied load is sustained by the matrix phase.

2. It protects the individual fibers from surface damage due to mechanical abrasion or chemical reactions with the environment.

3. It separates the individual fibers. By virtue of its relative softness and plasticity, the matrix prevents the propagation of brittle cracks from fiber to fiber, which may result in catastrophic failure. Thus, the matrix phase serves as a barrier to crack propagation. Even if some of the individual fibers fail, total fracture of the composite will not occur until large numbers of adjacent fibers fail and form a cluster of critical size.

The matrix material should be ductile. The elastic modulus of the matrix should be much lower than that of the fiber. In order to minimize the fiber pull-out, the adhesive bonding forces between the fiber and the matrix must be high. Indeed, it is the bonding strength that is the most important consideration in the choice of the matrix - fiber combination.

Since some ductility is essential, only metals and polymers are used as the matrix materials. Metals like Al and Cu, and commercial thermoplastic and thermosetting polymers are generally used as the matrix materials.

Matrix materials

Polymer matrix : Polymer is highly used as matrix material due to processibility, light weight and desirable mechanical properties. There are two kinds of polymer used as matrix materials viz. Thermosetts and thermoplastics.

i) Thermosetts : Thermosetts have crosslinked 3D molecular structure after curing they do not melt but decompose on heating. They can be retained in partly cured condition over a prolonged period of time. They are most suited for fiber composites and in structural engineering application. Main thermosetting polymers used are epoxy reisns (for high strength application and create smooth surface), phenolic resins (high temperature application), unsaturated polyester.

ii) Thermoplastic : Thermoplastics have one or two dimentional molecular structure which can melt. The use of thermoplastic matries avoides the presence of dangerous vapours in the workshops. TPMC are based on the use of trermoplastic polymber as matrix of the composites and this implies the reversibility of thermal action on the material during fabrication of the final element i.e. it is possible to pre-fabricate semi-finished items and later take them to final shape. Several thermoplastic polymers can be considered as possible matrices for composites. The most important for actual applications are polyetheleneimine, polyphenylene sulfide PPS and polyether ether ketone PEEK and polyetherketone ketone PEKK.

The main technological advantages of thermoplastic composites are. They have high toughness and impact resistance, good chemical resistance, it has high reparability and recyclability.

Metal matrix material :

Metal matrices offer high strength, high temperature resistance and ductility which increases toughness, and stiffness than those offered by their polymer counter parts. The matrix is usually a lighter metal such as Aluminium, Magnesium and Titanium and provides a complient support for the reinforcement. For high temperature application cobalt and cobalt nickel alloy are common.

Ceramic matrix material :

Ceramic materials exhibit ionic bonding, high melting point, good corrosion resistance stability at elevated temperature and high compressive strength is advantages of the ceramix

COMPOSITE MATERIALS

matrix materials. The ceramics mainly used are alumina, silica, zirconia etc.

Reinforcement

Anather constituent of composite which provide high strength, rigidity and enhance the matrix properties is reinforcement. Various reinforcement used in composites are as followes.

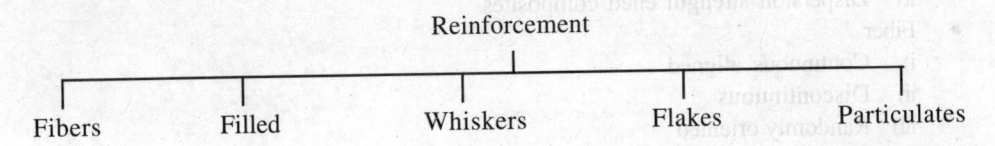

```
                    Reinforcement
        ┌─────┬─────────┬──────────┬──────────┐
     Fibers  Filled  Whiskers   Flakes   Particulates
```

Whiskers are single crystal fibres having large length to diameter ratios. Filamentary whiskers may have diameters of 0.1 to 1 μm and length of the order of 100 to 1000 μm. Due to their smaller size they show a high degree of crystalline perfection (almost free from defects) which is responsible for their exceptionally high strength. Whiskers are the strongest known materials. However extensive use of whiskers as a reinforcement medium is avoided due to their high cost and also because of the difficulty of incorporating them into a matrix.

Whiskers of metals, oxides, carbides and nitrides are frequently used as reinforcements. Variety of continuous fibers of glass, carbon, kevlar, silicon carbide, alimina, boron, tungsten etc. are used as reinforcement. These fibers exhibit different combinations of mechanical properties, physico chemical properties and electromagnetic properties.

Table 13: Characteristics of reinforcement materials

Material	Diameter filament, μm	Density, gm/cc	UTS, GPa	Melting / Softening / Decomposition
Fibers				
E-glass	9	2.6	3.45	700
Carbon (Graphite)	7-8	1.7-1.8	2.5-3.5	3650
Kevar - 49 (Aramides)	16	1.44	4.1	250
SiC	10.20	2.55	2.0	2700
Al_2O_3	3	3.3	2.0	>2000
Whiskers				
Al_2O_3	20	3.95	1.4	2045
SiC	0.1-1.0	3.19	3-14	2700
Metallic wires				
Steel	>15	7.75	4.1	1400
Molybdenum		10.2	5.1	
Tungsten		19.3	3.0	

Whisker : is a filament of material that is structure as a single defect free crystal. It has very high tensile strength (10-20) Gpa. Gold whisker are this filaments of elemental gold, silver whiskers are long filament of elemental silver.

10.3 CLASSIFICATION

Compositers can be classified into three categories
- Particle - reinforced,
  i) Large - particle composite
  ii) Dispersion strength ened composites
- Fiber
  i) Continuous aligned
  ii) Discontinuous
  iii) Randomly oriented
- Structural
  i) Laminates
  ii) Sandwich panels.

A) Particle Reinforced Composite :

i) **Large particle composites :** The particles in these composite are larger than in dispersion strengthed composites. The particle diameter is of the order of few microns. The particles carry a major portion of the load. The particles are used to increase the modulus and decrease the ductility of the matrix. viz. Automobile tire which has carbon black particles in the matrix of polyisobutylene elastomeric polymer.

ii) **Dispersion strengthened composites :** In this type of composites small particles of the order of 10 - 100 nm in diameter are added to the matrix material particle makes the composite harder and stronger These particles resist the matrix from deformation and matrix is the major load bering component.

Precipitation hardening. The strength and hardness of some metal alloys may be improved by the formation of extremely small uniformly dispersed particles of a second phase within the original phase matrix with the help of appropriate heat treatments. This process is called precipitation hardening because the small particles of the new phase are termed as "precipitates". This procedure is also called "Age hardening" because the strength develops with time or as the alloy ages. The strength of alloys of Al-Cu, Cu-Be, Cu-Sn, Mg-Al and some ferrous alloys are hardened by precipitation hardening process.

While the matrix bears the major part of the applied load, the small dispersed particles hinder or impede the motion of dislocations. Thus, the plastic deformation is restricted so as to improve the yield, tensile strength and hardness.

Metals and metal alloys may be hardened and strengthened by the uniform dispersion of high volume percent of very hard and inert materials, such as fine metallic or nonmetallic particies (oxide materials are generally used). The strength is achieved due to interactions between the particles and dislocations within the matrix, similar to what happens with precipitation hardening.

(B) Fiber-Reinforced Composites

Fiber-reinforced composites are those composites in which the dispersed phase is in the form of fibers. Technologically, these composites are very important as they provide high strength and stiffness on a weight basis. The characteristics which are responsible for this high strength/ stiffness are their high "Specific strength" (*i.e.*, the ratio of tensile strength to specific gravity) and high "Specific Modulus" (*i.e.*, the ratio of modulus of elasticity to specific gravity).

Fiber-reinforced composites having high specific strengths and high specific moduli have been produced by utilizing low-density fiber and matrix materials.

COMPOSITE MATERIALS 449

Fiber materials

Glass - glass is the most common and inexpensive fibre and is usually use for the reinforcement of polymer matrices. Glass has a high tensile strength and fairly low density (2.5 g/cc).

Carbon-graphite - in advance composites, carbon fibers are the material of choice. Carbon is a very light element, with a density of about 2.3 g/cc and its stiffness is considerable higher than glass. Carbon fibers can have up to 3 times the stiffness of steel and up to 15 times the strength of construction steel. The graphitic structure is preferred over the diamond-like crystalline forms for making carbon fiber because the graphitic structure is made of densely packed hexagonal layers, stacked in a lamellar style. This structure results in mechanical and thermal properties are highly anisotropic and this gives component designers the ability to control the strength and stiffness of components by varying the orientation of the fiber.

Polymer - the strong covalent bonds of polymes can lead to impressive properties when aligned along the fiber axis of high molecular weight chains. Kevlar is an aramid (aromatic polyamide) composed of oriented aromatic chains, which makes them rigid rod - like polymers. Its stiffness can be as high as 125 GPa and although very strong in tension, it has very poor compression properties. Kevlar fibers are mostly used to increase toughness.

Ceramic - fibers made from materials such as Alumina and SiC (Silicon carbide) are advantageous in very high temperature applications, and also where environmental attack is an issue. Ceramics have poor properties in tension and shear, so most applications as reinforcement are in the particulate form.

Metallic - some metallic fibers have high strength but since there density is very high they are of little use in weight critical applications. Drawing very thin metallic fibers (less than 100 micron) is also very expensive.

The mechanical characteristics of a fiber-reinforced composite depend upon the following factors :

1. Properties of the fiber.
2. The degree to which an applied load is transmitted to the fibers by the matrix phase. This depends on the magnitude of the interfacial bond between the fiber and the matrix phases.
3. Fiber length : For effective strengthening and stiffening of the composite, some critical fiber length is essential. The critical fiber length, l_c is dependent upon the fiber diameter, d its ultimate (or tensile) strength σ_f and on the fiber-matrix bond strength (or the shear yield strength) of the matrix, T_c as per the following equation :

$$l_c = \frac{\sigma_f \cdot d}{T_c}$$

For a number of carbon and glass fiber-matrix combinations, this critical length, l_c is of the order of 1 mm, which ranges between 20 to 150 times the fiber diameter.

If the fiber' length is much greater than l_c (*i.e.*, about 15 times the l_c) then the fibers are termed as "continuous". If the fiber length is considerably lesser than l_c, then the fibers are termed as "discontinuous" or "short" fibers.

4. Fiber orientation and concentration : The orientation and arrangement of the fibers relative to one another, their distribution and the fiber concentration have a significant role to play on the strength and other properties of the fiber - reinforced composites. As far as the orientation is concerned, two extreme cases are possible : (*i*) a parallel

alignment of the longitudinal axis of the fibers in a single direction and (*ii*) a totally random alignment. Continuous fibers are normally aligned, whereas discontinuous fibers may be aligned, randomly oriented or partially oriented. When the fiber distribution is uniform, superior overall composite properties can be obtained. The schematic representation of (*i*) continuous and aligned, (*ii*) discontinuous and aligned and (*ii*) discontinuous and randomly oriented fiber composites is shown in Fig. 10.1 reinforcing fibers must have high elastic modules, high strength, Low density, easily wetted by the matrix.

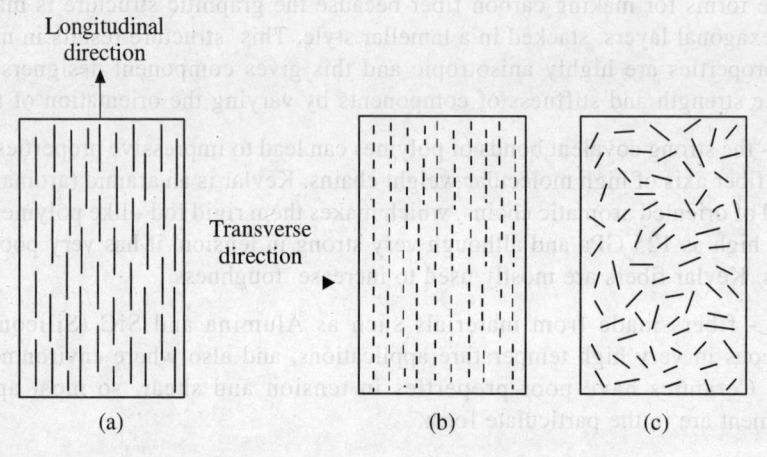

Fig. 10·1. Schematic representation of different types of Fiber - reinforced composites (*a*) Continuous and aligned (*b*) discontinuous and aligned, and (*c*) discontinuous and randomly oriented.

(*i*) Continuous and Aligned Fiber Composites :

The properties of **continuous and aligned composites** are highly **anisotropic,** that is, they depend on the direction in which they are measured. The maximum strength and reinforcement are achieved along the alignment (longitudinal) direction. However, in the transverse direction, fiber reinforcement is virtually non-existent and hence fracture usually occurs at relatively low tensile stresses. For other stress orientations, the strength of these composites lies between these extremes.

(*ii*) Discontinuous Fiber Composites

These are of two types viz., (*a*) Discontinuous and aligned fiber composites; (*b*) Discontinuous and randomly oriented fiber composites.

The reinforcement efficiency of discontinuous fibers is lower than that for continuous fibers. In spite of that, the discontinuous and aligned fiber composites are finding greater application in the commercial market. Chopped glass fibers are used most extensively while carbon and aramid discontinuous fibers are also employed.

Decisions regarding orientation and fiber length for a particular composite will depend on the level and nature of the applied stress as well as fabrication cost. Although the fabrication costs for continuous and aligned fiber composites are higher than those of discontinuous (short) fiber composites (both aligned and randomly oriented), rapid rates of production can be achieved in the case of latter. Further, in case of discontinuous (short) fiber composites, intricate shapes can be formed which are rather impossible with continuous fiber reinforcement. The fabrication techniques employed in case of short-fiber composite materials include compression moulding, injection

COMPOSITE MATERIALS

moulding and extrusion moulding.

SOME IMPORTANT TYPES OF FIBER - REINFORCED COMPOSITES

10.5. FIBER GLASS – REINFORCED COMPOSITES

Glass is a very popular fiber reinforcement material due to the following reasons :

(a) It is commonly available and can be economically fabricated into glass - reinforced plastic by the commonly available manufacturing techniques.

(b) High strength fibers can easily be drawn from molten glass. The fibers are very strong. When embedded in a plastic matrix, it produces a composite having a very high specific strength.

(c) When coupled with a plastic matrix, it exhibits chemical inertness thereby rendering the composite material so produced, useful in a variety of corrosive environments.

Fiber glass-reinforced composites can be produced by properly incorporating the continuous or discontinuous glass fibers within a plastic matrix. It is a very commonly used composite material. For obtaining a superior composite, the surface characteristics of the glass fiber employed are very important because even minute surface flaws can adversely affect the tensile properties of the product. Freshly drawn fibers are usually coated during the drawing process with a material that protects the surface from damage and other undesirable interactions and helps in promoting a better bond between the fibers and the matrix material.

Several plastic materials are used as matrix material for producing fiberglass - reinforced composites. Polyesters are the most commonly used matrix material. Glass fiber-reinforced composites of more recent commercial interest utilize glass fibers in a nylon matrix which yield a very high strength and high impact-resistance.

Fiberglass composites are widely used in automotive parts, marine bodies, storage tanks, plastic pipes and industrial floorings. Fiberglass reinforced plastics are extensively used in transportation industries to reduce vehicle weight and boost fuel efficiency.

10.6. OTHER FIBER - REINFORCED PLASTIC COMPOSITES

Carbon fibers have much higher specific modulus than glass fibers, and possess better resistance to temperature and corrosive chemicals. However, carbon fibers are more expensive and have only limited short-fiber utilization. The aircraft industry is currently using carbon-reinforced composites as structural components for the new-aircraft as a weight saving measure. The new structures thus fabricated are expected to be about 20 to 30% lighter than those fabricated from sheet metal parts.

Epoxy resins impregnated with boron fibers are being used for constructing helicopter rotor blades.

High-strength polymeric aramid fibers are being used in light-weight structural components e.g., airospace, aircraft, sporting and marine equipment. The possibility of using silicon carbide (SiC) and Silicon Nitride (Si_3N_4) fibers in plastic matrices is being currently investigated. Carbon-Carbon composites composed of carbon fibers imbedded within carbonized resin matrices are being considered mainly for use in high-temperature aerospace applications.

10.7. METAL MATRIX - FIBER COMPOSITES

A number of continuous fiber composites using alloys of Al, Mg, Cu and Ti matrices reinforced with 20 to 50% Volume of carbon, SiC, boron, borsic and metal fibers have been recently developed. Borsic-aluminium composite is prepared from continuous borsic fiber obtained by the vapour deposition of a layer of boron on a thin (10 μm diameter) tungsten wire followed by a thin coating of SiC on the fiber to retard undesirable interaction between Al and boron.

The metal matrix-fiber composites can be used at higher service temperatures than the polymer composites. Further, high specific strengths and high specific moduli are possible due to the relatively low densities of the base metals. This combination of useful properties renders them particularly attractive for some aerospace and new engine applications.

The high-temperature creep and rupture properties of some of the super-alloys (*e.g.*, Ni - and Co - based alloys) may be improved by fiber reinforcement using refractory metals like tungsten. Designs incorporating these composites permit higher operating temperatures and better efficiencies for turbine engines, while still maintaining excellent high-temperature oxidation - resistance and high impact strength.

10.8. HYBRID COMPOSITES

These are obtained by using two or more different types of fibers in a single matrix. They give better all-round combination of useful properties as compared to their single-fiber counterparts. A wide variety of fiber combinations and matrix materials can be used. In the common system, both carbon and glass fibers are incorporated into a polymeric resin. The glass-carbon hybrid composites provide the combination of strength, stiffness and low-density reinforcement of carbon fibers and cheapness of glass fibers. These are strong, tough and possess higher impact resistance. When the hybrid composites are stressed in tension, failure does not occur suddenly. In other words, failure is non-catastrophic.

The two different fibers in a hybrid composite may be all aligned and intimately mixed with each other or laminations may be constructed consisting of alternating layers of single fiber type. The properties of hybrid composites are an isotropic.

Hybrid composites are used in light-weight transport (land, water or air) structural components, light-weight orthopedic components and sporting goods.

10.9. PROCESSING OF FIBER - REINFORCED COMPOSITES

During the fabrication of fiber-rein forced composites, special attention is required towards the uniform distribution of the fibers within the matrix and their proper orientation. The various processing methods used for producing Fiber- Reinforced Plastics (FRPs) have been discussed under chapter 6.4. Some recent techniques of processing the fiber-reinforced composites are described below :

(*i*) Pultrusion

Pultrusion is a continuous process which can be easily automated. This is a cost-effective process because a wide variety of shapes can be fabricated at relatively high production rates and the stock length that can be manufactured is practically unlimited. Pultrusion is used generally with glass, carbon, and aramid fibers in concentrations of 40 to 70% volume with matrix materials such as epoxy resins, polyesters and vinyl esters. Pultrusion is used for the manufacture of components having continuous lengths and constant cross-sectional shape such as rods, tubes, beams, etc.

The schematic diagram of the pultrusion process is shown in Fig. 10.2.

Continuous fiber rowings or tows are initially impregnated with a thermosetting resin and then pulled through a steel die which performs to the desired shape establishing the desired resin to fiber ratio. The stock then passes through a curing die which is precision machined to impart the final shape. This die is also heated to initiate curing of the resin matrix. The stock is drawn through the dies at the desired speed with the help of a pulling device, which provides the driving force for the impregnated stock/strands to be forced through the die. Tubes and hollow sections can be made by using center mandrels or inserted hollow cores.

COMPOSITE MATERIALS

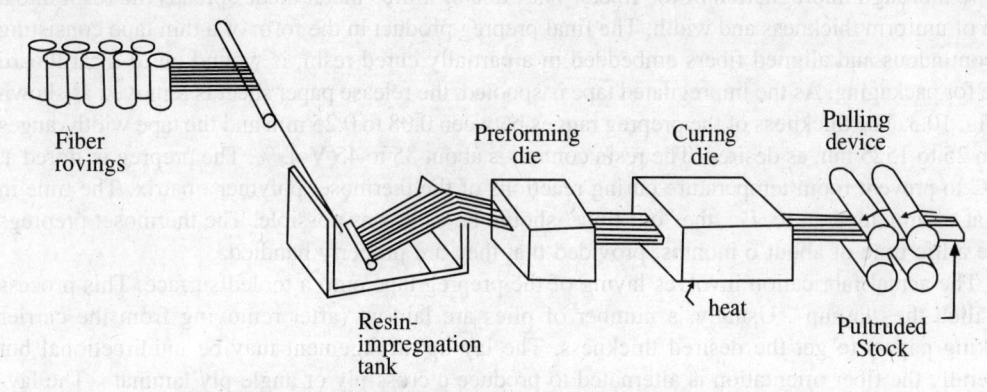

Fig. 10·2. Schematic diagram illustrating poltrusion process

(ii) Prepreg Production Process

Prepreg is the term used for continuous fiber reinforcement pre-impregnated with a partially cured polymer resin. This material is delivered to the manufacturers in the form of tape. The tape is molded and fully cured without adding any resin. This is the most widely used composite form for structural applications. Both thermoplastic and thermosetting resins are used in the prepreg production process while carbon, glass, and aramid fibers are employed as the reinforcements.

The schematic diagram illustrating the production of prepreg tape using thermosetting polymers is shown in Fig. 10.3.

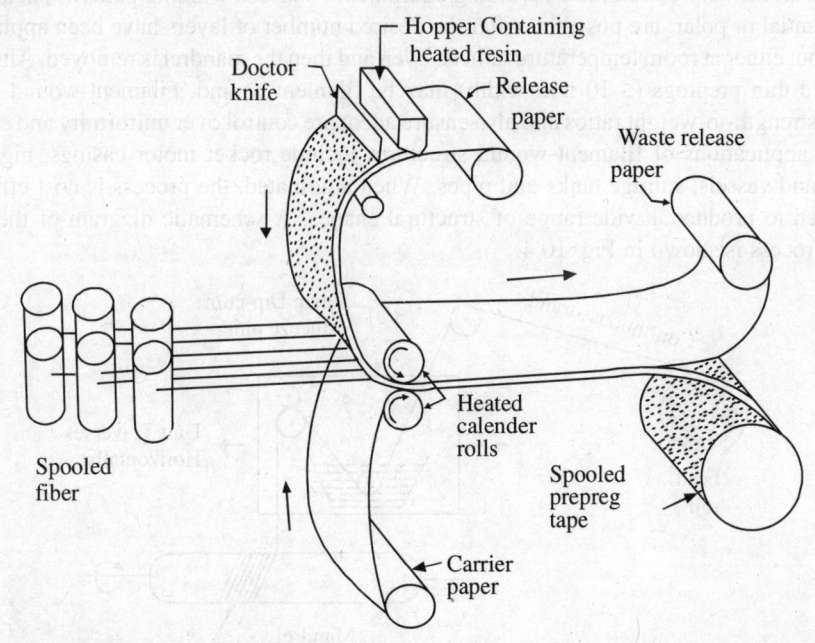

Fig.10·3. Schematic diagram showing the process of producing prepreg tape using thermoset polymers.

A series of spool-wound continuous fiber tows are sandwiched and pressed between sheets of release and carrier paper using heated rollers. This process is called "calendering." The release paper sheet was coated with a thin film of heated resin solution of relatively low viscosity so as to ensure thorough impregnation of the fibers. The "doctor knife" metal blade spreads the resin into a film of uniform thickness and width. The final prepreg product in the form of a thin tape consisting of continuous and aligned fibers embedded in a partially cured resin, is wound on to a cardboard core for packaging. As the impregnated tape is spooled, the release paper sheet is removed as shown in Fig. 10.3. The thickness of the prepreg ranges between 0.08 to 0.25 mm and the tape width ranges from 25 to 1525 mn, as desired. The resin content is about 35 to 45 Vol. %. The prepreg is stored at ≤0°C to prevent room temperature curing reactions of the thermoset polymer matrix. The time in use at room temperature, *i.e.,* the "out-time" should be as less as possible. The thermoset prepregs have a life time of about 6 months, provided that they are properly handled.

The actual fabrication involves laying of the prepreg tape onto a tooled surface. This process is called the "lay-up". Usually, a number of plies are laid-up (after removing from the carrier backing paper) to get the desired thickness. The lay-up arrangement may be unidirectional but generally the fiber orientation is alternated to produce a cross-ply or angle-ply laminate. The lay-up procedure may be carried out manually by cutting the lengths of tape and positioning them in the desired orientation on the tooled surface. Alternately, tape patterns may be cut with machines and then hand-laid. Automated methods are available for many applications of the composite materials to render them effective. Final curing of the laminates is accomplished by the simultaneous application of heat and pressure.

(*iii*) Filament Winding

Filament winding process comprises of accurate positioning of reinforcing fibers in a predetermined pattern to form a hollow (usually cylindrical) shape. The fibers, either in the form of individual strands or tows, are passed through a resin bath and then continuously wound onto a mandrel, with the help of automated winding equipment. Various winding patterns such as helical, circumferential or polar, are possible. After the desired number of layers have been applied, curing is carried out either at room temperature or in an oven and then the mandrel is removed. Alternatively, narrow and thin prepregs (≤ 10 mm width) may be filament-wound. Filament-wound parts have very high strength-to-weight ratios and also ensure adequate control over uniformity and orientation. Important applications of filament-wound structures include rocket motor casings, high-pressure cylinders and vessels, storage tanks and pipes. When automated, the process is cost-effective and can be used to produce a wide range of structural shapes. A schematic diagram of the filament-winding process is shown in Fig. 10.4.

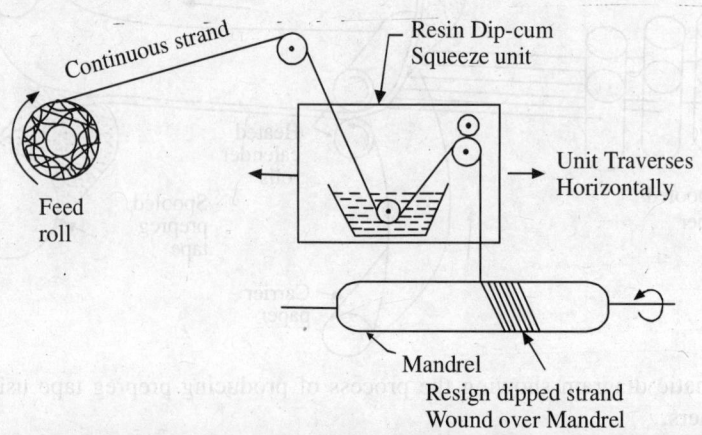

Fig. 10·4. Schematic diagram showing filament winding process.

COMPOSITE MATERIALS

10.10. STRUCTURAL COMPOSITES

A **Structural Composite** comprises of both homogeneous and composite materials, the properties of which depend, not only on the properties of the constituent materials but also on their geometrical design. Structural composites can be broadly classified as :
  (*i*) Laminar Composites, and
  (*ii*) Sandwich Panels.

(*i*) Laminar Composites

A Laminar composite consists of two-dimensional sheets or panels that have preferred high-strength direction as is found in wood and continuous and aligned fiber-reinforced plastics. Successive oriented fiber-reinforced layers are stacked and then cemented together in such a way that the orientation of the high-strength varies with each successive layer as shown in Fig. 10.5.

(a) Stacking of successive oriented fiber-reinforced layers

(b) Fabricated laminar composite

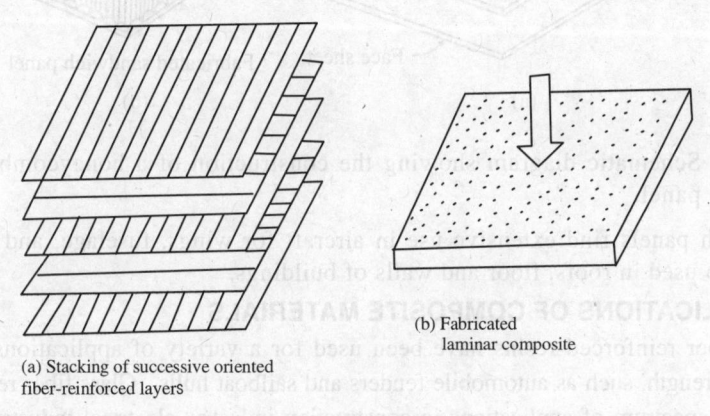

Fig.10·5. Successive stacking of oriented fiber-reinforced layers for producing a laminar composite.

For instance, adjacent wood sheets in plywood are aligned with the grain direction at right angles to each other. Laminar composites may also be constructed using fabric materials such as cotton, paper, or woven glass fibers embedded in a suitable plastic matrix. A Laminar Composite possesses relatively high-strength in a number of directions in the two-dimensional planes. Obviously, the strength in any given direction will be lower than what it would have been if all the fibers were oriented in that particular direction. Modern "SKi" is an example of a complex laminated structure.

(*ii*) Sandwich Panels

Sandwich panels usually consist of two strong outer sheets called "**faces**", separated by a layer of less-dense material called "**core**", which is of lower strength and lower stiffness. Typical "face" materials include fiber-reinforced plastics, plywood, titanium, steel and aluminum alloys. Typical "core" materials include synthetic rubbers, foamed polymers, balsa wood and inorganic cements.

The "faces' bear most of the in-plane loading, as well as any transverse bending stresses. The "core" serves the following two structural functions :

   (*a*) It separates the "faces" and resists any deformations perpendicular to the face plane.
   (*b*) It provides a certain degree of shear rigidity along planes which are perpendicular to the "faces".

A popular type of core material comprises of a "honeycomb" structure made up of thin foils that have been formed into interlocking hexagonal cells, with their axes oriented in a perpendicular direction to the "face" planes, as shown in Fig. 10.6.

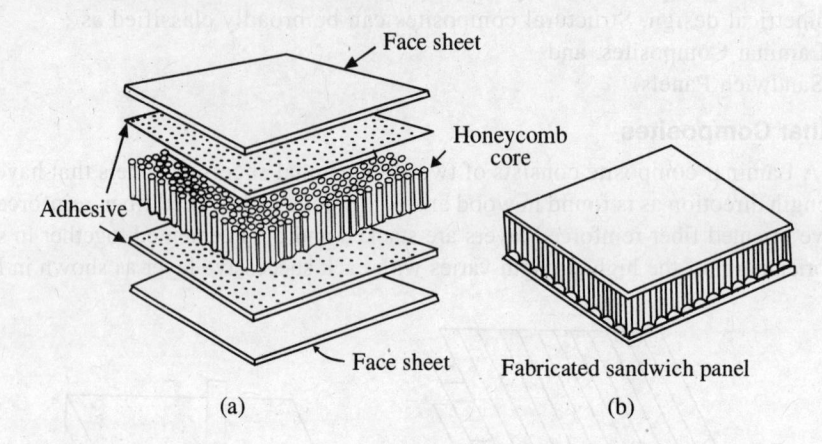

Fig.10·6. Schematic diagram showing the construction of a honeycomb core sandwich panel.

Sandwich panels find extensive use in aircraft for wings, fuselage, and tailplane skins. They are also used in roofs, floor and walls of buildings.

10.11 APPLICATIONS OF COMPOSITE MATERIALS

Glass fiber reinforced resins have been used for a variety of applications requiring high mechanical strength, such as automobile fenders and sailboat hulls. Glass fiber reinforced plastics have a broad spectrum of applications in construction industry, electrical industry, transportation, agriculture, consumer goods, sports and liesure activities, industry, agriculture, etc. Even automobile tyres can be considered as composite materials in that the rubber matrix and the textile reinforcing fibers are independent and either of them functions because of the other in a tire. Asbestos has been used as the reinforcing fiber in a variety of composite materials. For instance, asbestos-reinforced phenolic resins were used for automotive brake linings, but because of the carcinogenic properties of fine asbestos fibers, they have been replaced by fluorinated fibers and other substitutes. Some of the important fields of application of inorganic fibers are included in the Table given below :

Table : Areas of Applications of Inorganic Fibers

Area of application	Performance requirements	Fibers used
Cement reinforcement	Chemical resistance to cement and mechanical stabilty	Cement-resistant glass fibers, steel fibers, asbestos, etc.
Plastics reinforcement	High tensile strength, high elasticity modulus and low density.	Glass fibers, carbon fibers, boron fibers, SiC fibers, asbestos, Al_2O_3 fibers, etc.
Tire cord	Stability to alternating mechanical stress.	Steel fibers and very short metal fibers.
Frictional linings and seals	Compression strength, thermal stability, abrasion resistance and elastic deformation behaviour.	Asbestos, glass fibers, steel fibers, etc.

COMPOSITE MATERIALS

Composite materials have evolved as sophisticated materials. Examples include resins reinforced by carbon or graphite fibers for aircraft and aerospace applications ranging from aircraft wing surfaces to heat-resistant blocks used in aircraft brakes and nose cones for missiles. The industrial production of carbon fibers is based on the thermal degradation of non-melting organic polymers such as polyacrylonitrile (PAN).

The extent of mechanical properties that can be attained with carbon fibers depends on crystallinity and the structure of carbon formed, which in turn depend upon the quality and composition of the starting material and on the production technology. Reinforcing carbon fibers are relatively new entrants in industry. They were being used since the last 3 decades in military aircraft, spacecraft, and more recently in sports sector for the manufacture of sports cars, fishing rods, golf clubs, tennis rackets, skis, boots, yachts, masts, etc. Since 1980, they are used in the field of civil aviation such as tail units of air-bus. Further technological innovations and cost reduction is likely to extend their applications to weight savings in automobiles and components for light weight engines, robots and various other machineries.

QUESTIONS

1. What are composite materials? How are they classified? What are their advantages?
2. Write informative notes on the following :
    (a) Fiber glass-reinforced composites
    (b) Hybrid composites
    (c) Structural composites
3. Write short notes on the following:
    (a) Laminar composites
    (b) Sandwich panels
    (c) Poltrusion
    (d) Prepreg production process
    (e) Metal-fiber composites
    (f) Cermets

11

Thermodynamics Equilibrium and Kinetics

> *"Thermodynamics is a logical subject of great elegance. It is a powerful method for studying chemical phenomenon, which can be developed quite independently of the atomic and molecular theory and can be applied to systems of any complexity"*
>
> *"Goldon M. Barrow"*

The word thermodynamics is coined from two Greek words meaning *heat* and *powerful*. It is a science that deals with heat and work and those properties of substances that bear a relation to heat and work. It is an experimental science and its findings have been formalized into certain basic laws, which are known as the first, second and third laws of thermodynamics. In addition to these, the zeroeth law of thermodynamics is also logically developed which precedes the first law.

There are two approaches of study of thermodynamics. In the first approach, large scale (gross) behaviour of a substance is studied which is the macroscopic view point towards matter and its interactions, where the overall large-scale effect is considered. This study comes under the domain of *classical thermodynamics*.

Another approach to the study of thermodynamic properties and energy relationships is based on the statistical behaviour of large groups of individual particles. This method, founded on a microscopic viewpoint, is called *statistical thermodynamics*. It combines the computational techniques of statistical mechanics with the findings of quantum theory. In the present technology, where substances are employed under extreme ranges of temperature and pressure, the predictive methods of statistical thermodynamics are extremely important.

In this chapter emphasis will be placed upon the macroscopic viewpoint. The reasons for this are as following: First of all, the solutions of a great majority of thermodynamic problems require an analysis only in terms of macroscopic variables. Secondly, classical thermodynamics is an easier, more direct approach to the solution of engineering problems.

11.1. INTRODUCTION TO THE LAWS OF THERMODYNAMICS

The first law of thermodynamics, in its useful form, is a conservation of energy statement. It may be stated as follows :

(*i*) Although energy may be changed from one form to another, the total energy of a system and its surroundings remains constant.

(*ii*) In the light of Einstein relation $E = mc^2$, where E is the energy which can be produced by the destruction of mass m, the law may be stated in modified form as "the total mass and energy of an isolated system remains constant."

Mathematical statement of the first Law of Thermodynamics :

If the system changes from state A to state B, then it may be written as:
$$\Delta E = q + W \qquad \ldots(1)$$
where ΔE is the change in internal energy; q is the heat absorbed by the system, which apart from increasing the internal energy (ΔE) of the system, does mechanical work W.

This law has several applications. It is helpful to know the equivalence of different forms of energy in any process. Moreover it establishes the relationship between the amount of heat absorbed, change in its internal energy and work performed by the system.

The first law of thermodynamics says nothing about the character, possibility and direction of the processes by which various inter-conversions of energy are brought about. This creates the need for second law of thermodynamics.

Second Law of Thermodynamics

This law has many ramifications with respect to engineering processes. The law which specifies the conditions in which conversion of heat into work occurs, is called *Second Law of Thermodynamics*.

The importance of the law, so far as the chemical engineers and chemists are concerned, lies in the fact that it provides the basis of predicting feasibility of a particular reaction or process and if it is so, then to what extent? The law can be stated in many ways. This law deals with an important thermodynamic function-entropy, which will be discussed later.

Statements of the Second Law of Thermodynamics

1. All naturally occurring processes always tend to change spontaneously in a direction which will lead to equilibrium and can not be reversed on their own—

Ex. (*i*) Water flows from a high level to a lower level

(*ii*) Electricity flows from high point of potential to one of lower potential.

2. In a reversible process, the entropy of the universe is constant. In an irreversible process the entropy of the universe increases.

3. It is impossible for a self-acting machine, unaided by an external agency, to convey heat from a lower to a higher temperature.

4. Total heat absorbed by a system cannot be completely converted into work. In other words efficiency of any reversible engine is always less than one.

5. The energy of the universe is constant but entropy is continuously increasing.

Mathematical form of the Second Law of Thermodynamics

The second law of thermodynamics is mathematically defined as follows :
$$dS = \frac{dq_{rev}}{T}; \quad \Delta S = \frac{q_{rev}}{T}$$
where $\Delta S = (S_2 - S_1)$ which shows the change in entropy of the system. S_2 and S_1 are the actual entropy values in the final and initial states of the process. q_{rev} is the amount of heat absorbed by the system at constant temperature T.

Usefulness of the Second Law

The second law is extremely helpful to the engineers in the following ways :
  (*i*) It provides the means of measuring the quantity of energy
  (*ii*) It establishes the criteria for "ideal" performance of engineering devices.
  (*iii*) It determines the direction of change for processes
  (*iv*) It establishes the final equilibrium state for spontaneous processes.

In the study of the first and the second laws of thermodynamics, one comes across the following important thermodynamic functions
   (i) Internal Energy
   (ii) Enthalpy
   (iii) Entropy.

11.2. INTERNAL ENERGY (E)

As a consequence of the first law of thermodynamics, one can write that

$$W_{adi} = E_2 - E_1 = \Delta E \qquad \ldots(1)$$

where subscripts 1 and 2 indicate the initial and final states. W_{adi} is the work carried out for any adiabatic path between these two states. This equation is an operational definition of the change in the energy of a closed system between two equilibrium states. Equation 1 is a fundamental relation which enables one to evaluate the energy of a set of states in terms of adiabatic work interactions. The arbitrary value of E cannot be obtained but change in energy value of two states can be obtained. As a matter of fact, derivations of subsequent definitions of new forms of energy such as potential, kinetic, gravitational etc., are all special cases of equation 1. Thus the total energy E is the sum of all forms of energy associated with a closed system.

There are numerous forms of energy, which constitute the total energy of a macroscopic system. In general, we may write

Total energy = internal energy + kinetic energy + gravitational potential energy + electrostatic energy + magnetic energy + strain energy + surface energy, etc.

The proper inclusion of one or more of the terms on the right side of this expression in a thermodynamic analysis depends upon the type of system under study.

According to eqn. 1 where change in energy ΔE depends upon the initial and final states of the system, then this form of energy is called internal energy. It is usually denoted by E or U. It is not possible to ascertain its absolute value, but however, it is possible to measure the change in internal energy which is associated with a physical or chemical process. Then, change in energy is given by equation 1. It is dependent on the chemical nature, volume, temperature, pressure and mass of the system. It is determined by the state of the system or substance or system itself and is independent of the previous history of the system.

For the purpose of explaining the source of energy, it is believed that any kind of matter is a store of energy. It is difficult to experience or observe this energy as such. However, this energy may appear in the form of heat, work, electricity and light, which is observable. As mentioned above, the total energy possessed by a system is the sum total of all forms of energy.

The total contribution of kinetic energy $(\Delta E_{tran} + \Delta_{vib} + \Delta_{rot})$ towards the total magnitude of the energy change is very small. But the most important one which contributes towards breaking and making of the bonds in molecules is bonding energy which in turn consists of intermolecular and intramolecular energy.

Thus we have seen that internal energy E is an important thermodynamic function. Its value depends merely on initial and final states of the system and is independent of the path or manner in which change has been brought. It is extensive property and depends upon the mass of the system too.

11.3. ENTHALPY (H)

This is another important thermodynamic function. This can be evolved as follows :
Let us consider that
q_v = heat transferred to or from a system at constant volume, and
q_p = heat transferred to or from a system at constant pressure.

For chemical reactions that occur in a container of fixed volume, $P\Delta V = 0$ and then the first law equation (*i.e.*, $\Delta E = q - p\Delta V$) is reduced to the form

$$\Delta E = q_v \qquad \ldots(2)$$

However, chemical reactions occur more frequently in open vessels at constant barometric pressure, then we have

$$\Delta E = q_p - P\Delta V$$

This means, the change in energy is equal to the heat absorbed by the system at constant pressure, q_p, less the work done by the system at constant pressure, or

$$q_p = \Delta E + P\Delta V \qquad \ldots(3)$$

Since it is common practice to conduct reactions at constant pressure, we define a new quantity, called the Enthalpy, H, by the equation

$$H = E + PV$$

This new function enthalpy is also considered to be the measure of the total energy stored in a substance during its formation. Most of the chemical reactions are associated with heat (energy) changes. These heat changes are expressed in terms of this function enthalpy. Therefore heat effect in a reaction is the difference between 'H's of the products and 'H's of the reactants, and it is expressed as ΔH, as the heat of reaction, when state of a system is changed at constant pressure. Then,

$$\Delta H = H_2 - H_1 = \Delta E + \Delta(PV) = \Delta E + P\Delta V$$

i.e.,
$$\Delta H = \Delta E + P\Delta V$$

But from equation 3,

$$\Delta E + P\Delta V = q_v$$

and therefore

$$\Delta H = q_p \qquad \ldots(4)$$

In words, the change in enthalpy is equal to the heat absorbed or evolved when the process occurs at constant pressure and the only work is $P\Delta V$ (no electrical work).

Equations 2 and 3 yield the relationship between q_v and q_p

$$q_p = q_v + P\Delta V$$
$$= q_v + P(V_2 - V_1)$$
$$q_p = q_v + PV_2 - PV_1 \qquad \ldots(5)$$

Thus, for chemical reactions involving **only solids and liquids,** which undergo relatively small expansions and contractions. ΔV is negligible and q_v is practically equal to q_p, or $\Delta E = \Delta H$. However, if **gases are involved** in the chemical reaction, appreciable volume change may occur and the heat absorbed or evolved at constant pressure may differ appreciably from the heat absorbed or evolved at constant volume. Assuming ideal behaviour and constant temperature.

$$PV_2 = n_2 RT$$

and
$$PV_1 = n_1 RT$$

Then, substituting in eqn. (5)

$$q_p = q_v + n_2 RT - n_1 RT$$
$$= q_v + (n_2 - n_1)RT$$

∴
$$q_p = q_v + \Delta n\, RT$$

or
$$\Delta H = \Delta E + \Delta n RT \qquad \ldots(6)$$

where Δn is the total number of moles of *gaseous products* minus the total number of moles of gaseous reactants.

$$\Delta n = n_{products} - n_{reactants}$$

The value of R is 1·99 cal. per mole.

Thermochemical data are usually expressed in terms of enthalpy change, ΔH. For example, we can write

$$H_2O(g) + C(s) \longrightarrow CO(g) + H_2(g), \Delta H = 31\cdot 4 \text{ Kcal.}$$

A summary of the relations between the heat transfer for a reaction occurring at constant pressure or at constant volume is given below :

Constant pressure (Open vessel)	Constant Volume (Closed vessel)
$\Delta E = q_p - W$	$\Delta E = q_v - W$
$\Delta E = q_p - P\Delta V$	$\Delta E = q_v - P\Delta V$
$q_p = \Delta E + P\Delta V$	but $\Delta V = 0$
$q_p = \Delta H$	$\Delta E = q_v$

$$q_p = q_v + W$$
$$\text{or } \Delta H = \Delta E + W$$

Case 1. Reaction produces volume expansion

W is +ve, ($P\Delta V$ is + ve)

q_p is greater than q_v

ΔH is greater than ΔE

Reaction is more endothermic at constant pressure than at constant volume.

Case 2. Reaction produces volume contraction

W is —ve, ($P\Delta V$ is –ve)

q_p is less than q_v

ΔH is less than ΔE

Reaction is more exothermic at constant pressure than at constant volume.

Case 3. Reaction produces no volume change

W is 0, ($P\Delta V$ is 0)

$$q_p = q_v$$
$$\Delta H = \Delta E$$

Just as we have seen that the energy, E of a system depends on its state and is independent of the path taken to reach that state, the same is true of the product PV. Therefore, H, the sum of E + PV, like E, depends only on the state of the system. Consequently, ΔH, like ΔE, depends only on the final and initial states.

11.4. ENTROPY

A process or reaction which occurs under a given set of conditions without being forced, is known as **spontaneous** or **natural** process. If it is carried out reversibly, it yields maximum amount of work. In natural process the maximum amount of work is never obtained. Moreover all naturally occurring processes are spontaneous in nature.

The following are some examples of spontaneous processes :

(*i*) Distribution of solute throughout the solvent

(*ii*) Evaporation of water

(*iii*) Diffusion of gases or expansion of a gas

(*iv*) Flow of heat from a hot body to a cold body

(v) Flow of water from a hill to the ground

It was suggested that all spontaneous processes should be exothermic in nature (evolution of heat). But in practice many processes or reactions which are endothermic in nature occur spontaneously. For example, dissolution of potassium nitrate is accompanied by absorption of heat. Thus, sign of ΔH, that of enthalpy is not a perfect criterion for the prediction of spontaneity of the reaction. Here, existence of another driving force, which explains the spontaneity of the process, is conceived, which is nothing but entropy. Thus the concept of entropy has crept into thermodynamics.

ENTROPY (S)

Definition :

(i) Entropy, a thermodynamic state property, is a measure of degree of disorder or randomness or probability in a given system.

(ii) It is also considered as a measure of unavailable form of energy (TdS).

(iii) Entropy is a function which accounts for the irreversibility of the given change.

The entropy of a system and its surroundings together increases during all natural or irreversible processes. For a reversible process, the magnitude of the total entropy remains unchanged.

Mathematical Expression for Entropy

If any process is carried out reversibly between the limits of initial state 1 and final state 2, with absorption of heat q_{rev} at constant temperature, then the entropy change is given by

$$\Delta S = S_2 - S_1 = \frac{\text{Heat absorbed reversibly}}{\text{Temperature at which heat is observed}}$$

$$= \frac{q_{rev}}{T}$$

$$\Delta S = S_2 - S_1 = \frac{q_{rev}}{T}$$

where S_2 and S_1 are the actual entropy values in the final and initial states of the process. The total entropy change is given by

$$\Delta S = \int_{T_1}^{T_2} \frac{dq_{rev}}{T} = \frac{q_{rev}}{T} \qquad ...(7)$$

The entropy change being an extensive property, its value depends upon the quantity of substance expressed in moles. Thus unit of entropy is cal/deg/mole. This is usually called entropy unit (e.u.)

Entropy—Its Physical Significance

Entropy and Disorder

As per definition, the entropy is closely related with the disorder. Whenever disorder or disorganization takes piece in the process, entropy is bound to increase. According to one of the laws of nature, substances will tend to become as random as possible. Though this may seem strange, it is true. Any substance in the ordered position or state will have less entropy than the one which is in the disorderly state.

A substance at absolute zero is regarded to be in the highest ordered form (example— crystalline substance). The entropy value at absolute zero of temperature, of the crystalline substances, is taken as zero. With rise in temperature, crystalline substance gradually gets disorganized and thereby entropy also enhances. On this basis, melting of ice, vaporization of liquid, expansion of gas, dissolution of salts in water, breaking of molecules into atoms are the changes that involve increase in entropy.

In an attempt to give physical meaning to the entropy function Clausius interpreted TdS = dq "as degraded energy", that is, energy not available for work in a given process. It was Boltzmann and Plank, however, who gave the most useful picture of entropy by showing that a simple relationship existed between entropy and probability. Because probability and disorder are related, so are entropy and disorder. A schematic representation of these relationships is shown in Fig. 11·1.

Order
Low probability
Low Entropy

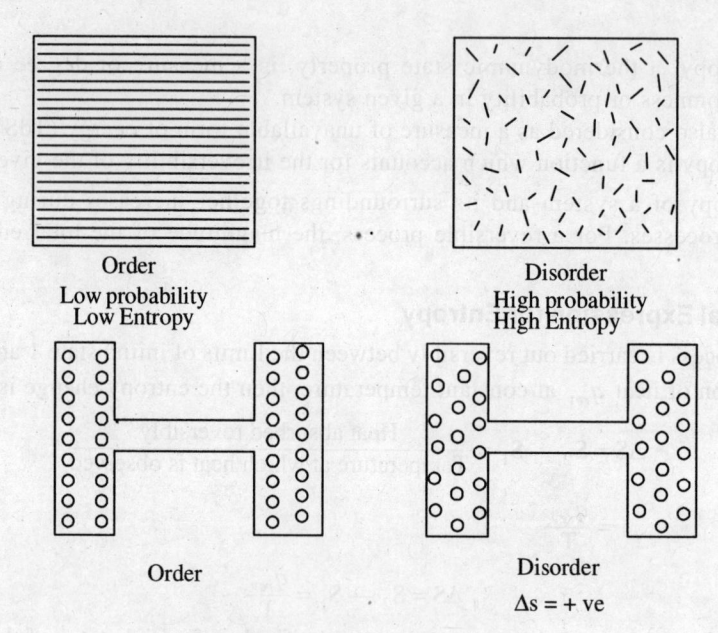

Disorder
High probability
High Entropy

Order

Disorder
Δs = + ve

Fig. 11·1. Order and Disorder

Entropy changes in different cases

(1) Entropy change in an isolated system :

For an isolated system, $q = 0$ and equation $dS = \dfrac{q_{rev}}{T} = 0$, and S = constant. For observable change $dS > 0$ as $dS > \dfrac{dq}{T}$. Thus for an isolated system any spontaneous change will tend to produce states of higher entropy until entropy reaches a maximum value. At this point the system will be in equilibrium and entropy will remain constant at its maximum.

(2) Entropy changes in Reversible Processes

A process is said to be *reversible* if conditions can be arranged so that the process can proceed in either direction *e.g.*, the melting of ice to give water at 0°C.

$$\text{Ice} \ (0°C) \rightleftharpoons \text{Water} \ (0°C)$$

This is just one example of an *equilibrium situation* which is obviously a dynamic one. Similarly in case of heat engine (carnot cycle) operation is reversible, for which the following relation can be written

THERMODYNAMICS, EQUILIBRIUM AND KINETICS

$$\frac{dq_1}{T_1} + \frac{dq_2}{T_2} = 0$$

For a complete reversible cyclic process involving a large number of carnot cycles, we can write

$$\frac{dq_{rev}}{T} = 0 \text{ or } \frac{q_{rev}}{T} = 0, i.e., \Sigma \Delta S = 0, \Delta S = 0$$

In words, we can write that net increase in entropy of the system will be zero in any reversible cyclic process.

(3) Entropy changes in Irreversible Process

The term *irreversible* is applied to reactions (or processes) that proceed spontaneously such as the sudden expansion of a gas into vacuum.

In these situations we note from the second law that $dS > \frac{dq}{T}$. This inequality provides a measure of the spontaneity of, or the tendency of, the reaction to proceed in a given direction. So long as $dS > \frac{dq}{T}$ the reaction will proceed.

Similarly we know that a reversible process absorbs more heat than the irreversible one, hence

$$\Delta S = \frac{q_{rev}}{T} > \frac{q_{irr}}{T} \text{ (when T is constant)}$$

But $\frac{q_{rev}}{T} = \Delta S_{sys}$

$$\therefore \quad \Delta S_{sys} > \frac{q_{irr}}{T}$$

Now consider that a small quantity of heat dq flows from a system at temperature T_2 to surroundings at temperature T_1. The total change in entropy is then given by

$$\Delta S_{net} = \frac{dq}{T_2} - \frac{dq}{T_1}$$

Since $\quad T_1 > T_2$
Therefore $\quad \Delta S_{net} > 0$

Hence it is clear that in an irreversible isothermal process, the entropy of an isolated system is greater than zero.

Similarly entropy changes in vaporization process, fusion and transition process can be obtained.

In *polymeric reactions*, such as conversion of ethylene into polyethylene, large, well ordered structures are formed. Such reactions result in a decrease in entropy.

Entropy of substances

Hard substances such as diamond, garnet, quartz, silicon carbide, etc., have generally small entropies because in these, individual atoms are bounded to each other in a most orderly way. On the contrary, soft substances, especially gases, have large entropy value because they contain large amount of thermal disorder.

Entropy of the Universe

The whole universe may be considered as an isolated system, and we have just seen that the tendency of an isolated system is to have maximum entropy. Similarly in case of spontaneous natural processes, $dS > 0$. This leads to the general statement : "The energy of the universe is constant but the entropy is continuously increasing."

11.5. ENTROPY AND EQUILIBRIUM

In case of an isolated system, entropy will remain constant at equilibrium or will increase if an observable change occurs. Observable changes will continue to occur until the 'entropy' attains a maximum value at which time the system will be in equilibrium. Thus in the process of expansion of the gas, it expands until it is uniformly distributed and then no further change occurs. This can be expressed as follows :

In a system at constant energy and volume (and which can do no work), the entropy is maximum at equilibrium.

$$(dS)_{E, V} = 0$$

With regard to entropy the following points are to be remembered.

The entropy depends on the various kinds of motions of all the particles (*i.e.*, molecules and atoms present in them). Similarly entropy is found to increase when the bonds between the atoms in the molecule are ruptured *i.e.*, when the molecule dissociates into atoms or groups. On the contrary, it decreases with formation of bonds.

11.6. GIBBS AND HELMHOLTZ FREE ENERGY

In most chemical processes, neither the energy nor the entropy of the system is held constant. It is necessary to find some way of determining how the *energy* and *entropy* factors act together to drive a system at equilibrium. Similarly criteria for reversible and irreversible processes are

$$\Delta S = 0 \text{ for reversible process}$$
$$\Delta S > 0 \text{ for irreversible process.}$$

In practice, the entropy change referred to is that of the *system* and *surroundings*. It is, however, more convenient to express the criterion of spontaneity in terms of the properties of the *system alone*.

To accomplish the above requirements, new thermodynamic functions are introduced—which are *free energy* (G) and work function (A). These functions are fundamental thermodynamic properties and depend upon the state of the system alone. These two functions show the compromise that is struck between both energy and entropy changes, whenever they are considered.

We know that a part of the total energy of a system is converted into work and the rest is unavailable. So, any kind of energy which can be converted into useful work is called 'available energy'. For instance, as in the operation of an engine or a motor. But the energy which cannot be converted into useful work is known as unavailable energy (represented by entropy function).

Total energy = Isothermally available energy
+ Isothermally unavailable energy

The isothermally available energy present in a system is called *free energy*. It is denoted by the symbol G. This function was introduced by J. Willard Gibbs in 1876 and is called the *Gibbs free energy G*, or *thermodynamic potential*.

It is also known that a part of internal energy of a system can be used at constant temperature to do useful work. So, this fraction of internal energy (E), which is isothermally available is called 'work function' of the system and is denoted by the symbol A. This function is also known as *Helmholtz free energy*.

Mathematical expression for free energy functions

The statement of the second law of thermodynamics indicates that a reaction or a chemical process will proceed so long as $dq < TdS$. The vast majority of reactions are carried out at constant temperature and pressure and under such conditions

$$dq = dH$$

i.e., the heat absorbed is equal to the change in enthalpy for the process. Substituting for dq in $dq < TdS$ we have

$$dH = TdS$$
or
$$dH - TdS < 0$$

So long as this condition is valid, the reaction will proceed. The importance of the quantity $(dH - TdS)$ as a measure of the extent to which a reaction will proceed is such that it is given a special symbol dG, *i.e.*,

$$dG = dH - TdS \qquad \ldots(8)$$

Thus if a process or reaction occurs, then $dG < 0$.

The equation (8) may be written also as

$$G = H - TS \qquad \ldots(9)$$

where H is enthalpy of the system.

Consider a change

$A \longrightarrow B$ at constant temperature T.

Let G_A, H_A and S_A be the free energy, enthalpy and entropy of A respectively.

Similarly G_B, H_B and S_B be the free energy, enthalpy and entropy of B respectively. According to equation (9).

For B: $G_B = H_B - S_B T$
For A: $G_A = H_A - S_A T$

$$\therefore \quad G_B - G_A = (H_B - H_A) - T(S_B - S_A)$$

$$\therefore \quad \Delta G = \Delta H - T\Delta S \qquad \ldots(10)$$

The above change will continue till an equilibrium is attained. The equation (10) gives the *relationship between entropy, enthalpy* and *free energy*.

At equilibrium, the Second Law states that

$$dq = TdS$$

so that under conditions of constant temperature and pressure

$$dH = TdS$$

or for a finite change

$$\Delta H = T\Delta S$$

Then at equilibrium, we have

$$\Delta G = \Delta H - T\Delta S = 0$$

Now it is easy to make a statement about possibility or feasibility of occurring of a process.

Thus, chemical reactions or processes will occur at constant pressure so long as the changes in G are negative *i.e.*, $dG < 0$. The equilibrium condition is that $\Delta G = 0$. The pictorial representation of the above general statement is given in Fig. 11·2.

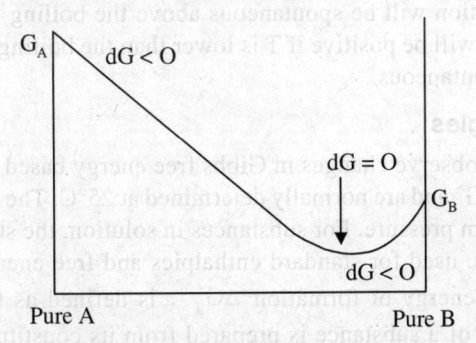

Fig. 11.2. The change in Gibb's free energy, as the reaction $A \rightleftharpoons B$ proceeds. At equilibrium G is minimum (*i.e.*, $dG = 0$).

Direction of a chemical change and significance of Free Energy

Any change in G at constant temperature ($\Delta T = 0$) is given by the equation

$$\Delta G = \Delta H - T\Delta S$$

The ΔG is seen to be made up of two terms, an energy term and an entropy term

$$\Delta G = \underset{\substack{\text{Energy} \\ \text{term} \\ \text{(1st law)}}}{\Delta H} - \underset{\substack{\text{Entropy} \\ \text{term} \\ \text{(2nd law)}}}{T\Delta S}$$

ΔG, i.e., change in free energy, is a measure of the work (or energy) that can be obtained from a reaction. In other words, free energy change which measures a balance between the two thermodynamic functions, does in fact give such a measure. Since most processes are carried at constant pressure, ΔH is the heat of reaction usually measured and therefore ΔG can be obtained in most of the chemical reactions.

The sign of the free energy change of a process is very important from the point of view of deciding the direction of change, feasibility and equilibrium state of the reaction. There are three possibilities as concerned to the sign of free energy.

$$A + B \longrightarrow C + D \quad \Delta G = -\text{ve (spontaneous reaction)}$$
$$A + B \rightleftharpoons C + D \quad \Delta G = 0 \text{ (zero) (reaction at equilibrium)}$$
$$A + B \longleftarrow C + D \quad \Delta G = +\text{ve (reaction is non-spontaneous)}$$

The arrows show the directions the reaction tends to follow spontaneously for the given sign of the free energy change.

A negative sign of free energy change for a process does mean that the process can occur provided the conditions are right. In fact, it is the sign of free energy change which determines whether the potentiality for a reaction to occur exists or not. Moreover the magnitude of the free energy change will tell us how large that potentiality is.

Thus we know that a reaction will proceed if $\Delta G < 0$ and will be at equilibrium if $\Delta G = 0$. When $\Delta G > 0$, the reaction cannot occur spontaneously.

To make these points clear, the example of vaporization of water can be considered.

(i) At its normal boiling point T_B (373 K) water is in equilibrium with its vapour. Under these conditions, $\Delta G = 0$, and therefore, $\Delta H = T\Delta S$. The entropy of vaporization at the boiling point is

$$\Delta S_{vap} = \frac{\Delta H_{vap}}{T_B}$$

(ii) For temperatures higher than the boiling point,

$$\Delta H_{vap} < T\Delta S_{vap}.$$

Since $T > T_B$ (It is assumed that both ΔH and ΔS are constant) it follows that ΔG will be negative, i.e., vaporization will be spontaneous above the boiling temperature.

(iii) Similarly, ΔG will be positive if T is lower than the boiling point; the vaporization will accordingly not be spontaneous.

Standard Free Energies

It is convenient to observe changes in Gibbs free energy based standard states; such values are given the symbol $\Delta G°$ and are normally determined at 25°C. The standard state of a substance is in stable form at 1 atm pressure. For substances in solution, the standard state is unit activity. Similar conventions are used for standard enthalpies and free energies.

The standard free energy of formation $\Delta G_f°$, is defined as the change in standard free energies, when 1 mole of a substance is prepared from its constituent elements. The standard free energy of formation of all elements is defined to be zero.

THERMODYNAMICS, EQUILIBRIUM AND KINETICS

The $\Delta G°$ for a reaction can be written as follows :
$$\Delta G° = G°_{products} - G°_{reactants}$$

Work function and maximum work

Work function 'A' is taken as
$$A = E - TS$$
or
$$\Delta A = \Delta E - T\Delta S$$

According to definition of entropy
$$\Delta S = \frac{q_{rev}}{T}$$

∴ The equation can be written as
$$\Delta A = \Delta E - q_{rev}$$

According to first law of thermodynamics
$$\Delta E = q_{rev} - W_{rev}$$

From the above two equations, we have
$$W_{rev} = -\Delta A$$

But W_{rev} is nothing but maximum work
∴
$$W_{max} = -\Delta A$$

Hence, in isothermal process maximum work can be obtained at the cost of Helmholtz free energy.

11.7. METASTABLE EQUILIBRIUM

In the study of phase rule, the concept of metastable equilibrium is dealt with. The phase rule is used to study hetereogeneous system in equilibrium by means of P—V—T diagrams or P—C—T or other kinds of diagrams, which are known as phase diagrams. Let us consider here an example of metastable equilibrium.

Point where Condensation would begin if equilibrium prevailed

Point where Condensation occurs very abruptly

(a)

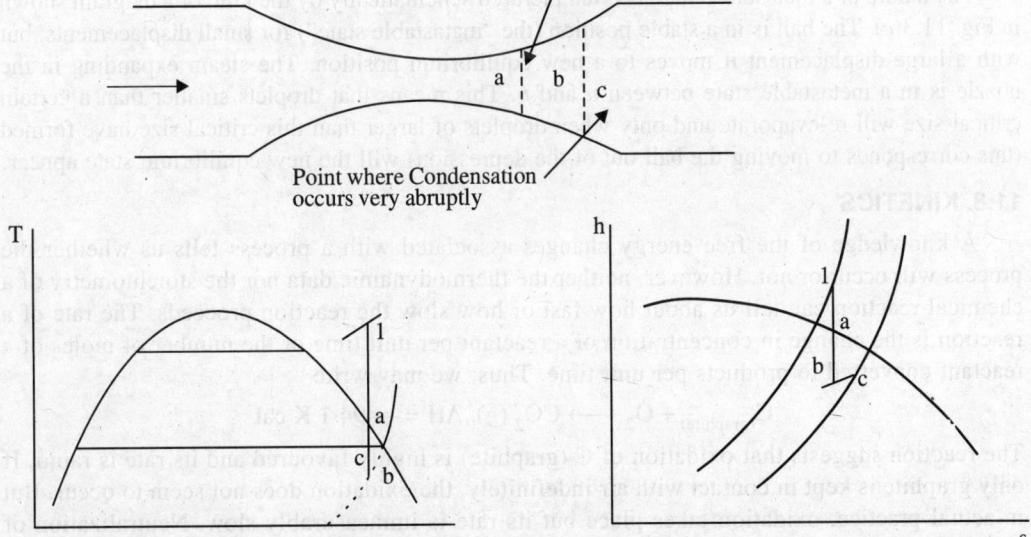

(b)

(c)

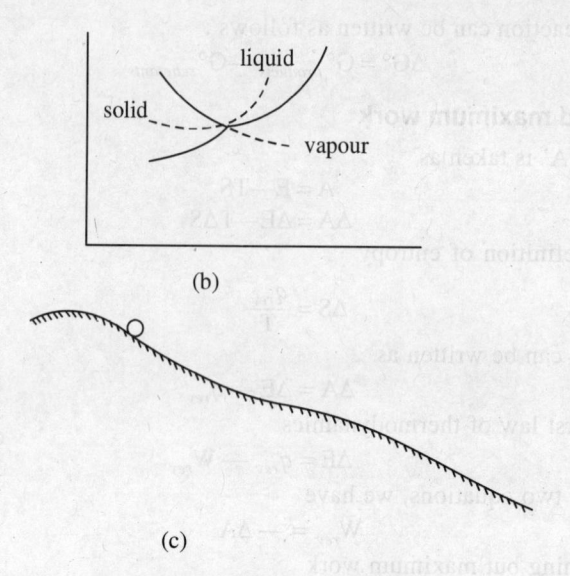

Fig. 11·3. Metalstable equilibrium.

Consider a slightly superheated vapour of water, such as steam, expanding in a convergent-divergent nozzle, as shown in Fig. 11·3.

Assuming that the process is reversible and adiabatic, the steam will follow path 1—a on the T—S diagram, and at a point a we would expect condensation to occur. However, if point a is reached in the divergent section of the nozzle, it is observed that no condensation occurs until point b is reached. At this point the condensation occurs very abruptly in what is referred to as condensation shock. Between points a and b the steam exists as a vapour, though the temperature is below the saturation temperature for the given pressure. The state is known as *metastable state*. The possibility of a metastable state exists with any phase transformation. The dotted lines on the equilibrium diagram of water system in Fig. 11.3(b) represent possible metastable states for solid liquid-vapour equilibrium.

The nature of a metastable state is often pictured schematically by the kind of a diagram shown in Fig. 11.3(c). The ball is in a stable position (the "metastable state") for small displacements, but with a large displacement it moves to a new equilibrium position. The steam expanding in the nozzle is in a metastable state between a and b. This means that droplets smaller than a certain critical size will re-evaporate and only when droplets of larger than this critical size have formed (this corresponds to moving the ball out of the depression) will the new equilibrium state appear.

11.8. KINETICS

A knowledge of the free energy changes associated with a process tells us whether the process will occur or not. However, neither the thermodynamic data nor the stoichiometry of a chemical reaction can tell us about how fast or how slow the reaction proceeds. The rate of a reaction is the change in concentration of a reactant per unit time or the number of moles of a reactant converted to products per unit time. Thus, we may write

$$C_{(graphite)} + O_2 \longrightarrow CO_2 (g), \Delta H = -94 \cdot 1 \text{ K cal}$$

The reaction suggests that oxidation of C (graphite) is highly favoured and its rate is rapid. If only graphite is kept in contact with air indefinitely, the oxidation does not seem to occur. But in actual practice, oxidation takes place but its rate is immeasurably slow. Neutralization of acid by a base represents also an example of rapid reaction. There is a class of reactions which take place in such a way that their rates can be measured in laboratory in minutes or hours.

THERMODYNAMICS, EQUILIBRIUM AND KINETICS

The rate of a reaction depends on a series of individual steps by which the reactants change to products. Some reactions proceed in one step, but more frequently the reaction occurs in sequence of steps. The single step or the sequence of steps is called the *mechanism of the reaction*. (A balanced equation such as

$$2\,NO\,(g) + Br_2(g) \longrightarrow 2\,BrNO\,(g)$$

tells us nothing about the mechanism of the reaction).

Thus the study of the rate and mechanism of chemical reactions is known as *chemical kinetics*.

Reaction Rates and Rate Laws

When a reaction proceeds, the general behaviour of the concentration of a reactant and product is shown in Fig. 11.4.

The rate of change of concentration of the reactant is given by the slope of the concentration-time curve. In addition it also decreases with time. The rate of change of concentration, also called the *rate of reaction*, is the change in concentration of a reactant or product per unit time. It is generally expressed as a time-derivative, $\dfrac{dC}{dt}$.

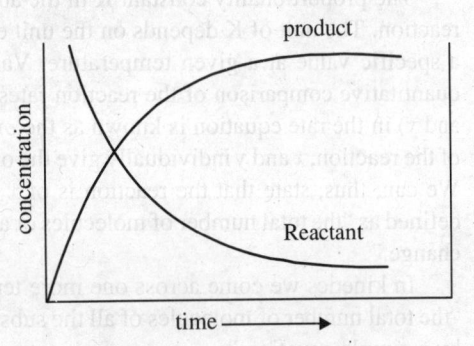

Fig. 11·4. Rate of reaction

Rate Expressions

Let us consider the simple gaseous reaction

$$N_2(g) + O_2(g) \longrightarrow 2\,NO\,(g)$$

The rate of this reaction is given by the rate of disappearance of the reactants N_2 or O_2. We can express this as

$$\text{Rate of reaction} = -\frac{d[N_2]}{dt} = -\frac{d[O_2]}{dt}$$

where $\dfrac{d[N_2]}{dt}$ and $\dfrac{d[O_2]}{dt}$ are both negative since the concentration decreases with time and the reaction rate itself will be positive as it should be. The reaction rate can also be expressed as the rate of formation of NO, $\dfrac{d[NO]}{dt}$, which is a positive quantity. Usually the rate of reaction is divided by the number of moles involved in the reaction. Thus for the above reaction we will have,

$$-\frac{d[N_2]}{dt} = -\frac{d[O_2]}{dt} = +\frac{1}{2}\frac{d[NO]}{dt}$$

Reaction Rate = Rate of disappearance of N_2 or O_2 = $\dfrac{1}{2}$ (Rate of formation of NO)

For a general reaction

$$aA + bB \longrightarrow cC + dD,$$

a, b, c and d are the number of moles of A, B, C and D species.

The reaction rate for this can be written as

$$\text{Reaction rate} = -\frac{1}{a}\frac{d[A]}{dt} = -\frac{1}{b}\frac{d[B]}{dt} = \frac{1}{c}\frac{d[C]}{dt} = \frac{1}{d}\frac{d[D]}{dt}.$$

In 1864, Guldberg and Waage established that the rate of a homogeneous reaction is proportional to some power of the concentrations of the reactants. The concentrations of reactants and products are expressed in moles per litre. For the general reaction

$$aA + bB \longrightarrow cC + dD$$
$$\text{Rate} \propto [A]^x [B]^y$$
or
$$\text{Rate} = K [A]^x [B]^y$$

where K is the proportionality constant. This manner of mathematically expressing reaction rates is called *rate law*. The numerical values of x and y can be determined experimentally, but cannot be deduced from the overall reaction. The values of x and y need not be the same as the stoichiometric coefficients a and b.

The proportionality constant K in the above equation is called the *specific rate constant* of the reaction. The unit of K depends on the unit employed for concentration and the unit of time. K has a specific value at a given temperature. Values of specific rate constants give us the basis for a quantitative comparison of the reaction rates. The sum of the powers of the concentration terms (x and y) in the rate equation is known as the *order of the reaction*. While $x+y$ gives the overall order of the reaction, x and y individually give the orders with respect to the reactants A and B respectively. We can, thus, state that the reaction is of x order with respect to A. The order of reaction is also defined as "the total number of molecules or atoms whose concentration changes during the chemical change."

In kinetics we come across one more term that is *molecularity* of the reaction. It is defined as "the total number of molecules of all the substances taking part in a chemical reaction as represented by a simple equation."

Example :
(i) Inversion of cane sugar (Molecularity = 2)

$$C_{12}H_{22}O_{11} + H_2O \xrightarrow{H^+ \text{ ions}} \underset{\text{Glucose}}{C_6H_{12}O_6} + \underset{\text{Fructose}}{C_6H_{12}O_6}$$

(ii) Dissociation of ammonia (Molecularity = 2)

$$2 NH_3 \rightleftharpoons N_2 + 3H_2$$

Reactions having molecularity one, two, three, etc. are known as *unimolecular, bimolecular, trimolecular* reactions respectively.

Orders of reactions

The orders of reactions provide the basis for classifying reactions. Generally, order of the reaction can be anywhere between zero and three. A *zero-order* reaction is one where the reaction rate does not depend on the concentration of the reactant, or the rate is proportional to the zeroeth power of the concentration of the reactant. Such reactions are not common. Decomposition of N_2O on a hot platinum surface is an example of a zero-order reaction.

$$N_2O \longrightarrow N_2 + \frac{1}{2} O_2$$

Rate $\propto [N_2O]^0$

Rate $= -d[N_2O]/dt = K[N_2O]^0 = K =$ constant

It is found that at a given temperature, a majority of reactions have rates that are proportional to the concentration of one, two, or possibly three of the reactants, with each reactant raised to a small integral power. Thus, if reactions are considered in which A, B, and C represent possible reactants, the rate equations for reactions with such concentration dependence would be as follows :

Rate = K[A]   (1st order reaction)

Rate = K[A]^2 or   K[A][B]   (2nd order reactions)

Rate = K[A]^3 or   K[A]^2[B]

THERMODYNAMICS, EQUILIBRIUM AND KINETICS

or   K[A] [B] [C]   (3rd order reactions)

The decomposition of N_2O_5 is a typical *first-order reaction*.

$$N_2O_5 \longrightarrow 2NO_2 + \frac{1}{2}O_2$$

$$\text{Rate} = -d[N_2O_5]/dt = K[N_2O_5]^1$$

The reaction of H_2 and I_2 to give HI is a *second-order* reaction

$$H_2 + I_2 \rightleftharpoons HI$$

$$\text{Rate} = K[H_2][I_2]$$

The decomposition of NO_2 to NO and O_2 is also a second order reaction.

The reaction of NO and oxygen to give NO_2 is a *third-order reaction* but the order with respect to NO is two and the order with respect to oxygen is one.

$$2NO + O_2 \longrightarrow 2NO_2$$

$$\text{Rate} = K[NO]^2[O]_2$$

It is mentioned above that there is no relation between the order (exponents in the rate equation) and the stoichiometric coefficients in the reaction. An example to illustrate this point would be the reaction between decaborane and ethyl alcohol to form triethyl borate and hydrogen.

$$B_{10}H_{14} + 30\,C_2H_5OH \longrightarrow 10\,B(OC_2H_5)_3 + 22H_2$$

$$\text{Rate} = K[C_2H_5OH][B_{10}H_{14}]$$

The overall order of this reaction is two, but the reaction is first order with respect to either reactant.

Order of reactions need not always have integral values. For example, the decomposition of acetaldehyde into CH_4 and CO under certain conditions is of 1·5 order.

$$CH_3CHO \longrightarrow CH_4 + CO$$

$$\text{Rate} = K[CH_3CHO]^{1.5}$$

The reaction between gaseous H_2 and Br_2 to form HBr looks very simple but it is found to obey the complex rate law.

$$\text{Rate} = \frac{d[HBr]}{dt} = \frac{K[H_2][Br]^{1/2}}{1 + K'[HBr][Br_2]}$$

It this case, there is no clear indication of reaction order.

QUESTIONS

1. Write short notes on
   (a) Second law of thermodynamics
   (b) Significance of entropy
2. What are the important thermodynamic functions :
   Explain at least one in detail.
3. What are the criteria, of thermodynamic equilibria ?
   Discuss with examples.
4. What do you understand by metastable equilibrium ?
5. Whet is meant by rate law and order of reaction?
6. How is the knowledge of thermodynamics and kinetics useful in the study of a reaction ?
7. Discuss the significance of the following terms :
   (i) Internal energy   (ii) Gibbs free energy
   (iii) Entropy   (iv) Enthalpy.
8. What is the relation between free energy and equilibrium ?

12

Crystal Structures

"Crystals are characterised by well-defined and, to some extent, symmetrically arranged planes. The study of crystals includes the examination of external form as well as the elucidation of the internal structure."

12.1. INTRODUCTION

In nature we observe many astonishingly ordered processes. In the cosmos, far away stars in the galaxy and nearby planets appear to follow the same orbits year after year. On earth, rhythmic sequence of tides, of lunar cycles beautiful natural patterns of many solids (garnet, quartz, diamond etc.) and myriad patterns of designs and colours in flowers and plants are some common examples. Here we will concentrate on the orderly arrangements of solids–which are *various* and *curious*. These adorn and beautify a large multitude of bodies. Among these solids, some crystals are commonly known to us.

The name *crystal* comes from the Greek word Krystallos meaning clear ice. This term was first applied to the beautiful transparent quartz "stones" found in the Swiss Alps. This choice of name was due to the belief that these stones were formed from water by the intense cold. Later, however, the name was used for solids bounded by many flat shiny faces symmetrically arranged.

In the gaseous state, molecules move randomly without exerting appreciable forces on one another. On the other hand, in the solid state, there is a regular order in the arrangement of ions, atoms or molecules constituting the solid and these are held together by fairly strong forces. These solids exhibit familiar macroscopic properties such as rigidity of definite shape, fixed volume and low compressibility. Solids diffuse very slowly as compared to liquids or gases and from various types of crystals (Fig. 12.1). All these properties lead to the fact that atoms or molecules constituting a solid occupy fixed positions with respect to each other.

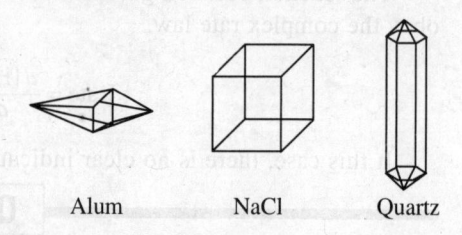

Alum   NaCl   Quartz

Fig. 12.1. Shapes of some crystals.

Crystallography is the branch of science which deals with geometry, properties and structure of crystals and crystalline substances.

A crystal is a solid figure having a definite geometrical shape, with flat faces and sharp edges. The study of crystals can be carried by two ways:

(*i*) the examination of external form
(*ii*) the elucidation of the internal structure.

With the help of X-ray diffraction technique, it is possible to elucidate the internal structure

474

CRYSTAL STRUCTURES

of crystals. X-ray scattering study provides a valuable tool for studying various crystal structures.

12.2. FUNDAMENTAL LAWS OF CRYSTAL STRUCTURES

Geometrical crystallography is concerned with the outward spatial arrangement of crystal planes and the geometrical shape of crystals. Thus, crystallography depends upon the following three fundamental laws:

1. Law of constancy of interfacial angles
2. Law of Rationality of indices
3. Law of Symmetry.

1. The Law of Constancy of interfacial angles

It states that the corresponding faces or planes forming the external surface of a crystal of a given substance always intersect at definite angle which remains constant at a given temperature no matter how the face develops.

A substance may crystallize under different conditions to produce crystals with faces of variable size and shape. The angle of intersection of any two corresponding faces, however, would always be found to be the same. For example, the interfacial angles, in all sodium chloride crystals are found to be 90°, irrespective of their size and shapes of faces.

2. The Law of Rational indices

According to this law, the ratio between intercepts on crystallographic axes for the different faces of a crystal can always be expressed by rational numbers.

For example consider a plane LMN in the crystal as shown in Fig. 12.2. This plane has intercepts OL, OM and ON along the X, Y, and Z axes at distances $2a$, $4b$ and $3c$ respectively, when $OA = a$, $OB = b$ and $OC = c$ are the chosen unit distances along the three coordinates. These intercepts are in the ratio of $2a : 4b : 3c$ wherein 2, 4, 3 are simple integral whole numbers.

The coefficients of a, b and c (2, 4 and 3 in this case) are known as the Weiss indices of a plane. It is to be noted that the Weiss indices are not always simple integral whole numbers as in this case. They may have fractional values as well as infinity. Weiss indices are, therefore, rather inconvenient in use and have

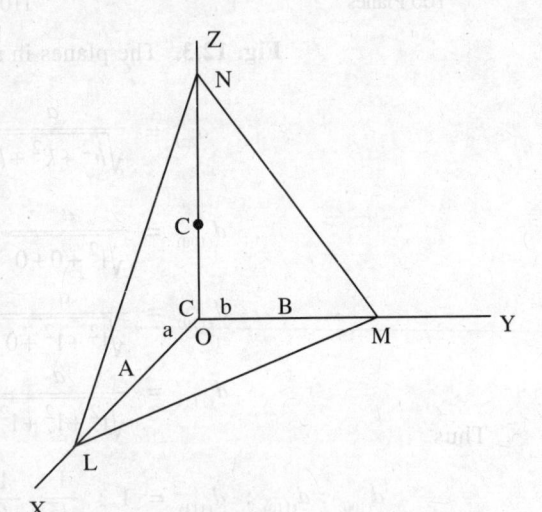

Fig. 12.2. Intercepts of crystallographic planes.

consequently been replaced by *Miller indices*. The Miller indices of a plane are obtained by taking the reciprocals of Weiss indices and multiplying throughout by the smallest number in order to make all reciprocals as integers. Consider a plane in which Weiss notation is given by $\infty a : 2b : c$.

Taking reciprocals of coefficients of a, b and c, we get the ratio $\dfrac{1}{\infty}, \dfrac{1}{2}, \dfrac{1}{1}$, *i.e.*, $0, \dfrac{1}{2}, 1$.

Multiplying by 2 in order to convert them into whole numbers, we get 0, 1, 2. These numbers are called *Miller indices* of the plane and the plane is designated as (0, 1, 2). In general, planes designated like this are termed (hkl) planes. In this case we have

$$h = 0, \ k = 1, \text{ and } l = 2; \ \begin{pmatrix} h, & k, & l \\ 0, & 1, & 2 \end{pmatrix}$$

Similarly the Miller's indices for the plane given in Fig. 12.2 can be obtained as follows:

(a) By taking reciprocals of Weiss indices, we get $\dfrac{1}{2}, \dfrac{1}{4}, \dfrac{1}{3}$

(b) Multiplying by 12, we get 6, 3, 4

The crystallographic plane is, therefore, designated as (634) plane.

The distances between the parallel planes in a crystal are designated as d_{hkl}. For various cubic lattices, these interplanar spacings are given by the general formula

$$d_{hkl} = \frac{a}{\sqrt{h^2+k^2+l^2}}$$

where a is the length of the cube side while h, k and l are the Miller indices of the plane.

For illustration purpose, let us consider the planes (100), (110) and (111) of a cubic lattice. Applying the above formula we can have (Fig. 12.3).

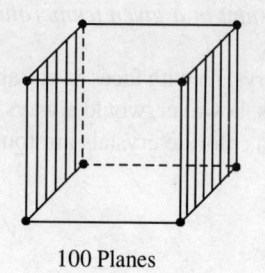

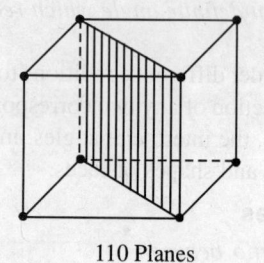

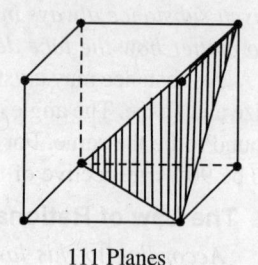

100 Planes   110 Planes   111 Planes

Fig. 12.3. The planes in a simple cubic crystal.

$$d_{hkl} = \frac{a}{\sqrt{h^2+k^2+l^2}}$$

$$d_{(100)} = \frac{a}{\sqrt{1^2+0+0}} = a$$

$$d_{(110)} = \frac{a}{\sqrt{1^2+1^2+0}} = \frac{a}{\sqrt{2}}$$

$$d_{(111)} = \frac{a}{\sqrt{1^2+1^2+1^2}} = \frac{a}{\sqrt{3}}$$

Thus

$$d_{(100)} : d_{(110)} : d_{(111)} = 1 : \frac{1}{\sqrt{2}} : \frac{1}{\sqrt{3}}$$

$$= 1 : 0.707 : 0.577$$

3. The law of Symmetry

This law states that all crystals of the same substance possess the same element of symmetry. Usually, there are three types of symmetry associated with a crystal.

(i) Plane of Symmetary

A crystal is said to have a plane of symmetry if it can be divided by an imaginary plane into two equal portions each of the which is mirror image of the other.

(ii) Axis of Symmetry

A crystal is said to have a line of symmetry if on rotation about an imaginary line it gives the same appearance more than once during one complete revolution. The crystal is said to have two, three or four-fold symmetry if it occurs two, three, or four times respectively during one complete revolution.

CRYSTAL STRUCTURES

(iii) Centre of Symmetry

Centre of symmetry of a crystal is such a point that any line drawn through it intersects the surface of the crystals at equal distances in both directions.

It may be pointed out that a crystal may have any number of planes or axes of symmetry but it has only one centre of symmetry.

Crystals can be described in terms of a regular three-dimensional array of points called *space lattice*. *The pattern of points which describes the arrangement of molecules or atoms in a crystal is known as a space lattice.* In other words a lattice is characterized by the distance between successive points along the three axes. Fig. 12.4 shows the space lattice of NaCl. Each of the points corresponds to the position of the centre of an ion. The solid circles locate the sodium ions and the open circles negative chloride ions. The circles do not represent sodium ions and chloride ions but only the positions occupied by their centres. In sodium chloride the ions are of different sizes and are practically touching each other, as shown in Fig. 12.5.

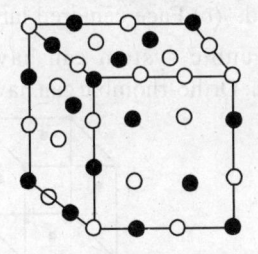

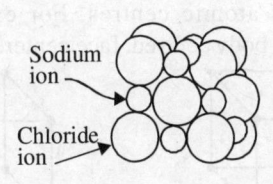

Fig. 12.4. Space lattice of NaCl.

Fig. 12.5. Model of NaCl

A space lattice is an infinitely extended regular distribution of points in space. While discussing the space lattice, it is sufficient enough to consider for it to represent the order of an arrangement. The small fraction of a space lattice which shows the pattern of the whole lattice, is called *unit cell*. It is defined as the smallest repeating unit, in three dimensions, which represents the entire crystal. The crystal may be considered to consist of a very large number of such unit cells, each one in direct contact with its nearest neighbours and all are similarly oriented in space. The unit cell of NaCl is the *cube*, as shown in Fig. 12.6.

If this cube is moved through its edge length in the 'a' direction, 'b' direction and the 'c' direction, many times, eventually the whole space is reproduced. Depending upon the various combinations of the lattice spacings along the three axes and angles between them, we can define *seven crystal systems* (Table 12.1). The simplest amongst these systems is the cubic, while triclinic is the most complex one.

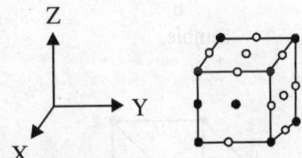

Fig. 12.6. Unit cell of NaCl

Table-12.1. Different Crystal Systems

System	Axes	Angles	Examples
Cubic	$a = b = c$	$\alpha = \beta = \gamma = 90°$	Cu, NaCl, CSCl, ZnS
Tetragonal	$a = b\,;\,c$	$\alpha = \beta = \gamma = 90°$	White tin, SnO_2, TiO_2 (rutile)
Ortho-rhombic	$a\,;\,b\,;\,c$	$\alpha = \beta = \gamma = 90°$	Rhombic sulphur TiO_2 (brookite)
Monoclinic	a, b, c	$\alpha = \gamma = 90°,\ \beta \neq 90°$	Monoclinic sulphur
Rhombohedral	$a = b = c$	$\alpha = \beta = \gamma \neq 90°$	Calcite ($CaCO_3$)
Hexagonal	$a = b\,;\,c$	$\alpha = \beta = 90°,\ \gamma = 120°$	Graphite, ZnS, BN
Triclinic	$a\,;\,b\,;\,c$	$\alpha, \beta\,;\,\gamma \neq 90°$	$K_2Cr_2O_7$, $CuSO_4, 5H_2O$

In the lattice, the points (atomic centres) may be assigned to corners, face-centres or body-centres of the seven types of crystal systems, listed in Table 12-1. In 1848, Bravasi showed that there can be *only fourteen lattices* associated with the seven crystal systems, depending on the

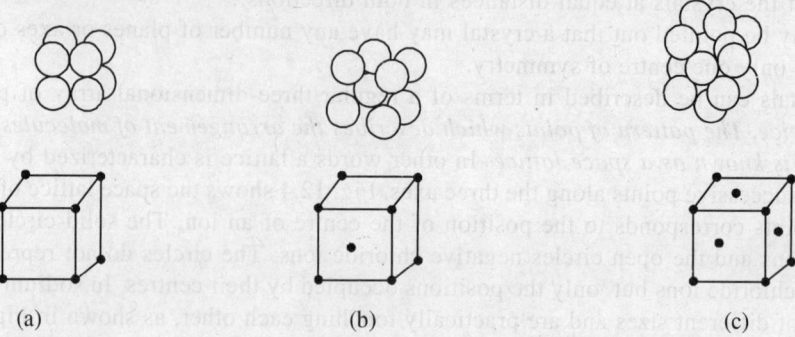

Fig. 12.7. (a) Simple, (b) Body centered, (c) Face centered lattice.

distribution of atomic centres. For example, the cubic system can have three Bravais lattices (simple, body centred, face centered – Fig. 12.7). Ortho-rhombic can have four, tetragonal

CUBIC: Simple, Body centered, Face centered

ORTHORHOMBIC: Simple, Body centered, End centered, Face centered

TETRAGONAL: Simple, Body centered

MONOCLINIC: Simple, End centered

Rhombohedral, Triclinic, Hexagonal

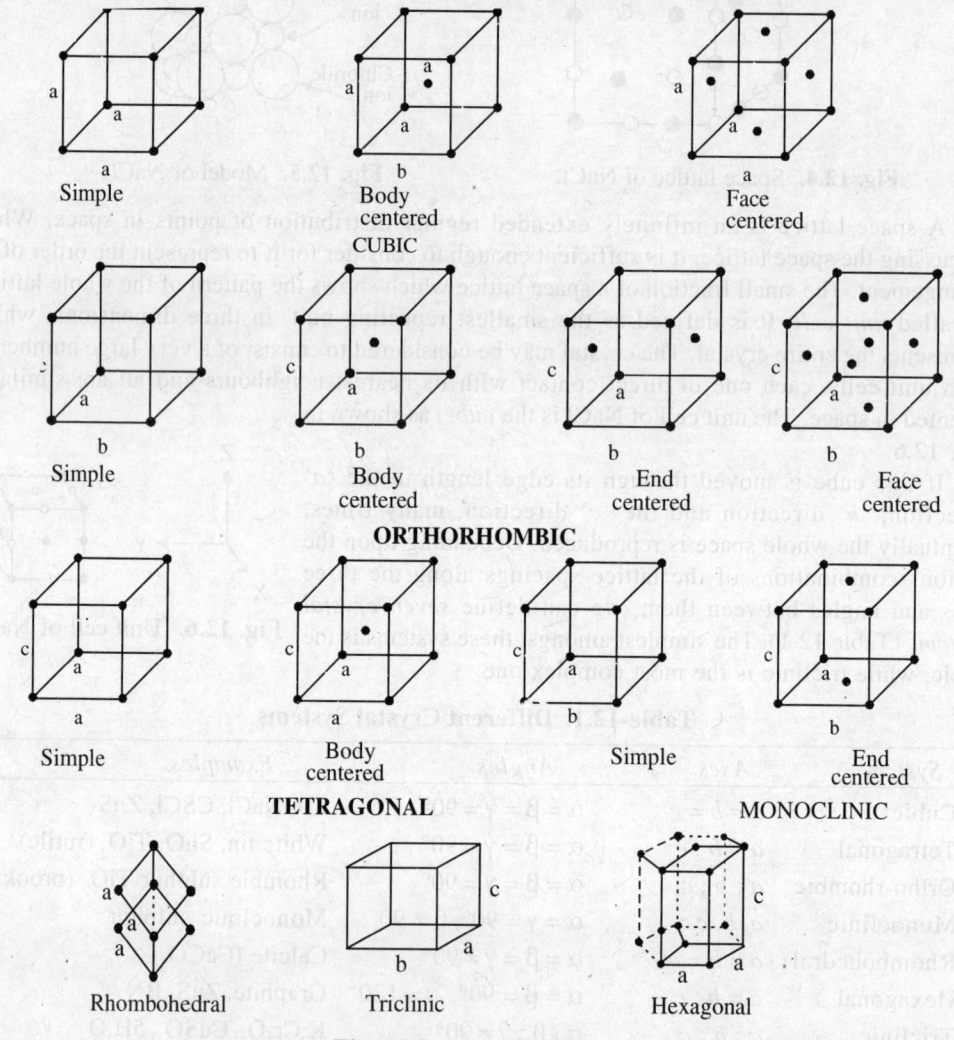

Fig. 12.8. Bravais lattices.

CRYSTAL STRUCTURES

system can have two and lastly three independent systems such as Rhombohedral, triclinic and hexagonal are left. In Fig. 12.8 all these Bravais lattices are shown. Every crystal possesses one of these lattice patters, but the detailed internal structure within the unit cell will vary in complexity from substance to substance. Whatever be the complexity, the unit cell is the basic repeating unit in the crystal, and X-ray diffraction patterns directly provide the interplanar spacings (from Bragg's law) and hence the unit cell dimensions.

12.3. X-RAYS AND CRYSTAL STRUCTURE

Crystals are made up of layers of atoms stacked in a regular fashion. Max Von Laue, in 1912, suggested that the orderly arrangement of atomic layers in the crystal should be spaced at a distance of a few angstroms. Since X-rays are electromagnetic radiations with wavelengths of about 10^{-8} cm or 1 Å, he suggested that a crystal should act as a natural and very fine three dimensional diffraction grating for the X-rays. Making use of the diffraction of X-rays, the structures of the crystals have been determined by various methods based on Bragg's law.

The following methods have been developed for the study of X-ray diffraction for the purpose of investigating the internal structure of crystals.
  (1) The Laue Photograph method
  (2) The Bragg's X-ray Spectrometer method
  (3) The powder method
  (4) The rotating crystal method and
  (5) Oscillating crystal method.

12.4. BRAGG'S LAW

A simple equation was developed by H. Bragg and his son William L.B. for the study of internal structure of crystals. According to them a crystal (composed of a series of equally spaced atomic planes) could be employed not only as a transmission grating, but also as reflecting grating. When X-rays are incident on a crystal plane (Fig. 12.9) these penetrate into the crystal and strike the atoms in successive planes. From each of these planes the X-rays are reflected like a beam of light from a bundle of glass plates of equal thickness. Based on this model they derived a simple relation between the wavelengths (λ), the angle of incidence θ and the distance (d) between the successive atomic planes. The relation is

$$n\lambda = 2d \sin \theta$$

(where $n = 1, 2, 3, ...$)
It is known as Bragg's law or Bragg's equation.
By measuring n, the order of reflection of the X-rays and θ, the angle of incidence, we can determine the ratio

$$\frac{d}{\lambda} = \frac{n}{2 \sin \theta}$$

From this, d can be calculated if λ is known and *vice-versa*.

Packing of atoms

As discussed above, we know that unit cells concern points that locate atomic or molecular centres. Atoms are space-filling entities and structures can be described as resulting from the packing together of representative spheres. The most efficient packing together of equal

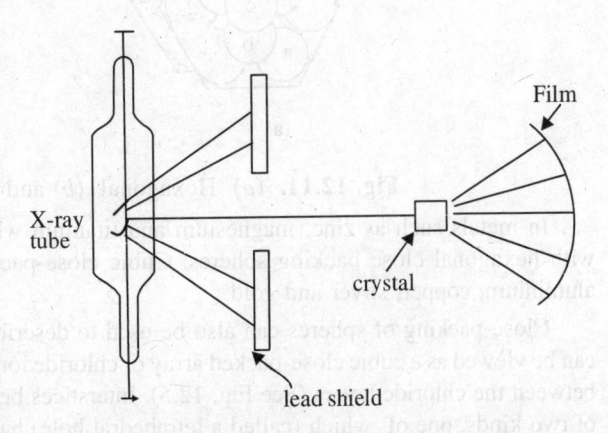

Fig. 12.9. X-ray determination of structure.

spheres, called *closed-packing,* can be obtained in two ways, which utilizes the same fraction (0.74) of total space. One of these close-packing arrangements is called *hexagonal close-packed*. Its build up can be considered as follows: Place a sphere on a flat surface. Surround it with six equal spheres as close as possible to the central sphere and in the same plane. If one looks down at this place, the projection is as shown in Fig. 12.10 (*a*). Now form a second layer of equally bunched spheres, staggered as shown in Fig. 12.10 (*b*) so that the second-layer spheres nestle into the depressions formed by the first-layer spheres. A third layer can now be added with each sphere directly above a sphere of the first layer. The fourth layer lies directly above the second layer, and so forth in alternating fashion until the hexagonal close-packed structure has been generated.

The other closed-packed structure, called *cubic-close-packed,* results if the build up of layers *a* and *b* is the same as that described above, but then a third layer *c* is added, as shown in Fig. 12.10(*c*). The spheres of layer *c* are not directly above those of either layer *a* or layer *b*. In generating this cubic close-packed structure the sequence of layers is *abcabc....* in contrast to the sequence *ababab....* for hexagonal close-packing.

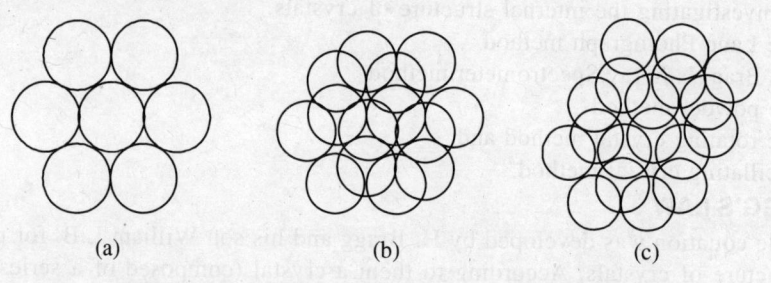

Fig. 12.10. (*a*) (*b*) (*c*) Close Packing of spheres.

These two structures are represented in terms of unit cells. For hexagonal close-packing, a unit cell is like that of Fig. 12.11(*a*) except that the *b* layer is inserted between the top and bottom faces, so as to add three lattice points at mid-height in the hexagonal prism shown. For cubic-packing the unit cell is a face-centred cube; the body diagonal is perpendicular to the stacking layers [Fig. 12.11(*b*)]. The difference between the two kinds of close-packing may be seen in Fig. 12.11.

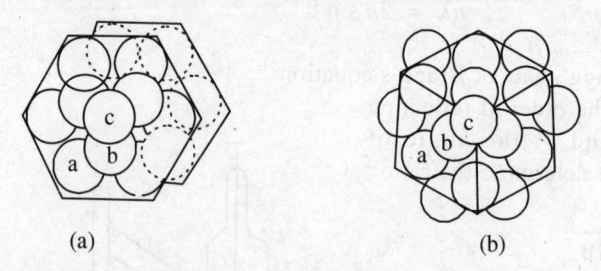

Fig. 12.11. (*a*) Hexagonal (*b*) and Cubic Close Packing

In metals such as zinc, magnesium and titanium while crystallization occurs, it takes place with hexagonal close packing spheres. Cubic close-packing is shown by other metals, such as aluminium, copper, silver and gold.

Close-packing of spheres can also be used to describe many ionic solids. For example, NaCl can be viewed as a cubic close-packed array of chloride ions with sodium ions, fitting into interstices between the chloride layers (See Fig. 12.5). Interstices between layers of close-packed spheres are of two kinds, one of which (called a tetrahedral hole) has four spheres adjacent to it and other of which (called an octahedral hole) has six spheres adjacent to it. The difference between the two kinds is illustrated in Fig. 12.12. The '*a*' layer of closed-packed spheres is represented by filled

CRYSTAL STRUCTURES

balls, and the '*b*' layer above it by open balls. On the left, marked by shades, is a grouping of four adjacent balls (three from layer *a*, and one from layer, (*b*) surrounding a tetrahedral hole. This is called a tetrahedral hole because a small atom inseted in the hole would have four neighbouring atoms arranged at the corners of a regular tetrahedron. Such a grouping of four-four spheres and its relation to a tetrahedron are shown at the bottom of the figure. It should be recalled that a regular tetrahedron is a triangle-based pyramid in which each of the four faces is an equilateral triangle.

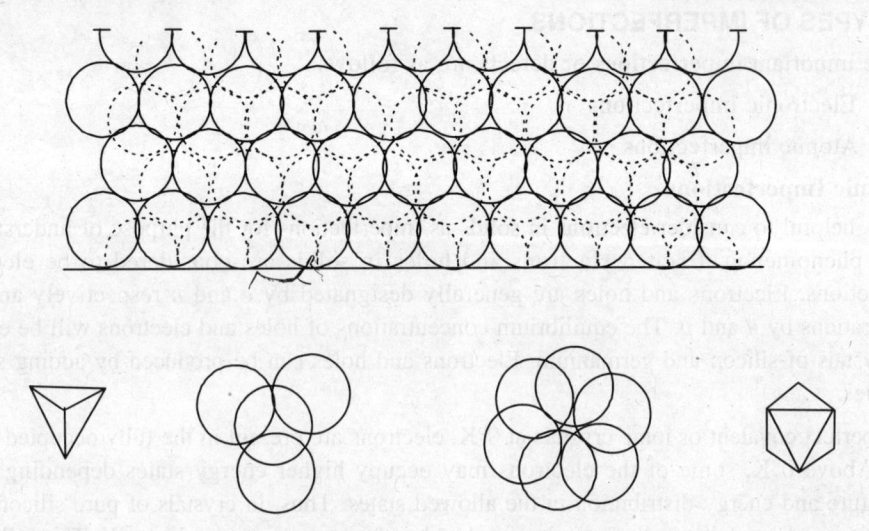

Fig. 12.12. Tetrahedral and Octahedral holes between layers of close-packed spheres.

On the right side in Fig. 12.12 there is also dashed area, which represents a grouping of six adjacent balls (three from layer '*a*' and three from layer '*b*') surrounding an octahedral hole. The octahedron, as shown at bottom of Fig. 12.12 is an eight-faced figure, all of whose faces are equilateral triangles. In the NaCl-structure (called the *rock-salt structure*), the Na^+ ions are regarded as being in the octahedral holes created between layers of Cl^- ions. The number of octahedral holes is just equal to the number of packed spheres, so that the number of Na^+ ions accommodated just equals the number of Cl^- ions.

Thus a close-packing is a way of arranging equi-dimensional objects in space so that the available space is filled very efficiently. A honeycomb is an example of a close packing found in nature. The bees form hexagonal cells to store their honey in order to utilize space more efficiently. Similarly, squirrels mound nuts in a close packing for the same reason.

The number of closed-packings possible in three dimensions are infinite. The various closed packing of spheres of equal size differ in the way that the hexagonal closed-packed layers comprising them are arranged.

Another important structure is the *zinc blende structure,* named after the mineral ZnS. It can be visualized as consisting of a cubic close-packed array of sulfur atoms with zinc atoms disposed in tetrahedral holes. These examples include the *technologically important group* III to V compounds, which are formed from an element of group III (for example, Al, Ga, In) plus an element of group V (for example P, As, Sb). They are important in solid-state electrical devices such as transistors and rectifiers.

Solid State Defects

In case of crystals, the state of complete order and of lowest energy is found at absolute zero of temperature. At any temperature above 0°K there will be some departure from complete order. In principle, any deviation from a completely ordered arranagment in a crystal constitutes native disorder. In addition, crystals may have disorder due to the presence of impurities. The term *imperfection* or *defect* is generally employed to denote a departure from a perfectly periodic array of atoms in a crystal, immaterial of whether the imperfection is intrinsic or extrinsic in origin. Some of the properties of solids, such as electrical conductivity, optical spectra or mechanical strength, cannot be explained on the basis of crystal structure alone. Imperfections (or defects) may not only alter the properties of a crystal, but also give rise to certain new properties of interest.

12.5. TYPES OF IMPERFECTIONS

The important imperfections or defects are as follows:

(1) Electronic imperfections

(2) Atomic imperfections

Electronic Imperfections

It is helpful to consider electrons in solids as imperfections for the purpose of understanding several phenomenon. Excited electrons and holes in solids are considered to be electronic imperfections. Electrons and holes are generally designated by e and h respectively and their concentrations by n and p. The equilibrium concentrations of holes and electrons will be equal in pure crystals of silicon and germanium. Electrons and holes can be produced by adding suitable impurities.

In perfect covalent or ionic crystals at 0°K, electrons are present in the fully occupied-energy states. Above 0°K, some of the electrons may occupy higher energy states depending on the temperature and energy distribution in the allowed states. Thus, in crystals of pure silicon, some electrons are released thermally from the covalent bonds at temperatures above °K. These electrons which are free to move in the crystal would be responsible for the electrical conductivity in silicon. The electron deficient bond produced by the removal of an electron is referred to as a *positive hole* or *hole*. Holes also gives rise to electrical conductivity, but the direction of motion of the holes in an electric field will be opposite to that of electrons.

Thus excited electrons and holes in solids are considered to be electronic imperfections.

Geermanium and silicon (elements of IV group) have a characteristic valence of four and form four bonds as in diamond. A large variety of solid state materials have been prepared by combination of elements of group III and V or II and VI to simulate the average valence of four, as in Ge or Si. Typical of groups III–V compounds are InSb, AlP and GaAs; whereas ZnS, CdS, Cd Se and Hg Te are examples of groups II–VI compounds. In these compounds the bonds are perfectly covalent and ionicity will vary depending on the relative electronegativities of the two component elements. The compounds of III–V groups and II–VI compounds exhibit interesting electronic (electrical and optical) properties which have been exploited in the electronics industry.

Atomic Imperfections

Ideally, a crystal consists of a perfectly periodic array of atoms whose arrangement conforms to the particular symmetry. The term *imperfection* or *defect* is generally used to describe any deviation from such an orderly array. If such deviation is found in the vicinity of an atom or a group of atoms, the imperfections are called *point defects*. If the deviations from periodicity extend over microscopic regions of the crystal, they are referred to as *lattice imperfections or defects*. Lattice imperfections may extend along lines (line defects) or surfaces (plane defects). Line defects are also called *dislocations*. The various types of atomic imperfections in crystals are given in Table 12.2.

CRYSTAL STRUCTURES

Table-12.2. Atomic Imperfections in Crystals

	Imperfection	Nature of Imperfection
1.	Point Defects	
	Schottky defect	Atom missing from the natural lattice site, creating a vacancy.
	Interstitual	Atom in a normally vacant interstitial site
	Antistructure	Atoms misplaced wrongly in sites assigned for some other atoms
	Frenkel defect	Atom in the lattice site displaced to an interstitial site, creating a vacancy.
2.	Line Defects	
	Edge dislocation	Dislocation line marks the edge of an extra plane of atoms which is inserted part way in a crystal.
	Screw dislocation	Displacement of atoms in one part of crystal relative to the crystal causes a spiral around the dislocation line.
3.	Plane Defects	
	Grain boundary	Boundary between two crystals in a polycrystalline solid
	Stacking faults	Boundary between two layers with different stacking sequences in closepacked structures.
	Shear structure	Structures where excess atoms are present in folded planes connecting slabs of the crystal of normal structure.

12.6. DISCUSSION OF SOME DEFECTS

One kind of lattice defect, known as *lattice vacancies,* arises if some of the lattice points are unoccupied. Another called *lattice interstitial,* arises if atoms occupy positions between lattice points. All crystals are imperfect to a slight extent and contain lattice defects. For example, in NaCl some of the sodium ions and chloride ions are missing from the regular pattern (Fig. 12.13a). In silver bromide (AgBr) some of the silver ions are missing from their regular positions and are found squeezed in between other ions (Fig. 12.13b). Lattice vacancies occur to some extent in all crystals. Their presence helps us to explain how *diffusion* and *ionic conductivity* can occur in the solid state. Lattice interstitials are considerably less probable. They are found to occur when small positive ions move into positions between the normal planes of larger negative ions.

```
+ − + − + − +         + − + − + − +
− + □ + − + −         − + − + − + −
+ − + − + − +         + − ⋮ − + − +
− + − + ⋮ + −         − + − + − + −
+ − + − + − +         + − + − + − +
− + − + − + −         − + − + − + −
+ − + − + − +         + − + − + − +
     (a)                   (b)
```

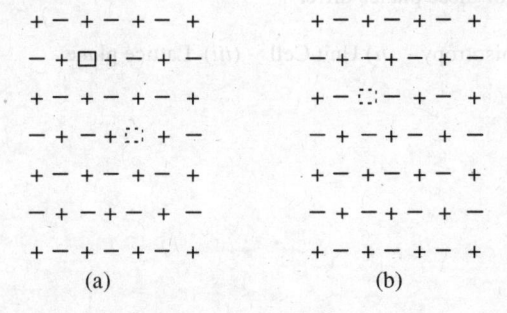

Fig. 12.13. Lattice defects (a) vacancies in NaCl, (b) misplaced Ag^+ in AgBr.

In addition to the defects arising from structural imperfections are defects of a *more chemical nature* in that they are associated with the presence of *chemical impurities.* Such impurities can drastically change the properties of materials; hence their introduction is being exploited in producing new materials with desirable combinations of properties. For example, the addition of less than 0.1% $CaCl_2$ to the NaCl can raise the conductivity by 10,000 times. This happens as follows. In the mixed crystal the Cl⁻ ion lattice is unchanged, but the Ca^{2+} ions, occupy positions in the Na^+ lattice. Because of the requirement for electrical neutrality, each insertion of Ca^{2+} for $2Na^+$ leads to the creation of lattice vacancy. The vacancy allows Na^+ ions to move and hence results in increased conductivity.

A more practical application and use of impurity defects involves germanium and silicon crystals. The electric conductivity of these group IV elements, in the pure state, is extremely low. However, on addition of trace amounts of elements from either group III or group V, the conductivity is greatly enhanced. Both germanium and silicon have the diamond structure. Each atom is bonded to four neighbours by four covalent bonds. These require all four outer electrons of each group IV atom. Thus extremely pure samples of germanium or silicon are, therefore, non-conductors. When a small number of foreign atoms (even one in a million) gets introduced into the crystal lattice, the situation changes drastically. For example, an atom of arsenic or antimony with five valency electrons can fit into the tetrahedral lattice of germanium or silicon only by setting one of the five electrons relatively free. Such electrons moving under the influence of an electric field, give rise to *n-type semi conductors* (named so because negative charge flows). However, when a trivalent boron atom is introduced, electron deficiency is created at the site of foreign atom. Such a defect is known as formation of a *positive hole* in the lattice. Under the influence of an electric field, electrons from a neighbouring atom fill this hole, leaving a positive hole around the atom which they leave. Such semiconductors are known as *p*-type semi-conductors due to the movement of a positive hole through the lattice in them.

A semi-conductor, in which there is junction between an electron rich and electron-poor region, acts as a rectifier, *i.e.,* can convert A.C. to D.C. The *n-p* junction can be formed by introducing a trace of boron on one side of a silicon disc impregnated with traces of arsenic.

QUESTIONS

1. State the law of rational indices. What are Miller indices ?
2. Write a short essay on various types of solids.
3. What is meant by space lattice of a crystal ? Draw a unit cell for space lattices of the following types:
    (*i*) Simple cubic (*ii*) Face centred cubic and (*iii*) Body centred cubic.
4. What are lattice planes of a crystal ? What is meant by (100), (110) and (111) planes of cubic lattice? How do the spacing of these planes differ ?
5. Explain the terms:
    (*i*) Isotropy and Anisotropy (*ii*) Unit Cell (*iii*) Lattice plane.

13

Structure of Solids and Magnetic Materials

"Although it is customary to use the expression 'Solid State' to substances which are of a crystalline nature (i.e., those having a regularity of the constituent atoms and molecules), even in the amorphous state, there appears to be some regularity which may be partial and transient. Nevertheless, it appears that there may be a gradual transition between the strict regimentation on the one hand and completely random distribution on the other, and a study of the solid state has also helped to elucidate the structure of amorphous solids."

In some respects the solid state seems simpler than the gaseous state. In the first place the volume and even shape of a solid is much less dependent on external conditions of temperature and pressure than is true for a gas. Secondly, whereas the gaseous state is characterized by disorder, in that molecules are in constant motion, the solid state consists of atoms arranged in fixed, ordered positions. This regularity of arrangements of atoms makes the detailed study of solids possible. These arrangements throw light upon many observed properties. There is a vast variety in solid structures. As a matter of fact, different solids have fewer properties in common than do gases. In this sense the solid state is not as simple as the gaseous state. The important macroscopic properties of a solid are rigidity, characteristic geometry and extremely low compressibility.

Another marked difference between solid and gaseous state is found in the diffusion rates. Any gas can diffuse through another very quickly, whereas many solids diffuse very slowly. One of the principal reason for slow diffusion of solids is that they are imperfect, having, for example, occasional vacancies where atoms or molecules should reside. Movement into these vacancies permits diffusion to occur. For many purposes, the existence of these defects can be ignored, but for certain properties such as diffusion, conductivity and mechanical strength, their presence can be decisive. A proper understanding of the nature and properties of solids provides the basis for developing new tailor-made materials with desired properties. Some special solids have industrially important applications in various fields of life.

In general, we come across two types of solids (*i*) crystalline solids and (*ii*) amorphous solids.

13.1. CRYSTALLINE SOLIDS

In these, atoms (or ions) or molecules are in highly ordered arrangement. This particular arrangement is responsible for their characteristic properties such as incompressibility, rigidity, definite volume, very slow diffusion, crystalline state and high melting points.

Crystals of a substance are bound by plane surfaces called *faces*. It has been said that the beauty of crystal lies in plainness of faces. These faces intersect at angles characteristic of the crystalline substance. Some of the crystals are shown in Fig. 13.1. On splitting up of a crystal, it has been found that it splits or cleaves along preferred directions. Many substances crystallise in more than one form and this phenomenon is referred to as *polymorphism or allotropy,* in Greek, *polymorph* means many forms. For example, sulphur exists in two crystalline forms (prismatic and rhombic octahedral crystals).

The properties of the crystals, such as electrical, mechanical and others, generally depend on the direction along which they are measured. Such a behaviour is termed *anisotropy*. Thus, crystals exhibit anisotropy and it is a clear indication of the presence of ordered lattices in crystals.

Sharp melting point is also one of the important properties of the crystalline solids.

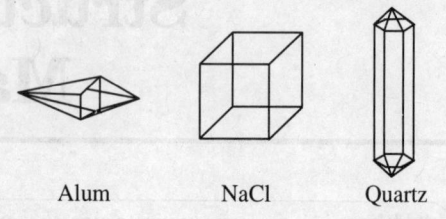

Alum   NaCl   Quartz

Fig. 13.1. Structures of some crystals.

13.2. AMORPHOUS SOLIDS

Amorphous solids resemble liquids in that they lack the extended ordering characteristic of crystalline solids. A good example of an amorphous solid is glass. The word glass is a general term used to describe many amorphous solids. It describes a rigid non-crystalline state. As might be expected, the X-ray picture of a glass is quite different from that of a crystal and resembles that of a liquid. Instead of a spot pattern, X-ray picture shows concentric rings. The existence of these rings indicates that there is a certain amount of order, but it is far from perfect. Another indication that glasses do not have a long-range ordering of atoms is their behaviour on being broken. Instead of showing cleavage with formation of flat faces and characteristic angles between faces, glasses break to give conchoidal or shell-like depression (as observed in the broken edges of glass).

An interesting class of amorphous solids are the plastics, such as bakelite, polyethylene, polystyene and nylon. These materials are rubber-like, fibery or brittle. Amorphous substances do not possess sharp melting points and do not have definite geometrical shapes though they are rigid and incompressible.

Amorphous solids do not melt sharply, but show gradual softening until they start to flow. Amorphous materials are generally isotropic *i.e.* their properties do not depend on the direction of measurement.

13.3. TYPES OF SOLIDS

On the basis of nature of binding forces present between units (ions or atoms or molecules) in crystalline solids, they are broadly divided into four types

1. Ionic solids
2. Covalent solids
3. Molecular solids
4. Metallic solid

This is an arbitrary classification. The important features of these four types of crystals are summarized in Table 13.1.

STRUCTURE OF SOLIDS

Table 13.1. Types of solids and their properties

	Type	Units present in the solid	Major binding force	Properties	Examples	App. Energy required to separate units (K-cals)
1.	Covalent	Atoms	Shared electrons	Hard, High melting, Non-conducting	C (diamond) Si	170 105
2.	Ionic	Positive and negative ions	Electrostatic attraction	Brittle, High melting, Insulating	LiF NaF NaCl AgCl	217 216 184 216
3.	Molecular	Molecules	Van der Waals, Dipole-dipole	Soft, Low Melting, Volatile, Insulating	Ar CO_2 CH_4 Cl_2	1.6 6.0 2.0 4.9
4.	Metallic	Positive ions in a 'sea' of electrons (electron gas)	Electrostatic attraction between positive ions and delocalised electrons	Low to high Melting, Highly Conducting	Li Al Fe W	38 77 99 200

Ionic Crystals

The units of this type of crystals are ions. The forces which hold these units are coulombic in nature. Some examples of ionic crystals are sodium chloride, caesium chloride and calcium chloride. In NaCl, each Na^+ is surrounded by six Cl^- ions and each Cl^- ion is surrounded by six Na^+ ions. The number of ions surrounding the oppositely charged ions is determined by the relative size and the charge of the ions. Sodium chloride in true sense is a giant molecule and it may rightly be called as an *ion pair molecule*. As the ions cannot move in the solid state, ionic compounds cannot conduct electricity. However, in aqueous solution or in the molten form, they are good conductors of electricity due to mobility of ions.

As the electrostatic forces are quite strong, the ionic solids are expected to be usually quite hard, having very low vapour pressure and high melting points. If a shearing stress be applied on an ionic crystal lattice, it causes a shift of one layer of ions over the other along a plane. In this situation two layers attain a position of mutual repulsion due to similar charges facing each other. Such a repulsion causes a break in the crystal.

Covalent Crystals

Atoms in a covalent molecule like methane or hydrogen are held by covalent bonds. In covalent solids, atoms are joined by covalent linkages forming a giant network. The units in these networks can either be atoms of the same elements or atoms of different elements with similar electronegativities. Carbon, silicon, silicon carbide (SiC), graphite, borazon (BN) are examples of covalent solids. In diamond (Fig. 13.2a), each carbon atom is linked to four other carbon atoms through the four tetrahedrally oriented covalent bonds (carbon having sp^3 hybridization), thus forming a three-dimensional network of carbon-atoms. Silicon also has four valencies directed to the vertices of a tetrahedron. SiC has the same geometrical arrangement as diamond, the only difference being that each silicon atom is surrounded by four carbon atoms and vice versa. Graphite is an allotropic form of carbon. In it, atoms are packed in a layer structure (Fig. 13.2b) and the layers are held together by weak van der Waals forces of attraction. Boron nitride, BN, has also the layer structure like that of graphite, but under pressure and temperature, the diamond form can be prepared. Since covalent bonds are strong and directional, this type of solids are fairly hard and resist deformation.

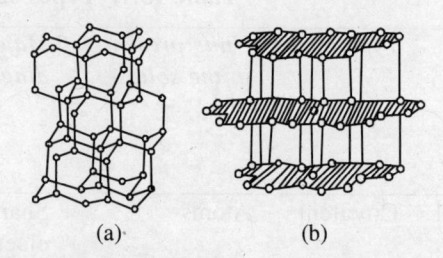

Fig. 13.2. Polymorphic forms of Carbon.
(a) Diamond structure.
(b) Graphite structure.

As mentioned above, in graphite, layer structure exists because of the weak forces between the layers, one layer may be made to slip past another layer. This property makes graphite quite soft and it can act as a lubricant. Cadmium chloride is another example of layer lattice structure.

Molecular Solids

In this type of solid, the molecule is the structural unit. In such solids a weak type of attraction, known as the Van der Waals force holds the atoms or molecules. The molecular crystals are of two types
  (i) Polar molecules
  (ii) Non-polar molecules

The polar molecular solids, e.g., ice, are formed from unsymmetrical molecules containing polar covalent linkage and have comparatively higher melting and boiling points than the non-polar molecular solids. This is so because in the former there are stronger dipole-dipole interactions. These solids have large thermal expansion and are also translucent but in the liquid form they are transparent to light. In many polymeric solids, strands or chains of polymers are held together by Van der Waals forces with other chains.

Molecular crystals are soft with low melting points and higher vapour pressures.

Metallic Solids

In this type, units are positive metal ions dispersed in electron gas. Since strong electrostatic attraction exists between the positive metal ions and electrons, these units are strongly bound. Consequently, their melting and boiling points are high. They are good conductors of heat and electricity because of the easy flow of electrons.

There is a close relationship between the physical properties of a solid and its structure and chemical composition. Some of these properties are exploited for various new innovations in electronic and magnetic devices such as transistors, computers, telephones etc. The information explosion in inorganic and organic solid state materials has been tremendous in recent years.

13.4. STRUCTURE OF SOLIDS (LEVELS OF STRUCTURE)

The cohesion of metals is due to an electrostatic attraction between the positive cores of the metal atoms and the negative fluid of mobile electrons. A simplified model of a one-dimensional structure can be considered as shown in Fig. 13.3. For this purpose, nuclei of sodium with charge of +11 can be taken up. The position of each sodium nucleus represents a deep potential energy well for the electrons. If these wells were fare apart, the electrons would fall into fixed positions on the sodium nuclei, giving rise to $1s^2\ 2s^2\ 2p^6\ 3s^1$ configurations, which are representative of isolated atoms. This situation is shown in Fig. 13.3(a). In the metals however, the potential wells are not far apart, and not infinitely deep. The actual situation in

STRUCTURE OF SOLIDS

this case is similar to that shown in Fig. 13.3(*b*). The electrons of sodium atoms tunnel through the barrier. An electron on one nucleus may slip through to occupy a position on a neighbouring nucleus. We are thus no longer concerned with levels of individual sodium atoms but with levels of the crystals as a whole. According to Pauli exclusion principle no more than two electrons can occupy exactly the same energy level. Therefore, once the possibility of electrons moving through the structure is accepted, we can no longer consider the energy levels to be sharply defined. The sharp 1s energy level in an isolated sodium atom is broadened in crystalline sodium into a bond of closely packed energy levels. A similar situation arises for the other energy levels, each becoming a band of levels as shown in Fig. 13.3(*b*).

Each atomic orbital contributes one level to a band. In the lower bands (1s, 2s, 2p), there are just enough levels to accommodate the number of available electrons, so that the bands are completely filled. If an external electric field is applied, the electrons in the filled bands cannot move under its influence. If these electrons are to be accelerated by the field, they would have to move into somewhat higher energy levels. This is impossible for electrons in the interior of a filled band, because all the levels above them are already occupied and the Pauli principle denies their accepting additional electrons. Nor can the electrons at the very top of a filled band acquire extra energy, since there are no higher levels for them to move into. It is a fact that occasionally an electron may receive a jolt of energy and be knocked completely out of its band into a higher unoccupied band.

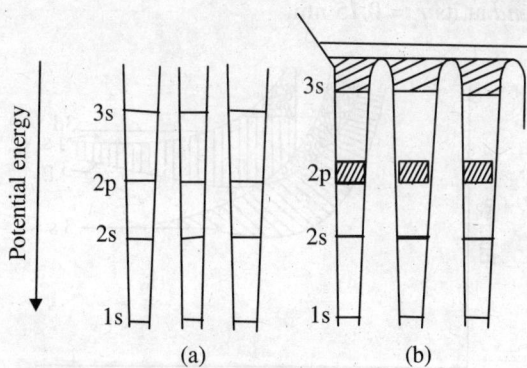

Fig. 13.3. Energy levels in sodium,
(*a*) Isolated atoms
(*b*) Section of a crystal. The sharp levels in the atom have become bonds in the crystal.

The situation is same for the electrons in the lower bands, but is different in the uppermost band, the 3s, which is only half-filled. An electron in the interior of the 3s band still cannot be accelerated because the levels directly above are already filled. Electrons near the top of the filled zone, however, can readily move up into unfilled (or empty) levels within the band. The topmost band has actually broadened sufficiently to overlap the peaks of the potential energy barriers, so that electrons in the top upper levels can move quite freely throughout the crystal structure. According to this idealized model, the nuclei are always arranged at the points of a perfectly periodic lattice. As a result there is no resistance to the flow of an electric current. The actual resistance is arised due to the departure from perfect periodicity. The loss of periodicity results from the thermal vibrations of the atomic nuclei. These vibrations destroy the perfect resonance between the electronic energy levels and thus cause a resistance to the flow of electrons. It is clear that the resistance from this mechanism increase with temperature. The resistance is found to increase when an alloying constituent metal is added to a pure metal and regular periodicity of the structure is decreased by the foreign atoms.

Thus for a univalent metal like sodium, an attractive picture is considered, but what of magnesium with its two 3s electrons and therefore completely filled 3s band? Why is it not an insulator instead of a metal? Detailed calculations have shown that in such cases, the 3p band is low enough to overlap the top of the 3s band, providing a large number of empty levels. Actually this happens for the alkali metals also. The way in which the 3s and 3p bands in sodium broaden as the atoms are brought together is shown in Fig. 13.4.

The interatomic distance in sodium at 1 atm pressure and 298 K is $r_e = 0.38$ nm. At this distance, there is no longer any gap between the 3s and 3p bands. In the case of diamond, on the other hand, there is a large energy gap between the filled *valence band* and the empty *conduction band* at its $r_e = 0.15$ nm.

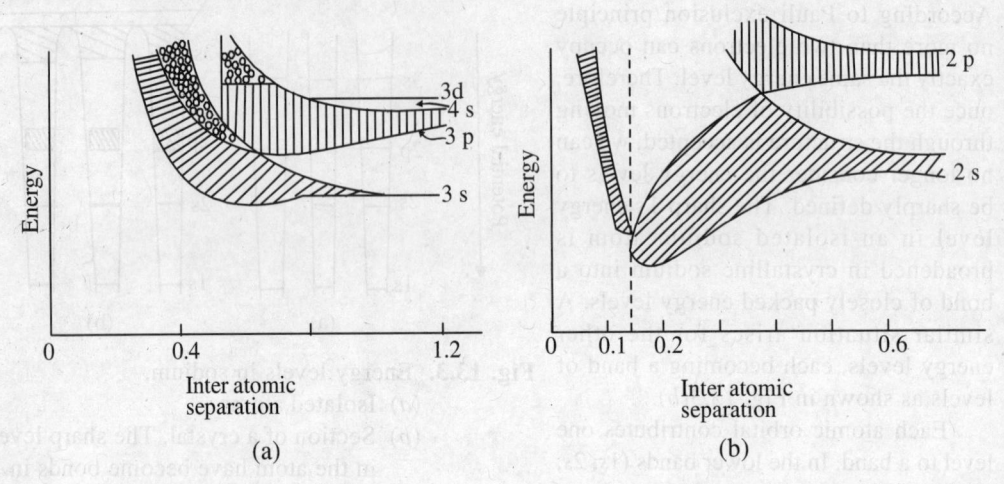

Fig. 13.4. Formation of energy bands, as atoms are brought together into a crystal (*a*) sodium, (*b*) diamond.

Thus, conductors are characterized either by partial filling of the bands or by overlapping of the topmost bands. Insulators have completely filled lower bands. These modes are represented in Fig. 13.5.

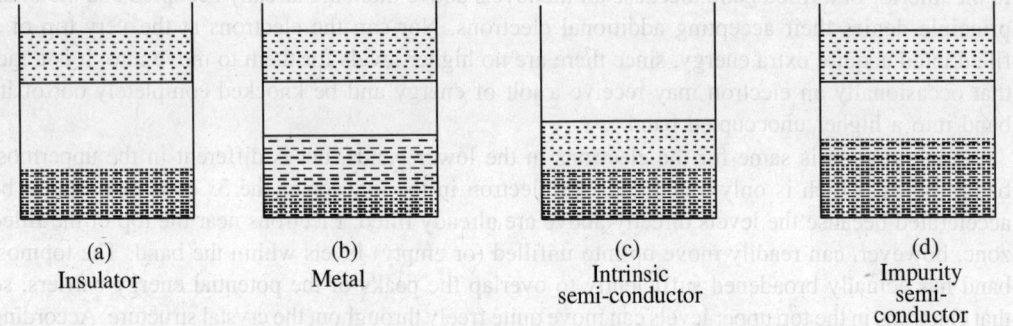

(a) Insulator (b) Metal (c) Intrinsic semi-conductor (d) Impurity semi-conductor

Fig. 13.5. Schematic bond models of solids classified according to electronic properties.

Semi-conductors

Solids are classified on the basis of their thermal conductivity into three types.

1. Metals or conductors

(which offer a low resistance to the flow of electrons when a potential difference is applied. The resistivities of metals lie in the range 10^{-6} to 10^{-8} Ω m at room temperature and increase with temperature).

2. Insulators

(which have high resistivities, from 10^8 to 10^{20} Ω m at room temperature).

3. Semiconductors

(The resistivities of these are intermediate between those of typical metals and typical insulators and decrease with rising temperature).

Electrical conductivity of solids may arise through the motion of electrons and holes (electronic conductivity) or other charged imperfections and ions (ionic conductivity). Substances like pure alkali halides (KCl, NaCl, etc.) where the conduction is only through ions, are generally insulators since the mobilities of ions are very much lower than those of holes or electrons. Conductivity of ionic solids increases because of presence of vacancies or other imperfections. Conductivity of metals generally decreases with increase in temperature while that of semiconductors increases with temperature.

At laboratory temperature, there is no effect of impurities present in metals on their conductivity. The electron concentration is mainly determined by the nature of metal (number of valence electrons in the metal) and the mobility by the lattice vibrations. At low temperatures, however, lattice vibrations are quite insignificant and the conductivity should be infinite. However, this is not true because of the presence of lattice imperfections and impurities. It is thus convenient to use the resistance ratio $\rho\ 300\ K/\rho 4.2\ K$ as a measure of purity of the metals.

Conductivity of *semiconductors* and insulators is prominently determined by the impurities and defects. Electrons and holes produced by the ionization of defects contribute to the electronic conduction in these solids. The behaviour of semiconductors and insulators can very well be explained in terms of *band model*. If the band structure of an element is such that the fully occupied band is separated by a large energy gap from the vacant band, then conduction is not possible. This situation is found in good insulators like diamond or MgO. On the contrary, if the energy gap is small, promotion of electrons can take place by raising the temperature – the higher the temperature greater is the number of electrons promoted and greater is the conductivity. This is the characteristic behaviour of semiconductors (*e.g.*, SnO_2, $SrTiO_3$). In fig. 13.6, the schematic band structures of metals, insulators and semiconductors are shown.

Empty conduction band
Forbidden energy gap
Filled valence band

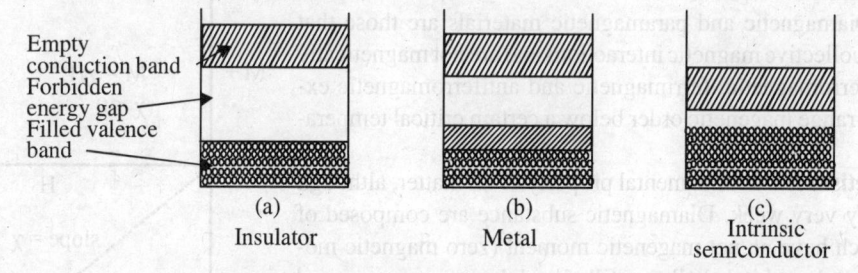

(a) Insulator
(b) Metal
(c) Intrinsic semiconductor

Fig. 13.6. Schematic band structure of (*a*) metals, (*b*) insulaters and (*c*) semi-conductors.

When the thermally produced electrons and holes are truely responsible for the conductivity of a semi-conductor (generally at higher temperature) rather than the electrons and holes produced by donors and acceptors (as in doped Ge or Si), the region is called the intrinsic region. At lower temperatures where the conductivity is mainly determined by the concentration of donors and acceptors, the region is called extrinsic. It should be remembered here that pure germanium or silicon by itself would be insulator; only by doping with a group III or V impurity can we increase the conductivity at ordinary temperature.

Apart from electronic semiconductors where band picture is applicable, there are various materials where the semiconduction is due to hopping of electrons from one site to another. Examples of these are oxides where cations of more than one valence state are present as in Pr_6O_{11}, NiO doped with Li_2O, etc.

Ionic conduction in a solid takes place through migration of atoms, ions or other charged species under applied field. This ion migration becomes possible because of the presence of vacancies or/and interstitials in the ionic solids. Conductivity measurements in ionic solids *e.g.*, NaCl give valuable information on the energy required for the creation of defects, as well as their migration energies.

13.5 MAGNETIC MATERIALS :

Magnetism is one of the phenomena by which materials exert attractive or repulsive force on other materials. Substance of transition metals and lanthanides which posses unpaired d and f electron shows magnetic behaviour.

When a substance is planed in an inhomogeneous magnetic field, it is either attracted toward the strong part of the field or repelled toward the weaker part. If it is attracted by the field, it is said to be paramagnetic if repelled, it is said to be diamagnetic. The force by which diamagnetic substance is repelled when placed in a field of strength H and gradient dH/dx is determined by

$$F = \chi VH \frac{dH}{dx}$$

Where χ is the magnetic susceptibility The degree of magnetization of a material in response to an applied field. and V is the volume of the substance. The magnetic susceptibility is a measure of the change in the magnetic moment of the atoms caused by an applied field.

Some materials are much more magnetic than others. The main distinction is that in some materials there is no collective interaction of magnetic moment, where as in other materials there is a very strong interaction between atomic moments. The magnetic behaviour of material can be classified into the following five groups

i) Diamagnetism
ii) Paramagnetism
iii) Ferromagnetism
iv) Ferrimagnetism
v) Antiferromagnetism

Diamagnetic and paramagnetic materials are those that exhibit no collective magnetic interactions and are not magnetically ordered. Ferromagnetic, ferrimagnetic and antiferromagnetic exhibit long -range magenetic order below a certain critical temperature.

Diamagnetism it is a fundamental property of all matter, although it is usually very week. Diamagnetic substance are composed of atoms which have no net magenetic moment (zero magnetic moment) i.e. all the orbital shells are filled and there are no unpaired electrons all the electrons are paired ex. He, N_2). However when exposed to a field, a negative magnetization is produced and thus the susceptibility is negative. When the field is zero the magnetization is zero. Diamagnetic substance include in unit of $10^{-8} m^3/kg$, are Quartz (SiO_2) - 0.62, Calcite ($CaCO_3$) 0.48, Water - 0.90.

Paramagnetism

In this class of materials, some of the atoms, ions or molecules in the materials have a net magnetic moment due to elctron spin A property of materials containing unpaired electron in partially filled orbitals.

These materials are attracted by magnetic field. They lose their magnetism in the absence of a magnetic field.

Individually magnetic moments do not interact magnetically the magnetization is zero when the field, is removed. In the presence of a field, there is a partial alignment of the atomic magnetic moments in the direction of the filed resulting in a net positive magnetization (the magnetic dipole moment per unit volume) and

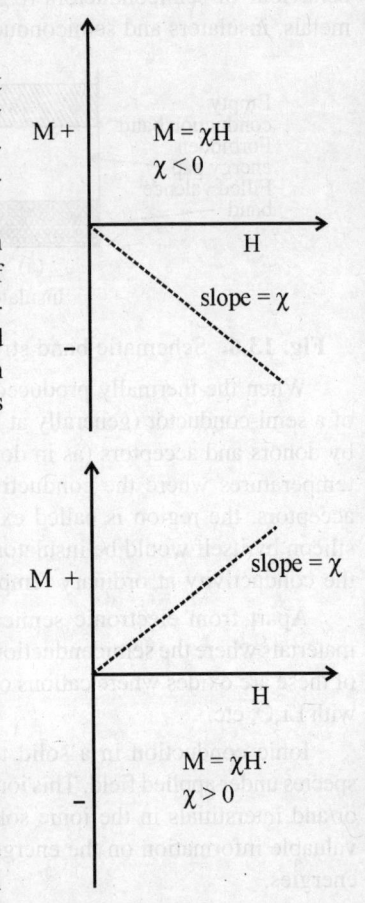

$M = \chi H$
$\chi < 0$

slope = χ

slope = χ

$M = \chi H$
$\chi > 0$

STRUCTURE OF SOLIDS

positive susceptibility.

The paramagnetism increases with the increase in the number of unpaired electrons. The paramagnetism of a substance is expressed in terms of its magnetic moment (μ). The magnetic moment is expressed in Bohr Magnetons (B.M.). Paramagnetic substance have net magnetic moment.

Magnetic moment can be calculated from the unpaired spin. Transition substance contains electron in their 3rd orbitals. The magnetic moment of one unpaired electron is equal to the sum of its magnetic moment due to its orbital motion called orbital magnetic moment (μ_s) and due to its spin motion called spin magnetic moment (μ_s).

1 B.M. = 9.2774×10^{-21} ergs/gauss = 9.274×10^{-24} JT^{-1} (Joules / Tesla)
Where, e - electron charge
      h - planks constant
      m - electron mass
      c - velocity of light

In many compounds the transition metal ions of 3d series, the electric fields of the surrounding atom may restrict the orbital motion of the electrons and hence the magnetic moment due to the orbital motion of the electron ((μ_l)) is partially vanishes $\mu_l = 0$, so the magnetic moment is entirely due to the spin of the parent ions (μ_s).

S for one electron is ½, and g is the gyromagnetic ratio = 2.0023 ~ 2.0

If a metal ion of 3d series element has 2, 3, 4 ... n unpaired electron, in its 3d Orbitals, the value of magetic moment for n unpaired electron called effective magnetic moment μ_{eff} will be obtained by putting the sum of the spin quantum numbers of n unpaired electrons.

The value of μ_{eff} depends on number of unpaired electrons present in 3d orbitals. Greater the number of unpaired electrons present in 3d orbitals, greater is the value of μ_{eff}.

	No. of unpaired electron	μ_{eff}, B.M.
$Zn^{+2} = 3d^{10}$	0	0
$Cu^{+2} = 3d^{9}$	1	1.79
$Ni = 3d^{8}$	2	2.84

A material is said to be ferromagnetic if all of its magnetic ions add a positive contribution to the net magnetization.

$$\mu = \mu_l + \mu_s$$

$$\mu_\ell = \sqrt{\ell(\ell+1)} \; \frac{eh}{4\pi \, mc}$$

$$\mu_s = g\sqrt{S(S+1)} \; \frac{eh}{4\pi \, mc}$$

$$\mu = \left[\sqrt{\ell(\ell+1)} + g\sqrt{S(S+1)}\right] \frac{eh}{4\pi \, mc}$$

$$\mu = \left[\sqrt{\ell(\ell+1)} + g\sqrt{S(S+1)}\right] B.M.$$

Where, B.M. is called Bohr Magneton Which is equal to eh / 4πmc

$$\mu = g\sqrt{s(s+1)} \ B.M.$$

$$\mu = 2\sqrt{\tfrac{1}{2}(\tfrac{1}{2}+1)} \ B.M.$$

$$\mu_{eff} = 2\sqrt{\tfrac{n}{2}(\tfrac{n}{2}+1)} \ B.M. \qquad \mu_{eff} = \sqrt{2(n+2)} \ B.M.$$

Ferromagenetism : It is like paramagnetism and it is also due to unpaired electrons. It is a bulk property of material. Ferromagnetic material are very important in technology. The property of ferromagnetism is due to the direct influence of two effects spin of an elctron and orbital angular momentum which gives the atom a resultant magnetic moment.

The atomic moment in these materials shows very strong interactions. These interactions are produced by electronic exchange forces and result in a parallel or antiparallel aligment of atomic moment. Ferromagnetic materials shows parallel alignment of moments resulting in large net magnetization even in the absence of magnetic field eg. the elements Fe, Ni, Co and their alloys are ferromagnetic. Ferromagnetic substances have vary large magnetic permeabilities which varies with the strength of the applied field. A given ferromagnetic substance loses its ferromagnetic properties at a certain critical temperature called Cuire temperature (Tc). Below the curie temperature the ferromagnet is ordered and above it disordered. The saturation magnetization goes to zero at the curie temperature.

Ferromagnets can retain a memory of an applied field once it is removed. This behaviour is called hysteresis and a plot of the variation of magnetization with magnetic field is called a hysteresis loop. Another hysteresis property is the coercivity of remanence (Hr.) This is the reverse field which, when applied and then removed, reduces the saturation remanence to zero. It is always larger than the coercive force. The various hysteresis parameters are not only intrinsic properties but are dependent on grain size, domain state, stresses and temperature.

Ferrimagnetism In ferromagnetic materials all of its magnetic ions add a positive contribution to the net magnetization. If some of the magnetic ions subtract from the net magnetization (if they are partially antialigned) then the materials is ferrimagnetic.

In these materials the magnetic moments of adjacent atoms are antiparallel, but of unequal strength i.e. some

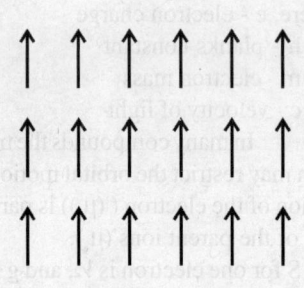

Ferromagnetic ordering
(Parallel alignment)

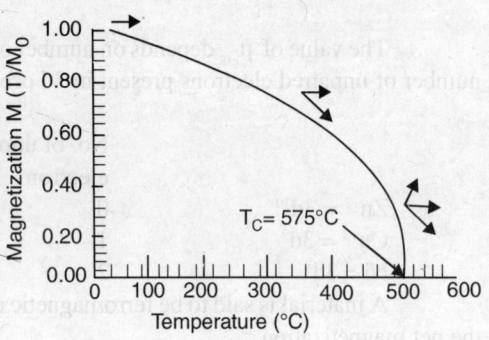

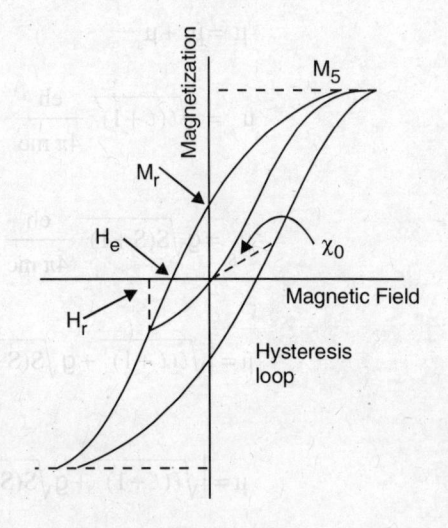

Hysteresis loop

STRUCTURE OF SOLIDS

magnetic moments oriented in one direction. Ferrimagnetic material therefore have a resultant magnetization similar to that of ferromagnetisms eg. ferrites. In ferrimagnetic materials the magnetic moment of the atoms on different sublattices are opposed and opposing moments are unequal and a spontaneous magnetization remains. This happens when the sublattices consist of different ions (such of Fe^{+2} and Fe^{+3}). Ferrimagnetism is exhibited by ferrites and magnetic garnets eg. Magnetite Fe_3O_4, Yttrium, iron garnet.

↑↑↓ ↑↑↓ ↑↑↓
↑↑↓ ↑↑↓ ↑↑↓
↑↑↓ ↑↑↓ ↑↑↓

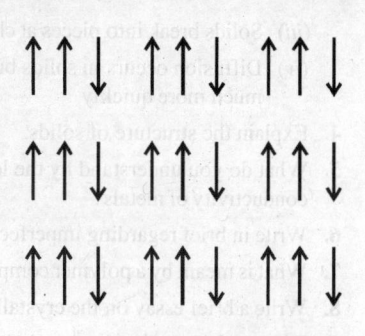

Ferrimagnetic ordering

The magnetic structure is composed of two magnetic sublattices (called A and B) separated by oxygens. The exchange interactions are mediated by the oxygen anions. When this happens, the interactions are called indirect or superexchange interaction. The strongest superexchange interaction result in an antiparallel aligment of spins between the A and B sublattice.

In ferrimagnets, the magnetic moments of the A and B sublattice are not equal and result in a net magnetic moment. Ferrimagnetism is similar to ferromagnetism. It exhibits all the standard mark of ferromagnetic behaviour spontaneous magnetization, curie temperature, hysteresis and remanence. However ferro and ferrimagenets have very different magnetic ordering. eg., Fe_3O_4 crystallizes with the spine structure. The large oxygen ions are close packed in a cubic arrangement and the smaller Fe ions fill in the gaps.

A-site : tetrahedral Fe - Fe ion is surrounded by four oxygens

B- site : Octahedral Fe - Fe ion is surrounded by six oxygens

The tetrahedral and octahedral site form the two magnetic sublattices, A and B respectively. The spins on the A sublattice are antiparallel to those on the B sublattice. The two crystal sites are different and result in complex forms of exchange interactions of the iron ions between and within the two types of sites.

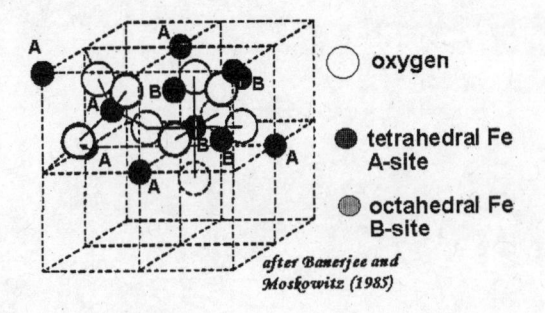

○ oxygen

● tetrahedral Fe A-site

◉ octahedral Fe B-site

after Banerjee and Moskowitz (1985)

The structural formula for magnetite is $[Fe^{+3}]$ A $[Fe^{+3}, Fe^{2+}]BO_4$

This arrangement of cations on the A and B Sublattice is called an inverse spinel structure. Fe^{+3} ions in tetrahedral and octahedral sites have antiparalled spins. The net magnetization is due to Fe^{2+} iron in octahectral sites.

Antiferromagnetism If the A and B sublattice moments are exactly equal but opposite the net moment is zero. This type of magnetic ordering is called antiferromagnetism.

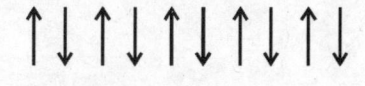

↑↓↑↓↑↓↑↓

QUESTIONS

1. What are the various properties of solids? Mention the types of solids.
2. What are the two arrangements for the closest packing in a system of uniform spheres?
3. Account for the following facts:
    (*i*) Solids are practically incompressible whereas gases can be easily compressed.
    (*ii*) Solids maintain their definite shape but the liquids flow.

(iii) Solids break into pieces at characteristic angles.
(iv) Diffusion occurs in solids but extremely slowly as compared with liquids and gases which diffuse, much more quickly.
4. Explain the structure of solids.
5. What do you understand by the levels of structure of solids. How it accoutns for conductivity and non-conductivity of metals?
6. Write in brief regarding imperfections in solids.
7. What is meant by a polymer compound? Discuss the importance of polymer compounds in day to day life.
8. Write a brief essay on the crystallinity of polymer compounds.
9. Discuss the various characteristic features of high polymers.
10. Explain:
    (a) Thermosetting polymers
    (b) Addition polymers
    (c) Condensation polymes.

14

Mechanical Properties

> *"The fabrication and applications of the various engineering materials depend mainly on their mechanical properties."*

A vast majority of engineering materials are in use in various walks of life. The fabrication and applications of such materials depend mainly on their mechanical properties such as strength, hardness, elasticity, etc. These properties in a large measure determine their behaviour under applied forces and loads. The mode of behaviour of the materials also depends on the type of bonding and structural arrangement of atoms or molecules in them. Manufacturing processes and operations, different types of stresses and mode of application bring variation in the properties, even in the materials of the same chemical composition.

A selection of material for a particular structure depends on its mechanical properties.

The engineering materials when subjected to applied forces undergo three basic types of deformation such as elastic, plastic and viscous. As per the nature of mechanism involved in their deformations, engineering materials are divided into three main types as

1. Elastoplastic materials — Structural metals
2. Viscoelastic materials — Plastics, rubbers, glasses, concrete and other amorphous bodies
3. Elastic materials — Ionic and covalent crystals.

Numerical data of mechanical properties may be obtained from some standard tests which are useful to designers, fabricators and research workers.

The capacity of a material to withstand a static load can be determined by testing that material in *tension* or *compression*. Information about its resistance to permanent deformation can be gained from *hardness tests*. *Impact* tests are used to indicate the *toughness* of a material under shock loading conditions. Over series of temperature measurements, transition from ductility to brittleness can be known. *Fatigue tests* measure useful time of a material under a cyclic load. Creep and *stress rupture* tests are carried to reveal the behaviour of a material when subjected to a load at high temperature for a long time.

In this chapter, some important mechanical properties of materials will be dealt with **Elastic behaviour**

14.1. STRESS AND STRAIN

Any force (or load) applied on the material establishes a stress and strain in it. Stress is measured by the force acting per unit area of a plane. It is expressed in lb/in^2. This is known as a unit stress (psi).

$$\sigma \text{(Stress)} = F/A$$

where F is the force and 'A' is the area.

Forces or loads acting on the body may be static or dynamic according to the mode of application. Static forces (stresses) remain essentialy constant or change slowly without exhibiting any repetitive characteristics.

Dynamic forces can be impact forces, alternating forces and reverse forces.

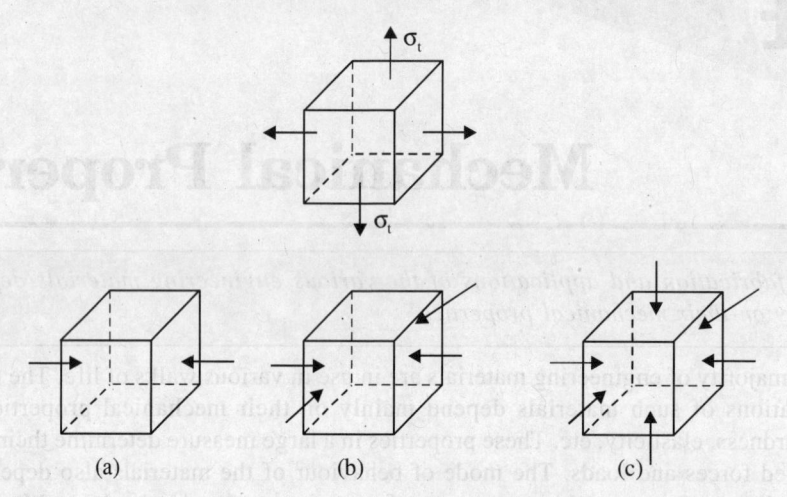

Fig. 14.1. States of Stress (*a*) uni-axial, (*b*) bi-axial, and (*c*) tri-axial.

The alternation in the shape or dimensions of a body resulting from stress is called *strain or deformation*. Strain is expressed in dimensionless units such as in/in or in percentage. Since there are three main types of stress, strain of tensile, compressive and shear character can be distinguished. Tensile strain is expressed as elongation per unit length.

$$\epsilon_t = \frac{DL}{L_0}, \quad \begin{array}{l} L_0 = \text{initial length} \\ L = \text{length after elongation} \\ \Delta L = \text{elongation produced} \end{array}$$

where $\Delta L = L - L_0$ and ϵ_t is the deformation in in/in of the original length. This is called the unit strain or simply strain.

Compressive strain is measured by the ratio of the contraction to the original length

$$\epsilon_t = \frac{L_0 - L}{L_0} = \frac{\Delta L}{L_0}$$

It can also be seen that tensile stress causes a contraction perpendicular to its own direction, whereas compressive stress cause an elongation perpendicular to its own direction.

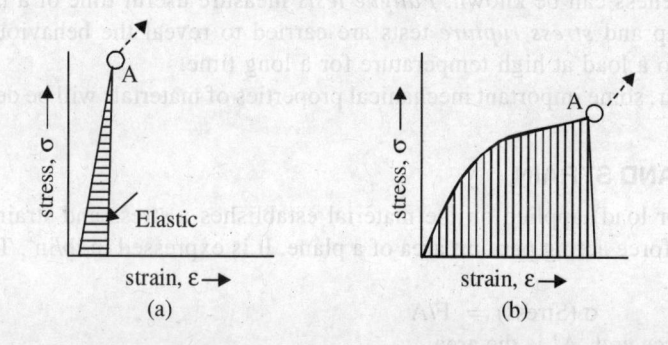

Fig. 14.2. (*a*) Linear, (*b*) non-linear elastic behaviour. The point A on each curve represents end of the elastic region.

MECHANICAL PROPERTIES

A change in shape, volume or both of a material takes place under the influence of an applied stress or a temperature change. This change is known as deformation. It is elastic if strain induced in them by a given stress (or by a temperature change) disappears when the stress is removed. The relationship between stress and strain in the elastic region is linear in some materials (crystalline) whereas non linear in others (non-crystalline, long chain molecular materials). This relationship between stress and strain may be correlated qualitatively with the structure and type of atomic bonding present. With regard to this, two graphs of stress (σ) versus strain (ϵ) are shown in Fig. 14.2.

The equation for linear portion of the tensile stress-strain (Fig. 14.2a) is $\sigma = E\epsilon$, where E is proportionality constant, *Young's modulus*. The value of Young's modulus may be determined by other methods. For example, if v is the velocity of sound in a material of density ρ, and Young's

$$v = \sqrt{\frac{E}{\rho}}$$

Several different elastic proportionality constants are in common use. They differ only in the type of stress and strain which they relate.

Young's module $\qquad E = \dfrac{\sigma}{\gamma}$

Shear modulus $\qquad G = \dfrac{\tau}{\gamma}$

Bulk modulus $\qquad K = \dfrac{\sigma_{Hyd}}{\Delta V / V_0}$

In the above equation, σ = uniaxial tensile or compressive stress

τ = shear stress
σ_{Hyd} = hydrostatic tensile or compressive stress
ϵ = normal strain
γ = shear strain
$\dfrac{\Delta V}{V_0}$ = fractional volume expansion or contraction

Fig. 14.3 below illustrates the geometrical relationship between stress and strain of the types described in the three equations above.

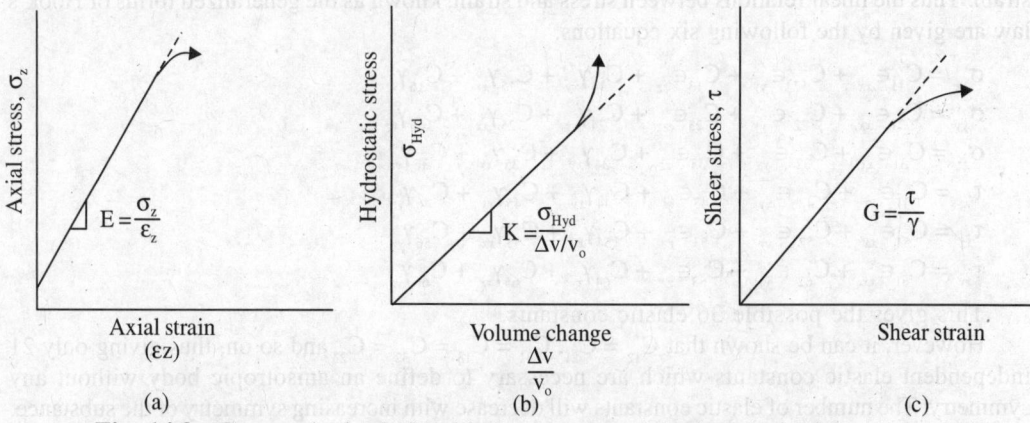

(a) Axial stress, σ_z vs Axial strain (ϵz): $E = \dfrac{\sigma_z}{\epsilon_z}$

(b) Hydrostatic stress σ_{Hyd} vs Volume change $\dfrac{\Delta v}{v_o}$: $K = \dfrac{\sigma_{Hyd}}{\Delta v/v_o}$

(c) Sheer stress, τ vs Shear strain: $G = \dfrac{\tau}{\gamma}$

Fig. 14.3. Geometrical relationship between stress and strain in three types.

14.2. HOOKE'S LAW

A material is termed elastic when the deformation produced in the body is wholly recovered after removal of the force. The relation between stress and the corresponding strain in the elastic range of the material is governed by Hook's law. It states that stress τ is proportional to strain ϵ and independent of time. Mathematically it is given by

$$\sigma \propto \epsilon$$

This law is generally applicable to most elastic materials for very small strains. In case of isotropic material (a material having same properties in one direction only) each stress will produce corresponding strain, but in case of anisotropic material a single stress component may produce more than one type of strain in the material.

Suppose a volume element as shown in Fig. 14.4 is subjected to a stress.

This stress can be resolved into nine force constants, three of each perpendicular to the three axis of the cube (Fig. 14.4). There are three normal forces perpendicular to the three faces of the cube; the other six are tangential forces or shear forces.

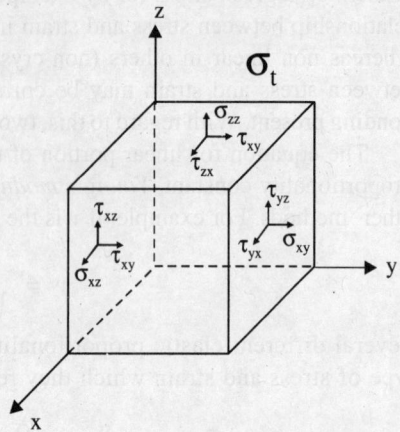

Fig. 14.4. The distribution of shear and normal stresses, acting on a volume-element.

$\sigma_{xx}\ \tau_{xy}\ \tau_{xz}$ stress components for forces acting in x direction
$\sigma_{yy}\ \tau_{yx}\ \tau_{yz}$ stress components for forces acting in y direction
$\sigma_{zz}\ \tau_{zx}\ \tau_{zy}$ stress components for forces acting in z direction

The force producing the shear stress τ_{yx} would tend to rotate the cube in one direction and the force producing the shear stress τ_{xy} would rotate the cube in the opposite direction, so that at equilibrium $\tau_{yx} = \tau_{xy}$. Similarly $\tau_{yz} = \tau_{zy}$ and $\tau_{xz} = \tau_{zx}$. Thus the stress can be completely determined (specified) by only six independent components. Out of these, three are normal stresses, $\tau_{xy}, \tau_{yz}, \tau_{zz}$. The normal stress is positive in tension and negative in compression.

For each stress component there is a corresponding strain component. Hence we have three normal strains, $\epsilon_{xx}, \epsilon_{yy}$ and ϵ_{zz} and three shear strains γ_{xy}, γ_{yz} and γ_{xz}. In anisotropic materials at pure stress σ_{xx} does not necessarily produce a pure strain ϵ_{xx} but it may also produce any other type of strain. Thus the linear relations between stress and strain, known as the generalized forms of Hook's law are given by the following six equations.

$$\sigma_{xx} = C_{11}\epsilon_{xx} + C_{12}\epsilon_{yy} + C_{13}\epsilon_{zz} + C_{14}\gamma_{xy} + C_{15}\gamma_{xx} + C_{16}\gamma_{yz}$$
$$\sigma_{yy} = C_{21}\epsilon_{xx} + C_{22}\epsilon_{yy} + C_{23}\epsilon_{zz} + C_{24}\gamma_{xy} + C_{25}\gamma_{xx} + C_{26}\gamma_{yz}$$
$$\sigma_{zz} = C_{31}\epsilon_{xx} + C_{32}\epsilon_{yy} + C_{33}\epsilon_{zz} + C_{34}\gamma_{xy} + C_{35}\gamma_{xx} + C_{36}\gamma_{yz}$$
$$\tau_{yz} = C_{41}\epsilon_{xx} + C_{42}\epsilon_{yy} + C_{43}\epsilon_{zz} + C_{44}\gamma_{xy} + C_{45}\gamma_{xx} + C_{46}\gamma_{yz}$$
$$\tau_{xz} = C_{51}\epsilon_{xx} + C_{52}\epsilon_{yy} + C_{53}\epsilon_{zz} + C_{54}\gamma_{xy} + C_{55}\gamma_{xx} + C_{56}\gamma_{yz}$$
$$\tau_{xy} = C_{61}\epsilon_{xx} + C_{62}\epsilon_{yy} + C_{63}\epsilon_{zz} + C_{64}\gamma_{xy} + C_{65}\gamma_{xx} + C_{66}\gamma_{yz}$$

This gives the possible 36 elastic constants.

However, it can be shown that $C_{12} = C_{21}, C_{31} = C_{13} = C_{32} = C_{23}$ and so on, thus giving only 21 independent elastic constants which are necessary to define an anisotropic body without any symmetry. The number of elastic constants will decrease with increasing symmetry of the substance. For example, for orthorhombic symmetry 9 elastic constants are necessary, for tetragonal symmetry 6, for hexagonal 5 and for cubic 3. For isotropic bodies only 2 independent elastic constants are required to specify the system completely. These elastic constants are also known as moduli of elasticity.

MECHANICAL PROPERTIES

14.3. MODULI OF ELASTICITY

From Hooke's law the relationship between stress and strain can be obtained in the form of a characteristic constant known as Young's modulus E.

Since Hook's law is

$$\text{Stress} \propto \text{Strain}$$
$$(\sigma) \quad (\epsilon)$$
$$\sigma = E \times \epsilon$$

where E = proportionality constant known as Young's modulus

The values of Young's modulus can be determined by other methods.

The important points regarding this relationship between stress and strain are

(i) It is linear in the region of materials, especially in crystalline materials and nonlinear in non-crystalline long chain molecular materials.

(ii) This relationship may be correlated qualitatively with the structure and type of atomic bonding present in a material.

There are three main types of stresses (i) tension, (ii) compression and (iii) shear. Consequently there will be three corresponding moduli of elasticity.

1. Young's modulus or the modulus of elasticity in tension, denoted by E is given by

$$E = \frac{\sigma_t}{\epsilon_t}$$

where σ_t = tensile stress
  ϵ_t = the tensile strain or longitudinal strain.

2. Bulk modulus K or modulus of compressibility is defined as the ratio of hydrostatic pressure to the relative change in volume. The resultant stress in the material is the effect of three stresses acting in three principal directions.

$$K = \frac{\text{Stress}}{\text{Volume Strain}} = \frac{\sigma}{\Delta V / V_0}$$

where σ is the stress
  V_0 = is the original volume
  ΔV = is the change in volume equal to $V_0 - V$

3. The modulus of rigidity or shear modulus, denoted by G, is given by

$$G = \frac{\tau}{\gamma}$$

where τ = shear stress, γ = shear strain.

4. An important elastic constant denoted by μ is known as Poisson's ratio. It is defined as the ratio of the lateral contracting strain to the elongation strain when a rod is stretched by a force applied at its ends.

$$\mu = \frac{\text{Lateral strain}}{\text{Longitudinal strain}}$$

14.4. RELATION BETWEEN E (Young's modulus) and K (bulk modulus)

The following equation gives the relation between E, K and μ.

$$K = \frac{E}{3(1-2\mu)}$$

The modulus of rigidity G is related to the Young's modulus E by the equation

$$G = \frac{E}{2(1+\mu)}$$

Young's modulus can also be calculated from the bulk modulus K and the modulus of rigidity G by the relation

$$\frac{1}{E} = \frac{1}{9K} + \frac{1}{3G}$$

For majority of the soft materials such as (colloidal substances) gels, pastes and putties, K is very large as compared to G. Hence it follows from above equation that E = 3G. Such materials are supposed to be incompressible because they do not compress to the extent to which they change their shape under the stress employed. For materials like metals, fibers and certain plastics, values of K *i.e.,* the bulk modulus must be taken into account. For liquids, which cannot sustain any pressure but a hydrostatic one for any length of time, only the bulk modulus K must be considered.

14.5. SIGNIFICANCE OF MODULUS OF ELASTICITY

The elastic behaviour depends upon interatomic and intermolecular forces, therefore, the

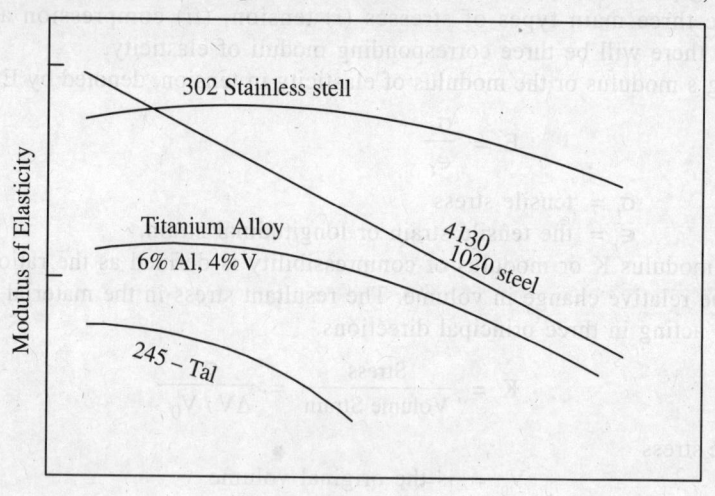

Fig. 14.5. Variation of modulus of elasticity with temperature.

value of elastic moduli vary directly with the magnitude of these forces. Thus, moduli of elasticity for covalent compounds such as diamonds are very high while they are lower for metallic crystals and ionic crystals. Relatively low values of moduli of elasticity are found for molecular amorpohous solids such as plastics, rubbers and molecular cyrstals.

14.6. VARIATION OF MODULUS OF ELASTICITY WITH TEMPERATURE

It has been observed that the moduli of elasticity decrease as the temperature increases. The elastic moduli of metals decrease at first only slowly with the increase of temperature, but near their melting points, they decrease rapidly (Fig. 14.5). The elastic moduli of plastic and other amorphous solids is affected by temperature more than those of metals. For highly anisotropic materials such as wood, nine elastic constants are required to define the system completely. Since the elastic properties of wood vary with the direction, which can be longitudinal, radial and tangential, we must distinguish three corresponding moduli of elasticity E, namely E_L (longitudinal), E_r (radial), as E_t (tangential), three Poisson's ratios and three moduli of rigidity G. The modulus E_L in longitudinal direction is much higher than in other directions, and its ratio to tangential modulus is commonly 20: 1, but it may be as high as 150: 1. Materials with laminated structure also show different properties in different directions. One important factor which influences elastic moduli and properties of wood is its moisture content.

MECHANICAL PROPERTIES

Table 14.1. Elastic moduli for various materials

Material	F 10^6 psi	G 10^6 psi	Poisson's Ratio, μ
Cast iron	16.0	7.4	0.17
Steel (mild)	30.0	11.8	0.26
Aluminium	10.0	3.6	0.33
Copper	16.0	6.4	0.36
Brass 70/30	14.5	5.3	–
Nickel (cold drawn)	31.0	11.5	0.30
Zirconium	13.6	5.2	–
Lead	2.6	0.9	0.4
Granite	6.7	2.8	0.2
Glass (soda-lime)	10.0	3.2	0.23
Alumina (sintered)	47.0	–	0.16
Concrete	1.5 – 5.5	–	0.2
Wood (longitudinal)	1.81	0.11	–
Nylon	0.4	–	–
Rubber, hard	0.4	–	–
P.V.C.	0.5	–	–

14.7. ATOMIC BASIS OF ELASTIC BEHAVIOUR

The potential energy V of a pair of atoms may be expressed as a function of distance of their separation r

$$V = -\frac{A}{r^n} + \frac{B}{r^m} \qquad \ldots(1)$$

where A and B are proportionality constants for attraction and repulsion and n and m are exponents giving the appropriate variation of V with r. Expressions for the forces of attraction and repulsion existing between the two atoms may be derived from the expression for potential energy; in the form

$$F = \frac{\delta v}{\delta r} = -\frac{nA}{r^{n+1}} + \frac{mB}{r^{m+1}} \qquad \ldots(2)$$

Letting $nA = a$, $mB = b$, $n + 1 = N$ and $m + 1 = M$,

$$F = -\frac{a}{r^N} + \frac{b}{r^M} \qquad \ldots(3)$$

Plots of the curves (Condon-Morse curves) of eqn. (1) and (3), which, clearly, have the same form, are shown in Fig. 14.6. The value of r corresponding to the minimum of potential energy is the equilibrium spacing d_0, of the two atoms. The net force is zero at d_0 and a displacement in either direction will call restoring forces into play. Although these curves describe the behaviour of an isolated atom pair, the same kind of behaviour is exhibited as a free atom approaches an existing crystal lattice: a net attractive force at first exists (potential energy decreases), which then reduces to zero (potential energy reaches a minimum) at a distance d_0, where the forces of attraction and repulsion are in balance. If the interatomic spacing were further reduced, a net repulsive force would act to restore the atoms to their equilibrium spacing. Atoms in crystal structure tend, therefore, to be arrayed in a definite pattern with respect to their neighbours.

Macroscopic elastic strain results from a change in interatomic spacing. The macroscopic strain $(l - l_0)/l_0$, in a given direction is equal to the average fractional change in interatomic spacing $(d - d_0)/d_0$, in that direction. It may easily be shown, then, that Young's modulus E is

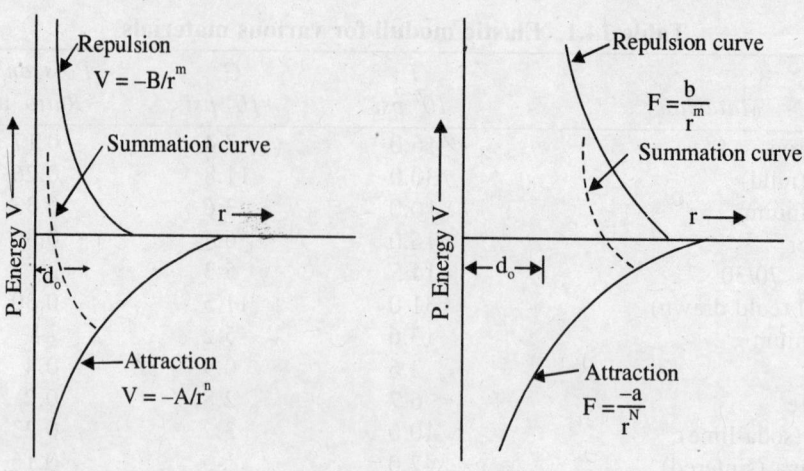

Fig. 14.6. Condon-Morse curves, showing qualitative variation of (a) energy, and (b) forces with distance of separation, r of atoms.

proportional to the slope, at d_0, of the Condon-Morse force curve or, alternatively, to the curvature of the Condon-Morse potential curve at d_0. The normal range of elastic strain in crystalline materials rarely exceeds $\pm^{1/2}$ per cent. Since, as may be seen in Fig. 14.7, the tangent $\frac{\partial F}{\partial r}$ very nearly coincides with the force curve in this area of strain, it is clear that for all practical purposes, stress is, as the theory of elasticity states, a linear function of strain. Maximum elastic strain in crystalline materials is usually very small, the stress necessary to produce this strain is, however, great. This stress-strain ratio is high because the applied stress works in opposition to the restoring forces of primary bonds (ionic, covalent, metallic). The elastic behaviour of such materials under the compression is the same as their behaviour under tension. In this case compressive stress-strain curve is merely an extension of the tensile stress-strain curve, as is shown in Fig. 14.8.

Certain non-crystalline materials, such as glass or cross-linked polymers also exhibit linear elasticity. This is due to the fact that their structure is such that distortion is opposed from the beginning by primary bonds. Other non-crystalline materials which contain intertangled long-chain molecules, such as rubber, may exhibit recoverable strains of several hundred percent. Such materials are called *elastomers*, and their elastic behaviour is usually called "high elasticity".

In elastomers, the straightening of chains in the direction of the applied

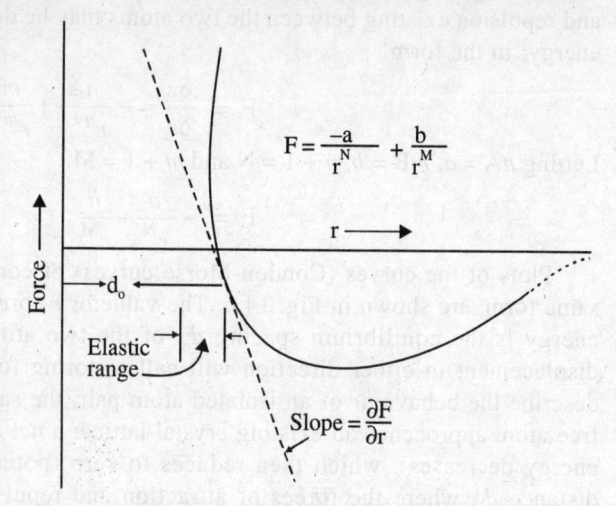

Fig.14.7. Summation curve and its tangents are for all practical purposes, coincident over the range of elastic strains, encountered in crystalline materials, thus, with virtually no error, stress may be considered proportional to strain, in the elastic range.

stress can produce appreciable macroscopic elastic strain at low stresses. Once the chains have been aligned, however, further elastic elongation requires the stretching of the chains in opposition to the primary bonding forces within them and to the secondary bonding forces between them. Elastomers therefore show the non-linear elastic tensile behaviour, as illustrated in Fig. 14.9(a) and (b). Compressive stress applied to elastomers (Fig. 14.8) initially causes a more efficient filling of space in the material. As the available space decreases, the resistance to further compression increases, until finally *primary bonding forces* within the chains begin to oppose the applied stress. The stress-strain curve in the compression thus increases in slope as deformation increases.

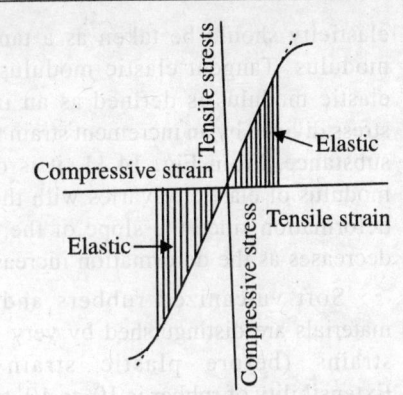

Fig. 14.8. Typical elastic behaviour of crystalline materials, in com-pression and tension.

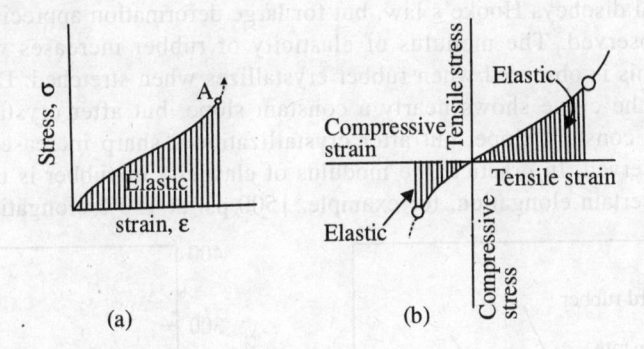

Fig. 14.9.(a) Non-elastic behaviour, (b) typical behaviour or elastomers, in compression and tension.

Certain cellular substances, such as wood, are fairly stiff in compression until the stress is sufficient to cause elastic buckling of the cell walls. At this point considerable strain may accumulate without much increase in stress. The stiffness may then increase again as the cells become compacted. Appreciable non-linear strains may be recoverable in such substances. However, if the stress becomes high enough, the cells will get crushed and the strain may not then be recovered. In tension the cell walls do not buckle elastically in the same way. A typical stress-strain curve is shown in Fig. 14.10.

14.8. ANELASTIC BEHAVIOUR

(Deviation from perfect elastic behaviour):

Very few materials behave as perfectly elastic bodies because of structural imperfections. Hooke's law is applicable in case of very small deformations. In case of larger deformation, the linear relationship between stress and strain no longer exists. Many engineering materials (such as cast iron, non-ferrous metals and concrete etc.,) yield a curve as shown in Fig. 14.11. According to the definition of modulus of elasticity, the stress versus strain curve should be linear. If the curve is not linear, the modulus of

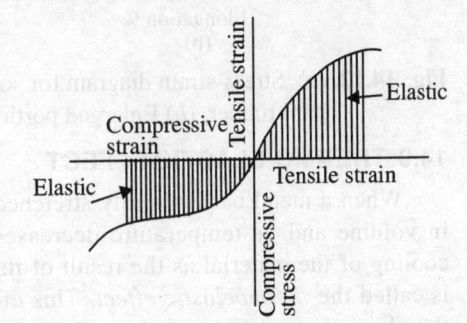

Fig.14.10. Typical elastic behaviour in compression and tension of cellular materials, that exhibit elastic buckling of cell-walls under compression.

elasticity should be taken as a tangent elastic modulus. Tangent elastic modulus. A tangent elastic modulus is defined as an increment of stress divided by an increment strain for an elastic substance. From Fig. 14.11, it is obvious that modulus of elasticity varies with the amount of deformation since the slope of the tangent line decreases as the deformation increases.

Soft vulcanized rubbers and other like materials are distinguished by very large elastic strains (before plastic strain appears). Extensibility of rubber is 10^3 to 10^4 times greater than metals or other materials (Fig. 14.12).

For very low deformation (Fig. 14.12b) the stress-strain curve is a straight line indicating

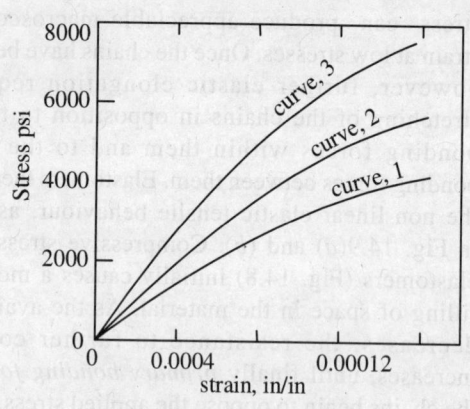

Fig. 14.11. Stress-strain diagram for concrete of different mixtures.

that the material disobeys Hooke's law, but for large deformation appreciable departure from curvature is observed. The modulus of elasticity of rubber increases with the amount of deformation. This is observed when rubber crystallizes when stretched. During the period of crystallization the curve shows nearly a constant slope, but after crystallization the curve shows nearly a constant slope, but after crystallization a sharp increase in the modulus of elasticity is observed. In practice, the modulus of elasticity of rubber is usually defined as a stress causing certain elongation, for example, 1500 psi at 150% elongation.

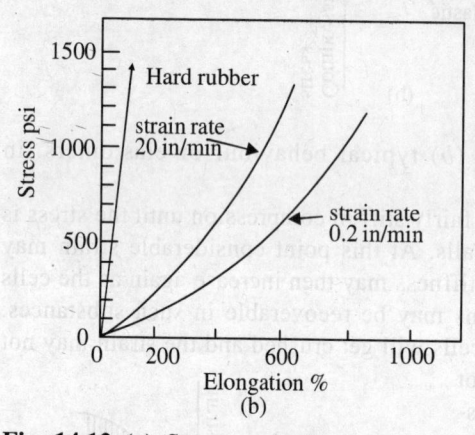

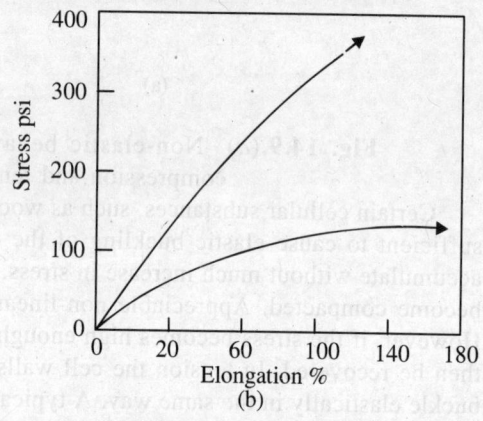

Fig. 14.12. (a) Stress-strain diagram for soft vulcanized acrylic rubber at two strain rates, and for a hard rubber, (b) Enlarged portion of stress-stain diagram for soft rubber (low strains).

14.9. THERMO-ELASTIC EFFECT

When a metal bar is rapidly stretched, it increases in volume and its temperature decreases. Heating or cooling of the material as the result of its deformation is called the *thermoelastic effect*. This effect is due to the phenomenon of retarded elasticity, called also the *elastic after effect*. The elastic after-effect can be best illustrated on the strain-time diagram (Fig. 14.13). When rapidly loaded, a material undergoes a certain instantaneous elastic deformation, ϵ_1, followed by a delayed or retarded elastic deformation ϵ_2 during a time

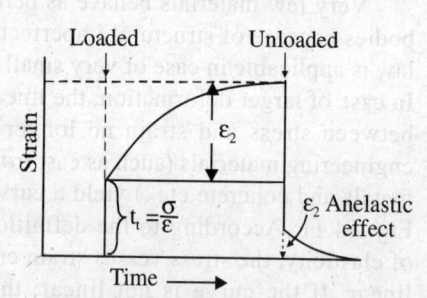

Fig. 14.13. Elastic after-affect.

MECHANICAL PROPERTIES

t, which approaches asymprotically a final value. If the load is removed, the instantaneous recovery, equal to ϵ_1, is followed by a delayed recovery in time t. *The term "anelasticity"* is applied to the stress and time dependence of elastic strain. Both thermoelastic effect and elastic after-effect are the aspects of "anelasticity" of the material, which is directly related to its internal friction. In polycrystalline metals the internal friction originates in the grain boundaries, whereas in polymeric materials it is due to specific molecular structure.

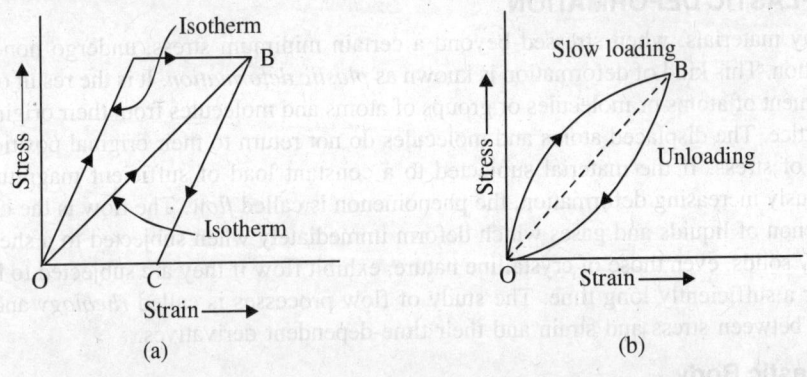

Fig. 14.14. (*a*) Thermo-elastic effect   (*b*) Elastic hysteresis

If the specimen is allowed to remain under the load for a sufficient length of time, it warms up to room temperature and expands (line AB in Fig. 14.14*a*). If unloaded at the original rate, it will adiabatically contract as shown by line BC; its temperature goes up and it decreases in volume. Allowed to stand, it cools to room temperature and it will accordingly contract (line CO). If a specimen is stretched at such a rate that its temperature will be constant, the isothermal modulus of elasticity will be represented by the line OB. It is seen from Fig. 14.14(*a*) that the adiabatic modulus (line OA) is greater than the isothermal odulus (line OB).

In practice the changes that take place are not adiabatic, since there is always some heat exchanged with the surroundings and the diagram will assume a shape as in Fig. 14.14(*b*). This is known as the hysteresis loop which represents the amount of energy dissipated as heat during loading and unloading of the material.

In high polymers and elastomers this effect is more pronounced. It may be attributed, to the specific structure and configuration of the molecules which account for a greater ease of movement under applied forces. When rubber is subjected to tension or compression, its volume remains essentially constant. In contrast to metals, rubber warms up when stretched and cools off when contracted. The stretching associated with straightening and orientation of the kinked rubber molecules, is accompanied by the evolution of heat. When the load is released the straightened molecules tend to return to their random arrangement, which is the most probable state from the thermodynamics point of view. This requires energy which is absorbed from the body, if the material contracts rapidly, thereby resulting in its cooling.

The time dependent component of elastic strain is described by a single number–the relaxation time T.

14.10. RELAXATION PROCESS

At low stress levels, the elastic distortion remaining in the crystal on removal of the load, may be sufficient to return the dislocations back to their original positions and thus with the passage of time the crystal reverts to its former dimensions. In a similar fashion, during stressing of long-chain polymers, adjacent chains are locally displaced and new secondary bonds between chains are formed. This local deformation may be recovered, however, when the elastic distortion remaining after removal of the load moves the displaced segments of the chains back to their initial position. This recovery is thermally activated.

The diffusion of interstitial atoms in a metal lattice can give rise to elastic hysteresis exactly as the diffusion of thermal energy in thermoelastic effect. In conclusion *Anelastic* effects also arise from the motion of substitutional solute atoms, grain boundary effects, motion of dislocations and intercrystalline and transcrystalline thermal currents. Such thermal currents have their origin in the elastic anisotropy typical of most crystalline solids.

The relaxation time is highly dependent on temperature.

14.11. PLASTIC DEFORMATION

Many materials, when stressed beyond a certain minimum stress, undergo non-recoverable deformation. This kind of deformation is known as *plastic deformation*. It is the result of permanent displacement of atoms or molecules or groups of atoms and molecules from their original positions in the lattice. The displaced atoms and molecules do not return to their original positions after the removal of stress. If the material subjected to a constant load of sufficient magnitude shows a continuously increasing deformation, the phenomenon is called *flow*. The flow is the characteristic phenomenon of liquids and gases which deform immediately when subjected to a shearing stress. But many solids, even those of crystalline nature, exhibit flow if they are subjected to high enough stress for a sufficiently long time. The study of flow processes is called *rheology* and deals with relations between stress and strain and their time-dependent derivatives.

Ideal Plastic Body

An ideal plastic body (which is also called St. Venant's solid) is represented on the stress-strain diagram, which shows a line parallel to the strain axis at a distance corresponding to the yield stress of the material (Fig. 14.15). Mathematically this can be expressed by the relation

$$\sigma = \sigma_y$$

where σ denotes the stress and σ_y the yield stress. It is seen that deformation proceeds continuously at constant stress equal to the yield stress. Closely connected with plastic deformation is the concept of *plasticity*, which can be defined as the ability of the material to be deformed continuously and permanently without rupture during the application of a force that exceeds the yield value of the material.

In case of majority of the materials the elastic deformation precedes the plastic strain yielding the curve shown in Fig. 14.16(a). A material obeys the law of elastic solids for stresses below the yield stress (AB), and this is

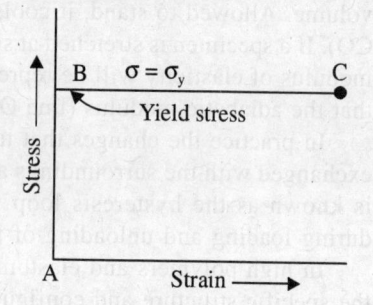

Fig. 14.15. Ideal plastic body.

followed by the ideal plastic deformation (BC) occurring at the yield stress value. Such materials are called *elastoplastic bodies* and represented by such an important group of structural materials as metals.

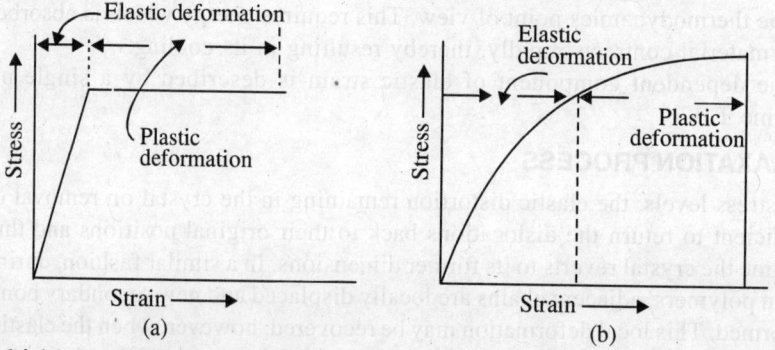

Fig. 14.16. (*a*) Ideal plastic deformation preceded by ideal elastic deformation.
(*b*) Elastic and plastic deformation of rigid bodies.

MECHANICAL PROPERTIES

A large number of engineering materials show deviations from both the perfect elastic and ideal plastic behaviour. Because of this, the relationship between stress and strain will not be linear and their stress-strain curve will be of the type shown in Fig. 14.16(b). It is seen from the Fig. 14.16(b) that there is a slightly curved line in the elastic range (AB) and a considerable increase in stress during the plastic deformation (BC).

The mechanism involved in plastic deformation of crystalline materials is different than that in amorphous materials. Crystalline materials undergo plastic deformation as the result of slip along definite crystalline plane, whereas in amorphous materials, sliding of individual molecules or group of molecules past one another occurs, resulting in a flow.

14.12. PLASTIC DEFORMATION OF A SINGLE CRYSTAL

Plastic deformation of a single crystal occurs either by slip or by twinning. As a matter of fact slip is most common. Although the crystal may be subjected to other stresses, both slip and twinning occur by pure shearing stress.

Slip

Slip represents a large displacement of one part of the crystal relative to another along particular crystallographic planes and in certain crystallographic directions. The particular, crystallographic planes are called the *slip* or *glide* planes and the preferable direction is called the slip direction (Fig. 14.17 and Fig. 14.18).

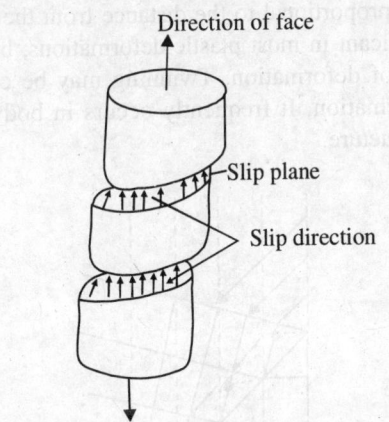

Fig. 14.17. Slip in a single crystal.

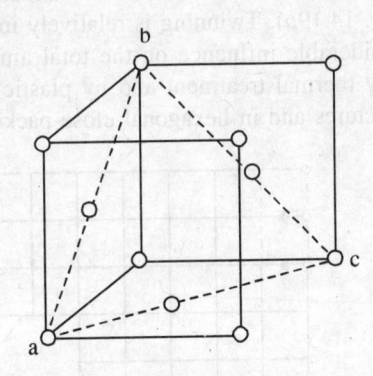

Fig. 14.18. A slip plane, for a cubic crystal. There are three slip directions *ab, bc* and *ac* in each slip plane.

In each crystal, there are one or more slip planes and one or more slip directions. For example, in face-centred cubic crystals like those of aluminium, copper and nickel, there are three octahedral slip planes and three dimensions of sliding for each plane which are parallel to one of the sides of the triangle *abc* (Fig. 14.18). Although slip planes are usually the most closely packed planes containing the maximum number of atoms, slip can be initiated on any other plane and in any other direction.

14.13. PLASTIC DEFORMATION OF POLYCRYSTALLINE METALS

The plastic deformation of polycrystalline metals is more complicated than that of single crystals because of the effect of neighbouring crystals and the character of grain boundaries.

As a result of grain boundary effects polycrystalline metals show much higher strength than single crystals, particularly at relatively low temperature. The strengthening effect is more pronounced in fine-grained metals than for the coarse-grained ones.

The mechanism of slip involves a translatory motion along the sliding planes and rotation of the specimen with respect to the axis of loading. As the amount of slip increases, further deformation becomes more and more difficult until finally the plastic flow cases. Slip may begin again only when a larger shear stress is applied. This phenomenon is known as *strain hardening* or *work hardening*.

Slip is a characteristic feature of plastic deformation in metal crystals. Hard crystals such as diamond, silicon carbide and hard metal carbides require such high values of dislocation stress that the crystal usually breaks before any measurable slip can take place. The crystals break because very strong directional covalent bonds between the atoms make very difficult the movement of any dislocations present in the crystal. In a metal crystal, relative movement of atoms under small initial stress is possible, owing to the non-directional character of the metallic bond. This is in agreement with the fact that the metals which lack a completely filled d shell so that certain directional forces exist, such as Vanadium, chromium, iron, cobalt, molybdenum, tantalum, tungsten, rhenium, osmium and uranium, are much harder than the alkali metals as well as copper silver and gold.

14.14. TWINNING

It is the process in which the atoms, in a part of a crystal subjected to stress, rearrange themselves so that one part of the crystal becomes a mirror image of the other part. Every plane of atom shifts in the same direction by an amount proportional to the distance from the twinning plane (Fig. 14.19*b*). Twinning is relatively insignificant in most plastic deformations, but it may have considerable influence on the total amount of deformation. Twinning may be caused by impact, by thermal treatment and by plastic deformation. It frequently occurs in body-centred cubic structures and in hexagonal close-packed structure.

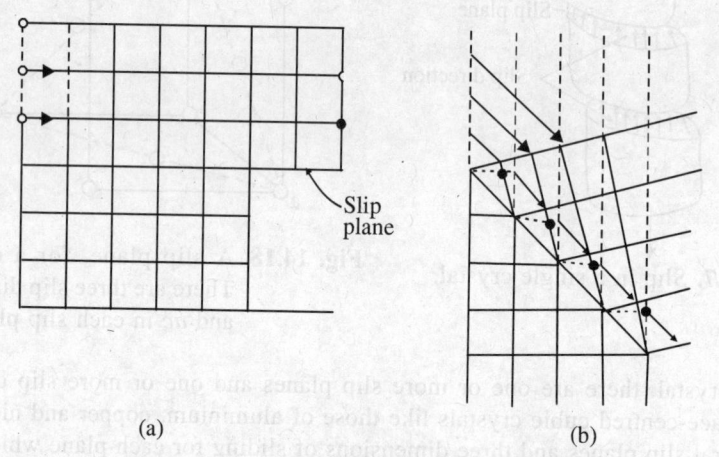

Fig. 14.19. Atom movement in twinning.

Every crystal contains imperfections from which slip can start at a low stress. The imperfections may be due to vacant lattice, interstitial atoms, foreign atoms in either interstitial or substitutional positions, dislocations and gross defects such as grain boundaries and voids. Furthermore, there is a group of transient imperfections such as electrons and holes, and phonons, which are intimately related to the translational energy within a crystal. Transient imperfections are generated by radiant energy in a proper crystal and account for its thermal, electrical and semiconducting properties.

All the above mentioned imperfections affect the physical and chemical properties of solids to a small or great extent. Dislocations are, however, considered to be mainly responsible for the plastic flow of crystals under relatively low stress at temperatures far below the melting point.

MECHANICAL PROPERTIES

14.15. DISLOCATIONS

Plastic or non-recoverable deformation of crystalline substances occurs primarily by the movement of crystal imperfections called *dislocations*. Dislocations in a crystal may occur during its growth from a melt or from a vapour or they may be produced during slip. A dislocation can be regarded as a linear disturbance of the atomic arrangements, which can move very easily on the slip plane, through the crystal. Although the concept of dislocation was first introduced primarily to explain deformation in metals, it is now established that dislocations play an important role in crystal growth, crystal transformation, electrical conductivity and other solid state phenomena. Dislocations in crystals have been observed directly with the aid of electron microscope.

There are two main types of dislocations. They are:

(*i*) Edge dislocations

(*ii*) Screw dislocations.

Many actual dislocations are truly the combinations of these two. An example of the edge dislocation is given in Fig. 14.20. It can be produced by applying a shear to the crystal. It can be seen that a linear disturbance of atoms appears at the edge of the crystal, which is represented in Fig. 14.20 as the area *abcde*. This line defect can be formed in crystal because a row of atoms is either removed from a lattice or displaced at a unit distance. This causes a part of the crystal above the *pp* plane to be under compression, whereas the lower part is under tension. It can also be seen that if the crystal is stretched in the direction to the right, only a partial plane of the atoms above *b* will move along the slip plane, thus accounting for a relatively easy movement of the dislocation. Edge dislocations are also called Taylor dislocations or Taylor-Orowan dislocations.

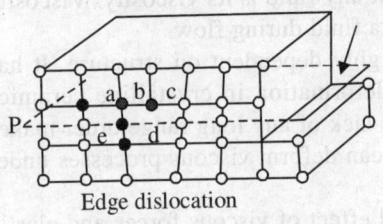

Fig. 14.20. Edge dislocation.

Screw dislocations, as described by Burgers, consist of a line of atoms with distorted coordination polyhedra. This is caused by a displacement of atoms in one part of the crystal with respect of the rest of the crystal, the displacement direction being parallel to the dislocation line, as shown in Fig. 14.21. The row of atoms marks the termination of the displacement in the dislocation. The polyhedra is not regular in the disturbed region and the distortion decreases progressively with increasing distance from the dislocation line.

A dislocation within a crystal can move either parallel to the slip plane or in the direction perpendicular to the slip plane. This latter movement requires a high activation energy and occurs mainly at elevated temperatures. Dislocations can also interact with each other or with impurities in the crystal, thereby accounting for a great diversity and variety of plastic phenomena. The impurities prevent the motion of dislocations. Therefore, a higher applied stress should be used to cause the slip. Dislocations may be held up by grain boundaries or other non-parallel dislocations. The mechanism of the strain hardening

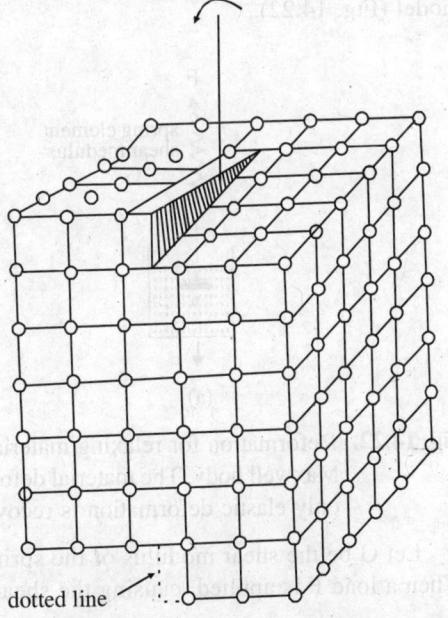

Fig. 14.21. Screw dislocations.

of a single crystal seems to be the result of the piling up of a group of dislocations at the end of the slip line. These piled-up dislocatons act as barriers or obstacles to further movements, considerably raising the stress of the metal.

The strength of commercial metals and alloys is found much lower than their theoretical cohesive strength. This may be accounted for due to the presence of dislocations in metal crystals. Experimentally it is estimated that a metal free from dislocations should be many times stronger than the commercial material, which may contain upto 10^8 dislocations per square centimeter in the annealed state. By means of special techniques, single crystals of metals such as iron, copper and tin can be produced in the form of very fine filaments, called **whiskers** (about 10^{-4} cm in diameter). Such whiskers do not have easily moved dislocations. These whiskers exhibit a strength many times greater than that of the normal metals, approaching the strength of a perfect crystal. The size for the whisker has a pronounced effect on its strength, which decreases as the diameter of the whisker increases. For example, iron whiskers usually fracture without only plastic deformation and obey Hooke's law upto 2% deformation, whereas silver whiskers exhibit some plastic deformation before the fracture.

14.16. VISCO-ELASTICITY

A characteristic property which determines the flow of any fluid is its viscosity. Viscosity may be defined as internal friction or as the resistance of a fluid during flow.

The mode of deformation in ceramic materials is highly dependent on structure. It has been discussed earlier that the nature of any possible deformation in crystalline ceramics results from plastic slip processes. In glassy ceramics, the lack of any long range order makes dislocation motion impossible. However, these materials can deform viscous processes under proper conditions of stress and temperature.

The deformation of any material caused by combined effect of viscous forces and elastic forces, is known as *viscoelastic* deformation. This can be represented by a model combining one spring and one dash pot element. (It is called *spring-dashpot model*). The spring illustrates purely elastic behaviour and the dash pot purely viscous behaviour. If the spring is Hookean and the dashpot Newtonian, the connection of these two elements in series gives the Maxwell model (Fig. 14.22).

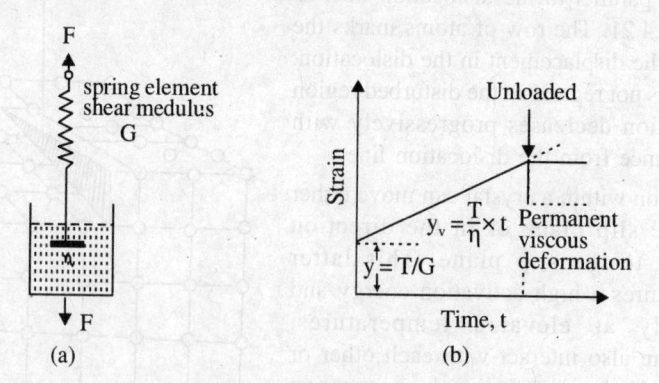

Fig. 14.22. Deformation for relaxing material (*a*) Maxwell element, (*b*) strain-time diagram for a Maxwell body. The material deforms continuously under constant stress. On unloading, only elastic deformation is recovered.

Let G be the shear modulus of the spring and η the viscosity of the dash not (Fig. 14.22). When a load F is applied, causing the shear stress τ the spring A instantaneously stretches by the amount $\dfrac{\tau}{G}$ and the dashpot begins to elongate steadily at the rate τ/η. The total deformation

MECHANICAL PROPERTIES

in the Maxwell model is therefore distributed between two elements A and B. The two elements are however subjected to the entire τ. Thus we can write

$$\tau = \tau_e = \tau_v \qquad ...(1)$$

and

$$\gamma_t = \gamma_e + \gamma_v \qquad ...(2)$$

here τ is the total stress causing the total deformation γ_t, τ_e is the stress causing elastic strain γ_e and τ_v is the stress causing viscous strain γ_t.

Since the element A behaves according to Hooke's law and the element B according to Newton's law

$$\gamma_e = \frac{1}{G}\tau \qquad ...(3)$$

and

$$\frac{d\gamma_v}{dt} = \frac{1}{\eta}\tau \qquad ...(4)$$

The rate of total strain with time is

$$\frac{d\gamma_t}{dt} = \frac{d\gamma_e}{dt} + \frac{d\gamma_v}{dt} \qquad ...(5)$$

Substituting eqn. 3 and 4 in eqn. 5, we obtain

$$\frac{d\gamma_t}{dt} = \frac{1}{G}\frac{d\tau}{dt} + \frac{1}{\eta}\tau \qquad ...(6)$$

Eqn. (6) states that the total rate change of γ_t with time is determined by the rate of flow plus the rate of change of the elastic deformation. If γ_t is constant, $\frac{d\gamma_t}{dt} = 0$ and eqn. 6 reduces to

$$\frac{1}{G}\frac{d\tau}{dt} + \frac{1}{\eta}\tau = 0 \qquad ...(7)$$

The solution of this simple differential equation is obtained by transferring the terms on both sides and integrating from time $t = 0$ to time t, and from γ_0 to τ.

$$\tau = \tau_0 e^{-(G/\eta)t} \qquad ...(8)$$

Eqn. (8) indicates that the stress τ gradually disappears exponentially, so that if the body is left to itself it will lose any internal stress. After a time t equal to η/G, the stress will have decayed to $\frac{1}{e}$ of its original value. This time t is the relaxation time T_{rel} for the material. Then

$$\tau = \tau_0 e^{-t/T_{rel}} \qquad ...(9)$$

If the stress is kept constant, then $d\tau/dt = 0$, and eqn. 6 becomes the Newtonian equation for flow of liquids, $d\gamma/dt = \left(\frac{1}{\eta}\right)\tau$. Thus under constant stress the Maxwell body is indistinguishable from a liquid, and its strain — time diagram is given in Fig. 14.23b.

14.17. CREEP

Creep can be defined as the slow and progressive deformation of a material with time under a constant stress. It is observed that metals, ionic and covalent crystals, and amorphous materials such as plastics, rubbers and the like, are very temperature-sensitive to creep.

The common method of carrying out a creep test is to subject the specimen to a constant tensile stress, while maintaining a constant temperature and measuring the extent of deformation with the elapsed time. Creep is also determined in compression, shear and bending. Although simple in principle, the test requires in practice elaborate laboratory apparatus and great care and precision in its performance. The time for each test may be a matter of hours, weeks or months, or even years. Some creep tests on metals have been run for more than 10 years. The customary testing time in Japan is from 1000 to 3000 hours and the general method for creep testing is covered by the American Society for Testing Material (ASTM) standard specifications, E – 22 – 71. The data are represented by plotting the creep curve as deformation versus elasped time at constant temperature and stress as in Fig. 14.23 and 14.24.

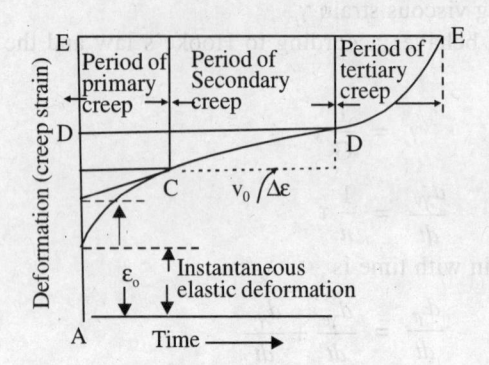

Fig. 14.23. Creep curve at constant temperature and stress. The minimum creep rate $\Delta \epsilon / \Delta t$ is determined by the slope v_0 of the curve in the period of secondary creep (CD).

Total creep at any time 't' can be found from the relation $\epsilon_z = \epsilon_0 + v_0 t$, where ϵ_0 is the intercept of the line CD with creep-strain axes.

When load is applied at the beginning of a creep test, the instantaneous deformation (elongation) AB is followed by the primary or transient creep BC, then by secondary or steady-state creep CD and finally by the tertiary or accelerated creep DE (Fig. 14.23).

The primary creep or transient creep has a decreasing creep rate, because of the work hardening process due to deformation. The primary creep is similar in its mechanism to delayed elasticity or elastic after-effect and as such is recoverable by unloading the specimen. Because of the absence of grain boundaries, single metal crystals do not show recovery to such an extent as polycrystalline metals do.

The secondary creep or the steady-state creep (also called the minimum creep rate) is due to the result of an equilibrium attained between work-hardening process and the annealing effect. The steady-state creep may be viscous or plastic in character. It depends on the stress level and temperature. Generally, the higher the temperature and stress level, the more the viscous creep. It differs however, from pure viscous flow in that the creep rate is highly temperature-sensitive. Moreover it can be related to the temperature by the same type of equation as viscosity or diffusion.

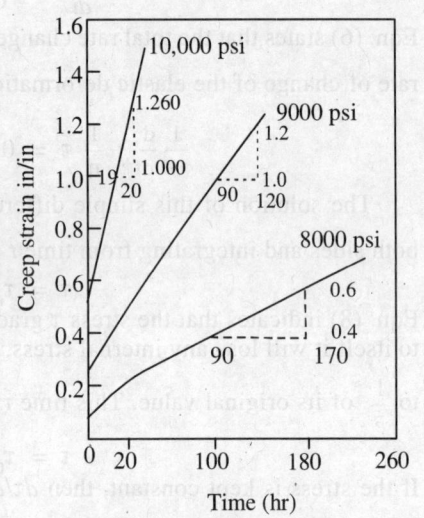

Fig. 14.24. Creep curves of 24S–T4 aluminium alloy at a constant temperature of 350°F and for various stresses.

$$\frac{\Delta E}{\Delta t} = A e^{-E/RT}$$

where E is the activation energy, A is the constant and T, R and e are as previosuly defined.

Similarly the relation between the minimum creep rate and stress at constant temperature can be represented by the exponential equation

MECHANICAL PROPERTIES

$$\frac{\Delta \tau}{\Delta t} = B \sigma^n$$

where B and n are constants, characteristic for each material.

The tertiary creep occurs at an accelerated rate; it actually represents a process of progressive damage resulting in an imminent fracture of the material through inter-crystalline or other causes.

The relation between the time to rupture and the stress a constant temperature can be represented by the following equation

$$t_r = a \sigma^n$$

where t_r is the time to rupture under initially applied stress and a and n are material constants. On plotting logarithm stress versus logarithm time, a linear relation is obtained.

14.18. CREEP IN METALS

Pure metals show good creep resistance at high temperature. For example, molybdenum with a high melting point and high density shows high creep resistance, whereas titanium with high melting point, has relatively poor creep resistance because of its low elastic modulus. Metals having low melting points such as aluminium will tend to creep extensively at increased temperatures. However, lead with a very low melting point exhibits creep at room temperature. The creep resistance of pure metals can be considerably increased by alloying them with suitable elements. Therefore, at high temperature and severe stresses, various types of creep-resistant alloys are used.

The function of alloying elements is to increase the temperature of recrystallization of the alloy or to produce finely dispersed precipitate such as of carbides, nitrides and intermetallic compounds along the grain boundaries. The precipitates of solute element in form of submicroscopic particles impede the movement of dislocations, thereby increasing creep resistance of alloys.

However, extensive precipitation from a super-saturated solution followed by subsequent coagulation of the particles will decrease the creep resistance of the alloy.

The extent of the creep in metals is also affected by grain size, micro structure and cold work. Generally, coarse-grained materials show better creep resistance than fine grained ones. For this reason single crystals exhibit a higher creep resistance than polycrystalline material. The tungsten filaments of electrical bulbs are made of single crystals, for they must withstand very high temperature.

Another important factor is the thermal stability of the microstructure of alloys and its resistance to oxidation at high temperature.

Another important factor is the thermal stability of the microstructure of alloys and its resistance to oxidation at high temperatures.

14.19. CREEP IN AMORPHOUS MATERIALS

Amorphous materials such as high polymers, asphalts etc., show high rates of deformation under relatively low stresses and temperature, because of their less rigidity and more temperature sensitiveness than metals, and ionic and covalent crystals. This puts the limitations of the use of plastics for structural purposes. Thermosetting plastics are much more creep resistant than thermoplastic polymers. However, at high temperatures and stresses, they undergo deformation.

The addition of fillers and various reinforcing materials to the plastics, rubbers and asphalt-paving materials such as glass fibres and glass cloth increases considerably the creep resistance.

In industry the steady-state creep of plastics is often referred to as cold flow, whereas the term creep is used to define the primary creep which has its maximum value immediately when the elastic or rapid deformation ceases and when cold flow (steady-state creep) commences. Creep resistance of plastic is also adversely affected by moisture which gradually penetrates the material and acts as plasticizer.

Creep in concrete is due to the viscous flow of the hardened cement paste. A high proportion of rigid solid ingredients imparts rigidity to the concrete, whereas a large amount of cement paste enhances its viscous properties. The viscosity of cement paste increases with time by the process called hardening, thereby causing the creep of concrete to decrease with time. Hardened cement paste is essentially a Maxwell body of very high viscosity and cohesion. The aggregate forms a noncohesive, granular mass, whose resistance to irrecoverable deformation by shear is the result of friction between grains. High-grade concretes, which are rich in cement paste, deform essentially like a viscoelastic material with a viscosity of about 10^{16} poises. The larger the amount of aggregate, the more pronounced the effects of internal friction. Various mixes of concrete will show behaviour varying between that of purely viscoelastic and that of plastic material.

Creep in concrete can be considerably reduced by reinforcing it with steel rods. The viscous deformation of the cement paste is thus coupled in parallel with the elastic deformation of steel which governs the total deformation of the reinforced concrete. In contrast to that of plastics, creep of concrete is not very sensitive to temperature.

14.20. EFFECT OF PRECIPITATION PARTICLES ON DISLOCATION MOTION

Precipitation-hardening is nothing but strengthening of a material by the formation of dispersion of hard particles within the matrix. This impedes the motion of dislocations through the material. A dislocation may, in passing through the matrix, arrive at a hard, high modulus particle which does not shear as easily as the matrix. The dislocation on being arrested in its motion at precipitate particle, starts to bulge through between them, as shown in Fig. 14.25. After bulging through between the particles, the dislocation line reforms, leaving a dislocation loop around the particles (See Fig. 14.22). Each additional dislocation leaves another dislocation loop around the particle, effectively increasing the spacing between nearby particles, so that an ever-increasing stress is required to push successive dislocation through.

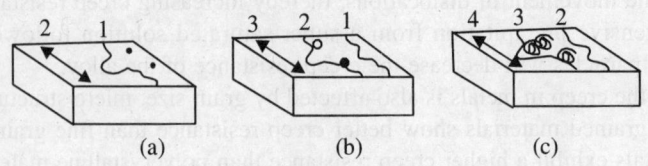

Fig. 14.25. The formation of dislocation loops around precipitate particles.

Precipitation-hardening is most effective when the particles are small and coherent with the matrix lattice.

A sufficient number of dislocation loops may be created around a particle after the passage of many dislocations that it will shear. This creates a crack nucleus within the matrix which often produces sudden fracture at high stress and diminishes the strain to fracture.

QUESTIONS

1. How many independent elastic constants are sufficient to define completely a diamond crystal, zinc crystal, steel, and wood? Give reasons.
2. What is meant by mechanical properties of solids? Account for atleast one property in details.
3. Explain deviation from perfect elastic behaviour in case of materials used for engineering purposes.
4. By what mechanism does the plastic deformation occur in a metal crystal?
5. Show how the true-stress true-strain diagram helps to explain the extent of plastic deformation and strain hardening of metals.
6. What do you understand by the terms—
    (*i*) Plastic deformation   (*ii*) Slip and   (*iii*) Creep
7. Describe the behaviour of viscoelastic materials.
8. Explain the effect of precipitate particles on dislocation phenomenon.
9. Write about
    (*a*) Plastic deformation, and   (*b*) Mechanism of Creep
10. How you will explain an elasticity of materials ?

15

Glass and Ceramics

> *"Glass & ceramics include a variety of materials suck as clay and clay products, refractories, glasses, enemels, cements, abrasives, porcelains, insulators etc. Ceramic materials of wide range of properties can be prepared for different applications."*

15.1 GLASS

A glass is an inorganic substance hard, brittle, amorphous mixture of silicate of calcium, sodium or other metal. It may be colorless or colored and transparent or opaque. It has no definate composition. The major ingredients are silica, soda ash, limestone. The other ingredients are borax, litharge, potash, fluorspar, zinc oxide, barium carbonate etc.

Structure of glass - Many solids have a crystalline structure on microscopic scales. The molecules are arranged in a regular lattice. As the solid is heated molecules vibrates about their position in the lattice until, at the melting point, at the melting point, the crystal breaks down and the molecules start to flow. There is a sharp distinction between the solid and the liquid state which are separated by a first order phase transition i.e. a discontinuous change in the properties of the material such as density. Freezing is marked by a release of heat called the heat of fusion. As the liquid cooled its viscosity increase, but viscosity also has a tendency to prevent crystallization. Normally when a liquid cooled below its melting point, crystals form and it solidifies, but sometimes it can becomes supercooled and remain liquid below its melting point because there are no nucleation sites to initiate the crystallization. If it is cooled further the viscosity rises, it may not crystalise. However as viscosity rises continuously form a thick fluid and finally to amorphous solid. The molecules have a disordered arrangement but have sufficient cohesion to maintain some rigidity, this state is called glass.

15.1.1 Manufacturing of glass

Glass manufacturing has essentially three steps. Batching, melting and forming.

Batching - In the batching operation the raw materials are weighted, mixed and milled. The common raw ingredients are given below.

Row materials	Chemical composition
Sand	SiO_2
Soda ash	Na_2CO_3
Limestone	$CaCO_3$
Borax	$Na_2B_4O_7 \cdot 10H_2O$
Boric acid	$B_2O_3 \cdot H_2O$
Litharge	PbO
Potash	$K_2CO_3 \cdot 1.5H_2O$

Fluorspar	CaF_2
Zinc oxide	ZnO
Barium carbonate	$BaCO_3$

The ingradients are separately crushed to a proper size and then mixed in proper proportion. The nixture of ingradient is called batch material. Trace impurities can be a serious problem to the glass maker. The most common impurity is iron oxide which give a greenish shade to the glass.

Melting - Melting is performed in a furnace Glass furnace are of two types. Pot furnaces and tank furnaces. Both furnaces are regenerative type to utilize the heat of waste gases for heating the incoming gas and air used for burning.

Pot Furnace - The glass is melted in either closed or open pots, places inside the combustion chamber of the furnace in a circle. A pot is a large monkey shaped crucible made from high alumina fine clay such as mullite ($3Al_2O_3 \cdot 2SiO_2$). A pot is of about 2 tonnes capacity. The number of pots varies from 6 to 20 except in the case of optical glass where only one pot is used.

Tank Furnace - In a tank furnace, batch materials are charged into one end of a large rectangular tank built of refractory block, heated by producer gas. In tank furnaces, the process is continuous and they are used where the output expected is large and are used in the manufacture of sheet glass and bottle glass.

Silica is the glass having the wide variety of useful application. It has low thermal expansion, it is resistant to attack by water and acids. Pure silica (SiO_2) has a glass melting point of 2300°C. Pure silica can be made into glass for special application, to simplify the processing other substance such as sodium carbonate (Na_2CO_3) are added which lower the melting point to about 1500°C. Soda make the glass water soluble, $CaCO_3$ is added to increase the resistance to moisture. Alumina is added to enhance durability and to lower the coefficient of expansion. Potash (K_2O) on substitution to Soda (Na_2O) increase durability and reduces the tendency to devitrification.

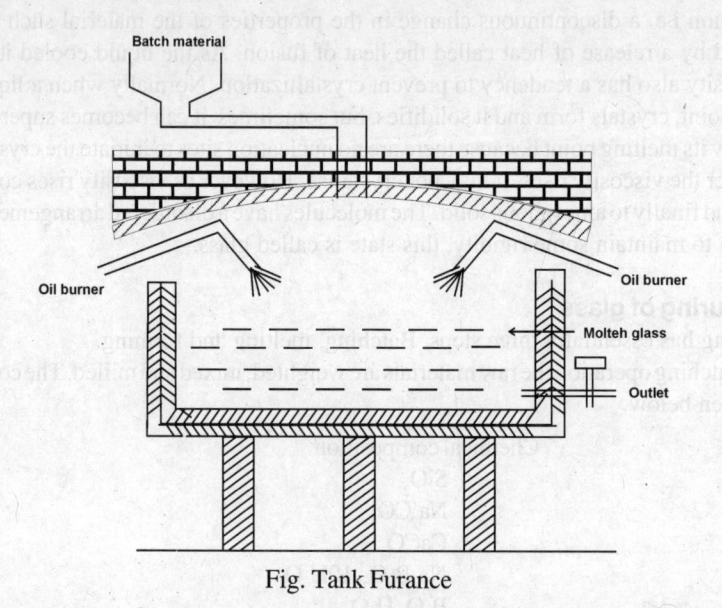

Fig. Tank Furance

Silica heated with soda ash, sodium silicate formed known as water glass (soluble in water)

$$Na_2CO_3 + SiO_2 \xrightarrow{720\text{-}900°C} Na_2SiO_3 + CO_2 \uparrow$$
Soda ash Silica   Water glass

Silica heated with limestone calcium
silicate formed which is insoluble in water but reactive towards acids.

$$CaCO_3 + SiO_2 \xrightarrow{1000°C} CaSiO_3 + CO_2$$
limestone   Calcium silicate

$$Pb_3O_4 \rightarrow 3bPO + O$$
$$PbO + SiO_2 \rightarrow PbSiO_3$$

Soda-lime glass can be obtained by heating silica with Na_2CO_3 and $CaCO_3$
$$SiO_2 + Na_2CO_3 + CaCO_3 \rightarrow Na_2O \cdot CaO \cdot 6SiO_2 + 2CO_2 \uparrow$$

During melting gases are produced and it will be remain in the molten mass and get solidify and defects were created. The process of removing these defects is called refining.

Forming - Forming is nothing but the manufacturing of different types of glass wares from the molten glass.

Annealing - During manufacturing when a glass is cooled down suddenly from the melt, strain were introduced due to unequal cooling rates in various parts of the body. Cracks were developed on the articles on heating or even by keeping for some time, the change of temperature cannot be withstand. In order to avoid this problem the articles were heated at annealing temperature and it cooled very slowly the process is called annealing.

The process of annealing is a thermal treatment designed to reduce these residual strain to an acceptable level. For different types of glass, there is a fixed annealing temperature. Annealing is a two-step process in which the glass is first raised to a sufficiently high temperature such that the viscosity is sufficiently low that all the strain is removed and than cooled slowly through various hot chembers in a controlled manner.

Annealing point for typical glasses.
Borosilicate glass   -   518-550°C
Lime glass   -   472-523°C
Lead glass   -   419-451°C

15.1.2 Types of Glasses

Soda - lime glass - The soda lime glass (for containers) has the following chemical composition SiO_2 -74%, Na_2O - 13%, CaO - 10.5%, Al_2O_3 - 1.3%, K_2O - 0.3%, MgO - 0.2%, TiO_2 - 0.01%, Fe_2O_3 - 0.04%. It has density of 2.52 g/cm^3 at 20°C and refractive index 1.518, glass transition temperature 573°C, coefficient of thermal expansion 9 x 10^{-6} K^{-1}. It is used in the manufacture of all kinds of containers, plate glass, window sheet, electric lamp, bulbs etc.

Borosilicate glass - The chemical composition of borosilicate glass is SiO_2 - 81%, B_2O_3 - 12.5%, Na_2O - 4%, Al_2O_3 - 2.2%, CaO - 0.02%, K_2O - 0.06%. It has low coefficient of thermal expansion 3 x 10^{-6} K^{-1} making more resistant to thermal shock and high chemical resistance. It has glass transition temperature 536°C, density 2.235 g/cm^3, refractive index 1.473. It is used in the manufacture of chemical laboratory glassware, equipments, pipelines, insulators and washers. It is also used in the processing of high-level nuclear - waste.

Special optical glass - It has chemical composition SiO_2 - 41.2%, PbO - 34.2%, BaO - 12.4%, ZnO - 6.3%, K_2O - 3.0%, CaO - 2.5%, Sb_2O_3 - 0.35% and As_2O_3 - 0.2%. The coefficient of thermal expansion is 7×10^{-6} K^{-1}. Glass transition temperature 540°C. density 3.86 g/cm³ and refractive index 1.65.

Glass wool - Glass wool is a fibrous wool like material. Its chemical composition is SiO_2 - 63%, Na_2O - 16%, CaO - 8%, B_2O_3 - 3.3%, Al_2O_3 - 5.1, MgO - 3.5%, K_2O - 0.8%, Fe_2O_3 - 0.3%. It has high coefficient of thermal expansion 10×10^{-6}, K^{-1}, density is 2.55, refractive index 1.531, glass transition temperature 551°C. It is chemically resistive and it has low thermal conductivity. Glass wool is used for heat insulation purpose in ovens, motors, vacuum cleaners. It is also used in the filtration of corrosive liquids such as acids and alkalies due to its chemical resistance.

Fused silica glass - Fused silica glass is also called as quartz glass. Its chemical composition is only SiO_2. Glass transition temperature is 1140°C, density - 2.203 glcm³, refractive index - 1.459, coefficient of thermal expansion is very less 0.55×10^{-6} K^{-1}. It is very resistant thermally as well as chemically, It is transparent to ultraviolet light and is used for applications that require transparency in this region.

15.2 CERAMICS

A ceramic is an inorganic non-metallic solid prepared by the action of heat and subsequent colling . Ceramic materials may have a crystalline or partly crystalline structure, or may be amorphous i.e. a glass

The word "ceramics" comes from to the Greek word "Keramikos" which means "burnt material". Now, the term "ceramics" is broadly used for variety of materials such as refractories, glasses, structural clay products, enamles, abrasives, cements, ferroelectrics and some non-matalic materials. Ceramic materials are characterized by high temperature resistance, good electrical resistance, resistance to chemical attack and weathering, high compressive strength and good tensile strength. Ceramics are classfied into three categories :

Oxides : Alumina, zirconia

Non-oxides : Carbides, borides, nitrides, siliçides

Composites : Particulate reinforced, combinations of oxides and non-oxides.

15.2.1 Clays

Clay is a general term used for certain naturally occurring mineral aggregates or earths which essentially consist of silica and alumina. Clays are formed by ageing and weathering of granites and other igneous rocks containing quartz, feldspar and mica. Clays remaining at the place of their origin are called "primary clays". *e.g.* : Kaolin, (China clay) etc. These are low in iron content and so they "burn" white. On the other hand, if the clays are transported from their original location by wind or water, they are called "sedimentary clays". They usually contain lime stone powder, hydrated oxide mud and some organic impurities.

Clay has several unique properties that enable it to be used for ceramic art.

i. Clay has a property of plasticity. which makes it possible to give it a permanent shape.
ii. Clay shrinks when it is fired in kilns. This is a setting process, which allows the clay to become hard and retain the shape that is given to it.
iii. Clay molecules can remain attached to each other. Also, clay molecules stick very close to each other without leaving many pores, due to which clay is an ideal material to make ceramic vessels, Depending on the composition, there are different types of clays. The main types of clays used in ceramics are as follows.

i. **China clay (kaolin)** - Kaolin is a silicate clay, with the chemical composition of Al_2O_3 $2SiO_2$ $2H_2O$. It is considered to be the richest form of clay used in the ceramics clay industry. This clay is pure white in color and it retains that color when the article is made. The degree of purity of china clay ceramics is adjudged on the basis of its whiteness. However a pinkish or orange red hue can be obtained in kaolin by the addition of some amount of iron oxide.

ii. **Ball Clay** - Ball clay is a further metamorphosis of kaolin which deposits in swampy areas. Here, the acids formed due to decomposition of vegetal and animal matter can break the kaolin to still finer particles. The result is, the cohesion properties increase, but the whiteness gets reduced. Ceramics made of ball clay are much harder than those made of china clay.

iii. **Fireclays** - Fireclays are ceramics clay that can withstand high temperatures. They are commonly used to make utility wares such as those which are used in chemical laboratories. This is actually a name given to different kinds of clays whose composition consists of high amounts of aluminum and silicates, but they have little amounts of iron in them. Feldspar and calcium are absent i n all kinds of fireclays.

iv. **Earthenware clays** - These are similar to fireclays in their composition, but they cannot withstand as high temperatures. Earthenware clays are used to make utensils and utility items, such as water pots and urns. They are attached by cohesion at low firing temperatures in kilns, and thus even small scale potters can create them. They are also called as potters' clay. However, earthenware clays could require several fluxes for their formation into ceramics, such as iron oxides. This makes the surfaces of this ceramics clay quite rough and highly porous. Smooth vitrification does not occur in earthenware clays.

v. **Stoneware clays** - These are the cheapest and the most abundantly found varieties of clay used in ceramics clay range. They are gray to brown in color. These clays are highly impure with iron, calcium and feldspar, due to which they require a higher hardening temperature.

vi. **Porcelain clays** - Porcelain is heat-resistant and the most good-looking in the ceramics clay family. Thus its sheer beauty it is widely used in making crockery like dinner plates. It is pure white clay and keeps that color after the firing process. It is a daunting task to make ceramic ware from porcelain since it has a tendency to crack when drying. For this reason, hand building is difficult with this type of clay.

Zirconium dioxide ZrO_2 known as zirconia is a white crystaline oxide of ziroconium. Pure ZrO_2 has a monoclinic crystal structure at room temperature and transitions to tetragonal and cubic at increasing temperatures. The volume expansion caused by the cubic to tetragonal to monoclinic transformation induces very large stresses, and will cause pure ZrO_2 to crack upon cooling from high temperatures. Several different oxides are added to zirconia to stabilize the tetragonal and/or cubic phases: magnesium oxide (MgO), yttrium oxide, (Y_2O_3), calcium oxide (CaO), and cerium (III) oxide (Ce_2O_3), amongst others.

Zirconia is very useful in its 'stabilized' state. In some cases, the tetragonal phase can be metastable. If sufficient quantities of the metastable tetragonal phase is present. then an applied stress, magnified by the stress concentration at a crack tip. can cause the tetragonal phase to convert to monoclinic, with the associated volume expansion. This phase transformation can then put the crack into compression, retarding its growth, and enhancing the fracture toughness. This mechanism is *known* as *transformation toughening,* and significantly extends the reliability and lifetime of products made with stabilized zirconia. A special case of zirconia is that of tetragonal zirconia polycrystalline or TZP, which is indicative of polycrystalline zirconia composed of only the metastable tetragonal phase.

The cubic phase of zirconia also has a very low thermal conductivity, which has led to its use as a thermal barrier coating or TBC in jet and diesel engines to allow operation at higher temperatures. Thermodynamically the higher the operation temperature of an engine, the greater the possible efficiency. It is used as a refractory material, in insulation, abrasives, enamels and ceramic glazes. Stabilized zirconia is used in oxygen sensors and fuel cell membranes because it has the ability to allow oxygen ions to move freely through the crystal structure at high temperatures. This high ionic conductivity (and a low electronic conductivity) makes it one of the most useful electroceramics. ZrO_2

bandgap is dependent on the phase (cubic, tetragonal, monoclinic or amorphous) and preparation methods, estimates from 5-7 eV.

Aluminium oxide Al_2O_3 is an amphoteric oxide of aluminium with the chemical formula. It is also commonly referred to as **alumina.** It is produced by the Bayer process from bauxite. Its most significant use is in the production of aluminium metal, although it is also used as an abrasive due to its hardness and as a refractory material due to its high melting point.

Aluminium oxide is an electrical insulator but has a relatively high thermal conductivity (40 W/mk) In its most commonly occurring crystalline form, called corundum or α- aluminium oxide, its hardness makes it suitable for use as an abrasive and as a component in cutting tools.

Aluminium oxide is responsible for metallic aluminium's resistance to weathering. Metallic aluminium is very reactive with atmospheric oxygen, and a thin passivation layer of alumina quickly forms on any exposed aluminium surface. This layer protects the metal from further oxidation. The thickness and properties of this oxide layer can be enhanced using a process called anodising A number of alloys, such as aluminium bronzes, exploit this property by including a proportion of aluminium in the alloy to enhance corrosion resistance.

Crystal structure - The most common form of crystalline alumina, a-aluminium oxide, is known as corundum. Corundum has a trigonal Bravais lattice with a space group of R - 3c. The primitive cell contains two formula units of aluminium oxide. The oxygen ions nearly form a hexagonal close-packed structure with aluminium ions filling two-thirds of the octahedral interstices. Alumina also exists in other phases, namely eta, chi, gamma, delta and theta aluminas. Each has a unique crystal structure and properties. The so-called *p*-alumina proved to be $NaAl_{11}O_7$.

Production - Aluminium hydroxide minerals ore the main component of bauxite, the principal are of aluminium. The bauxite ore is made up of a mixture of the minerals gibbsite ($Al(OH)_3$), boehmite (γ - AlO(OH)), and diaspore (α - AlO(OH)) along with iron oxides and hydroxides, quartz and clay minerals. Bauxite is purified by the Bayer process:

$$Al_2O_3 + 3 H_2O + 2 NaOH \rightarrow 2NaAl(OH)_4$$

The Fe_2O_3 does not dissolve in the base. The SiO_2 dissolves as silicate $Si(OH)^{2-}_6$ Upon filtering, Fe_2O_3 is removed. When the Bayer liquor is cooled, $Al(OH)_3$ precipitates. leaving the silicates in solution. The mixture is then calcined -(heated strongly) to give aluminium oxide.

$$2Al(OH)_3 + heat \rightarrow Al_2O_3 + 3H_2O$$

The formed Al_2O_3 is alumina. The alumina formed tends to be multi-phase; i.e., constituting several of the alumina phases rather than solely corund um The production process can therefore be optimized to produce a tailored product. The type of phases present affects, for example, the solubility and pore structure of the alumina product which. in turn, affects the cost of aluminium production and pollution control.

Ball clays are secondary clays which are sticky and burn "white". They possess high plasticity due to their fine grain size and possess high dry strength. They assume light colour when fired and exhibit long range of vitrification. They are usually associated with considerable amount of organic matter and other impurities. This is the reason why they have low refractoriness and their slight colour.

The following are some important general properties of clays.

Finely pulverised clays, when wet, are plastic and mouldable but, when dry, are rigid. At high temperatures, they are vitreous. These properties enable the clays to be moulded into suitable and even complex shapes when mixed with water and then, on firing, yield rigid and glassy mass unaffected by water. When moist clay is dried at 100°C, it becomes hard and brittle due to loss of free water. But if it is soaked in water, it once again becomes soft and plastic. When heated to 400 to 700°C, the combined water is expelled and a hard porous mass, which does not soften with water, is obtained. When heated further to 900 to 1000°C, the hardness and porosity increases. At 1400°C to 1800°C, it melts to give a glassy mass.

The plasticity of clays can be increased by ageing, that is, by storing it under wet condition or by weathering. This is due to hydration accompanied by gelation and also due to further subdivision of the particles by bacterial action.

The impurities associated with clays also influence their properties to some extent. For instance, presence of iron oxide imparts red colour to the burnt material. Presence of silica increases its refractory nature and porosity and reduces the extent of air-shrinkage. Presence of CaO, MgO, and Fe_2O_3 act as flux and lower the fusion point of the clay. Lime reduces the vitrification range. Clays containing high percentage of silica and very low percentage of Fe_2O_3 are known as "fire-clays" which have important industrial applications due to their high fusion point. Clays, in general, are very important raw materials for the manufacture of pottery, earthenware, bricks, tiles, terracotta, conduits for electrical cables and drain pipes.

15.2.2 Silica

Silica occurs commonly in nature as sandstone, silica sand or quartzite. It is the starting material for the production of silicate glasses and ceramics. Silica is one of the most abundant oxide materials in the earth's crust. It can exist in an amorphous form (vitreous silica) or in a variety of crystalline forms.

There are three crystalline forms of silica; quartz, tridymite, cristobalite. A high purity grade of silica, fused silica (which is around 99.4-99.9% SiOz) is produced by carbon arc, plasma arc, gas fired continual extrusion or carbon electrode fusion. Fused is primarily used in the electronics industry.

Silica has good abrasion resistance, electrical insulation and high thermal stability. It is insoluble in all acids with the exception of hydrogen fluoride (HF).

Table No. 1 Properties of quartz and fused silica

Material	Quartz	Fused silica
Density (g/cm^3)	2.65	2.2
Thermal expansion coeff. ($10^{-6}K^{-1}$)	12.3	0.4
Permittivity (ε')	3.8-5.4	3.8
Loss factor (ε'')	0.0015	
Dielectric field strength (kV/mm)	15.0 - 25.0	15.0 - 40.0
Resistivity (Ωm)	$10^{12} - 10^{16}$	$> 10^{18}$

When silicon dioxide is heated various tranformations take place as follow

$$\alpha - \text{quartz} \underset{}{\overset{575°C}{\rightleftharpoons}} \beta - \text{quartz} \underset{}{\overset{867°C}{\rightleftharpoons}} \text{tridymite} \underset{}{\overset{1470°C}{\rightleftharpoons}} \text{Cristobalite}$$

d - 2.65 g/cc

d - 2.2 g/cc

↓ 1750°C

The quartz has close packed structure and high density whereas the tridymite and cristobalite forms are comparatively open structured. The great changes in density between quartz and tridymite is responsible for the large expansion that occure during the fomration of tridymite.

Table 2. Differences between the different crystal structures of silica.

Phase	Density (g/cm^2)	Thermal expansion (10^{-6} K^{-1})
Quartz	2.65	12.3
Tridymite	2.3	21
Cristobalite	2.2	10.3

Fused silica has extremely good dielectric and insulating properties. For these reasons it is used as an inert, low expansion filler material for epoxy resins in electronic circuits.

Applications

Semiconductors - The electrical conductivity of the semiconductors is not as high as that of metals. in spite of that they have unique electrical characteristics that render them especially useful. The electrical properties of these materials are sensitive to the presence of impurities. Silicon is an intrinsic semiconductor, which means that its electrical behaviour is based on its inherent electronic structure. Silicon semiconductors are used in integrated circuits applications.

Piezoelectrics : Piezoelectricity is a property of a material where polarisation is induced and an electric field is established across a specimen by the application of an external force. Reversing the direction of the external force (i.e. tension to compression) reverses the direction of the electric field. Quartz has this property as it has a complicated crystal structure with a low degree of symmetry. Piezoelectric materials such as quartz are used in transducers such as phonograph pickups, ultrasonic generators, buzzers, alarms, strain gauges and igniters in cookers.

Refractory Materials : Silica has considerably low thermal expansion, a fairly high melting point and is resistant to creep aking it a good refractory material. It tends to be used in acid environments if **used** on its own or used as a starting material for the synthesis of other refractory products. Due is insoluble in the majority of acids, it is used as a refractory material in acidic environments. Silica is classified as an acid refractory as it behaves like an acid at high temperatures reacting with bases. The majority of its applications are in the glass industry.

Glass Refractories : Silica is an important material for use as a refractory in the production of glass. There are essentially two types of silica commonly used in glass refractories. The crystallised silica, which is composed of pure crystalline quartz. This material is crushed, graded and pressed into bricks. Upon heating up to high temperatures the quartz will transform to tridymite and cristobalite. In bricks tridymite tends to be the most favourable of all crystalline forms of silica, as it has a smooth, predictable and low thermal expansion up to 600°C. Once beyond this temperature its thermal expansion is practically zero. Bricks of this type are used in the melter crown of the glass furnace as they provide good resistance to creep and good mechanical strength at the operating temperature.

15.3 METHODS FOR FABRICATION OF CERAMIC WARE

The most common methods of formation of ceramic ware include :

(*a*) **Soft mud process.** This is the oldest method of forming clay on the Potter's wheel and is still used today particularly in art pottery. Jiggering is the most popular soft mud process particularly used in whiteware manufacture to form plates and high voltage insulators. In mechanical jiggering process, a column of de-aired clay is fed to the jigger machine and spread on the mould where the jiggering tool forms the profile surface. It is then dried to remove the ware from the mould.

(*b*) **Stiff mud process.** In this process, the stiff mud containing 12 to 20% water is forced through a steel die of any desired shape in a pug mill. The column thus obtained can be cut into desired length. Steam or oil lubrication helps in reducing the imperfections due to die friction and decrease the force required. Vacuum treatment helps to improve the workability of clays and to yield dense and homogeneous columns. This method is particularly used for quicker manufacture of bricks, hollow tiles, electrical insulators and other technical ceramic wares.

(c) **Dry Pressing.** In this method, the ceramic ware is shaped under pressure while restricting the moisture content to 5-15%. Hydraulic or automatic pressing is usually employed. This method is used for shaping floor tiles, wall tiles, high grade refractories, electrical insulators, special types of magnetic ceramics etc.

(d) **Hot pressing.** Hot pressing achieves accelerated sintering of solid particles below their melting point. The die and the plunger used for compressing are made up of graphite and the heating is done electrically by induction or by resistance. Shaping and firing take place together in this process. This process is particularly used for preparing dense shaped pieces of refractory oxides and carbides.

(e) **Slip Casting.** In this process, a 30 to 40% suspension of ceramic material in water is poured into a porous mould made up of plaster of pairs which sucks the water from the contact area. Thus a hard layer of clay is formed on the mould surface. This process is continued until the entire inner part of the mould is filled. This method is known as solid casting. In the other method known as drain casting, the excess liquid is pooured out after the desired wall thickness is built up in the mould. Suitable quantities of deflocculants e.g., Na_2CO_3, Na_3PO_4 and Na_2SiO_3 are usually added to reduce the water content and to give a firm cast. The slip is allowed to age and then is agitated to attain equilbrium. Although this process is slow, it is advantageous in case of intricate shapes.

Drying of ceramic ware. The shaped ceramic wares obtained by the above processes must be first dried carefully before firing. Direct firing of the shapes in the kiln results in cracking of the ware. The drying rate should be carefully controlled by regulating the humidity, flow rate of the air used for drying and its temperature. Excess shrinkage during drying leads to crack formation. Shrinkage can be reduced by controlling the particle size and the initial water content.

The method of drying used depends on the quality, size of the ware and the process used for preparing the ware. Hand-made ware requires slower rate of drying as compared to the ware formed by automatic processes.

Drying sheds are used for drying hand-made wares. Hot floor drying rooms are used for large shapes and refractories that require moderately slow drying rate. Infrared drying is used for thin ceramic wares which require rapid drying.

Firing of ceramic ware. The dried ceramic ware is fired at temperatures in the range of 700-2,000°C (depending on the composition and the properties desired) to impart hardness, durability and strength. Unglazed ceramic ware is usually fired once only whereas glazed ware is fired twice i.e., biscuit firing and glost firing. However, the recent trend is to apply glaze to the unfired ware itself and then subject it to one-fire process only. The kiln used for firing may be:

(a) periodic or batch kiln (e.g., up-draft kiln, down-draft kiln and scove kiln).

(b) Continuous or counter-flow kiln (e.g., chamber kiln or moving fire zone kiln and tunnel kiln or fixed fire zone kiln).

Obviously, continuous kilns offer advatanges such as more thermal efficiency, lesser pollution due to smoke and ash, lesser contamination, lesser firing time, lesser fuel and labour charges. Some of these kilns are described under the chapter on "Refractories".

15.4. CERAMIC PRODUCTS

Important ceramic products fall under the following three types : (1) Structural clay products (2) whitewares and (3) earthenware and stonewares.

(1) Structural clay products :

These include bricks, tiles, blocks, terracotta (glazed or unglazed decorative products made from fire-clay e.g., flower vases) etc. Many of the structural products are made from relatively low-grade clays, shales and "grog" (i.e., old rejected bricks). The following steps are usually involved in the manufacture of the structural clay products.

(*a*) *Grinding and Screening* :- The clay or shale ground and screened using vibratory screens.

(*b*) *Tempering* :- The raw materials are mixed in dry or wet pans with small quantity of water and some organic matter to increase plasticity.

(*c*) *Forming* :- The sticky and plastic clay is then moulded to the desired shapes using stiff-mud process or soft-mud process or dry-pressing process described earlier depending upon the nature of the clay, its plasticity and the water content.

(*d*) *Drying* :- The shapes i.e., bricks or tiles are extruded through the dies and dried outdoors, or in sheds or in tunnel driers as required.

(*e*) *Firing* :- The bricks or tiles are then fired in a kiln for about 7 days at 875°C to 1100°C depending upon the properties of the raw material used and of the product desired. Some finely divided "grog" is usually incorporated to prevent undue shrinkage during firing. To avoid cracking of the bricks, slow cooling to room temperature over 7 days is generally adopted.

(2) White wares

The term "whiteware" or "white pottery" usually refers to glazed or unglazed ceramic materials which have white or cream colour after firing and have fine structure. Whitewares are usually prepared from suitable proportions of China clay (or any white burning clay free from iron compounds), feldspar and quartz (or flint). The mixture is made into paste with suitable quantity of water and subjected to any of the forming processes described earlier to produce the desired shape. This is followed by careful drying and then firing at 1350 to 1500°C when partial vitrification takes place.

Whiteware products consist of a refractory body with a glossy coating called the "glaze". These are usually made by any of the following two process ":

(*a*) In the "porcelain process", the body and the glaze are developed in one firing only.

(*b*) In the "China process", the glaze is developed in the second stage and hence is superior to that obtained by the "porcelain process".

The following sequential steps are involved in the manufacture of white pottery :

(*a*) *Preparation of the body of the ware* : The finely powdered ingredients are mixed with suitable quantity of water. The desired shape is produced by any of the forming processes described earlier e.g., hand moulding or "throwing" on a potter's wheel or slip casting or jiggering or pressing. The article thus produced is carefully dried and fired in a biscuit oven (called biscuit firing). This produces porous ware called "bisque".

(*b*) *Glazing* :- The porous ware has to be coated with a glaze to render it water-tight. This is done as follows : The glaze constituents viz., quartz, feldspar, boric oxide and lead oxide are finely powdered and mixed with water to form a slurry ("slip"). The "bisque" is dipped into the "slip" and then fired once again (called the "glost firing") until the glaze is matured.

(*c*) *Decoration* :- In order to decorate the article, the desired design may be painted on the body before glazing (which is called "under-glazed") or the design may be painted upon the glaze followed by another firing to melt the pigment into the glaze (which is known as "over-glazed"). Metallic oxides are used as pigments for "under-glazed" whereas coloured glasses are generally used for "over-glazed" designs.

Attractive and coloured floor and wall tiles are produced by adding chromium oxide or cobalt

GLASS AND CERAMICS

oxide or magnese oxide to the white burning clay bodies to get the colour shades. They are biscuit fired, glazed and re-fired.

Porcelain articles are prepared by the wet process or the dry process or the casting process depending on the properties of the end-product desired and its application. If fine grained and highly glazed insulation is required for high voltage service, wet process is employed. If open-textured low voltage pieces are to be produced quickly on large scale, then dry process is employed. If very big and intricate articles are to be prepared, the casting process is employed. Although the raw materials used are same, the above three processes differ in the forming and drying operations.

Chinaware articles are produced by more complicated procedures. Small and round object such as cups, saucers and plates are produced in large scale by jiggering. Other articles are either cast from clay slip in moulds made of plaster of paris or they are shaped by throwing on the potter's wheel by expert hands to obtain the desired shapes.

Whitewares are characterised by low-porosity, translucency and good strength. A fully vitrified ware will have zero porosity and zero absorption. White wares find extensive use as electrical insulators, spark plugs, crucibles, dishes and other laboratory equipment and decorative pottery.

(3) Earthenwares and stonewares

Clay products which are hard and strong like stone come under this category. Relatively softer type of clay products which are prepared by burning at lower temperatures are called "earthenwares" while the clay products, which are denser and harder and are prepared by firing at higher temperatures are known as "stoneware". Earthenwares and stonewares are usually glazed to render them strong, compact and impervious to water.

Earthenwares :

Earthenwares are of two types, the "coarse" (e.g, terracotta) and the "fine". Coarse earthen wares are prepared from crude clay containing about 50% of sand or chalk and about 10% iron compounds with small quantities of associated mineral matter. Articles of coarse earthenware are generally unglazed but for some special purposes, an interior or exterior coating is applied. Fine earthenwares are made of clays which burn white e.g., China clay and ball clay. Hence fine earthen wares are white or almost white in colour. China clay imparts texture and translucency whereas ball clay gives strength to the unfired shapes. Fine earthenwares are invariably glazed. The mechanical strength of fired earthenware bodies depends on their porosity and the nature of the glassy phase present.

Chemical stonewares

These are generally coarse and robust. They can be produced with plastic clays (a) containing relatively higher percentage of iron oxide; (b) which undergo uniform shrinkage during drying and firing, and (c) which sinter well at relatively low temperatures (1100-1200°C). Chemical stonewares are generally glazed with hard porcelain type of glass and are single fired. Sanitary wares are constituted of glazed vitreous body of high strength. Sanitary wares require a relatively higher firing range than earthenwares.

A typical formulation for producing stoneware is as follows : clay - 50%, feldspar – 20%, flint – 25%, kaolin – 5% and grog – 10%. Sanitary wares are made from raw material compositions such as feldspare (23-35%), ball clay (10-25%), China clay (20-35) and quartz (20-30%). The shaping is done by casting, moulding or throwing on a potter's wheel. The shapes are dried and then fired in a tunnel kiln at high temperatures, above 1000°C. They are then subjected to salt glazing in which common salt is thrown on the fired articles at about 1100°C. The salt vapour produces sodium aluminium silicate which being more fusible fills the surface pores effectively.

The stonewares have high chemical resistance, high impermeability to liquids, long service life and high mechanical strength. Stonewares with very high strength, very high degree of impermeability, high thermal stability, high dielectric strength, high thermal conductivity and low coefficient of thermal expansion are prepared for special applications. By virtue of the above properties, stonewares are widely used in industry for pipes, condensers, retorts, stills, chlorinating vessels, acid pumps, stopcokcs, absorption towers, valves, and pumps used in chemical industry. Stonewares are extensively used for sanitary fixtures *e.g.*, sinks, bath tubs, water closets, drain pipes, sewerage pipes, hume pipes and underground cable sheathings. Stonewares are relatively cheap but are fragile and become totally useless when once broken.

15.5. GLAZES

Glazes are low melting glasses of different compositions, which form a firm and continuous layer on a ceramic surface to render it smooth, nonporous, non-permeable, hard, attractive, and easy to clean. A glaze comprises of a mixture of glass-forming materials of proper composition such as lead silicates, borosilicates, etc. A glaze is applied on a ceramic body to achieve the following objectives : smoothness, durability, glossy surface, non-permeability to liquids, hardness, facelift and decorative effect.

Glazes are classified as follows :-

(1) Classification based on composition :-

   (*a*) Raw glazes containing insoluble raw materials e.g., porcelain glazes, glazes containing Pb or Zn etc.

   (*b*) Fritted glazes containing a small quantity of glass before firing.

   (*c*) Vapour glazes which are deposited from the vapour phase e.g., salt glaze, smear glaze etc.

(2) Classification based on optical properties :-

   (*a*) Transparent (*b*) Opaque (*c*) Crystalline, etc.

(3) Classification based on surface characteristics :-

   (*a*) Glossy   (*b*) Mat   (*c*) Semimat, etc.

Glazing compositions are usually ground in ball mills lined with porcelain to a fineness of about 10 μm. The glaze is usually applied in the form of a suspension of the powder ingredients in water, which dries on the surface of the ceramic body as a layer. The common methods of application of glazes on ceramic bodies are brushing, pouring, dipping and spraying. Spraying is most commonly used commercially. On firing, the ingredients react and form a thin layer of glass on the surface of the ceramic body.

Ordinary glazing mixtures free from iron salts and colouring pigments produce colourless glaze. Pigments and metal oxides mixed in proper proportions yield coloured glazes as follows :

             Iron oxide – Red or brown

             Copper oxide – Blue

             Cobalt salt – Blue

             (Iron oxide + lime) – Cream.

Salt glazing is used for impairing glossy film over earthenwares. In this process, common-salt or NaCl is thrown into the furnace containing the ceramic shapes in red hot condition. The salt vapours react with silica present on the article to form glossy and impervious film of sodium silicate.

In liquid glazing, which is a superior method, the powdered glazing mixture and the desired colouring pigments are mixed with water to get a colloidal solution called "glaze slip". The ceramic shapes which are to be glazed are burnt in a kiln carefully at a low temperature. They are then dipped momentarily in the "glaze slip" and then fired once again in the of soot and dust during firing.

Glazes are strong in compression and weak in tension. Therefore, cracks can develop if the glaze has a high coefficient of expansion than that of the ceramic body. Defective glazing may result in pinholes, crawling and peeling of the surface.

15.6. PORCELAIN AND VITREOUS ENAMELS

Porcelain or vitreous enamels are fused silicate coatings applied to metals to give protective coating as well as decorative effect. Vitreous enamels with a wide range of physical and mechanical properties can be formulated. The metal is first given a "ground coat" and then the "cover coat". The composition of a typical ground coat "frit" is : Silica (20%), feldspare (30%), borax (30%), soda as (8%), fluorspar (6%), $NaNO_2$ (40%), MnO_2 (1.2%), NiO (0.3% and cobalt oxide (0.5%). The composition of a typical cover coat "frit" is as follows : flint (17.1%), feldspar (17.1%), borax (17.1%), soda ash (30.8%), $NaNO_2$ (2.7%), ZnO (3.4%), Sb_2O_3 (2.4%) and cryolite (9.3%). Opacifiers such as tin oxide, TiO_2 and zirconium silicate may also be added during pebble milling of the frit with clay and water to produce the workable suspension called "slip". After applying a layer of the slip on the metallic article, it is dried and then fired in a muffle furnace for about 2-3 minutes at about 800°C. Then, the cover coat is sprayed on the ground coat and dried. Then it is fired again for about 2-15 minutes at about 800°C.

QUESTIONS

1. What are ceramics ? How are they clasified ? What are the major constituents of ceramics ?
2. What are the important raw materials for the manufacture of whiteware products ? How are white wares prepared ?
3. Describe the various steps in the manufacture of whitewares. Discuss the various characerstics of the cermic whiteware products and their industrial applications.
4. Describe the various methods available for the fabrications of ceramic ware.
5. How are structural clay products manufactured ?
6. Describe the manufacture and uses of porcelain.
7. Write an informative note on any three of the following :
   (a) Glazes   (b) Vitreous enamels
   (c) Firing of ceramic ware   (d) Clays   (e) Feldspars

16

Refractories

Refractories are such inorganic materials which can withstand very high temperatures without softening, melting or deformation. The essential function of a refractory is to serve as structural material and maintain its mechanical functions at high temperatures under the service conditions. There is no well-defined line of demarcation for separating refractories from non-refractories. However, according to ASTM classification, a fire-clay refractory brick must be thermally stable atleast upto 1515°C. With increasing use of high temperature processes, the demand for various types of ceramic refractories is constantly growing in mechanical engineering as well as in the metallurgical, chemical and power industries. Refractories are very widely used for providing high-temperature resistant lining for furnaces, kilns, ladles, crucibles, etc., in various industries such as ferrous and non-ferrous, glass, ceramics, power-generation, oil refining and cement. They are also used in the manufacture of rocket nozzles and launch pads on one hand and for domestic heating on the other. The technical utility of a refractory depends upon the judicial balancing of the properties and behaviour of the refractories on one hand and the destructive conditions likely to be encountered in the actual service conditions on the other.

16.1. REQUISITES OF A GOOD REFRACTORY

The requisites of a good refractory material are as follows :
(1) They should possess good refractory properties *i.e.*, their physical, chemical and mechanical properties should not undergo substantial changes at high temperatures.
(2) They should be chemically stable under the service conditions in which they are employed *i.e.*, they should not react with corrosive agents such as acidic or basic molten slags, hot gases etc.
(3) They should possess good thermal strength *i.e.*, they should be able to withstand thermal shock due to rapid and repeated temperature fluctuations.
(4) They should possess good resistance to abrasion by dusty gases and erosion by molten metals.
(5) They should be able to withstand the charge load at the working temperature and other severe operating conditions.
(6) They should possess low permeability.

16.2. CLASSIFICATION OF REFRACTORIES

Refractories can be broadly classified into the following three categories :
(1) Acid refractories which are not attacked by acid slags, *e.g.*, silica and fire clay refractories.
(2) Basic refractories which are not attacked by basic or neutral media, *e.g.*, magnesite,

REFRACTORIES

dolomite, chrome-magnesite.
(3) Neutral refractories which are not attacked by slightly acidic or neutral media, *e.g.* Carbon and graphite, chromite, silicon carbide etc. In addition, special refractories are also available which possess superior properties :
   (*a*) Single oxide refractories *e.g.*, alumina, magnesia and zirconia.
   (*b*) Mixed oxide refractories *e.g.*, zircon, spinel, mullite.
   (*c*) Non-oxide refractories, *e.g.*, carbides, nitrides, silicides and borides.

Some other classifications are also in vogue which are based on service conditions, mineral constituents, applications or manufacturing methods.

Refractories are available in different shapes and sizes as follows : (*a*) bricks (*b*) crucibles and tubes (*c*) granules (*d*) castables and (*e*) cements.

Physical terms

Refractories are supplied to the trade in several physical forms which include bricks, finely ground cementing materials, plastics, castables, and granular materials in bulk.

The principal refractory product is a brick or other preformed shape. However, there are numerous sizes and shapes to fit all types of construction: rectangular forms, wedges, arches, keys, skews, jambs, feather-edges, necks, bung arches and segmental shapes including circle brick, cupola and rotary kiln blocks. These are used in coke ovens, runners, tuyeres, burners, muffles, crucibles, saggers, glass pots, stoppers, nozzles, tubes, feeder parts for glass tanks, spark plug cores and highly specialized items of laboratory ware.

Refractory materials are also available as mortars or cements, for laying-up, coating or patching brickwork. They are supplied either dry or wet in a ready-mixed form for immediate application. They may be air-setting at ordinary temperatures, or heat-setting during furnace operations.

Plastic refractories are essentially moist unformed brick mixes supplied for forming special shapes and solid jointless (monolithic) furnace sections at the installation point. They can be rammed into place with relatively low pressure and fired by the heat of the furnace in which they are installed.

Castables are refractory concretes and the aggregates now comprise of almost all common refractory materials. Highly porous refractory aggregates are used for insulating castables.

Bulk products are prepared from refractories like grain magnesite, dolomite, chrome ore, fire-clay, sand, and ganister and are supplied in different grain sizes for use in making bottoms, banks and fills of furnaces, as well as for other miscellaneous applications.

Behaviour of refractories

Refractory bricks and other products are generally compounded from several raw materials designed to control the refractoriness and other physical properties and behaviour. The chemical composition is thus influenced by the proportions of the raw materials employed and their relative availability in different producing centres. A knowledge of the relative behaviour of the various refractories available is helpful for the judicious selection of the right type of the refractory material for use in presence of numerous destructive forces encountered in high temperature processes. The behaviour or resistance to these forces can be determined by standard laboratory tests designed for this purpose and correlating these results with actual service observations.

Chemical attack, slugging or melting, compressive and tensile stresses at elevated temperatures, thermal and mechanical shock and abrasion can destroy a refractory material. The specific heat and thermal conductivity are of vital importance. If the refractory is designed to act as an insulator also, a low thermal conductivity is desirable. On the contrary, if the refractory is used for muffle walls and recuperative systems, a high thermal conductivity is desirable. A high heat capacity is desirable for a refractory used in the checkers of regenerative systems.

The performance of a refractory product is influenced by its composition, by its mechanical processing variables, and to a great extent by the thermal treatment received during its manufacture and the subsequent use.

Phase equilibria

Some physical and chemical characteristics of refractories can be expected from the phase equilibrium data of the refractory oxide systems involved. During the thermal treatment received in manufacture, most of the refractories do not attain complete chemical equilibrium. In some cases, this may proceed at an accelerated rate when the refractory is later exposed to high temperature under service conditions. Although many refractories contain a mixture of oxides, their actual behaviour depends mostly on two or three-oxide-component systems. Under the actual service conditions, they may interact with other reactive materials in the form of furnace charges, slags or vapours which may complicate the physical and chemical reactions. The crystal-liquid relationships in such systems are of vital significance in understanding the ability of the refractory to maintain its identity when subjected to high temperature or its stability under stresses. The viscosity of the liquid phase and the degree of solubility of the crystalline components of the system also play a vital role.

16.3. PROPERTIES OF REFRACTORIES

In order to select a refractory, which is best suited for a particular operation, the following aspects should be thoroughly assessed :

(1) the working temperature of the furnace
(2) the nature of the materials to be handled
(3) the magnitude of temperature fluctuations and their frequency
(4) the load applied, and
(5) the chemical reactions likely to be encountered.

Generally, for constructing a furnace, several types of refractories may be required because no single refractory can withstand the different conditions prevailing in different parts of the furnace. The following are the important characteristics of the refractories :

(1) Chemical properties

As stated earlier, commercial refractories are classified as acid, basic and neutral, although in many cases, a sharp distinction cannot be made. However, an acid refractory such as a silica brick should not be used in contact with an alkaline product. Similarly, a basic refractory such as magnesite brick should not be used in contact with an acidic product. The acceptability of a refractory should be determined on the basis of both the chemical and physical properties. Chemical reactions of refractories may occur due to contact with furnace gases, slags, fuel ashes, steel, glass etc.

(2) Refractoriness

It is the property of a material by virtue of which it can withstand high temperatures without appreciable softening or deformation under working conditions. Refractoriness is usually measured by the softening or fusion temperature of the material. Obviously, a refractory material should have softening temperatures higher than the operating furnace temperatures. Most of the commercial refractories soften gradually over a wide range of temperatures and do not exhibit sharp melting points because they are composed of several minerals, both crystalline and amorphous in nature. Fusion temperatures of some pure and commercial refractories are listed in Table 1.

REFRACTORIES

Table 1. Fusion Temperature of some Refractories

Refractory Material	Fusion Temperature, °C
Silica (SiO_2)	1710
Silica brick	1700
Fire clay brick	1600-1750
Kaolinite ($Al_2O_3 \cdot 2SiO_2 \cdot 2H_2O$)	1785
Bauxite brick	1732-1850
High alumina clay brick	1802-1880
Alumina (Al_2O_3)	2050
Magnesia brick	2200
Magnesia	2830
Chromite ($FeO \cdot Cr_2O_3$)	1770
Chromite brick	1950-2200
Spinel ($MgO \cdot Al_2O_3$)	2135
Silicon Carbide (SiC)	2700
Forsterite ($2MgO \cdot SiO_2$)	1890
Sillimanite (Al_2SiO_5)	1816
Zirconia (ZrO_2)	2710
Zirconia brick	2200-2700
Boron nitride	2720
Boron Carbide (B_4C)	2450
Zirconium boride (ZrB)	3040
Titanium boride	2940
Silicon nitride (Si_3N_4)	1900
Molybdenum disilicide ($MoSiO_2$)	2100
Lime	2570
Carbon (C)	3500

The softening behaviour (fusion point or refractoriness) is commonly determined by means of the standard PCE (Pyrometric Cone Equivalent) test. In this test (A.S.T.M. C_{24-46}), the finely ground refractory is molded into small cones in the form of slim tetrahedrons and heated at a prescribed rate. Their bending or softening behaviour is compared on the same test plaque with standard pyrometric cones calibrated for testing the refractories. Standard data are available pertaining to the bending intervals and the end points (temperature, °C) for each of the standard pyrometric cones at various heating rates (*e.g.*, 20°C/hr, 100°C/hr, 150°C/hr and occasionally, 600°C/hr). The bending interval is the interval between the temperature at which the cone starts to bend and the end point (which is the temperature at which the cone finally touches the plaque), as shown in Fig. 16.1. The

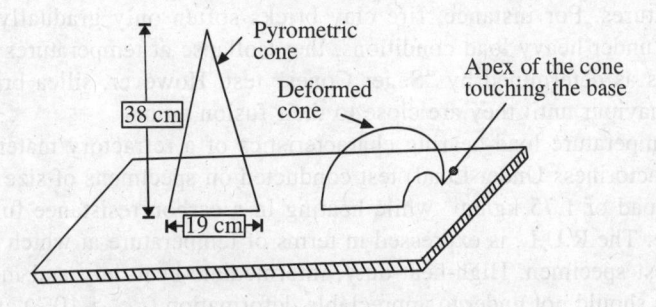

Fig. 16.1. Pyrometric Cone Equivalent Test.

P.C.E test is used to measure the refractoriness on all forms and types of refractories. In case of bricks this test is supplemented by a hot compressive or load test (A.S.T.M.C$_{16-49}$). This is carried out by applying a static load of 25 p.s.i. throughout a specified heating period of a 9 inch straight brick standing on end. Five heating schedules are used depending on the type of refractory brick being tested.

The maximum service temperatures recommended for fire-clay bricks are generally somewhat lower than the PCE temperature values. Silica bricks can be used at temperatures very close to the PCE temperature. However, ladle bricks are an exception and they are normally exposed to temperatures several hundred degrees higher than the PCE temperature to take advantage of the softening of the brick.

The PCE test is also used to check the uniformity of composition of refractory raw materials and finished products, to classify fire-clay refractories, and to determine the contamination of the refractories from the fluxes and other materials encountered in service conditions.

(3) Porosity

Porosity of a material is given by the ratio of its pores volume to that of its bulk volume. The higher the porosity of a refractory brick, the more easily it is penetrated by gases and molten fluxes. For a particular class of refractory brick, the one with the lowest porosity may be taken as the best because it will have the greatest strength, heat capacity, and thermal conductivity. Further, it will have greater resistance to abrasion and corrosion although the resistance to thermal spalling decreases with the decrease in porosity. Thus, porosity is an important property of a refractory material as it is related directly to many other physical properties.

(4) Thermal Spalling

This is a property by virtue of which a refractory brick suffers fracturing, flaking, peeling or cracking at high temperatures or under rapid fluctuations in temperature due to the development of uneven stresses and strains or compression. Spalling may also occur due to penetration of molten slag into the pores of refractory brick which leads to cracking due to the differences in their coefficient of expansion and contraction. Bricks with the highest expansion and least uniformity are more susceptible to thermal spalling. Spalling can be minimised by employing refractory bricks having low coefficient of expansion, good thermal conductivity, uniformity and relatively high porosity within the permissible limits. Further, avoiding sudden temperature fluctuations, overfiring the refractory bricks during manufacture and improved furnace design to minimise the stresses and strain can also help in reducing the spalling tendency.

(5) Strength

Commercial refractories that are used for lining high temperature furnaces are expected to withstand varying loads of the charge. Hence, refractory materials must possess high mechanical strength under the strenuous service conditions. Cold strength has not much relation to strength at high temperatures. For instance, fire clay bricks soften only gradually over a range of temperature but under heavy load conditions, they collapse at temperatures much lower than the fusion points as determined by "Seger Cones" test. However, silica bricks exhibit good load-bearing behaviour until they are close to their fusion points.

The high-temperature load-bearing characteristics of a refractory material is determined by R.U.L. (Refractoriness Under Load) test conducted on specimens of size 5 cm^2 and height 75 cm under a load of 1.75 kg/cm^2 while heating in a carbon resistance furnace at a rate of 10°C per minute. The R.U.L. is expressed in terms of temperature at which 10% deformation occurs on the test specimen. High-heat-duty, intermediate-heat-duty and moderate-heat-duty refractory bricks should not undergo appreciable deformation (*i.e.*, ≯ 10%) at 1350°C, 1300°C and 1100°C respectively under a load of 3.5 kg/cm^2.

REFRACTORIES

(6) Thermal Conductivity

Thermal conductivity is an important property of a refractory because it determines the amount of heat that flows through a furnace wall under given service conditions. Insulating refractories with low thermal conductivity will not allow much loss of the furnace heat and hence they are used in blast furnace, open hearth furnace, etc. On the other hand, refractories with high thermal conductivity are used in the construction of muffle walls, retorts and recuperators where efficient heat transfer from the outer surface to the charge is needed. The densest and least porous bricks possess the highest thermal conductivity due to the absence of air in the voids. If the brick is porous, the air entrapped in the voids (or pores) provides an insulating effect. Porous bricks can be prepared by mixing copious amount of a carbonaceous material with the refractory mix before moulding. When the moulds are burnt, the carbonaceous material burns off leaving behind minute voids which provide the insulating effect.

(7) Heat Capacity

The heat capacity of a furnace depends upon the following three factors: (*i*) the thermal conductivity (*ii*) the specific heat and (*iii*) the specific gravity of the refractory.

Lightweight refractory bricks have low heat capacity and hence are 'suitable for intermittently operated furnaces because the working temperature of the furnace can be achieved in lesser time and with lesser consumption of fuel. Conversely, the dense and heavy fire-clay bricks have higher heat capacity and as such are best suited for regenerative checkerwork systems used in coke ovens, stoves for blast furnaces, glass furnaces etc.

(8) Resistance to temperature changes

Refractory bricks having coarsest textures and lowest thermal expansions are most resistant to rapid thermal changes because the strain developed in them will be relatively less. However, such bricks are more susceptible for corrosion and abrasion and they also possess relatively low crushing strength. Conversely, fine and dense-textured refractory bricks, on account of their lower porosity and greater crushing strength are more resistant to sudden temperature changes and are also less susceptible for corrosion and abrasion.

(9) Dimensional stability

It is the resistance of a refractory material to changes in volume due to its prolonged exposure to high temperatures. These dimensional changes could be reversible or irreversible. Irreversible changes may lead to contraction (as in the case of fire-clay bricks) or expansion (as in the case of silica bricks). These changes take place because of variations in the extent of vitrification or because of transformation from one crystalline form into other having different density.

(10) Electrical Conductivity

Refractories are poor conductors of electricity with the exception of graphite. Refractories used for lining of electric furnaces should also have low electrical conductivity.

(11) Bulk density

This influences many other important properties. A high bulk density will improve strength, volume stability, heat capacity and spalling resistance. However, for insulating refractories, a porous structure is needed which is provided by a low density of the refractory.

16.4. RAW MATERIALS FOR THE REFRACTORIES

Following are the raw materials used for the production of bulk refractories :

(A) Minerals

(1) Fire clay. It is a sedimentary clay having low percentage of alkali metal salts and iron which is responsible for its refractoriness. Rich deposits of fire clay occur in Bihar, Orissa, West Bengal, Gujarat, M.P., Maharashtra and Rajasthan.

(2) Bauxite. It can be represented by the formula $Al_2O_3.nSiO_2.nH_2O$ and is majorly used in the alumino-silicate refractories. Good bauxite deposits occur in Bihar, M.P., Gujarat, Maharashtra, Karnataka, Tamil Nadu, Orissa and Jammu & Kashmir.

(3) Kyanite and Sillimanite. Both these minerals have the same chemical composition, $(Al_2O_3.SiO_2)$. Kyanite occurs in long, bladed crystals having light green or blue colour whereas sillimanite occurs as fibrous masses comprising of thin slender crystals having greyish colours. India produces large quantities of these two minerals. The best deposits of kyanite occur in Singhbhum (Bihar) while the richest deposits of sillimanite occur in Jaintia and Khasi hills (Meghalaya).

(4) Diaspore. It has the chemical composition, $AlO(OH)$, and it occurs in lamellar masses with pearly luster or in prismatic crystal form. The best deposits of diaspore are in Hamirpur and Jhansi (U.P.) and Shimpuri and Tikamgarh (M.P.).

(5) Magnesite. It is nothing but magnesium carbonate but it generally contains small quantities of iron carbonate. Richest deposits of magnesite occur in Salem (Tamil Nadu) while notable deposits are located in Karnataka and U.P.

(6) Dolomite. Limestone containing more than 10% $MgCO_3$ are called dolomitic. True dolomites contain more than 45% $MgCO_3$. Dolomite deposits are widely distributed in India and are mainly used in refractory industry and metallurgy.

(7) Quartzite. It mainly contains microcrystals of quartz which have been firmly cemented with secondary silica. Refractory grade quartzites occur in Bihar, Karnataka, Orissa, M.P., A.P., Rajasthan, West Bengal, Gujarat, Tamil Nadu and Kerala.

(8) Graphite. It occurs in thin inelastic flexible laminae or in massive deposits. Both the crystalline and amorphous forms of graphite are used in refractory industry. Best deposits of graphite occur in A.P., Orissa, Kerala, Tamil Nadu and Karnataka.

(9) Chromite. It is a double oxide of Cr and Fe. Notable deposits of chromite occur in Orissa, Karnataka and Bihar.

(10) Zircon. It is $ZrSiO_4$ and occurs in granites, pegmatites and nepheline syenites. It is also commonly found in beach sands of coastal India along with monazite and ilmenite.

(11) Mica. India is the largest producer of mica in the world. It has low thermal conductivity. Its chemical composition is complex as it consists of a number of minerals. Chief mining centres of mica in India are located in Bihar, A.P. and Rajasthan.

(12) Silicon Carbide. It is prepared by heating sand and coke at 2000°C in an electric oven. It is known as carborundum.

(13) Diatomite. It is also known as diatomaceous earth or kieselgurh. Its name is derived from the fact that it is composed of skeletons of diatoms which are microscopic organisms. Its closed cell structure and high porosity are responsible for its low density and low thermal conductivity. It is used as a heat insulator. Workable deposits of diatomite are rather scarce in India.

(14) Vermiculite. It is a hydrated magnesium aluminium silicate resembling Biotite, the micaceous mineral. It can be split into thin laminae which are soft, pliable and inelastic. When heated in the absence of air, it changes its colour to silvery grey and when heated in the presence of air, it turns to golden brown. On heating to 150°C, it expands or exfoliates. The expanded granules entrap air producing light weight aggregates having about six times its original volume. This contributes to its insulating property. Best deposits of vermiculite occur in A.P., Bihar, Rajasthan and Karnataka.

REFRACTORIES

(B) Fuels

The commonly used fuel for refractory manufacture in India is non-coking steam coal of selected A and Grade I types. The coal is mixed in 1:1 proportion for firing alumina refractories and in 3:1 proportion for firing silica and basic refractories.

(C) Binders

The fluxes required to bind together the particles of the refractories are generally kept at a minimum to reduce the degree of vitrification. The binders used may be organic or inorganic in nature. The organic binders include molasses, starch, saw dust, shellac, gum, dextrin, cellulose, ethyl silicate etc. The inorganic binders in common use include bentonite, sodium silicate, lime, calcined gypsum, plastic clay and sulphite lye.

The possibility of shaping articles made from bodies without clay or even without natural plasticity paved the way for the manufacture of single-component ceramics (*e.g.*, pure oxide refractories) which exhibit superior properties. Such refractories are monocrystalline and self-bonded in contrast to the conventional vitreous-bonded types.

16.5. MANUFACTURE OF REFRACTORIES

The manufacture of refractories includes several physical operations and chemical conversions which are essentially same for different products although the raw materials involved are variable. For the manufacture of bulk refractories, the following operations are usually involved :

(1) Preliminary treatment

(*a*) **Calcination.** Raw materials like dolomite, magnesite, bauxite and fire clay are calcined at temperatures ranging from 1200-1700°C in downdraught, shaft tunnel or rotary kilns. This operation removes the volatile constituents in the respective raw materials and renders them dimensionally stable by minimising shrinkage in the subsequent firing operation.

(*b*) **Crushing and grinding.** The raw materials are crushed in jaw crushers or gyratory crushers and then finely ground in ring rolls, hammer mills, ball mills or tube mills.

(*c*) **Sizing.** The ground raw materials are sized into coarse, medium and fine fractions with the help of multideck vibratory screens. Finer fractions are sized with the help of air classifiers. The coarse fractions are recycled to the grinding mills for further grinding to the specified fineness.

(*d*) **Beneficiation.** If the raw materials contain any undesirable impurities, they should be beneficiated by processes like magnetic separation, electrostatic separation and froth floatation. Beneficiation is usually needed for enriching chromite and magnesite from impurities like silica.

(2) Blending and mixing

This is the stage at which the raw materials of the correct size distribution are blended with bond additions and water so as to obtain the final product of desired porosity. Low porosity refractories can be produced by selecting proper particle sizes of the raw materials so that the interstices between the particles are filled by the finer ones. The type and percentage of the bond additions depends upon the type of refractory (ordinary or porous) that is produced. The quantity of water added depends upon the type of moulding to be followed later.

>3 to 7%-dry pressing
>10 to 15%-extrusion and repressing
>10 to 20%-hand moulding

Blending is usually carried out with the help of devices such as paddle mills, pug-or pan-mills, enrich mixers, etc. The following are the main functions of blending and mixing :

(*a*) Controlling the porosity of the final product by reducing the air entrapped in the interstices.

(*b*) Producing a homogeneous mixture of the batch ingredients.

(*c*) Distribution of the plastic material to coat thoroughly the non-plastic constituents thereby developing plasticity.

(3) Forming or moulding

The forming methods usually employed are similar to those used for preparing ceramic ware. These are described briefly as follows :

(a) **Soft mud process.** In this process, plastic mixes containing 10 to 20% water are used for hand moulding large and intricate shapes in wooden moulds lined with metal. Sand or water are used as lubricant to facilitate the release of the formed shape to leave the mould. Pneumatic ramming is sometimes adopted to achieve close packing. Automatic moulding presses are employed for large-scale production of standard shapes.

(b) **Stiff process.** In this process, semiplastic mixes having 10 to 15% water content are horizontally extruded under moderate pressures through a die of desired shape by a de-airing auger. The friction is reduced with the help of steam or oil. The column emerging from the die is cut into desired lengths with a wire cutter. The wire-cut bricks are passed through a machine repress of de-airing type for final shaping and embossing.

Mechanical moulding gives high strength and high density to the refractory brick as compared to hand moulding. In order to increase the strength and density of the refractory by mechanical moulding further, de-airing is recommended. De-airing decreases laminations and cracking of the brick when pressure is released after pressure moulding and increases the density. De-airing is carried out by one of the following techniques : (a) Removing the air in the voids of the refractory by decreasing the rate of pressure application and release, (b) Double pressing (Repressing) in which the material is first pressed and allowed to crack and then it is repressed to close the voids. (c) Application of vacuum through vents in the moulds and (d) sending a gas like butane to displace the air followed by application of pressure when the gas is absorbed by the clay.

(c) **Dry pressing.** The great demand for refractory bricks of high strength, density and uniformity has resulted in the use of dry-press method of moulding with mechanically operated presses. This method is particularly suitable for non-plastic mixes containing 3 to 7% water content. Measured quantities of the raw mix are placed in a die and are pressed into shapes under high pressure (30 to 150 MN/m^2) in a dry press e.g., toggle-, rotating table-, hydraulic- or friction-press preferably of de-airing type. Dry-pressing can produce large quantities of refractory bricks (2000 per hour) of standard shape having uniformity, good warp resistance, load resistance and spalling resistance.

(d) **Slip Casting.** The process is particularly suited for intricate shapes with thin sections although the process is slow. A stable suspension of the finely powdered mix in water, just fluid enough to pour (called "slip") is prepared. About 0.02% of a deflocculant e.g., sodium polyphosphate or a mixture of sodium carbonate and sodium silicate is added to the slip, and it is agitated and poured into a plaster of paris mould. The agitation is sometimes done in vacuum to remove any entrapped air. The plaster of paris mould absorbs the water from the contact surface and a uniform deposit of the material is obtained on the inner side of the mould. After the shape becomes somewhat hard, the mould is dismantled and the cast is subjected to further processing viz., drying and firing.

(4) Drying

The main purpose of drying is to remove the moisture added to develop plasticity in the mix before forming (moulding). The rate of drying should be carried out very slowly under selected set of humidity and temperature so that no voids are left in the refractory nor cracks are produced due to internal stresses developed because of shrinkage. The following three types of driers are usually employed :

REFRACTORIES

(a) **Hot Floor Dryers.** These dryers are used for soft mud refractories and for large shapes. The hot floor dryer comprises of a heated concrete floor on which the green shapes are placed and a roof through which the water evaporated escapes. The heating is done by burning coal or oil or by exhaust steam from engines or by the waste heat from kilns.

(b) **Tunnel Dryers.** These dryers are used for bricks and small shapes. The green shapes are set on trolleys which move in the tunnel from one end to the other. A stream of hot gas passes through the tunnel counter-current or co-current with the direction of movement of the trolleys.

(c) **Controlled Humidity Dryers.** The green shapes set on trolleys are moved through a continuous tunnel at a uniform speed. The tunnel is set at carefully controlled temperature and humidity in different sections, depending upon the type of refractory.

(5) Firing or Burning

This is the final stage in the manufacture of refractories. The important functions of firing are (i) Development of a permanent bond by partial vitrification of the mix which imparts mechanical strength to the refractory, and (ii) Development of stable mineral forms to give a permanent set at dimensions that will not change appreciably in future service. The chemical changes that take place include removal of water of hydration, calcination of carbonates and oxidation of ferrous iron. During these changes, a shrinkage in volume upto 30% may occur which may set up severe strains in the refractory. This shrinkage may be eliminated by prestabilization of the materials used e.g., by appropriate sizing and pressing.

The firing techniques and schedules vary with the type of refractory being manufactured. For the firing of refractories, the following three types of kilns are commonly used :

(i) Round down drought periodic kilns

In this type of kilns, the fire boxes are arranged around the kilns on which the shapes are set and heating is continued to attain the firing temperature at a pre-determined rate. The products of combustion go up inside the walls to the dome-shaped roof of the kiln and then are pulled down through the shapes with the help of flues connected to an external stack. Down drought kilns are generally used for large and intricate shapes. A typical down drought kiln is illustrated in Fig. 16.2.

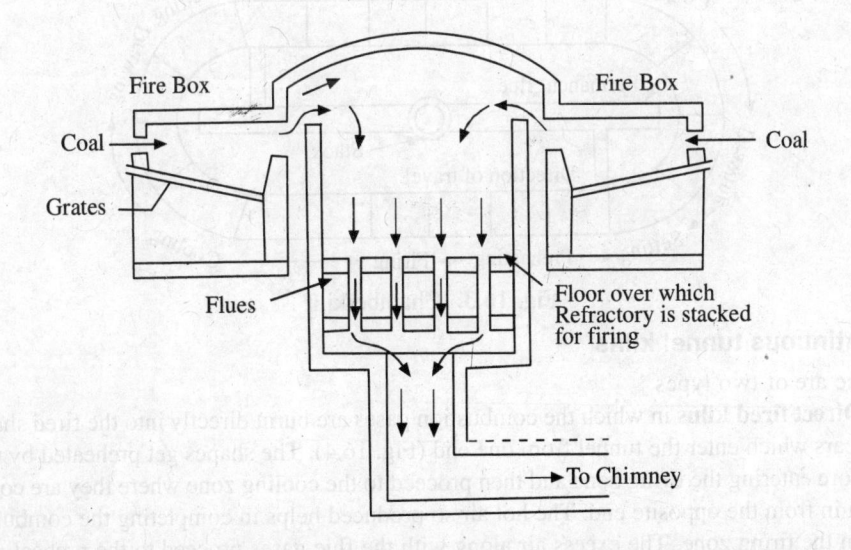

(a) Fig. showing the Down Drought Kiln.

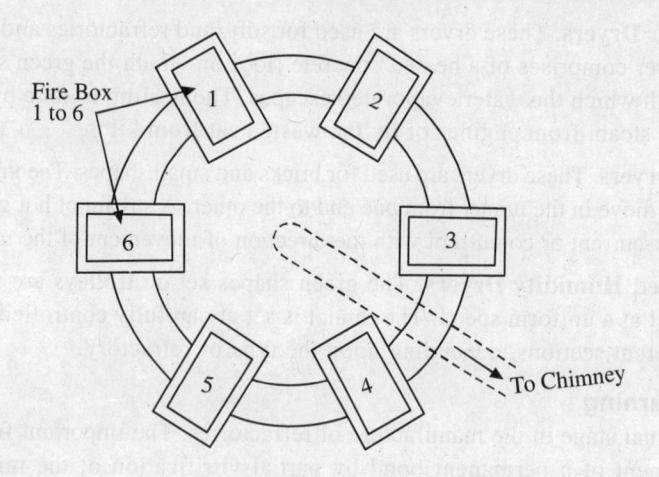

(b) Fig. showing the plan of Down Drought Kiln.
Fig. 16.2

(ii) Continuous chamber kilns

A continuous chamber kiln is shown in Fig. 16.3.

This type of kilns consist of a number of inter-connected chambers each of which is similar to the Down Drought Kiln excepting that the gases pass from one chamber to the next one counter-current to the shapes instead of passing out to the stack. The heat from the combustion gases is utilized to pre-heat the bricks in the chamber before firing while the heat from the cooling bricks is used to pre-heat the air used for combustion. The process is continuous and some chambers will be getting cooled as the others are fired while some others will be simultaneously getting heated by the waste heat of the other chambers. Thus, the fuel consumption in these kilns is much less than that in the case of Down Drought Kilns.

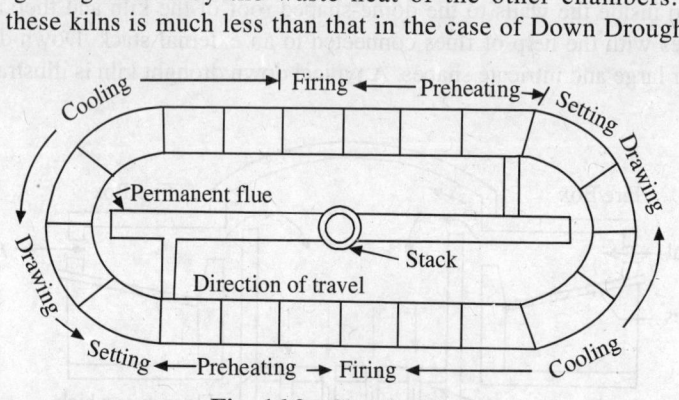

Fig. 16.3. Chamberkiln

(iii) Continuous tunnel kilns

These are of two types :

(a) **Direct fired kilns** in which the combustion gases are burnt directly into the fired shapes set on open cars which enter the tunnel from one end (Fig. 16.4). The shapes get preheated by the flue gases before entering the firing zone and then proceed to the cooling zone where they are cooled by air blown in from the opposite end. The hot air so produced helps in completing the combustion of the fuel in the firing zone. The excess air along with the flue gases proceed to the preheating zone from which they are then withdrawn with the help of a suction fan at the entrance end of the tunnel

REFRACTORIES

kiln, for further use in an adjacent dryer. Coal, oil or natural gases are used for firing. Continuous tunnel kiln has several advantages over the Down Drought Periodic Kiln such as higher thermal efficiency, better operating control, lower and shorter labour costs.

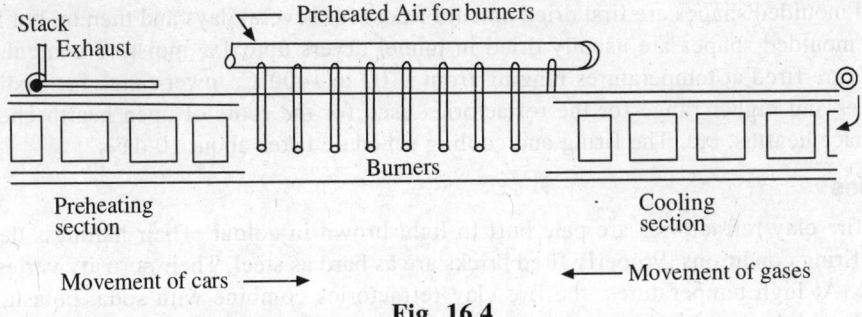

Fig. 16.4

(*b*) **Muffle kilns** in which the flue gases are not allowed to come into contact with the shapes being fired. As these kilns are expensive, they are used for manufacture of special refractories which are very sensitive to the presence of external impurities.

Quality Control

A strict quality control is essential to ensure that the refractories produced are mechanically strong, compact, homogeneous in texture and free from cracks, voids, soft corners and other flaws. This is achieved by subjecting the products at various stages of their manufacture to standard testing procedures as per (*a*) IS: 1528-1974, 1527-1972, 1335-1959 (*b*) BS: 1902, 1966-70 and ASTM Part 17-1974. The important properties tested include chemical composition, pyrometric cone equivalent (PCE), refractoriness under load (RUL), apparent porosity, true density, spalling resistance, cold crushing strength, permanent linear change on reheating, built density, reversible thermal expansion and carbon monoxide disintegration. Study of the phase diagrams of the relevant constituents in the refractory helps in determining the proper batch composition. Similarly, study of thermal properties (*e.g.*, differential thermal analysis, DTA and thermogravimetric analysis, TGA) helps in determining the firing schedules for the particular refractory being manufactured. Depending upon the specific application of the refractory being manufactured, the requirements of the various properties are judiciously determined.

16.6. TYPES OF REFRACTORY PRODUCTS

I. Acid Refractories.

(A) Fire Clay Refractories

Fire clays are the most widely used refractory materials and are well suited for a variety of applications. These contain 25 to 44% of Al_2O_3 and upto 70% SiO_2. The chief raw materials used for the manufacture of this type of alumino-silicate refractory are raw and calcined fire clays, and since these clays can withstand high temperatures, these are called fire clay refractories. Calcined fire clay, known as the "Grog", accounts for 50% or more of the batch mix. Grog is prepared by calcining fire clay at about 1450°C in rotary kilns, followed by crushing, sieving and grading. The graded fractions are suitably blended. Plastic clay is the usual bonding material used. For high grog bricks, additional bonding material like molasses, dextrin or sulphite may also be required. Addition of water may vary from 3 to 30% depending upon the method of forming proposed to be used subsequently. Moulding is done by hand or by machine. Hand moulding is done for non standard sizes and shapes. In this the mixture containing 15 to 20% water is rammed and beaten into mould taking care to see that air pockets, voids etc., are absent. Bricks of low porosity and high slag resistance are generally produced by dry pressing the mix containing 3 to 74 water in de-airing type mechanical or hydraulic toggle presses at a pressure of 75 to 150 MN/m^2. Hollow bricks like sleeves, pipes and tuyers are prepared in specially designed vertical auger machines using semi-plastic mixes having 10 to 12% water.

Hand moulded shapes are first dried on cold floors for several days and then on hot floors. Machine moulded shapes are usually dried in tunnel dryers upto 1% moisture content. Then the bricks are fired at temperatures ranging from 1100 to 1400°C, lower range for casting pit refractories and higher range for the refractories used for the roofs of open hearth checkers, blast furnace hearths, etc. The firing and cooling schedule takes about 10 days.

Properties

The fire clay refractories are pale buff to light brown in colour. Their hardness depends upon the firing conditions. Properly fired bricks are as hard as steel. Their porosity varies from 8 to 30%. At high temperatures, the fire clay refractories combine with soda, potash, lime, MgO, FeO, sulphates, chlorides and carbonates to form fusible salts. Their crushing strength in cold is about 950 kg/cm^2 which goes down with increasing temperature. Their refractoriness depends upon the softening point of the fire clay used. Their safe working temperature is about 1545°C. They have softening point of 1350°C under a load of 50 lbs sq.in. Their spalling tendency can be decreased by using a coarse grog, by increasing the porosity, and decreasing the coefficient of expansion. The specific heat of fire clay brick is as low as 0.25 which increases with temperature and reaches 0.264 at 1300°C. Its thermal conductivity varies from 0.8 to 0.95 (in CGS units) from 300°C to 1100°C. Though they are bad conductors of heat, their radiation power is high and hence fire chambers made of fire clay can be heated to very high temperature. They possess lower porosity, lower refractiveness and higher resistance for thermal spalling as compared to silica bricks. Their spalling resistance makes them particularly suitable for making checker work of regenerating furnaces which are susceptible to temperature fluctuations. Fire clay refractories are classified as high duty (> 40% Al_2O_3), moderate heat duty 'A' (> 30% Al_2O_3 and < 65% SiO_2) and moderate heat duty 'B' (> 25% Al_2O_3 and < 70% SiO_2) types. The specifications for general purpose refractories are covered in IS : 6, 7, 8-1967. Similarly, specifications covering steel plant refractories, glass tank refractories are also available. The ranges for the various specifications are as follows : PCE - ASTM Cone No. 27-32; RUL : t_a : 1300-1450°C and t_e : 1500-1600°C; Apparent porosity : < 17-30% vol; cold crushing strength : > 17-39 MN/m^2; permanent linear change after preheating : < 0.5-1.5%.

Uses

The steel industries are the largest consumers of fire clay refractories. They are used for lining of blast furnaces, open hearths, stoves, ovens, flues, crucible furnaces etc. Fire clay refractories are also widely used in foundries, lime kilns, regenerators, pottery kilns, continuous ceramic and metallurgical kilns, glass furnaces, cupolas, brass and copper furnaces, boiler settings, gas-generators, etc.

(B) Silica Refractories

In silica brick, the two oxides of interest are SiO_2 and CaO. Owing to the predominance of silica as a separate phase the thermal behaviour of the refractory is mostly influenced by the performance of the various polymorphous crystalline forms of different temperature stability.

The quartz from which all silica refractories are made is converted during the firing step to a mixture of cristobalite and tridymite, which have good high temperature stability. The CaO added (about 2%) to the silica brick mixes as a "mineralizer" accelerates the formation of cristobalite and tridymite, and also helps to form a bond during firing. Cristobalite when heated to about 230°C inverts suddenly from its low-temperature-crystalline form to its high-temperature crystalline form accompanied by a large increase in volume. The reverse process occurs on

REFRACTORIES

cooling. In order to obviate this problem, silica refractories are generally heated and cooled slowly in this temperature range.

Silica bricks have particular ability to withstand compressive loads almost to their complete fusion point (1700°C) and hence they are particularly suitable for constructing furnace roofs. This rigidity is due to the interlocking of the crystalline silica and the fact that the amount of liquid (actually, two immiscible liquids) formed is comparatively small and quite viscous. The presence of even small percentages of alumina and alkalies eliminates the immiscibility and thus reduces the refractory properties; further, these compounds form lower-melting eutectics than what SiO_2 and CaO do alone. This is the reason why quartzites are beneficiated to lower the percentage of alumina and alkalies for producing silica bricks for high temperature use.

Silica refractories contain at least 94% silica. The chief raw material for silica refractories is quartzite which is used without precalcination. The grading is finer than that used for fire clay refractories. The bond additions are lime (1-2%) and Fe_2O_3 and/or TiO_2 (1%) which are so small that they require thorough blending. Sometimes, 1 to 1.5% of molasses or sulphite lye (1-1.5%) is used for green strength. For non-standard shapes and sizes, hand moulding with pneumatic ramming is used. For standard shapes, dry pressing is used. Drying is usually carried out in tunnel dryers at 90-120°C for about 24 hours until the moisture is reduced to 0.2%. Dry pressed bricks can be set directly on the kiln cars and passed through tunnel dryers.

Firing of the silica bricks is usually done in round down drought periodic kilns or continuous tunnel kilns. Firing is the most vital step in the manufacture of silica bricks. As stated earlier, on heating and cooling, quartz (which is the mineralogical constituent of quartzite undergoes a number of different crystalline allotropic transformations, some of which are stable and others unstable

$$\alpha\text{-quartz} \xrightleftharpoons{573°C} \beta\text{-quartz} \xrightleftharpoons{867°C} \text{Tridymite} \xrightleftharpoons{1470°C} \beta\text{-cristobalite}$$
$$\text{Sp. gr.} = 2.65 \quad \text{Sp. gr.} = 2.20 \quad \text{Sp. gr.} = 2.32$$

The inversion points are 573°C for quartz, 260°C for cristobalite, and 120°C, 160°C and 250°C for tridymide. After firing, the silica bricks must consist of cristobalite and tridymite only and quartz should be absent. If during firing of silica brick, quartzite is not converted into cristobalite and tridymite, the bricks will expand during use in the furnace, which may lead to collapse of the structure. The conversion of quartz to cristobalite is accelerated by lime, whereas the formation of tridymite is accelerated by Fe_2O_3 and TiO_2. Since these reactions are rather slow, the temperature during firing is gradually from 90°C to the final firing temperature of 1400-1500°C in about 7 days. The bricks are soaked at this temperature for 2 days. The cooling rate, particularly below 300°C, is very vital for the inversions of cristobalite and tridymite. The cooling takes about 6 days.

Properties

Silica bricks have homogeneous texture, free from air pockets and moulding defects. They possess a low porosity (17 to 25%). These are the properties desirable for resistance to slag penetration for a refractory used in furnaces. The silica bricks can withstand a load of about 3.5 kg/cm^2 upto about 1600°C. They are resistant to thermal spalling below 800°C. They possess low permeability to gases. Their physical strength when heated is much higher than that of fire clay bricks. Furnaces using silica bricks must be heated and cooled gradually to minimize spalling and cracking.

The specifications for various silica refractories as per IS and ASTM standards are as follows : pyrometric cone equivalent (ASTM Cone No.) min. 31; Refractoriness under load (RUL) min. t_a 1650-1670°C; Cold crushing strength, min. 24.5 to 49 MN/m^2; apparent porosity, max. 17 to 25% vol; reversible thermal expansion at 1000°C, max. 0 to 13%; permanent linear change after reheating at 1450°C for 4 hrs, max. 0.4 to 1%; true relative density, max. 2.35 to 2.37.

Uses

Silica bricks are used for arches in large furnaces because of their high physical strength. Silica bricks are used for main arch, side walls, port arches and built-heads of open hearth furnaces, coproduct coke-ovens, gas retorts, glass furnaces, electric furnace, roofs, copper stove domes, acid converter linings, etc.

(C) High-alumina Refractories

When the alumina content in fire clay brick reaches above 47.5%, then it is called high alumina refractory brick. These are made from clays rich in bauxite and diaspore. The refractoriness and the temperature of incipient vitrification increase with the alumina content, as can be seen from the phase diagram illustrated in Fig. 16.5.

The essential oxides of both fire clay and high-alumina refractories are Al_2O_3 and SiO_2. Points A and D indicate the melting points of the two essential oxides at 1723 and 2050°C. In fired refractories, at or near these points, silica is in the form of cristobalite and alumina is in the form of corundum. Point B indicates the melting point of the eutectic composition in this system at 1545°C. Paint C locates the temperature of 1810°C at which mullite, the only compound in the system, melts incongruently to form corundum and a liquid. This point also shows the composition of the liquid. Points 1 to 6 represent several commercial refractory material compositions in the following order : Siliceous fire clay; Kaolin (dehydrated); the

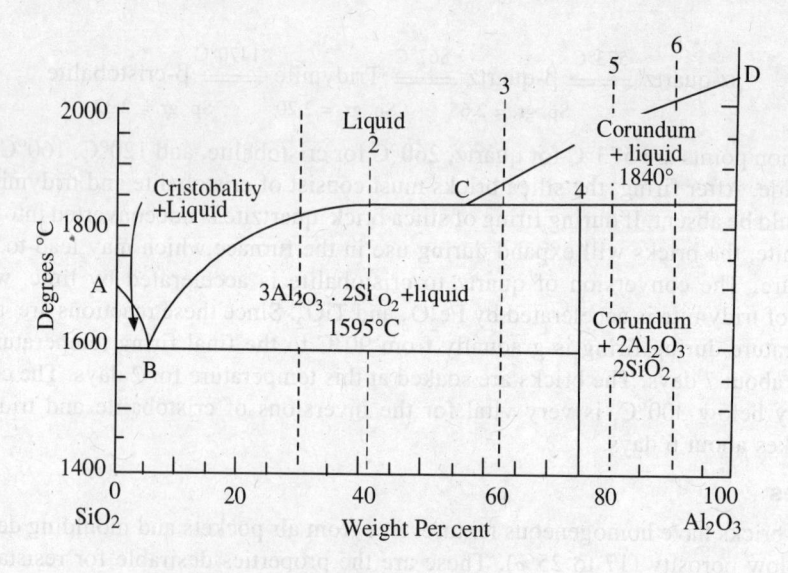

Fig. 16.5. Phase diagram of $MgO-SiO_2$ System

sillimanite minerals (and alusite, kaynite and sillimanite); mullite; diaspore; and bauxite (both dehydrated). Even though fire clay refractories, which are predominantly kaolin, may contain an appreciable proportion of siliceous liquid at elevated working temperatures, they are good refractories, because the liquid is so viscous. Refractories in this system containing enough alumina to be near or above mullite in composition are mostly crystalline at furnace temperatures and thus perform as super refractories, High-alumina refractory bricks are superior to fire clay bricks and Silica-bricks as they have more refractoriness, better slag resistance, inertness to carbon monoxide and stability to natural gas environment upto 1000°C. Special high-alumina refractories are made from silimanite and kyanite which are anhydrous aluminosilicate minerals : $Al_2O_3.SiO_2$. They are bonded with plastic clay and the mixture is moulded into bricks, followed by drying and firing at 1600°C. The extensively used silimanite refractories have the approximate composition of about 63% alumina and 34% silica. Higher percentages of alumina are classed as super-refractories and almost pure alumina refractories are classed under pure oxide refractories.

Uses

High-alumina refractory bricks are employed in cement industry, paper industry, for boiler settings, in the linings of glass furnaces, oil-fire furnaces, high-pressure oil stills, in the roofs of lead softening furnaces and in regenerator checkers of blast furnaces.

II. Basic Refractories

Basic bricks are usually made from magnesia, dolomite, magnesite, chromite and forsterite. In order to achieve the strength and other physical properties, the basic bricks are generally power-pressed and are either hard-burned or chemically bonded. The drawbacks of lack of bond and volume stability in unburnt basic or other bricks have been overcome by the following innovations in the manufacturing process: (*a*) Maximum interfitting of grains was achieved by using only selected particle sizes mixed in proper proportions to fill all the voids (*b*) Increasing the forming pressures to the order of 10,000 *psi* and utilizing de-airing equipment to reduce the air voids between the grains and (*c*) Use of a refractory chemical bond.

(A) Magnesite Refractories

These are made from magnesites. These bricks do not withstand much higher load at elevated temperatures. This drawback is overcome by blending with chrome ores.

The magnesite bricks are generally made from dead-burnt magnesite grains which is prepared by burning magnesite in a rotary kiln or a shaft at about 1700°C using 3 to 4% iron oxide or mill scale powder to reduce the sintering temperature. It is then crushed, sized and graded. Molasses or sulphite lye is used as a binder. 2 to 6% of alumina is added to impart thermal shock resistance of the refractory. The ingredients are blended after adding requisite quantity of water and the mix is aged for 1 to 10 days to ensure complete hydration of any free lime present. Then the mix is moulded by dry pressing at pressures above 70 MN/m^2. Then the shapes are dried first in open sheds and then in controlled humidity dryers. Then they are subjected to firing which should be slow at the beginning and then at the final temperature range of 1400-1600°C. The ceramic bond is usually developed by forsterite ($2MgO.SiO_2$). However, the current trend is to prepare the bricks from purer materials and firing them at 1700°C or more so that direct bonding takes place between the periclase grains (Periclase is a stable form of magnesia). The cooling rate also should be slow. Since magnesite brick cannot bear heavy load during firing in kilns, they are boxed in load bearing silica bricks.

Fire bonded magnesite bricks are susceptible to spalling. This drawback is removed by chemical bonding or by metal casing. For preparing chemically bonded magnesite refractories, the mechanical strength is developed by adding a chemical bonding agent *e.g.*, oxychloride or oxysulphate, instead of by firing. In this case, higher forming pressures are needed to minimize voids. Curing and drying are carried out in specially designed tunnel dryers.

Metal cased magnesite refractories are prepared by ramming dead burnt magnesite into soft steel containers; which welds together at the heated end in high temperature service forming a practically monolithic structure.

Properties

As per IS standards (IS : 1747-1972), the magnesite refractories should have the following composition : MgO : 85% min. SiO_2 : 5.5% max; and CaO : 2.5% max. Refractoriness under load : min. t_a 1550°C; apparent porosity : max. 2.5% vol; Relative density min. 3.53, Cold crushing strength for pressed bricks : min. 34.3 MN/m^2.

Magnesite bricks have low resistance to acid slag but high resistance to basic slag. They have high thermal conductivity and low permeability. Their resistance to spalling for normal bricks is very poor. However, chemically bonded and metal cased bricks have better spalling resistance. Magnesite bricks can be used upto 2000°C without load and upto 1,500°C under load of 3.5 kg/cm^2. Their abrasion resistance is poor. They have a tendency to combine with water and CO_2.

Uses

Magnesite refractories are preferred where basic materials in molten state are to be heated at high temperatures. These refractories are used in open-hearth and electric-furnace walls in the roofs of non-ferrous reverberatory furnaces *e.g.*, those used for Cu, Pb and Sb. They are also used for lining of basic converters in steel industry, hot mixer linings, copper converters, refining furnaces for Ag, Au and Pt. etc.

(B) Forsterite Refractories

Forsterite ($2MgO.SiO_2$) is used both as a bonding material as well as a base for high-temperature refractories. When forsterite is used as a base, the refractories are usually made from olivine 2 (Mg, Fe) $O.SiO_2$ which is characterized by its high refractoriness. In the manufacture of forsterite refractories, dead burned magnesite is usually added to convert some accessory

$$MgO.SiO_2 \; + \; MgO \longrightarrow 2MgO.SiO_2$$
$$\text{Enstatite} \quad \text{Magnesia} \quad \text{Forsterite}$$

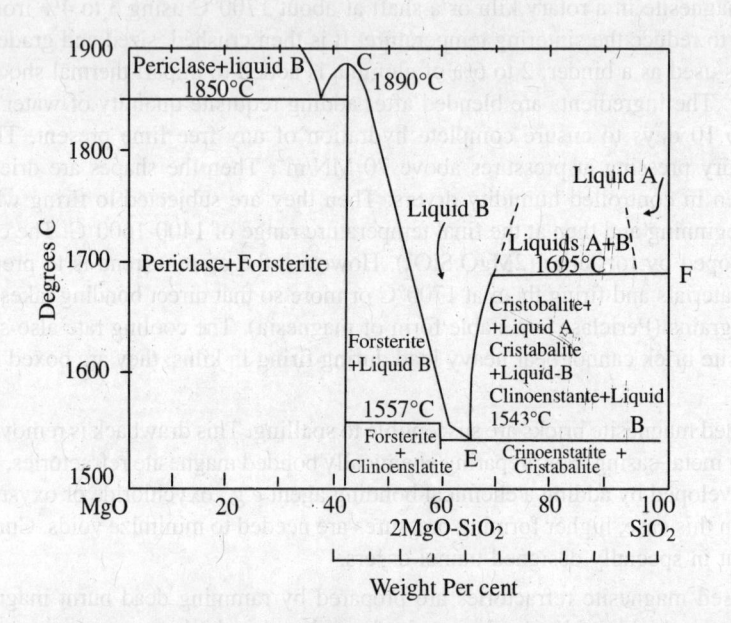

Fig. 16.6

minerals also to forsterite, which is the most stable silicate at high temperatures, as can be seen from the phase diagram of $MgO.SiO_2$ system shown in Fig. 16.6. For instance, enstatite or clinoenstatite which occurs in mined olivine rocks, is converted into forsterite:

Such materials are called super-refractories which possess high melting point, good volume stability and stability against transformations during heating. Their preparation does not require calcining.

Uses

Forsterite is used in glass tank super-structures and checkers because they possess high chemical resistance to the fluxes used and good strength at high temperatures allow increased tank output. These refractories are also used in open-hearth end walls and copper refining furnaces.

(C) Magnesite-chrome and Chrome-magnesite Refractories

Magnesia bricks do not stand much load at elevated temperatures. In order to overcome this difficulty, they are blended with chrome ores. If the blend contains more of magnesite, it is called magnesite-chrome and if chromite predominates, it is called chrome-magnesite. These refractories are made from dead-burnt magnesite and chrome ore mixed in proper proportions. This may be fire bonded, chemically bonded or metal cased as in the case of magnesite bricks described earlier. A coarse grading provides better thermal shock resistance.

Their resistance to acid slag is low and resistance to basic slag is high. They have high resistance to abrasion, erosion and spalling. Their load-bearing capacity and volume stability at elevated temperatures is high. Chrome-magnesite refractories have a tendency to bursting in contact with iron oxide.

(D) Dolomite Refractories

Dolomite is a mixed carbonate of calcium and magnesium, $CaMg(CO_3)_2$. For making dolomite bricks, the mineral is washed, crushed and calcined when dolomite decomposes to give CaO and MgO. This is then mixed with a binder (such as silicate) and water. The mixture is allowed to stand for sometime and then moulded into bricks. The dried bricks are then fired at 1500°C in a kiln for about 24 hours.

Properties

Dolomite bricks are more soft, more hygroscopic, more porous and possess less strength as compared to magnesia bricks. They have greater shrinkage and lesser resistance to thermal shock.

The properties of dolomite bricks can be improved by mixing the dolomite with serpentine, $MgO.SiO_2$ and calcining the mixture which brings about the formation of di- and tri-calcium silicates. Then it is blended with a silicate binder and moulded into bricks followed by firing at 1500°C for a day or two to give "stabilized" dolomite bricks. These are more stable towards basic slags.

Uses

Dolomite refractories are quite cheap and used in granular form to pitch the bottom of open-hearth furnace and for repairing works. Stabilized dolomite bricks are used in Bessemer convertors, ladle linings and for basic electric furnace lining.

(III) Neutral Refractories

(A) Carbon Refractories

The raw materials are low ash content (0.15%) carbon *e.g.*, petroleum coke (for electrodes) or coke derived from coal (for bricks and blocks). The crushed and finely graded coke is dried in a rotary dryer and hot blended at 58 to 80°C with 15 to 18% tar or pitch in a dry pan mill. Forming of shapes is done by dry pressing of the hot mix at a pressure of 30 to 40 MN/m^2 or by pneumatic ramming in case of large shapes. Hardening takes place within 1 to 6 days, depending upon the size. Firing is carried out in a reducing atmosphere at 800 to 1000°C packing the shapes in carbon. Specially designed down drought periodic kiln or chamber kiln or tunnel kiln may be used. A tunnel kiln car takes a load upto 1600 kg in which bricks are set edgewise into muffles built of fire clay bricks, all the interstices, being filled with crushed coke and a layer of coke is laid on the top on which another layer of fire clay bricks is embedded in fire clay or refractory mortar. The heating and cooling cycle may take 4 to 6 weeks in a periodic kiln but the schedule can be minimised if tunnel kilns are used.

Properties

Carbon refractories have high refractoriness, high resistance to acid slag, high resistance to thermal shock and spalling and high load bearing capacity at high temperatures. However they have low resistance to basic slag, low resistance to oxidising atmosphere and possess low thermal expansion. Their thermal and electrical conductivity is wide ranging. They react with water above 600°C and CO_2 above 700°C. Carbon refractories are practically infusible, close textured and can withstand temperature fluctuations and chemical attack in neutral or reducing conditions.

Uses

Carbon refractories are used for making electrodes and for linings in chemically reactive equipments.

(B) Graphite Refractories

Graphite is a natural allotropic form of carbon having dark colour, bright metallic lustre. In addition to various other industrial applications, graphite is also used as a refractory material. Graphite refractories are produced by one of the following two methods : (1) by graphatising of pear shaped carbon products by prolonged heating (3 to 5 weeks) at about 2850°C in an electric furnace (2) shaping of graphite flakes bonded with fire clay or carbon. This method is particularly used for preparing crucibles. The shapes are dried at 80 to 100°C for one day in controlled humidity driers and fired at 900 to 1200°C in down draught periodic kilns or continuous muffle kilns. In order to prevent oxidation, the crucible surface is coated with boric acid, borax, soda ash, sodium silicate, $CaCl_2$ or $MgCl_2$. The fired crucibles are slowly cooled. In India, clay or carbon bonded graphite crucibles upto 100 kg capacity are produced in small scale sector.

The properties of graphite refractories are more or less similar to the carbon refractories excepting that they have higher oxidation resistance than the latter.

Graphite crucibles are widely used in industry. Graphite bricks are used for construction of electrodes, linings of chemically resistant equipments, atomic reactors, electric furnaces and in non-ferrous metal smelting furnaces.

(C) Silicon carbide (Carborundum) Refractories

Silicon carbide is a very hard abrasive having a very high melting point. It is made by heating a mixture of 60% sand and 40% coke together with some saw dust and salt in an electric furnace at 1500°C. Saw dust helps in increasing the porosity and salt helps in the removal of iron etc. in the form of volatile chlorides. The SiC so formed is in the form of interlocking crystals. This is crushed, sized and suitably graded so as to get a dense packing. The bonding agents used are less than 10% of plastic fire clay, graphite, ethyl silicate or even finely divided SiC itself. Recently, silicon nitride or oxynitride is preferred as bonding agent because it imparts superior oxidation resistance to the product. The final firing of the bricks is carried out in a reducing atmosphere at about 1500°C. The silicon carbide refractories generally contain 85-90% SiC. However, for less severe service conditions, SiC content may be as low as 60%.

REFRACTORIES

Self-bonding type of silicon carbide refractory bricks can be prepared by mixing SiC particles with a temporary binding agent like glue, then pressing and firing at 2000°C when inter-crystalline bond develops.

Properties

Silicon carbide super-refractories are characterised by their chemical resistance, high refractoriness and ability to withstand sudden temperature fluctuations. The bricks are extremely refractory and possess high resistance to spalling and abrasion. They have a high thermal conductivity, low thermal expansion, high mechanical strength and can withstand loads in furnaces even at 1650°C and higher. Their resistance to acid slag and to reducing atmosphere is high, whereas their resistance to oxidising atmosphere is medium and their resistance to basic slag is low. Self-bonded silicon carbide refractory bricks have superior properties than silicon nitride-bonded bricks which in turn are superior to the clay bonded bricks. Silicon carbide bricks have a tendency to oxidize to silicon in oxidizing atmosphere around 950°C. This tendency can be counteracted by coating with a thin layer of zirconium.

Uses

Silicon carbide refractories on account of their good thermal conductivity, are mainly used in muffles. Their ability to absorb and release heat rapidly and their resistance to spalling under repeated temperature fluctuations make them an ideal choice for recuperators. Silicon carbide bricks are also used for partition walls of chamber kilns, coke ovens and furnace floors. Owing to their high electrical conductivity, they are used as heating elements in furnaces in the form of rods and bars known as globars. Silicon carbide bonded with tar are excellent for making high conductivity crucibles. Clay-bonded bricks are preferred for high conductivity bricks while lime-bonded bricks are used for muffle and electric furnace lining.

(D) Chromite Refractories

These are made by firing chromite ($Cr_2O_3.FeO$) blended with clay at a temperature of 1500 to 1700°C. These are neutral refractories having good slag resistance. They possess high density, moderate spalling resistance and moderate thermal conductivity. They can be used upto 1800°C. Their crushing strength is about 350 to 550 kg/cm^2 and their refractoriness under a load of 3.5 kg/cm^2 is 1430°C. They are used in furnace linings, sodium carbonate recovery furnaces and bottoms of soaking pits.

(IV) Special Refractories

(A) Single or Pure Oxide Refractories

The refractory industry has been constantly facing demands for products having very high refractoriness and which can withstand very severe operating conditions. Pure oxide refractories are developed to meet these demands. Refractory oxides of interest under this category in the increasing cost per unit volume are: alumina, magnesia, zirconia, beryllia and thoria. All these have been developed commercially for light refractory products. Alumina, magnesia and zirconia are available in high purity (greater than 97%) and are composed of electrically fused grains. Beryllia is not used commercially in heavy wear because of its exorbitant cost and volatalizing tendency above 1610°C in presence of water vapour. Thoria has a disadvantage of its being radio-active and hence is under strict controls. Among the pure oxide refractories, sinter alumina has the widest application. It was successfully used upto 1870°C. Magnesia is a basic refractory and is reduced at high temperatures. Its applications are limited to oxidizing atmospheres upto about 2,200°C. Pure zirconia undergoes crystalline transformation from monoclinic to tetragonal form at about 980°C, accompanied by drastic volume change on inversion. Hence stabilization of the crystal structure of the cubic form (which does not undergo inversion) is necessary. This can be accomplished by adding certain metallic oxides e.g., CaO and MgO. Processing temperatures with the fused stabilized zirconia are of the order of 2600°C. The pure zirconia refractories are used in kilns needed for firing barium titanate resistors.

(i) Pure Alumina Refractories

These are made from calcined or fused alumina having a purity greater than 97%. The usual additions are 2% of Kaolin (which reduces the sintering temperatures and improves pressing characteristics) and 2% of talc or magnesite (which controls the grain size and improves pressing characteristics). The binding agent used depends upon the method adapted for forming. For dry pressing, organic binders e.g., cellulose or polyvinyl alcohol are used. Firing temperatures are in the range of 1700-1800°C depending upon the purity of alumina. Forming can also be done by hot pressing in graphite moulds. Pure alumina refractories belong to neutral type having good resistance to oxidising and reducing atmospheres even upto 1900°C. High temperature kilns furnish alumina bricks which approach pure corundum in properties which have high slag resistances. Such refractories are used where severe slagging occurs.

(ii) Magnesia Refractories

These are prepared from fused magnesia grains. Cellulose is normally used as binder. Firing temperatures are in the range 1900-1950°C. Care must be taken in their setting to ensure that they do not react with the kiln material at this high temperature. Forming by hot pressing is similar as in the case of pure alumina refractories. This is a basic type of refractory having refractoriness of the order of 2300°C.

(B) Mixed Oxide Refractories

(i) Mullite ($3Al_2O_3.2SiO_2$) Refractories

These are prepared from high alumina minerals and fire clay in proper proportions. The grog is first prepared by fusing or sintering the raw material mix at 1820°C or by calcining it at 1600°C. After blending the fine powder with a binder such as plastic clay or sulphite lye, the shapes are formed by the usual methods followed by firing at about 1750°C. Addition of a small quantity of aluminium fluoride during blending might reduce the mullitisation temperature.

Electrocast or corhart mullite refractory blocks are prepared from electrically fused mullite. The manufacturing procedure consists of introducing a mixture of diaspore clays of high alumina content (furnishing the proportion of $3Al_2O_3.2SiO_2$) into a top of an electric furnace. Molten aluminium silicate at a temperature of about 1870°C is tapped from the furnace periodically into moulds built from sand slabs. They are annealed for 6 to 10 days before the blocks are usable. The refractory blocks so obtained have a vitreous non-porous structure which exhibits a linear coefficient of expansion of about one-half of that of good fire-brick. The electrocast has only 0.5% voids as compared to 17 to 29% for fire clay blocks. Cast refractories are used in glass furnaces, as linings of hot zones of refractory kilns, modern boiler furnaces exposed to severe duty and in metallurgical equipment such as forging furnaces. The higher initial cost of these refractories is balanced by the long life and minimum wear. A still superior type of fused cast refractories, with 83 to 95% alumina is also available now.

Mullite refractories belong to neutral type of refractories. They possess high refractoriness and high resistance to acid slag, basic slag, oxidising atmosphere, reducing atmosphere and to thermal and structural spalling.

(ii) Zircon ($ZrO_2.SiO_2$) Refractories

These are prepared from micronised zircon mixed with finely ground clay or alumina. The grog is prepared by the partial sintering of zircon. Firing temperature is 1650-1700°C. These refractories belong to acid type and possess high resistance to acid slag. Their resistance to thermal shock or spalling is high. Their thermal expansion is low and resistance to molten metals is high. Their resistance to oxidising atmosphere is low.

REFRACTORIES

(C) Non-oxide Refractories

These include silicon nitride, boron carbide, molybdenum disilicide, zirconium boride and titanium boride. They can be formed into dense shapes by hot-pressing or sintering in controlled conditions. Molybdenum disilicide exhibits oxidation resistance even at 1800°C, while others tend to oxidize from 1000°C. Non-oxide refractories are expensive, and are mostly used at present in nuclear and space research programmes.

(D) Insulating Refractories

These are characterised by low thermal conductivity and low bulk density. They are of two types :

(1) Backing up insulations which are made from highly porous inorganic refractory materials like asbestos, mica, vermiculite and diatomite.

(2) Hot face insulations which are similar in composition to dense refractories. The raw materials used for this type of insulating refractories are chromite, magnesite, fire clay, silica and alumina. The insulating value of this type of refractories is due to the method of their preparation. For example, waste cork is ground and sized and is mixed with fire clay. This is then moulded and burnt in a kiln. The cork burns out leaving a highly porous and light weight refractory brick.

The following methods are used for rendering the insulating refractories porous :

(i) by adding to the mix organic materials like rice husk, saw dust, ground cork, coke or pitch, which are burnt out during firing leaving a porous mass.

(ii) by adding to the mix a highly porous inorganic refractory material like porous grog, expanded mica, calcined diatomite or vermiculite.

(iii) by adding to the mix a granular volatile substance like naphthalene flakes which sublimes at high temperature leaving behind a porous product. The fired product will be flux-free unlike in the other methods. The sublimed naphthalene can be cooled and reused.

(iv) by converting the mix to the consistency of a thick slip and introducing gas bubbles (by chemical effervescence or by inclusion of a preformed foam) which are so stable that they remain in it until the product is dried and fired. Chemical effervescence can be obtained by Zn or Al and an alkali which produces hydrogen bubbles or by adding finely divided lime and acid which produces CO_2 bubbles. Preformed foam can be obtained by stirring into the slip soap, sodium resinate or saponin in water. In both the methods, stabilizers like plaster of paris, glue or gum arabic are added which prevent coalescence of the bubbles.

(E) Monolithic Refractories

These are made from a volume stable refractory (*e.g.*, calicined fire clay, SiC, silica, magnesia, alumina, calcined diatomite, vermiculite or expanded mica depending on the refractory type being made) and a bonding agent (*e.g.*, high alumina, cement, phosphate, sulphate, sodium silicate, ethyl, silicate, polyvinyl alcohol or phenolic resins depending upon the application). Monolithic refractories are in the form of unfired loose mixture which can be made into castables, mortars, gunning compounds or plastic and ramming mixes before application. IS specifications are available for all these products.

Industrial outlets for the Refractories

The various types of refractories described above are widely used in iron and steel, non-ferrous metals, cement, gas, power, ceramics, glass and chemical industries as well as nuclear and space research programmes.

QUESTIONS

1. What are the requisites of a good refractory material ?
2. What are refractories ? How are they classified ? State some important industrial applications of refractories.
3. Discuss the important properties of refractories which have a direct bearing on their industrial use.
4. Write informative notes on
   (a) Fire clay refractories
   (b) Silica refractories
   (c) Magnesite refractories
5. (a) Define the term "refractory". How are refractories classified ? Give two examples of each type.
   (b) What are the causes for failure of a refractory?
6. Write short notes on the following :
   (a) Pyrometric Cone Equivalent
   (b) Thermal spalling
   (c) Graphite refractories.
7. (a) Discuss the essential characteristics of a good refractory.
   (b) What type of refractories are used for each of the following and why?
   (i) Bessemer converter   (ii) Regenerative furnace
   (iii) Blast furnace   (iv) Open-hearth furnace.
8. Give an account of the preparation, properties and uses of the following:
   (a) High alumina refractories
   (b) Dolomite refractories
   (c) Fire clay refractories
9. Write informative notes on:
   (a) Single oxide refractories
   (b) Mixed oxide refractories
   (c) Non-oxide refractories.
10. (a) What are the raw materials for refractories ?
    (b) What are the different steps in the manufacture of refractories ?
    (c) What do you mean by super-refractories ?
11. Discuss the significance of the following properties of a refractory brick :
    (a) Refractoriness   (b) Dimensional stability
    (c) Spalling   (d) Thermal conductivity
12. Discuss the various physical and chemical factors which affect the industrial uses of refractories.
13. Write briefly on the preparation, properties and uses of the following:
    (a) Carborundum bricks
    (b) Zirconia bricks
    (c) Chromite bricks
14. Write short notes on :
    (a) Insulating refractories
    (b) Monolithic refractories
    (c) Silica refractories
15. Fill up the blanks in the following:
    (a) Refractoriness of a material is defines as ....
    (b) Acid refractories should not be used in contact with .... whereas basic refractories should not be used in contact with ....
    (c) Porosity of a refractory material is given by the ratio of ....
    (d) Thermal spalling means ....
    (e) The heat capacity of a furnace depends upon the following three factors namely ....
    (f) Dimensional stability of a refractory indicates ....

17

Electroplating

Electroplating is the electrodeposition on metals, alloys, and non-metallics. Electroplating is done for decoration, surface protection and engineering performance at costs that are lower than those of comparable articles prepared by other methods. The first documented electroplating work was done in 1840 and the process remained only as an art until 1910. With the spectacular advances made in the knowledge of physical chemistry of solutions, with the improvements made in electrical machines and measuring instruments and with the increasing availability of very pure chemicals, the "art" of electroplating was transformed into a process of science and technology since 1930.

The main purpose of electroplating is to alter the surface properties of metals or non-metals in order to achieve improved appearance, improved resistance to corrosion, wear or chemical attack, improved frictional and galling behaviour and increased hardness.

The electroplating of common metals includes the processes in which ferrous or non-ferrous base material is electroplated with nickel, chromium, copper, zinc, lead, iron, cadmium, aluminium, brass, bronze or suitable combinations thereof. In electroplating involving precious metals, a ferrous or non-ferrous base material is plated with gold, silver, platinum, palladium, indium, iridium, rhodium, ruthenium, osmium or combinations thereof.

In electroplating, metal ions from acidic, alkaline or neutral media are reduced on the workpieces being plated which act as cathodic surfaces. The metal ions in solution are usually replenished by the dissolution of metal from anodes or small pieces contained in inert wire or metal baskets. Replenishment by metal salts is also in practice as in the case of chromium plating. In such a case, an inert material must be used as anode. Several types of electroplating solutions have been used commercially although only some of them have been adopted widely. For instance, cyanide solutions are popular for Cu, Cd, Zn, Ag, Au and brass. However, noncyanide alkaline solutions containing pyrophosphate have come into vogue recently for Cu and Zn. Zn, Cu, Sn and Ni are plated with acid sulphate solutions particularly for plating relatively simple shapes. Zn and Cd are sometimes electroplated from slightly acidic or neutral solutions. The most common methods of plating are in barrels, on racks and continuously from a spool or coil.

17.1. APPLICATIONS OF ELECTROPLATING

(1) Plating for decorative purposes

Chromium, nickel, silver, gold, copper, brass and rhodium are the most widely used metals for decorative plating. Zinc, cadmium, tin, lead, platinum and palladium are also used for special decorative effects.

The base metals and alloys that are the most commonly subjected to electroplating are steel, brass, copper, nickel, silver, white-metal alloys of lead, zinc, or tin base, and aluminium alloys.

The type of the plating required is determined on the basis of the appearance desired, the intended use of the finished article and the nature of the brass metal being plated. Chrome plating is most commonly used for decorative purposes because of its durability and resistance to chemical attack, abrasion and tarnish.

Electroplating is widely used in the manufacture of aircraft, automobiles, refrigerators, electrical appliances such as irons, fans, hot plates and toasters, builders hardware such as door knobs, locks, hinges and strike plates, Jewellery, radios, cameras, typewriters, watches, purses, umbrellas, etc.

(2) Plating for protection

Decorative plating necessarily imparts some protection. However, when plating is done mainly for protection, appearance may not be a factor under consideration.

Steel, being the most commonly used construction material, is the most widely used basic material which must be protected from corrosion and chemical attack. The protective plates that are commonly used on steel are Zn, Cd and Sn. These are applied on stamped, spun or cast articles at some stage in their manufacture. Zinc plating is preferred in industrial and rural atmospheres whereas cadmium plating is a better choice in sea-shore or salty atmospheres. Tin does not provide the galvanic protection offered by Zn and Cd. However, in case of food containers and food handling equipment, tin plating is preferred as it not only protects steel but also does not contaminate the food. In applications where protection as well as decorative effects on steel are needed, copper, nickel and chromium plates are used. Copper and brass are generally plated with Ni and Cr. When metals are plated for protection against their environments, due consideration must be given to the galvanic effects of the metal couple to avoid undesirable effects.

(3) Plating for special surface and engineering effects

Electroplating occupies an important place in engineering. A composite part can be designed in which the basis metal is an alloy which possesses the required mechanical and fabrication properties although it may have poor resistance to surface attack under service conditions. The required surface stability can be provided by means of a plated coating of a metal. Thus special alloys that have good surface properties but are difficult to work in massive form can be combined with alloys that are relatively easy to work. In engineering applications, preparation and cleaning of the basis metal surface and selection of the appropriate plate are of paramount importance in producing a satisfactory product.

Electroplates are applied for temporary use in metal treatment. Many steel parts are copper-plated prior to carburizing to prevent carburization at undesired locations. Tin and copper-tin alloy plates are used to prevent nitriding of steel at surface areas where hardening is not desired.

Improved performance and service life of cast-iron or steel glass moulds is achieved by the application of low-contraction chromium plate. Chromium plate on the important surfaces of molds for plastics improves the life of the mold and also the appearance of the molded product. Chromium plate on the surface of dies for drawing or extruding copper, brass and steel brings about considerable reduction of the wear on the die and thus improves the appearance of the drawn metal.

(4) Electroforming

The radio, radar, aircraft, automobile, glass, rubber, steel, printing and munition industries use several products made either partly or fully by electroforming. This method helps in the manufacture of articles where other methods have been totally ineffective.

(5) Plating on non-metallic materials

Non-metallics such as synthetic resins, cloth, paper, wood, porcelain, leather etc., are usually

ELECTROPLATING

plated for decoration, preservation or to prepare light-weight materials with metal surface characteristics, surface conductance and strength. Non-metallics are also plated in connection with electroforming and electroplating.

Preparation of Basis Material

Electroplating involves more than just a formulation for a solution. Before a metal or non-metal can be electroplated, the surface must be physically, chemically and mechanically clean.

Physical cleanliness means freedom from oils, greases, superficial dirts associated with polishing, buffing and atmospheric dust.

Chemical cleanliness means freedom from oxides or other surface-formed compounds of the basis metal.

Mechanical cleanliness means freedom from a surface skin of damaged metal (*e.g.*, loose fragments, scratches and strains) produced by mechanical operations.

Electroplating involves a series of operations each of which is vital in determining the quality of the final product. Following are the minimum number of steps involved in succession :

1. Removal of scale or tarnish (pickling)
2. Mechanical preparation of the surface
3. Cleaning for physical cleanliness
4. Rinsing
5. Acid dipping for chemical cleanliness
6. Rinsing
7. Electroplating
8. Rinsing, and
9. Drying.

Pickling

This involves the removal of relatively heavy scale of oxides from hot rolling, casting and heat treating, and also the removal of rust or tarnish due to exposure to atmosphere. This operation usually precedes polishing and buffing. Pickling consists of immersing the article in dilute HCl or H_2SO_4. For cleaning copper, nickel or brass article the pickling bath consists of a dilute solution of HNO_3 or a mixture of dilute HNO_3 and dilute H_2SO_4.

Mechanical Preparation

Mechanical preparation of the basis metal surface is generally required to achieve better appearance, better protection and superior engineering properties. In order to achieve optimum results from electroplating, special attention must be paid in the techniques adopted for polishing, buffing and colouring or electropolishing.

The rough surfaces of freshly prepared articles are rendered smooth by mechanical operations such as grinding with a grinding stone or grinding wheel, sand blasting and scratch rubbing by a wire brush or a abrasive stone or paper. For finer work, decorative polishing is carried out by polishing with rouge.

Cleaning for physical cleanliness

This involves the removal of oils, greases, polishing and buffing compound and other forms of superficially attached dirt. This is usually accomplished by cleaning with organic solvents (*e.g.*, trichloroethylene) and perchloroethylene or by aqueous cleaning agents (*e.g.*, sodium carbonate, hydroxide, or silicate) with or without electric current. To remove the surface dirt, simple immersion in an aqueous cleaner may not be adequate. Better results may be obtained by discharging gas on the surface of the metal by rendering it as cathode or anode, which provides scrubbing action.

Rinsing

Rinsing with clean water is essential between the various steps in the electroplating sequence. This is accomplished by dipping the article in clean running water or by spray rinsing or both depending upon the shape and drag-out characteristics of the article being plated.

Chemical Cleanliness

Dipping in 10-30% (by volume) of HCl or H_2SO_4 after cleaning is an important step particularly if the parts are to be first plated in an acid solution. This operation removes any tarnish film produced during cleaning step and also neutralizes any alkaline film formed which cannot be completely removed by rinsing. This is essential to ensure that the pH of the plating bath is not affected. Subsequent to the acid dip and the rinse following it, neutralization by dipping the workpiece in a dilute cyanide solution may be desirable to prevent rusting.

Preparation of Al and Mg

Al and Mg require special treatment before subjecting to electroplating. For successful plating on Al and Mg, the oxide film is first replaced by a thin film of zinc by immersing in a solution containing (ZnO + NaOH) or ($ZnSO_4$ + KF + $Na_4P_2O_7$ + K_2CO_3). A copper plate from a low pH Rochelle salt or pyrophosphate bath usually precedes a plate from any other bath, in order to prevent attack of Al through pores in the thin zinc film.

Preparation of Non-metallics

Non-metallics must be rendered clean, smooth and conducting before subjecting to electroplating. Synthetic plastics such as Vinyl chloride, Vinylidine chloride, bakelite, cellulose derivatives and laminated phenolics can be polished by methods comparable to those used in case of metals. Leather, paper, cloth etc. are cleaned to remove oils, greases and waxes and are then rendered non-porous by coating with shellac or lacquer. A conducting surface is produced by depositing a continuous coating of Cu or Ag by chemical reduction or metal spraying and by sputtering at high voltages.

The acid $CuSO_4$ plating bath is generally most suitable for plating on the conductive film. The plating solutions used for metals are also applicable for non-metallics. If a different metal plating is needed, the acid $CuSO_4$ bath should be used as a first coating followed by the desired metal plate.

17.2. ELECTROPLATING EQUIPMENT AND OPERATING ONDITIONS

The electroplating equipment (Fig.205) essentially consists of a plating tank (made of steel or rubber-lined or lead-lined or synthetic-lined steel) into which anodes are hung from the outside bars, connected to the positive bus bar conducting current from a generator or rectifier. The workpiece to be plated is hung on racks from the center or

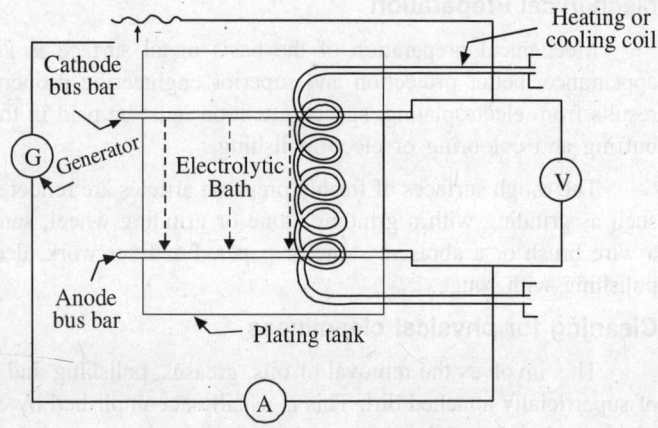

Fig. 17.1. Electroplating Assembly

cathode bar (which is also called work bar). The racks must possess the necessary current carrying capacity and enable uniform current distribution. A low-rpm drive motor helps in providing oscillating motion to bring about cathode agitation. The plating bath can be heated with steam or cooled with water through coils or pipes. Immersion heaters or external heat exchangers may also be used for heating or cooling.

In addition to the factors already discussed, the properties of the plate *e.g.,* smoothness, porosity, uniformity of coverage, hardness, condition of stress, current deficiency and crystal size and orientation of the plate depend upon the operating variables such as the bath composition, the acidity or alkalinity of the bath, the temperature, the current density of deposition, the agitation of the bath and the presence or otherwise of addition agents. For instance, too high a current density gives a rough, spongy plate.

Stress in electroplates produces the same kind of distortion in the position of the metal atoms in the crystal structure as occurs in wrought or cast metals. Stress in plates can be relieved by proper heat treatment. If stress is not properly relieved, the plate might crack upon ageing in service as a result of which some plates *e.g.,* nickel will lose protective action against corrosion. Bright metal plates are more susceptible to stress.

In commercial electroplating applications, special attention is given to factors like cathode and anode current densities, cathode and anode current efficiencies, rate of deposition and throwing power. Current density is commonly expressed as amperes/sq. ft. of anode and cathode surface. The cathode current efficiency refers to the amount of metal plated to the amount that could theoretically be deposited by the same amount of current according to Faraday's law : 96,500 coulombs could theoretically deposit one equivalent of the metal. Anode current efficiency refers analogously to the amount of metal dissolved at the anode. The product of the cathode current density and efficiency gives the rate of deposition at the cathode. Throwing power refers to the uniformity of plate thickness that can be expected on a shaped article over the surface of which the current density will vary because of current distribution. The distribution of the plating current is influenced by the relative distance of any given part of the surface from the anodes. The anodes employed are generally of the same metal as that to be deposited, so that the dissolving anode maintains the metal ion concentration in the bath. Such anodes are referred to as "soluble" anodes which constitute the main raw material consumed. However, in some cases, "insoluble' anodes are used.

17.3. ELECTROPLATING BATHS :

Electroplating baths contain acids, alkalis, metal salts and various additives used as bath control compounds. The constituents of some commonly used electroplating baths are given in Table 1.

Table 1.

Electroplating process	Bath composition
Copper cyanide	Copper cyanide, sodium cyanide, sodium carbonate, sodium hydroxide and Rochelle salt.
Gold cyanide	Metallic gold, sodium phosphate and potassium cyanide
Iron	Ferrous sulphate, ferrous chloride, ferrous fluoborate, ammonium chloride, calcium chloride, sodium chloride and boric acid.
Silver	Silver cyanide, potassium or sodium cyanide, potassium or sodium carbonate, potassium hydroxide, potassium nitrate and carbon disulphide.
Chromium	Chromic acid and sulphuric acid. Sometimes with fluoride catalyst.

Cadmium cyanide	Cadmium cyanide, sodium cyanide, sodium hydroxide and cadmium oxide.
Cadmium fluoborate	Cadmium fluoborate, fluoboric acid, boric acid, ammonium fluoborate and licorice.
Zinc cyanide	Sodium cyanide, zinc oxide, sodium hydroxide and zinc cyanide.
Acid-zinc	Zinc sulphate, ammonium chloride, ammonium sulphate or sodium acetate and glucose or licorice
Tin-zinc alloy	Potassium stannate, zinc cyanide, potassium hydroxide and potassium cyanide.
Copper-fluoborate	Copper fluoborate and fluoboric acid
Acid-copper sulphate	Copper sulphate and sulphuric acid
Copper pyrophosphate	Copper pyrophosphate, potassium hydroxide and ammonia.
Brass and Bronze	Copper cyanide, zinc cyanide, sodium cyanide, sodium carbonate, ammonia and Rochelle salt.
Acid-tin	Tin fluoborate, fluoboric acid, boric acid, stannous sulphate, sulphuric acid, cresol, sulphonic acid, napthol and gelatin.
Stannate-tin	Sodium stannate, sodium hydroxide, sodium acetate and hydrogen peroxide.
Lead-tin	Lead fluoborate, tin fluoborate, boric acid, fluoboric acid, glue and hydroquinone.
Nickel (watts)	Nickel sulphate, nickel chloride, nickel fluoborate, boric acid, phosphoric acid, phosphorus acid and "stress-reducing additives"
Nickel-Acid fluoride	Nickel chloride, hydrofluoric acid, citric acid, sodium lauryl sulate as wetting agent.
Black nickel	Nickel ammonium sulphate, nickel sulphate, ammonium Sulphate, zinc sulphate and sodium thiocyanate.
Tin-Nickel Alloy	Nickel chloride, stannous chloride, ammonium fluoride, ammonium bifluoride, sodium fluoride and hydrochloric acid.

17.4. CHARACTERISTICS OF ELECTROPLATING WASTES

All the constituents of the electroplating baths contribute to the wastewater stream either through part dragout, batch dump or floor spill. Electroplating baths may contain Cu, Ni, Au, Ag, Zn, Cd, Cr, Pt, Sn, Pd, Fe, Pb, K, Na and ammonia. The anionic components likely to be present include carbonate, borate, fluoborate, cyanide, fluoride tartrate, phosphates, chloride, sulphide, sulphate, nitrate, and sulphamate. Further, many plating solutions contain various other metallic, organic and metallo-organic additives to in duce grain refining, levelling of the plating surface and deposit brightening. Mo, Se, As, Co, saccharin, aldehydes are some such constituents. All these materials contribute to the wastewater stream from electroplating units. Typical non-metallic contaminants such as oil, grease, biodegradable mass, suspended solids, invert solids, toxic and organic compounds may also be present. A complete survey of all the metallic and non-metallic contaminants, anticipated flow rates and the concentration of the constituents which are influent to the waste treatment system must be done before finalizing its design.

Waste-treatment Technologies

Some specific applications of waste-treatment technologies used for electroplating wastes are summarized in Table 2.

Table 2. Waste Treatment Technologies

Waste treatment problem	Treatment technology
(1) Removal and/or recovery of dissolved salts	Hydroxide precipitation, sulphide precipitation, ion exchange, Membrane filtration, insoluble starch xanthates, peat adsorption, carbon adsorption, electrolytic recovery, high pH precipitation (for complexed metals)
(2) Removal of dissolved salts for reuse of water	Reverse osmosis
(3) Recovery of process baths	Electrodialysis
(4) Removal of organics	Aerobic decomposition, carbon adsorption, resin adsorption.
(5) Destruction of cyanides	Electrochemical oxidation, oxidation by chlorine, oxidation

	and cyanates	by ozone, oxidation by ozone with u.v. radiation, oxidation by H_2O_2 (for destruction of cyanides only).
(6)	Suspended solids removal	Flotation, centrifugation, ultra filtration, sedimentation, diatomaceous earth filtration.
(7)	Sludge dewatering	Centrifugation, pressure filtration, diatomaceous earth filtration. Sludge bed drying, vacuum fittration.
(8)	Reduction of chromium (VI) (from metal finishing and cooling tower blow-downs)	Electrochemical reduction
(9)	Concentration and recovery of process chemicals	Evaporation
(10)	Oil removal	Centrifugation, coalescing, flotation, skimming, ultrafiltration (oil breakdown can be accomplished by aerobic decomposition).

17.5. SAFETY PRECAUTIONS

Most of the cleaning, pickling and plating operations may give rise to toxic gases, vapours, sprays, fumes, etc., which have to be removed by suitable ventilation Removal of sprays and gases is particularly vital for cyanide and chromic acid solutions. Special precautions should be taken in handling cyanides, fluoborates and chromic acid solutions so as to ensure that they do not come into contact with the skin nor enter the mouth or nasal passages. Cyanide solutions should not be allowed to come into contact with acids which would react with them forming gaseous hydrogen cyanide. Cyanide wastes are destroyed by chlorination or by electrolytic oxidation.

QUESTIONS

1. What is electroplating ? Why it is done ? Discuss the various steps involved in electroplating.
2. Discuss the various applications of electroplating.
3. Discuss the characteristics of electroplating wastes and the waste treatment technologies commonly employed.
4. Write information notes on any three of the following :—
    (a) Safety precautions involved in electroplating operations.
    (b) Electroplating equipment and operating conditions
    (c) Sequential steps involved in plating on non-metallic materials.
    (d) General constituents of any two commonly used electroplating baths and their functions.

18

Environmental Chemistry and Control of Environmental Pollution

"In a fragile biosphere, the ultimate fate of humanity may depend on whether we cultivate a deep sense of self-restraint, founded on a widespread ethic of limiting consumption and finding non-material enrichment".

18.1 ENVIRONMENTAL CHEMISTRY

Environmental Chemistry deals with the study of the various chemical phenomena taking place in the environment. In a broader perspective, it comprises of the study of the chemical species existing in the various segments of the environment, their sources, pathways, reactions and their consequences on the activities of human beings and other life-forms. Thus, environmental chemistry may be considered as a multi-disciplinary study, involving physical and life-sciences, meteorology, agriculture, public health, engineering etc. The basic concepts of environmental chemistry are interesting not only to the scientists engaged in various scientific and engineering activities but also to the personnel involved in resource planning and material management. It is now universally realised that any future development activity has to be viewed in the light of its ultimate environmental impact. The tremendous increase in industrial activity during the last few decades and the release of obnoxious industrial wastes into the environment, have been of considerable concern in recent years from the point of view of environmental pollution. Environmental pollution on one hand and deforestation and population explosion on the other, are threatening the very existence of life on the earth. This situation can improve only if people from all walks of life realize the importance of environmental protection. Hence, environmental education (which includes basic concepts of environmental chemistry) at all levels of non-formal and formal education, is of paramount importance.

Environmental Segments

The environment may broadly be considered to comprise of the following four segments.

(1) Lithosphere   (2) Hydrosphere
(3) Atmosphere   (4) Biosphere.

(1) Lithosphere. The mantle of rocks constituting the earth's crust is called lithosphere. For all practical purposes, the soil covering the rocks (which results from physical, chemical and biological processes during weathering) is also considered to be an important part of the lithosphere. The soil mainly consists of complex mixture of inorganic and organic matter and water. The inorganic mineral constituents include complex mixture of silicates of Na, K, Ca, Al and Fe; oxides of Fe, Mn and Ti; and carbonates of Ca and Mg. The organic matter, which constitutes not more than about 5% of the soil mainly determines the productivity of the soil. It consists of biologically active components such as polysaccharides, nucleotides, organo-phosphorous and organo-sulphur compounds, sugars and humic materials. The clay minerals and humus present in the soil possess a very high cation exchange capacity and thus help in supplying essential trace metals to the plants as nutrients (The cations in the soil are exchanged for H^+ from the carbonic acid present in the soil and thus supply the trace metals to the plants through the roots).

(2) Hydrosphere. This includes all the surface and ground water resources *viz.*, oceans, seas, rivers, streams, lakes, reservoirs, glaciers, polar ice caps, ground-water and the water locked in rock-crevices and minerals lying deep below the earth's crust. Earth is called the blue planet, because about 80% of its surface is covered by water. But, however, about 97% of the earth's water resources is locked-up in the oceans and seas, which is too saline to drink and for the direct use for agricultural and industrial purposes. About 2.4% is trapped in giant glaciers and polar ice-caps. Thus not even 1% of the total world's water resources is available for exploitation by man for domestic, agricultural and industrial purposes.

Water exhibits unusual properties, as compared to the hydrides of VI Group elements, due to the presence of hydrogen bonding. Without hydrogen bonding, water would have boiled at $-100°C$ and in that case life on the earth would have been impossible. Because of hydrogen bonding, water is a liquid at room temperature, with melting point of $0°C$ and boiling point of $100°C$ at atmospheric pressure[*].

(3) Biosphere. This is the region of the earth where life exists and includes a global girdle extending from about 10,000 m below sea level to 6,000 m above sea level. Thus, the biosphere covers the entire realm of living organisms and their interactions with the other segments of the environment, namely the lithosphere, the hydrosphere and the atmosphere.

(4) Atmosphere. The atmosphere comprises of a mixture of gases (*e.g.*, N_2, O_2, CO_2, Ar etc) and it extends upto about 500 kms above the surface of the earth. A constant exchange of matter takes place between the atmosphere, biosphere and hydrosphere. Their relative weights are of the following order :

Biosphere : Atmosphere : Hydrosphere
1 : 300 : 69,100

The weight of the atmosphere is approximately $(4.5 \text{ to } 5) \times 10^{15}$ metric tons. The atmospheric temperature, pressure and density vary considerably with altitude. Thus, the atmospheric temperature varies from $-100°C$ to $+1200°C$, depending upon the altitude. The atmospheric pressure at sea-level is 1 atmosphere, whereas at 100 km above sea-level, it drops to 3×10^{-7} atmosphere. The atmospheric density at the surface of the earth is about 0.0013 g/cu.m. which decreases sharply with increasing altitude and gradually thins out into space. At about 600 km and above, the atoms and molecules describe free elliptical orbits in the earth's gravitational field.

The atmosphere surrounding the earth acts as a gaseous blanket, protecting the earth from dangerous cosmic radiations from outer space and helps in sustaining life on the earth. The atmosphere screens the dangerous u.v. radiations from the sun (< 300 nm) and transmits only the radiations in the range 300 nm to 2500 nm, comprising of near ultra-violet, visible and near infra-red radiations and radio-waves $(0.01 - 4 \times 10^5 \text{ nm})$.

The atmosphere plays a vital role in maintaining the heat balance on the earth by absorbing the i. r. radiation received from the sun and re-emitted by the earth. In fact, it is this phenomenon,

---

[*]It has a density of 1.0 gm/cm³; surface tension of 73 dynes per cm at 20°C; viscosity of 0.01 poise at 20°C; dielectric constant of 80; and specific heat of 1 cal per gram per °C.

called "the Green-house effect", which keeps the earth warm enough to sustain life on the earth. Apart from this, the important gaseous constituents of the earth, viz., O_2, N_2 and CO_2 play important roles in sustaining life on earth. Oxygen supports life on earth, Nitrogen is an essential macro-nutrient for plants (via nitrogen fixation and fertilizer manufacture) and Carbon dioxide is essential for photosynthetic activity of plants. Moreover, atmosphere is a carrier of water from oceans to land, which is so vital for the hydrologic cycle.

Any major disturbance in the atmospheric composition, either by extra-ordinary or anthropogenic activities, may lead to disastrous consequences and may even endanger the very survival of life on the earth.

Composition of the atmosphere

The composition of clean, dry air, near sea-level in given in Table1.

Table 1 - Composition of clean, dry air, near sea-level

Components	Content per cent by volume	ppm
(a) Major components		
Nitrogen (N_2)	78.09	7,80,900
Oxygen (O_2)	20.94	2,09,400
Water Vapour (H_2O)	0.1-5	1,000-50,000
(b) Minor Components		
Argon (Ar)	0.934	9,340
Carbon dioxide (CO_2)	0.032	320
(c) Trace Components		
Neon (Ne)	0.00182	18.2
Helium (He)	0.000524	5.24
Methane (CH_4)	0.00018	1.8
Krypton (Kr)	0.00011	1.1
Nitrous oxide (N_2O)	0.000025	0.25
Hydrogen (H_2)	0.00005	0.5
Xenon (Xe)	0.0000087	0.087
Sulphur dioxide (SO_2)	0.0000002	0.002
Nitrogen dioxide (NO_2)	0.0000001	0.001
Ammonia (NH_3)	0.000001	0.01
Carbon monoxide (CO)	0.000012	0.12
Ozone (O_3)	0.000002	0.02
Iodine (I_2)	Traces	Traces

The proportions of the various components of the atmosphere more or less remain constant upto a height of about 16 km from the earth's surface. Above this height, gravitational separation begins, although this becomes significant above 130 km. The weight of the atmosphere is about $4.5 - 5 \times 10^{15}$ metric tonnes, which is about one millionth of the total weight of the earth.

Atmospheric Structure

The atmosphere, which extends upto about 500 km, can be broadly divided into four major regions, with widely varying temperatures, even within each region. Some important characteristics of the four major atmospheric regions viz., Troposphere, Stratosphere, Mesosphere and Thermosphere are summarised in Table-2 and Fig. 18.1.

Table 2 - Characteristics of the major regions of the atmosphere

Name of the region	Height above the earth's surface, km.	Temperature range °C	Major chemical species present
Troposphere	0- 11	15 to -56	O_2, N_2, CO_2, H_2O
Stratosphere	11- 50	-56 to -2	O_3
Mesosphere	50- 85	- 2 to -92	O^+_2, NO^+
Thermosphere	85-500	-92 to 1200	O^+_2, O^+, NO^+

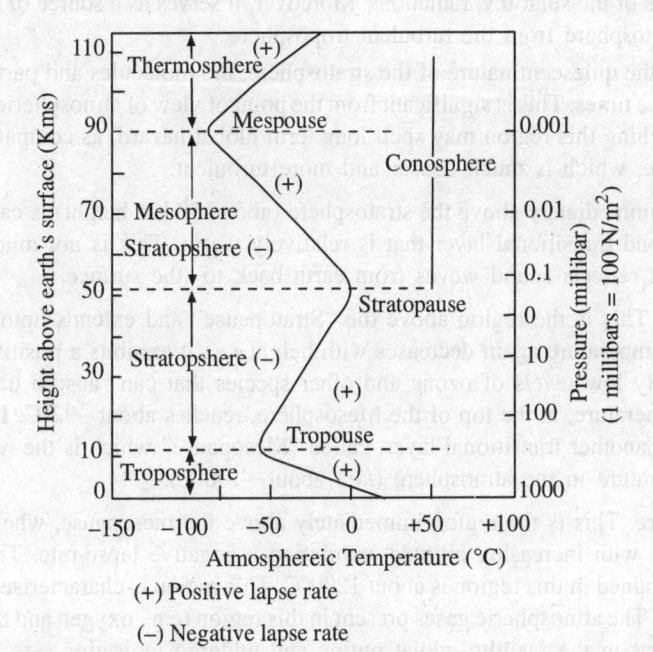

Fig. 18.1. Major Atmospheric Regions with temperature and Pressure Profile.

(+) Positive lapse rate
(-) Negative lapse rate

(1) Troposphere. This is the region nearest to the earth's surface and extends upto an altitude of 11 km. The upper limit may vary by a few kilometers, depending upon temperature, nature of the terrestrial surface and some other factors. The troposphere accounts for over 70 % of the atmospheric mass. The composition of air in this region remains more or less constant in the absence of any significant air pollution. This is mostly due to the turbulence and constant circulation of air masses, as a result of convection currents arising from differential heating and cooling rates between the equator and the poles. The density of air, in this region, decreases exponentially with increasing altitude. The troposphere contains most of the water, cloud and particulate matter of the atmosphere.

The temperature of air in the troposphere decreases fairly steadily with increasing altitude from the ground temperature to a temperature of about − 56°C. It can be seen from Fig. 18.1. that the temperature-altitude curve then changes its slope (*i.e.*, the temperature starts increasing with increasing altitude) rather suddenly in a narrow transitional layer at the top of the troposphere, known as the "Tropopause", which is usually at an altitude of 10 km to 20 km. The temperature of the tropopause is the least at the equator.

The change of temperature with height is called the "lapse rate". The decrease of temperature with increasing altitude in the troposphere is called positive lapse rate. The transition from positive lapse rate to negative lapse rate at the tropopause marks what is called the "temperature inversion".

(2) Stratosphere. The region above the tropopause is called the stratosphere. In this region, the temperature-altitude curve shows a warming trend with increasing altitude *i.e.*, it exhibits a negative lapse-rate. The temperature in this stratospheric region continues to increase with height, until 50 kms, where the temperature attains a maximum of $-2°C$. This warming up tendency in the stratosphere is due to the absorption of solar u.v. radiation by ozone, whose concentration in this region is in the range of 1 to 5 ppm by volume and this is responsible for the negative lapse-rate. The air in this region is very dry and the clouds and convection currents from the troposphere normally do not penetrate into it.

The presence of ozone in the stratosphere serves as a shield to protect life on the earth from the harmful effects of the solar u.v. radiations. Moreover, it serves as a source of heat for separating the quiescent stratosphere from the turbulent troposphere.

Because of the quiescent nature of the stratosphere, the molecules and particles in the region have long residence times. This is significant from the point of view of atmospheric pollution because any pollutant reaching this region may spell long term global hazard, as compared to their impact in the troposphere, which is much denser and more turbulent.

The region immediately above the stratosphere (above 50 km height) is called "Stratopause" which is the second transitional layer that is relatively warm. This is not much cooler than the earth's surface. It reflects sound waves from earth back to the surface.

(3) Mesosphere. This is the region above the "Stratopause" and extends upto 85 km height. In this region, the temperature again decreases with height *i.e.*, it exhibits a positive lapse-rate. This is due to relatively low levels of ozone and other species that can absorb u.v. radiations from the sun. The temperature, at the top of the Mesosphere, reaches about $-92°C$. Immediately above the Mesophere is another transitional layer, called "Mesopause" which is the region of minimum or coldest temperature in the atmosphere (*i.e.*, about $-100°C$).

(4) Thermosphere. This is the region immediately above the mesopause, where the temperature rises very rapidly with increasing altitude, exhibiting a negative lapse-rate. The maximum temperature that is attained in this region is about $1200°C$. This region is characterised by low pressures and low densities. The atmospheric gases present in this region (e.g., oxygen and nitric oxide) absorb the solar radiations in the far ultra-violet region and undergo ionisation.

The region above the stratosphere, in the altitude range of 50 km to 100 km, is called "Ionosphere". In this region, positive ions e.g., O_2^+, O^+, NO^+ etc and electrons exist at significant levels. These charged species persist for long periods of time, without mutual neutralization, due to the rarefied conditions existing in the region.

The troposphere, the stratosphere and the mesosphere are fairly uniform in composition. From air pollution point of view, troposphere is of particular significance.

Radiation balance of the earth

The earth continuously receives energy from the sun, a part of which is absorbed, while the remaining is emitted back into space. The earth maintains its heat balance within narrow limits, due to complex mechanisms of energy-transfer. Hence, optimum climatic conditions are maintained for supporting life.

The sun radiates energy like a black body at $6000°K$. Out of the solar radiations reaching the earth, 92% consists of radiations in the range 315 to 1400 nm. 45% of this is in the visible region (400 to 700 nm). The earth absorbs radiations mainly in the visible region and emits radiation in the infra-red region (2-40 μ, with maximum at 10 μ).

The quantity of solar energy per unit time passing through unit area at right angles to the direction of the solar beam measured just outside the earth's atmosphere, is called the "solar-constant" (S). The value of the solar flux reaching the earth's upper atmosphere is estimated to be about 1400 watts m^{-2} min^{-1}. The heat equivalent of the solar radiation reaching the earth is estimated to be 2.68×10^{24} joules per year. Out of this energy, 50% is reflected before it reaches the earth's surface, 15% is reflected by the earth's surface, 5.3 % is absorbed by the soil, 27.8 % is utilized for heating and evaporation of water and the remaining 1.9 % is absorbed by the marine and land vegetation. The energy transport plays a vital role in maintaining the earth's radiation balance and this proceeds through various mechanisms, such as the following :

(1) Conduction of energy through the interaction of atoms, molecules and other species.
(2) Convection of energy via massive air circulation.
(3) Radiation of energy from the earth in the infrared region (2 to 40 μ).
(4) Re-absorption of most of the outgoing infra-red radiation (2 to 40 μ) by water vapour (4-8 μ), CO_2 (12 to 16.3 μ) and other gases such as CH_4 and re-emitting a part of this radiation to the earth's surface. This process, called the "green-house effect" (discussed later) is mostly responsible for maintaining the climatic features and temperature of the earth.

Further, the particulate matter in the atmosphere, which is released by natural forces (*e.g.*, wind, sea sprays and volcanoes) or by anthropogenic activities (*e.g.*, agricultural and industrial activities which release dust, fumes, smoke) also exert cooling or heating effect, depending upon the nature of the particles which may reflect, scatter or absorb radiations.

The fraction of the incident solar radiation that is reflected and scattered back into the space is called "albedo". Deforestation and the consequent soil erosion, industrial operations and other such activities also influence the earth's radiation balance by altering the `albedo'.

Chemical Species and Particulates present in the Atmosphere

The various chemical species present in atmosphere have been summarised in Table–1 and Table–2.

Ions. That region of the atmosphere which is at an altitude of 50 km to 100 km, is called "Ionosphere" because appreciable levels of electrons and positive ions, such as O_2^+, O^+, NO^+, etc exist in this region with considerable residence times. The u.v. radiations from the sun are mostly responsible for the formation of the ionic species in ionosphere. During night times, when the u.v. radiations are not present, the positive ions tend to recombine with the free electrons, yielding the neutral species from which they are originated initially. This process occurs more rapidly in the lower regions of the ionosphere where the concentration of these species is relatively high.

Radicals. Apart from the ionic species, the atmosphere also consists of highly reactive free radicals, generated by photo-chemical reactions *e.g.*, $HO^\bullet$, $CH_3^\bullet$, $SO_2^\bullet$, $ROO^\bullet$, $NO_2^\bullet$, and $HCO^\bullet$. The free radicals, which may be organic or inorganic, comprise of atoms or groups of atoms with impaired electrons having short half-lives. Owing to their high reactivity, they interact with other chemical species available, to propagate chain reactions until chain termination takes place due to destruction of one of the free radicals in the chain, by any of the various possible mechanisms. Free radicals play significant role in the formation of photo-chemical smog.

Particles. Several types of particles exist in the atmosphere in varying sizes from 0.1 μ to 10 μ. Even pure air may contain several hundreds of particles per c.c., whereas highly polluted air may contain about a lakh of particles per c.c. Particles of colloidal dimensions are called "aerosols". The word "particulates" is generally used to describe the particles present in aerosols. Aerosols of natural origin having diameters < 0.2 μ are called "Aitkin" particles. The various particles existing in nature include dust, smoke, fog, pollen grains, volcanic ash and bacteria. Dusts, mist, smoke, smog and fumes may also result from anthropogenic activities. Inorganic particulates, such as iron oxide, calcium oxide etc., may result from combustion of coal and metallurgical operations. Particulate lead results from automobile exhausts. Aerosol mists are generated from oxidations of SO_2 to SO_3 which in presence of water vapour forms droplets of H_2SO_4.

Organic particulate matter arises from automobile exhausts, combustion of fuels and evaporation of organic matter from vegetation. Certain organic particulates contain polycyclic aromatic hydrocarbons (PAH) which are carcinogenic. They are generated from the pyrolysis of higher paraffins present in fuels and in some vegetable matter.

Atmospheric particulates may enter the respiratory tract and cause serious health hazards. Removal of obnoxious particulates from gaseous emissions is an important step in controlling air pollution.

The size and chemical characteristics of the particles in atmosphere are more significant than their concentration. The particles in the range of $0.1\ \mu$ to $1\ \mu$ manifest in several important effects in the atmosphere, such as the following :

(1) They play a vital role in maintaining the radiation balance and heat balance of the earth.
(2) They provide nuclei for condensation of water vapour.
(3) They are responsible for fog and cloud formation.
(4) They may bring about heterogeneous gas-phase reactions. The particles may absorb gases and catalyze some reactions. For example, the decomposition of ozone is catalyzed by a solid surface.
(5) Particles absorb and destroy free radicals, thereby reducing the rate of free radical chain reactions.
(6) The absorption of gases on some solid surfaces may change the absorption spectrum of the gas for sun light. For instance, oxygen absorbed on carbon particles might absorb sunlight more strongly than what free oxygen would do.
(7) Particles would help in several types of reactions in the atmosphere such as
   (a) Catalyzing the oxidation of SO_2 by O_2 or O_3 in aqueous droplets.
   (b) Neutralization reactions between sulphuric acid droplets and lime stone dust.
   (c) Reaction between O_3 or NO_2 with salt particles in water droplets.
   (d) Photochemical reactions involving dusts and aerosols.

Reactions in the atmosphere

The various chemical and photochemical reactions taking place in the atmosphere, mostly depend upon the temperature, composition, humidity and intensity of sun light. Thus the ultimate fate of a chemical species in the atmosphere depends upon these parameters. Photochemical reactions take place in the atmosphere by the absorption of solar radiations in the u.v. region. Absorption of photons by chemical species gives rise to electronically excited molecules, which can bring about certain reactions, which are not possible under normal laboratory conditions, excepting at higher temperatures and in presence of chemical catalysts. The electronically excited molecules produced by the absorption of a photon may undergo any of the following changes :

   (a) Reaction with other molecules on collision
   (b) Polymerisation
   (c) Internal rearrangement
   (d) Dissociation
   (e) De-excitation by fluorescence or de-activation to return to the original state.

Any of the first four changes stated above may serve as an initiating chemical step or a primary process. The three steps involved in an overall photochemical reaction are : (1) Absorption of radiation, (2) Primary reactions, and (3) Secondary reactions.

The various chemical species that can undergo photo-chemical reactions in the atmosphere include NO_2, SO_2, HNO_3, N_2, Ketones, H_2O_2, organic peroxides and several other organic compounds and aerosols such as metal oxides.

Reactions involving oxides of nitrogen. The oxides of nitrogen present in the atmosphere are N_2O, NO and NO_2. Nitrous oxide mainly originates from various microbiological processes whereas NO and NO_2 mainly originate from the combustion of fossil fuels and other anthropogenic activities. The NO + and NO_2 together are referred to as NO_x, which is an important constituent of polluted air.

In the stratosphere, N_2O undergoes photochemical decomposition to NO, which in turn, depletes the protective ozone layer :

$$N_2O + h\nu \longrightarrow N_2 + O$$

$$N_2O + O \longrightarrow NO + NO$$

$$NO + O_3 \longrightarrow NO_2 + O_2$$

The most important primary photochemical reaction is the dissociation of NO_2.

$$NO_2 + h\nu \longrightarrow NO + O$$

The NO so formed may be oxidised by O_3 or more slowly by O_2, thus leading to a cyclic chain reaction. The chain may be broken only when the NO_2 is completely converted into HNO_3 by hydration and catalytic oxidation in presence of aerosols, fog or photochemical smog. The reactions may take place in the following sequence :

$$2\ NO + O_2 \longrightarrow 2\ NO_2$$

$$NO + O_3 \longrightarrow NO_2 + O_2$$

$$2NO_2 + O_3 \longrightarrow N_2O_5 + O_2$$

$$4\ NO_2 + O_2 + 2H_2O \longrightarrow 4HNO_3$$

$$N_2O_5 + H_2O \longrightarrow 2HNO_2$$

HNO_2 and HNO_3 may also undergo photochemical dissociation as follows :

$$HNO_2 + h\nu \longrightarrow NO_2 + H^\bullet$$

$$HNO_2 + h\nu \longrightarrow NO + HO^\bullet$$

$$HNO_3 + h\nu \longrightarrow NO_2 + HO^\bullet$$

In the stratosphere, NO_2 may react with the HO free radical, forming HNO_3 :

$$NO_2 + HO^\bullet \longrightarrow HNO_3$$

The HNO_3 so formed is removed as acid rain or is converted into particulate nitrates due to neutralisation by NH_3 or particulate lime.

Oxidation of Sulphur dioxide. Sulphur dioxide, present in the atmosphere, absorbs solar radiation in the range of 300-400 nm and produces electronically excited states of SO_2. This undergoes oxidation to SO_3 and in presence of water vapour, this is converted to H_2SO_4.

$$SO_2 + h\nu \longrightarrow SO_2^*$$

$$SO_2^* + O_2 \longrightarrow SO_4^*$$

$$SO_4^* + O_2 \longrightarrow SO_3 + O_3$$

$$SO_3 + H_2O \longrightarrow H_2SO_4$$

The overall reaction in presence of sunlight and relative humidity 30-90% may be represented as

$$SO_2 + \tfrac{1}{2} O_2 + H_2O \longrightarrow H_2SO_4$$

This photochemical oxidation of SO_2 leading to the formation of H_2SO_4 aerosol is greatly accelerated in the presence of olefinic hydrocarbons and oxides of nitrogen which are usually present in "photochemical smog".

"Photochemical Smog" is initiated by the photochemical dissociation of NO_2 and the consequent secondary reactions involving unsaturated hydrocarbons, other organic compounds and free radicals, leading to the formation of organic peroxides and ozone. This phenomenon takes place during sunny days with low winds and low level inversion. The photochemical smog and the consequent formation of aerosols reduce the visibility, cause irritation to eyes and damage plants and rubber goods.

The oxidation of SO_2 can also take place by interaction with the free radical $HO^{\bullet}$ present in photochemical smog

$$SO_2 + HO^{\bullet} \longrightarrow HOSO_2^{\bullet}$$
$$HOSO_2^{\bullet} + O_2 \longrightarrow HOSO_2 O_2^{\bullet}$$
$$NO + HOSO_2 O_2^{\bullet} \longrightarrow HOSO_2 O^{\bullet} + NO_2$$
$$\text{(sulphate)}$$

Chemical oxidation of SO_2 may also take place in water droplets present in aerosols. This reaction is accelerated in presence of NH_3 and catalysts e.g., oxides of Mn, Fe, Cu, Ni.

Solid particles, such as soot, bring about catalytic oxidation of SO_2 by providing a heterogeneous phase for contact. Soot is formed during combustion of solid and liquid fuels in domestic and industrial operations and automobile emissions.

Sulphur dioxide is a pollutant responsible for smog formation, acid rains, and corrosion of metals and alloys.

Oxidation of Organic Compounds. Organic compounds such as hydrocarbons, aldehydes and ketones absorb solar radiation and undergo various photochemical and chemical reactions involving free radicals. Some of these reactions are catalysed by particulate matter such as soot and metal oxides. Some of the intermediates and final products formed contribute to photochemical smog formation.

$$CH_4 + O_2 \longrightarrow H_3C^{\bullet} + HO^{\bullet}$$
Methane
$$CH_4 + HO^{\bullet} \longrightarrow H_3C^{\bullet} + H_2O$$
$$H_3C^{\bullet} + O_2 + M \text{ (Third body)} \longrightarrow H_3COO^{\bullet} + M$$
$$CH_3CH = CH_2 + HO^{\bullet} \longrightarrow CH_3 - CH^{\bullet} - CH_2OH$$
Propylene
$$RCHO + h\nu \longrightarrow R^{\bullet} + HCO$$
Aldehyde
$$\begin{matrix} R \\ R \end{matrix} \!\! > \!\! CO + h\nu \longrightarrow R^{\bullet} + R^{\bullet}CO$$
Ketone

where R is an alkyl or aryl radical or even a hydrogen atom. The alkyl or aryl radicals may react with oxygen to form a peroxyl radical, which in turn may react with another oxygen atom to give O_3. The peroxyl radical may react with NO_2 to give peroxyacyl nitrate (PAN), formaldehyde (which is an irritant) and various polymeric compounds which reduce visibility.

$$\text{R COO}^\bullet + \text{NO}_2 \longrightarrow \text{R COO NO}_2$$
$$\text{(PAN)}$$

Greenhouse effect

The earth is heated by sunlight and some of the heat that is absorbed by the earth is radiated back into space. However, some of the gases in the lower atmosphere, acting like glass in a greenhouse, allow the solar radiations (in the range 300 to 2500 nm, *i.e.*, near u.v., visible and near I.R. region, while filtering the dangerous u.v. radiations, *i.e.* <300 nm) but do not allow the earth to re-radiate the heat into space. In other words, these gases in the atmosphere are transparent to the sun-light coming in but they strongly absorb the infra-red radiation, which the earth sends back as heat. A part of the heat so trapped in these atmospheric gases is re-emitted to the earth's surface. The net result is the heating of the earth's surface by this phenomenon, called the "Greenhouse effect". The gases that are responsible for this Greenhouse effect are CO_2, water vapour, CH_4 and man-made chloroflourocarbons (CFC's). Water vapour strongly absorbs I.R. radiations in the range 4000 to 8000 nm and CO_2 in the range 12,000 to 16,300 nm. The radiations in the range 8000 to 12,000 nm escape unabsorbed and this is known as the region of atmospheric window.

Carbon dioxide is released by volcanoes, oceans, decaying plants as well as human activities such as deforestation and combustion of fossil fuels. Automobile exhausts account for 30% of CO_2 emissions in developed countries.

Methane is released from coal mines, decomposition of organic matter in swamps, rice paddy cultivation, guts of termites in forest debris and stomachs of ruminants.

Chloroflourocarbons (CFC's) are used as coolants in refrigerators, propellants in aerosol sprays, plastic foam materials like "Thermocoles" or "Styrofoam and in automobile air-conditioners.

In fact, the "Greenhouse gases" (particularly CO_2 and water vapour) are responsible for keeping our planet warm and thus sustaining life on the earth. If the Greenhouse gases are very less or totally absent in the earth's atmosphere, then the average temperature on the earth would have been at sub-zero levels. But, however, if the concentration of Greenhouse gases is larger, they may trap too much of heat, which may threaten the very existence of life on the earth. For instance, the CO_2 present in the atmosphere of the planet Venus is about 60,000 times more than that in the earth's atmosphere. Hence the average temperature of Venus is about 425°C, thus making the existence of life impossible there.

Oceans and bio-mass are the major sinks for the atmospheric CO_2. Oceans convert CO_2 into soluble bicarbonates. The photosynthetic activity in the green plants increases with increase in CO_2 level in the atmosphere. Forests are the places where lot of photo-synthetic activity occurs. They also act as vast reservoirs of fixed but readily oxidisable carbon in the form of vegetation, wood and humus. Hence, forests maintain a balance in the atmospheric CO_2 level. Therefore, deforestation definitely upsets this balance and increases the atmospheric CO_2 level.

It is estimated that the atmospheric CO_2 content has increased by 25% during the last two centuries. This is mostly attributed to the industrial revolution over these two centuries. This is one of the reasons for the slight increase in the global temperature (about 0.5°C). Since the concentrations of the Greenhouse gases have been continuously increasing because of deforestation, industrialization, increased burning of fossil fuels, mining, exhausts from increasing number of automobiles and other anthropogenic activities, there is an increasing concern about the possible "global warming". Some scientists fear that if proper precautions are not taken, the concentration of the green-house gases in the atmosphere may double within the next 50–100 years. If this happens, the average global temperature may increase by 4 to 5°C. This will increase the evaporation of surface waters, which may influence climatic changes depending upon the pattern of cloud formation. For instance, low-level dense clouds may exert cooling effect whereas high level thin cloud formation may exert heating effect due to increased green-house effect.

The projections from computer modelling regarding the climatic changes that could be triggered off due to "global warming" reveal alarming scenarios. Even 1.5°C raise in surface temperature can adversely affect the food production in the world. Thus, the wheat growing zones in the northern latitude may be shifted from the USSR and Canada to the polar regions *i.e.*, from fertile soils to poor soils near the north pole. The biological productivity of the ocean would also decrease due to warming of the earth's surface layer, which in turn, may reduce the transport of the nutrients from deeper layers to the surface by vertical circulation. The computer modelling also indicate the following effects due to "global warming": melting of the polar ice caps; dry areas becoming drier; humid areas like the Amazon, suffering more intense tropical storms; drastic drop in food production, particularly in lands within 35 degrees north and south of the Equator; increased breeding of pests and diseases due to more humid conditions; shorter, wetter and warmer winters; and longer, hotter and drier summers, particularly in mid-continental areas. Global warming may also trigger increased thermal expansion of oceans and melting of glaciers, which results in lifting up of the sea-level by 20 cm to 1.5 meters by the later part of the 21st century. Thus, cities like Bombay, Miami, London, Venice, Bangkok and Leningrad may become extremely vulnerable. Defences against the raising sea-levels and expanding oceans are very difficult and expensive, which many nations of the world cannot afford. Further, a global temperature raise, even about 1.5°C, is likely to cause floods, hurricanes, tornadoes, apart from raising of the sea-level due to melting of the polar ice caps and inundating coastal cities like Madras, Sydney, New York and Boston.

There are differences of opinion among experts regarding the dynamics and effects of "global warming" due to the complexity of neutral phenomena that might be operating simultaneously. More accurate future climatic projections will be possible with better super-computer models, based on greater understanding of the complex natural climatic forces involved. But until that time, the possible devastating effects due to "global warming" by the "greenhouse effect" cannot be underestimated. Some of the steps suggested to minimise the "Greenhouse effect" include reduction in the use of fossil fuels, encouraging the use of alternative sources of energy (*e.g.*, solar, geothermal, wind, bio-gas, etc). Conservation of forests, extensive afforestation, encouraging community forestry, reduction in the use of automobiles, research in development of more efficient automobile engines, ban on CFC's and nuclear explosions, development of environmentally compatible technologies with the help of intensive inter-disciplinary research, effective check on the growth of population and imparting of non-formal and formal environmental education.

Formation and Depletion of Ozone in the Stratosphere

Ozone is an important chemical species present in the stratosphere. At an altitude of about 30 km, its concentration is about 10 ppm. The ozone layer present in the stratosphere acts as a protective shield for the life on earth. It strongly absorbs ultra-violet radiations from the sun in the region 220-330 nm and thereby protects the life on earth from severe radiation damage, such as DNA mutation and skin cancer. Thus only a small fraction of u.v. radiation reaches the lower atmosphere and the earth's surface.

Ozone is formed in the stratosphere by photochemical reaction :

$$O_2 + h\nu \ (242 \text{ nm}) \longrightarrow O + O$$

$$O + O_2 + M \text{ (third body, such as } N_2 \text{ or } O_2) \longrightarrow O_3 + M$$

The third body absorbs the excess energy liberated by the above reaction and thereby the ozone molecule is stabilized. Thus, ozone is constantly formed in the stratosphere. However, it is also destroyed by chlorine, released due to volcanic activity and also by reaction with (a) Nitric oxide, (b) Atomic oxygen and (c) Reactive hydroxyl radical, which are also present in the atmosphere[*], by the following reactions :

(a) $O_3 + NO \longrightarrow NO_2 + O_2$

(b) $O_3 + O \longrightarrow O_2 + O_2$

(c) $O_3 + HO^{\bullet} \longrightarrow HO_2 + HOO^{\bullet}$

$HOO^{\bullet} + O \longrightarrow HO^{\bullet} + O_2$

Ozone, in the stratosphere, is also found to be destroyed by man-made chloroflourocarbons (CFC's), which are used as coolants in refrigerators, air-conditioners, propellants in aerosol sprays and in plastic foams, such as "Thermocole" or "Styrofoam". The CFC molecules, escaping into the atmosphere, decompose to release chlorine in the ozone layer (by photo-dissociation) and each atom of chlorine, thus liberated, is capable of attacking several ozone molecules.

$Cl + O_3 \longrightarrow ClO + O_2$

This reaction is followed by

$ClO + O \longrightarrow Cl + O_2$

which regenerates Cl atoms, so that a long chain process is involved, which conserves Cl atoms. The environmental hazards of CFC's were recognised as early as 1970. In fact, temporary thinning in the stratospheric ozone layer, leading to the formation of "Ozone hole" was actually detected over the Antarctica during September to November in 1985. Reported increase in cases of skin cancer in South Australia are also attributed to u.v. radiations reaching the earth, due to depletion of ozone layer, over that part of the world temporarily.

The detection of the "Ozone hole" over Antarctica in 1985 attracted the attention of scientific community in the world. The U.S. immediately banned the use of CFC's in spray cans. Further, in the year 1987, twenty four nations of the world signed the Montreal Protocol, which aims at 35% reduction in the global production of the CFC's by the year 1999. Simultaneously, efforts to produce chlorine-free substitutes have also started. In fact, synthesis of a product called HFC-134a has already been reported as an effective substitute for CFC. Hydrofluorocarbons (HFC), Hydrofluorocarbons (HCF) and methyl cyclohexane (MCH) are envisaged as substitutes for CFC's.

Lapse Rate and Temperature Inversion

The change of atmospheric air temperature with altitude is called the "lapse rate". As we go high in the atmosphere, the pressure decreases. A gas, that is free to expand, has a tendency to move from high pressure region to low pressure region. While doing so, the gas expands and if there is no opportunity to exchange its heat with the surroundings, the gas undergoes adiabatic cooling. This adiabatic lapse rate can be calculated with the help of pressure-volume-temperature relationship, the adiabatic expansion and the change of pressure with the altitude. The value for adiabatic lapse rate has been calculated to be about—10°C/km for dry air and about—6°C to—7°C/km for humid air, the higher values occurring at greater altitudes where the water content is less.

Under conditions of adiabatic lapse rate, a smoke plume rises upwards into the atmosphere by virtue of its low density (because of its higher temperature) until it reaches air of similar temperature and similar density. However, in many practical situations, the lapse rate is found to be either greater

---

[*] In atmosphere, NO comes from chemical and photo-chemical reactions in the atmosphere, supersonic jets, nuclear explosions etc; Cl_2 comes from CFC's and volcanoes; and OH• comes from bio-mass burning and from natural water systems.

or lesser than that of the adiabatic lapse rate due to external heating and cooling effects. From air-pollution point of view, the following two situations are of great importance :

(i) "Super adiabatic lapse" where the lapse rate is considerably higher than that of adiabatic. On a clear summer day, the air near the earth's surface gets heated rapidly to reach a condition where the lapse rate is super adiabatic. In such a situation, the atmosphere is said to be in an unstable equilibrium. It is under such conditions when the atmospheric pollutants are rapidly dispersed due to considerable vertical mixing of air.

(ii) "The negative lapse rate" or "temperature inversion" when the temperature decreases with increasing altitude. Inversion occurs in the atmosphere mainly due to the following two reasons :

(a) Radiation inversion :- Such a situation occurs when the air near the earth's surface gets cooled because of the loss of heat by the earth at night by emitting longwave radiations. Under the conditions of temperature inversion, the air near the ground is denser than the air above and hence very little mixing or turbulence takes place.

(b) Subsidence inversion :- Such a situation occurs when the upper layer of air "subsides" (i.e. descends) during a developing anticyclone (i.e. high pressure centre). This subsiding air warms as it contracts. But, the extent of warming will be greater at the upper part than at the lower part near the ground.

Under the conditions of inversion, the denser cold air at the bottom cannot rise up due to lack of any driving force and the atmosphere is said to be stable. Such a situation is dangerous from the atmospheric pollution point of view, because the pollutants in the air cannot be dispersed. Development of inversion conditions generally occur on still and clear nights, when the ground is strongly cooled. Hence, the air near the ground level is also cooled, while the air above is relatively warm. In such a condition, if a pollutant is discharged at or near the ground level, the pollutant tends to concentrate rather than getting dispersed in the air. This increasing intensity of air pollution may lead to disastrous consequences. Inversion is generally destroyed by the next morning due to the solar heating of the ground and the buildings which enable upward current of the warmed air.

Simultaneous occurrence of mist or fog may prolong the duration of inversion by reducing the sunlight reaching the ground.

The problem of air pollution assumes serious dimensions under conditions of inversion and other meteorological conditions which are adverse to the effective dispersion of pollutants. Inversion is responsible for many air-pollution episodes in the world. The notorious London smog of 1952 that lasted for five days killing four thousand people is an example.

In towns and villages which are situated in valleys, inversion conditions are particularly disastrous, because the dense air charged with pollutants descends down to the valleys and forms a sort of "lake of polluted air".

Fortunately for India, the climatic conditions in many parts are favourable for dispersion for most of the time. But still, every effort should be made to prevent improper and indiscriminate release of pollutants into the atmosphere.

18.2 AIR POLLUTION CONTROL

Introduction

According to U.S. Public Health Service, "Air pollution may be defined as the presence in the outdoor atmosphere of one or more contaminants or combination thereof in such quantities and of such duration as may be, or may tend to be injurious to human, plant or animal life, or property, or which unreasonably interfere with the comfortable enjoyment of life, or property, or the conduct of business."

The atmospheric structure and composition have been already discussed in the earlier sections of this chapter. The atmosphere is a dynamic system, which steadily absorbs various pollutants from natural as well as man-made sources, thus acting as a natural sink. Gases such as CO, CO_2, H_2S, SO_2 and NO_x as well as particulate matter, such as sand and dust, are continually released into the atmosphere through natural activities such as forest fires, volcanic eruptions, decay of vegetation, winds and sand or dust storms. Man-made pollutants e.g., NO_x, SO_2, CO, hydrocarbons, particulates etc. are also released into the atmosphere. These have surpassed the pollutants contributed by nature thousand-fold. The magnitude of the problem of air-pollution has increased alarmingly due to population explosion, industrialization, urbanisation, automobiles and other human proclivities for greater comfort. The pollutants travel through the air, disperse and may interact with other substances in the atmosphere before they reach a sink, such as an ocean or a human receptor. If the pollutants enter the atmosphere at a faster rate than are absorbed by the natural sinks, then they gradually accumulate in the air. Such a disturbance in the dynamic equilibrium in the atmosphere by the air pollutants released by anthropogenic activities resulting in considerable accumulation in the atmosphere may affect the very life on earth and its environment. Further, the dilution and dispersion of the gaseous pollutants in the atmosphere depend upon the meteorological conditions prevailing at a given time.

Classification of air-pollutants

The air pollutants may be classified in different ways as follows :

(a) According to origin :

(i) Primary pollutants which are directly emitted into the atmosphere and are found as such, e.g., CO, NO_2, SO_2 and hydrocarbons.

(ii) Secondary pollutants which are derived from the primary pollutants due to chemical or photo-chemical reactions in the atmosphere, e.g.,-Ozone, Peroxy- acyl nitrate (PAN), Photo-chemical smog, etc.

(b) According to chemical composition :

(i) Organic pollutants, e.g.,-Hydrocarbons, aldehydes, ketones, amines and alcohols

(ii) Inorganic pollutants. C

Carbon compounds (e.g., CO and carbonates

Nitrogen compounds (e.g., NO_x and NH_3)

Sulphur compounds (e.g., H_2S, SO_2, SO_3 and H_2SO_4)

Halogen compounds (e.g., HF, HCI and metallic fluorides)

Oxidising agents (e.g., O_3)

Inorganic particles (e.g., fly ash, silica, asbestos and dusts from transport, mining, metallurgical and other industrial activities).

(c) According to state of matter :

(i) Gaseous pollutants which get mixed with the air and do not normally settle out, e.g.,- CO, NO_x and SO_2.

(ii) Particulate pollutants which comprise of finely divided solids or liquids and often exist in colloidal state as aerosols, e.g.,- smoke, fumes, dust, mist, fog, smog and sprays.

Air Pollutants and their Effects

The most common air pollutants are : (1) Carbon monoxide (2) Oxides of nitrogen (3) Sulphur dioxide (4) Hydrocarbons and (5) Particulates.

(1) Carbon monoxide (CO). The atmospheric air contains 0.1 to 0.12 ppm of carbon monoxide. The natural processes which contribute to carbon monoxide in the atmosphere are volcanic activity, natural gas and marsh gas emissions, electrical discharges in the atmosphere during storms, seed germination etc. However, most of the CO in the atmosphere is due to human activities such as (1) Automobile exhausts (which accounts for 60% of CO in the atmosphere), (2) Forest fires and agricultural burning (*i.e.* burning of forest debris, crop residues, bushes, weeds and vegetation, which contributes to about 17% of the CO in the atmosphere), and (3) Industrial operations such as electric and blast furnaces in iron and steel industry, petroleum refining, paper industry, gas manufacture and coal mining (which constitutes about 9.6% of CO in the atmosphere). The concentration of CO in city air is about 55 ppm. Soil micro-organisms act as a major sink for CO from ambient atmosphere. It has been demonstrated that 2.8 kg of a soil sample removed 120 ppm of CO from ambient air in 3 hours. Although soil sinks can take care of atmospheric CO, still it exists in significant concentrations in the atmosphere, because the largest CO-producing areas have the least amount of soil sink available around them.

Since transportation sector accounts for about 74% of the entire global CO-emissions, out of which major contribution comes from gasoline-fed internal combustion engines, lot of attention and efforts have been directed towards controlling CO-emission from this sector. Considerable amount of research work is going on along the following directions :

(a) Modification of engine design Designing of more efficient internal combustion engines to minimise the emission of pollutants formed during combustion of gasoline.

The automobile emissions consist of CO, NO_x, hydrocarbons and particulates. If some control measures are applied to reduce one of these pollutants, the amount of other pollutants are also affected simultaneously. For instance, a low air-fuel ratio reduces NO_x emissions considerably but increases CO and hydrocarbon emissions. The stoichiometric air-fuel ratio is the ratio at which the right proportion of oxygen is maintained for complete oxidation of C and H_2 to CO_2 and H_2O respectively. However, automobile carburettors are set to the fuel rich side of the stoichiometric ratio (i.e. greater % of fuel in the air-fuel mixture) because the engine tends to stall on the lean-fuel side (*i.e.* lessor % of fuel in the air-fuel mixture).

Using leaner air-fuel ratios which result in improved fuel distribution and retarded ignition timing and the use of lower compression ratios are some of the modifications in engine design that have proved successful in checking pollution but with some sacrifice in fuel and power economy.

(b) Fuel modification and development of substitute fuels. Efforts are continuing to develop fuel substitutes for gasoline which, on combustion may release lower concentrations of pollutants. Natural gas, methane, steam reformed gasoline, blends of light hydrocarbons etc., have been recommended. With such fuels, emissions due to incomplete combustion are eliminated thereby reducing noxious pollutants in the exhaust gases. However, further technological progress is necessary to enable their wider use. Alcohols and blends of gasoline and alcohol have also been used as fuels. Some of their combustion products such as aldehydes are eye irritants.

Some fuel additives such as barium salts have been also reported to produce excellent results and further studies are in progress in the field of fuel additives too. Extensive toxicological studies are also needed before introducing any new fuel additives.

Tetraethyl lead (TEL), which is widely used as anti knocking additive to gasoline, is the major source of lead pollution in the environment. Lead-free gasoline is being marketed in USA, India and some parts of Europe in an attempt to curb lead pollution. However, thorough toxicological and other studies are needed while releasing these compositions for general use. Also, cheap and efficient catalytic converter must be used to control CO, HC and NO_x emissions.

(c) Treatment of exhaust gases. Treatment of exhaust gases with the help of thermal or catalytic reactors in the exhaust pipe has been a topic for extensive research. In two stage catalytic converters, NO_x are reduced to N_2 and NH_3 in the first converter at elevated temperatures in presence of catalysts such as Pt, Pd and Ruthenium or some base metal alloys in the presence of reducing gas e.g., CO and hydrocarbons. For oxidising CO and hydrocarbons in the second converter, oxidising catalysts of noble metals (e.g., Pt, Ru and Ir) or oxides of CO, Cr, Cu, Ni, Fe and Mn supported on ceramic materials are used in conjunction with suction of additional air to ensure complete oxidation. However, in case of noble metal catalysts, lead-free gasoline will have to be used to prevent poisoning of the catalysts. Further, catalysts suitable for use in automobiles should be cheap, harmless, long-lasting and widely available. As an alternative to catalytic converters, "stratified charge engines" are being developed in Japan. These engines provide for an additional combustion chamber, where a fuel-rich mixture is introduced and ignited with a spark. This sets in combustion at a relatively low temperature, due to which formation of NO_x is minimised. Then the burning mixture enters into the larger main chamber where it gets mixed with a lean-fuel mixture (i.e., higher % of air and lesser % of fuel in the air-fuel mixture). This ensures complete combustion of CO and hydrocarbons without stalling the engine and, at the same time, build up of NO_x is limited because of the lower temperature maintained in the chamber. However extensive research has to be carried out before introducing such engines on a large scale in automobiles.

(d) Exhaust gas recirculation. It is another approach that is being studied for the purpose of automobile pollution abatement.

(2) Oxides of Nitrogen (NO_x). Out of the eight possible oxides of nitrogen, only N_2O, NO and NO_2 are the important constituents of the atmosphere. The atmospheric background concentrations of N_2O, NO and NO_2 are 0.25 ppm; 0.1 to 2 ppm; and 0.5 to 4 ppm respectively. Although the concentration of N_2O is more in the atmosphere, NO and NO_2 are more significant from air pollution point of view and they are usually represented together as NO_x. The formation of NO from N_2 and O_2 is favoured at high temperatures (~1210 to 1765°C) which are usually attained in the combustion proceses involving air. Rapid cooling of the combustion products (quenching) prevents the dissociation of NO. The oxidation of NO to NO_2 is also favoured at high temperatures (~1100°C) but the amount of NO_2 formed is usually not more than 0.5% of the total NO_x present. NO_2 is also formed by photolytic reactions in the atmosphere, as discussed earlier. Some of the reactions involved are as follows :

$$[1210\text{-}1765\ °C]$$
$$N_2 + O_2 \longrightarrow 2NO$$
$$2NO + O_2 \longrightarrow 2NO_2$$
$$NO_2 + h\nu \xrightarrow{398nm} NO + O$$
$$O_2 + h\nu \xrightarrow{242nm} O + O$$
$$O + O_2 + M \longrightarrow O_3 + M$$
$$\text{(third body)}$$
$$O_3 + NO \longrightarrow NO_2 + O_2$$
$$2NO_2 + O_3 \longrightarrow N_2O_5 + O_2$$
$$4NO_2 + 2H_2O + O_2 \longrightarrow 4HNO_3$$

Alternative Mechanism for Nitric Acid formation
a) $O_3 + NO_2 \longrightarrow NO_3 + O_2$
b) $NO_3 + NO_2 \longrightarrow N_2O_5$
c) $N_2O_5 + H_2O \longrightarrow 2HNO_3$

Oxides of nitrogen may be formed either by the natural or artificial fixation of nitrogen from the atmosphere or from nitrogen compounds present in organic matter. The annual global release of NO_x from man-made sources is about 5×10^7 tonnes, which is only slightly less than that discharged by natural bacterial activity. Oxides of nitrogen are produced by the combustion of coal, oil, natural gas and other organic matter. Thus, NO_x is introduced into the atmosphere from automobile exhausts, incinerators, furnace stacks, coal based power plants and other similar sources. When the fuels are burnt in air, some of the N_2 in the air is oxidised to NO. The amount of NO formed depends on the temperature of the flame and the rate of cooling or quenching of the combustion products. Higher flame temperatures and rapid cooling of the combustion products favour the formation of NO.

The average residence times of NO and NO_2 in the atmosphere are 4 days and 3 days respectively. They undergo various photochemical and chemical reactions in the atmosphere, leading to the formation of HNO_3 which gets precipitated as nitrates during rainfall or as dust.

The NO_x from man-made sources may be 10 to 100 times more in urban areas, as compared to rural areas. Even in urban areas, the ambient NO_x levels vary with sunlight and traffic density at any given point of time.

Chemical and photo-chemical reactions involving NO_2 and hydrocarbons induced by sunlight are responsible for the formation of photochemical smog.

As discussed in the earlier section, use of two stage catalytic converters can minimize the NO_x from automobile emissions. Similarly, NO_x from power plant emissions can be reduced by 90% by using a two-stage combustion process. The fuel can be first fired at a relatively high temperature using only about 90% of the stoichiometric air required so that only a minimum quantity of NO is formed under these conditions. Then the combustion of the fuel may be completed at a relatively low temperature, in the excess air. NO is not formed under these conditions.

(3) Sulphur dioxide (SO_2). Combustion of any sulphur-bearing materials produces SO_2 accompanied by a small quantity of SO_3. This mixture is usually denoted as SO_x. Nearly 67% of the global SO_x pollution is due to volcanic activity and other natural sources, over which we have no control. The remaining 33% SO_2 emission is because of human activities such as combustion of fuels, coal-fired power stations, transportation, refineries, metallurgical operations such as smelting of sulphide ores and chemical plants *e.g.*, manufacture of sulphuric acid. Most of the man-made SO_x pollution is concentrated in urban and industrial areas.

Almost all the sulphur present in liquid and gaseous fuels and about 80% of sulphur present in the solid fuels appears as SO_x in the flue gases. Depending on the sulphur content of the fuel burnt and the conditions of combustion (*e.g.*, % of excess air used), the concentration of SO_x in flue gases varies from 0.05 to 0.4%. However, in metallurgical operations such as smelting of sulphide ores, the SO_2 concentration in stack gases may be 5 to 10%.

SO_2 is oxidised to SO_3 in atmospheric air by photolytic and catalytic processes involving ozone, NO_x and hydrocarbons, giving rise to the formation of photochemical smog. Oxidation of SO_2 can take place in presence of catalysts such as NO_x, metal oxides, soot and dust. Under normal humid conditions of the atmosphere, SO_3 reacts with water vapour to produce droplets of H_2SO_4 aerosol which give rise to the so-called "acid rain" discussed later. The sulphuric acid and sulphate aerosols present in the urban air are smaller than 2μ and hence can easily reach the pulmonary region of lungs, causing serious respiratory problems, particularly in old people.

$$SO_2 + O_3 \rightarrow SO_3 + O_2$$
$$SO_3 + H_2O \rightarrow H_2SO_4 \rightarrow (H_2SO_4)_n \text{ aerosol}$$

$$SO_2 + \tfrac{1}{2}O_2 + H_2O \xrightarrow{\text{Catalyst such as metal oxide, soot etc.}} H_2SO_4 \rightarrow (H_2SO_4)_n \text{ aerosol}$$

Control of SO_x emissions from the anthropogenic activities is contemplated on the following lines :

(1) Removing SO_x from flue gases before letting them out into the atmosphere : Chemical scrubbers such as (a) Lime stone or (b) Citric acid are suggested to absorb SO_2 from the flue gases.

(a) $2\,CaCO_3 + 2SO_2 + O_2 \rightarrow 2\,CaSO_4 + CO_2$
(b) $SO_2 + H_2O \rightleftharpoons HSO_3^- + H^+$
$HSO_3^- + H_2\,cit \rightleftharpoons (HSO_3 \cdot H_2\,cit)^{-2}$

(2) Removing sulphur from the fuels used for combustion : Pyritic sulphur in coal can be removed by grinding and washing in coal washeries. However organically bound sulphur cannot be easily removed from coals. Research is in progress to synthesise special type of micro-organisms using bio-technology, which are capable of converting organically bound sulphur into soluble form.

(3) Utilizing low-sulphur fuels.

(4) Generation of power by alternative energy sources and discouraging fossil-fuel based thermal power-plants.

Acid rain

Rain has always been valued by mankind, because good crops and abundant water supplies are possible only by timely and plentiful rainfall. Summer rains refresh people. Spring rains recharge the aquifers and cleanse the groundwaters. Autumn rains and winter snow help cleansing the air. Rain in general, brings with it a sense of hope, vitality and a promise for the future.

However, over the last few decades, simple rainfall has taken on a threatening complexity in some parts of the world. In these locales, the rain must pass through an atmosphere polluted with oxides of sulphur (SO_x) and oxides of nitrogen (NO_x). The falling rain and snow react with these oxide pollutants to produce often a mixture of sulphuric acid, nitric acid and water. This is known as acid precipitation or acid rain.

Rain tends to be naturally acidic with a pH of 5.6 to 5.7 due to the reaction of atmospheric CO_2 with water to produce carbonic acid. This small amount of acidity is, however, sufficient to dissolve minerals in the earth's crust and make them available to plant and animal life; yet not acidic enough to inflict any damage. Other atmospheric substances from volcanic eruptions, forest fires and other similar natural phenomena also contribute to the natural sources of acidity in rain. Still, even with the enormous amounts of acids created by nature annually, normal rain-fall is able to assimilate them to the point where they cause little, if any, known damage. But, it is the contributions of SO_x, NO_x, etc. from anthropogenic activities that disturb this acid balance and convert natural and mildly acidic rain into precipitation with far-reaching environmental consequences.

It has been stated in the earlier sections that SO_x and NO_x emissions into the atmosphere are eventually converted into H_2SO_4 and HNO_3 droplets due to a series of photochemical reactions and chemical reactions catalysed by other species present in the atmosphere. These acid droplets are partly neutralised with bases e.g., particulate lime, NH_3 etc. These salts and the remaining H_2SO_4 and HNO_3 droplets along with HCl released into the atmosphere by man-made and natural HCl emissions give rise to acidic precipitation, which is popularly known as "Acid Rain". Analysis of acid rain samples showed the following species : H^+, NH_4^+, Na^+, K^+, Ca^{2+}, Mg^{2+}, SO_4, NO_3^- and Cl^-. The contributions from the three acids in the "acid rain" are mostly in the order : $H_2SO_4 > HNO_3 > HCl$.

Acid rainfall may occur at a place far away, may be even 500 km to 1000 km from the sources of the pollutants. Some events of acid rainfall in Sweden and Canada have been traced to be due to large scale SO_x emissions from densely populated industrial areas of UK and USA.

Acid rain represents one of the major consequences of air pollution, because of large SO_x and NO_x emissions from big industrial areas into the atmosphere. The longer the SO_x and NO_x remain in the atmosphere, the greater the chances of their oxidation of H_2SO_4 and HNO_3 by the various photochemical and catalytic chemical reactions. Acid rains may cause extensive damage to materials and terrestrial ecosystems such as water, fish, vegetation, stone, steel, paint, soil and mankind as follows:

(1) Damage to building and structural materials as well as valuable ancient sculptures, carved

from marble, limestone, sand-stone etc., because of pitting and mechanical weakening due to attack by the acidic components. Deformation and degeneration of stone statues in countries like Greece and Italy have been reported to be due to damage by the acid rains. Architectural monuments like Taj Mahal in our country will face a similar threat if proper measures are not taken.

(2) Acidification of soils with the consequent effects on microbial and soil fauna and fixation of nitrogen. Also, indirect effects due to alternation in soil chemistry because of acidification of soils leading to reduced forest productivity.

(3) Foliar damage to crops and forests, leaching of nutrients from leaves and alternation of seed germination characteristics. Damage to young growing plant tissues and the process of photosynthesis, thus hindering the development of plants and threatening their very survival.

(4) Potential effects on aquatic systems, such as acidification, decreased alkalinity and mobilization of metals like aluminium.

(5) Other biological effects on aquatic biota, such as altered species composition among plankton, vegetation and invertebrates, decline in productivity of fish and amphibian, skeletal deformities and increased fish mortality.

(6) Corrosive damage to steel, zinc, oil-based paints and automobile coatings.

(7) Possible effects on humans. Lungs, skin, hair may be affected. The heavy metals released by acid rain also may cause potential threat to human health. Acidification of drinking water reservoirs and concurrent increases in heavy metal concentrations may exceed public health limits and may cause injurious effects.

The extent of damage due to acid rain depends upon factors such as climate, topography, geology, biota and human activity.

The phenomenon of acid rains is a highly interactive problem and remedial measures to control it are very expensive. The potential probability of acid rains is one of the major arguments against installation of coal-based thermal power stations. Keeping in view the potential ill-effects of acid rains, it is highly desirable to control the pollutants (causing acid rains) at source, rather than treating the undesirable consequences of acid rains.

The only practical approach to counter the problem of acid rain is to reduce SO_x and NO_x emissions. The following three general options are considered for this purpose :

(1) **Energy conservation** resulting in reduced fuel consumption and hence slower emissions of SO_x and NO_x. Conservation via more efficient fuel use and through improved thermal insulation is also being studied.

(2) **Desulphurization** and denitrification of fuels of stack gases and increased use of fuels naturally low in sulphur content or use of technologies that reduce the SO_x and NO_x emissions. Desulphurization and use of low-NO_x-producing technologies are the only viable control options today and will perhaps continue to be so for some more time.

(3) **Substitutions** for fossil fuels by other alternative energy forms may offer future solutions to this problem.

Reduction of SO_x emissions can be accomplished by (a) removing the sulphur content before the fuel is burnt with the help of techniques such as coal cleaning, coal gasification and desulphurization of liquid fuels (b) removing the sulphur content during combustion, as in fluidized-bed combustion, and (c) removal of sulphur emissions after combustion, as in stack or flue gas desulphurization systems or scrubbers. The future of SO_x control from traditional fuel sources lies in the perfection of these techniques.

Reduction of NO_x emissions from stationary combustion sources can be achieved by modification of furnace and burner design and/or modification of operating conditions. The combustion modification techniques available now include using 2-stage combustion, precisely controlling air, injecting water during combustion, recirculating flue gases, and/or by altering design of firing chambers. NO_x emissions from mobile combustion sources may be achieved by lowering the combustion temperatures in the engine and catalytic removal of NO_x from exhaust gases using devices such as a 3-way system that reduces carbon monoxide, hydrocarbons and NO_x simultaneously.

Some techniques used for controlling SO_x and NO_x emissions have already been described in earlier sections. More extensive research efforts are necessary to develop economical and effective technologies to control SO_x emissions at the source, to discern the mechanism of long-distance dispersion of SO_x emissions and to monitor and study the various effects of acid rains on plants, animals and humans in greater details.

It is interesting to note that some scientists feel that SO_2, an air-pollutant resulting from burning fossil fuel and a major contributor to "acid rain", may also counteract the dangerous phenomenon of "global warming due to the Greenhouse effect", by modifying the cloud formation in such a way as to bounce back the sun-light into space and cool the planet.

(4) Hydrocarbons

Hydrocarbons are emitted into the atmosphere by natural biological activity as well as anthropogenic sources such as automobile exhausts; burning of coal, oil, wood and refuse; and solvent evaporation. Anthropogenic sources account for about 15% of the total hydrocarbon emissions into the atmosphee. The annual global hydrocarbon emission to the atmosphere by man-made sources is roughly estimated to be 57×10^7 tonnes per year. Methane (CH_4) is the major hydrocarbon emitted into the atmosphere by natural activities such as anaerobic decomposition of organic matter in water, soil and sediments by micro-organisms. About twenty other hydrocarbons have been identified in ambient atmosphere in areas of heavy vehicular traffic which include ethane, n-butane, n-petane, isopenetane, isobutane, m-xylene, propane, ethylene, acetylene and toluene. The Los Angeles basin (USA) has been extensively studied for hydrocarbon pollution because of its heavy automobile traffic and also because of its topography and climatic features which result in heavy accumulation of hydrocarbons. These hydrocarbons being thermodynamically unstable tend to get oxidised in the atmosphere by a series of chemical and photochemical reactions. This gives rise to the formation of various end-products such as CO_2, solid organic particulates (which settle down) and water-soluble acids and aldehydes which are washed down by rain. It is some of these products from photochemical reactions of hydrocarbons that have harmful effects on human beings rather than the hydrocarbons themselves. The hydrocarbons have a definite role to play in the formation of "photochemical smog" which is characterised by reduced visibility, eye irritation, peculiar odour and damage to vegetation and accelerated cracking of rubber products.

Photochemical Smog

When the atmosphere is loaded with large quantities of automobile exhausts during warm sunny days with gentle winds and low level inversion, then the exhaust gases are trapped by the inversion layers with stagnant air masses and simultaneously exposed to intense sunlight. Then, a number of photochemical reactions involving NO_2, hydrocarbons and other organic compounds and free radicals take place leading to the formation of ozone, peroxides and other photochemical oxidants in the atmosphere (Fig. 18.2). This gives rise to the phenomenon of "photochemical smog" which is characterised by the formation of aerosols that reduce visibility, generation of brown hazy fumes that irritate the eyes and lungs, and which cause extensive damage to vegetation and rubber goods. Photochemical smog was observed in some parts of Los Angeles and Denver in USA and that is why it is sometimes referred to as "Los Angeles Smog". Photochemical smog is an oxidising smog and it should be clearly distinguished from the usual reducing "smog" which forms due to the combination of smoke and fog.

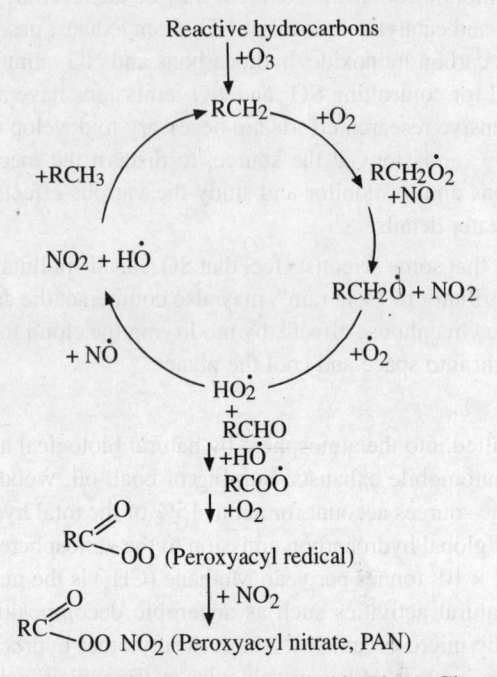

$$\text{Reactive hydrocarbons} \xrightarrow{+O_3} RCH_2 \xrightarrow{+O_2} RCH_2\dot{O}_2 \xrightarrow{+NO} RCH_2\dot{O} + NO_2$$

Fig. 18.2. Typical free-radical reactions, involved in Photo-Chemical Smog formation.

Cycle shown:
- Reactive hydrocarbons $+O_3 \rightarrow RCH_2$
- $RCH_2 + O_2 \rightarrow RCH_2\dot{O}_2$
- $RCH_2\dot{O}_2 + NO \rightarrow RCH_2\dot{O} + NO_2$
- $RCH_2\dot{O} + O_2 \rightarrow HO\dot{2} + RCHO$
- $RCHO + H\dot{O} \rightarrow RC\dot{O}$
- $RC\dot{O} + O_2 \rightarrow RC(=O)OO$ (Peroxyacyl radical)
- $+ NO_2 \rightarrow RC(=O)OO\,NO_2$ (Peroxyacyl nitrate, PAN)
- $HO\dot{2} + N\dot{O} \rightarrow NO_2 + H\dot{O}$
- $H\dot{O} + RCH_3 \rightarrow RCH_2$

For controlling the formation of noxious photochemical pollutants in the atmosphere, the emission of primary pollutants *viz.*, hydrocarbons and NO_x should be controlled.

(5) Particulates

Some details about particulates have already been presented in a preceding section. Particulates (solid or liquid) are important constituents of the atmosphere. About 2000 million tonnes of particulate matter per year are released from natural agencies such as volcanic eruptions, wind and dust storms, salt spray etc. Man-made activities such as burning of wood, coal, oil and gaseous fuels, industrial processes, smelting and mining operations, fly-ash emissions from power plants, forest fires, burning of coal refuse and agricultural refuse etc., release about 450 million tonnes of particulates per year. The diameter of particulates may range from $2 \times 10^{-4} \mu$ to $5 \times 10^2 \mu$ with varying life times depending upon the size and density of the particles and turbulence of air which control their settling rate. Their number per cubic centimeter may vary from several hundred to several thousands. In industrial and urban areas, the number of particulates may be more than a lakh per cubic centimeter with mass varying from 60 μg to 2mg/m³. However, the size of the particulates and their chemical nature are more vital than their number. Owing to their large surface areas, particulates provide excellent sites for absorption of various organic and inorganic species which encourage heterogeneous phase reactions in the atmosphere. The particles are also capable of scattering light and reducing visibility. Particulates include Fe_3O_4, V_2O_5, CaO, $PbCl_2$, $PbBr_2$, fly ash, aerosols, soot etc. Polycyclic aromatic hydrocarbons (PAH) are important constituents of several organic particulates which are carcinogenic. Soot is a highly condensed product of PAH compounds and can itself adsorb many PAH compounds and toxic trace metals *e.g.*, Be Cd, V, Cr, Ni and Mn as well as carcinogenic organics such as benzo-α-pyrenes.

Fine particulates having size of < 3μ (such as airborne toxic metals like Be and airborne asbestos) which can penetrate through nose and throat, reach the lungs and cause breathing problems and irritation of the lung capillaries. Similarly, pulmonary fibrosis in asbestos mine workers, black-lung disease in coal miners and emphysema in urban populations are attributed to the particulate pollution.

Further, airborne particulates such as dust, mist, fumes and soot can cause damage to various materials. Particulates may accelerate corrosion of metals and cause damage to paints and sculptures. The extent of damage depend upon the physical and chemical properties of the particulates. Particulates present in the atmosphere may influence the climate through formation of clouds and snow. They also may absorb solar radiation and reduce visibility.

Particulate matter can be removed from gaseous effluents from industries with the help of devices such as cyclone collectors, settling chambers, wet scrubbers and electrostatic precipitators.

The effect of some important air-pollutants and their sources are summarised in Table 3.

Table 3. – Sources of some important air pollutants and their effect on human beings and animals

Pollutant	Major Sources	Typical Effects
Carbon monoxide (CO)	Incomplete combustion of fuels, automobile exhausts, jet engine emissions, blast furnaces, mines and tobacco smoking	Toxicity, blood poisoning, increased proneness to accidents, CNS impairment, CO combines with haemoglobin, forming carboxy heamoglobin, which is useless for respiratory purposes and hence leads to death.
Sulphur dioxide (SO_2)	Combustion of coal, combustion of petroleum products, burning of refuse, petroleum industry, oil refining, power houses, sulphuric acid plants, metallurgical operations and from domestic burning of fuels.	Increased breathing rate and feeling of air-starvation, suffocation, aggravation of asthma and chronic bronchitis, impairment of pulmonary functions, respiratory irritation, sensory irritation, of throat and eyes.
Oxides of Nitrogen (NO_x)	Automobile exhausts, coal-fired and gas-fired furnaces, boilers, power stations, explosives industry, fertilizer industry, manufacture of HNO_3, combustion of wood and refuse.	Respiratory irritation, headache, bronchitis, pulmonary emphysema, impairment of lung defences, oedema of lungs, lachrymatory effect, loss of appetite, corrosion of teeth.
Hydrogen sulphide (H_2S)	Coke ovens, kraft paper mills, petroleum industry, oil refining, viscose rayon manufacturing plants, manufacture of dyes, tanning industry and sewage treatment plants.	Headaches, conjunctivitis, sleeplessness, pain in the eyes, irritation of respiratory tract, respiratory paralysis, asphyxiation, malodorous. In high concentrations, it may lead to blockage of oxygen transfer, poisoning cell enzymes and damaging nerve tissues.
Chlorine (Cl_2)	Accidental breakage of chlorine cylinders, electrolysis of brine, bleaching of cotton pulp and other process industries using chlorine.	Irritation to eyes, nose and throat, toxicity, respiratory irritation, lachrymatory effects. In large doses it may cause oedema, pneumonitis. emphysema and bronchitis.
Hydrogen fluoride (HF)	Glass fibre manufacture, chemical industry, fertilizer industry, aluminium industry, ceramic industry, phosphate rock processing.	Irritation, respiratory diseases, fluorosis of bones, mottling of teeth.
Carbon dioxide (CO_2)	Combustion of fuels, automobile exhausts, jet engine emissions.	Toxic in large quantities, Hypoxia.
Hydrocarbons (HC)	Organic chemical industries, petroleum refineries, automobile exhausts, rubber manufacture.	Some hydrocarbons have carcinogenic effects, lachrymatory effect. Irritation of lungs, eyes and respira-

Oxidants (*e.g.*, O$_3$)	Photochemical reactions in atmosphere involving organic materials and NO$_2$ etc. Reactions induced by silent electrical discharge and intense u.v. radiations in the atmosphere.	tory tract. Accumulation of fluids in lungs and damage to lung capillaries. These biochemical effects of O$_3$, mostly arise from generation of free radicals which attack the —SH groups present in the enzymes.
Dusts	Asbestos factories, mining activities, power stations, metallurgical industries, ceramic industry, factory stacks, glass industry, cement industry, foundries.	Respiratory diseases, toxicity from metallic dusts, silicosis and asbestosis from the specific dusts. Asbestos dust causes pulmonary fibrosis, pleural calcification and lung cancer.
Ammonia (NH$_3$)	Chemical industries, coke oven refineries, stocks yards, fuel incineration.	Damage to respiratory tracts and eyes, corrosive to mucous membranes. Irritation to eyes, skin and respiratory tract.
Formaldehyde (HCHO)	Waste incineration, automobile exhausts, combustion of fuels, photochemical reactions.	Inflammation and ulceration of upper respiratory tract, clouding of cornea, coughing and choking while inhalation.
Hydrochloric acid (HCl)	By-product from chlorination of organic compounds, combustion of gasoline in which ethylene dichloride is present, burning of coal, paper and chlorinated plastics.	
Radioactive gases and dusts	Radioactive gases and suspended dusts from natural and artifical radio isotope sources.	Somatic effects such as leukaemia and other types of cancer, cataracts and reduction in life expectancy. Genetic effects such as mutations in human gametes.
Arsenic (As)	Arsenic containing fungicides, pesticides and herbicides, metal smelters, by-product of mining activities, chemical wastes.	Inhalation, ingestion or absorption through skin can cause mild bronchitis, nasal irritation or dermatitis. Carcinogenic activity also is suspected. Attack-SH groups of enzymes, coagulate proteins.
Beryllium (Be)	Coal, nuclear power and space industries, production of fluorescent lamps, motor fuels and other industrial use.	Damage to skin and mucous membranes, pulmonary damage, perhaps carcinogenic.
Boron (B)	Boron producing units, production and use of petroleum fuel and additives, burning coal and industrial wastes, detergent formulations.	Ingestion or inhalation as dust causes irritation and inflammation. Boron hydrides can damage CNS and may result in death.
Cadmium (Cd)	Cadmium producing industries, electroplating, welding. Byproduct from refining of Pb, Zn and Cu, fertilizer industry, pesticide manufacture, cadmium nickel batteries, nuclear fission plants, production of TEL used as additive in petrol.	Inhalation of fumes and vapours causes kidney damage, bronchitis, gastric and intestinal disorders, cancer, disorder of heart, liver and brain. Chronic and acute poisoning may result. Renal disfunction, anaemia, hypertension, bone-marrow disorder and cancer. Toxic to body tissues, can cause irritation, dermatitis, ulceration of skin,

Chromium (Cr)	Metallurigcal and chemical industries, processes using chromate compounds, cement and asbestos units.	perforation of nasal septum. Carcinogenic action suspected.
Lead (Pb)	Automobile emissions, lead smelters, burning of coal or oil, lead arsenate pesticides, smoking mining and plumbing.	Absorption through gastrointestinal and respiratory tract and deposition in mucous membranes, cause liver and kidney damage, gastro-intestinal damage, mental retradation in children, abnormalities in fertility and pregnancy.
Zinc (Zn)	Zinc refineries, galvanizing processes, brass manufacture, metal plating, plumbing.	Zinc fumes have corrosive effect on skin and can cause irritation and damage mucous membranes.
Manganese (Mn)	Ferromanganese production, organo-manganese fuel additives, welding rods, incineration of manganese containing substances.	Poisoning of CNS, absorption, ingestion, inhalation, or skin contact may cause manganic pneumonia.
Nickel (Ni)	Metallurgical industries, using nickel, combustion of fuels containing nickel additives, burning of coal and oil, electroplating units using nickel salts, incineration of nickel containing substances, vanaspati manufacture.	Respiratory disorders, dermatitis, cancer of lungs and sinus.
Vanadium (V)	Vanadium refining, production of vanadium containing alloys, power plants, burning of oil rich in vanadium.	Gastro-intestinal and respiratory disorders, inhibition of synthesis of cholesterol, heart disease and cancer in case of chronic exposures.
Selenium (Se)	Burning of fuels and residual oils, fumes and gases from refinery wastes, incineration of paper and other wastes, natural sources.	Irritation of gastro-intestinal and respiratory tracts, irritation of eyes, nose and throat, damage to lungs, liver and kidneys.
Mercury (Hg)	Mining and refining of mercury, organic mercurials used in pesticides, laboratories using mercury.	Inhalation of mercury vapours may cause toxic effects and protoplasmic poisoning. Organo-mercurials are highly toxic and may cause irreversible damage to nervous system and brain.

Effects of Air Pollutants on Man and his Environment

(1) Damage to materials. The materials that may be affected by air pollutants include metals, building materials, rubbers, elastomers, paper, textiles, leather, dyes, glass, enamels and surface coatings. The types of possible damage to these materials by air pollutants include corrosion, abrasion, deposition, direct chemical attack and indirect chemical attack. The intensity of damage depends upon factors such as moisture, temperature, sunlight, air movement and of course the nature and concentration of the pollutant. Table 4 summarizes the possible damage to various materials because of air pollution.

Table 4. Manifestation or Air Pollution on Various Materials

Materials	Type of damage	Principal air pollutants	Other Environmental factors Influencing the rate of attack
Metals	Corrosion, soiling and tarnishing	SO_x, other acid gases	Moisture, temperature, air, salt
Building Materials	Surface erosion, leaching, corrosion discoloration.	SO_x, other acid gases acid mist, sticky particulate matter.	Moisture, freezing, temperature fluctuations
Paints	Discoloration, softened finish, surface erosion.	SO_2, H_2S, O_3, sticky particulates	Moisture, sunlight, fungus and other micro-organisms.
Textiles	Soiling, spotting, reduced tensile strength	SO_x, NO_x, other acid gases, particulates	Moisture, sunlight, physical ware.
Leather	Weakening, powdered surface	SO_x, other acid gases	Physical ware
Paper	Embrittlement	SO_x, other acid gases	Moisture, sunlight physical ware
Dyes	Fading, colour change	NO_x, ozone. oxidants	Sunlight, Moisture.
Ceramics	Change in surface appearance	HF, acid gases	Moisture
Rubber	Craking, Weakening.	Ozone, oxidants	Sunlight.

(2) Damage to Vegetation. Air pollutants, such as sulphur dioxide, HF, particulate fluorides, smog, oxidants like ozone, ethylene (from automobiles), NO_x, chlorine and herbicide and weedicide sprays exert toxic effects on vegetation. The damage usually manifests in the form of visual injury such as chlorotic marking, banding, silvering or bronzing of the underside of the leaf. Retardation of plant growth may also occur in some cases. The extent of damage to a plant depends upon the nature and concentration of the pollutant, time of exposure, soil and plant condition, stage of growth, relative humidity and the extent of sunlight.

(3) Damage to farm animals. Arsenic, lead and fluorides are the main pollutants which cause damage to livestock. These air-borne contaminants accumulate in vegetation and forage and poison the animals when they eat the contaminated vegetation.

Arsenic occurs as an impurity in coal and many ores. It is also used in insecticides. Livestock near smelting and other industrial operations suffer arsenic poisoning with symptoms like salivation, thirst, liver necrosis, inflammation and depression of central nervous system.

Lead is emitted from metallurgical smelters, coke-ovens and coal combustion operations, lead arsenate sprays and automobile exhausts. Lead poisoning occurs in horses and other animals with symptoms such as depression, lethargy, gastritis, paralysis and breathing troubles.

Cattle and sheep are particularly susceptible to fluorine toxicity which may cause fluorosis of teeth and bones.

(4) Darkening of sky and reduction in visibility. Sky darkening may be caused by heavy smoke and fog or by dust storms. The reduction in visibility may be due to smoke, fog and industrial fumes which contain particulates in the size range of 0.4 to 0.9 μm that scatter light. The intensity of these effects depends upon the particle size, the angle of the sun, aerosol density, thickness of the affected air mass and meteorological factors such as inversion height, wind speed and humidity.

(5) Effect on human health and human activities. The effects of air pollution on humans, animals and vegetation has already been discussed in earlier sections. Air pollution can effect the health of workers within the industrial premises, causing absenteeism, sickness and drop in production. Industrial hygiene measures are being taken by many industrial managements to combat these occupational disease. However, apart from the effects on industrial workers, air pollution also affects larger segments of general population. The notorious London smog of 1952, which lasted for 5 days causing 4000 deaths, is an example. Epidemiological and toxicological studies indicate a link between air pollution and respiratory conditions like chronic bronchitis, bronchial asthma, pulmonary emphysema and lung cancer. The vulnerability to air pollution depends upon age, sex, general health status, nutrition, pre-existing diseases, concurrent exposures, concentration and nature of the pollutants involved, extent of exposure, temperature and humidity at the time of exposure. People who are very young or very old and infirm, people of poor health, smokers, people with asthma, bronchitis and coronary heart disease are usually more vulnerable. Irritation of nose, eyes and throat and bad odours due to air pollutants cause annoyance, allergy and health hazards.

Air pollution may cause sickness, absenteeism among workers and general lethargy, which naturally result in decrease in efficiency in all facets of human activity.

Interdependence of Human Activities, Meteorology and Air-Pollution

The dispersion of air-pollutants at a particular place are influenced by meteorological factors, such as wind speed, wind direction and turbulence. Precipitation and humidity also influence the pollution potential.

Wind speed and Wind Direction. The drift and diffusion of air pollutants near the ground level are profoundly influenced by the wind speed and wind direction. The higher the wind speed near the point of discharge of the air pollutants, the more rapidly the pollutants are dispersed and diluted with air. On the other hand, the pollutants tend to be concentrated near the area of discharge when the wind speeds are low. However, in rough terrain, the wind speed and direction near the source may not fully govern the movement of the pollutants. For instance, hills may deflect the flow of air contaminated by pollutants either vertically and/or horizontally. The extent of deflection depends on the vertical atmospheric stability. In valleys, the winds carrying the pollutants may flow up or down the valley. These meanderings or channeling effects are more pronounced with increasing depth of the valley.

Atmospheric stability and temperature inversion. The influence of atmospheric stability and temperature inversion on the upward dispersion and dilution has been already discussed earlier. The decrease of air temperature with height is called the lapse rate. Under conditions of super-adiabatic lapse rate, the atmosphere is unstable and the pollutants are rapidly dispersed, because of the vertical mixing of the air. On the other hand, under conditions of negative lapse rate or temperature inversion, (*i.e.*, when the temperature increases with height), the pollutants in the air do not disperse vertically, because the air near the ground is denser than the air above and hence the turbulence or mixing is very less (since the atmosphere is stable). Thus, air pollution problem may assume alarming dimensions when the climatic conditions are adverse to effective dispersion of the pollutants. Fortunately, climatic conditions in most parts of our country are favourable to effective dispersion of pollutants, as compared to the cold countries.

Plume characteristics under different lapse conditions. Chimneys are used with boilers, furnaces etc., to create a natural draft, to provide the necessary air for combustion and to disperse the gaseous pollutants efficiently. The chimney design, particularly the stack height, should be carried out with due consideration of the meteorological factors, such as wind speed and wind direction, to ensure that the population near the factory is not affected. The diffusion profile of the stack gases mainly depends on the stability of the atmosphere. The various types of plume characteristics, along with the corresponding temperature profiles are shown in Fig. 11.3.

586   A TEXTBOOK OF ENGINEERING CHEMISTRY

It can be seen from the figure that when the atmosphere is very stable, the gaseous effluent forms a "fanning" plume. Clear sky with light winds during the night are favourable for "fanning" and under these conditions the plume will spread horizontally. When an inversion occurs with

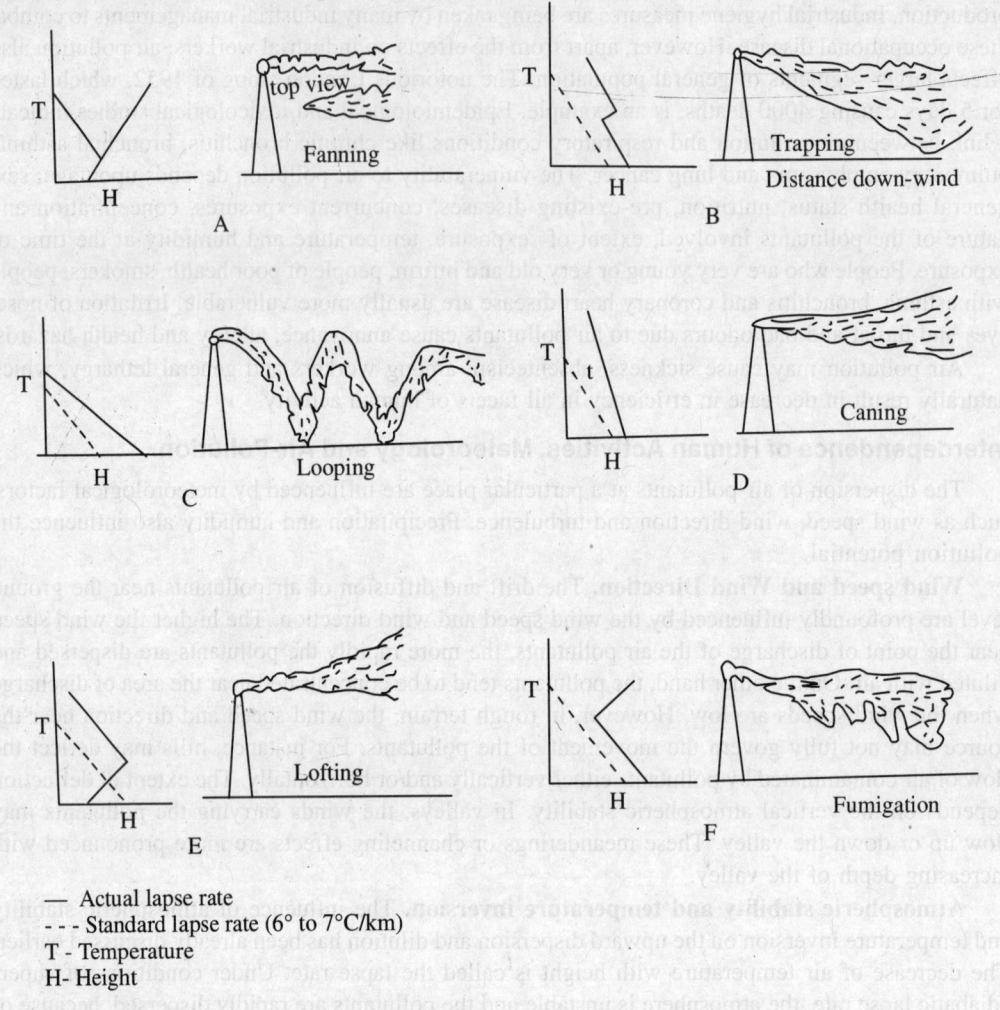

Fig. 18.3. Types of smoke plume under different conditions of atmospheric stability.

— Actual lapse rate
- - - Standard lapse rate (6° to 7°C/km)
T - Temperature
H - Height

light winds aloft and moderate turbulence below a "trapping" plume may be released. When super-adiabatic lapse rates occurs, "looping" plumes are observed because of the development of large thermal eddies in the unstable air, resulting in bringing high concentrations of the plume gases to the ground for short periods. A clear hot day, accompanied by light winds may produce a "looping" plume. When the atmosphere is cloudy or strongly windy and the turbulence is more mechanical than thermal, "coning" plume is observed. The cone shaped plume tends to reach the ground at greater distances than that in case of "looping". When there is a strong lapse rate above a surface inversion, "lofting" is observed. Under this condition, rapid upward diffusion takes place and emissions will not reach the surface because downward diffusion does not penetrate the inversion layer. When strong lapse rate occurs at the lower layers and inversion in upper layers, "fumigation" is observed. Under these conditions, high concentration of the plume gases reach the ground along with full length of the plume due to thermal turbulence. "Fumigation" is usually prevalent with clear skies and light winds during summer.

Precipitation and Humidity. The cleaning of pollutants may take place by the three natural processes, namely, fall out, rain out and wash out. Gravitational settling of larger pollutant particles is called "fall out". Pollutants present in cloud particles act as nuclei of condensation and grow to rain drops, which reach the surface by "rain out". When the contaminants in air are swept out by precipitation during its descent to the surface, the process is called "wash out".

The knowledge of local and regional meteorological factors is helpful in location of an industrial plant, the design and use of pollution control equipment and in establishment of air quality criteria.

Just as the extent of air pollution depends on weather conditions, air pollution, in turn, also affects the weather to some extent. For instance, air pollution may reduce visibility, increase the duration and frequency of fog and reduce the incoming solar radiation. These may affect, to some extent, the human activities. On the other hand, human activities, such as deforestation, massive shifting of surface and ground waters, burning of fossil fuels, automobile emissions etc., may also change the climatic conditions on the earth.

Air Quality Standards

Air Quality standards indicate the levels of pollutants that cannot be exceeded during a specified time period in a specified geographic area, with due reference to the method of measurement, units of measurement, concentration and time of exposure. These are derived from air quality criteria, which are in turn derived on the basis of effects of ambient air pollution on human health, vegetation, animals, materials, visibility etc. The "sanitary standards" adopted in USA are worked out with consideration of the local conditions, technical and economic feasibility of controlling the particular pollutant to the desired level etc. The "hygienic standards" adopted in USSR permit only such pollutant concentrations which do not cause any harmful or unpleasant effects directly or indirectly on man, which may adversely affect his capacity to work or his physical or mental well-being. Hence these standards represent the ideal requirements which every country should strive to attain.

Table 5 summarizes some national ambient air quality standards for the U.S. proposed in 1971. The primary air-quality standards define the levels judged necessary to protect the public health with an adequate safety margin.

Table 5. Ambient Air Quality Standards for U.S.

Pollutant	Primary Air Quality Standards
SO_x	(1) 80 μg/m^3 (0.03 ppm) annual arithmetic mean
	(2) 365 μg/m^3 (0.14 ppm) maximum 24 h concentration which should not be exceeded more than once a year.
NO_2	100 μg/m^3 (0.05 ppm) annual arithmetic mean
CO	(1) 10 μg/m^3 (9 ppm) maximum 24 h concentration.
	(2) 40 μg/m^3 (35 ppm) maximum 1 h concentration.
Particulate	(1) 75 μg/m^3 (0.03 ppm) annual geometrical mean.
	(2) 260 μg/m^3 (0.1 ppm) 24 h maximum concentration which should not be exceeded more than once a year.
Oxidants	160 μg/m^3 (0.08 ppm) max 1 h concentration not to be exceeded more than once a year.

The standards prescribed for air pollution control in India are as follows:

Sr. No.	Area category	Pollutant concentration, $\mu g/m^2$			
		SO_2	NO_x	CO	Suspended particulate matter
1.	Industrial and mixed-use areas	120	120	5000	500
2.	Residential and rural areas	80	80	2000	200
3.	Sensitive areas	30	30	1000	100

Emission standards for industrial environments are also being finalised.

Air Monitoring

The harmful effects of air pollution can be assessed only on the basis of adequate data collected at properly selected sampling stations, using well defined sampling procedures and analytical techniques. It is preferable to conduct both physico-chemical and biological monitoring and correlate the results to evolve an integrated approach for air pollution control. The primary objective of the air monitoring programme is to collect basic data on the quantity and quality of the pollutants present in air and the influence of topography, population and meteorological factors on air pollution.

Atmospheric sampling and Analysis

The principal objectives of atmospheric sampling are measurement of pollution, collection of base-line information and control data suitable for source detection and trend evaluation. These are helpful to develop air quality criteria, which eventually form the basis for setting air quality standards. Sampling procedures are developed on the basis of the pollutant being sampled, the collection technique and the device being employed and the sensitivity, specificity and selectivity of the analytical method being used.

In general, a sample of 10 m^3 is collected at a rate of 0.003 m^3/min to 3.0 m^3/min over 1 to 3 hr period. The pollutant concentration is expressed in $\mu g/m^3$.

The various collecting devices and techniques employed for the particulate and gaseous air pollutants are summarised below :

Collecting techniques employed for Air pollutants

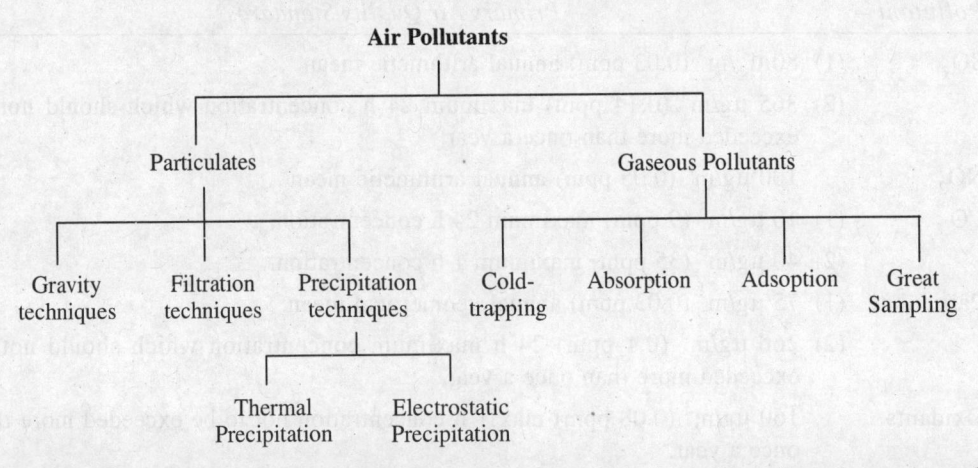

```
                        Air Pollutants
                       /              \
                Particulates        Gaseous Pollutants
              /     |     \         /    |      |      \
         Gravity Filtration Precipitation  Cold-  Absorption Adsoption Great
       techniques techniques techniques  trapping                    Sampling
                            /      \
                      Thermal    Electrostatic
                   Precipitation  Precipitation
```

Gravity techniques are used to collect settleable particulates (*e.g.*, fly-ash, soot, smoke etc.) in air. Devices used include dust fall sampling instruments, such as dust fall bucket, dust fall jar, etc. Dust fall jar is an open-mouthed glass or polyethylene container with a little water at the bottom. It is exposed to the atmosphere for a month, in a bird-shield on a stand kept 1 m above the sampling floor. The mass of the dust deposits per unit area of jar water surface area is determined after evaporating the water to dryness and weighing. The result is reported in mg/cm^2/month or tonnes/km^2/month. Sedimentation technique is used for collecting particulates of diameters >5μ. Filtration techniques are used for collecting suspended particulates that do not settle out early. Devices used include High Volume Sampler, Paper tape sampler, etc. The filter media used are glass fibre (for organic compounds), filter paper (for metals and anions) and silica felts (for trace organic and inorganic species). The pore size of these filters is usually 0.45 μ. The particulate content is usually expressed in terms of μg/m^3 of air. High volume air samplers are most commonly used for sampling suspended particles. These are used to pump large volumes of air (upto 2000 m^3) at a rate of about 1.7 m^3/min and retain particles down to ~ 0.1 μ size on the glass fibre filters.

In a special technique, called "inertial technique", the polluted air sample is drawn into a sampler provided with obstacles which cause the air-stream to change direction, while the particulates travelling in the initial direction collide with the obstacle. In the dry impinges, the obstacles have an adhesive surface on which the particles are impacted. In the dry impinges of the cascade impactor type, particles can be differentiated with size by using progressive decreasing orifice size. In the wet impinges, the obstacle is immersed in a fluid in which the particulates are collected by impingement.

Precipitation Techniques

Thermal precipitation. When exposed to a high temperature gradient, suspended particles tend to move to lower temperature region. This principle may be applied for collecting aerosol particles in the range 0.001 μ to 10 μ. The technique can also be used efficiently for collecting radio-active particulates.

Electrostatic precipitation. In this technique, electrical charge is used to force radioactive or other particulates in the range of 0.001 to 10 μ to migrate out of the air stream onto a collecting surface.

Cold trapping or Freeze out or Condensation sampling

In this technique, the gaseous pollutants from the air stream are trapped in different collecting chambers, each maintained at progressively decreasing temperature, ranging from 0°C (in an ice bath) to –196°C (in liquid nitrogen bath). This technique is used to collect insoluble or non-reactive vapours, hydrocarbons, radioactive gases, etc., simultaneously using freeze-out sampling trains containing chambers maintained at progressively decreasing temperature.

Absorption sampling. In this technique, the desired gaseous contaminant in air is collected by closely contacting with or bubbling through the corresponding absorbent solution. Maintaining suitable bubble size and the requisite residence time are essential for efficient sampling. The absorbents used include water (for absorbing gases *e.g.*, HF) oils (for absorbing hydrocarbons), alkali (for absorbing acidic gases), and acids (for absorbing alkaline gases). The absorbing devices used include glass scrubbers, packed columns, impinges and counter-current scrubbing systems. A typical fritted glass scrubber used for gaseous air pollutants is shown in Fig. 18.4.

Adsorption sampling. In this technique, gases and vapours in the polluted air are adsorbed on suitable adsorbents *e.g.*, activated charcoal, activated carbon, activated alumina, silica gel and molecular sieves made from sodium or calcium zeolites. A typical charcoal adsorption tube is shown in Fig.18.5. The amount of the pollutant adsorbed is proportional to the surface area of the adsorbent, the physico chemical characteristics of the adsorbent, the temperature and pressure maintained in the sampling train. When activated charcoal is used as adsorbent, the pollutant adsorbed can be leached with a suitable solvent such as carbon disulphide and the solution is analysed by gas chromatography. In some cases, the sample from the collecting device is directly injected into a gas chromatograph for analysis.

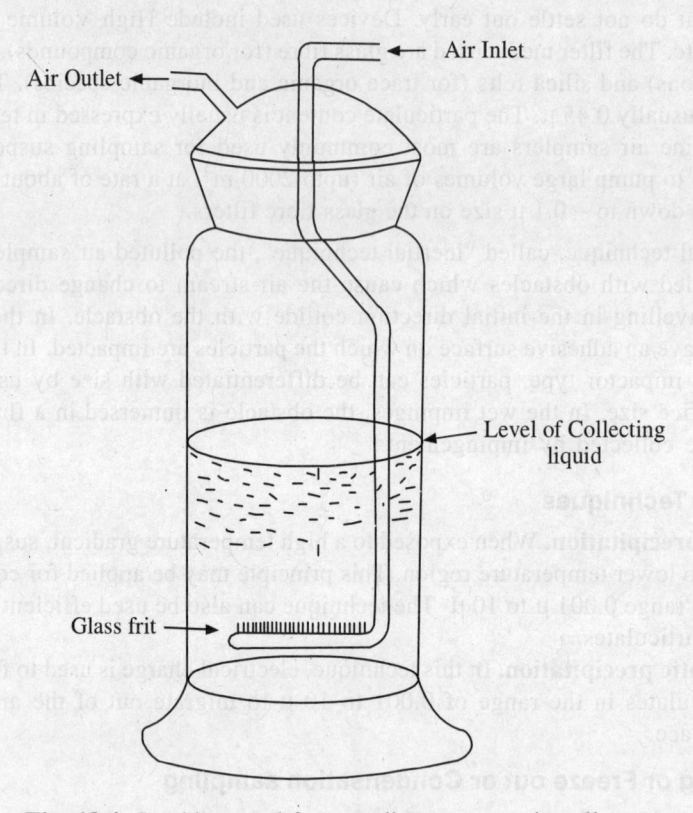

Fig. 18.4. Scrubber used for sampling gaseous air-pollutants.

Great Sampling: In this technique, a great sample is taken out between suitable intervals and is allowed to come into contact with a suitable absorbing solution in the sampling apparatus. A liquid or gas-displacement collector or a deflated plastic bager evacuated flask are commonly employed.

Analytical and Instrumental Techniques used in the Estimation of Atmospheric Pollutants

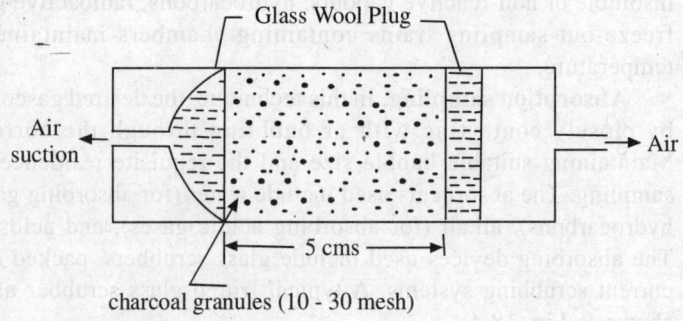

Fig. 18.5. Adsorption tube, packed with charcoal.

The sampling equipment and analytical methods used for the estimation of some common pollutants are summarized in Table 6.

Table 6. Sampling equipment and analytical methods used for the estimation of some pollutants

Sl. No.	Name of the Pollutants	Sampling equipment	Method for estimation
1.	Dust	Dust fall jar	Gravimetric
2.	Suspended particulates	High Volume Sampler	Gravimetric
3.	Sulphation rate	Lead Candle	Gravimetric
4.	Hydrogen sulphide	Air sampling kit	Methylene Blue Method
5.	Sulphur dioxide	Air sampling kit	West & Gaeke Method, Conductometry; Amperometry
6.	Oxides of Nitrogen	Air sampling kit	Jacob & Hochneisser Method; (Spectrophotometry); Chemiluminiscense.
7.	Carbon monoxide	Special sampling techniques	I.R. Spectroscopy; Gas Chromatography.
8.	Hydrocarbons	Special sampling techniques	Gas chromatography (in conjunction with mass spectrograph for low level of hydrocarbons)

1. Dust fall jar. This is used to determine the settleable particles in air. The dust fall jar comprises of an open-mouthed glass or polythene container with some water in the bottom. The container is usually placed in a bird-shield on a stand 1 m above the sampling floor and is exposed to the atmosphere for a month. The mass of the dust deposited per unit area of jar water surface area is then determined. A dust fall jar and lead candle fixed on a tripod stand are shown in Fig. 18.6.

2. Determination of suspended particulates with a high volume sampler. The high volume sampler (Fig. 18.7) is used to determine the suspended particulates in air that remain for extended periods. This is used for estimating the total levels of particulate matter. The sampler is fixed in a wooden shelter in such a way that air is drawn from the vertically upward flow through a 20 cm × 25 cm glass fibre filter at a flow-rate of about 1.7 m^3/min. Particles in the size range of 0.1 to 100 microns diameter are usually collected on the filter. Sampling is usually done for 24 hours. From the mass of the particulates collected and the volume of the air sampled, the mass concentration of the suspended particulates in the ambient air can be computed as μg/m^3 of air. Particulate matter collected on the filter can be incinerated or extracted with a suitable solvent or digested with an oxidizing acid mixture for the subsequent chemical analysis. Characterization of atmospheric particles is best done with the help of simple or electron microscopy.

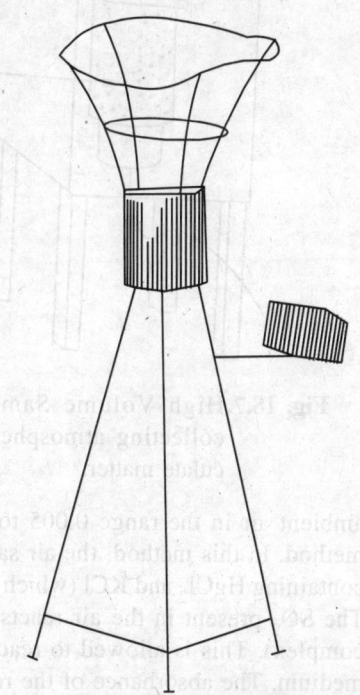

Fig. 18.6. Dust-Fall Jar and Lead Candle fixed on a Tripod stand.

3. Determination or sulphation rate. Sulphation rate is a cumulative index useful for evaluating SO_2 and other sulphur compounds in the atmosphere. It is determined by lead peroxide method. The principle involved in this method is that when sulphur compounds in gaseous state come into intimate contact with solid lead peroxide, lead sulphate is formed, which can be estimated by gravimetry or by turbidimetry.

A cylindrical polyethylene or glass tube of area 100 sq. cm. is wrapped with a piece of tapestry cloth or stockinette or gauze bandage (10 cm × 10 cm). A paste containing gum tragcanth mucilage in alcohol and lead peroxide is applied evenly on the fabric, covering the cylinder (only over 100 sq. cm. area) with the help of a spatula and is allowed to dry in a desiccator. The "lead candle" so prepared is installed in a louvered box which keeps the air off but allows the air to come into contact with the fabric surface. After exposure for a month, the $PbSO_4$ formed is estimated in the laboratory by gravimetry or turbidimetry.

Sulphation rate = [Total $BaSO_4$ formed in mg due to sulphation] × 0.343 mg of SO_3 per 100 cm^2 per day.

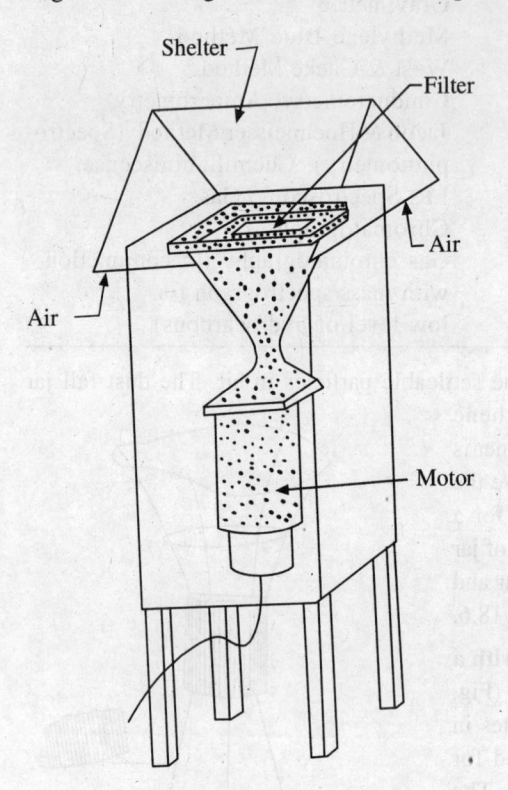

(1 mg of $BaSO_4$ = 0.343 mg of SO_3).

Fig. 18.7. High Volume Sampler, for collecting atmospheric particulate matter.

4. Estimation or hydrogen sulphide. A measured volume of air at a suitable flow rate is passed through an absorbing solution containing $CdSO_4$ and NaOH, taken in an impinger or an absorption bottle. The solution is then treated with sulphamic acid solution, followed by sulphuric acid, N-N-dimethyl P-Phenylene diamine and ferric chloride. The absorbance of the methylene blue formed is determined on a spectrophotometer at 670 nm. The H_2S present can be calculated with the help of a calibration curve prepared with different volumes of standard Na_2S solution. The H_2S concentration is expressed as $\mu g/m^3$ of air passed.

5. Estimation of sulphur dioxide SO_2 in ambient air in the range 0.005 to 5 ppm can be determined by the modified West & Gaeke method. In this method, the air sample containing SO_2 is drawn through a scrubbing solution containing $HgCl_2$ and KCl (which yields $HgCl_4$) from potassium tetra-chloromercurate formed. The SO_2 present in the air reacts with $HgCl_4$ giving $HgCl_2.SO_3^-$ (dichloro-sulphitomercurate complex). This is allowed to react with formal-dehyde, followed by para-rosaniline in acidic medium. The absorbance of the resulting red-violet dye is measured at 548 nm.

The collection efficiency of the scrubbing solution for SO_2 is about 95%. If more than 2 ppm of NO_2 is present in the air, its interference in the above method of estimating SO_2 can be eliminated by the addition of sulphamic acid (H_2NSO_3H) which reduces NO_2 to N_2.

Commercial continuous SO_2 analysers are based on conductometry wherein the SO_2 is absorbed in H_2O_2 solution and the increased conductivity of the solution, due to the formation of H_2SO_4, is measured to determine the SO_2 content. SO_2 monitors, using amperometry, are also available, wherein the changes in electrode potential due to reaction between SO_2 and Br_2 are measured to determine the SO_2.

6. Estimation of oxides of Nitrogen. An aqueous solution of NO_2 yields equal quantities of nitrate and nitrite ions. The nitrite concentration is determined on the basis of the formation of a reddish purple azo dye produced at pH 2.0 to 2.5 by the coupling of diazotized sulphanilic acid with N-(1-napthyl)-ethylene diaminedihydro-chloride. The colour system obeys Beer's law upto 180 μ/1 N with a 1-cm light path at 543 nm.

NO_x analysis can be best done by chemiluminescence. Interaction of NO with O_3 generates electronically excited NO_2 molecule. This emits radiation in the range 600-3000 nm. This can be measured by a photomultiplier, since the intensity is proportional to concentration.

$$NO + O_3 \longrightarrow NO_2^* + O_2$$
$$NO_2^* \longrightarrow NO_2 + h\nu$$

The air sample containing NO and NO_2 is passed through a thermal converter, which converts NO_2 to NO. Analysis of this sample by chemiluminescence gives total NO_x. Analysis of another sample without passing through the thermal converter gives only NO. The difference of the above two analyses corresponds to NO_2 in the air sample.

7. Analysis or carbon monoxide. Carbon monoxide strongly absorbs I.R. radiation at certain wavelength. Hence it can be determined by non-dispersive infra-red spectrometry.

Carbon monoxide at levels < 10 ppm can be conveniently determined by gas chromatography, using flame ionization detector.

8. Analysis of hydrocarbons. Hydrocarbons present in air can be collected in an absorption column packed with a suitable absorbent such as porous styrene-divinyl benzene polymer, by sampling about 10 litres of air at a rate of 4 litres/minute. The sample can be desorbed by heating the absorption common for the subsequent gas chromatographic determination, using a flame ionization detector or a mass spectrometer (for low levels).

Apart from the above pollutants, ammonia can be determined by spectro-photometry or potentiometry; volatile pesticides can be determined by gas chromatography; heavy metal ions such as Cu, Cd, Cr, Fe, Mn, Pb, Zn etc., can be determined by Atomic Absorption Spectrometry or Emission Spectroscopy or X-Ray Fluorescence or Neutron Activation Analysis (for Cu, Cr, Fe, Mn, etc). Techniques such as chromatography (for silicates and polycyclic hydrocarbons), potentiometry (for fluorides), Electron spectroscopy (for sulphates) are also used for analysing particulate matter.

Air-Pollutants from some major industrial and other sources

The following table (Table 7) summarizes some major air-pollutants from some typical industries and other sources:

Table 7

Source	Typical pollutant emissions
Petroleum refineries and petrochemical industry	SO_x, NO_x, hydrocarbons, CO, particulate matter, aldehydes NH_3, odours of mercaptans, smoke.
Iron and steel industry	SO_2, CO, Dust, NO_2, particulates, acid fumes, oxide fumes, oil and solvent fumes, odours, smoke.
Fertilizer industry	SO_x, acid mist, NO_x, fluorides, NH_3, urea dust, fertilizer dust, smoke.
Thermal Power Plants	Flyash, SO_x, NO_x. particulates such as soot, CO, unburnt coal dust.
Chemical industries	SO_x, NO_x, H_2S, fluorides, hydrocarbons and CO (from organic

	chemical industries); mercury and Cl_2 (from chlor-alkali plants); and particulates, acid fumes (from acid plants, synthetic fibre plants, etc.).
Cement industry	Cement, lime dust, SO_x, NO_x, fly ash, smoke.
Tanneries and leather industry	Mercaptans, sulphides.
Paints, pigments and dyes	Nitrobenzene, aniline, thinners, solvents, base materials
Piper and paper products	H_2S, mercaptans
Automobiles	SO_x, NO_x, HCHO, hydrocarbons, smoke

Assessment of pollutant emissions from industries

The assessment of pollutant emissions from various industries can be made on the basis of the following :

(1) Calculation of material balance based on inputs and outputs,

(2) Calculation of emission factor, which is the statistical average of the mass of a pollutant emitted per each source of pollution per unit quantity of material processed or used.

(3) Carrying out stack sampling to actually determine the various pollutants and their quantities emitted from each source. The dust from the gas streams is retained in a suitable filtering media (*e.g.*, filter paper thimble, alundum thimble, membrane filters, fibre glass filters, bubblers, etc.) and the gas passed through is analysed for its constituents.

Air Pollution from Automobiles

The principal types of automobiles used in our country are : (*i*) Two-wheelers and three-wheelers, powered by two-stroke petrol engines (*ii*) passenger cars, powered by four-stroke petrol engines, and (*iii*) trucks and buses, powered by four-stroke diesel engines. The extent of pollution by these automobiles depends upon the engine design, fuel composition, operating conditions (*e.g.*, idling, cruising, acceleration, deceleration), etc. The emissions from petrol engines are classified as follows :

(*a*) **Exhaust emissions.** These are mostly comprised of CO, NO_x, particulates containing lead compounds and unburnt hydrocarbons. The various approaches that are made to control these emissions are : (1) Modification of engine design (2) Modification of operating conditions (3) Treatment of exhaust gases with devices such as absorbers, adsorbers, catalytic converters and after-burners. (4) Modification or alteration of fuels. Some of these have already been discussed earlier.

(*b*) **Crank case emissions.** These include the engine blow-by which leak past the piston, particularly during the compression stroke and the oil vapours produced in the crank case. Obviously, crank case emissions mostly contain hydrocarbons. The quantity of blow–by mostly depends on the design of the engine, condition of the engine and operating conditions. For instance, worn out cylinder and piston rings enhance the blow–by. The crank case emissions can be controlled by installing "positive crank case ventilation systems" which recycle the blowby gases and crank case ventilation air to the engine intake, instead of allowing them to escape into the atmosphere.

(*c*) **Evaporative emissions.** It is estimated that an average evaporative emission of hydrocarbons from a passenger car in India is about 20 kg per year. The various methods that are being tried to control the evaporative emissions include (*i*) Storing the fuel vapours in the crank case or absorbing them in a charcoal canister for subsequent recycling to the engine (*ii*) Modifying the fuel by replacing C_4 and C_5 olefinic hydrocarbons by the corresponding paraffinic hydrocarbons (*iii*) Introducing suitable mechanical devices to combat the problem.

The emission norms for petrol and diesel vehicles for ECE countries (Euro norms) are given below:

Emission Norms for Petrol Vehicles

Emmission Norm	Pollutant (gm/Km)	Four Wheeler T/A	Four Wheeler COP	Two Wheeler T/A	Two Wheeler COP	Three Wheeler T/A	Three Wheeler COP
CMVR 2000 Norm (W.E.F. 1st April 2000)	CO HC+NOx PM	2.72 0.97 ---	3.16 1.13<>---	2.00 2.0 ---	2.40 2.40 ---	4.0 2.0 ---	4.80 2.40 ---
EURO (W.E.F. 1.7.92 IN ECE COUNTRIES)	CO HC+NOx PM	2.72 0.97 0.14	3.16 1.13 0.18	NOT APPLICABLE		NOT APPLICABLE	
EURO II (W.E.F. 1st Jan. 1996 IN ECE COUNTRIES)	CO HC+NOx PM	2.20 0.50 -----	2.20 0.50 ----	NOT APPLICABLE		NOT APPLICABLE	
EURO III (W.E.F. 1st Jan. 2000 IN ECE COUNTRIES)	CO HC NOx HC+NOx PM	2.20 0.20 0.15 ----- -----	2.20 0.20 0.15 ---- ----	NOT APPLICABLE		NOT APPLICABLE	
EURO III (W.E.F. 1st Jan. 2005 IN ECE COUNTRIES)	CO HC NOx HC+NOx PM	1.00 0.10 0.08 ----- -----	1.00 0.10 0.08 ---- ----	NOT APPLICABLE		NOT APPLICABLE	

NOTE:
AS PER SUPREME COURT ORDER, ALL NON-COMMERCIAL PETROL VEHICLES REGISTERED IN NATIONAL CAPITAL REGION (i.e. NEW DELHI) SHOULD MEET EURO 1 NORMS EFFECTIVE FROM 1st JUNE 1999 AND EURO II NORMS FROM 1st APRIL 2000.

Abbreviations
T/A : Type Approval
COP : Confirmation of Production
CMVR : Central Motor Vehicle Rules
ECE : European Committee on Environment

Emission Norms for Diesel Vehicles

Emmission Norm	Pollutant (gm/Km)	R < 1250 Kg T/A	R < 1250 Kg COP	1250 < R > 1700 T/A	1250 < R > 1700 COP	1700 < R T/A	1700 < R COP
CMVR 2000 NORMS (W.E.F. 1st April 2000)	CO	2.72	3.16	5.17	6.0	6.9	8.0
	HC+NOx	0.97	1.13	1.4	1.6	1.7	2.0
	PM	0.14	0.18	0.19	0.32	0.25	0.29
EURO (W.E.F. 1.7.92 IN ECE COUNTRIES)	CO	2.72	3.16	◄·········		◄·········	
	HC+NOx	0.97	1.13				
	PM	0.14	0.18				
EURO II (W.E.F. 1st Jan. 1996 IN ECE COUNTRIES)	CO	1.0	1.0	◄·········		◄·········	
	HC+NOx	0.7	0.7				
	PM	0.08	0.08				
EURO III (W.E.F. 1st Jan. 2000 IN ECE COUNTRIES)	CO	0.64	0.64	◄·········		◄·········	
	HC	----	----				
	NOx	0.5	0.5				
	HC+NOx	0.56	0.56				
	PM	0.05	0.05				
EURO IV (W.E.F. 1st Jan. 2005 IN ECE COUNTRIES)	CO	0.5	0.5	◄·········		◄·········	
	HC	-----	-----				
	NOx	0.25	0.25				
	HC+NOx	0.3	0.3				
	PM	0.025	0.025				

NOTE:
AS PER SUPREME COURT ORDER, ALL NON-COMMERCIAL DIESEL VEHICLES REGISTERED IN NATIONAL CAPITAL REGION (i.e. NEW DELHI) SHOULD MEET EURO 1 NORMS EFFECTIVE FROM 1st JUNE 1999 AND EURO II NORMS FROM 1st APRIL 2000.

Indian Driving Cycle (IDC)

Total time per cycle : 108 seconds
6 cycles makes a test sequence = 108 x 6 = 648 seconds
Average speed = 21 Km/hr.
Max. speed = 42 Km/hr.
Typical distance covered = 3.9 Km
Now IDC is modified as IDC + EUDC 90
EUDC 90 = Extra Urban Driving Cycle upto 90 Km/hr.

Air-pollution control

The following two basic approaches are used for controlling air-pollution :
- (i) Controlling or confining the pollutants at source. This can be achieved by
    - (a) Modifying the process in such a way that pollutants do not form at all beyond permissible concentrations.
    - (b) Reducing the pollutant concentrations to tolerable levels before they are released to the environment, by use of suitable equipment to destroy, alter or trap the pollutants formed.
- (ii) Dilution of the pollutants in the atmosphere to permissible levels before they can reach the receptor. This can be achieved by using tall stacks, controlling the process parameters, with due regard to the local meteorological conditions and proper community planning to prevent accumulation of dangerous ground level concentrations within the designated areas.

Methods and equipment used to control gaseous pollutants

(1) Combustion. This technique is used when the pollutant contains gases or vapours, which are organic in nature. Flame combustion or catalytic combustion of these pollutants converts them into water vapour and relatively innocuous products, such as CO_2. The equipment used for flame combustion include fume incinerators, steam injection or venturi flares and after-burners. Catalytic combustion is resorted to in situations where lower operating temperatures are desirable e.g., burning or waste cracking gases, fumes from paint or enamel baking ovens and coffee roasting processes.

(2) Absorption. In this technique, the gaseous effluents are passed through scrubbers or absorbers containing a suitable liquid absorbent to remove or modify one or more of the pollutants present in the gas stream. The efficiency of gas absorption process depends upon: (a) The chemical reactivity of the gaseous pollutant in the liquid phase (b) The extent of surface contact between the liquid and the gas (c) The contact time and (d) The concentration of the absorbing medium. The equipment used include plate towers, spray towers, packed towers, bubble-cap plate towers and liquid jet scrubber towers. The gas absorption technique is widely used for removing pollutants like NO_x, H_2S, SO_2, SO_3 and fluorides from gaseous effluents. The various absorbing liquids commonly employed are as follows:

Pollutant	Absorbent
NO_x —	H_2O, aq. HNO_3
HF —	H_2O, NaOH
H_2S —	Ethanol arnines, NaOH + Phenol (in mole ratio of 3:2), sodium alamine soda ash, tripotassium phosphate, ammonia liquor from coke ovens, sodium thioarsenate, etc.
SO_2 —	Water, alkaline water, suspension of $Ca(OH)_2$, sulphites of Ba or Ca or Na, ethanolamine, dimethyl aniline, 1:1 mixture of water and xylidine, aluminium sulphate, etc.

(3) Adsorption. In this technique, the gaseous effluents are passed through porous solid adsorbents taken in suitable containers. The organic and/or inorganic constituents of the effluent gases are held at the interface of the solid adsorbent by physical adsorption or chemisorption, The efficiency of adsorption depends upon the surface area per unit weight of the adsorbent, other physical and chemical characteristics of the adsorbent and nature and concentration of the gas being adsorbed. The common adsorbents used for various gaseous pollutants are as follows :

Pollutant	Adsorbent
NO_x	Silica gel, commercial zeolites
HF	Lump lime stone, porous pellets of NaF.
H_2S	Iron oxide
SO_2	Pulverised lime stone or dolomite, alkalised alumina (Al_2O_3 + Na_2O)
Organic solvent vapours	Activated carbon
Petroleum fractions	Bauxite
Vapours associated with gases	Alumina, Silica gel, Bauxite

The preferential adsorption characteristics of some adsorbents render some adsorbents selective for certain applications. For instance, Silica gel-6, activated alumina, and synthetic zeolite or silicate molecular sieves preferentially adsorb water vapour from a mixture of the water vapour and organic pollutants. Desorption of the sorbed gases is usually achieved by increasing the temperature or reducing the pressure.

When the waste gas streams contain higher concentrations of the gases e.g., NO_x, SO_2, etc., the gases can be recovered economically and used for the manufacture of HNO_3, H_2SO_4, etc. SO_x from power plants can be removed by injecting pulverised limestone into the boiler furnace. The CaO formed reacts with the SO_x to form calcium sulphite and calcium sulphate. Thus, the SO_x emission into the atmosphere is prevented.

Odorous gases can be controlled by masking, counter-action, sorption in a suitable solvent or adsorption on activated carbon.

Methods and equipment used for controlling particulate emissions

Particulate materials in ambient air may originate from stationary as well as mobile sources. The particulate collection devices are based on the size, shape, electrical properties and hygroscopic properties of the particulates concerned. The various devices used may be classified as follows :

(*i*) **Mechanical devices.** These devices mostly operate on the basis of the following two mechanisms :

(*a*) Gravity settling in which the velocity of the horizontal carrier gas is reduced adequately so that the particles settle by gravitational force.

(*b*) Sudden change of direction of the gas flow causes the particles to separate out due to their greater momentum.

Settling chambers, buffer chambers and cyclone separators are the commonly used mechanical devices to separate particulates from gases. Settling chambers collect particulate matter by gravity or centrifugal force. These are used in power plants and industries dealing with rock products.

In cyclone collectors, the velocity of the incoming gas stream is transformed into a confined vortex, from which the centrifugal forces drive the suspended particles to the walls of the collector structure. Cyclone collectors are used to remove particulate impurities from rock product industries, iron and steel plants, mining and metallurgical industries.

(*ii*) **Filtration Systems.** Dust-laden gases are forced through a porous medium such as woven or filled fabric. The particles are trapped and collected in the filters and the gases devoid of the particles are discharged out. Fibrous or deep-bed filters and cloth bag filters are commonly used. Fabric filter media made from cotton, wool, nylon, dacron, asbestos, silicone coated glass cloth etc., are in use. Cloth and nylon filters are used upto 80-90°C whereas asbestos and silicone covered glass cloth filters can be used upto temperature of 250-350°C.

Wool, orlon and vinyon filters are good for acidic gases whereas cloth, nylon and asbestos filters are good for alkaline media. Glass fibre filters have superior chemical resistance. Fabric filters are used in industries dealing with rock products, pigments, etc.

(iii) **Electrostatic precipitators.** When a gas or an air stream containing aerosols *e.g.*, dust, fumes or mist is passed between two electrodes which are electrically insulated from each other and between which appreciable difference in electrical potential exists, then the aerosol particles get precipitated on the electrode that is at a lower potential. An electrostatic precipitator comprises of the following parts:

(*a*) A source of high voltage (*b*) A high voltage discharge electrode (usually negative) of small cross-sectional area *e.g.*, a wire, and a collecting electrode (usually positive and at ground potential) of large surface area *e.g.*, a tube or plate. (*c*) A device for disposing the collected material and (*d*) An outer housing enclosing the electrode. The following four basic steps are involved in the operation of an electrostatic precipitation : (1) Electrically charging the particles by ionisation (2) Transporting the charged particles by the force exerted upon them in the electric field to the collecting surface (3) Neutralising the electrically charged particles precipitated on the collecting surface and (4) Removing the precipitated particles from the collecting surface by rapping or washing.

In one-stage electrostatic precipitators, ionisation and collection are performed in a single step. In two-stage precipitators, a pre-ionising step is followed by collection. The latter are unsuitable for use with heavy dust concentrations. These are mostly used in air-conditioning plants.

Electrostatic precipitators are the devices of choice when (1) Very large volumes of gases are to be handled (2) Valuable dry material is to be recovered (3) Very high collecting efficiency for the removal of fine particulates are essential and (4) When the gas temperatures are very high. Electrostatic precipitators are widely used in power plants, paper and pulp industries, chemical industries such as sulphuric acid plants, iron and steel plants, mining and metallurgical industries, rock products, refineries, carbon black manufacturing industries etc.

(iv) **Wet Scrubbers.** Wet scrubbers are used (*a*) When fine particles have to be efficiently removed (*b*) When particulates as well as gaseous contaminants have to be removed (*c*) When the gases to be treated are combustible (*d*) When the volume of the gases being treated is low (*e*) Where cooling is desired and addition of water is not objectionable (*f*) When large variations in process flow-rates have to be accommodated, and (*g*) Where the temperature of the gases to be treated is as high as 300°C or even above.

Wet scrubbers are classified according to the method of particle collection as follows :

(*a*) Liquid carriage type where the gas stream containing the particles is allowed to strike a liquid surface within the collector and the liquid carrying the trapped gas particles moves to a location outside the collector for ultimate disposal.

(*b*) Particle conditioning type where the dust particles in the gas stream is brought into intimate contact with water so that the effective size of the particles is increased due to the formation of heavier water-particulate agglomerates. These can be more easily separated from the gas stream by any of the collection mechanisms.

A large variety of wet scrubbers are in use in air pollution control which include ventury scrubber, gravity spray scrubber, wet impinger scrubber, cyclone spray chambers, wet centrifugal scrubber, etc.

Wet scrubbers are used in chemical, mining and metallurgical industries to trap SO_2, NH_3, metal fumes, etc.

The effective removal of gaseous pollutants and particulates from gaseous effluents from anthropogenic activities is of vital significance. If the air-pollutants are indiscriminately discharged into the atmosphere, the dynamic equilibrium existing in the atmosphere will be disturbed thereby causing adverse effects on man and his environment.

Micro, Meso and Macro Air Pollution

Air pollution problems may occur in micro-, meso- and macro-scale. Micro-scale problems range from those covering less than a centimeter to those of the size of a house or somewhat larger. Meso-scale air pollution problems are those covering a few hectares up to the size of a city or a cluster of cities, or a state. Macro-scale problems extend from nations, countries and, in the broadest sense, to the entire globe. Indoor air pollution is of a micro-scale, whereas acid rain, ozone layer depletion, and greenhouse effect are macro-scale problems.

Indoor Air Pollution

Indoor air quality is equally important as ambient air quality. People spend more time indoors than outside. The air we breathe indoors is sometimes more polluted than outdoor air. For example, many pollutants such as cigarette smoke or radon gas, if they are emitted outdoors have plenty of air for dilution which protects them from exposure to hazardous levels of these pollutants. However, these pollutants tend be concentrated inside the houses, leading to harmful exposure levels.

People living in cold climates spend 70 to 90% of their time indoors. Most of the dwellings in underdeveloped countries in Asia are made of tree branches with roofs and sides made of leaves and floors made of bamboo strips or twigs. These dwellings are cool in summer, comfortably hot in winter and are free from indoor air pollution. In these houses, there are free interchanges between the inside of the thatch and ambient air through holes in the dwelling envelope. Quite in contrast, people in the developed / advanced countries build their houses as airtight as possible to conserve energy for air-conditioning. Because of this, pollutants released inside their houses remain practically inside. This is one of the reasons why more people in advanced countries die of cancer as compared to those in Asian countries.

Combustion occurs in homes daily to cook food, heat water and in winter to provide space heating. This can produce elevated levels of carbon monoxide (CO) and nitrogen oxides (NO_x).

Important indoor sources of NO_x and CO include range oven, range-top burner, pilot light, gas space heaters, gas dryers, cigarette smoke and kerosene space heaters. SO_2 may come from kerosene space heaters. NO_2 levels have been found to range from 70 µg / m^3 in air-conditioned houses with electric range to 182 µg / m^3 in non-air-conditioned houses with gas stoves. The latter value is quite high in comparison to the ambient air quality limits. SO_2 levels were found to be within the limits whereas respirable particulate matter (RSP) was found to increase considerably with one smoker and to rise dramatically with increasing number of smokers.

Some photocopying machines convert oxygen to ozone. Tobacco smoke contains suspected carcinogens such as benzene, hydrazine, benzo (α-) pyrene and nickel. The average particle size of tobacco smoke particles are of the order of 0.2 µm, which go deep into the lungs. A single cigarette gives off about 10^{12} particles, most of which are released while the cigarette is simply smoldering in the air (sidestream smoke) rather than when a smoker takes off a puff (mainstream smoke). It is important to note that while the smokers voluntarily expose themselves to the large concentrations of these particles, nonsmokers too face the danger of exposure to significant concentrations. Respirable particulate matter (RSP) was found to increase with one smoker and to rise dramatically with two.

Tobacco smoke can also release other indoor pollutants like carbon monoxide, nitrosoamines, nicotine, acrolein, and other aldehydes. If pets are kept at home, the condition of asthmatic patients will be aggravated. Further, if forced-air re-circulation is used as a method of heating, feathers can accumulate in air cleaners, practically, anything that gasifies will accumulate in airtight houses.

Wood-burning stoves and fire-places represent another important source of indoor air pollution because of CO, NO_x, hydrocarbons, respirable particles and other carcinogenic emissions such as of benzo--α-pyrene.

Another important indoor air pollutant that received considerable attention is radon. Radon is an inert gas which is radioactive and emits α-radiation. It emanates from natural geologic formations and from some construction materials. Radon gas and its daughter products (e.g., polonium, lead and bismuth) can attach themselves to inhaled particulates and can be lodged deep in the alveoli. Continuous lung irradiation with these materials causes a disease which is believed to lead to lung cancer followed by death. Radon is considered as the most serious indoor air pollutant. Radon gas and its decay products are beleived to cause 5,000 to 20,000 lung cancer deaths per year in the U.S. Radon-222 is a radioactive gaseous isotope of radon having a half-life of 3.8 days. It is a part of a natural radioactive decay series (chain) called the uranium series. It begins with uranium-238 and ends with the inactive lead isotope, Lead-206.

$$^{238}_{92}U \xrightarrow{-\alpha} {}^{234}_{90}Th \xrightarrow{-\beta} {}^{234}_{91}Pa \xrightarrow{-\beta} \cdots \cdots \rightarrow {}^{226}_{88}Ra$$
$$(t_{\frac{1}{2}}=1.62\times10^3 \text{ yrs})$$
$$\downarrow -\alpha$$
$$^{214}_{83}Bi \xleftarrow{-\beta} {}^{214}_{82}Pb \xleftarrow{-\alpha} {}^{218}_{84}Po \xleftarrow{-\alpha} {}^{222}_{86}Rn$$
$$(t_{\frac{1}{2}}=19.7 \text{ minutes}) \quad (t_{\frac{1}{2}}=26.8 \text{ minutes}) \quad (t_{\frac{1}{2}}=3.1 \text{ minutes}) \quad (t_{\frac{1}{2}}=3.66 \text{ days})$$
$$\downarrow -\beta$$
$$^{214}_{84}PO \xrightarrow{-\alpha} \cdots \cdots \xrightarrow{-\alpha} {}^{206}_{82}Pb$$
$$(t_{\frac{1}{2}}=1.6\times10^4 \text{ seconds}) \qquad\qquad \text{stable end product}$$
$$(t_{\frac{1}{2}}=\text{Half-life period})$$

The uranium series

Although radon is chemically inert, its short-lived decay products viz., Polonium-218, Lead-214 and Bismuth-214 are chemically active and easily become attached to inhaled particles that can lodge on the lungs. The α-emitting polonium causes the greatest lung damage.

Radon emerging out of the soil can accumulate in houses. Radon can also exist in groundwater since it is one of the intermediate products when uranium undergoes natural radioactive decay to its stable end product, lead. Radon may get released and inhaled in high concentrations when shower is used for bathing. The U.S. Environmental Protection Agency (US EPA) suggests a radon guideline concentration of 8 p Ci/L (pico curies per litre of Radon-222 in equilibrium with its progeny) as a level at which residents should take remedial action.

The problem can be mitigated by dilution of radon to below harmful levels. If ambient air is blown in from outside, the concentrations of all pollutants inside including radon get diluted. Warm air tends to rise due to the so called stack effect. As the warm air rises, it leaves a void at the bottom. The void acts as a partial vacuum which cause radon from the outside soil to be sucked into the house. One of the engineering controls therefore is to plug holes such as those in the ceiling, electric fixture holes, openings around ducts, basement walls etc. to eliminate the stack effect. The third method is what is called subslab suction in which the basement slab is penetrated with a pipe and sucking the radon from under the slab by installing a blower.

Other pollutants that are somewhat unique to the indoor air pollution are (i) formaldehyde emissions from particle boards, plywood, urea-formaldehyde foam insulation, textiles, various adhesives and other building materials (ii) asbestos used for insulation and fire proofing and (iii) volatile organics released from household cleaning products.

Formaldehyde and radon are not regulated as ambient air pollutants. However, they have been found indoors in alarmingly high concentrations. In one study, formaldehyde concentrations detected indoors was found to be in the range 0.0455 to 0.19 ppm.

The sources and exposure guidelines of some important indoor air contaminants are summarised in Table-8

Table-8 : Sources and exposure guidelines for indoor air pollutants

S.No.	Pollutant	Indoor sources	Fuidelines, average concentrations.
1	Carbon monoxide	Kerosene and gas space heaters, wood stoves, gas stoves, fire places and smoking.	10 mg/m^3 for 8 hrs ; 40 mg/m^3 for 1 hr.
2.	Nitrogen dioxide	Kerosene and gas space heaters, and gas stoves.	100 μg/m^3 (annual)
3.	Sulphur dioxide	Kerosene space heaters.	800 μg/m^3 (annual) ; 365 μg/m^3 for 24 hrs.
4.	Formaldehyde	Particle board, paneling, plywood, carpets, ceiling, tiles, urea-formaldehyde foam insulation, and other construction materials.	120 μg/m^3
5.	Ozone	Photocopying machines and electrostatic air cleaners.	235 μg/m^3/hr (once a year)
6.	Asbestos and other fibrous aerosols.	Fireproofing, thermal and acoustic insulation, friable asbestos, decorations, hard asbestos, vinyl flooring and cement products.	0.2 fibers/mL for fibers longer than 5 μm.
7.	Radon and radon progeny	Diffusion from soil, groundwater, and building materials.	0.01 working levels annual (The unit of working level is defined as 100 P Ci/L of Radon-222 in equilibrium with its progeny) (PCi = Pico-Curies)
8.	Volatile organics.	Cooking, smoking, room deodarizers, paints, varnishes, cleaning sprays, solvents, carpets, furniture, and draperies.	None available
9.	Inhalable particulate matter.	Smoking, vacuum cleaning, fire places and wood stoves.	55 to 110 μg/m^2 (annual) ; 365 μg/m^3 for 24 hrs.

Source : Environmental Engineering - A Design Approach by A.P. Sincero and G.A. Sincero, Prentice Hall of India Pvt. Ltd., New Delhi (1999).

As in the case of outdoor air, in case of indoor air also, the amount of air available to dilute pollutants is an important indicator of the possible concentration of the contaminants. Indoor air can be exchanged with outdoor by one or more of the following three mechanisms :
(i) Infiltration
(ii) Natural ventilation
(iii) Forced ventilation

Infiltration is the natural air exchange that occurs between a building and its environment through cracks and holes, leakage areas created by plumbing, ducts, ceilings floors,, exhaust vents etc., existing in the building envelope even when the doors and windows are closed.

Natural ventilation is the air exchange that occurs when the doors and windows are purposely opened to increase air circulation.

Forced ventilation occurs when mechanical air handling systems induce air exchange with the help of blowers or fans.

It is generally assumed that increasing the infiltration rate enhances indoor air quality. However, in case of radon which is emitted from soil, this may not happen. Although wind-driven infiltration helps reduce indoor concentrations by allowing relatively radon-free fresh air to blow into the building, stack-driven infiltration, which draws air through the floor, may actually cause new radon to enter the building, negating the cleaning that is usually caused by infiltration.

Large amount of energy is lost when conditioned air (heated or cooled) that leaks out of buildings is replaced by outside air by infiltration, because outside air entering the house has to be once again heated or cooled, as the case may be, to maintain the desired inside temperature. One method to get extra ventilation with minimum heat loss is by using mechanical heat-recovery ventilator (HRV). In this system, the warm outgoing stale air transfers much of its heat to the cold, fresh air being drawn into the house. Another simpler and cheaper approach is to provide mechanical ventilation systems that can be used intermittently in the immediate vicinity of concentrated sources of pollutants. Exhaust fans in bathrooms and range hoods over gas stoves, for instance, can considerably reduce indoor pollution, with minimum heat loss and thus with minimum loss of power.

Indoor air quality modelling can be done by applying box model concepts. The building can be treated as a single, well-mixed box, with sources and sinks for the pollutants under consideration. If necessary, the simple model can be expanded to include several boxes, each characterized by uniform pollutant concentrations. For instance, a two-box model is sometimes used for radon estimates, where one box is used to model radon concentrations within the living space of a dwelling while the other is used to model the air space beneath the house. (For further details the students may refer Introduction to Environmental Engineering and Science by Gilbert M. Masters, Prentice-Hall of India Pvt. Ltd., New Delhi, 1994).

18.3 WATER POLLUTION CONTROL

Water is essential for the survival of any form of life. On an average, a human being consumes about 2 litres of water everyday. Water accounts for about 70% of the weight of a human body. About 80% of the earth's surface (*i.e.*, 80% of the total 50,000 million hectares in area) is covered by water. Out of the estimated 1,011 million km^3 of the total water present on earth, only 33,400 m^3 of water is available for drinking, agriculture, domestic and industrial consumption. The rest of the water is locked up in oceans as salt water, polar ice-caps and glaciers and underground. Owing to increasing industrialization on one hand and exploding population on the other, the demands of water supply have been increasing tremendously. Moreover, considerable part of this limited quantity of water is polluted by sewage, industrial wastes and a wide array of synthetic chemicals. The menace of water-borne diseases and epidemics still threatens the well-being of population, particularly in under-developed and developing countries. Thus, the quality as well as the quantity of clean water supply is of vital significance for the welfare of mankind.

India receives about 1400-1800 mm of rainfall annually. It is estimated that 96% of this water is used for agriculture, 3% for domestic use and 1% for industrial activity. An analysis conducted in 1982 revealed that about 70% of all the available water in our country is polluted. In appreciation of this situation, several steps are being taken to control water pollution.

Classification of Water Pollutants

The various types of water pollutants can be broadly classified into the following five major categories :

(1) Organic Pollutants. The organic pollutants may be further categorized as follows :

(a) **Oxygen-demanding wastes.** These include domestic and animal sewage, bio-degradable organic compounds and industrial wastes from food-processing plants, meat-packing plants, slaughter-houses, paper and pulp mills, tanneries etc., as well as agricultural run-off. All these wastes undergo degradation and decomposition by bacterial activity in presence of dissolved oxygen (DO) This results in rapid depletion of DO from the water, which is harmful to aquatic organisms. The optimum DO in natural waters is 4-6 ppm, which is essential for supporting aquatic life. Any decrease in this DO value is an index of pollution by the above mentioned oxygen-demanding wastes. Many aquatic organisms cannot survive at lower DO levels in water.

(b) **Disease-causing wastes.** These include pathogenic microorganisms which may enter the water along with sewage and other wastes and may cause tremendous damage to public health. These microbes, comprising mainly of viruses and bacteria, can cause dangerous water-borne diseases such as cholera, typhoid, dysentery, polio and infectious hepatitis in humans. Hence, disinfection is the primary step in water pollution control.

(c) **Synthetic Organic Compounds.** These are the man-made materials such as synthetic pesticides, synthetic detergents (syndets), food additives, pharmaceuticals, insecticides, paints, synthetic fibres, elastomers, solvents, plasticizers, plastics and other industrial chemicals. These chemicals may enter the hydrosphere either by spillage during transport and use or by intentional or accidental release of wastes from their manufacturing establishments. Most of these chemicals are potentially toxic to plants, animals and humans. Some bio-refractory (*i.e.*, resistant to microbial degradation) organics such as aromatic chlorinated hydrocarbons may cause offensive colours, odours and tastes in water, even when present in traces and makes the water (or fish present in it) unacceptable from aesthetic point of view. Non-degradable chemicals, such as alkyl benzene sulphonate from synthetic detergents often lead to persistent foams. Volatile substances, such as alcohols, aldehydes, ethers and gasoline may cause explosion in sewers.

(d) **Sewage and agricultural run-off.** Sewage and run-off from agricultural lands supply plant nutrients, which may stimulate the growth of algae and other aquatic weeds in the receiving water body. This unwieldy plant-growth results in the degradation of the value-of the water body, intended for recreational and other uses. Further, the water body loses all its DO in the long run due to the natural biological process of eutrophication and ends up as a dead pool of water.

(e) **Oil :** Oil pollution may take place because of oil spills from cargo oil tankers on the seas, losses during off-shore exploration and production of oil, accidental fires in ships and oil tankers, accidental or intentional oil slicks (as in the Gulf War between Iraq and U.S.-led allied forces in the year 1991) and leakage from oil pipelines, crossing waterways and reservoirs. Oil pollution results in reduction of light transmission through surface waters, thereby; reducing photo-synthesis by marine plants. Further, it reduces the DO in water and endangers water birds, coastal plants and animals. Thus, oil pollution leads to unsightly and hazardous conditions which are deleterious to marine-life and sea-food. Oil pollution in seas has been increasing in recent years due to the increase in oil-based technologies, massive oil shipments, accidental oil spillages and intentional oil slicks during international hostilities.

(2) Inorganic Pollutants

Inorganic pollutants comprise of mineral acids, inorganic salts, finely divided metals or metal compounds, trace elements, cyanides, sulphates, nitrates, organometallic compounds and complexes of metals with organics present in natural waters. The metal-organic interactions involve natural organic species, such as fulvic acids and synthetic organic species, such as EDTA. These interactions are influenced by or influence redox equilibria, acid-base reactions, colloid formation and reactions involving micro-organisms in water. Algal growths in water and metal toxicity in aquatic eco-systems are also influenced by these interactions.

Various metals and metallic compounds released from anthropogenic activities add up to their natural background levels in water. Some of these trace metals play essential roles in biological processes, but at higher concentrations, they may be toxic to biota.

The most toxic among the trace elements are the heavy metals, such as Hg, Cd and Pb and metalloids, such as As, Sb and Se. The heavy metals have a great affinity for sulphur and attack the -SH bonds in enzymes, thereby immobilizing the latter. Protein carboxylic acid groups (-COOH) and amino-groups ($-NH_2$) may also be attacked by the heavy metal ions. The heavy metals that may be bound to the cell membranes interfere with the transport phenomena across the cell wall. Heavy metals also tend to precipitate phosphate biocompounds or catalyse their decomposition. Water pollution by heavy metals occurs mostly due to street dust, domestic sewage and industrial effluents.

Polyphosphates from detergents serve as algal nutrients and thus are significant as water pollutants.

(3) Suspended solids and sediments

Sediments are mostly contributed by soil erosion by natural processes, agricultural development, strip mining and construction activities. Suspended solids in water mainly comprise of silt, sand and minerals eroded from the land. Soil erosion by water, wind and other natural forces are very significant for tropical countries like India. It is estimated that out of the total land area of 328 million hectares, 175 million hectares are susceptible to degradation by soil erosion. It is also estimated that the continents are losing 5.8 cm of surface soil every 1000 years. About 6000 metric tonnes of soil are washed away into the sea every year, which means that about 5.37 million tonnes of NPK (nitrogen, phosphorous and potassium) fertilizers are washed away into the sea. This erosion leads to qualitative and quantitative degradation of soil in land area. Thus, soil may be getting removed from agricultural land to areas where it is not at all required, such as water reservoirs. Soil particles eroded by running water ultimately find their way into water reservoirs and such a process is called 'siltation'. Reservoirs and dams are filled with soil particles and other solid materials, because of siltation. This reduces the water storage capacity of the dams and reservoirs and thus shortens their life. Such problems are faced with our reservoirs such as Ram Ganga, Hirakud, Nizamsagar, Bhakra and Maithan, and the resultant reduction in live storage capacity of the reservoirs may lead to severe loss of irrigation potential of our country by the end of this century. Apart from the filling up of the reservoirs and harbours, the suspended solids present in water bodies may block the sunlight required for photosynthesis by the bottom vegetation. This may also smother shell fish, corals and other bottom life forms. Deposition of solids in quiescent stretches of streams impair the normal aquatic life in the streams. Further, sludge blankets containing organic solids decompose, leading to anaerobic conditions and formation of obnoxious gases. The tremendous problem of soil erosion can be controlled by proper cultivation practices and efficient soil and forest management techniques.

The organic matter content in sediments is generally higher than that in soils. Sediments and suspended particles exchange cations with the surrounding aquatic medium and act as repositories for trace metals such as Cu, Co, Ni, Mn, Cr and Mo. Suspended solids such as silt and coal may injure the gills of the fish and cause asphyxiation.

(4) Radioactive Materials

The radioactive water pollutants may originate from the following anthropogenic activities:

(*a*) Mining and processing of ores, *e.g.*, Uranium tailings.

(*b*) Increasing use of radioactive isotopes in research, agricultural, industrial and medical (diagnostic as well as therapeutic) applications, *e.g.*, I^{131}, P^{32}, Co^{60}, Ca^{45}, S^{35}, C^{14}, Rb^{86}, Ir^{132} and Cs^{137}.

(*c*) Radioactive materials from nuclear power plants and nuclear reactors, *e.g.*, Sr^{90}, Cs^{137}, Pu^{248}, Am^{241}.

(*d*) Radioactive materials from testing and use of nuclear weaponry, *e.g.*, Sr^{90}, Cs^{137}.

The radioactive isotopes found in water include Sr^{90}, I^{131}, Cs^{137}, Cs^{141}, Co^{60}, Mn^{54}, Fe^{55}, Pu^{239}, Ba^{140}, K^{40}, Ra^{226}.

These radioactive isotopes are toxic to life-forms. For instance, Sr^{90}, which emanates from testing of nuclear weapons, accumulates in bones and teeth and causes serious disorders in human beings. The maximum permissible level of Sr^{90} in water is 10 pico curies per liter (1 pico curie = 10^{-12} curie).

(5) Heat

Waste heat is produced in all processes in which heat is converted into mechanical work. Thus, considerable thermal pollution results from thermal power plants, particularly the nuclear-power-based electricity generating plants. In such industries, where the water is used as a coolant, the waste hot water is returned to the original water bodies. Hence the temperature of the water body increases. This rise in temperature decreases the DO content of water, which adversely affects the aquatic life. Moreover, any rise in temperature may increase the susceptibility of aquatic biota to the toxic effects of some chemicals, such as methyl mercury and some polycyclic aromatic hydrocarbons. Reduction of DO in water may alter the spectrum of organisms that can adopt to live at that temperature and that DO level. Suspended solids in water may also cause bad odours and tastes and also may promote conditions favourable for growth of pathogenic bacteria.

If the pollutant concentration in the receiving waters is not within the acceptable limits, adequate steps must be taken to minimize or remove them by suitable treatment techniques *e.g.*, sedimentation, filtration, biological oxidation, chemical precipitation or adsorption by activated carbon.

Characterisation of Waste Waters

Waste waters are characterised on the basis of various physical, chemical and biological characteristics apart from flow data details:

(1) Physical Characteristics. Colour, Odour, Dissolved Oxygen (DO), Insoluble Substances (settleable solids, suspended solids), Corrosive properties, Radio-activity, Temperature range, Foamability, etc.

(2) Chemical Characteristics. Chemical oxygen demand (COD), pH, Acidity or Alkalinity, Hardness, Total Carbon, Total dissolved solids, chlorine demand, known organic and inorganic components such as Cl^-, S^{2-}, SO_4^{2-} N, P, Pb, Cd, Hg, Cr, As, surfactants, phenols, hydrocarbons oils and greases.

(3) Biochemical Characteristics. Biochemical oxygen demand (BOD), presence of pathogenic bacteria etc., and toxicity to man, aquatic organisms, plants and other life forms.

The actual methods used for the treatment of a waste depend upon the characteristics of the particular waste.

Suspended solids. The suspended solids are determined by filtering an aliquot of the sample through a previously weighed, sintered crucible or a tared Gooch crucible and drying the crucible at 103°C to 105°C to constant weight. The difference in weight indicated as mg/l gives the suspended solids content of the sample.

Settleable solids. The settleable solids content of a sample is obtained by allowing 1 litre of the sample to settle for about 1 hour at 20°C in an Imhoff cone, which is a tapered conical tube. The volume of settleable matter in the cone is recorded as ml/1. The settleable solids may also be expressed in mg/1 which can be calculated by the difference between mg/l suspended soids minus mg/l non-settleable matter determined by the procedure described as above.

Total Solids. The total solids content of a sample is determined by evaporating a known volume of the sewage or waste water sample, and drying the residue for 24 hours at 103°C to 105°C, followed by weighing. This gives the total solids content of the sample, which includes the dissolved as well as suspended solids.

Dissolved oxygen (DO). The measurement of DO gives a ready assessment of purity of water. The determination of dissolved oxygen is the basis for BOD (Biochemical Oxygen Demand) test, which is commonly used to evaluate the pollution strength of waste waters. The determination of DO content is also essential for maintaining aerobic conditions in the receiving waters and also in the aerobic treatment of sewage and industrial waste waters.

The DO content of a water sample is determined iodometrically by the modified Winkler's method. The principle involved in this method is that when manganous sulphate is added to the water sample containing alkaline potassium iodide, manganese hydroxide is formed. This is oxidised to basic manganic oxide by the DO present in the water sample. When sulphuric acid is added, the basic manganic oxide liberates iodine, which is equivalent to the DO originally present in the water sample. The liberated iodine is titrated with a standard hypo solution, using starch as indicator.

Interference due to nitrite can be eliminated by adding sodium azide to the alkaline potassium iodide solution used above.

$$MnSO_4 + 2\ KOH \longrightarrow Mn(OH)_2 + K_2SO_4$$

$$2Mn(OH)_2 + O_2 \text{ (from DO)} \longrightarrow 2MnO(OH)_2$$
Basic manganic oxide

$$MnO(OH)_2 + 2H_2SO_4 \longrightarrow Mn(SO_4)_2 + 3H_2O$$
Manganic sulphate

$$Mn(SO_4)_2 + 2\ KI \longrightarrow MnSO_4 + K_2SO_4 + I_2$$

$$2Na_2S_2O_3 + I_2 \longrightarrow Na_2S_4O_6 + 2NaI$$
Sodium thiosulpate   Sodium tetrathionate

The DO is usually expressed as mg/l (or ppm)

Biochemical Oxygen Demand (BOD). Biochemical oxygen demand represents the quantity of oxygen required by bacteria and other micro-organisms during the biochemical degradation and transformation of organic matter present in wastewater under aerobic conditions. BOD test is a very valuable test in the analysis of sewage, industrial effluents and grossly polluted waters. In spite of the inherent limitations, the BOD test is still valued as the best test for assessing the organic pollution. BOD is considered as the major characteristic used in stream pollution control. It gives very valuable information regarding the purification capacity of streams and serves as a guide-line for the Regulatory Authorities to check the quality of effluents discharged into such water bodies.

The BOD test essentially consists of measurement of Dissolved Oxygen Content of the sample, before and after incubation at 20°C for 5 days. If the sample does not contain any oxygen, it is supplied with oxygen and the depletion caused is calculated as the measure of BOD. While carrying out the BOD test, microbial organism (called "seed") may also have to be provided if necessary. The BOD is usually expressed as mg/l. (5 days at 20°C).

Chemical Oxygen Demand (COD). The chemical oxygen demand (COD) is a measure of the oxygen equivalent to that portion of organic matter present in the waste water sample that is susceptible to oxidation by potassium dichromate. This is an important and quickly measured parameter for stream, sewage and industrial waste samples to determine their pollutional strength.

According to the American Society of Testing and Materials (ASTM), COD is defined as the amount of oxygen (expressed in mg/l) consumed under specified conditions in the oxidation of organic and oxidisable inorganic matter, corrected for the influence of chlorides.

The principle involved in the determination of COD is that when the waste water sample is refluxed with a known excess of potassium dichromate in a 50% H_2SO_4 solution in presence of $AgSO_4$ (as catalyst) and $HgSO_4$ (to eliminate interference due to chloride), the organic matter of the sample is oxidised to water, CO_2 and ammonia. The excess dichromate remaining unreacted in the solution is titrated with a standard solution of ferrous ammonium sulphate. The COD of the sample is calculated as follows :

$$COD \text{ in } mg/l = \frac{(V_1 - V_2) N \times 8 \times 1000}{X}$$

where V_1 and V_2 are the volumes of ferrous ammoniun sulphate (of normality, N) run down in the blank and test experiments respectively and X is the volume of the sample taken for the test.

Since in the COD test, both the biologically oxidisable and the biologically inert matter are oxidised, the COD value for a sample is always higher than BOD value.

Methods and Equipment used in Waste Water Treatment

The various methods used in sewage and industrial waste water treatment are as follows :

(i) Preliminary Treatment

The principal objectives of preliminary treatment are the removal of gross solids (*i.e.*, large floating and suspended solid matter, grit, oil and grease if they are present in considerable quantities.

Large quantities of floating rubbish such as cans, cloth, wood and other larger objects present in waste water are usually removed by metal bars, acting like stainers as the waste water moves beneath them in an open channel. The velocity of the water is then reduced in a grit-settling chamber of a larger size than the previous channel.

Removal of gross solids is generally accomplished by passing waste water through mixed or moving screens. Different types of these screens are available, which include bar screens (described above), hand raked or mechanical raked screens, drum screens and wire rope screens.

The modern mechanical screens cum filters include rotary, self-cleaning, gravity type units and circular overhead fed vibratory units. These are costlier, as compared to the conventional bar screens, but are very effective in reducing the suspended solids and BOD. Sometimes; instead of screening, the gross solids in the sewage are cut into small pieces with the help of macerators or comminutors.

Grit (or detritus) is removed in the early stages of treatment in grit channels or tanks to safeguard against any damage to pumps and other equipment by abrasion and also to avoid settling in pipe bends and channels. Grit, being heavier than organic solids, can be separated from organic solids by careful regulation of the flow velocity in the grit tanks. The grit settling chambers are periodically disconnected from the main system to remove the grit manually, for possible use in land-filling, road making and on sludge drying beds. If the waste waters contain appreciable quantities of oil and grease, then it is advisable to remove as much of these as possible, in the preliminary treatment itself to avoid adverse effects on the rest of the plant. This is achieved by passing the waste water through skimming tanks where oil and grease are skimmed off. This process can be rendered more efficient by aeration, chlorination or vacuum flotation.

If the oil and grease are in emulsified condition, as in wool-scouting wastes, ordinary skimming methods are ineffective. In such cases, they may be removed with the help of chemical reagents in primary sedimentation tanks.

(*ii*) Primary Treatment

After the removal of gross solids, gritty materials and excessive quantities of oil and grease, the next step is to remove the remaining suspended solids as much as possible. This step is aimed at reducing the strength of the waste water and also to facilitate secondary treatment.

Sedimentation. The suspended matter can be removed efficiently and economically by **sedimentation**. This process is particularly useful for treatment of wastes containing high percentage of settleable solids or when the waste is subjected to combined treatment with sewage.

The sedimentation tanks are designed to enable smaller and lighter particles to settle under gravity. The most common equipment used include horizontal flow sedimentation tanks and centre-feed circular clarifiers. The settled sludge is removed from the sedimentation, tanks by mechanical scrapping into hoppers and pumping it out subsequently. In a well-designed continuous flow sedimentation tank, about 50% of the suspended solid matter is settled out within two hours of detention time. An efficient sedimentation system is expected to remove about 90% of the suspended solids and 40% of organic matter (thus reducing the BOD).

In waste waters containing larger proportion of industrial wastes, a long detention time helps in mixing and balancing (or equalizing) of the various wastes and safeguards against unduly heavy loads being passed on to the biological purification plants subsequently.

Sedimentation aids. Finely divided suspended solids and colloidal particles cannot be efficiently removed by simple sedimentation by gravity. In such cases, mechanical flocculation or chemical coagulation is employed.

In *Mechanical flocculation,* the wastewater is passed through a tank with a detention time of 30 minutes and fitted with paddles rotating at an optimum peripheral speed of 0.43 m/s. Under this gentle stirring, the finely divided suspended solids coalesce into larger particles and settle out. Specialised equipment such as *Clariflocculator* is also available, wherein flocculating chamber is a part of a sedimentation tank.

In *chemical coagulation,* the sewage or other waste water is treated with certain chemicals which form a floc (flocculent precipitate) that absorbs and entrains the suspended and colloidal particles present. The coagulants in common use are (1) Hydrated lime (2) Alum, $Al_2(SO_4)_3.18H_2O$ (3) Copperas $FeSO_4.7H_2O$ (4) Feric chloride and (5) Chlorinated copperas, $FeSO_4.Cl$ (mixture of ferric sulphate and chloride). Alum is the most popular coagulant used both in water and waste water treatment. For best results, the chemicals used for coagulation are well mixed with the wastewater in baffled channels followed by mechanical flocculation before sedimentation. Pre-aeration for about 10 minutes before sedimentation is also found to help in the removal of entertained gases like CO_2 and H_2S and improved flocculation and separation of oil and grease.

Coagulation is the most effective and economical means to remove impurities.

Sometimes, in addition to the coagulants, other chemicals called *"coagulant aids"* are also used in very small quantities to promote the formation of large and quick settling floc and thereby enhancing coagulation. Activated silica and polyeletrolytes (such as polymers of cyanamide, acrylic or methacrylic acids and their derivatives, and hydrolysed high molecular weight polymers having molecular mass of 10^4 to 10^6 of acrylamide or acrylonitrile) are the most commonly used coagulant aids. Polyelectrolytes are generally used in dilute solutions (~0.25 ppm). Owing to their selective property, care should be taken in selecting the most suitable polyelectrolytes.

The synthetic coagulant aids work by the following two mechanisms:

(i) The coagulant aids, having long chain molecular structure, are absorbed on two or more particles, thus drawing them together.

(ii) By reducing the charge on the particles and thus reducing the repulsive power of the like charges on the particles.

The sequence of operations in the chemically aided coagulation are (a) Addition of lime, if there is no sufficient alkalinity, (b) Addition of coagulants, followed by rapid mixing for 4 to 6 minutes, (c) Addition of coagulant aids, followed by gentle agitation or slow stirring for about 40 minutes.

Equalization. Some industries produce different types of wastes, having different characteristics at different intervals of time. Hence, uniform treatment is not possible. In order to obviate this problem, different streams of effluents are held in big holding tanks for specified periods of time. Each unit volume of waste is mixed thoroughly with other unit volumes of other wastes to produce a homogeneous and equalized effluent. Aeration or mechanical agitation with paddles usually give better mixing of the different unit volumes of effluents.

Neutralization. Highly acidic or highly alkaline wastes should be properly neutralized before being discharged. Acidic wastes are usually neutralized by treatment with lime stone or lime-slurry or caustic soda, depending upon the type and quantity of the waste. Alkaline wastes may be neutralized by treatment with sulphuric acid or CO_2 or waste boiler flue gas.

If both acidic and alkaline wastes are produced in the same plant or at nearby plants, storing them in separate holding tanks and mutual neutralization by mixing them in appropriate proportion is the cheapest method.

(iii) Secondary Treatment

In secondary treatment, the dissolved and colloidal organic matter present in waste waters is removed by biological processes involving bacteria and other micro-organisms. These processes may be aerobic or anaerobic. In aerobic processes, bacteria and other micro-organisms consume organic matter as food. They bring about the following sequential changes:

(i) Coagulation and flocculating of colloidal matter,

(ii) Oxidation of dissolved organic matter to CO_2 and

(iii) Degradation of nitrogenous organic matter to ammonia, which is then converted into nitrite and eventually to nitrate.

Thus, secondary treatment reduces BOD. It also removes appreciable amounts of oil and phenol. However, commissioning and maintenance of secondary treatment systems are expensive.

The effluent from primary sedimentation tanks is first subjected to aerobic oxidation in systems, such as aerated lagoons, trickling filters, activated sludge units, oxidation ditches or oxidation ponds. Then the sludge obtained in these aerobic processes, together with that obtained in the primary sedimentation tanks, is subjected to anaerobic digestion in the sludge digesters.

Certain micro-organisms, in presence of dissolved oxygen and in proper environmental conditions, utilise organic waste as their food, and convert into simpler compounds such as CO_2, H_2O, nitrates and sulphates, which are non-pollutants. This process, therefore, can be used to remove organic substances from wastes. Almost all organic substances, with a few exceptions, such as hydrocarbons and ethers, can be oxidised by aerobic biological treatment. Complex cell tissues and protein materials are also synthesised during this process, which are then agglomerated and removed from the waste by settling. Germicidal and resistant organics, such as cyanides and phenols also can be destroyed by special types of micro-organisms, after prolonged acclimatization periods.

Under anaerobic conditions (*i.e.*, in the absence of dissolved oxygen or gaseous oxygen), certain groups of micro-organisms *e.g.*, hydrolyte and methane forming organisms, can carry out the digestion of complex organic wastes. The hydrolyte organisms convert complex organic compounds to simple and low-molecular weight organic acids and alcohols. These are then converted by methane bacteria to CO_2 and CH_4. Anaerobic treatment process can be carried out in depth without the need for large surface area. It can take place in mixed or enriched cultures and can, therefore, be maintained easily on large scale. The process can be, applied to most types of substrates excepting a few like lignin and mineral oil. The process is less expensive but the final effluent is less satisfactory, as compared to that from aerobic treatment, because of the dark colour, odour and higher residual BOD.

Anaerobic treatment is mainly employed for the digestion of sludges. However, organic liquid wastes from dairy, slaughter house etc., were treated by this method economically and effectively. The efficiency of this process depends upon pH, temperature, waste loading, absence of oxygen and toxic materials.

Some of the commonly used biological treatment processes are described below :

(*i*) **Aerated Lagoons.** These are large holding tanks or ponds having a depth of 3-5 m and are lined with cement, polythene or rubber. The effluents from primary treatment processes are collected in these tanks and are aerated with mechanical devices, such as floating aerators, for about 2 to 6 days. During this time, a healthy flocculent sludge is formed which brings about oxidation of the dissolved organic matter. BOD removal to the extent of 90% could be achieved with efficient operation. The operation and maintenance are relatively simple. The major dis-advantages are the larger space requirements and the bacterial contamination of the lagoon effluent which necessitates further biological purification in maturation pond or by secondary sedimentation and sludge digestion.

(*ii*) **Trickling Filters.** The trickling filters usually consist of circular or rectangular beds, 1 m to 3 m deep, made of well-graded media (such as broken stones, PVC, coal, coke, synthetic resins, gravel or clinkers) of size 40 mm to 150 mm, over which wastewater is sprinkled uniformly on the entire bed with the help of a slowly rotating distributor (such as a rotary sprinkler) equipped with orifices or nozzles. Thus, the waste water trickles through the media. The filter is arranged in such a fashion that air can enter at the bottom, counter current to the effluent flow and a natural draft is produced. A gelatinous film, comprising of bacteria and aerobic micro-organisms known as "Zooglea", is formed on the surface of the filter medium, which thrive on the nutrients supplied by the sewage or the waste water. The organic impurities in the waste water are adsorbed on the gelatinous film during its passage and then are oxidised by the bacteria and the other micro-organisms present therein. When the thickness of the film on the medium increases, a part of it gets detached and carried away along with the effluent. Hence, the effluent from the trickling filters is allowed to settle in a settling tank to retain the sludge particles and is then discharged. The sludge is then pumped to the sludge digestion unit.

A Schematic representation of a typical Trickling filtration process is given in Fig. 18.8.

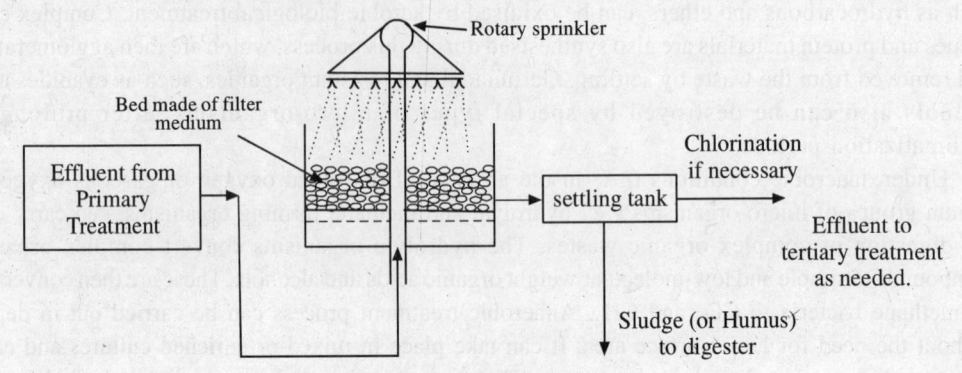

Fig. 18.8. Trickling Filter.

Although trickling filtration is classified as an aerobic process, it is indeed a facultative system*. Aerobic bacterial species *e.g.*, spore forming bacteria and bacillus are mostly present in the upper layer of the filter, whereas anaerobic species, such as Desulfo vibrio, are present in the interfaces of the stones. Facultative bacteria, such as Pseudomonas, Alcaligens, Flavo bacterium, Enterobactericeae and Micrococcus are also present in the trickling filters. Algae, fungi, ciliates, protozoan, worms, snails, insect larvae that feed on micro-organisms are also present.

By and large, smaller media give better results but they tend to choke easily. Synthetic plastic media are found to be particularly useful for industrial wastes of higher loading. Moreover, they can afford maximum surface area for the formation of microbial film and their light weight helps in greater economy in laying the underdrain of the filter.

The microbial film formed is very sensitive to temperature. The metabolic activity is proportional to the temperature of the waste water passing through the filter. Thus, the efficiency of the filter decreases in winter season.

The efficiency of the filter depends upon the composition of the waste, strength of hydraulic loading, temperature, pH, depth of the filter, the size and uniformity of the filter medium, uniformity of waste water distribution over the filter and proper air supply.

The trickling filter has greater resistance to toxic waste, as compared to the "Activated Sludge Process" and can recuperate more promptly from an overdose of toxic materials. However, shock loads or sudden surges should be avoided lest the efficiency of the filter might be impaired temporarily or even permanently.

Trickling filters are simple to operate and can produce BOD removal to the extent of 65 to 85%, depending upon the rate of filtration. Moreover, constant manual attention is not needed for this process. Trickling filters produce effluents of consistent and better quality.

The disadvantages of the process include the cost of construction and the need for ventilation ducts for the under drain system. The efficiency decreases with increased loading of the waste water. To overcome this problem, the waste water may be diluted with the effluent from the previous treatment.

Trickling filters are effectively used for the treatment of industrial wastes from dairy, distillery, brewery, cannery, food processing, pulp and paper mills, pharmaceuticals, petrochemicals, slaughter house and poultry processing industries.

---

* "Facultative bacteria" can grow with or without the presence of oxygen.

(***iii***) **Activated Sludge Process.** This is the most versatile biological oxidation method employed for the treatment of waste water containing dissolved solids, colloids and coarse solid organic matter. In this process, (Fig. 18.9), the sewage or industrial waste water is aerated in a reaction tank in which some microbial floc is suspended. The aerobic bacterial flora bring about biological degradation of the waste into CO_2 and H_2O, while consuming some organic matter for synthesising bacteria. The bacterial flora grows and remains suspended in the form of a floc, which is called "Activated sludge". The effluent from the reaction tank is separated from the sludge by settling and is discharged. A part of the sludge is recycled to the same tank to provide an effective microbial population for a fresh treatment cycle. The surplus sludge is

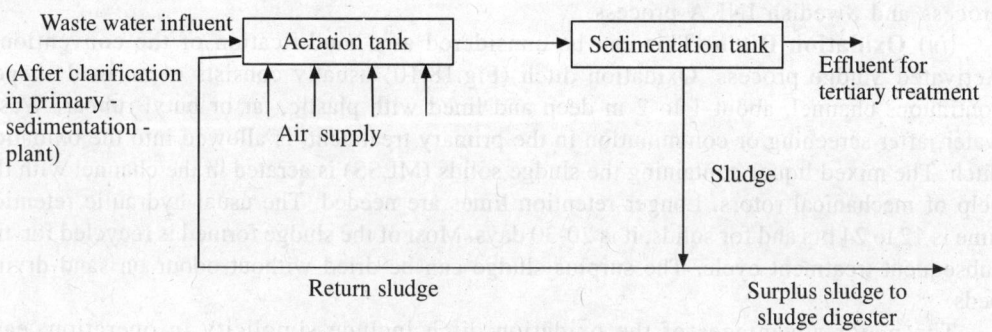

```
Waste water influent  →  [Aeration tank]  →  [Sedimentation tank]  →  Effluent for tertiary treatment
(After clarification
in primary -
sedimentation -
plant)
        ↑ Air supply                              ↓ Sludge
        ←─────── Return sludge ───────────────────┘
                                                  → Surplus sludge to sludge digester
```

Fig. 18.9. Activated Sludge Process

digested in a sludge digester, along with the primary sludge obtained from primary sedimentation. An efficient aeration for 3 to 6 hours is adequate for sewage, whereas for industrial wastes, 6 to 24 hours of aeration is required for this process. A BOD removal to the extent of 90-95% can be achieved.

The microbial flocs formed in this process comprise of Zoogleal masses of living organisms, embedded with their food and slime material and act as active centres for biological oxidation and that is why it is called "Activated Sludge". A young light sludge is preferred to old heavy sludge, because the latter would be mineralised and become devoid of oxygen. For this process to be efficient, at least 0.5 ppm oxygen must be present all the time. Oxygen is supplied either by mechanical aeration or by diffused aeration systems.

The micro-organisms should be provided with essential nutrients, such as N and P, which are supplied in the form of urea and mono-or di-ammonium hydrogen phosphate. Other nutrients *e.g.*, K, Mg, Ca, etc. are generally present in the waste. Other important factors which determine the efficiency of the activated sludge are pH, temperature and oxidation-reduction potential. The optimum pH range for the process is 6.5. to 9.0. Low temperature slows down the rate of metabolism, while high temperature increases the metabolic activity to such an extent that the oxygen is consumed fast, leading to anaerobic conditions.

The presence of synthetic detergents, such as alkyl benzene sulphonates (ABS) and polyethylene glycols, are not susceptible for microbial degradation. They lead to foam formation and make the process difficult and dangerous. Antifoaming compounds are used in such situations.

The performance of the activated sludge process can be assessed with the help of microbial indicators, as observed by regular microscopic examination of the activated Sludges. A good activated sludge contains a relatively high population of free swimming (*e.g.*, stylonichia) and stalked ciliates (*e.g.*, vorticella) apart from a few rotifers. A very high population of dispersed bacteria indicates a poor activated sludge system. The presence of filamentous micro-organisms indicates the deficiency of N and/or P, low pH and low oxygen levels. It may be noted that filamentous bacterial growths' retard floc compaction and settling and result in turbid effluents.

Activated sludge process produces a high quality effluent with relatively small areas. With efficient systems, about 95% BOD removal is possible. The disadvantages include high cost of operation and maintenance, need for careful attention and sensitivity to shock loads of toxic and organic substances.

Activated sludge process was used satisfactorily for the treatment of effluents from food processing, sugar, textile processing, antibiotic manufacturing industries, etc.

Several modifications of the conventional activated sludge process have come into vogue and they mostly differ in the method of air supply. These include : compressed air process, tapered aeration process, extended aeration process, dispersed aeration process, contact stabilization, high rate aerobic treatment process, simplex process, aeration rotor process and Swedish INKA process.

(iv) Oxidation Ditch. This can be considered as a modification of the conventional Activated Sludge process. Oxidation ditch (Fig.18.10) usually consists of an oval shaped continuous channel, about 1 to 2 m deep and lined with plastic, tar or butyl rubber. Waste water, after screening or comminution in the primary treatment, is allowed into the oxidation ditch. The mixed liquor containing the sludge solids (MLSS) is aerated in the channel with the help of mechanical rotors. Longer retention times are needed. The usual hydraulic retention time is 12 to 24 hrs and for solids, it is 20-30 days. Most of the sludge formed is recycled for the subsequent treatment cycle. The surplus sludge can be dried without odour on sand drying beds.

The major advantages of the oxidation ditch include simplicity in operation, easy maintenance, low cost of construction, operation and maintenance, overall efficiency and flexibility. This process is generally used for wastes having low BOD.

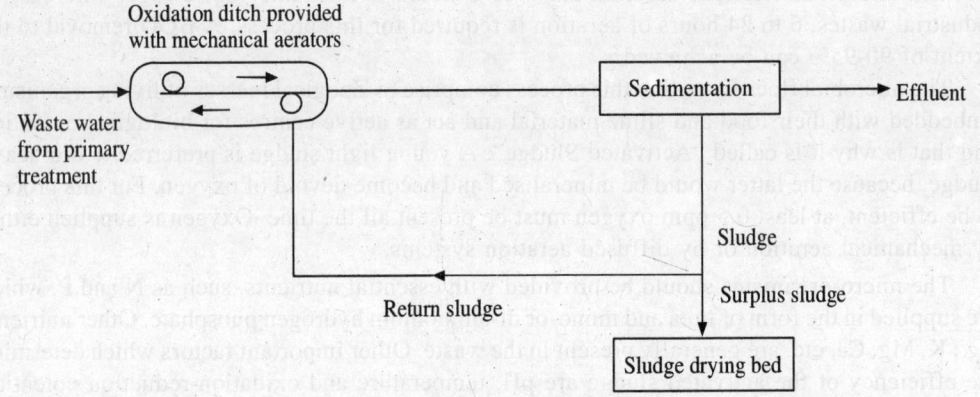

Fig. 18.10. Schematic representation of Oxidation Ditch Process.

Oxidation ditch process is used effectively for the treatment of waste water from beet-sugar manufacture, vegetable and fruit canning industry, slaughter house and meat packing industry and edible oil refineries.

(v) Oxidation Pond. An oxidation pond is a large shallow pond (0.5 m to 1.5 m depth) with arrangements to measure the inflow and outflow. The wastes enter the pond at one end and the effluent is removed at the other end. Stabilization of organic matter in the waste is brought about mostly by bacteria, such as Pseudmonas, Flavo bacterium and Alcaligenes, and to some extent by flagellated protozoa. The oxygen requirement for their metabolism is provided by algae present in the pond. The algae, in turn, utilize the CO_2 released by the bacteria for their photo synthesis. Oxidation ponds are also called waste stabilization ponds.

For efficient waste water treatment by this process, an adequate natural aeration, good mixing by wind and proper penetration of sunlight required for photosynthesis of algae are essential. This is the reason why oxidation ponds should be shallow.

Addition of nutrients may be necessary to enhance the growth of algae. This will enhance the amount of oxygen released, which in turn increases the rate of purification of waste water. Any deficiency of oxygen may lead to anaerobic conditions and consequent release of bad odours due to putrefaction of wastes. This is the reason why mechanical aeration is also provided at some places in addition to natural aeration.

Although waste water treatment in oxidation ponds is generally considered as an aerobic process, the purification is performed by a combination of aerobic, facultative and anaerobic processes. The waste water present in the upper part of the pond (which constitutes the major portion of the waste) undergoes aerobic oxidation to CO_2 and H_2O. Solids present in the waste, which settle as a layer at the bottom, act as anaerobic phase. Here, the organic matter is oxidised by anaerobic bacteria to CH_4, CO_2, and NH_3. The facultative zone exists near the anaerobic phase.

Waste treatment by oxidation pond is cheap and the operation and maintenance are simple. The process can be used for all types of wastes and any degree of purification can be obtained. The process can withstand organic and hydraulic shock loads. The heavy metal ions present in the waste water are precipitated as hydroxides (due to the high pH of the waste water in the oxidation pond) which settle as sludge. However, oxidation ponds require larger space. The effluent from the oxidation ponds may require disinfection or further treatment in a separate maturation pond before final discharge.

(v) Anaerobic digestion. This treatment is mainly used for sludge digestion. Sludge is the watery residue from the primary sedimentation tank and humus tank (from secondary treatment). The constituents of the sludge undergo slow fermentation or digestion by anaerobic bacteria in a sludge digester, wherein the sludge is maintained at a temperature of 35°C at pH 7-8 for about 30 days. CH_4, CO_2 and some NH_3 are liberated as the end products. The schematic representation of the anaerobic sludge digestion process is shown in Fig. 219.

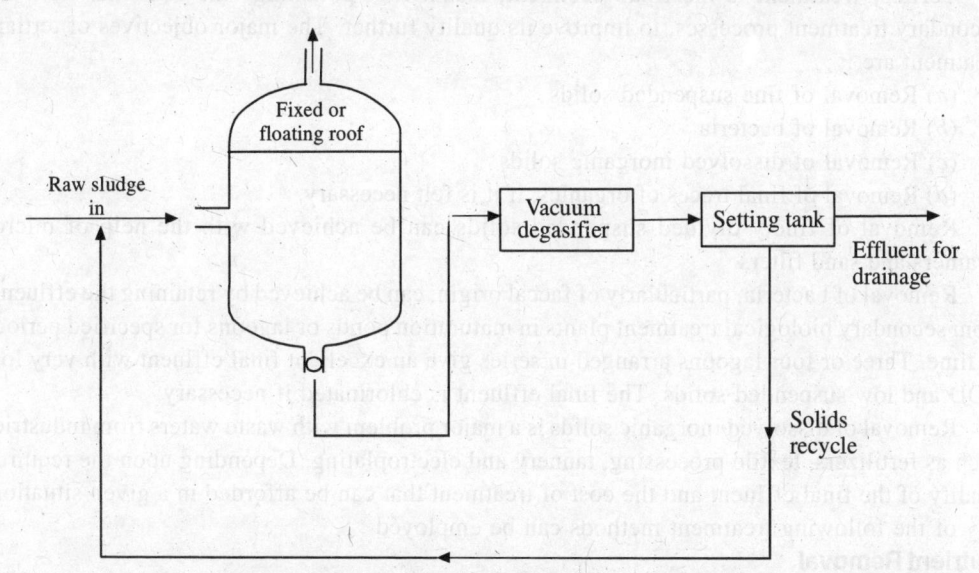

Fig. 18.11. Anaerobic Sludge Digestion Process.

Species of Pseudomonas, Flavo bacterium, Aerobacter, Alcalagenes etc., convert complex organic compounds to low molecular weight organic acids and alcohols. Methane bacterium, Methanosarcina and Methanococcus types of bacteria are responsible for the generation of CH_4 in this process. Desulfo Vibrio bacteria reduce sulphates to sulphides and thus, H_2S is released.

The anaerobic digestion process can be accelerated at higher temperatures.

The advantages of anaerobic digestion process are as follows :-
(1) Reduction in volume of the waste by about 65%.
(2) The digested sludge is safer to be used as manure than the undigested sludge.
(3) The digester gas obtained has the following percentage composition by volume : CH_4- 65 to 80%; CO_2- 5 to 30%, N_2, H_2; H_2S and CO together - about 5%. About 0.6 to 1.25 m^3 of the gas is produced per kg of the organic matter destroyed. The calorific value of the gas is about 26 MJ/m^3 (*i.e.*, about 700 BTU/ft^3). The gas can therefore be used as a fuel to provide the heat required to warm the digestion tanks. In large installations, it can be used for power generation.
(4) Although anaerobic treatment is a slow process, it is useful for treating small quantities of wastes, containing readily oxidisable dissolved organic solids in liquid form or in finely divided form. The operation and maintenance costs are lesser with this treatment. That is why some liquid wastes containing soluble organics from dairy, slaughter house and paper mill industries have been economically and effectively treated by this process.

Sludge treatment and disposal

The sludge from the digester may contain about 90 to 93% water. The sludge is de-watered in drying beds, filter presses or vacuum filters. The de-watered sludge, after chlorination, can be sent for ultimate disposal. The various methods used for ultimate disposal include dumping in land-fills, incineration, dumping at selected sites in sea, or utilizing as a low grade fertilizer.

Tertiary Treatment

Tertiary treatment is the final treatment, meant for "polishing" the effluents from the secondary treatment processes, to improve its quality further. The major objectives of tertiary treatment are :
(*a*) Removal of fine suspended solids
(*b*) Removal of bacteria
(*c*) Removal of dissolved inorganic solids
(*d*) Removal of final traces of organics, if it is felt necessary.

Removal of finely divided suspended solids can be achieved with the help of micro-strainers and sand filters.

Removal of bacteria, particularly of faecal origin, can be achieved by retaining the effluents from secondary biological treatment plants in maturation ponds or lagoons for specified periods of time. Three or four lagoons arranged in series give an excellent final effluent with very low BOD and low suspended solids. The final effluent is chlorinated if necessary.

Removal of dissolved inorganic solids is a major problem with waste waters from industries such as fertilizers, textile processing, tannery and electroplating. Depending upon the required quality of the final effluent and the cost of treatment that can be afforded in a given situation, any of the following treatment methods can be employed :

Nutrient Removal

The quantity of wastewater produced increases with the increase of population and increase of industrialization. However, the capacity of the streams to assimilate the waste is finite. This is the reason why a given stream is not only effluent limited but also water quality limited, thus necessitating advanced wastewater treatment systems.

One of the main reasons for removing nutrients from waste-water is to prevent eutrophication.

ENVIRONMENTAL CHEMISTRY & CONTROL OF ENVIRONMENTAL POLLUTION

From this point of view, the nutrients of concern are phosphorous and nitrogen.

Removal of nitrogen

Nitrogen is generally removed from wastewater by the following two methods :

(1) Physical method utilizing stripping columns : The reaction of NH_3 with water may be represented as :

$$NH_3 + H_2O \rightleftharpoons NH_4^+ + OH^-$$

From this reaction, it is clear that by increasing the pH, the reaction will be driven to the left and the concentration of NH_3 is increased. This facilitates easy removal of NH_3 by stripping. In practice, the pH is increased to about 10 to 11 by using lime. Stripping is done by introducing the wastewater at the top of the column and allowing it to flow down concurrent to the flow of air introduced at the bottom. The stripping medium inside the column may comprise of packings or fillings such as Raschig rings and Berr Saddles, or sieve trays and bubble caps. The liquid flows in thin sheets around the medium, thereby allowing more intimate contact between liquid and the stripping air.

(2) Biological method utilizing the process of nitrification and denitrification : Nitrification is the process of oxidizing nitrogen to nitrate and denitrification is the process of converting the nitrate to nitrogen gas, thus removing the nitrogen from the wastewater. Both these steps are involved in the biological nitrogen removal (BNR).

In the oxidation step (i.e. nitrification step), nitrogen present in the wastewater is first oxidized to the nitrite form by *Nitrosomonas,* followed by the oxidation of the nitrite to nitrate by *Nitrobacter.*

In the second step (i.e. denitrificatin step), the process takes place under anaerobic conditions utilizing the normal heterotrophic organisms, resulting in the reduction of nitrate to gaseous N_2.

Removal of Phosphorus

As stated earlier, nutrients are removed from wastewater being discharged to control eutrophication of the receiving streams. In most cases, removal of phosphorous is targeted.

Domestic wastewater and agricultural return water have been identified as the principal sources of phosphorous. Phosphorous in wastewater may be present as orthophosphate, polyphosphate and organic phosphorus. Typically, the phosphorus enters the wastewater from human body wastes, from food wastes discharged to the sewers from kitchen, and from the condensed inorganic phosphate compounds used in various household detergents. Commercial washing and cleaning compounds are also a source of phosphates.

Two general methods are used to remove phosphorus :
(i) by biological uptake by incorporation into cell tissues.
(ii) by chemical precipitation using lime and salts of iron (e.g., ferric chloride or ferric sulphate) and of aluminium (e.g., alum).

The precipitates usually formed are metal phosphates. If lime is used, hydroxyapatite, $Ca_5(OH)(PO_4)_3$ is formed.

$$Fe^{3+} + HPO_4^{2-} \rightarrow FePO_4 + H^+$$
$$Al^{3+} + HPO_4^{2-} \rightarrow AlPO_4 + H^+$$
$$5\,Ca(OH)_2 + 3\,HPO_4^{2-} \rightarrow Ca_5(OH)(PO_4)_3 + 6\,OH^- + 3H_2O$$
$$H^+ + OH^- \rightarrow H_2O$$

Alum and iron reactions are buffered between pH 5.5 to 7. Lime is normally used at pH > 1.

Evaporation. This is an expensive process. It is used only when the recovered solids or the concentrated solutions are reused, *e.g.*, some electroplating wastes. This method is also used when the volume of the waste water to be treated is less. This method is also employed for concentrating radio-active liquid wastes.

Ion-exchange. The use of ion-exchange for de-mineralisation of water is well known. It is widely used for obtaining de-ionised water for use in high-pressure boilers. This process is now extended to waste water treatment for the removal and recovery of toxic materials from waste water. Ion-exchange process is economical only when the recovered salts are reused in the process, as in electroplating industry. Despite the simplicity of its operation, the method may not be economical if the objective of the treatment is only the removal of dissolved solids from waste water. Special ion-exchangers are available for the retrieval of toxic metal ions from industrial waste water.

Adsorption. Adsorption by activated carbon is advantageous to remove small quantities of organic contaminants from waste-water. Special adsorbents are commercially available for the removal and retrieval of toxic heavy metal ions from industrial waste water. Activated carbon treatment is particularly useful for the removal of pesticides (*e.g.*, DDT) and carbamate insecticides.

Reverse osmosis. When a waste water containing dissolved solids is allowed to pass through a semi-permeable membrane, at a pressure over and above the osmotic pressure of the waste water, only the water from the waste permeates through the membrane, leaving behind a concentrated liquor, containing the dissolved solids. This process is particularly suitable and effective for the removal of dissolved solids from waste water. The cost of the membranes and the fouling of the membranes are the major limitations of this process.

Chemical precipitation. The dissolved solids in the waste water, particularly the heavy metal ions, can be removed by precipitation as their hydroxides with cheap precipitating agent like lime. Chromates in electroplating waste water are highly toxic and can be removed by treatment with $FeSO_4$ first to reduce the chromates to Cr(III), followed by precipitation with lime.

Several other processes are available for removing dissolved impurities from waste water. These include electrodialysis, solvent extraction fertilization, freeze purification, removal of dissolved solids by algae, etc. Although their effectiveness in removing dissolved solids from waste water is unquestionable, economic feasibility often becomes the limiting factor.

Unconventional Methods: Materials like Chitin, Chitosan, modified Cotton, modified Wool, Keratin, Poly-electrolytes, Starch xanthates, cellulose xanthates, fly ash, agricultural byproducts [*e.g.*, Onion skin, peanut skin and bagasse], and tree barks [*e.g.*, Redwood bark, Accacia arabica bark and Techtona grandis bark] have been found to be capable of removing/retrieving of heavy metal ions from waste water. These processes may be used for local use in case of small scale industries which cannot afford conventional waste treatment techniques due to commercial and technical limitations.

18.3-A. POLLUTION DUE TO SOME TYPICAL INDUSTRIES SOURCES, CHARACTERISTICS, EFFECT AND TREATMENT OPTIONS

(1) Textile Industry

(A) Introduction

The basic raw materials in a textile industry are cotton, wool or synthetic fibers such as rayon, nylon, acrylic, cellulose acetate or polyester. The textile industry comprises of preparation of the yarn, weaving, knitting and processing.

(*i*) **Cotton textile industry :** The various operations involved in a cotton textile mill are combing, spinning, sizing, weaving and knitting. All these processes except sizing are dry processes. The "grey cloth" thus obtained after the above operations is subjected to the various wet treatment processes such as desizing, scouring, bleaching, mercerizing, dyeing or printing and finishing. All these processes generate considerable volumes of effluents and hence form the major sources of wastewaters.

(*ii*) **Synthetic textile industry** : The fabrics of rayon, cellulose acetate, nylon, acrylic, Dacron, etc. do not require much of dry processing but certain treatments are required for removal of size, antistat and lubricating oil (*e.g.*, removal of PVA, resins, gelatin etc.) used in weaving operations. The wet processes are more or less similar *e.g.*, scouring, dyeing, rinsing, bleaching, finishing, etc. The characteristics of effluents naturally depend upon the type of fiber and the process followed.

The cotton textile industry produces relatively larger effluent volumes and higher pollution loads as compared to the synthetic textile industry.

(B) Wastewater sources and characteristics

(*i*) **Cotton textile industry** : In cotton textile processing, the characteristics of the wastewaters produced in some major operations involved are as follows :

Sizing : (Sizing is the process by which the warp thread is sized with starch to give the necessary tensile strength and smoothness required for weaving).

The wastewater is generally coloured and contains starch, polyvinyl alcohol and softeners. It has a high BOD and high content of dissolved and suspended solids.

Desizing, Scouring and Mercerizing :

(Desizing is the process used for removing the sizing materials present in the grey cloth with the help of 0.5% H_2SO_4 or with the help of enzymes).

(Scouring is the process used to remove natural impurities like greases, waxes and fats by boiling with NaOH, soda ash, sodium silicate, NaO_2, with small amount of a detergent).

(Mercerization is the process of boiling the cloth with 20% caustic soda solution followed by washing with water. This process gives luster and strength to the fabric).

The wastewater contains starch, acids, alkali silicates and enzymes.

Bleaching : (*Bleaching* is used to remove natural colouring materials. It is done using alkaline hypochlorite or chlorine or Na_2O_2. Bleaching is usually followed by washing with water and then by scouring treatment with sodium bisulfite to remove traces of alkali and Cl_2).

The effluent contains chlorine, hypochlorites and peroxides.

Dyeing, printing and finishing : The effluents contain dyes, alkalies, chromium, phenolics, oils and waxes.

(*ii*) **Synthetic textile industry:** The wastewater is generally coloured and contains alkalis, organic solvents, resins, PVA, etc. The effluent has high values of COD and BOD. The origin of wastes is more or less same as above.

(*iii*) **Wool industry** : The origin of wastewater and the characteristics are given below :

Scouring : The effluent is hot, highly coloured and contains greases, soaps, alkalis, detergents and suspended solids. Possess high BOD and COD.

Oiling : Olive oil or other vegetable oils.

Sizing : Starch.

Filling : Soda ash, soaps and detergents.

Dyeing : Dyes, phenolics, chromium salts, alkalis, etc.

(C) Environmental effects of the wastewaters

(*i*) The dyes present impart persistent colour to the receiving streams and interfere with photosynthesis of phytoplankton.

(*ii*) The high pH is deleterious to aquatic life.

(*iii*) The colloidal and suspended impurities cause turbidity in the receiving waters.

(*iv*) The oil present interferes with the oxygenation of the receiving water streams.

(*v*) The dissolved minerals increase the salinity of the water, thus rendering it unfit for irrigation purposes.

(vi) The toxic chemicals such as Cr, aniline, sulphides, etc. destroy fish and other microbial organisms responsible for self-purification of water streams.

(vii) The immediate oxygen demand due to the impurities such as starch, sulphides, nitrites etc., depletes the dissolved oxygen content and adversely affects the aquatic life.

(viii) The dissolved solids form incrustations on the surface of the sewers.

(ix) The dissolved impurities present cause corrosion in the metallic parts of the sewage treatment plants.

(D) Treatment Options

Segregation of wastes, screening to remove coarse suspended matter, grease removal (in case of scouring effluents), equalization, neutralization, chemical coagulation to remove colour, suspended and colloidal impurities, aerobic biological treatment (*e.g.*, trickling filtration, activated sludge process, oxidation pond, oxidation ditch or aerated lagoons) and finally tertiary treatment (*e.g.*, reverse osmosis or electrodialysis) to remove dissolved solids, as and when required or affordable.

(2) Pulp and paper Industry

(A) Introduction

The paper manufacturing process mainly consists of the following three steps :

(*i*) **Raw material preparation :** In this the cellulosic raw materials *e.g.*, wood, bamboo, cotton liners, bagasse, rags, straw, jute and hemp are slashed and cut into small chips.

(*ii*) **Pulping :** In this, the raw materials are digested with chemicals under high temperature and pressure so as to free the cellulose fibers from the binders *viz.*, lining, resins etc. Several types of pulping processes are available of which the following three are more common.

(*a*) **Sulfate or craft pulping process :** This is a widely used process suitable for any kind of wood. Since the pulp produced is coloured and difficult to bleach and since long and strong fibers are produced in this process, it is used for preparing strong wrapping papers, bag making papers, wall papers, paper boards for cartons, corrugated boards, etc. The process consists of digesting the raw materials with Na_2S, $NaOH$ and Na_2CO_3 at about 175°C and 120 psi pressure for 2 to 5 hours. During this process, most of the linings are hydrolysed to alcohols and acids. Toxic chemicals such as mercaptans and dimethyl sulphide are also formed. After digestion, the pressure is allowed to drop and the charge (called brown stock) is blown into a pit where the fibers are separated. The remaining waste cooking liquor, known as "black liquor" is sent to another unit for recovering chemicals such as Na_2S, Tall oil etc.

(*b*) **Acid sulfide pulping process :** This is used for the manufacture of high grade paper such as bond paper, tissue paper, bread wrap papers etc. In this process, the raw material containing mostly coniferous woods is digested with calcium bisulphite or magnesium bisulfite at 125 to 160°C with steam at a pressure of about 100 psi for about 6 to 12 hours. The pulp produced is dull white in colour and easy to bleach but the fibers are weaker than those obtained in case of sulphate kraft pulping process.

(*c*) **Semi-chemical pulping process :** In this process, lesser amounts of $NaOH$ and Na_2S are employed for pulping and Na_2SO_4 is used as the make-up chemical. This process produces high quality bleaching pulp suitable for manufacturing writing paper, bond paper, tissue, offset, newsprint papers and corrugated boards.

After digestion by any of the above processes, washing of the pulp followed by bleaching (with Cl_2 or ClO_2, $NaOH$ and $CaOCl_2$ in successive stages) must be employed. These steps along with the chemical recovery step contribute to the bulk of the effluents.

(*iii*) **Paper making :** The pulp is disintegrated in a beater along with fillers (*e.g.*, clays, $CaCO_3$, $BaSO_4$ or TiO_2), rosin or wax emulsion and $Al_2(SO_4)_3 \cdot 18H_2O$ (for providing a gelatinous film on the fiber), dyes and other chemicals as required. Then the pulp is refined. The fibers are cut to the desired size, passed on to stuffing boxes to get the proper consistency and then passed over a moving belt of fine wire. The water passes through the screen while the fibers form a mat on the surface of the screen. This is then sent on to rolls and then to calendering machine for final finishing.

(B) Wastewater Sources and Characteristics

All the above steps *viz.*, raw material preparation, pulping, washing, bleaching, chemicals recovery, screening of pulp and paper making contribute to the waste effluents.

The important characteristics of combined effluent of integrated pulp and paper mills with chemicals recovery system are dark brown colour, characteristic odour, high content of suspended and dissolved solids (about 1350 mg/l and 1650 mg/l respectively), high COD (about 1500 mg/l) and resistant to biological oxidation.

(C) Environmental Effects

(*i*) The dark colour of the effluent is due to the lining compounds which are not easily biodegradable and hence it imparts persistent colour to the receiving water streams and inhibits photosynthesis and other natural self-purification process of the water streams.

(*ii*) The immediate oxygen demand of the effluent brings about depletion of oxygen of the receiving stream with the concomitant adverse effects to the aquatic life.

(*iii*) The chemicals present in the effluent, *e.g.*, sulfites, phenols, free chlorine, methyl mercaptan, pentachlorophenol are harmful to fauna and flora of the receiving waters. The settleable materials present may sink to the bottom and interfere with aquatic life.

(D) Treatment Options

Recovery of by-products, preliminary screening, coagulation, sedimentation, flocculation and flotation to remove suspended matter, chemical treatment to remove colour, activated sludge treatment to remove dissolved organics, lagooning for storage and biodegradation of organic matter.

(3) Electroplating industry

(A) Introduction

In this process, a ferrous or non-ferrous base material is electroplated with Ni, Cr, Cu, Zn, Pb, Cd, Al, Ag, Au etc., to alter the surface properties of the base metal in order to achieve corrosion resistance, wear resistance, improved or decorative appearance etc. The important steps involved in this process are

(*i*) **Surface Cleaning :** The surface of the base metal is freed from oils and greases by saponification with warm alkali. Greases of mineral oil origin are removed by emulsification with alkali mixture containing NaOH, Na_2CO_3, Na_3PO_4 and Na_2SiO_3, etc, or with the help of solvents such as benzene or trichloroethylene. The articles are then washed. The rinse water forms an alkaline effluent.

(*ii*) **Pickling or stripping :** The decreased base metal is then treated with H_2SO_4 or HCl to remove scales and rust. The pickled articles are then washed with water. This wash water and the spent pickle liquor (containing ferrous sulphate and unused acids) form the waste effluents.

(*iii*) **Plating :** The pickled articles are electroplated in electrolytic cells containing plating baths as desired. The plating baths may be acidic or occasionally alkaline. Cyanides are generally used in plating baths as they are good oxide solvents and yield brighter and less porous plates. Several types of chemical additives are also used in the plating baths.

Among the plating wastes, cyanide concentrates and spent chromate bath wastes have received particular attention due to their toxicity. These wastes are usually segregated and stored for separate treatment. Apart from them, spent plating bath liquors from Cd, Pb, Zn, Cu, and Ni plating operations, spill overs, floor washes containing the above toxic chemicals and other compounds also form waste effluents from electroplating industry.

(B) Sources and Characteristics of the Plating Wastes

(i) Cleaning : These operations contribute to alkali wastes containing NaOH, carbonates, silicates, wetting agents and organic emulsifiers. Whenever cleaning is performed with organic solvents, the effluent from this operation consists of the solvents such as trilene, benzene, petrol etc. as well as emulsifiers.

(ii) Stripping or Pickling : These effluents contain unused acids (*e.g.*, HCl, H_2SO_4 and HNO_3) and ferrous sulfate.

(iii) Electroplating : All the constituents of the plating baths contribute to the wastewater stream either through part drag-out, batch dump or floor spill. Electroplating baths may contain Cu, Ni, Ag, Au, Zn, Cd, Cr, Sn, Pb, Fe, ammonia, etc. The anionic components likely to be present include borate, carbonate, fluoborate, cyanide, fluoride, tartrate, phosphate, chloride, sulfide, sulfate, sulfamate, nitrate, etc. Further, many other additives to induce grain refining, deposit brightening, surface levelling, etc. are also added to the plating baths. These include Mo, Se, As, Co, saccharin, aldehydes etc., all of which contribute to the waste streams. Apart from these, contaminants like oil, grease, biodegradable mass, suspended solids, etc. may also be present in the wastewaters.

(C) Effects

(*i*) Plating effluents are highly toxic and corrosive.

(*ii*) Cyanide, chromic acid, chromates, salts of heavy metals, *e.g.*, Cd, Pb, Ni, Zn and Cu present are toxic to aquatic life. Their toxicity to micro-organisms inhibits self-purification property of the streams.

(*iii*) Fe, Sn, etc. impart colour to the receiving stream.

(*iv*) Phosphates and nitrates present in the effluent help in excessive algal growth which is undesirable.

(*v*) Colloidal and suspended impurities impart unaesthetic appearance to the stream.

(*vi*) Owing to the toxic nature of the effluents, they are not disposed into rivers or waste courses. They are generally discharged into sewers. If cyanide is not completely removed, the HCN gas formed may affect the workers in the sewage treatment plant and sewer system. The organic solvents present may cause explosion in the sewer system. Oils and greases present may interfere with the biological treatment of the sewage. Acidic or alkaline plating effluents may corrode the concrete structures. Suspended impurities present may clog the municipal sewer system.

(D) Treatment Options

(*i*) Segregation of cyanide wastes, chromium wastes and other toxic metal bearing wastes.

(*ii*) **Treatment of cyanide waste by alkaline chlorination :** The pH of the waste is raised to 11 by adding lime or NaOH. Chlorine gas is then passed into the wastewater while agitating it thoroughly for several hours. The cyanide is decomposed under these conditions as per the following chemical reaction :

$$2NaCN + 12NaOH + 5Cl_2 \longrightarrow 2Na_2CO_3 + 10NaCl + 6H_2O + N_2$$

(*iii*) **Treatment of chromium bearing effluents :** The wastewater is to be brought to pH 3 or below by adding a mineral acid if necessary and is then treated with a reducing agent such as $FeSO_4$, SO_2 or sodium bisulphite. The Cr (III) formed is then precipitated as the hydroxide by adding lime slurry. The reactions taking place may be represented as follows:

$$H_2Cr_2O_7 + 6FeSO_4 + 6H_2SO_4 \longrightarrow Cr_2(SO_4)_3 + 3Fe_2(SO_4)_3 + 7H_2O$$

$$Cr_2(SO_4)_3 + 3Ca(OH)_2 \longrightarrow 2Cr(OH)_3 + 3CaSO_4$$

$$3Fe_2(SO_4)_3 + 9Ca(OH)_2 \longrightarrow 6Fe(OH)_3 + 9CaSO_4$$

(*iv*) Mixing the effluents obtained after treatment of cyanide and chromium wastes with the other metal-bearing wastes, followed by adjustment of pH to about 9 and settling the precipitates obtained. The sludge may be dried on sand beds and disposed in landfills.

(*v*) Scraping of the floating oils and greases from holding tank and removal of the emulsified oils by pretreatment with alum and caustic soda before entering into the holding tank.

(*vi*) The rinse waters may be reused after removing the ionic impurities by passing through cationic and anionic exchange resins. Oil is removed from the rinse waters by passing through activated carbon filters before passing through the ion-exchange resins. The effluents from the cation- and anion-exchange resins may be treated separately to remove the toxic impurities.

The various treatment technologies available for dealing with electroplating waste problems are given in the following table.

Table 8 : Electroplating waste treatment technologies

Waste treatment problem	*Treatment technology*
(1) Removal and/or recovery of dissolved metals.	Hydroxide precipitation, sulfide precipitation, ion exchange, Membrane filtration, insoluble starch xanthates, peat adsorption, carbon adsorption, electrolytic recovery, high pH precipitation (for complexed metals).
(2) Removal of dissolved salts for reuse of water.	Reverse osmosis
(3) Recovery of process baths.	Electrodialysis
(4) Removal of organics.	Aerobic decomposition, carbon adsorption, resin adsorption.
(5) Destruction of cyanides and cyanates.	Electrochemical oxidation, oxidation by chlorine, oxidation by ozone, oxidation by ozone with u.v. radiation, oxidation by H_2O_2 (for destruction of cyanides only).
(6) Suspended solids removal.	Flotation, centrifugation, ultrafiltration, sedimentation, diatomaceous earth filtration.
(7) Sludge dewatering.	Centrifugation, pressure filtration, diatomaceous earth filtration, sludge bed drying, vacuum filtration.

(8) Reduction of chromium (VI) (from metal finishing and cooling tower blowdowns). Electrochemical reduction

(9) Concentration and recovery of process chemicals. Evaporation.

(10) Oil removal. Centrifugation, coalescing, flotation, skimming, ultrafiltration (Oil breakdown can be accomplished by aerobic decomposition).

(4) Leather Tanning Industry

A. Introduction

In tanning industry, the animal skins and hides are treated to convert them to non-putrescible and tough leather. The following two processes are generally used :

(i) **Beam house processing :** The dry skin is made up of 85% collagen (protein) with small quantities of impurities such as albumin, lipids, globulin and carbohydrates. In this process the following steps are used to remove the impurities and prepare the collagen for tan-yard processing.

(a) **Soaking** : The skins and hides are salted and soaked in water in pits containing wetting agents for a day to remove salts, blood, dirt etc. and to restore the moisture content.

(b) **Liming** : The skins and hides are washed with water and again soaked in lime and Na_2S to soften and remove the hairs and other trace protein impurities. The hides are again washed with water.

(c) **Deliming** : The hides are treated with NH_4Cl or $(NH_4)_2SO_4$ and washed again.

(ii) **Tan-yard processing:** This is done by the following two types of tanning processes:

(a) **Vegetable tanning** : This process is used to produce heavy leathers which can be easily tooled and embossed. In this process, the delimed skins and hides are soaked for about 15 days in pits containing vegetable tan liquor (which comprises of extracts of bark, wood, nuts, etc.) and pyrogallol or catechol base. Some types of vegetable tannins and syntans can reduce this tanning time to a few hours only. The tanned materials are then soaked in myrobalan liquor. Then, vegetable oil such as pungam oil is applied to render the hides soft, pliable and tear-resistant. The leather is then dried and dyed.

(b) **Chrome tanning** : This process is used to prepare light and tough leather. The process involves more number of operations but the time required is much less.

Bating : The delimed hides are treated with a bating agent (which is a mixture of proteolytic enzymes such as typsin and chymotypsin). This process reduces the pH and swelling, peptizes the fibers, removes the protein degradation products and renders the material smooth, porous and silky.

Pickling : The material is treated with H_2SO_4 and common salt for a day in a drum.

Tanning: The material (leather) in then soaked in chrome tan liquor containing chromium sulphate for 6 hours. Then, Na_2SO_4 is added to the leather in the same drum to fix the chrome. It is then taken out of the drum and kept aside for a day to complete the chrome fixation.

Dyeing : The tanned leather is washed, neutralised and dyed as required.

Flat liquoring : An emulsion of sulphonated oil is applied to the leather to render it soft and pliable.

B. Sources and characteristics of tannery wastes

(i) **Soaking** : The effluent contains lot of NaCl, dirt, dung, soluble proteins like albumin, organic matter, suspended solids. This putrefies quickly giving bad odour.

(ii) **Liming** : The effluent contains alkali, sulphides, lime, $CaCO_3$, colloidal proteins and their degradation products. It has a high BOD value (4000 to 9001 mg/l).

(*iii*) **Dehairing and deflushing :** The effluent contains hairs, flesh, sulphides and suspended solids.

(*iv*) **Deliming :** Effluent is slightly acidic and has BOD values of 1000 to 2000 mg/l.

(*v*) **Vegetable tanning :** Effluent contains large amount of organic matter, suspended solids, dark colour, bad odour. BOD value is about 12000 mg/l.

(*vi*) **Bating :** Effluent contains ammoniacal and organic nitrogen, ammonium salts and soluble proteins.

(*vii*) **Pickling :** Effluent is acidic. Contains excess NaCl.

(*viii*) **Chrome tanning :** Effluent is acidic, toxic and contains 100 to 200 mg/l of Cr(III). BOD is about 1000 mg/l.

(*ix*) **Dyeing :** Effluent contains various dyes as used.

(*x*) **Fat-liquoring :** Effluent contains oils and greases.

C. Effects

(*i*) Tannery effluents contain several constituents which are deleterious, irrespective of the fact that where they are discharged *viz.,* into river, stream, sewer, land or sea.

(*ii*) It imparts persistent dull brown colour to the receiving water causing aesthetic and other problems described earlier.

(*iii*) Highly repulsive odour is imparted to the receiving water. The dissolved constituents like proteins are putrefiable.

(*iv*) The acidic or alkaline effluents are corrosive to concrete and metal pipes.

(*v*) Excess NaCl in the effluent is also corrosive and renders the receiving water unsuitable for irrigation.

(*vi*) The effluents may contain pathogenic bacteria.

(*vii*) The dissolved chromium present is toxic to fish and aquatic life and thus affects the natural self-purification property of the stream.

(*viii*) The suspended solids such as hair, flesh, $CaCO_3$ etc. interfere with aeration and photosynthetic activities of the aquatic flora.

(*ix*) If the wastewater is discharged into sewer, the suspended impurities such as $CaCO_3$, hairs etc. may choke the sewerage pipes. The sulphides present in the wastewater cause "crown corrosion" to the concrete structures, etc.

(*x*) The chromium and sulfides present in the wastewater being toxic to microorganisms disrupt the biological treatment operation such as trickling filtration. The suspended lime etc. also interfere with the biological activities in the sewage treatment plants.

(*xi*) The presence of excessive salt and Cr in the wastewaters may deteriorate the quality of the ground water in the affected areas.

D. Treatment Options

Primary treatment includes screening to remove hairs, fleshings etc. and then sedimentation. The secondary treatment includes processes such as chemical coagulation and biological treatment. Removal of chromium (by precipitation with lime) followed by activated sludge process gives an effluent of better quality. Oxidation pond and anaerobic lagoon are recommended for small and isolated tanneries. Since excessive quantities of NaCl still remain in solution after any of the above treatments, disposal of the treated effluent still poses a problem.

(5) Fertilizers industry

Fertilizer Raw materials : P, K, S.

Fertilizer intermediates : H_2SO_4, H_3PO_4, HCl and HNO_3

Fertillzer products : Urea, ammonium sulphate, ammonium nitrate, normal super phosphates, triple super phosphate, monoammonium and diammonium phosphate, ammonia and other liquid or slurry formulations.

Sources of waste effluents : Spill overs from manufacture of acids used as raw materials : spill over of the final fertilizer products; boiler blow-down, cooling waters, etc.

Characteristics and effects : Effluent from ammonia production is highly alkaline. Contains excess NH_3 from gas scrubbing and gas cleaning operation. Effluent from phosphoric acid manufacture is acidic and contains high amount of phosphates and suspended solids. When discharged into inland surface waters, fertilizer effluents cause eutrophication due to algal bloom. High amounts of fluoride (over 1000 mg/l) present in phosphatic fertilizer effluent enrich the fluoride content of the receiving waters causing dental and skeletal fluorosis to humans, abnormal calcification of bones in animals and adverse effects on plants. Presence of Cr, cyanide, ammonia are harmful to aquatic life.

Treatment Options : Sedimentation, neutralisation, recycling, segregation of wastes, physico-chemical or biological methods to remove NH_3, removal of urea by thermal or enzymatic hydrolysis, removal of fluoride (present as fluosilicic acid) and phosphate by precipitating with chalk, lime or double lime treatment, removal of hexavalent chromium by reduction with SO_2 in acid medium followed by precipitation of chromium hydroxide by lime treatment, destruction of cyanides by alkaline chlorination, removal of arsenic from segregated CO_2-scrubber liquid for isolation or recycling, removal of oils and grease by mechanical oil separation followed by settling and filtration through a bed of activated carbon or coke.

(6) Dairy

Sources of waste effluents : Spillage, washings of cans, equipments and floor waste from butter and cheese producing units.

Waste Characteristics : High BOD, high dissolved organics content, easily putrescibe. Foul odour. Amenable for biological treatment.

Treatment options : Equalization, skimming, oil and grease removal by grease traps, biological treatment, *e.g.*, aeration lagoon, oxidation pond or stabilization pond. Oxidation ditches with aeration and sludge return arrangements are recommended. Anaerobic lagoons many also be used. The treated effluent may be disposed for spray irrigation.

(7) Rubber industry

Source of effluents : Coagulation of rubber and latex, washings, etc.

Characteristics : High BOD, high suspended solids, characteristic odour, variable pH.

Treatment methods : Aeration, activated carbon adsorption treatment or chlorination to remove odour and phenolic substances, trickling filtration or activated sludge process.

(8) Soap industry

Source of effluents : Processing wastes, spill-overs of chemicals and floor washings.

Waste characteristics : Alkali, unsaponified oils, suspended solids. High BOD and COD.

Treatment methods : Skimming of floating fatty acids with help of trap tanks, Recovery of glycerine; Primary settlement, neutralization, Secondary treatment before discharge.

(9) Detergent Industry

Sources of effluents: Spill-overs of constituent chemicals, flocr washings.

Waste characteristics: Usually alkaline pH, contains alkyl benzene sulphonate or other surface active reagents, builders such as phosphates, borax, etc. Not amenable for biological oxidation. Affect aeration and reoxygenation of the stream, causes fronthing, toxic to flora and fauna.

Treatment methods: Flotation, skimming, precipitation with $CaCl_2$, equilization, pH adjustment, chemical coagulation, settling, activated sludge treatment.

(10) Cane Sugar Industry

A. Introduction

In sugar mills, sugar cane is washed, chopped and shredded. The juice is extracted by passing it through crushers containing a series of roller mills. The residue left is called bagasse, which is used as a boiler fuel or for manufacture of paper and boards. The juice is screened to remove floating impurities and milk of lime is added to increase the pH from 7.6 to 7.8 thereby preventive the inversion of sucrose to glucose and fructose and also to help in clarification by coagulation of the colloidal impurities by adding brontonite or some other coagulant aid. The mixture is heated to about 102°C using high pressure steam and allowed to settle. The clarified juice is bleached by treatment with sulpher dioxide and is sent for evaporation. The residue is sent to filter press or rotary drum vacuum filters and the juice in it is reclaimed. The mud from the filter press is known as press mud which is used as a manure. The clarified juice, which contains about 85% water, is evaporated to approximately 40% water in multiple effect evaporators to thick, pale yellow juice. This syrup is then fed into a single effect evaporator where the sugar is crystallized. This latter process is called "pan boiling". Sugar crystal nuclei are added at this stage to improve crystallization. This syrup now contains 10% water. The mixture of syrup and crystals known as "massecuite" is passed into centrifugal basket crystallisers to separate the crystallised sugar from the mother liquor. The spent mother liquor is called "mollasses", which is drained away separately. The sugar from the centrifugal basket is dried in a drier, graded and bagged.

B. Sources of Effluents

(i) **Cane wash water:** Cane is washed only if it is mechanically harvested. The wash-water contains suspended solids and some sugar.

(ii) **Mill House effluent:** This comprises of water used for cooling the bearings of milling machines which pick-up lubricants, it also includes spill-overs and floor washes.

(iii) **Wastewater from Washing Filter Press-cloth:** The cloth used in filter press has to be washed periodically to remove the mud that clogs the pores of the cloth. This wastewater possesses high BOD and contains lot of suspended solids.

(iv) **Effluent from evaporators:** Cooling and condenser waters contain sugar particles which are collected due to concentration of juice at multiple evaporators.

(v) **Effluent from boiler house:** This is generated by leakages from centrifuges and by periodic floor washing. This contains high BOD. The boiler house effluent also contains blow-down water.

(vi) **Spray overflow:** The cooling water is recycled from the pond. Excess water is sometimes let out as waste. This wastewater possesses low BOD and suspended solids.

(vii) **Molasses effluents:** Leakage and overflow from molasses storage tanks are responsible for the high BOD of sugar factory effluents.

(viii) **Off-season cleaning waters:** Off-season cleaning of juice heaters and other heating surfaces to remove scales is done with hot caustic soda solution, followed by dilute hydrochloric acid, followed by rinsing with water. These operations produce a large volume of waste effluents.

C. Characteristics of Effluents

The effluents from sugar factories are generally coloured, possess disagreeable odour, oil and grease, high BOD and suspended solids, and high dissolved carbonaceous matter.

D. Effects of Effluents

The immediate oxygen demand of sugar factory effluents results in depletion of dissolved oxygen content of the receiving water bodies and cause foul odours. The degradation of dissolved and suspended solids also contributes to obnoxious odours. The suspended solids block the drainages. The excess oil and grease also prevents aeration.

E. Treatment Options

Conventional biological treatment methods are not very effective for sugar factory effluents due to high concentration of carbohydrates. These techniques are not economical because of the seasonal nature of the sugar industry. Volatile organic acids produced during biological oxidation inhibit the biological activity and hence these acids have to be neutralised by lime.

Treatment of sugar factory effluents by lagooning or in oxidation ponds has been found to be most practical. However, a 2 stage oxidation was found to be most effective and economical. After equalisation (with a detention period of one day), the waste is first subjected to anaerobic digestion in an open pond of over 2.5 m for about 6 days. Then, in the second stage, the waste-effluent is subjected to aerobic oxidation in open shallow ponds of over 1 m depth. A detention period of 10 to 15 days is recommended.

F. Important parameters to be determined

Colour, odour, pH, temperature (in the field), BOD, COD, suspended solids, TDS, Sulfides, oil and grease, Boron, Sodium.

(11) Steel Industry

An integrated steel mill consists of the following major units:

A. Coal Washery

Coal is crushed, screened and washed with water to remove soil, dirt, soluble silica, and other unwanted impurities. Although most of the water used for washing is recycled, considerable quantity of water is discharged as waste. The impurities present in the coal washery effluent are suspended coal fines, shale, clay, Kaolin, calcite, pyrite, gypsum, etc.

B. Carbonization of Coal for manufacture of coke and byproduct recovery

Coke is produced by heating bituminous coal at about 900 to 1100°C in the absence of air in coke ovens. It is passed through various recovery units to recover ammonia, benzene, toluene, xylene, coal tar, phenol, napthalene, etc. The coke produced is quenched with cold water. About 30% of the quench water gets evaporated, while the remaining water contaminated with coke dust (known as breeze) is sent for recovery of the breeze, recirculated and then discharged as waste. This waste effluent from the quench tower contains toxic materials *e.g.* tar, ammonia, H_2S, ammonia, Phenols, Oils, Pyridine, benzole etc, and hence special attention is required for its treatment.

C. Production of Iron

A mixture of iron ore, coke and lime stone (or dolomite) is charged from the top of a blast furnace, while a blast of hot air is blown from the bottom. Iron ore is reduced to metallic iron, which in molten condition passes down to the bottom. This iron is called pig iron. Gaseous products formed are allowed to pass through dry dust catchers, then through wet scrubbers to clean the upflowing gases from dust, and then finally to electrostatic precipitators to remove fine dust particles.

Waste effluents are generated during (*i*) cooling furnace (*ii*) wet scrubbing of fuel gases and (*iii*) flushing of electrostatic precipitators.

The furnace cooling water is usually recycled. The wet scrubbing effluent contains suspended solids comprising of iron oxide, alumina, silica, carbon, lime, magnesia, coke and sinter. Flushing of electrostatic precipitator contributes only small quantity of effluents, but the finest dust particles are collected in appreciable quantity.

D. Steel Making

Depending on the final product required and the availability of raw materials, steel is produced from the pig iron using open hearth furnace or basic oxygen furnace or electric furnace.

Scrubbing water, used to separate dust particles from hot gases evolved during loading or heating the charge in the furnace, is the major effluent from steel making. This effluent is usually acidic and contains lot of fine suspended solids. Large quantity of water is used for cooling the furnace but it is uncontaminated and hence reused.

To improve the physical properties, molten steel produced is degassed under vacuum. The effluent from vacuum degassing contains, iron, manganese and fluoride.

The excess heat produced in the furnace is used by waste heat boilers. Consequently, waste effluents arise from boiler blow-down and from regeneration of ion-exchange resigns used to obtain boiler feed water.

E. Finishing

(*i*) **Rolling:** In rolling mills, steel ingots are converted to blooms, billets and slabs. During this process, the scales dropped into the scale flume, situated below the roll tables. The scales are then flushed to a "scale-pit". Further, during rolling, the rolls get heated and are cooled by water. This wastewater, which also is contaminated by oil and grease, also joins the "scale pit". Cold rolling effluents also contain oils in the form of emulsified rolling oils which get mixed with the water sprayed on the metal during rolling.

(*ii*) **Pickling:** This is an important treatment given to steel products before the final finishing step and is intended to remove dirt, greases and iron oxide scale which accumulate on the metal during fabrication, so as to enable good finish. Pickling comprises of dipping the steel products in dilute sulfuric acid or hydrochloric acid (about 20%) by weight). As the pickling process is continued, the acid becomes weak and there will be progressive build up of ferrous sulfate or ferrous chloride. After a particular stage, the pickle liquor gets unsuitable for use and is rejected as waste. This spent pickle liquor contains 5 to 15% of free acid and 2 to 10% of ferrous salts. This is a very strong waste with considerable pollution potential and its volume is 100 to 1000 liters per ton of steel.

(*iii*) **Rinsing:** About 250 to 20,000 liters of water are required for rinsing 1 ton of pickled steel. The rinse waters contain about 0.02 to 0.5% free acid.

(*iv*) **Other finishing operations:** Tinning, plating, galvanizing etc., are among the other final finishing operations. Each of these operations generate effluents contaminated with acids, alkalies, salts of Cr, Zn, Sn etc. along with fluorides, cyanides, etc.

F. Effluents from auxiliary and service units

(*i*) **Thermal Power Plant:** The major effluent from thermal power plant is the boiler blow-down water which is usually alkaline (pH about 11) and contains Phosphates, antifoaming additives and other organic compounds.

Wastes arising out of regeneration of ion-exchange resins, cleaning and flushing of boilers are also important. They contain brine, alkalies (*e.g.* NaOH), acids (*e.g.* HCl, H_2SO_4), Phosphates, and hydrazine.

Burning of pulverised coal in boiler furnace releases significant quantities of flyash which is scrubbed with water. The effluent therefore contains high concentration of suspended solids and some soluble salts of rare and heavy metals.

(*ii*) **Oxygen Plant:** Large quantities of oxygen are produced from air for use in blast furnace and other steel making furnaces. CO_2 and moisture from air are removed by passing through 0.5 N sodium, hydroxidesolution, which is ultimately disposed as waste.

(*iii*) **Foundry:** For controlling the dust, the air in the foundry is scrubbed with water spray or filtered through bag filters. Effluents from these operations comprise of excess suspended solids.

(*iv*) **Slab Plant:** Blast furnace slag is quenched in water. Scrubbing of the air to control dust and washing the conveyer belt generate waste effluents rich in suspended solids.

EFFECTS OF EFFLUENTS

1. The acidic spent pickling wastes destroy micro-organisms and inhibit self-purification of water streams. Fish and other aquatic animals are adversely affected by sudden change of pH conditions of the receiving river bodies. The spent pickle liquor being acidic is corrosive and causes damage to metallic materials, concrete structures, Pumps, etc.
2. The suspended and colloidal impurities, though not toxic, interfere with the self-purification of streams by reducing the photosynthetic activity of water plants and by smothering benthic organisms. Abrasive materials present in the suspended matter may choke the gills of fishes and thereby cause damage to fisheries. The black and opaque coal washery effluents discolour the stream and render it unfit for most uses.
3. The phenolic substances present in Coke Oven effluents impart unpleasant taste and odour to the receiving waters. They also taint the fish by imparting phenolic taste to the flesh of fish. Higher concentration of phenol is toxic to fish.
4. The biodegradable organic matter-present in Coke Oven effluents deplete the dissolved oxygen in the receiving water courses and cause obnoxious odours and effect fish and other aquatic life forms.
5. The oils and grease present in the effluents interfere with aeration, affect natural purification of water courses and produce unsightly appearance by the formation of ugly oil slike and irridiscent colours.
6. The coal tar present in Coke Oven affluents choke the sewers and interfere with treatment processes.
7. The hot effluents arising from cooling processes decrease the dissolved oxygen content of water, cause excess algal growth, increase the virulence of pathogens and decrease the viscosity of water thereby accelerating the settlement of suspended particles.
8. Other impurities present in steel plant effluents such as ammonia, cyanide, sulfide, iron and manganese salts, etc., cause undesirable or deleterious effects.

In view of the above, the deleterious impurities from different steel mill effluents have to be properly treated.

Treatment options:
1. **Effluents from coal washery:** High suspended solids present in coal washery effluent can be best removed by clariflocculation using alum, ferric sulfate or ferric chloride as coagulants and starch or polyelectrolytes as coagulant aids.
2. **Effluents from coal carbonization:** After removing ammonia and phenols, the Coke Oven effluent is subjected to biological treatment such as activated sludge process or trickling filtration.
3. **Effluents from Blast Furnace operations for iron production:** Fuel dust from this effluent is removed by chemical coagulation (using alum, $FeCl_3$ or $Fe_2(SO_4)_3$ followed by sedimentation. The sludge is dewatered in vacuum filters and disposed for land fill. The clarified effluent is recycled for cooling and other purposes.

4. **Effluents from steel making:** These are settled in primary settling tanks and subjected to coagulation and flocculation in clariflocculators. The clarified effluent is reused.
5. **Effluents from metal finishing operations:** The coarse scale particles from wastewaters from roughing stand are settled in a primary settling tank. The effluent is then passed to secondary settling tank with 2 to 3 compartments having oil skimming facility. The settled mass and skimmed oils are sent to furnace for burning. Effluents from finishing stand and scarfer machines contain fine scales. This is sent to settling tanks and then to clariflocculators for coagulation and flocculation. Further purification is achieved by passing through sand filters.

   Waste effluents from cleaning lines are passed through primary gravity separators for removing oils and settleable matter and then subjected to coagulation using $FeCl_3$ or $Fe_2(SO_4)_3$. The cooling water from reheating furnaces is relatively clear but hot. This is cooled and reused.
6. **Pickling Waste:** This is neutralised by treating with lime and aerated to form Ferric hydroxide, which is removed as sludge.

   The pickling rinse water is usually treated with other cold mill wastes contaminated with oils so as to remove the oils by the ferric hydroxide flee formed.

   In some steel mills, the pickle liquor is concentrated by evaporation and placed in a reactor into which HCl gas is bubbled. $FeCl_2$ or H_2SO_4 are thus recovered. The $FeCl_2$ is roasted to produce iron oxide and HCl gas. The HCl gas is scrubbed and reused.
7. **Effluents from auxiliary and service units:** The power plant wastes such as brines, acids and alkalis are collected in equalization tanks and neutralised if necessary before disposal. Other wastes are subjected to plain sedimentation and then recycled or disposed.

Parameters to be determined

Colour, odour, pH; total suspended solids, BOD, oils and greases, free ammonia, cyanide, sulfide, phenols, thiocyanate, B, Na.

(12) Cement Industry

The raw materials required for the manufacture of portland cement are

(*a*) Calcareous materials (which supply lime) *e.g.*, lime stone, cement rock, chalk, mart, marine shells and waste calcium carbonate from industrial processes.

(*b*) Argillaceous materials (which supply silica, alumina and iron oxide) *e.g.*, clay, shale, blast furnace slag, ash and cement rock.

(*c*) Pulverised coal as fuel.

Portland cement is manufactured by Dry Process, Wet Process or Semi-Dry Process.

In the Dry process, the calcareous and argillaceous materials are crushed, dried, mixed in proper proportion, pulverised in tube mills and homogenised in a mixing mill with the help of compressed air. This dry mix is introduced into the upper end of a rotary kiln while a blast of burning coal dust is blown from the lower end of the rotary kiln. The raw materials react at a temperature of 1400 to 1600°C to form cement clinkers. The clinkers are cooled and pulverised with about 2% gypsum. The cement is then stored in concrete silos.

In the Wet process, the raw materials are finely ground and blended in the desired proportion. The mix is converted into a slurry containing 30-40% water. The slurry is introduced from the upper end of the rotary kiln while it meets blast of burning coal dust from the lower end. Cement is formed at a temperature of 1400-1600°C in the kiln.

In Semi-dry process, the raw materials are ground dry and nodulised with 10-14% water. The nodules are fed on a travelling grate where they get dried and preheated before entering a short rotary kiln where they are burnt to form cement clinkers. Cement industry contributes much to air pollution whereas liquid effluents do not cause much of problem.

The cement clinkers, after grinding to a fine powder are cooled by water-jacketed heat exchangers. This cooling water is usually recycled.

Wet scrubbing of kiln dust yields an effluent that possesses a high pH value, alkalinity, suspended and dissolved impurities. This effluent is clarified and the sludge obtained is recycled to the kiln.

The treatment systems used for cement industry effluents are majorly
   (i) cooling towers or ponds to reduce the temperature of the cooling water, and
   (ii) clarifiers for separation of solids from wet scrubber effluents or from dry dust leaching.

The air pollution due to fuel gases should be controlled by using suitable chimneys and efficient dust catching devices.

18.4. SOIL POLLUTION

Soil is a very important constituent of lithosphere. It plays a vital role in the production of food for the sustainance of human beings and animals. Soil mainly comprises of a mixture of organic and inorganic matter and water. The inorganic mineral constituents of soil include complex mixture of silicates of Na, K, Ca, Al and Fe; oxides of Fe, Mn and Ti; and carbonates of Ca and Mg. The common mineral constituents of soil include finely divided quartz (SiO_2), othroclase ($KAlSi_3O_8$), albite ($NaAlSi_3O_8$) epidote ($4CaO.3(AlFe)_2O_3.FeO(OH)$) and magnetite ($Fe_3O_4$). The clay minerals in soil, which are essentially hydrates, aluminium silicates and iron silicates, which bind cations e.g., Na^+, K^+, Ca^{2+} and Mg^{2+}. Since these cations are not leached out by water, they serve as plant nutrients. The organic matter in the soil, which hardly constitutes 5% of it, mainly determines the productivity of the soil. It consists of biologically active components such as polysaccharides, nucleotides, organosulphur compounds, sugars and humic materials. The clay minerals and humus present in the soil possess a very high cation-exchange capacity and thus help in supplying essential trace metals to the plants as nutrients. The humus is present in soil in colloidal form. The water present in the soil helps in the biological activity. The important micro-organisms present in soil are algae, protozoa and actinomycetes which render the soil productive. These micro-organisms play an important role in mineral and fertilizer break-down. Some types of fungus readily produce citric acid in soil and then solubilizes Si, Al, Fe, and Mg in basalt. Some types of fungus convert organic nitrogen to inorganic nitrogen.

The essential macro-nutrients required for plant growth are C, H, O, N, P, S, K, Ca and Mg. C, H and O are supplied by soil, water and atmosphere. Nitrogen may be obtained by some plants, directly from the atmosphere, with the help of nitrogen-fixing bacteria. However, nitrogen, phosphorous and potassium (N, P, K) are usually supplied to soil as fertilizers.

The essential micro-nutrients for plants are boron, Na, Cu, Zn, Fe, Mn, V, Mo etc. Most of them are important components of essential enzymes while some of them contribute for photosynthetic activity of the plants. Although these elements are essential in trace levels, they may prove to be toxic at higher levels.

Pollution of soil takes place due to the waste products from industrial, domestic and agricultural activities. High levels of Zn and Pb in this soil around Zinc and lead mines and smelters, accumulation of sulphate in the soil due to the combustion of fuels containing sulphur, and accumulation of lead particulates from automobile exhausts on soil around the high-traffic density high-ways are examples of soil pollution due to anthropogenic activities. Indiscriminate release of untreated or inadequately treated industrial waste-water into nearby water sources and land also cause soil-pollution.

Sewage sludge containing high levels of toxic metals may also contribute to soil-pollution.

Excessive use of synthetic fertilizers may lead to excessive release of nitrates and phosphates in the soil, which may cause water pollution. Nitrates and phosphates may be leached out of soil by run-off waters and pollute the nearby surface water or groundwater sources. High concentrations of nitrates are objectionable in drinking water because they may be reduced to nitrites in the stomach and cause a disease called methemoglobinemia (i.e. blue babies). Fertilizers and pesticides applied to crops are retained in the soil in considerable quantities. They enter into cyclic environmental processes such as absorption by soil, leaching by water, etc. Pesticides contaminate not only the lithosphere but also the atmosphere. Pesticides, including insecticides, herbicides, fungicides and rodenticides, are persistent pollutants. Due to interaction of lithosphere and biosphere, they may enter the food chain and pose serious health hazards. They are cumulative poisons. Some of them undergo metabolic formation and biodegradation. The degradation products of pesticides so formed are more dangerous than the parent compounds. Some of these pesticide residues are carcinogenic. The metabolic products too are toxic. Some arsenic pesticides if used indiscriminately, may render the soil permanently infertile. The rate of degradation of pesticides depends upon their nature and chemical structure. The pesticide residues in soil may be taken up by plants and cause phytotoxicity. They may be added to the aquatic environment and enter the food-chain.

The pesticides of common use may be classified as :

(a) Chlorinated hydrocarbons (e.g., Aldrin, DDT, Dieldrin, Lindane, BHC, etc.).

(b) Organo-phosphorous compounds (e.g., methyl or ethyl parathion, malathion, guthion, etc.).

(c) Carbamate compounds (e.g., carboryl or sevin, zectrion, etc.)

(d) Inorganic compounds (e.g., $CuSO_4$, $NiCl_2$, As_2O_3, PbO_2, etc.).

(e) Miscellaneous compounds (e.g., Organo-mercurials, 2,4D; 2,4,5T etc.).

Analysis of pesticides is usually done by Thin layer chromatography (TLC), Gas-liquid chromatography (GLC), High performance liquid chromatography (HPLC), U.V.-I.R. spectrophotometry, NMR spectrometry, GC-MS technique, and Polarography. Carbamate pesticides are best analysed by TLC while chloro-organic pesticides are analysed by GLC using electron capture detectors.

Soil pollution is receiving greater and greater attention due to its direct impact on public health.

Soil pollution was originally defined as the contamination of the soil system by considerable quantities of chemical or other substances, resulting in the reduction of its fertility or productivity with respect to the qualitative, and quantitative yield of the crops. However, if some of the contaminants are such that if they are taken up by the plants (with or without any detrimental effect on them), and enter into the food chain and impart detrimental or toxic effects on the consumers, then that also should be treated as soil pollution.

The important sources of soil or land pollution include (a) indiscriminate dumping of untreated or inadequately treated domestic, mining and industrial wastes (solids or liquids) on land (b) indiscriminate or un-regulated use of chemical fertilizers and/or pesticides (c) fall-out of gaseous and particulate air-pollutants from mining and smelting industries, smoke-stacks, automobile exhausts, etc. (d) soil-erosion due to deforestation, unplanned irrigation and unscientific agricultural practices (e) disposal of huge quantities of fly-ash and bottom ash from coal-based thermal power plants. It is estimated that two lakh tons of ash is being deposited per day by the thermal power plants in our country, thereby rendering large areas of agricultural land unfit for cultivation due to pollution and degradation of soil. The toxic trace metals leached out of the ash dumps pollute the nearby surface and ground waters.

Immediate attention should be paid for controlling land/soil pollution. The obvious approaches to achieve this objective include (1) population control (2) launching extensive afforestation and community forestry programmes (3) stringent and deterrent measures against deforestation (4) stringent pollution control legislations and more effective and powerful administrative machinery for their implementation (5) banning the use of highly toxic and persistent synthetic chemical pesticides or at least regulating their use only for special purposes under adequate monitoring (6) encouragement to use bio-pesticides rather than highly toxic chemical pesticides (7) construction of suitable sanitary landfills by municipal corporations for disposal of municipal wastes (8) construction of security land-fills for permanent disposal of hazardous and recalcitrant industrial wastes (9) imparting public awareness regarding bad effects and health hazards due to environmental pollution and measures to be taken to combat them with the help of mass media and curricular measures at all levels of formal and informal education, and (10) involving the public at large in environmental protection measures.

18.5. HAZARDOUS WASTES AND TREATMENT TECHNOLOGIES :

According to the working definition given by WHO, UNEP and World Bank given in 1987, a "Hazardous Waste" is defined as any waste, excluding domestic and radioactive wastes, which because of its quantity, physical, chemical and infectious characteristics can cause significant hazard to human health or the environment when improperly treated, stored, transported or disposed. Hazardous wastes included solids, liquids, gases and sludges that may arise from a wide range of industrial, commercial and agricultural sources.

As per the classification made by the United States Environmental Protection Agency (USEPA), wastes are designated as hazardous if they exhibit the following four characteristics viz., ignitability, corrosivity, reactivity and toxicity as shown in Table 8 given below. The Table also gives the methods recommended for measurement of the four characteristics. Toxicity is evaluated in terms of leachable chemical constituents present in the waste. For regulatory control, USEPA recommended toxicity characteristic leaching procedure (TCLP) and prescribed concentration limits on the leachate for heavy metals and organic chemicals having proven chronic toxicity. The levels prescribed for these chemical constituents in the leachate are 100 times the concentration of constituents known to exhibit chronic toxicity.

Table 8 : Parameters of Hazard Potential and Their Measurement

Parameter	Method for measurement	Defining characteristics
Ignitability	ASTM Standard : D-3278-1978	Wastes which spontaneously ignite in dry or moist air at or below 60 C.
Corrosivity	ASTM Standard : pH G-1-1972	An aqueous solution having less than 2 or greater than 12, or a liquid that corrodes steel at rate greater than 6.35 mm per year at 55° C.
Reactivity	U.S. Federal Register, Rules and Regulations, 45, No. 98, Sec. 261. 23, 1980	A Solid waste: - that is unstable and readily undergoes violent change without detonating. - reacts violently with water - generates toxic gases HCN, H_2S, etc. when exposed to favourable pH conditions.
Toxicity	U.S. Federal Register, March 29, 1990	Wastes which release toxic materials on leaching in excess of the given concentration to pose a

substantial hazard to human health or environment as measured by the Toxicity Characteristics Leaching Procedure (TCLP). The list includes selected volatile and semivolatile organics, pesticides, herbicides and heavy metals.

Table 9 : Toxicity Characteristics of Leachate Constituents and the Methods for their Measurement

EPA HW NO.	Constituent (mg/l)	CAS No.	Chronic Toxicity Reference	Regulatory Level
D004	Arsenic	7440-38-2	0.05	6.0
D005	Barium	7440-39-3	1.0	100.0
D018	Benzene	71-34-2	0.005	0.5
D006	Cadmium	7440-43-9	0.01	1.0
D019	Carbon Tetrachloride	56-23-5	0.005	0.5
D020	Chlordane	57-74-9	0.0003	0.03
D021	Chlorobenzene	108-90-7	1.0	100.0
D022	Chloroform	67-66-3	0.06	6.0
D007	Chromium	7440-47-3	0.05	5.0
D023	o-Cresol	95-48-7	2.0	200.0
D024	m-Cresol	108-39-4	2.0	200.0
D025	p-Cresol	106-44-5	2.0	200 0
D026*	Cresol		2.0	200.0
D016	2, 4-D	94-75-7	0.1	10.0
D027	1, 4-Dichlorobenzene	106-46-7	0.075	7.5
D028	1, 2-Dichloroethane	107-06-2	0.005	0.5
D029	1, 1-Dicholoroethylene	75-35-4	0.007	0.7
D030	2, 4-Dinitrotoluene	121-14-2	0.0005	0.13
D012	2, Endrin	72-20-8	0.0002	0.02
D031	Heptachlor	76-44-8	0.00008	0.008
D032	Hexachlorobenzene	118-74-1	0.0002	0.13
D033	Hexachloro-1, 3-butadiene	87-68-3	0.005	0.5
D034	Hexachloroethane	67-72-1	0.03	3.0
D008	Lead	7439-92-1	0.05	5.0
D013	Lindane	58-89-9	0.004	0.1
D009	Mercury	7439-97-6	0.002	0.2
D014	Methoxychlor	72-43-5	0.1	10.0
D035	Methyl ethyl ketone	78-93-3	2.0	200.0
D036	Nitrobenzene	98-95-3	0.02	2.0
D037	Pentachlorophenol	87-86-5	1.0	100.0
D038	Pyridine	110-86-1	0.04	5.0
D010	Selenium	7782-49-2	0.01	1.0
D011	Silver	7440-22-4	0.05	5.0

*If o–, m–, p-Crespol cannot be differentiated, total Cresol is to be used.

D039	Tetrachloroethylene	127-18-4	0.007	0.7
D015	Toxaphene	8001-35-2	0.005	0.5
D040	Trichloroethylene	79-01-6	0.005	0.5
Do41	2, 4, 5-Trichlorophenol	95-95-4	4.0	400.0
D042	2, 4, 6-Trichlorophenol	88-06-2	0.02	2.0
D017	2, 4, 5-TP (Silvex)	93-72-1	0.01	1.0
D043	Vinyl Chloride	75-01-4	0.002	0.2

Table 10 : Constituents Considered Hazardous for land disposal

Hazardous constituent	TCLP extract concentration, mg/l
Acetone	2.0
n-Butyl alcohol	2.0
Carbon disulfide	2.0
Carbon tetrachloride	0.1
Chlorobenzene	2.0
Cresols	2.0
Cyclohexanone	2.0
Ethyl acetate	2.0
Etyly benzene	2.0
Ethyl ether	2.0
HxCDD (all hexachlorodibenzo-p-dioxins)	0.001
HxCDF (all hexachlorodibenzofurans)	0.001
Isobutanol	2.0
Methanol	2.0
Methylene chloride	1.2
Methyl ethyl ketone	2.0
Methyl isobutyl ketone	2.0
Nitrobenzene	0.09
PeCDD (all pentachlorodibenzo-p-dioxins)	0.001
PeCDF (all pentachlorodibenzofurans)	0.001
Pentachlorophenol	1.0
Pyridine	0.7
TCDD (all tetrachlorodibenzo-p-dioxins)	
TCDF (all tetrachlorodibenzofurans)	0.001
Tetrachloroethylene	0.015
2, 3, 4, 6-Tetrachlorophenol	2.0
Toluene	2.0
1, 1, 1-Trichloroethane	2.0
1, 2, 2-Trichloro-1, 2, 2-trifluoroethane	2.0
Trichtoroethylene	0.1
Trichlorofluoromethane	2.0
2, 4, 5-Trichlorophenol	8.0
2,4, 6-Trichlorophenol	0.04
Xylene	2.0

A total of 8 heavy metals and 31 volatile and semi-volatile organic chemicals including pesticides and herbicides are included in the list shown in Table 9. The fortieth chemical added to the list is cresol and to be used only if its isomers cannot be separated. USEPA has also proposed the use of TCLP for land disposal restictions of hazardous waste. Table 10 gives a list of 34 organic constituents which have been designated or considered hazardous for landfill disposal if their concentrations in TCLP extract are equal to or grater than the limit indicated in Table 10.

Proximate Analysis :

The proximate analysis includes determination of
(a) moisture, volatile solids and ash content
(b) elemental composition (C, H, N, S, P, F, Cl, Br, I)
(c) heating value of the waste
(d) viscosity or physical form

The proximate analysis data provides valuable information regarding the physical form of the waste and an approximate mass balance on its composition. This also helps in deciding the most appropriate treatment procedure for the waste. It also helps in predicting the likely combustion products (e.g. NO_x, SO_x, P_2O_5, helogens, and hydrogen halides) which are of significance.

Survey Analysis :

This provides an overall description of the sample in terms of major types of organic compounds and inorganic elements present in the waste and the overall chemistry of the waste sample. This information also helps in identifying as to which of the hazardous constituents listed under 40 CFR part 264 and 270 Appendix VIII (us Federal Register, 1986) dealing with groundwater monitoring and hazardous waste delisting, are present in the particular waste. This also helps to predict the major POHs present in the waste and to identify the hazardous byproducts or products of incomplete combustion (PICs) or other possible emissions that may require sampling and analysis.

Directed Analysis :

The directed analysis is mainly meant for the measurement of designated POHCs in the waste samples. This information is helpful to decide which organic compounds should be analysed in the ash and which PICs are to be analysed in stack emissions.

Table 11 : Methods for chemical Analysis of Water and Wastes

Priority pollutants	EPA method No.	Method of analysis
Purgeable halocarbons	601	Purge-and-trap (PAT), GC, detection with a Hall (electrolytic) detector
Purgeable aromatics	602	PART, GC, photo ionization detector
Acrolein and acrylonitrile	603	PAT, GC, flame-ionization detector
Phenols	604	Extraction, Kuderna-Danish (K-D) concentration, GC, flame-ionization-electron-capture detection
Benzidines	605	Extraction, concentration, HPLC, electro-chemical detection
Phthalate esters	606	Extraction, Florisil or alumina cleanup K-D concentration, GC, flame-ionization-electron-capture detection
Nitrosamines	607	Extraction, Florisil or alumina clean up K-0 concentration, GC. flame-ionization-electron capture detection
Organochlorine pesticides	608	Extraction, Florisil or Alumina and PCDs

Nitroaromatics and isophorone	609	cleanup, K-D concentration, GC, detection Extraction, K-D concentration, GC, flame ionization-electron-capture detection
Polynuclear aromatic	610	Extraction, K-D concentration, GC, hydrocarbons flame-ionization detection or HPLC/ UV fluorescence detection
Haloethers	611	Extraction, solvent exchange, K-D concentration, GC, electron-capture detection
Chlorinated hydrocarbons	612	Extraction, solvent exchange, K-D concentration, GC, electron-capture detection
2, 3, 7, 8-Tetrachlorodibenzo 8-TCDD)	613	Spiking with labelled 2, 3, 7, 8- -p-dioxin (2, 3, 7, TCDD, extraction, solvent exchange, K-D concentration, analysis by GC/MS
Yolatile organic compounds (purgeables)	624	Purge-and-trap, analysis by GC/MS
Semivolatile organic compounds (base-neutral and acid-extractable compounds)		Extraction, K-D concentration, analysis by GC/MS

Table 12 : Recommended AAS Analysis Methods for trace Metals

Element	SW 846 method No.	Description
Ag	7760	Direct aspiration
	7761	Graphite furnace
As	7060	Graphite furnace
	7061	Gaseous hydride method
Ba	7080	Direct aspiration
	7081	Graphite furnace
Cd	7130	Direct aspiration
	7131	Graphite furnace
Cr	7190	Direct aspiration
	7191	Graphite furnace
	7195	Hexavalent Cr: coprecipitation
	7196	Hexavalent Cr: colorimetric
	7197	Hexavalent Cr: chelation-extraction
	7198	Hexavalent Cr: differential-pulse polarography
Hg	7470	Hg in liquid waste (Manual cold-vapor technique)
Pb	7420	Direct aspiration
	7420	Graphite furnace
Se	7740	Graphite furnace
	7741	Gaseous hydride method

Analytical Methods :

USEPA (1980) document listed 717 hazardous substances (HS) composed of 611 unique chemical compounds and 06 waste streams. Over 300 POHCs are considered among the toxic and hazardous categores of organic compounds. Owing to the complexity of the hazardous wastes, analysis and quantification of specific organic compounds is rather difficult. Mass spectrometer (MS) coupled with gas chromatograph (GC)/ liquid chromatograph (LC), Atomic Absorption Spectorscopy (AAS) and Inductively coupled Atomic Emission Spectroscopy are the techniques of choice used for the analysis of various pollutants. Table 11 and 12 respectively indicate the methods used for chemical analysis of wastewaters and AAS analysis of trace metals.

Hazardous Waste Treatment Technologies :

The following treatment technologies are available for the treatment of hazardous wastes :
- (a) **Physical methods** : Physical treatment processes include gravity separation, phase change systems such as air and steam stripping of volatiles from liquid wastes, adsorption, reverse osmosis, ion-exchange, electrodialysis.
- (b) **Chemical methods** : Chemical methods usually aim at transforming the hazardous waste into less hazardous substances using techniques such as pH neutralization, oxidation or reduction, and precipitation.
- (c) **Biological methods** : Biological treatment methods use micro-organisms to degrade organic pollutants in the waste stream.
- (d) **Thermal Methods** : Thermal destruction processes that are commonly used include incineration and pyrolysis. Incineration is becoming a more preferred option. In pyrolysis, the waste material is heated in the absence of oxygen to bring about chemical decomposition.
- (e) **Fixation/Immobilization /Stabilization** techniques involve dewatering the waste and solidifying the remaining material by mixing it with a stabilizing agent such as portland cement or a pozzolanic material, or vitrifying it to create a glassy substance. For hazardous inorganic sludges, solidification process is generally used.

a) Physical Methods :

- (i) **Sedimentation** : - Gravity settling and flotation (natural or by employing finely devided air bubbles) are the simplest physical treatment systems used to achieve solid - liquid separation. Several types of special sedimentation tanks and clarification tanks are designed to encourage solids to settle at the bottom which can be collected as a sludge. In flotation, some solids can be floated with the help of tiny air bubbles which can be skimmed from the surface. The sludges separated by any of the above two methods can be further concentrated by evaporation, filtration or centrifugation.
- (ii) **Adsorption** :- Small quantities of dissolved organic hazardous wastes can be removed by adsorption. Granualr activated carbon (GAC) which has an enormous surface area (about 1000 m^2/g) is most commonly used as an adsorbent. Contaminated water is allowed to trickle down through the GAC packed in a series of vessels. The hazardous organics in polluted water are adsorbed on the porous GAC matrix. The GAC filter after some time gets clogged with the adsorbed contaminats and hence must be replaced or regenerated or disposed properly. Regeneration is usually done by burning the contaminats from the surface of the adsorbent granules or using a solvent.
- (iii) **Aeration** : This technique is used to drive volatile contaminants out of waste solution. Contaminated waste solution is sprayed downward through a suitable packing material in a tower while air is blown upward which carries away the volatile materials with it. Such packed - tower air - stripper can remove over 95% volatile organic compounds (VOCs) such as trichloroethylene, tetrachloroethylene, benzene, toluene, trichloroethane, and other organics derived from solvents.

In the induced - draft strippen contaminated water is sprayed through a series of nozzles horizontally from the sides of a chamber, while air passing through it draws off the volatile.

By passing contaminated water first through an air stripper and then through granulated activated carbon adsorption system, many volatile and non-volatile organic contaminants can be removed from waste water to undetectable levels.

(iv) **Ion-exchange:**- This technique is often used to remove toxic metal ions from solution.

(v) **Electrodialysis :** - This technique uses ion - selective membranes and an electric field to separate anions and cations in solution. This technique which was mostly used for desalination of brackish water, is now used in the field of hazardous waste treatment. Removal of metal salts from plating rinses is an example.

(vi) **Reverse osmosis:** - This device uses pressure to force contaminated water through the pores of a semipermeable membrane while restricting the passage of larger contaminant molecules. This process is particularly effective for removing dissolved solid contaminants from waste waters.

(b) Chemical Methods :

Chemical methods for treating hazardous waste have the dual advantages of converting it to less hazardous forms and also to recover useful by-products in some cases which may offset a part of treatment costs.

(i) **Neutralization :** As per Resource Conservation and Recovery Act (RCRA), hazardous wastes comprise of such wastes which are corrosive and having a pH of less than 2 or more than 12.5. Such wastes can be rendered less hazardous by chemical neutralization. Acidic wastewaters are usually neutralized with slaked lime [$Ca(OH)_2$] in a continuously stirred chemical reactor provided with a feed back control system to monitor pH and control the rate of addition of slaked lime. Alkaline wastewaters may be neutralized by adding acid direcly or by bubbling in gaseous CO_2 which forms carbonic acid (H_2CO_3). CO_2 has an advantage that it is generally available readily in the exhaust gas from any combustion process at the treatment site. Simultaneous neutralization of acidic and caustic waste can be accomplished in the same reaction vessel.

(ii) **Chemical precipitation :** The solubility of toxic metals in a waste stream can be decreased, leading to the formation of a precipitate that can be removed by settling and filtration. Precipitation of toxic metals as their hydroxides using lime is the most common metal removal process used.

$$M^{2+} + Ca(OH)_2 \longrightarrow M(OH)_2 + Ca^{2+}$$

Each metal has its own optimum pH for precipitation as hydroxide. Each metal hydroxide has its own pH at which its solubility is minimum. Therefore, it is tricky to control precipitation of metals from a hazardous waste containing a mixture of toxic metals.

Precipitation of metal ions in hazardous waste as sulfides has an advantage to achieve much lower metal ion concentrations in the effluent because metal sulfides have much lower solubilities as compared to the corresponding metal hydroxides. However, sulfide precipitation has a disadvantage because of the possibility of formation of odorous and toxic H_2S gas.

(iii) **Chemical reduction-oxidation :** Redox reactions offer another important alternative for chemical treatment of hazardous wastes. Trivalent chromium ions are far less toxic and more easily precipitated than hexavalent chromium. Therefore, chromium (VI) from electroplating effluents is reduced to Cr (III) using a reducing agent such as SO_2 :

$$3SO_2 + 3 H_2O \longrightarrow 3 H_2SO_3$$
$$2CrO_3 + 3H_2SO_3 \longrightarrow Cr_2(SO_4)_3 + 3 H_2O$$

Another important redox treatment system commonly used in metal finishing industry is the oxidation of cyanide wastes. The cyanide is oxidized to less toxic cyanate using alkaline chlorination. Further, chlorination oxidizes the cyanate to CO_2 and N_2, thereby accomplishing total destruction of the hazardous cyanide.

$$NaCN + Cl_2 + 2\,NaOH \longrightarrow NaCNO + 2\,NaCl + H_2O$$
$$2NaCNO + 3\,Cl_2 + 4\,NaOH \longrightarrow 2\,CO_2 + N_2 + 6\,NaCl + 2\,H_2O$$

Wastes that can be successfully treated via oxidation include benzene, phenols, cyanide, As, Fe, Mn and most organics while those which can be treated via reduction include Cr (VI), Hg, Pb, Ag, Chlorinated organics like Polychlorinated biphenyls (PCBs), and unsaturated hydrocarbons.

(c) Biological Treatment :

As described earlier, biological treatment systems use micro-organisms, mainly bacteria, to metabolize organic matter, converting it to CO_2, water and new bacterial cells. The microbes need a source of carbon and energy, which they can get from the organics that they consume, as well as nutrients such as N and P. They are sensitive to temperature and pH. Some of them need oxygen. Although, living organisms are susceptible to toxic substances, it is surprising that most hazardous organics are amenable to biological treatment, provided that the proper distribution of the proper organisms can be established and maintained. For any given organic substance, there may be some organisms that accepet that substance as an acceptable food supply, while others may find it toxic. Further, organisms that flourish with the substance at one concentration may die when the concentration is increased beyond some critical level. Moreover, even though a mocrobial population has been established to handle a particular type of organic waste, it may be destroyed if the characteristics of the waste are changed too rapidly. However, if changes are made slowly enough, selection pressures may allow the microbial consortium to adjust to the new conditions and thereby retaining their effectiveness.

Biological treatment of a hazardous wastewater (e.g. leachates from hazardous waste landfills) is just one of the steps in an overall treatment system, as shown in the flowsheet given below :

General flowsheet for the treatment of liquid hazardous waste

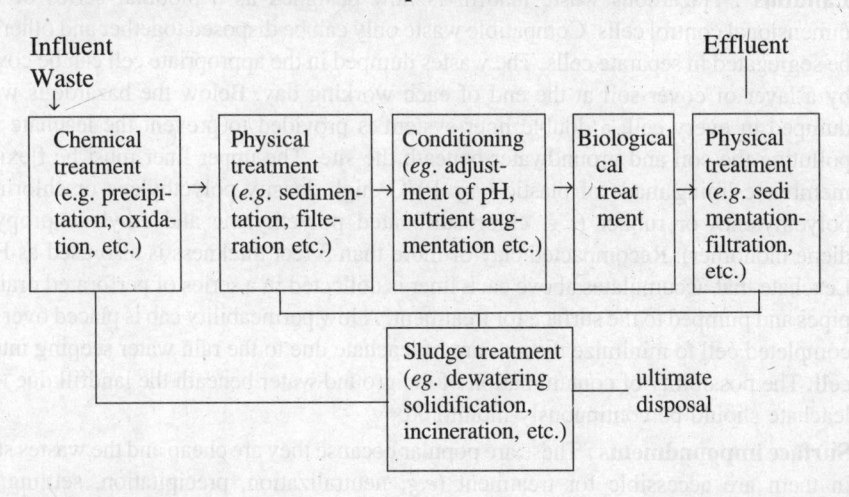

Influent Waste → Chemical treatment (e.g. precipitation, oxidation, etc.) → Physical treatment (e.g. sedimentation, filteration etc.) → Conditioning (e.g. adjustment of pH, nutrient augmentation etc.) → Biological treatment → Physical treatment (e.g. sedimentation-filtration, etc.) → Effluent

Sludge treatment (e.g. dewatering solidification, incineration, etc.) → ultimate disposal

(ii) "In Situ" Bio-degradation :

To clean a contaminated aquifer, the following techniques are available: One is to pump out the ground water, treat it, and then either reinject it back into the ground or find some acceptable way to reuse it, or dispose off the treated water in a suitable way.

Another method is to treat the contaminated soils above the aquifer by removing huge quantities of the soil from the site, then treating and disposing them elsewhere.

A relatively new promising alternative to the above two approaches is to move the treatment system to the site of contaminated soil and water. This can be accomplished by "In situ biodegradation" wherein suitable bacteria are used to degrade the organic contaminants in the soil and groundwater on the contaminated site itself. The following two approachers are used to achieve this process : First one is to enhance the environment of existing microbial populations by supplying necessary nutrients to the contaminated aquifer. Oxygen supply can be enhanced if necessary be injecting an oxidant such as H_2O_2 or by forcing air through wells with diffusers. The second approach is to after the underground microbial population by seeding with new micro-organisms that have acclimated to the pollutants to be degraded. These new microbes can be selected on the basis of laboratory studies. Genetically altered microbes may be of great potential for this purpose. "In situ" biodegradation has been successfully used to treat soil contaminated with gasoline and diesel oil. It also shows great promise to treat trichloroethylene, tetrachloro-ethylene and 1, 2 - dichloroethylene which are the most commonly found contaminants in underground water supplies.

(d) Incineration :

This is a technology of choice for treating many types of hazardous organic wastes present in soils and other solids, liquids, gases, slurries and sludges. However, this process is not capable of destroying inorganic wastes, although it can concentrate them in ash, which can be transported and disposed more easily. A number of types of hazardous waste incinerators are available. However, liquid injection incinerator and the rotary kiln incinerator are the two designs which are most popular. In spite of numerous controls, hazardous waste incinerators may emit noxious gases, products of incomplete combustion, odors, and particulates.

(e) Land disposal :

The various land disposal techniques used for containing hazardous wastes included landfills, surface impoundments, injection wells, and waste piles.

(*i*) **Landfills :** Hazardous waste landfill is now designed as a modular series of three dimensional control cells. Compatible waste only can be disposed together and others can be segregated in separate cells. The wastes dumped in the appropriate cell can be covered by a layer of cover soil at the end of each working day. Below the hazardous wastes dumped in every cell, a double liner system is provided to prevent the leachate from polluting the soil and groundwater beneath the site. The upper liner must be flexible - membrane lining made of plastic (*e.g.* PVC, high density polyethylene or chlorinated polyethylene) or rubber (*e.g.* chlorosulfonated polyethylene and ethylene propylene diene monomer). Recompacted clay of more than 3 feet thickness is also used as FML. Leachate that accumulates above each liner is collected in a series of perforated drainage pipes and pumped to the surface for treatment. A low permeability cap is placed over each completed cell to minimize the amount of leachate due to the rain water seeping into the cell. The possibility of contamination of the ground water beneath the landfill due to the leachate should be continuously monitored.

(*ii*) **Surface impoundments :** These are popular because they are cheap and the wastes stored in them are accessible for treatment (*e.g.* neutralization, precipitation, settling, and biodegradation) during storage. Surface impoundments are nothing but excavated or diked areas used to store liquid hazardous wastes. Storage in these facilities is usually temporary unless the impoundment is designed to be eventually closed as a landfill.

Modern surface impoundment facilities are provided with liners, leachate collection system and monitoring programs similar to those of landfills.

(iii) **Underground injection :** The most popular way for disposal of liquid hazardous wastes is to force them underground atleast below 700 m from the surface through deep injection wells. However, stringent regulations have to be followed to ensure prevention of contamination of ground water.

QUESTIONS

1. Give a detailed account the various environment segments.
2. Describe the characteristics of major atmospheric regions.
3. What are the various chemical species and particulates present in the atmosphere ?
4. Write informative notes on any two of the following :
   (a) Earth's radiation balance, (b) Lapse rate and temperature inversion, (c) Photochemical ssmog and (d) Particulates in atmosphere.
5. What are the major gases responsible for causing Green-house effect and how are they released into the atmosphere ? What are the possible consequences of uncontrolled global warming ?
6. How is ozone formed and depleted in nature? What are the consequences of depletion of ozone layer in the atmo-sphere ?
7. How are air-pollutants classified? What are the major effects of CO, NOx, SO_2, Hydrocarbons and particulates, on human health ?
8. What are the sources of NOx and SO_2 in the atmosphere? Give an account of the global efforts to control the release of these two pollutants into the atmosphere by antropogenic source.
9. What are the pollutants responsible for causing acid rain ? What are its deleterious effects on materials and terrestrial ecosystems?
10. Discuss the effects of air pollutants on man and his environment.
11. Discuss the various techniques used for atmospheric air sampling.
12. Discuss the various methods and equipment used for controlling particulate emission into atmosphere.
13. How do you estimate the following atmospheric pollutants?
    (a) Suspended particulates, (b) H_2S, (c) SO_2 and (d) NOx
14. Write a detailed account on water-pollutants.
15. Discuss the principles involved in the operation of Trickling filters. What are the advantages and limitations of this process ?
16. Write informative notes on any two of the following :
    (a) Oxidation pond (b) Anaerobic digestion (c) Activated sludge process.
17. Write a detailed account on Soil pollution.

19

Non-Conventional Energy Sources

Increasing industrialization and unsustainable consumption patterns are escalating the environmental problems due to depletion of resources and energy. The unsustainable use of renewable resources and the generation of toxic materials during industrial operations are creating problems to biodiversity, environment and human health. Industrial production by using environmentally sound technologies is considered to be one of the best strategies to provide the paradigm to put our society on the path of sustainability.

Energy is a primary input in any industrial operation. Energy is also a major input in sectors such as commerce, transport, telecommunications, etc., besides the wide range of services required in the household and industrial sectors. Owing to the far-reaching changes in the forms of energy and their respective roles in supporting human activities, research and training on various aspects of energy and environment have assumed great significance.

19.1. Sources of Energy

Energy is the capacity for doing useful work. Energy resources are broadly classified as primary and secondary.

(*i*) **Primary energy resources** are those which are mined or otherwise obtained from the environment.

Ex. (*a*) **Fossil Fuels** : Coal, Lignite, Crude oil, Natural gas, etc.

(*b*) **Nuclear Fuels** : Uranium, Thorium, Deuterium, other Nuclei used in fission or fusion reactions.

(*c*) **Hydroenergy** : The energy of falling water used to turn turbines or mill wheels.

(*d*) **Geothermal** : The heat from the underground streams or the heat stored in the hot rocks beneath the earth's surface.

(*e*) **Solar Energy** : Electromagnetic radiation from the sun.

(*f*) **Wind energy** : The energy from moving air used by wind mills.

(*g*) **Tidal energy** : The energy associated with the rise and fall of the tidal waters.

However, the sun is the source for most of our energy resources; circulation of the atmosphere due to the sun's heat is the basis of solar energy and wind energy. The evaporation of water thereby causing rain and water flow is the basis of hydroenergy. Photosynthesis because of solar radiation forms the basis for chemical energy in the plants. The chemical energy stored in fossil fuels is also due to the solar heat.

The primary energy resources can be further classified as *renewable resources and non-renewable resources*. Renewable resources are those which are not exhaustible and which can hence provide continuous supply *e.g.*, wood, tidal energy, solar energy, wind energy, geothermal energy etc. Non-renewable resources are those which are finite and exhaustible *e.g.*, fossil fuels (viz., coal, petroleum, natural gas), nuclear fuels etc.

(*ii*) **Secondary energy resources** are those which do not occur in nature but are derived from primary energy resources.

Ex. : Petrol or gasoline, electrical energy from coal burning, hydrogen obtained by electrolysis of water etc.

Fossil Fuels

During the early parts of the biological history of the earth, photosynthesis outpaced the activity of the consumers and docomposers. Therefore, large amount of organic matter accumulated, particularly on the bottoms of swamps and shallow seas. Gradually, this material was buried under sediments eroding from the land, and, over millions of years, was converted to coal, crude oil and natural gas. These three are called Fossil fuels because all of them once were living matter.

Formation of additional quantities of fossil fuels by natural processes may be still continuing even today. However, we are using fossil fuels outrageoulsy faster than their formation. Hence, there is no hope of replenishment of the fossil fuels. Thus, sooner or later, we shall run out of the fossil fuels.

Almost the entire transportation in the world is dependent on crude oil. Nuclear energy, coal and hydroenergy are exclusively used for the production of electrical power. Natural gas and oil are more versatile energy sources but they should be used as efficiently as possible. For instance, using natural gas to generate electrical power and then using the electricity for space heating is extremely inefficient because of the 60–70% thermodynamic loss that occurs in the production of the electricity. The natural gas can be used more efficiently in a modern furnace, in which more than 90% of the energy content of the gas goes into heating the home and the rest 10% goes up to chimney.

As the supplies of fossil fuels in nature are finite, we shall be running out of them in near future. Therefore, we should make every effort to accommodate ourselves to the decreasing supplies of fossil fuels and we cannot afford to wait until the crisis develops. There are only two options :

(1) Conservation

(2) Developing non-fossil fuel energy alternatives.

While exploratory drilling in recent years has not proved any appreciable reserves of crude oil, it has shown considerable reserves of natural gas. With the installation of a tank for compressed gas in the trunk and some modifications of the engine fuel intake system, cars can run perfectly well on natural gas. Such cars are widely used in Buenos Aires, where service stations are equipped with compressed gas to refill the tank.* Natural gas is a clean-burning fuel, free from hydrocarbon emissions. However, using natural gas as a substitute for total oil demand is not a permanent solution. However, natural gas can provide an interim relief until non-fossil fuel energy alternatives are developed.

As against the limited crude oil reserves, several countries are endowed with coal reserves which can meet the energy needs for over 150 years. However, coal cannot be used directly as fuel in vehicles. However, coal can be converted to a liquid or gaseous fuel by chemical processes. However, these coal-based synthetic fuels are too expensive atleast for the present.

These processes may also generate some pollutants. Further, mining and using coal as fuel are accompanied by adverse environmental impacts. Many coal deposits can be exploited practically only by strip-mining. In underground mining, atleast 50% of the coal must stay in place to support the mine roof. In strip mining, gigantic power shovels turn aside the rock and soil above the coal seam and then remove the coal. Obviously, this procedure results in total destruction of the ecosystem. Strip-mined areas may be turned into permanent deserts. Furthermore, erosion and acid reaching from the disturbed earth may have adverse effects on surface and ground water of the area.

---

*Consequent to the directives from the Supreme Court of Indias in the year 2001, taxies and auto-rickshaws in Delhi are using CNG or LPG as fuel. Mumbai and other cities are also using CNG or LPG as automobile fuel since 2001.

Burning of coal produces considerable quantity of CO_2, SO_2 and other pollutants.

An important problem with all fossil fuels is that their combustion yields carbon dioxide which can cause global warming through the greenhouse effect.

There are only two alternatives to fossil fuels. These are nuclear power and solar energy.

Nuclear Fission and Nuclear Fusion

Nuclear fission reactors are based on the fission of U—235 nuclei by thermal neutrons:

$$_{92}U^{235} + _{0}n^{1} \rightarrow _{51}sb^{133} + _{41}Nb^{99} + 2.5 \, _{0}n^{1} + 200 \, meV$$

The energy from these nuclear reactions is used to heat water in the reactor and generate steam to drive a steam turbine.

High Temperature Gas-cooled reactors and Fast-Breeder reactors convert non-fissionable pu^{238} and Th^{232} to fissionable pu^{239} and U^{233}.

A major problem in the widespread use of nuclear fission power reactors is the generation of large quantity of radioactive fission-waste products, some of which may remain lethal for thousands of years. The safe disposal of the radioactive wastes also poses some problems.

Nuclear fusion power reactors are based on deuterium-deuterium reactions and deuterium-tritium reactions:

$$_{1}H^{2} + _{1}H^{2} \rightarrow _{2}He^{3} + _{0}n^{1} + 3.3 \, meV$$

$$_{1}H^{2} + _{1}H^{3} \rightarrow _{2}He^{4} + _{0}n^{1} + 17.6 \, meV$$

The deuterium-deuterium reactions promise an unlimited source of energy. However, harnessing of fusion energy will take several more years due to the technical problems involved.

Solar Energy

Solar energy originates from the thermonuclear fusion reactions taking place in the sun. The earth continuously receives energy from the sun, a part of which is absorbed while the remaining is emitted back into space. Out of the solar radiations reaching the earth, 92% consists of radiations in the range 315 nm to 1400 nm. 45% of this radiation is in the visible region (400 nm to 700 nm). The earth absorbs radiations mainly in the visible region and emits radiation in the infra-red region (to, with maximum at 10μ). The value of the solar flux reaching the earth's upper atmosphere is estimated to be about 1400 watts $m^{-2} min^{-1}$. The heat equivalent of the solar radiation reaching the earth is estimated to be about 2.68×10^{24} Joules per year.

Light, through photosynthesis, is the energy source for all major ecosystems. Solar energy, being non-polluting and non-depletable, is considered, as renewable energy and thus fits into the principle of sustainability. Only 0.2 to 0.5% of the solar energy reaching the earth is trapped by photosynthesis. Thus, only a tiny fraction of the solar energy reaching the earth drives all our ecosystems. Similarly, a small percentage of the solar energy reaching the earth could sustainably supply all the energy needs of human societies without altering the biosphere in any way. In spite of these advantages, large scale use of solar energy is limited by the following:

(i) Problem of economically collecting solar energy over large areas and converting it to other forms that can be conveniently transported, stored and used in existing equipment.

(ii) Problems of designing facilities that can utilize diffused sun-light. Recent technological advances have tackled these problems to a considerable extent. Solar energy is now beginning to be used as an economically feasible and sustainable energy source.

Space Heating and Water-heating by Solar Energy

About 25% of the U.S. energy budget is used for heating buildings in cold weather and for providing hot water for washing purposes. Temperature of about 20°-22°C are used for space heating and 50°-60°C are used for hot water. Using an oil or gas flame to provide such low-temperature heat is considered as thermodynamically inefficient use of energy. Utilizing electrical power for this purpose is even more wasteful. Solor energy is ideally suited for providing low-temperature heat for such purpose because sunlight falling on any black surface is readily absorbed and converted to heat in the desired temperature range. Therefore, complex collection or conversion equipment is not required when solar energy is to be used for obtaining low-temperature heat. A simple flat-plate collector is sufficient. There are several designs of flat-plate collectors but all of them basically consist of a black surface covered by a clear plastic or glass "window" (Fig. 19.1). The black surface absorbs sunlight and converts it to heat, and the window prevents the heat from escaping out. Air can be heated by passing it between the window and the black surface whereas water heating can be done by passing it through tubes embedded in the surface. Thus, minimum cost is involved in collecting and converting solar energy to heat.

However, beyond the collector, solar heating system may be "Active" or "Passive" and may or may not include a means of heat storage. *An active solar heating system* uses either pumps or blowers to circulate the air or water from the collector to the desired location. On the other hand, *a passive solar heating system* depends upon natural convection currents (created when hot water or air rises due to lower density) to move water or air. Passive systems are relatively inexpensive and practically free from maintenance. Several plants are available for passive solar heating of buildings. The basic concept involves using of large, sun-facing windows due to which the building itself acts as the collector. In winter, sun-light beams through the window, and heats the interior while at night, insulated drapes or shades are pulled down to trap the heat inside. In summer, excessive heat load can be avoided by shielding the windows from the high summer sun using suitable awnings or overhangs or deciduous plantings or vines. There is no need of additional heat-storage facilities. Good insulation also serves to keep the buildings cool in summer, reducing the use of air-conditioning. Along with improved insulation, appropriate landscaping can contribute to the heating and cooling efficiency of both solar and non-solar designs. Particularly, deciduous trees or vines on the sunny side of a building will block much of the excessive summer heat while letting the desired winter heat pass through. Thus, suitable architecture can provide heating and cooling virtually free of cost.

A common criticism regarding passive or active solar heating is that a backup heating system is still needed during inclement weather. While this is true, it misses the point that even if solar heating and improved insulation provides only 20% of the overall heating needs, still it reduces the demand for conventional fuels by 20%. The fuel costs are also reduced by 20%. The same holds true for solar water heating.

On a per capita basis, cyprus is

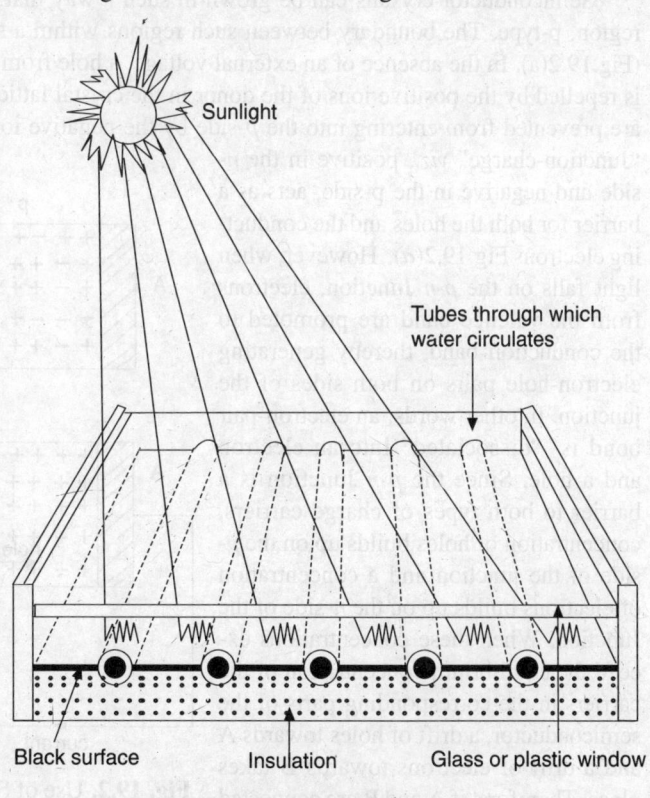

Fig. 19.1. Flat-plate solar collector.

the world's largest solar energy user; 90% of the homes and several hotels and apartment buildings have solar water heaters. In Israel, 65% of domestic hot water is provided by solar energy. In the U.S., about 7 lakh solar hot water systems are operating although it represents only 0.5% of the total requirement.

Thus it is clear that there is tremendous opportunity to make use of solar energy in conjunction with improved insulation for space heating and water heating in a cost-effective manner.

Production of Electricity using Solar Energy

Solar energy can be used to generate electrical power, thereby providing an alternative to thermal and nuclear power. At present, the following two methods seem to be cost-effective :

(*i*) Photovoltaic cells (Solar Cells).

(*ii*) Solar trough collectors.

(*i*) **Photovoltaic cells**— These are commonly known as "*Solar Cells*", which were primarily developed for providing power supply for space satellites in 1950 s. A typical solar cell consists of two very thin layers of material. The lower layer has atoms with single electrons in the outer orbital, which are easily lost. The upper layer has atoms lacking one electron from their outer orbital and hence can readily gain electrons. The kinetic energy of light striking this "Sandwich" dislodges electrons from the lower layer, which are trapped into the upper layer, thus creating an electric potential between the two layers. This potential provides the electric current through the rest of the circuit, which connects the upper side through a motor or other electrical device back to the lower side. Thus without any moving parts, solar cells convert light energy directly to electrical power. Since there are no moving parts in them. Solar cells do not wear out. However, their current life span is about 20 years because they deteriorate due to exposure to weather.

Semiconductor crystals can be grown in such a way that one region is n-type and an adjacent region, p-type. The boundary between such regions within a single crystal is called a *p-n* Junction (Fig.19.2(a). In the absence of an external voltage, a hole from the *p*-side penetrating into the-n-side is repelled by the positive ions of the donor in the crystal lattice and similarly, conducting electrons are prevented from entering into the *p*-side by the negative ions of the acceptor in the lattice. This "Junction-charge" *viz*., positive in the *n*-side and negative in the p-side, acts as a barrier for both the holes and the conducting electrons Fig.19.2(*a*). However, when light falls on the *p-n* Junction, electrons from the valence band are promoted to the conduction band, thereby generating electron-hole pairs on both sides of the junction. In other words, an electron-pair bond is "dissociated" into an electron and a hole. Since the *p-n* Junction is a barrier to both types of charge carriers, concentration of holes builds up on the *p*-side of the junction and a concentration of electrons builds up on the *n*-side of the Junction. When these concentrations exceed the equilibrium concentration of the carriers in the corresponding parts of the semiconductor, a drift of holes towards A and a drift of electrons towards B takes place. Therefore, if A and B are connected with a conductor, there occurs a flow of

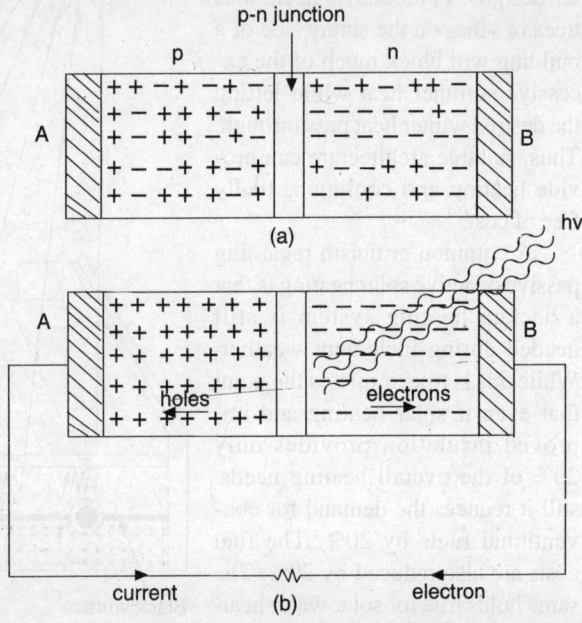

Fig. 19.2. Use of Semiconductors as Solar Cells cells.

current from A to B and flow of electrons from B to A. (Fig.19 2.(b). The electrons recombining with the holes ensures an uninterrupted flow of electrons as long as light is falling on the p-n Junction.

The solar cells made of silicon have been used as source of Power in space-craft and satellites. Other solar cells developed are CdS(n-type)/Cu$_2$ S(p-type) gallium arsenide and indium phosphide.

A part of the electrical energy derived from solar energy by using solar cells can be stored by charging Ni-Cd batteries or lead-acid batteries. Alternatively, it can be used to produce hydrogen by electrolysis of water. The liberated hydrogen can be used in hydrogen-oxygen fuel cells.

The huge capital cost is the major limitation for the large– scale use of solar cells.

The largest solar cell manufactured in large scale is about 5 cm in diameter, which in full sunlight can provide power output similar to a standard flash-light battery. However, several cells can be connected together to obtain large amount of power. The present solar cells can achieve a conversion efficiency of light energy to electrical energy in the order of 10 to 20%. This low efficiency is immaterial in one respect because the primary energy source, namely sun light, is available free of cost. However, the cost of producing the number of solar cells required to obtain a given amount of power is considerable. As is common with any other new technology, the first cell produced was very expensive. In the early 1970's, photovoltaic power was costing about 30 dollars per killowatt-hour. With improvements in production techniques, the cost came down to about 30 cents per killowatt-hour in 1991. However, even this cost is still 4 to 6 times the cost of power from traditional sources. In spite of this, even at this cost photovoltaic cells are economically practical for many applications such as in pocket calculators where they replace batteries. Panels of solar cells are the most economical way of providing power at points that are far from utility lines *e.g.*, rural homes, light-houses, irrigation pumps, radio-transmitters, offshore oil drilling platforms, traffic signals etc.

Recent breakthroughs promise to cut the costs by 50 to 80%. Rural electrification projects based on photovoltaic cells are becoming popular in the Third World Countries. In the near future, photovoltaic power plants may be producing power nearly at the same cost as traditional power plants. The Japanese are developing solar-powered air-conditioners to be marketed soon. Experiments with vehicles run on photovoltaic cells are under way using ultra-efficient designs.

Solar Trough Collectors

The solar trough concept was invented by Charles Abbott in the 1930's. Solar trough collection systems represent another economically practical method for producing electrical power from sunlight. Sunlight hitting the solar trough collector is reflected onto a pipe running down the centre of the trough, and heats fluid circulating through the pipe. The heated fluid is then used to boil water, thereby generating steam to run conventional turbogenerator.

Lux International Company is reported to have built a large number of solar trough collection systems in southern California with a total capacity of 350 megawatts, which is about 30% of the capacity of a large nuclear power plant. The recent facility built by Lux International achieved 22% conversion of incoming sunlight to electrical power and produces power at a cost of 8 cents per kilowatt-hour. This is slightly higher than the cost of nuclear or coal–fired power plants. This solar power may be considered cheaper if one considers the hidden costs of air-pollution, strip mining, and nuclear waste disposal in the production of thermal and nuclear power.

The drawback that solar power system require large stretches of land to "harvest" the sunlight may not be treated as serious if we consider the thousands of acres which are being wasted by strip-mining of coal every year or the huge area of land which was rendered uninhabitable due to nuclear accidents. Further, photovoltaic systems can utilise existing rooftops without using additional land.

Since about 65% of the electrical power demand occurs in daytime hours when industries and offices are in operation, the problem of sunlight being not available during the night is not very serious. Atleast, solar power will conserve the traditional fuels to a considerable extent for future use.

Further, in the long run, the night time power demands can be met by different forms of indirect solar energy such as water power or wind power discussed later. Also, the excess solar power generated during day time may be stored by using it to pump water from a low-elevation reservoir to a high elevation reservoir. When additional power is required, the water from the high elevation reservoir can be allowed to run down through turbo-generators. Alternatively, the excess solar power can be used to pump air into underground caverns and let this compressed air to drive turbines whenever the power is needed.

Two more approaches of producing electrical power from sunlight are receiving attention :

(*i*) **"Power Tower"** : In this method, an array of suntracking mirrors is used to focus the sunlight falling on a large area of land (several acres) onto a boiler mounted on a tower in the centre of the land. The intense heat produces steam in the boiler which drives a turbogenerator. This system is expected to generate power more economically than the solar trough system.

(*ii*) **"Solar Pond"** : In this method pioneered in Israel, an artificial pond is partially filled with brine and fresh water is placed over the brine. Brine, being denser than fresh water, remains at the bottom without any appreciable mixing. Sunlight passes through the fresh water but is absorbed and converted to heat in the brine. This fresh water then acts as an insulating blanket and holds the heat in. The hot brine solution can be circulated through buildings for heating. Alternatively, it can be converted to electrical power by vaporizing the fluids with low boiling points and using the vapours to drive a low–power turbogenerator. Since the solar pond also acts as an efficient heat storage unit, it supplies power continuously.

Solar Energy for Driving Vehicles

While considering energy sources to sustain our transportation system, conservation by increasing vehicle mileage through better engine design, proper use and maintenance still remains the choicest option for reducing demand for crude oil in the short term. Use of solar heating facilities reduces the demand for natural gas in commercial and residential sectors thereby transfering the conserved natural gas to the transportation sector.

Use of solar energy for running electric vehicles, though feasible, still suffers from the following limitations :

(*a*) Storage batteries are expensive and heavy relative to the amount of power stored. For instance, General Motors' newest prototype electric car, which is likely to be marketed shortly, carries 360 Kg of batteries to store the equivalent of about 4 litres of petrol.

(*b*) The range of even an ultra-efficient car is less than 140 km.

(*c*) Lack of a breakthrough in battery technology.

Solar energy can be used to produce hydrogen which in turn, can be used as fuel for existing vehicles. Hydrogen is a clean-burning gas without any pollution because the product of combustion is only water vapour.

$$2H_2 + O_2 \rightarrow 2H_2O + energy$$

Present vehicles could easily be adapted to run on hydrogen with some modifications in the fuel tank, fuel intake system and the type of service stations. The service stations should dispense hydrogen gas into vehicle "fuel tanks", which would consist of materials that would adsorb large amounts of hydrogen and release it only slowly, on being heated. This slow release would ensure safety against the hazard of explosion.

However, hydrogen does not exist in free state on earth and it should be produced by electrolysis of water. Solar energy can be used to produce hydrogen by electrolysis of water at places where land is cheap and sunlight is plenty.

Power from Indirect Solar Energy

Hydropower, wind and biomass are considered as indirect forms of solar energy because water cycle, atmospheric circulation and photosynthesis are driven by solar energy. These indirect forms of solar energy also provide substantial sources of energy, just like direct solar energy. Mankind has been utilizing these indirect forms of energy from ancient times.

(*a*) **Hydropower**— Water power has been used since ancient times by diverting water from natural streams or rivers over various kinds of paddle wheels or turbines. The power output from waterwheels being low, people started building high dams from the last century to obtain a substantial head of hydrostatic pressure. Thus, the water under high pressure, flows through the base of the dam and drives turbo–generators producing hydroelectric power. In U.S., about 300 large dams generate 9.5% of its total electrical power production.

Although hydroelectric power is basically a non-polluting renewable energy source, it is still associated with serious problems :-

(1) Dams have drowned out beautiful stretch of rivers, wildlife habitat, forests, productive farmlands, and areas of historic, archeological, and geological significance. The construction of big dams have also rendered several farmers and tribals homeless and without any livelihood.

(2) The reservoir behind the Aswan High Dam in Egypt has caused the spread of parasitic worms which caused a debilitating disease. Further, the increase in humidity over a large area because of the reservoir is causing rapid deterioration of ancient monuments and artefacts which were existing over many centuries.

(3) Since water flow from the dam is regulated as per the requirement of power, dams play havoc down-stream because water levels may changes from extremes of near flood levels to virtual dryness and back to flood even in a single day. Other ecological factors are also affected because sediments rich in nutrients settle in the reservoir and only small amounts reach the river's mouth.

(4) Devastating earthquakes, observed near Koyana in India, are attributed to the Koyana dam (Maharashtra) by some Scientists.

Many developing countries have great potential for large hydel power projects but due to the above problems, there is lot of opposition from people as well as from Environmental protection organisations.

(*b*) **Wind Power**— Wind mills have been used since ancient times. A large number of different designs were tested, but most practical one seems to be the age-old concept of airplane type propeller blades turning a generator geared to the shaft. Modestly sized "wind turbines", comprising of machines with blade diameter of about 17 m which can generate about 100 Kilowatts, have provided to be most practical. "Wind Frams" consisting of arrays of 50 to several thousand such machines, are now producing power in a number of places around the world. California, with 17,000 machines generating 1500 megawatts is the world's largest producer of wind-generated power. This supplants the need for two nuclear power plants. By the turn of this decade, European countries like Britain, Netherlands, Germany, Italy and Denmark will have a combined wind generated power capacity of over 3,000 megawatts. With gradual improvements in design and reliablity, wind farms are now generating power at as little as 6 cents per kilowatt-hour, which is certainly competitive with traditional energy sources. Power-generating wind turbines are now installed in 95 countries right from the tropics to the Arctic. However, still there is lot of potential which is untapped. Many regions of the world have areas where winds are constant enough to render wind turbines practical. Wind farms in different locations can be connected to the already existing electrical grid so that it can provide backup for each other since the wind is invariably flowing somewhere or the other. Wind farms can also provide a sustainable complement and backup to direct solar power facilities.

Wind power is a non-polluting, renewable and hence sustainable source of energy. However, it has the following drawbacks :

(i) Location of wind farms on migratory routes could spell hazard to birds and disaster for some avian populations.

(ii) Their appearance on the landscape and their continual whirring and whistling can be irritating.

*Wind energy has proved to be economically competitive in certain parts of India, which include coastal regions in Gujarat, Kachchh, Saurashtra; gaps in the eastern ghats in Tamil nadu, Tuticorin and Kayathar; and plains in Rajasthan, U.P., hill tops in M.P., etc. Total wind energy resource potential in India is estimated to be about 25,000 MW.

Biomass Energy

Biomass energy or Bioconversion refers to the direct burning of wood, waste paper, manure, agriculture or any form of biomass or converting them to a fuel. Certain microorganisms when they digest biomass in the absence of air, produce either alcohol or methane gas, which themselves give energy on combustion. Since biomass is obtained through the process of photosynthesis, biomass energy is considered to be another form of indirect use of solar energy.

Firewood can be considered as a sustainable energy resource where forests are plenty relative to population, which was the case in ancient times. However, any idea of using firewood as a large-scale energy resource must be considered in the context of maximum sustainable yields and preservation of the bio-diversity of eco-systems. Many Third World countries where millions of people still depend on firewood as fuel for cooking food and getting warmth, are already suffering from the severe ecological effects of deforestation and human deprivation caused by non-availability of wood. In this back-drop, any blanket recommendation to burn wood because it is renewable, is environmentally irresponsible.

Similarly, proposals for growing trees or other crops solely from the point of view of biomass energy, either for direct burning or for conversion to alcohol or methane – must be considered in the light of possible soil erosion and requirement of fertilizers, pesticides, etc. for growing such crops and their possible environmental impact. Further, the possibility of depriving the land for growing food crops in favour of energy crops should also be kept in view because both may be grown on the same land.

Biomass energy should be preferred wherever energy can be produced as a byproduct of waste disposal (e.g., Sawmill waste, Sugar refinery waste, Municipal refuse, etc.). However, even while doing this, advantages of recycling, refuse or composting of the wastes should be kept in view.

Use of alcohol as fuel is being vehemently promoted in some grain-growing regions of developed countries such as U.S. Alcohol is produced by fermentation of grains, starches, sugar or similar food products. However, developing and under-developed countries should be very cautious in entertaining such ideas. Since many parts of the world suffer from shortage of grains, production of alcohol from grains may lead to shortage of food for their rapidly growing population, thereby leading to widespread malnutrition. Further, while alcohol, by itself, is clean-burning, we should remember that producing fuel grade alcohol requires distillation for which cheap, dirty-burning fuels such as soft coal are used as fuels. The net energy yield of alcohol is therefore modest, because fuel equivalent of about 0.5 gallon of alcohol is used for every gallon of alcohol produced.

Anaerobic digestion of sewage sludge and animal manure is a biomass utilizing method which creates a valuable synergism between recycling and energy production. Such a mehtod yields biogas and a nutrient-rich compost which is a good organic fertilizer which can be recycled back to the land in order to maintain the fertility of the fields growing forage for the animals. Further, we can thus get cheap energy which is sustainable and is free from adverse environmental consequences. In China, millions of small farmers maintain a simple digester in the form of a sealed pit into which they put agricultural wastes. The biogas generated is used as domestic fuel. This is certainly a better alternative than using wood as fuel.

Ocean Thermal Energy Conversion (OTEC)

In oceans, a thermal gradient (i.e., temperature difference) of about 20° C exists between surface water heated by the sun and colder deep water. This temperature difference can be harnessed to produce power. This concept is known as Ocean Thermal Energy Conversion (OTEC). An OTEC power plant can be built on a brage (*i.e.*, a sailing vessel) that could travel anywhere in the ocean. It uses the warm surface water to heat and vaporise a low boiling liquid such as ammonia. The increased pressure of the vaporized liquid would drive turbo generators. The ammonia vapour leaving the turbines would then be condensed by cold deep water which is about 100 m below the surface and is returned back to start the cycle again. The electrical power so generated could be used to produce hydrogen and shipped to the shore. Alternatively, an energy intensive industry can be located on factory ships that would anchor alongside the OTEC Plant. A few OTEC Plants have been tested.

Owing to the small temperature difference between the surface water and deep water, the conversion efficiency is as low as 2 – 3%. This low efficiency by itself is immaterial since the primary energy source *viz.*, the temperature difference between the surface and deep waters over most of the ocean, is freely available. However, this low efficiency coupled with other drawbacks such as high capital costs, persistent maintenance problems and fouling of pipes and pumps due to marine organisms, results in meager energy yields thereby rendering OTEC power uneconomical at the present state of the OTEC technology.

19.2. OTHER RENEWABLE ENERGY SOURCES

(*i*) **Geothermal Energy**— Heat from the molten core of the earth offers a huge and sustainable energy source. This geothermal energy can be used for space heating or for boiling water and generate steam to drive turbogenerators. In volcanic regions of the earth, the hot rock is relatively close to the surface. Also, in such regions sometimes, the natural ground water comes in contact with the hot rock and the heated water or steam may find its way to the surface through natural steam vents. In such regions, holes can be drilled into the hot rock-ground-water structure and make the rising steam to drive turbogenerators to produce electric power. The world's largest Geothermal energy production facility exists at a location known as "The Geysors" near San Francisco in U.S. The electrical output from this facility in the year 1988 was about 2000 megawatts, which is equivalent to the power produced by two large nuclear power plants. Similar geothermal facilities exist in the Philippines, Mexico, Japan, Italy and Iceland, generating a total power of about 3,000 megawatts.

However, the problems with the Geysers was that while the source of heat is unlimited, the amount of ground water is not. This is the reason why power output from the Geysers has been steadily declining since 1988. What is happening at the Geysers is similar to the "Pot" boiling dry. Attempts are underway to inject water into it but the results are not conclusive.

Natural steam vents occur only in a few regions whereas hot dry rock structures are available in almost all places if we can drill deep enough (about 5 to 6 km) to reach them. Two parallel holes can be drilled into the hot rock and fractures created between the two holes. Water is forced down one of the holes which gets heated as it seeps through the fractures and the steam so formed comes out from the other hole. This steam can be used to drive turbines to produce electricity. Attempts are in progress to solve the technological problems involved in such an approach.

Geothermal power production may be accompanied with the following problems :

(i) Hot steam and water coming to the surface are usually contaminated with salts and other pollutants such a sulphur compounds. Some of these contaminants are highly corrosive to turbines and other equipment. They may also cause air pollution if the steam escapes into the atmosphere. SO_2 pollution from a geothermal plant may be as much as that of a high- sulphur-coal based thermal power plant.

(ii) Hot brine released into surface waters may be ecologically hazardous.

Tidal Power

A lot of energy is inherent in the twice-a-day rise and fall of the tides. Scientists have tried to utilize this tidal energy which is eternal and pollution-free. One of the simple schemes is to construct a dam across the mouth of a bay and mount turbines in the structure. The incoming tide flowing through the turbines generates power. As the tide shifts, the blades may be reversed so that that outflowing water continues to generate power. At present, tidal power plants are in operation in Russia, France and Nova Scotia.

For any tidal power project to be of practical use, a fluctuation of atleast 6 meters is needed between the high tide and low tide. Such a situation exists only at about 15 locations in the world. The Bay of Fundy in North America is one of such locations and a large tidal power plant is being developed.

Tidal power plants are accompanied by the adverse environmental effects because of the dams which may trap sediments, impede the migration of marine organisms, change water circulation and cause mixing of fresh water with salt water.

Conclusion

The potential availability of electrical power from traditional and alternative fuels does not mean that we should overlook the importance of conservation. If our energy use continues to grow with the growth of population as in the past, it will be difficult to keep pace with the growing power demands even with massive installations of solar and other alternative energy sources. Hence, energy conservation should from an important element in the energy policy, irrespective of the source of power. Conservation of energy should be achieved by making heating, lighting and transportation systems more efficient without sacrificing our comfort. Relentless pursuit of energy conservation through greater energy efficiency and lesser wastage will help in supporting our growing economy without exerting large power demands.

QUESTIONS

1. Discuss the use of solar energy for space heating, water heating and for production of electricity.
2. Write informative notes on any two of the following :
   (a) Flat plate solar collectors (b) Photovoltaic cells (c) Solar trough collectors.
3. Write short notes on any three of the following :
   (a) Solar energy for producing fuel for vehicles   (b) Power towers
   (c) Solar pond   (d) Solar production of electricity.
4. Discuss the use of indirect solar energy for generation of electrical power.
5. Discuss the principle involved in producing hydropower and the advantages and environmental consequences involved in hydropower generation .
6. Write informative notes on any two of the following :
   (a) Wind power   (b) Biomass energy   (c) Tidal power.
7. Discuss the principles involved in Ocean Thermal Energy Conversion (OTEC). Discuss its merits and demerits.
8. How is Geothermal energy used for generation of electrical power? Discuss its advantages and disadvantages.
9. Write short notes on any three of the following :
   (a) Conservation of energy   (b) Advantages and limitations of nuclear power
   (c) Bio-conversion   (d) Problems associated with the use of fossil fuels

20

Powder Metallurgy and its Industrial Applications

"Owing to the better understanding of the under-pinning scientific principles, technical innovations within the industry and those emanating from R & D studies and the exacting requirements in high-tech applications, there has been a considerable growth of powder metallurgy industry during the past few decades".

Terms and Definitions

Activated sintering :	The use of additives, such as chemical addition to the powder or additions to the sintering atmosphere to improve the densification rate.
Activator :	The additive used in activated sintering, also called a dopant.
Adhesion :	The force of attraction between the atoms or molecules of two different phases.
Agglomerate (noun):	An assembly of powder particles of one or more constituents clustered closely together.
Agglomerate (Verb) :	To develop an adherent cluster of particles.
Aggregate (noun) :	A mass of particles.
Alloy Powder :	A metal powder consisting of atleast two constituents that are partially or completely alloyed with each other.
Amorphous Powder :	A powder that consists of particles that are substantially non crystalline in character.
Annealing :	A generic term denoting a treatment that consists of heating to and holding at a suitable temperature followed by cooling at an appropriate rate, primarily for the softening of metallic materials. Steels may be annealed to facilitate cold working or machining, to improve mechanical or electrical properties or to promote dimensional stability.
Annealed Powder :	A powder that is heat treated to render it soft and compactible.
Atomisation :	The disintegration of a molten metal into particles by a rapidly moving gas or liquid stream or by other means.
Axial loading :	The application of pressure on a powder or compact in the direction of the press axis.

655

Ball mill :	A machine in which powders are blended or mixed by ball milling.
Ball milling :	Grinding blending or mixing in a receptacle of rotational symmetry that contains balls of a metal or non metal harder than the material being milled.
Billet :	A compact (green or sintered) that will be further worked by forging, rolling or extrusion, sometimes called an ingot.
Binder :	(1) A substance added to the powder to increase the strength of the compact.
	(2) Cement together powder particles that alone would not sinter into a strong object.
Blend (noun):	Thoroughly intermingled powders of the same nominal composition.
Blending :	Thorough intermingling of powder fractions of the same nominal composition to adjust physical characteristics.
Can :	A sheathing of soft metal that encloses a sintered metal billet for the purpose of hot working (hot isostatic pressing, hot extrusion) without undue oxidation.
Cermet :	A material consisting of ceramic particles bonded with a metal.
Compact (noun):	The object produced by compression of a metal powder, generally while confined in a die.
Compact (verb) or Compacting or Compaction :	The operation or process of producing a compact sometimes called pressing.
Deformation :	Change of shape resulting from application of force.
Degassing :	Specifically, the removal of gases from a powder by a vacuum treatment at ambient or at elevated temperature.
Die :	The part or parts making up the confining form in which a powder is pressed or a sintered compact is repressed or coined. The term is often used to mean a die assembly.
Dopant :	A substance added in small quantity in metallic powder to prevent or control recrystallization or grain growth during sintering or during use of the resultant sintered object (Also see activator).
Doping :	Addition of small amount of an activator to promote sintering.
Dust :	Specifically a superfine powder having predominantly sub micron size particles.
Ejection :	Removal of the compact after completion of pressing, whereby the compact is pushed through the die cavity by one of the punches. Also called knock out.
Extrusion :	Shaping metal powder into a chosen form by forcing it through an orifice or die of the appropriate shape. The powder may either be mixed with a plasticizer or contained in a can which is extruded with the contained powder.

POWDER METALLURGY AND ITS INDUSTRIAL APPLICATIONS 657

Exudation :	The action by which all or a portion of the low melting constituent of a compact is forced to the surface during sintering, sometimes referred to as bleed out or sweating.
Forging :	(1) Reshaping a billet or ingot by hammering. (2) The process of placing a powder in a container, removing the air from the container and sealing it, followed by the conventional forging of the powder and container to the desired shape.
Forming :	A generic term in powder metallurgy describing the first step in changing a loose powder into a solid specific configuration.
Gap :	The clearance between a moving punch and the die cavity.
Grain :	An individual crystal within a polycrystalline metal or alloy. Sometimes used for particle.
Green :	Unsintered.
Green Compact:	An unsintered compact
Green density :	The density of a green compact.
Green strength :	The ability of a green compact to maintain size and shape during handling and storage prior to sintering.
Hard metal:	A collective term that designates a sintered material with high hardness, strength and wear resistance, and is characterised by a tough metallic binder phase and particles of carbides, borides or nitrides of the refractory metals.
Heavy alloy :	A sintered tungsten alloy with Ni, Cu and or Fe, the tungsten content being atleast 10% by weight and the (heavy metal) density being atleast 16.5g/cm^3.
High energy-rate compacting :	Compacting of a powder at a very rapid rate by the use of explosives in a closed die.
Hot densification :	Rapid deformation of a heated powder preform in a die assembly for the purpose of reducing porosity. Metal is usually deformed in the direction of the punch travel.
Hot forging :	The plastic deformation of a pressed and/or sintered compact in atleast two directions at temperatures above the recrystallisation temperature.
Hot isostatic pressing :	A process for simultaneously heating and forming a compact in which the powder is contained in a sealed flexible sheet metal or glass enclosure and the so contained powder is subjected to equal pressure from all directions at temperature high enough to permit plastic deformation and sintering to take place.
Hot pressing:	Simultanceous heating and forming of a compact (see also pressure sintering).
Hydrostatic pressing :	A special case of isostatic pressing that use a liquid such as water or oil as a pressure transducing medium and is therefore limited to near room temperature operation.
Impact sintering :	An instantaneous sintering process during high energy rate

	compacting that causes localised heating, welding or fusion at the particle contacts.
Neck :	The contact area between abutting particles in a compact undergoing sintering.
Net Shape :	The shape of a sintered part that conforms closely to specified dimensions.
Network structures:	Structures in which one constituent occurs primarily at the grain boundaries thus enveloping the grains of the other constituent(s). A desirable feature in cemented carbides, as in the system cobalt/tungsten carbide, where the cobalt phase forms a ducitle network surrounding the brittle carbide grains.
Open Porosity (or interconnected porosity or interlocking porosity):	A network of connecting pores in a sintered object that permits a d flui or gas to pass through the objects.
Particle:	A minute portion of matter. A metal powder particle may consist of one or more crystals.
Particle morphology :	The form and structure of an individual particle.
Permeability :	The ability of a compact to permit liquid or gas to flow through it as measured under specific conditions.
Plasticizer :	A substance added to a powder or powder mixture to render it more formable during cold pressing or extrusion.
Platelets :	Flat particles of metal powder having considerable thickness. The thickness, however is smaller as compared to the length and width of the particles.
Impregnation :	The process of filling the pores of a sintered compact with a non metallic material such as oil, wax, or resin.
Indirect Sintering :	A process whereby the heat needed for sintering is generated outside the body and transferred to the compact by conduction, convection, radiation, etc. Contrast with direct sintering in which the heat needed for sintering is generated in the body itself, such as by induction or resistance heating.
Infiltration :	The process of filling the pores of a sintered or unsintered compact with a metal or alloy of lower melting temperature.
Interface :	A surface that forms the boundary between phases in a sintered compact.
Keying :	The deformation of metal particles during compacting to increase interlocking and bonding .
Knockout (verb) :	Ejecting a compact from a die cavity.
Knockout punch:	A punch used for ejecting compacts.
Liquid phase sintering:	Sintering of a compact or loose powder aggregate under conditions where a liquid phase is present during part of the sintering cycle.

POWDER METALLURGY AND ITS INDUSTRIAL APPLICATIONS

Loose powder sintering:	Sintering of uncompacted powder using no external pressure.
Lower punch:	The lower member of a die assembly which form the bottom of the die cavity.
Lubricant :	A substance mixed with a powder to facilitate compacting and subsequent mold ejection of compact, often a stearate or proprietary wax. It may also be applied as a film to the surfaces of the punches or the die cavity wall, such as by spray coating.
Matrix metal :	The continuous phase of a polyphase alloy or mechanical mixture.
Metal powder :	Elemental metal or alloy particles, usually in the size range of 0.1 to 1000 nm.
Milling :	Mechanical comminution of a metal powder or a metal powder mixture, usually in a ball mill, to alter the size or shape of the individual particles, to coat one component of a mixture with another or to create uniform distribution of components.
Porosity :	The amount of pores (voids) expressed as a percentage of the total volume of the powder metallurgy part.
Powder Metallurgy :	The technology and art of producing metal powders and of the utilization of the metal powders for the production of massive materials and shaped objects.
Powder metallurgy part:	A shaped object that has been formed from metal powders and sintered by heating below the melting point of the major constituent. A structural or mechanical component made by the powder metallurgy process.
Preform :	The initially pressed compact to be subjected to repressing.
Pulverisation :	The process of reducing mental powder particle sizes by mechanical means also called comminution or mechanical disintegration.
Punch :	The part of a die assembly which transmits pressure to the powder in the die cavity.
Pyrophorecity :	The property of a substance with a large surface area to self-ignite and burn when exposed to oxygen or air.
Reaction Sintering :	The sintering of a powder mixture consisting of atleast two components that chemically react during the treatment.
RZ-Powder :	The reduced iron powder made in Germany from the scale of pig iron.
Sedimentation :	The settling of particles suspended and dispersed in a liquid through the influence of an external force, such as an external gravity or centrifugal force.
Segregation :	Separation of a blend or mixture into fine and coarse portions as a result of overmixing or vibration.
Sizing :	The pressing of a sintered compact to secure desired dimensions.
Slip-casting :	A process used in the production of refractory products. In this process,

Tap density : The density of a powder when the volume receptacle is tapped or vibrated under specified conditions while being loaded.

Theoretical density : The density of the same material in the wrought condition.

Whiskers :
(1) Metallic filamentary growths, often microscopic in size, that attain very high strengths.

(2) Oxide whiskers such as Saphire which because of their strength and inertness at high temperatures are used as reinforcement in metal matrix composites.

[Continuation from previous page:] ground material is mixed with water to form a creamy liquid, which is poured into plaster moulds where the surplus water is absorbed and a solid replica of the inside of the mould is obtained.

20.1. INTRODUCTION

As per the powder metallurgy committee of the American Society for Metals, Powder Metallurgy is defined as "the art of producing metal powders and objects shaped from individual, mixed or alloyed metal powders, with or without the inclusion of non metallic constituents, by pressing or molding objects which may simultaneously or subsequently heated to produce a coherent mass, either without fusion, or with fusion of a low melting constituent only".

Powder metallurgy is essentially the art of producing metal powders by processes which do not involve the fusion of the metal. The metal powder may contain non-metallic admixtures as well as lower melting metal additions, which may undergo melting during processing. However the major portion of the starting material remains in the solid state. The standard procedure of powder metallurgy comprises of shaping metal powder or powder mixtures by pressure and subsequently heating the pressed powder compact to temperatures well below the melting point. The shaping by pressure is called "compaction" and the heat treatment process is called "sintering". Sometimes, the processes of compaction and sintering can be carried out simultaneously and this procedure is called "hot pressing" or "pressure sintering". In special cases, powders are sintered without previous compaction while in other cases; compacted powders are used without subsequent sintering.

Powder metallurgy comprises of the art of producing of not only metal and alloy powders intended for processing by compaction and sintering, but also powders used in other applications such as catalysts, paints and pyrotechnics. The words "metal powders" many include not only metal and alloy powders but also powders of compounds such as nitrides, carbides, and silicates, which are characterized by metallic properties such as metallic bonding as indicated by their thermal and electrical conductivity and metallic appearance.

Metal powders find extensive and diverse application in various industries such as catalysts, paints, explosives, printing inks, etc. In all such applications, the powder particles retain their identities. On the other hand, traditional powder metallurgy process comprises of conversion of a solid metal, alloy or ceramic in the form of a mass of dry particles, usually less than 150 microns in diameter, into an engineering component of pre-determined shape and possessing such properties which enable them to be used in most applications without further processing. The traditional powder metallurgy process involves the following three basic steps:
(1) Production of metal powder
(2) Compaction of the powder usually in rigid dies, to a handleable pre-form and
(3) "Sintering" which involves heating the perform to a temperature near the melting point of the metal, when the powder particles lose their identities through inter diffusion-processes and the desired properties are developed. The sintered product retains certain degree of porosity. Some variations of the process were necessitated to

achieve improvements in densification and other properties. For instance, if the sintered component is brought into contact with an appropriate metal of lower melting point, the latter is absorbed to fill the interconnected porosity. Similarly, in multicomponent systems, if one of the minor components becomes molten at the sintering temperature, rapid densification takes place through the liquid phase, thereby eliminating the porosity.

Since powders do not flow other than in the direction of applied pressure, the shapes which can be produced in rigid dies, will be limited. However isostatic pressing in flexible moulds provides greater shape flexibility and improved densification, particularly when the pressing is combined with the sintering operation. This is also true with the conventional hot pressing techniques.

Conventional metal fabrication processes may also be employed for improving densification, including the mechanical working of sintered billets and the hot forging of sintered preforms. Powders may be vacuum-sealed into metal cans and rolled or extruded into semi-finished products. Metal strip in certain alloys is produced by the roll forming of either lubricated powder or of a metallic paste followed by sintering and final consolidation by cold rolling. Another recent development in this field is to spray molten metal on to a substrate to produce intermediate products which are then fabricated by more conventional methods. Production of complex shapes in materials of high intrinsic value can be done using injection moulding techniques.

Archeological evidences indicate the use of powder metallurgy processes by early civilizations even before man learnt to extract metals. Certain metals in powder form were produced by carbonaceous reduction of locally available oxide ores, consolidated by sintering and forged to meet the requirements of some early civilizations, much before man discovered ways and means to melt the metals. The iron pins keying the marble blocks of the parthenon in Athens are believed to have been made in this way. The Delhi pillar in its massive form constructed around 300 A.D. is supposed to have been made out of about 6 tons of sponge iron. However, after the development of melting practices for iron and copper, powder metallurgy as an industrial process, remained more or less dormant for several centuries. However, For a short period in the beginning of the 19th century, there arose a need for producing malleable platinum for chemical laboratory ware. The platinum ware was fabricated by pressing/compacting of fine platinum powder into a cake followed by hot working (hot forging). However, this process was abandoned in the mid 19th century in favour of fusion process.

Further significant development of powder metallurgy as an industrial process occurred only in the earlier part of the 20th century when the invention of the electric lamp inspired a search for a filament material having high melting point, adequate conductivity and low rate of evaporation. Tungsten was found to meet these requirements but the sintered metal was brittle. This difficulty was resolved in the year 1910 when it was observed that tungsten bars, produced by sintering at 3000°C by direct electric resistance heating when worked in hot condition developed sufficient ductility to allow continued working at progressively lower temperatures until a filament of required diameter could be cold drawn.

The pattern for the future industrial development of powder metallurgy processes was set in mid 1920 with the invention of the hard, wear-resistant cemented carbide and the porous bronze self-lubricating bearing. The former product made from a mixture of tungsten carbide and cobalt demonstrated the principle of liquid phase sintering, whereas the latter product provided the basis for fabricating net-shaped engineering components.

The cemented carbide was produced by compacting and sintering a mixture of powders containing tungsten carbide with upto 15% cobalt. At about 1400°C a molten phase is formed between the cobalt and some of the tungsten carbide which promotes rapid and complete densification yielding a final structure of carbide particles in a tungsten carbide /cobalt alloy matrix. Cemented carbides originally developed for wire drawing dies, were found to have extensive application in metal cutting, rock drilling and hot working dies. Several new compositions have been developed later on with addition of other refractory carbides such as titanium carbide and tantalum carbide.

The principle of sintering in presence of a liquid metal phase described above was later on extended to heavy alloys such as those containing about 90% of tungsten with about 10% nickel-iron or copper-nickel. These alloys were originally developed for radium therapy screening but were later used for mass balancing in aircraft and gyroscopes. Another remarkable extension of this process is that in which a porous sintered metal is brought into contact with a molten metal of lower melting point to rapidly absorb and fill the pores. Heavy duty electrical contacts comprising of tungsten/silver and tungsten/copper were produced by this techniques.

The second remarkable powder metallurgy product developed in 1920s referred to above, was the porous bronze self-lubricating bearing made from a mixture of powders containing about 90% copper, 10% tin and a little graphite. The mixture was pressed and sintered to form a tin – bronze alloy. The final close dimensional tolerances required in the bearing were achieved by pressing the sintered compact in a sizing die followed by impregnation with oil. When in use, the bronze reservoir supplies oil to the bearing interface. Porous products such as filters for separation processes which are in use today are made from slightly compacted and sintered powders of bronze, stainless steel, etc.

This process used for producing the porous bearing formed the basis for the sintered iron structural parts produced during 1930s. These developments were highly significant because the powder metallurgy process which was so far used to produce products that could not be made by any other technique, now posed a challenge to the conventional fusion methods. The earlier products of powder metallurgy, which still retained about 10% porosity, were required to meet only modest engineering specifications. The major commercial attraction of the powder metallurgy process was its ability to produce components having good dimensional tolerances and at competitive prices. These features remained as the primary commercial advantages for several years even as the product base was gradually broadened. Although the major tonnage growth has been in iron and alloy steel structural parts, products of aluminum and metals of higher intrinsic value such as Ni, Cu, Ti and other refractory metals as well as cemented carbides have considerably contributed to the overall market value. Soft magnetic parts for pole pieces, relay cores etc, and high energy magnets including the more recent rare earth types are important products of the industry and are manufactured in substantial quantities. The versatility of the process has been demonstrated by lending itself for the production of a variety of industrially important metal/non-metal products such as metal/diamond abrasive wheels; copper/ graphite brushes for electric motors, friction materials for vehicle brake linings comprising of a dispersion of compounds such as silicates in a metal matrix; and silver/cadmium oxide light duty electrical contacts. A special class of products known as "cermets" were also manufactured by this process which include nuclear materials such as the oxide or carbide of uranium dispersed in a metallic matrix, and molybdenum / Zirconium oxide which is used for hot working dies. Metal powders were also processed by more traditional fabrication processes which include rolling of a lubricated powder or a metallic paste into strip, mechanical working of sintered preforms, and extrusion of powder which is sealed in metal cans into shapes.

Technological innovations emerging out of industrial experience as well as from scientific research have contributed to the overall growth of the powder metallurgy industry during the past 3 decades. Better understanding of the under-pinning scientific principles involved has achieved further progress in manufacturing technology with better control of processing parameters and automation of the pressing and sintering operations. Further, better coordination between the design engineer and powder metallurgy technologist resulted in improved densification of the Sinter Component and attainability of a better level of mechanical properties. Alloying has also helped in producing products which can be heat treated and with properties such as homogeneity and reproducibility. Thus the products of powder metallurgy offered a challenging alternative to those produced by fusion process.

Powder metallurgy process has several advantages. It is neither labour intensive nor energy intensive. It is ecologically clean. It conserves material. It can produce net shape components with homogeneous and reproducible properties. These attributes have attracted further studies and attention for possible use of this process in high technology applications.

Further accumulation of scientific knowledge of the process and of related processes and product developments and the challenging demands for new and improved processes and materials for advanced engineering application such as aerospace applications over the past 15 years contributed to the further development of traditional as well as as new sectors of the powder metallurgy industry. Important products emerging out of this technological upsurge include sintered and dispersion strengthened alloys wherein a very fine dispersion of an oxide in a metal matrix such as nickel or lead, conferred improved micro structural stability and enhanced creep resistance. One of the recent developments in this field is what is called "mechanical alloying" process in which mixture of metallic and non-metallic powders are subjected to high energy milling. The non-metallic oxide or carbide powder particles get coated with the softer metal and are re-distributed in the form of a very fine dispersion through repeated fracture and re-welding of the composite powder particles. This process was originally developed for producing alloys for high temperature service but was later extended to light alloys used for aero-space application and for creep resisting steels and other engineering materials.

Novel microstructures are also produced by the consolidation of very fine powders generated by the "gas atomization" of molten metal. During quenching of the atomizing gas, the metal droplets solidify so rapidly that they can exhibit non-equilibrium behaviour such as high solubility extensions of certain alloying elements, together with micro crystalline or even amorphous structure. Several Al-alloys processed like this have reached commercial stage while efforts are in progress on alloys of Cu, Mg and Ti. An important extension of this techniques is the osprey spray-forming process in which the atomized metal stream is directed onto a substrate on which it solidified and is allowed to build up to the required form. This process is applicable to a wide range of engineering materials including steels and light alloys.

In the techniques described above, metal matrix composites are produced using small fiber whiskers or particulates of SiC, Al_2O_3 etc. to reinforce the metallic base. The reinforcement together with the refined microstructure contributes to enhanced stiffness, strength and other high temperature properties. Besides the conventional powder metallurgy techniques and mechanical alloying, other manufacturing process such as spray-forming have come into vogue. For instance, non metallic fibers or particulates can be injected into the molten metal as it is sprayed onto substrate. This technique is already in use for producing aluminum alloy matrix composites whereas it is under development for those based on Mg, Ti and some other high temperature alloys.

New class of ceramic products obtained from powders based on oxides, nitrides, or carbides find application in wear-resisting materials in heat-exchangers, pump parts in corrosive environments, special components of petrol and diesel engines, dielectrics, piezoelectric, electrical and thermal insulators, magnetics, ion conductors, sensors and high temperature ceremic super conductors.

20.2. TECHNIQUES FOR PRODUCING METAL POWDERS:

Powders production occupies an important place in metal industry. Powders of iron, steel, copper and copper base, aluminum, tin, nickel, molybdenum, tungsten and tungsten carbide have several industrial applications. Therefore production of metal powders is a vital step in the field of powder metallurgy. Powders are also used for non-powder metallurgy applications such as welding electrodes, submerged arc-welding (as in case of iron), paints, pyrotechnics, etc.

The various mechanical, chemical and electrochemical powder production techniques for producing metallic powders and powders of carbides, borides, nitrides, ceramics and some special processes used for producing whiskers and short fibres are described in the following sections:

20.2.1 MECHANICAL PROCESS :

The mechanical process of powder production comprises of size reduction performed in solid state or in molten metal. The former involve grinding and milling whereas the latter involves atomisation. Atomisation is of particular significance for mass production of metal powders.

(a) Grinding and milling

Grinding and milling are the traditional processes used for producing particulate materials. These processes are widely used for size reduction of brittle materials in ceramic and cement industries. However, their use in powder metallurgy is rather limited because metallic materials exhibit considerable degree of plasticity which makes the process less effective. All the same, materials such as ferroalloys and inter metallic compounds can be effectively comminuted by mechanical means. Mechanical disintegration processes involving high energy milling procedures cause severe embrittlement of the metal. However, this can be used to advantage in the process of "mechanical alloying" described later in this section. The major phenomena taking place in grinding and milling operations are the nucleation of cracks followed by crack propagation and fracture. The limit of the minimum obtainable particle size depends on the conditions of the mechanical process and on the nature of the material whose size is being reduced. This so-called "grinding or milling equilibrium" has been determined for several metallic and non metallic powders and ranges between 0.1 nm to 1 nm.

A wide variety of grinding and milling equipment is available which includes the following:

(i) *Ball mills*: The milling balls in action in a conventional ball mill vessel is shown in Fig. 20.1.
Grinding/milling is accomplished by symmetrical rotation of the material in a vessel containing metallic or non-metallic balls which are harder than the material being milled.

(ii) *Vibration mills*: In vibration mills the design of the vessel is similar to that in the rotation all mills, but the vessel is mounted on special steel springs. The amplitude and frequency of vibration can be adjusted according to the vibration

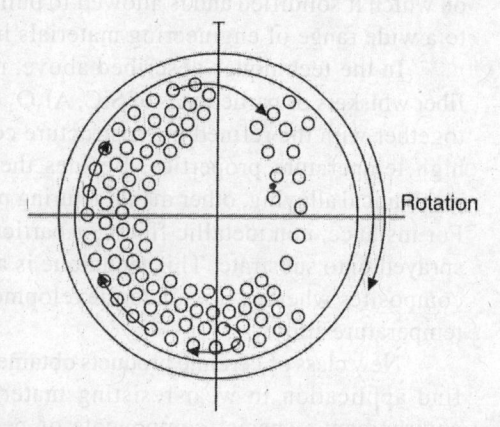

Fig. 20.1. Steel balls in milling action in a conventional ball mill vessel.

mechanism and the characteristics of the springs used. Vibration mills are generally used only when the milling time is short.

(iii) *Planetary mills*: In this, the single vessel rotation is superposed by the rotation of the table supporting the fixed vessels, thereby accelerating the movement of the milling balls.

(iv) *Attritor Mills*: An attritor is a ball mill system wherein both the balls as well as the materials being milled are set in motion together with the help of a shaft with stirring arms which rotate upto 2000 rpm. The cylindrical vessel container is usually water cooled because considerable heat is generated in the process. Attritor millling can be performed under dry or wet condition. The balls used in attritor milling are of 0.5 to 2 mm which are smaller than in the conventional ball mills. The milling liquid (water or organic liquid) can be recycled with the help of a device like distillation apparatus. The milling intensity of an attritor is generally much higher than that of the conventional ball and hence are faster by at least one order of magnitude for getting similar results. The processes for wet grinding by batch attritor milling are shown in Fig. 20.2.

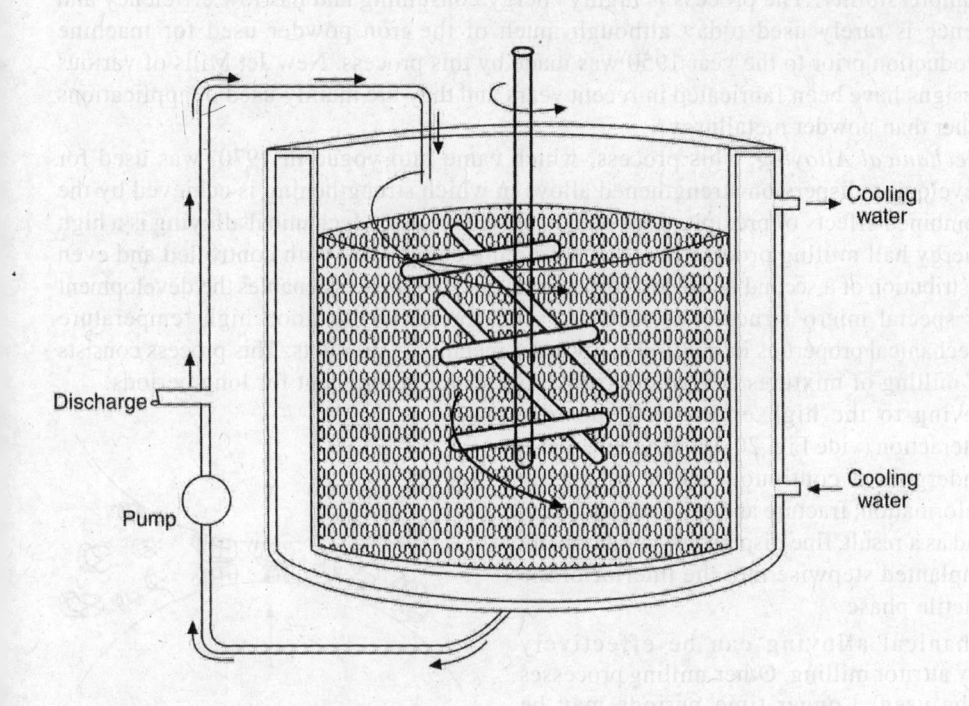

Fig. 20.2. Batch attritor for wet grinding.

Attritors (dry and wet) are widely used in paint and pigment industry, hard metal industry and for production of oxides and lime stone powder. Many metals and alloys may become amorphous during extended milling in an attritor. This has resulted in the development of tailor made amorphous alloys.

(v) *Roller Mills*: These mills are used for size reduction of brittle materials. High compression roller mills operate with profiled rollers in the pressure range of 50 to 500 M pa. Products with narrower particle range can be obtained by this techniques. The service life of roller mills is about 10 to 20 times than that of ball mills.

(vi) *Cold stream process*: In this process, coarse particles (<2mm) at high pressure (about

70 bar) and in a high velocity gas stream (mach 1) are allowed to strike a highly wear-resistant target material through a venturi-nozzle. As the gas stream passes through the venturi-nozzle before entering the collision chamber, there occurs a pressure drop and a rapid cooling of the particles occurs in the nearly adiabatic system. This results in considerable embrittlement of the particles thereby enabling easy fracturing of the particles during collision with the target materials. The process is continuously repeated and the fine particles are extracted and collected in special chambers while the coarse particles stay in the repeating cycles of operation. By using argon as the inert gas, the process produces fine powders (a few nm) having very low oxygen content. The surface oxide can be spalled off and separated during the first few cycles of operation. As this process involves high energy consumption, it is economically feasible for producing powders of high value materials such as super alloys and cemented carbides.

(vii) *Eddy Mills and jet mills*: In the Hametag (Eddy mill) process, ductile material (such as fine iron scrap, wires or sheets) is comminuted by rotating propellers in a vessel under an inert protective gas. The plate like particles produced have only a moderate compressibility. The process is highly energy consuming and has low efficiency and hence is rarely used today although much of the iron powder used for machine production prior to the year 1950 was made by this process. New Jet Mills of various designs have been fabricated in recent years and they are mainly used in applications other than powder metallurgy.

(viii) *Mechanical Alloying*: This process, which came into vogue in 1970, was used for developing dispersion strengthened alloys in which strengthening is achieved by the combined effects of precipitations and dispersed oxides. Mechanical alloying is a high energy ball milling process used for producing composites with controlled and even distribution of a second phase in a metallic matrix. This process enables the development of special micro structures which are vital for achieving good high temperature mechanical properties in multi-phase powder metallurgy products. This process consists of milling of mixtures containing the ductile major component for long periods.

Owing to the high energy ball powder interaction (vide Fig. 20.3) the ductile phase undergoes a continuous cycle of plastic deformation, fracture and rewelding process and as a result, fine dispersoids are gradually implanted stepwise into the interior of the ductile phase.

Mechanical alloying can be effectively achieved by attritor milling. Other milling processes may also be used. Longer time periods may be required with low energy ball mills. Mechanical alloying may be used with almost all combinations of brittle phases (*e.g.* Oxide, nitrides, carbides, carbon and intermetallics) and ductile metallic powders and hence is applicable for the development of a broad range of composite materials.

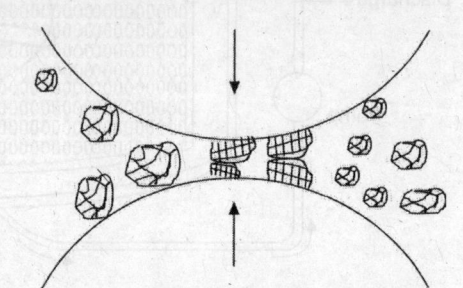

Fig. 20.3. Ball powder interation during mechanical alloying including fragmentation and mini-forging.

(ix) *Processing of metal chips*: This may be of greater significance in future in the context of the need for conservation of raw materials. Powder production from metal chips involves cleaning from impurities, followed by high energy ball vibration milling. The cold deformed powders are annealed in order to improve their compactability. This process may be economical in case of expensive non-ferrous metals (*e.g.* Cu) and alloys (*e.g.* bronze and brass).

The process of manufacturing of metal powder starting from metal chips involves the following steps.
 (i) Cleaning to remove impurities
 (ii) Fragmentation using high-energy ball milling
     The cold deformed powders are usually annealed to enhance their compactibility. From the point of view of conservation of raw materials, this process is expected to be more important in future.

20.2.2. Atomisation

Melt atomisation is the most important method for producing metal powders.

The major stages in this process are
 (i) Melting.
 (ii) Disintegration of the melt into droplets (atomisation).
 (iii) Solidification and cooling.
     In order to impart the necessary properties on the powders, additional processing steps *e.g.* reduction of surface oxides, size classification, degassing, etc. are generally required.
     The classification of atomisation techniques is usually done on the basis of the way in which energy is introduced to achieve disinteg-ration of the melt. These variants are summarised in Table 20.1

Table 20. 1: Classifications of Atomisation Process

Sr.No.	Mode in which the Energy for dis-integration is introduced in the melt	Name of the process
1.	Capillary forces	Melt drop process
2.	Mechanical impact	Impact disintegration
3.	Electrostatic forces	Electro dynamic atomisation
4.	Liquid/gas/ streams or jets	liquid/gas/ atomisation
5.	Centrifugal forces	Centrifugal atomisation
6.	Ultrasonics	Ultrasonic atomisation
7.	Gas super saturation of the melt.	Vacuum atomisation

 (i) *Melting*: In the melting stage, the following two important criteria exist :
   (a) Whether crucible system is required for the melting and melt distribution. (Crucible system is a main source of contamination of the atomized powder product.)
   (b) The type of heating source employed. All the melting techniques employed in metallurgical operations such as induction, plasma, arc and electron beam melting can be used. However, some of them such as arc melting may lead to contamination of the atomized powder product.
 (ii) *Disintegration of the Melt* : In this processing step, the main classification criterion is the manner in which the energy for disintegration is introduced into the melt.
 (iii) *Solidification and cooling* : The rate of cooling is the controlling parameter in this step which in its turn depends upon the dimensions the liquid droplets or the solid powder particles and also on the type the surround medium *i.e.* by radiation, convection and/or contact cooling. The microstructure of the powder particles is determined by
   (a) Cooling rate
   (b) Under-cooling prior to nucleation
        These two factors also are important to determine the required dimensions of the atomisation unit, which has to provide a path, on which the droplets can solidify without touching the unit walls or structure.

The Various types of melt disintegration methods are described below.

(i) Methods based on capillary drop formation:

(a) *Melt drop process*: In this process, the liquid metal flows vertically through the outlet capillary of a tundish, which controls the melt stream diameter. The melt stream breaks up into droplets of a small diameter. In view of the relatively large particle size achievable and the small production capacity, this process is useful only for laboratory scale production of metal powders.

(b) *Impact disintegration process*: This is based on mechanical impact. A smaller particle size can be achieved by impact disintegration of the resulting droplets from the capillary. This process has potential application for the production of rapidly quenched powders.

(c) *Eectrodynamic atomisation*: In this process a DC voltage of 3-20 KV is applied between the capillary orifice and a perforated electrode plate placed in front of it. Melt droplets emitting from the orifice attain a strong positive charge. Particle size in the range of 0.1 to 10 nm can be obtained by using a capillary diameter of 76 nm. Although fine powders can be obtained by this process, the production capacity of the existing units is only in the order of grams per hour and the process is yet to grow to commercial scale.

(ii) Liquid Atomisation

These are among the most important processes for industrial production of metal powders.

(a) Atomisation by liquids: Water atomisation is mainly used for producing iron base powders. A typical water atomisation unit is schematically illustrated in Fig. 20.4.

The starting material is first melted and metallurgically treated in a separate furnace. It is then fed into a tundish which is a device to provide a uniformly flowing vertical melt stream, which is disintegrated into droplets with the help of several water jets, arranged in the focal areas. The impact from the high pressure water stream brings about disintegration of the flowing molten metal. Configuration of the water jets may vary in the number of jets, the angle between the jets and the metal stream, and the focussing of the jets into a point or line. The usual pressure range of the water stream is 6 to 21 M pa., resulting in a velocity of 70-250m^3 per second of the water jets. Throughput is 10 to 100 kg per minute of metal for 0.1-0.4m^3 per inch of water consumed. The overall process efficiency attainable is much better than in the mechanical disintegration, although it still remains below 1%.

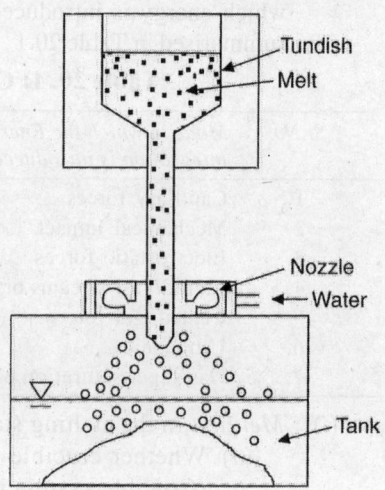

Fig. 20.4. Water atomisation unit.

Since atomisation takes place under unsteady state of turbulent flow conditions, theoretical treatment of the process is rather complicated.

However, on the basis of emperically derived relationships, the most important parameters which control the average size of the powder particles are the water pressure, the water velocity the angle between the water jets and the metal stream, the melt stream diameter, the viscosity of the melt, the density of the melt, the surface tension of the melt and the ratio of flow rate of the melt to that of the water.

The mean particle size obtained by water atomisation process ranges between 30 nm to 1000 nm. Different particle shapes can be produced depending upon the process parameters. With decreasing superheating of the melt, increasing jet velocity and decreasing flow rate ratio, the shape of the particle may vary from nearly spherical to irregular. Irregular particle shapes confer a good green strength of the cold compacted powders and hence are preferred for producing iron and steel powders used for structural parts.

The capital and operating costs of water atomisation are low. However, the possible reaction of the atomised metal with water is a limiting factor. Water atomisation is therefore restricted to metals and alloys of low oxygen affinity, which pick up only small amounts of oxygen or which can be easily reduced in subsequent processing step.

In order to counteract the problem of oxidation, synthetic oils have been used in place of water. However, oil atomisation leads to carbon pickup in the metal powder produced. This can be removed by decurburization. Oil atomisation is industrially used for the production of low alloy Mn-Cr Steel powders.

(b) Gas atomisation: This is the second most important atomisation process of industrial value. The gases used are nitrogen, helium, argon or air, depending upon the requirements of the metal being atomised. Gas atomisation process may be performed either in vertical or horizonal units. A typical scheme of inert gas atomisation is illustrated in Fig. 20.5.

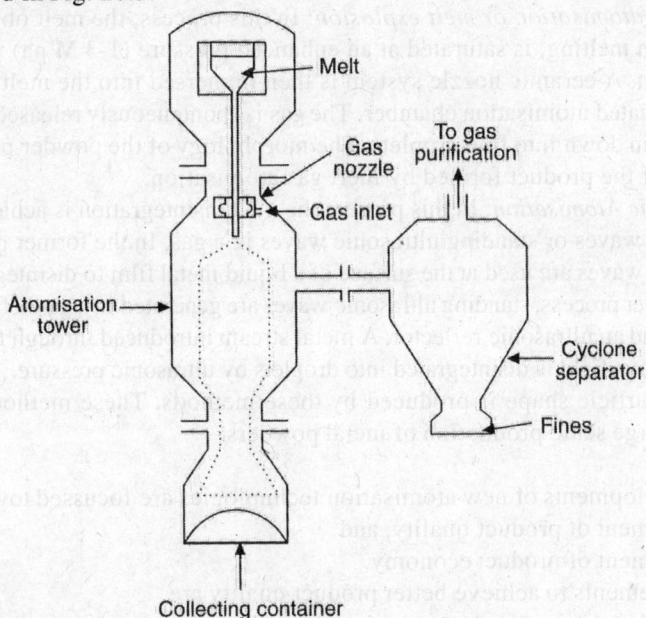

Fig. 20.5. Schematic representation of an inert gas atomisation process.

The design of the nozzle which forms the gas stream or Jets is vital for the success of this process. The overall process efficiency is comparable to that of water atomisation (*i.e.* £ 1%). However, the process costs are higher. The throughput of a single melt stream installation is about 50 kg/min. Gas pressures upto 12 Mpa are used which result in supersonic velocities of the order of Mach 2 and a gas flow rate of about 40 m_3 min^{-1}. The gas exits from cyclone, where the fine powder particles are separated before the gas is recycled into the process. Since the cooling rates are relatively low, the atomisation chambers required are large. The mean particle size of the metal powders produced is in the range of 20-300 nm.

Air atomisation is used for the production of powders of aluminium, aluminium alloys copper, copper alloys, tin, lead and precious metals.

Inert gas atomisation is used for producing high alloy products such as stainless steel, tool steels, iron, nickel, super alloy powders based on nickel, cobalt and aluminium alloy powders.

(vi) *Centrifugal atomisation* : In this process centrifugal force is used for the disintegration of the molten metal. The process can be undertaken in one or two process steps. In the single step process, liquid film is produced on the surface of a rapidly rotating consumable alloy bar by an arc (rotating electrode process), plasma, or electron beam source. The melt cannot be highly super heated because the molten material in that case quickly runs off from the electrode. This problem is solved by the two step process comprising of melting and melt disintegration stages. This is performed with the help of a rapidly revolving cooled wheeled, cup or disc.

(vii) Centrifugal atomisation process yields particle sizes in narrow range.

In single step centrifugal atomisation, the melt does not come into contact with ceramic crucibles or liners. Hence this process is useful for producing refractory powders of tungsten and highly reactive powders of titanium alloys etc. Two step centrifugal atomisation is used for the production of aluminium, titanium alloys, super alloys and powders of refractory metals.

(viii) *Vacuum atomisation or melt explosion*: In this process, the melt obtained by vacuum induction melting, is saturated at an enhanced pressure (1-3 M pa) with a gas such as hydrogen. A ceramic nozzle system is then immersed into the melt and connected to the evacuated atomisation chamber. The gas is spontaneously released thereby breaking the stream down into fine droplets. The morphology of the powder produced is similar to that of the product formed by inert gas atomisation.

(ix) *Ultrasonic Atomisation*: In this process the melt disintegration is achieved by ultrasonic capillary waves or standing ultrasonic waves in a gas. In the former process, stationary capillary waves are used at the surface of a liquid metal film to disintegrate into droplets. In the latter process, standing ultrasonic waves are generated in a gas between an ultrasonic source and an ultrasonic reflector. A metal stream introduced through the pressure points of this wave field is disintegrated into droplets by ultrasonic pressure.

Spherical particle shape is produced by these methods. These methods are yet to find application for large scale production of metal powders.

Future trends:

Future developments of new atomisation technologies are focussed towards :

(1) Improvement of product quality, and
(2) Improvement of product economy.

The major requirements to achieve better product quality are

(a) Ultra high purity
(b) Ultra fine particle size and
(c) Ultra fine cooling rate

Future development are targeted to achieve one or more of the above requisites. Impurities can be reduced by avoiding contact of the melt with electrodes, crucibles or linings. From this point of view, electron beam skull melting may be preferred. For producing ultra fine powders, "explosive evaporation" techniques seem to be preferable. These techniques comprise of introducing very large amount of energy in very short time, thereby causing melting and explosive separation of the material into fine droplets. The energy required may be supplied by electric arc discharge, laser or electron beam heating or capacitor discharging by direct electric current through thin wires.

Ultra rapid cooling rates can be best achieved in combination with ultra fine particle sizes.

20·2.3. Chemical Processes

(A) **Reduction:** This is the most important chemical process in powder metallurgy. Metal compounds such as oxides, carbonates, nitrates or halogenides can be reduced with hydrogen or carbon or highly reactive metals. In most cases, the metal compounds being subjected to reduction are in solid state. However, hydro metallurgical processes have been developed for the reduction of nickel and cobalt solutions using hydrogen at high pressure. Obviously, the reduction processes are governed by the free energy of the reaction involved. The more stable the compound, the stronger should be the reducing media. Alkali and Alkaline earth metals, which form very stable oxides, are the very strong reducing agents.

(i) **Reduction by Hydrogen:** This process is generally employed for producing pure and fine powders of refractory metals (*e.g.* tungsten, molybdenum), ferrous metals and copper. The reduction is carried out at temperatures well below the melting point of the respective metal. The process is often carried out in tube furnaces, in which the powdered metal oxide is moved in flat crucibles which meets the hydrogen stream coming in the opposite direction. The overall chemical reaction may be represented as follows.

$$MeO + H_2 \rightarrow Me + H_2O$$
metal oxide　　　　　　metal

In the actual process, several intermediate reaction steps may be involved, depending on the stability of the lower oxides and the degree of Me-O solid solution. Further, the properties of the resultant metal powder and the reproducibility depend upon the actual conditions in which the reduction process is carried out. Some general rules are given below.

Conditions or reduction	Properties of the product
(a) Higher reduction temperature and longer time	Longer particle size, lower specific surface, lower residual oxygen content and possibly the formation of sintered cake
(b) Lower reduction temperature and shorter time	Smaller particle size, higher specific surface, higher residual oxygen content and possibly the formation of pyrophoric powders.
(c) High flow rate of hydrogen with a low dew point	Higher reduction rate, lower residual oxygen content and no or low reoxidation during cooling

Chemical equilibria are generally defined by the equilibrium constant, K in accordance with the law of mass action.

This process is used for the production of tungsten powder from WO_3 and molybdenum powder from MoO_3. For technical production of tungsten filaments for electric bulbs, tungsten powder of particle size 3-5 nm is required. This is done by adding aqueous solutions of potassium silicate, potassium chloride and aluminium trichloride each in the concentration range of 0.2 to 0.6% to WO_3.

Powders of other metals such as Cu, Ni, Co and Re can be produced by hydrogen reduction of their respective suitable chemical compounds, at appropriate conditions.

(i) Hydro chemical reduction

Powders of metals such as cobalt and nickel can be directly reduced from aqueous or organic solutions by reduction with gaseous hydrogen under appropriate temperature and pressure in an autoclave. This process was developed first by Sherrit Gordon Mines Ltd., Canada for production of cobalt and nickel powder from sulfide ores containing cobalt, nickel, copper, iron and sulfur. They are subjected to froth flotation and the fraction rich in cobalt and nickel is separated. This is leached in autoclaves with ammonium hydroxide under air pressures of 7-9 bar. Nickel and cobalt (as well as some residual Cu) form readily soluble ammines while most of the other impurities are insoluble and remain in solid phase along with the Fe $(OH)_3$ precipitate. A few metals like cadmium and tin enter the solution. The clean solution is then treated with gaseous hydrogen under a pressure of 28 to 35 bar at 180 to 220°C in an autoclave, when the metals are precipitated. Powders of precious metals such as platinum, palladium and gold were produced by hydro chemical reduction of the solutions of their compounds with hydrogen. Gold can be obtained by hydro-chemical reduction process only forom organic solvents. Fine powders of precious metals *e.g.* gold are used in the electronic industry for producing multilayer ceramic capacitors, for conductor plates in thick film applications and conductive tracks on ceramics and glass.

(ii) Reduction by Carbon

This is an important process for producing sponge iron powder, which is the starting material for the mass production of sintered iron and steel parts. In this process, high grade magnetite ores (Fe_3O_4) are subjected to direct reduction by carbon in a tunnel furnace heated by natural gas. The iron sponge formed is crushed, classified and subjected to a final treatment process to adjust carbon and oxygen contents and to reduce residual stresses. The powder is milled to get the required particle size for producing powder metallurgy parts. Powders with particle size > 150nm are sold for the manufacture of welding electrodes.

The reduction process occurs, after an early solid state reaction, mainly via gaseous phase. The overall reaction is

$$Fe_3O_4 + 4\,CO \rightleftarrows 3Fe + 4CO_2$$

The sponge iron powder has some internal porosity, good compressibility and is of irregular shape.

(iii) Reduction by metals

Metallo-thermic reduction using metals like Na, Ca and Mg as reducing agents is a standard procedure in metallurgy for the production of some refractory and highly reactive metals like Ti, Ta, Zr, Th, and U.

Ti Cl_4 + 2Mg → Ti + 2 Mg Cl_2

K_2 Ta F_7 + 5 Na → 2KF+5Na F+Ta (ΔH = – 2985 KJ Per Kg)

(In argon atmosphere or in vacuum at 800 to 900°C)

Zr O_2 + 2Ca → Zr + 2 CaO

Zr Cl_4 + 2Ca → Zr + 2 Ca Cl_2

At suitable temperatures, the metal melts down in situ nascendi, forming a sponge or sometimes an ingot. All metallo-thermic reactions in principle, can be adjusted (*e.g.* by inert additives, batch size and geometry etc.) in such a way that the metal is produced in a powdered form Powders of Be, Ta, Nb. and Th are also of interest industrially.

(B) Carbonyl Process

Metals like iron and nickel react with carbon monoxide under certain conditions to form metal carbonyls, which on thermal decomposition yield fine powders of the respective metal.

POWDER METALLURGY AND ITS INDUSTRIAL APPLICATIONS

$$Fe + 5\,CO \underset{\substack{\text{Decomposition}\\(1\text{ bar, }250°C)}}{\overset{\substack{(200\text{ bar, }2000°C)\\ \text{Reaction}}}{\rightleftarrows}} Fe(CO)_5$$

$$Ni + 4\,CO \longrightarrow Ni(CO)_4$$

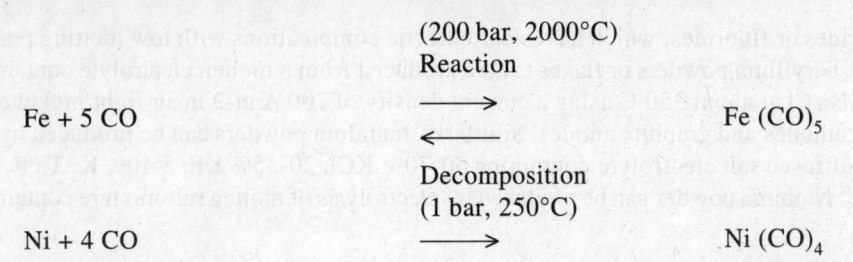

Carbonyl iron powders are used for electronic industries (*e.g.* in soft magnetic alloys) and powder metallurgical applications while carbonyl nickel finds use in several powder metallurgy applications only.

(C) Hydride-Dehydride processes

The hydrides of Zr, Hf, Ti, Ta, U, Th, Y and Mg are of interest for powder metallurgy applications. Very fine and reactive powders of such metals can be produced via hydride formation, which takes place when the metal is reacted with molecular hydrogen at elevated temperatures under a suitable partial pressure. The brittle hydrides can be powdered easily and on dissociation in vacuum at similar or somewhat higher temperatures than those at which they are formed.

Metal powders prepared from hydrides are highly active and often pyrophoric, depending on the conditions of dissociation. Special precautions, such as inert gas absorption on the powdered surfaces, are often needed to avoid their possible instantaneous ignition on exposure to air.

20·2.4. Electro chemical processes

In the electrochemical processes for producing metal powders, the electrolyte consisting of the metal salt either in aqueous solution or in molten form as the case may be, is subjected to electrolysis under appropriate conditions. The positive charged ions migrate to the cathode, get discharged and deposited. The cathodic potential necessary for the deposition of the metal depends upon its position in the electromotive series and on the various inhibiting electrode processes. Powdered metal can be obtained directly or indirectly by milling the deposit at the cathode. Important process parameters include metal ion concentration in the electrolyte, conductivity of the electrolyte, current density, voltage, temperature, bath circulation and wherever necessary, the addition of colloids to inhibit the growth of nuclei.

(a) Electrolysis of aqueous solutions

For electrolysis in aqueous solutions, lower temperature and high current density facilitate the production of metal powders rather than formation of larger deposits. Electrolysis of aqueous solutions is generally conducted at about 50-60°C This process is particularly important for the production of powdered copper and powdered iron, while powders of Co, Ni, and Zr also can be produced. For production of copper, the electrolyte consisting of 5 to 35 g per litre of Cu^{++} and 120-250 g per litre of sulphuric acid operates at 50°C. The electrolysis is performed in a group of plastic lined steel tanks forming a parallel working unit with circulating electrolyte system. The current is 7500 – 10000 A. The cathodic current density is about 4000 A per square meter, the anodic density being 1/10th of this. The total surface area of the negative electrodes is about 1/10th of that of the positive electrodes. The anode and cathodes are made of dense, pure electrolytic copper. The copper powder obtained is washed and dried.

(b) Fused salt electrolysis

Highly reactive metals *e.g.* Be, Ta, Nb and Th which form highly stable oxides, may be deposited by electrolysis of molten salt electrolytes. The electrolyte consists of salt mixtures, mostly chlorides or fluorides, which have near eutectic compositions with low melting points. For example, beryllium powders or flakes can be produced from a molten electrolyte containing Be Cl_2-KCl-Na Cl at about 350°C using a current density of 700 A m-2 in air tight nickel cells with nickel cathodes and graphite anodes. Similarly, tantalum powders can be produced by the electrolysis of fused salt electrolyte containing 50-70% KCl. 20-35% KF, 5-10% $K_2 Ta F_7$ and 4 -5% $Ta_2 O_5$. Niobium powder can be produced by electrolysis of molten salt mixture containing $K_2 Nb F_7$.

20·2.5. Metal Carbides

Carbides of transition metals, particularly those of groups IV A, V A and VII A of the periodic table such as WC, Nb C and Ta C are the basic materials for producing the so-called cemented carbides. These are generally prepared by the following methods.

(a) By direct high temperature reaction between the transition metal and carbon in a graphite furnace in hydrogen atmosphere at 1700⁰C

$$W + C \longrightarrow WC$$

(b) By reaction of the transition metal with hydrocarbons.

$$W + CH_4 \longrightarrow WC + 2H_2$$

(c) Direct carburisation of the transition metal oxides.

$$Ti O_2 + 3C \longrightarrow Ti C + 2 CO$$

Apart from the single metal carbides, solid solutions of two or three metal carbides, nitrides etc. are also of great interest, particularly for use in cemented carbides for cutting or milling purposes. The following methods are used for producing mixed carbides.

(i) Carburisation of mixed oxide or metal powders with soot or graphite powders in a graphite tube furnace at about 1700°C in presence of hydrogen

(ii) Metal carbide powders at 1600-2000°C to enable the formation of solid solution.

20·2.6. Metal Nitrides

Nitrides of metals e.g. Ti, Zr, V, Nb, Ta and W can be prepared by reaction of the respective powder with nitrogen or ammonia at 1200°C or higher, depending on particle size of the powder. Very pure nitrides can be prepared by this method provided the starting materials are pure Nitridation of the metal oxide, in presence of carbon, with nitrogen or ammonia at 1250-1400°C offers a more economic route.

$$2 Ti O_2 + 4C + N_2 \rightarrow 2 TiN + 4CO$$

20·2.7. Metal borides

Pure metal borides can be produced in laboratory by reaction of metal halogenides with boron halogenides as follows

$$Me x_4 + 2Bx_3 + 5H_2 \rightarrow MeB_2 + 10Hx$$

However, industrial production of metal borides can be done by carbothermal reduction of mixtures of metal oxides, $B_2 O_3$ and C at 1700 to 2000°C or by boro thermal reduction of metal oxides and boron at 1500 - to 1700°C

$$TiO_2 + B_2O_3 + 5C \rightarrow TiB_2 + 5CO$$
$$TiO_2 + 4B \rightarrow TiB_2 + 2BO$$

Another type of carbo-thermal reaction is the boron carbide method

$$2TiO_2 + B_4C + 3C \longrightarrow 2TiB_2 + 4CO \ (1700\text{-}2000°C)$$

The boro-thermal method is relatively more expensive because of high price of boron. However the nitrides produced by carbo-thermal method contain 0.1 % free carbon and oxygen as the major impurities.

20·2.8. Metal silicides

Many of the methods given above for the production of carbides and borides can be used for producing metal silicides.

20·2.9. Ceramic powders

The traditional and commonly used silicate ceramics are the naturally occuring raw materials such as clay, kaolin, etc. However, the processing of advanced ceramis requires pure and well defined powders which are generally made synthetically. Further, the manufacture of ceramic parts is more or less similar to powder metallurgy processing. Therefore, advanced ceramics are also usually considered under powder metallurgy products.

Ceramic powders are synthesised by employing a wide variety of techniques which yield powders of varying particle size distribution and varying properties *e.g.* Chemical and phase composition, purity, compactibility and sinterability. Sinter grade powders for producing dense fine grained parts must be produced in the lower micron or even sub micron range coarser powders are used in abrasives (Al_2O_3, SiC) and in refractories (Al_2O_3, ZrO_2, mullite). Some general methods of producing ceramic powders are as follows:

(a) *Solid State reactions:* Many of the oxide ceramics are produced by thermal decomposition of hydroxides, carbonates, oxalates, sulfates and other suitable compounds by the conventional techniques. Depending on temperature and reaction time, loose or agglomerated powder or a sintered cake is produced, which can be milled to obtain the desired particle size.

(b) *Solid gas reactions:* This method is used for synthesising oxides, carbides and nitrides by the reaction of metals with oxygen, hydrocarbons, and nitrogen or ammonia respectively.

(c) *Gas phase reactions:* This method is useful for producing ceramic powders having high specific surface area (100 m^2 g^{-1} and higher). The process comprises of vapour phase decomposition or hydrolysis in a flame and is specially used to produce SiO_2 and TiO_2 from Ti-Cl_4 and Si-Cl_4 respectively.

$$SiCl_4 + 2H_2O \longrightarrow SiO_2 + 4HCl$$

The process can be made continuous.

(d) *Melting:* Ceramic oxides such as Al_2O_3 and ZrO_2 are manufactured by arc melting with subsequent milling to produce the desired particle size. This process in mainly employed for producing coarse grade powders required for the manufacture of refractory materials.

(e) *Reactions from solution:* Aqueous or non-aqueous solution of salts is the precursor in several techniques for producing ceramic powders. This process yields very fine powders and uniformly distributed powder mixtures. This process can be carried out by any of the following techniques :

    (i) precipitation and filtration of Single compounds or mixtures.

    (ii) Hydro thermal reaction at elevated temperature and pressure. This process is particularly suitable for manufacturing pure and doped ZrO_2 based powders starting from $ZrOCl_2$ solution.

    (iii) Solution combustion wherein alcohol based or organo metallic solutions are burnt in air or oxygen to produce oxide particulates.

    (iv) Solvent vaporisation or dehydration by direct evaporation or spray drying or spray roasting or freeze drying of the solution.

    (v) *Sol-Gel process:* The sol-gel technique is receing greater attention for producing advanced ceramics. The technique comprises of the formation of a three dimensional network of inorganic matter (gel) from colloidal or molecular solutions of the precursor (sol). This reaction may take place in aqueous or non-aqueous media such as, by diminishing the water content by changing the PH or the surface charge (Zeta potential) of the sol or by any other procedure that leads to gelation of the liquid precursor.

Manufacture of some important oxide and non-oxide ceramic powders

(a) **Alumina (Al_2O_3):** This is the most important oxide ceramic and is produced in large scale in powder form and to a wide variety of specifications. Very high purity aluminas (>99.99%) are produced by the decomposition of high purity aluminium based salts (such as sulfates, chlorides and nitrides). Powders of lesser purity are produced from bauxite via the Bayer process. After separating Al_2O_3 from Fe_2O_3 and other oxides by converting into sodium aluminate, the latter is converted into $Al(OH)_3$ by hydrolysis which is then calcined to Al_2O_3. It is used in various ceramic applications such as cutting tools, spark plugs, machine parts, electronic parts, bio ceramics, Chinaware and special glasses. 99.99% pure Al_2O_3 is used in sodium vapour lamp tubes to provide the necessary translucency. Electronic applications require high purity, low Na_2O content and easy sinterability.

(b) **Zirconia (ZrO_2):** This is a highly refractory material with a melting point of 2950 K. It is manufactured from Baddeleyite Crude ZrO_2) or zircon sands ($ZrSiO_4$). At 1370 K, transformation of the tetragonal form to the monoclinic form takes place with considerable volume change, due to which it is not possible to sinter pure ZrO_2 to dense, crack free bodies. It is for this reason that ceramic grade ZrO_2 powder must be produced with some additions (e.g. Ce O, MgO, and Y_2O_3) for ensuring fully or partially stabilised materials.

The Zr SiO_4 is converted into sodium Zirconate ($Na_2Zr O_3$) by strong heating with $Na_2 CO_3$ or Na OH.)

$$Zr SiO_4 + 2Na_2 CO_3 \xrightarrow{1270 \ K} Na_2 ZrO_3 + Na_2 SiO_3 + 2CO_2$$

$$Zr SiO_4 + 4 Na OH \xrightarrow{870 \ K} Na_2 Zr O_3 + Na_2 SiO_3 + 2H_2O$$

After removing the $Na_2 Si O_3$ by rinsing with water, the Na_2, $Zr O_3$ is treated with HCl to obtain water-soluble $ZrOCl_2$. At this stage, chlorides of the necessary additives mentioned above are added as required and the solution is hydrolysed to $Zr (OH)_4$ followed by calcination to yield ZrO_2. Sub-micron size particles of the ceramic powder with homogeneous dispersion of the doping elements can be prepared using milling and spray drying steps. It is from these powders that oxide ceramics of highest strength are manufactured.

Silicon carbide (SiC): This is the most important non oxide ceramic which is widely used for abrasives, clay-bonded or porous recrystallised high temperature-or wear-resistant materials and for manufacturing dense sintered parts in conjunction with sintering additives e.g. Carbon or boron. Silicon carbide exists as hexagonal α-SiC and cubic β-SiC. The SiC which is the stable form, was first manufactured in 1891 by Acheson process.

$$SiO_2 + 3C \rightarrow SiC + 2CO$$

The starting material is pure sand, pertoleum coke, some saw dust to decrease packing density and 1-3% of NaCl for purification. This is heated with graphite electrodes and a temperature of 2000-2300°C is reached during a 30 hour cycle which is essential for good crystallisation of the SiC.

β-SiC is obtained by gas phase reaction of $SiCl_4$ with hydrocarbons or by pyrophysis of trimethyl silane,

[$(CH_3)_3$ SiH] or trimeltyl chlorosilane [$(CH_3)_3$ SiCl],

POWDER METALLURGY AND ITS INDUSTRIAL APPLICATIONS 677

at a temperature range of 800-1500°C with argon or H_2 as carrier gas.

Silicon Nitrides ($Si_3 N_4$): It occurs in two polymorphous forms, α- and β- both of which are hexagonal. The C-axis in the α-form is nearly double the spacing of the β-form.

Silicon nitrides powder is produced by reaction of Si or Si compounds with N_2 or NH_3 at 1000-1400°C when both α- and β- form of the compounds are formed

$$3 Si + 2 N_2 \rightarrow Si_3 N_4$$
$$3 Si + 4 NH_3 \rightarrow Si_3 N_4 + 6H_2$$

Very fine and pure powder of silicon nitrides can be produced commercially by Diimid process by the reaction of $SiCl_4$ with ammonia

$$Si Cl_4 + 6 NH_3 \rightarrow Si (NH)_2 + 4 NH_4Cl$$
$$3 Si (NH)_2 \rightarrow Si_3 N_4 + 2 NH_3$$

Sintering to high density can be achieved by using special additives eg. $Al_2 O_3$, MgO and $Y_2 O_3$

Boron Carbide: It is synthesised by the reduction of boric oxide with carbon at temperatures over 2500°C (ESK electro furnace process), followed by milling and purification.

$$2B_2 O_3 + 7C \rightarrow B_4C + 6CO$$

20·2.10. NANO CRYSTALS

These have particle size in the range between 50nm to 20nm, and are also known as ultra-fine powders. They were obtained by various methods such as

(a) Gas phase reactions in DC, R F or microwave plasma (ex: $Si_3 N_4$, metallic powders)
(b) Gas phase reaction powered by laser (ex :- SiC)
(c) Hydrogen reduction of metal chloride vapours (ex.:- powders of Ag, Ni, Cu, W etc.)
(d) Evaporation and condensation techniques (ex:- metallic powders of Fe and Fe Ni)

The powders react with oxygen and may be pyrophoric. However, under controlled exposure to oxygen, they can be passivated.

Nano-crystalline powders are used for high density sintered products but more often in the form of bonded powders for use as catalysts (e.g. Ni, Pd), conductive pastes (ex: Ag, Cu, Ni), magnetic recording media, (Fe-Co-Ni) alloy powders and micro porous filters (ex. Ni).

20·2.11. WHISKERS

These are short metallic or non-metallic fibres, which are generally mono crystalline. Whiskers are usually deposited by vapour phase processes. Whiskers of copper and silver were known for a long time as naturally grown minerals. The conditions for whisker formation (instead of formation of crystal or crystallites) is a strongly anisotropic unidirectional growth conditions within a particular range of temperature and concentration. Whiskers of many metals and ceramic materials (e.g. SiC, $Al_2 O_3$ etc.) can be easily prepared Industrial scale production of whisker is limited because of safety considerations about their handling.

Whiskers are mostly manufactured from volatile compounds from which deposition take place on solid or liquid substrates. Metallic whiskers can be prepared by the reduction of metal halide vapour with hydrogen in a carrier gas.

$$2 FeCl_3 + 3H_2 \rightarrow 2Fe + 3HCl$$

Non-metallic whiskers such as those of SiC, $Al_2 O_3$ and $Si_3 N_4$ find greater application in the development of high performance materials as compared to metallic to whiskers. Cubic β-SiC whiskers, which are widely used commercially are obtained by vapour liquid solid (VLS) process, in which a hydrocarbon (e.g. CH_4) is allowed react with a volatile silicon compound (e.g. Si Cl_4, Si O) at 1400°C in presence of liquid iron alloy as catalyst. A cheaper process is the pyrolytic treatment of rice husk which contains SiO_2 which reacts with the organic matter to yield whiskers of SiC.

Whiskers can be prepared in a wide range of diameters (from 0.1 nm to several nm) and lengths (10 to several hundred nm, and in some cases upto several mm) depending on the type of material and process conditions). The cross section of whiskers may be hexagonal, rectangular or even irregular, depending upon crystal structure and growth conditions. The extremely high strength of whiskers depends on their diameter and length, the smaller the diameter the greater the strength. Whiskers can be incorporated into normal powder metallurgy process such as mixing compacting and sintering. This is the major advantage of whiskers in their use for reinforcement.

20.3. CHARACTERISTICS AND PROPERTIES OF METAL CERAMIC POWDERS

The properties of the final powder metallurgy parts depend to a large extent on the properties of the starting powders. The powders are considered as dispersed systems which consist of atleast one disperse phase and continuous surrounding medium. The dispersed phase, in its turn, consists of a number of individual elements namely the particles. Dispersed systems are generally characterised by qualitative description of the individual disperse elements and by the quantitative particle and structural characteristics of the system. In metal or ceramic powders, a variety of disperse elements or particles can exist. The powder metallurgical terminology used for describing these elements is illustrated in Fig. 20.6.

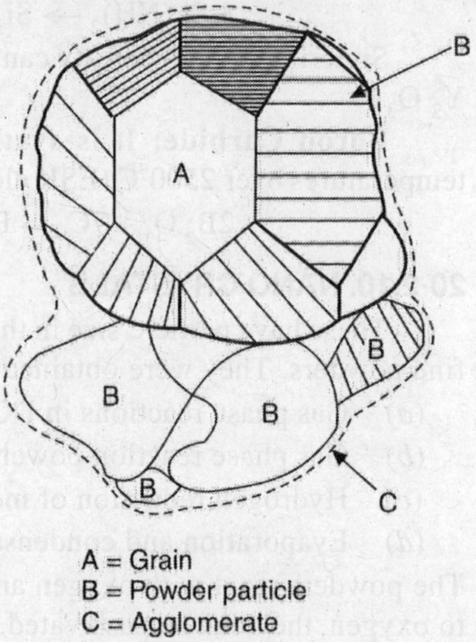

The term powder particle denotes only the primary particles that are formed as individual particles during the production process and not those developed by agglomeration of other particles. These powder particles can be crystalline or amorphous. Crystalline particles can be either single crystals or polycrystalline (grains). Polycrystalline powder can be single or multi phased. The powder particles can form secondary particles or agglomerates (vide Fig 20.6 above) Agglomeration generally occurs unintentionally in powder metallurgy process, but it can also take place intentionally by controlled stages.

A = Grain
B = Powder particle
C = Agglomerate

Fig. 20.6. Powder elements in powder metallurgy terminology.

The measurable properties of disperse elements are called the particle characteristics which include fineness, particle size, particle shape, etc. The particle shape of the powder is usually characterised qualitatively as dendritic, acicular, fibrous, flaky, spheroidal, granular, nodular, angular, irregular, etc. some of which are shown in Fig. 20.7.

However quantitative characterisation requires application of shape factors that represent relationship of linear 2- or 3-dimensional parameters.
Particle size analysis can be carried out by a variety of methods such as counting methods, sedimentation methods, classification methods, light scattering methods, diffraction methods, photon correlation spectroscopy, field flow fractionation and hydro dynamic chromatography.

Surface analysis of powder particles can be done by methods such as permeametry, gas adsorption, etc.

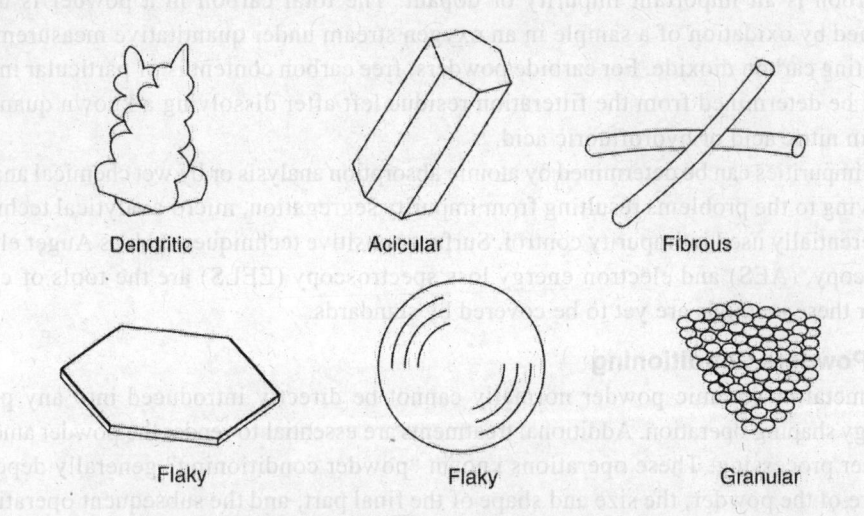

Fig. 20.7. Some typical powder particle shapes.

20.4. TECHNOLOGICAL PROPERTIES OF POWDERS

The behaviour of powders during powder metallurgy operations such as powder handling, conveying, compaction, etc. though related to the powder characteristics in a complex manner, is difficult to predict on that basis. Therefore, test procedures such as powder flowability, apparent density, tap density and compact ibility, have been developed which are useful for comparative characterisation of powders with respect to powder metallurgy processing. These test procedures are also covered by standards.

Powder flowability is measured as the time required for 50 g of a powder to leave the flowmeter (*e.g.* Hall flowmeter) under the influence of gravity.

The apparent density of a bulk powder is defined as the powder mass divided by the bulk powder volume. Standard test procedures for its measurement are based on the Hall flowmeter or the Scot volumeter. The tap density is the apparent density of a powder when packed vertically, by vibration into a calibrated measuring cylinder manually or by a standard tapping apparatus at frequency of 1.7 to 5 Hz and an amplitude of 3 mm. Compactability characterises the densification behaviour of bulk powder in terms of compact pressure and compact density. This is determined with the help of a standard floating die apparatus.

20.5. IMPURITIES IN POWDERS

Owing to the high specific surface of powders, they pick up impurities from the surrounding medium by adsorption or chemisorption during different stages of powder production and processing. Even small quantities of these impurities can sometimes completely change the properties of the final material which may render it defective and unacceptable. The most important impurities are oxygen, carbon, and metallic impurities.

Oxygen is the very important impurity in metal and non oxide ceramic powders. Standards are available for the determination of the oxygen content of the metal powder by reduction methods and it is possible to distinguish between hydrogen loss, hydrogen reducible oxygen, and total oxygen by the procedures given. The hydrogen loss of powder is defined as the mass loss divided by the total mass during hydrogen annealing of a sample.

Carbon is an important impurity or dopant. The total carbon in a powder is usually determined by oxidation of a sample in an oxygen stream under quantitative measurement of the resulting carbon dioxide. For carbide powders, free carbon content is of particular interest. This can be determined from the filteration residue left after dissolving a known quantity of powder in nitric acid or hydrofluoric acid.

Metallic impurities can be determined by atomic absorption analysis or by wet chemical analyses.

Owing to the problems resulting from impurity segregation, micro analytical techniques are preferentially used in impurity control. Surface sensitive techniques such as Auger electron spectroscopy, (AES) and electron energy loss spectroscopy (EELS) are the tools of choice. However these methods are yet to be covered by standards.

20.6.1. Powders Conditioning

A metal or ceramic powder normally cannot be directly introduced into any powder metallurgy shaping operation. Additional treatments are essential to render the powder amenable for further processing. These operations known "powder conditioning" generally depend on the nature of the powder, the size and shape of the final part, and the subsequent operations to be carried out.

Metal powders, during their processing, have to be handled, conveyed and filled into moulds or dies, and often cold-or hot-compacted. Therefore, the powders must be rendered amenable for such operations and also impart such characteristics which provide the basis for the desired properties for the green compact and final sintered components. This can be done by adding temporary or permanent additives. Temporary additives remain during some process steps and are removed in later processing whereas permanent additives like alloying additions remain in the material during the entire process. Introduction of these additives and behaviour of the conditioned powder during its transport, compaction or forming are the fundamental problems of powder conditioning.

Temporary additives are used for lubrication, plasticizing and binding. Lubricants and plasticizers help in the movement of the powder particles under externally applied forces by reducing inter-particle and die wall friction forces. Binders help in enhancing the strength of the green compacts, so as to withstand the stresses arising during subsequent handling.

The particle sizes of powders vary widely, ranging from a few nanometres upto some hundred micrometers. The handling and processing behaviour of such powders is totally different at the fine and the coarse ends of this range. This is mainly due to the relation between surface attractive forces and inertial forces, which increases strongly with decreasing particle size. This causes problems in handling and transportation of fine powders, which tend towards incontrolled agglomeration, thereby losing their flowability. Fine powders may also cause dust emissions and severe damage to the compaction tools due to adhesive wear if they are entrapped in the gap between the die and the punch. Therefore very fine particles or dust have to be removed by classification. Further, for fine powders used in producing hard metals or ceramics, the inertial forces have to be enhanced and surface attractive forces have to be reduced. The former is usually achieved by enlarging particle size by controlled agglomeration (granulation) of the primary particles whereas the latter can be accomplished by wet processing, *i.e.* by dipersing the powders. Therefore in powder condition, special attention is required towards problems such as mixing or dispersing, segregation and agglomeration.

Blending is the process of inter-mixing of powders of the same nominal composition to achieve a desired particle size distribution.

Mixing is the process of inter-mingling powders of different chemical composition.

A variety of equipment is available for blending, mixing and dispersing are available.

Tumble mixers utilize gravitational forces for causing motion of the particles. Shear-agitated mixers employ paddles or other moving components in a stationary container for shearing planes within the bulk powder. Centrifugal mixers utilize centrifugal forces to control the particle motion.

In fluidised bed blenders the contribution of convection and diffusion to particle motion and mixing is greater as compared to the above mixers. Blending by fluidised bed blenders is very efficient and is free from contamination by abrasion products. However, it is sensitive to segregation caused by differences in particle size, shape or density. However, tumble mixers and low shear mixers are generally used for dry mixing and blending of metal powders, due to possible degradation of particles in higher shear equipment.

The tendency of segregation in a powder or powder mixture mainly depends on the differences in particle density, particle size and particle shape.

Agglomeration (granulation) allows enlargment of very fine powder particles so as to render them free flowing and safe. The bonding between the particles in either controlled or uncontrolled agglomeration process can be due to the following mechanishms;

(*i*) solid bridges formed by crystallized salts or sintering contacts.
(*ii*) Immobile liquid bridges formed by viscous binders, adhesives and absorption layers.
(*iii*) Mobile liquid bridges (capillary forces).
(*iv*) Van der waals forces.
(*v*) Mechanical interlocking

The forces of adhesion increase with decreasing size of the primary particles.

The following are the methods of granulation :

(*a*) Layering agglomeration which involves fragmentation and agglomeration which occur simultaneously, with greater probability of the latter, thereby resulting in increasing agglomerate size with processing time. This process enables in getting a relatively homogenous arrangement of primary particles and uniform distribution of the binding agent added for granulation. Granules with diameters upto several millimeters can be produced by this method.

(*b*) Press agglomeration which takes place under high pressure and results in agglomerates having high strength and low porosity and of well defined shape and size.

(*c*) Spray drying in which powder suspensions are atomized into droplets, which are dried by evaporation of the liquid during free fall of the droplets.

In powder metallurgy practice, granulation is necessary for particles only in the sub sieve range (10-5 nm downwards). Granulation is applied mainly for systems which are dispersed in a suspension in a previous processing step (eg. Sol – Gel methods). In such cases, spray drying is the method of choice

Agglomerate strength, its shape, size distribution and porosity are the important parameters which control flowability and tap density. The granules formed must be strong enough to withstand handling and transportation, but weak enough to be completely deformed during shaping. Otherwise, it may cause inhomogeneities in the green compact thereby resulting in agglomerate related defects due to differential sintering.

20·6.2. Heat Treatment

The various objectives for performing heat treatment of metallic powders include chemical reduction, decarburisation, annealing, degassing and size enlargement. The chemical reduction and decarburisation processes of heat treatment have been already discussed earlier under the section on powder production techniques.

Raw metallic powders produced from various production processes generally exhibit reduced compressibility due to rapid cooling during atomisation, work hardening from mechanical size reduction, residual interstitial impurities such as carbon, oxygen or nitrogen, and also due to oxide layers formed during production or storage. Annealing of metal powders helps in attainment of good subsequent compressibility. During annealing, residual stresses are relieved by recovery and recrystallisation. Alloying elements in supersaturation can be precipitated in the form of relatively coarse particles, thereby reducing the work hardening tendency during powder compacting. Apart from these micro structural effects, annealing also helps in reducing the levels of impurities *e.g.* Oxygen, carbon, and nitrogen. In many cases, decarburisation is accomplished simultaneously with deoxidation by the reaction of carbon and oxygen present in the powder to form carbon monoxide. Further, if a reducing atmosphere is provided by using hydrogen or dissociated ammoina, further lowering of oxygen content and prevention of reoxidation can be accomplished. A very striking example is carbon iron powder containing 0.8% C and 2% O in the initial powder which after hydrogen annealing were reduced to 0.02% of C and 0.15% O. Martensite and ferritic stainless steel powders are annealed after atomisation to enhance compactibility.

Superalloy or high performance spherical powders such as titanium or gas atomised tool steel powders are usually consolidated by hot compaction in sealed claddings. Absorbed gases (eg. O.A or N) from atomisation and subsequent handling can be done by evacuation degassing methods at ambient or elevated temperatures. Unintentional size enlargement by heat treatment results from sintering of powder particles. This necessitates a subcequent crushing of the annealed product agglomerates. That is why heat treatment processes are conducted at as low temperatures as possible to minimize sintering and to allow for a subsequent mild comminution without significant strain heardening.

20.7. SAFETY ASPECTS DURING HANDLING OF METAL POWDERS

20.7.1 Toxicity: The toxic effects of metal and alloy powders to the humans depend upon their size, density, the biochemical characteristics of the material, the route of its entry into the body (*i.e.* by inhalation, oral ingestion, or absorption through skin) and its dosage. Larger powder particles (> 12nm) mostly get lodged in the upper respiratory tract but are cleared in a short time. The particles are expelled by coughing or are swallowed and either reabsored in gastro-intestinal tract or excreted in the faecies. Smaller particles tend to pentrate further down into the aveolar region of the lungs, with half lives ranging from a few months to years. The presence of particulates in the respiratory tract may cause allergic, fibrogenic and immunological responses. The particles may also be absorbed to other organs by the bloodstream. The permissible exposure limits of the metal and alloy powders are regulated in many countries in the form of threshold limit values (TLV) or maximum exposure limits (MEL).

Toxicity due to inhalation of Ni powder by workers can be prevented by masks during its handling. Nickel Carbonyl powder is extremely toxic. "Berylliosis" may be caused by inhalation of fine particulates or dust containing Be. Toxicity due to Cd and Pb is a health risk during manufacture of bearings using fine powders of these metals. Similarly cobalt toxicity is signisficant in the hard metal industry. Chromium powders are also toxic. SiC Whickers

(< 3 nm) have been found to be carcinogenic after inhalation.

20.7.2 Pyrophoricity and Explosivity: Finely divided metal powders can be hazardous due to their thermal instability in ambient air. If the heat of reaction is generated more rapidly than its dissipation into the environment, metal powder may become self-igniting at ambient temperatures. This property is called "pyrophoricity" and it depends directly on the particle size, surface area and the heat of formation of the respective metal oxide. In powder metallurgy, metals like Al, Cr, Co, Cu, Fe, Ni, Ta, Ti and compounds like TiB_2 are known to be pyrophoric. The risk of pyrophoricity increases with decreasing size of the powder particles. For instance, Al powder of particle size of about 0.03 nm and Zr powder of particle size of about 3 nm are highly pyrophoric. However porous powders of Ni having Fe or Cu having much larger particle size, when prepared from their respective oxides at low reduction temperatures may also exhibit pyrophoricity if the surface area approaches 1 square meter per gram. The presence of moisture tends to increase pyrophoricity because the hydrogen released during the reaction often lowers the ignition temperature. Protective films of thin oxide or stearate reduce the pyrophoric tendency of metal powders.

Pyrophoricity should be clearly distinguished from "autogeneous" ignition. The latter refers to the lowest temperature at which combustion begins and continues in a metal powder when it is heated in air.

Explosivity of metallic dust clouds is a more important hazard encountered during processing and handling of metal powder. The dust clouds may be generated in pulverizers, mixers or in operations such as powders pouring, stirring or in powder elevators. Flames and mechanically or electrically generated sparks are the major sources of ignition for metallic dust clouds. The explosivity of a metal powder depends on physical factors (*e.g.* Small particle size and high specific surface area), chemical factors (*e.g.* Chemical protective layer such as a stearate or thin oxide films) and atmospheric conditions such as oxygen content of the ambient air. The table given below shows the classification of metal powders on the basis of their explosivity.

Table 20(2) Conditions of Explosivity of Some Metals

	Highly explosive	Moderately explosive	Slightly Explosive
Powders	Zr, Mg, Al	Cu Fe, Mn, Zr, Sn, Si,	Pb, Co, Mo
Percentage of oxygen in ambient air	<3%	>3%	>10%
Ignition Temperature	<600°C	300-800°C	>700°C and strong ignition source
Explosive limit	20 to 50 g/m³	100 to 500 g/m³	very high powder concentration

The destruction potential of powder explosions is very high if the explosirity conditions are favourable. The following precautions will go a long way for minimising powder dust explosions.

(*i*) Installation of venting systems in areas where protection of critical tube or vessel systems by the explosion exists
(*ii*) Protection of critical tube or vessel systems by careful design to make them strong enough to bear the explosion pressure.

(iii) Protection of critical tube/vessel systems by bursting disks
(iv) In case of fire, only dry powder extinguishers should be used
(v) Minimising the dust cloud generation or dust accumulation
vi) Proper awareness about explosive characteristics of metal powders.

20.8 COMPACTION AND SHAPING

Compacting and Shaping are the processes by which bulk powders are transformed into preforms of desired shape and density with or without applying external pressure. However there are some process variants which combine shaping operation with sintering step and in such a case, the end product is a sintered part instead of a preform.

1. Pressure assisted shaping

The pressure assisted forming operation can be sub-divided into cold compaction methods and hot compaction methods.

(A) **Cold compaction methods:** Cold compaction takes place in a temperature range (mostly at ambient temperature) within which high temperature deformation mechanisms like dislocation or diffusion creep are negligible several methods of cold compacting are available.

(i) **Cold pressing**

This is the most important compaction method in powder metallurgy. The starting material is the bulk powders without or with very small amounts of lubricant or binder additions. The cold pressing process may involve axial die pressing or isostatic pressing principles, which are illustrated in Fig. 20.8.

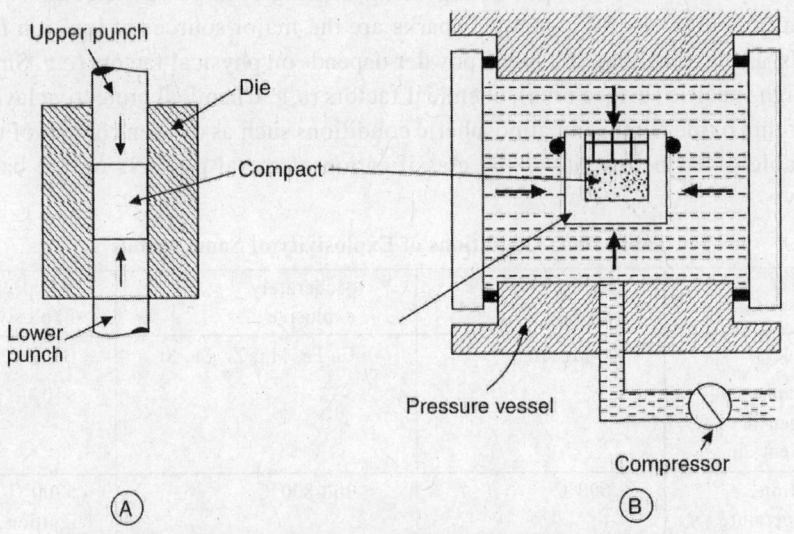

Fig. 20.8. Schematic representation of cold pressing principles:
(A) Axial pressing (B) Isostatic pressing.

In axial pressing, the powder is compacted in rigid dies by axially loaded punches. The main process variable in this technique is the axial compaction pressure, which is defined as the ratio of the punch load and the punch surface area. On the other hand, in isostatic pressing, the powder is sealed in an elastic mould and exerted to the hydrostatic pressure of a liquid pressure medium.

Axial pressing is the most important practical forming method which is very economical for mass production of precision parts. In this process, the powder is compacted between the punch faces and die walls, which undergo only very limited elastic deformation. Therefore, the compacts can be fabricated to very close geometrical tolerances. The compaction sequence of die filling, compaction and ejection of the compact can be carried out in both hydraulic or mechanical presses or as mixed mode presses at high production rates. In axial pressing compaction pressures beyond 600 M pa are feasible and sometimes used in practice, but give rise to increased tooling costs. Depending on shape complexity, axial pressing technique requires large quantity output to compensate for the tooling costs and hence is competetive only in mass production.

Isostatic cold pressing is perfomed either by "wet bag" or "dry bag" methods. In wet bag tooling, the flexible bar or the mould is filled outside the pressure vessel sealed outside the pressure vessel. The bag is fully surrounded by the pressure exerting fluid. In dry bag tooling, the bag is permanently sealed within the pressure vessel and powder filling takes place without removing it out of the vessel. Wet bag tooling is more versatile but requries manual handling of the filled bag. Dry bag tooling is more amenable for mass production, though limited in shape flexibility. Liquids such as special oils or water with anti-corrosive and lubricating additives are generally used for exerting pressure. The major criteria for selecting pressure medium include its price, availability and compressibility. Tooling for isostatic pressing differs considerably from axial pressing tools. Elastomers such as natural and synthetic rubber, silicone rubber, PVC and polyurethane are generally suitable as bag materials. The criteria for selecting a bag material are compatibility with the pressure fluid, the required stability of the mould (thin or thick wall) and the quantity of parts to be produced within the life span of the mould. Rubber and PVC moulds, which can be produced by simple dipping methods, are often preferred for single use tooling. Polyurethane is the most commonly used bag material for multi use toolings. However, polyurethane moulds are relatively expensive due to complicated casting techniques required for mould productions.

Isostatic pressing offers some advantages over axial die pressing. Isostatic pressing provides better shape capability and more homogeneous density distribution within the compacts.

Isostatic cold pressing method is used for manufacturing spark plug insulators, semi finished products as well as net shaped parts from metal and ceramic powders, other semi finished products such as blocks of high speed steels or aluminium alloys, shaped articles such as filter elements, tubes and crucibles, balls for use in bearings or as milling media, high speed steel tool preforms and special structural parts with threads or undercuts, which are difficult to produce by die compaction.

(*ii*) **Powder injection moulding (PIM)**

This process is wellknown in the manufacture of parts from thermoplastic polymer materials and injection moulding is a very important forming method applied for large scale production. In the context of powder metallurgy, this process is called metal injection moulding. Fig. 20.9 illustrates the flow chart with the major steps involved in process.

The moulding operation requires a preconditioning step in which the powder is transformed into a mouldable state, the so called 'feedstock'. This is accomplished by mixing the powder with binders (mostly organic), which renders the powders to flow under moulding conditions, but at the same time provide good shape stability and green strength of the moulded parts. These requisites are satisfied by modifying the rheological behaviour during the moulding operation. The particle size of PIM base powder is generally chosen to be <25 nm. Fine powders decrease the size of moulding defects and increase simterability. However the lower size limit of the base powder is determined by the solids volume fraction and the acceptable debinding time. The most widely used binder systems are thermoplastics such as polymers and waxes.

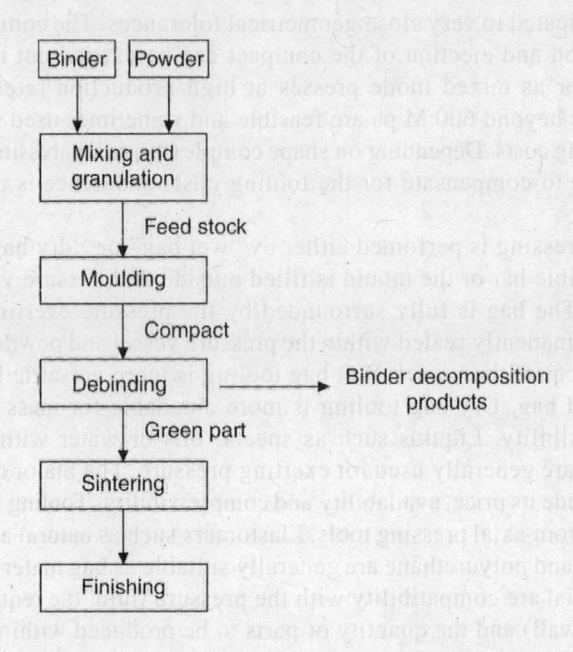

Fig. 20.9. Flow-chart of the major steps involved in the powder injection moulding (PIM) process.

Moulding of a thermoplastic feedstock consists of heating it above the melting temperature of the binder, forcing a sprue and runner channel system into the mould cavity, and then cooling the mould and solidifying the powder-binder mixture. Prior to sintering, the binder has to be removed from the compact without particle disruption and without contaminating the compact. At the same time care should be taken to see that the compact shape is retained during this operation. Two major debinding techniques are used, namely thermal de binding and solvent debinding. Thermal debinding is carried out by evaporation, thermal decomposition or by extraction of the liquid binder by a wick substrate or powder bed. Solvent extraction is carried out by immersing the compact in a solvent, which partially dissolves the binder.

The major components of injection moulding equipment are shown in Fig. 20.10.

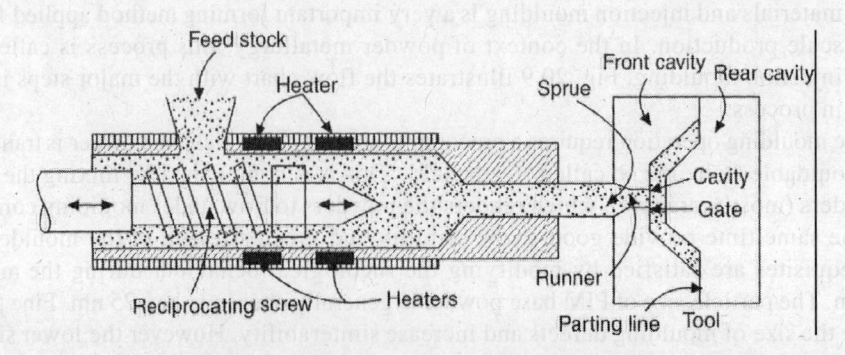

Fig. 20.10. Schematic representation of injection moulding equipment.

PIM enlarges the shape capability of the powder metallurgy process and is mainly used for parts with very complex shapes. This process competes favourably with forming techniques like precision casting or machining of powder metallurgy preforms. Its main advantage is the possibility of production of net shape complex parts via the powder route as with powder materials like cemented caribides, ceramics, tungsten heavy alloys and superalloys. PIM process is superior to other forming techniques particularly for producing short fibre or whisker reinforced metal matrix composites (MMC).

(*iii*) **Cold Powder extrusion**

This is a plastic forming method (similar to injection moulding), where a plasticised powder system is forced by a piston or screw unit into a forming die. A typical powder extrusion equipment is schematically illustrated in Fig. 20.11.

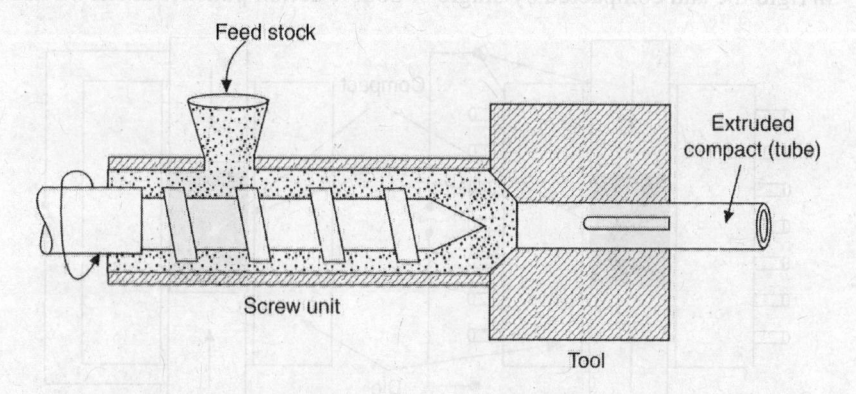

Fig. 20.11. Powder extrusion equipment (Schematic representation)

It differs from injection moulding mainly in tooling, which limits extrusion to products of constant cross section, such as tubes and rods. Also in this method, the possibility of changing the rheology of the extrudate during the process by cooling or heating is very limited, because it is a semi-continuous process. Therefore the powder binder systems must provide flow under the extrusion pressure, and also the shape stability of the extruded compact under gravity and handling stresses. Cold powder extrusion is an established technology for the manufacture of traditional ceramic products like bricks, tiles, stoneware tubes etc. In powder metallurgy, this technology is used in producing rod shaped preforms for hard metal drilling tools, etc.

Industrial extrusion pressures lie in the range of 4-15 MPa. An extrusion velocity of l M per minute is common. Tooling can be very sophisticated, as seen in ceramic honeycomb catalyst substrates with upto 200 channels per sq. cm. and channel walls as thin as 0.2 mm. Plasticizers employed are organic binders with medium and high viscosity. However, in case of oxide ceramics where extrusion is widely applied, aqueous binder solutions are used. In case of metal powders non aqueous extrusion based on solvents like alchol, petroleum and liquid wax are used.

(*iv*) **Roll Compaction**

In this method, metal powders are compacted in the gap between two rollers. This is an alternative technique for producing semi-finished products like strips and sheets. The most notable commercial application of this process is the production of high purity strips and sheets of Co and Ni. The stirp is used for making coin blanks for the mints, electronic tubes, batteries, thermostats and other bimetal applications.

(B) Hot Compaction

In this process, the deformation mechanisms of the powder are activated by the simultanous application of higher processing temperature and extermal pressure. The following are the major hot compaction techniques:

(i) Axial hot pressing
(ii) Isostatic hot pressing
(iii) Hot forging
(iv) Hot extrusion

In most of these techniques, final forming, shaping and sintering of the powder compact are included.

(i) **Axial hot pressing:** In this process, the powder or a precompacted preform, is placed in rigid die and compacted by single or double action punches as shown in Fig. 20.12.

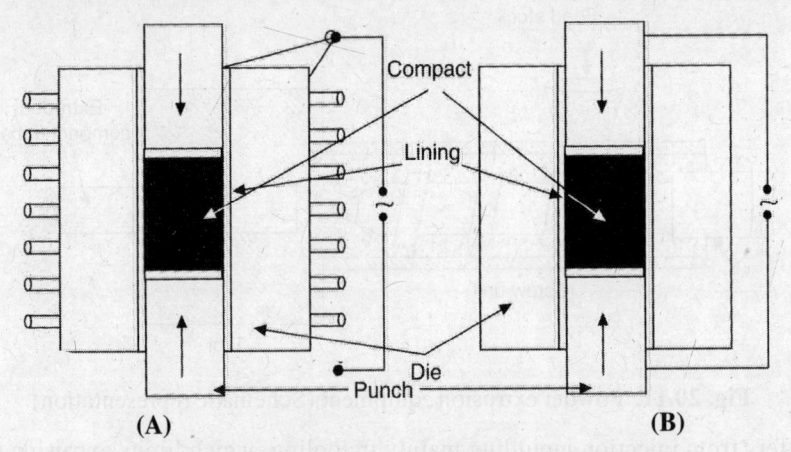

Fig. 20.12. Tooling for axial hot pressing
(A) By induction heating (B) By direct current heating.

The tooling material has to be chosen on the basis of strength and compatibility with the compact under the hot pressing conditions. Graphite is most commonly used as it enables pressing temperatures of even 2500°C. other materials that are used include carbon fibre reinforced carbon (CFC), dispersion hardened molybdenum (TZM), tungsten, cabalt alloys (stellites), heat-resistant alloy steels and ceramics. The compact can be heated by direct resistance heating or by heat transfer from the tool which itself can be resistance heated by direct or induction currents, or by convection or radiation from an external heat source. For most metal and alloy powders, an atomosphere of inert or reducing gases or vacuum is used keeping in view the adaptability of the compact material. Compaction pressures upto 50 M pa are used in graphite dyes and upto 100 M Pa are used in steel alloy dies. Lower pressures are used for liquid phase systems to avoid squeezing of the liquid out of the compact. The cycle times vary from several seconds upto an hour.

Axial hot pressing technique of fabrication is used for many non oxide ceramics (*e.g.* SiC, Si_3N_4, B_4C, BN), for metal matrix composites like metal bonded diamond tools, beryllium pieces and blocks and cemented carbide parts.

(II) Isostatic Hot Pressing

Hot isostatic pressing (HIP) is a hot compaction technique under isostatic pressure conditions. The three main HIP processes variants are give below:

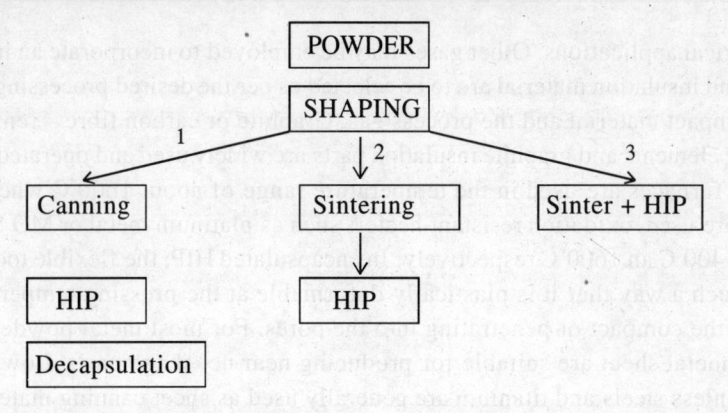

```
            POWDER
              │
            SHAPING
         ┌────┼────┐
         1    2    3
         ▼    ▼    ▼
      Canning Sintering Sinter + HIP
         │    │
         ▼    ▼
        HIP  HIP
         │
         ▼
     Decapsulation
```

Encapsulated HIP, using a suitable tooling or canning system, is the only processing route which permits pressure aided densification over the complete range from green compact density to theoretical density. Containerless HIP has two sub-variants Viz post-HIP and sinter HIP. Post HIP is the hot isostatic compaction of preforms which are sintered in a separate cycle to the density level of 90-93% T.D. (theoretical density) whereby the pores are closed. Sintered HIP is a continuous process combining pressureless sintering to allow pore closure and subsequent hot isostatic pressing without intermediate cooling of the compact. A schematic representation of a HIP unit is given in Fig. 20.13.

It consits of a cooled-wall pressure verssel contaning a furnace and the workload. The process gas can be pressurised with the help of a gas compressor system. A sophisticated insulation is provided to prevent heat losses to the cooled vessel walls. The commonly used process gases are the inert gases such as helium and argon. In sintering HIP consolidation of $Si_3 N$, nitrogen is used as an inert as well as is reactive process

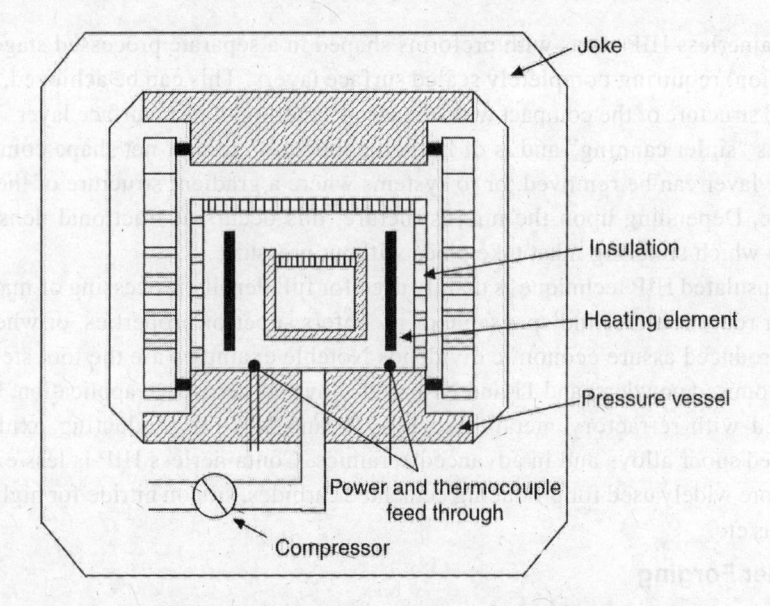

Fig. 20.13. Schematic overview of a HIP unit.

Atmosphere where it suppresses the high temperature decomposition of the nitride. As per recent developments, oxygen atmosphere is used for hot compaction of oxide ceramic materials used for electrical applications. Other gases may be employed to incorporate an in situ coating. The furnace and insulation material are to be selected as per the desired processing temperature, capsule or compact material and the process gas. Graphite or carbon fibre – renforced carbon (CFC) heating elements and graphite insulation parts are widely used and operated upto 2200°C. Molybdenum furnaces are used in the temperature range of about 1600°C wherever oxygen atmoshperes are used, oxidation resistant heaters such as platinum metal or MO Si_2 are used to operate upto 1400°C an 1600°C respectively. In encapsulated HIP, the flexible tooling has to be designed in such a way that it is plastically deformable at the pressing temperature without reacting with the compact or penetrating into the pores. For most metal powders, containers made of thin metal sheet are suitable for producing near net shape parts. Low carbon steel, austenitic stainless steels and titanium are generally used as sheet canning materials. Steel or titanium capsules may be used at the temperatures upto 1400°C and 1650°C respectively for hot compaction of refractory materials.

Other practical encapsulation methods include ceramic investment moulds and glass encapsulation. Ceramic investment moulds are prepared in the same manner as in the investment casting technique. A wax pattern is coated with a ceramic slurry, followed by drying, dewaxing and firing. The ceramic shell is then filled with the powder to be compacted, and packed into a pressure transmitting ceramic granule bed inside a steel can which is evacuated and sealed. This method is applied for producing complex alloy parts used in aerospace industry. Glass encapsulation in mainly used in hot consolidation of ceramics and refractory compounds, where metallic capsule materials cannot be used because of high procesing temperature. In this method, shaping of compact is first carried out by a cold compaction method such as cold pressing or injection moulding. A layer of glass-frit is applied to the surface of the compact. After degassing of the compact, the temperature is raised to melt the glass powder, which forms a dense surface layer.

Containerless HIP starts with preforms shaped in a separate processed stage (as in glass encapsulation) requiring completely sealed surface layers. This can be achieved, in principle, by a graded structure of the compact with enhanced sintering of near surface layer. This methods is known as "sinter canning" and is of limited application to non-net shape compacts, where the surface layer can be removed, or to systems where a gradient structure of the final part is appropriate. Depending upon the microstructure, this occurs at fractional density levels of 85–95% to which sintering must take place without pressure.

Encapsulated HIP technique is mainly used for full density processing of materials where the powder route is either indispensable or it confers superior properties, or where net shape products produced assure economic dividends Notable examples are the tool steels produced from gas atomised powders, and Ti-and Ni-based alloys for aerospace application. Encapsulated HIP is used with refractory metals like Mo, W and Nb, for producing oxide-dispersion strengthened super alloys and in advanced ceramics. Containerless HIP is less expensive and hence is more widely used for producing cemented carbides, silicon nitride for high temperature applications etc.

(III) Powder Forging

This process is specially used for producing net shape structural steel parts and semi-finished products from special powder materials. This process enables to achieve mechanical properties, comparable or even superior to those obtained from conventional forging techniques. The process is more economical because it requires lesser machining operations. Powder forging is used to achieve higher densities and hence higher strength and ductility than those obtainable from cold compaction followed by sintering. The process is used commercially for producing highly stressed parts such as gears, roller bearing races, stator cams and connecting rods. A flow chart of operations involved in powder forging is given on below.

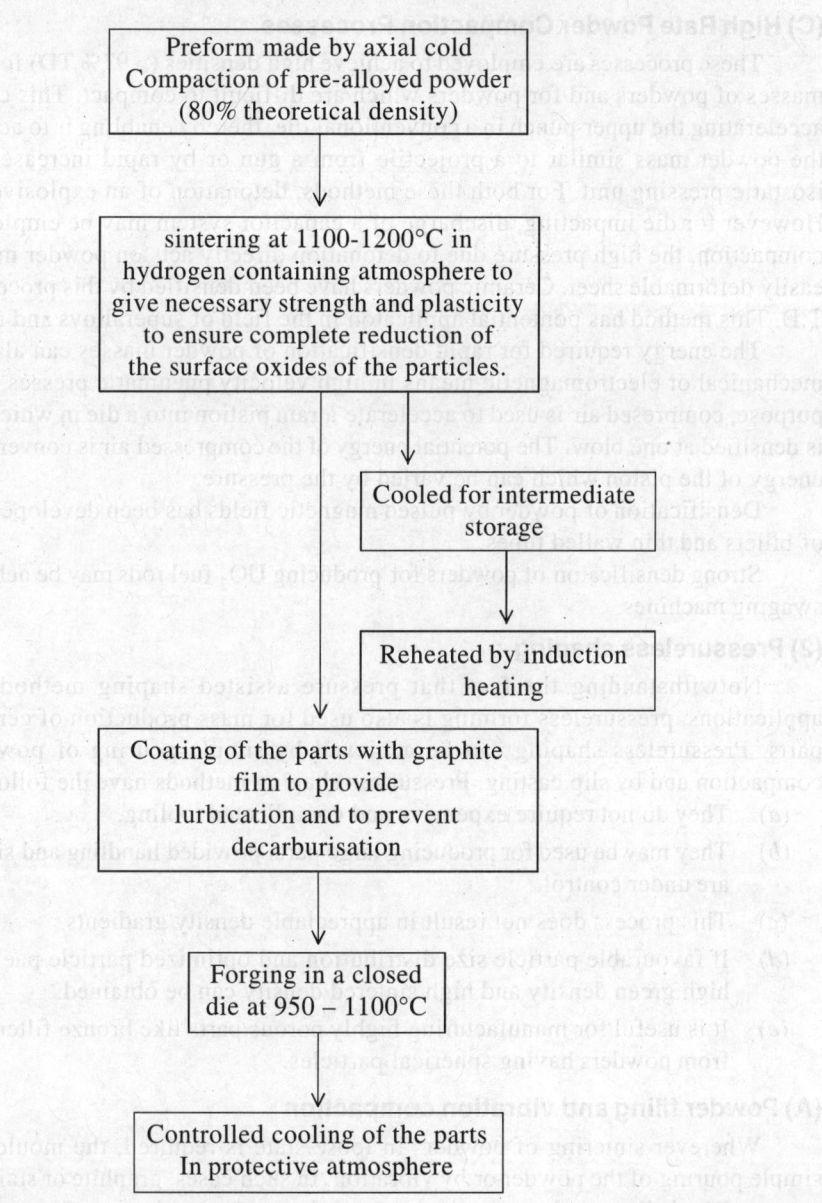

```
┌─────────────────────────────────┐
│   Preform made by axial cold    │
│  Compaction of pre-alloyed powder│
│      (80% theoretical density)   │
└─────────────────┬───────────────┘
                  ▼
┌─────────────────────────────────┐
│     sintering at 1100-1200°C in  │
│   hydrogen containing atmosphere to│
│   give necessary strength and plasticity│
│       to ensure complete reduction of│
│   the surface oxides of the particles.│
└─────────────────┬───────────────┘
                  ▼
┌─────────────────────────────────┐
│     Cooled for intermediate      │
│             storage              │
└─────────────────┬───────────────┘
                  ▼
┌─────────────────────────────────┐
│      Reheated by induction       │
│             heating              │
└─────────────────┬───────────────┘
                  ▼
┌─────────────────────────────────┐
│  Coating of the parts with graphite│
│       film to provide            │
│  lurbication and to prevent      │
│        decarburisation           │
└─────────────────┬───────────────┘
                  ▼
┌─────────────────────────────────┐
│     Forging in a closed          │
│       die at 950 – 1100°C        │
└─────────────────┬───────────────┘
                  ▼
┌─────────────────────────────────┐
│  Controlled cooling of the parts │
│      In protective atmosphere    │
└─────────────────────────────────┘
```

Tooling is kept at about 250-300°C. Powder forging tools are completely closed, in contrast to conventional forging dies.

Forging, rolling and extrusion are directional hot working operations in which the material undergoes shear deformation. These operations are carried out starting from powders or from powder preforms which are already consolidated to near theoretical density. Examples for the latter type are components made from nickel and titanium alloys for aerospace applications.

Hot powder extrusion is generally carried out with vibrated or cold precompacted powders enclosed in a metal capsule. This technique is used mainly for producing dispersion strengthened aluminium, superalloys, composite materials, seemless stainless steel tubing and high speed tool steels.

(C) High Rate Powder Compaction Processes

These processes are employed to achieve high densities (> 97% TD) for compacting large masses of powders and for powders which are difficult to compact. This can be achieved by accelerating the upper punch in a conventional die, thereby enabling it to act as an impactor to the powder mass similar to a projectile from a gun or by rapid increase of pressure in an isostatic pressing unit. For both these methods, detonation of an explosive material is used. However for die impacting, discharge of a capacitor system may be employed. In explosive compaction, the high pressure due to detonation directly acts on powder mass enclosed in an easily deformable sheet. Ceramic powders have been densified by this process upto about 95% T.D. This method has pontential applicaton in the field of superalloys and refractory metals.

The energy required for rapid densification of powder masses can also be generated by mechanical or electromagnetic means in high velocity pneumatic presses developed for this purpose, compressed air is used to accelerate a ram pistion into a die in which the powder mass is densified at one blow. The potential energy of the compressed air is converted into the kinetic energy of the piston which can be varied by the pressure.

Densification of powder by pulsed magnetic fields has been developed for densification of billets and thin walled tubes.

Strong densificaton of powders for producing UO_2 fuel rods may be achieved using rotary swaging machines.

(2) Pressureless shaping

Notwithstanding the fact that pressure assisted shaping methods have important applications, pressureless forming is also used for mass production of ceramic and metallic parts. Pressureless shaping can be achieved by simple pouring of powders by vibration compaction and by slip casting. Pressurless shaping methods have the following advantages.

(a) They do not require expensive and complicated tooling.

(b) They may be used for producing large parts provided handling and sintering procedures are under control.

(c) This process does not result in appreciable density gradients.

(d) If favourable particle size distribution and optimized particle packing are provided, high green density and high sintered density can be obtained.

(e) It is useful for manufacturing highly porous parts like bronze filters and diaphragms, from powders having spherical particles.

(A) Powder filing and vibration compaction

Wherever sintering of powders in loose state is required, the mould may be filled by simple pouring of the powder or by vibration. In such cases, graphite or stainless steel moulds are employed. Pressureless shaping is used for preparing bronze filters and porous nickel membranes. Vibration compaction was studied for the development of nuclear fuels from coarse spherical UO_2 powders in order to obtain high vibration densities in stainless steel tubes without subsequent sintering. This technique was investigated as an alternative for produceing pellets. Vibration compaction is also an important step in encapsulated HIP or hot extrusion of powder materials.

(B) Slip Casting

This technique is conventionally used for shaping large and complicated clay ceramic parts such as high voltage insulators. This can be used for metal powders and oxide and non oxide powders. The mould is made out of porous plaster of paris. The slip consists of an aqueous or alcholic suspension of ceramic or metal powder in70 to 80% wet concentration. The slip is poured into the mould, where it loses some of its water content by capillary forces and forms a semi hard layer on the surface of the mould. The casting which still contains 15% of water is further dried and then sintered.

Pressurized slip casting process is used for porcelain and sanitary ware in order to increase the rate of water removal and therely accelerating the production rate. The most important requisites of the slip are its stability against sedimentation during processing, its castability and the resultant green density after drying. The properties are mainly controlled by the nature of the powder, its concentration, the stabilizing additves, the PH and the viscosity. Stablity against sedimentation can be achieved by electrostatic methods (by using surface active ions or polarized molecules in the liquid) or by steric methods (by absorbing surface active neutral organic compounds.)

Slip casting is an important method in the ceramic industry and for producing short fibre and whisker containing composites and also for coatings. However this process is unsuitable for mass production of small parts.

Continuous slip casting is a special process developed for casting thin sheets, strips and foils with thickness range of 0.2 to 1.5 mm for use in the manufacture of electronic pars such as ceramic housings, capacitors, heat exhangers catalyst carriers,etc. The slip constitutes a slurry containing a plasticizer and is cast on a hot, stainless steel conveyor belt where it loses most of its moisture and remains like a plastic or rubber like strip. The tape can be cut or punched and coiled and stacked. It is converted into rigid ceramic foil after dewaxing and sintering. This process is used in mass production of ceramic substrates for use in electronic devices.

"Wet pouring" is a novel modified slip casting technique. In this process, a solvent mixture containing a binder is poured into a mould. After the volatile solvent is evaporated, the green body is mechanically stabilised by the binder. After thermal debinding, it is subjected to sintering.

(C) Electrophoretic Forming

With the help of electrophoresis phenomenon extremely fine powders of Al_2O_3 or SiC can be compacted by this process into simple geometric forms which have good sinterability.

(3) Spray Forming

Surface coating of metals, alloys, compounds and composites by spray deposition has been in practice for many decades to enhance corrosion resistance, oxidation-resistance, wear-resistance, for thermal or electrical insulation and to impart special optical, electrical or chemical properties. Flame spraying and plasma spraying are the most popular thermal spray technologies used for free standing near net shaped parts, semi finished products like big rods and tubes as well as clad billets.

Osprey process is a commercially important process which enables direct conversion of molten metal into a fine grained semi-finished product by means of a combined gas atomising and deposition process. The process involves the following five stages which determine the quality and integrity of the preforms:

(*i*) Metal delivery (*ii*) atomisation (*iii*) transfer of the droplets (*iv*) consolidation (*v*) preform osprey process is used for producing stinless steel tubes upto 8 m in length, high speed steel preforms upto 250 kg, super-alloy billets, rod/wire mills, coper or aluminium alloys and composites. Aluminium billets are further processed by extrusion whereas high speed steel preforms are subjected to forging by conventioanl methods. The process involves lesser number of processing steps and offers improved economy as compared to normal powder routes.

Vacuum plasma spraying technique which directly uses powders as raw materials, is preferred where higher deposition rates are required and where the materials involved are costly and difficult to process by other methods.

20.9. SINTERING

Almost all metal or ceramic powder parts are sintered at adequate temperatures before use. During sintering, the interparticle contacts increase due to the formation of bondings between the atoms or ions which are comparable with the bonding strength of a regular crystal lattice. (the bonding within a green compact is mainly by adhesive forces and by formation of large contact areas due to hooking and clamping to each other and these are responsible for the reasonable green strength). In pure single components, sintering takes place completely in solid state. However in multicomponent systems, a liquid phase may be involved only to the extent that the solid skeleton ensures geometrical stability of the part being sintered. While coarser particles undergo sintering without siginificant change in dimensional stability, fine powders may sinter with some shrinkage leading to densification. Sintering process may be visualised as a thermally activated material transport in a powder mass or a porous compact, decreasing the specific surface due to increase in inter partical contacts, shrinkage of pore volume and change of pore geometry. From practical point of view sintering may be considered as a heat treatment of a powdered mass or a porous compact in order to change their properties towards the properties of a pore-free body.

The driving force of any solid state sintering process is the decrease in free energy of the system due to the following reasons :

(i) Decrease of the specific surface area due to to initiation or growth of particle contact areas

(ii) Decrease in pore volume

(iii) Spheroidisation of the pores

(iv) Elimination of the residual (from powder manufacturing processes) non-equilibrium lattice defect concentrations, such as point defects and dislocation, in the powder mass.

(v) In case of multi-components system, elimination of non-equilibrium states due to mutual solid solubility (homogenisa-tion of concentration gradients) characterised by the free energy of solution formation.

(vi) In case of multi-component systems, elimination of non-equilibrium states due to chemical reactivity characterised by the free energy of compound formation.

Phenomenologically, the sintering process is sub-divided into the following three stages.

(i) **First Stage:** The particle contacts are transformed to sintered bridges called necks. During this stage, the powder particles remain discrete and grain boundaries are generally formed between two adjacent particles in the place of contact. Very little shrinkge may occur.

(ii) **Intermediate Stage:** Single particles begin losing their identity. A coherent network of pores and grain growth occur, which result in a new microstructure. Most of the shrinkage takes place.

(iii) **Final Stage:** Densification upto 90-95% of theoretical density takes place. Increasing Spheroidisation of isolated pores occurs. Relative proporation of closed pore space increases.

Single component sintering: The various mechanisms that might be involved during sintering process are as follows :

- Adhesion   } (Whitout material transport)

- Surface diffusion
- Grain boundary diffusion
- Volume diffusion via vacancies or Interstitials
- Vaporisation and Recondensation

} (With material transport involving movements of individual atoms)

- Viscous flow
- Plastic flow
- Particle rotation
- Grain boundary Sliding

} (With material transport involving collective movement)

The contact area and contact quality within a pressed or unpressed powdered mass and consequently, the sintering behaviour is dependent upon internal parameters (*e.g.* Praticle size, particle shape, surface roughness and morphology) and external parameters (*e.g.* Temperature, time and sintering atmosphere). The lower the temperature and the coarser the powder, lesser is the densification. The easier sintering of fine powders is because of the larger number of particle contacts and the higher driving force for sintering.

When a high density is reached during sintering the grain size generally increases. This fact should be kept in view particularly when a sintering cycle has to be optimised to provide high density and small grain size. Though difficult to achieve, both these properties are essential for developing adequate strength in high steength metallic materials.

When too high temperatures or too long times are used, oversintering may take place giving rise to undesired properties such as coarsening of grains, decrease in density and strength, newly created gas filled pores and dimensional inaccuracies.

Appearance of interconnected (open) porosity and closed porosity in a sintered body, dependent on total porosity, is another important property. When the total porosity is in the range of 15% to 12%, open pores are predominant. When the total porosity reaches below 7%, open pores completely disappear whereas when the total porosity is in the range 7% to 12% both open and closed pores exist in comparable volumes.

In cases where green compacts are inhomogeneous, as in axially pressed compacts with large height to diameter ratio the density gradient present may lead to differential sintering. This is because dense areas undergo higher sintering rate and larger shrinkage than the more dense areas, thereby resulting in dimensional inaccuracies and distortions.

Some parameters may influence the material transport during sintering process and thereby the properaties of the sintered product. Activated sintering may result from an increased driving force and/or increased mobility induced by physical or chemical means. Activation energy of the overall process may be reduced by special compositons of interfaces (grain or phase boundaries) that can be achieved with the help of additives or by the activation of gas phase transport, etc. Inhibition of sintering may result from reduced mobility of the atoms and also due to the presence of inactive films at sintering contacts. Activating and inhibiting phenomena are also encountered during reactions with sintering atmospheres.

Minor sintering additives, called dopants sometimes result in substantal enhancements of sintering (usually called activated sintering). For instance, about 0.1% of nickel lowers the sintering temperature of tungsten or molybdenum particles, enters the grain boundaries formed during sintering and thereby increasing the grain boudary self diffusion of W or Mo by a factor upto 5000 at 1300°C such paths of high diffusivity at the interfaces may provide very rapid sintering, provided that the activating element remains concentrated at the grain boundaries during sintering. Metals like Fe and Co have similar effects, though smaller.

Chemical reactions with sintering atmospheres causing in situ formation of fresh particle surfaces also may cause activated sintering effects. For example incorporation of hydrogen chloride gas or halide into the sintering atmosphere facilitates mass transport by the formation of vapur phase molecules (*e.g.* Fe Cl_2 during sintering of iron). Enhanced pore spheroidisation as well as improved sintering may confer improved mechanical properties (such as higher ductility) on the products.

Metallic surfaces other than those of noble metals in contact with ambient air is coated with an oxide film. The higher the specific surface of a powder, the larger is the oxide formed and this may influence the sintering behaviour substantially. Oxides with high energy of formation (such as MgO, Al_2O_3 and Cr_2O_3) are stable under normal protective gas - or vacuum –sintering conditions and they strongly inhibit sintering process because of the reduced diffusional exhange between the metallic surfaces. When a low melting point *e.g.* Mg or Al, is covered with its high melting point oxide, the actual sintering temperature for the metal as compared to its oxide film, is relatively so low that practically no transport of the atoms takes place in the oxide. That is why it is impossible to produce parts of aluminium powder by conventional cold pressing and sintering. The practical procedure is to introduce a liquid phase to destroy the oxide films or to apply a strong deformation process by extrusion, which removes the oxide films mechanically and provides fresh metallic surfaces formed in situ. Metal or alloy powder with highly reactive consistuents (*e.g.* Superalloys, titanium alloys, hard magnets) have to be manufactured with low oxygen content. For instance, the oxygen specification for superalloy powder is < 100 ppm. The oxygen content not only inhibits sintering but also degrades phyical and mechanical properties including reliability, because of oxide inclusion.

Oxide films with low energy of formation (such as those of Ni, Cu, Fe, Mo, W etc.) are not stable during sintering. The oxide is either reduced (eg. Fe or Ni oxides), evaporates (eg. Mo O_3), or is partly soluble in the matrix metal (as in the case of copper oxides). The oxide so disappeared may leave behind a highly active metallic surface.

Additives are often used in oxide ceramics mainly to activate sintering and to suppress grain growth at high sintered densities. Producing fine grained micro-structures in highly dense ceramics is of particular value for optimisation of mechanical, magnetic or dielectric properties. Soluble dopants (*e.g.* CaO in ZrO_2 or Th O_2) may improve sintering by changing defect chemistry and thereby diffusibility in the matrix. MgO in Al_2O_3 inhibits grain growth and increase pore mobility and shrinkage.

Activated sintering was also observed with the help of mechanical shocks (*e.g.* by explosives) and neutron irradiation.

Ultrafine particle (nano sized) powders have to be processed while taking special precautions.

Multicomponent sintering

Sintering of multi component powder mixtures takes place in solid stae or in presence of a liquid phase. During solid state sintering, solid solubility may or may not take place. Sinterability of powder mixtures, particularly in the solid state, may be influenced by the generation of sintering stresses, which results from factors, such as the following :

(*a*) Mismatch of thermal expansion between phases.
(*b*) Themal expansion anisotropy in non cubic phases.
(*c*) Inhomogeneous sintering due to long range density gradients within the mixture caused by compacton.
(*d*) Differential sintering resulting from short range density gradients within the mixture caused by agglomeration.
(*e*) Differential sintering caused by phase inhomogeneities in the mixture.

Sintering stresses may lead to micro or even macro cracks. These are particularly important in case of metal ceramic composities containing an essentially un-deformable ceramic component, *e.g.* SiC whiskers or fibres. This may be relevant in ceramic-ceramic composites too. Presure assisted sintering helps to minimise this type of damage. Multiphase and ceramic parts are generally subjected to residual stresses after cooling. Liquid phase sintering is one of the widely used fabrication processes in powder metallurgy for both metallic as well as ceramic products. It has the following advantages:

(1) increase in the sintering rate
(2) achieving of almost full density
(3) short sintering times as long as good wetting by the liquid phase occurs. Even small amounts of liquid can bring about appreciable enhancement of sintering in many systems

Sintering can take place with transient or permanent liquid phase. Transient liquid phase sintering is generally employed to introduce and homogenize smaller amounts of alloying elements such as in the systems Fe-P and Fe-Cu. Permanent liquid phases occur during sintering of high speed steels and ceramic systems such as porcelain and Si_3N_4 with different additives.

Infiltration is a two step liquid phase sintering process, in which the pores of green or pre-sintered compacts is filled with a molten metal or alloy usually of low viscosity. The molten metal or alloy is often deposited in the solid state at the surface of the comapct, melts during heating and fills the pores due to capillary forces. The process is easy and reproducible when the surfaces are free of any contamination. This technique was found to be of limited practical application in systems such as W-Cu, WC–Cu, W-Ag, WC-Ag, Mo-Ag etc. Large high-voltage electrical contacts have been manufactured by this process.

Reaction or Reactive sintering is a sintering process of powder mixtures taking place simultaneously with exothermic chemical reactions, so that a different phase composition results after sintering. The reactions involved may be of solid, liquid or gas phase formation of intermetallic phases in systems like Ni-Al or Ti-Al are typical examples. However in such cases, an extensive swelling takes place instead of shrinkage. The driving force of the chemical reactions involved is generally orders of magnitude higher than that of pure solid phase sintering where only a decrese in surface energy occurs. Reaction sintered (bounded) silicon nitride (technically known as RBSN) is obtained by the following exothermic reaction taking place below the melting point of Si which is 1410°C.

$$3\,Si + 2\,N_2 \rightarrow Si_3N_4 + Q$$

Reactive sintering may be of intrest in producing ceramic-ceramic and metal-ceramic composites.

Pressure sintering or hot compaction or hot pressing are synony-mous terms used for processes involving simultaneous application of final shaping and sintering.

Induction sintering involves sintering by direct coupling of pressed compacts in a high frequency induction field of the order of 4 KHZ. The process has the advantages of low energy consumption due to the high heating rates, short sintering times, lesser amounts of insulation materials to be heated, simplicity and low investment costs. However, a critical heating-up rate must not be exceeded lest local melting or cracking may take place. Induction sintering process has been developed only on a laboratory scale for simple shaped parts of hard metals and some alloyed and unalloyed steels.

Plasma sintering consists of sintering in plasmas performed as zone sintering in a hallow cathode discharge. This process has been developed to produce small diameter rods of MgO doped Al_2O_3 and thin walled tubes. The processe achieves high densities (of the order of 99% theoretial density) within low sintering times.

However special precuations are needed for practical applications. *Microwave sintering* which is expected to have several advantages, is in development stage.

Cold sintering prcess has been used for powders of irregular shapes with high plasticity *e.g.* Fe, Al and Cu which could be densified by high pressure compaction to 97 – 99% TD at a compacting pressure of about 3 G Pa. With a suitable mechanical design, dies and punches made from tool steel or Cemented carbide may be used at this pressure. Under such conditions, extensive plastic deformation within the powder particles and shear deformation occur, leading to fresh surfaces formed in situ, which are free of oxide and other contaminations. This facilitates bringing the surfaces upto such atomic distances, resulting in strong bonding, which are similar to those achieved in high temperature sintering. Cold sintering has potential application in consolidation of rapidly solidifiable powders (*e.g.* Fe based or Al based) and for composite materials such as metal-bonded diamond or carbide composites.

Rate controlled sintering (RCS) process allows optimised temperature schedule with optimised shrinkage rates over a wide range of densification.

Sintering equipment used in industrial practice is conveyor belt furnace, walking beam furnace and vacuum furnaces.

The conveyour belt furnace consists of a stainless steel network, which carrys the parts of to be sintered, directly in case of large parts, or stacked in plain sheet iron boxes in case of small parts. The temperature attained by such a furnace is about 1150°C by the heat resistance of the belt, which is in partial contact with hot gases. This type of furnaces are mainly employed for sintering unalloyed or alloyed iron and steel and also for copper alloy parts. The furnace consits of dewaxing zone, sintering zone and cooling zone.

Walking beam furnace is an improvised version of push through type furnaces. In this furnace, the parts to be sintered, which are stacked in sheet iron boxes, are heated by molybdenum elements in the range of 1350°C to 1600°C as required. Reducing atmosphere is maintained in the furnace to protect the heating elements. This type of furnace is used for sintering alloyed steel parts and some special products.

Vacuum furnaces are used in case of hard metal parts and special products. They are integral parts of the sinter hot-isostatic pressing (HIP) cycle. The heating elements are made of W,Ta or Mo which yield temperature upto 2400°C, 1700°C and 1500°C respectively. For producing hard metal parts, graphite heating elements are used and special dewaxing valves are provided. The final vacuum is about 10^{-2} to 10^{-3} m bar. Single or multi-chamber furnaces are available. The walls of the vacuum vessel are water-cooled and are shielded against radiation from the heating elements with the help of graphite felt.

Sintering atmospheres are provided during sintering to achieve mainly the following objectives.

(1) To eliminate or control chemical reactions such as reduction, oxidation, decomposition, nitridation, carburisation and decarburisation as well as to eliminate impurities.
(2) To remove volatile admixtures and their decomposition products such as lubricants or plasticizers from injection moulding process.
(3) To prevent undesirable evaporation of the main component or alloying elements.
(4) To facilitate heat transfer by convection in the furnace
(5) To provide external hydrostatic pressure required in hot isostatic pressing (HIP).

The sintering atmosphere should be chosen in such a way that it is compatible with the construction material of the furnace and also with the heating elements that are in contact with the gas. In practice, sintering atmospheres are provided using pure hydrogen gas, pure nitrogen gas, noble gases such as argon and helium, burned gases such as endothermic gas (endo gas) or exothermic gas (exo gas) obtained from partial combustion of natural gas, propane or other hydrocarbons.

Though expensive, vacuum sintering is practised for special products because of the availability of sophisticated equipment which enable the production of high quality products.

Manufacture of wide variety of powder metallurgy products including special ceramics require many types of processing technologies such as single sinter process, double sinter process, indirect shaping, pressureless sintering, hot pressing, hot isostatic pressing, powder forging, infiltration, powder rolling, shock compaction, ultra rapid sintering, reaction sintering etc. several post-sintering or secondary operations are used in mass production of powder metallurgy parts (including sintered machine parts). These operations include finishing, sizing, infiltration, machining, heat treatment, surface hardening, protection from corrosion etc.

20.10. IMPORTANT POWDER METALLURGY PRODUCTS AND THEIR INDUSTRIAL APPLICATIONS

20·10.1. Iron and Steels

Powder metallurgy parts made of sintered iron and steels are produced in large quantities. Their properties mostly depend upon their composition, manufacturing route, porosity and finishing. The applications include bearings, filters and forged parts.

20·10.2. Low alloy steels

High temperature sintered parts are finding increasing use in high performance applications in car industry because of their improved strength as a consequence of more complete reduction of residual oxides. Powder metallurgy parts are used in electric and other engines, business equipment and domestic appliances. Sinter-forged parts consisting of Mn, Cr, Ni and Mo alloyed steel with 0.2 to 0.7% carbon generally exhibit higher fatigue resistance as compared to conventionally sintered steel. Hence they are employed for heavy duty applications such as for gear parts and connecting rods for car engines.

20·10.3. High alloy steels

Austenitic, ferritic and martensitic stainless steels having composition of 12% Ni, 18% Cr. 2% Mo and and 68% Fe are manufactured by powder metallurgy. The pre-alloyed powders are produced by water gas atomisation with sintering atomoshpere of vacuum or high purity hydrogen. Large quantity of stainless steel tubes are produced by cold isostatic pressing of gas atomised powder followed by hot extrusion. Metal injection moulding (MIM) is used in the mass production of very small filters with controlled porosity.

20.10.4. High speed steels

These contain 10 to 20% of Cr, W, V, Mo and Co. Excepting cobalt, these metals are mostly present as carbide, depending upon the carbon content. These high speed steels (HSS) are partly manufactured by the powder metallurgy route, starting from atomised powders. Semi-finished or finished products are produced by die or cold isostatic pressing followed by sintering in vacuum. However, the sintering operation should be carried out under close temperature control. The powder metallurgy route enables increased carbide content as compared to cast and wrought HSS products. Moreover, the carbides are highly dispersed, which is generally impossible to obtain by melting and casting, particularly with high carbon content. This gives rise to better isotropic properties, better grindability and for several applications, to increased tool life as compared with products manufacture by conventional processes. Some sintered steels based on a plain iron or iron alloys and containing 10 to 20% by volume of hard phases (*e.g.* TiC, Al_2O_3 or NbC) are introduced by mechanical alloying, together with sintering aids such as boron and phosphorous. These sintered products possess good mechanical and wear properties and they can even compete with HSS for some special application.

20·10.5. Copper and its alloys

Copper based structural parts are widely used. Limited quantities of pure sintered copper are used for some special high conductivity applications. Sintered bronze parts with controlled porosity are used as bearings, filters etc. Lead bronzes are used as bearings supported with steel shells. Brass with 10 to 40% Zn is used as engine parts by using water atomised alloyed powders.

20·10.6. Superalloys of Nickel and Cobalt

Nickel and cobalt alloys having heat-resistance and oxidation-resistance, often called "superalloys", are available in many compositions and varied properties. Many of these superalloys have been produced earlier by melting and casting. Powder metallurgy route has been found to offer some advantages in the production of superalloys. However, micro structural optimization and minimising oxygen content are essential for the superalloy production by powder metallurgy. Powder handling and processing must be done in clean environment under protective gas, particularly for applications such as production of gas turbine disks. It includes encapsulated HIP consolidation, followed by die forging, machining and heat treatment.

20·10.7. Metals having high melting point

(*a*) **Chromium:** Owing to its excellent properties such as corrosion resistance against reactive gases, acids, alkalis and slags, chromium has been recognised as a structural material in chemical engineering. Powder metallurgy route to produce high purity chromium as finished or semi-finished products involves cold pressing of the electrolytic powder, sintering in hydrogen atmosphere and subsequent canning in vacuum tight steel containers followed by forming to obtain fully dense products.

(*b*) **Molybdenum and Tungsten:** Although molybdenum, used for producing large sheets is partly manufactured by conventional metallurgical routes, tungsten (melting point, 3420°C) is exclusively produced by powder metallurgy. For the manufacture of rods of W and Mo, which are the intermediate products for fabrication of wires, the metal powder is die pressed or isostatically pressed, presintered and then finally sintered at about 80% of the melting temperature in furnaces with cylindrical tungsten heaters under pure hydrogen. This indirect sintering procedure has replaced sintering under direct low voltage current at about 3000°C in order to avoid strong density gradients at the rod ends. The sintered rods are further processed to produce fine wires for lamp bulbs.

A liquid phase sintered tungsten alloy having composition of 1.5 to 5.5% Ni and 1.5 to 4% Cu or 3% Fe, which attains full density (about 18 g/Cm^3) after sintering at 1200 to 1400°C, can be machined by conventional tools. It is used for producing vibration dampeners, counter weights, penetrators for rotating masses required for ordnance applications, isotope containers and for radioactive shielding.

(*c*) **Niobium and Tantalum:** Nb and Ta are processed by vacuum electron beam melting or arc melting or by the powder metallurgy route. The metal powders are die pressed, presintered and finally sintered at low voltage and at current intensities up to 50,000 A. The residual impurities *viz.*, H, halogens, C, N, and O are progressively removed as the temperature increases from 800°C to 2000°C during the sintering of Ta, Both Ta and Nb are processed mainly to wires and sheets. Ta is used as a corrosion resistant material in chemical industries, for high temperature heating elements and in electrolytic capacitors, as electrodes which are covered by anodic oxidation with Ta_2O_5 as the dielectric. High purity Ta and Nb containing less than 100 ppm of interstitial impurities (C,N, and O) are ductile at room temperature. Nb in the form of Nb_3Sn is the base for metallic superconductors.

20·10.8. Light metals
(a) Aluminium alloys

Compacts of pure aluminium powder cannot be sintered to high densities because of the dense and stable layer of Al_2O_3 on the powder particles. At the limiting sintering temperature of 630°C (which is near to the melting point of Al), no appreciable atom mobility occurs in the Al_2O_3 layer. Therefore alloyed aluminium powders (mainly those produced by atomisation) are used. These alloyed powders form a liquid phase at the sintering temperature which penetrates the oxide films at Al - Al_2O_3 phase boundaries and promotes the sintering process. Powders of aluminium alloys such as Al - Cu - Mg, Al – Zn – Cu - Mg etc. are easily compressible and are sintered under vacuum or under pure nitrogen, giving tensile strengths in the range 120 to 200 Mpa, which increase after treatment to 230 – 320 Mpa. They are used as engine parts. High strength aluminium alloys such as Al – Mg –CO, Al – Li, Al – Fe – CO, Al – Fe – Ce are used in aircraft and space applications. Ternary and quaternary Al-Fe alloys are corrosion-resistant and are used for high temperature applications. Owing to their good strength: density ratio, corrosion-resistance, amenability for surface treatment and high ductility, aluminium alloys can compete to some extent with sintered steel parts and high strength polymers in their use in the car industry.

(b) Beryllium

Fine grained beryllium with some degree of ductility is produced by powder metallurgy process. The flake type powders are consolidated by hot pressing in vacuum followed by extrusion, rolling or machining. Owing to its low density coupled with high strength and high modulus, it is used for application in space technology as well as for gears and control parts in ordnance industry.

(c) Titanium and its Alloys

Although titanium and its alloys are generally manufactured by melting and investment casting, powder metallurgy route is employed for producing complicated parts which require lot of machining. The metal powders are produced by vacuum rotational atomisation and consolidated by hot isostatic pressing in capsules, at 920°C and at pressure of 2 K bar for about 2 hours, to almost 100% theoretical density. This technology can be used for producing Ti and its alloys. By virtue of their high strength: density ratio, titanium alloys are used in aerospace industry, especially where precise specifications, regarding long term strength and endurance limits are warranted.

20·10.9. Hard metals (cemented carbides)

Hard metals are among the most important powder metallurgy products. They comprise of at least one hard compound and a binder metal or alloy. The basic metallic hard compounds are carbides of metals belonging to IV A, VA and VIA groups of the periodic table. The most important among them are tungsten carbide (WC) and titanium carbide (TiC). Cobalt (3 to 25%) is the most widely used binder WC-Co system is the largest in the field of hard metals. The binder metal fulfills the following two important functions :

(i) It provides a liquid phase during sintering which helps in a high densification rate and high final density.

(ii) It results in the development of a bi-or multiphase microstructure with high bend strength and fracture toughness. Tungsten caribide (WC) behaves like metal and possesses excellent thermal conductivity and limited plasticity under service conditions. Its strength and hardness increase with increasing metal content, whereas the hardness decreases and crack propagation is effectively inhibited by the ductile binder phase. It is this feature which makes WC-Co system as a unique hard metal system.

TiC is the most important alloying carbide for WC. Minor amounts of TaC or NbC are also added together with TiC. It enhances the high temperature properties viz; hardness, wear resistance, oxidation resistance and decreases the welding tendency with chips formed in situ. TiC, however, decreases the bend strength and toughness.

Hard metals with special wear-resistant coatings (5 to 10 nm thick) of TiC, Ti N, Ti (CN), Al_2O_3, etc are of particular significance for applications of hard metals, particularly for cutting purposes. The combination of relatively tough matrix with a very hard surface layer confers additional advantages, particularly for cutting purposes. Multi-layer coating provides better toughness as compared to single layers of comparable thickness. The coating is done by physical vapour deposition (PVD) or chemical vapour deposition (CVD). Plasma assisted CVD was found to give the combined advantages of PVD and CVD. Working rate and life time of the coated tools are increased by a factor of 2 to 6, depending on the nature of applicaton. Coatings of diamond and boron nitrides, which have been developed recently seem to further promote the use of coated tools in industry.

One of the recent developments in this field is the manufacture of micrograined tungsten carbide hard metals using ultrafine WC powders having particle size of < 0.1nm. They exhibit improved cutting performance on hardened steel, high temperature alloys and under interrupted cutting conditions. Nano-sized powders are successfully used nowadays for developing micrograined hard metals, by introducing adequate grain growth inhibition. These hard metals are harder and more wear-resistant and are expected to find wider use in future.

Tungsten free hard metals such as TiC, Ti (CN) and Cr_3C_2 with Ni-binder are of special interest because of their hardness, chemical stability and wear resistance. The performance gap between cemented carbides and high alloy tool steels has been bridged by materials containing 50 wt% TiC or Ti N embedded in a heat treatable steel matrix. These materials are used for special milling and drilling purposes.

Successful efforts have been made recently to replace Co-binder with Fe/Ni or Fe/NI/Co alloys in WC hard metals to achieve improved hardness and improved abrasive properties as compared to WC-Co.

$Ti B_2$ - Fe hard metals have been developed using fine grained raw materials and subjecting to vacuum sintering. Sintered densities > 99% TD could be achieved with good strength and toughness. These may find future applications.

Hard metals find extensive use for cutting, milling and drilling tools (for metals, plastics, wood, etc.) chisels, saws, rods, drawing dies for wires, deep drawing and cold working, etc. Most of the applications are for cutting purposes. Hard metals are widely used in mining industry, construction industry and for working rocks for constructing tunnels.

20·10.10. Materials of high porosity

A variety of parts with controlled interconnected porosity is manufactured by powder metallurgy route for different applications.

Products of low density are processed for producing, gas filters and liquid filters and for other applications. Bronze filters are produced from spherical particles by loose sintering in graphite dies. Filters made of nickel and titanium based alloys and austenitic steels are manufactured from non-spherical and atomised powders by cold pressing followed by sintering. Monosized spherical particles give good sintered strength and a good control of pore sizes and also avoid closed pores. However irregular shaped particles provide greater amount of porosity. These filters are widely used in oil, gasoline, gas, aqueous liquid and molten plastic filtration, as flame arresters, distributors, separators, conveyor systems, sound absorbers and air bearings. The porosity ranges from 25 to 60% and pore sizes from 1 to 2000 nm. For filters made from sintered fibers, the porosity may be even upto 90%.

Bearings in which the porosity serves as a lubricant reservoir are produced by powder metallurgy process. Bearing of bronze, plain iron and Fe-Cu containing 15 to 35% of interconnected porosity are examples of this type. After pressing sinterng and sizing, they are infiltrated with oil or special solid polymers. Graphite or Mo S_2 admixed to the powder before processing serve as lubricants particularly for low sliding speeds. Sintered bearings retain their lubricant mostly during their entire life and hence they are self-lubricating. The rate of supply of the lubricating oil depends on temperature and the speed of rotations, the greater the speed, greater is the oil supply. For heavy duty appllctions, non porous bearings with steel backs are employed. Cu-Pb or Cu-Pb-Sn powder mixtures are sintered at atmospheric pressure on a steel strip and rolled to the desired density and thickness.

Electrodes for batteries and fuel cells with high internal surface area (about 75% porosity) are used as carriers for the electrochemically active material. For example, for Ni-Cd batteries, the Ni powders are sintered on nickel plates using loose powder, slurry or powder rolling technology. The resultant porous structure is impregnated by a nickel salt, which is converted to $Ni(OH_2)$ forming the positive electrode. $Cd(OH)_2$ is used as the negative electrode. Highly porous metals are also used as catalyst carriers in chemical industry.

Another interesting area of application of high porosity materials is metallic foams, with pore volumes of $\geq 90\%$. Aluminium foams are produced from cold compacted conventional aluminium powders with homogeneously distributed gas evoluting media such as metal hydrides. On heat treatment, the material forms a foam, forming essentially closed porosity, despite this high pore volume. Conventional shaping operations *viz*, Rolling, extrusion, forging, etc, at sufficiently low temperature before heat treatment makes it possible, to produce a variety of geometries. The porosity of the final product depends on the amount of gas evoluting media and the heating rate. Foamed metals are used as fire retardents, energy absorbents, barriers and light weight structures.

20·10.11. Magnetic materials

Although the conventional melting and casting technologies are available particularly for producing soft magnetic alloys and Alnico permanent magnets, soft and hard magnets are also manufactured by powder metallurgy. Important metallic soft magnetic materials *e.g.* Fe, Fe-P, Fe-Ni, Fe-Co, Fe-P, Fe-Si-P and Fe-Co-V alloys are produced by pressing the respective elemental powder mixtures or pre alloyed powders (*e.g.* Fe-Si) and sintering under very pure hydrogen. Homogeneity of the solid solutions produced and a stress-free condition are necessary for good magnetic properties. Powder metallurgy processing offers an advantage of providing a product, free from segregations. Considerable quantities of soft magnetis are also produced by pressing metal carbonyl powders embedded in polymer matrix. These materials are used in unsintered form despite the fact that a thermal treatment for stress release is beneficial for obtaining optimised magnetic properties. Sintered soft magnets are widely used in computers, communication technology and electro-engineering. Permalloy is used as laminated components in transformers. The powder composites are used as magnetic cores in electrical convertors.

The three important physical properties required for a good, hard magnetic material are.
- (*i*) Multiphase micro-structure with fine and homogeniously distributed ferromgnetic precipitates.
- (*ii*) Single domain particles, preferably without Bloch walls.
- (*iii*) Non cubic phases with high magnetic anisotropy.

The important reasons for using powder metallurgy in preference to the conventional melting and casting technologies are as follows:
- (*i*) More suitable microstructures and higher strength are achievable by sintering.
- (*ii*) Very little or no machining is required for sintered materials.

(*iii*) Better economy, particularly in mass production of smaller pieces.

Alnico magnets are produced by pressing of powder mixtures of Fe, Co, Ni and an Al-Fe master alloy and sintered in hydrogen atmoshpere at 1300°C. Anisotropic magnetic property can be conferred on the material by cooling in a magnetic field thereby causing texturised precipitates.

Cobalt rare earth (Co-RE) magnets. ESD (Elongated single domains) powders and Fe Nd B magnets are also produced by powder metallurgy. They find application in electromagnetic devices and machines, some of which are useful in communication technology.

COMPOSITES: Composites may be considered as multiphase materials, whose single phases generally belong to different groups of materials such as metals, polymers or ceramics.

On the basis of shapes of at least one phase of the composite, a subdivision is usually made as particulate, whisker-and continuous-fiber-containing composites. Whiskers and fibers are generally used for mechanical reinforcement, as is the case with particulate matter found in dispersion strengthened alloy. Composition, concentration, and shape of the phases involved and the microstructures obtainable, can be varied to produce a variety of composites. In this, powder metallurgy plays a vital role. Powder metallurgy is mostly used in the manufacture of materials reinforced by particles, whiskers and short fibers.

In ceramic metal composites, the matrix may be metallic or ceramic. Accordingly, they are subdivided into
 (*a*)   Metal matrix composites (MMC).
 (*b*)   Ceramic matrix composites (CMC).

(a) Cermets:

Cermets consist of atleast one ceramic and one metallic component. They are usually manufactured by mixing of the components, followed by compaction and consolidation. As far as the conventional technology of mixed powders is concerned, the phase dominating in Vol % acts as matrix while the other acts as dispersed phase. However, by special type of processing *viz.* CVD of metals on larger ceramic particles followed by consolidation, it is possible to provide a minor metal constituent as a matrix. At similar concentration, both the phases provide a three-dimensionally inter connected net works in such a way that no matrix is defined.

Cermets based on ZrO_2-Mo are used as protective tubes for thermocouples and also as extrusion dies for alloys of Al and Cu. U_3Al-Al and U_3O_8-Al with aluminum claddings are used all over the world as dispersion-type fuel elements in high-flux, low-temperature atomic research reactors. Ag-based cermets with 10-15 % CdO are employed as electrical switching contacts for moderate and heavy loading. However, in view of toxicity of cadmium, CdO has been recently replaced by SnO_2 and other oxides.

(b) Whisker/ Short Fiber-reinforced Metals

Powder Metallurgy techniques can be used to reinforce metals, alloys, polymers, ceramics, implantation of whiskers (*e.g.*, SiC) or short or continuous fibers (*e.g.* SiC, Al_2O_3, C) of diameters less than a few nm. In view of the hazards involved in handling of whiskers they are being replaced by particle or chopped fiber for reinforcement of metals such as Al.

(c) Dispersion strengthened metals and alloys

The basic principle involved in dispersion strengthening/hardening is that the finely dispersed particles can inhibit the dislocation motion in a ductile matrix. Dispersion hardening is generally based on incoherent particles. This is in contrast with precipitation hardening wherein the precipitates are coherent or semicoherent with the matrix. Since the particles are stressed by the dislocations in motion, they must possess adequate strength lest they may be mechanically destroyed during deformation. Further, they should possess high melting point, microstructural stability, chemical stability and solid solubility, otherwise their efficiency may be reduced by partial disappearance of the dispergent. These specifications are met by oxides of thermodynamic stability such as ThO_2 and Y_2O_3. Dispersion strengthening may be applied for many common metals and alloys.

Sintered aluminum powder (SAP) was the first dispersion hardened material invented in 1950. This contained Al with 8-15% Al_2O_3. A recent development is Al reinforced with a few percent Al_4C_3 and Al_2O_3 particles. Al reinforced by SIC particles received much attention among the high strength aluminum alloys.

Nickel with 2% ThO_2 was the first among oxide dispersion strengthened (ODS) alloys. This was first produced by Du point using coprecipitation as Ni-and Th-salts with further processing and sintering under reducing atmosphere. Recently, Ni-alloys with a few % Y_2O_3 are preferred, as they satisfy all the strengthening requirements and are resistant to creep and oxidation. These are mainly used in gas turbines for military aircraft. ODS-iron-chromium alloys with minor additives of Al and Ti and 0.5 wt % Y_2O_3 offer an economic alternative to the alloys based on austenitic nickel.

Oxide dispersion strengthened (ODS) Pt-or Pt-Rh-alloys with less than 0.5 wt % ZrO_2 possess high long term rupture strength. They are used in modern glass industry (as crucible material for glass melts upto 1400°C) for the production of glass fibers and optical glasses. Oxide dispersion strengthened-Nb-and Ta-alloys were found to have excellent high temperature properties.

(d) Infiltrated alloys

Powder metallurgy products such as W-Cu, W-Ag, Mo-Ag and Mo-Cu were prepared by the infiltration process described earlier. These alloys are used as heavy duty electrical contacts. W or Mo provides matrix resistance against heat, wear and arc erosion whereas Ag or Cu phase provides good conductivity. Fe-Pb and Fe-polymer composites are used for bearings. Copper infiltrated Fe and Fe-C compacts are used as high strength parts with good machinability.

(e) Miscellaneous Composites

Composites based on Cu, Fe or bronze with adequate quantity of non-metallic components, are manufactured by powder metallurgy Processes as friction anti-friction materials in tribological application. Oxide and graphite containing composites are used as high performance friction materials for brakes and clutches in heavy vehicles, excavators, aeroplanes and heavy engines as well as in military equipments. Higher service temperatures of 1200°C are achieved by high strength carbon fibre reinforced-carbon (CFC) which is used in aircraft brakes. Composites without mineral oxide additives are used as anti-friction materials containing solid or liquid lubricants. Composites based on graphite with metal powder *e.g.* Cu-powder upto 30 wt % are used for carbon brushes in electric motors. Silver-based composites with graphite or Ni are used as electrical contact materials.

Diamond tools are composites of 2-35 vol % of natural or synthetic diamond with grain sizes in the range 10 to 500 nm which are embedded in a metal matrix *e.g.*, bronze, Cu or Co-alloys. Coated diamonds are used to improve the bonding to the metal matrix. They are manufactured by hot pressing to nearly full density. They are used for grinding and cutting of hard alloys and ceramics for concrete and rock-working. Cubic boron nitride composites are preferred for grinding of hardened steels.

20·10.12. Advanced Ceramic Materials

Ceramics were produced in the past for bricks, refractory materials, corrosion-resistant products, sanitary ware, insulation, etc. These classical ceramics are mainly silicates with complicated chemical composition and lattice structures with ionic bonding and multi phase, coarse and have porous structure. Modern ceramics are used as high tech materials in various industrial applications, They are oxide or non-oxide materials having simple chemical composition and lattice structure, and possess fine-grained and dense microstructure. The non-oxide ceramics contain covalent bonding and possess superior high temperature properties Classical, refractories and insulators are generally not considered under the category of advanced ceramics. On the basis of their properties and applications, the advanced ceramic materials are sometimes categorized as Functional ceramics and Structural ceramics. Functional ceramics are used in electronic, electrical, and telecommunication engineering industries. On the other hand, structural engineering ceramics are used as engineering components, wear parts, high temperature parts and in chemical devices. The mechanical and physical properties of ceramics depend upon their chemical composition and more strongly on their microstructural properties *e.g.* grain size disturbution, grain boundary phases (*e.g.* glassy phase and inclusions) micro-cracks, residual porosity, etc.

Some important industrial applications of advanced ceramic material in various fields are Summarised in the Table 20(3) given below.

20·10.13. Miscellaneous applications

Powder or pastes of noble metals such as Au, Pt and Pd are used for metallisation of ceramics for use in electronic parts, for conducting tracks on ceramics and glass. They are also used as sintered parts for special applications such as ODS – Pt and Pt-Rh-alloys with ZrO_2, and as cermets.

Mixing Ag-Sn alloy powder with Hg leads to the formation of Ag_3Hg_4 and $Sn_7 Hg_4$ by liquid phase room temperature reaction sintering. The solid and corrosion resistant amalgams so formed are used as dental amalgams. However, they are now replaced by polymers and ceramics. Intermetallic compounds such as TiAl, Ti_3Al, Mo SiO_2 represent important powder metallurgy products with potential future applications as high temperature parts *e.g.* car engines, gas turbines, heating elements, because of their thermal stability, corrosion resistance, high oxidation resistance and good mechanical properties.

Pure fine grained high density Al_2O_3 –based ceramics are used as medical implants particularly as parts of hip joints and artificial teeth.

Table 20.(3). Industrial Application of Advanced Ceramics

Field of Application	Requisite Properties	Examples of the Ceramic products used and remarks
(1) Magnetic, Electronic and Electrical Applications: Magnets, sensors, heating elements, IC packages, ferro-electrics, piezo-electrics, varistors (variable resistors), solid electrolytes, super-conductors, communication technology, etc.	Low electrical conducti-vity for magnetic appli-cations. High electrical conductivity for electrical applications, suitable semiconducting, piezo thermoelectricity, and dielectrical properties as required for the specific electronic applications, as the case may be.	• Magnetic ceramics (Hard and soft ferrites) • Soft ferrites have the general formula of MeO Fe_2O_3 (where Me = Fe, Ni Co, Zn, Mn) having a cubic structure of inverse spinel. • Hard ferrites have the general formula of Me O.X Fe_2O_3 (Where x is about 6 and Me = Ba, Sr.) having a hexagonal structure and high magnetic crystal anisotrophy. • Soft ferrites have low eddy current losses. Hence they are

		used in communication technology, *e.g.* cores for radio and TV
• Thin plates of Al_2O_3 are widely used as substrates for electronic circuits
• Ceramics with special electrical and optical properties have the general formula of ABO_3 (Where A = Ca^{2+}, Sr^{2+}, Ba^{2+} or Pb^{2+}, and B = Ti^{4+}, Sn^{4+} or Zr^{4+}). These compounds have perowskite (Ca Ti O_3) structure. The perowskite ceramics may also exhibit elecro-optical effects. Their optical refraction constant is influenced by electrical fields. This effect is used for light switching within microseconds. Doped pb (Ti Zr) O_3 gives adequate transparency. These materials are electrical insulators.
• Semiconducting ceramics are based on ZnO. Its properties can be changed by doping with Bi, Mn, Co or Sb. Which yield high resistivity in the low voltage range and low resistivity at high voltages. These variable resistors (varistors) are used for over voltage protection.
• Ba TiO_3, which is an insulator, can be transformed into a semiconductor, which exhibit strong positive temperature coefficients. Hence they are used in self-regulating heating elements and temperature sensors.
• Y_2O_3-doped ZrO_2 is used as oxygen sensor for determining oxygen concentration in auto-mobile exhaust gases.
• Complex oxides such as (La Sr) CuO_4 and $YBa_2Cu_3O_7$ are examples of superconducting ceramics. Such compounds have great future potential.
(2) Nuclear Application: Fuel and breeding elements, shields, absorbers and matrix for waste conditioning	Irradiation resistance, high temperature resistance, corrosion resistance, and high absorption coefficient	• UO_2 and UO_2-Pu O'_2 ceramics are used as fissionable as well as breeding materials.

(3) **Mechanical Applications:** Cutting tools, bearings, seals, wear parts, motor and gas turbine parts, thermal barrier layers for gas turbine parts.	Long term high temperature resistance, wear resistance, thermal shock resistance.	• Engineering ceramic parts being used are mostly monolithic and comprise of oxides or non oxides. Al_2O_3 and ZrO_2 are example of oxide ceramics. Non oxide ceramics parts are based on SiC and Si_3N_4. Al_2TiO_5 is used for thermal insulating applications • Increases in strength and fracture toughness have been achieved by reinforcement of the ceramics by SiC-whiskers or by fibres or by transformation toughening. Ceramic coatings have been successfully used in various engineering applications. For instance, wear resistant coatings are used for cutting tools whereas corrosion resistant and thermal barrier layers are used for gas turbine parts.
(4) **Thermal:** Burner tubes, nozzles, heating elements heat exchangers, high temperature components, insulating parts, thermal barrier coatings, non-ferrous metallurgy.	Resistance for high temperatures thermal shock, high or low thermal conductivity as required.	• Al_2O_3- and Si_3N_4-based ceramics are used as engine parts, wear resistant materials eg. Cutting tools.
(5) **Chemical and biological Applications:** Corrosion protection, catalyst carriers, environmental protection, organ implants, sensors, filters, membranes fuel cell materials, etc.	Corrosion resistance, bio-compatibility	• Honey comb ceramics (cordierite) are used for catalyst carriers
(6) **Optical properties:** Lamps, windows IR optics and fibre optics	Low absorption coefficient	

20.11. CONCLUSION

Powder metallurgy has to compete with the conventional technologies *e.g*, machining of wrought semi-finished products, forging, fine casting, arc melting, etc. The larger the number of pieces produced in a single run, the more economical will be the powder metallurgy route of production. The future of powder metallurgy products depends upon improvement of new technologies, diversification of product base, economy and quality assurance.

QUESTIONS

1. Describe different processes for producing metal powders.
2. Discuss the various chemical methods used for preparing metal powders

3. What are the various ceramic powders of industrial importance? How are they manufactured?
4. Write informative notes on any two of the following :
   (a) Nano crystals (b) Whiskers (c) Transitional metal carbides
5. Write short notes on any three of the following :
   (a) Pyrophoricity and explosivity of metal powders
   (b) Powder injection moulding
   (c) Powder forging
   (d) Slip casting
6. Describe various methods of powders conditioning.
7. Discuss different processes of compaction and shaping of metal powders.
8. Discuss the sintering processes used in powder metallurgy practice with special reference to Sintering atmospheres and sintering equipment.
9. Discuss important powder metallurgy products of industrial importance giving specific examples and their applications.
10. Write short notes on any three of the following powder metallurgy products:
    (a) Cemented carbides   (b) Composites
    (c) Advanced ceramics   (d) Magnetic materials

21

Batteries and Battery Technology

"Batteries are storehouses of electrical energy on demand." They provide well contained energy conversion devices which greatly contributed to the needs of mankind. Zero emission vehicles of the future will be battery powered only. Many non-polluting energy conversion devices such as photovoltaic systems require the concomitant use of rechargeable batteries for energy storage.

Terms and Definitions

Anode : It is the negative electrode of a primary cell. It is always associated with oxidation or the release of electrons into the external circuit. In a rechargeable cell, the anode acts as negative pole during discharge and as positive pole during charge.

Battery : This term is generally used to denote one or more electrically connected galvanic cells.

Cathode : It is the positive electrode of a primary cell and is always associated with reduction or taking of electrons from the external circuit. In a rechargeable cell, the cathode is the positive pole during discharge and the negative pole during charge.

Charge : This is the operation of a cell when an external source of current reverses the electrochemical reactions of the cell to restore the battery to its original charged state.

Closed-circuit voltage: This is the voltage measured across the terminals of the cell or battery when current is flowing into the external circuit.

Discharge : This is the operation of a cell when current flows spontaneously from the battery into an external circuit.

Electrochemical couple: It is the combination of the electrode reactions of the anode and cathode to form the complete galvanic cell. The number of electrons given up by the anode to the external circuit must be identical with the number of electrons withdrawn from the external circuit by the cathode.

Electrolyte : This is the material which provides ionic conductivity between the positive and negative electrodes of the cell.

Impedance or Internal resistance : The impedance or internal resistance of the battery is the resistance to the flow of current, which operates in addition to the resistance of the external load.

Open Circuit voltage : This is the voltage measured across the terminals of the cell or battery when no external current is flowing. When measured on a single cell, it is usually close to the thermodynamic electro-motive force (emf).

Separator : It is a physical barrier between the positive and negative electrodes to prevent direct shorting of the electrodes. Separators must be permeable to ions, but must not conduct electrons. They must be inert in the total environment.

21.1. INTRODUCTION

In the year 1786, Galvani found that the effect of dissimilar metals on the nerves of frogs electrical. However he erroneously thought that the animal tissue itself was the source of electricity. It is significant that his name was used to denote the galvanic current of the battery. Further a unit cell of a battery is generally called as a galvanic cell. Around the year 1800, Allessandro Volta observed that a stack of alternating dissimilar metals with a layer of paper between the metal bi layers gave rise to a source of electricity. He then believed that the simple fact of contact of dissimilar metals in a pile was the source of perpetual power. Volta's name has been associated to the intensive property describing the ability of the battery to do work (i.e. the voltage of the battery). Further a series assembly of cells is sometimes referred to as "Voltaic pile". Later Michael Faraday after intensive work for years together from 1830, discovered that the chemical reaction at each electrode was the source of electricity produced by a battery. It is for this reason that his name is used to describe electrode operation as a Faradaic process. Faraday was the first to use the words "electrode" as a general term for a pole of a battery, "anode" for the negative pole of a battery, and "cathode" for the positive pole of a battery. He also coined the words "anion" for the charge that migrates to the anode, "cation" for the charge that migrates to the cathode "electrolyte" for the solution containing anions and cations, and "electrolysis" for the procss occurring during passage of electrical current through an electrolyte, in touch with anode and cathode.

For the rest of the 19th century, batteries were almost always used for experiments in chemistry and physics laboratories. The batteries used in early periods were voltaic piles. They consisted of a series of zinc and copper plates fastened back to back to form a high voltage, low current source. Initially there was only a moistened paste-board (with low conductivity because of the adventitious ions that might be present) between the plates. The 19th century saw the development or invention of several types of batteries. The invention of the lead acid battery in 1859 by plante and the zinc-manganese dioxide cell by Leclanche in 1866 represent landmarks.

The so-called "dry cell" was invented in the first part of the 20th century by Gassner. In this cell, the electrolyte was immobilised by starch or any other gelling agent. Since it was easier to seal such a cell, it paved the way for the development of portable power which eventually led to the present day batteries. Rechargeable batteries were dominated by the lead acid system. Simultaneously, nickel-cadmium and nickel-iron cells were also invented and developed as storage cells during the same period.

The nickel cadmium system was developed further in Germany for war-time requirements in 1950s. This led the way for the development of small rechargeable battery systems for consumer applications. Further it was the 1950s that saw the change over from zinc-manganese dioxide cells to alkaline electrolyte system. The Ruben cell, which is nothing but a zinc-mercuric oxide battery with alkaline electrolyte, was also developed for the electronic circuits used for military applications during the second world war.

Many new battery systems came into vogue in the latter part of the 20th century. Several new batteries now are based on lithium, used as anode material. Important rechargeable batteries are based on metal hydrides as negative electrodes. A recent special rechargeable battery is based on carbon intercalated by lithium, as the negative electrode such systems seem to be very promising for future development.

Batteries may be considered as storehouses for electrical energy on demand. The size of battery ranges from a tiny coin to that of a large house. Tiny coin and button sized cells are used for electronic applications requiring only small capacity. Liter-container sized batteries are commonly used in motor vehicles for starting, lighting and ignition purposes. Very large house sized batteries are used for utility storage of electrical energy. The basis for battery technology is that the chemical energy derived from the chemical reactions in the battery is converted into electrical energy. This is in contrast to what happens in heat engines, where intermediate thermal or combustion processes bring about the conversion of chemical energy into electrical energy. Therefore due to direct conversion of chemical energy into electrical energy in a battery, all the free energy of the chemical system is available for conversion. In heat engines such a conversion efficiency is limited by the carnot cycle. Further, the conversion of electrical to mechanical energy is quite efficient because of the inherent simplicity and low friction of electrical motors. This is the basis for many applications of batteries. Conversion of electricity to light or sound is also efficient and easily controlled, thereby enabling development several applications of batteries for lighting or creation and reproduction of sound.

As is the case with any energy conversion device, in batteries too there are losses, and there has been constant efforts to minimize the same.

Batteries are broadly classified into the following three categories :

(a) **Primary battery**, which is designed to be discharged only once and then discarded.

(b) **Secondary battery**, which is rechargeable and can be used like the primary battery, then recharged and used again, the cycle is repeated until the capacity fades or is lost suddenly due to internal short-circuit

(c) **Reserve battery**, in which active materials are kept separated by a special arrangement. When it has to be actually used, an activation device makes it ready. Such a battery is designed for long storage before use.

21·2. THEORETICAL PRINCIPLES

On the basis of investigations on the electrolysis of aqueous solutions, Faraday (1833) established a relationship between the quantity of electricity passed and the amount of chemical change (i.e. the amount of any material deposited on or dissolved from electrodes) that occurs. His investigations are summarised in the form of the following two laws :

First Law: The amount of any substance deposited or dissolved as a result of passage of an electric current is proportional to the quantity of electricity passed. If 'W' grams of a chemical material is deposited or dissolved by passing I amperes of current for t seconds, then

$$W \propto I \times t \text{ ampere seconds or coulombs}$$
$$W = e \times I \times t \qquad \ldots(1)$$

Where e is called electrochemical equivalent (ECE)

When I and t each is equal to unity, then

$$W = e$$

Thus the electrochemical equivalent (ECE) is defined as the weight of the substance deposited when one coulomb of electricity (3.0×10^9 esu; 1 ampere second) is passed.

Second Law: When the same quantity of electricity is passed through different electrolytes, the amounts of different substances deposited or dissolved are proportional to their respective chemical equivalent weights.

If W_1 and W_2 are the amounts of two substances deposited or dissolved by passing q coulombs of electricity these are in the ratio of their respective equivalent weights E_1 and E_2

$$\frac{W_1}{W_2} = \frac{E_1}{E_2} = \frac{e_1}{e_2} \qquad ...(2)$$

Since one coulomb of electricity is needed to deposit or dissolve one ECE for 1 gram equivalent of any substance the quantity of electricity required is E/e coulombs

$$\frac{E_1}{e_1} = \frac{E_2}{e_2} = F \qquad ...(3)$$

where F is called the faraday. Its significance can be visualised by combining the first and scond laws.

$$W = \frac{1 \times t \times E}{F} \qquad ...(4)$$

If $I \times t = F$, then $W = E$

Thus when one faraday of electricity is passed, the weight deposited or dissolved is the gram equivalent weight. The value of F has been found to be 96,500 coulombs.

$F = 96,5000 \times C = 26.8$ A-hr (ampere hours)

Thus $\qquad W = M \times I \times t /nF \qquad ...(5)$

where M is the molecular weight of the material, n is the number of electrons involved in the electrode reaction.

An important relation derived form Faraday's law (Eqn 5) and thermodynamics shows that the Gibbs free energy (ΔG) for the cell reaction is found to be directly related to the electromotive force (emf) or voltage (E) of the system.

$$\Delta G = -n\,FE = \Delta H - T\,\Delta S \qquad ...(6)$$

where n is the number of electrons involved in the cell reaction as it is written to compute the Gibbs free energy; ΔH and ΔS are the enthalpy change and entropy change respectively for the reaction and T is the absolute temperature in degrees kelvin.

A standard emf is defined by Eqn 6 if the standard free energy at 25^0 C and unit activities for all reactants are used. This quantity is widely tabulated for electrode reactions, and the standard potential of combinations of couples can be calculated from them. The standard potentials are referred to the standard states for the reactions of hydrogen gas to form hydronium ions as the zero of potential. From Eqn. 6, the more negative the value of ΔG, the more useful is the work obtained from the reaction (from the definition of the free energy) and the larger is the voltage of the cell. Further development of the thermodynamic relationship leads to the following important reactions.

$$\Delta S = nF\,(dE/dT)_p \qquad ...(7)$$

and $\Delta H = nF\,[E - T\,(dE/dT)_p] \qquad ...(8)$

which enables the determination of the reaction enthalpy, ΔH and the reaction entropy, ΔS from the emf and its temperature dependence. Conversely, if the enthalpy and entropy are known from the thermodynamic tables, the temperature behaviour of emf can be accurately predicted. The Nernst equation emerges from the combination of the Vant Hoffs isotherm with the relationship for the standard and total emf's and Eqn (6).

$$E = E^\circ - (RT/nF)\,\ln\,[\Pi\,A^s\,(products)/\Pi\,A^s\,(reactants) \qquad ...(9)$$

Where $\Pi\,A^s$ (products) is the product of the activities of the products, each raised to its stoichiometric coefficient, and ΠA^s (reactants) is the corresponding product of activities of the reactants and R is the gas constant. This equation enables the determination of activities for the reactants and products and also correspondingly to determine the cell emf, if the activities are known.

For comparing various battery types and couples, battery efficiency is an important parameter. Thermodynamic efficiency (E) is considered as the most fundamental efficiency which is given by the following equation.

$$E = \Delta G°/\Delta H° = I - T\Delta S°/\Delta H° \qquad ...(10)$$

The thermodynamic efficiency (E) is considered as the highest efficiency that can be obtained from a given electrochemical couple and this can be determined from thermodynamic tabulations of enthalpy and free energy. Many electrochemical couples employed in commercial batteries have thermodynamic efficiencies of 90% or more, as against the heat engine efficiencies based on the Carnot cycle which are mostly lesser than 40%. However an electrochemical cell in actual use performs with less than thermodynamic efficiency due to irreversible losses involved in the electrode reactions, such as polarization. The overall electrochemical efficiency Eff (electrochem) is given as per the following.
Equation

$$\text{Eff}_{(electrochem)} = {}_{t=0}^{t=t}\!\int E' 1\, dt / \Delta G \qquad ...(11)$$

where E' = the closed circuit termical voltage when the net current I is flowing at time, t. Depending on the type of discharge occurring, both E and I may be functions of time.

A battery is usually designed with one of the electrodes of smaller capacity than the other in order to achieve special properties, such as overcharge protection, which ensures safe operation. Hence it is necessary to know the coulombic efficiency of the active material:

$$\text{Eff}_{(coulombe)} = {}_{t=0}^{t=t}\!\int 1\, dt / Q \qquad ...(12)$$

where q is the quantity of coulombs (I x t) calculated from Faraday's laws (vide Eqn 5) this efficiency depends on the current, design of the electrode, cut off voltage and other properties.

The reversible heat of the reaction if given by the difference between the enthalpy and the free energy (i.e. T x entropy) of the cell reaction. As indicated earlier in the discussion on thermodynamic efficiency, this is quite small for practical batteries. An irreversible heat due to polarization of the electrodes is always present. The total amount of heat released during operations of the battery is given by the following relationship:

$$Q = T \Delta S + \int I(E - E')\, dt \qquad ...(13)$$

where 1 and E are functions of time. This heat is released inside the battery at the reaction site. Heat release in a battery can cause serious problem for high rate applications. Battery designers have to make adequate provision to accommodate battery heat. Failure to do this may result in thermal runaways and other catastrophic situations.

In aqueous batteries, an ionogenic salt, a Bronsted acid, or Bronsted base dissolved in water is used as an electrolyte. (According to Bronsted and Lowry, an acid is defined as the substance which has a tendency to lose a proton whereas a base is a substance with a tendency to accept a proton). In non-aqueous batteries, a strong electrolyte salt is usually dissolved in a non- aqueous liquid to form the electrolyte. The ions formed due to dissociation of the electrolyte carries the current through the electrolyte phase from one electrode to the other.

In solid electrolytes, which are ionic solids, the current is carried by ions or vacancies, depending on the nature of the solid. It is very important that no electronic current is carried by the electrolyte. This is normally not a problem for liquid electrolytes, because only a few unusual liquid solutions are electronic conductors. Electrons which flow internally in this manner are not available for the external circuit to do work, and represent a loss of efficiency of the cell. In time, during storage, the entire battery capacity may be lost through this kind of internal short circuit.

Ionic conduction in the electrolyte obeys Ohms law

$$I = kV \qquad \ldots(14)$$

where I is the current in amperes, and

V is the voltage drop (in volts) across the electrolyte of conductivity k in S/cm.

The conductivity is determined by the degree of ionic dissociation and the mobility of each kind of ions present, which is closely related to the viscosity of the medium and the availability of special conduction mechanisms. Conductivity ranges from about 1 S/cm in concentrated H_2SO_4 in lead acid batteries and 9 N potassium hydroxide in alkaline batteries to as low as 1 mS/cm for organic solvent-based electrolytes in lithium batteries. Conductivity can be as low as 1 nS/cm in case of some solid electrolyte batteries.

The high conductivity of aqueous acids and bases is due to the special conductance mechanism which enables rapid transfer of proton from water molecules to another water molecule or to hydroxyl ion in the case of bases. Further anions and cations are also strongly solvated by water. However is case of organic media, the cations are generally solvated strongly but the anions are not. The ions can approach each other closely and form ion pairs that act as entities that do not move in the electric field of the electrolyte. Only a few solids are capable of conducting ions sufficiently well for a battery operating at or mean room temperature. Even these special solids have to be used in very thin layers to minimize the ohmic loss. The time response to the conductance of the electrolyte is almost instantaneous. This is also true for other ohmic contributions to loss or polarization, such as resistance of the electrodes and connectors, resistance of leads to external circuits. Also, this effect steadily increases with increase of current because the change in ohmic polarization depends on the total current. An even greater increase in ohmic polarization is observed if there is an increase in resistance due to precipitation of product to block the electrolyte conduction or the formation of insulating products from the conducting ones.

Activation polarization is another source of loss, though it has a slower response than the ohmic loss described above. The source of this loss is the kinetic limitation of transferring charge across the electrode interface with the electrolyte. Activation polarization has an intermediate response to changes in current, which is in the range of microseconds to milliseconds. The variation of activation polarization with current is logarithmic and hence the effect is most pronounced at low to intermediate levels of current. Since it is the current density, not the total current, which is the independent variable, the activation polarization is dependent on surface area of the active material. Therefore, near the end of discharge at a constant current or resistance, the activation polarization may become significant even at higher current, because the surface area of the active material drops down to low levels since it is exhausted by the discharge reaction.

Another type of polarization is the concentration polarization, which is related to the transport properties of the cell. Since the batteries have restricted fluid flow in most designs this problem is often treated in terms of the diffusion of the ions within the electrolyte or in the solid phases of the active materials.

When the electrode acts as a sink for the ions, a depletion of the respective ions occurs in the vicinity of the electrode. When the electrode produces the ion in question, the electrode acts as a source of the ion. In either case, the polarization caused by the limiting transport process can be analyzed with the Nernst equation:

$$n = (RT/nF) \ln (C_e/C) \qquad ...(15)$$

Where C_e is the concentration of the ion at the electrode surface and C is the concentration in the bulk electrolyte. When the electrode reaction is a sink for the ion, the concentration at the electrode surface tends to zero as the current is increased. This results in a limiting current situation in which the polarization tends to infinity as the concentration tends to zero, which is evidenced by a sharp drop in the cell potential. The current density (A/cm2) at which the sharp drop in the cell potential occurs is called the limiting current density, i.e. in terms of this measurable current, the concentration polarization is given by

$$n_c = (RT/nF) \ln (1 - i/i_l) \qquad ...(16)$$

When the electrode is a source of the ion as per the eqn. (15), the concentration polarization gradually increases with time until a steady state is reached. The steady state polarization too increases uniformly with an increase in current in this case as per eqn. (16), unless any other process such as precipitation takes place. This behaviour is difficult to see in many batteries because the other electrode polarization increases so much when it approaches the limiting current situation. This situation generally represents the ultimate limitation to the passage of current in many batteries. As stated earlier, the onset of concentration polarization is the slowest among the three types of polarization. In case of an ideal stirred electrolyte, a steady state can be reached within milliseconds, but in a battery containing a separator and porous electrodes, the time required to reach a steady state may be several minutes or even hours.

The complex voltage-current-time behaviour of batteries can often be facilitated experimentally by the inclusion of a reference electrode within the cell. This electrode, having a stable potential, is allowed to remain at its open circuit value with the help of a very high-impedance electrometer. Each operating electrode is then independently measured against the reference electrode while current passes through the cell, and many of the individual electrode effects can be sorted out.

Recently, development of modelling methods for simulation of porous electrodes of the type used in batteries helped the battery designers to understand the implications of the porous electrode parameters e.g. porosity, pore size, and tortuosity and their interaction with the local impedance of the electrodes. On the basis of these models, it is gathered that the current is not produced uniformly throughout the electrode except at very low current, but has a complex distribution that depends on time, current and local properties such as electrolyte concentration, resistance etc.

Earlier, battery scientists used to compare the battery performance in terms of hours of service. However, the present practice is to compare energies per unit of volume, called the energy density and measured in wh/L, and per unit of weight, called specific energy and measured in wh/kg. A full comparison is often made of the variation of specific energy with specific power (w/kg) and of the variation of energy density with power density (W/L). Plots of this type, called Ragone Plots, provide information regarding the power at which the energy begins to fall off rapidly, due to approach of a limiting current or other modes of abrupt failure.

Other market driven criteria include cost per unit of energy or power, capital cost of rechargeable cell and charger versus the cost of a primary cell, and suitability of the cell for application. The active materials and the cell components must be stable for at least 5 years in the operating environment to provide adequate value and safety. Rechargeable cells may suffer self-discharge during storage and are returned to their starting state by the charging process. There are some other criteria driven by customer preference or convenience, such as primary versus rechargeable, difficulty and sophistication for a given charging regime, or large primary

BATTERIES AND BATTERY TECHNOLOGY

cell capacity versus the necessity for frequent recharging but overall longer life of a rechargeable cell. Another criterion is that some battery materials (such as lead in lead-acid batteries) are considered by some legislative bodies as undesirable unless proper collection and special disposal or recycling techniques are available. Other important considerations for the practical use of batteries are the temperature range of operation and storage life under the particular climatic conditions. For general consumer use, the temperature range of operations is from –10ºC to +50ºC, whereas for military use, it is –50ºC to +75ºC. In case of rechargeable battery systems, the additional requirements include the cycle life, the capability to maintain its capacity from cycle to cycle, over charge ability, cell reversal, and special charger requirements. Further, the safety of the battery is of primary importance. The battery should not leak, it should not vent any hazardous materials, or undergo any violent reactions under the conditions of actual use or even likely abuse by an uninformed or ignorant user.

The voltage available from a current producing chemical cell depends on the magnitude of the electrode potentials, which is generally a few volts only. If we want to produce higher voltages, a number of cells can be combined in series. Such an assembly is called "battery". Commercial cells which are used as a source of electrical energy belong to the following types : primary cells, secondary cells, reserve cells and fuel cells.

In a primary cell, a chemical reaction proceeds spontaneously and its free energy is converted into electrical energy. The production of electrical energy at the expense of the free energy of the cell is called "discharging of the cell". In the secondary cells, electrical energy is passed into the cell when a chemical reaction is induced and the products of the reaction remain on the electrodes. These products later react in the backward direction at our choice and liberate free energy in the form of electrical energy. Such cells serve to accumulate the electrical energy in the form of some chemical reaction and later on, the reaction is reversed at our will to release the electrical energy. This process is called "charging the cell". The primary cells have the inherent disadvantage that once they are run down, they cannot be used again. Further, manufacturers of primary cells strongly warn that consumers should not attempt to recharge them because of possible safety hazards, such as leakage or gas generation causing venting.

The cathode (i.e. the electrode at which reduction occurs), during the "discharge" of primary or secondary cells is designated "positive". In secondary cells, the electrode which functioned as "cathode" during "discharge" becomes the "anode" during charging.

The examples of primary cells include the simple voltaic cells, Daniel Cell, Lechlanche Cell, Bichromate Cell, Grove-Bunsen Cell, Clarke Standard Cell and the Weston Cadmium Cell. From commercial point of view, the main categories of primary cells are carbon-zinc cells, (including Lechlanche cell and Zinc chloride cells), alkaline cells and lithium cells.

21·3. PRIMARY CELLS

(A) The simple voltaic cell: It comprises of copper and Zinc plates that are separately dipped in sulphuric acid contained in glass vessel. The chemical action of zinc with dilute sulphuric acid furnishes chemical energy which forces the electricity from Zn to Cu inside the cell. The emf of the cell is the potential difference between Zn and Cu plates when they are disconnected. The emf is independent of the size and shape of the elements but depends only on the nature of the elements used and the concentration of the solution. Conventional current flows from Cu to Zn in the external circuit, and hence from Zn to Cu in the internal circuit or inside the cell.

Apart from Cu and Zn, any two metals having different reduction potentials or different oxygen affinities may be combined to constitute a cell. The order of oxygen affinities of some elements is as follows.

$Zn > H_2 > pb > Sn > Fe > Cu > Hg > Ag > Pt > C$.

Zinc has the highest oxygen affinity and carbon the least. Thus if Zn is used as the negative element any other element from the above series can be used as the positive element.

The layers of hydrogen that may accumulate on the positive element tends to increase the resistance for the passage of current inside the cell and also causes a back e.m.f. This is called polarization.

The impurities present in the negative element may combine with negative element to form local cells. Thus, the impure elements present in Zinc may lead to unnecessary dissolution of Zinc. This can be prevented by amalgamation of Zn. The Zinc amalgam covers the impurities and thereby the surface of pure zinc only is presented to the sulphuric acid.

(B) Daniel cell: This cell may be represented as follows

$$Zn-|\ Zn\ SO_4\ \|\ Cu\ SO_4\ |^+ Cu$$

Anode reaction: $Zn \rightarrow Zn^{2+} + 2e^-$
Cathode reaction: $Cu^{2+} + 2\ e^- \rightarrow Cu$
Cell reaction: $Zn + Cu\ SO_4 \rightarrow Zn\ SO_4 + Cu$, emf 1.1 V

(C) Grove Bunsen cell: This cell may be represented as follows

$$Zn\ |\ 8\%\ H_2\ SO_4\ |\ 66\%\ HNO_3\ |\ Pt$$

Nitric acid acts as depolarizer. The liquid Junction potential that might be generated between the two acids is negligible.

$$Zn + H_2\ SO_4 \rightarrow Zn\ SO_4 + H_2$$
$$2\ HNO_3 + H_2 \rightarrow 2H_2O + 2NO$$

The Voltage may vary between 1.86 to 1.96 V as the concentration of HNO_3 varies from 45% to the furming nitric acid.

(D) Bichromate Cell: This cell may be represented as follows:

$Zn\ |\ H_2\ Cr\ O_4;$	$H_2\ SO_4;$	$H_2O\ |\ C$
18 parts	22 parts	100 parts

$$3\ Zn \rightarrow 3\ Zn^{2+} + 6\ e^-$$
$$6e^- + Cr_2\ O^{2-}_7 + 14\ H^+ \rightarrow 2Cr^{3+} + 7\ H_2O$$

Cell reaction: $Zn + H_2\ SO_4 \rightarrow Zn\ SO_4 + H_2$ The chromic acid is prepared by reacting $K_2\ Cr_2\ O_7$ with $H_2\ SO_4$. The chromic acid acts as a depolarizer.

The voltage of the cell is about 2 V. However, due to intense polarization, this cell is suited only for intermittent use.

(E) The Clark Standard cell: The cell may be represented as follows:

Zn (amalgam 10%) | Zn $SO_4.7H_2O$ (satd.) | Hg_2SO_4 (satd.) | Hg

The cell reaction is

$$Zn + Hg_2\ SO_4 \cdot 7\ H_2O \rightarrow Zn\ SO_4.7H_2O + 2\ Hg$$

The emf of the cell is 1.433 V

(F) The Weston – Cadmium Cell :- The cell may be represented as follows:

Cd(amalgam 12.5%) | 3 Cd $SO_4 \cdot 8H_2O$ | Hg_2SO_4 (satd.)| Hg

$$Cd \rightarrow Cd^{2+} + 2e^-$$
$$Hg_2\ SO_4 + 2e^- \rightarrow 2\ Hg + SO_4^{2-}$$
$$Cd + Hg_2\ SO_4 + 8/3 \cdot H_2O \rightarrow Cd\ SO_4 \cdot 8/3\ H_2O + 2\ Hg$$

The emf of the cell is 1.0183 V

(G) Carbon – Zinc Cells: There are two basic versions of Carbon – Zinc Cells: the Leclanche Cell and Zinc Chloride or heavy duty cell. The original Leclanche cell consisted of an amalgamated zinc rod as anode and a carbon plate surrounded by a mixture of granular carbon and Mn O2 as cathodes dipping into a 20% solution of NH_4Cl as electrolyte. The cell reaction is

$$Zn^- | NH_4 Cl | MnO_2 + Carbon^+ |.$$

In the "dry cell", which is a modified version of Leclanche cell, the electrolyte is immobilized by using the electrolyte in the form of a paste. Such a cell is christened as "dry cell" because of the absence of any liquid or mobile phase. The electrolyte is soaked up by some absorbent material like cardboard. The dry cell consists of a zinc anode which is shaped as a container for the electrolyte and a carbon cathode surrounded by MnO_2, and a paste of NH4 cl and Zn as a cathodic depolalizer, and facilitates the H^+ - discharge reaction by removing the adosrbed hydrogen atoms.

Cathode reaction: $2 MnO_2 + 2H_2O + 2e^- \rightarrow 2 MnO (OH) + 2OH^-$

Anode reactions:
(Primary reaction) $\quad Zn - 2 e^- \rightarrow Zn^{2+}$
(Secondary reactions) $\quad 2 NH_4 Cl + 2 (OH^-) \rightarrow 2 NH_3 + 2 Cl^- + 2 H_2O$
$\quad\quad Zn^{2+} + 2NH_3 + 2 Cl^- \rightarrow [Zn (NH_3)_2] Cl_2$

Overall cell reaction:

$$Zn + 2 Mn O_2 + 2 H_2 O \rightarrow Zn^{2+} + 2 OH^- + 2 Mn O (OH)$$

The voltage of the cell, which is due to the primary electrode reactions, is about 1.5 V. However, the secondary reactions consume the Zn^{2+} and OH^- as shown above, and once the cell is discharged, it cannot be charged again.

A Leclanche type primary cell employing Mg as anode and carbon as cathode has also been developed. The Eo values of Mg and Zn are respectively –2.38 V and –0.76 V. The electrolyte used was gelled magnesium bromide containing lithium chromate to inhibit the corrosion of the anode. The cathodic depolarizer comprises of a mixture of barium chromate and MnO_2. The cell voltage is 1.9 V.

Anode reaction: $Mg + 2 (OH^-) - 2e^- \rightarrow 2 Mg (OH)_2$
Cathode reaction: $2 Mn O_2 + 2 H_2O + 2e^- \rightarrow 2 Mn O (OH) + 2 (OH^-)$
Overall cell reaction: $Mg + 2MnO_2 + 2 H_2O \rightarrow 2 Mn O (OH) + Mg (OH)_2$

Carbon-Zinc Cells are the most commonly and the most widely used primary cells all over the world. They traditionally contain a carbon rod (for cylindrical cells) or a carbon coated plate (for flat cells) to collect the current at the cathode with a zinc anode, which is also used as the primary container of cylindrical cells and it is this combination that has led to the name "carbon – zinc cells". As stated earlier, the two basic versions of this type of cells are the Leclanche cell and the zinc chloride or heavy – duty cell. Both the types have zinc anodes and Mn O_2 cathodes and include Zn Cl_2 in the electrolyte. The Leclanche cell also contains an electrolyte saturated with NH_4 Cl (some additional undissolved NH_4 Cl is usually added to the cathode) whereas in the Zn Cl_2 cell, only a very small amount of NH_4 Cl is added to the electrolyte. Both types are considered as dry cells because of the absence of any excess liquid electrolyte in the system. The Zn Cl_2 cell is often made with synthetic Mn O_2 and gives higher capacity than the Leclanche cell, which uses inexpensive natural Mn O_2 for the active cathode material. Since Mn O_2 is only a moderate conductor, the cathodes of both the types of cells contain 10 to 30% carbon black in order to distribute the current.

Leclanche cells usually contain paste separators, in which the electrolyte solution and some cereals are cooked until thick. (In some designs, a cold set paste is used). The paste is metered into the cell and the pre-pressed cathode body is inserted, so as to force the paste in a separating layer. A cutaway view of a typical pasted cylindrical cell is shown in Fig. 21.1.

In paper lined cells, a paper – separator coated with starch or modified starch is used, which is much thinner and more conductive than the starch separator. This type of separator is used in premium Leclanche cells and zinc chloride cells. In such cells, the separator is inserted into the zinc can, followed by insertion of the carbon rod into the cathode mix. Fig.21.2 shows a cutaway view of a typical paper lined cell.

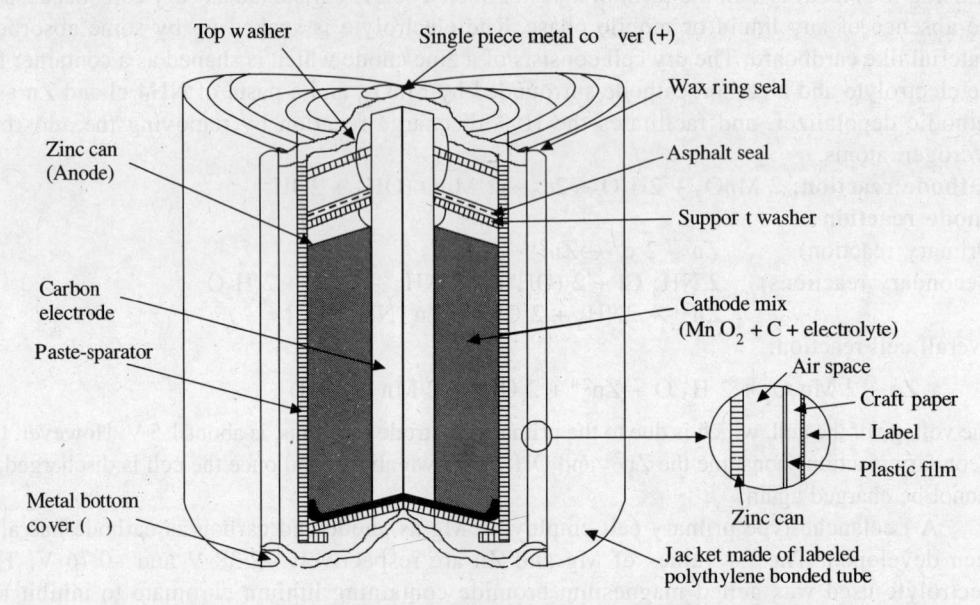

Fig. 21.1. Cut-away view of a typical pasted cylindrical Leclanche cell.

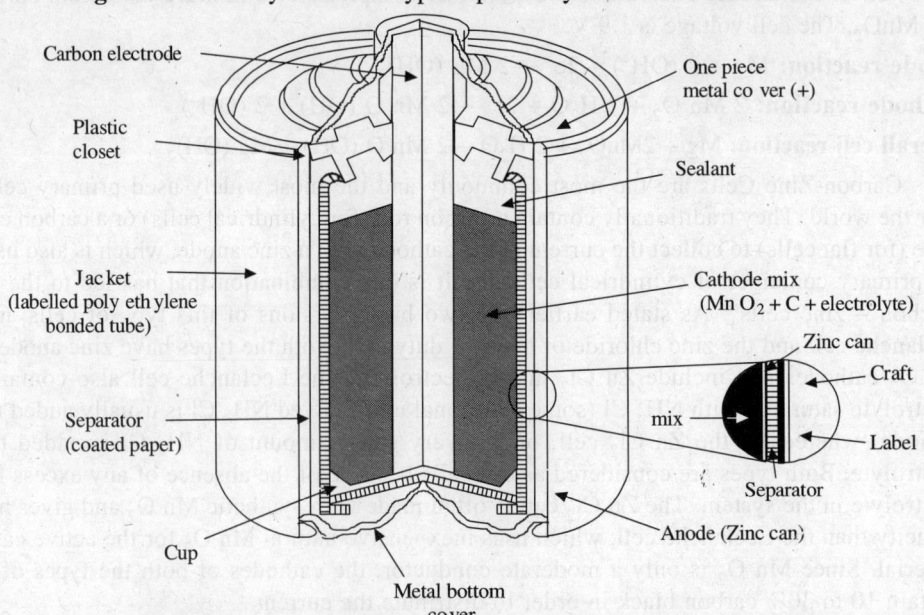

Fig. 21.2. Cut-away view of a paper-lined carbon-zinc cell.

Flat cells are used to stack in multicell batteries, most commonly for 6-Volt or 9-Volt batteries. Fig. 21.3 shows a cutaway of a typical flat cell, as used in a 9-Volt battery. In such a cell, the zinc plate is coated with carbon to serve as the collector for the cathode of the cell

BATTERIES AND BATTERY TECHNOLOGY

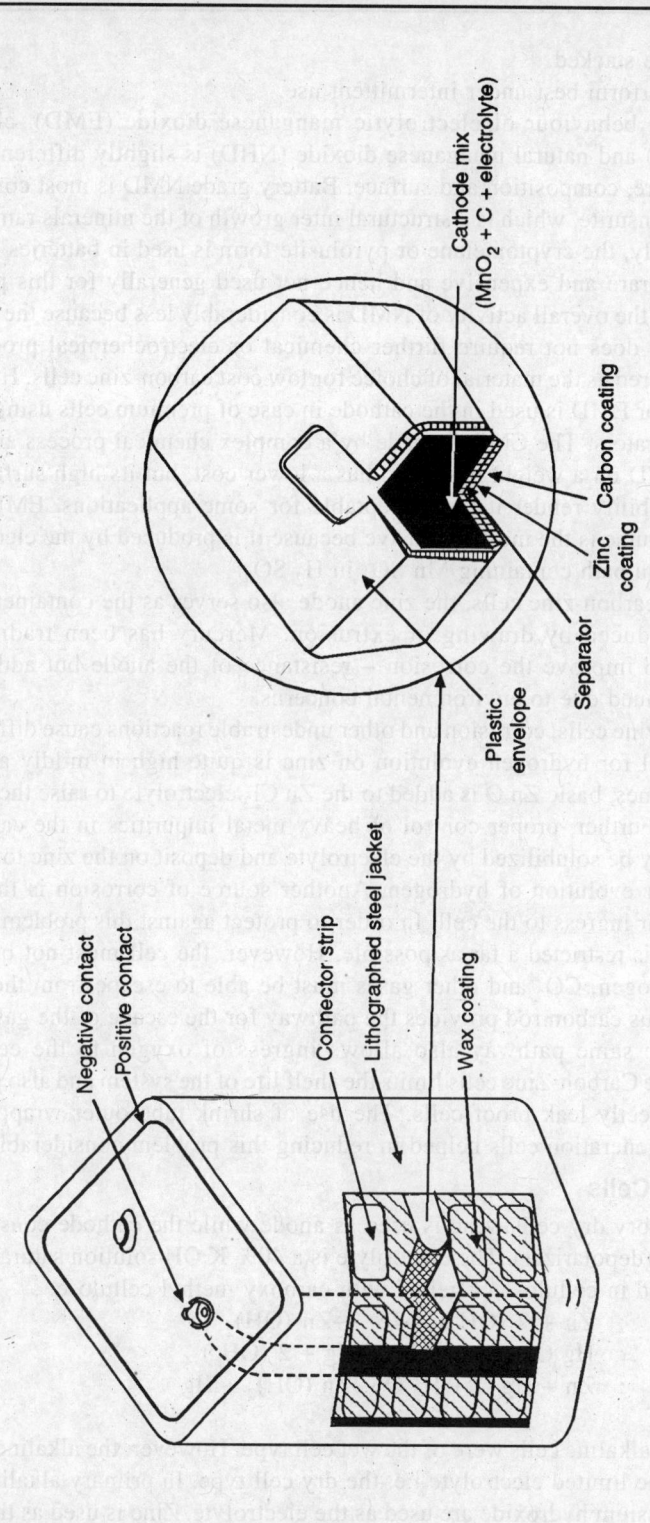

Fig. 21.3. A cut-away view of a typical flat cell used in multi-cell batteries.

Labels: Cathode mix (MnO_2 + C + electrolyte); Carbon coating; Zinc coating; Separator; Plastic envelope; Wax coating; Lithographed steel jacket; Connector strip; Positive contact; Negative contact.

beneath, when the cells are stacked.

Carbon-Zinc cells perform best under intermittent use.

The electrochemical behaviour of electrolytic manganese dioxide (EMD), chemical manganese dioxide (CMD) and natural manganese dioxide (NHD) is slightly different due to their differences in structure, composition and surface. Battery grade NMD is most commonly in the form of the mineral insutite, which is a structural inter growth of the minerals ramsdellite and pyrolusite. Occasionally, the cryptomelane or pyrolusite form is used in batteries because ramsdellite form is rather rare and expensive and hence not used generally for this purpose. The theoretical capacity or the overall activity of NMD is considerably less because the mineral is relatively abundant and does not require further chemical or electrochemical processing. Therefore, NMD is considered as the material of choice for low cost carbon-zinc cells. However, some proportion of CMD or EMD is used in the cathode in case of premium cells using $ZnCl_2$ electrolyte and paper separators. The CMD is made by a complex chemical process and has a comparable activity to EMD on a weight basis and has a lower cost, but its high surface area and poor packing compatibility render it less acceptable for some applications. EMD is the most active form and of course is the most expensive because it is produced by the electrolytic deposition of MnO_2 from a bath containing $MnSO_4$ in H_2SO_4.

In many cylindrical carbon-zinc cells, the zinc anode also serves as the container for the cell. The zinc can is produced by drawing or extrusion. Mercury has been traditionally incorporated in the cell to improve the corrosion – resistance of the anode but addition of mercury is being discontinued due to environmental concerns.

In designing carbon-zinc cells, corrosion and other undesirable reactions cause difficulties. Luckily, the over potential for hydrogen evolution on zinc is quite high in mildly acidic or alkaline solutions. Sometimes, basic ZnO is added to the $ZnCl_2$ electrolyte to raise the PH and lower the corrosion rate. Further, proper control of heavy metal impurities in the cathode is essential, because they may be solubilized by the electrolyte and deposit on the zinc to become low overpotential sites for evolution of hydrogen. Another source of corrosion is the direct reaction of oxygen from air ingress to the cell. In order to protect against this problem, the cell is sealed and access of air is restricted a far as possible. However, the cell must not be sealed hermetically because hydrogen, CO_2 and other gases must be able to escape from the cell as they are formed. The porous carbon rod provides the pathway for the escape of the gases from the cell but however, the same pathway also allows ingress of oxygen to the cell. This compromise situation in the Carbon-Zinc cells limits the shelf life of the system and also prevents the construction of a perfectly leak proof cells. The use of shrink tube outer wrapping and other devices in the new generation cells helped in reducing this problem considerably.

(H) Ruben-mallory Dry Cells

In the Ruben – Mallory dry cells, zinc is used as anode while the cathode consists of a paste of carbon and HgO (depolarizer). The electrolyte is a 40% KOH solution saturated with potassium zincate absorbed in cellulose or gelled with carboxy methyl cellulose.

Anode reaction : $Zn + 2(OH^-) \rightarrow 2e^- \rightarrow Zn(OH)_2$

Cathode reaction : $HgO + H_2O + 2e^- \rightarrow Hg + 2(OH^-)$

Overall cell reaction : $Zn + HgO + H_2O \rightarrow Zn(OH)_2 + Hg$

The cell voltage is 1.3 V.

(I) Alkaline Cells: Early alkaline cells were of the wet cell type. However, the alkaline cells of the 1990s are mostly of the limited electrolyte i.e. the dry cell type. In primary alkaline cells, sodium hydroxide or potassium hydroxide are used as the electrolyte. Zinc is used as the anode material while a variety of materials can be used as cathode. Depending on the type of construction, alkaline cells are divided into the following two classes:

BATTERIES AND BATTERY TECHNOLOGY

(i) The larger cylindrical shaped batteries
(ii) The miniature, button type cells.

Cylindrical alkaline batteries are mainly produced using Zinc – MnO_2 chemistry whereas miniature cells are produced using a variety of systems to meet the requirements of particular applications.

Cylindrical alkaline cells are mostly Zinc – MnO_2 cells using an alkaline electrolyte and are constructed in standard cylindrical sizes. They can be used in the same type of devices as ordinary leclanche cells and zinc chloride cells. Further, their high level of performance makes them ideally suited for applications like cameras, toys and audio devices.

A cross – sectional representation of a typical cylindrical alkaline cell is shown in Fig. 21.4.

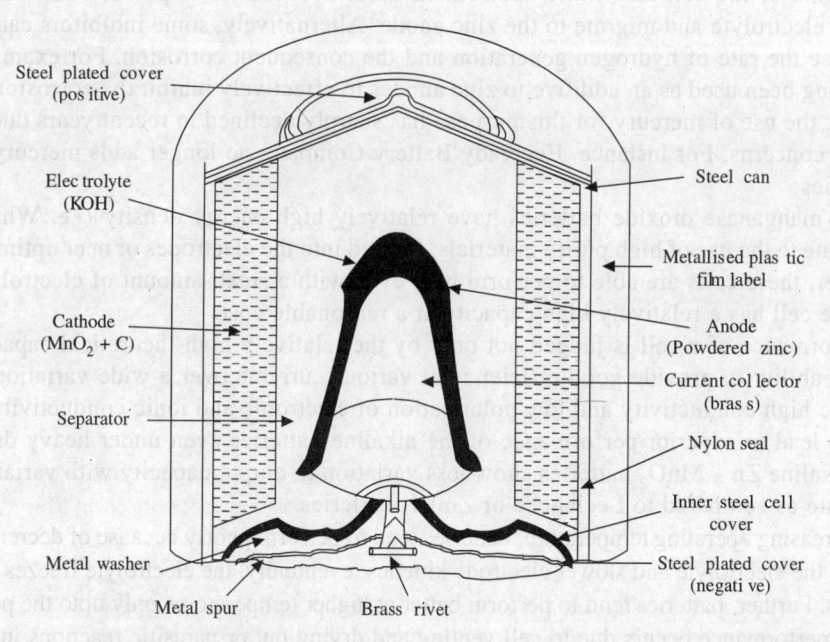

Labels (left): Steel plated cover (positive); Electrolyte (KOH); Cathode (MnO_2 + C); Separator; Metal washer; Metal spur
Labels (right): Steel can; Metallised plastic film label; Anode (Powdered zinc); Current collector (brass); Nylon seal; Inner steel cell cover; Steel plated cover (negative); Brass rivet

Fig. 21.4. Cross-sectional view of a typical cylindrical alkaline cell.

The battery is housed in a steel can, which also serves as the current collector for the cathode. The can contains dense, compacted cathode material consisting of MnO_2, carbon and at times, a binder. The cathode has a hollow center, lined with a separator to isolate the cathode from the anode. Within the separator lined, cavity, the anode mix is placed. The anode mix consists of alkaline electrolyte, zinc powder and a small quantity of gelling material to immobilize the electrolyte and suspend the zinc powder. A metal leaf or pin inserted into the anode mix provides the contact to the anode. A plastic seal assembly is provided to the cell to keep the electrolyte to prevent the leakage of electrolyte out of the cell and to prevent the air from getting into the cell. It is through this seal that the contact to the anode collector is provided for. The seal contains a safety vent, which is activated if the internal pressure of the battery exceeds a particular limiting value.

The alkaline cell derives its power from the reduction of the MnO_2 cathode and the oxidation of the zinc anode.

Anode reaction : $Zn + 2 OH^- \rightarrow ZnO + H_2O + 2e^-$
Cathode reaction : $2e^- + 2 MnO_2 + 2 H_2O \rightarrow 2 MnOOH + 2 (OH)^-$

However, the actual anode and cathode reactions may be much more complicated than what are represented in the above equations.

Alkaline batteries having high output capacity and high current carrying ability are being manufactured by many companies all over the world, which are competing with each other to improve the performance of the cylindrical alkaline batteries.

Since the electrolyte in the alkaline batteries is corrrosive to human tissue as well as materials in the devices where they are used, it is imperative to provide a good seal against any possible leakage in the alkaline batteries than in case of carbon – zinc cells. However, providing for a good seal is incompatible with the formation of a non-condensable gas such as hydrogen and hence it is important to maintain only a low level of gas in the battery. Further, certain impurities present in zinc can catalyze the generation of hydrogen, thereby greatly increasing the corrosion rates. Hence, the zinc used in alkaline cells must be of high purity. Similarly, other components of the cell also must not contain harmful levels of impurities that might dissolve in the electrolyte and migrate to the zinc anode. Alternatively, some inhibitors can be used to decrease the rate of hydrogen generation and the consequent corrosion. For example, mercury has long been used as an additive to zinc anodes to effectively inhibit the corrosion of zinc. However, the use of mercury for this purpose has sharply declined in recent years due to environmental concerns. For instance, Eveready Battery Company no longer adds mercury to alkaline batteries.

Alkaline manganese dioxide batteries have relatively high energy density (i.e. Wh/L). This is partly due to the use of high purity materials, formed into the electrodes of near optimum density. Further, these cells are able to perform well even with a small amount of electrolyte. This is why the cell has a relatively high capacity at a reasonable cost.

The performance of a cell is judged not only by the relatively high theoretical capacity but also by its ability to provide good efficiency at various currents over a wide variation of conditions. The high conductivity and low polarization of electronic and ionic conductivity in the electrodes, lead to superior performance of the alkaline batteries even under heavy drain conditions. Alkaline Zn – MnO_2 batteries show less variation in output capacity with variation in discharge rate as compared to Leclanche or Zn Cl_2 batteries.

With decreasing operating temperature, batteries tend to perform poorly because of decreased conductivity of the electrolyte and slower electrode kinetics; eventually, the electrolyte freezes and the battery fails. Further, batteries tend to perform better at higher temperatures only upto the point that decline in performance occurs due to cell venting and drying out or parasitic reactions in the cell. On the whole, it may be summarized that alkaline cells show less performance loss at low and high temperatures as compared to Leclanche Cells.

Alkaline Zinc – MnO_2 cells exhibit good capacity retention on long term storage. The use of high purity materials ensures very low rates of parasitic reactions in the cells.

Miniature alkaline cells are small, button shaped cells, which use zinc anodes, alkaline (Na OH or K OH) electrolyte and a variety of cathodic materials eg. Mn O_2, Hg O, Ag_2 O, or even air (in case of zinc – air batteries). While the chemistry of the anode is essentially identical as in the larger cylindrical alkaline cells, the cathode chemistry varies with the type of cathode material used. Miniature alkaline cells are used in watches, calculators, cameras, hearing aids and other miniature devices. The construction of a typical miniature alkaline cell is illustrated in Fig. 245. The cathode mix is placed into the can and then the separator layer is inserted. The type of the separator materials used and the number of layers depend upon the type of cathode used. The anode mix is then placed in contact with the anode cup. The anode cup and the can are electrically separated with the help of an insulating, sealing gasket, which also helps in preventing the leakage of the electrolyte from the cell. However, the cathode material, the type of separator and the electrolyte used vary with the nature of application.

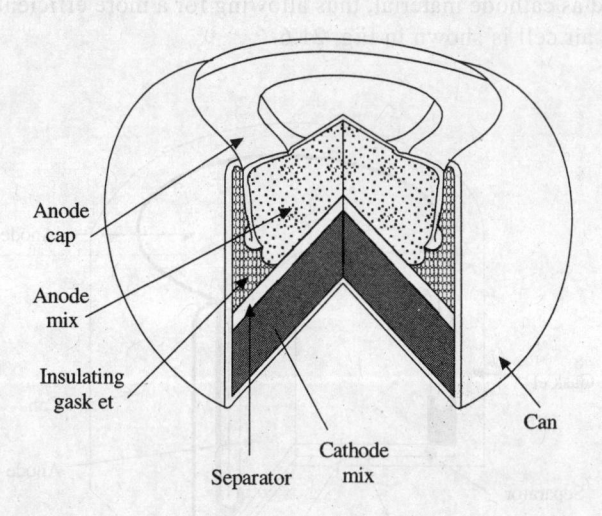

Fig. 21.5. Cut-away view of a miniature alkaline cell.

The zinc – MnO_2 miniature alkaline cells are used where economical power source is desired. The chemistry is identical with that in case of the Zinc – MnO_2 cylindrical battery.

Miniature zinc – HgO alkaline batteries have higher capacity that the Zin – MnO_2 batteries. The cathode reaction is

$$HgO + H_2O + 2e^- \rightarrow Hg + 2(OH^-).$$

The toxicity of mercury causes serious disposal problem.

Miniature zinc – silver oxide batteries have high energy density, almost as high as that of mercury cells. They operate at higher voltages than mercury cells but for somewhat lesser time. The cathode reaction is

$$Ag_2O + H_2O + 2e^- > 2Ag + 2(OH^-).$$

Miniature zinc – silver oxide batteries are made with KOH or NaOH as electrolyte. However, KOH – containing batteries operate more efficiently at high current drains, whereas NaOH - containing batteries are more resistant to leakage and easier to seal. Miniature zinc – silver oxide batteries are generally used in electronic watches and in such applications where high energy density, a flat discharge profile and higher operating voltage than a mercury cell are needed. They have good storage life too.

If Ag_2O is replaced by AgO as the cathode in the miniature zinc – Ag_2O battery, higher capacity and higher energy density can be obtained. The cathode reaction in this case is –

$$AgO + H_2O + 2e^- \rightarrow Ag + 2(OH^-).$$

The air depolarized cell or air cell :- In this type of cell, the zinc anode and the porous carbon cathode are immersed in 20% solution of NaOH.

Anode reaction : $Zn + 4(OH^-) - 2e^- \rightarrow ZnO_2^{2-} + 2H_2O$

Cathode reaction : $H_2O + [O] + 2e^- \rightarrow 2(OH)^-$

Overall cell reaction : $Zn + 2(OH^-) + [O] \rightarrow ZnO_2^{2-} + H_2O$

Since OH^- ions take part in the electrode reaction, the alkali gets diluted. The voltge of the cell is 1.45 V. The zinc – air batteries offer the possibility of obtaining very high energy densities.

In place of a cathode material placed in the battery during manufacture, oxygen from the atmosphere is used as cathode material, thus allowing for a more efficient design. The cutaway view of miniature air cell is shown in Fig. 21.6.

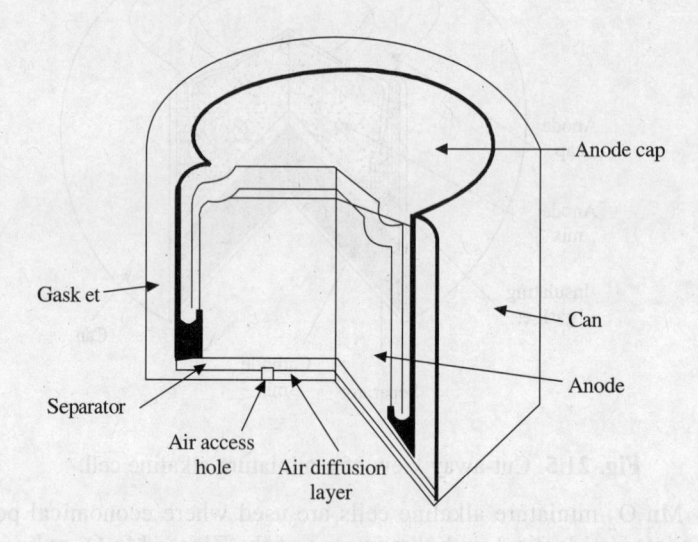

Fig. 21.6. Cross-sectional view of a miniature air cell battery.

It resembles a typical miniature cell, except for the air access holes in the can. The anode occupies more internal volume of the cell. However, in place of a thick cathode pellet, it is provided with a thin layer containing the cathode catalyst and air distribution passages. Air enters the cell through the holes in the can and the oxygen reacts at the surface of the cathode catalyst. The air access holes are generally covered with a protective tape, which can be removed when the cell is used.

The miniature air cell cathode usually contains special type of carbon to provide a surface for the initial reduction of oxygen and also to catalyze to peroxyl decomposition. Small amounts of metal oxides can also be used as catalysts.

The cathode reaction shown above requires several steps. Initially, oxygen is reduced or it may be catalytically decomposed, producing hydroxide and oxygen. The performance of the air cathode largely depends on its ability to catalyze this decomposition reaction. If this reaction is very slow, large amounts of peroxide will build up in the cathode during discharge, and this result in large polarization and low operating voltage. If the reaction is well catalyzed, the battery can operate at a higher and more useful voltage.

Air cells are ideal for use in such applications where the usage is largely continuous, and where the discharge level is relatively constant and well defined. Air cells are not general purpose cells. They have to be carefully designed for particular application. Miniature air cells are mainly used in hearing aids, where they are required to produce a relatively high current for a relatively short period (i.e. for a few weeks). Miniature air cells can be operated for continuous service in the temperature range of -10^0 C to $+ 55°$C.

The air cells are widely used in Railways and the locations where the electric supply by cables is either expensive or impossible.

Important characteristics of some aqueous primary batteries are listed in Table 21(1).

Table 21(1) – Important Characteristics of Some Aqueous Primary Batteries

Sr.No.	Battery System	Working Voltage, V	Energy Density, Wh/L	Specific Energy Wh/Kg	Temperature range, °C
1.	Carbon-Zinc (Zn/MnO_2)	1.2	70 to 170	40 to 100	–40 to +50
2.	Alkaline-Manganese dioxide (Zn/MnO_2)	1.2	150 to 250	80 to 95	–40 to +50
3.	Zinc-Mercuric Oxide (Zn/HgO)	1.3	400 to 600	100	–40 to +60
4.	Zinc-Silver Oxide (Zn/Ag_2O) (Zn/Ag_2O)	1.55	490 to 520	130	–40 to +60
5.	Zinc/Air (ZnO_2)	1.25	700 to 800	230 to 400	–40 to +50

Lithium Cells

Conventionally, cells having lithium anodes are called Lithium Cells irrespective of the type of the cathode used. Lithium cells can be broadly categorized into the following two types:

(*i*) **Cells with solid cathodes:** These may have solid or liquid electrolytes. Solid electrolyte systems, other than lithium-iodine systems, are yet to develop into commercial level.

(*ii*) **Cells with liquid cathodes:** These have liquid electrolytes. Further, at least one component of the electrolyte solvent and the cathode active material are identical.

The lithium cells take advantage of the high energy film – forming property. This property allows for lithium compatibility with solvents and other materials with which it is thermodynamically unstable, yet permits high electrochemical activity when the external circuit is closed. This is possible because the film formed is conductive to lithium ions, but not to electrons. Hence, corrosion reactions occur only to the extent of formation of thin, electronically insulating films on lithium which are coherent and also adherent to the base metal. This thinness of the film is vital for allowing reasonable rates of transport of lithium ions through the film when the circuit is closed. Water, alcohol and some other compounds which are thermodynamically unstable with lithium, do not form such a possivating film. Organic compounds such as propylene carbonate ether, butyrolactone, dimethoxy ethane, tetrahydrofuran and dioxolane are among the best solvents that were found to be useful in lithium batteries Similarly, the most successful electrolyte salts include lithium perchlorate, lithium tetrafluoroborate, lithium trifluoromethane sulfonate, and lithium hexafluorophosphate. Lithium hexafluoroarsenate which possesses good properties as electrolyte but is toxic because of arsenic is at present used mostly in military applications only.

Organic electrolytes have relatively low conductivity and also lead to slow kinetics of the cathode reactions. It is mostly these reasons that compelled the development of designs such as thin electrodes, and very thin separators for lithium batteries. This usage, in its turn, led to the development of coin cells instead of button cells for miniature batteries and Jelly – roll or spiral wound designs instead of bobbin designs for cylindrical cells. Despite the high cost associated with glass-to-metal hermetic seals or compression seals designed to minimise the ingress of water and oxygen and egress of volatile solvent in the cylindrical lithium cells, their superior energy densities offer considerable economic advantages as compared to aqueous systems.

The lithium-manganese dioxide is emerging as the most widely used 3-Volt solid cathode lithium primary battery. The electrolytic MnO_2 should be heated to > 300°C to effectively remove water, before incorporating it in the cathode. This is very important to get good performance of the cell. The anodic and cathodic reactions taking place in the cell are as follows:

Anodic reaction : $Li \rightarrow Li^+ + e^-$

Cathodic reactions : **Step-I**

$X Li^+ + xe^- + MnO_2 \rightarrow Li_x MnO_2$

This first step is a homogeneous reaction in which a partially lithiated material is formed. This step is further followed by a heterogeneous process to a new phase, the structure of which is yet to be conclusively determined.

Step-II

$(1 - x) Li^+ + (1 - x) e^- + Li_x MnO_2 \rightarrow Li MnO_2$

The energy density of the system depends on the type of the cell and the current drain. The Li – MnO_2 batteries are competitive with other systems. The cylindrical cells are used in fully automatic cameras. The coin cells are widely used in electronic devices such as calculators, watches, etc. The lithium iron disulfide battery with nominal voltage of 1.5 V was first manufactured in button cell size and can be used as replacement for zinc – silver oxide cell. The performance of the cylindrical cell is superior to alkaline cells at all rates of discharge. The cell is widely used as power source for electronic camera flash guns, which yield 3 to 5 times more flashes than those obtained with alkaline cells.

Lithium-iodine cells are widely used in medical applications. Iodine forms a charge transfer complex with poly (2 – Vinyl pyridine) (PVP), which constitutes the cathode for these cells. The Li – Iodine battery systems are highly stable and dependable and hence are universally used for heart pacers which can be used for even 10 years as compared to 1.5 to 2 years of operation for alkaline batteries. The limitation to low current densities are intrinsic with the Li – Iodine systems. Researchers involved in development of medical electronics, in their efforts to manufacture of equally stable batteries with higher current capabilities, succeeded in developing lithium – silver vanadium oxide battery which holds promise for defibrillator applications. Less expensive coin – type cells have also been developed for consumer electronics applications, but the use of such cells are restricted because of severe current limitations.

Among the lithium cells with liquid cathode, the first successful example is lithium – sulfur dioxide cells. They use either acetonitrile (AN) or propylene carbonate (PC) or a mixture of the two as cosolvents with the SO_2, usually 50 Vol %. The cell reaction is as follows :-

$2 Li + 2SO_2 \rightarrow Li_2 S_2 O_4$

Lithium – thionyl chloride cells possess very high energy density; mostly due to the nature of the cell reaction, namely –

$4 Li + 2 SOCl_2 \rightarrow 4 LiCl + S + SO_2$

The reaction involves two electrons per thionyl chloride molecule. Further, one of the products, SO_2, is a liquid under the internal pressure of the cell, facilitating a more complete use of the reactant. Also, no co-solvent is needed for the solution, because thionyl chloride is a liquid having only a modest vapoar pressure at room temperature. Li Al Cl$_4$ is generally used as the electrolyte salt. The system has an open circuit voltage of 3.6 V. Because of the high activity of the cathode and the good mass transfer of the liquid catholyte, the cell is capable of very high rate discharges. Owing to the excellent voltage control, lithium – thionyl chloride batteries are increasingly used on electronic circuit boards for supplying a fixed voltage for memory protection and other stand-by functions. Since the cells are designed in low rate configuration, it is possible to obtain maximum energy density and cell stability. Because of the wide range of temperature and performance capability coupled with high specific energy, these cells are also attracting military and space applications. These cells also find application in medical devices such as neuro-stimulators, drug delivery systems, etc. Characteristics of some lithium primary batteries are given in Table 21(2).

Table 21(2) – Important Characteristics of Some Lithium Primary Batteries

Sr. No	Cell Systems	Working Voltage, V	Energy Density, Wh/L	Specific Energy, Wh/Kg	Temperature range of operation, °C
(A) Solid Cathode Systems					
1.	Li / CF	2.5 – 2.7	400	200	– 20 to + 60
2.	Li / Mn O$_2$	2.7 – 2.9	400	200	– 20 to + 55
3.	Li / I$_2$	2.4 – 2.8	920	250	37
4.	Li / Fe S$_2$	1.3 – 1.7	460	250	– 20 to + 60
(B) Liquid cathode systems					
5.	Li / SO$_2$	2.7 – 2.9	440	250	– 55 to + 70
6.	Li / SO Cl$_2$	3.3 – 3.5	950	300	– 50 to + 70

21·4. SECONDARY BATTERIES

In secondary batteries the cell reaction can be made to proceed in any direction by withdrawing or supplying the current. These secondary or rechargeable batteries may be broadly divided into the following three categories:
(1) Lead Acid accumulators or Acid Batteries
(2) Alkaline storage Batteries
(3) Others including lithium/lithium ion batteries. Each of these systems are discussed below.

(1) Lead Acid Accumulators

It comprises of a grid of lead-antimony alloy coated with lead dioxide as positive pole (cathode) and spongy lead as negative pole (anode). The electrolyte is a 20% solution of H$_2$SO$_4$ (Specific gravity 1.15 at 25°C) in which a number of the above electrode pairs, containing inert porous partitions in between, are dipped. The electrode reactions which take place during the discharge of the cell (*i.e.* when the current is drawn from the cell) are as under

Anode
$$Pb - 2\,e^- \rightarrow Pb^{2+}$$
$$Pb^{2+} + SO_4^{2-} \rightarrow Pb\,SO_4$$
$$Pb + SO_4^{2-} - 2\,e^- \rightarrow Pb\,SO_4$$

Cathode
$$Pb\,O_2 + 4H^+ + 2\,e^- \rightarrow Pb^{2+} + 2H_2O$$
$$Pb^{2+} + SO_4^{2-} \rightarrow Pb\,SO_4$$
$$\overline{PbO_2 + 4H^+ + SO_4^{2-} + 2e^- \rightarrow Pb\,SO_4 + 2H_2O}$$

Since the lead sulfate gets precipitated on the cathode and in the solution, the cell reaction may be written as follows.
Pb | Pb SO$_4$ (S) | H$_2$SO$_4$ (aq), Pb SO$_4$ (s) | Pb O$_2$ (s) | Pb
The net cell reaction for the passage of two faradays is

$$Pb\ O_2 + Pb + 2H_2SO_4 \underset{\text{charging}}{\overset{\text{discharging}}{\rightleftarrows}} 2\ Pb\ SO_4 + 2\ H_2O$$

During the discharge process, the consumption of H$_2$SO$_4$ is replaced by an equivalent quantity of water. Thus as the cell produces electric current, the concentration of the sulfuric acid decreases. However, during the charging process, the reverse reaction takes place and hence sulphuric acid is generated while an equivalent water is consumed. It is because of this that the original strength of the acid is restored. Since both these changes are associated with variations in the specific gravity of the acid, the extent of charge or discharge of the cell at any time can be determined by testing the specific gravity of the acid. The cell voltage lies in the range of 1.88 V to 2.15 V as the concentration of the sulphuric acid varies from 5% to 40%.

(2) Rechargeable Alkaline storage batteries

Several types of rechargeable alkaline batteries were manufactured from time to time since the latter part of the 19th century. In 1881, Lalande and Chaperson identified the benefits of an alkaline electrolyte in zinc-copper oxide cell which eventually resulted in the development of a secondary cell. Junger and Edison independently developed nickel-iron and nickel-cadmium batteries after discovering nickel oxyhydroxide and its high positive potential. Nickel-cadmium battery systems are still used in several applications. Several other alkaline rechargeable battery systems such as nickel-hydrogen, nickel-metal hydride, manganese dioxide, zinc and silver-zinc have been produced recently and are gaining importnace. The salient features of some of these systems are incorporated in Table 21(3).

Several other new types of secondary battery systems are under various stages of development and some of them such as the lithium and lithium ion cells have already been produced in limited hit the market.

The salient features of some such systems are tabulated in Table 21(4).

The nickel-iron batteries (the Edison cells) were widely used as a traction battery and is again under study for such application. Otherwise it is not considered today as an important commercial battery.

The nickel iron cell may be represented as follows.

Fe | FeO (S) KOH (Aq) | Ni O (S) or Ni$_2$ O$_3$ (S) | Ni

It comprises of steel grid cotnaining iron powder as an anode and Ni$_2$ O$_3$ or Ni O mixed with Ni and supported on steel grid as cathode. The electrolyte is a 20 to 25% solution of KOH taken in an inert steel container. The overall cell reaction for the passage of two faradays is

Fe + 2 Ni (OH)$_3$ →Fe (OH)$_2$ + 2 Ni(OH)$_2$

The cell voltage is about 1.34 V.

The nickel-cadmium cell is essentially similar to the nickel-iron cell excepting that cadmium is used instead of iron. The overall cell reaction for the passage of two faradays may be represented simplistically as follows.

Cd+2 Ni (OH)$_3$ →Cd (OH)$_2$ + 2 Ni (OH)$_2$

The knowledge on basic electrochemistry, design principles and technology of nickel cadmium batteries has developed significantly in recent times. This battery is usually

Table 21(3) SALIENT FEATURES OF SOME ALKALINE RECHARGEABLE BATTERY SYSTEMS

Battery System	Battery Types	Battery Applications	Salient features, advantages, limitations, and remarks
Nickel-cadmium	a) prismatic cells with pocket electrodes	Stationary/remote power requirements, load leveling	Easy to produce. Relatively low efficiency low energy per volume and weight. Open circuit voltage cell capacity 1.3 volts. Energy density(Wh/L) is 1.36. Specific energy is 1.21 Wh/kg. Toxicity due to to cd.
	b) Vented prismatic cells having sintered electrodes	Air craft batteries portable	High electrode efficiency, high cycle life, high rate capability, open circuit voltage cell capacity is 1.315 Ah. Energy density is 2.72 Wh/L. specific energy is 2.31 Wh/kg. Toxicity of Cd.
	c) Button cells having pressed powder electrodes	lowpower applications	Small size. Amenable for mass production. Relatively low current capability. Toxicity due to cd.
	d) Sealed cylindrical cells with Jelly roll electrodes and prismatic cells with stacked electrodes	portable highpower and highenergy applications	Relatively low electrode efficiency cadmium toxicity
Nickel-Hydrogen	Pressure vessel, thin electrode stacks	High profile applications in space missions; As power systems for civil and miltiary satellite applications.	High cycle life. High electrode efficiency. Easily charageable. Resistant against possible abuse. Expensive. Difficult to manufacture, poor energy density and poor charge resistance. Open circuit voltage 1.3 V. Energy density 60 Wh/L. specific energy 50 Wh/kg.
Nickel-Metal hydride	Sealed cylindrical, Jelly roll electrodes	Portable high-power and high-energy applications	High energy density, high cycle life, high negative electrode efficiency, Expensive metals are used and hence costly, poor charges retention as compared to Ni-Cd cells; cell capacity 1.3V. Energy density is 184 Wh/L. specific energy is 55 Wh/kg.
Silver-Zinc	Prismatic cells with flat plate electrodes	Military applications in aerospace, sub-marine and communications	High rate and high energy density. poor cycle life. Expensive because of use of silver; cell capacity 1.6V; Energy density is 190 Wh/L; specific energy is 90 Wh/kg.
MnO_2-Zinc	Sealed cylindrical cells with concentric cylinder electrodes	Portable, low to moderate power applications.	Relatively cheap. Absence of pb and Cd and hence environmentally safe relatively, Alternatve for primary alkaline batteries. Very poor cycle life. Rapid fall of capacity. possible mercury contamination in some designs. cell capacity 1.5V; Energy density is 125 or 240 Wh/L for 10 to 1 cycle as the case may be; specific energy is 90 Wh/kg.

Table 21(4) RECHARGEABLE BATTERY SYSTEMS UNDER DEVELOPMENT

Battery System	Nominal Voltage	Remarks
Lithium-Manganese dioxide	3.0 V	Room temperature operation, positive electrode material is air-sensitive, problems due to plating of lithium
Lithium-V_6O_{13}	2.5 V	Solid polymeric electrolyte cross linked using radiation polymerisation, lithium plating problems
LithiumTi-S_2	2.3 V	Room temperature operation, air-senstive positive electrode material, problems due to lithium plating.
Aluminium-Air	1.6 V	Circulating electrolyte, low efficiency negative electrode, replaceable negative electrode
Lithium-Aluminum-FeS	1.33 V	Fused salt electrolyte, problem due to corrosion
Nickel Zinc	1.65 V	High rate capability, low cycle life
Zinc-Bromine	1.85 V	Circulating liquid phase contains bromine complex and the related containment problems
Sodium-Nickel Chloride (or iron Chloride)	2.5 (or 2.3)V	Solid positive electrode, molten sodium is used as negative electrode, high temperature molten salt as catholyte and solid β-Al_2O_3 separator
Sodium-Sulfur	2.1 V	β-Al_2O_3 separator, high temperature liquid electrodes and containment problems.

manufactured in the discharged state. The active materials are mainly Ni $(OH)_2$ or CdO. It is widely used in the sealed version for high current applications (eg. Power tools) and for high cycle life applications (*e.g.* Computer power supply units). The sealed, jelly roll type cell system is also designed for high rate charging. The charging time is as short as 15 minutes by using special chargers. Larger sealed cells have been widely used in space application where excellent system reliability and high cycle life are required for synchronous low earth orbit applications in conjunction with solar cells. The battery supplies power on the dark side of each orbit and gets recharged on the light side by photovoltaic solar cells. The nickel-cadmium cell is also manufactured as a wet cell in two versions; (a) the sintered cell, mainly used for stand-by power and aircraft starting batteries and (b) the pocket cell mainly used for starting diesel engines, and for emergency lighting. Though expensive as compared to acid batteries, these cells have excellent performance capability.

The chemistry involved in all the above three types of nickel-cadmium batteries is same. The cadmium electrode is discharged and charged in the overall reaction as follows:

$$Cd + 2\ OH^- \underset{\text{charge}}{\overset{\text{discharge}}{\rightleftarrows}} Cd\ (OH)_2 + 2e^-$$

The normal reaction at nickel electrode is

$$\beta\text{-Ni OOH} + H_2O + e^- \rightarrow \beta\text{-Ni (OH)}_2 + OH^-$$

However this reaction is complicated because of the existence of two additional phases that undergo electrochemical transformation under various conditions into the original phases.

$$Y - \text{Ni OOH} + (1-x)H_2O + e^- \rightarrow \alpha - \text{Ni (OH)}_2 \text{ X. } H_2O + OH^-$$

The Y – Ni OOH phase appears in the system because of possible overcharging of the β–Ni OOH electrode. This material is then reduced during discharged to the α–form, which is slowly converted to the β–phase in the strongly alkaline electrolyte. These reactions interfere with the smooth functioning of the nickel electrode.

Adequate care is taken while designing the sealed nickel cadmium batteries to provide for adequate porosity within the electrodes and the separator to permit oxygen gas transport, to accommodate the heat generated in the cell and to provide appropriate venting system which can be activated as and when pressure build-up occurs in the cell. The electrodes in the sealed nickel- cadmium batteries have also undergone considerable development in recent years to achieve higher energy density. Originally, sintered nickel strips of relatively high porosity were impregnated by solutions of nickel or cadmium salts and the corresponding hydroxides were precipitated in place by adding NaOH or KOH solution. Alternatively, a water electrolysis reaction at the sintered surface causes as local change of PH to alkaline condition resulting in the precipitation of nickel hydroxide on the sintered surface. These sintered electrodes possess high mechanical and electrochemical stability and are capable of high currents during charge as well as discharge and give very high cycle life. Another approach to increase the energy density of the negative electrode is by using fibrous nickel mat of very fine fibers and high porosity. A recent development for the positive electrode involves posting a blend of nickel hydroxide and conductor into a very high porosity (about 95%) nickel metal foam structure. Special precipitation processes have been developed to prepare high density nickel hydroxide for these electrodes to achieve higher energy densities. Additives like cobalt oxide, Zn and Cd were found to reduce swelling of the positive electrode, evolution of oxygen and minimizing the formation of undesirable phases. Under special conditions, cells, with energy density over 140 Wh/L and specific energy over 50 Wh/kg have been produced.

(3) Nickel hydrogen batteries

These batteries are widely used in satellites for civilian and military purposes. The principle involved in the nickel-hydrogen cell is that the hydrogen electrodes are reversible when properly catalyzed and hence can be used to generate hydrogen from the water present in the cell during charging.

$$H_2 + 2 OH^- \underset{\text{charge}}{\overset{\text{discharge}}{\rightleftharpoons}} H_2O + 2 e^-$$

In this, the negative electrode is the limiting electrode so that oxygen is not evolved in the charging process, thereby avoiding a potentially dangerous situation in which hydrogen and oxygen could be present in an explosive mixture. The pressure vessel made of Inconel or stainless steel can withstand upto 200 bar. The operation is usually designed at about 10^0C to facilitate the dissipation of heat, to a safe design level of 50 bar. Since the reaction between the nickel electrode with gaseous hydrogen is slow, such a cell is ideally suited to the short delays between charging and discharging and the short discharge times of the low earth orbit. The specific energy of the battery system is quite high (50 Wh/kg) and that is a good feature for space applications. Further, the cycle life behaviour is remarkably good and more than 40,000 cycles are obtained. Hence over 10 years of low earth orbit can be obtained from this type of battery system with 10 charge-discharge cycles taking place in a day.

(4) Nickel-Metal Hydride Cells

The cell has been developed more recently and is produced in two different versions using different alloy types.

(a) A B$_2$ types where A is group IV metal such as titanium and B is a group VIII metal such as nickel

(b) A B$_5$ type where A is a rare earth metal such as lanthanum and B is a group VIII metal such as nickel.

Many additional alloying elements have been incorporated into both the above two base formulations to improve the hydride formation properties. The voltage of this cell is comparable to that of nickel-cadmium cells, whereas its energy density is comparable to the most advanced types of nickel-cadmium cells (i.e. 110 – 150 Wh/L).

The shelf life of this cell is better than that of hydrogen cells because the nickel alloy is designed to reduce the pressure of hydrogen gas to only a few bars as compared to 50 bars in case of nickel- air battery. Further the negative electrode is the limiting electrode to forestall the evolution of oxygen from the electrode during charge, thereby avoiding any explosive mixture of hydrogen and oxygen in the cell. These batteries are used for portable high power and high energy applications. Further improvement in cycle life and shelf properties may enable potential use of this cell as source of power for portable computers, portable telephones, etc.

(5) Manganese dioxide-zinc battery

This is nothing but an improved version of well known primary manganese dioxide-zinc battery system with respect to battery design and containment of zinc electrode eg. by using laminated cellophane separator, to restrict the growth of the dendritic zinc which reduces the battery life. This battery is of low cost and is suitable for low to moderate power applications.

(6) Silver-Zinc Alkaline Battery

The electode reactions of this cell are similar to the primary cell described earlier, excepting that the high voltage plateau of divalent silver oxide is purposely sought to raise the energy density of the system. The high cost of the system because of silver and the fact that when the system is fully discharged, only 80% of the initial capacity is recoverable because of shape change and dendritic growth on zinc electrode, are the two factors which are detrimental for non-military application. The high energy density of the system and the charge retention of the cell during storage are the factors for the military applications in the fields of aerospace, submarine and communications.

(7) Rechargeable Lithium-ion Batteries

A group of battery systems known as lithium-ion rocking chair, or swing batteries are receiving great attention and interest recently. The positive electrode in such systems is a lithiated transitional metal oxide such as $Li Co O_2$, $Li Mn_2 O_4$ or $Li NiO_2$ mixed with a conductor and binder and coated on a metal foil. $Li Co O_2$ is at present used in commercial cells. This material is in the discharged state. The negative electrode consists of carbon or graphite and a binder coated on another metal foil. This is also in the discharged state. The cell is charged after production. The cell reaction on charge involves oxidation of the positive electrode in a homogeneous reaction to form a lithium – deficient transition metal oxide with the transition metal in a higher oxidation state :

$$Li Co O_2 \rightarrow Li\, xCo O_2 + (1-x) Li^+ + (1-x) e^-$$

The cell reaction at the negative electrode depends on the nature of the carbon, to some extent. If the negative electrode is made of amorphous carbon, it seemed to undergo a homogeneous reaction with lithium to form a lithium insertion compound of variable stoichiometry. If graphitic carbon is used as negative electrode, it undergoes a stepwise reaction leading to the final formation of C_6 Li. The reaction may be represented generically as follows :

$$C + Y\ Li^+ + Y\ e^- \rightarrow C\ Li_y$$

The charged open circuit voltage of the cell is about 4.1 V. The energy density is about 250 wh/L while the specific energy is 115wh/Kg. The cycle life is quoted at 1200 cycles with reasonably high current capability and a self-discharge rate of 10% per month. The unit cells are cylindrical, jelly roll designs, generally of the same size as Ni – Cd cells. The batteries were used for portable telephones, computers and camcorders. These battery systems are under further development, particularly for use in electric vehicles and for energy storage services.

(8) Rechargeable Lithium Batteries

These cells had a lithium foil as negative electrode and a specially treated Mo – S_2 as positive electrode. A lot of developmental work is in progress on rechargeable lithium batteries with liquid electrolytes because of the high specific energies and high energy densities possible with this technology. A number of positive electrodes such as Mn O_2, Ti S_2 and $V_2 O_5$ are under study.

Another approach under investigation with rechargeable lithium ion and lithium batteries is to utilise an immobilised polymer electrolyte as the ionically conductive medium instead of a liquid electrolyte with a polymer separator. Some lithium salts dissolve in certain polymers such as polythylene oxide and the conductivity is sufficiently high so as to enable the battery operation at elevated temperatures eg., 50°C. But, if a stable solvent is incorporated to plasticize the polymer, the conductivity increases to such a level to enable room temperature operation. The positive electrode materials being investigated for polymer electrolytes include Li $Mn_2 O_4$, Mn O_2, Ti S_2 and $V_6 O_{13}$.

(9) High Temperature Batteries

Some elevated temperature battery systems have been studied for possible use as power source for electric vehicles. They include (a) iron sulfide – lithium aluminum battery with a molten chloride electrolyte and (b) sodium polysulfide – sodium battery with a solid beta alumina separator. The iron sulfide battery operates at about 400° C. (c) Sodium – Ni Cl_2 (or Fe Cl_2) battery using molten salt electrolyte in the positive electrode compartment and molten sodium in the negative electrode compartment. The two compartments are separated by beta aluminium oxide tubes. The temperature of operation is about 250° C.

21·5. RESERVE BATTERIES

In recent years, there has been a requirement of cells with special characteristics for the use in proximity fuses, artillery, mines, pilot balloons, guided missiles, radars, bazookas, night-viewing equipment, sub-marine torpedo propulsion, walkie-talkies, and special electronic equipment. Such batteries must withstand vibration, shock, extreme variations in temperatures and also must have capability of high voltage, long inactive shelf life, rapid activation to the ready state as and when required, and must be available in different sizes and shapes. As a result of these requirements, considerable R & D work has been done with promising outcome. The so - called "Reserve Batteries" are the outcome of such R & D efforts. These batteries are generally classified by the mechanism of activation employed, such as

    (*a*) by mechanical means
    (*b*) by water activation using sea water or tap water
    (*c*) by electrolyte activation using only the solvent or the complete electrolyte
    (*d*) by gas activation, using a gas as an active cathode
    (*e*) by thermal activation, using a solid salt electrolyte that is activated by melting on application of heat.

Activation is usually done by adding a missing component just before use, which should be done in a simple way such as pouring the water into an opening in the cell in case of water – activated cell or in a more complicated manner by using valves, pistons, heat pellets activated by gravitational or electric signals in case of thermally – activated or electrolyte – activated cells. Some of them are described below :

(1) Perchloric Acid and Fluoboric Acid Cells

These cells were developed shortly before and during the second world war. They were specially designed for radio-sonde batteries but have found some applications in electronic equipment. They are relatively inexpensive, light – weight, stable, shock – proof and have good low – temperature characteristics. They can be stored for long periods. The cell may be represented as follows:

Pb | HCl O_4 or HBF_4 (aq) | Pb O_2, inert material.

The electrodes used are similar to those used in lead – sulphuric acid storage cell but the reaction products formed in this case are readily soluble in water or acid solutions. The cell can be activated at the time of use and can give several hours of service.
Reaction at lead electrode (negative electrode) is :

$$Pb \rightarrow Pb^{2+} + 2 e^-$$

Reaction at lead dioxide electrode (positive electrode) is :

$$Pb\ O_2 + 4H\ Cl\ O_4 + 2\ e^- \rightarrow Pb\ (Cl\ O_4)_2 + 2\ Cl\ O_4^- + 2\ H_2O$$

The overall cell reaction is:

$$Pb\ O_2 + Pb + 4\ H\ Cl\ O_4 \rightarrow 2\ Pb\ (Cl\ O_4)_2 + 2\ H_2O$$

When fluoboric acid is used, the reaction is same excepting that lead fluoborate is formed on discharge. The cell cannot be charged nor can it be stored in the assembled state for more than a couple of days.

Perchloric and fluoboric acid cells are assembled in various ways. For instance, an A – B battery is made of polyvinyl plastic. The electrodes are of hair pin type with metallic lead plated on one prong and lead dioxide on the other. The battery can be evacuated by a special device and the electrolyte is drawn into the battery by suction. In other types, the electrolyte is supplied by a reservoir which is broken by means of a button or by other mechanical means at the time of activation.

The e.m.f. of the cell varies with changes in concentration or activity of the perchloric acid. For normal temperatures, an electrolyte of 60% perchloric acid is preferred, while for low temperatures, 50% perchloric acid is satisfactory. In the latter case, an e.m.f. of 1.92 V is obtained at –50° C. With fluoboric acid, 1.86 V is obtained.

Prchloric acid is hazardous in presence of organic matter. Fluoboric acid is safe. Fluosilicic acid and some sulfonic acids may be used as electrolyte, but they yield lower voltage and lower capacity.

(2) Silver Chloride Cell

This reserved cell is activated by sea – water or even by fresh water. It was also developed during the second world war to fulfil the demand for a light - weight high - powered reserve battery. This battery must be preserved in a dry state in hermetically sealed containers with silica gel (used as a desiccant) until activated by mere immersion in sea water or fresh water at the time of use. The cell may be represented as:

Mg | water | Ag Cl | Ag

It is constructed with long thin strips of silver foil covered on both sides with silver

BATTERIES AND BATTERY TECHNOLOGY 737

chloride by electroplating and wound in a spiral with strips of commercially pure magnesium of like dimensions. The strips of magnesium and silver chloride are separated by a strip of absorbent paper. The roll of alternate strips of magnesium and silver chloride is placed in plastic containers open on the sides to admit water for activation of the cell. The cell warms up as it operates and gives a working voltage of 1.5 volts under moderate loads. Sea water activates the cell faster than fresh water. When sea water is used to activate the cell, a voltage of 1.42 volts is obtained within 2 minutes. The cell has good low temperature properties. These cells are made in different sizes and shapes. They are ideally suited for radiosonde and other applications where a light high-powered cell is desired. For high – voltage applications, batteries are made with flat – type cell. Batteries are made up by stacking plates with absorbent paper between the electrodes of each cell and insulation between adjacent cells. The magnesium – silver chloride cells are used for powering emergency communication devices for air-plane crews whose planes come down in the sea.

(3) The Gordon – Magnesium Battery

This cell consists of magnesium as the negative electrode and an oxygen (from air) depolarizer as the positive electrode. Tap water is used as the electrolyte. The cell may be represented as follows :

$$Mg \mid Water\ or\ KBr\ (Aq) \mid O_2\ (gas)\ from\ air, C$$

The reaction at the magnesium electrode is the formation of magnesium ions whereas at the carbon electrode, oxygen is consumed with the formation of water presumably through the formation of hydrogen peroxide at the intermediate stage.

The cell actually consists of a carbon tube in which magnesium rod is placed. The space between the two is filled with a wick of wood wool. Four such units are joined in parallel. The wicks of each unit protruding from their lower ends are bundled together to form a pad, which is fully immersed in the electrolyte made of K Br solution. Water is added before the cell is used. The four carbon tubes are tied together tightly by a wrapping of copper gauze to ensure good contact. The electrolyte does not cover the electrodes.

The cell has an open – circuit voltage of 1.6 volts. About half an hour is required to attain the peak voltage and the battery warms as it attains its full voltage.

(4) Silver Peroxide – Zinc Alkaline Cell

This cell was designed during the second world war to meet the requirement for a high rate primary cell. Howerver, during investigations, it was found that this cell was best suited for "one shot" purposes because of its shorter shelf life, arising out of the high solubility of silver oxide in Na OH or KOH used as electrolyte. Thus, this cell became a reserve – type cell which can be activated at the time of use. The cell may be represented as follows:

$$Ag \mid Ag\ O \mid K\ OH\ (25\%\ aq.\ Soln) \mid Zn$$

The reactions at the silver oxide electrode (positive electrode) occurs in two reduction steps:

$$2\ Ag\ O + H_2\ O + 2\ e^- \rightarrow Ag_2\ O + 2\ OH^-$$

$$Ag_2\ O + H_2\ O + 2\ e^- \rightarrow 2\ Ag + 2\ OH^-$$

The reactions at the zinc electrode (negative electrode) are as follows:

$$Zn \rightarrow Zn^{2+} + 2\ e^-$$

$$Zn^{2+} + 2\ OH^- \rightarrow Zn\ (OH)_2 \rightleftharpoons H_2\ Zn\ O_2$$

$$H_2\ Zn\ O_2 + 2\ K\ OH \rightleftharpoons K_2\ Zn\ O_2 + 2\ H_2\ O$$

Then, the overall reaction at the zinc electrode is as follows :

$$Zn + 2 K^+ + 4 OH^- \rightarrow K_2 Zn O_2 + 2 H_2 O + 2e^-$$

A solution of sodium hydroxide may also be used as electrolyte but the conductivity of KOH is higher. On the other hand, zincates are more soluble in NaOH than in KOH.

This cell cannot be used as a secondary cell.

At high – rate discharges, the working voltages are about 1.36 V, 1.15 V, 1.05 V and 0.85 V respectively, for discharge rates of 5, 10, 20 and 30 amperes at room temperature.

The limitations of this cell include limited shelf life, high cost because of use of Ag, suceptibility of silver oxide for reduction by organic impurities, reduced capacity at low temperatures, enhanced corrosion at high temperatures and diffusion of soluble silver salts from the positive electrode to the negative electrode. However, these limitations are not considered too serious for reserve type cells.

(5) Fused electrolyte cells

These cells are activated by heat. They usually employ a stable solid electrolyte of low melting point. These cells have zero voltage at normal temperature but attain their maximum voltage at the melting point of the solid electrolyte. A cross sectional view of such as cell is shown in Fig.21.7.

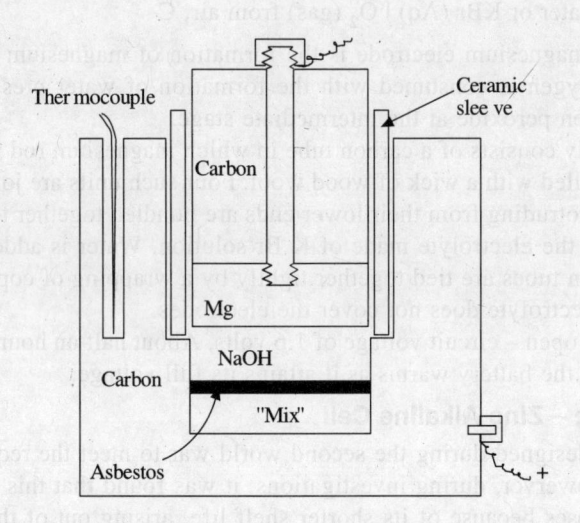

Fig. 21.7. Cross-sectional view of a fused-electrolyte reserve-type cell.

The cell may be represented as follows.

Mg | molten electrolyte | Mn O$_2$, C

In this cell, sodium hydroxide, potassium hydroxide, sodium nitrate potassium nitrate, lithium nitrate, sodium nitrite or sodium hydrogen sulfate were used as the solid or molten electrolyte. A mixture of manganese dioxide and acetylene black called the "mix" is placed in a reservoir made of carbon block. The electrolyte is then packed over the "Mix" and a carbon rod inserted in magnesium disc in such a way that the disc and mix are separated by about 2mm. The carbon rod along with the magnesium disc is insulated from the carbon block by ceramic sleeve, which extends to a ledge above the reservoir of "mix". Lead wires are fastened to the carbon block and the rod. A thin strip of soft absbestos is placed above the "mix" to prevent Mn O$_2$ from diffusing through the molten electrolyte to the magnesium electrode. For experimental purposes, the temperatures of the cell is determined with a thermocouple, the hot junction of which is placed in a well in the carbon block of the cell. The cell is heated by an external source of heat. When fused Na OH is used as electrolyte, 750 minutes of service are obtained at a working voltage of 1.3 volts to a cut off voltage of 1 volt. Longer service times may be obtained with larger cells. The cell reaction is as follows:

$$Mg + 2\ MnO_2 + 2\ NaOH\ (fused) \rightarrow$$
$$Mg(OH)_2 + Mn_2O_3 + Na_2O$$

(6) Gyuris Cell

This cell was orginally proposed as a storage cell but it may also serve as a reserve cell. The cell consists of sodium nitrate as electrolyte and MnO_2 and molten sodium as electrodes. The electrodes are separated by a diaphragm of copper gauze. The cell operates at the temperature of about 450°C and gives an average closed circuit voltage of an 2.5 volts. The cell maintains a suitable temperature when once it has been heated.

(7) Lithium cells

Many battery types, including many of those described under, primary batteries, can be used as reserve cells. For instance, lithium-thionyl chloride and lithium-sulfur dioxide systems are often used in "reserve" configurations, in which the respective electrolyte stored in a sealed compartment, is released for activation with the help of a piston or any other means to force it into the inter-electrode space. These high energy systems have taken over some of the applications of the older liquid ammonia reserve batteries which usually employ magnesium as an anode and metadinitrobenzene as cathode. Another new reserve battery is the lithium-vanadium pentoxide battery which performs at high rates and high energy. Such batteries are employed in mines and fuses used in military ordnance.

Another interesting liquid cathode reserve battery is the lithium water battery in which water serves not only as the electrolyte but also as the liquid cathode. The reaction product is the soluble lithium hydroxide. In some cases, a solid cathode such as silver oxide, or another liquid reactant such as hydrogen peroxide, is used in combination with the lithium anode and aqueous electrolyte to improve the rate or to decrease the gaseous emissions in the system. These cells are generally used in the marine environment where water is available in plenty or compatible with the cell reaction product. These cells are generally used for applications such as torpedo propulsion and powering sono-buoys and submersibles.

Thermally activated "reserve cells" of the older type use calcium or magnesium anodes, but the newer type cells use lithium alloys as anodes. Lithium alloys with aluminium, boron and silicon are of intrest for this purpose. Lithium can diffuse rapidly within the alloy phase, thereby allowing high currents to flow. The electrolyte is usually the eutectic composition of lithium chloride and potassium chloride which melts at 352^0C. This electrolyte has temperature dependent conductivities with an order of magnitude higher than the best aqueous electrolytes. The high conductivity, the enhanced kinetics and the enhanced mass transport allow the battery to be discharged at a very high rate. The cathodes used with the older calcium anode cells are metal chromates such as calcium chromate. The anode reaction product is calcium chloride while the reaction products at cathode is a mixed calcium chromium oxide of uncertain composition. Iron sulfide is one of the best cathodes for lithium alloy cells. The pellet used for thermal activation is generally a reactive metal such as iron or zirconium and an oxidizing agent such as potassium perchlorate. An electrical or mechanical signal ignites a primer, which in its turn ignites the heat pellet, which melts the electrolyte. Adequate heat is evolved by the high current which is capable of sustaining the necessary temperature during the life time of the application. Millions of these batteries have been manufactured for military ordnance, for use in bombs, rockets, missiles etc.

Reserve batteries can be stored for 10 to 20 years while awaiting use. They are usually manufactured in large numbers for military use. However, they are also occassionally employed for industrial and safety purposes.

21·6. FUEL CELLS

(A) Introduction: A fuel cell is an electrochemical cell which can convert the chemical energy contained in a readily available fuel-oxidant system into electrical energy by an electrochemical process, in which the fuel is oxidised at the anode. Similar to any other electrochemical cell, the fuel cell consists of an electrolyte and two electrodes. However, the fuel and the oxidizing agent are continuously and separately supplied to the electrodes of the cell, at which they undergo reactions. These primary cells are capable of supplying current as long as the reactants are supplied.

Fuel cells work at high efficiency and the resulting emission levels are far below the permissible limits. Fuel cell systems are modular and hence can be built in a wide range of power requirements from a few hundred watts upto multi kilo watt and even megawatt sizes. This enables construction of highly efficient power plants at any specified location. Further, due to low emission levels fuel cell power plants can be installed on site where energy is actually required for consumption even in densely populated area. Hence the power transimission lines are more economical and transmission losses are minimum.

The basic principles of a fuel cell are identical to those of the well known electrochemical batteries. The only difference is that in fuel cell, the chemical energy is provided by a fuel and an oxidant stored outside the cell in which the chemical reactions takes place. Electrical power therefore, can be obtained as long as the cell is supplied with the fuel and the oxidant.

Despite the fact that the history of fuel cell dates back to 1839 with the experiments of the British scientist, Grove and the subsequent R & D efforts particularly since 1950, fuel cells were used relatively for a few application only. The most important applications have been their use as the main source of electrical energy for the manned space-crafts of the Gemini, the Apollo and the space shuttle programs of the National Aeronautics and Space Adminstration (NASA) of the United States of America. Attempts to develop fuel cell systems for terrestrial application were made in the 1980s. Medium to higher power range (200 kw to 10 Mw) applications were aimed for load levelling for energy utilities, for remote power plants, cogeneration, industrial waste utilization and emergency power supply. In the lower power range, (kilowatt range), studies were focussed on the development of fuel cell systems for propulsion of electric vehicles and for military use.

Given the most important charactrstics, namely high efficiency, low emission and low noise levels, the fuel cells are expected to play a vital role in the future energy scenario, particularly when hydrogen is projected as the main energy source of the 21st century.

(B) Historical Development

Sir william Grove in 1839 electrolysed sulphuric acid using two platinum electrodes and observed that the gases evolved namely, hydrogen and oxygen, were electrochemically active and provided a potential difference of about one Volt between the electrodes.

Ostwald in 1894 realised on the basis of thermodynamic considerations, that an electrochemical cell can provide a higher energy conversion efficiency than heat engines, which are limited by Carnot's law.

Nernst and Haber directed their efforts on the study of direct carbon oxidizing fuel cells but met with only a limited success because of the difficulty at that time to procure the appropriate materials.

Baur in 1933 Proposed for the first time that an electrochemical system can work at room temperature, with alkaline electrolyte, using hydrogen as fuel. Bacon, a British engineer in

1950 carried this idea forward and successfully developed porous nickel electrodes. He had built the first high pressure medium temperature alkaline fuel cell system in the kilowatt range. It was on this basis that the fuel cell for the Appolo program of NASA (USA) was developed.

Justi, Bishoff, spenger and Winsel of Braunschweig, technical university, Germany around 1965, prepared highly active porous electrodes made of pressed Raney nickel and Raney silver powder. The hydrogen and oxygen electrodes developed later at Varta and siemens were based on this work.

Broers and Ketelaar of Holland in 1958 improved the high temperature molten carbonate fuel cell to achieve longer life for the cell. This work led to the development of the present power plant systems which can oxidize hydrocarbon fuels directly without a reforming step.

In order to achieve low temperature air operation with low cost carbonaceous materials, several other scientists further continued their studies on gas diffusion electodes by schmid (1923), studies on peroxide mechanism of carbon-oxygen electrodes by Berl (1943) and the work on carbon-air electrodes by Heise and schumacher (1932 – 1947).

Kordesch and Marko of the university of Vienna in 1951 prepared special metal oxide catalysts based on cobalt aluminium spinels. The carbon plate and tube electrodes were activated by steam to enhance their active surface. The carbon hydrogen electrodes contained small amounts of platinum catalyst which could oxidize fuels like alchols and aldehydes directly at low temperature. These new carbon electrodes provided high current densities in alkaline electrolytes at low temperature, thereby confirming the predictions of Baur (1933).

Fuel cells received worldwide attention in the 1950s and the 1960s because of the space programmes. Hydrogen-oxygen fuel cells were chosen as the best for use in manned space flights not only because of their high energy density but also because the reaction product generated is pure water which can be used by the crew for drinking and also for cooling purposes.

After a brief full due to reduction in the number of space projects, a new wave of R & D programmes on fuel cells started emerging once again in 1973 due to the oil crisis. Further investigations were directed to achieve more efficiency in energy conversion schemes and better enviornmental controls. Dedicated efforts have been directed towards development of technology of the five most promising systems, viz alkaline fuel cell system, molten carbonate fuel cells, phosphoric acid fuel cells, ion exchange membrane fuel cells, and solid oxide fuel cells.

(C) Theoretical Principles

The conventional process of utilizing the chemical energy of a fuel may be depicted as follows:

$$\text{chemical energy} \xrightarrow{(1)} \text{heat} \xrightarrow{(2)} \text{mechanical energy} \xrightarrow{(3)} \text{electrical energy.}$$

The losses in energy conversion in steps (1) and (3) can be minimized but it cannot be done in step (2) because the extent to which heat can be converted into mechanical work is limited by the second law of thermodynamics. Notwithstanding the improvements in design, about 50 to 95% of the chemical energy of the fuel is wasted. In contrast to heat engines, in a fuel cell, the chemical energy is directly converted into electrical energy without going through an intermediate conversion into heat. Hence the limitation imposed by the second law of thermodynamics is not applicable to such a conversion. Therefore, this direct conversion of chemical energy into electrical energy in a fuel cell can theoretically have an efficiency of 100%.

The basic arrangement in a fuel cell can be represented as follows.

Fuel | Electrode | Electrolyte | Electrode | oxidant.

At the anode, the fuel undergoes oxidation liberating electrons and the oxidation products of the fuel. The electrons so liberated from the oxidation process taking place at the anode reduce the oxidant at the cathode.

The electrons liberated at the anode can perform useful work when passing through the external circuit. In a fuel cell, the electrical work done corresponds to a change in free energy of the reacting system.

$$\Delta G = -n\, FE$$

where n is the number of electrons involved, F is the Farady constant, and E is the theoretical potential.

If the enthalpy change of the cell reaction is ΔH the difference in energy ($\Delta H - \Delta G$) can be converted only to heat but not to work.

$$\Delta H - \Delta G = T.\Delta S$$

where ΔG is the Gibbs free energy of reaction, T is the absolute temperature and ΔS is the entropy of the reaction.

The thermodynamic efficiency (eff_{th}) of the fuel cell is determined by the ratio of the work obtained from the cell and the heat effect of the reaction.

$$\left(Eff_{th}\right) = \frac{\Delta G}{\Delta H} = 1 - \frac{T.\Delta S}{\Delta H}$$

The thermodynamic efficiency of the fuel cell depends on the sign and magnitude of ΔS. If the gaseous substances are consumed in the reaction so that Δn, the change in the number of molecules of the system is Zero, the efficiency is unity.

(Δn, the change in the number of the molecules of the system = total number of molecules of gaseous products − total number of molecules of gaseous reactants).

If Δn is $< O$, then ΔS is $< O$ and ΔH is negative. Then the eff_{th} is <1.

It is possible to have $eff_{th} > 1$ if Δn is > 0 so that ΔS is $>O$ However, the actual efficiency realised is about 50 to 80% and this is certainly greater as compared to the maximum practical efficiency of heat engines.

The electrochemical $\left(eff_{el}\right)$ efficiency of a cell in operation is defined as:

$$\left(eff_{el}\right) = V/E$$

Where V is the terminal cell voltage and E is the cell potential.

In case of hydrogen-oxygen fuel cell having porous electrodes and potassium hydroxide as electrolyte, the following reactions take place.

Cathode

$$O_2 + 2H_2O + 4\,e^- \rightarrow 4\,OH^-$$

This reaction occurs through the intermediate peroxy ion formation, which is followed by the merely catalytic peroxide decomposition reaction with no net flow of electons.

$$O_2 + H_2O + 2\,e^- \rightarrow HO_2^- + OH^-$$

$$HO_2^- \rightarrow OH^- + \tfrac{1}{2} O_2$$

The cathode reaction given above is the sum of the above intermediate reactions occuring at the cathode)

At anode

$$2H_2 + 4\,OH^- \rightarrow 4H_2O + 4\,e^-$$

Overall reaction

$$2H_2 + O_2 \rightarrow 2H_2O$$

Hydrogen enters the pores of the anode and reaches the reaction zone, where the gas, the liquid electrolyte (i.e aqueous potassium hydroxide solution) and the solid conducting structure meet. The hydrogen gas diffuses to the electrochemically active site (e.g. a surface coated with a platinum catalyst) where it is absorbed and dissolved by the electrolyte. It then dissociates and ionizes to 2 H^+. Two protons react with the hydroxyl ions of the electrolyte to form water, which dilutes the KOH electrolyte. Two electrons are made available to the external circuit. The hydroxyl ions which are thus used up are "replenished" from the cathode reaction, in which ½ O_2 reacts with water to produce 2 OH^- ions, taking up the two electrons from the outer circuit. If the reaction is related to one molecule of oxygen, the process produces a total of four electrons. These facts are reflected in the cell reactions given above.

For the successful functioning of a fuel cell, the following basic requirements have to be fulfilled.

(a) The fuel cell electrode design should provide for promoting a high rate of electrode processes by creating a stable interface between the three phases, viz. the solid electrode, the liquid electrolyte and the gaseous fuel. Some of the electrodes used are platinum, porous PVC, teflon coated with silver, nickel boride, raney nickel, etc.

(b) The fuels and the oxidants used must be relatively cheap and readily available. The liquid fuels that are now used include methanol, ethanol, hydrazine, formaledehye, etc. the gaseous fuels that are used include hydrogen, alkanes, carbon monoxide etc. The oxidants used include oxygen, air hydrogen peroxide and nitric acid, Oxygen is used as oxidant in many fuel cells employing aqueous electrolytes because it is cheap and readily available.

(c) The reactions proceed at a faster rate at increased temperatures and also in presence of suitable catalysts. The emf of a fuel cell falls appreciably when current is drawn from the cell, due to polarization effect, especially over-voltage. These effects are minimised and the rates of the electrode processes are increased at higher temperatures. Molten salt and oxide systems such as an eutectic mixture of Li_2CO_3 and Na_2CO_3 or K_2CO_3 as electrolytes and fuels such as hydrogen or carbon monoxide are used for this reason.

(d) In order to have an appreciable conductivity and to minimize the effect of concentration polarization, fairly concentrated aqueous solutions of the electrolytes should be used as far as possible. Whenever high temperatures are employed, the loss of water due to evaporation should be minimized by the application of pressure. The electrolytes used include KOH, sulphuric acid or phosphoric acid solutions. However, aqueous systems are not suitable if the temperatures are over 250^0C. In such cases, fused carbonates (eg. Li_2CO_3, Na_2CO_3, K_2CO_3) or oxides are used. Since the electrode reactions are fast at these high temperatures, high activity catalysts are not required.

(D) Advantages of Fuel cells

1. High efficiency of energy conversion (75 to 82.8%), from chemical energy to electrical energy.
2. No emission of gases and pollutants are within permissible limits
3. Fuel cells offer excellent method for efficient use of fossil fuels.
4. Hydrogen-oxygen systems produce drinking water of potable quality.

5. Low noise pollution and low thermal pollution.
6. Modular and hence parts are exchangeable.
7. Low maintenance costs.
8. Fast start-up time of low temperature systems
9. The regenerative hydrogen-oxygen system is an energy storage system for space applications.
10. Low cost fuels can be used with high temperature systems.
11. The cogeneration of heat will increase the efficiency of high temperature systems.
12. Fuel cell automotive batteries can render electric vehicles efficient and refillable.
13. Fuel cells are suitable for the future nuclear solar-hydrogen economy.
14. Hydrogen and air electrodes are useful in other battery systems eg. Nickel-hydrogen, Zinc-air, aluminium-air, etc.
15. Fuel cells hold promise in the energy scenario where hydrogen is projected as the main energy carrier of the 21st century.
16. Saves fossil fuels.

(E) Limitations

1. High initial cost.
2. Large weight and volume of gas-fuel storage systems.
3. High cost of pure hydrogen.
4. Lack of infrastructure for distributing hydrogen
5. Liquifaction of hydrogen requires 30% of the stored energy.
6. Necessity of further R&D efforts for direct oxidation of hydrocarbons in fuel cells.
7. Most alkaline cells suffer from carbon dioxide degradation and hence CO_2 should be removed from the fuels and the air.
8. Life-times of the cells are not accurately known.

Some efforts have already been made towards minimizing the above disadvantages. Indeed, considerable progress has already been made regarding cost reduction of catalysts by using small amounts at higher utilization rate. Mass production of fuel cell components by improved techniques reduced the costs. The use of carbon-based catalyst substrates and other construction components gave promising results in increasing the life expectancy of low and medium temperature cells. New discoveries in the manufacture of metal hydrides and for the liquifaction of hydrogen have been made. Pressure cylinders have already decreased in weight by 50% over the last few decades. Fuel reformer units are already getting smaller and more efficient. Alcohols and ammonia appear promising as hydrogen carriers. Further R & D efforts may pave the way for more extensive use of fuel cells.

(F) Fuel cell Technology

(i) Terminology

A fuel cell is an energy conversion device or a reactor in which electrical energy is generated by special type of electrochemical cells. A "battery" may comprise of an arrangement of such cells in parallel or series (modules/stack of cells). The term "fuel cell system" commonly refers to a fuel cell battery together with all the necessary accessories for its operation. The main components of a fuel cell system are shown in Fig. 21.8.

BATTERIES AND BATTERY TECHNOLOGY 745

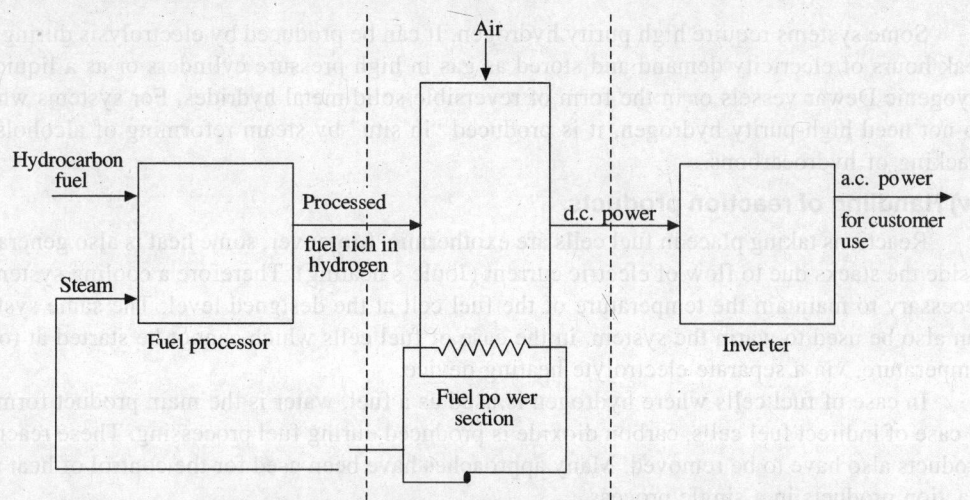

Fig. 21.8. Fuel cell system (stack and accessories)

(ii) Electrodes

Most of the reactants used in fuel cells are gaseous, and hence the development of the porous gas diffusion electrodes represent a significant break-through in fuel cell technology. Today, some fuel cells such as the monolythic solid oxide fuel cell can yield power densities almost similar to those produced in combustion engines.

The main function of a porous gas diffusion electrode is to provide a large surface area for reaction zone with minimum hindrance for the mass transport thereby facilitating smooth access for the reactants and prompt removal of products. Use of metal powders with a very high specific area ($100 m^2/g$) or carbon (specific area $1000 m^2/g$) for the manufacture of the porous gas diffusion electrode ensures large surface area for the three-phase contact zone where the gaseous reactant, liquid electrolyte and solid catalyst meet and interact with each other. Sometimes, fillers are mixed with the electrode components during manufacture and are removed at the end of the manufacturing process, leaving voids that enhance the porosity of the structure.

Depending on the basic structure (i.e. carbon or metal), the binders and/or impregnation material used during their fabrication, the electrodes have a hydrophobic (carbon) or a hydrophillic (metal powder) surface. Hydrophobic gas diffusion electrodes are made of fine carbon powder bonded with teflon (polytetrafluroethylene). Hydrophilic electrodes are made of sintered metal powders such as of Raney metal. Porous metal structures are heavy, but have a high conductivity.

The voltage of a single fuel cell is less than 1 volt, and hence many cells are packed up into multi-cell stacks or batteries, connected in series or parallel, according to the voltage or current requirements. The stack construction depends also on the configuration of the system. A sequential stack design out of prefabricated units is essential for the necessary circulation of reactants and coolants in an immobile (matrix) electrolyte system and for the provision of electrolyte loops in mobile electrolyte systems.

(iii) Reactant supply systems

The reactant supply system of a fuel cell can be a simple storage tank or even a complex fuel processing tank. For instance, hydrogen which is the most commonly used fuel in practical systems, can be supplied directly from a hydrogen cylinder or indirectly from other fuels through a fuel processing unit. The most commonly used oxidants are oxygen or air.

Some systems require high purity hydrogen. It can be produced by electrolysis during off peak hours of elecricity demand and stored as gas in high pressure cylinders or as a liquid in cryogenic Dewar vessels or in the form of reversible solid metal hydrides. For systems which do not need high-purity hydrogen, it is produced "in situ" by steam reforming of alcohols or cracking of hydrocarbons.

(*iv*) Handling of reaction products

Reactions taking place in fuel cells are exothermic. Moreover, some heat is also generated inside the stacks due to flow of electric current (Joule's heating). Therefore a cooling system is necessary to maintain the temperature of the fuel cell at the designed level. The same system can also be used to warm the system, in the case of fuel cells which cannot be started at room temperature, via a separate electrolyte heating device.

In case of fuel cells where hydrogen is used as a fuel, water is the main product formed. In case of indirect fuel cells, carbon dioxide is produced during fuel processing. These reaction products also have to be removed. Many approaches have been used for the control of heat and reaction products in a single process.

Fuel cells generate direct current (DC). However, in many cases, the fuel cell cannot provide electrical energy in the required voltage and current. In such situation, power conditioner is also included in the system.

(*v*) Fuel cell systems

In "direct fuel cell system", the fuel is directly oxidized. In "indirect fuel cell system", the fuel is processed by chemical means prior to its oxidation. Excepting on a laboratory scale, there are no fuel cells which directly oxidize hydrocarbons or alcohols at ambient or medium temperatures. Direct hydrocarbon fuel cell systems are possible only at very high temperatures. Conversion of alcohols to hydrogen lowers the efficiency of the system to 40%. If diesel oil has to be cracked or reformed, the overall efficiency drops to 30%.

Fuel cell systems are also classified on the basis of the temperature of operation as follows:

(*a*) **Low temperature fuel cells:** These cells operate below 100^0C. These are simple in design. However, the electrode processes have to be accelerated by using special type of catalysts. Since these catalysts are easily poisoned, only high-purity reactants must be used. The fuels used are hydrogen (produced by reforming of steam or cracking of ammonia) or methanol or hydrazine. Pure oxygen is used as oxidant. The electrolyte used for hydrogen-oxygen fuel cells or hydrazine–oxygen fuel cells is potassium hydroxide solution. Dilute sulfuric acid is used as the electrolyte in case of methanol- oxygen fuel cell.

(*b*) **Medium or moderature temperature fuel cells:** These are operated at temperature range of 100^0 to 250^0C. The fuel usually employed is pure hydrogen. However, cells employing reformed hydrocarbons as fuels are also known. The oxidants employed are oxygen or air free from carbon dioxide and other impurities. The electrolyte used generally is an aqueous potassium hydroxide. However, in case of hydrocarbon fuels, phosphoric acid is used as electrolyte. In Bacon cell, Ni and NiO are used as anode and cathode respectively. In case of hydrocarbon fuels, platinum electrodes are used. If aqueous electrodes are used in these cells, operation under pressure is required.

(*c*)**High temperature fuel cells:** These operate generally at temperatures $>500^0C$. These cells require fused salts (e.g. alkali metal carbonates or hydroxides) or solid electrolytes with appreciable electrical conductivity. Generally, hydrocarbon fuels are used.

Fuel cells are also classified sometimes on the basis of the operating pressure. Accordingly, the systems are referred to as high pressure, medium pressure or atmospheric pressure systems.

Monopolar and bipolar current collection designs are available for use. For current collection, monopolar electrodes have tabs and are connected with one another on their edges.

BATTERIES AND BATTERY TECHNOLOGY

In monopolar cells, the current distribution of high current densities is an important factor. On the other hand, bipolar current collection is practiced while stacking up of many cells in series in order to obtain a high stack voltage.

From the practical point of view, fuel systems are distinguished on the basis of the type of electrolyte used :-
(1) Alkaline fuel cells (AFC)
(2) Molten Carbonate Fuel Cells (MCFC)
(3) Phosphoric acid fuel cells (PAFC)]
(4) Solid oxide fuel cell (SOFC)
(5) Proton exchange membrane fuel cells (PEMFC)

(i) Alkaline Fuel Cells (AFC)

In these cells, 30 – 45 wt.% aqueous solution of potassium hydroxide is used as electrolyte, depending on the system. Acidic impurities such as CO_2 should not be present in the reactants. Removal of CO_2 from air can be accomplished by passing it through an absorption tower containing soda lime. Alkaline systems generally use high purity hydrogen from electrolysis plants as fuel and oxygen or air as oxidants. Alkaline systems operate well at room temperature. They yield the highest voltage (at comparable current densities) of all fuel cell systems. The cell and electrodes can be built from low cost carbon and plastics. Because of

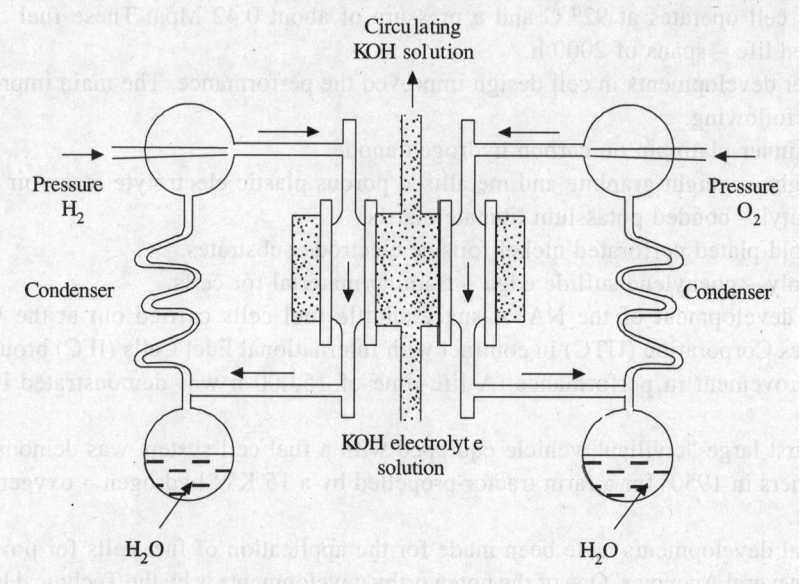

Fig. 21.9. Alkaline Fuel Cell equipped with dual-loop water removal system with circulating jets.

good compatibility with many construction materials, alkaline fuel cells have long operating life. A wide choice of catalysts is available for these cells (e.g. platinum, silver, gold, metals oxides etc) in contrast to acidic cells which are limited to platinum group metal and tungsten carbide. Alkaline fuel cells with circulating electrolyte have several advantages such as removal of accumulated impurities, working as a cooling liquid and usefulness as a water removal vehicle. A hydrogen-oxygen fuel cell system with dual loop water removal system with circulating. Jets is shown in Fig. 21.9.

Bacon Fuel cell system is a developed alkaline fuel system which consists of porous nickel anodes and lithiated porous nickel oxide cathodes separated by a circulating 30% aqueous

KOH solution. Operating temperature was 200°C. The cell is operated under pressure (upto 5 Mpa) to prevent the electrolyte from boiling. Pure hydrogen and oxygen were used as reactants.

The alkaline fuel cell system developed for the NASA Apollo Space Program (1960 to1965) used the structure of Bacon's double porosity sintered nickel electrodes but activated them further with platinum metal catalysts to boost the performance despite lowering the operating pressure to about 0.33 Mpa. The electrolyte used was 85% KOH which was solid at room temperature, but liquifies on heating to 100°C. Operating temperature was 200-230°C. The electrodes were circular with diameter of 200mm and were 2.5 mm thick. Each individual cell was packed between nickel sheets and 31 such were stacked together and electrically connected in series.

Further development of the NASA space shuttle fuel cells was carried out at the United Technologies Corporation (UTC).

In the space shuttle fuel cell system of NASA (1974), silver – plated magnesium foils are used for the light – weight bipolar plates, which help in the heat transfer too. The electrolyte is 35 – 45% KOH solution in water, which is totally immobilized by a reconstituted asbestos separator. The anode is of carbon bonded with PTFE (polytetrafluoroethylene) carrying 10 mg/cm^2 platinium – palladium catalyst pressed on a silver – plated screen. The cathode catalyst consists of gold (90 wt %) and platium (10 wt%) and is deposited on gold – plated nickel – screen. The cell operates at 92^0 C and a pressure of about 0.42 Mpa. These fuel cells have demonstrated life – spans of 2000 h.

Further developments in cell design improved the performance. The main improvements include the following:

(a) thinner platinum on carbon hydrogen anode
(b) light – weight graphite and metallised porous plastic electrolyte reservoir plates.
(c) butyl – bonded potassium titanate matrix
(d) gold plated perforated nickel foils as electrode substrates.
(e) Poly – phenylene suflide edge – framing material for cells.

Later development of the NASA space shuttle fuel cells carried out at the Universal Technologies Corporation (UTC) in contract with International Fuel Cells (IFC) brought about further improvement in performance. A life-time of 15,000 h was demonstrated in smaller units.

The first large "civilian" vehicle equipped with a fuel cell system was demonstrated by Allis Charmers in 1950^s for a farm tractor propelled by a 15 KW hydrogen – oxygen fuel cell power plant.

Several developments have been made for the application of fuel cells for power plants and cars by several Agencies. One of the noteworthy developments is by the Technical University of Graz (Austria) using teflon – bonded carbon fuel cells to be operated with air and hydrogen, in 1985. Extruded carbon – polypropylene plates are used as bipolar plates and manifolds. The major objectives of these efforts were to develop low – cost fuel cell systems for remote locations, for the replacement of rechargeable batteries and for the propulsion of electric vehicles.

(2) Molten Carbonate Fuel Cells (MCFC)

The first technical – size molten carbonate fuel cells were built in 1951 by Broers and Ketelaar using a mixture of sodium carbonate and tungsten trioxide as electrolyte. The electrodes for the fuel and oxidant were nickel powder and silver powder. They were the first to use mixtures of lithium, sodium and potassium carbonates in a sintered porous magnesium oxide disk.

In 1960, Douglas of General Electric Co. (USA) designed molten carbonate cells with gas diffusion electrodes consisting of porous silver and nickel electrodes attached to porous alumina tubes.

Presently, mixtures of molten lithium – potassium – carbonates immobilized in lithium aluminate matrix are preferred as electrolytes. Anodes consist of a porous structure of nickel treated with oxides to prevent sintering. Cathodes are made from lithiated sintered nickel oxide. The cell hardware is made of stainless steel or other high temperature steel alloys. The stack configuration is bipolar. Since the operating temperatures (600 to 700° C) are high, noble metal catalysts are not needed. Internal steam reforming (in the manifolds behind the electrodes) is possible. Thus, some hydrocarbons can be directly fed to the cell, where they are reformed into hydrogen rich gas, carbon monoxide and carbon dioxide. Typical cell reactions are as follows :

Anode $H_2 + CO_3^{2-} \rightarrow H_2O + CO_2 + 2e^-$
Cathode $\frac{1}{2} O_2 + CO_2 + 2e^- \rightarrow CO_3^{2-}$

In the presence of carbon monoxide, the shift reaction will occur :

$$CO + H_2O \rightarrow H_2 + CO_2$$

A great advantage of MCFC is that the waste heat is useful for co-generation. Therefore, most MCFC development programs are aimed at medium to high power range (100 KW to 10 MW) plants for electric utility or industrial use.

Coal – fueled MCFC power plants have been developed by International Fuel Cells (IFC) for electric utility and dispersed power generation.

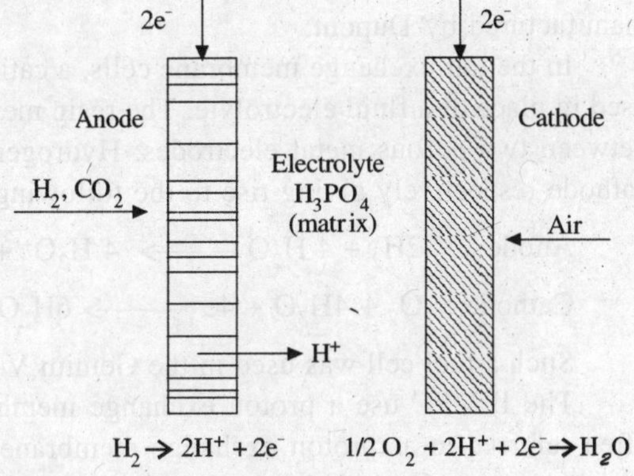

Fig. 21.10. Reactions occuring in a Phosphoric Acid Fuel Cell with a matrix electrolyte.

Various research institutes and universities in various countries such as Japan, USA, UK, Italy, etc. are carrying out R & D efforts on internal gas mainfolding and external gas manifolding and new materials to improve the efficiency and cell life.

(3) Phosphoric Acid Fuel Cells (PAFC)

These are considered as the most advanced fuel cells after the alkaline fuel cells used in space-crafts. Phosphoric acid fuel cells have been operated in a wide range of power out-put (1 KW to 5 MW).

By the end of 1969, the matrix phosphoric acid fuel cell operating between 170 to 200° C with reformed fuels was considered as the best concept for this system. The high temperature improves the conductivity of the phosphoric acid. The oxygen reduction reaction is the rate determining factor for the performance of the cell. The reactions occurring in a PAFC with a matrix electrolyte are represented in Fig. 21.10. Phosphoric acid fuel cell electrodes contain a noble metal (eg. Pt) catalyzed onto a PTFE (teflon) bonded carbon applied on a graphite – cloth support. The electrolyte is contained in a silicon carbide matrix deposited on or placed between the electrodes. The bipolar plates are hot-pressed carbon or graphite plastic mixtures.

PAFC can operate with reformed hydrocarbons or alcohols. Reforming of fuel into a hydrogen – rich gas takes place in separate reforming units which are coupled with the fuel cell stacks. Catalyst poisoning due to CO is minimum because of the high operating temperature. Air is used as oxidant and also for cooling. Some systems use liquid cooling system.

United Technology Corporation, USA developed PAFC plants producing 40 KW, 100 KW and 500 KW were intended for generation of electricity in remote areas or in city centers with the co-production of heat.

PAFC Power plants using methanol as fuel were developed for U.S. army. Systems producing 20 KW on that basis were used for electric vehicle power plants.

(4) Ion Exchange Membrane Fuel Cells (IEMFC)

The General Electric Company, USA in 1960 demonstrated the first practical application of an ion-exchange membrane fuel cell with a 1 KW fuel cell (Solid Polymer Electrolyte, SPE) System in the Gemini space flights. Initially Polystyrene Sulfonates were used as membrane materials. They were later replaced by more stable Sulfonated Poly-tetrafluoroethylene, manufactured by Dupont.

In the ion-exchange membrane cells, a cation-exchange resin membrane in the H-form is used in place of a fluid electrolyte. The resin membrane is about 1mm thick and is sandwiched between two porous metal electrodes. Hydrogen and Oxygen are fed into the anode and the cathode respectively giving rise to the following reactions.

Anode : $2H_2 + 4H_2O \longrightarrow 4H_3O^+ + 4e^-$

Cathode : $O_2 + 4H_3O + 4e^- \longrightarrow 6H_2O$

Such a fuel cell was used in the Gemini V-rocket in 1965.

The IEMFC use a proton exchange membrane as electrolyte and hence these cells are often referred to as proton exchange membrane fuel cells (PEMFC). Thus the cell electrode reactions correspond to those of acid fuel cell systems. The polymer membranes provide an effective barrier against leakage of gases between the electrodes. The membranes are very thin and hence reduce ohmic losses in the electrolyte, thereby producing high current densities. The carbon electrodes contain platinum as catalyst and are pressed onto both sides of the membrane. Since the membrane possesses acidic characteristics, hydrocarbon reformates can be used as fuel. However the fuel should be free from carbon monoxide to protect the catalyst from getting poisoned at the low operating temperature, viz. 80°C.

The Gemini SPE power plant contained a stack of 32 cells which were built into a power module. Each cell consisted of an electrode-electrolyte assembly, associated components for gas distribution, current collection, heat removal and water removal. Heat removal was accomplished by circulation of a coolant through tubes placed in each cell assembly. Water removal was accomplished with the help of wicks placed in the Cathode gas chamber. The Gemini fuel cell power plant had a voltage of about 25V and was rated at 1 KW.

The General Electric Company has also developed a regenerative SPE-PEM (Solid Polymer electrode-proton exchange membrane) electrolysis – fuel cell system for space station applications. While in orbit, during shady periods, the fuel cell provides energy for the space station and during sunny periods, the output of the solar cells electrolyze the water formed in the fuel cells. The hydrogen and oxygen thus generated are supplied to the fuel cells again.

(5) Solid Oxide Fuel Cells (SOFC)

In these cells, a solid electrolyte *e.g.* a mixture of Yttrium dioxide and Zirconium dioxide is used. The cells operate at a temperature of about 1000°C. Charge transfer in the electrolyte is done by oxygen ions. The anodes are made of nickel/Zirconium oxide cermet whereas the cathodes are made of Lanthanum manganate (La Mn O_3). Reformate gases (H_2 + CO) are used as fuel and oxygen is used as oxidant. The electrode reactions may be summarized as follows.

Anode reaction

$$X (H_2(g) + Y CO (g) + (X + Y) O^{2-} \rightarrow X H_2O (g) + Y CO_2 (g) + 2 (X+Y) e^-$$

Cathode reaction :

$$1/2 (X + Y) O_2 (g) + 2(X + Y) e^- \rightarrow (X+Y)O^{2-}$$

$$\tfrac{1}{2} (X+y) O_2 (g) + x H_2 (g) + x H_2(g) + Y CO (g) \rightarrow x H_2O (g) + Y CO_2 (g)$$

Since the cathode reaction uses only oxygen (or air) as oxidant, re-circulation of Carbon dioxide from the anode exhaust is not needed, thus simplifying the system. The reaction rate is enhanced because of the high operating temperature (about 1000°C) and noble metal catalysts are not required. Carbon monoxide does not poison the electrodes and indeed it is also used as fuel. Internal fuel reforming is possible. The essential features of a solid oxide fuel cell are shown in fig. 251.

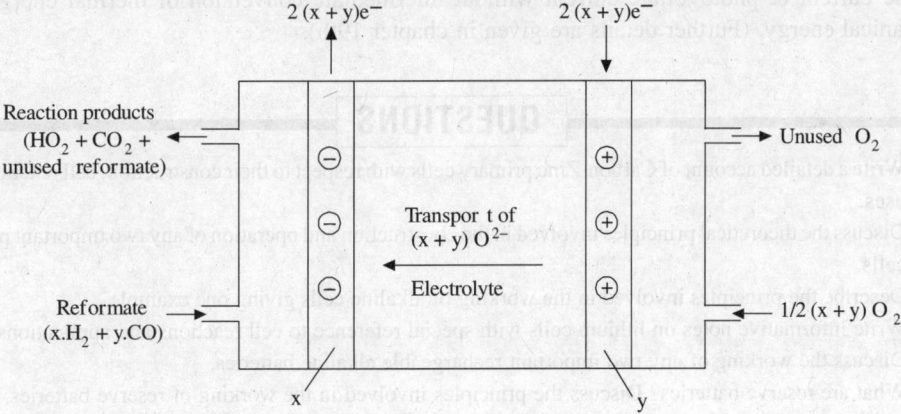

Fig. 21.11. Schematic representation and reactions taking place in a solid oxide Fuel Cell.

Since only a solid electrolyte is used, the problems associated with liquid handling and corrosion are avoided.

Some R & D efforts are made to design medium temperature solid oxide fuel cells. The cells tested are hydrogen-oxygen cells. The solid material used is hydrogen-exchanged ß-alumina. The operating temperature for this type of solid proton conductor is in the range of 150-200°C.

(6) Biochemical Fuel Cells

In 1911, Potter demonstrated a bio-cell in which disintegration of organic compounds by micro-organisms could generate electricity.

The corrosion of the oil pipelines submerged under sea water was considered to be due to the electricity generated by marine bacteria.

In biochemical fuel cells, organic substrates (eg. Urea and glucose) act as a fuel which is oxidised with the help of living organisms such as bacteria or enzymes derived from bacteria. These function as bio-anodes. It is also possible to accomplish direct reduction of oxygen or some other oxidant at the cathode (biocathode). In the indirect approach, the biological system can be used to generate reactants that are consumed at the electrodes. For instance, pseudomonas mathraniea bacteria can utilize the carbon from methanol or methane to liberate hydrogen which can be subsequently used as a fuel. Such systems hold promise for controlling pollution due to organic waste and at the same time generating electricity.

(7) Regenerative Fuel Cells

In such cells, the fuel, after it has been consumed to produce electricity, is regenerated and reused. For instance, in a photogalvonic cell, a chemical like nitrosyl chloride (NOCl) is photochemically dissociated into NO and Chlorine, which are used as fuels in fuel cells. The fuels are regenerated photochemically. Thus, in effect, the light energy is converted into electrical energy.

Notwithstanding several advantages of fuel cells in producing power with minimal pollution, their use is still limited only for special applications mostly because of the economic and power-output considerations. Intensive R & D efforts must be encouraged, particularly for innovative processes and designs as well as cheaper and more effective catalysts to produce electrical energy directly from hydrocarbons or carbon using fuel-cell technology.

21.7 SOLAR CELLS

Direct or diffuse solar energy can be converted with the help of solar cells directly into electric current & photovoltaic current without intermediate conversion of thermal energy into mechanical energy. (Further details are given in chapter 19th).

QUESTIONS

1. Write a detailed account of Carbon-Zinc primary cells with respect to their construction, cell reactions and uses.
2. Discuss the theoretical principles involved in the construction and operation of any two important primary cells.
3. Describe the principles involved in the working of alkaline cells giving one example
4. Write informative notes on lithium cells with special reference to cell reactions and applications
5. Discuss the working of any two important rechargeable alkaline batteries.
6. What are reserve batteries? Discuss the principles involved in the working of reserve batteries, giving some examples.
7. What are the different types of fuel cells available? Discuss the principles involved in the working of hydrogen – oxygen fuel cell.
8. Write informative notes on any two of the following :
    (a) Fuel cell technology
    (b) Advantages and limitations of fuel cells
    (c) Alkaline fuel-cell systems
9. Write short notes on any three of the following
    (a) Phosphoric acid fuel cells
    (b) Ion-exchange membrane fuel cells
    (c) Solid-oxide fuel cells
    (d) Molten carbonate fuel cells
    (e) Solar cells

22

Instrumental Techniques in Chemical Analysis

22.1. COLORIMETRY AND VISIBLE SPECTROSCOPY

Introduction

The variation of the colour of a system with change in concentration is the basis of colorimetry. In colorimetry, the concentration of a substance is determined by measurement of the relative absorption of light with respect to a known concentration of the substance. In visual colorimetry, natural or artificial white light is generally used as a source of light, and determinations are made with the help of a simple instrument called a calorimeter or color comparator. If the eye is replaced by a photoelectric cell, the instrument is called a photoelectric colorimeter. This instrument makes use of relatively narrow range of wavelengths of light obtained by passing white light through filters (e.g., plates of coloured glass, gelatin, etc.) and hence such an instrument is called a filter photometer. If a source of visible radiation extending up to the UV region of the spectrum and if radiations up to bandwidth of less than 1 nm (1 nm = 10^{-7} cm) can be obtained with a suitable device (e.g., such as a spectrometer) and if it contains a device (e.g., photometer) with which measurement of intensity of transmitted radiation can be made, then such an instrument is called a spectrophotometer.

Colorimetric and spectrophotometric methods offer simple means for determining minute quantities of substances.

Theoretical principles

Lambert's Law: If a monochromatic light passes through a transparent medium, the rate of decrease in intensity with the thickness of medium is proportional to the intensity of the incident light. In other words, the intensity of the emitted light decreases exponentially as the thickness of the absorbing medium increases arithmetically. The law may be expressed in the form of a differential equation;

$$\frac{-dI}{dl} = KI$$

where I is the intensity of the incident light of wavelength λ, l is the thickness of the medium, and K is proportionality factor. On integration, and equating I with I_0 if l = 0, we get

$$\ln \frac{I_0}{I_t} = Kl \qquad \ldots(1)$$

or

$$I_t = I_0 . e^{-Kl} \qquad \ldots(2)$$

where it is the intensity of transmitted light, I_0 is the intensity of the incident light, l is the thickness of the absorbing medium and K is the constant for the wavelength and the absorbing medium used. By changing from natural to common logarithms, we get

$$I_t = I_0 \cdot 10^{-0.4343 Kl} = I_0 \cdot 10^{-Kl} \qquad ...(3)$$

where K = k/2.3026, which is known as the absorption coefficient.

The ratio I_t/I_0 is called transmittance, T.

The ratio I_0/I_t is called opacity, and

log I_0/I_t = A, the absorbance of the medium or optical density, D or the extinction, E.

Beer's Law: The intensity of a beam of monochromatic light decreases exponentially as the concentration of the absorbing substance increases arithmetically. This may be expressed in the form

$$I_t = I_0 \cdot e^{-K'c}$$
$$= I_0 \cdot 10^{-0.4343 K'c}$$
$$= I_0 \cdot 10^{-k'c} \qquad ...(4)$$

where c is the concentration, and K' and k' are constants.

Combining equations 3 and 4, we get

$$I_t = I_0 \cdot 10^{-acl} \qquad ...(5)$$

or

$$\log I_0/I_t = a\,c\,l \qquad ...(6)$$

This expression is known as Beer-Lambert's Law, which is the fundamental equation for colorimetry and sprectrophotometry.

If c is expressed in mole L^{-1} and l in cm, 'a' is given a symbol 'e', which is known as the molar absorption coefficient or molar absorptivity.

Since I_0/I_t is called absorbance, A, we arrive at the following inter-relationship:

$$A = e\,c\,l = \log I_0/I_t = \log 1/T = -\log T$$

If l is constant (as in the case of matched cells used in a colorimeter or spectrophotometer), the Beer-Lambert's may be written as

$$c \propto \log I_0/I_t$$
$$c \propto \log 1/T$$

or

$$c \propto A$$

Hence, by plotting A or log 1/T as ordinate, against concentration as abscissa, straight line will be obtained and this will pass through a point (c = 0, A = 0 i.e., T = 100%). This calibration line can be used for determining unknown concentrations of the same material by measuring their respective absorbances, and obtaining their corresponding concentrations from the calibration line.

Deviations from or exceptions to the Beer-Lambert's Law

The law does not hold when coloured solute ionises, associates, dissociates, or undergoes complexation, because the nature of the coloured species will vary with concentration.

Additivity of absorbances

According to Beer's law, the absorbance at any particular wavelength is directly proportional to the number of absorbing molecules. If a solution contains more than one type of absorbing species, the total absorbance will be the sum of the absorbances of all the species, provided they do not interact chemically.

Instrumentation

1. **Radiation Source:** Tungsten filament lamp is most widely used for producing visible light in the wavelength range 400–750 nm. If colorimetric analysis is carried out in the UV range, (i.e., down to 200 nm) hydrogen discharge lamp or deuterium is used as radiation source, whereas for work in IR region, Nernst Glower is used.

INSTRUMENTAL TECHNIQUES IN CHEMICAL ANALYSIS

2. **Dispersing Device:** The selection of a narrow band of wavelength, which is required for colorimetry and spectrophotometry, is accomplished with the help of a monochromator. Monochromators used in various spectral regions are:

(a) filters (b) prisms, and (c) gratings

Plates of coloured glass can be used as filters in the visible region. From the stability point of view, gelatin filters are better. The gelatin filter consists of a layer of gelatin admixed with suitable organic dyes and sandwiched between two sheets of glass. Liquid filters are solutions of appropriate absorbing components. Interferance filters contain a transparent dielectric material (e.g., calcium fluoride or magnesium fluoride) sandwiched between two glass plates whose inner surfaces have been coated with a semi transparent material like silver.

A filter is not considered as a real monochromator. A good monochromator should provide much narrower wavelength bands than those obtained by filters. Glass prisms are used for work in the visual range, quartz prisms are used for UV and alkali halide prisms are generally used for IR range.

Gratings are of two types: (a) Transmission gratings (b) Reflection gratings. A transmission grating comprises of a series of closely spaced parallel grooves ruled on a piece of glass or other transparent material. The greater the number of lines per square inch, shorter will be the wavelength dispersed by the grating and greater will be the dispersing power. A grating suitable for visible and UV work should have 15,000–30,000 lines per square inch. The dispersion through a grating is based on the principle of diffraction. When a transmission grating is eliminated by radiation from a source through a slit, each of the grooves on the grating acts as a new light source, and interference among the multitude of beams produced brings about dispersion of radiation into component wavelengths.

Reflection gratings are used in most of the instruments. These are prepared by ruling closely spaced grooves on a polished metal strip or thin metallic film deposited on a glass strip, which acts as a mirror. The grating can be mounted on a spectrograph in many ways but the following two are commonly used. In Littrow mounting, the grating is put in a place of prism, and a concave mirror is used (instead of lens) for collimation. In Ebert-Fastie mounting, a single, large, special mirror serves the purpose of both collimation and focusing. Use of two small spherical mirrors mounted systematically can be cheaper alternative.

3. **Slits:** A slit system is used for selecting the desired wavelengths from the light dispersed by a monochromator. An entrance slit and an exit slit are placed each side of the prism or grating. The slit jaws are made of metal in the shape of knife-edges, and they can be moved with respect to each other to control the width. The entrance slit chooses a small parallel beam of incident light, while preventing stray radiations entering the optical path. The light, after being dispersed (by the prism or grating) goes through the exit slit and travels through the sample or reference cell and finally reaches the detector system.

4. **Sample holders:** To hold the sample solution to be analysed and the reference solution, optically matched colour-corrected fused glass cells are used in the wavelength range 300–2500 nm. Corex glass or quartz cells are used in the wavelength range 210–300 nm, and fused silica cells are used for measurements at somewhat lower wavelengths.

5. **Detectors:**

(a) **Photovoltaic or barrier layer cells:** It consists of a metallic base plate (e.g., Fe or Al), which acts as an electrode. On the surface of this electrode, a thin layer of a semiconductor material like Selenium is deposited. This surface of selenium is then covered by a very thin layer of silver or gold, which acts as a collector electrode. When radiation falls upon the surface of selenium, electrons are generated at the selenium - silver interface. These electrons are collected by the silver. Accumulation of these electrons on the silver surface creates a voltage difference between the silver surface and the base plate of the cell. If the external circuit has a low resistance, a photocurrent will flow and this is directly proportional to the intensity of the incident radiation. If the cell is connected to a galvanometer, a current will flow and this is proportional to the intensity of the incident radiation.

These cells have the following limitations:

(i) These cells are less sensitive in the blue region as compared to other regions of the visible spectrum.

(ii) The current output of the cells depend upon the wavelength of the incident light.

(iii) These cells show tendency to fatigue.

(iv) The current produced by the photovoltaic cells cannot be readily amplified by the conventional electronic circuits because of the low internal resistance.

In view of the above, photovoltaic cell detectors are used only in simple filter photometers but not in sophisticated spectrophotometers.

(b) **Photo emissive cells or phototube:** This comprises of an evacuated glass bulb, coated internally with a thin photosensitive layer, such as caesium oxide (or potassium oxide) and silver oxide, which acts as cathode. (This light sensitive layer emits electrons when illuminated). A metal ring inserted near the center of bulb forms the anode, which is maintained at a high voltage by means of a battery. The interior of the bulb is either evacuated or filled with inert gas at low pressure (e.g., argon at 0.2 mm). When light, penetrating the bulb, falls on the photosensitive layer, electrons are emitted, thereby causing a current to flow through an outside circuit. This current may be amplified by electronic means, and this is taken as a measure of the amount of light striking the photosensitive surface.

(c) **Photomultiplier Tube:** The sensitivity of a photoemissive cell can be considerably increased with the help of a photomultiplier tube. This consists of an electrode covered with a photoemissive material. The tube is also provided with a series of positively charged plates, known as dynodes. Each of the plate is also covered with a material, which emits 2–5 electrons for each electron that strikes its surface. Further, each dynode is charged at a successively higher potential. when light radiation is incident upon the surface of the photoemissive cathode, electrons are ejected. these electrons are accelerated to the sensitive surface of a dynode where secondary electrons are emitted in greater number than those which were incident on it. These secondary electrons, in turn, are accelerated on to the second dynode, which was maintained at a higher potential, where still more number of secondary electrons are emitted, the increase of electrons is by a factor of 4 to 5. This process goes on and on, at every dynode thereby producing a very large number of electrons. The number of electrons thus reaching the collector is a measure of the intensity of light incident on the cathode surface. Generally, about 10 dynodes are used, each of which is maintained at about 80 V more positive than the preceding dynode. Thus, the overall amplification factor of about 10^6 can be achieved by a photomultiplier. The output of a photomultiplier tube is limited to several milliamperes. Hence only low incident radiant energy intensities can be employed. It can measure intensities about 200 times weaker than those measurable with an ordinary photoelectric cell and amplifier. For scanning the complete spectral range, two different phototubes. one 'red' sensitive cell (600 - 800 nm) and one 'blue' sensitive cell (200 - 600 nm) will have to be used.

(d) **Silicon Diode or Photodiode:** This detector consists of a strip of p-type silicon on the surface of silicon chip (n-type silicon). When a biasing potential is applied with the silicon chip connected to the positive pole of the biasing surface, electrons and holes are caused to move away from the p-n junction. This creates a depletion region in the neighbourhood of the junction, which in effect becomes a capacitor. When light falls on the surface of the chip, free electrons and holes are generated which migrate to discharge the capacitors; the magnitude of the resultant current is a measure of the intensity of the incident light. This detector has a greater sensitivity than a single phototube but lesser than that of a photomultiplier.

With the help of modern technology, it is possible to form a number of photodiodes on the surface of a single silicon chip. This chip also contains an integrating circuit which can scan each photodiode in turn to give a signal which is transmitted to a microprocessor. Each photodiode can be programmed to respond to a particular narrow band of wavelengths so as to enable scanning of the complete spectrum almost instantaneously.

Colorimetric Methods

The fundamental principal for colorimetric analysis is to compare, under well designed conditions, the colour produced by the substance in unknown amount with the same colour produced by a known amount of material being determined. The following methods are available for the quantitative comparison of colours of known and unknown solutions:

(*i*) **Methods using visual comparators**

(*a*) **Standard series method:** In this method, Nessler tubes are employed, which are nothing but colourless glass tubes of uniform cross section and flat bottoms. The solution of the substance being determined is made up to a known volume, and the color is compared with that of a series of standards in the same way starting from known amounts of the substance being determined. 100 ml of the solutions of the unknown and each of the standard solutions are placed in Nessler tubes, and the solutions are viewed vertically through the length of the liquid columns. The concentration of the unknown solution is equal to that of the standard solution having the same colour. LOVIBOND–2000 comparator works on this principle.

(*b*) **Duplication method:** In this method, a known volume (say x ml) of the test solution is treated in Nessler cylinder with a measured volume (say y ml) of appropriate reagent so as to develop a colour. Now, x ml of distilled water is placed in a second Nessler cylinder together with y ml of the reagent. A concentrated standard solution of the substance under test is now added to the second Nessler cylinder from a micro burette until the colour thus developed exactly matches the colour in the first Nessler cylinder already developed with unknown solution.

(*c*) **Dilution method:** In this method, a colour developed from the sample and standard solution are taken in two identical glass tubes of same diameter and are observed horizontally through the tubes. The more concentrated solution is diluted until the colours in the tubes are of identical intensity when observed horizontally through the same thickness of solutions. The relative concentrations of the two original solutions (known and unknown) are then proportional to the heights of the matched solutions in the tubes.

(*d*) **Balancing method:** In this method, comparison is made in two tubes, and the height of the liquid in one tube is so adjusted that when both the tubes are observed vertically, the colour intensities in the tubes are equal. If the concentration of solution in one of the tubes is known then the concentration in the other can be calculated from the relation

$$c_1 l_1 = c_2 l_2$$

where c_1 and c_2 are the known and unknown concentrations and l_1 and l_2 are their respective lengths of the two liquid columns.

$$c_2 = c_1 l_1 / l_2$$

Duboscq colorimeter works on this principle.

(*e*) **Photoelectric photometer method:** In this method, the human eye is replaced by a photoelectric cell. Instruments which incorporate photoelectric cells measure the light absorption and not the colour of the substance, hence the term **'photoelectric colorimeters'** is a misnomer. They may be better called as **photoelectric photometers** or **photoelectric comparators** or **absorptiometers**.

These instruments essentially consist of a light source, a suitable light filter to provide nearly monochromatic light, a glass cell for the sample/standard solution, a photoelectric cell to receive the radiation transmitted by the solution, and a measuring device to determine the response of the photoelectric cell. The various types of each of these components used in visible spectrophotometry and colorimtery have been already discussed in earlier sections. The photometer is first calibrated by measuring the absorbance of a series of solutions of known concentrations (at the wavelength at which the coloured species has maximum absorbance). A calibration curve is plotted connecting the concentration with the readings of the measuring device employed (optical density). The concentration of the unknown solution is then determined by noting the cell response and referring it to the calibration curve.

The photoelectric photometers are available in different type incorporating one or two photocells. In the one-cell type of instruments, the absorption of light by the solution is usually measured directly by determining the current output of the photoelectric cell in relation to the value obtained with the pure solvent. In instruments of the one-cell type, it is essential to ensure that the light source is of constant intensity and the photocell used doesn't exhibit fatigue effect.

The two-cell type of photoelectric photometer is considered to be more trustworthy because any fluctuation of the intensity of light source will affect both cells alike as they are matched for their spectral response. In this type of instrument, the two photocells, which were illuminated by the same source of light, are balanced against each other through a galvanometer. The test solution is placed before one cell and the pure solvent before the other, and the difference in current output is measured.

(f) **Spectrophotometric method:** This is the most accurate method for determining the concentration of a substance. However, it is more expensive. A spectrophotometer may be considered as a refined photoelectric photometer which provides a continuously variable and more nearly monochromatic light. The essential parts of a spectrophotometer are a source of radient light, a monochromator to produce narrow bands of radient energy, glass or silica cells for the

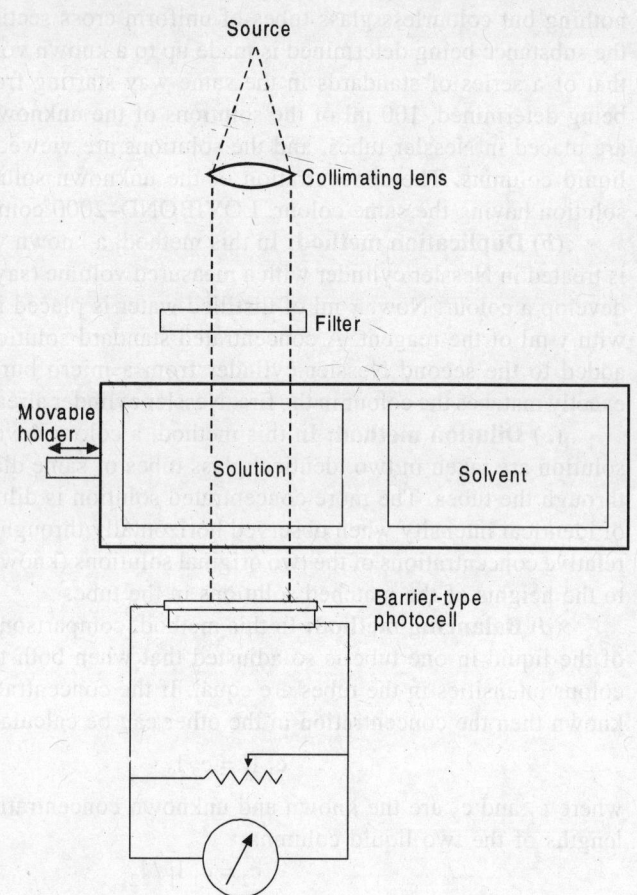

Fig. 22.1. Optical diagram of a simple photoelectric photometer.

solvent and for the test solution, a photocell or a photomultiplier, an amplifier and a recorder. Single-beam and double beam spectrophotometers are commercially available. The essential features of a *Single-beam spectrophotometer* are illustrated in Fig. 22.2.

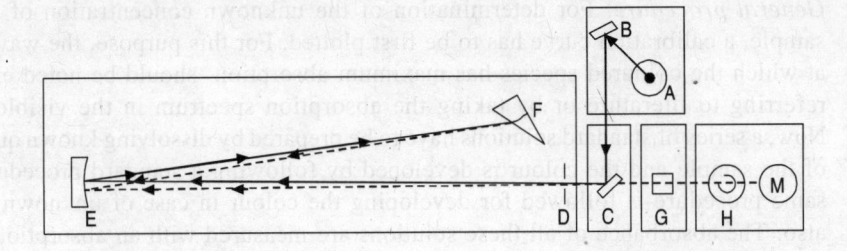

INSTRUMENTAL TECHNIQUES IN CHEMICAL ANALYSIS

Fig. 22.2. Essential features of a single-beam spectrophotometer.

A - Light source, B - dondensing mirror, C - diagonal mirror, D - entrance slit,
F - quartz prism, G - absorption cell, H - photocell, M - meter

An image of the light source, A is focussed by the condensing mirror, B and the diagonal mirror, C on the entrance slit, D. The light beam falls on the collimating mirror, E where it is rendered parallel and reflected to the quartz prism, F. The back surface of the prism is aluminised, so that the light refracted at the first surface is reflected back through the prism. This light beam undergoes further refraction as it emerges from the prism. The collimating mirror focuses the spectrum in the plane of the slit system, D. The light of the wavelength for which the prism is set then passes out of the monochromator through the exit slit, through the absorption cell, G and finally to the photocell, H. The response of the photocell is amplified and registered on the meter. In modern versions of the instrument, the prism is replaced by a diffraction grating.

Double beam spectrometers are the most modern general-purpose UV-visible instruments covering the spectral region from 200-800 mm by a continuous automatic scanning process and producing the absorption spectrum as a pen trace on calibrated chart paper.

Salient features of colorimetric analysis:

(a) It gives accurate results at low concentrations.

(b) It enables the analysis of such substances for which gravimetric and titrimetric procedures are not available (*e.g.* In case of some biological substances).

(c) It is ideal for quick and routine analysis of components of a number of similar samples. When once the calibration curve is plotted, a large number of samples of the same component in different concentrations can be rapidly analysed.

Criteria of a satisfactory colorimetric analysis:

(a) The colour reaction should be specific or atleast selective for the particular substance being analysed under the chosen experimental conditions.

(b) Beer's law should be obeyed in the desired concentration range. In other words, the colour should be proportional to concentration.

(c) The colour should be sufficiently stable to enable accurate analysis

(d) The colorimetric procedure should give reproducible results under the specific experimental conditions.

(e) The colour reaction should be highly sensitive and the reaction product should strongly absorb in the visible region.

(f) The coloured solution should be clear and free from turbidity.

(g) *General procedure*: For determination of the unknown concentration of a given sample, a calibration curve has to be first plotted. For this purpose, the wavelength at which the coloured species has maximum absorption should be noted either by referring to literature or by taking the absorption spectrum in the visible range. Now, a series of standard solutions have to be prepared by dissolving known quantities of the sample and the colour is developed by following a standard procedure. The same procedure is followed for developing the colour in case of unknown sample also. The absorbance of all these solutions are measured with an absorptiometer at the optimum wavelength λ_{max}. The absorbance of each of the standard solutions is plotted against the respective concentration. A straight-line plot is obtained if the Beer's Law is obeyed. The concentration of the unknown sample solution can directly be referred from the calibration curve.

The absorption spetcra of 0.001 M potassium dichromate solution if 1F H_2SO_4 taken in 1 cm cell is given in Fig. 22.3

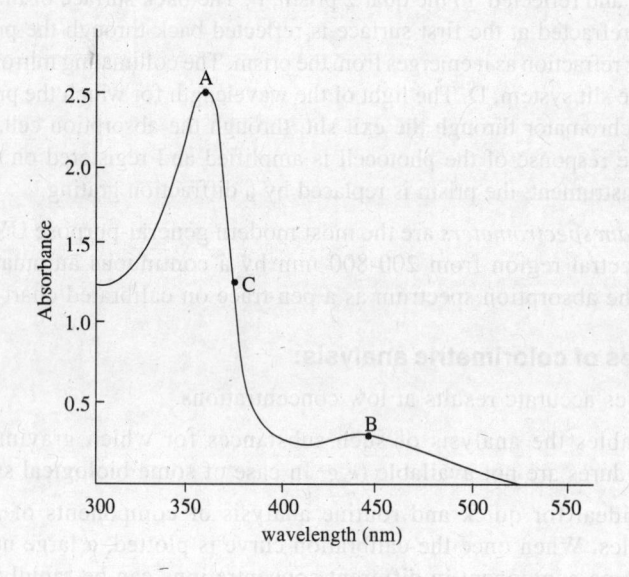

Fig.22.3. Absorption spectrum of 0.001 M $K_2Cr_2O_7$ solution in 1 F. H_2SO_4 in 1 cm cell.

A standard calibration curve is shown in Fig. 22.4.

It shows how the concentration of an unknown sample can be inferred from the calibration plot.

Applications of Visible Spectroscopy

(1) It is useful in identifying chemical substances. According to Hartlay's rule, compounds having similar structures would have similar absorption spectra. By comparing the spectrum of an unknown sample with standard spectra of known compounds, the unknown sample can be identified.

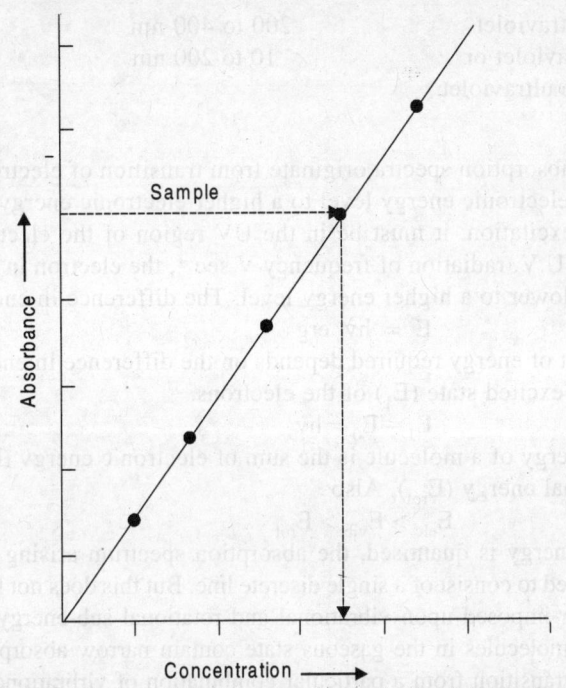

Fig. 22.4. Standard calibration plot for colorimetric analysis.

(2) It is useful in the determination of concentration of solutions as low as 10^{-7} M, which cannot be determined by the conventional volumetric of gravimetric methods. This technique is successfully used for quantitative analysis of a large number of substances.

Ex:-
  (a) Determination of iron with o-phenanthroline, batho-phenanthroline or thiocyanate.
  (b) Determination of ammona using Nessler's reagent.
  (c) Determination of phosphate using ammonium molyb-date.
  (d) Determination of aluminium using Eriochrome Cyanine or Aluminon.
  (e) Determination of arsenic by 'molybdenum blue' method.
  (f) Simultaneous determination of Cr and Mn in steel, etc.
  (g) Analysis of ores, minerals, alloys and other industrial raw materials and finished products.
  (h) Analysis of environmental samples.

(3) In determination of molecular composition of complexes.
(4) For determination of pk value of an indicator.
(5) For studying instability constants of metal complexes.
(6) For studying cis-trans isomerism.

22.2. ULTRAVIOLET SPECTROSCOPY

Introduction

The UV region of the electromagnetic spectrum is subdivided into two spectral regions as follows:

(i) Near ultraviolet   200 to 400 nm
(ii) Far ultraviolet or   10 to 200 nm
Vacuum ultraviolet

Origin

Ultraviolet absorption spectra originate from transition of electrons within a molecule or ion from a lower electronic energy level to a higher electronic energy level. For a radiation to cause electronic excitation, it must be in the UV region of the electronic spectrum. When a molecule absorbs U.V. radiation of frequency V sec^{-1}, the electron in that molecule undergoes transition from a lower to a higher energy level. The difference in energy in given by

$$E = h\nu \text{ erg}$$

The actual amount of energy required depends on the difference in energy between the ground state (E_0) and the excited state (E_1) of the electrons.

$$E_1 - E_0 = h\nu$$

The total energy of a molecule is the sum of electronic energy (E_{elec}), vibrational energy (E_{vib}) and rotational energy (E_{rot}), Also

$$E_{elc} > E_{vib} > E_{rot}$$

When UV energy is quantised, the absorption spectrum arising from a single electronic transition is expected to consist of a single discrete line. But this does not happen because electronic absorption is super-imposed upon vibrational and rotational sub-energy levels. That is why, the spectra of simple molecules in the gaseous state contain narrow absorption peaks, wherein each peak represents a transition from a particular combination of virbrational and rotational levels in the electronic ground state to a corresponding combination in the excited state. However, in case of polyatomic complex molecules, broad absorption bands are obtained due to coalescence of discrete bands.

When energy is absorbed by a molecule in the U.V. region, it brings about some changes in the electronic energy of the molecule resulting from transitions of valence electrons. The following three types of electrons are involved in organic molecules.

(i) π Electrons: The electrons responsible for double bonds are called Π electrons. These are involved in unsaturated hydrocarbons like trienes and aromatic compounds. In unsaturated systems, Π electrons predomi-nantly determine the energy state of the electron sheaths which are excited by the absorption of U.V. or visible light.

(ii) σ-electrons: Electrons forming single bonds are called σ- electrons according to moecular orbital notation. These are involved in saturated bonds such as those between C and H in paraffins and such bonds are known as s bonds. Since the energy requiring to excite electrons in σ-bond is much more than that obtained by UV radiation, compounds containing σ-bonds do not absorb UV radiation.

(iii) η-electrons: These are the unshared or non-bonded electrons and are not involved in the bonding between atoms in moleculars. Example: Organic compounds containing N, O or halogens. However, η-electrons can be excited by UV radiation and hence compounds containing atoms like N, O, S, halogen compounds or unsaturated hydrocarbons may absorb UV radiation.

The η-electrons in the case of the elements of the first two rows of the periodic table are the P electrons.

The above three types of valence electrons are illustrated by formaldehyde molecule shown below.

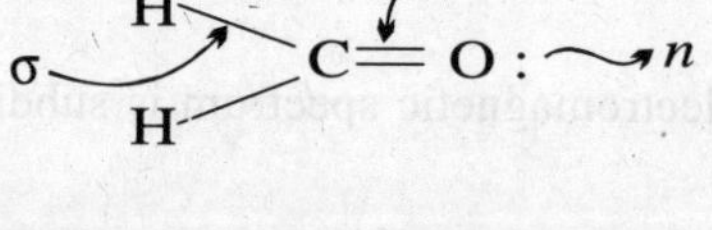

INSTRUMENTAL TECHNIQUES IN CHEMICAL ANALYSIS 763

The interactions between h and P or P and P electrons are considerable, whereas the interaction between s and P electrons may be ignored. In the bonding electrons, the s-electrons are more strongly bound than the P electrons, while in the antibonding levels, the s* level has higher energy compared to the P* level. These facts are represented in the electronic energy levels given below.

Energy absorbed in the UV region by complex organic molecules results in transitions of valence electrons in the molecule. These transitions are

$$\sigma \rightarrow \sigma^*, n \rightarrow \sigma^*, n \rightarrow \pi^*, \text{ and } \pi \rightarrow \pi^*$$

As per the selection rules, $n \rightarrow \pi^*$ transition is forbidden, while the remaining three are allowed.

The electronic energy levels in simple organic molecules are shown below.

Level	
Ryderg	R —
Antibonding σ	σ* —
Antibonding π	π* —
Non-bonded n	p —
Bonding π	π —
Bonding σ	σ —

Energy (not to scale)

From the figure, it can be seen that the order of change in energy values for different transitions are as follows.

1. **$n \rightarrow \pi^*$ transition:** These transitions are shown by unsaturated molecules which contain atoms like N, O and S. These transtions show a weak band in their absorption spectrum. In aldehydes and ketones (having no C ≡ C or C = C bonds), the band due to $n \rightarrow \pi^*$ transition generally occurs in the range 270-300 nm. On the other hand, carbonyl compounds having double bonds separated by 2 or more single bonds exhibit the bands in the range 300 to 350 nm due to $n \rightarrow \pi^*$ transitions.

2. **$\pi \rightarrow \pi^*$ transition:** These type of transitions are related to the promotion of an electron from a bonding Π orbital to an anitbonding π^* orbital. Selection rules based on symmetry concepts determine whether transitions to a particular π^* orbital is allowed or forbidden. For example,

$$n \rightarrow \pi^* < \pi \rightarrow \pi^* < n \rightarrow \sigma^* << \sigma \rightarrow \sigma^*$$

the UV spectrum of ethylence exhibits an intense band at 174 nm and a weak band at 200 nm, both of which are due to $\pi \rightarrow \pi^*$ transitions. According to the selection rules, only the band at 174 nm represents an allowed transition. The intensity of absorption by ethylene is independent of the solvent on account of non-polar nature of the olefinic bond. Alkyl substitution of the olefins moves the absorption to a longer wavelength. This is known as **bathochromic effect (or red shift)** This effect is progressive with increasing number of alkyl groups.

In the study of some substituted olefins exhibiting cis-trans-isomerism, the trans-isomer absorbs at the longer wavelength with greater intensity than the cis-isomer.

The UV absorption spectrum of benzene exhibits three bands, two intense bands at 180-200 nm and a weak band at 260 nm.

3. **n-σ* transitions:** The energy required for $n \rightarrow \sigma^*$ transition is generally less than that required for $\sigma \rightarrow \sigma^*$ transition and their corresponding absorption bands appear at longer wavelengths in the near untraviolet region (180 to 200 nm). Saturated compounds with lone pair (non-bonding) electrons undergo $n \rightarrow \pi^*$ transitions apart from $\sigma - \sigma^*$ transitions.

4. **σ → σ* transitions:** Such transitions occur in case of saturated hydrocarbons, which do not contain lone pairs of electrons. The energy required for this type of transitions is very large and the absorption band occurs in the far ultraviolet region (126 to 135 nm). For example, methane has λ_{max} at 121.9 nm and ethane at 135 nm correspond to this transition. These transitions cannot be observed in commercial spectrophotometers which generally do not operate at wavelengths below 180 nm.

Instrumentation for UV spectroscopy

The following are the important components of a UV spectrometer:

1. **Source of radiation:** The following are the most common radiation sources used in UV spectrometers:

(a) Hydrogen discharge lamps; (b) Deuterium lamps; (c) Xenon discharge lamps; (d) Mercury arcs.

2. **Monochromators:** Monochromators are used to disperse the radiation according to the wavelength. The essential components of a monochromator are an entrance slit, a dispersing element (eg., a prism or a grating) and an exit slit. The prisms are generally of quartz or fused silica. The dispersing element disperses the heterochromatic radiation into its component wavelengths, whereas the exit slit allows the nominal wavelength to pass through.

3. **Detectors:** The following three types of detectors are commonly used.

(a) **Photovoltaic or Barrier-layer cell:** In this cell, the light striking the surface of a semiconductor such as selenium (mounted upon an iron base plate) leads to the generation of electric current. The magnitude of the current generated is proportional to the intensity of the light beam falling on it. Though widely used, these cells have low sensitivity for low light levels and has a tendency to become fatigued.

(b) **Photocell or Photoemissive cell:** In the simplest form, this cell consists of a glass bulb coated internally with a thin, sensitive layer of caseum or potassium oxide and silver oxide. This layer emits electrons when light falls on it. This layer is the cathode. A metal ring insert near the centre of the bulb forms the anode. This is maintained at a high voltage with the help of a battery. The interior of the bulb is either evacuated, or less preferably, filled with an inert gas (e.g., argon) at low pressure (about 0.2 mm). When light penetrating the bulb falls on the sensitive layer, electrons are emitted, thereby causing a curent to flow through an external circuit. This current can be amplified by electronic means, and is taken as a measure of the amount of light striking the photosensitive surface. In other words, the emission of electrons leads to a potential drop across a high resistance in series with the cell and the battery, the fall in potential may be measured by a suitable potentiometer, and is related to the amount of light falling on the cathode.

(c) **Photomultiplier tube:** It consists of an electrode covered with a photo-emissive material (such as caesium or potassium oxide and silver oxide, which emits electrons when illuminated) and a series of positively charged plates, each charged at a successively higher potential. The plates (called dynodes) are covered with a material such as Be – Cu or Cs – Sb which emits several (2 to 5) electrons for each electron collected on its surface. When the electrons hit the first plate, secondary electrons are emitted in greater number than those which had initially struck the plate. These emitted electrons are then attracted by a second dynode where more secondary electrons are generated. The process is repeated progressively over all the dynodes present (9 to 16) in the photomultiplier. The net result is the large amplification (up to 10^6) in the current output of the cell. The number of electrons reaching the collector is a measure of the intensity of light falling on the detector. The photomultiplier can measure light intensities which are 200 times weaker than those measurable with an ordinary photoelectric cell and amplifier. This is extremely sensitive and very fast in response.

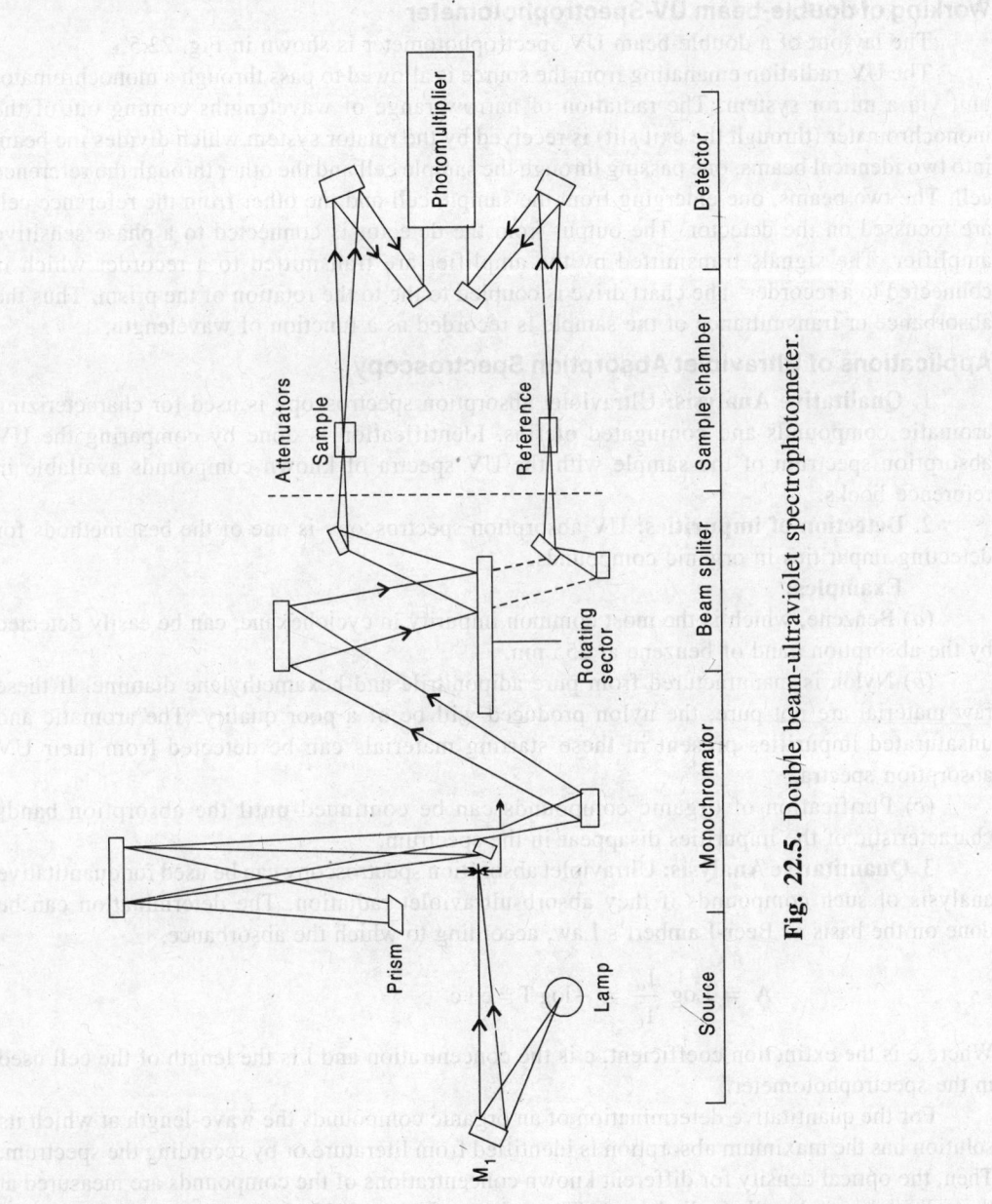

INSTRUMENTAL TECHNIQUES IN CHEMICAL ANALYSIS

Fig. 22.5. Double beam-ultraviolet spectrophotometer.

(*d*) **Recorders:** The signal from the detector is received by the recording system provided with a recorder pen.

(*e*) **Sample and reference cells:** Matched pair of cells made of quartz or fused silica are used. Single-beam and double –beam UV-spectro-photometers are available commercially.

Working of double-beam UV-Spectrophotometer

The lay-out of a double-beam UV spectrophotometer is shown in Fig. 22.5.

The UV-radiation emanating from the source is allowed to pass through a monochromator unit via a mirror system. The radiation of narrow range of wavelengths coming out of the monochromater (through the exit slit) is received by the rotator system which divides the beam into two identical beams, one passing through the sample cell and the other through the reference cell. The two beams, one emerging from the sample cell and the other from the reference cell are focussed on the detector. The output from the detector is connected to a phase sensitive amplifier. The signals transmitted by the amplifier are transmitted to a recorder which is connected to a recorder. The chart drive is coupled to the to the rotation of the prism. Thus the absorbance or transmittance of the sample is recorded as a function of wavelength.

Applications of Ultraviolet Absorption Spectroscopy

1. **Qualitative Analysis:** Ultraviolet absorption spectroscopy is used for characterizing aromatic compounds and conjugated olefins. Identification is done by comparing the UV absorption spectrum of the sample with the UV spectra of known compounds available in reference books.

2. **Detection of impurities:** UV absorption spectroscopy is one of the best methods for detecting impurities in organic compounds.

Examples:

(*a*) Benzene, which is the most common impurity in cyclohexane, can be easily detected by the absorption band of benzene at 255 nm.

(*b*) Nylon is manufactured from pure adiponitrile and hexamethylene diamine. If these raw material are not pure, the nylon produced will be of a poor quality. The aromatic and unsaturated impurities present in these starting materials can be detected from their UV absorption spectra.

(*c*) Purification of organic compounds can be continued until the absorption bands characteristic of the impurities disappear in the spectrum.

3. **Quantitative Analysis:** Ultraviolet absorption spectroscopy can be used for quantitative analysis of such compounds if they absorb ultraviolet radiation. The determination can be done on the basis of Beer-Lambert's Law, according to which the absorbance,

$$A = \log \frac{I_o}{I_t} = -\log T = e\,l\,c$$

Where e is the extinction coefficient, c is the concentration and l is the length of the cell used in the spectrophotometer.

For the quantitative determination of an organic compounds the wave-length at which its solution has the maximum absorption is identified from literature or by recording the spectrum. Then, the optical density for different known concentrations of the compounds are measured at the selected wavelength (called l_{max}). The values of optical density are then plotted against their respective concentrations, to obtain a calibration curve. If Beer's law is obeyed over the concentration range, a straight line graph must be obtained. The solution of unknown concentration is put in the spectrophotometer and its optical density is measured. The concentration of the unknown solution can be directly inferred from the calibration curve.

If a solution contains several absorbing compounds, the absorbance of the solution measured by the UV spectrophotometer will be the sum of all the individual absorbing species present in the solution, provided they do not interact with each other. On this basis, multicomponent mixtures can be analysed. This procedure is used for the determination of xylenes.

Tellurium can be determined as its iodide complex, $(Te\ I_6)^{2-}$ by absorption measurement at 335 nm.

Cobalt and Uranium can be determined as their thiocyanate complexes by absorption measurements at 312 nm and 375 nm respectively.

Nickel can be determined as its complex with β mercapto propionic acid by measuring its absorbance at 330 nm

Aluminium can be determined by making absorption measurements of its complex with 2- methyl 8- hydroxyquionline at 390 nm.

Twelve elements of the lanthanide series have been determined simultaneously with the help of recording spectrophotometer by measuring their characteristic absorption bands in the visible and UV regions.

4. Studying kinetics of chemical reactions: Ultraviolet spectroscopy can be used to study the kinetics of chemical reactions by following the change in concentration of a product or a reactant with time during the reaction.

5. Determination of dissociation constants of weak acids or bases: UV spectroscopy can be used to determine the dissociation constants of acids or bases. For instance, the dissociation constant (pk_a) of an acid HA can be determined by determining the ratio $[HA]/[A^-]$ spectrophotometrically from the graph plotted between absorbance and wavelenghts at different pH values. This value can be substituted in the equation.

$$Pk_a = pH + \log [HA]/[A^-]$$

6. **Determination of molecular weight:** UV spectroscopy can be used for determining the molecular weight of a compound if it can be converted into a suitable derivative which shows an absorption band in its spectrum. For instance the molecular weight of an amine can be determined by converting it into picrate. One litre of the solution of the amine picrate is prepared by dissolving a known weight of the amine picrate in a suitable solvent and determining its optical density at 380 nm. The concentration of the amine picrate can be determined using the formula.

$$C = \frac{\log (I_o / I_t)}{\epsilon_{max} \times 1}$$

7. **Study of tautomeric equilibria:** UV spectroscopy can be used to determine the % of keto and enol forms present in compounds such as ethyl acetyl acetate by measuring the strength of the respective absorption bands.

8. **Determination of Calcium in blood cerum:** Calcium in the blood can be indirectly determined by converting the Ca present in 1 ml of the serum as its oxalate, redissolving it in sulphuric acid, and treating it with dilute ceric sulphate solution. The absorption of the excess ceric ion is measured at 315 nm. The amount of Ca in the blood serum can thus be indirectly calculated.

9. **Determination of ozone in the environment:** The ozone concent-ration present in smoke-fog (smog) in the environment can be determined by measuring its absorption at 260 nm.

10. **Miscellaneous applications:** (*a*) UV spectroscopy has been used for confirmation of the structure of chloral.

(*b*) This technique has also been used to study charge-transfer transitions, such as the one observed with iodine and benzene in heptane.

22.3. INFRARED SPECTROPHOTOMETRY

Introduction

Infrared spectrophotometry is a powerful analytical technique useful for chemical identification. When coupled with intensity measurements, this technique can be used for quantitative analysis. Infrared spectra originate from the absorption of energy by a molecule in the infrared region and the transitions occur between two vibration levels. By measuring molecular vibrational frequencies, useful information regarding molecular structure can be obtained. The vibrational spectra, therefore are considered as **molecular finger prints**.

The infrared region of the electromagnetic spectrum which extends from the red end of the visible spectrum to the microwave region, may be divided into the following three regions:

1.	Near-infrared (overtone region)	0.8 to 2.5 µm (12,500 to 4000 cm^{-1})
2.	Middle-infrared (vibration-rotation region)	2.5 to 50 µm (4000 to 200 cm^{-1})
3.	Far-infrared (rotation region)	50 to 1000 µm (200 to 10 cm^{-1})

The important region of interest from analytical point of view is 2.5 to 25µm (4000 to 400 cm^{-1}). Infrared spectra originate from the different modes of vibration and rotation of molecule. At wavelengths below 25µm the radiation has sufficient energy to cause changes in vibrational levels of the molecule, which are accompanied by changes in the rotational energy levels. The pure rotational spectra of molecules occur in the far-infrared region and these are used for determining molecular dimensions.

The near-infrared region is accessible with quartz optics, and this is coupled with greater sensitivity of near-infrared detectors and more intense light sources. The near-infrared region is often used for quantitative work. Near-infrared spectrometry is a valuable tool for analyzing mixtures of aromatic amines.

The mid-infrared region may be further divided into

(*i*) the "group frequency" region, 2.5 to 8 µm (4000 to 1300 cm^{-1})

(*ii*) the "finger print" region, 0.8 to 15.4 µm (1300 to 450 cm^{-1})

In the group frequency region, the principal absorption bands may be assigned to the vibration units consisting of only two atoms of a molecule; that is, units which are more or less depend only on the functional group giving the absorption and not on the complete molecular structure. However, structural influences do reveal themselves as significant shifts from one compound to another. While eliciiting information from an infrared spectrum, prominent bands in this region are noted and assigned first. In the range of 2.5 to 4 µm (4000 to 2500 cm^{-1}), the absorption is characteristic of hydrogen streching vibrations with elements of mass 19 or less. When coupled with heavier masses, the frequencies overlap the triple-bonded region. The intermediate frequency region, viz., 4 to 6.5 µm (2500 – 1540 cm^{-1}) is called the unsaturated region. Triple bonds appear from 4 to 5 µm (2500 to 2000 cm^{-1}), whereas double bond frequencies fall in the region, 5 to 6.5 µm (2000 to 1540 cm^{-1}). With the help of empirical data, one can distinguish between C = O, C = C, C = N, N = O, and S = O bands. The major factors in the spectra between 7.7. to 15.4 µm (1300 to 650 cm^{-1}), which fall in the "finger print" region, are single bond stretching frequencies and bending vibrations (skeletal frequencies) of polyatomic systems, which involve motions of bonds linking a substituent group to the remainder of the molecule. The absorption bands help in their identification.

The far-infrared region contains, the bending vibrations of carbon, nitrogen, oxygen and fluorine with atoms heavier than mass 19, additional bending motions in cyclic or unsaturated systems. The low-frequency molecular vibrations found in far-infrared are particularly sensitive to changes in the overall structure of the molecule. While studying the conformation of the molecule as a whole, the far-infrared bands differ often in a predictable manner for different isomeric forms of the same basic compound. The region is well suited for the study of coordination bonds and organometallic compounds.

In the case of simple diatomic molecules, we can calculate the vibrational frequencies by treating the molecule as a harmonic oscillator. The frequency of vibration is given by the following relation:

$$\nu = \frac{1}{2\pi}\sqrt{f/\mu} \ S^{-1}$$

Where ν = the frequency (vibrations per second),

f = force constant (the streching or restoring force between two atoms) in newtons per meter and m is the reduced mass per molecule (in kilograms) defined by the following relationship:

$$\mu = \frac{m_1 \times m_2}{(m_1 + m_2)} = \frac{A_{r1} \times A_{r2}}{(A_{r1} + A_{r2}) \times L \times 1000} \ kg$$

Where m_1 and m_2 are masses of the individual atoms, and A_{r1} and A_{r2} are the relative atomic masses, L is Avogadro's constant.

The absorption bands are usually quoted in units of wave numbers $(\tilde{\nu})$ which are expressed in reciprocal centimeters, cm^{-1}. In some cases, wavelengths (λ) measured in micrometers (nm) are also used. The inter-relationship between these units is given by.

$$\tilde{\nu} = \frac{1}{\lambda} = \frac{V}{C}$$

Therefore,

$$\tilde{\nu} = \frac{1}{2\pi C} \times (f/\mu)^{1/2} \ cm^{-1}$$

Usually, reasonably good agreement is found between calculated and experimental values for wave numbers.

However, this simple calculation has not taken into consideration any possible effects arising from other atoms in the molecule. More sophisticated methods of calculation have been developed to take these interactions into account.

Vibrational Spectra of Polyatomic molecules

A diatomic molecule has only one vibrational mode and hence it yields a rather simple system. But for a polyatomic molecule, several vibrational modes are possible and, therefore, it gives a complicated IR spectrum. The vibration of atoms in a polyatmoic molecule may be visualised from a mechanical model of the system. The atoms in a molecule can be seen as resembling a system of balls by varying masses and arranged in accordance with actual space geometry of the molecule. These balls are connected with the mechanical springs whose forces are proportional to the bending forces of the chemical bonds. These forces keep the balls in position of balance. From such a model, it can be visualised that a molecule has two types of fundamental vibrations; (*i*) in one type of vibrations, the distance between two atoms increases or decreases but the atoms remain in the same bond axis. This is known as **stretching vibration.** When the stretching and compressing occurs in a symmetric fashion, it is called **symmetric** stretching. On the other hand, when one bond is compressing while the other is stretching, then it is called asymmetric stretching (ii) The other type of vibration is known as bending or deformation in which the position of the atom changes relative to the original bond axis. This bending involves oscillation of the atoms perpendicular to its bond axis. Four types of deformations may be distinguished:

(a) **Scissoring:** When the two atoms joined to a central atom deformation produced is known as scissoring.
(b) **Rocking:** When the two atoms joining a central atom move back and forth in the plane of the molecule, the resulting deformation is called rocking.
(c) **Wagging:** In this type of deformation, the structural unit moves back and forth, out of the plane of the molecule.
(d) **Twisting:** In this type of deformation, the structural unit rotates about the bond which joins the rest of the molecule.

It is not mandatory that all the fundamental vibrations should exist in the infrared spectrum. Some of the vibrations may be inactive in the infrared region or the symmetry of the molecule may be such that two or more fundamental frequencies are exactly similar. Such are called degenerate bands.

For a vibrational mode to appear in the infrared spectrum for absorbing energy from the incident radiation, it is essential that a change in dipole moment occurs during the vibration. Vibration of two similar atoms against each other (for example O_2 or N_2 molecules), will not result in a change of electrical symmetry or dipole moment of the molecule, and hence such molecules will not absorb in the IR region.

The vibrational modes are characteristic of the groups in the molecule. They are very useful in the identification of a compound or establishing the structure of an unknown substance. IR spectra may be used for detection or identification of impurities in a substance. Some typical group frequencies are given Table 22(1).

Many bands of weak intensity may occur at shorter wavelengths and these are called overtone bands or combination bands. These should not be confused with the intense fundamental bands originating from normal vibration modes. The IR spectrum of a compound is regarded as a molecular "finger-print" of the compound. For instance, the IR spectrum of polystyrene is given in Fig.22.6.

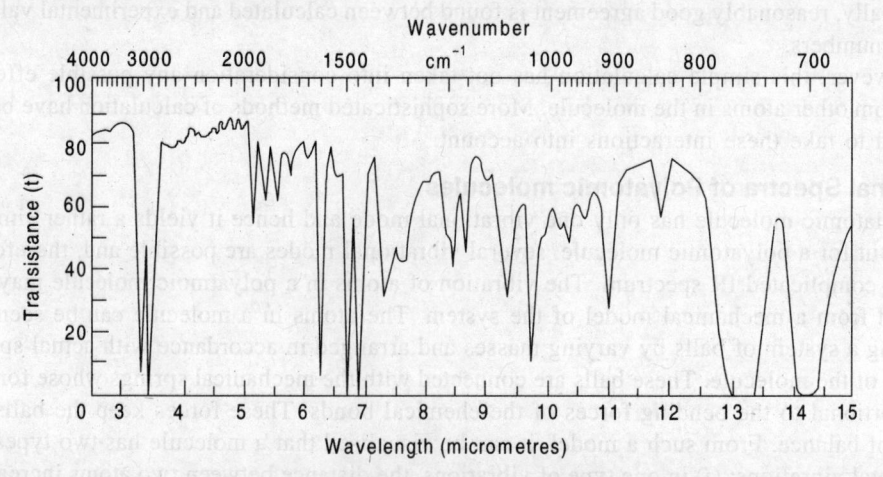

Fig. 22.6. IR Spectrum of Polystyrene.

The IR spectrum of an unknown substance is compared with spectra of possible substances as indicated by other properties. When the spectra match, the identity of the unknown substance can be established. This procedure is particularly useful for distinguishing between structural isomers.

Table 22(1). Approximate Group Positions of Some Infrared Absorption Bands

Group	Mode	Wavelength (microns)	Wave number (cm^{-1})
C—H	Bending (in-phase)	6.8 – 7.7	1300 – 1500 (m,s)
C—H	Streching	3.0 – 3.7	2700 – 3300 (m,s)
C—H	Bending (out-of-phase)	12.0 – 12.5	800 – 830 (w)
C—H	Rocking	11.1 – 16.7	600 – 900 (w)
O—H	Bending	6.9 – 8.3	1200 – 1500 (m,w)
O—H	Streching	2.7 – 3.3	3000 – 3700 (m)
N—H	Bending	6.1 – 6.7	1500 – 1700 (s,m)
N—H	Rocking	11.1 – 14.0	600 – 900 (s,m)
C—C	Streching	8.3 – 12.5	800 – 200 (m,w)
C = C	Streching	5.9 – 6.3	1600 – 1700 (m)
C ≡ C	Streching	4.2 – 4.8	2100 – 2400 (m,w)
C = N	Streching	5.9 – 6.3	1600 – 1700 (m,s)
C ≡ N	Streching	4.2 – 4.8	2100 – 2400 (m)
C—O	Streching	7.7 – 11.1	900 – 1300 (m,s)
C = O	Streching	17.0 – 21.0	480 – 600 (s)
C—Br		15.0 – 20.0	500 – 670 (s)
C—Cl		13.0 – 14.0	710 – 770 (s)
C—F		7.4 – 10.0	1000 – 1350 (s)

s – Sharp   m – Medium   w – Weak

Instrumentation

An infrared spectrometer contains the following three major components:

(1) **Source of radiation:** The main sources of mid-infrared radiation are (1) Nichrome wire wound on a ceramic support (2) Nernst glower, which is a filament containing oxides of zirconium, thorium yittrium and cerium held together with a binder (3) Globar, which is a bonded Silicon carbide rod. When heated electrically at 1200 to 2000°C, they glow and produce mid-IR radiation.

(2) **Monochromator:** Prisms and gratings are commonly used for this purpose. The prism materials used commonly are as follows:

Prisms and materials

Sodium chloride (rock salt)	4000 to 650 cm^{-1}
Potassium bromide	1000 to 400 cm^{-1}
Caesium iodide	1000 to 260 cm^{-1}
Calcium fluoride	50,000 to 1100 cm^{-1}
Lithium fluoride	16,666 to 1670 cm^{-1}

The most commonly used prism materials, NaCl or KBr are hygroscopic. Middle-infrared region normally necessitates the use of two different prisms for obtaining adequate dispersion over the whole range. For these reasons, diffraction gratings have replaced prisms as the main means of monochromation in the IR region. Further, gratings provide greater resolving power than prisms and can be designed to operate over a wider spectral range.

Advanced infrared spectrophotometers produce the IR spectra by a procedure based upon interferometry. This is known as **Fourier Transform Infrared Spectroscopy** (FT – IR).

(3) **Detectors:** The IR detectors generally convert thermal radiant energy into electrical energy. Two types of IR detecors are in use; (*a*) **selective and** (*b*) **non-selective. In selective detectors,** the response depends upon the wavelength of the incident radiation. **Example:** Photo cells, Photoconductive cells, Photographic plates, and IR phosphors.

Photoconductivity cell: This is a non-thermal detector of greater sensitivity. It comprises of a thin layer of lead sulphide or lead telluride supported on glass and enclosed in an evacuated glass cover. When IR radiation falls on lead sulphide or lead telluride, its conductance increases, thereby causing more current to flow. This detector has high sensitivity and an excellent response time (about 0.5 m sec) in IR detection. But, its range at room temperature is limited to near infrared. However, the range can be broadened by drastic cooling.

In non-selective detectors, the response is relatively independent of the wavelength of incident radiation but is directly proportional to the incident energy. **Example:** Thermocouples, bolometers, pneumatic cells, etc. These detectors are best suited for spectroscopic work.

Thermocouples are generally made by evaporating metals like Bi, Sb or semiconductor alloys, on a thin film of cellulose nitrate or any other suitable support material. It is placed in an evacuated chamber with a KBr or CsI window to minimise loss of energy by convection. The incident I.R. radiation is absorbed by blackened junction of dissimilar metals, thereby causing a rise in temperature at the junction. This gives rise to an increase in emf developed across the leads at the junction. Thermocouple detectors usually show a response time of about 60 m sec. and sensitivities of the order of 6 to 8 microvolts.

A **bolometer** is essentially a thin blackened platinum strip or a thin film of a noble metal deposited on a nonconducting support material, placed in an evacuated glass vessel with a KBr or CsI window which is transparent to IR radiation. This is connected to one arm of a wheatstone bridge. Any absorbed radiation raises the temperature of the strip and thereby alerts the resistance. Two identical elements are usually placed in the opposite arms of the wheatstone bridge, one of the elements is in path of the IR beam and the other compensates for variations in ambient temperature. Both the above receptors give a small direct current, which may be amplified to drive a recorder. Bolometers have faster response time bacause of their small thermal capacity. Recently, **thermistor bolometers** have been introduced. These have shown increased sensitivity.

The **Golay Pneumatic detector** consists of a gas-filled chamber. When the IR radiation falls on the gas, its pressure increases due to increase in temperature. These small pressure changes cause deflections of the movable wall of the chamber, which also functions as a mirror and reflects a light beam directed upon it to a photocell. The amount of light reflected bears a direct relation to the expansion of the gas chamber, and hence to the radiant energy of the light from the monochromator.

The **Pyroelectric detectors** fitted in many modern IR spectrophoto-meters use ferroelectric materials operating below their curie temperatures. When an IR radiation is incident on the detector, the change in polarisation produced gives an electrical signal. This varies with the intensity of the incident radiation. These detectors are particularly useful in FT-IR.

Sample Preparation

(*a*) **Solid Sampling:** The following four techniques are generally used for preparing solid samples:

(*i*) **Solids run in solution:** If a suitable solvent which is transparent in the IR region can be found to dissolve the solid, then the solution can be run in one of the cells for liquids.

(*ii*) **Solid films:** If the solid sample is amorphous, it can be deposited on the surface of a KBr or NaCl cell by evaporation of a solution of the solid sample. This technique is suitable for quick quantitative analysis but not for quantitative work.

(*iii*) **Nujol mull technique:** This is the most convenient and routine method. The finely powdered sample is mixed with a small quantity of a heavy paraffin oil, (usually a medicinal grade **nujol**) and mulled to form a thick paste. The paste is allowed to spread between IR trannsmitting windows. This is then mounted in the path of the IR beam and the spectrum is run. Nujol is transparent in the IR region, but it has absorption bands near 2857 cm^{-1}, 1449 cm^{-1} and 1389 cm^{-1}, where C—H streching and C—H bending absorption bands occur. In order to have the complete spectrum, a second mull has to be prepared using perfluorokerosene, fluorolube or halocarbon oil, and a separate spectrum is run. The mull technique is useful for quantitative analysis.

(*iv*) **Pressed pellet technique:** In this method, a very small quantity of finely ground solid sample is intimately mixed with about 100 times its weight of pure and desiccated KBr (or less commonly with KI or CsBr), and then pressured in an evacuated die under high pressure. The resultant transparent disc is inserted into the sample holder of the spectrophoto-meter, while a blank KBr pellet of identical thickness is kept in the path of the reference beam.

Advantages of pallet techniques

(*a*) This method is free from problem of bands due to the mulling agent, appearing in the IR spectrum.

(*b*) KBr pellets can be stored for long periods.

(*c*) The resolution of the spectrum is superior

(*d*) Since the concentration of sample taken in the pellet can be suitably adjusted, this technique is useful for quantitative analysis.

Limitations

(*i*) Owing to the traces of moisture present in the sample, a band at 3450 cm^{-1} always occurs due to OH- group from the moisture. Care should, therefore, be taken while investigations are carried out in this region.

(*ii*) The high pressure involved in pressing of the disc may bring about polymorphic changes in the crystallinity in the samples (particularly inorganic complexes) which may cause complications in the IR spectrum.

(*iii*) This method is not suitable for some polymeric substrances which cannot be powdered easily. In such cases, grinding is done at liquid nitrogen temperatures, at which the material becomes brittle.

(*b*) **Liquid sampling:** Liquid sample are directly taken into rectangular cells of NaCl, KBr or for work in double beam spectrophotometers, matched cells of identical thickness are used. Absorption cells for liquids or solutions are available commercially.

(*c*) **Sampling of gases:** Gas samples are directly introduced into 10 cm long special cells provided with IR transparent windows. For analysing dilute gases, long path cells (20 cm or 30 cm long) are employed.

Single-beam spectrophotometers

The lay-out of a single-beam infrared spectrophotometer is shown in Fig.22.7.

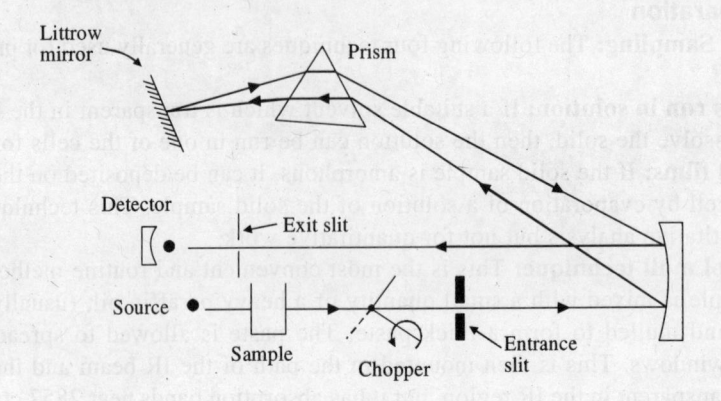

Fig.22.7. Lay-out of a single beam infrared spectrophotometer.

In this system, the radiation emitted from the source passes through the sample entrance slit, collimating mirror and then through a fixed prism and a littrow mirror. The prism and the littrow mirror select the desired wavelength and allows it to pass on to the detector, with the help of the collimating mirror The detector measures the intensity of radiation. On comparing this with the original intensity of radiation, one can measure the fraction of radiation that has been absorbed by the sample. The absorption spectrum can be obtained by measuring the degree of absorption of radiation at different wavelengths in the desired range.

The single beam instruments have the following disadvantages.

(*i*) The intensity of the emission of radiation source may change with wavelength from time to time during the analysis, which results in sloping of the base-line and deformation of the spectra

(*ii*) When the sample is analysed in solution, the bands of the solvent appear in the spectrum. This may lead to problems in interpreting the bands.

These difficulties can be overcome by double beam spectrophotometers.

Double-beam spectrophotometers

These are constructed in such a way that the radiation emitted by the source is split into two identical beams having equal intensity, one of the beams passes through the sample, whereas the other passes through the reference (air or pure solvent) for compensation. The two half beams are recombined on to a common axis, and are alternately focussed on to the entrance slit of the monochromator.

When there is no sample in the sample cell, the half-beam travelling along the sample cell is equal to that travelling through the reference cell, when these two identical half beams recombine, a steady signal reaches the detector.

However, when the sample is present in the sample cell, the half-beam travelling through it becomes less intense (depending on the nature of the sample). When the two half beams (one coming from the reference and the other from the sample) recombine, an oscillating signal is produced, which is measured by the detector. The signal from the detector then passes through a servomotor to the recorder.

Applications of Infrared Spectrophotometry

(*i*) **Quantitative analysis**

According to Bear-Lambert's law the absorbance, A is given by the following relation

$$A = ecl = \log \frac{I_0}{I_t} = \log \frac{1}{T} = -\log T$$

Where I_0 = Intensity of the incident light falling upon an absorbing medium of thickness, 1 cm

I_t = Intensity of transmitted light
T = Transmittance
c = Concentration of the absorbing species expressed moles per liter.
e = Molar absorption coefficient (or molar absorptivity or molar extinction coefficient).

For a mixture of compounds, the observed absorbance at a particular wavelength, will be the sum of the absorbances for the individual constituents of the mixture at that wavelength.

$$A_{observed} = A_1 + A_2 + A_3 = e_1 c_1 l + e_2 c_2 l + e_3 c_3 l$$

(Since the path length is constant)

In spite of the fact that the molar absorption coefficients in IR region are relatively low (smaller by a factor of 10 than those in the electronic region), innovative instrumental designs and improvements with signal to noise ratio recording and FT-IR (Fourier Transform Infrared spectroscopy) have overcome the accuracy, limitations and instrumental limitations. Therefore, quantitative infrared procedures are now much more widely used and are frequently applied in quality control and material investigations

(*a*) **Measurements using Beer Lamberts Law:** By comparing the magnitude of the absorbance (A_u) of the unknown concentration (C_u) with the absorbance (A_s) of the standard solution of known concentration using a cell of measured path length, quantitative measurement of unknown sample can be achieved.

$$A_u = e C_u l$$
$$A_s = e C_s l$$
$$\frac{A_s}{C_s} = \frac{A_u}{C_u} = e l$$
$$C_u = \frac{A_u \times C_s}{A_s}$$

Provided that linear absorbance/concentration relationship exists over the concentration range concerned.

(*b*) **Measurement using calibration graph:** In this procedure, all standards and samples are measured in the same fixed-path length cell, although the dimensions of the cell and molar absorption coefficient for the chosen absorption band are not needed as these are constant throughout the measurement. Unknown concentration can be determined from the calibration graph prepared under the same conditions as with the series of standards.

(*c*) **Standards addition method:** These are used in determinations of low concentration components in multi-component mixtures. In this procedure, a series of solutions of pure analyte in increasing concentration are prepared in a suitable solvent which does not have any absorbance in the region of chosen absorption band. Now, to each of these solutions, a known amount of the sample containing unknown concentration is added. All the solutions are diluted to a fixed volume and their absorbances measured in a fixed path length cell by scanning over the chosen absorption band. A plot of the absorbance against concentration of the pure analyte does not pass through the origin, because equal quantities of the sample of unknown concentration has been added to all the solutions. Extrapolation of the graph back to the X-axis gives the concentration of the unknown as a negative value.

(d) Internal standard method: In this procedure, a substance (e.g. KCNS or Na_3N) having a prominent isolated absorption band is intimately ground with KBr and this is used for preparing a disc. The KSCN concentration is 0.1 to 0.2% in the KBr. The KBr/KSCN disc gives a characteristic absorption band at 2125 cm^{-1}. For quantitative measurement of an organic compound, a series of standards have to be prepared using different known quantities of the pure organic compounds with the KBr and KSCN mixture and preparing the discs in each case. When the IR spectra of each of the standard discs have been obtained, the calibration curve is prepared by plotting the ratio of intensity of the selected band of the organic compound and the KSCN peak at 2125 cm^{-1} against the concentration of the organic compound in the respective discs. The peak ratio of measured absorbances obtained with the KBr disc prepared by incorporating the sample in it by following similar procedure, is referred to the above calibration curve to get the amount of the organic compound in the sample.

This techniques is successfully applied for the determination of the purity of commercial benzoic acid, etc.

(e) Quantitative analysis of gases: This can be done by applying Beer- Lambert's law:
$$(\text{Absorbance}), A = e\ p\ l$$
Where e = molar absorptivity, l = sample thickness, and p = partial pressure of the gaseous component

(f) Quantitative analysis of solid samples: This is done by preparing a solid sample in nujol or KBr and measuring only band absorption ratio, then calibrating the absorption ratios using absorption coefficient ratios which can be determined from the standards, and finally determining the sample concentration on this basis.

(g) Multicomponent analysis: If the various components present in the sample mixture do not react with each other, dissociate, associate or form complexes, quantitative determination of the components can be carried out on the principle that the absorbances of each component in the mixture are additive.

(h) Qualitative analysis: IR spectroscopy is particularly useful in the qualitative identification of compounds or components in a mixture. By referring to the enormous spectral data available, it is possible to identify compounds, or in some cases, even mixture of compounds by comparison of the spectra of the unknown compounds with those of the reference spectra. IR spectroscopy was successfully used in identification of various organic compounds (such as carboxylic acids amines, alkanes, alkines, aromatics), inorganic ions (such as carbonates, nitrates, nitrites, sulphates, phosphates, silicates, cyanides, cyanates, thiocyanates, ammonium ion, etc), coordination compounds (eg. cyanocomplexes, metal carbonyls etc.)

(3) IR spectroscopy has been useful to study inter-molecular and intra-molecular hydrogen bonding with the help of X-H stretching frequency in the molecules such as b phenyl ethanol and O-halogen phenols.

(4) IR spectroscopy is used for identifying cis-and trans-isomeric forms of o-chloro phenol, etc.

(5) IR spectroscopy is used in conformational studies of some compounds by studying the C-X stretching frequency present in equitorial and axial positions. It is also used in determining conformational equilibrium constant, k which is given by
$$k = C_e / C_a$$
Where C_e and C_a are the integrated intensities of the C − X stretching peaks in the equitorial and axial positions.

(6) IR spectroscopy is used in determining the purity of samples e.g., commercial benzoic

acid by studying the intensity of carbonyl band at 1695 cm^{-1}.

(7) IR spectroscopy can be used for studying the progress, of a chemical reaction. For instance during the oxidation of a secondary alcohol to a ketone, by examining the IR spectrum of aliquots withdrawn from the reaction mixture from time to time. As the reaction proceeds the O-H stretching band (at 3570 cm^{-1}) of secondary alcohol slowly disappears and the C = O stretching band (at 1725 cm^{-1}) due to the formation of ketone appears.

(8) IR spectroscopy is useful to study the progress of chromatographic separations.

(9) Determination of aromaticity: The difference in the wavelengths of overtones of C-H bands in different environments can be used to determine the relative proportions of unsaturated and saturated rings present in hydrocarbons and also to determine the percentage of aromatics or olefines present in the mixture.

(10) In determining the shape or symmetry of a molecule such as NO_2, which shows 3 peaks as per (3n-6) formula whereby it is confirmed that it is not a linear molecule.

(11) In calculation of force constants of molecules. The force constant is a measure of the force (in dynes/cm) required to deform a bond.

(12) In studying tautomeric equilibria such as Keto-enol, lactum-lactum and mercaptothioamide tautomerism by examining the characteristic frequencies of groups such as C = O; O-H, H-H or C = S in the respective IR spectra.

(13) Environmental monitoring of carbon monoxide using non-dispersive IR spectrometry: When IR radiation passes through a long cell (100 cm) containing carbon monoxide, a part of the energy of radiation is absorbed by CO, which can be correlated with its concentration.

In the non-dispersive IR spectrometer, the radiation from the source is not dispersed according to wavelentgh using a prism or grating as is done in a standard IR spectrometer. Instead, it is made very specific for a given compound by using the material under investigation as per the detector. In this procedure, radiation from an IR source is "Chopped" by a rotating device so that it alternately passes through a sample cell and a reference cell. The beams emerging out of the sample cell and the reference cell separately fall on the two ends of a detector which is filled with CO and divided into two equal compartments with a flexible diaphragam. Obviously, the IR beam emerging out of the sample cell and the reference cell will have different intensities. Greater the concentration of CO in the sample, greater is the absorption of radiation (at 4.2 nm) and less intense is the beam reaching the sample cell side of the detector compartment. Owing to the difference in intensity of the IR beams falling the two sides of the detector compartment, the temperature of the two compartments will be different. This temperature difference of the two compartments results in bulging of the separating diaphragm towards one side. Even the slightest movement of the diaphragm is picked up by the detector and recorded. By this method, upto 150 ppm of CO can be measured with a relative accuracy of ±5%. Such analysers are called "Dedicated Process Analysers", and they are used for industrial monitoring and for environmental studies.

(14) Another dedicated application of quantitative infrared spectrometry is its application for measurement of ethanol in the breath of motorists who are suspected of consuming alcoholic drinks before starting to drive the motor vehicles. Such devices are called "Intoximeters".

(15) Industrial Applications: (*a*) IR spectroscopy is used to determine bulk structure and incidental structure of industrial polymers. Bulk structure results from the normal polymerisation of the monomers, whereas the incidental structures arise from impurities in the monomers, side reactions, etc. IR studies helped in assigning bulk structure of ploymers such as butadiene polymers and incidental structures of polythylene, etc.

(*b*) The degree of crystallinity of nylon-66 has been studied by IR spectra. Absorption band at 934 cm^{-1} is a measure of crystallinity while the band at 1238 cm^{-1} is used as a measure of the amorphous content.

(*c*) IR spectroscopy has been used to determine molecular weight of polymers by measuring end group concentrations.

22.4. CHROMATOGRAPHY

Chromatography is a techniques used for the separation of a mixture of solutes brought about by the dynamic partition or distribution of dissolved or dispersed materials between two immiscible phases, one of which is moving past the other.

Chromatographic processes may be conveniently classified broadly as follows:

(i) Partition chromatography; (ii) Adsorption chromatography.

(i) **Partition Chromatography:** In this technique a mixture of substances is separated by means of partition between a moving solvent phase (mobile phase) and a stationary liquid phase which is held on a suitable solid support.

When the mobile phase is liquid, the technique is called liquid. liquid chromatography (LLC). This technique is useful in the separation of water soluble substances.

The liquid – liquid separations are carried out on cellulose or moist silica gel, which may be prepared in the form of thin chromatographic grade paper strips (paper chromatography PC) or thin layers (thin layer chromatography, TLC) or packed into columns (partition column chromatography). The medium in each case acts as a support for the stationary phase (generally water).

When the mobile phase is a gas, the technique is called **Vapour Liquid Chromatography** or **gas liquid** chromatography. In this technique, the stationary phase is a thin layer as a non-volatile liquid which is coated on porous solid and packed in a column. The components of the mixture are separated due to partition between this stationary phase and the mobile gaseous phase.

(ii) **Adsorption Chromatography:** In this technique small differences in the adsorption behaviour of substances, between a moving solvent (liquid or gas) and a stationary solid phase bring about their separation. If the moving phase is a liquid, it is called **liquid solid chromatography,** or **adsorption column chromatography.** If the moving phase is a gas it is called gas-solid chromatography.

Chromatographic processes may also be classified under (I) Liquid chromatography and (II) Gas chromatography.

(i) **Liquid Chromatography:**

The term liquid chromatography covers a number of separation techniques:

(a) Liquid-solid chromatography

(b) Liquid-liquid chromatography

(c) Ion exchange chromatography

(d) Bonded phase chromatography

(e) Size exclusion chromatography etc.

Classical liquid column chromatography is characterised by the use of wide diameter glass columns packed with finely divided stationary phase. The mobile phase, percolates through the column under gravity. Many important and useful separations have been achieved by this process. However, the separations are rather slow and the analysis of the separated species (e.g., by chemical or spectroscopic techniques) at times, may be cumbersome. Development of modern high-performance liquid chromatography (HPLC) since 1969 provided a highly efficient and versatile technique with excellent features, surpassing even the Gas Chromatography (GC) techniques in some respects.

Important features of some major types of liquid Chromatography are discussed below:

(A) Liquid – Solid Chromatography (LSC):

This process, generally referred to as **adsorption chromatography,** is based on interactions between the solute and fixed active sites on a finely divided solid adsorbent used as the stationary phase. The molecules of the solvent, used as mobile phase, compete with the solute molecules for the polar adsorption sites. The stronger the interaction between the mobile phase and the stationary phase, the weaker will be the adsorption of the solute on the stationary phase.

The adsorbent (stationary phase) used may be an active solid having high surface area, such as charcoal, alumina or silica gel.

The adsorbant may be packed in a column or spread on a plate. Adsorbents of widely different particle size from 5 mm for HPLC upto 40 mm for TLC are commercially available. Liquid-Solid Chromatography is particularly efficient for separation of compounds which are non-ionic and are soluble in organic solvents.

(B) Liquid-liquid (Partition) Chromatography (LLC):

This process is based on the partition (or distribution) of different solute molecules between two immiscible liquid phases according to their relative solubilties. The principle involved is akin to that of a solvent extraction process. The separating medium consists of a finely divided inert support such as silica gel or Kieselguhr, which holds the fixed (stationary) liquid phase. Separation of the solutes is accomplished by passing a suitable mobile phase over the stationary phase. The stationary phase may be in the form of a packed column, a paper strip, or a thin layer on glass.

Depending on the relative polarities of the stationary and mobile phase, liquid-liquid (partition) chromatography techniques can be of the following two types:

(*i*) **Normal liquid-liquid chromatography:** This technique is employed when the mobile phase is non-polar, whereas the stationary phase is polar. In such a situation, the non-polar solutes prefer the mobile phase and elute first, whereas the polar solutes show preference to the stationary phase and elute later.

In the normal LLC, the stationary phases used are carbowax, β–β'– oxydipropionitrile, ethylene glycol, cyanoethyl silicone, etc., whereas the mobile phases employed include hexane, heptane, benzene, xylene, saturated hydrocarbons mixed with 10% dioxan, methanol, ethanol, chloroform, etc.

(*ii*) **Reverse – Phase Chromatography (RPC):** This technique is employed when the mobile phase is polar and the stationary phase is non-polar. In such a case, the polar compounds show preference to the mobile phase and elute first, whereas the non-polar solutes elute later. This technique is particularly suitable for the separation of polar substances which are either insoluble in organic solvents or bind too strongly to solid absorbents for successful elution. The versatility of RPC arises from the fact that almost all organic molecules contain hydrophobic parts in their structure and hence are capable of interacting with the non-polar stationary phase. However, with the advent of hydrophobic bonded phases, RPC technique is recently becoming less preferred.

In RPC, the stationary (reverse) phases used include squalane, Zipax-HCP and cyanoethyl silicone; whereas the mobile phases used include water and alcohol-water mixtures, acetonitrile and acetonitrile-water mixtures.

Ion-pair chromatography is another partition-type separation process similar to the ion-association systems used in liquid-liquid extraction (solvent extraction). In this process, the species of interest associates with a large counter ion of opposite charge, the latter being selected in such a way that the resulting ion-pair is rendered soluble in an organic solvent. A typical example of such a process is the separation of sulpha drugs on microparticulate silica using 0.1 M tetrabutyl ammonium sulphate buffered at pH 9.2 as the stationary phase and a hexane-butanol (75:25) mixture as mobile phase.

(C) Ion-Exchange Chromatography (IEC)

In this process, separation of two or more different cations can be achieved by passing the solution containing the mixture of the cations through an ion exchange column. Depending upon the affinity of the cations to the exchange sites on the ion-exchange resin column, the cations are absorbed at different levels on the resin column, provided that the quantities of the ions are quite small as compared to the total ion-exchange capacity of the column. The absorbed ions can be separately and consecutively eluted by employing a suitable eluting agent(regenerating solution). Typical examples of this process are the separation of amino-acids, lanthanides and complex mixtures of colsely related substances. Ion-exchange chromatography is a method of choice for compounds with ionic or ionizable functional groups.

(D) Bonded-Phase Chromatography (BPC)

Conventional liquid-liquid chromatography has some problems such as loss of stationary phase from the support material. This may be overcome by ensuring that the stationary phase is chemically bonded to the support material. This type of liquid chromatography, in which both the monomeric and polymeric phases have been bonded to the support materials used, is called "bonded phase chromatography (BPC)". Silylation reactions are widely used to prepare bonded phases. In these reactions, silanol groups, at the surface of the silica gel are reacted with substituted chlorosilanes.

$$\equiv Si-OH + Cl-\underset{\underset{H_3C}{|}}{\overset{\overset{CH_3}{|}}{Si}}-R \longrightarrow Si-O-\underset{\underset{CH_3}{|}}{\overset{\overset{CH_3}{|}}{Si}}-R + HCl$$

For instance, reaction of silica with a dimethyl chlorosilane produces a monomeric bonded phase; whereas use of di-or tri-chlorosilanes in the presence of moisture produces a polymeric layer formed at the silica surface, thus yielding a polymeric bonded phase. The nature of the main chromatographic interaction can be altered by changing the chracterictics of the functional group, R on the silane molecule. In analytical high-performance liquid chromatography (HPLC), the most important bonded phase is the non-polar C-18 type, in which the modifying functional group, R is an octadecyl hydrocarbon chain. The siloxane phases are stable under the physico-chemical conditions used in most chromatographic conditions. They are stable in the pH range 2 to 9. A wide variety of bonded-phase packings are available commercially for use in HPLC.

(E) Gel-Permeation Chromatography (GPC) or Size Exclusion Chromatography

This type of chromatography is used for the separation of substances on the basis of their molecular size and shape. The stationary phases used in GPC are porous materials with a closely controlled pore size. The primary mechanism of retention of solute molecules is the differential penetration (or permeation) of solute molecules into the interior of the gel particles. Smaller molecules can penetrate into the interior of the gel particles, depending upon their size and the available pore size distribution, and are more strongly retained. However, larger molecules cannot penetrate (or permeate) through the openings in the gel network, and hence pass through the column mainly by way of the interstitial liquid volume.

The stationary phase materials originally used in GPC were the xero-gels of polyacrylamide (Bio-gel) and Cross-linked dextran (sephadex) type. However, these semi-rigid gels were found to be incapable of withstanding the high pressures used in HPLC. The modern stationary phases used in GPC comprise of micro-particles of styrene divinyl benzene copolymers, silica, or porous glass. GPC can be used for simple organic molecules, inorganic molecules, and studies on complex biochemical and highly polymerised molecules.

Choice of Separation techniques

The selection of the appropriate column type depends on the physical characteristics of the sample. For separation of compounds with molecular weights higher than 2000, the method of choice would be size exclusion or gel permeation chromatography (GPC). For separation of compounds of molecular weight lesser than 2000, the broad selection of the separation mode is made on the following basis:

(a) For water-soluble ionic compounds → Ion Exchange Chromato-graphy (IEC) and Ion pair chromatography (IPC) are preferred.

(b) For water-soluble non-ionic componds → Liquid-Solid chro-matography (LSC), Bonded-Phase Chromatography (BPC) and Reverse Phase Chromatography (RPC) are preferred.

(c) For organic solvent-soluble polar compounds → Bonded-Phase Chromatography (BPC) and Reverse Phase Chromatography (RPC) are preferred.

(d) For organic solvent-soluble non-polar compounds, Reverse Phase Chromatography (RPC) and Liquid-Solid Chromatography (LSC) are preferred.

(e) For complex samples, a combination of the chromatographic techniques may be required. Computer aided methods for optimization of separation conditions in High Performance Liquid Chromatography (HPLC) are available.

High-Performance (or High-Pressure) Liquid Chromatography (HPLC)

The modern HPLC, developed around 1969, enabled liquid chromatography to be transformed into a versatile technique, which in some respects proved to be more versatile than Gas Chromatography (GC). The success of HPLC is mainly due to its following features :

(i) High resolving power
(ii) Rapid separation
(iii) Accuracy in determination
(iv) Repetitive and reproducible analysis using the same column
(v) Continuous monitoring of the column effluent
(vi) Automation of the analytical procedure and data handling
(vii) Wide choice of stationary and mobile phases, and
(viii) Applicability to wide variety of samples.

Equipment For HPLC

Important components of a HPLC system are as follows:

(a) Solvent Delivery System: This includes a high-pressure pump, associated pressure and flow controls and a filter on the inlet side. The pump is one of the most important components of HPLC because its performance directly affects the retention time, reproducibility and detector sensitivity. For analytical applications in which columns of 25-50 cm in length and 4 to 10 mm internal diameters and packed with particles as small as 5 to 10 nm are used, the pump should be capable of delivering the mobile phase at the rate of 1 to 5 ml per minute. Most of the analytical work in HPLC is performed using pressures in the range 400 to 1500 psi (1 bar = 10^5 pascals = 14.5 psi). The mobile phase should be supplied to the column in a constant, reproducible and pulse-free manner. Single-or multi-head reciprocating high-pressure pumps are generally used in HPLC.

The **choice of mobile phase** is vital in HPLC. The solvent used as mobile phase should be carefully selected keeping in view the nature of the sample components, the polarity of the stationary phase, the type of separation process (*i.e.*, normal phase or reverse phase), boiling point, viscosity, flammability, detector-compatibility, toxicity, etc. HPLC grade solvents are available commercially.

(*b*) Sample Injection System : Sample introduction is done using a syringe injection or through a sampling valve. Automatic sample injectors are available. Valve injection is preferred for quantitative work.

(*c*) The Column: The HPLC columns are made from precision-bore polished stainless steel tubing, 10-30 cm long and 4 to 5 mm internal diameter. The retention of the stationary phase or package at each end in the column is accomplished with the help of thin stainless steel frits with a mesh of 2 mm or less.

The packings used in modern HPLC columns consist of small rigid particles having a very narrow particle range. The following three types of packing are generally used:

(*i*) Porous polymeric beds based on styrene-divinyl benzene copolymers. These are used for ion-exchange and size exclusion chromatography. However, off late, they are being replaced for many applications by silica-based packings which are more efficient and mechanically stable.

(*ii*) Totally porous silica particles having diameters lesser than 10 nm and narrow particle size range. Many commercially important column packings for analytical HPLC are based on this material because of its enhanced column efficiency, sample capacity and analytical rapidity.

(*iii*) Porous-layer beads having diameters in the range 30 to 55 mm. They comprise of a thin silica shell or modified silica or some other suitable material on an inert spherical core (*e.g.*, glass beads). These pellicular type packings are still used for some IEC applications. However, their general use in HPLC has declined with the development of totally porous microparticulate packings, refered to under (*ii*) above.

The development of bonded phases (discussed above) for liquid-liquid chromatography on silica gel columns is of major significance. For instance, the widely used C-18 type bonded-phases are excellent for the separation of moderately polar mixtures and is used for the analysis of drugs, pharmaceuticals and pesticides.

The useful life of an analytical HPLC column can be increased by introducing a short "guard column" introduced between the injector and the HPLC column to protect the latter from damage or loss of efficiency caused by particulate matter or strongly absorbed substances in samples or solvents.

(*d*) Detectors: Detectors are used to monitor the mobile phase as it emerges out of the HPLC column. Two types of detectors are often used.

(*i*) Bulk property detectors measure the difference in some physical property of the solute in the mobile phase as compared to the mobile phase alone. **Example:** Refractive index detectors (*e.g.*, deflection refractometer, Fresnel refractometer) and conductivity detectors, which are universally used for ionic species.

(*ii*) Solute property detectors *e.g.*, spectrophotometric detectors, flurescence detectors and electrochemical detectors (*e.g.*, amperometric or coulometric detectors)

(*e*) Strip-chart recorder.

(*f*) Data handling device and microprocessor control.

Quantitative analysis by HPLC requires a clear-cut relationship established between the magnitude of the detector signal and the concentration of a particular solute in the sample, the former being measured by either the corresponding peak area or the peak height. Manual methods can be used for calculating peak areas but computing integrators are preferred for data handling in HPLC and are a part of the instrumental package now-a-days. The percentage of each constituent of the mixture may then be calculated on the basis of area normalisation, *i.e.*, by expressing each peak area as a percentage of total area of all the peaks in the chromatogram.

HPLC is used for the separation and/or analysis of water pollutants e.g., phenol, cresol, DDT, pesticide residues in food, caffeine, steroids, chlorinated hydrocarbons, polyaromatic hydrocarbons (PAH). Coupled with graphite furnace atomic absorption spectroscopy, it is used for analysing arsenial pesticides and their metabolites in soil.

Gas Chromatography (G.C.)

This is a process by which a mixture is separated into its constituents by a moving gaseous phase passing over a stationary sorbent. Gas chromatography may be of the following two types :

(*i*) Gas-Liquid chromatography (GLC), in which the separation takes place by partitioning a sample between a mobile gas phase and a thin layer of a non-volatile liquid coated on an inert support. This is more important technique than GSC.

(*ii*) Gas-Solid Chromatograophy (GSC), in which a suitable solid material with a large surface area such as granular silica, alumina or carbon is used as a stationary phase, while employing a mobile gas phase.

Gas chromatography technique was originally developed in 1941 by A.J.P. Martin and R.L.M. Synge for which they were awarded Nobel prize in 1952. Today this technique is the most important and extensively used analytical tool for the determination of number of components in a mixture, the presence of impurities in a substance, and identification of a compound. This technique is also becoming important for process control in chemical industries and refineries.

Although gas chromatography is limited to volatile materials, the applicability of this technique has been further extended because of

(*i*) availability of column temperatures upto 450^0C

(*ii*) availability of pyrolytic techniques, and

(*iii*) the possibility of converting non-volatile materials into volatile derivatives.

Theoretical Principles

Gas chromatography process is controlled by three physical transport phenomena namely, flow, diffusion and more importantly, the partition of the solutes between the stationary phase and the mobile phase. As the solute is introduced into the column, the molecules are distributed (or partitioned) between the stationary phase and the mobile phase (the carrier gas) and a dynamic equilibrium is soon established. The process of distribution of the solutes between the two phases continues further as the fresh mobile phase (carrier gas) passes over the column, thereby establishing a fresh dynamic equilibrium.

The process goes on and on until a final equilibrium is established and at this stage, the concentration of molecules of each solute in the two phases is constant.

Partition ratio,

$$K = \frac{\text{Concentration of solute molecules in the stationary phase}}{\text{Concentration of the solute molecules in the mobile phase}}$$

Partition ratio depends upon

(*a*) the nature of the solute

(*b*) the nature of the solvent (i.e., the stationary liquid phase)

(*c*) the concentration of the liquid phase, and

(*d*) temperature.

The solute components of the sample mixture travel down the column at their own rate depending upon their respective partition ratios and the extent of their band spreading, thereby allowing a clear and clean separation of the components and their subsequent detection and determination.

Sequence of Gas Chromatographic process steps: A sample mixture containing different solutes is injected into a heating block where it is immediately vaporized and swept by the mobile phase (carrier gas) stream into the column inlet. The solutes are absorbed at the head of the column by the stationary phase and then desorbed by fresh carrier gas. This sorption-desorption process occurs repeatedly as the sample is moved by the carrier gas down towards the column outlet. Each solute band will travel at its own rate through the column. Their bands will separate to a degree that is determined by partition ratios of the individual solutes present in the sample, and the extent of their band spreading. The solutes are then eluted sequentially in the increasing order of their partition ratios and enter a detector attached to the column exit. If a recorder is used, the signals appear on the chart as a plot of time versus the composition of the carrier gas stream. The time of emergence of a peak is characteristic of each component. The peak area is proportional to the concentration of the component in the sample mixture.

Equipment: A schematic representation of a typical laboratory gas chromatographic apparatus is shown in Fig.22.8.

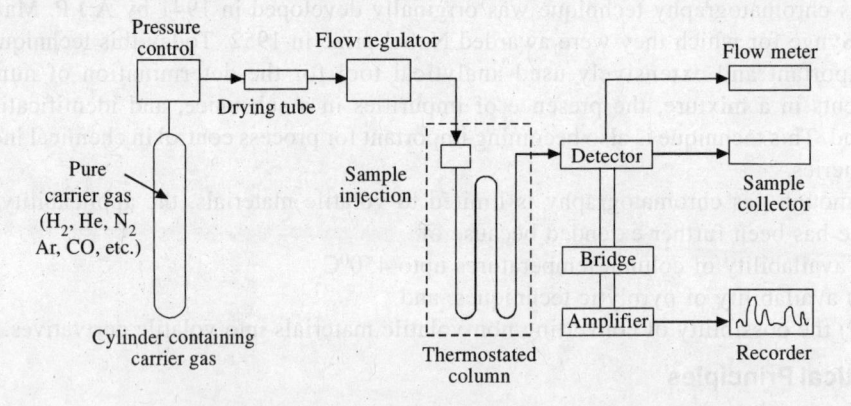

Fig. 22.8. Schematic diagram of a gas chromatograph.

The essential parts of a typical gas chromatographic assembly are as follows :

(*i*) **High pressure gas cylinder to supply pure carrier gas (mobile phase):** The operating efficiency of a gas chromatographic process depends on the maintenance of a constant flow of pure carrier gas. The carrier gases commonly used are pure helium, nitrogen, hydrogen and argon. The choice of a carrier gas depends upon (*a*) the nature of the sample (*b*) the type of detector system employed and (*c*) the column efficiency. Hydrogen and helium are most suited for use with thermal conductivity detectors because of their high thermal conductivity (as compared to the vapours of most organic compounds) and low density. Hydrogen is seldom used as carrier gas because of fire hazard and its reactivity towards unsaturated or reducible sample components.

The high pressure supply of carrier gas is associated with pressure regulators and flow meters to monitor and control the flow rate of the carrier gas.

(*ii*) **Sample Injection System:** The quantity of sample required for gas chromatographic analysis depends upon the nature and concentration of the solutes being analyzed, the size of the column and sensitivity of the detector. Generally, 0.1 to 50 microlitres are used for gases and liquid samples, whereas a fraction of a milligram is used for solid samples.

The sample is introduced into the carrier gas using (*i*) a specially prepared micro-syringe having a hypodermic needle (*ii*) a glass ampoule (used for viscous liquids or solids) or (*iii*) a gas sampling valve.

Analysis of non-voltile organic compounds can be done by first converting them into a volatile derivative before introducing into the carrier gas. For example, amino acids can be converted into their acetyl or amyl esters and analysed on a carbon column. Fatty acids may be analysed as their methyl esters. Similarly, non-volatile organic compounds containing functional groups such as $-OH$, $-COOH$, $-SH$, $-NH_2$, or $= NH$ groups can be analysed by converting them into volatile derivatives using salylation reagents (salylation is the term used for introducing tri-methyl salyl (TMS) group, $-Si(CH_3)_3$ or dimethyl salyl (DMS) group, $= Si(CH_3)_2$ in the organic compound by replacing active hydrogen).

Some non-volatile inorganic compounds can be analysed by gas chromatography by converting them into volatile, neutral metal chelates. For example, β– diketone ligands e.g., acetylacetone, and the fluorinated derivatives e.g., trifluoroacetylacetone and hexafluoro acetylacetone form stable volatile chelates with Al, Be, Cr (III), etc., thereby enabling their analysis by gas chromatography.

High molecular weight compounds may be subjected to pyrolysis to break the large molecules into smaller and more volatile fragments and then analysed by gas – liquid chromatography. Such a technique is called **pyrolysis gas chromatography.**

(*iii*) **The column:** Actual separation of sample components takes place in the column. Important factors for obtaining the desired separation are the nature of the solid support, type and amount of the stationary phase, method of packing, length of the column and temperature. The column is enclosed in a thermostatically controlled oven to ensure reproducible conditions.

The basic types of columns generally used in GC are (*a*) packed columns and (*b*) open tubular or capillary columns. Packed columns are made of stainless steels, copper, cupronickel, glass or even plastic (nylon or polythene) which may be coiled in U- or W-shape. Metal columns are preferred, though they are expensive. The lengths of the column may be anywhere from 120 cm to 5 m.

Adsorption columns are used in gas-solid chromatography (GSC) with suitable adsorbing materials such as:

Activated carbon	(for H_2, N_2, O_2, NO, CO, CH_4)
Silica gel	(for CH_4, C_2H_4, C_2H_6, CO_2)
Alumina	(for CH_4, C_2H_4, C_2H_6)
Linde Sieve 5A	(for H_2, N_2, O_2, CH_4, C_2H_6, CO)

Partition columns are used in gas liquid chromatography (GLC). They are packed with an inert support carrying a non-volatile liquid phase. The support material consists of a finely divided celite, ground fire-brick or glass beads, which can hold adequate amount of the liquid phase. The liquids used include:

Silicone oils or greases	— (for general purpose)
Apieson grease, apieson oil, and squalane	— (for separation of hydrocarbons, ethers, olefines, Ketones, halides, etc.)
Dinonyl phthalate	— (for hydrocarbons and halides)
Polyethylene glycols	— (for alcohols, amines, ethers, aromatics)
2-dimethyl formanide	— (for olefines and other low boiling hydrocarbons)

Various types of porous polymers have been developed commercially. These include:

Poropak series	— (developed by Waters Associates)
Chromosorb Series	— (developed by Johns Manville)
Tenax GC	— (based on 2,6- diphenyl phenylene oxide which can withstand operating temperatures upto 400°C. Used for concentration and determi-nation of trace level volatile organics present in gases and biological fluids)

The selection of the most suitable liquid phase for a particular separation is quite crucial and depends upon the nature of the substances to be separated. Further, the partitioning liquid should be non-volatile at the working temperature, thermally and chemically stable, and should exhibit different affinities for the different components present in the sample.

Open tubular or capillary columns are increasingly used in GLC because of their superior resolving power for complex mixtures. These are made of glass, or high purity fused silica or stainless steel and are 25 to 300 meters long and 0.1 to 1.0 mm in internal diameter. The inner walls of the tubes of these capillary columns are coated with the stationary phase either directly (wall-coated open tubular, WCOT) or after depositing a finely divided layer of a solid support (support–Coated Open Tubular, SCOT).

(*iv*) Thermal Compartments

Gas chromatograms are generally obtained with the column maintanined at a constant temperature. This isothermal operation mode results in the following two disadvantages :

(*i*) The early peaks are sharp and closely spaced (due to poor resolution), whereas the later peaks tend to be low, broad and widely spaced (due to excessive resoulution).

(*ii*) Compounds of high boiling points at times go undetected, particularly with mixtures of totally unknown composition and wide boiling point range.

These undesirable consequences of operation in isothermal mode can be minimised by using the technique of programmed-temperature gas chromatography (PTGC), in which the temperature of the entire column is raised during the sample analysis. The type of instrumentation required for temperature programming is different from that required for isothermal operation. Linear and non-linear temperature programming of sample and reference columns are available. Vapour jackets containing benzene (B.pt.80.1°C) toluene (B.pt. 110°C), cyclohexane (B.pt. 161°C), tetralin (B.pt. 207.5°C) and bromonaphthalene (B.pt. 281°C) or liquid baths may be used to maintain the column temperature. The columns are seldom operated at room temperature except in case of extremely volatile sample mixtures.

(*v*) **Detectors:** The detector is situated at the exit end of the separation column. The main functions of the detector are to sense and measure the small quantities of the separated components present in the carrier gas stream leaving the coluimn. The output of the detector is fed to a recorder which produces a pen-trace, called **chromatogram.** The choice of the detector depends upon the nature and concentration level of the components being measured.

The fundamental requisites of a good GC detector are

(*i*) **Reasonably high sensitivity:** (Sensitivity is defined as the detector response (mv) per unit concentration of the analyte (mg/ml).

(*ii*) Linearity (i.e., the detector response must be directly proportional to the concentration of the analyte).

(*iii*) Stability (i.e., reasonable constancy of the signal output with time, when the input is constant).

(*iv*) Universal response (i.e., it should respond to all the components present in the analyte, which are being measured. Sometimes, selective response indicators are used).

The detectors most widely used in gas chromatography are as follows:

(*i*) Thermal Conductivity detectors (TCD) : This is an important bulk property detector which is universal, non-destructive and concentration-sensitive. This is generally used for the detection of permanent gases, light hydrocarbons, metal chelates and other compounds which respond poorly to the flame-ionisation detector.

(*ii*) **Ionisation Detector:** Ionisation detectors in current use include:
   (*a*) Flame ionisation detector (FID).
   (*b*) Photoionisation detector (PID).

(c) Electron capture detector (ECD).
(d) Thermionic Ionisation Detector (TID).

FID and ECD are widely used ionisation detectors. Flame ionisation detector (FID) works on the principle that when the effluent from the column is mixed with hydrogen and burnt in air, the flame produced can ionise solute molecules. The ions thus produced are collected at the electrodes and the resulting ion current is measured. FID is a universal detector for organic compounds. This is the most popular detector because of its stability, high sensitivity, fast response and wide range of linear response.

Electron Capture detector (ECD): In this detector, a b-ray source (generally a foil containing ^{3}H or 6/3 Ni is used to generate slow electrons by ionisation of the carrier gas (N_2 is preferred) flowing through the detector. These electrons migrate to the anode under a fixed potential and give raise to a steady base line current. When an electron capturing gas (i.e., eluate molecules) emerges from the column and reacts with an electron, the net result is the replacement of an electron by as negative ion of much greater mass with a corresponding reduction in current flow. The response of the detector is related to the electron affinity of the eluate molecules. The ECD is the second widely used detector due to its sensitivity to a wide range of compounds. It is used for trace analysis of pesticides, herbicides, drugs, biologically active compounds and metals as their chelates.

Element sensitive detectors such as thermionic ionisation detectors are selectively used for phosphorous and/or nitrogen containing compounds; whereas flame photometric detectors are used for the determination of compounds containing sulphur and phosphorous.

Atomic absorption spectrometry is used in conjuction with gas chromatography to separate and determine organo-metallic compounds like tetraethyl lead in petroleum.

Gas chromatography interfaced with mass spectrometry (GC-MS), Fourier transform infrared spectrometry (GC-FTIR), and optical emission spectroscopy (GC-OES) can provide excellent information about a sample.

(vi) Recorder

Almost all the detectors give small and weak electrical signal. Hence it is necessary to pass these signals through an amplifier before going to the recorder. The recorder comprises of a mobile recording pen activated by the signal and a recording chart strip moving at a pre-selected speed. The amplified signals drive the pen on the moving strip of paper and trace out a chromatogram containing a series of Gaussian (i.e., bell shaped) peaks.

The time required for the sample components to travel through the column is compared with that for known compounds, which serves as a means of identification. The area of the peak is a direct measure of the concentration of the corresponding compound present in the sample.

Applications of gas chromatography

(a) Identification and separation of compounds: petroleum products, waxes, nitrogen and sulphur compounds, saturated and unsaturated hydrocarbons have been separated by GC and identified by IR, UV, NMR and mass spectrometric techniques.

GC was used to analyse fatty acids, steroids, herbicides, pesticides, esters, esterogens, perfumes, plastics, cosmetics, beverages, fertilizers, rubber products and detergents.

(b) Quantitative determination of compounds: The respective peak area in a chromatogram is the product of peak height and peak width at half peak height. The peak area is usually obtained with the help of planimeter or electrical integrators, commercial chromatographs are provided with intergrators which directly print, along the edge of the chart, a series of spikes proportional in frequency to the area beneath the curve. The spikes associated with a peak are counted, which directly gives the amount of the material in the sample.

(c) Preparative gas chromatography: Though originally developed as an analytical tool, GC has recently been proved to be a preparative tool too.

(d) Determination of carbon monoxide: Trace levels of CO (less than 10 ppm) can be determined by gas chromatography. The sample is subjected to catalytic reduction by hydrogen over a nickel catalyst at 360°C and the resulting CH_4 is measured by gas chromatography using flame ionization detector.

$$CO + 3H_2 \longrightarrow CH_4 + H_2O$$

(e) Gas Chromatography interfaced with Mass Spectrometry (GC/MS), Fourier Transform Infrared Spectrometry (GC/FTIR) and Optical Emission Spectroscopy (GC/OES) can detect nonogram quantities of materials. These techniques proved to be powerful tools for organic environmental analysis.

22.5. NUCLEAR MAGNETIC RESONANCE (NMR) SPECTROSCOPY

Principle

The nucleus of hydrogen (proton) spins about its own axis. Owing to its positive charge, the spinning proton generates a small magnetic moment and thus behaves like a tiny bar magnet. If the proton is placed in a strong magnetic field, it either aligns along the field or against it. The protons can be compelled to change their orientation by absorbing a discrete quantum of energy (hv) in the radio frequency range of the electromagnetic spectrum. The frequency of radiation to achieve this depends on the strength of the external magnetic field. The strength of the magnetic field and the radio frequency are adjusted in such a way that a quantum of energy is absorbed by the proton. In such case, the system is considered to be in resonance. This phenomenon is the basis of nuclear magnetic resonance spectroscopy.

Introduction

The isotopes of most of the elements possess gyromagnetic properties. i.e., they behave like tiny spinning bar magnets. In NMR spectroscopy, the characteristic absorption of energy by certain spinning nuclei in a strong magnetic field when irradiated by another much weaker rotating radiofrequency (r.f.) magnetic field perpendicular to it, helps in identification of atomic configurations in molecules. When these nuclei undergo transitions from one alignment in the applied field to an opposite one, energy is absorbed. The amount of energy required to cause a particular nucleus to re-align depends upon the following factors : field strength, electronic configuration around the particular nucleus, type of molecule, anisotropy and intermolecular interactions. The spectra obtained helps in identification and characterization of molecules: (a) the number of signals in the absorption spectra gives information about the number of different kinds of protons in the molecule, (b) the positions of the signals (chemical shifts) tell about the electronic environment of each kind of protons, (c) the intensity of the signals indicates the number of each kind of protons present and (d) the splitting of a signal into several peaks reveals the environment of proton with respect to other nearby protons.

Theory

The following two properties of nuclear particles are important in understanding the NMR spectroscopy: (1) Net spinning of the proton and neutron (both having a spin quantum number, $I = 1/2$) and (2) distribution of positive charge. If the spins of all the particles are paired, there would not be any net spin (i.e., $I = 0$). The distribution of positive charge is spherical, so that the nuclear quadrupole moment of the nucleus (eQ) is zero, where 'e' is the unit of electrostatic charge and Q is the measure of the deviation of the charge distribution from spherical symmetry (which is zero in this case). For the spherical non-spinning nucleus,

INSTRUMENTAL TECHNIQUES IN CHEMICAL ANALYSIS

both I and eQ are zero. A nucleus may have I = 1/2 and eQ is still zero. These nuclei, because of spin, will have nuclear magnetic moment. If I is greater than or equal to 1, the nuclei will not only have spin associated with them but the distribution of protonic charge will be non-spherical. If Q is positive, the protonic charge will be oriented along the direction of the applied field vector. If Q is negative, the charge would be perpendicular to the principal axis.

Nuclear magnetic energy levels: For a nucleus to be magnetic, it must possess spin angular moment, whose magnitude is $(h/2\pi)\sqrt{I(I+1)}$, where I is the spin quantum number of the particular nucleus and h is planck's constant. Nuclei with I = 0 are non-magnetic and hence is not important from NMR point of view. Nuclei with I = 1/2 give the best resolved NMR spectra. Important examples are ^1H, ^{13}C, ^{19}F, and ^{31}P nuclei. Nuclei with I > ½, which are of interest from NMR point of view, include ^2H (I = 1), ^{14}N (I = 1) and ^{11}B (I = 3/2). The magnetic axis of the nucleus can assume (2I + 1) orientations with respect to the external magnetic field. Each orientation corresponds to a discrete energy level, given by the following relationship:

$$E = \frac{m\mu}{I} \times \beta H_o$$

Where E is the energy of transition, m is the magnetic number, μ is the magnetic moment of nucleus expressed in nuclear magnetons, I is the spin quantum number, and H_o is the external magnetic field strength in gauss. The spectrum of allowed values, in terms of spin quantum number, is I, (I – 1),...., - (I – 1), –I. Each value corresponds to a discrete orientation (or energy level). Therefore, a nucleus with spin 1 has three orientations, and so on.

When a nucleus is placed in a system where it absorbs energy, it gets excited. It then loses energy and reverts to the unexcited state. It absorbs energy again and enters the excited state. Such a nucleus, which becomes excited and unexcited alternately, is said to be in a state of resonance. For determining the resonance frequency, the energy absorbed by the nuclei is measured while the magnetic field is varied. As the magnetic field is increased, the processional frequency of the nucleus increases. When this frequency becomes equal to the frequency of the oscillation field, transitions occur between the nuclear energy states. The energy absorbed in this process produces a signal at the detector, which after amplification is recorded as a band in the spectrum.

An NMR spectrum is then plotted between absorption signal at the detector and the strength of the magnetic field.

Instrumentation

Two types of NMR spectrometers are in use :

(1) Wide line NMR spectrometers.
(2) High resolution NMR spectrometers.

The high resolution instruments can resolve the fine structure that is associated with the absorption pak for a particular nucleus, the chemical environment of which

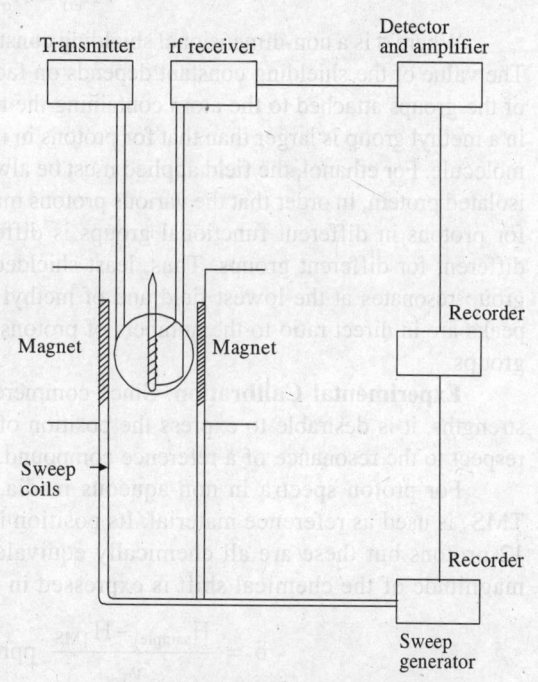

Fig. 22.9. Schematic diagram of NMR spectrometer.

reveals the nature of this fine structure. Wide line instruments are relatively inexpensive and are useful for quantative elemental analysis and for studying physical environment of a nucleus. A schematic diagram of an NMR spectrometer is given in Fig. 22.9.

Spectrometers can be of absorption type or induction type. The essential components of an NMR spectrometer are:

(a) A strong magnet to provide the principal part of the magnetic field, H_o
(b) A sweep circuit consisting of a set of Helmholtz coils superimposing the main magnetic field, to provide the additional field required to bring the total field to the resonance condition.
(c) A transmitter to supply the desired radio frequency (r.f.) energy
(d) A detector-amplifier circuit to pick-up and amplify the resonance signal.
(e) A probe which serves to hold the sample between the pole pieces.
(f) A device to receive and record the signal.

For recording NMR spectra, the sample is placed between the poles of the strong magnet. The sample is then irradiated with radiowaves. At certain value of the magnetic field, absorption of r.f. energy occurs. A sensitive detector monitors the absorption energy which is recorded as a peak on the graph. The spectra can be plotted at low resolution or high resolution as desired. The peak areas are measured automatically by the modern NMR spectrometers.

NMR spectra can be described in terms of chemical shifts and coupling constants.

Chemical shifts: Chemical shift is an important feature of high-resolution NMR spectra. In different chemical environments, the same type of nucleus will be shielded slightly from the applied field surrounding electrons. For a fixed external field, H_o, different screening factors cause slightly different frequencies.

The magnitude of the effective field experienced by each group of nuclei can be expressed as

$$H_{eff} = H_o (1 - \sigma)$$

Where σ is a non-dimensional shielding constant, which may be a positive or negative number. The value of the shielding constant depends on factors such as hybridization and electronegativity of the groups attached to the atom containing the nucleus. Thus, the shielding constant for protons in a methyl group is larger than that for protons in methylene, and it is zero for an isolated hydrogen molecule. For ethanol, the field applied must be always greater than the field for the resonance of an isolated protein, in order that the various protons may resonate. Since the value of shielding constant for protons in different functional groups is different, the required applied field would also be different for different groups. Thus, least shielded (i.e., low shield constant) proton of hydroxyl group resonates at the lowest field and of methyl group at the highest field. The areas under the peaks are in direct ratio to the numbers of protons, e.g., 1:2:3 on hydroxyl, methylene and methyl groups.

Experimental Calibration: Since commercial NMR spectrometers employ different field strengths, it is desirable to express the position of resonance, in field-independent units and with respect to the resonance of a reference compound.

For proton spectra in non-aqueous media, tetramethyl silica, $(CH_3)_4 Si$ (abbreviated at TMS) is used as reference material. Its position is assigned as 0.0 on the δ scale. TMS contains 12 protons but these are all chemically equivalent and hence give rise to a sharp signal. The magnitude of the chemical shift is expressed in ppm as follows:

$$\delta = \frac{H_{sample} - H_{TMS}}{v_1} \text{ ppm}$$

Where H_{sample} and H_{TMS} are the positions of the absorption lines for the sample and reference respectively, expressed in frequency units (hertz) and v is the operating frequency of the spectrometer. A positive δ value represents a greater degree of shielding in the sample than in the frequency. Chemical shifts are also expressed tau (τ) units, with t = 10 − δ.

Values for the chemical shifts (δ) of protons (hydrogen atoms) in some chemical groups are given in Table 22(2) given below.

Table 22(2). Chemical shifts for protons in ppm relative to TMS = 0

Type of proton	δ, ppm
R — C — H ‖ O	9.7
R — OH	5
R — OCH$_3$	3.8
R — CH$_2$ — R	1.3
R — CH$_3$	0.9

Materials recommended for other nuclei include CS_2 or TMS for ^{13}C, trichlorofluoromethane (C Cl$_3$F) for ^{19}F, and phosphoric acid for ^{31}P.

NMR spectra are displayed on charts having the magnetic field strength versus energy absorption. For obtaining exact positions of the absorption due to the protons in the given organic compound, a small amount of the TMS (standard) is mixed with the sample and the NMR spectrum is recorded. The data is interpreted as per the following guidelines :
1. The number of peaks (NMR signals) indicates the number of protons (or H atoms) in the molecule.
2. The value of (chemical shift) indicates the type of group containing the protons (hydrogen atoms); (e.g., methyl group, ethyl group, methylene olefins, ethers etc.).
3. The relative areas of peaks indicate the number of protons present in each group.
4. The spin-spin splitting or multiplicity reveals the possible arrange-ments of groups in the molecule. At high resolution, the main peak for each group may split into two or more peaks, which indicates the number of protons present on the adjacent carbon.

To illustrate a typical example, the NMR spectrum of ethanol, (CH$_3$CH$_2$OH) is shown in Fig. 22.10. Wherein (a) is the spectrum of low resolution and (b) is the spectrum of high resolution.

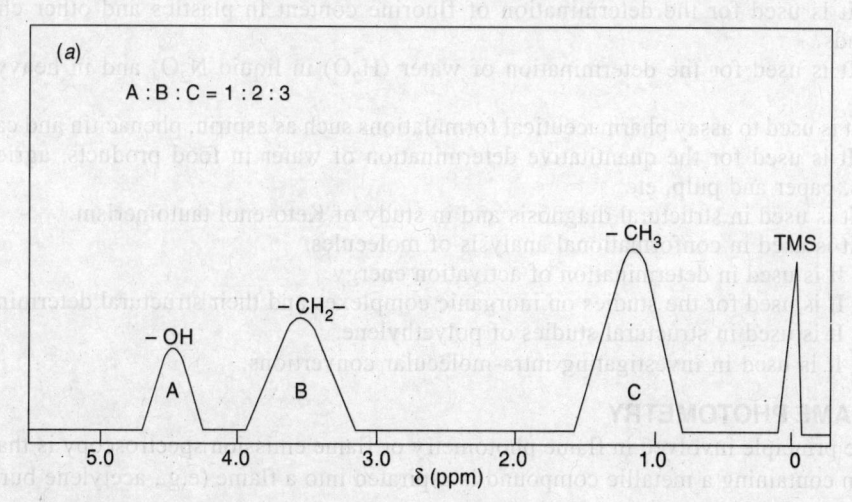

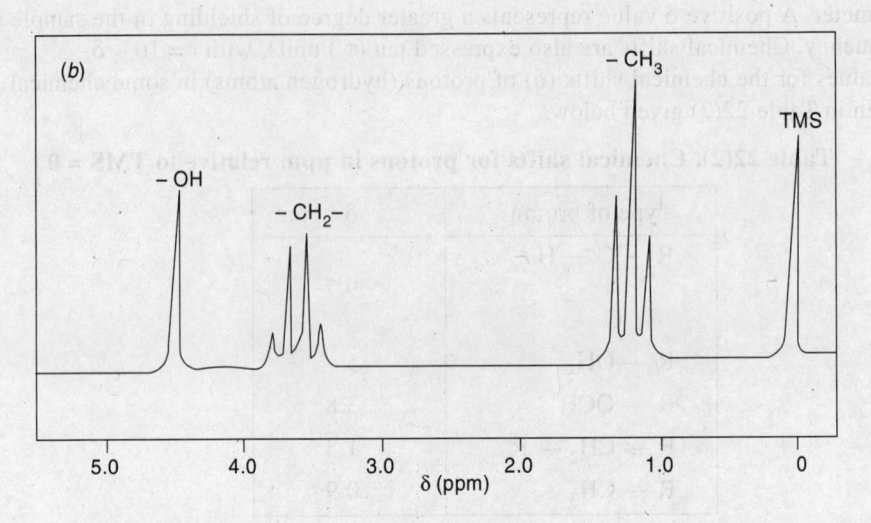

Fig. 260. NMR spectrum of ethanol at low and high resolution.

On the basis of the above standard guidelines, the spectra of ethanol can be interpreted as follows:

(1) The spectrum shows 3 peaks. They correspond to the protons in –OH, –CH$_2$–, and CH$_3$ respectively.

(2) The spectrum shows 3 peaks at the values 0.9, 3.7 and 4.8, which shows the presence of the groups –CH$_3$, –CH$_2$ – and –OH respectively.

(3) The areas under the 3 peaks are in the ratio of 1:2:3 which correspond to the number of protons in OH, CH$_2$ and CH$_3$

(4) At high resolution, the main peak for –CH$_2$– splits into 3 sub-peaks, thereby indicating the presence of 3 protons on the adjacent carbon.

Applications of NMR Spectroscopy

1. It is used for identification of atomic configurations in molecules.
2. It is used for quantitative analysis of materials for particular isotope content from the integrated area under the NMR absorption band.
3. It is a rapid, non-destructive method for analyzing proton content of fats and oils.
4. It is used for the determination of fluorine content in plastics and other chemical compounds.
5. It is used for the determination of water (H$_2$O) in liquid N$_2$O$_4$ and in heavy water (D$_2$O).
6. It is used to assay pharmaceutical formulations such as aspirin, phenacitin and caffeine.
7. It is used for the quantitative determination of water in food products, agricultural materials, paper and pulp, etc.
8. It is used in structural diagnosis and in study of Keto-enol tautomerism.
9. It is used in conformational analysis of molecules.
10. It is used in determination of activation energy.
11. It is used for the studies on inorganic complexes and their structural determinations.
12. It is used in structural studies of polyethylene.
13. It is used in investigating intra-molecular convertions.

22.6. FLAME PHOTOMETRY

The principle involved in flame photometry or flame emission spectroscopy is that when a solution containing a metallic compound is aspirated into a flame (e.g., acetylene burning in air), a vapour containing the metal atoms will be formed. Some of these metal atoms in gaseous state, may be raised to an energy level, which is sufficiently high to permit the emission of radiation, which is characteristic to the metal under investigation. Flame photometers are generally used for the analysis of sodium, potassium, lithium and calcium, because these elements have an easily excited flame spectrum of sufficient intensity for detection by a photocell.

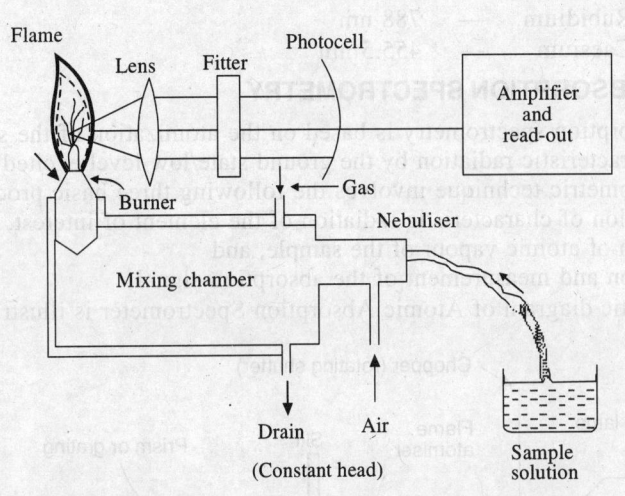

Fig. 22.11. Layout of a simple flame photometer.

The layout of a simple flame photometer is shown in Fig. 22.11

Air, at a given pressure, is passed into an atomizer and the suction thus produced draws a solution of the sample into the atomizer, where it joins the air stream in the form of a fine mist and passes into the burner system. Here, the air meets the fuel gas supplied under pressure, in a small mixing chamber, and the mixture is burnt in the burner. The radiations resulting from the flame pass through an optical filter which permits only the radiation characteristic of the element under investigation to pass through the photocell. The output from the photocell is measured on a digital read-out system. Procedure A calibration curve has to be plotted, first using solutions of known concentrations. This is done by aspirating into the flame known concentrations of the element to be determined, measuring the respective readings and plotting them against the respective concentrations of the standard solutions used.

If necessary, the test solutions may be suitably diluted to get readings in the range of 0.1 to 0.4. When once the calibration curve is ready, it is very easy to determine the concentration of the unknown samples by simple interpolation. The calibration curve should be checked periodically by making measurements with the standard solutions, and if necessary a new calibration curve must be drawn.

Some of the sophisticated instruments, now available, include a microcomputer which stores the calibration curve and allows a direct read-out of concentration.

Generally, stock solutions containing about 1000 mg ml^{-1} are prepared and then the working standard solutions are prepared by suitable dilution of the stock solution. Solutions which contain less than 10 mg ml^{-1} are often found to deteriorate on standing owing to adsorption of the solute on to the walls of the glass vessels. Hence, standard solutions of such low solute concentrations should not be stored for more than a day or two. The stock solutions should be prepared from the pure metal or from the pure metal oxide by dissolution in a suitable acid of high purity.

The characteristic wavelengths of the emitted radiations

Sodium	—	589 nm
Lithium	—	671 nm
Potassium	—	766 nm
Calcium	—	623 nm
Barium	—	553.6 nm
Strontium	—	460.6 nm
Rubidium	—	788 nm
Caesium	—	455.5 nm

22.7. ATOMIC ABSORPTION SPECTROMETRY

Atomic absorption spectrometry is based on the atomization of the sample followed by absorption of characteristic radiation by the ground state/low level excited atoms. The atomic absorption spectrometric technique involves the following three basic processes:

(i) Generation of characteristic radiation of the element of interest,
(ii) Creation of atomic vapour of the sample, and
(iii) Detection and measurement of the absorption signal.

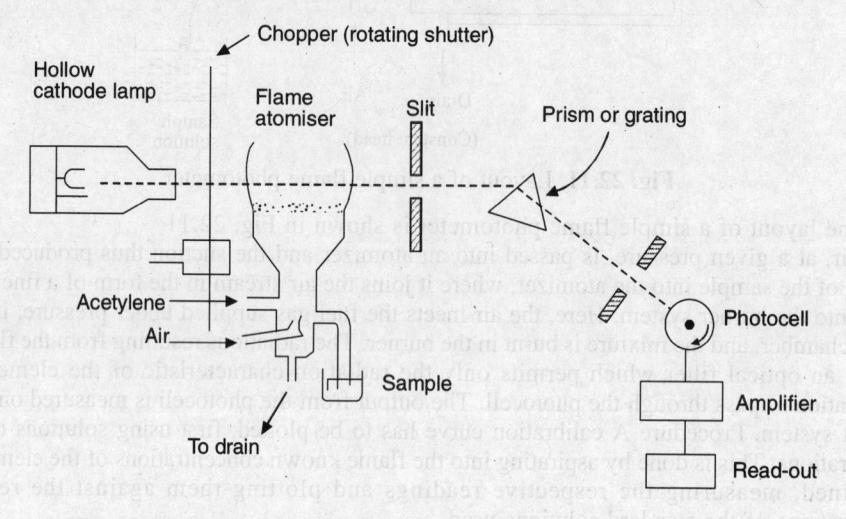

The schematic diagram of Atomic Absorption Spectrometer is illustrated in Fig. 22.12.

Fig. 22.12. Schematic diagram of Atomic Absorption spectrometer.

The characteristic radiation is obtained with the hollow cathode lamp which consists of a glass tube containing noble gases, primarily Argon (Ar), at several mm pressure, an anode and a hollow cathode which is either fabricated out of the analyte metal or has a coating of a suitable compound of the metal to be analysed. A high potential is applied across the electrodes producing an electrical current of several milliamperes flowing through the lamp when it is in operation. Ar^+ ions produced inside the tube impinge upon the cathode with a very high energy, leading to sputtering of metal atoms from the cathode atoms and emit radiation with a very narrow wavelength characteristic of the metal. The characteristic radiation thus obtained from the hollow cathode lamp passes through a flame into which the sample is aspirated. The metallic compounds are decomposed in the flame and the metal is reduced to the elemental state, forming a cloud of atoms. The atoms absorb a fraction of radiation in the flame. The decrease in radiant energy increases with the concentration of the metal being analysed in the sample according to the Beer's Law. The resultant radiant energy (light beam) passes through a monochromator to eliminate extraneous light resulting from the flame and then to detector, amplifier and recorder/read-out system.

A modification of the Atomic Absorption technique is the replacement of the flame as an atomizer and the sample holder, for certain elements, by a graphite furnace. The furnace consists of a hollow graphite cylinder placed in such a way that the light beam passes through it. About 100 microliters of the sample are placed in the graphite tube through a hole in the tip. An electrical current is passed through the tube for heating to incandescence in a programmed manner. The volatile metals are vaporized in the hollow graphite tube and the absorption of the light by the metal atoms is recorded as a spike shaped signal.

By using graphite furnace, the detection limits of many elements are improved by about 1000 times as compared to those obtained by flame atomization.

The detection limits of various elements by the two techniques are given in Table 22(3) given below:

Table 22(3). Detection limits (in parts per billion, PPb) of various Elements by Atomic Absorption Spectrometry

Element	Detection limit in parts per billion	
	Flame AAS	Electrothermal Atomisation (Graphite furnace) AAS
Al	20	0.004
As	100	0.06
Cu	2	0.008
Cr	3	0.005
Cd	1	0.008
Co	5	0.03
Fe	5	0.003
Mn	3	0.004
Zn	0.6	0.0007
Ni	8	0.02
Pb	10	0.03
Se	100	0.1
V	20	0.15
Ti	50	0.003

The Electrothermal Atomisation (ETA) – AAS offers better sensitivity due to improved atomization efficiency and higher attainable temperatures and enables picogram level determinations in some favourable cases while flame – AAS techniques are more precise and can offer precision levels of 1–3% R.S.D.(Relative Standard Deviation) in many cases. In view of the small sample size requirement, high atomization efficiency, prevention of open flame within the critically important facilities and compatibility with solid sampling, ETA – AAS has found better acceptability in many fields. The drawback of the technique lies in the strong interaction of analyte atoms with the carbon walls of the atomizer leading to poor precision of determinations as compared to that attainable by flame – AAS technique. Platform technique and tungsten wire heating are some of the recent innovations in ETA – AAS and are aimed at radiational mode of heating of the sample.

Special Techniques Used in AAS

(1) Solid sampling for metallurgical and geological samples.

(2) Standard addition technique for miscellaneous types of samples where sample – Standard matching cannot be achieved perfectly.

(3) Determination of As, Se, Ge, Te, Sb, and Sn by converting them to their hydrides and aspiration into Argon – Hydrogen flame.

$$\text{As (V) Solution} \xrightarrow[\text{(H}^+\text{)}]{\text{Na BH}_4} \text{As H}_3 \xrightarrow[\text{Gas}]{\text{Heat in flame}} \text{As}^0 + \text{H}_2$$

(4) **Flameless (Cold Vapour) Atomic Absorption technique for the determination of Hg**

This procedure is confined to the determination of mercury, which in the elemental state has an appreciable vapour pressure at room temperature so that the gaseous atoms exist without the need for any special condition. In this technique, Hg^{2+} is reduced to elemental Hg^0 by sodium borohydride or more usually by $SnCl_2$. The resulting Hg vapour is swept into an absorption cell with air, where it is subjected to a beam of radiant energy from a mercury vapour lamp. Mercury is analysed at 253.7 nm. Nanogram (10^{-9} gm) quantities of Hg can be estimated by this technique.

(5) Determination of Al, Ba, Be, Si and V by direct aspiration into N_2O–C_2H_2 flame.

(6) Determination of low concentrations of Pb, Cd and Cr by chelation with ammonium pyrrolidine dithiocarbamate, extraction into methyl isobutyl ketone (MIBK) and aspiration into air – acetylene flame.

(7) Determination of low concentration of Al and Be by chelation with 8-oxylquinolinol, extraction into MIBK and aspiration of the organic extract into N_2O – acetylene flame.

Operation

In practice, the meter of the AAS assembly is adjusted to zero absorbance or 100% transmittance when a blank solution is aspirated into the flame, when the light of the hollow cathode lamp passes on to the photomultiplier tube. Now, the solution of the sample is aspirated into the flame. A part of the light from the hollow cathode lamp is absorbed by the atomic cloud of the analyte, resulting in a proportionate decrease of light intensity falling on the photomultiplier tube thereby producing a deflection in the meter needle.

Standard solutions of the elements to be determined are used to construct a calibration curve with the help of which the elemental content of the test solution can be measured. A typical calibration curve for copper at 34.28 Å is illustrated in Fig.22.13.

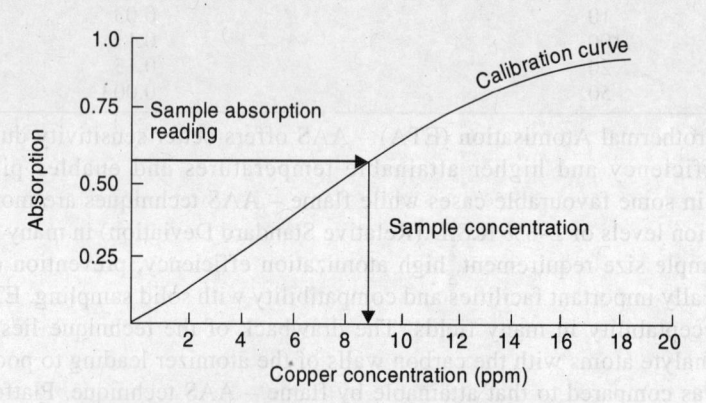

Fig 22.13. Calibration curve for copper at 34.28 Å

The slope of the linear calibration curve is noted and then the concentrations of the unknown solutions can be calculated by using the following equation:

$$A = mc$$

where A = absorbance, m = slope of the calibration curve and c = concentration.

It is important that the experimental conditions during atomization and measurement of absorbance should be identical while preparing the calibration curve and while analyzing the unknown sample.

Applications of Atomic Absorption Spectrometry

(1) Rapid and accurate determination of Ca, Mg, Cr, Cu, Co, Cd, Fe, Mn, Ni, Pb, Ag and Zn by aspiration into air – C_2H_2 flame.

INSTRUMENTAL TECHNIQUES IN CHEMICAL ANALYSIS

(2) Determination of Hg by flameless AAS technique.

(3) Determination of As, Se, Ge, Bi, Te and Sn by converting them to their respective hydrides and aspiration into $N_2O - C_2H_2$ flame.

(4) Determination of vanadium in lubricating oils.

(5) Determination of trace quantities of lead in alloys.

(6) Determination of trace elements in contaminated soils, environmental samples and industrial materials.

QUESTIONS

1. State Beer-lambert's law and discuss its applications and limitations.
2. Sketch the optical diagram of a photoelectric photometer and explain its working.
3. What are the essential parts of a spectrophotometer? What are its advantages over a colorimeter?
4. Discuss the origin of ultraviolet spectra and explain the working of UV-Spectrometer.
5. Discuss the essential components of a UV-spectrometer. Give four important applications of UV-Spectroscopy.
6. How do IR-spectra originate? Describe the working of an IR spectrometer.
7. What is the instrumentation required for IR spectroscopy? Give five important applications of IR spectroscopy.
8. Describe the essential components of HPLC system. State some important advantages of HPLC.
9. Discuss the principles involved in Gas Chromatographic. Describe the working of a gas chromatograph.
10. What are the essential parts of a Gas Chromatographic assembly? Give five important applications of gas chromatography.
11. Discuss the principle involved in the working of a flame photometer. State some applications of flame photometry.
12. What is the instrumentation required for atomic absorption spectrometry? Give some important applications of atomic absorption analysis.
13. Discuss the principles involved in the MNR spectroscopy. State some important applications.

23

Green Chemistry for Clean Technology

23.1. INTRODUCTION

An ideal chemical reaction should have a number of attributes such as
(a) Safety
(b) Simplicity
(c) Selectivity
(d) High yield
(e) Energy efficiency
(f) Use of renewable or recyclable raw materials and reagents.
(g) Absence of hazardous byproducts or at least minimizing or containing them.

In practice, it is impossible to achieve all these attributes simultaneously. Indeed it is a challenge for chemists and engineers to identify environmentally preferable reaction pathways that optimize the balance of all the desirable attributes.

The concept of green chemistry was coined by Paul Anestas of America. He enunciated 12 principles of Green Chemistry in 1994 towards ideal synthetic methods to save natural resources. Green chemistry is the use of chemistry for pollution prevention by environmentally-conscious design of chemical products and processes that reduce or eliminate the use or generation of hazardous substances. The basic principles of green chemistry are given in Appendix-2.

23.2. Goals of Green Chemistry

The goals of "green chemistry" perspective include the following :
(1) To reduce adverse environmental impacts by appropriate and innovative choice of materials and their chemical transformations.
(2) To develop processes based on renewable (plant-based) rather than non-renewable (fossil carbon-derived) raw materials.
(3) To develop processes that are less prone to obnoxious chemical releases, fires and explosions.
(4) To minimize byproducts in chemical transformations through redesign of reactions and reaction sequences. In other words, to achieve better "Atom economy".

$$\% \text{ Atom economy} = \frac{\text{Formula weight of the product}}{\text{(Sum of formula weights of all the reactants)}} \times 100$$

(Good atom economy means most of the atoms of the reactants are incorporated in the desired products and only small amounts of unwanted byproducts are formed and hence lesser problems of waste disposal or waste treatment).

(5) To develop products that are less toxic or which require less toxic raw materials/feedstocks.
(6) To develop products that degrade more readily/rapidly in the environment than the current products.

(7) To reduce the requirements for hazardous or environmentally persistent solvents and extractants in chemical processes.
(8) To improve energy efficiency by developing low temperature and low pressure processes by using new/improved catalysts.
(9) To develop efficient and reliable methods to monitor processes (e.g. monitoring reactions and releases) for improved control.

23.3. SIGNIFICANCE OF GREEN CHEMISTRY

The chemical industry releases about 5 billion tons of chemical wastes annually to the environment. It also spends over 300 billion dollars annually for treatment, control and disposal of these chemical wastes. This poses a formidable challenge for chemists and chemical engineers to review the design, manufacture and use and then to further accomplish improvements or changes in the design of new products and processes. The innovations in technology must be planned and implemented in such a way that they are sustainable both economically and environmentally. From the stand point of environmental protection, many industrial sectors have made significant studies in reducing emissions. Yet, even with these improvements, the impact of industrial processing, use and disposal of chemicals on environment is staggering. The environmental protection in many countries was based on promulgation of statutes and regulations and implementing them using command and control approach. In order to implement the environmental mandates, industries had to install a variety of waste handling, treatment, control and disposal systems, which require new equipment with high capital costs. On a societal level, this means increase in the price of the consumer goods. Further, the true costs of the environmental impacts due to the manufacture, processing, use and disposal have not been fully incorporated into the price of the goods produced. These costs include site remediation, ecosystem destruction and health care. It is precisely in this context that green chemistry provides the best opportunity for manufacturers, processors and users of chemicals to carry out their work in the most economical and environmentally beneficial way.

Many countries have been contemplating to levy environmental tax on industries towards payment of environmental impacts ostensibly caused because of their operations. By using the principles of green chemistry in the design and manufacture of chemical products and processes, many industries in the world have been successful in lowering the overall costs associated with environmental safety and health. Environmental expenditures are now being thought of as the cost of doing business. Green chemistry is already demonstrating the potential and promise to develop new techniques and methodologies to enable industries to pursue their traditional innovations and at the same time minimizing environmental impacts. The funds which they might have to pay towards environmental tax (as and when levied) may now be diverted for R & D efforts.

23.4 BASIC COMPONENTS OF GREEN CHEMISTRY RESEARCH

The major research efforts in green chemistry may be broadly classified into the following four areas :
(1) Alternative feedstocks or starting materials,
(2) Alternative reagents or transformations,
(3) Alternative reaction conditions,
(4) Alternative final products or target molecules.

These four areas are inter-related and at times overlapping. Although incremental innovations may be made in any of these areas to achieve a more environmentally benign process, the final judgement should be made on the basis that whether this specific change will result in the net improvement in the overall synthetic process from the standpoint of environmental impact. Examples of some important innovations made in each of the above areas are given below :

23.4.1. Alternative feedstocks or starting materials

Utilizing environmentally benign feedstocks reduces the risk to human health and environment. This can be done by different methods as follows :

(a) reducing the amount of the feedstock used
(b) reducing the intrinsic toxicity of the feedstock through structural modification
(c) replacing the feedstock

Eventhough the risk can be reduced through protective gear and control technologies, they are often very expensive. Green chemistry offers economically viable solutions, as illustrated by the following examples.

Ex :

(i) Synthesis of aromatic amines : Chlorinated aromatics are used in the synthesis of a variety of aromatic amines. However, chlorinated aromatics are known to be environmentally hazardous and are known to be persistent bioaccumulators. Stern and coworkers at the Monsanto Corporation successfully proved that by using nucleophilic substitution for hydrogen, the use of chlorinated hydrocarbons can be avoided. This research on feedstock replacement helped in removing the environmental concerns in the commercial preparation of aromatic amines.

(ii) Synthesis of Adipic acid : Significant advances have been made by Frost at Michigan State University in using alternative feedstocks coupled with biosynthetic methodologies. New synthetic pathways have been reported for the manufacture of adipic acid, catechol and hydroquinone using glucose as the starting material in place of traditionally used benzene which is carcinogenic. Traditional and alternative synthetic pathways for the manufacture of adipic acid are given in Fig.23.1.

(A) Traditional pathway

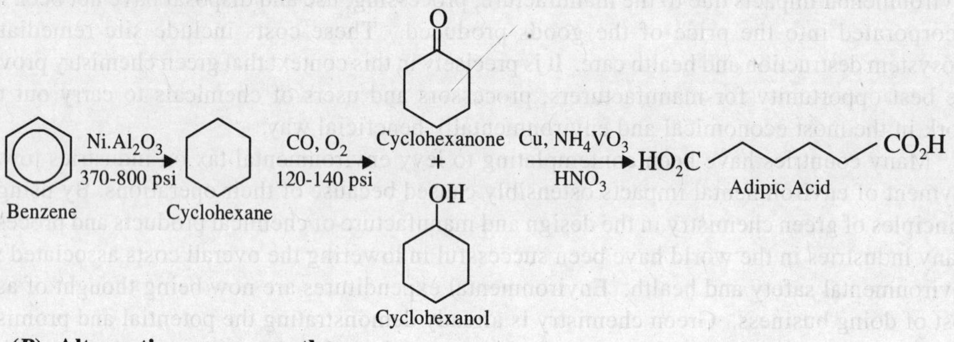

(B) Alternative greener pathway

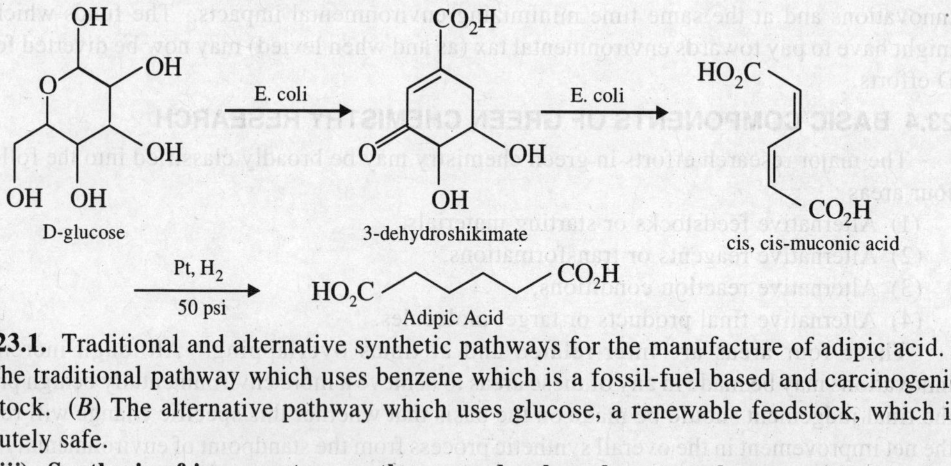

Fig. 23.1. Traditional and alternative synthetic pathways for the manufacture of adipic acid. (A) The traditional pathway which uses benzene which is a fossil-fuel based and carcinogenic feedstock. (B) The alternative pathway which uses glucose, a renewable feedstock, which is absolutely safe.

(iii) Synthesis of isocyanates, urethanes and polycarbonate polymers without using hazardous phosgene : Riley, Mc Ghee and coworkers at Monsanto Corporation have been successful in eliminating phosgene in the manufacture of isocyanates and urethanes by designing a process for direct reaction of CO_2 with amines.

The largest industrial use of phosgene is in the preparation of isocyanates. Isocyanates are mostly used for the manufacture of polyurethanes and pesticides. Polyurethanes are used in the manufacture of cushions, mattresses, insulation, car bumpers, paints, adhesives, etc. Hence, isocyanates are produced on the billion pounds scale annually and are mostly produced so far via phosgene technology. The reaction of an amine with phosgene has several problems such as high toxicity of phosgene and generation of H Cl. The use of CO_2 as a phosgene replacement has been explored. Experimental conditions for the highly selective synthesis of isocyanates from carbamate anions (generated by the reaction between amine and CO_2) have been found using various "dehydrating agents" (e.g., acetic anhydride, benzoic anhydride, trifluoroacetic anhydride, benzene sulphonic acid anhydride, etc). The initial work was based on the use of electrophilic dehydrating agents for the generation of isocyanates from amines and CO_2.

$$R\,NH_2 + CO_2 + Base \rightarrow R\,NH\,CO_2^{-}\,{}^{+}H\,Base$$
$$R\,NH\,CO_2^{-}\,{}^{+}H\,Base + Dehydrating\ agent \rightarrow RNCO + Salts$$

These reactions were shown to proceed with high selectivity and high yield using phosphorous containing electrophiles (e.g., $PO\,Cl_3$, $P\,Cl_3$ and P_4O_{10}) and also offer a relatively low cost and mild reaction route to obtain new materials. However, a large amount of salt waste is generated. To solve this problem, the possible use of non-halide and potentially low to no waste reagents capable of achieving the same high selectivities and yields has been explored. It is during these efforts that the use of O-sulphobenzoic acid anhydride as the dehydrating agent has been found to be successful. This reagent has not only provided selective conversion of amines and CO_2 into their corresponding isocyanates (thereby eliminating the need to use toxic phosgene) but also provided a process by which the dehydrating agent can be recycled (thereby eliminating the huge amount of salt waste). This reaction is supposed to proceed via nucleophilic attack of the carbamate anions at the anhydride giving rise to the formation of a mixed anhydride, which then undergoes base-induced elimination to form the corresponding isocyanate.

This reaction may also pave the way for generating isocyanates in situ followed by immediate conversion of the isocyanate into urethane materials.

Menzer at Du pont developed a catalytic process where the use of phosgene in the traditional isocyanate process is eliminated. In Menzer's process, the amine is directly carbonylated through the use of carbon monoxide in a proprietory system. The process is reportedly commercialised.

Komiya et al from Asahi Chemicals, Japan reported the successful synthesis of polycarbonates without using phosgene. The process utilizes a molten state reaction between a dihydroxy compound such as bis-phenyl A and a diaryl compound such as diphenyl carbonate. This new process clearly accomplishes the following goals of green chemistry : (*a*) It eliminates the use of phosgene which is hazardous, (*b*) The reaction is carried out in molten condition thereby eliminating the use of methylene chloride as solvent, which is a suspected carcinogen.

802   A TEXTBOOK OF ENGINEERING CHEMISTRY

$$COCl_2 \quad HO-\!\!\langle\text{C}_6H_4\rangle\!\!-\!\!\underset{CH_3}{\overset{CH_3}{C}}\!\!-\!\!\langle\text{C}_6H_4\rangle\!\!-OH \quad NaOH$$

Phosgene   Bis-A

Solvent H_2O CH_2Cl_2 → **Interfacial Polymerization**

Washing with water → Water, NaCl (CH_2Cl_2)

Solvent removal → CH_2Cl_2
                            Difficult to separate

$$\left[-O-\langle C_6H_4\rangle-\underset{CH_3}{\overset{CH_3}{C}}-\langle C_6H_4\rangle-O-\overset{O}{\underset{\parallel}{C}}-\right]_n$$

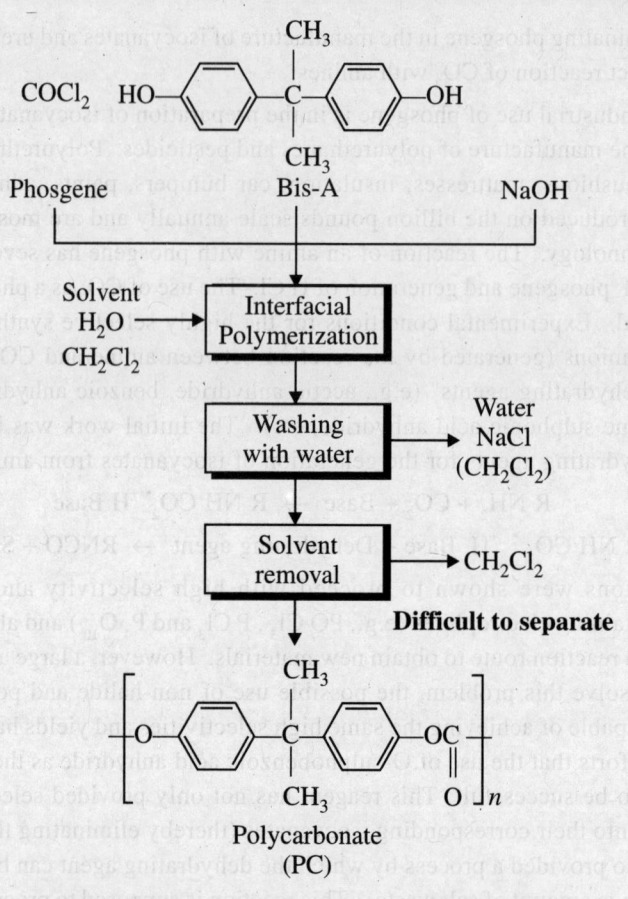

Polycarbonate (PC)

┌─ Problems ─────────────────────────┐
│ 1. Use of dangerous Phosgene   ☠ Poison │
│ 2. Use of a very large amount of CH_2Cl_2 │
│ 3. Presence of Cl-impurities in PC │
└────────────────────────────────────┘

Phosgene process of polycarbonate and its problems

Fig. 23.2. Synthesis of polycarbonate
(A) Phosgene process (Hazardous traditional route)

GREEN CHEMISTRY FOR CLEAN TECHNOLOGY 803

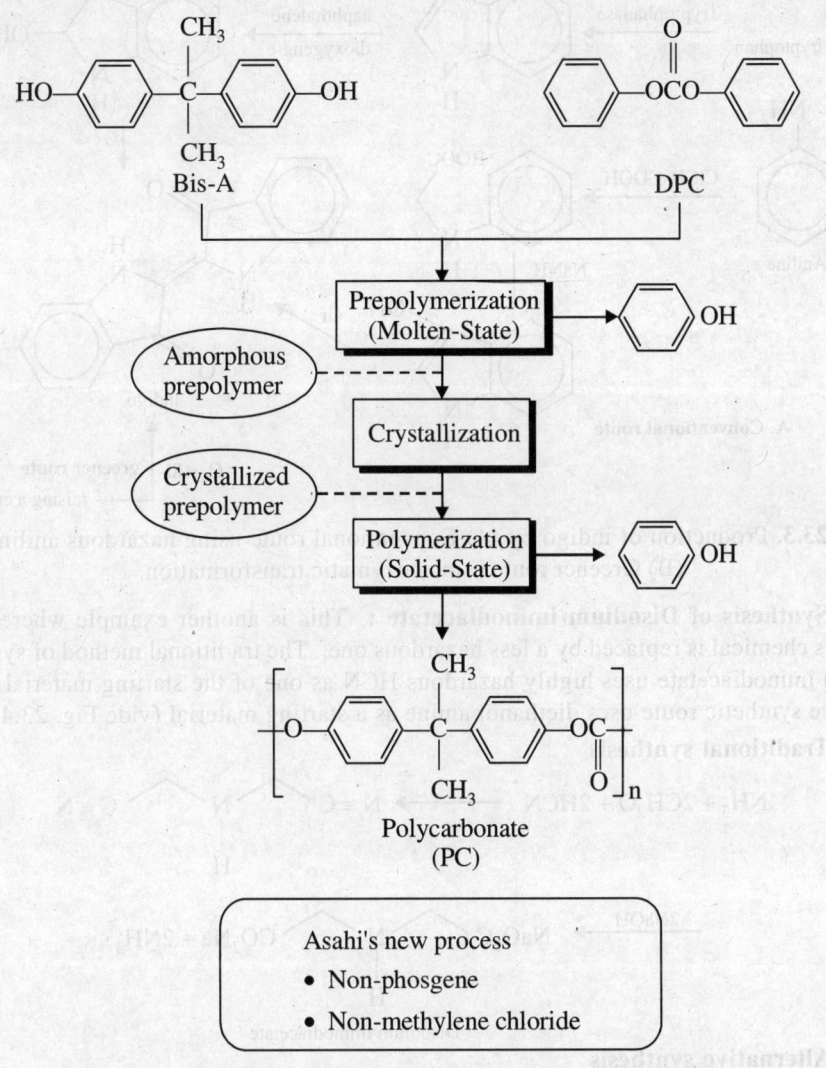

Fig.23.2. Solid-state polymerization process of polycarbonate

(*B*) Asahi's solid state polymerization process without using phosgene and methylene chloride as solvent (Benign green chemistry route).

(*iv*) **Synthesis of indigo :** Indigo, the dye used to colour blue Jeans, can be made enzymatically by removal of the side chain of tryptophan to give indole, which can be dehydroxylated enzymatically, and then oxidized with oxygen to indigo. The presently used commercial process starts with aniline which is highly toxic. Moreover, it produces considerable amount of waste salts, thereby causing disposal problem.

An even greener way to produce indigo would be to raise it is a crop as practised in colonial America. A new commercial strain yields 5 times more indigo than the traditional plant.

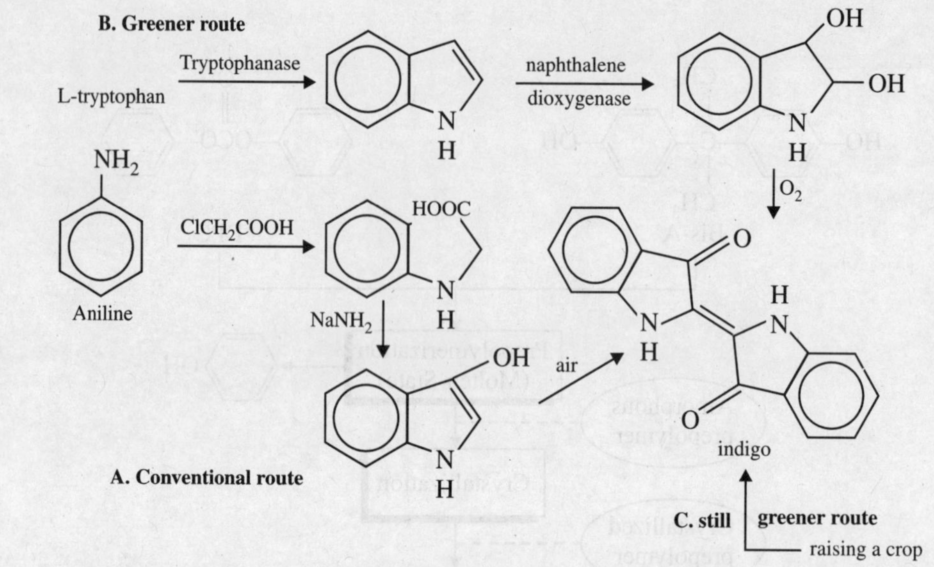

Fig. 23.3. Production of indigo by (A) Conventional route using hazardous aniline and (B) Greener route using enzymatic transformation.

(v) **Synthesis of Disodium iminodiacetate :** This is another example where a highly hazardous chemical is replaced by a less hazardous one. The traditional method of synthesis of disodium iminodiacetate uses highly hazardous HCN as one of the starting material. Now, an alternative synthetic route uses diethanol amine as a starting material (vide Fig. 23.4).

(A) Traditional synthesis

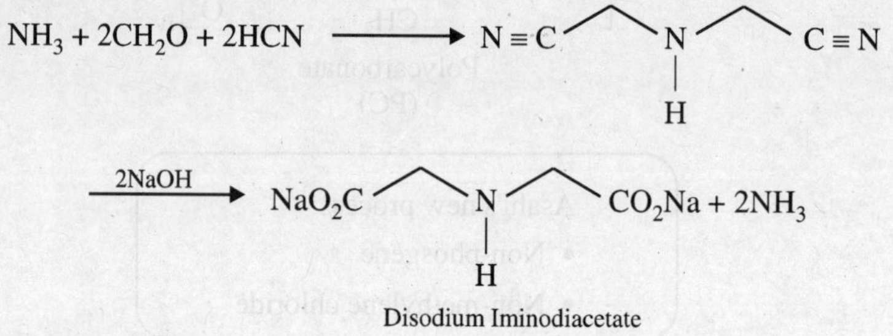

$$NH_3 + 2CH_2O + 2HCN \longrightarrow N\equiv C\text{-}CH_2\text{-}NH\text{-}CH_2\text{-}C\equiv N$$

$$\xrightarrow{2NaOH} NaO_2C\text{-}CH_2\text{-}NH\text{-}CH_2\text{-}CO_2Na + 2NH_3$$

Disodium Iminodiacetate

(B) Alternative synthesis

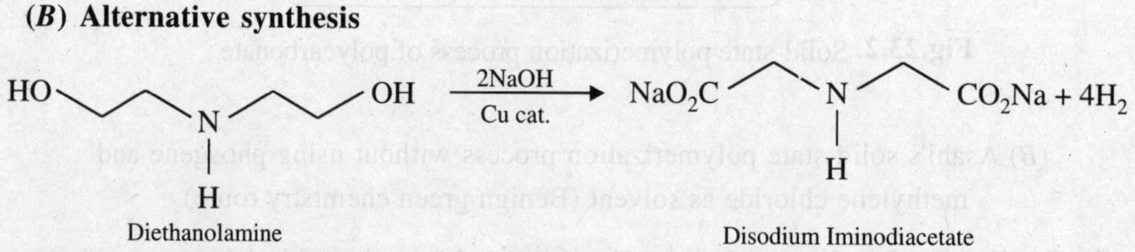

$$HO\text{-}CH_2CH_2\text{-}NH\text{-}CH_2CH_2\text{-}OH \xrightarrow[\text{Cu cat.}]{2NaOH} NaO_2C\text{-}CH_2\text{-}NH\text{-}CH_2\text{-}CO_2Na + 4H_2$$

Diethanolamine   Disodium Iminodiacetate

Fig. 23.4. Synthetic pathways for disodium imminodiacetate. (A) Traditional route using hazardous hydrogen cyanide. (B) Alternative route using diethanol amine with copper catalyst.

(vi) **Synthesis of polymeric substances from biological/agricultural wastes :** Gross at the university of Masachusettes prepared new polymeric substances from biological/agricultural wastes such as polysaccharides using biocatalytic transformation. This process simultaneously addresses several green chemistry objectives and environmental concerns such as

(a) use of environmentally benign feedstocks

(b) use of biocatalytic transformation rather than the conventional monomers and harsh experimental conditions

(c) addresses the environmental concerns of persistent polymers and plastics by paving the way for the manufacture of biodegradable gross polymers.

23.4.2. Alternative reagents or transformations

There can be several opportunities to substitute more benign chemicals for the toxic reagents used at present to carry out some transformations. The following examples illustrate how such efforts lead to reduce environmental and human health risks.

(a) Paquetta of Ohio State university reported the use of Indium, instead of metals of environmental concern being used so far, to bring out several transformations. Such transformations also demonstrated increased selectivities. Further, these transformations could be carried out in aqueous media instead of volatile organic solvents.

(b) In the synthesis of several important chemicals, methylation is a vital step. However, many strong methylating agents such as dimethyl sulphate are acutely toxic and often corcinogenic. Fundo at the university of Venice reported the use of dimethyl carbonate for methylation of acrylonitriles and methylacrylacelates, which are important reagents in the synthesis of a class of antidepressant drugs. This advance in green chemistry serves dual purposes — (i) It eliminates the use of toxic dimethyl sulphate. (ii) It also eliminates the problem of high salt production during the transformation because this process requires only catalytic amount of the base to proceed.

However, dimethyl carbonate is traditionally synthesized by the reaction of highly toxic phosgene with methanol. Therefore, the overall benefit accruing from replacement of dimethyl sulphate with dimethyl carbonate becomes questionable unless the latter is synthesized by using a more benign reaction pathway. Indeed, such an effort resulted in the successful work of Rivetti and coworkers of Enichem Chemical Company. These workers synthesized dimethylcarbonate by the oxidative carbonylation of methanol using carbon monoxide

(c) Epling and coworkers at the university of connecticut reported alternative methodology, using visible light as a "reagent", for the cleavage of a variety of dithiane and oxathiane ring systems commonly used as protecting groups. Traditionally, rings of this type are cleaved by toxic heavy metal-catalyzed reduction. Epling's work has the potential to accomplish important transformations useful in dye and pharmaceutical industries. This methodology addresses the health and environmental concerns associated with the use of toxic heavy metals for such reactions.

(d) Kraus from Jowa State University studied the widely used Friedel Craft's acylation reaction which is catalyzed by the Lewis acids. He found that by using quinolic moieties and aldehydes, formal synthesis of a number of target molecules (such as diezapam and analogues), which are of vital importance in the pharmaceutical industry, has been achieved.

Alternative reaction conditions

During the manufacture, processing, formulation and use of chemical products, a variety of associated substances contribute to environmental pollution. Examples are the solvents used in reaction media, separations and formulations. Many solvents, particularly the volatile organic solvents (VOC's) came under severe scrutiny and regulatory restriction due to their toxic contributions to air and water pollution. Therefore, much of the R & D in green chemistry reaction conditions is focussed on alternative solvents. Some of the main alternatives to traditional solvents include super-critical fluids, aqueous solvents, polymerized or immobilized solvents, ionic liquids, solventless systems and reduced hazard organic solvents.

(a) Super-critical fluid (SCF) systems :

Solvent systems such as supercritical (SC) carbon dioxide and supercritical CO_2 / H_2O mixtures are under investigation for use in various types of reactions. In general, super-critical fluids (SCFs) appear to offer promise for low cost, innocuous, solvents that can supply "tuneable" properties.

Super-critical CO_2 (SC CO_2) is non-toxic, non-inflammable and inexpensive. Moreover, since the solubility of most of the solutes in super-critical fluids changes considerably near the critical point, it is often possible to recover the solute from super-critical CO_2 merely by reducing the pressure to below the critical point. In this manner, several applications of SC CO_2 such as de-caffeinating coffee (where supercritical CO_2 has now replaced the earlier solvent viz. methylene chloride), strictly depend upon the physical, solvating properties of this solvent. However, all materials are not soluble in super-critical CO_2. For materials such as high molecular weight hydrocarbons, which are not highly soluble in supercritical CO_2, the advantages of super-critical fluids can be obtained by adding a surfactant to SC CO_2. Addition of the appropriate surfactant generates a micelle phase in which materials that are not normally soluble in SC CO_2 can also be suspended. The following reaction illustrates this concept.

Catalytic copolymerization of CO_2 with epoxides :

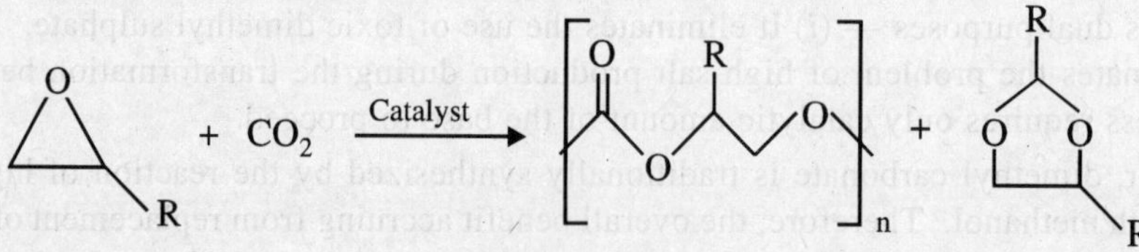

$$\overset{O}{\underset{R}{\triangle}} + CO_2 \xrightarrow{\text{Catalyst}} \left[\underset{O}{\overset{O}{\parallel}}C-O\underset{R}{\overset{}{\wedge}}O \right]_n + \underset{R}{\overset{O}{\underset{O}{\square}}}$$

Fig. 23.5. Example of the use of supercritical CO_2 to replace conventional solvents. A co-solvent is used along with SC CO_2 to allow a polymerization reaction to take place.

In this case, a polymerization reaction, which produces high molecular weight materials, is conducted in a surfactant-super-critical CO_2 system. Thus, the need for conventional solvents is eliminated.

Tanko, at Virginia polytechnic Institute demonstrated successfully some free-radical halogenations in SCF. He proved that when compared with standard bromination reactions of alkylated aromatics as the model system, the yields and selectivities of the reactions of SCFs are equal or even superior to those conducted in conventional solvent systems.

De Simone at the university of North Carolina demonstrated that a variety of polymer types can be synthesized with different monomeric systems in super-critical fluids. His studies in methyl methacrylate polymers showed many definite advantages in using SCFs as a solvent system over conventional halogenated organic solvents.

Tumas et al at the National Laboratory at Los Almos showed reactions such as polymerization of epoxides, oxidation of olefins and asymmetric hydrogenations in SCF systems were comparable and even superior to the conventional halogenated organic solvent systems.

In many cases, solvents are used strictly for their physical, solvating properties. However, in some situations, such as chemical reactions occurring in solvents, the solvents play a specific role in the chemical synthesis. In such cases where a reaction takes place in a solvent, supercritical fluids (SCFs) may enhance or inhibit the desired reaction. The effect of SCFs on reaction chemistry is an active area of research. A particular type of reactions, known as asymmetric catalytic reductions have been studied in super-critical CO_2. It was found that the selectivities achieved in these reactions are comparable or even superior to those obtained in conventional solvents.

GREEN CHEMISTRY FOR CLEAN TECHNOLOGY

(b) Aqueous solvent systems :

Several projects are underway on the use of aqueous solvent systems in place of organic solvent systems in chemical manufacturing. Many of these show great promise and efficacy.

Although water is non-toxic, non-inflammable, renewable and inexpensive solvent, the limited solubility of many hydrocarbon reactants in water has been a limitation in its utility. However, Anastas and Williamson (1998) and Li and Chan (1997) reported some case studies showing the innovative use of water as a reaction medium. Water with an alcohol co-solvent is used in Diels Alder reactions.

Aqueous conditions for the Diels-Alder Reaction

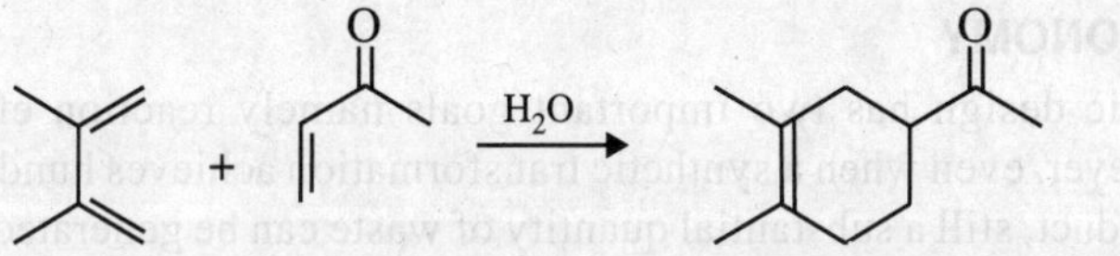

Fig. 23.6. Use of water as a solvent for certain Diels Alder reactions to increase reaction rate.

Some Diels Alder reactions, such as the dimerization of 1,3 cyclopentadiene, were found to be accelerated in water. This may be due to favourable packing of hydrophobic surfaces in the transition state of the reaction.

Many other organic reactions that were traditionally carried out in organic solvents have now been carried out in aqueous media. Examples of such reactions include Barbier-Grignard reaction, pericyclic reactions, and transition metal-catalyzed reactions (Li 1998).

(c) Derivatized / Immobilized solvent materials :

Hurter and Hatton, 1992 and U.S. EPA 1996 used a novel concept of using derivatized / immobilized solvent materials with the twin-objectives of reducing emissions and promoting the recovery of hazardous solvents. In this case, the hazardous substance under consideration, viz. tetrahydrofuran (THF) is attached to a polymeric backbone using chlorinated styrene derivative. The THF then remains relatively mobile, but because it is attached to a polymeric backbone, it is less likely to volatilize and is easily recoverable using methods such as ultrafiltration.

Derivatized/Polymeric solvent replacement for THF

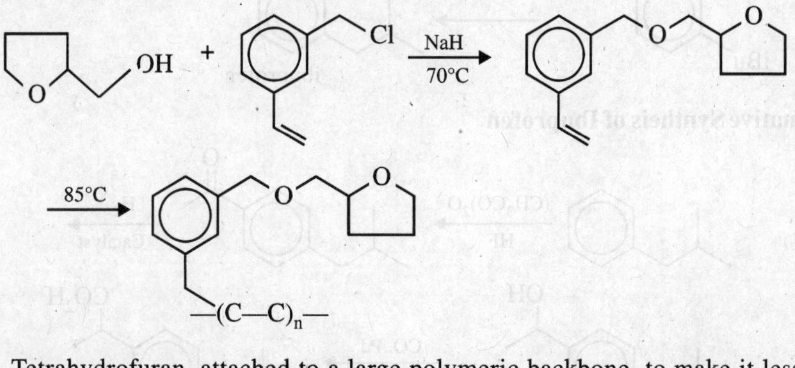

Fig. 23.7. Tetrahydrofuran, attached to a large polymeric backbone, to make it less volatile.

4. Alternative products and Target Molecules

In many cases, the part of a molecule which provides its intended activity is separate from the part responsible for its toxicity. Thus it is possible to greatly reduce the hazards associated with certain chemical products while maintaining their performance and innovation. This is particularly of interest in pharmaceutical and pesticide industries.

(a) De Vito at the United States Environmental Protection Agency (US EPA) found that by blocking the alpha position, the ability of nitriles to form an alpha radical is prevented. Thus the toxicity of nitrile can be decreased by several orders of magnitude without adversely affecting the ability of the nitrile to function as a cross-linking agent. The nature of the hazard posed by the release of cyanide depends upon the ability of nitrile to form an alpha radical.

(b) De Pompei at Tremco, Inc. designed alternatives to isocyanates as a sealant by maintaining their function without toxicity. In this work, acetylacetate esters are used as alternatives to hazardous isocyanates. Acetylacetate esters do not have any environmental or health hazards.

23.5. ATOM ECONOMY

Any synthetic design has two important goals namely reaction efficiency and product selectivity. However, even when a synthetic transformation achieves hundred percent selectivity to the desired product, still a substantial quantity of waste can be generated, if the transformation is not "atom economical". Atom economy is defined as the ratio of the formula weight of the target molecule to the formula weights of all the starting materials and the reagents. It indicates the intrinsic efficiency of the desired transformation. The example given below illustrates the atom economy achieved in the synthesis of ibuprofen by the traditional and alternative routes.

Traditional Syntheis of Ibuprofen

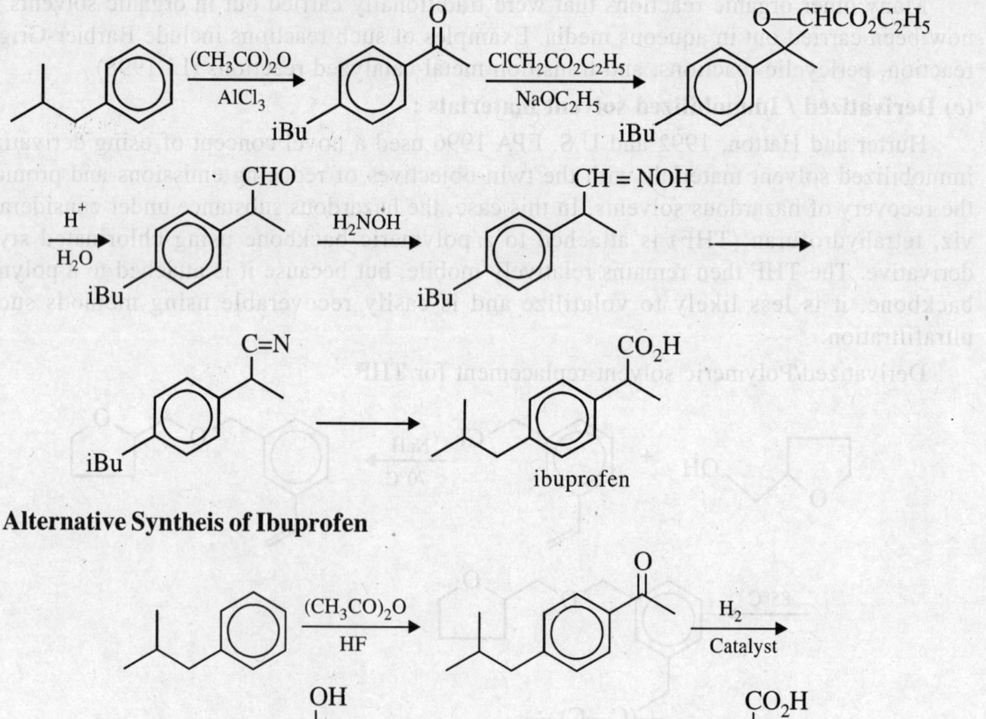

Alternative Syntheis of Ibuprofen

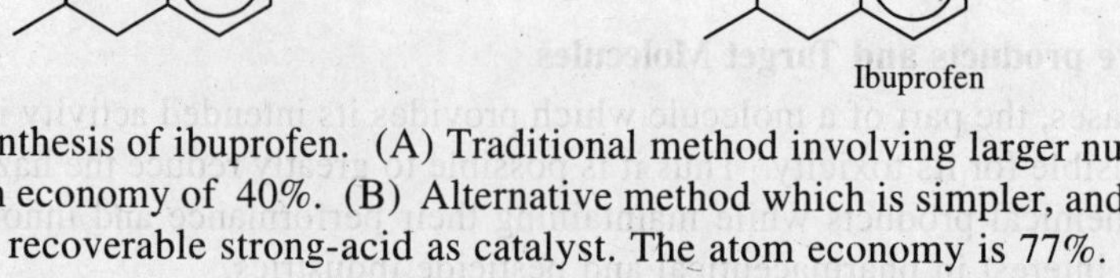

Fig. 23.8. Synthesis of ibuprofen. (A) Traditional method involving larger number of steps with atom economy of 40%. (B) Alternative method which is simpler, and employs recoverable strong-acid as catalyst. The atom economy is 77%.

23.6. FUNCTIONAL GROUP APPROACHES TO GREEN CHEMISTRY

There are several tools in the design of more environmentally benign chemistry, which include the following :

23.6.1. Structure - Activity Relationship

Today, about 15,000 chemicals are commercially produced and every year, a thousand or more new chemicals are developed. Every chemical in use may possess a number of potential risks to human health and the environment. It is difficult to evaluate all possible environmental impacts rigorously and precisely. However, preliminary screening of the potential environmental impacts of chemicals is possible. It is also essential to identify potential risk reduction opportunities with the limited amount of available information about the chemical structure. Table 23.1 identifies the chemical and physical properties that may influence each of the processes that determine environmental exposure.

Table 23.1 : Physical and chemical properties required to perform Environmental Risk Screenings.

S.No.	Environmental process	Relevant chemical and physical properties
1.	Toxicity and health effects	Dose-response relationship
2.	Persistence in the environment	Atmospheric oxidation rate, aqueous hydrolysis rate, rate of microbial degradation, rate of photolysis, adsorption, etc.
3.	Uptake by humans	Transport across dermal layers, rate of transport across lung membranes, degradation rate within human body.
4.	Uptake by organisms	Volatility, lipophilicity, degradation rate in organisms, molecular size, etc.
5.	Dispersion and fate	Density, volatility, melting and boiling point, water solubility, effectivensss of waste water treatment, etc.

Melting point, boiling point, vapour pressure and solubility describe the partitioning of the chemical between solid, liquid and gaseous phases. Additional molecular properties used in assessment of environmental fate of chemicals include octanol-water partition coefficient, soil sorption coefficients, Henry's Law constants, bioconcentration factors. Once the basic physical and chemical properties are defined, the properties that influence the fate of chemicals in the environment can be estimated. These include estimates of the rates at which chemicals will react in the atmosphere, the rates of reaction in aqueous environments and the rate at which the compounds get metabolized by organisms. If we are able to estimate the environmental concentrations based on release rates and environmental fate properties, then the human exposures to the chemicals can be estimated. Finally, if exposures and hazards are known, the risk to humans and the environment can be estimated.

Each of the above properties can be estimated on the basis of the structure of the chemical under consideration. These methods are based on the reasonable presumption that a molecule is composed of a set of functional groups or molecular fragments and that each of these fragments contributes in a well-defined manner to the properties of the molecule. These methods are therefore known as group contribution methods, structure activity relationships (SARs), or Quantitative Structure Activity Relationship (QSAR).

Although the precise mechanism of action of a chemical may not be known, many a times, the structure activity relationships can be used to identify structural modifications that may improve the safety of the chemical under consideration. For example, if we know that the methyl-substituted analog of a substance has a very high toxicity and that the toxicity decreases as the substitution moves from ethyl to propyl, then we can try to reduce the toxicity of a chemical by increasing the alkyl chain length. Thus, structure-activity relationship can be used as a powerful tool to design safer chemicals.

23.6.2. Elimination of Toxic Functional Group

Chemicals of any particular class are defined by their structural features, such as alcohols, aldehydes, ketones, nitriles or isocyanate functional groups. In case that the toxicity of any chemical or the mechanism by which it generates toxicity are not known, one can reasonably assume that certain reactive functional groups will react similarly in the body or in the environment. Thus if the toxicity regarding other compounds in a chemical class is known, one may suspect toxic effect of any chemical having the same functional group. Thus, we can design a safer chemical by removing the toxic functionality (*i.e.* the functional group that is suspected to produce toxicity). In cases where the functionality of toxicity is the same as the functionality of the desired activity (*i.e.* if the functional group responsible to cause toxicity and the functional group that provides the necessary properties required for a specific application are the same), then, we can mask the functional group to a non-toxic derivative form. The parent functionality can be released or regenerated as and when necessary.

This technique can be best illustrated by the following example. The vinyl sulfone functionality is highly electrophilic and reacts with cellulosic fibers, making it an effective component of dyes. However, several toxic effects are attributed to this functionality. The sulfones can be made safer by masking the functional group by synthesizing a relatively less hazardous hydroxyethylsulfone. This will make the manufacturing, storing and transporting the chemical safer. The hydroxy ethyl sulfone can be readily converted to vinyl sulfone as and when required.

$$R-\underset{\underset{O}{\|}}{\overset{\overset{O}{\|}}{S}}-CH_2CH_2OH \xrightarrow{H_2SO_4} R-\underset{\underset{O}{\|}}{\overset{\overset{O}{\|}}{S}}-CH_2CH_2OSO_3 \xrightarrow{\text{Strong base}} R-\underset{\underset{O}{\|}}{\overset{\overset{O}{\|}}{S}}-CH=CH_2$$

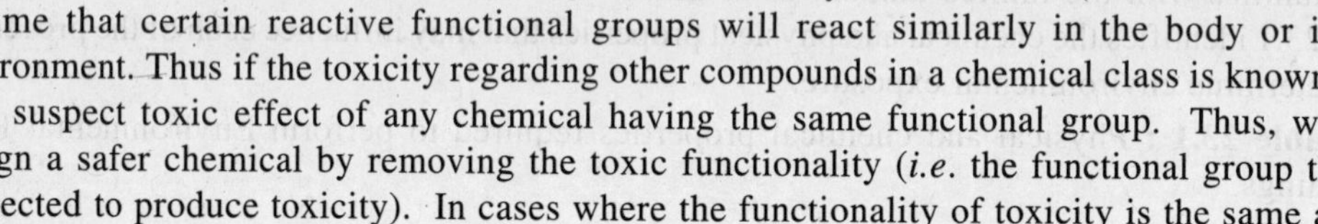

Hydroxy ethyl sulfone　　　　　　Not Reactive　　　　　　Vinyl Sulfone
　　　　　　　　　　　　　　　　　　　　　　　　　　　Highly Reactive

Fig. 23.9. Masking of highly reactive and hazardous vinyl sulfone as safer hydroxy ethyl sulfone. The latter can be converted to vinyl sulfone as and when required.

23.6.3. Reducing the bioavailability

If a potentially toxic substance cannot reach the target of toxicity due to some features of structural design, then for all practical purposes, it may be considered as innocuous. If we do not know what structural features of the molecule have to be modified to render it less hazardous, still we can exercise the option to minimize its bioavailability. This can be achieved by manipulation of the water solubility / lipophilicity relationships that govern the ability of a substance to pass through biological membranes such as skin, lungs or the gastro-intestinal tract.

The same principle also applies to designing safer chemicals for the environment such as ozone depleting substances (ODS). For any chemical to exhibit a considerable ozone depleting potential, it must satisfy two basic conditions : (1) It should be able to reach the necessary altitudes, and (2) It should have adequate life time in those altitudes so as to cause the damage. Keeping this in view, several substances are now being designed which possess the same useful properties, for their applications as those of ozone depleting substances but at the same time are not available to the target of hazard, viz. stratospheric ozone layer.

23.6.4 Designing for Innocuous Fate

It is generally desirable to design substances that are robust and long lasting but at the same time they should not be toxic and bioaccumulating. They should be designed to degrade after their useful life. They should not persist in the environment or in a landfill for ever. Therefore, safer chemicals have to be designed not only from the point of manufacture and use of the chemicals but also keeping in view of their disposal and ultimate fate at the end of their life cycle.

23.7. OPTIMIZATION OF FRAMEWORKS FOR THE DESIGN OF GREENER SYNTHETIC PATHWAYS

In order to design green chemical synthesis pathways, identifying the alternatives is not only vital but also challenging. Combinatiorial approaches are generally used for this purpose. The following steps are usually followed :

(a) Select a set of molecular or functional group building blocks from which the target molecule can be constructed.

The following guidelines are suggested by Buxton, et al (1997) for selecting a group of starting materials :

(i) The building blocks should include the groups that are present in the product.

(ii) Groups present in any available industrial raw materials or byproducts or coproducts may be included.

(iii) Such groups which provide the basic building blocks for the functionalities of the product or of similar functionalities may be included.

(iv) Sets of groups associated with the general chemical pathway employed (e.g., cyclic, acyclic or aromatic) may be selected.

(v) Groups that violate the property restrictions should be rejected.

The functional group building blocks thus selected can be used to identify a set of potential molecular reactants. This exercise will generate a very large number of potential reactants. To restrict this number of potential reactants, constraints based on chemical principles and intution are generally imposed. For example, if the product contains aromatic monosubstituted group, it is better to start with a monosubstituted aromatic molecule. Similarly, such reactants for which, the carbon skeleton will have to be altered (by ring condensation, etc.) to get the product should be avoided as far as possible. By using such criteria and assumptions, one can arrive at a limited set of reactants for further consideration.

(b) Having selected a limited set of potential reactants, a set of stoichiometric constraints have to be applied to describe how the reactants can interact to form the desired molecules. For instance, if the desired product molecule contains 7 aromatic carbons bound to hydrogen and 2 aromatic carbons bound to other aromatic carbons, the reactants that we select must provide sufficient number of aromatic carbons of various types, to synthesize the desired product molecule. Simultaneously, thermodynamic constraints too have to be identified for selectivity. These constraints help to reduce the number of possibilities that might be considered further.

(c) Finally, a set of criteria have to be used to identify the reaction pathways followed by economic and environmental ranking.

Such approaches have helped in designing alternative synthetic procedures which are relatively safer as compared to the traditional methods. Synthesis of carbaryl provides an example.

The Bhopal tragedy occurred on Dec. 3, 1984 was attributed to the catastrophic release of methyl isocyanate. The Union Carbide was manufacturing a pesticide called carbaryl (1-naphthyl-methyl carbamate) by the traditional method using 1-naphthol and methylisocyanate as reactants. Now, an alternative reaction pathway is suggested to minimize the use of hazardous materials. In this method, 1-naphthol and phosgene are used as reactants to produce 1-naptholenyl chloroformate, which when treated with methyl amine yields carbaryl. Although the reactants used are same, this alternative process eliminates the formation of methyl isocyanate, which is highly hazardous. Further insights and innovations in synthetic organic chemistry are needed to avoid the use of phosgene which is also toxic. The traditional and alternative synthesis pathways are illustrated in Fig. 23.10.

Traditional Synthesis of Carbaryl

$$CH_3NH_2 + COCl_2 \longrightarrow CH_3-N=C=O + 2HCl$$ 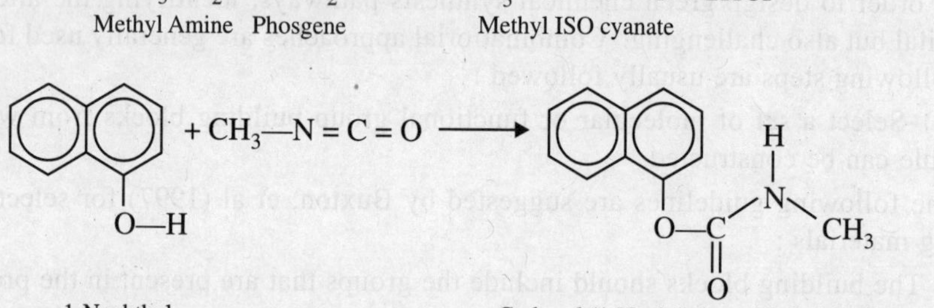
Methyl Amine Phosgene Methyl ISO cyanate

[1-Naphthol + CH₃—N=C=O → Carbaryl (1-Naphthaleny, Methyl Carbonate)]

Alternative Synthesis of Carbaryl

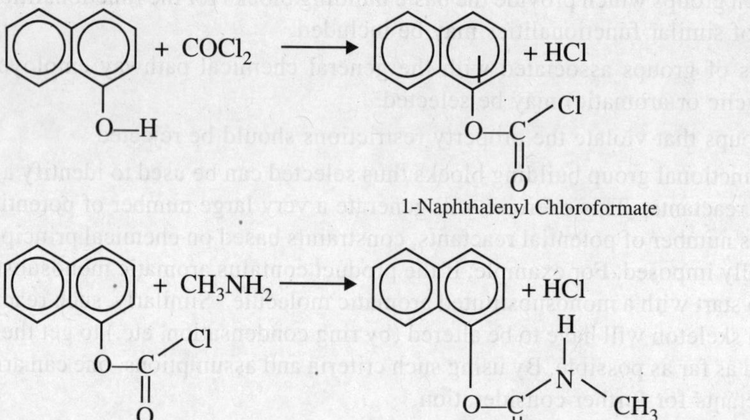

[1-Naphthol + COCl₂ → 1-Naphthalenyl Chloroformate + HCl]

[1-Naphthalenyl Chloroformate + CH₃NH₂ → Carbaryl + HCl]

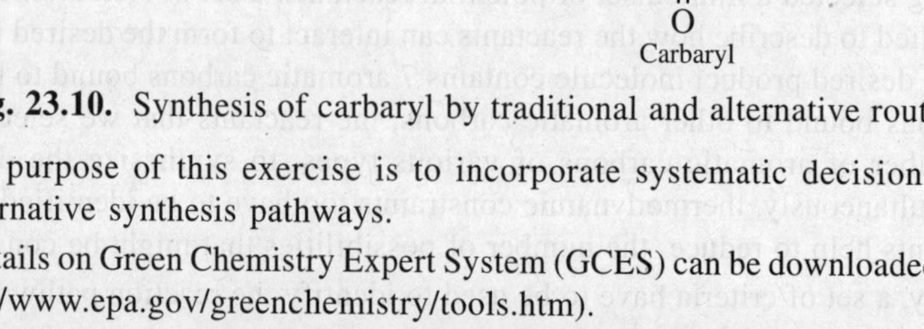

Fig. 23.10. Synthesis of carbaryl by traditional and alternative routes.

The basic purpose of this exercise is to incorporate systematic decision rules into the search for alternative synthesis pathways.

Further details on Green Chemistry Expert System (GCES) can be downloaded from US EPA website (http://www.epa.gov/greenchemistry/tools.htm).

23.8 INDUSTRIAL APPLICATIONS OF GREEN CHEMISTRY

Nature produces globally around 170 billion tonnes of plant biomass annually in the form of wood, cereals and other crops. However, not more than 4% of this is exploited by the economy. As per the predictions of the U.S. National Research Council, the market share of renewable resources will reach 25% by the year 2020.

Products from natural materials

The oleochemicals branch of Cognis Company (U.S.A.) offers complete range of chemical products produced from natural materials to its customers all over the world. Using palms, coconuts, palm seeds, soy oils etc., as starting materials, the company produces a variety of products such as fatty acids, their methyl esters, glycerol and long-chain alcohols. These products, in turn, are used as starting materials for various items used in bodycare, pharmaceutical, and food industries.

Palm oils can now compete with mineral oils even in the production of petrol and natural gas. The new drilling fluid based on plant-derived esters of palm oil is an example. The resulting rock powder can be returned to the seas without any concerns.

The biotechnological process for the production of adipic acid, which is the precursor of nylon, from glucose is already described earlier in this chapter.

The Bioactive polymer systems (Biopos) at Teltow is developing one of the first biorefineries in Germany. Their initial target is to couple this refinery to an existing drying facility for green materials, where the biomass is separated into a liquid extract and solid residue. This solid residue is processed to produce animal feed, composite fibre materials and a starting material for producing lerulinic acid. The liquid extract is used for producing proteins and fermentation media.

The Cargill-Dow Company (U.S.) is producing 1,40,000 tonnes of eco-plastics annually from corn waste in its biorefinery in Nebraska. These eco-plastics can be used for routine packaging like foil or plastic pots and even T-shirts. For this purpose, the starch is degraded to a glucose syrup which is eventually converted to lactic acid. This is polymerised from which the thread is spun.

Green Solvents : Supercritical media are fast emerging as promising alternatives for organic solvents in various processes. The initial industrial application was the removal of caffeine from coffee and tea. Encouraged by this technology other applications such as extraction of other natural products, decaffeination via hop extraction, fat reduction in cocoa, extraction of peanut oil, etc. are also being actively pursued. Supercritical fluids are also being used in the production of well-defined polymer particles. For example, polyamides used in paints and varnishes for automobile industry must be possessing precise characteristics such as particle diameter. If the particles are too fine, the spray painting requires special efforts to exude dust. On the other hand, if the particles are large, they will appear as defects in the finished lacquer. The traditional procedure followed for achieving well- defined distributions of grain sizes is the repeated sieving and sorting, which is expensive and cumbersome. However, such a narrow distribution of grain size can be achieved with supercritical fluid media by rapidly changing the solubility by suddenly relaxing the system (*i.e.* by reducing the pressure to below critical point) to end supercriticality, which leads to precipitation of the solvated particles. This is because the solubility of most solutes in supercritical fluids dramatically change around the critical point and hence by just reducing the pressure below the critical point, desired materials can be recovered.

Dupont has developed a process for producing Teflon using supercritical CO_2 instead of trichlorotrifluoroethane.

Supercritical fluids are also finding application in nano-electronics too. In nano-electronics, supercritical CO_2 can replace ultra-high purity water because it does not affect nanometric structures. Moreover, the controllable evaporation rates observed when supercritical CO_2 is gradually moved out of supercriticality render it possible to obtain extremely thin homogeneous films of photo varnishes. These coating methods are found to be superior to the traditional methods.

Ionic liquids : Ionic liquids are coming to be used as "green solvents". Ionic solvents have the advantage of low vapour pressures. Researchers at the university of Carolina are studying the use of chloroaluminate compounds for electrophilic alkylation, acylation and cationic polymerisation.

The first commercial process based on ionic liquids has been developed by Ludwig Maase and coworkers at BASF, Ludwigshaten, Germany for the production of alkoxyphenylphosphine, which is a precursor for photoinitiators. This process has the advantage that the ionic liquid is generated in the course of the reaction itself, and it absorbs HCl, which is an unwanted byproduct. The product is claimed to give higher yields and it has the advantage of recycling and hence is cost-effective.

Green fuels and E-Green propellants : The use of Biodiesel as an environmentally friendly fuel is already established. Now, the space industry is also looking for alternative environment friendly fuels known as "Green propellants". The challenges are to replace the present engines with those that can work with environmentally benign fuels, and to understand the relevant parameters of the selected green propellants.

Among the alternative fuels under consideration, hydrogen peroxide (H_2O_2) is the current favourite. It decomposes without producing any harmful substances and it also qualifies as the "green propellant" from the point of view of energy balance. Some companies have already developed the expertise to produce ultrapure H_2O_2, which is suitable for this application. Green propellants and green rocket engines offer great promise and potential to replace the existing ones in near future.

Zeolites : Zeolites are crystalline inorganic polymers made of aluminosilicates and have tetrahedral structures of the type XO_4 in which each X atom is linked by shared oxygen ions. X may be Al, B, Ga, Ge, Si or P.

Zeolites have been extensively used in heterogeneous acid-catalysis involving hydrocarbon transformation. Their high acidity is derived from protons that are required to maintain electrical neutrality as shown below :

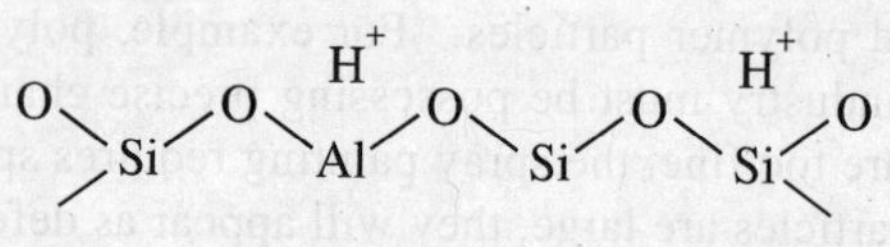

Fig 23.11. Super acid sites in zeolites used as solid acid catalysts.

Zeolites have dramatically improved the performance of fluidized catalytic cracking unit. For example, in the alkylation of butene, large amounts of corrosive acids like HF and H_2SO_4 are used conventionally. In the alternative greener process, the corrosive acids are replaced by *Solid Acid Catalysts* such as Zeolites.

Base-catalyzed and bifunctional (acid / base) applications of zeolites have been reported for manufacturing bulk chemicals. Base catalysis may play major role in the synthesis of fine and speciality chemicals. For example, in the synthesis of 4-methyl thiazole (a systemic fungicide), the basic sites of the caesium zeolite catalyst were found to work very efficiently without using Cl_2, CS_2 or NaOH.

Chemical modifications of zeolite allow selectivity based on shape. Thus, these catalysts offer promising and versatile technology for the future.

Biocatalysis : Biocatalysis mainly includes enzyme catalysis, antibody catalysis and biomimetic catalysis. Among them, enzyme catalysis has been more widely studied. Enzyme catalysts are no longer confined to biochemical reactions only. They are having impact on complex organic synthesis of bioactive molecules and in biotechnology in general. Enzyme catalysts exhibit chemoselectivity, regioselectivity and sterioselectivity. Enzyme catalysis is applicable not only in aqueous media but also in non-aqueous solvents, including supercritical fluids. However, enzymes require water to function as catalysts. Usually, a small amount of water, corresponding to a monolayer of the enzyme molecule is sufficient.

Enzymes having broad substrate selectivity, while still retaining other features of selectivity can prove to be powerful and versatile catalysts. The remarkable properties of enzyme catalysts and their selectivity make them desirable in many synthetic organic pathways. Synthesis of acrylamide is an example :

Conventional method

$$CH_2=CH-CN + H_2O \xrightarrow[2.NH_3]{1.H_2SO_4} CH_2=CH.C(=O)NH_2 + (NH_4)_2SO_4$$
$$\text{acryl amide}$$

Alternative method

$$CH_2=CH-CN \xrightarrow[\text{hydritase}]{\text{nitril}} CH_2=CH.C(=O).NH_2$$
$$\text{acryl amide}$$

In the year 1981, Degussa developed the concept of an enzyme membrane reactor. It is a milestone in the development of modern biocatalysis because it combined the advantages of both enzyme reactor and homogeneous catalysis.

The "projektaus katalyse" centre established by Degussa has been working on the entire spectrum of new catalyst development. In an effort to discover enzymatic methods to the existing synthetic pathways, scientists of that centre developed several methods. One such example is the synthesis of 1-tert-lencine (1-Tle). This amino acid does not occur in nature but serves as a chiral auxiliary compound in asymmetric synthesis and as a building block for synthesis of drugs against cancer, inflammation and viral infection. The redox enzyme leucin dehydrogenase reduces a keto-carbonic acid precursor to form the 1-Tle. The problem is that the natural cofactor NADH is required as a hydrogen donor in stoichiometric amounts, and the reactor does not include the NADH regeneration system that is in place in the cell. The high cost of such cofactors came in the way of its industrial application so far. However, Maria-Regina Kula at the Institute for Enzyme Technology of the university of Dusseldorf was able to find an enzyme which can generate the precious redox cofactors during the reaction. She isolated the enzyme formate dehydrogenase from the yeast "candida boidinii" and discovered that it is suitable for a large number of redox systems requiring NADH as a cofactor.

Another important enzymatic method that can compete with a chemical process is the production of L-Dopa; this drug is used for the treatment of parkinson's disease. In this process, catechol, pyruvate and ammonia are allowed to react in presence of the enzyme L-Tyrosine phenol lyase. Dehydroserine is formed as an intermediate during the course of a reversible α, β–elimination, which is subsequently converted to L-Dopa by stereoselective addition of catechol. The enzyme accepts catechol as a mimic of its natural substrate phenol. The process is now used for industrial production by Ajinomoto Company, Japan.

Biocatalysis, by virtue of its merits such as selectivity, sustainability and environmental safety, holds a strong promise to discover new and novel production pathways for very useful chemical compounds.

23.9 CONCLUSION

There have been several occasions in the history of chemistry when the chemistry community focussed on goals (such as synthesizing natural products), developing different reaction types (such a stereospecific reactions) to specific applications (such as anti-cancer agents). Now, it is the turn of green chemistry towards sustainable development via pollution prevention and resource conservation. Goals merely provide an objective on which energy, knowledge and resources are focussed. More often than not, pursuit of a goal provides the avenues for several other accomplishments. The same is expected to be true for green chemistry as well. Every project on green chemistry may not achieve its goal of innocuous feedstocks or reagent or benign conditions or products, but striving for this sublime goal will certainly result in excellent chemistry. What else can be more satisfying for the chemistry community as a whole ?

QUESTIONS

1. What is green chemistry and what are its principles ?
2. What are the goals of green chemistry perspective ? What are the various approaches that lead towards those goals ?
3. What are the basic components of green chemistry research ? Discuss with suitable examples.
4. Discuss the significance of green chemistry towards cleaner production and sustainable development. Give suitable examples.
5. Write informative notes on any two of the following from the point of view of green chemistry approach :
   (a) Supercritical carbon dioxide
   (b) Ionic liquids
   (c) Atom economy
   (d) Structure-activity relationship.
6. Discuss with the help of suitable examples the various Functional Group Approaches to Green Chemistry.
7. Write short notes on any three of the following :
   (a) Optimization of frameworks for the design of greener synthetic pathways
   (b) Green solvents
   (c) Solid acid catalysts
   (d) Biocatalysis.
8. Discuss the various applications of green chemistry for achieving sustainable development.

24

Mechanism of Organic Reactions

24.1. INTRODUCTION

A chemical equation is a symbolic representation of a chemical change. It indicates the initial reactants and final products involved in a chemical change. Reactants generally consist of two species :

(i) One which is being attached; it is called a substrate.

(ii) The other which attacks the substrate; it is referred to as a reagent. These two interact to form products.

$$\text{Substrate} + \text{Reagent} \rightarrow \text{Products}$$

It is important for us to know not only what happens in a chemical change but also how it happens. Most of the reactions are complex and take place via intermediates which may or may not be isolated. The intermediates are generally very reactive. They react readily with other species present in the environment to form the final products.

The detailed step by step description of a chemical reaction is called its mechanism.

$$\text{Substrate} \rightarrow \text{Intermediates} \rightarrow \text{Products}$$
$$\text{(Transitory)}$$

It would be seen that most of the attacking reagents bear either a positive or a negative charge. Naturally these would not attack the substrate successfully unless the latter somehow possessed oppositely charged centres in the molecule. In other words, the substrate molecule although as a whole electrically neutral must develop polarity on some of its carbon atoms and substituents linked together.

This is made possible by the displacement of the bonding electrons (partially or completely) resulting in the development of polarity in the molecule. Such changes or effects involving the displacement of electrons in the substrate molecule are often referred to as Electron displacement effects. These displacement effects are of great significance in understanding reaction mechanisms.

24.2. ELECTRON DISPLCEMENT EFFECTS

There are four types of electron displacement mechanisms frequently observed in organic molecules.

(1) Inductive effect
(2) Electromeric effect
(3) Mesomeric effect
(4) Hyperconjugative effect

1. Inductive effect

A covalent bond is formed on account of the equal contribution of electrons by two atoms. In case two atoms are similar, the electron pair occupies a central position between the two nuclei of the atoms and such a bond is known as non-polar covalent bond. For example, the bond between H_2 molecule (H : H), chlorine molecule (Cl : Cl), Carbon and Carbon —C—C— (—C—C—) are non-polar covalent bonds as the influences of the electron pair is same on both the atoms.

In a covalent bond between two dissimilar atoms (having different electronegativities), the electron pair does not remain in the centre but is attracted towards the more electronegative atoms. The bond becomes somewhat polar due to unequal sharing of the electron pair. For example, in the bond —C—X, if X is more electronegative than C, the electron pair is attracted towards X. With the result of this shifting, X acquires a partial negative charge denoted by δ^- and C attain a partial positive charge denoted by δ^+.

—C—X or —C : X or $—\overset{\delta+}{C}—\overset{\delta-}{X}$ or —C→X

The polarity thus produced in the molecule as a result of higher electron activity of one atom compared to another is termed as "inductive effect".

It is important to note that the electron pair, although permanently displaced, remains in the same valency shell. This inductive effect is always transmitted along a chain of carbon atoms. However, its intensity decreases as distance from the source atoms increases.

$$—C_4 \rightarrow —C_3 \rightarrow —C_2 \rightarrow —C_1^{\delta+} \rightarrow —Cl^{\delta-}$$

The positive charge on C_1 attracts the electron pair between C_1 and C_2. This will cause C_2 to acquire a small positive charge but this charge on C_2 will be smaller than on C_1. Similarly, C_3 will acquire positive charge that will be still smaller. This effect is still relayed further. In fact, inductive effect tends to be insignificant beyond the second carbon atom. It is a permanent effect in the molecule and can be observed practically in the form of dipole moment. This type of electron displacement along a carbon chain due to the presence is a source is called inductive effect or transmission effect.

The carbon-hydrogen bond is used as a standard for comparing the tendency of electron attraction and repulsion. On this basis inductive effect is classified into two classes.

1. Positive Inductive effect (+ I effect) : When the atom or group which has less power to attract the electron than hydrogen it is said to have positive inductive effect.

+ I effect : Group (Electron-releasing or repelling)

$(CH_3)_3 C > (CH_3)_2 CH > CH_3 CH_2 > CH_3 > H$

2. Negative Inductive effect (− I effect) : When the atom or group which has more power to attract electron in comparison to hydrogen, it is said to have − I effect.

− I Effect : Atoms or Groups (Electron-attracting or withdrawing)

$NO_2 > CN > COOH > F > Cl > Br > I > OR > OH > C_6H_5 > H$

Important Features of Inductive effect

(i) It is a permanent effect in the molecule or ion.

(ii) The shared pair of electrons although permanently shifted towards more electronegative atom, yet remains in the same valence shell.

MECHANISM OF ORGANIC REACTIONS

(*iii*) As a result of electron shifting, the more electronegative end acquires partial negative charge and the other acquires partial positive charge.

(*iv*) The inductive effect is not confined to the polarization of one bond but is transmitted along a chain of carbon atoms through σ-bonds. However, the effect is insignificant beyond second carbon in the chain.

(*v*) Inductive effect brings changes in physical properties such as dipole moment, solubility, etc. It affects the rate of the reaction.

(*vi*) Carbon-hydrogen bond is taken as a standard of inductive effect. Zero effect is assumed for this bond. Atoms or groups which have a greater electron-withdrawing capacity than hydrogen are said to have –I effect whereas atoms or groups which have a greater electron releasing power are said to have +I effect.

Applications of Inductive Effect

The phenomenon of inductive effect is very important in organic chemistry as it is helpful in explaining a number of facts. Some of them are discussed below.

(i) Reactivity of Alkyl halides

The presence of halogen atom in the molecule of alkyl halide creates a centre of low electron density which is readily attacked by the negatively charged reagents. For example CH_3Cl is more reactive than CH_4 as inductive effect is present in CH_3Cl and it is not present in methane.

$$H_3C \rightarrow Cl \quad \text{or} \quad H_3C^{\delta+} - Cl^{\delta-}$$
$$\uparrow$$
site for the attack by electron–rich reagent.

The activity also increases from primary to secondary and from secondary to tertiary halides as the +I effect of methyl groups enhances –I effect of the halogen atom.

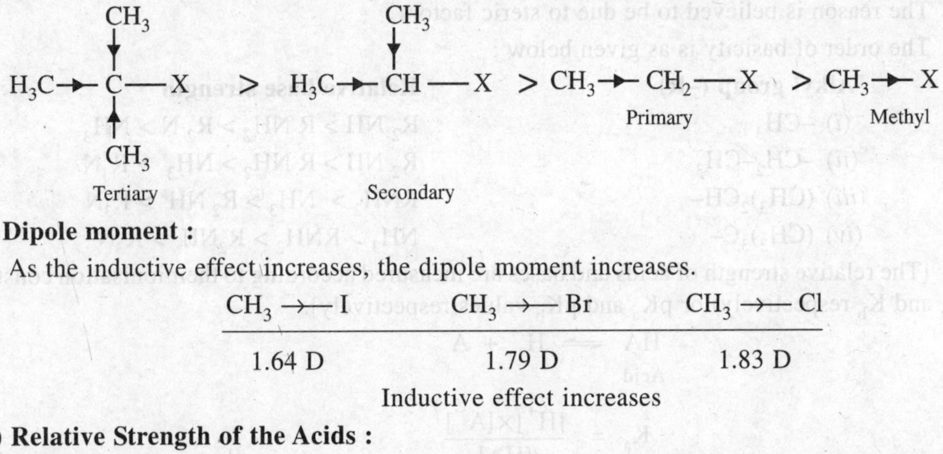

$$\underset{\text{Tertiary}}{\overset{CH_3}{\underset{CH_3}{H_3C \rightarrow \overset{|}{\underset{|}{C}} - X}}} > \underset{\text{Secondary}}{\overset{CH_3}{H_3C \rightarrow \overset{|}{CH} - X}} > \underset{\text{Primary}}{CH_3 \rightarrow CH_2 - X} > \underset{\text{Methyl}}{CH_3 \rightarrow X}$$

(ii) Dipole moment :

As the inductive effect increases, the dipole moment increases.

$CH_3 \rightarrow I$	$CH_3 \rightarrow Br$	$CH_3 \rightarrow Cl$
1.64 D	1.79 D	1.83 D

Inductive effect increases

(iii) Relative Strength of the Acids :

Strength of the acids depends on the tendency to release proton when the acid is dissolved in water. Any group or atom which helps in the release of proton increases the strength of the acid. Thus, the group or atom having –I effect increases the strength as it decreases the negative charge on the carboxylate ion.

$$CH_3COOH < I\,CH_2COOH < BrCH_2COOH < ClCH_2COOH < Cl_2CHCOOH < Cl_3CCOOH$$

–I, effect increases, so acid strength increases

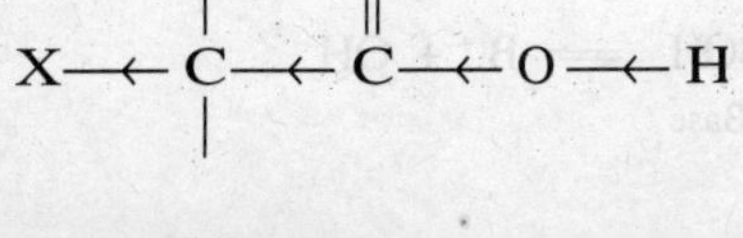

$$X \leftarrow \underset{|}{\overset{|}{C}} \leftarrow \overset{O}{\overset{\|}{C}} \leftarrow O \leftarrow H$$

Any group or atom showing +I effect decreases the acid strength as it increases the negative charge on the carboxylate ion which holds the hydrogen firmly.

$$HCOOH > CH_3COOH > C_2H_5COOH > C_3H_7COOH > C_4H_7COOH$$

+I, effect increases, so acid strength decreases

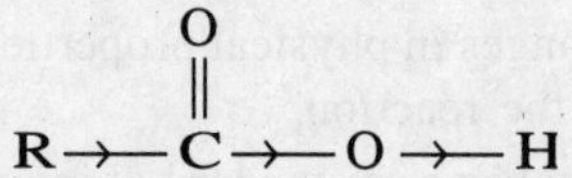

$$R \rightarrow C(=O) \rightarrow O \rightarrow H$$

(iv) Strength of Bases :

Base strength is defined as the tendency to donate an electron pair for sharing. The difference in base strength in various amines can be explained on the basis of inductive effect. The +I effect increases the electron density while −I effect decreases it. The amines* are stronger bases than NH_3 as the alkyl groups increase electron density on nitrogen due to +I effect while $Cl\,NH_2$ is less basic due to −I effect. So, greater the tendency to donate electron pair for coordination with proton, greater is the basic nature, i.e. greater the negative charge on nitrogen atom (due to +I effect of alkyl group), greater will be the basic nature.

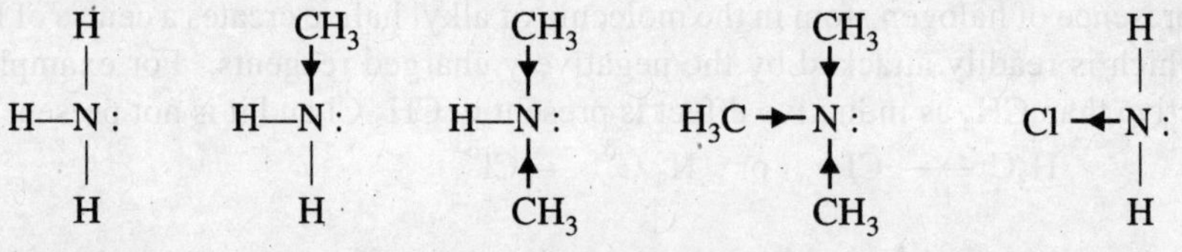

* It is important to note that the relative basic character of amines is not in total accordance with inductive effect (T > S > P) but it is in the following order.

Secondary > primary > tertiary

The reason is believed to be due to steric factors.

The order of basicity is as given below :

Alkyl group (–R)	Relative base strength
(i) –CH_3	$R_2NH > RNH_2 > R_2N > NH_3$
(ii) –CH_2–CH_3	$R_2NH > RNH_2 > NH_3 > R_3N$
(iii) $(CH_3)_2CH$–	$RNH_2 > NH_3 > R_2NH > R_3N$
(iv) $(CH_3)_3C$–	$NH_3 > RNH_2 > R_2NH > R_3N$

[The relative strength of acids and bases are measured according to their ionisation constants (K_a and K_b respectively) or pK_a and pK_b values respectively].

$$HA \rightleftharpoons H^+ + A^-$$
Acid

$$K_a = \frac{[H^+] \times [A^-]}{[HA]}$$

$$pK_a = -\log K_a$$

Greater the value of K_a or lower the value of pK_a, stronger will be the acid.

Acid	pK$_a$	Acid	pK$_a$
HCOOH	3.77	NO_2CH_2COOH	1.68
CH_3COOH	4.76	$CH\,CH_2\,COOH$	2.47
CH_3CH_2COOH	4.88	$ClCH_2COOH$	2.86
C_6H_5COOH	4.17	$Cl_2CHCOOH$	1.29

$$BOH \rightleftharpoons B^+ + OH^-$$
Base

$$K_b = \frac{[B^+][OH^-]}{[BOH]}$$

$$pK_b = -\log K_b$$

Greater the value of K_b or lower the value of pK_b, stronger will be the base.

Amine	$(CH_3)_2 NH$	$CH_3 NH_2$	$(CH_3)_3 N$
pK_b	3.23	3.32	4.2

2. Electromeric effect (Electro = electron; meros = part)

It is a temporary effect involving the complete transfer of a shared pair of electrons to one or other atom joined by a multiple bond such as a double or triple bond.

A multiple bond (double or triple) consists of Sigma and π-bonds. The electrons of the π-bond are loosely held and easily polarisable. When a compound having a multiple bond is approached by a charged reagent, the π-electrons of the bond are completely polarised i.e. shifted towards one of the constituent atom. Thus,

$$>C=C< \xrightarrow{\text{Polar reagent}} >\overset{+}{C}-\overset{=}{C}<$$

The atom which acquires the electron pair becomes negatively charged while the other atom gets the positive charge.

"The effect involving the complete transfer of a shared pair of electrons to one of the atoms joined by a multiple bond (double or triple) at the requirement of attacking reagent is known as electromeric effect."

It is indicated by E and represented by a curved arrow showing the shifting of electron pair.

Direction of the shift of electron pair

The direction of the shift of electron pair can be decided on the basis of the following points.

(i) When the groups linked to a multiple bond are similar, the shift can occur to either direction. For example, in ethylene the shift can occur to any one of the carbon atoms.

$$H_2C = CH_2 \longrightarrow H_2\overset{+}{C}-\overset{=}{C}H_2$$
$$H_2C = CH_2 \longrightarrow H_2\overset{=}{C}-\overset{+}{C}H_2$$

Both are similar

(ii) When the dissimilar groups are linked on the two ends of the double bonds, the shift is decided by the direction of inductive effect. For example, in propylene the shift can be shown in the following ways :

$$CH_3-CH=CH_2 \longrightarrow CH_3-\overset{+}{C}H-\overset{=}{C}H_2 \qquad ...(i)$$

$$CH_3-CH=CH_2 \longrightarrow CH_3-\overset{=}{C}H-\overset{+}{C}H_2 \qquad ...(ii)$$

Due to electron repelling nature of methyl group, the electronic shift occurs as shown in (i) and not as according to (ii) shown above.

In the case of carbonyl group, the shift is always towards oxygen i.e. towards the more electronegative atom.

$$>C=O \longrightarrow \overset{+}{C}-\overset{=}{O}$$

[In cases where inductive effect and electromeric effect simultaneously operate, usually electromeric effect predominates.]

Since the electromeric effect takes place only at the requirements of the attacking reagent, it always facilitates the reaction and never inhibits it. This effect is of common occurence during addition of polar reagents on >C=C< and >C=O bonds.

Types of electromeric effect

Electromeric effects are of two types :

(i) Positive Electromeric (+E) Effect :

When the electrons are displaced away from an atom or group in multiple bond compound, it is known as +E effect.

e.g. $\ddot{X} \curvearrowright C = \overset{\curvearrowright}{C}$ + E effect
(Displacement away from X)

(ii) Negative Electromeric (–E) Effect :

When the electrons displacement is towards an atom or group in multiple bond compound, it is known as – E effect.

e.g. $Y = \overset{\curvearrowleft}{C} — \overset{\curvearrowleft}{C} = C$ – E effect
(Displacement towards Y)

The electromeric effect differs from inductive effect in the following respects.

Electromeric Effect	Inductive Effect
1. Shown by substrate molecule containing double (C=C) or triple (C≡C) bonds.	1. Shown by molecules containing single bonds.
2. Takes place when the molecule is exposed to an attack by an electrophilic reagent.	2. Takes place under the influence of a substituent (electron-attracting or electron pumping) linked to the terminal carbon atom.
3. Polarity caused by complete transfer of an electron-pair to one of the two joined by multiple bond.	3. Polarity caused by the displacement of bonding electron-pair from one atom towards the other.
4. The charge acquired by the atom that gains the electron pair is +1 while the other gets –1.	4. The charge developed by the carbon linked to the substituent is small fractional δ^+ or δ^- according as the substituent is electron attracting or electron pumping.
5. Temporary effect : disappears with the removal of the attacking reagent.	5. Permanent effect : depending on the structure of the substrate molecule.

3. Mesomeric effect

If the single and double bonds are present alternatively in a molecule, it is said to contain double bonds in conjugation and the molecule is called a conjugated molecule. For example butadiene is a conjugated molecule.

$$CH_2 = CH — CH = CH_2$$

In such a compound, the π-electrons are delocalised and polarity in the molecule is developed. The same thing also happens when an atom or group with lone pair of electrons is situated in conjugation with a conjugated double bond.

"The Mesomeric effect refers to the polarity produced in a molecule as a result of interaction between two π-bonds or a π-bond and a lone pair of electrons."

MECHANISM OF ORGANIC REACTIONS

The effect is transmitted in the chain in a similar manner as the inductive effect.

In a carbonyl group (>C=O), the oxygen atom is more electronegative than the carbon atom. The π-electrons of the double bond get displaced towards the oxygen atom. (The shifting of electron is shown by a curved arrow (↷). This will give an ionic structure. The actual structure seems to lie in between the structure (I) and (II) which can be best represented as structure (III) in which π-electrons are drawn perferentially towards oxygen.

$$>C=O \quad\quad C=\ddot{O} \longleftrightarrow C^+-\ddot{\ddot{O}}^- \quad\quad >C\cdots\cdots O$$
$$(I) \quad\quad\quad\quad (II) \quad\quad\quad\quad\quad\quad\quad (III)$$

Now if the carbonyl group is conjugated with C=C type system, the above polarisation is transmitted further through π-electrons.

$$H_3C-CH=CH-CH=O \longleftrightarrow H_3C-CH=CH-\overset{+}{C}H=\ddot{\ddot{O}}^-$$
$$(I)$$

$$H_3C-CH-CH-CH-O \quad\text{or}\quad H_3C-CH-CH=CH-\ddot{\ddot{O}}^-$$
$$(III) \quad\quad\quad\quad\quad\quad\quad\quad\quad (II)$$

The polarisation also occurs when an atom or group having lone pair of electrons is present in conjugation with the double bond (conjugated system) as in bromo benzene.

Like inductive effect, mesomeric effect also classified as follows :

(i) **Positive Mesomeric effect (+M)** : When the transference of electrons is away from the atom or group in a conjugated system it is known as M effect.

$$\ddot{X}-C=C-$$
$$(+M\,\text{effect})$$

+ M effect Groups : –Cl, –Br, –I, NH$_2$, –NR$_2$, –OH, –OCH$_3$

(ii) **Negative Mesomeric effect (–M)** : When the transference of electrons is towards atom or group in a conjugated system are known by –M effect.

$$-C=C-C=Y$$
$$(-M\,\text{effect})$$

– M effect Groups : —NO$_2$, —CN, —C(=O)—

Salient features of Mesomeric Effect

(i) π-electrons and lone pair of electrons are involved. This effect operates through conjugative mechanism.

(*ii*) It is a permanent effect present in the molecule in the ground state.

(*iii*) It affects the physical properties such as dipole moment solubility, etc. Rate of reaction of the substance is also affected.

(*iv*) It is denoted by M. Groups which have the capacity to increase the electron density of the rest of the molecule are said to have +M effect. Such groups possess lone pair of electrons. Groups which decrease the electron density of the rest of the molecule by withdrawing electron pair are said to have –M effect.

Similarities between Inductive and Mesomeric effect :

(*i*) Both are permanent effects.

(*ii*) Both affect the physical properties of the molecule.

(iii) Both can either hinder of facilitate a particular reaction.

(*iv*) Both can affect the rates of reaction, the strength of acids and bases, the reactivity of halides and the substitution of different aromatic species.

The two effects differ in the following aspects :

Inductive Effect	Mesomeric Effect
(1) It is operative in saturated compounds.	(1) It is operative in unsaturated compounds especially having conjugated systems.
(2) It involves electrons of sigma bonds.	(2) It involves electron of π-bonds or lone pair of electrons.
(3) The electron pair is slightly displaced from its position and thus partial charges are developed.	(3) The electron pair is completely transferred and thus full positive and negative charges are developed.
(4) It is transmitted over a quite short distance. The effect becomes negligible after second carbon atom in the chain.	(4) It is transmitted from one end to the other end of the chain provided conjugation is present.

4. Hyperconjugation

Baker and Nathan suggested that alkyl groups with at least one hydrogen atom on the α-carbon atom, attached to an unsaturated carbon atom, are able to release electrons by a mechanism similar to that of the electromeric effect, e.g.

$$-\overset{H}{\underset{}{C}}-C=C \longrightarrow -\overset{H^+}{C}=C-\bar{C}$$

Note that the delocalisation involves α and π bonds orbitals (or π orbitals in case of free radicals); thus it is also known as $\alpha - \pi$ conjugation. This type of electron release due to the presence of the system $H—C—C = C$ is known as hyperconjugation.

$$H-\underset{H}{\overset{H}{C}}-CH=CH_2 \longleftrightarrow H-\underset{H}{\overset{H^+}{C}}-CH=\bar{C}H_2 \longleftrightarrow H^+ \; \underset{H}{\overset{H}{C}}=CH-\bar{C}H_2$$

Hyperconjugation in propene

$$H-\underset{H^+}{\overset{H}{C}}-CH=\bar{C}H_2$$

MECHANISM OF ORGANIC REACTIONS

[Resonance structures of toluene showing hyperconjugation]

Hyperconjugation in toluene

More the number of H-C bonds attached to the unsaturated system more will be the probability of electron release by this mechanism. Thus the electron release by this mechanism will be greater in methyl (possessing three hyper conjugated H-C bonds), less in ethyl (having two such bonds) and iso-propyl (one) and essentially zero in tert-butyl (no hyperconjugated H-C bond) group.

[Structure: H—C(H)(H)—C=C]
Methyl compound
(containing three H—C hyperconjugated bonds)

[Structure: H₃C—C(H)—C=C]
Ethyl compound (containing two H—C hyperconjugated bonds)

[Structure: H₃C—C(H)(CH₃)—C=C]
Isopropyl compound
(containing one H—C hyperconjugated bond)

[Structure: H₃C—C(CH₃)(CH₃)—C=C]
Tert-Butyl compound
(containing no H—C hyperconjugated bond)

It is important to note that although hyperconjugation like inductive effect clauses the release of electrons and thus the two effects reinforces each other in this respect, the magnitude of the two effects changes in opposite disreaction as use pass along a series of alkyl groups.

$$H_3C—, \quad CH_3—CH_2—, \quad (CH_3)_2 CH—, \quad (CH_3)_3 C—$$

(Increasing inductive effect, Decreasing hyperconjugation)

The phenomenon of hyperconjugation can also be applied to group Cl — C — C = C (cf. H — C — C = C) where the effect operates in the reverse direction.

$$Cl — C — C = C \leftrightarrow Cl^- \; C = C — C^+$$

Effects of Hyperconjugation

(i) **Stability of Alkenes :** Hyperconjugation explains the stability of certain olefins over other alkenes. For example, propylene is more stable than ethylene because in propylene there are three H—C hyperconjugated bonds and thus the σ-electrons of C—H bond can delocalise over three different structures.

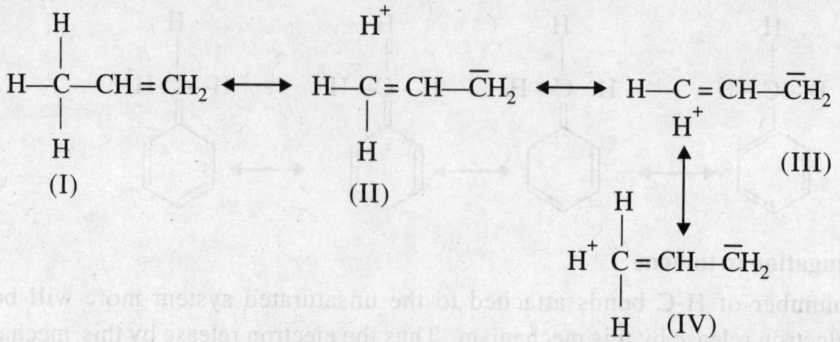

$$\underset{(I)}{\overset{H}{\underset{H}{H-C-CH=CH_2}}} \longleftrightarrow \underset{(II)}{\overset{H^+}{\underset{H}{H-C=CH-\bar{C}H_2}}} \longleftrightarrow \underset{(III)}{H-C=CH-\bar{C}H_2}$$

$$\overset{H^+}{\underset{(IV)}{\underset{H}{H^+\,C=CH-\bar{C}H_2}}}$$

Note that in the resonating structures II, III and IV, there is no definite bond between one of the carbon atoms and one of the hydrogen atoms, hence hyperconjugation is also known as no-bond resonance.

Further greater the number of alkyl groups attached to the doubly-bonded carbons, the more stable the alkene is. Thus 2-Methyl propene and butene-2 are more stable than propylens.

$$\underset{\text{2-Methyl propene}}{\overset{CH_3}{\underset{}{CH_2-C=CH_2}}} \qquad\qquad \underset{\text{Butene-2}}{CH_3-CH=CH-CH_3}$$

(Six H–C hyperconjugated bonds)   (Six H — C hyperconjugated bonds)

(*ii*) **Stability of Alkyl free-radicals :** The concept of hyperconjugation can also be extended to explain the following relative stabilities of alkyl radicals.

$$\text{tert.Alkyl} > \text{sec-Alkyl} > \text{prim-Alkyl} > \text{Methyl radical}$$

e.g., $\qquad (CH)_3\overset{\bullet}{C} > (CH_3)_2\overset{\bullet}{CH} > CH_3-\overset{\bullet}{CH_2} > \overset{\bullet}{CH_3}$

In general, more the number of hyperconjugative structures of a species higher is its stability. Thus ethyl radical may be regarded as a hybrid of the following hyperconjugative structures.

$$\underset{H\;\;H}{\overset{H\;\;H}{H-C-C^\bullet}} \longleftrightarrow \underset{H\;\;H}{\overset{H\;\;H}{H^\bullet\;C=C}} \longleftrightarrow \underset{H\;\;H}{\overset{H^\bullet}{H-C=C-H}} \longleftrightarrow \underset{H^\bullet\;H}{\overset{H\;\;H}{H-C=C}}$$

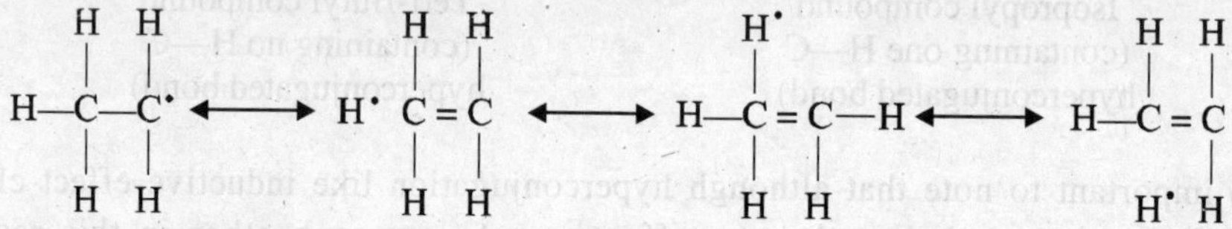

(*iii*) Orienting influence of alkyl group in aronautic systems of o, p–positions and of —CCl_3 group is m–position.

24.3. Reaction Mechanism

By the mechanism of a chemical reaction, we mean the actual series of discrete steps which are involved in the transformation of reactants into products. Now since organic compounds are covalent, organic reactions involve :

(A) The breaking of the old covalent bond, and

(B) Making of the new covalent bond

(A) Breaking of a Covalent Bond :

The first step, *i.e.* breaking of a covalent bond between two atoms can take place mainly in two alternative ways, viz homolytic or heterolytic fission depending upon the relative electronegativity of the two concerned atoms, e.g.

MECHANISM OF ORGANIC REACTIONS

(1) Homolytic fission takes place when the two atoms (say A and B) are usually of similar electronegativity.

$$A:B \longrightarrow \overset{\bullet}{A} + \overset{\bullet}{B}$$
Free radicals

(2) Heterolytic fission takes place when the two atoms (A and B) are of different electronegativities. It may again take place in two different ways.

(a) Where A is more electronegative than B :

$$A:B \longrightarrow A:^{\ominus} + B^{\oplus}$$
Ions

(b) When B is more electronegative than A :

$$A:B \longrightarrow A^{\oplus} + :B^{\ominus}$$
Ions

It is important to note that homolytic fission requires much less energy (e.g. 67.2 Kcal/mole for C—Br bond in H_3C—$^{\bullet}CH_2$—$^{\bullet}Br$ into H_3C—CH_2 amd Br) than the heterolytic fission (e.g. 183 Kcal/mole for C—Br bond in CH_3—CH_2—Br into CH_3—CH_2^+ and Br^-).

(1) **Homolytic fission or Homolysis** : As clear from the above representation, in homolytic bond fission one electron of the bonding pair goes with each of the departing atom or group resulting in two electrically neutral fragments or atoms generally known as free radicals.

e.g. $$H_3C:X \longrightarrow H_3\overset{\bullet}{C} + \overset{\bullet}{X}$$

(i) **Free Radicals** : A free radical may be defined as an atom or group of atoms having an odd or upaired electron. These results on account of homolytic fission of a covalent bond and are denoted by putting a dot (•) against the symbol of atom or group of atoms.

$$\overset{\bullet}{Cl} \quad ; \quad \overset{\bullet}{H} \quad ; \quad H_3\overset{\bullet}{C} \quad ; \quad H_3C-\overset{\overset{\displaystyle H}{|}}{\underset{\underset{\displaystyle H}{|}}{\overset{\bullet}{C}}}$$

Chlorine   Hydrogen   Methyl free   Ethyl free
free radical   free radical   radical   radical

The formation of free radical is initiated by heat, light or catalyst.

(a) $Cl:Cl \xrightarrow[\text{Sunlight}]{\text{Energy}} \overset{\bullet}{Cl} + \overset{\bullet}{Cl}$

(b) $H_3C:H \xrightarrow{102 \text{ Kcal}} H_3\overset{\bullet}{C} + \overset{\bullet}{H}$

(c) $H_3C-CH_2-H \xrightarrow{97 \text{ Kcal}} H_3\overset{\bullet}{C}-CH_2 + \overset{\bullet}{H}$

(d) $\begin{matrix} CH_3 \\ \\ CH_3 \end{matrix}\Big\rangle CH-H \xrightarrow{94 \text{ Kcal}} \begin{matrix} CH_3 \\ \\ CH_3 \end{matrix}\Big\rangle \overset{\bullet}{CH} + \overset{\bullet}{H}$

(e) $\begin{array}{c} CH_3 \\ CH_3 \\ CH_3 \end{array}\!\!>\!C\!-\!H \xrightarrow{91\ Kcal} \begin{array}{c} CH_3 \\ CH_3 \\ CH_3 \end{array}\!\!>\!\overset{\bullet}{C} + \overset{\bullet}{H}$

Characteristics of free radicals :

(i) Free radicals are formed by homolytic bond fission.
(ii) They are short lived reaction intermediates.
(iii) They show paramagnetic character due to presence of unpaired electrons.
(iv) They show the following three types of reactions :

   (a) The mutual combination of free radicals forms neutral molecules.

$$Cl^\bullet + Cl^\bullet \longrightarrow Cl_2 \quad (Cl:Cl)$$
$$^\bullet CH_3 + ^\bullet CH_3 \longrightarrow CH_3-CH_3 \quad (H_3C:CH_3)$$

   (b) The reaction between a free radical and a neutral molecule gives a new radical.

$$CH_4 + Cl^\bullet \longrightarrow {}^\bullet CH_3 + HCl$$

   (c) A free radical can lose a neutral molecule to form a new radical.

$$\underset{\text{Acetate radical}}{CH_3COO^\bullet} \longrightarrow \underset{\text{Methyl radical}}{^\bullet CH_3} + CO_2$$

Relative stabilities of free radicals

The tiertiary alkyl free radicals are most stable and methyl free radical is least stable, i.e. the free radical formed easily has greater stability.

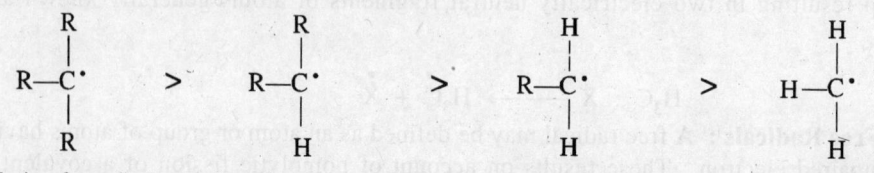

$$\underset{\text{Tertiary free radical}}{R-\underset{R}{\overset{R}{|}}\overset{|}{C^\bullet}} > \underset{\text{Secondary free radical}}{R-\underset{H}{\overset{R}{|}}\overset{|}{C^\bullet}} > \underset{\text{Primary free radical}}{R-\underset{H}{\overset{H}{|}}\overset{|}{C^\bullet}} > \underset{\text{Methyl radical}}{H-\underset{H}{\overset{H}{|}}\overset{|}{C^\bullet}}$$

The stability of carbon (alkyl) free radicals is not influenced by inductive effect because they have no charge (difference from carbonium ions and carbanions). However, they are stabilised by hyperconjugation (no-bond resonance). The relative order of stabilities of some common free radicals is given below :

Triphenylmethyl > benzyl > allyl > tertiary > secondary > primary > methyl > vinyl

Extra stability of the aromatic acid allyl radicals is due to resonance. The relative stability order of ter., sec. and pri-free radicals is explained on the basis of hyperconjugation.

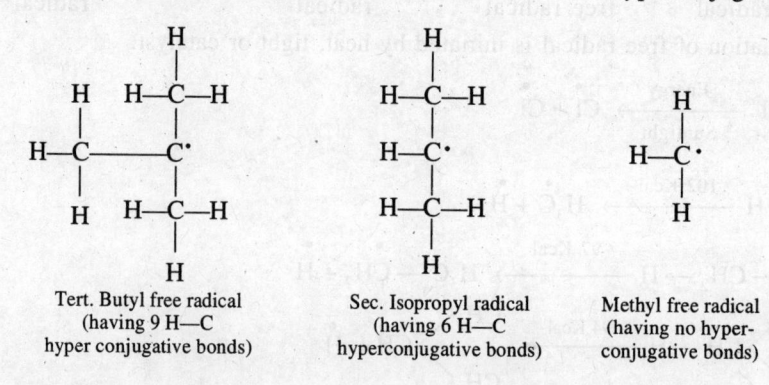

Tert. Butyl free radical
(having 9 H—C
hyper conjugative bonds)

Sec. Isopropyl radical
(having 6 H—C
hyperconjugative bonds)

Methyl free radical
(having no hyper-
conjugative bonds)

MECHANISM OF ORGANIC REACTIONS

Structure of alkyl free radicals

The carbon atom of alkyl free radicals which is bonded to only three atoms or groups of atom is sp² hybridized. Thus, free radicals have a planar structure with odd electron situated in the unused p-orbital at right angles to the plane of hybrid orbitals.

120°

(2) Heterolytic fission or heterolysis

This involves the breaking of a covalent bond in such a way that both the electrons of the shared pair are carried away by one of the atoms.

This type of fission occurs when the two atoms differ considerably in their electronegativities. The electron pair is carried away by the atom which is more electronegative in comparison to the other.

$$A:B \begin{cases} A:^{\ominus} + ^{\oplus}B \quad \text{(A is more electronegative than B)} \\ ^{\oplus}A + :B^{\ominus} \quad \text{(B is more electronegative than A)} \end{cases}$$

Heterolytic fission leads to the formation of charged or ionic species, one having a positive charge and the other a negative charge. This type of fission occurs most readily with polar compounds in polar solvents like water or alcohol and is influenced by the presence of ions due to acid and base catalyst. Quite often the ionic species formed by heterolytic fission bear the positive or negative charge on carbon atom. Such ionic species are known as carbonium ions or carbanions according to the charge which the carbon atom carries (positive or negative).

(i) Carbonium ions (carbocations)

When a covalent bond, in which carbon is linked to a more electronegative atom or group, breaks up by heterolytic fission, the more electronegative atom takes away the electron pair while carbon loses its electron and thus acquires a positive charge.

Such organic ions carrying a positive charge on carbon atom are known as carbonium ions or carbocations.

$$>C:X \longrightarrow >\overset{\oplus}{C} + :X^{\ominus}$$
$$\text{Carbonium ion}$$

Carbonium ions are named by adding the words 'Carbonium ion' to the parent alkyl group. These are also termed as primary, secondary, tertiary, depending upon the nature of the carbon atom bearing positive charge.

H–C⁺H₂ (H)	H₃C–C⁺H₂ (H)	(CH₃)₂–C⁺H	(CH₃)₃C⁺
Methyl carbonium ion	Ethyl carbonium ion	Isopropyl carbonium ion	Tert. butyl carbonium ion
(Primary)	(Primary)	(Secondary)	(Tertiary)

Formation of carbonium ions

(i) **By Heterolysis :** They are formed by heterolysis of halogen compounds.

$$(H_3C)_3 C-Cl \rightarrow (CH_3)_3 \overset{\oplus}{C} + \overset{\ominus}{Cl}$$

(ii) **By protonation of alkenes or alcohols :**

$$CH_2 = CH_2 \overset{H^+}{\rightleftharpoons} \overset{+}{CH_2} - CH_3$$

$$R-O-H \overset{H^+}{\rightleftharpoons} R-\overset{+}{O}H_2 \overset{-H_2O}{\rightleftharpoons} \overset{+}{R} + H_2O$$

(iii) **By decomposition of Diazo compounds :**

$$C_6H_5-N_2-Cl \overset{-Cl}{\longrightarrow} C_6H_5\overset{+}{N_2} \overset{-N_2}{\longrightarrow} \overset{+}{C_6H_5} + N_2$$

Characteristics of carbonium ion

(i) Its formed by heterolytic bond fission.

(ii) Carbon atom carrying positive charge has six electrons in its valence shell, *i.e.* 2 electrons less than octet.

(iii) It is a short lived reaction intermediates.

(iv) Carbonium ion is diamagnetic in nature.

(v) Due to electron deficiency, it behaves as a Lewis acid.

Structure of carbonium ion

The positively charged carbon atom in the carbonium ion is in sp^2 state of hybridization. The three hybridized orbitals which lie in the same plane or involved in the formation of three bonds with other atoms while the unhybridized p-orbitals remains vacant. The carbonium ion has a planar structure.

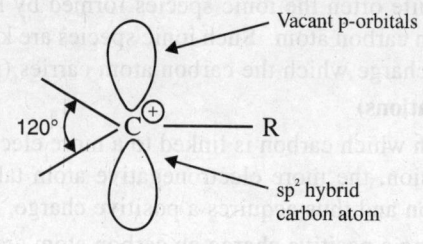

Stability of carbonium ion

The stability of carbonium ions is influenced by both resonance and inductive effects. An alkyl group has an electron releasing inductive effect. An alkyl group attached to the positively charged carbon of a carbonium ion tends to release electrons towards that carbon. In doing so it reduces positive charge on the carbon. In other words, the positive charge gets dispersed as the alkyl group becomes somewhat positively charged itself. This dispersal of the charge stabilises the carbonium ion.

More the number of alkyl groups, the greater the dispersal of positive charge, and therefore, more the stability of carbonium ion is observed.

$$R \rightarrow \underset{R}{\overset{R}{\underset{|}{C^{\oplus}}}} \quad > \quad R \rightarrow \underset{H}{\overset{R}{\underset{|}{C^{\oplus}}}} \quad > \quad R \rightarrow \underset{H}{\overset{H}{\underset{|}{C^{\oplus}}}} \quad > \quad H_3C^+$$

Tertiary   Secondary   Primary   Methyl

MECHANISM OF ORGANIC REACTIONS

Stability decreases as +I decreases (dispersal of positive charge decreases)

$$CH_3CH_2\overset{+}{C}H_2CH_2 > CH_3CH_2\overset{+}{C}H_2 > CH_3\overset{+}{C}H_2 > \overset{+}{C}H_3$$

Stability decreases as molecular mass decreases or +I effect decreases.

Allyl and benzyl carbonium ions are much more stable as these are stabilized by resonance.

$$CH_2=CH-\overset{+}{C}H_2 \longleftrightarrow \overset{+}{C}H_2-CH=CH_2 \text{ (Allyl)}$$

Benzyl carbonium ion

The group like —NO$_2$ and —Br which have –I effect reduce the stability of carbonium ions.

(ii) Carbanions

When a covalent bond, in which carbon is attached to a lesser electronegative atom, breaks up by heterolysis the atom leaves without taking away the bonding pair of electrons and thus the carbon atom acquires a negative charge due to an extra electron.

$$\geqslant C : Y \longrightarrow \geqslant \overset{\ominus}{C} : + \overset{\oplus}{Y}$$

Such organic ions which contain a negatively charged carbon atom are called carbanions. These are named after the parent alkyl group and adding the word carbanion.

Methyl carbanion   Ethyl carbanion

These are also termed as primary, secondary and tertiary depending on the nature of carbon atom bearing the negative charge.

Organic compounds which possess a labile or acidic hydrogen have the tendency to produce carbanions as in the case of reactive methylene compounds which lose proton in presence of sodium ethoxide (C$_2$H$_5$ONa).

$$CH_2\begin{matrix}COOC_2H_5\\COOC_2H_5\end{matrix} + C_2H_5O^- \longrightarrow \overline{C}H\begin{matrix}COOC_2H_5\\COOC_2H_5\end{matrix} + C_2H_5OH$$

Characteristics of carbanions

(i) It is a product of heterolytic bond fission.
(ii) It is a short lived reaction intermediates.
(iii) Carbon atom carrying negative charge has eight electrons in its valence shell.
(iv) It behaves as a Lewis base.

Structure of carbanions

The negatively charged carbon is in a state of sp^3 hybridization. The hybrid orbitals are directed towards the corners of a tetrahedron. Three of the hybrid orbitals are involved in the formation of single covalent bonds with other atoms while the fourth hybrid contains a lone pair of electrons. Thus, carbanions have a pyramidal structure similar to NH_3 molecule.

sp^3 orbital containing lone pair of electrons

Stability of Carbanions : The stability of carbanions is influenced by resonance, inductive effect and s-character of orbitals. The group having +I effects decrease the stability while groups having –I effect increase the stability of the carbanions.

Stability decreases as +I effect increases (Methyl > 1° > 2° > 3° carbanions). Alkyl and benzyl carbanions are stabilised due to resonance.

$$CH_2 = CH—\overset{\ominus}{C}H_2 \longleftrightarrow \overset{\ominus}{C}H_2—CH = CH_2 \quad (Allyl)$$

Benzyl carbanion

Stability of carbanion increases with increase in s-character of orbitals.

$$R—C \equiv \overset{\ominus}{C} \quad > \quad R_2C = \overset{\ominus}{C} \quad > \quad R_3C—\overset{\ominus}{C}H_2$$
$$\qquad\qquad\qquad\qquad\qquad\quad |$$
$$\qquad\qquad\qquad\qquad\qquad\quad R$$
50% s-character $\qquad$ 33% s-character $\qquad$ 25% s-character

Reagents : Most of the attacking reagent carry either a positive or a negative charge. The positively charged reagents attack the regions of high electron density in the substrate molecule while the negatively charged reagents will attack the regions of low electron density in the substrate molecule. The fission of the substrate molecule to create centres of high or low electron density is influenced by attacking reagents. Most of the attacking reagents can be classified into two main groups.

(1) Electrophiles or Electrophilic Reagents

Electrophiles or electrophilic reagents are electron loving species (electro = electron, philic = loving). These species carry either positive charge or electron deficient molecule. So a reagent which can accept an electron pair in a reaction, is called an electrophile. These contain, generally, two electrons less than, the octet. On this basis - electrophiles are classified into two

MECHANISM OF ORGANIC REACTIONS

subclasses.

(a) **Charged Electrophiles**: The electrophiles which have positive charge are known as charged electrophiles e.g.,

H^+ (H_3O^+), Cl^+, Br^+, I^+, NO_2^+, R_3C^+, NH_4^+, NO^+, etc.

(b) **Neutral Electrophiles**: The electrophiles which have electron deficiency are known as neutral electrophiles e.g.,

SO_3, BF_3, $AlCl_3$, $ZnCl_2$, $FeCl_3$, $RCOCl$, $(RCO)_2O$, $R-MgX$, etc.

Electrophiles act as Lewis acids. The reactions involving the attack of electrophiles are known as electrophilic reactions.

(2) Nucleophiles or Nucleophilic Reagents:

Nucleophiles or nucleophilic reagents are electron donating species. The name nucleophile means nucleus loving. Nucleophiles are electron rich, *i.e.* they normally possess an unshared electron pair which they can donate. They are either negative ions or neutral molecules with free electron pairs to donate. They attack regions of low electron density (positive centres) in the substrate molecule. On this basis nucleophiles are classified into two subclasses.

(a) **Charged Nucleophiles**: The nucleophiles which have negative charge are known as charged nucleophiles. eg.,

Cl^-, Br^-, I^-, OH^-, OR^-, CN^- etc.

(b) **Neutral Nucleophiles**: The nucleophiles which are electron rich, i.e. they normally possess an unshared electron pair are known as neutral nucleophiles e.g.,

NH_3, RNH_2, H_2O, ROH, ROR, RSR etc.

(B) Bond formation: The first step of bond fission is followed by the next step of bond formation for which there are again two possibilities, viz.(a) either it may occur in a next and separate step or (b) it may occur simultaneously with the step of bond fission. For example, an ion : OH^- may react in substitution reaction with the molecule CH_3Cl in either of the following two ways.

(a) $H_3C / :Cl \xrightarrow[\text{bond fission}]{\text{First step}} H_3C^{\oplus} + :Cl^{\ominus}$

then $H_3C^{\oplus} / :OH^{\ominus} \xrightarrow[\text{bond formation}]{\text{second step}} CH_3 : OH$

So, in this case, the intermediate species has a zero life time and therefore is not formed. In such case reaction proceeds through an intermediate 'transitory state' in which the old bond is cleaved and new bond is formed at the same time (simultaneously). Such transitory state can best be represented as below.

$$OH :\xrightarrow[\text{bond making}]{\delta-} H_2C \xrightarrow[\text{bond breaking}]{\delta+} Cl$$

(A transition state formed during reaction of methyl chloride and alkali solution.)

Difference between transition state and intermediate

A transition state refers to an imaginary state and cannot be isolated. On the other hand, an intermediate is a stable real species and can be isolated under appropriate conditions. The three important reaction intermediates are free radicals, carbonium ion and carbanions are present in organic reactions.

Reaction Mechanisms - Their Types: From the mechanistic viewpoint described earlier in this chapter, organic reactions could be divided into two classes :

(i) Homolytic reactions involving homolytic bond fission;

(ii) Heterolytic reactions involving heterolytic bond fission. The reaction mechanism in the two cases is entirely different.

Free-radical Mechanism : In homolytic reactions, the first step is the production of a free radical (R') from a normal molecule by the application of energy (heat or light). The free radical then attacks the substrate to bring about homolytic fission.

$$\overset{\bullet}{R} + X—Y \longrightarrow R—\overset{\bullet}{X} + Y$$
substrate

This type of reaction mechanism when the attacking species is a free radical, is called free-radical mechanism.

The free-radical mechanism may be illustrated by the chlorination of methane in sunlight. In the first step, illumination causes a molecule of chlorine to break homolytically into two chlorine free radicals. These then attack a molecule of methane forming hydrogen chloride and methyl radical (H$_3$C). In turn, the methyl free radical attacks a molecule of chlorine giving methyl chloride (CH$_3$Cl) and a fresh chlorine radical.

$$Cl:Cl \xrightarrow{\text{light}} 2\,Cl^{\bullet}$$

$$\left. \begin{array}{l} Cl^{\bullet} + CH_4 \longrightarrow HCl + H_3C^{\bullet} \\ H_3C^{\bullet} + Cl_2 \longrightarrow CH_3Cl + Cl^{\bullet} \end{array} \right\} \text{Chain reaction}$$

Once the free radical H$_3\overset{\bullet}{C}$ is produced, the reaction proceeds by the last two steps in a sort of endless chain. Such a self-propagating reaction is often referred to as a chain reaction.

Since the free radical carries no charge, it can attack any part of the substrate molecule regardless of the electronic distribution of the latter.

Polar or Ionic Mechanism :

This type of mechanism applies to organic reactions in which heterolytic fission takes place. Here the substrate molecule develops negative and positive charge centres, or partial ionic character, by the displacement of electrons due to inductive effect, electromeric effect etc. These negative and positive poles of the molecules are so to say activated for attack by an electrophilic or nucleophilic reagent. The mechanism of reactions involving the attack of electrophilic or nucleophilic reagents on the polar or ionic molecules of the substrate is referred to as polar or ionic mechanism.

The cardinal principle of electrostatics that 'the unlike charges attract while like charges repel each other' forms the basis of polar or ionic mechanism. The negatively charged nucleophilic reagents are anionic in character and attack the substrate molecule at the positive centre causing a nucleophilic reaction. On the other hand, the positively charged electrophilic reagent would attack the molecule at the negative centre causing electrophilic reaction.

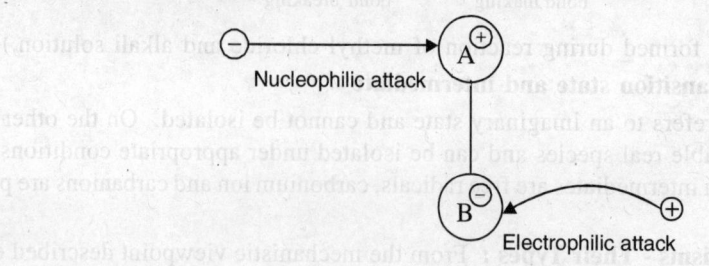

⊖ ⟶ A⊕
Nucleophilic attack

B⊖ ⟵ ⊕
Electrophilic attack

MECHANISM OF ORGANIC REACTIONS 835

The polar or ionic mechanism may be beautifully illustrated as in above Fig. The imaginary molecule AB of the substrate has a positive end A and a negative end B. This would be expected to be attacked by anionic nucleophilic reagent at A and cationic electrophilic reagent at B.

An example of the above reaction mechanism is afforded by the addition of HCN (H^+ + CN^-) to acetone.

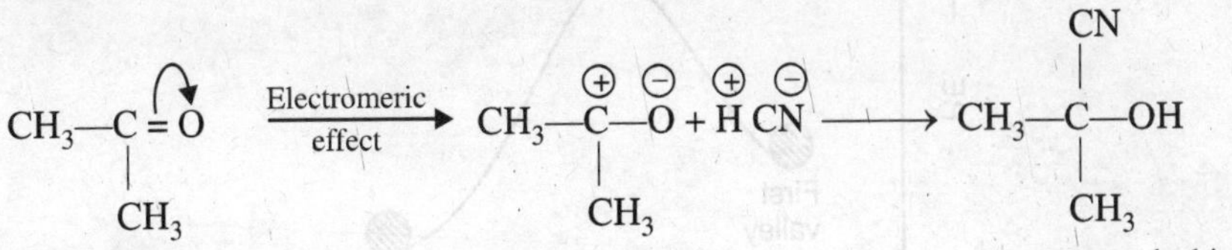

$$\underset{\text{Acetone}}{\underset{\underset{CH_3}{|}}{CH_3-C=O}} \xrightarrow{\text{Electromeric effect}} \underset{\underset{CH_3}{|}}{CH_3-\overset{\oplus}{C}-\overset{\ominus}{O}} + \overset{\oplus}{H}\overset{\ominus}{CN} \longrightarrow \underset{\text{Acetone cynohydrin}}{\underset{\underset{CH_3}{|}}{\overset{\overset{CN}{|}}{CH_3-C-OH}}}$$

24.4. Energy Requirements of a Reaction

In an organic reaction, some bonds break while some new ones are formed. Energy is required for the breaking of bonds and also it is liberated during the formation of bonds. Let us consider, for example, the reaction between methane and chlorine to give chloromethane (CH_3Cl) and hydrogen chloride. Here one C–H bond and one Cl–Cl bond are broken, requiring a total energy of 157 Kcal/mole. At the same time one Cl–Cl bond and one H–Cl bond are formed, liberating total energy of 184 Kcal/mole. The net result is that the reaction occurs with a liberation of energy equal to 27 Kcal/mole and is, therefore, said to be exothermic. Thus,

$$\underset{99}{CH_3-H} + \underset{58}{Cl-Cl} \longrightarrow \underset{81}{CH_3-Cl} + \underset{103}{H-Cl} \quad \text{(Bond energy Kcal/mole)}$$

Conversely, the reverse reaction would require the consumption of 27 Kcal/mole energy and will be Endothermic. Thus an exothermic reaction could be considered like a ball rolling down a hill side and would be expected to occur of its own accord. On the other hand an endothermic reaction which is like a ball rolling uphill, would not occur normally, unless energy is supplied to the system. The total 'chemical energy' stored in the molecules before and after the reaction could be described as the 'potential energy' of the system.

Activation Energy : We have discussed above that a reaction proceeds by breaking of bonds in the reacting molecules. Therefore, whether a reaction is exothermic or endothermic, to start with, the energy must come from some source to break the bonds.

We know from our knowledge of physical chemistry that molecules of the reactants are in a state of rapid motion and possess kinetic energy. The reaction occurs when the reacting molecules approach in proper alignment with respect to one another and collide. On such collisions, the kinetic energy possessed by the molecules is transformed into 'chemical energy' or 'potential energy' of the system. Thus to start a reaction, the required energy is suplied by the collisions of the reacting molecules, whose rate could be enhanced if necessary, say by heating.

Not all collisions between the reacting molecules are fruitful. Rather, it is only the collisions of such molecules as possessing a certain minimum energy which brings about the reactions while others do not. The molecules that come to possess higher potential energy through collisions are said to have been 'activated' to enter into reaction.

"The minimum amount of 'chemical' or 'potential' energy that must be provided by collisions of the reacting molecules for the reaction to occur is termed as the "Activation energy."

It is true that for an exothermic reaction, the collisions of the reactant molecules readily supply the required initial energy and the reaction takes place spontaneously. On the other hand, for an endothermic reaction the molecules need to be activated by supplying energy, say in the form of heat to make the reaction go.

For illustration, imagine some balls attempting to climb over a hill from one valley to another only those will cross over which possess a certain minimum of energy to reach the hill-top from where they would get over the barrier or the 'energy cliff' (vide Fig. 24.1 below).

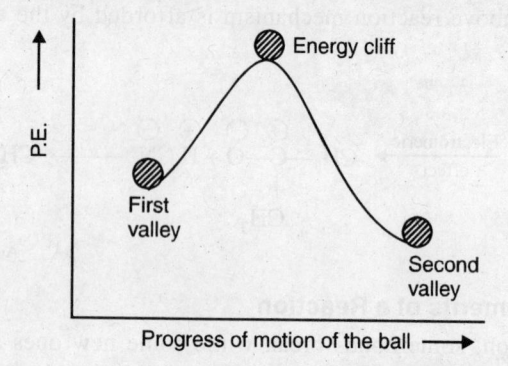

Fig. 24.1 : Illustration of energy cliff concept.

From analogy, only such reacting molecules that possess a certain minimum energy could change into products but before doing so, these would have to cross over the 'activation energy barrier'.

Let us consider the energy changes during the course of the reaction.

$$X + Y-Z \rightarrow X-Y + Z$$

In the beginning both X and Y–Z possess certain potential energy represented by the point (*a*) on the curve (vide Fig. below). The reacting molecules also possess, kinetic energy which on collision is transformed into potential energy.

Exothermic reaction **Endothermic reaction**

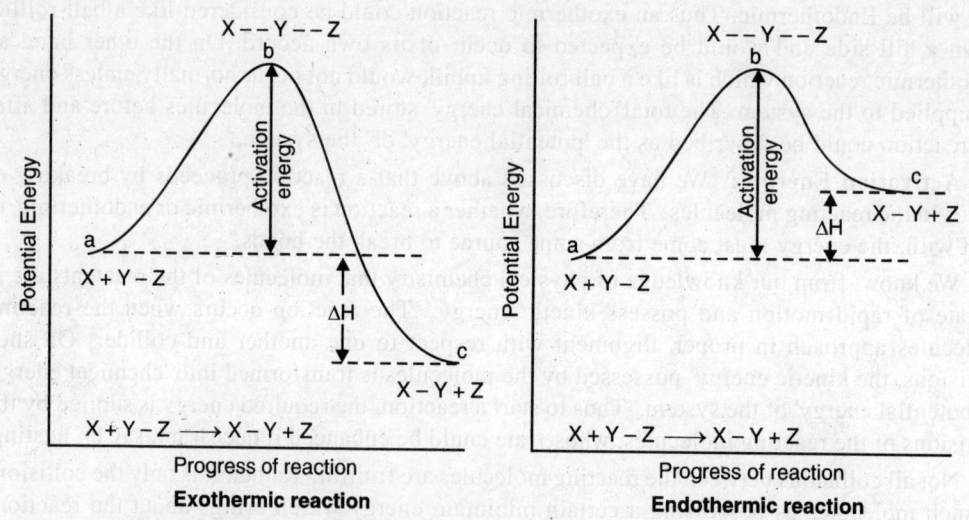

Fig. 24.2 : Energy change in Exothermic and Endothermic reactions.

This results in the increase of potential energy and the system moves up along the curve till cliff (b) is reached. The 'energy cliff state' is a sort of temporary phase and leads to the product (c), when the potential energy of the system is again changed into kinetic energy and then into heat or any other form of energy.

Fig. 24.2 illustrates that for an exothermic reaction the system originally possesses more potential energy than the product and the excess energy (ΔH) is liberated as heat. For an endothermic reaction, the system to start with has less potential energy (PE) than at the end (–ΔH) and, therefore, it absorbs heat from the surroundings.

The reactions X + Y – Z → X – Y + Z could be visualised to take by the following steps. The molecule X approaches Y - Z from a direction remote from Z (proper alignment) while X draws nearer to Y, Z starts being repelled from Y until a stage is reached when X and Z are rather loosely attached to Y and are approximately equidistant from it. Thus an 'activated complex X...Y...Z is formed in which X to Y and Y to Z distances are slightly more than the normal bond lengths. This is the least stable arrangement and is called the transition state or activated complex. The sequence of events may be represented by the following equation.

$$\underset{\text{Reactants}}{X + Y - Z} \rightarrow \underset{\substack{\text{Transition state} \\ \text{(Activated complex)}}}{X...Y.....Z} \rightarrow \underset{\text{Products}}{X - Y + Z}$$

The activated complex is not a true molecule, the bonds being partial. In this state, the system possesses maximum energy and is most unstable. Hence the transition state of a system could be described as an extremely transitory specific arrangment of atoms and groups through which a reaction system must pass on its way to the products. In other words, the activated complex has infinitesimally short life time and at once decomposes to give the products.

Transition State (T.S.) : An Intermediate :

A transition state or an 'activated complex' refers to an imaginary molecule and hence cannot be isolated. While on the other hand, an intermediate is quite a stable chemical entity and can be isolated under the given experimental conditions. A reaction that proceeds through an intermediate has to surmount two energy barriers; one for the conversion of the reactants to the intermediate and the other for the conversion of the intermediate into the products as depicted in the energy diagram (Fig. 24.3).

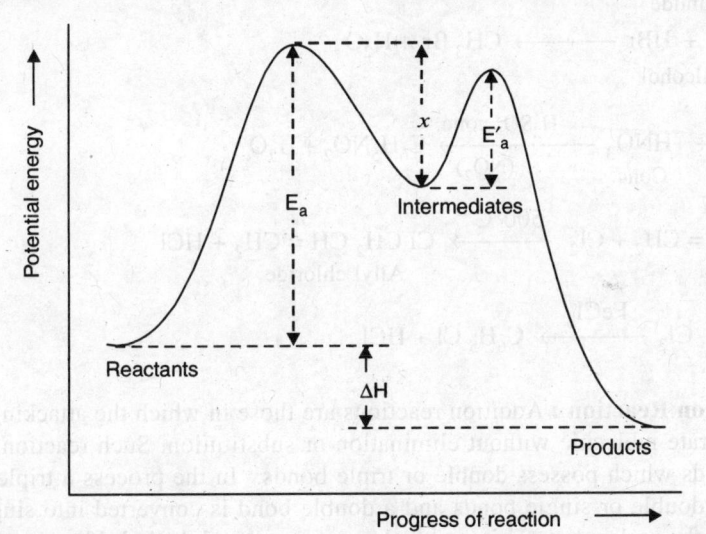

Fig. 24.3 : Energy diagram of a reaction proceeding through an intermediatee.

For a true intermediate, the energy of activation, E_a for the conversion of reactants into products is necessarily higher than the energy of activation, E'_a for conversion of the intermediate into the products. Had the energy of activation E_a' been higher than the energy of activation E_a then the very idea of an intermediate would have been forfeited ? The greater the dip 'X', the more stable will be the intermediate. Consequently, the shallower the dip 'x' the less stable will

24.5. TYPES OF ORGANIC REACTIONS

The reaction of organic compounds can be classified into four main types.

(i) Substitution or Displacement reactions
(ii) Addition reactions
(iii) Elimination reactions
(iv) Rearrangement reactions

1. Substitution or Displacement Reaction : Substitution or displacement reactions are those reactions in which an atom or group of atoms attached to a carbon atom in a substrate molecule is replaced by another atom or group of atoms. During the reaction no change occurs in the carbon skeleton *i.e.*, no change in the saturation or unsaturation of the initial organic compound. Depending on the mechanism, the substitution reactions are further classified into three types.

(a) Free radical substitution reactions
(b) Electrophilic substitution reactions
(c) Nucleophilic substitution reactions.

Some of the examples of substitution reactions are :

(i) $CH_4 + Cl_2 \xrightarrow{\text{UV light}} CH_3Cl + HCl$
    Methane

(ii) $CH_3CH_2Br + NaOH \longrightarrow CH_3CH_2OH + NaBr$
    Ethyl bromide

(iii) $CH_3OH + HBr \longrightarrow CH_3Br + H_2O$
    Methyl alcohol

(iv) $C_6H_6 + HNO_3 \xrightarrow[+ (NO_2)]{H_2SO_4 \text{ conc.}} C_6H_5NO_2 + H_2O$
    Benzene   Conc.

(v) $CH_3CH=CH_2 + Cl_2 \xrightarrow{500°C} ClCH_2CH=CH_2 + HCl$
    Propylene   Allyl chloride

(vi) $C_6H_6 + Cl_2 \xrightarrow{FeCl_3} C_6H_5Cl + HCl$
    Benzene

2. Addition Reaction : Addition reactions are those in which the attacking reagent adds up to the substrate molecule without elimination or substitution. Such reactions are given by those compounds which possess double or triple bonds. In the process a triple bond may be converted into double or single bonds and a double bond is converted into single bond. For each π-bond of the molecule two sigma bonds are formed and the hybridization state of carbon atoms changes from sp to sp^2 and sp^2 to sp^3.

Like substitution reactions, addition reactions are also of three types.

(a) Free radical addition reactions
(b) Nucleophilic addition reactions
(c) Electrophilic addition reactions.

Some of the examples of addition reactions are :

MECHANISM OF ORGANIC REACTIONS

(i) $HC \equiv CH + H_2 \xrightarrow{Ni, 140°C} CH_2 = CH_2 \xrightarrow{H_2/Ni, 200°C} CH_3-CH_3$
    Acetylene → Ethylene → Ethane

(ii) $CH_2 = CH_2 + Br_2 \longrightarrow CH_2Br-CH_2Br$
    Ethylene → 1,2-Dibromoethane (Ethylene bromide)

(iii) $CH_3-\underset{H}{\overset{H}{C}}=O + HCN \longrightarrow CH_3-\underset{CN}{\overset{OH}{C}}-H$
    Acetaldehyde → Cyanohydrin

(iv) $CH_3C \equiv N + H_2O \xrightarrow[H_2O_2]{Acid} CH_3-\overset{O}{\underset{\|}{C}}-NH_2$
    Acetamide

(v) $CH \equiv CH + HBr \longrightarrow CH_2 = CHBr \xrightarrow{HBr} CH_3-CHBr_2$
    Acetylene → Vinyl bromide → 1,1-Dibromoethane (Ethylidine bromide)

3. Elimination Reactions: The reverse of addition reactions are termed as elimination reaction. In these reactions generally atoms or groups from two adjacent carbon atoms in the substrate molecule are removed and multiple bond is formed. In the process two sigma bonds are lost and a new π-bond is formed, *i.e.* state of hybridization of carbon atoms changes from sp^3 to sp^2 and sp^2 to sp.

Some examples are:

(i) $H-\underset{H}{\overset{Br}{C}}-\underset{H}{\overset{Br}{C}}-H + Zn \xrightarrow[-Br_2]{Heat} H-\underset{H}{\overset{}{C}}=\underset{H}{\overset{}{C}}-H + ZnBr_2$
    1,2 dibromoethane → Ethylene

(ii) $H-\underset{H}{\overset{H}{C}}-\underset{H}{\overset{Br}{C}}-H + Alc. KOH \xrightarrow{-HBr} H-\underset{H}{\overset{}{C}}=\underset{H}{\overset{}{C}}-H + KBr + H_2O$
    Ethyl bromide → Ethylene

(iii) $H_3C-\underset{H}{\overset{H}{C}}-OH \xrightarrow[-H_2]{Cu, 800°C} H_3C-\overset{}{C}=O$
    Ethyl alcohol → Acetaldehyde

(iv) $H-\underset{H}{\overset{H}{C}}-\overset{O}{\underset{\|}{C}}-NH_2 + P_2O_5 \xrightarrow[-H_2O]{Heat} H_3C-C \equiv N + H_2O$
    Acetamide → Methyl cyanide

4. Rearrangement Reactions : The reactions which involve the migration of an atom or group from one side to another within the molecule (nothing is added from outside and nothing is eliminated) resulting in a new molecular structure are known as rearrangement reactions. The new compound is actually the structural isomer of the original one.

Some of the examples are :

(i) $CH_3 . CH_2 . CH_2 . CH_3 \xrightarrow[\text{Heat}]{\text{Anhydrous AlCl}_3} CH_3-CH-CH_3$
　　　n-butane　　　　　　　　　　　　　　　　　　　　　　 |
　　　　　　　　　　　　　　　　　　　　　　　　　　　　　　CH_3
　　　　　　　　　　　　　　　　　　　　　　　　　　　　　iso-butane

(ii) $CH_3 . CH_2 . CH = CH_2 \xrightarrow[\text{Heat}]{\text{Al}_2(SO_4)_3 \text{ or AlCl}_3} CH_3CH = CHCH_3 + CH_3-C = CH_2$
　　　|
　　CH_3
　　　n-butane　　　　　　　　　　　　　　　　　(2-butane)　　(2-methyl propene)

(iii) $NH_4CNO \xrightarrow{\text{Heat}} NH_2CONH_2$
　　　Ammonium　　　　　　　Urea
　　　 cyanate

(iv) Phenyl acetate $\xrightarrow[\text{Fries rearrangement}]{\text{Anhydrous AlCl}_3}$ Ortho-hydroxy acetophenone + Para-hydroxy acetophenone

Mechanism of Substitution Reactions

As already defined, a substitution reaction involves the replacement of an atom or group attached to a carbon atom by another atom or group. These reactions may follow free radical, nucleophilic or electrophilic mechanism. Some typical examples are considered to explain the three types of mechanism.

1. Free Radical Substitution Reactions : Such substitution reactions in which attacking species are free radicals are known as free radical substitution reactions.

Ex. : (a) Chlorination of Methane : The chlorination of methane in the presence of ultraviolet light is an example of free radical substitution.

$$CH_4 + Cl_2 \xrightarrow[\text{light}]{UV} CH_3Cl + HCl$$
　Methane　　　　　　　Methyl chloride

The reaction does not stop with the formation of methyl chloride (CH_3Cl) but the remaining hydrogen atoms are replaced one by one with chlorine atoms to give rise to chain reaction.

$$CH_3Cl + Cl_2 \rightarrow CH_2Cl_2 + HCl$$
$$CH_2Cl_2 + Cl_2 \rightarrow CHCl_3 + HCl$$
$$CHCl_3 + Cl_2 \rightarrow CCl_4 + HCl$$

Mechanism : The reaction is initiated by the breaking of chlorine molecule into chlorine-free radicals in presence of UV light.

$$Cl_2 \xrightarrow[\text{light}]{UV} Cl^\bullet + Cl^\bullet \quad \text{(chain initiation)} \quad ...(i)$$

MECHANISM OF ORGANIC REACTIONS

The chlorine-free radicals attack methane molecules ($Cl^\bullet$ is a substituent).

$$CH_4 + Cl^\bullet \rightarrow {}^\bullet CH_3 + HCl \quad \text{(chain propagation)} \qquad \ldots(ii)$$

Each of the methyl-free radicals, in turn reacts with chlorine molecule to form methyl chloride and at the same time chlorine-free radical is produced.

$$^\bullet CH_3 + Cl_2 \rightarrow CH_3Cl + Cl^\bullet \quad \text{(chain propagation)}$$

The chlorine-free radical can react with fresh methane molecule as in step (ii) or methyl chloride.

$$^\bullet Cl + CH_3Cl \rightarrow {}^\bullet CH_2Cl + HCl \quad \text{(chain propagation)}$$

The CH_2Cl radical would react with another molecule of chlorine to form dihalogenated methane. This process may extend further till all the replaceable hydrogen atoms in the methane have been substituted by chlorine atoms.

$$\begin{aligned}
^\bullet CH_2Cl + Cl_2 &\rightarrow CH_2Cl_2 + Cl^\bullet &&\text{(chain propagation)} \\
Cl^\bullet + CH_2Cl_2 &\rightarrow {}^\bullet CHCl_2 + HCl &&\text{(chain propagation)} \\
^\bullet CHCl_2 + Cl_2 &\rightarrow CHCl_3 + Cl^\bullet &&\text{(chain propagation)} \\
Cl^\bullet + CHCl_3 &\rightarrow {}^\bullet CCl_3 + HCl &&\text{(chain propagation)} \\
^\bullet CCl_3 + Cl_2 &\rightarrow CCl_4 + Cl^\bullet &&\text{(chain propagation)}
\end{aligned}$$

The chain of reaction initiated and propagated as shown above may be terminated if free radicals combine amongst themselves without giving rise to any new radicals.

$$\begin{aligned}
Cl^\bullet + Cl^\bullet &\rightarrow Cl_2 &&\text{(chain termination)} \\
H_3C + Cl^\bullet_3 &\rightarrow CH_3Cl &&\text{(chain termination)} \\
^\bullet CH_3 + {}^\bullet CH_3 &\rightarrow CH_3\text{---}H_2 &&\text{(chain termination)}
\end{aligned}$$

Evidence in favour of free Radical Mechanism :

(i) The reaction does not occur in dark but requires energy to initiate the process.

(ii) It has been observed that addition of substances that are sources of free radicals can initiate the reaction in the dark and even at low temperature. For example, in the presence of 0.02% of tetraethyl lead [$(C_2H_5)_4$Pb], the chlorination of butane can take place in dark at 140°C.

(iii) It has been observed that oxygen acts as an inhibitor in this reaction. This is due to the fact that methyl radical combines with oxygen to form much less reactive peroxy methyl radical $CH_3OO^\bullet$.

(b) Arylation of Aromatic Compounds (Gomberg reaction) :

The reaction of benzene diazonium halide with benzene gives diphenyl by a free radical substitution reaction.

$$\underset{\text{Benzene}}{C_6H_5\text{---}H} + \underset{\substack{\text{Benzene} \\ \text{diazonium halide}}}{C_6H_5N_2X} \xrightarrow{\text{Alkali}} \underset{\text{Diphenyl}}{C_2H_5\text{---}C_6H_5} + N_2 + HX$$

The mechanism of the reaction is as follows :

Initiation step : $C_6H_5\text{---}N=N\text{---}X \longrightarrow C_6H_5^\bullet + N_2 + X^\bullet$
(in neutral or basic medium)

Propogation step :

$C_6H_5^\bullet + C_6H_6 \longrightarrow$ [cyclohexadienyl radical resonance structures with C_6H_5 and H substituents]

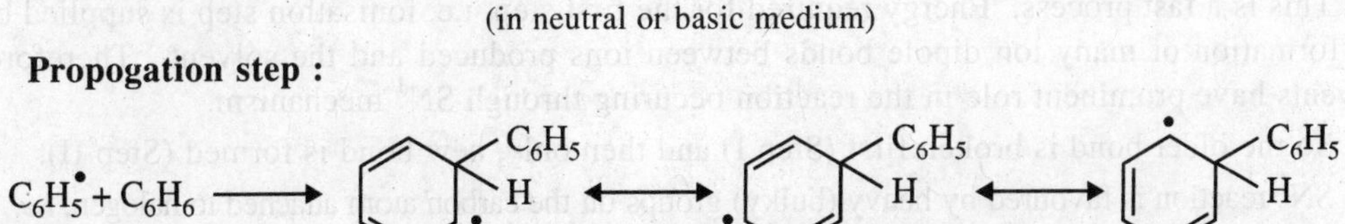

Termination step :

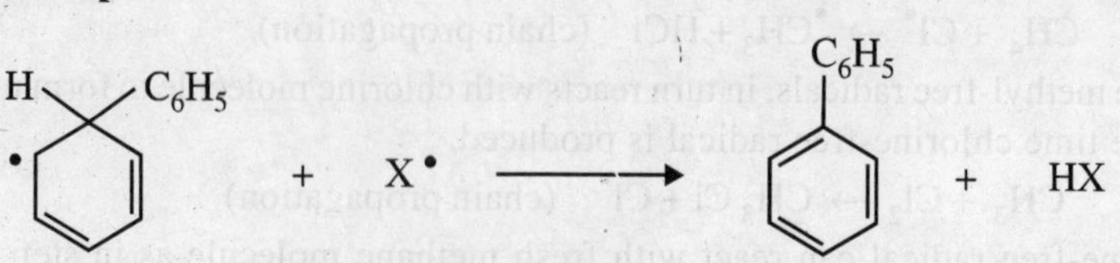

[Diagram: cyclohexadienyl radical with H and C₆H₅ + X• → biphenyl-type product (C₆H₅ on ring) + HX]

(c) Wurtz Reaction : Ethyl bromide on treatment with metallic sodium forms butane, ethane and ethylene by involving free radical mechanism.

$$C_2H_5Br + Na \longrightarrow {}^{\bullet}C_2H_5 + NaBr$$

$$^{\bullet}C_2H_5 + {}^{\bullet}C_2H_5 \longrightarrow C_2H_5 - C_2H_5 \text{ (Butane)}$$

$$2\,{}^{\bullet}CH_3CH_2 + C_2H_5 \xrightarrow{\text{Disproportionation}} \underset{\text{Ethane}}{CH_3 - CH_3} + \underset{\text{Ethylene}}{CH_2 = CH_2}$$

(2) Nucleophilic Substitution Reaction : The substitution reactions in which attacking reagents are nucleophiles are known as nucleophilic substitution reactions.

The nucleophilic substitution reactions are divided into two classes.

(a) Unimolecular Nucleophilic Substitution Reaction or SN^1 - Reaction.

(S = Substitution, N = Necleophilic, 1 = Unimolecular)

The nucleophilic substitution reaction in which rate of reaction depends only on the concentration of the substrate molecule is known as SN^1-reaction.

It shows first order kinetics. The reaction rate is proportional to the concentration of the substrate.

$$\text{Rate} \propto [\text{substrate}]$$

The hydrolysis of tert-butyl bromide is an example of SN^1 -reaction. The reaction consists of two steps.

Step I :

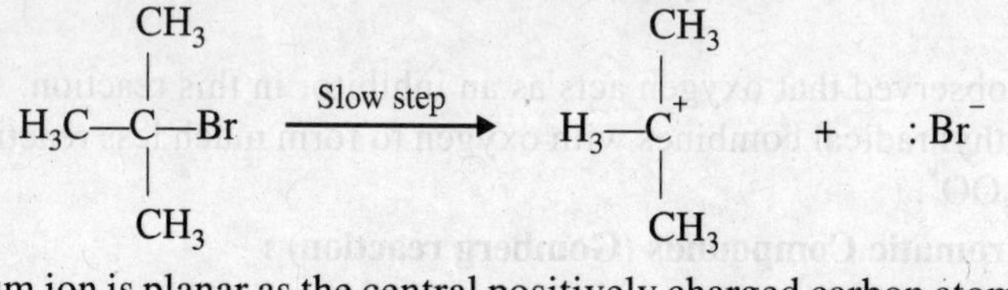

$$H_3C-\underset{\underset{CH_3}{|}}{\overset{\overset{CH_3}{|}}{C}}-Br \xrightarrow{\text{Slow step}} H_3C-\underset{\underset{CH_3}{|}}{\overset{\overset{CH_3}{|}}{C^+}} + :Br^-$$

The carbobnium ion is planar as the central positively charged carbon atom is sp^2 hybridised.

Step II : The nucleophile can attack the planar carbonium ion from either side to form tert-butyl alcohol.

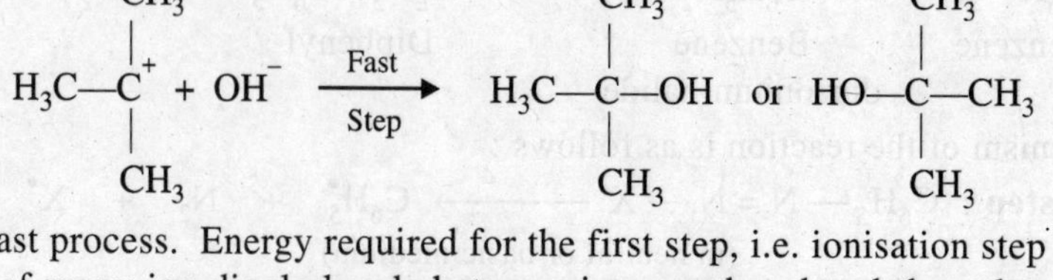

$$H_3C-\underset{\underset{CH_3}{|}}{\overset{\overset{CH_3}{|}}{C^+}} + OH^- \xrightarrow[\text{Step}]{\text{Fast}} H_3C-\underset{\underset{CH_3}{|}}{\overset{\overset{CH_3}{|}}{C}}-OH \text{ or } HO-\underset{\underset{CH_3}{|}}{\overset{\overset{CH_3}{|}}{C}}-CH_3$$

This is a fast process. Energy required for the first step, i.e. ionisation step is supplied by the formation of many ion dipole bonds between ions produced and the solvent. Therefore, solvents have prominent role in the reaction occuring through SN^1 mechanism.

So the older bond is broken first (Step I) and then only, new bond is formed (Step II).

SN^1 reaction is favoured by heavy (bulky) groups on the carbon atom attached to halogen, *i.e.*

$$\underset{\text{Tertiary}}{R_3C-X} > \underset{\text{Secondary}}{R_2CHX} > \underset{\text{Primary}}{RCH_2X} > CH_3-X$$

and the nature of carbonium ion in substrate is SN^1, order : Benzyl > Allyl > tertiary > secondary > primary > methyl halides.

(b) **Biomolecular Nucleophilic Substitution Reaction : Or SN^2 (S-substitution, N-Nucleophilic; 2= Biomolecular)**

The nucleophilic substitution reaction in which the rate of reaction depends on the concentration of both substrate and the nucleophile are known as SN^2 reaction. Hence it is a 2nd order reaction. The rate of reaction is proportional to the concentration of the substrate multiplied by concn. of nucleophile.

$$\text{Rate} \propto [\text{substrate}] [\text{nucleophile}]$$

Hydrolysis of methyl chloride is an example of SN^2 - reaction. The chlorine atom present in methyl chloride is more electronegative than the carbon atom. Therefore C—Cl bond is partially polarized.

$$H-\overset{\overset{H}{|}}{\underset{\underset{H}{|}}{C}}{}^{\delta+}-Cl^{\delta-}$$

When the methyl chloride is attacked by OH$^-$ nucleophile from the opposite side of the chlorine atom, a transition state results in which both OH$^-$ and Cl$^-$ are partially bonded to carbon atom.

$$HO^- + \underset{\underset{H}{|}}{\overset{\overset{H}{\diagup}}{C}}-Cl \longrightarrow \underset{\underset{H}{|}}{HO^{\delta-}\text{---}\overset{\overset{H}{|}}{C}\text{---}Cl^{\delta-}} \longrightarrow \underset{\underset{H}{|}}{HO-\overset{\overset{H}{|}}{C}-H} + Cl^-$$

$$\qquad\qquad\qquad\qquad\text{Transition state} \qquad\qquad \text{Alcohol}$$

In transition state chlorine starts taking hold of the electron pair through which it is bonded to carbon and the OH$^-$ ion offers a pair of electrons for the formation of bond with carbon. Finally, chlorine leaves the molecule as a chloride ion and OH$^-$ ion forms a covalent bond with the carbon giving alcohol as a reaction product. In this reaction configuration of carbon is changed i.e. inverted.

SN^2-mechanism is followed in the case of primary and secondary alkyl halides, i.e., SN^2 reaction is favoured by small groups on the carbon atom attached to halogens. Therefore,

$$CH_3-X > R-CH_2-X > R_2CH-X > R_3-C-X$$

or primary is more reactive than secondary and then tertiary alkyl halides, and nature of carbonium ion in substrate is :

SN^2-order : Methyl > ethyl > isopropyl > tert-butyl > allyl > benzyl chloride

(Benzyl and allyl carbonium ion are stabilised due to resonance)

3. Electrophilic Substitution Reaction : The substitution reaction in which attacking reagents are electrophile are known by electrophilic substitutions. It is represented by SE

(S = Substitution; E = Electrophilic).

The bromination of benzene in the presence of FeBr$_3$ is an example of electrophilic substitution reaction.

$$C_6H_6 + Br_2 \xrightarrow{FeBr_3} C_6H_5Br + HBr$$

Mechanism :

Step I : Formation of Electrophile takes place.

$$Br:Br + FeBr_3 \rightarrow Br^+ + FeBr_4^-$$

Step II : The electrophile takes (Br^+) attacks the benzene ring to form a resonance stabilised carbanium ion.

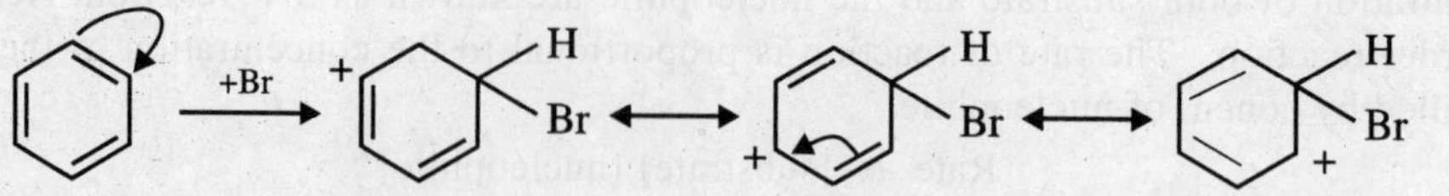

[Resonance structures showing +Br attack on benzene ring producing carbanium ion with H and Br on the same carbon, with positive charge delocalized around the ring]

Step III : Elimination of proton occurs and the substitution product is formed.

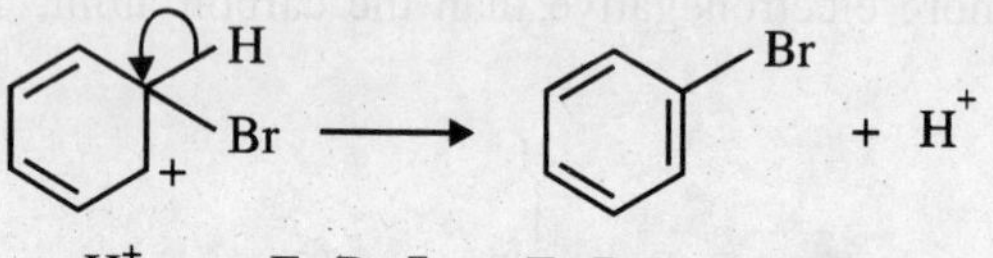

[Structure showing loss of H+ from carbanium intermediate to give bromobenzene + H+]

$$H^+ + FeBr_4^- \rightarrow FeBr_3 + HBr_4$$

Mechanism of Addition Reactions :

The addition reactions are the reactions of the double or triple bonds. These reactions may be initiated by electrophiles, nucleophiles or free radicals. The molecule having >C=C< or –C≡C– are readily attacked by electrophilic reagents while molecules having >C=O or –C≡N are readily attacked by nucleophilic reagents.

(a) Electrophilic Addition Reactions : In electrophilic addition reactions an electrophile approaches the double or triple bond and in the first step forms a covalent bond with one of the carbon atoms resulting in the formation of carbonium ion which then take up a nucleophile to result in an addition product.

$$\underset{\text{Olefin}}{>C=C<} + \underset{\text{Electrophile}}{E^+} \xrightarrow{\text{Slow}} \underset{\text{Carbonium ion}}{>\overset{+}{C}-C-E}$$

$$\underset{\text{Carbonium ion}}{>\overset{+}{C}-C-E} + \underset{\text{Nucleophile}}{X^-} \xrightarrow{\text{Fast}} \underset{\text{Addition product}}{>C-C-E \atop |\atop X}$$

The addition of HBr on ethylene is an example of electrophilic addition. Ethylene is a symmetrical olefin.

$$\underset{\text{Ethylene}}{CH_2=CH_2} + HBR \longrightarrow \underset{\text{Ethyl bromide}}{\overset{H}{\underset{|}{C}H_2}-\overset{Br}{\underset{|}{C}H_2}}$$

Mechanism :

Step I : Hydrogen bromide gives a proton and bromide ion.

$$HBr \rightarrow \underset{\text{(Electrophile)}}{H^+} + \underset{\text{(Nucleophile)}}{:Br^-}$$

Step II : The electrophile attacks the double bond to form a carbonium ion.

$$CH_2=CH_2 + H^+ \longrightarrow \underset{\text{Carbanium ion}}{H_2\overset{+}{C}-CH_3} \text{ or } CH_3-\overset{+}{C}H_2$$

MECHANISM OF ORGANIC REACTIONS

Step III : The nucleophile (Br⁻ ion) now attacks the carbonium ion to form the addition product.

$$CH_3-\overset{+}{C}H_2 + :Br^- \longrightarrow CH_3-CH_2Br$$

$$H_2\overset{+}{C}-CH_3 + :Br^- \longrightarrow Br-CH_2-CH_3$$

Both the products are identical.

In case both alkene and the adding reagent are unsymmetrical, two different products are expected.

$$CH_3-CH=CH_2 + H-Br \longrightarrow \begin{cases} CH_3-CH_2-CH_2-Br \quad \text{(n-propyl bromide)} \\ CH_3-\underset{\underset{Br}{|}}{CH}-CH_3 \quad \text{(isopropyl bromide)} \end{cases}$$

Experimentally, it is observed that isopropyl bromide is the major product. This can be explained on the basis of following mechanism.

Consider the addition of HBr to propene which is unsymmetrical in nature.

Step I : Hydrogen bromide gives a proton (H⁺) and a bromide ion (Br⁻).

$$HBr \rightarrow H^+ + :Br^-$$

Step II : The proton (H⁺) attacks the π-bond to give a stable carbonium ion.

$$CH_3-CH=CH_2 + H^+ \longrightarrow \begin{cases} CH_3-\overset{+}{C}H-CH_3 \quad \text{(More stable)} \\ CH_3-CH_2-\overset{+}{C}H_2 \quad \text{(Less stable)} \end{cases}$$

Step III : The nucleophile bromide ion attacks the more stable carbonium ion to give isopropyl bromide (major project).

$$CH_3-\overset{+}{C}H-CH_3 + :Br^- \longrightarrow CH_3-\underset{\underset{Br}{|}}{CH}-CH_3$$

After studying a large number of addition reactions, the Russian chemist Markownikoff presented a rule.

Markownikoff Rule :

"When an unsymmetrical reagent adds to an unsymmetrical double bond, the negative part of reagent gets attached to that carbon atom which is joined to lesser number of hydrogen atoms."

Under normal conditions always Markownikoff's rule is applied.

If the H atoms linked to a double bond are equal in number, then negative part of the reagent to be added (addendum) goes to larger (higher) alkyl group. For example.

$$CH_3 \rightarrow CH_2 \rightarrow CH=CH-CH_3 \xrightarrow[\text{(AR)}]{\overset{\delta+ \; \delta-}{H-Br}} CH_3-CH_2-\overset{+}{C}H-\overset{-}{C}H-CH_3$$

$$\downarrow \overset{\delta+}{H}----\overset{\delta-}{Br}$$

$$CH_3-CH_2-\underset{\underset{Br}{|}}{CH}-CH_2-CH_3$$

3-bromo pentane

Addition in alkenes and alkynes (in solution) are electrophilic additions, because the positive end of the adding molecule is more closer to $\bar{c}$ centre (carbanion) and thus addition of electrophile, E^+, occurs in first step.

Deviation from Markownikoff's rule or peroxide effect or Kharaseh's effect or Anti-Markownikoff's rule : In presence of peroxide, propene reacts with HBr to form n-propyl bromide as the major product and not isopropyl bromide as expected from Markownikoff's rule.

$$\underset{\text{propene}}{CH_3-CH=CH_2} \xrightarrow[+HBr]{\text{Peroxide}} \underset{\text{n-propyl bromide}}{CH_3-CH_2-CH_2-Br}$$

[However, this effect is not observed in the case of addition of HCl or HI]. The addition of HBr has been explained via free radical mechanism.

Chain initiation :

$$\underset{\substack{\parallel \ \parallel \\ O \ \ O}}{C_6H_5COOCC_6H_5} \longrightarrow 2C_6H_5\overset{\bullet}{\underset{\substack{\parallel \\ O}}{C}O}$$

Chain Propagation :

$$\overset{\bullet}{Br} + CH_3CH=CH_2 \longrightarrow \underset{\substack{\text{Secondary free radical} \\ \text{which is more stable}}}{CH_3\overset{\bullet}{C}HCH_2Br}$$

$$CH_3-\overset{\bullet}{C}HCH_2Br + HBr \longrightarrow CH_3CH_2CH_2Br + \overset{\bullet}{Br}$$

Chain termination :

$$\overset{\bullet}{Br} + \overset{\bullet}{Br} \longrightarrow Br_2$$

$$\overset{\bullet}{Br} + \overset{\bullet}{C}H_3 \longrightarrow CH_3Br$$

(By disproportionation)

In HBr, both the chain propagation steps are exothermic while in HCl, first step is exothermic and second step is endothermic, while in HI, first step is endothermic. On account of this, the addition of HBr is easier.

(b) Nucleophilic Addition Reaction :

When the addition reaction occurs on account of the initial attack of nucleophile, the reaction is said to be a nucleophilic addition reaction. Due to presence of strongly electronegative oxygen atom the π-electrons of the carbon oxygen double bond in carbonyl group (>C=O) get shifted towards the oxygen atom and thereby such bond is highly-polarised. This makes carbon atom of the carbonyl group electron deficient.

$$>C=O \longleftrightarrow >\overset{+}{C}=\overset{-}{O} \equiv >\overset{\delta+}{C}-\overset{\delta-}{O}$$

The addition of HCN to acetone (>C=O) is an example of nucleophilic addition.

$$\underset{\text{Acetone}}{\overset{CH_3}{\underset{CH_3}{>}}C=O} + HCN \longleftrightarrow \underset{\text{Acetone cyanohydrin}}{\overset{CH_3}{\underset{CH_3}{>}}C\overset{OH}{\underset{CN}{<}}}$$

The mechanism of the reaction involves the following steps.

Step I : HCN gives a proton (H^+) and a nucleophile, namely cyanide ion (CN^-).

$$HCN \rightarrow H^+ + CN^-$$

Step II : The nucleophile (CN^-) attacks the positively charged carbon to form an anion [H^+ does not attack the negatively charged oxygen as anion is more stable than the cation.]

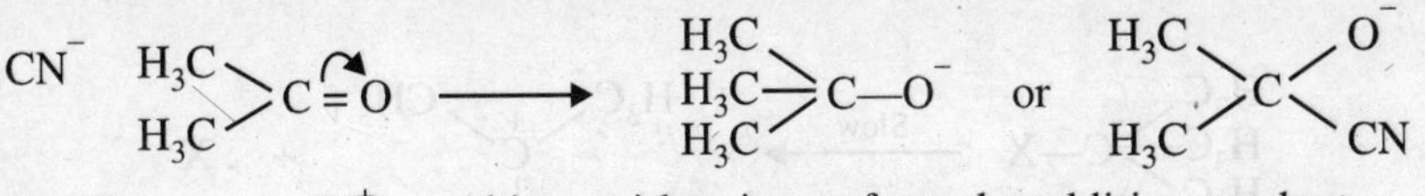

MECHANISM OF ORGANIC REACTIONS

$$CN^- + \underset{H_3C}{\overset{H_3C}{>}}C=O \longrightarrow \underset{H_3C}{\overset{H_3C}{>}}C-O^- \quad \text{or} \quad \underset{H_3C}{\overset{H_3C}{>}}C\underset{CN}{\overset{O^-}{<}}$$

Step III : The proton (H$^+$) combines with anion to form the addition product.

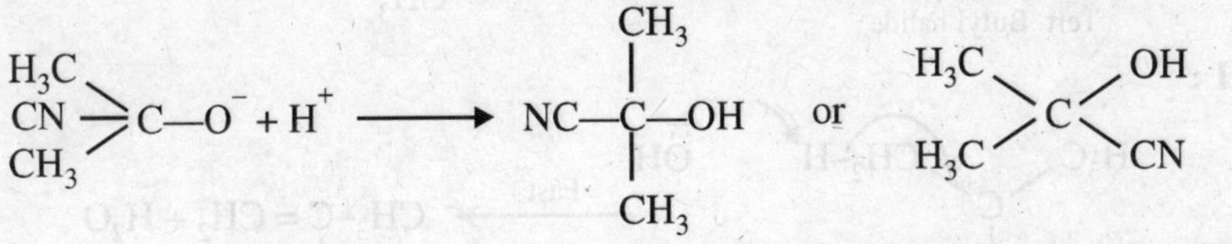

$$\underset{CH_3}{\overset{H_3C}{\underset{|}{>}}}\overset{CN}{C}-O^- + H^+ \longrightarrow NC-\underset{CH_3}{\overset{CH_3}{\underset{|}{C}}}-OH \quad \text{or} \quad \underset{H_3C}{\overset{H_3C}{>}}C\underset{CN}{\overset{OH}{<}}$$

In >C=O compounds, the addition of liquid HCN gives cyanohydrin and the addendum is CN ion (addition is catalysed by bases and retarded by acids) and not HCN directly.

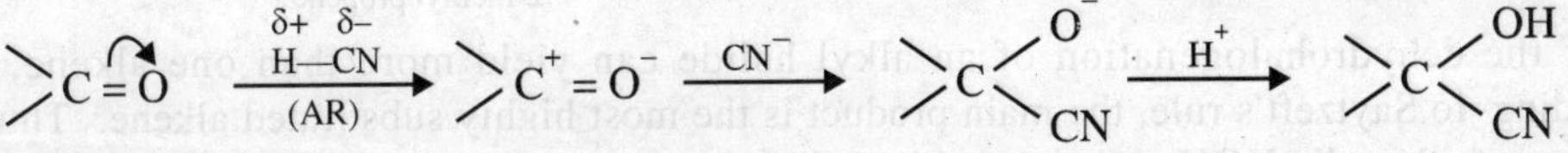

$$>C=O \xrightarrow[(AR)]{\overset{\delta+\;\delta-}{H-CN}} >C^+ = O^- \xrightarrow{CN^-} >C\underset{CN}{\overset{O^-}{<}} \xrightarrow{H^+} >C\underset{CN}{\overset{OH}{<}}$$

3. Mechanism of Elimination Reactions :

An elimination reaction, generally, involves loss of atoms or groups from adjacent carbon atoms resulting in the formation of a π-bond between these carbon atoms; so they are reverse of addition reactions.

On the basis of mechanism, elimination reactions may be two types.

(i) α-Elimination Reaction : When the two groups are eliminated from the same carbon atom, the process is known as α-elimination e.g.,

$$CHCl_3 \longrightarrow :CCl_2 + HCl$$
$$\text{Chloroform} \quad\quad \text{Dichloro carbene}$$

In general,

$$>C\underset{X}{\overset{H}{<}} \longrightarrow C: + Hx$$

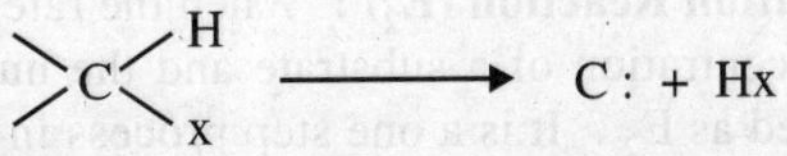

However, this type of elimination reaction is very rare.

(ii) β-Elimination Reaction : When the two groups or atoms are removed from the two adjacent (alpha and Beta) carbon atoms of the molecule, the process is known as β-elimination reaction e.g.,

$$CH_3-CHBr-CH_3 \xrightarrow{\text{Alc. KOH}} CH_3-CH=CH_2 + HBr$$

In general,

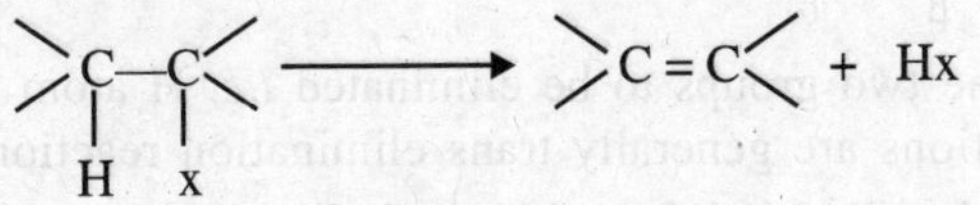

$$\underset{H\;\;X}{\overset{|\;\;|}{>C-C<}} \longrightarrow >C=C< + Hx$$

This type is the most common and hence commonly it is also known as elimination reaction. These reaction may proceed either by unimolecular (E$_1$) or biomolecular (E$_2$) mechanism.

(a) Unimolecular Elimination Reaction (E$_1$) : When the rate of an elimination reaction is dependent only upon the concentration of the substrate and is independent of the concentration of the nucleophile, the reaction is of first order and is designated as E$_1$.

It is a two step processes. The first step involves the slow ionisation of the alkyl halide to give the carbonium ion and the second step involves the fast abstraction of a proton by the base from the adjacent β-carbon atom leading to the formation of an alkene e.g.

Step I :

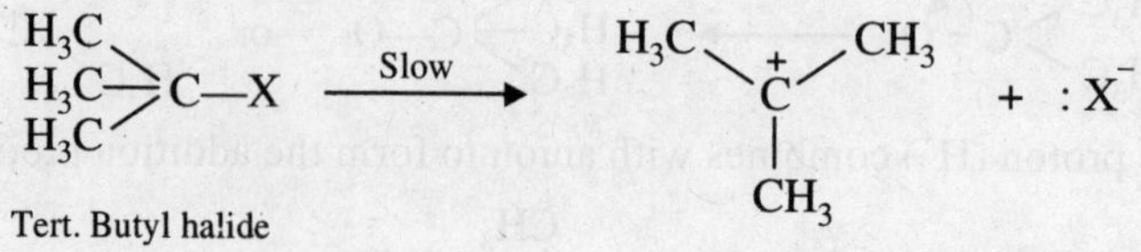

$$\underset{\text{Tert. Butyl halide}}{\begin{array}{c}H_3C\\H_3C\\H_3C\end{array}\!\!\!\!\!\!\!\!\!\!\!>\!C\!-\!X} \xrightarrow{\text{Slow}} \underset{CH_3}{\overset{H_3C}{\underset{|}{\overset{|}{C^+}}}\!\!\!\!\!\!\!\!\!\!\!\!\!\!CH_3} + :X^-$$

Step II :

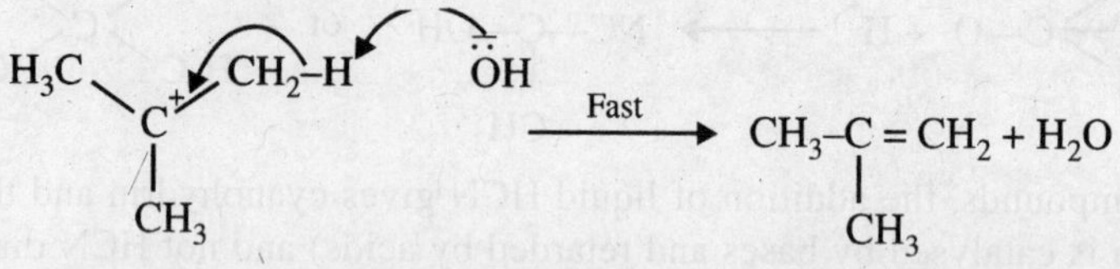

$$\underset{CH_3}{\overset{H_3C}{\underset{|}{\overset{|}{C^+}}}}\!\!\!\!\!\!\!\!\!\!CH_2\!-\!H \quad \overset{..}{O}H \xrightarrow{\text{Fast}} \underset{\underset{\text{2-methyl propene}}{CH_3}}{CH_3\!-\!C=CH_2} + H_2O$$

If the dehydrohalogenation of an alkyl halide can yield more than one alkene, then according to Saytzeff's rule, the main product is the most highly substituted alkene. Thus the reaction of ethanolic KOH on sec-butyl bromide can, in principle, yield two isomeric alkenes, 1-butene and 2-butene as shown below.

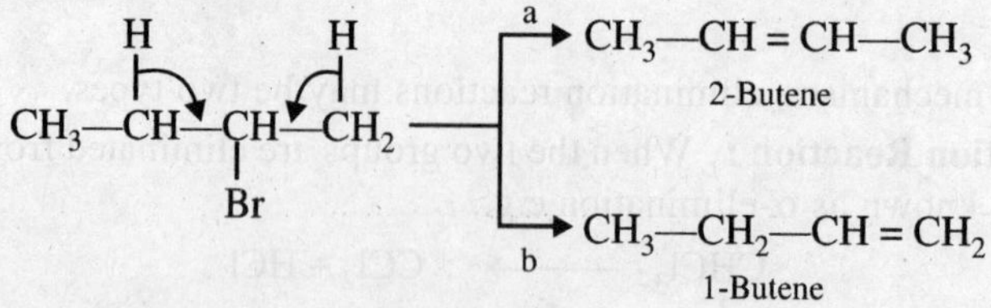

$$CH_3\!-\!\underset{}{\overset{H}{\underset{|}{CH}}}\!-\!\underset{\underset{}{\overset{|}{Br}}}{\overset{H}{\underset{|}{CH}}}\!-\!CH_2 \begin{array}{c}\xrightarrow{a} CH_3\!-\!CH=CH\!-\!CH_3 \\ \text{2-Butene} \\ \xrightarrow{b} CH_3\!-\!CH_2\!-\!CH=CH_2 \\ \text{1-Butene}\end{array}$$

In accordance with the Saytzeff's Rule, the main product is the disubstituted alkene, 2-butene, rather than the monosubstituted, 1-butene.

(b) Bimolecular Elimination Reaction (E_2) : When the rate of an elimination reaction is dependent both upon the concentration of a substrate and the nucleophile, the reaction is of second order and is represented as E_2. It is a one step process in which the abstraction of the proton from the β-carbon and the expulsion of the leaving group *i.e.*, halide ion etc. from the α-carbon atom occur simultaneously. The mechanism for such a reaction may be represented as follows :

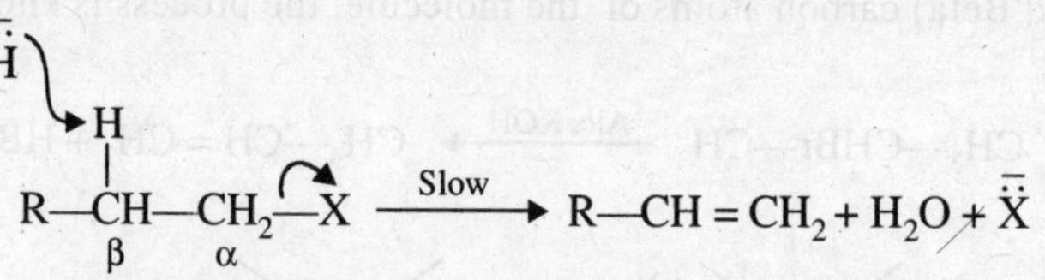

$$\overset{\overset{\displaystyle \overset{..}{\overset{..}{O}}H}{\downarrow}}{R\!-\!\underset{\beta}{CH}\!-\!\underset{\alpha}{CH_2}\!-\!X} \xrightarrow{\text{Slow}} R\!-\!CH=CH_2 + H_2O + \overset{..}{\underset{..}{X}}$$

In these reactions, the two groups to be eliminated *i.e.* H atom and X are trans to each other and hence E_R reactions are generally trans elimination reactions.

The dehydration of the alkyl halides with alcoholic alkali is an example of this type of reaction.

$$\underset{\text{1-Bromopropane}}{CH_3\!-\!CH_2\!-\!CH_2\!-\!Br} + C_2H_5O^- \longrightarrow \underset{\text{Propene}}{CH_3\!-\!CH=CH_2} + Br^- + C_2H_5OH$$

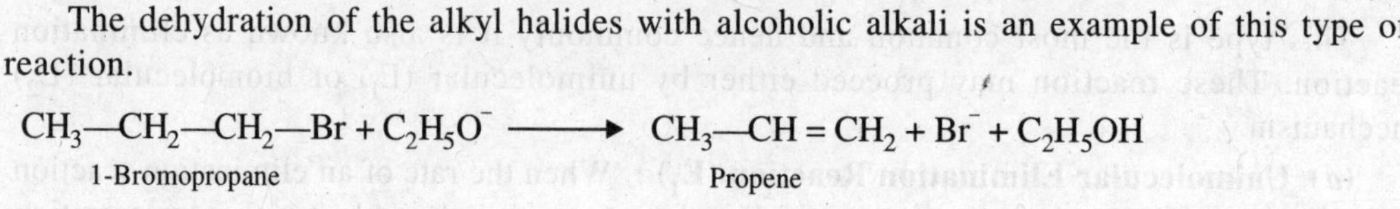

$$C_2H_5O^- + H\!-\!\underset{\underset{H}{|}}{\overset{\overset{CH_3}{|}}{C}}\!-\!\underset{\underset{H}{|}}{\overset{\overset{H}{|}}{C}}\!-\!Br \xrightarrow{\text{Slow}} \underset{\text{Transition state}}{C_2H_5O\cdots H\cdots\underset{\underset{H}{|}}{\overset{\overset{CH_3}{|}}{C}}\!-\!\underset{\underset{H}{|}}{\overset{\overset{H}{|}}{C}}\cdots Br} \xrightarrow{\text{Fast}} \underset{\underset{H}{|}}{\overset{\overset{CH_3}{|}}{C}}=\underset{\underset{H}{|}}{\overset{\overset{H}{|}}{C}} + C_2H_5OH + Br^-$$

MECHANISM OF ORGANIC REACTIONS   849

4. Rearrangement Reactions

In this type of reactions, some atoms/groups shift from one position to another within the substrate molecule itself, giving a product with a new structure. The reaction which proceed by a rearrangement or reshuffling of the atoms/groups in the molecule to produce a structural isomer of the original substance are called rearrangement reactions.

These rearrangement reactions may proceed either by an intramolecular or intermolecular change. Those rearrangements in which the migrating group is never fully detached from the system during the process of migration, are called Intramolecular Rearrangements, whereas those in which the migrating groups get completely detached and is later on re-attached are called Intermolecular Rearrangements.

Intramolecular Rearrangements : Some examples of intramolecular change are given below.

(1) Isomerization : Butane changes to isobutane on heating in presence of aluminium chloride.

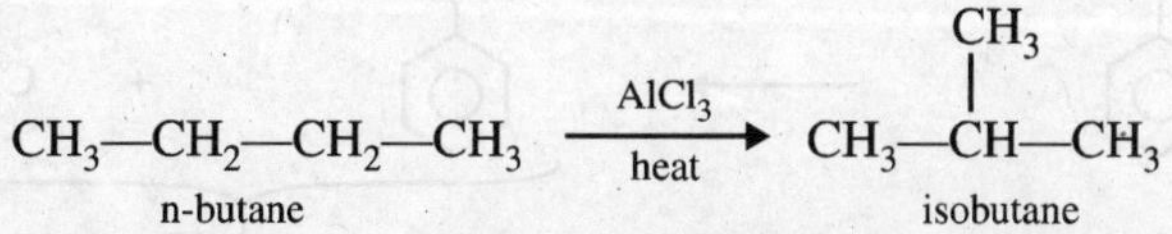

$$CH_3-CH_2-CH_2-CH_3 \xrightarrow[\text{heat}]{AlCl_3} CH_3-\underset{\underset{CH_3}{|}}{CH}-CH_3$$

n-butane   isobutane

Such reactions which produce chain isomers of the original substrates are called isomerization reactions.

Similarly n-propyl bromide on heating with $AlBr_3 + HBr$ gives isopropyl bromide. This change may be visualized to be taking place via the following steps.

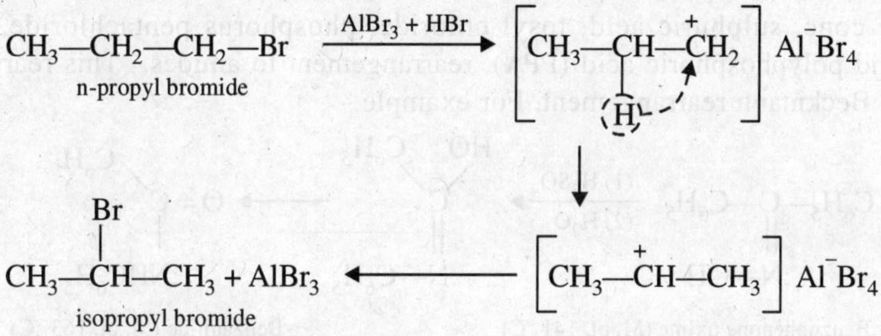

$$CH_3-CH_2-CH_2-Br \xrightarrow{AlBr_3 + HBr} \left[CH_3-\overset{}{CH}-\overset{+}{CH_2} \right] Al\bar{B}r_4$$

n-propyl bromide

$$CH_3-\underset{\underset{Br}{|}}{CH}-CH_3 + AlBr_3 \longleftarrow \left[CH_3-\overset{+}{CH}-CH_3 \right] Al\bar{B}r_4$$

isopropyl bromide

(2) Beckmann Rearrangement : When a ketoxime is treated with an acidic catalyst such as H_2SO_4, polyphosphoric acid, thionyl-chloride, phosphorous pentachloride, benzene sulphonyl chloride etc., it gets converted into a substituted amide. The main mechanistic features of the rearrangements are given below :

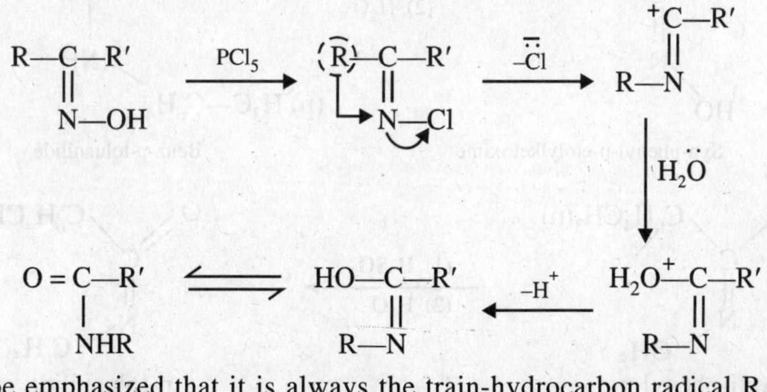

$$\underset{\underset{N-OH}{\|}}{R-C-R'} \xrightarrow{PCl_5} \underset{\underset{N-Cl}{\|}}{(R)-C-R'} \xrightarrow{-\bar{C}l} \underset{\underset{R-N}{\|}}{\overset{+C-R'}{}}$$

$$\downarrow H_2\ddot{O}$$

$$\underset{\underset{NHR}{|}}{O=C-R'} \rightleftharpoons \underset{\underset{R-N}{\|}}{HO-C-R'} \xleftarrow{-H^+} \underset{\underset{R-N}{\|}}{H_2\overset{+}{O}-C-R'}$$

It should be emphasized that it is always the train-hydrocarbon radical R with respect to the hydroxyl group that migrates. Furthermore, R radical or group never gets completely detached from the remainder of the molecule during the course of transformation.

Intermolecular Rearrangements : In these rearrangement the atom or group undergoing migration becomes completely free from the rest of the molecule and later on gets reattached at some site of the rest of the molecule producing thereby a structural isomer of the original substance. Some examples of such inter-molecular changes are mentioned below.

Orton Rearrangement : In this rearrangement, N-chloroacetanilide gets converted into a mixture of o and p-chloroacetanilide on treatment with dil HCl. Thus,

$$CH_3-CO-N(Cl)-C_6H_5 \xrightarrow{\text{Warm dil. HCl}} Cl-C_6H_4-NH-COCH_3 \text{ (p-chloroacetanilide)} + C_6H_4(Cl)-NH-COCH_3 \text{ (o-chloroacetanilide)}$$

A probable mechanism for this change is as under :

$$C_6H_5-N(Cl)-CO-CH_3 \longrightarrow C_6H_5-NHCOCH_3 + Cl_2 \xrightarrow{SF} \text{O- \& P- Chloroacetanilide}$$

24.6. Mechanisms of some Reactions

1. Beckmann Rearrangement : The ketoximes, and aldoximes, when treated with acidic reagents like conc. sulphuric acid, tosyl chloride, phosphorus pentachloride, phosphorus pentaoxide and polyphosphoric acid (PPA), rearrangement to amides. This rearrangement is referred to as Beckmann rearrangement. For example -

$$C_6H_5-\underset{\underset{:N-OH}{\|}}{C}-C_6H_5 \xrightarrow[(2) H_2O]{(1) H_2SO_4} \underset{\underset{:N-C_6H_5}{\|}}{\overset{HO\;\;C_6H_5}{C}} \longrightarrow O=\underset{:NHC_6H_5}{\overset{C_6H_5}{C}}$$

Benzophenone oxime (M. pt. 141 °C)   Benzanilide (M. pt. 163 °C)

This rearrangement leads to different products in the case of syn and anti ketoximes, the groups which are anti to each other changing takes place e.g.,

Syn-phenyl-p-etolylketoxime $\xrightarrow[(2) H_2O]{(1) H_2SO_4}$ Benz-p-toluanilide

Anti-phenyl p-tolylketoxime $\xrightarrow[(2) H_2O]{(1) H_2SO_4}$ p-toluanilide

MECHANISM OF ORGANIC REACTIONS

Mechanism : This rearrangement involves following sequence of reactions as illustrated bellow :

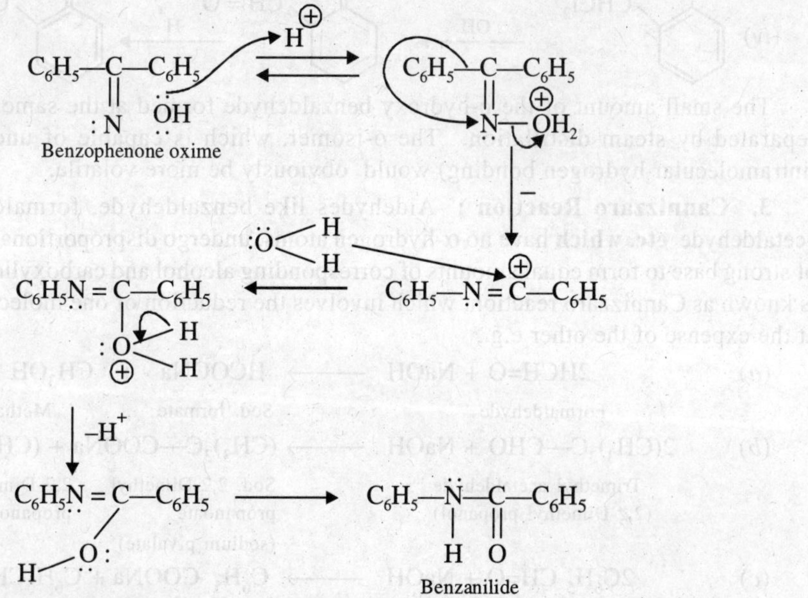

$$C_6H_5-C-C_6H_5 \quad \underset{:N \; OH}{\|} \quad \text{Benzophenone oxime} \xrightarrow{H^+} \quad C_6H_5-C-C_6H_5 \quad \underset{N-OH_2}{\|}$$

$$C_6H_5N=C-C_6H_5 \quad \underset{:O}{|} \quad \xleftarrow{} \quad C_6H_5-N=C-C_6H_5$$

$$\downarrow -H^+$$

$$C_6H_5N=C-C_6H_5 \quad \underset{O:}{|} \quad \xrightarrow{} \quad C_6H_5-\underset{H}{\overset{|}{N}}-\underset{O}{\overset{\|}{C}}-C_6H_5$$

Benzanilide

2. Reimer-Tiemann Reaction : An aldehyde group is introduced, generally ortho to the —OH group, when a phenol is treated with chloroform and sodium hydroxide solution. For example,

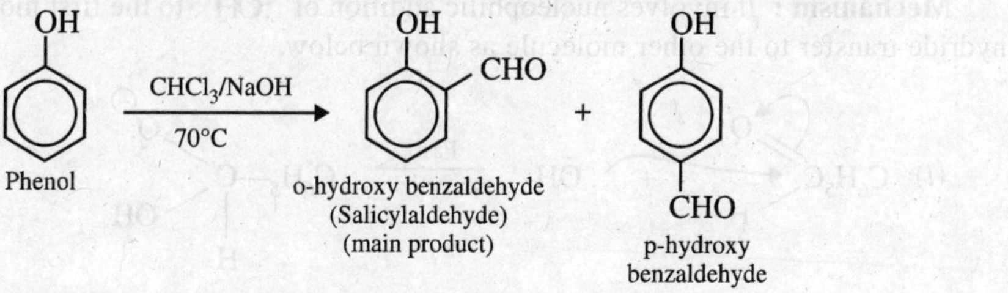

Phenol $\xrightarrow{\text{CHCl}_3/\text{NaOH}, \; 70°C}$ o-hydroxy benzaldehyde (Salicylaldehyde) (main product) + p-hydroxy benzaldehyde

Mechanism : It involves electrophilic substitution of the aromatic nucleus the electrophile being dichlorocarbene :CCl$_2$ (carbon has only a sextet of electrons).

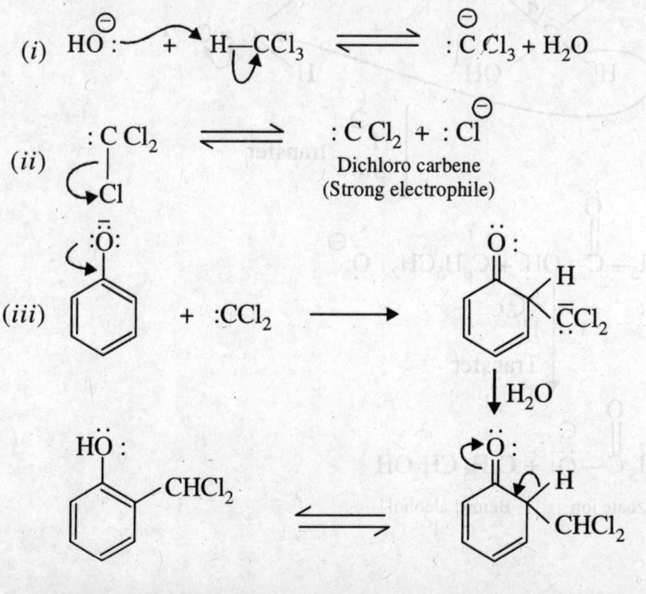

(i) $HO^{\ominus} + H-CCl_3 \rightleftharpoons :CCl_3^{\ominus} + H_2O$

(ii) $:CCl_2-Cl \rightleftharpoons :CCl_2 + :Cl^{\ominus}$
    Dichloro carbene (Strong electrophile)

(iii) [phenoxide] + :CCl$_2$ $\longrightarrow$ [cyclohexadienone intermediate with CCl$_2$]

$\downarrow H_2O$

[o-HO-C$_6$H$_4$-CHCl$_2$] $\rightleftharpoons$ [cyclohexadienone with CHCl$_2$]

(iv) [reaction scheme: o-dichloromethyl phenol with :OH⁻ → intermediate → H⁺ → o-hydroxybenzaldehyde]

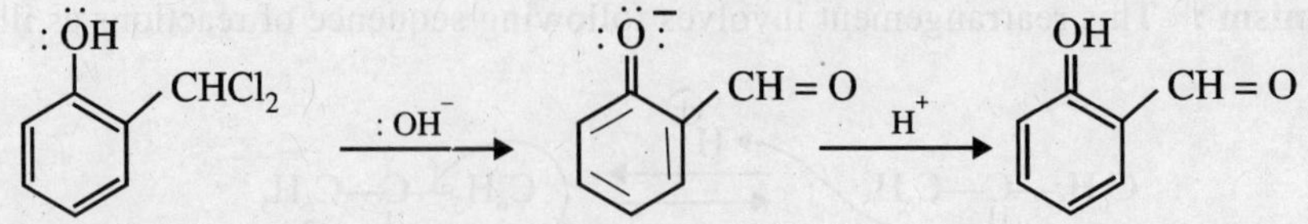

The small amount of the p-hydroxy benzaldehyde formed at the same time can be easily separated by steam distillation. The o-isomer, which is capable of undergoing chelation (intramolecular hydrogen bonding) would obviously be more volatile.

3. Cannizzaro Reaction : Aldehydes like benzaldehyde, formaldehyde, trimethyl-acetaldehyde **etc. which** have no α-hydrogen atoms, undergo disproportionation in the presence of strong base to form equal amounts of corresponding alcohol and carboxylic acid, this reaction is known as Cannizzaro reaction, which involves the reduction of one molecule of the aldehyde at the expense of the other e.g.,

(a) $\quad$ 2HCH=O + NaOH $\longrightarrow$ HCOONa + CH$_3$OH
$\qquad$ Formaldehyde $\qquad\qquad\qquad$ Sod. formate $\quad$ Methanol

(b) $\quad$ 2(CH$_3$)$_3$C—CHO + NaOH $\longrightarrow$ (CH$_3$)$_3$C—COONa + (CH$_3$)$_3$C—CH$_2$OH
$\qquad$ Trimethyl acetaldehyde $\qquad\qquad$ Sod. 2,2-Dimethyl $\quad$ 2,2-Dimethyl-1-
$\qquad$ (2,2-Dimethyl propanol) $\qquad\qquad$ propanoate $\qquad\qquad$ propanol
$\qquad\qquad\qquad\qquad\qquad\qquad\qquad\qquad$ (sodium pivalate)

(c) $\quad$ 2C$_6$H$_5$CH=O + NaOH $\longrightarrow$ C$_6$H$_5$COONa + C$_6$H$_5$CH$_2$OH
$\qquad$ Benzaldehyde $\qquad\qquad\qquad\qquad$ Sod. benzoate $\quad$ Benzyl alcohol

Mechanism : It involves nucleophilic addition of :OH$^{\ominus}$ to the first molecule followed by hydride transfer to the other molecule as shown below.

(i) C$_6$H$_5$C(=O)H + :OH⁻ $\underset{}{\overset{Fast}{\rightleftharpoons}}$ C$_6$H$_5$—C(O⁻)(H)(OH)

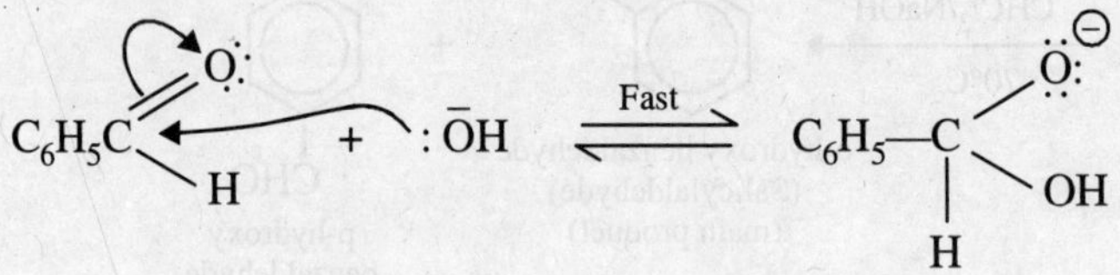

[Second step: two molecules — C$_6$H$_5$—C(H)(OH)(O⁻) + C$_6$H$_5$—C(H)=O: with hydride (H:⁻) transfer, Slow]

↓

C$_6$H$_5$—C(=O)—OH + C$_6$H$_5$CH$_2$—Ö:⁻

(ii) $\qquad\qquad\qquad\qquad\qquad$ ↓ H⁺ Transfer

$\qquad\qquad\qquad$ C$_6$H$_5$C(=O)—Ö:⁻ + C$_6$H$_5$CH$_2$OH
$\qquad\qquad\qquad$ Benzoate ion $\qquad$ Benzyl alcohol

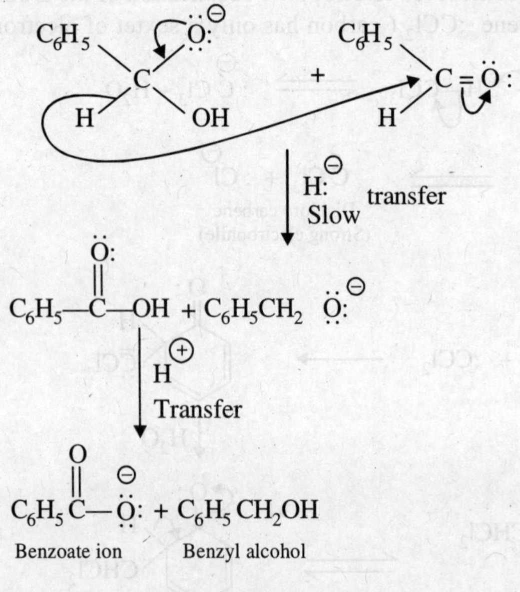

MECHANISM OF ORGANIC REACTIONS

It is evident from above mechanism, that one molecule of the aldehyde acts as a hydride (:H) donor and the other acts as a hydride acceptor. In other words, Cannizzaro reaction is an example of self oxidation and reduction.

Crossed Cannizzaro Reaction : The reaction between two different aromatic aldehydes will give all possible products and therefore, such a reaction is not of much synthetic value. However, a crossed Cannizzaro reaction between an aromatic aldehyde and formaldehyde is an important synthetic reaction due to the greater susceptibility of formaldehyde to oxidation and we get mainly the sodium formate and alcohol corresponds to aromatic aldehyde.

$$\text{Anisaldehyde (p-OCH}_3\text{-C}_6\text{H}_4\text{-CHO)} + HCH=O \xrightarrow{NaOH} \text{p-methoxy benzyl alcohol (p-OCH}_3\text{-C}_6\text{H}_4\text{-CH}_2\text{OH)} + HCOONa$$

4. Skraup Synthesis : It consists in heating a mixture of aniline, nitrobenzene (or arsenic acid), glycerol, conc. sulphuric acid and ferrous sulphate. The overall process involves following steps.

(i) $HOCH_2CH(OH)CH_2OH \xrightarrow[-2H_2O]{\text{conc. }H_2SO_4} H_2C=CH-CH=O$
    Glycerol   Acrolin

(ii) Aniline + Acrolein $\xrightarrow{\text{1, 4-Addition}}$ [intermediate] $\rightleftharpoons$ [intermediate]

$\xrightarrow{\text{Cyclisation}}$ [intermediate] $\xrightarrow{-H^+}$ [intermediate] $\xrightarrow[-H_2O]{\text{conc. }H_2SO_4}$ [dihydroquinoline] $\rightarrow$ [quinoline]

(iii) 1,2-dihydroquinoline $\xrightarrow{C_6H_5NO_2}$ Quinoline + H_2O

5. Diels-Alder Reaction : The Diels-Alder reaction consists in the 1,4-addition of the double bond of a suitable vinyl derivative (called the dienophile) to the two ends of a conjugated diene system. Thus 1,3-butadiene combines with acrolein at 100°C to form tetrahydrobenzaldehyde.

1,3-butadiene + CH₂=CHCHO $\xrightarrow{100\ °C}$ 1,2,3,6-tetrahydrobenzaldehyde

This reaction is highly stereospecific and occurs exclusively in the cis–fashion.

Mechanism : The exact mechanism of Diels-Alder reaction is still uncertain. It has been suggested that a six-membered transition state compound is formed. That is,

[Diene] + CH₂=CH—CHO ⟶ Transition state ⟶ 1,2,3,6-tetrahydrobenzaldehyde

6. Hoffmann Degradation Reaction : When an amide is treated with bromine (or chlorine) in alkali solution, it is converted to a primary amine that has one carbon less than the starting amide.

$$R-\overset{O}{\underset{\|}{C}}-NH_2 + Br_2 + NaOH \xrightarrow{H_2O} R-NH_2$$
(1°-amine)

This reaction which was discovered by Hoffmann and results in the formation of a primary amino by elimination of carbonyl group (≡ one carbon) from an amide, is often referred to Hoffmann-Degradation. The overall reaction of the amide with bromine in the presence of three equivalents of alkali may be represented by the equation.

$$R-CO-NH_2 + Br_2 + 3NaOH \xrightarrow{H_2O} R-NH_2 + 2NaBr + NaHCO_3 + H_2O$$
1°-amine

MECHANISM OF ORGANIC REACTIONS

Hoffmann degradation is particularly useful for stepping down any homologous series by one carbon atom less.

Mechanism : The mechanism of Hoffmann degradation involves the following steps.

(i) Base-catalysed bromination of amide produces N-bromoamide

$$R-\underset{\underset{O}{\|}}{C}-NH_2 + Br_2 + NaOH \xrightarrow{H_2O} R-\underset{\underset{O}{\|}}{C}-\underset{\underset{H}{|}}{N}-Br + NaBr + H_2O$$

(ii) The N-bromoamide then reacts with alkali (NaOH) to give acyl nitrene ion.

$$R-\underset{\underset{O}{\|}}{C}-\underset{\underset{H}{|}}{\overset{..}{N}}-Br + OH^- \longrightarrow R-\underset{\underset{O}{\|}}{C}-\overset{..}{N}: + Br^- + H_2O$$

Acyl nitrene (unstable)

The nitrogen atom of nitrene ion has only six electrons and the ion is highly unstable.

(iii) R group of nitrene then migrates (or rearranges) to form a relatively stable intermediate, called isocyanate.

$$R-\underset{\underset{O}{\|}}{C}-\overset{..}{N} \longrightarrow R-\overset{..}{N}=C=O$$

Isocyanate

(iv) The isocyanate finally reacts with water to give unstable carbamic acid which eliminates a molecule of CO_2 to yield the primary amine.

$$R-\overset{..}{N}=C=O + H_2O \longrightarrow \left[R-\underset{\underset{H}{|}}{\overset{..}{N}}-\underset{\underset{O}{\|}}{C}-OH \right] \longrightarrow R-\overset{..}{N}H_2$$

Carbamic acid (unstable)   1°-amine

Since the key step (iii) in the above mechanism involves rearrangement, the overall reaction resulting in a primary amine is also called Hoffmann Rearrangement. This reaction is often refered to as Hoffmann Bromoamide reaction because 'bromoamide' is an important intermediate as shown in the first step of the mechanism.

QUESTIONS

1. Discuss the various electron displacement effects with examples.
2. Write informative notes on any two of the following :
   (a) Inductive effect   (b) Electromeric effect
   (c) Mesomeric effect   (d) Hyperconjugative effect
3. Write short notes on any three of the following :
   (a) Activation energy   (b) Structure and stability of carbonium ion
   (c) Structure and stability of carbanion   (d) Nucleophilic reagents
   (e) Electrophilic reagents
4. How are the organic reactions classified ? Discuss the mechanism of substitution reactions with a few examples.
5. Explain the mechanism of any two of the following with examples :
   (a) Beckman rearrangement   (b) Reimer – Tiemann reaction
   (c) Cannizaro reaction   (d) Diels–Alder reaction
   (e) Hoffmann degradation

25

Reaction Dynamics & Catalysis

25.1. INTRODUCTION

For the study of chemical processes, the following two areas of chemistry are of vital importance:

(i) *Chemical thermodynamics,* which determines whether a reaction is feasible or not, and

(ii) *Chemical kinetics,* which provides information regarding the rate of reaction, the factors that influence a reaction and the conditions under which a reaction take place and the mechanism or the sequential steps involved in a reaction.

25.2. RATE OF A REACTION OR REACTION VELOCITY

The rate of a reaction may be defined as the change in concentration of any of the reactants and products per unit time.

Let us consider the following simple reaction:

$$R \text{ (reactant)} \longrightarrow P \text{ (product)}$$

As the reaction proceeds with time, the concentration of the reactant R decreases and that of P increases. Hence, the rate of this reaction will be equal to the rate of disappearance of R or the rate of appearance of P.

That means,

$$\text{the rate} = \frac{-d[R]}{dt} = \frac{+d[P]}{dt}$$

where $[R]$ and $[P]$ are the respective concentrations in moles per liter of the reactant and product while d represents infinitesimally small change in concentration. The negative sign shows that the concentration of the reactant R decreases whereas the positive sign shows that there is an increase in concentration of the product, P, as the time passes.

Similarly, if we consider the following reaction

$$A + B \longrightarrow C + D,$$

the rate of reaction $\quad \dfrac{dx}{dt} = \dfrac{-d[A]}{dt} = \dfrac{+d[B]}{dt} = \dfrac{+d[C]}{dt} = \dfrac{+d[D]}{dt}$

Similarly, for a more general reaction

$$aA + bB \longrightarrow cC + dD,$$

the rate of reaction $\quad = -\dfrac{1}{a} \times \dfrac{d[A]}{dt} = -\dfrac{1}{b} \times \dfrac{d[B]}{dt}$

$$= +\frac{1}{c} \times \frac{d[C]}{dt} = +\frac{1}{d} \times \frac{d[D]}{dt}$$

25.3. REACTION RATE AND TIME

According to Guldberg and Waage's law of Mass Action, the rate of a chemical reaction is directly proportional to the product of the active masses of the reactants. Thus, as the chemical reaction proceeds, the molecular concentrations of the reactants are progressively reduced and the rate of the reaction proportionally decreases with time. Figure 5.1 given below shows a graph drawn between the reaction velocity and time.

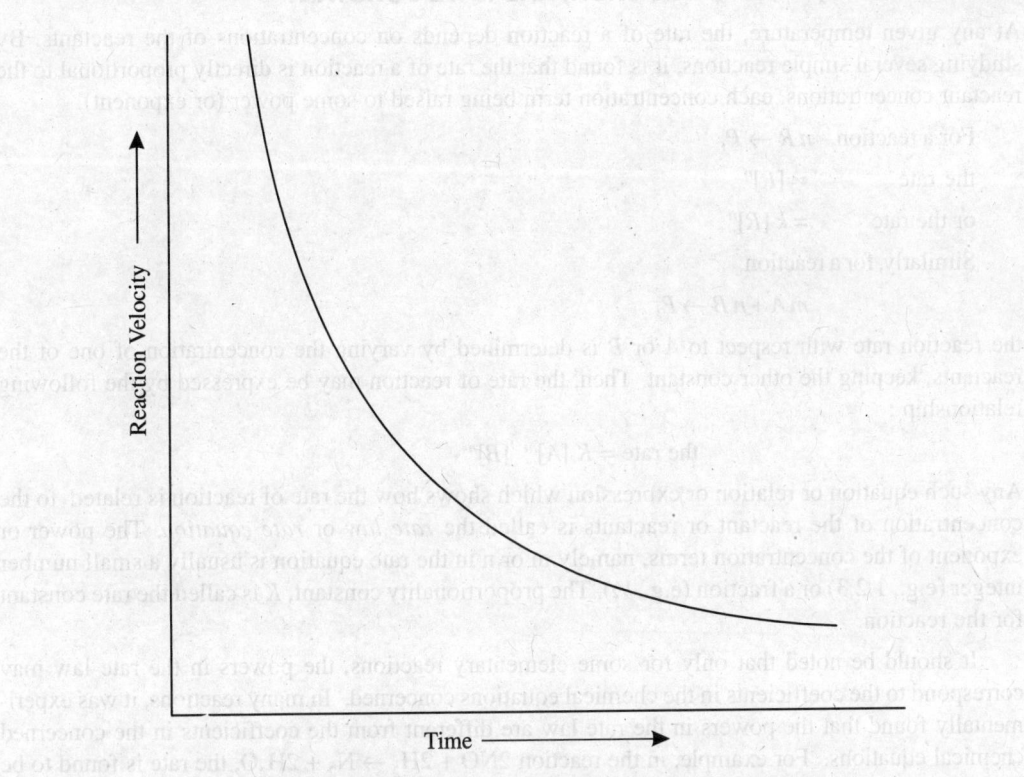

Fig. 25.1. Graph between the reaction velocity and time.

It can be seen from the graph that the reaction velocity rapidly decreases initially and in due course, it decreases very slowly with time. Thus, for completion of the reaction, it takes a very long time.

Units of reaction rate. The reaction rate has the units of concentration (*i.e.* moles per liter) divided by time (*i.e.* seconds, minutes, hours, and so on as the case may be). Thus the units of rate of reaction may be mole/litre/sec (mole l^{-1} s^{-1}) or mole/litre/min (mole l^{-1} min^{-1}) or mole/litre/hour (mole l^{-1} hr^{-1}) as the case may be.

25.4. FACTORS INFLUENCING THE REACTION RATE

1. **Concentration of the reactants.** The rate of a chemical reaction decreases with the decrease in their concentration.

2. **Temperature.** In general, the rate of reaction increases with temperature. In case of some reactions, the rate of reaction may double or even trible at every 10°C of temperature. However, there may be some exceptions. Some reactions which do not take place at room temperature may take place at elevated temperatures.

3. **Catalyst.** The rate of reaction may be increased with the help of a catalyst (Ex : Decomposition of KCl_3 increases in presence of MnO_2 as a catalyst) or decrease in presence of a negative catalyst (Ex : The decomposition of H_2O_2 is retarded in presence of traces of acetanilide).

4. **Surface area of reactants.** The finer the particle size, greater will be the surface area and higher will be the rate of reaction.

5. **Radiation.** The rate of reaction sometimes may be enhanced by visible or U.V. radiation Ex : Reaction between H_2 and O_2 in sun-light.

25.5. RATE LAW (OR RATE EQUATION) AND RATE CONSTANT

At any given temperature, the rate of a reaction depends on concentrations of the reactants. By studying several simple reactions, it is found that the rate of a reaction is directly proportional to the reactant concentrations, each concentration term being raised to some power (or exponent).

For a reaction, $nR \to P$,

the rate $\propto [R]^n$

or the rate $= k[R]^n$

Similarly, for a reaction,

$$mA + nB \to P,$$

the reaction rate with respect to A or B is determined by varying the concentration of one of the reactants, keeping the other constant. Then, the rate of reaction may be expressed by the following relationship :

$$\text{the rate} = K[A]^m[B]^n$$

Any such equation or relation or expression which shows how the rate of reaction is related to the concentration of the reactant or reactants is called the *rate law* or *rate equation*. The power or exponent of the concentration terms, namely m or n in the rate equation is usually a small number integer (e.g., 1,2,3) or a fraction (e.g., ½). The proportionality constant, K is called the rate constant for the reaction.

It should be noted that only for some elementary reactions, the powers in the rate law may correspond to the coefficients in the chemical equations concerned. In many reactions, it was experimentally found that the powers in the rate law are different from the coefficients in the concerned chemical equations. For example, in the reaction $2NO + 2H_2 \to N_2 + 2H_2O$, the rate is found to be proportional to $[H_2]$ although the coefficient of H_2 in the equation is 2.

∴ In this case, the rate of reaction $= K[H_2][NO]^2$

Now, consider the reaction,

$$aA + bB \to cC + dD$$

It is found experimentally that the rate of reaction depends upon only α concentration terms of A and β concentration terms of B. Then, we may write that

The reaction rate $\alpha [A]^\alpha [B]^\beta$

∴ The reaction rate $= K \cdot [A]^\alpha [B]^\beta$

where $[A]$ and $[B]$ are the respective molar concentrations of the reactants A and B, and K is called the *rate constant* or *velocity constant*. The above equation is called the *rate law or rate equation*.

If $[A] = [B] = 1$ mole/litre, then

the reaction rate $= K$. That is the reason why the rate constant, K is also called *specific reaction rate*.

REACTION DYNAMICS & CATALYSIS

25.6. MEASUREMENT OF RATE OF REACTION

The rate of reaction can be measured by determining the concentration of one of the reactants or products at different intervals of time. The usual procedure followed is to withdraw an aliquot of the reaction mixture at different time intervals, cool it to 0°C to freeze the reaction and determine the concentration of the desired reactant or product as the case may be, by a suitable technique such as titrimetry, colorimetry or spectrophotometry. The values of concentration of the reactant are plotted against the corresponding time intervals, when a graph of the type shown in Fig. 5.2 is obtained.

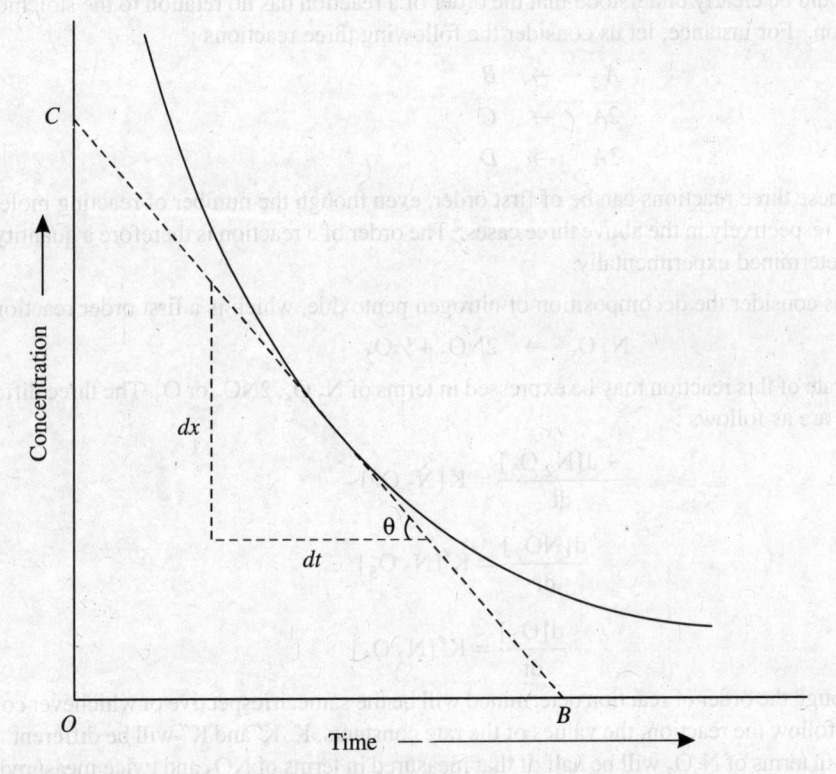

Fig. 25.2. Plot of concentration of the reactant against the corresponding time interval.

If the rate of reaction at the time B is required, a tangent corresponding to that time is drawn on the curve as shown in the figure. The slope of the tangent gives the rate of reaction *i.e.*

the rate of reaction at time $B = \tan \theta = \dfrac{dx}{dt} = \dfrac{OC}{OB}$

Sometimes, instead of concentration, any other suitable property related to concentration (e.g., pressure, volume, optical rotation, absorbance etc.) at different time intervals may be measured to study the reaction kinetics.

25.7. ORDER OF A REACTION

According to the rate law, for a reaction of the type

$$aA + bB \rightarrow cC + dD,$$

if it is experimentally found that the rate of reaction depends only on α conentration terms of A and β, concentration terms of β, then

the rate of the reaction $= K[A]^\alpha [B]^\beta$

(For simple and elementary reactions, α and β may be the same as a and b respectively).

The order of a reaction is defined as the sum of the exponents to which the concentration terms in the rate law are raised to express the rate of reaction.

Thus, the order of reaction in the above case = $(\alpha + \beta)$.

Depending upon the values of $(\alpha + \beta)$ as equal to 0, 1, 2 or 3, the reactions are said to be of zero order, first order, second order or third order respectively. Further α and β are called the orders of the reaction with respect to the reactants A and B respectively.

It should be clearly understood that the order of a reaction has no relation to the stoichiometry of the reaction. For instance, let us consider the following three reactions :

$$A \rightarrow B$$
$$2A \rightarrow C$$
$$3A \rightarrow D$$

All these three reactions can be of first order, even though the number of reacting molecules are 1, 2 and 3 respectively in the above three cases. The order of a reaction is therefore a quantity that can only be determined experimentally.

Let us consider the decomposition of nitrogen pentoxide, which is a first order reaction :

$$N_2O_5 \rightarrow 2NO_2 + \tfrac{1}{2}O_2$$

The rate of this reaction may be expressed in terms of N_2O_5, $2NO_2$ or O_2. The three different rate equations are as follows :

$$\frac{-d[N_2O_5]}{dt} = K[N_2O_5]$$

$$\frac{d[NO_2]}{dt} = K'[N_2O_5]$$

$$\frac{d[O_2]}{dt} = K''[N_2O_5]$$

Although the order of reaction determined will be the same, irrespective of whichever component is used to follow the reaction, the values of the rate constants, K, K' and K'' will be different. The rate measured in terms of N_2O_5 will be half of that measured in terms of NO_2 and twice measured in terms of O_2 because for one mole of N_2O_5 decomposing, two moles of NO_2 and half mole of O_2 are formed,

$$\therefore \quad 2\frac{-d[N_2O_5]}{dt} = 2\frac{d[NO_2]}{dt} = 4\frac{d[O_2]}{dt}$$

$$\therefore \quad 2K = K' = 4K''$$

Hence, while stating the value of rate constant in such cases, it is necessary to specify that with respect to which component, the reaction rate was determined.

EXAMPLE:

Chemical reaction	Rate equation	Order of the reaction
$2N_2O_5 \rightarrow 4NO_2 + O_2$	rate = $K[N_2O_5]$	1
$H_2 + I_2 \rightarrow 2HI$	rate = $K[H_2][I_2]$	2
$2NO + 2H_2 \rightarrow N_2 + 2H_2O$	rate = $K[NO]^2[H_2]$	3
$CHCl_3 + Cl_2 \rightarrow CCl_4 + HCl$	rate = $K[CHCl_3][Cl_2]^{1/2}$	1½

25.8. ZERO ORDER REACTION

A reaction is said to be of zero order when the reaction rate is independent of concentration. That means the rate of reaction remains same irrespective of the concentration.

Example:
$$NO_2 + CO \rightarrow NO + CO_2$$

At 200° C, the rate of this reaction = $K[NO_2]^2$. Obviously, the rate does not depend on the [CO], and that is why it is not included in the rate equation, and the power of [CO] is considered to be zero. Thus, the reaction is of zero order with respect to [CO] whereas with respect to [NO_2], it is of second order. Accordingly, the order of the overall reaction is $(2 + 0) = 2$.

The rate equation for a zero order reaction is : $dx/dt = K_o$, where K_o is the rate constant.

The kinetic equation for a zero order reaction is : $K_o = x/t$.

The half-life period of a zero order reaction, $t_{1/2} = a/2K_o$.

This shows that for a zero order reaction, the half life period is directly proportional to initial concentration.

Zero order reactions have come across mostly in heterogeneous reactions.

In some reactions, the order with respect to a reactant can be rendered zero by taking the reactant in sufficiently large concentration so that the change in its concentration due to its involvement in the reaction is negligible.

The derivation of the integrated rate equation and examples are given in latter sections.

25.9. MOLECULARITY OF A REACTION

The stoichiometry of a reaction gives information as to the minimum number of molecules of reactants which lead to the formation of products. This is known as molecularity of the reaction. Thus, molecularity is defined as the sum total of the number of molecules of various reactants that take part in the chemical reaction as represented by the balanced chemical equation.

On the other hand, the sum of the powers to which the concentration terms are raised in the rate law is known as the order of the reaction.

Chemical reactions may be broadly considered under two types :
(i) Elementary reactions which take place in a single step only.
(ii) Complex reactions which take place in two or more steps.

Example: For the reaction,
$A + B \rightarrow$ products,
reaction rate α $[A]$ $[B]$
reaction rate $= K$ $[A]$ $[B]$

Hence, in this case, both the molecularity as well as order is 2.

Molecularity Vs. order of a reaction

	Molecularity of a reaction	Order of a reaction
1.	It is the sum total of the number of molecules of the various reactants which take part in the chemical reaction as represented by the balanced chemical equation.	It is the sum total of the powers of the concentration terms in the rate law expression.
2.	In case of simple one step reactions, it is inferred from the stoichiometric chemical reaction represented by the balanced equation.	It is determined directly by kinetic experiments.

3. It is always a whole number.	It can be a whole number or it can have a fractional value
4. It can never be zero or negative.	It can be zero or negative also.
5. It doe snot vary with conditions such as temperature, pressure or concentration.	It can vary with conditions such as temperature pressure or concentration.

In complex reactions or such reactions which involve more than one step, the molecularity can be determined only after knowing the reaction machanism, *i.e.* knowledge about the various elementary reactions that constitute the overall chemical reaction. These elementary reactions may not proceed at the same rate. According to the *"bottle-neck principle"*, the rate of the *slowest of all the elementary reactions* determines the overall rate of the reaction, and hence it becomes the rate-determining step.

Hence, in case of complex reactions, where the reactions may take place in 2 or more stepes, the molecularity is taken as the number of reacting species taking part in the rate-determining step in the overall reaction mechanism.

25.10. PSEUDO-ORDER REACTIONS

If one of the reactants in a chemical reaction is present in a large excess, small changes in concentration of that reactant does not seem to have any effect. In such cases the experimentally determined order of the reaction will be different from the actual value. Such reactions are called pseudo-order reactions.

Example

$$CH_3COOC_2H_5 + H_2O \rightarrow CH_3COOH + C_2H_5OH$$
ethyl acetate (large excess) acetic acid ethyl alcohol

rate = K [CH$_3$COOH] [H$_2$O]
    = K' [CH$_3$COOH]

Because of the large excess of H$_2$O present, its concentration remains practically unchanged during the course of the reaction. Therefore, although the reaction is actually of second order, in experimental practice it appears to be of first order. Thus, this reaction is said to be a pseudo-first-order reaction.

Another example of pseudo-first-order reaction is provided by the hydrolysis of sucrose in the presence of a dilute mineral acid :

$$C_{12}H_{22}O_{11} + H_2O \rightarrow C_6H_{12}O_6 + C_6H_{12}O_6$$
surface (large excess) glucose fructose

reaction rate = K [C$_{12}$H$_{22}$O$_{11}$] [H$_2$O]
            = K' [C$_{12}$H$_{22}$O$_{11}$]

25.11. INTEGRATED RATE EQUATIONS

(A) Zero Order Reactions

As stated earlier, for any zero order reaction, the rate of reaction is independent of concentration of the reactants. Hence the rate of reaction remains constant through out the reaction. Let us consider a general zero order reaction of the following type :

(a) Integrated rate reaction

$$A \longrightarrow products$$

If the initial concentration of A is 'a' moles/liter and after time t, let x moles per liter of it have reacted or decomposed. Then, the rate of equation for zero order reaction can be written as follows:

$$\frac{dx}{dt} \propto a$$

or

$$\frac{dx}{dt} = K_0$$

where K_0 is the rate constant for the zero order reaction,

$$\therefore \quad dx = K_0 \cdot dt$$

on integration of this equation, we get,

$$x = K_0 t + I$$

where I is the constant of integration.

When $t = 0$, $x = 0$, and hence $I = 0$.

$$\therefore \quad x = K t \qquad \qquad \ldots(1)$$

That is, the amount of A reacted $\propto$ time. Also, the rate constant is equal to the rate of reaction at all concentrations.

(b) **Half-life period of a reaction** is defined as the time taken for half of the reaction to complete. For a zero order reaction, the half life period can be calculated as follows:

When $\qquad x = a/2$, $t = t_{1/2}$

Substituting these values in the equation (1) above, we get

$$a/2 = K\, t_{1/2}$$

or $\qquad t_{1/2} = a/2K \qquad \therefore t_{1/2} \propto a.$

It means that the half-life period of a zero order reaction is proportional to the initial concentration of the reactant.

(c) **Characteristics of zero-order reactions**

1. The rate of reaction is independent of the concentration of all the reactants.
2. The rate constant is equal to the rate of reaction at all concentrations.
3. The half-life period is proportional to the initial concentration of the reactants -

$$t_{1/2} = a/2K$$

4. The unit of the rate constant of a zero order reaction is mole lit^{-1} time^{-1}.
5. The concentration of the product increases linearly with time. Therefore, if a graph is drawn between the moles of product formed (x) and time (t), a straight line passing through the origin is obtained.

(d) **Examples of zero-order reactions**

(i) The reaction between hydrogen and chlorine in presence of light leading to the formation of hydrogen chloride. This reaction can be studied in an experimental assembly shown in Fig. 5.3A. The H_2 ↑ and Cl_2 ↑ enclosed in the tube (which is inverted in a trough of waters react with each other in the presence of light forming HCl ↑. As this dissolves in water and the level of water in the tube rises (Fig. 5.3B). In this manner, although the quantities of H_2 and Cl_2 decrease as the reaction proceeds, their quantities (or concentration) per unit volume remain the same. Thus, the same rate of reaction is maintained in spite of the decrease in the initial quantities of H_2 and Cl_2.

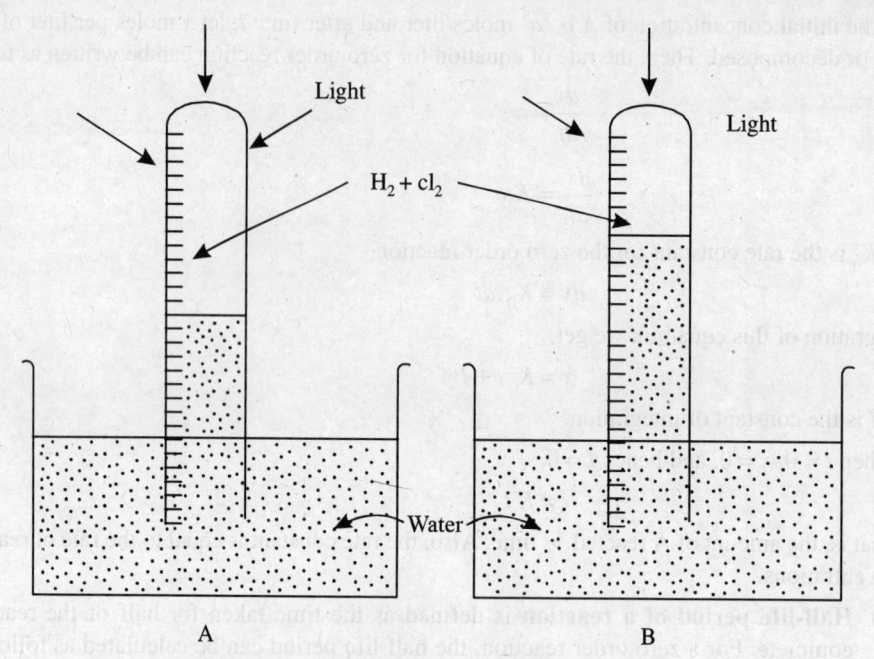

Fig. 25.3. Photochemical reaction between H_2 and Cl_2.

(ii) Decomposition of hydrogen iodide on the surface of gold.

(iii) Decomposition of ammonia on the surface of catalyst such as tungsten, platinum or molybdenum at high temperatures. The gaseous reactant molecules get absorbed on the catalyst surface. As soon as some of these reactant molecules are converted into products and escape out, new reactant molecules occupy their position, thereby maintaining the constant concentration of the reacting molecules sorbed on the catalyst. Thus the rate of reaction remains constant irrespective of the quantity of the gaseous reactants initially taken.

(iv) Some enzymatic reactions.

(B) First Order Reactions

A reaction is said to be of first order if the rate of the reaction depends upon concentration term of one reactant only. In other words, the reaction rate of a first order reaction is determined by the change in concentration term of one of the reactants only.

(a) Integrated rate equation

Let us consider a reaction

$$A \longrightarrow B + C$$

Initial concn. a o o
Concn. after time, $a-x$ x x

Let us suppose that the concentration of A at the beginning of the reaction (*i.e.* when $t = o$) be 'a' moles per liter. If after time, t seconds, let x moles of A have reacted. Then, the concentration of A becomes $(a - x)$. Thus amounts of B and C formed after time, 't' seconds will accordingly be x moles/liter each.

Since we know that for a first order reaction, the rate of reaction is directly proportional to the concentration of the reactant at any given time,

$$\frac{dx}{dt} \propto (a-x) \quad \therefore \quad \frac{dx}{dt} = K(a-x) \qquad \ldots(1)$$

where K is the rate constant for the reaction.

$$\therefore \quad \frac{dx}{(a-x)} = K \cdot dt \qquad \ldots(2)$$

On integration of this rate equation, we get

$$\int \frac{dx}{(a-x)} = \int K \cdot dt \qquad \ldots(3)$$

or
$$-\log_e (a-x) = Kt + I \qquad \ldots(4)$$

where I is the constant of integration. The constant K may be evaluated by considering that when $t = o$, x will also be $= 0$. Substituting these values in the above equation 4, we get

$$-\log_e a = I$$

Substituting this value of I in eqn. (4), we have

$$-\log_e (a-x) = K.t - \log_e a$$

$$\therefore \quad K = \frac{1}{t} \log_e \frac{a}{(a-x)} \qquad \ldots(5)$$

converting into common logarithms, we get

$$K = \frac{2.303}{t} \log_{10} \frac{a}{(a-x)} \qquad \ldots(6)$$

The values of K can be obtained by substituting the values of a and $(a-x)$ determined experimentally at different intervals of time.

When the values of $\log_{10} \frac{a}{(a-x)}$ is plotted against corresponding values of time, t, a straight line passing through the origin is obtained. The slope of such a straight line will be $\frac{2.303}{k}$

Sometimes, the integrated rate law is used in the following form :

$$K = \frac{2.303}{(t_2 - t_1)} \log \frac{(a - x_1)}{(a - x_2)} \qquad \ldots(7)$$

where x_1 and x_2 are the amounts of the reactant decomposed time intervals t_1 and t_2 respectively from the beginning of the reaction.

(b) Half-life period of a first order reaction

Half life period of a reaction is defined as the time required for completion of half of the reaction. In other words, it is the time required for half of the reactant to convert into products. It is denoted by $t_{\frac{1}{2}}$. At half-life period *i.e.* at time $t_{\frac{1}{2}}$, the concentration of A will be $a/2$. Hence from eqn. (6), we get,

$$t_{1/2} = \frac{2.303}{K} \log_{10} \frac{a}{(a/2)} = \frac{2.303}{K} \log_{10} 2$$

$$\therefore \quad t_{1/2} = \frac{0.693}{K} \qquad \ldots(8)$$

Thus, it can be seen that for a first order reaction, the half life period is independent of the initial concentration of the reactant.

(c) Characteristics of first order reactions :-

(i) In a first order reaction, the rate of reaction depends upon one concentration term only or any property depending upon the concentration of optical rotation, pressure, absorbance etc.

(ii) A first order reaction should obey the following rate equations with usual notations :

$$K = \frac{2.303}{t} \log_{10} \frac{a}{(a-x)}$$

$$K = \frac{2.303}{(t_2 - t_1)} \log_{10} \frac{(a - x_1)}{(a - x_2)}$$

(iii) The units of K are independent of the units in which concentrations are expressed as long as both 'a' and $(a - x)$ are expressed in the same concentration units. The units of K depends only upon the unit of time, usually second^{-1}, minute^{-1} or hour^{-1}.

(iv) The half-life period of a first order reaction is independent of initial concentration of the reactant. Also, the half-life period is inversely proportional to the rate constant

$$t_{1/2} = \frac{0.693}{K}$$

In fact, the time taken for any fraction of the reaction to be completed is independent of the initial concentration.

(v) The first order rate expression

$$K = \frac{2.303}{t} \log \frac{a}{(a-x)}$$

can also be written as follows :

$$(a - x) = a \cdot e^{-Kt} \qquad ...(9)$$

It is evident from this expression that $(a - x)$ will become zero only at infinite time.

Examples

Radioactive decay of radium (Ra) whose half-life period is 1620 yrs. and also, the decay is independent of the initial quantity of Ra.

Examples of First Order Reaction

1. Inversion of cane sugar (sucrose) in presence of an acid which acts as a catalyst.

$$\underset{\text{Sucrose}}{C_{12}H_{22}O_{11}} + H_2O \xrightarrow[\text{as catalyst}]{H^+ \text{ ions}} \underset{\text{Glucose}}{C_6H_{12}O_6} + \underset{\text{Fructose}}{C_6H_{12}O_6}$$

The reactant sucrose rotates the plane of polarisation to the right (dextro-rotatory) whereas one of the products glucose is dextro-rotatory (+ 52.5° C) and the other product, fructose is leavo-rotatory i.e., it rotates the plane of polarisation to the left (– 92°). Thus, the change in rotation is proportional to the decomposition of sucrose. Therefore, as the sucrose disappears during the reaction, the dextro-rotation goes on decreasing progressively due to the predominant laevo-rotation of the progressively increasing amount of fructose formed. The order of the reaction as being first order can be confirmed by substituting the values of the rotation of plane of polarised light measured at different time intervals using a polarimeter in the following first order rate expression :

$$K = \frac{2.303}{t} \log \frac{r_o - r_\infty}{r_t - r_\infty}$$

where r_o = initial value of rotation of polarised light due to sucrose solution,

r_t = value of rotation after time, t

r_∞ = value of rotation at the end of the inversion experiment.

2. Hydrolysis of an ester such as ethyl acetate in presence of a mineral acid :

$$\underset{\text{ethyl acetate}}{CH_3COOC_2H_5} + H_2O \xrightarrow{H^+} \underset{\text{acetic acid}}{CH_3COOH} + \underset{\text{ethyl alcohol}}{C_2H_5OH}$$

As the hydrolysis of ethyl acetate proceeds, there will be proportional increase in the concentration of acetic acid formed. Therefore, the kinetics of the reaction can be studied by taking a known quantity of ethyl acetate and mixing it with a relatively large quantity of N/2 HCl. An aliquot of the reaction mixture is withdrawn at different intervals of time and titrated against a standard alkali. Obviously, as the reaction proceeds, the value of alkali required to neutralise the acid (HCl present as catalyst + CH_3COOH produced by hydrolysis of the ester) progressively increases.

The fact that this is a first order reaction is established by substituting the results in the first order rate expression;

$$K = \frac{2.303}{t} \log \frac{a}{(a-x)}$$

and verifying the constancy of the value of rate constant, K.

or in this case, $(a-x) \propto (v_\infty - v_t)$ and $a \propto (v_\infty - v_o)$

$$\therefore \quad K = \frac{2.303}{t} \log \frac{(v_\infty - v_o)}{(v_t - v_o)}$$

where, v_o = initial titre value,
v_∞ = final titre value at the end of the experiment, and
v_t = titre value at the various time intervals chosen.

3. Decomposition of N_2O_5 in carbon tetrachloride :

$$N_2O_5 \xrightarrow{\text{in CCl}_4} 2NO_2 + \frac{1}{2}O_2$$

The kinetics of the reaction can be studied by measuring the volume of oxygen evolved at different time intervals (vide Fig. 5.4)

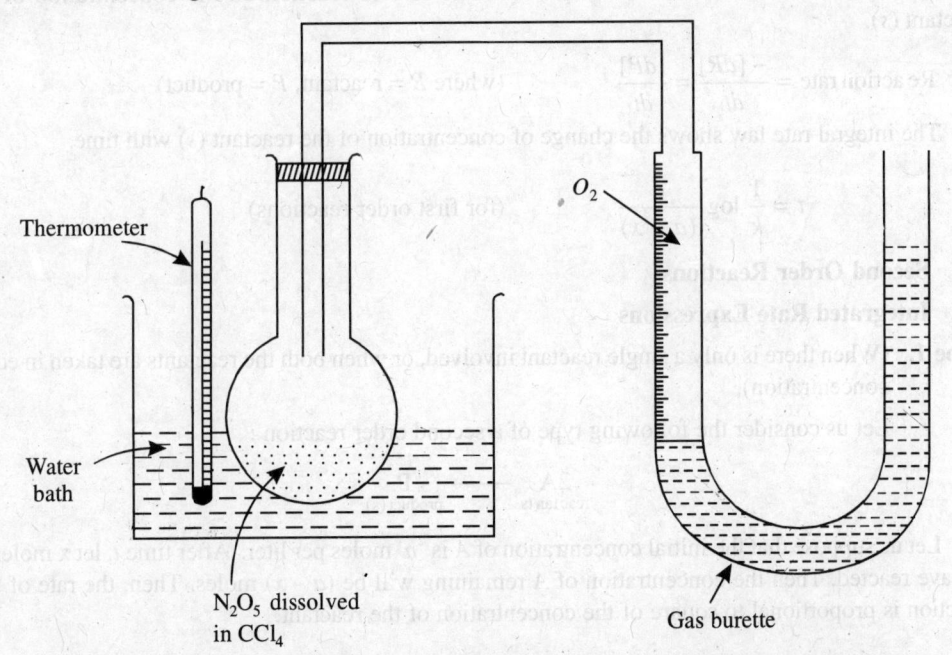

Fig. 25.4. Experimental assembly to monitor the kinetics of decomposition of N_2O_5 in CCl_4.

The fact that this reaction is of first order can be established by substituting the volumes of oxygen evolved at various time intervals selected (V_t) and the volume of oxygen evolved at the completion of the reaction, in the first order rate expression :

$$K = \frac{2.303}{t} \log_{10} \frac{v_\infty}{(v_\infty - v_t)},$$

where v_∞ is proportional to 'a' and $(v_\infty - v_t)$ is proportional to $(a - x)$ in the original first order rate equation.

4. **Radioactive decay of radioisotopes :**

The spontaneous decay of radioactive isotopes follow first order reaction. The number of radioactive atoms N present in a system at any time t is related to the number of the radioactive atoms initially present, N_o. Also, K is the decay constant.

As stated in eqn. (9) earlier, the first order rate expression can be stated as

$$(a - x) = a\, e^{-kt}$$

Substituting N_o for 'a' and N for $(a - x)$, we get

$$N = N_o \cdot e^{-kt}$$

we know that $K = \dfrac{0.693}{t_{1/2}}$ where $t_{1/2}$ = half life period of a radioisotope.

$$\therefore \qquad N = N_o \cdot e^{\frac{0.693}{\lambda} t}$$

The radioactivity of the material is proportional to the number of radioactive atoms present. Hence, this equation is useful for the safe application of radioisotopes in medicine and in carbon dating.

Difference Between Differential and Integral Rate Laws

The differential rate law shows the dependence of rate of reaction on the concentration of the reactant (s).

$$\text{Reaction rate} = \frac{-[dR]}{dt} = \frac{[dP]}{dt} \qquad \text{(where } R = \text{reactant, } P = \text{product)}$$

The integral rate law shows the change of concentration of the reactant (s) with time.

$$t = \frac{1}{k} \log \frac{a}{(a - x)} \qquad \text{(for first order reactions)}$$

(C) Second Order Reactions

(a) Integrated Rate Expressions

Type 1. (When there is only a single reactant involved, or when both the reactants are taken in equal concentration).

Let us consider the following type of a second order reaction :

$$\underset{\text{reactants}}{2A} \longrightarrow \underset{\text{product(s)}}{P}$$

Let us suppose that the initial concentration of A is 'a' moles per liter. After time t, let x moles of A have reacted. Then the concentration of A remaining will be $(a - x)$ moles. Then, the rate of this reaction is proportional to square of the concentration of the reactant.

i.e.
$$\frac{dx}{dt} \propto (a-x)^2$$

∴
$$\frac{dx}{dt} = K(a-x)^2 \qquad ...(10)$$

where K is the rate constant

On rearrangement of this equation, we have,

$$\frac{dx}{(a-x)^2} = K \cdot dt \qquad ...(11)$$

on integration, we get

$$\frac{1}{(a-x)} = Kt + I \qquad ...(12)$$

where I is the integration constant.

If $x = o$ and $t = o$,

$$I = \frac{1}{a}$$

Substituting this value of I in eqn. (12), we have

$$\frac{1}{(a-x)} = Kt + \frac{1}{a} \qquad ...(13)$$

$$Kt = \frac{1}{(a-x)} - \frac{1}{a}$$

∴
$$K = \frac{1}{t} \cdot \frac{x}{a(a-x)} \qquad ...(14)$$

This is the integrated rate equation for a second order reaction of Type-1.

The units of the rate constant, K for a second order reaction are mole^{-1} L time^{-1}

Type-2. (When different concentrations of the reactants are taken)

Let us consider a second order reaction of the following type :

$$\underset{\text{reactants}}{A + B} \longrightarrow \underset{\text{product(s)}}{P}$$

Let us suppose that the initial concentration of A = 'a' moles/litre, and the initial concentration of B = 'b' moles/litre. Let x molecules of A and B each have reacted after time, t. Then, the respective concentrations of A and B after time, t would be $(a-x)$ moles/litre and $(b-x)$ moles / litre.

∴
$$\frac{dx}{dt} \propto (a-x)(b-x) \qquad ...(15)$$

$$\frac{dx}{dt} = K(a-x)(b-x) \qquad \text{where K is the rate constant.}$$

on rearrangement,

$$\frac{dx}{(a-x)(b-x)} = K\,dt \qquad ...(16)$$

On resolving the left hand side of this equation into partial fractions, we can rewrite this equation as follows :

$$\frac{1}{(a-b)}\left[\frac{1}{(b-x)} - \frac{1}{(a-x)}\right]dx = K\,dt \qquad ...(17)$$

on integration of this equation, we get

$$\frac{1}{(a-b)}\left[\int\frac{dx}{(b-x)} - \int\frac{dx}{(a-x)}\right] = \int K\,dt$$

or

$$\frac{1}{(a-b)}[-\log_e(b-x) + \log_e(a-x)] = Kt + I \qquad ...(18)$$

where I is the constant of integration.

If $x = 0$ and $t = 0$,

$$I = \frac{1}{(a-b)} \cdot \log_e \frac{a}{b} \qquad ...(19)$$

Substituting this value of I in eqn. (18), we get

$$\frac{1}{(a-b)} \log_e \frac{(a-x)}{(b-x)} = Kt + \frac{1}{(a-b)} \log_e \frac{a}{b}$$

$$\therefore \quad Kt = \frac{1}{(a-b)} \log_e \frac{b(a-x)}{a(b-x)}$$

$$\therefore \quad K = \frac{1}{t(a-b)} \log_e \frac{b(a-x)}{a(b-x)} \qquad ...(20)$$

converting it to logarithm to the base 10, we get

$$K = \frac{2.303}{t(a-b)} \log_{10} \frac{b(a-x)}{a(b-x)} \qquad ...(21)$$

Eqn. (21) is the integrated rate equation for a second order reaction of Type 2. The units of the rate constant, K for a second order reaction will be

$$\frac{1}{\text{sec}} \cdot \frac{1}{\text{mole/litre}} = \text{mole}^{-1}\,\text{L time}^{-1}$$

(b) Characteristics of Second Order Reactions

1. A second order reaction follows one of the following integrated rate equations :

For Type-1 reactions

$$K = \frac{1}{t} \cdot \frac{x}{a(a-x)}$$

(when there is a single reactant or when both the reactants are taken in equal concentrations)

For Type 2 reactions

$$K = \frac{2.303}{t(a-b)} \cdot \log_{10} \frac{b(a-x)}{a(b-x)}$$

(when different concentrations of the reactants are taken)

2. It is not mandatory that in a second order reaction, there should be only two reactants always. Indeed, there could be only one reactant but the reaction rate may be proportional to the square of its concentration. On the other hand, there could be more than two reactants, but the reaction rate may be proportional to the product of concentrations of any two of the reactants only.

For example, the following reaction :
$$K_2S_2O_8 + 2KI \longrightarrow 2K_2SO_4 + I_2$$
appears to be a third order reaction, but the rate of reaction is given as :
$$\frac{dx}{dt} = K.[S_2O_8^{2-}][I^-]$$
Hence, this is a second order reaction.

3. The units of second order rate constant, K are
   (time^{-1} × concn^{-1}) or (mole/litre^{-1} sec^{-1})
4. The units of K depend upon the units of concentration as well as the units of time used.
5. The time taken for the completion of any fraction of the reaction is inversely proportional to the initial concentration of the reactant.

For instance, let us calculate the time required for the completion of half of the reaction, denoted by $t_{1/2}$, when $x = a/2$, $t = t_{1/2}$. Substituting these values in the second order eqn. of Type-1, viz.,

$$K = \frac{1}{t} \cdot \frac{x}{a(a-x)}$$

or
$$t = \frac{1}{k} \cdot \frac{x}{a(a-x)}$$

we get
$$t_{1/2} = \frac{1}{k} \cdot \frac{a/2}{a(a-a/2)} = \frac{1}{K} \cdot \frac{1}{a}$$

$$\therefore \quad t_{1/2} \propto \frac{1}{a}$$

Similarly, we can prove that the time taken for the completion of any fraction of time is inversely proportional to the initial concentration of the reactants.

6. The rate equation of the second order Type-1 reaction can also be verified by plotting time vs. $\frac{1}{(a-x)}$ or $\frac{x}{(a-x)}$. A linear plot is obtained (Fig. 5.5). K, the second order rate constant can be calculated from the slope of the plot.

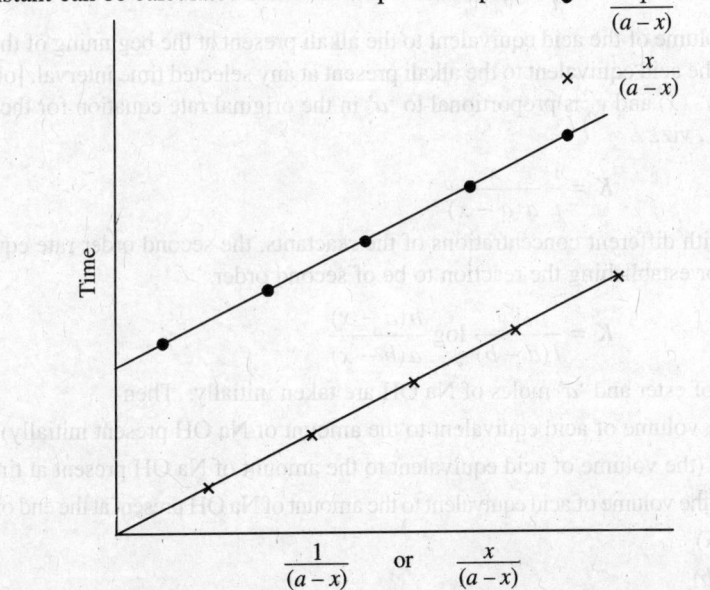

Fig. 25.5. Evaluation of the second order rate constant.

7. If one of the reactants is present in a large excess, the second order rate expression approximates to the first order rate expression.

Thus, if we take very large excess of 'a', the values of 'b' and 'x' in the second order rate equation given below, can be neglected because they will be insignificantly small in comparison with 'a'.

$$K = \frac{2.303}{t(a-b)} \log_e \frac{b(a-x)}{a(b-x)} \quad \text{(Second order rate expression)}$$

This second order rate expression, accordingly, approximates to the first order rate expression as follows:

$$K = \frac{2.303}{t \cdot a} \log \frac{ba}{a(b-x)}$$

or

$$K' = \frac{2.303}{t} \log_{10} \frac{b}{(b-x)} \quad \text{(First order rate expression)}$$

Under these conditions, such reactions are called pseudo first order reaction, although originally they are second order reactions

(c) Examples of Second Order Reactions

1. Saponification of esters with sodium hydroxide.

$$CH_3COOC_2H_5 + NaOH \longrightarrow CH_3COONa + C_2H_5OH$$

As the reaction proceeds, the concentration of the reactants decrease. If we start with equal concentrations of the reactants, the concentration of NaOH can be conveniently followed at different time intervals by withdrawing an aliquot from the reaction mixture and determining the NaOH present in it by direct or indirect titration with a standard solution of an acid.

The fact that this reaction is of second order can be established by the constancy of the values of K determined by substituting the titration results at different time intervals in the second order rate equation as follows :

$$K = \frac{1}{t} \cdot \frac{(v_o - v_t)}{v_o \cdot v_t}$$

where v_o is the volume of the acid equivalent to the alkali present at the beginning of the expt., and v_t is the volume of the acid equivalent to the alkali present at any selected time interval. [obviously, v_t is proportional to $(a - x)$ and v_o is proportional to 'a' in the original rate equation for the second order reaction of type 1, viz.,

$$K = \frac{1}{t} \cdot \frac{x}{a(a-x)}$$

If we start with different concentrations of the reactants, the second order rate equation Type-2 should be used for establishing the reaction to be of second order.

$$K = \frac{1}{t(a-b)} \log \frac{b(a-x)}{a(b-x)}$$

If 'b' moles of ester and 'a' moles of NaOH are taken initially. Then

'a' $\propto v_o$ (the volume of acid equivalent to the amount of NaOH present initially).

$(a - x) \propto v_t$ (the volume of acid equivalent to the amount of NaOH present at time, t).

$(a - x) \propto v_\infty$ (the volume of acid equivalent to the amount of NaOH present at the end of the reaction).

$x = a - (a - x)$

$b = a - (a - b)$

$$\therefore \quad K = \frac{1}{t \cdot v_\infty} \log \frac{(v_o - v_\infty) v_t}{v_o (v_t - v_\infty)}$$

2. Decomposition of hydrogen iodide in gaseous state

$$2HI\ (g) \rightarrow H_2(g) + I_2\ (s)$$

3. Reaction between potassium persulfate and potassium iodide

$$K_2S_2O_8 + 2KI \rightarrow 2K_2SO_4 + I_2.$$

This reaction takes takes place in the following three stages

$$S_2O_8^{2-} \rightleftharpoons 2SO_4^- \quad \text{................(1st stage)}$$

$$SO_4^- + I^- \rightarrow SO_4^{2-} + I \quad \text{............(2nd stage)}$$

$$I + I \rightarrow I_2 \quad \text{............(3rd stage)}$$

while stage 1 and stage 3 are fast, stage 2 is slow and hence the rate determining one. Accordingly, this reaction is of second order.

4. Conversion of ammonium cyanate to urea.

$$NH_4CNO \rightarrow CO\ (NH_2)_2$$

The rate of reaction is found to be proportional to the square of the concentration of ammonium cyanate. Hence, the reaction is of second order.

(D) Third Order Reactions

(a) Integrated Rate Equation

Let us consider a simple type of third order reaction such as

$$3A \rightarrow \text{products}$$

Let the initial concentration of A be 'a' moles/liter and after time t, let x moles of A have reacted, leaving behind (a − x) moles of A. Then, the reaction rate

$$\frac{dx}{dt} \propto (a - x)^3 \quad ...(22)$$

$$\frac{dx}{dt} = K (a - x)^3 \quad ...(23)$$

where K is the rate constant.

By rearranging the equation, we get

$$\frac{dx}{(a - x)^3} = K\,dt \quad ...(24)$$

on integration, we get

$$\frac{1}{2(a - x)^2} = Kt + I \quad ...(25)$$

where I is the constant of integration.

'I' can be evaluated by considering that when $t = o, x = o$. By substituting these values in eqn. (25), we get

$$I = \frac{1}{2a^2}$$

on substituting this value of I in eqn. (24) and rearranging, we get,

$$kt = \frac{1}{2(a-x)^2} - \frac{1}{2a^2}$$

$$\therefore \quad K = \frac{1}{2t} \left[\frac{x(2a-x)}{a^2(a-x)^2} \right] \quad \ldots(26)$$

This is the integrated rate equation for a third order reaction.

(b) Units of K.
The units of the third order rate constant are

$$\frac{1}{(mol/litre)^2} \times \frac{1}{time} = mole^{-2}\, L^2\, time^{-1}$$

(c) Half Life Period, $t_{1/2}$.
At $t_{1/2}$, $x = a/2$. Substituting these values in eqn. 26, we get

$$t_{1/2} = \frac{1}{2k} \left[\frac{a/2(2a - a/2)}{a^2(a - a/2)^2} \right]$$

on simplification, we get

$$t_{1/2} = \frac{1}{2k} \cdot \frac{3}{a^2} = \frac{3}{2k} \cdot \frac{1}{a^2}$$

$$\therefore \quad t_{1/2} \propto \frac{1}{a^2}$$

(d) Examples of Third Order Reactions
(i) $2NO\,(g) + O_2\,(g) \rightarrow 2\,NO_2\,(g)$
(ii) $2FeCl_3\,(aq) + SnCl_2\,(aq) \rightarrow 2FeCl_2\,(aq) + SnCl_4\,(aq)$
(iii) $2NO + Cl_2 \rightarrow 2NOCl$
(iv) $2NO + H_2 \rightarrow N_2O + H_2O$
(v) $2NO + D_2 \rightarrow N_2O + D_2O$
(vi) $2NO + Br_2 \rightarrow 2NOBr$

25.12. REACTIONS INVOLVING MORE THAN THREE MOLECULES

Reactions involving four or more molecules (*i.e.* reactions with high molarity) are extremely rare. According to the kinetic theory, the rate of a chemical reaction is proportional to the number of colisions taking place between all the reacting molecules. Obviously, the possibility of simultaneous collision of more than 3 molecules is very rare. However, such reactions can take place through a series of steps involving lesser number of molecules. Such reactions, which do not take place in one-step, but take place in a series of steps or stages, are known as complex reactions. In such cases, the slowest of the reaction steps becomes the rate determining step.

25.13. METHODS FOR DETERMINATION OF ORDER OF A REACTION

(1) Methods Using Integrated Rate Equations
The reaction under investigation is conducted with different initial concentrations of the reactant, 'a' and determining the concentration $(a - x)$ after regular time intervals, (t). Then the experimental values of t, a, and $(a-x)$ are substituted into the integrated rate equations of the first, second and third order reactions. The rate equation which gives a constant value for K corresponds to the order of the reaction being studied.

(2) Fractional Changes Method or Method Based on Half-Life Period

We know that the half-life period ($t_{1/2}$) for reactions of different orders are related to the initial concentration (A) of the reactants as follows.

1st order reaction → $t_{1/2}$ is independent of concentration.

2nd order reaction → $t_{1/2} \propto \dfrac{1}{[A]}$

3rd order reaction → $t_{1/2} \propto \dfrac{1}{[A]^2}$

n th order reaction → $t_{1/2} \propto \dfrac{1}{[A]^{n-1}}$

For evaluating the order of a reaction under study, two separate experiments are performed by taking different initial concentrations of a reactant and the progress of the reaction is observed. When the initial concentration is reduced to half, the time is noted in each case. Let the two initial concentrations selected be $[A_1]$ and $[A_2]$ and the corresponding half-life periods be t_1 and t_2 respectively.

Substituting the values, we have

$$\frac{t_2}{t_1} = \left[\frac{A_1}{A_2}\right]^{n-1}$$

∴ $\quad (n-1)\log\left[\dfrac{A_1}{A_2}\right] = \log\left[\dfrac{t_1}{t_2}\right]$

Then, the order of reaction, n is obtained by the following relation:

$$n = 1 + \frac{\log[t_1/t_2]}{\log[A_2/A_1]}$$

(3) Vant Hoff's Differential Method

The rate of a reaction of nth order is proportional to the nth power of concentration.

$$\frac{-dc}{dt} \propto c^n$$

$$\frac{-dc}{dt} = Kc^n \qquad \ldots(1)$$

Experiments with two different initial concentrations (c_1 and c_2) of a reactant are carried out. Then,

$$\frac{-dc_1}{dt} = Kc_1^n \qquad \ldots(2)$$

$$\frac{-dc_2}{dt} = Kc_2^n \qquad \ldots(3)$$

Taking logarithms of the eqns. 2 & 3, we get

$$\log\left[\frac{-dc_1}{dt}\right] = \log K + n\log c_1 \qquad \ldots(4)$$

$$\log\left[\frac{-dc_2}{dt}\right] = \log K + n\log c_2 \qquad \ldots(5)$$

Eqn. (4) - Eqn. (5) gives

$$n = \frac{\log\left[\frac{-dc_1}{dt}\right] - \log\left[\frac{-dc_2}{dt}\right]}{\log c_1 - \log c_2} \qquad ...(6)$$

The different graphs are drawn by plotting c_1 and c_2 against time (t) in each case. The slopes of the tangents in each graph, $\frac{-dc_1}{dt}$ and $\frac{-dc_2}{dt}$ when substituted in eqn. (6) gives the value of n, which is the order of the reaction.

(4) Graphical Method

(a) For First Order Reactions

We have seen that for a first order reaction of the type

$$A \rightarrow \text{products},$$

the rate eqn. is given as follows :

$$\log_e \frac{a}{(a-x)} = Kt$$

$$\log_e(a-x) = -Kt + \log_e a$$

which is in the form of : $y = mx + c$.

The two variables in this first order rate equation are $\log_e \frac{a}{(a-x)}$ and t. Therefore, if a plot of $\log_e \frac{a}{(a-x)}$ against t yields a straight line, the reaction is of first order. If a curve is obtained, the reaction is not of the first order.

(b) For Second Order Reactions

For a second order reaction of the type :

$$2A \rightarrow \text{products},$$

the rate of reaction is given by the expression :

$$K = \frac{1}{at} \cdot \frac{x}{(a-x)}$$

This can be written as

or

$$\frac{1}{(a-x)} = Kt + 1/a$$

which is in the form of

$$y = mx + c$$

The two variables in the above second order rate expression are $\frac{1}{(a-x)}$ and t.

Thus, when $\frac{1}{(a-x)}$ is plotted against t, a second order reaction should yield a straight line. If a curve is obtained, the reaction is not of second order.

(5) Ostwald's Isolation Method

This method is used to determine the order of complicated reactions. This method comprises of "isolating" one of the reactants as far as its influence on the rate of reaction is concerned. This is done

REACTION DYNAMICS & CATALYSIS

by taking excessive concentrations of all other reactants expecting the one which is to be "isolated". The reactant which is not taken in excess is said to be "isolated". Then the order of the reaction with respect to the "isolated" reactant is determined. Similarly, the order of the reaction with respect to other reactants is determined by "isolating" the other reactants one at a time. Then, the order of the reaction is given by the sum of the orders determined with respect to each of the "isolated" reactants.

Let us consider the reaction,

$$A + B + C \rightarrow \text{products}$$

In this method, the order of the reaction with respect to A is first determined by isolating A by taking B and C in excess. Let this order be n_A. Likewise, the order of the reaction with respect to B and C is separately determined. Let these be n_B and n_C respectively. Then the order of the reaction, n is calculated as follows :

$$n = n_A + n_B + n_C.$$

This method offers an additional advantage of following the behaviour of each reactant separately under the given experimental conditions.

25.14. Complex or Simultaneous or Composite Reactions

Sometimes, the main chemical reaction may be accompanied by some side reactions. If these side reactions occur simultaneously, they are called parallel reactions. If they occur successively they are called consecutive reactions.

Complex reactions proceed in a series of steps and the rate of the overall reaction will be according to the stoichiometric equation.

(A) Parallel or Concurrent Reactions

In such reactions, the reactant follows two or more different paths simultaneously giving two or more different products. The rate of such reaction may be changed by varying the experimental conditions such as temperature, pressure or catalyst as per our preference. The reaction which gives maximum yield of the product is called the main reaction while others are called side reactions.

Let us consider a reactant, A which simultaneously gives two different compatible products B and C by following different chemical reactions with rate constants K_1 and K_2.

$$A \begin{array}{c} \xrightarrow{K_1} B \\ \text{path-1} \\ \text{path-2} \\ \xrightarrow{K_2} C \end{array}$$

If $K_1 > K_2$, then the reaction $A \rightarrow B$ will be the major reaction and $A \rightarrow C$ will be the side reaction.

If we start with 'a' moles per liter of A and in time t, let x moles per liter of A reacted to form the products, B and C. Let y moles per liter of B and Z moles per liter of C are formed in time t. Then, the rate of formation B (from A) = dy/dt and the rate of formation of C (from B) = dz/dt.

$$\frac{-dx}{dt} = \frac{dy}{dt} + \frac{dz}{dt} = (K_1 + K_2)(a - x)$$

If the two reactions are of the same order, then

$$\text{(The amt. of } B \text{ formed at any stage)} = \frac{dy}{dt} = K_1(a - x)^n$$

(The amt. of C formed at any stage) = $\dfrac{dz}{dt} = K_2 (a-x)^n$

$\therefore \quad \dfrac{dy}{dz} = \dfrac{K_1}{K_2}$

Therefore, by analysing the products B and C formed at the end of the reaction and plotting the concentration of one against the other, a linear plot is obtained, the slope of which gives the ratio of rate constants, K_1/K_2. Further, according to Wegscheider's principle, the ratio of the amounts of the products formed in two side reactions (not consecutive or opposing reactions) is independent of time.

If we consider the reaction

$$A + B \longrightarrow \begin{cases} C \quad (y) \\ D \quad (z) \end{cases}$$
$(a-x) \quad (b-x)$

If the two reactions are of the same order, then

$$\dfrac{dy}{dt} = K'(a-x)^n (b-x)^m$$

$$\dfrac{dz}{dt} = K''(a-x)^n (b-x)^m$$

$$\dfrac{dz}{dt} = \dfrac{K'}{K''}$$

The two products, C and D can be determined at the end of the reaction. The plot of the concentration of C versus concentration of D gives a straight line, the slope of which gives the ratio of rate constants, K'/K''. We can also plot the concentration of one product against the concentration of reactants consumed to get a linear plot. Since both the parallel reactions have the same order, analysis of the disappearance of the reactants will show that the data corresponds to the overall order of the reaction n or $(n + m)$, as the case may be. Thus, if each of the parallel reactions are of second order, the analysis based on reactant disappearance would also respond to second order.

However, if the parallel reactions are of different orders, the analysis is more difficult.

If we consider the following type of reaction, where one of the reactants react with two different reactants, we will be confronted with competetive parallel reactions :

$$A \longrightarrow \begin{cases} + B \longrightarrow D \\ + C \longrightarrow E \end{cases}$$

In such cases too, suitable rate equations can be formulated.

Examples of Parallel Reactons

(i) Nitration of phenol (where o – and p – nitrophenols are formed)
(ii) Bromination of bromobenzene (where o – and p – dibromobenzene are formed)

(iii) Reaction of ethyl bromide with caustic potash (where ethylene and ethyl alcohol are formed)

(iv) Removal of H Br from alkyl halides by alcoholic potash (where butene – 1 and butene – 2 are formed)

(B) Consecutive Reactions

In consecutive or sequential reactions, the final product is formed through one or more intermediate steps. In these reactions, the product of the first elementary reaction forms the reactant for the second elementary reaction, and so on until the final product is formed.

Let us consider the following reaction :

$$A \longrightarrow B \longrightarrow C$$

Initial concn. a o o

Concn. after time, t $(a-x)$ y z

Assuming the reactions to be of first order, we get

$$-\frac{d[A]}{dt} = \frac{dx}{dt} = K_1[A] = K_1(a-x) \qquad \ldots(1)$$

$$\frac{d[C]}{dt} = \frac{dz}{dt} = K_2[B] = K_2 y \qquad \ldots(2)$$

$$\frac{d[B]}{dt} = \frac{dy}{dt} = K_1[A] - K_2[B]$$

$$= K_1(a-x) - K_2 y \qquad \ldots(3)$$

We have seen earlier that the rate equation for a first order reaction is :

$$K_1 = \frac{2.303}{t} \log \frac{a}{(a-x)} \qquad \ldots(4)$$

$$\therefore \qquad (a-x) = a \cdot e^{-kt} \qquad \ldots(5)$$

Substituting eqn. (5) in eqn. (3), and integrating, we get

$$y = \frac{K_1 a}{(K_2 - K_1)} = [e^{-K_1 t} - e^{-K_2 t}] \qquad \ldots(6)$$

$$z = a \left[1 - \frac{K_2(e^{-K_1 t}) - K_1(e^{-K_2 t})}{(K_2 - K_1)} \right] \qquad \ldots(7)$$

The way in which the concentrations of A, B and C vary with time is depicted in Fig. 5.6.

From Fig. 5.6, it is clear that the concentration of A decreases exponentially with time, while the concentration of B first increases and then decreases, whereas the concentration of C increases (from zero) with time and finally attains the value of initial concentration of A i.e. $[A_0]$ when all of A has changed to C, the final product, at the end of the reaction.

Examples of Consecutive Reactions

(i) **Removal of oxygen in a pollutant stream.** Oxygen [species B] is dissolved from air by a process similar to A giving B, and is consumed for the biological oxidation of organic matter at a rate proportional to the amount of organic matter present,

$$A \longrightarrow B \longrightarrow C$$

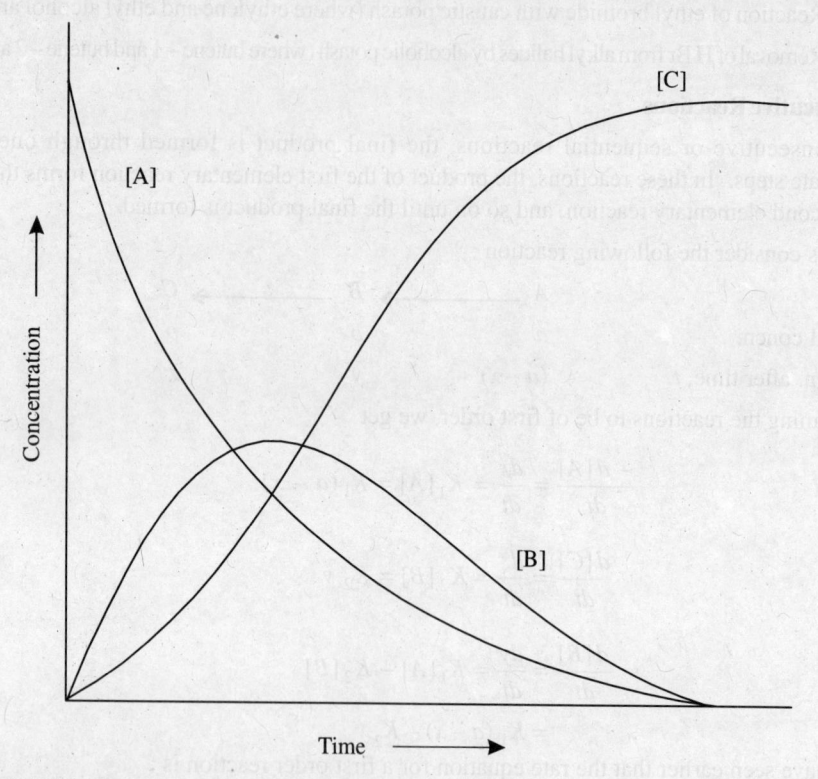

Fig. 25.6. Variation of concentration of reactants and product (s) in a consecutive reaction

(*ii*) **Bacterial nitrification of ammonia.** Ammonia is oxidized by nitrosomonas bacteria to nitrite, which is then oxidized to nitrate by nitrobacter bacteria.

$$NH_3 \xrightarrow[\text{(nitrosomonas)}]{O_2} NO_2^- \xrightarrow[\text{(nitrobacter)}]{O_2} NO_3^-$$

(*iii*) **Decomposition of ethylene oxide**

$$\begin{array}{c} CH_2 \\ \end{array}\!\!\!>\!\!O \longrightarrow (CH_3CHO)^* \longrightarrow CH_4 + CO$$
$$\begin{array}{c} CH_2 \end{array}$$

(*iv*) **Gaseous phase decomposition of dimethyl ether**

$$CH_3OCH_3 \longrightarrow CH_4 + HCHO \longrightarrow H_2 + CO$$

(*v*) **Several radioactive decay systems such as uranium disintegration series.**

$${}^{226}_{88}Ra \xrightarrow{-\alpha} {}^{222}_{86}R_n \xrightarrow{-\alpha} {}^{218}_{84}PO \xrightarrow{-\alpha} {}^{214}_{82}Pb \xrightarrow{-\beta} {}^{214}_{83}Bi \xrightarrow{-\beta} ...$$

(*vi*) **BOD Reaction**

The BOD reaction is only an apparent first order reaction. Indeed it is a consecutive reaction (vide example 1). The rate of reaction is proportional to the amount of oxidizable organic matter remaining at any time. This concentration is modified by the population of the active organisms. When once the population of the organisms reaches a more or less stable level the reaction is controlled by the amount of organic matter available as food for the organisms.

In the BOD test, O_2 is not allowed to dissolve in the sample continuously from air, thereby eliminating one step in the consecutive reaction scheme. When the population of the organisms reaches a steady value, an apparent first order reaction is observed with respect to oxygen expressed in terms of oxygen demand and the organic matter.

(C) Reversible Reactions (or Opposing Reactions or Equilibrium Reactions)

Reversible reactions are such reactions wherein the reactants react to form the products and the products also react to give back the original reactants. That means, the reactions take place both in the forward direction and also in the backward direction under the same conditions and oppose each other.

Let us consider the following reversible first order reaction :

$$A \underset{K_b}{\overset{K_f}{\rightleftharpoons}} B$$

Initial concn. → 'a' mols/L —
Concn after time t → $(a-x)$ moles/L x mols/L
Concn. at equilibrium → $(a-x_e)$ mols/L x_e mols/L

K_f = rate constant for forward reaction.
K_b = rate constant for backward reaction.
x = fraction of 'a' transformed after time, t.
x_e = value of x at equilibrium.

$$\frac{-d[A]}{dt} = K_f[A] - K_b[B]$$

$$= K_f(a-x) - K_b x \qquad \ldots(1)$$

At equilibrium, $\qquad \dfrac{-d[A]}{dt} = 0, \, x = x_e,$

and the net rate $\qquad \dfrac{dx}{dt} = 0.$

∴ $\qquad K_f(a - x_e) - K_b x_e = 0 \qquad \ldots(2)$

or $\qquad K_f(a - x_e) = K_b x_e \qquad \ldots(3)$

∴ $\qquad \dfrac{x_e}{(a - x_e)} = \dfrac{K_f}{K_b} = \text{say } K \qquad \ldots(4)$

where x_e is the equilibrium concentration of B and has a definite value for a given temperature and for a given value of 'a'.

$$K_b = \frac{K_f(a - x_e)}{x_e} \qquad \ldots(5)$$

Substituting this value of K_b from eqn. (5) in eqn. (1), we have

$$\frac{-d[A]}{dt} = \frac{dx}{dt} = K_f(a-x) - \frac{K_f(a-x_e)}{x_e} \cdot x$$

$$\therefore \quad \frac{dx}{dt} = \frac{K_f}{x_e} \cdot a(x_e - x) \qquad ...(6)$$

$$\therefore \quad \frac{dx}{(x_e - x)} = \frac{K_f a}{x_e} \cdot dt$$

On integration, and applying the condition that when $t = 0$, $x = 0$, we get

$$= \frac{K_f a}{x_e} \cdot t = \log_e \frac{x_e}{(x_e - x)} \qquad ...(7)$$

Although this reaction is of reversible first order reaction, it behaves like a simple first order reaction with x_e (i.e. the equilibrium concentration of the product) instead of the initial concentration of the reactant, and the rate constant being $K_f a/x_e$ instead of K_1. We can see from below that

$$\frac{K_f a}{x_e} = (K_f + K_e).$$

From eqn. (5), we have

$$K_b = \frac{K_f(a - x_e)}{x_e} = \frac{K_f a}{x_e} - K_f \qquad ...(8)$$

$$\therefore \quad \frac{K_f a}{x_e} = K_f + K_b \qquad ...(9)$$

From Eqn. (4) we get the value of $\dfrac{K_f}{x_b}$

From Eqn. (9) we get the value of $(K_f + K_b)$

From these above two values determined experimentally as per eqn. (4) and (9), K_f and K_b can be evaluated.

The rate expressions can similarly be obtained for reversible second order reactions.

Examples of Reversible Reactions

(i) Reaction between gaseous carbon monoxide and nitrogen peroxide.

$$CO\,(g) + NO_2\,(g) \rightleftharpoons CO_2\,(g) + NO\,(g)$$

(ii) Conversion of ammonium cyanate into urea where both forward and backward reactions are of first order.

$$\underset{\text{ammonium cyanate}}{NH_4\,CNO\,(aq)} \rightleftharpoons \underset{\text{urea}}{NH_2 - CO - NH_2\,(aq)}$$

(iii) Dissociation of hydrogen iodide, where both forward and backward reactions are of second order.

$$2HI \rightleftharpoons H_2 + I_2.$$

(iv) Oxidation of NO with oxygen to NO_2, where the forward reaction is of third order and the backward reaction is of second order.

$$2NO + O_2 \rightleftharpoons 2NO_2$$

(v) Hydrolysis of ethyl acetate in presence of acid and in excess of water. Here K_f is of first order and K_b is of second order.

$$CH_3COOC_2H_5 + H_2O \rightleftharpoons CH_3COOH + C_2H_5OH$$

(vi) Mutarotation of α-glucose in isolation is partially transformed to β-glucose, where the forward first order reaction is opposed by backward first order reaction *i.e.* both forward and backward reactions are of first order.

25.15. THEORIES OF REACTION RATES

The major objective of any theory of reaction rates is to calculate the rate constant of a reaction from the fundamental properties of the reactant molecules. Two important theories have been proposed towards this objective.

(a) The collision theory by Trautz and Lewis (1918) and Lindemann (1922).

(b) The transition state theory or the absolute reaction rate theory by Eyring and Polamji (1935).

(a) The Collision Theory

According to this theory, the reactant molecules, considered as hard spheres, will have to "approach" or collide with each other before their bonds can rearrange and re-group to form products. The number of times the reactant molecules approach each other in unit time is the number of collisions taking place.

According to the collision theory; (*i*) Chemical reactions take place by collisions between reacting molecules. (*ii*) All collisions are not fruitful. Only a small fraction of the colliding molecules are effective to bring about a reaction. For example, it was shown that in the decomposition of HI (hydrogen iodide) at 556 K at a concentration of 1 mole / L, only about one in 10^{17} collisions were effective in producing the reaction. (*iii*) The colliding molecules must possess sufficient kinetic energy to cause a reaction.

For example, the change of energy of the colliding molecules of the reactants with the progress of the exothermic reaction:

$$AB + C \longrightarrow AC + B$$

is shown in Fig. 5.7.

A chemical reaction takes place by breaking bonds between the atoms of the reacting molecules and forming new bonds in the molecules of the product. The energy required for this is provided by the kinetic energy of the reacting molecules before collision. To cause a reaction between the colliding molecules, the molecules should attain the so called activation energy, E_a, which is the minimum energy necessary to cross the energy barrier, (between the reactants and products) and bring about an effective and fruitful reaction between the colliding molecules. If the molecules are colliding with kinetic energy less than E_a, they fail to surmount the energy barrier and hence the collisions become unproductive and the molecules are simply bounced off by each other.

(c) The reactant molecules must collide with proper orientation which ensures direct contact between the atoms involved in the breaking and forming the bonds (*i.e.* transforming and regrouping of bonds).

On the basis of the above two postulates of the collision theory, the reaction rate for an elementary reaction is given by

$$\text{rate} = f \times p \times z$$

where f = fraction of molecules which possess sufficient energy to react,

p = probable fraction of effective collisions with proper orientation, and

z = collision frequency.

Limitations of Collision Theory

(i) This theory is applicable to simple gaseous reactions and for solutions in which the reacting species exist as simple molecules.

(ii) In this theory only the kinetic energy of the colliding molecules is supposed to contribute towards the energy required to surmount the energy barrier. There doesn't seem to be any valid reason to ignore rotational and vibrational energies of molecules.

(iii) There is no method for determining the probability factor, P for a reaction whose rate constant has not been determined experimentally.

(iv) For reactions involving complex molecules, the rate constants determined experimentally do not agree with calculated values.

(v) The theory does not throw light on the clevage and bond formation involved in the reaction.

(vi) Though the theory is logical and rational, it appears that it is over-simplified.

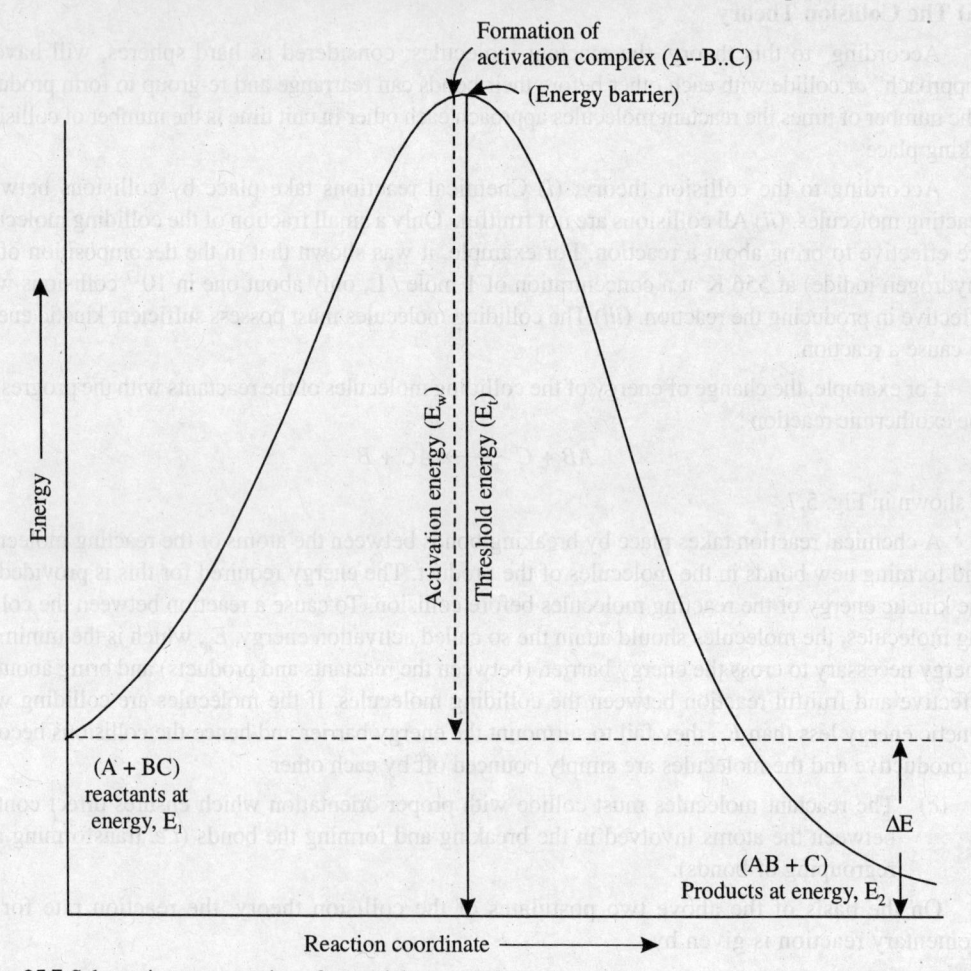

Fig. 25.7. Schematic representation of potential energy of a system of reactants and products as a function of the reaction coordinate.

E_a = Activation energy

E_t = Threshold energy

$\Delta E = (E_1 - E_2)$ = Energy change accompanying the reaction $(E_1 - E_2)$.

A...B...C = Activation complex, where the bond A – B is in the process of being broken and the bond B – C is in the process of being formed.

REACTION DYNAMICS & CATALYSIS

Collision Theory for Unimolecular Reactions (Lindermann's Hypothesis)

A number of unimolecular reactions such as :

$$2N_2O_5 (g) \rightarrow 4NO_2 (g) + O_2 (g)$$

are found to be first order reactions. If two molecules must collide in order to provide the necessary activation energy, a second order rate law is normally expected to be obeyed. This anomaly is explained by Lindemann in 1922 by assuming that a time lag exists between activation and reaction of molecules. During this time lag, the activated molecules may either react or get deactivated to ordinary molecules.

Step-1. $\underset{\text{ordinary molecules}}{A + A} \longrightarrow A + \underset{\text{activated molecule}}{A^*}$ (activation)

Step-2. $A + A^* \longrightarrow A + A$ (deactivation)

Step-3. $A^* \longrightarrow \text{products}$ (reaction)

If the time lag is long, step-3 is slow, the reaction follows first order kinetics.

If A reacts as soon as it is formed, step-2 is the slow step, and hence the reaction follows second order kinetics.

Thus, the rate of reaction will not be proportional to all the molecules that are activated but only to those which remain active.

$$\underset{\text{(ordinary molecules)}}{A + A} \underset{K_2}{\overset{K_1}{\rightleftharpoons}} \underset{\text{(activated molecules)}}{A^* + A} \xrightarrow{K_3} \text{Products}$$

where K_1, K_2 and K_3 are rate constants of the different reactions involved as above.

(B) Transition state theory of absolute reaction rates or theory of activated complex formation.

According to this theory, the reacting molecules first form an intermediate activated complex before forming the products. This activated complex possesses higher energy than both the reactants and products. The activated complex, which is in transition state, will be in equilibrium with the reacting molecules and the products, and the rate of reaction is controlled by the rate of decomposition of the activated complex to form the products :

$$\underset{\text{reactants}}{A + B} \rightleftharpoons \underset{\text{activated complex}}{X^*} \xrightarrow{K} \text{products}$$

(The activated complex is regarded as being situated at the top of the energy barrier lying between the initial and the final states. The rate of the reaction is controlled by the rate at which the complex travels over the top of the barrier and decompose to form the products).

The various steps involved in a chemical reaction, as per the transition state theory are:

(a) In a collision, the fast approaching reactant molecules slow down due to gradual repulsion between their electron clouds. In this process, the kinetic energy of the two molecules is converted into potential energy.

(b) As the molecules come closer to each other, the interaction of the electron clouds lead to the rearrangement of the valence electrons.

(c) An activated complex or transition state is formed. This is highly transient and decomposes to give rise to the products.

Henry Eyring showed that the rate constant, K is given by the following relationship:

$$K = \frac{RT}{Nh} \cdot K' \qquad ...(1)$$

where R = Gas constant
  T = absolute temperature
  N = Avogadro's number
  h = Planck's constant, and
  K' = Equilibrium constant for the formation of the activated complex from the reacting molecules, i.e.

$$K' = \frac{[X^*]}{[A][B]}$$

The free energy ΔG^* of activation for a reaction of the type:

$$A + B \longrightarrow X^*$$

is given by thermodynamic considerations as:

$$\Delta G = -RT \log_e K'$$

$$\log K' = \frac{-\Delta G^*}{RT}$$

$$\therefore \qquad K' = e^{-\Delta G^*/RT} \qquad ...(2)$$

Substituting this value of K' in eqn. (1), we get

$$K = \frac{RT}{Nh} \cdot e^{-\Delta G^*/RT} \qquad ...(3)$$

This shows that at any given temperature, greater the value of free energy of activation for a reaction, lower will be the rate of reaction. We also know from thermodynamics that

$$\Delta G^* = \Delta H^* - T \Delta S^* \qquad ...(4)$$

where ΔH^* and ΔS^* are the changes in enthalpy and entropy of activation respectively.

Substituting this value for ΔG^* in eqn. (3), we get

$$K = \frac{RT}{Nh} \cdot e^{(-\Delta H - T\Delta S)/RT}$$

$$= \frac{RT}{Nh} \cdot e^{\Delta S^*/R} \cdot e^{-\Delta H^*/RT} \qquad ...(5)$$

The transition state theory of reaction rates offers the following advantages over the collision theory:

(i) In collision theory, the probability factor, P was introduced rather arbitrarily (or empirically), whereas in the transition state theory, its inclusion in terms of entropy of activation, ΔS^* seems to be more reasonable from thermodynamics point of view.

(ii) The concept of formation of intermediate actiated complex offers more convincing explanation for formation of products from reactants, than assuming that only the collisions of molecules lead to the formation of the products.

25.16. EFFECT OF TEMPERATURE ON RATES OF REACTION - ARRHENIUS EQUATION

It has been stated earlier that, in general, the rate at which a reaction takes place increases with increase in temperature. The ratio of specific rate constants at two temperatures separated, by 10° C (usually 25° C and 25° C) is known as temperature coefficient.

$$\text{Temperature coefficient} = \frac{K_{t+10}}{K_t}$$

It was noted that by an increase of 10° C, the reaction rate in case of many reactions was almost doubled or even trebled in some cases.

It was found that the variation of rate constant, K with abolute temperature, T could be expressed by the following empirical relationship :

$$\log K = A - \frac{B}{T} \qquad ...(1)$$

where A and B are constants.

Van't Hoff (1884) suggested with the help of his reaction isochores that the logarithm of the specific reaction rate is a linear function of the reciprocal of absolute temperature. This theory is extended by Arrhenius (1889).

According to Arrhenius hypothesis, all molecules of a system cannot take part in a chemical reaction. Only a certain fraction in molecules called active molecules undergo reaction. There will be an equilibrium existing between the number of active molecules and passive molecules, when the temperature is increased, some of the passive molecules absorb heat energy and become active and thereby increasing the rate of reaction.

On the basis of his studies, Arrhenius gave the following equation showing the relationship between the rate constant and temperature :

$$K = A \cdot e^{-E_a/RT} \qquad ...(2)$$

where K = rate of constant

A = frequency factor for the reaction or pre-exponential factor or collision number

E_a = energy of activation, which is characteristic of a reaction

R = gas constant

T = absolute temperature

Taking logarithms of both sides of eqn. (2), we get

$$\log_e K = \log_e A - \frac{E_a}{RT} \qquad ...(3)$$

If K_1 and K_2 are the values of rate constant at absolute temperatures T_1 and T_2, we have

$$\log_e K_1 = \log_e A - \frac{E_a}{RT_1} \qquad ...(4)$$

$$\log_e K_2 = \log_e A - \frac{E_a}{RT_2} \qquad ...(5)$$

Eqn. (5) – eqn. (4) gives

$$\log_e K_2 - \log_e K_1 = \frac{-E_a}{RT_2} - \left(\frac{-E_a}{RT_1}\right)$$

$$= \frac{E_a}{RT_1} - \frac{E_a}{RT_2}$$

$$\therefore \quad \log_e \frac{K_2}{K_1} = \frac{E_a}{R}\left(\frac{1}{T_1} - \frac{1}{T_2}\right)$$

$$= \frac{E_a}{R}\left(\frac{T_2 - T_1}{T_1 T_2}\right)$$

or $$\log_{10} \frac{K_2}{K_1} = \frac{E_a}{2.303\,R}\left(\frac{T_2 - T_1}{T_1 T_2}\right) \qquad \ldots(6)$$

Thus, if we know the values of rate constants K_1 and K_2 at the corresponding temperatures T_1 and T_2, the value of the energy of activation, E_a can be calculated.

Similarly, if we know the values of E_a and K_1 at a temperature, T_1, then it is possible to calculate the rate constant K_2 at any other temperature, T_2.

Further, eqn. (3) can be also written as :

$$\log_e K = \frac{-E_a}{RT} + \log_e A$$

or $$\log_e K = \frac{-E_a}{2.303\,RT} + \log A$$

Since this equation is of the form : $y = mx + c$, this represents an equation for a straight line. Therefore, a plot of $\log K$ vs. $\frac{1}{T}$ gives a straight line with a slope equal to $\frac{-E_a}{2.303\,R}$. (Vide Fig. 5.8) From the value of the slope, the value of the energy of activation, E_a can be evaluated.

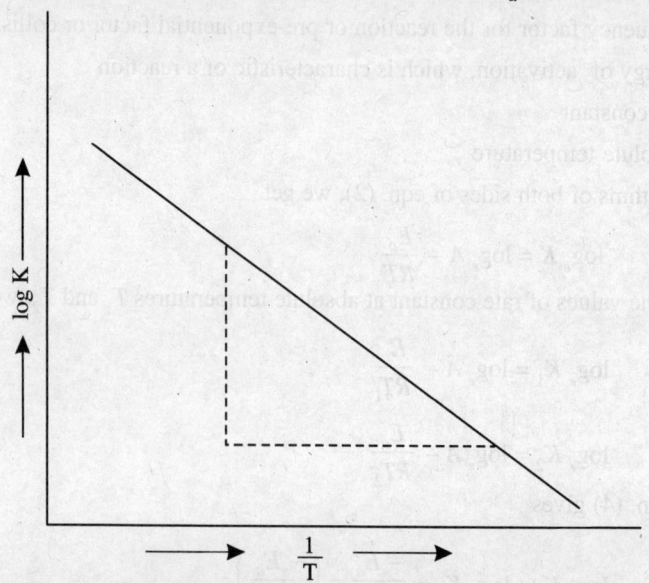

Fig. 25.8. Graphical determination of energy of activation E_a from a plot of $\log K$ vs. $1/T$.

$$\text{Slope} = -\frac{-E_a}{2.303\,R} \qquad \log K = -\frac{-E_a}{2.303\,RT} + \log A$$

REACTION DYNAMICS & CATALYSIS

25.17. ACTIVATION ENERGY AND CATALYSIS

As stated earlier, for a reaction to take place, a certain energy barrier must be surmounted. For this, the reactant molecules must acquire the activation energy. If the entropy of activation remains constant, the rate of a reaction is determined by the energy of activation. A catalyst enables the same overall process to take place by an alternative path involving a smaller free energy of activation. Even then, the energy difference, ΔE between the reactants and products remain the same.

25.18. EXAMPLE 1.

The following data is obtained in the studies on decomposition of oxalic acid as per the reaction :

$$(COOH)_2 \longrightarrow CO_2 + CO + H_2O$$

Time in minutes	Volume of $KMnO_4$ required for titration of residual oxalic acid at the corresponding time interval, in ml.
0	22
300	17
450	15
600	13.4
1200	7.9

Determine the value of the velocity constant and demonstrate that the reaction is of the first order.

Solution

For any reaction of the first order, the integrated rate equation is as follows :

$$K = \frac{2.303}{t} \log \frac{a}{(a-x)}$$

where K = velocity constant ,

a = initial concentration of the reactant ,

$(a-x)$ = concentration at time, t.

If the value of K determined at different time intervals is constant, then we can conclude that the reaction is of first order.

In the example given above, the initial concentration 'a' of oxalic acid corresponds to (or equivalent to) 22 ml of $KMnO_4$. The data given and the calculations can be fitted in the following tabular form :

t (minutes)	a (ml)	$(a-x)$ (ml)	$K = \frac{2.303}{t} \log \frac{a}{(a-x)}$ (min^{-1})
300	22	17	$K = \frac{2.303}{300} \log \frac{22}{27} = 0.00086$
450	22	15	$K = \frac{2.303}{450} \log \frac{22}{15} = 0.00085$

600	22	13.4	$K = \dfrac{2.303}{600} \log \dfrac{22}{13.4} = 0.00087$
1200	22	7.9	$K = \dfrac{2.303}{1200} \log \dfrac{22}{7.9} = 0.00086$
			Average, $K = 0.00086$ min^{-1}

The near constancy of the value of K in the last column demonstrates that the reaction is of the first order.

The value of the velocity constant = 0.00086 minute^{-1}

$= 8.6 \times 10^{-4}$ minute^{-1}.

This implies that at any instant, 0.86% of the reactant molecules present will decompose per minute.

Example 2.

The following data was obtained while studying the decomposition of N_2O_5 in CCl_4 at 48°C. Show that the reaction is of first order. Also, calculate the rate constant for the reaction.

Time, in minutes	Volume of O_2 evolved, in ml
10	6.30
15	8.95
20	11.40
25	13.50
∞	34.75

Note:

N_2O_5 is a volatile solid which decomposes in gaseous state as well as when present in the form of solution in an inert solvent eg., $CHCl_3$ or CCl_4. The reaction can be represented as :

$$N_2O_5 \longrightarrow N_2O_4 + \dfrac{1}{2} O_2$$
$$\Downarrow$$
$$2NO_2$$

When the reaction is carried out in solution, N_2O_5 and NO_2 remain in solution. The volume of O_2 gas collected at different time intervals is measured to follow the reaction kinetics.

Solution

Any reaction of first order must obey the following rate equation :

$$K = \dfrac{2.303}{t} \log \dfrac{a}{(a-x)} \qquad ....(1)$$

REACTION DYNAMICS & CATALYSIS

In the present case, the rate equation can be written as :

$$K = \frac{2.303}{t} \log \frac{V_\infty}{(V_\infty - V_t)}$$

where, V_∞ = volume of oxygen gas collected at infinite time, which is proportional to the initial quantity of N_2O_5 (i.e. 'a', in the eqn. 1).

Time, t (minutes)	V_t (ml)	$(V_\infty - V_t)$ (ml)	$K = \frac{2.303}{t} \log \frac{V_\infty}{(V_\infty - V_t)}$ (min^{-1})
10	6.30	(34.75 − 6.30)	$K = \frac{2.303}{10} \log \frac{34.75}{28.45} = 0.01996$
15	8.95	(34.75 − 8.05)	$K = \frac{2.303}{15} \log \frac{34.75}{25.80} = 0.01988$
30	11.40	(34.75 − 11.40)	$K = \frac{2.303}{20} \log \frac{34.75}{23.35} = 0.01989$
			Average, $K = 0.01992$ min^{-1}

The nearly constant value of K in the three experiments show that the reaction is of first order. The average value of the rate constant, $K = 0.01992$ minute^{-1}.

Example 3.

A first order reaction has been found to take 20 minutes for 15% of the reaction to complete. How much time it would take for 60% of the reaction to be completed.

Solution

The rate equation for a first order reaction is as follows :

$$K = \frac{2.303}{t} \log_{10} \frac{a}{a-x}$$

It follows that for times t_1 and t_2, the rate equations will be :

$$K_1 = \frac{2.303}{t_1} \log \frac{a_1}{a_1 - x_1}, \text{ and}$$

$$K_2 = \frac{2.303}{t_2} \log \frac{a_2}{a_2 - x_2}$$

V_t = Volume of oxygen gas collected at time, t; which is proportional to the amount of N_2O_5 decomposed at time, t (i.e. x in the eqn. 1).

i.e. $x \propto V_t$

$a \propto V_\infty$

$(a - x) \propto (V_\infty - V_t)$.

Since $K_1 = K_2$, we have

$$\frac{2.303}{t_1} \log \frac{a_1}{a_1 - x_1} = \frac{2.303}{t_2} \log \frac{a_2}{a_2 - x_2}$$

In the present problem,

$$x_1 = \frac{15}{100} \times a_2; \quad t_1 = 20 \text{ minutes}$$

$$x_2 = \frac{60}{100} \times a_2; \quad t_2 \text{ is to determined.}$$

$$\therefore \frac{2.303}{20} \log \frac{a_1}{\left(a_1 - \frac{15}{100} \times a_1\right)} = \frac{2.303}{t_2} \log \frac{a_2}{\left(a_2 - \frac{60}{100} \times a_2\right)}$$

$$\therefore \frac{1}{20} \log \frac{100}{85} = \frac{1}{t_2} \log \frac{100}{40}$$

$$\therefore t_2 = \frac{20 \log \frac{100}{40}}{\log \frac{100}{85}} = 112.7 \text{ min utes}$$

∴ This first order reaction takes 112.7 minutes to reach 60% completion.

Example 4.

The following is a first order reaction with a rate constant of 2.2×10^{-5} sec^{-1} at 302° C :

$$SO_2Cl_2 \xrightarrow{K} SO_2 + Cl_2$$

Calculate the percentage of SO_2Cl_2 that would decompose in 90 minutes at the same temperature.

Solution

We know that the rate constant for a first order reaction is given by the rate equation :

$$K = \frac{2.303}{t} \log \frac{a}{a-x}$$

In the present problem,

$$K = 2.2 \times 10^{-5} \text{ sec}^{-}$$

$$t = 90 \times 60 = 5400 \text{ sec}$$

$$\therefore 2.2 \times 10^{-5} = \frac{2.303}{5400} \log \frac{a}{a-x}$$

$$\therefore \log \frac{a}{a-x} = 0.0516$$

Taking antilog of both sides, we have

$$\frac{a}{a-x} = \text{antilog } (0.0516) = 1.127$$

$$\therefore \quad a = 1.127 (a-x) = 1.127 a - 1.127 x$$

REACTION DYNAMICS & CATALYSIS

$$\therefore \quad 1.127\, x = (1.127\, a - a) = 1.127\, a$$

$$\therefore \quad \frac{x}{a} = \frac{0.127}{1.127} = 0.113$$

Thus, it is clear that 11.3% of SO_2Cl_2 would decompose in 90 minutes at the given temperature.

Example 5.

The reaction involving inversion of cane sugar as follows is a pseudo-unimolecular reaction obeying the first order rate law :

$$\underset{\text{sucrose}}{C_{12}H_{22}O_{11}} + H_2O \longrightarrow \underset{\text{glucose}}{C_6H_{12}O_6} + \underset{\text{fructose}}{C_6H_{12}O_6}$$

The kinetics of this reaction are followed by measuring the optical rotation of the solution from time to time as the sucrose and invert sugar (equimolar mixture of fructose and glucose) produced will have different values of specific rotation. In a particular experiment the following data are obtained:

Time, t (minutes)	0.0	7.2	36.8	46.0	68.0	∞
Rotation of polarised light (r°)	+ 24.1	+ 21.4	+ 12.4	+ 10.0	+ 5.5	− 10.9

(a) Demonstrate that the reaction is of first order.
(b) Calculate the value of the rate constant for the reaction.
(c) What will be the time taken for the solution to become optically inactive.

Solution

Any first order reaction follows the rate equation given below :

$$k = \frac{2.303}{t} \log \frac{a}{a-x}$$

where 'a' is the initial concentration and $(a-x)$ is the concentration of the reactant at time, t.

Since the reaction kinetics are followed by measuring the optical rotation of the solution from time to time, we have :

$a \propto (r_\infty - r_o)$, and
$(a - x) \propto (r_\infty - r_t)$

where r_o, r_t and r_∞ are the values of optical rotation at initial stage, at time t, and at infinite time (i.e. final stage) of the reaction.

Therefore, the rate equation in this case can be written as :

$$k = \frac{2.303}{t} \log \frac{(r_\infty - r_o)}{(r_\infty - r_t)}$$

The data given and the calculations can be fitted into tabular form as follows :

$r_o = + 24.1; \quad r_\infty = -10.9$
$\therefore (r_\infty - r_o) = -35.0$

Time, t (mins)	rotation observed (r_t)	$(r_\infty - r_t)$	$K = \dfrac{2.303}{t} \log \dfrac{(r_\infty - r_o)}{(r_\infty - r_t)}$ (min^{-1})
0.0	+24.1 (r_o)	—	—
7.2	+21.4	−32.3	$K = \dfrac{2.303}{7.2} \log \dfrac{-35.0}{-32.3} = 0.113$
36.8	+12.4	−23.3	$K = \dfrac{2.303}{36.8} \log \dfrac{-35.0}{-23.3} = 0.112$
46.0	+10.0	−20.9	$K = \dfrac{2.303}{46.0} \log \dfrac{-35.0}{-20.9} = 0.113$
68.0	+5.5	−16.4	$K = \dfrac{2.303}{68.0} \log \dfrac{-35.0}{-16.4} = 0.113$
∞	−10.9	—	—
			Average value of K = 0.1128 min^{-1}

(a) The constancy of the value of rate constant, K shows that the reaction is of first order.

(b) The average value of rate constant for the reaction = 1.13×10^{-2} min^{-1}.

(c) When the solution becomes optically active, $r_t = 0$.

$$K = \dfrac{2.303}{t} \log \dfrac{(r_\infty - r_o)}{(r_\infty - r_t)}$$

∴

$$t = \dfrac{2.303}{K} \log \dfrac{(r_\infty - r_o)}{(r_\infty - r_t)}$$

Substituting the values for K, r_∞, r_o and r_t, we have

$$t = \dfrac{2.303}{0.113} \log \dfrac{-35.0}{-10.9} = 103.3 \text{ minutes}$$

∴ The solution becomes optically inactive after 103.3 minutes

Example 6.

How many atoms C^{14} would produce an average of 10 beta emissions per minute? Given that the half life of C^{14} is 5600 years.

Solution

The rate of decay of a radioactive isotope is proportional to the number of the radioactive nuclei that undergo decay. If n is the number of radioactive nuclei present at time, t, then the rate of decay, $-\dfrac{dn}{dt} = Kn$ where K is the radioactive decay constant. Further, we also know that the half-life period of a radioactive isotope ($T_{0.5}$) is related to the decay constant, K as follows :

REACTION DYNAMICS & CATALYSIS

$$\frac{0.693}{K} = T_{0.5}$$

$$\therefore \quad K = \frac{0.693}{K} = T_{0.5}$$

In this case, the half life period of C^{14} is given as 5600 years which is equal to $(5600 \times 365 \times 24 \times 60)$ minutes

$$= 2.95 \times 10^9 \text{ minutes.}$$

$$\therefore \quad K = \frac{0.693}{T_{0.5}} = \frac{0.693}{2.95 \times 10^9} = 2.33^{-10} \times 10^{-10} \text{ min}^{-1}$$

Now since $-\dfrac{dn}{dt} = Kn$, we have

$$n = \frac{-dn/dt}{K}$$

$$= \frac{10}{2.35 \times 10^{-10}}$$

$$= 4.255 \times 10^{10} \text{ atoms.}$$

Thus, 4.255×10^{10} atoms of C^{14} are required to give an average of 10 beta emissions per minute.

Example 7.

Equal concentrations of an ester and NaOH are used to study the kinetics of saponification. 5 ml of the reaction mixture were withdrawn at different time intervals and titrated against an acid solution. The data obtained are as follows :

Time, t (minutes)	Titre value, V_t (in ml)
0	16 (V_o)
5	10.3
15	6.1
25	4.3

Demonstrate that the reaction is of the second order. Also, calculate the average value of the rate constant for the reaction.

Solution

We know that any second order reaction should obey the following rate equation :

$$K = \frac{1}{t} \cdot \frac{x}{a(a-x)}$$

In the present case,

$$x \propto (V_o - V_t)$$

$$a \propto V_o$$

$$(a-x) \propto V_t$$

$\therefore$ The rate equation may be written as :

$$K = \frac{1}{t} \cdot \frac{(V_o - V_t)}{(V_o \cdot V_t)}$$

In this experiment,

$$V_o = 16 \text{ ml}$$

The data and calculations may be tabulated as follows :

Time, t (minutes)	V_t (ml)	$V_0 - V_t$ (ml)	$K = \frac{1}{t} \cdot \frac{(V_0 - V_t)}{(V_o - V_t)}$ min^{-1}
5	10.3	$(16 - 10.3) = 5.7$	$K = \frac{1}{5} \times \frac{5.7}{(16 \times 10.3)} = 0.00692$
15	6.1	$(16 - 6.1) = 9.9$	$K = \frac{1}{15} \times \frac{9.9}{(16 \times 6.1)} = 0.00678$
25	4.3	$(16 - 4.3) = 11.7$	$K = \frac{1}{25} \times \frac{11.7}{(16 \times 4.3)} = 0.00682$
			Average value of $K = 0.00684$ min^{-1}

The near constancy of the value of K, as calculated according to the second order rate equation shows that this reaction is of second order.

The average value of the rate constant is 0.00684 min^{-1}.

MODEL QUESTIONS

Chemical Kinetics

1. Explain clearly the significance of "molecularity of a reaction" and "order of a reaction" with suitable examples. Distinguish between the two.
2. Discuss the significance of the term "order of a reaction" with examples. Derive an expression for the rate constant (or velocity constant or specific reaction rate).
3. Write explanatory notes on the following :
   (a) An isolated chemical reaction.
   (b) Velocity constant of a reaction.
   (c) Pseudo-unimolecular reactions.
4. What do you mean by the term "order of a reaction" ? Derive an expression for the rate equation of a second order reaction when the concentrations of both the reactants are same.
5. Define the term "order of a reaction". Derive an expression for the velocity constant (rate constant) of the reaction : $CH_3 COOC_2 H_5 + Na OH \rightleftharpoons CH_3 COO Na + C_2 H_5 OH$ when the initial concentration of each of the reactant is 'a' moles / litre. Under what conditions would such a reaction behave kinetically as a first order reaction ?
6. How does the half-change period depend on the initial concentration in case of first order and second order reactions ? Derive the corresponding expressions.
7. State the various methods available for the determination of velocity constant of a reaction in the laboratory.
8. What are the units of the rate constant in case of first order and second order reactions. Derive expression for half-change periods in both the cases.
9. Describe a method for the laboratory determination of order of a reaction.
10. How do you determine the rate of hydrolysis of ethyl acetate by an alkali and the order of the reaction ?

REACTION DYNAMICS & CATALYSIS

11. (a) How is the velocity of a reaction influenced by variation of temperature?
    (b) Why do we not generally come across reactions of third and higher order?

12. (a) Derive the rate equation for a first order reaction.
    (b) A 20% solution of cane sugar having a dextro-rotation of 34.5° is inverted by 0.5 N lactic acid at 25° C and the following data was recorded while studying the kinetics of the reaction:

Time (minutes)	0	1435	11365	∞
Rotation (degrees)	34.5	31.0	14.0	−10.8°

    Establish that this reaction follows the kinetics of a unimolecular reaction.

13. What are the characteristics of a first order reaction?

    On the basis of the following kinetic data obtained during the decomposition of H_2O_2 in aqueous solution, show that the reaction follows first order kinetics:

Time in minutes	Moles of $KMnO_4$ required to react with a fixed aliquot of the reaction mixture.
0	25.40
15	9.83
30	3.81
40	2.03

14. In a study to follow the kinetics of decomposition of N_2O_5 in an inert solvent, the following data was obtained:

Time in seconds	Volume of O_2 evolved in ml
0	0
300	3.42
600	6.30
900	8.95
∞	34.75

    Prove that the reaction is of first order.

15. In the kinetic study of thermal decomposition of gaseous acetaldehyde, the increase of gaseous acetaldehyde, the increase of pressure (ΔP) recorded with the progress of time was as follows:

Time in seconds	42	242	480
ΔP in mm	34	134	194

    The initial pressure of the reactant in the vessel was 363 mm.
    Show that the reaction is of second order.

16. In a kinetic study during the hydrolysis of methyl acetate at 25°C, using N/20 HCl as catalyst, the following results were recorded:

Time, t	Volume of alkali used while titrating equal quantity of the reaction mixture at the corresponding time (in ml)
0	24.36
1200	25.85
4500	29.32
7140	31.72
∞	47.15

Determine the rate constant of the reaction and show that the reaction follows first order kinetics.

(**Ans.** 5.51×10^{-5} sec^{-1}; First order)

17. N-chloroacetanilide changes slowly in presence of an acid into p-chloroacetanilide. But of the above two, only N-chloroacetanilide liberates iodine from KI. Therefore, the progress of the reaction can be followed by titrating equal quantities of the reaction mixture from time to time with hypo solution. In such an experiment, the following results were recorded :

Time, t (in hours)	0	1	2	4	6	8
Volume of hypo run down (in ml)	45	77	67.5	56.3	50.7	47.9

Find the average value of K.

Show that this reaction is mono-molecular.

Also, determine the fraction of N-chloroacetanilide decomposed after 3 hours.

(**Ans.** $K = 0.3441$; fraction : 64.31%)

18. The variation in the partial pressure of azomethane with time at 600 K is determined experimentally and the data is recorded as follows :

Time (seconds)	0	1000	2000	3000	4000
Partial pressure (10^{-2} Torr)	8.20	5.72	3.99	2.78	1.94

The decomposition of azomethane takes place according to the following reaction :

$CH_3 N_2 CH_3 (g) \longrightarrow CH_3 CH_3 (g) + N_2 (g)$

Show that the reaction is first order with respect to azomethane. Also, calculate the value of K.

(**Ans.** $K = 3.6 \times 10^{-4}$ s^{-1})

19. In a particular experiment, the concentration of N_2O_5 in liquid bromine varied with time as follows :

Time (seconds)	0	200	400	600	1000
Conc. of N_2O_5 (mol L^{-1})	0.110	0.073	0.048	0.032	0.014

Show that the decomposition reaction follows first order kinetics in N_2O_5 and determine the rate constant.

(**Ans.** $K = 2.1 \times 10^{-3}$ s^{-1})

REACTION DYNAMICS & CATALYSIS 899

20. Benzene diazonium chloride in aqueous solution decomposes as follows :

$$C_6H_5N_2Cl \longrightarrow C_6H_5Cl + N_2$$

Starting with an initial concentration of 10 g / litre, the volume of N_2 gas obtained at 50° C at different time intervals was found to be as under :

Time (minutes)	6	12	18	24	30	∞
Volume of N_2 collected (ml)	19.3	32.6	41.3	46.5	50.4	58.3

Determine the rate constant and show that the reaction follows first order kinetics.

(**Ans.** $K = 6.76 \times 10^{-2}$ min^{-1})

21. Show that in case of a first order reaction, the time required for completion of 99.9% of the reaction is about 10 times of that required for completion of 50% of the reaction.

22. The half-life period for the thermal decomposition of N_2O_5 at 100° C is 4.6 seconds and it is independent of the initial pressure of the gas. Calculate the specific rate constant of the reaction under these conditions.

(**Ans.** 15.07×10^{-2} sec^{-1})

23. Discuss the kinetics of saponification of esters.

24. The saponification of ethyl acetate was carried out starting with equal concentrations of ethyl acetate and sodium hydroxide. Equal aliquots of the reaction mixture were withdrawn at different time intervals and titrated with an acid. The results obtained were recorded as under :

Time (minutes)	0	5	15	25
Titre Value (ml)	16	10.24	6.13	4.32

Establish that the reaction is of second order.

25. Why are the reactions of a higher order rare ? A second order reaction in which the initial concentrations of the reactants are identical is 25% complete in 600 seconds. Calculate the time required for completion of 75% of the reaction. (**Ans.** 5400 seconds)

26. A second order reaction involving decomposition of a gas with initial concentration of 4×10^{-2} moles L^{-1} takes 40 minutes for 40% of the reaction to complete. Calculate the specific reaction rate. (**Ans.** $K = 0.4167$ litre mol^{-1} min^{-1})

27. Describe the fractional change (half-life period) method for the determination of order of a reaction.

28. The half-life period of a decomposing compound is 50 minutes at a certain concentration, C. If we start with an initial concentration of $C/2$, the half-life period is 25 minutes. Find the order of the reaction. (**Ans.** Zero order)

29. Write informative notes on any two of the following :
    (a) ostwald's isolation method for the determination of the order of a reaction.
    (b) chain reactions
    (c) Arrhenius equation for the effect of temperature on rate of reaction.

30. Discuss transition state theory (or theory of absolute reaction rates or theory of activated complex formation). What are its advantages over the collision theory ?
31. Write short notes on any three of the following :
    (a) consecutive reactions
    (b) parallel or side reactions
    (c) opposing or reversible reactions
    (d) chain reactions
    (e) zero order reactions
32. Write short notes on any three of the following :
    (a) Energy of activation
    (b) Temperature coefficient of a reaction
    (c) Transition state
    (d) Probability factor
    (e) Arrhenius equation
    (f) Half change time
    (g) molecularity Vs. order of a reaction.
33. The following data is corrected for the decomposition of ammonium nitrite in aqueous solution :

Time (in minutes)	10	15	20	15	∞
Volume of N_2 (in ml)	6.25	9.00	11.40	13.65	35.05

Show that the reaction follows first order kinetics.

34. In a saponification reaction of ethyl acetate using identical concentrations of the ester and an alkali, the following data was collected : (The alkali reacted and hence the progress of reaction at different time periods is determined by titrating with an acid).

Time (in minutes)	0	4.9	10.1	23.7	∞
Titre value (in ml)	47.6	38.9	32.7	22.6	11.8

Show that the reaction is of second order

35. The $t_{1/2}$ of a reaction became half when the initial concentration of the reactant is doubled. Show that the reaction is of second order.
36. The rate constant of a reaction increased by three times when the temperature is raised from 27° C to 37° C. Calculate the activation energy of the reaction. (**Ans.** 82.18 KJ mol^{-1})
37. Two different reaction have their activation energies of 25 KJ mol^{-1} and 52 KJ mol^{-1}. Which of the reactions would be more susceptible to changes in temperature and why ?
38. (a) What do you mean by a "Zero order reaction" ? What is meant by the activation energy for a reaction ?
    (b) Consider a gaseous decomposition reaction :

    A → products,

    at 500° C and at an initial pressure of 350 torr sec when 5% of the decomposition was over and and 0.76 torr sec^{-1} when 20% of the decomposition was over. Determine the order of the reaction.

(c) For the reaction :

$$2NO + Cl_2 \rightleftharpoons 2NOCl$$

the following mechanism was proposed.

Step -1 : $NO + Cl_2 \underset{K_{-1}}{\overset{K_1}{\rightleftharpoons}} NOCl_2$

Step -2 : $NO + NOCl_2 \longrightarrow 2 NOCl$

show that the overall rate of reaction is :
$K [NO]^2 [Cl_2]$, where, $K = K_1 K_2 / K_{-1}$
Assume that $K_2 [NO] \ll K_{-1}$

(d) Consider the following esterification reaction :

$$CH_3COOH + C_2H_5OH \rightleftharpoons CH_3COOC_2H_5 + H_2O$$

1 mole of the acid and 1 mole of the alcohol are mixed at a temperature of 25° C. At equilibrium 0.667 moles acid have reacted. Calculate the equilibrium constant Kc. How much ester would be obtained if 2 moles of the acid were mixed with 1 mole of the alcohol under identical conditions ?

(Biju Patnaik University of Technology, Bhuvaneswar, 2003)

CATALYSIS

25.19. INTRODUCTION

In the year 1835, J. J. Berzelius reviewed a number of observations where the rate of a reaction was increased by the presence of some substances that remained unchanged at the end of the process. The examples quoted include.

(a) decomposition of alkaline hydrogen peroxide in presence of certain metals,
(b) conversion of starch into sugar in presence of acids, and
(c) combination of H_2 and O_2 on spongy platinum.

Berzelius thought that such substances help to loosen the bonds which hold the atoms in the reacting molecules together. Therefore, he coined the term "catalysis" (Greek : *Kata* = wholly, *lien* = to loosen) to describe such phenomena. The substance that brought about this change is called "catalyst".

However, a number of cases are known where the rate of a reaction is retarded (*i.e.* slowed down) by the presence of some substances.

According to the present concept, *"A catalyst is defined as a substance which alters the rate of a chemical reaction, while itself remaining chemically unchanged at the end of the reaction"*.

Thus, there can be two types of catalysts. A catalyst which enhances the rate of a reaction is called a "positive catalyst" or simply "catalyst", and the phenomenon is called "positive catalysis" or simply "catalysis". The word catalyst, when used alone, is almost invariably taken to imply acceleration of the chemical process.

Example : Increased rate of decomposition of $KClO_3$ in presence of MnO_2.

If the rate of a reaction is retarded by the added substance, it is said to be a "negative catalyst" and the process is called "negative catalysis" or "inhibition".

Example : Decreased rate of decomposition (retardation) of H_2O_2 in presence of dilute acids or glycerol.

25.20. ACTION OF A CATALYST

The catalyst does not initiate a reaction. It merely accelerates a reaction that is already taking place at a slow pace. As stated in the chapter on "Chemical Kinetics", a catalyst functions by providing an alternate path which involves lower energy of activation for the reaction.

25.21. CHARACTERISTICS OF CATALYTIC REACTIONS (OR CRITERIA OF CATALYSIS)

(1) A catalyst remains unchanged chemically at the end of the reaction.

It must be emphasized that the amount and chemical composition of a catalyst remains unaltered at the end of the process. However, it may undergo a physical change.

Example : (a) Granular MnO$_2$ employed to catalyze the decomposition of KClO$_3$ changes to a fine powder at the end of the reaction.

(b) Platinum gauze used as catalyst for the oxidation of NH$_3$ becomes rough after some time.

The physical alteration of catalyst at the end of the reaction shows the involvement of catalyst in the reaction mechanism involved in the catalytic process.

(2) A small amount of catalyst is often sufficient to bring about a considerable extent of reaction.

Since the catalyst is not consumed in the reaction and is regenerated at the end, even a small amount of it often causes large quantities of reactants to combine. Some catalysts are particularly effective for certain reactions.

Example : (a) Cupric ions even at a concentration of 1 g ion in 10^6 litres cause considerable oxidation of Na$_2$SO$_3$ by oxygen.

(b) Colloidal platinum present at a concentration of 1 g. atom in 10^6 liters catalyze the decomposition of H$_2$O$_2$.

However, in homogeneous processes and where no chain mechanism is involved, the reaction velocity increases with the amount of catalyst employed. This is also true to some extent for many heterogeneous reactions involving gases on the surface of a solid, where the rate of reaction is proportional to the effective area of the catalytic surface. This is the reason why finely divided materials are employed as catalysts to give the best results.

(3) A catalyst does not affect the position of equilibrium in a reversible reaction.

Since the catalyst remains unchanged chemically at the end of the reaction, it cannot contribute

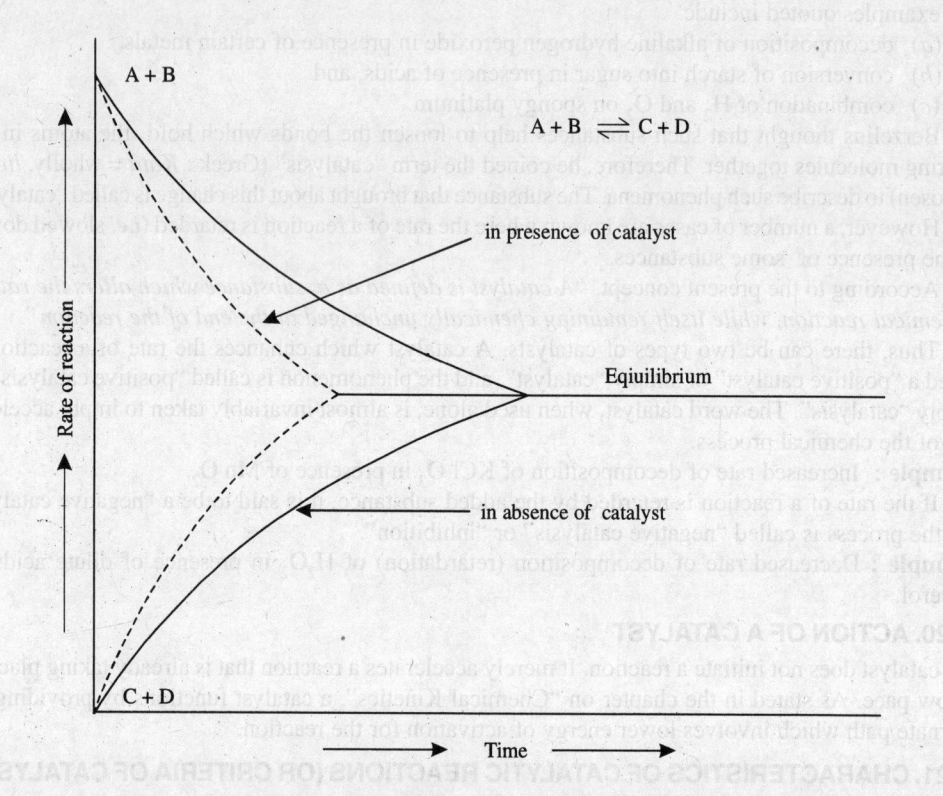

Fig. 25.9. Effect of catalyst on time taken for establishment of equilibrium.

any energy to the system. Hence, according to the second law of thermodynamics, the same position of equilibrium should be attained ultimately, whether a catalyst is used or not. However, a catalyst affects the forward and backward reactions in a reversible process to the same extent. Therefore, a catalyst helps in attaining the equilibrium more quickly, although the position of equilibrium and the equilibrium constant remain unchanged. These concepts are best illustrated in Fig. 5.9.

Example : (a) The same position of equilibrium was reached in the reaction of SO_2 and O_2 using three different catalysts, namely V_2O_5, Pt and Fe_2O_3.

(b) In the Haber's process for the manufacture of ammonia, iron is used as catalyst. The reaction between N_2 and H_2 is very slow but in presence of catalyst, the equilibrium is reached quickly. However, the percentage yield remains unchanged.

$$N_2 + 3H_2 \xrightleftharpoons{Fe} 2NH_3$$

(c) Many enzymes, which fall into the category of catalysts, are known to accelerate both the forward and reverse reactions in an equilibrium process.

(4) A catalyst, in general, cannot initiate a reaction

As stated earlier, a catalyst functions by providing an alternative path involving smaller free energy of activation. Thus, it is reasonable to state that a catalyst can only change the speed of a reaction which is already taking place, even if it is extremely slowly but it does not initiate the reaction.

Example : (a) The reaction between H_2 and O_2 in presence of platinum black or finely divided platinum as catalyst takes place very quickly even at room temperature, but in the absence of the catalyst, the reaction is so slow that no appreciable amount of water is formed even after several years.

(5) A catalyst is specific in its action

While a catalyst is effective in altering the rate of a particular reaction, it is not necessary that it is effective for another reaction. For a particular reaction, only a particular catalyst is effective. In fact, it is known that different products are obtained in presence of different catalysts although the reactants are same.

Example : (a) Ethanol is converted into ethanal in presence of hot Cu as catalyst while it gives ethene when passed over hot Al_2O_3.

$$\underset{\text{ethanol}}{C_2H_5OH} \xrightarrow{Cu} \underset{\text{ethanal}}{CH_3CHO}$$

$$\underset{\text{ethanol}}{C_2H_5OH} \xrightarrow{Al_2O_3} \underset{\text{ethene}}{C_2H_4} + H_2O$$

(b) Hydrogen and carbon monoxide react at high temperature and pressure, to give different products in presence of different catalysts, as given below :

(i) $CO + 2H_2 \xrightarrow{ZnO, Cr_2O_3} CH_3OH$

(ii) $CO + 3H_2 \xrightarrow{Ni} CH_4 + H_2O$

(other volatile hydrocarbons may also be formed)

(iii) $CO + H_2 \xrightarrow{Co}$ mixture of higher paraffins

(c) Enzymes are highly specific in their action. An enzyme that breaks down carbohydrates has no effect on proteins of fats in the digestive system.

(6) The efficiency of a catalyst depends on its physical state.

In heterogenious catalysis, it is found that a catalyst is far more effective when it is present in a fine state of subdivision as compared to the catalyst when used in bulk or in lump form.

Example : (a) Finely divided nickel has been found to be more effective than pieces or lumps of nickel.

(b) Platinum black, platinised asbestos or finely divided platinum have been proved to be far more effective than using lumps of platinum.

(7) Change of temperature alters the rate of a catalytic reaction.

Just as in case of uncatalysed reactions, the efficiency of a catalyst is altered with change of temperature. In some reactions where the catalyst is physically altered, there may be change in catalytic activity. The temperature at which the rate of reaction is maximum is called the optimum temperature.

(8) The activity of a catalyst can be altered by the presence of a foreign substance. If this substance increases the activity of catalyst, it is called a promoter. On the contrary, if the presence of the foreign substance inhibits or destroys the catalytic activity, then it is called an anti-catalyst or catalytic poison.

Example : (a) Molybdenum acts as a promoter for iron used as catalyst in the manufacture of NH_3 by Haber's process.

(b) Arsenic oxides poison the platinum used as catalyst for oxidation of SO_2 to SO_3 in contact process for manufacture of H_2SO_4.

25.22. TYPES OF CATALYSIS

Catalysed reactions are broadly classified into the following two groups on the basis of the mechanism of reactions and the phases involved :

(1) Homogeneous catalytic reactions

In these reactions, the reactants and catalyst are in the same single phase and the reaction involves molecules, ions or free radicals as intermediates.

Homogeneous catalysis can occur in gaseous phase as well as in the liquid (solution) phase.

(a) Examples of homogeneous catalysis in gas phase

(i) Oxidation of sulfur dioxide to sulfur trioxide by the oxides of nitrogen (as in chamber process for the manufacture of sulfuric acid.

$$[NO]_{gas} + 1/2\, O_{2\,gas} \longrightarrow NO_{2\,gas}$$

$$SO_{2\,gas} + NO_{2\,gas} \longrightarrow SO_{3\,gas} + [NO]_{gas}$$

(ii) Molecular iodine is an efficient catalyst for many gas phase thermal decompositions such as pyrolysis of acetaldehyde. A free radical mechanism is involved in this reaction.

$$CH_3CHO_{vapour} + I_{2\,vapour} \longrightarrow CH_{4\,gas} + CO_{gas} + I_{2\,vapour}$$

REACTION DYNAMICS & CATALYSIS

(iii) Decomposition of ozone in presence of nitric oxide

$$2O_3(g) \xrightarrow{NO(g)} 3O_2(g)$$

(b) Examples of homogeneous catalysis in solution phase

(i) Hydrolysis of esters eg. ethyl acetate catalysed by an acid or an alkali.

$$CH_3COOCH_3(aq) + H_2O(l) \xrightarrow[\text{(aq)}]{H^+/OH^-} CH_3COOH(aq) + CH_3OH(aq)$$

methyl acetate — acetic acid — methanol

(ii) Decomposition of H_2O_2 catalysed by iodide ion.

$$2H_2O_2(aq) \xrightarrow[\text{(aq)}]{I^-} 2H_2O(l) + O_2(g)$$

(iii) Inversion of cane sugar catalysed by hydrogen ion from a mineral acid

$$C_{12}H_{22}O_{11}(aq) + H_2O(l) \xrightarrow{H^+} C_6H_{12}O_6(aq) + C_6H_{12}O_6(aq)$$

cane sugar — glucose — fructose

(iv) Condensation of acetone to diacetone catalysed by hydroxyl ions

(v) Mutarotation of sugar catalysed by hydroxyl ions.

(vi) Hydrolysis of starch to glucose in presence of an acid.

$$(C_6H_{10}O_5)_n(aq) + nH_2O(l) \xrightarrow[\text{(aq)}]{H^+} nC_6H_{12}O_6(aq)$$

(2) Heterogeneous Catalytic Reactions

In these reactions, the reactants and the catalyst are of different phases and the reactions proceed at an interface between the two discrete phases.

(a) Heterogeneous catalysis with solid reactants.

Example : (i) Decomposition of potassium chlorate in presence of manganese dioxide as catalyst.

$$2KClO_3 + [MnO_2] \longrightarrow 2KCl + 3O_2 + [MnO_2]$$

solid — solid

(ii) Heterogeneous catalysis with liquid reactants

Example : (i) Reaction between benzene and ethanoyl chloride catalysed by anhydrous aluminium chloride.

$$C_6H_6(l) + CH_3COCl(l) + [AlCl_3](s) \longrightarrow C_6H_5COCH_3 + HCl + [AlCl_3]$$

phenyl methyl ketone

(ii) Decomposition of aqueous solution of hydrogen peroxide in presence of MnO_2 as catalyst.

$$2H_2O_2(l) + [MnO_2](s) \longrightarrow 2H_2O + O_2 + [MnO_2]$$

(iii) Hydrogenation of animal and vegetable oils using nickel as catalyst.

(iv) Manufacture of synthetic petrol by cracking of heavy oils in presence of aluminium silicate as catalyst.

(c) Heterogeneous catalysis with gaseous reactants (contact catalysis)

Example: (i) Reaction between carbon monoxide and hydrogen catalysed by ZnO and Cr_2O_3 for the manufacture of methyl alcohol.

$$CO(g) + 2H_2(g) + [ZnO, Cr_2O_3](s) \longrightarrow CH_3OH + [ZnO, Cr_2O_3](s)$$

(ii) Reaction between sulfurdioxide and oxygen using finely divided platinum or Vanadium pentoxide as catalyst, used in the manufacture of sulfuric acid by contact process.

$$2SO_2(g) + O_2(g) + [Pt \text{ or } V_2O_5](s) \longrightarrow 2SO_3(g) + [Pt \text{ or } V_2O_5](s)$$

(iii) Oxidation of ammonia in presence of platinum gauze as catalyst used in the manufacture of nitric acid.

$$4NH_3(g) + 5O_2(g) + [Pt](s) \longrightarrow 4NO(g) + 6H_2O(l) + [Pt](s)$$

(iv) Synthesis of ammonia by Haber's process in which nitrogen and hydrogen combine in presence of finely divided iron as catalyst.

$$N_2(g) + 3H_2(g) + [Fe](s) \longrightarrow 2NH_3 + [Fe](s)$$

(v) Use of aluminium silicates as catalyst for cracking (*i.e.* conversion of high molecular weight hydrocarbons into low-molecular hydrocarbons.

(vi) Use of $Mo_2O_3 \cdot Al_2O_3$ mixed catalyst as catalyst for reforming (*i.e.* conversion of cycloalkanes into aromatic hydrocarbons).

(vii) Use of catalytic converters to convert harmful gases (*eg.* CO) in automobile exhaust into less harmful ones, to control atmospheric pollution.

25.23. CATALYTIC PROMOTERS

Certain substances were found to increase the activity of a catalyst, although they themselves are not considered as catalysts. These substances are called catalytic promoters or activators.

Thus, a catalytic promoter or activator is defined as a substance, which when added in small quantities, promotes the activity of a catalyst.

Example : (i) In the Haber's process for the manufacture of ammonia from N_2 and H_2, molybdenum (Mo) or aluminium oxide and potassium oxide ($Al_2O_3 + K_2O$) is used as a promoter to increase the activity of iron catalyst.

$$N_2 + 3H_2 \xrightleftharpoons{Fe + Mo} 2NH_3$$

(ii) In the hydrogenation of oils for manufacture of Dalda (or vegetable ghee), tellurium (Te) is added as promoter to increase the activity of nickel catalyst.

(iii) In the isomerization of paraffins, the activity of $AlCl_3$ as catalyst, is increased by the presence of HCl or H_2O as promoter.

(iv) In the Bosch's process for the manufacture of hydrogen from water gas, finely divided iron is used as catalyst and metallic copper is used as promoter.

(v) A mixture of catalysts, viz. ($Zn + Cr_2O_3$) is used for the synthesis of methanol from carbon monoxide and hydrogen to achieve better catalytic efficiency.

Mechanism for the Action of Promoter

No solid surface can be ideally smooth. The *"residual field"* of atoms or any catalytic surface (or any solid surface to that matter), which is responsible for adsorption and catalysis will not be uniform. An atom on a flat surface is attached to similar atoms over a solid angle of 180°C, but an atom at the apex of a peak, or at the edge or corner of a crystal, will have a much smaller fraction of its

REACTION DYNAMICS & CATALYSIS

electrostatic attraction shared by others, and therefore, its *"residual field"* will be large.. Such *"unsaturated"* atoms generally constitute the *"active centres"* involved in catalysis. *"Discontinuities"* such as edges of crystals and grain-boundaries, or cracks or imperfections in crystals, all of which represent positions of unsaturation, function as active centres on a catalyst surface.

The mechanism of the action of a promoter in increasing the efficiency of a catalyst is not clearly understood. However, the increase of catalytic activity in presence of a promoter may be due to the following reasons :

(a) Alteration in lattice spacing to provide for the enhanced spacings between the catalyst particles. This results in further weakening and cleavage of the bonds between reactant molecules. This enables further faster reaction to form the products. (Fig. 5.10). This phenomenon is common in promoting catalytic activity in heterogeneous catalysis.

(b) Increase of peaks, cracks and corners on the catalyst surface (because of addition of promoter) thereby increasing the number of reactant molecules on the catalyst, which contributes to the enhanced rate of reaction.

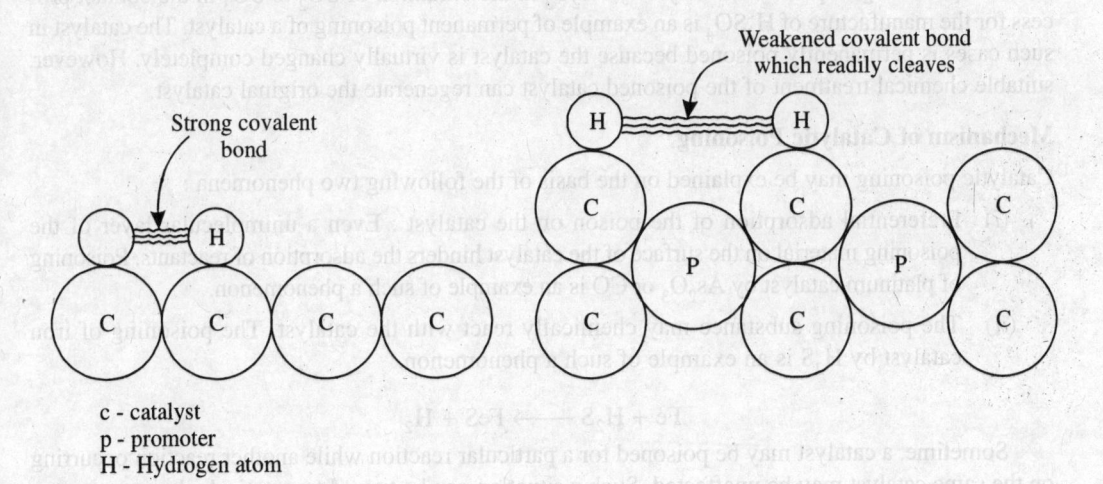

Fig. 25.10. Action of promoter in altering the lattice spacing of a catalyst thereby making the reaction faster.

25.24. CATALYTIC POISONS

A heterogeneous catalyst is often rendered ineffective due to the presence of minute quantities of extraeous impurities, thereby inhibiting the catalytic reaction.

Any substance which inhibits or destroys the catalytic activity to accelerate a reaction is called a catalytic poison. The process is called poisoning of catalyst.

Example : (i) The activity of platinum catalyst in the reaction between hydrogen and oxygen is inhibited by H_2S, CS_2 and CO.

$$2H_2 + O_2 \xrightarrow[\text{poisoned by CO}]{\text{Pt catalyst}} 2H_2O$$

(ii) The platinum catalyst used in the oxidation of SO_2 to SO_3 (in the contact process of manufacture of sulfuric acid) is poisoned by the presence of arsenic trioxide.

$$2SO_2 + O_2 \xrightarrow[\text{poisoned by As}_2\text{O}_3]{\text{Pt catalyst}} 2SO_3$$

However, the new V_2O_5 catalyst used in contact process is not susceptible for poisoning.

(iii) In the Haber's process for the manufacture of ammonia, the iron catalyst used to catalyse the reaction between N_2 and H_2 is poisoned by the presence of H_2S.

$$N_2 + 3H_2 \xrightarrow[\text{poisoned by } H_2S]{\text{Fe catalyst}} 2NH_3$$

(iv) In the reaction between ethylene and hydrogen, the copper catalyst used is poisoned by Hg or CO.

Temporary and Permanent Poisons

In the reaction between N_2 and H_2 catalysed by Fe catalyst in Haber's process, water vapour and O_2 act as temporary poisons by combining with Fe to form an oxide of iron. If the catalytic poisons, viz., water vapour or oxygen is removed from the reactant gases, the iron oxide is reduced to iron by the hydrogen present, and thus the catalytic activity of Fe is restored.

The poisoning of platinum catalyst by As_2O_3 in the oxidation of SO_2 to SO_3 in the contact process for the manufacture of H_2SO_4 is an example of permanent poisoning of a catalyst. The catalyst in such cases is permanently poisoned because the catalyst is virtually changed completely. However, suitable chemical treatment of the poisoned catalyst can regenerate the original catalyst.

Mechanism of Catalytic Poisoning

Catalytic poisoning may be explained on the basis of the following two phenomena :

(i) Preferential adsorption of the poison on the catalyst : Even a unimolecular layer of the poisoning material on the surface of the catalyst hinders the adsorption of reactants. Poisoning of platinum catalyst by As_2O_3 or CO is an example of such a phenomenon.

(ii) The poisoning substance may chemically react with the catalyst. The poisoning of iron catalyst by H_2S is an example of such a phenomenon.

$$Fe + H_2S \longrightarrow FeS + H_2$$

Sometime, a catalyst may be poisoned for a particular reaction while another reaction occurring on the same catalyst may be unaffected. Such a situation can be turned to practical advantage sometimes. When vapours of ethyl alcohol are passed over a copper catalyst at about 300°C, acetaldehyde is produced. The acetaldehyde first formed tends to decompose into CH_4 and CO. If some water is present in ethyl alcohol only this latter reaction is inhibited by partial poisoning of the catalyst. Thus, the yield of acetaldehyde is increased.

25.25. NEGATIVE CATALYSIS AND INHIBITION

When the rate of a reaction is diminished due to the presence of a substance, it is called a *negative catalyst* or an *inhibitor*.

Example : (i) Oxidation of sodium sulfite solution by oxygen gas is inhibited by small amounts of benzyl alcohol, aniline, benzaldehyde, mannitol, brucine, etc.

(ii) Decomposition of organic acids (*eg.*, formic acid, oxalic acid, citric acid and malic acid) takes place in presence of H_2SO_4 as catalyst. This reaction is inhibited by the presence of small quantities of water, ammonium sulfate, potassium sulfate and some organic oxygen compounds.

(iii) Addition of *antiknock* materials such as tetraethyl lead, $(C_2H_5)_4$Pb known as TEL inhibits the rapid and explosive combustion of the petrol. Thus, knocking in the engine is prevented.

(iv) Chloroform is used as a general anaesthetic in surgery. It undergoes aerial oxidation forming carbonyl chloride, COCl$_2$ which is highly toxic. When 2% ethanol is added to chloroform, it acts as a negative catalyst to suppress the formation of poisonous carbonyl chloride.

$$4CHCl_3 + 3O_2 \longrightarrow 4COCl_2 + 2Cl_2 + 2HO \text{ (Ethanol inhibits this reaction)}$$

(v) The decomposition of H$_2$O$_2$ is retarded by the presence of glycerol or a dilute acid which act as negative catalyst.

Mechanism of Negative Catalysis

Negative catalysis, may be caused by any of the following mechanisms :

(i) By poisoning or removing or counteracting the existing positive catalyst.

Example : (a) Even a minute concentration of cupric ions catalyze the reaction between a sulfite solution and oxygen. Addition of a negative catalyst such as a cyanide or mannitol counteracts the effect of the positive catalyst and retards the reaction.

(b) The decomposition of H$_2$O$_2$ is catalyzed by traces of alkali dissolved from the glass container.

Addition of an acid will destroy the alkali catalyst and retard the decomposition of H$_2$O$_2$.

(ii) By breaking a chain reaction. A negative catalyst functions by preventing the formation or propagation of reaction chains.

Negative catalysis in gas reactions is generally considered to be due to the breaking of reaction chains.

Example : (a) The phenomenon of "knocking" in the internal combustion engines is connected with the propagation of non-stationary branched chains. The "antiknock" materials such as tetraethyl lead (TEL) or nickel carbonyl, Ni(CO)$_4$ produce metal compounds such as peroxides which destroy the chain carriers.

(b) Retardation of thermal decomposition of vapours of acetaldehyde and ethyl ether by nitric oxide which removes the free radicals.

(c) Iodine acts as negative catalyst for the homogeneous combination of hydrogen and oxygen by destroying the chain carriers and thereby inhibiting the reaction. Nitrogen trichloride, NCl$_3$ also functions as negative catalyst for the reaction between H$_2$ and O$_2$ by destroying the chain reaction.

$$\left.\begin{array}{l} Cl_2 \longrightarrow Cl^\bullet + Cl^\bullet \\ H_2 + Cl^\bullet \longrightarrow HCl + H^\bullet \\ H + Cl_2 \longrightarrow HCl + Cl^\bullet \end{array}\right\} \text{chain propagation}$$

$$NCl_3 + Cl^\bullet \longrightarrow 1/2\, N_2 + 2Cl_2 \} \text{ chain termination}$$

(d) Stabilizers like acetanilide are added to prevent decomposition of H$_2$O$_2$ solution. The stabilizers are preferentially adsorbed on the glass walls of the container on which reaction chains are normally initiated. The stabilizers also may act as chain breakers by absorbing light which causes photochmical decomposition of H$_2$O$_2$.

(iii) By altering the surface of the walls of the reaction vessel.

Example : The oxidation of liquid benzaldehyde by oxygen gas is believed to be a wall reaction and the walls of the reaction vessel catalyze the reaction. Negative catalysts affect the nature of the wall and retard the reaction velocity.

25.26. AUTOCATALYSIS

In some reactions, one of the products formed itself catalyzes the reaction. Such a phenomenon is called autocatalysis, and the substance (*i.e.* reaction product) which catalyzes the reaction is called an autocatalyst. For an autocatalytic reaction, the rate of reaction increases with time, because the concentration of the catalytic product (*i.e.* the autocatalyst) increases. Therefore, a plot of the progress of reaction (or the reaction rate or the product formed) with time gives a sygmoid curve which reaches a maximum when the reaction is complete (vide Fig. 5.11).

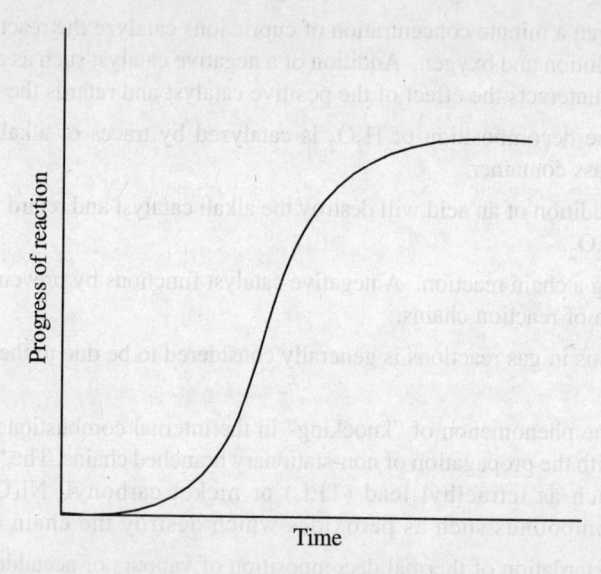

Fig. 25.11. Plot showing the progress of an autocatalytic reaction with time (in arbitrary units).

Example : (*i*) In the reaction between an oxalate and acidified potassium permanganate solution, the manganese (II) ions generated by the reduction of the permanganate catalyse the reaction.

$$2\ MnO_4^- + 5\ C_2O_4^{2-} + 16H^+ \longrightarrow 2\ Mn^{2+} + 10\ CO_2 + 8H_2O$$

(*ii*) In the hydrolysis of esters, the acid formed as one of the products, itself catalyzes the reaction.

$$\underset{\text{ethyl acetate}}{CH_3COOC_2H_5} + H_2O \longrightarrow \underset{\substack{\text{acetic acid}\\\text{(acts as autocatalyst)}}}{CH_3COOH} + \underset{\text{ethanol}}{C_2H_5OH}$$

(*iii*) The reaction between Cu and nitric acid is slow in the beginning, but the nitrous acid produced in the reaction, itself catalyses the rate of dissolution of the metal in the acid. Subsequently, the rate of reaction decreases again. This is because the nitrous acid itself decomposes when its concentration exceeds a certain amount.

$$3\ Cu + 8\ HNO_3 \longrightarrow 3\ Cu(NO_3)_2 + 4H_2O + 2\ NO$$

$$2\ NO + H_2O + O \longrightarrow 2H^+ + 2\ NO_2^-$$

25.27. INDUCED CATALYSIS

If the rate of a reaction, which does not take place under normal conditions, is influenced by another reaction, such a phenomenon is called induced catalysis. For instance, sodium arsenite solution is not easily oxidised by air where as sodium sulfite is readily oxidised. However, when air is passed through a mixture of sodium arsenite and sodium sulfite solutions, both of them undergo oxidation.

25.28. ACTIVATION ENERGY AND CATALYSIS

According to the collision theory of reaction rates, a chemical reaction takes place only by collisions between the reacting molecules. However, all the collisions are not effective and only a small fraction of the collisions bring about a reaction. A collision will be fruitful only when colliding molecules collide with sufficient kinetic energy and with proper orientation.

A chemical reaction takes place by breaking bonds between the atoms of the reactant molecules to enable the formation of new bonds to produce the molecules of the product. The energy required for this purpose comes from the kinetic energy associated with the reacting molecules before collision. At ordinary temperatures, the molecules do not possess adequate energy for effective collisions. It has been found that to bring about a fruitful reaction between the colliding molecules, they should acquire some minimum energy, known as the *activation energy*. Only such molecules that collide with a kinetic energy greater than the activation energy are able to surmount the energy barrier for the particular reaction. The activated molecules crossing the energy barrier, on collision, form an activated complex or a transition state momentarily which then decomposes to give the product.

One of the methods to increase the kinetic energy of the molecules is to increase the temperature.

Alternatively, the reaction can be made to take place by the use of a suitable catalyst, which lowers the activation energy of the reaction by providing an alternative new pathway. Thus, it is possible to increase the number of effective collisions in a reaction in presence of a catalyst than what would have occurred at the same temperature in the absence of the catalyst. Thus, under otherwise identical conditions, the presence of a catalyst renders the reaction faster.

The above principles are illustrated in Fig. 25.4 and 25.5.

Approaching molecules of the reactants

Activated complex or Intermediate state

Molecules of the product

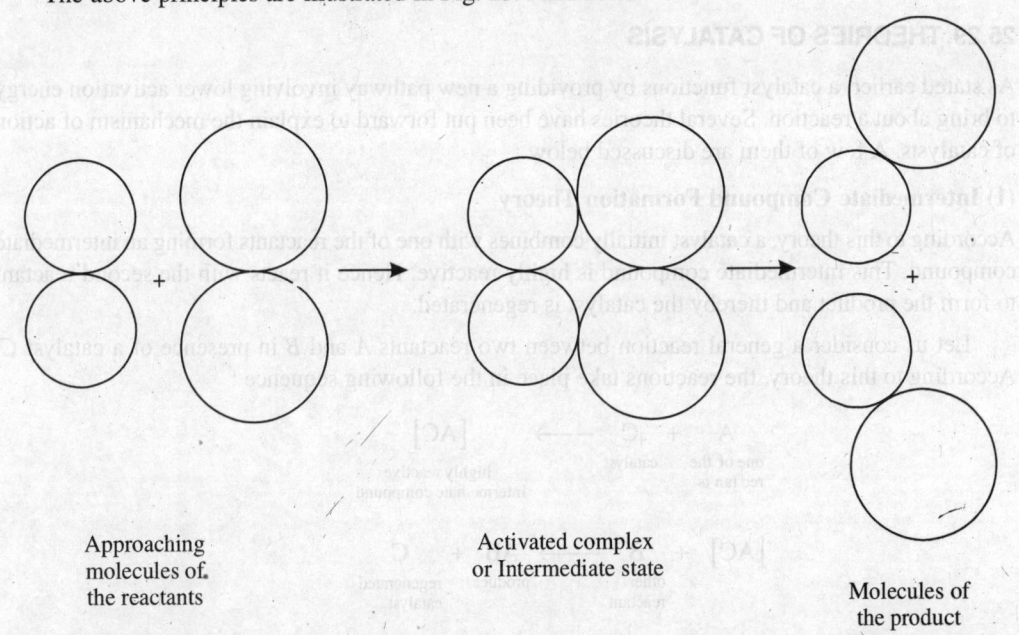

Fig. 25.12. Mechanism of a molecular reaction involving the formation of a transient intermediate state of activated complex

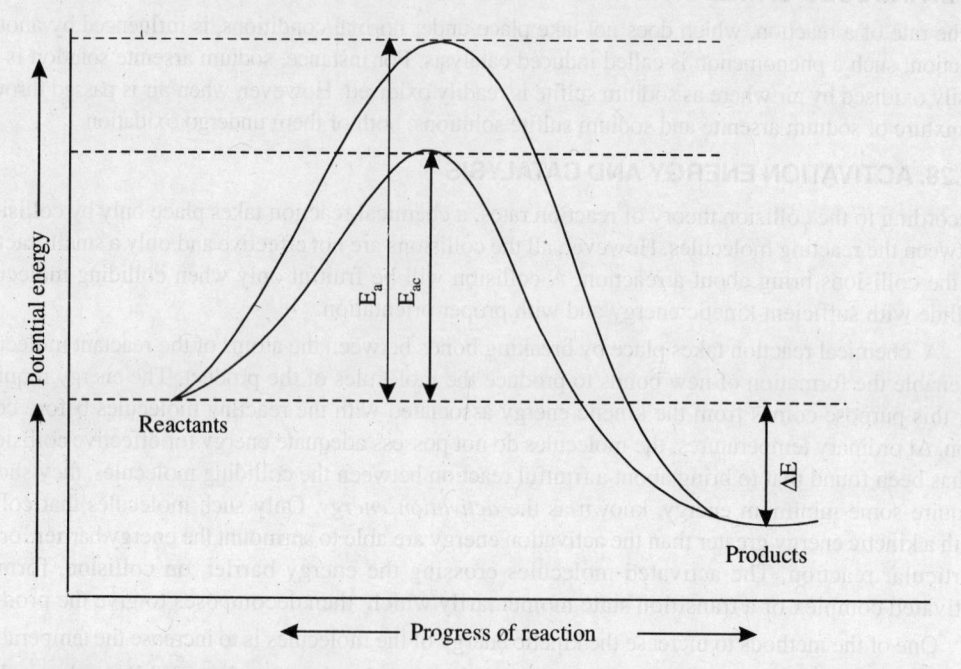

Fig. 25.13. Energy diagram depicting the lowering of activation energy in presence of a catalyst by providing a new pathway for the reaction.

E_a = activation energy for uncatalyzed reaction.

E_{ac} = activation energy for catalyzed reaction.

ΔE = energy change accompanying the reaction

25.29. THEORIES OF CATALYSIS

As stated earlier, a catalyst functions by providing a new pathway involving lower activation energy to bring about a reaction. Several theories have been put forward to explain the mechanism of action of catalysts. A few of them are discussed below :

(1) Intermediate Compound Formation Theory

According to this theory, a catalyst initially combines with one of the reactants forming an intermediate compound. This intermediate compound is highly reactive. Hence it reacts with the second reactant to form the product and thereby the catalyst is regenerated.

Let us consider a general reaction between two reactants A and B in presence of a catalyst C. According to this theory, the reactions take place in the following sequence :

$$A + C \longrightarrow [AC]$$
one of the reactants, catalyst, highly reactive intermediate compound

$$[AC] + B \longrightarrow AB + C$$
other reactant, product, regenerated catalyst

Overall Reaction

$$A + B + [C] \longrightarrow AB + [C]$$

REACTION DYNAMICS & CATALYSIS 913

Thus the rate of reaction is increased because of the involvement of the catalyst.

Alternatively, one can envisage the formation of a highly reactive and transient intermediate state, which requires low activation energy.

$$\underset{\text{(Reactants)}}{A + B} + \underset{\text{Catalyst}}{C} \longrightarrow \underset{\substack{\text{(Reactive and transient activated} \\ \text{complex or intermediate state)}}}{A \ldots B \ldots C}$$

This complex readily and spontaneously breaks up to form the product and thereby regenerating the catalyst.

$$A \ldots B \ldots C \longrightarrow \underset{\text{product}}{AB} + [C]$$

This is because, the energy of AB is far lower than that of the intermediate complex. Thus the rate of reaction is increased because of the presence of the catalyst.

It is extremely difficult to isolate the intermediate complex because it is very reactive and transient.

Homogeneous catalytic processes in solution appear to be generally characterized by the formation of intermediate compound followed by its decomposition to yield the product and regenerate the catalyst.

Example : (a) Catalytic oxidation of SO_2 in presence of [NO] as catalyst in the chamber process for the manufacture of sulfuric acid.

$$2 SO_2 + O_2 + [NO] \longrightarrow 2 SO_3 + [NO]$$

Mechanism

$$\underset{\text{catalyst}}{[NO]} + 1/2 O_2 \longrightarrow \underset{\text{(intermediate)}}{NO_2}$$

$$NO_2 + SO_2 \longrightarrow \underset{\text{product}}{SO_3} + \underset{\text{catalyst regenerated}}{[NO]}$$

(b) Decomposition of formic acid catalyzed by the presence of hydrogen ions as catalyst.

$$HCOOH + \underset{\text{catalyst}}{[H^+]} \longrightarrow \underset{\text{intermediate}}{[HCOOH_2]^+}$$

$$[HCOOH_2]^+ \longrightarrow CO + H_2O + \underset{\substack{\text{catalyst} \\ \text{regeneration}}}{[H^+]}$$

(c) Formation of diethyl ether from ethanol in presence of H_2SO_4 as catalyst.

$$2C_2H_5OH + [H_2SO_4] \longrightarrow C_2H_5OC_2H_5 + H_2O + [H_2SO_4]$$

Mechanism

$$C_2H_5OH + \underset{\text{catalyst}}{[H_2SO_4]} \longrightarrow \underset{\text{intermediate}}{C_2H_5HSO_4} + H_2O$$

$$C_2H_5 \;\boxed{HSO_4 + H}\; OC_2H_5 \longrightarrow \underset{\text{product}}{C_2H_5OC_2H_5} + \underset{\text{catalyst regenerated}}{[H_2SO_4]}$$

(d) Reaction between benzene and methyl chloride in presence of aluminium chloride as catalyst (Friedel-Crafts reaction)

$$C_6H_6 + CH_3Cl + [AlCl_3] \longrightarrow C_6H_5CH_3 + HCl + [AlCl_3]$$
catalyst methyl benezene catalyst regenerated

Mechanism

$$CH_3Cl + [AlCl_3] \longrightarrow [CH_3]^+ [AlCl_4]^-$$
methyl chloride catalyst int ermediate

$$C_6H_5 + [CH_3]^+ [AlCl_4]^- \longrightarrow C_6H_5CH_3 + HCl + [AlCl_4]$$
product catalyst regenerated

(e) The reaction between ceric and thiosulfate ions is rather slow but it is accelerated by the presence of a small amount of iodide. This enables the reaction rapid enough to be used for volumetric analysis.

$$Ce^{+++} + I^- \longrightarrow Ce^{+++} + I$$
catalyst int ermediate

$$I + I \longrightarrow I_2$$

The iodine molecules (or even iodine atoms) react rapidly with the thiosulfate ions, regenerating the iodide.

$$2S_2O_3^{--} + I_2 \longrightarrow S_4O_6^{--} + 2I^-$$
catalyst regenerated

In heterogeneous gaseous reactions catalyzed by solid catalysts also, the reacting molecules adsorbed on the surface of the catalyst, may also be regarded as an intermediate activation state.

The intermediate compound formation theory cannot fully explain the various heterogeneous catalytic processes. Further, this theory cannot explain the action of promoters, catalytic poisons, etc.

(2) The Adsorption Theory

This theory has been enunciated mainly to explain the mechanism of heterogeneous catalytic processes, particularly those reactions between gaseous reactants catalyzed by a solid catalyst. In such processes, the reaction is initiated by adosrption of the reactant molecules on the surface of the catalyst. The adsorption process results from the residual forces on the catalyst surface. These forces may lead to the formation of a chemical bond or just to polarization of the adsorbed molecule. In the process of bond formation with the surface, the adorsbed molecule may also undergo dissociation. This adsorbed species is therefore a more reactive form, and hence it can undergo reactions more readily than when the catalyst is not present. The catalytic selectivity may thus be considered as a consequence of the specific manner in which bond formation, polarization, etc. takes place.

It can therefore be considered that there is a continuous formation and decomposition of unstable intermediates which offer a new pathway involving a lower energy of activation for the formation of final reaction products. Such a formation of unstable intermediates is a valid pathway for homogeneous catalysis reactions too. For catalysis involving solid surfaces, chemisorption of reactant molecules on the catalytic surface is a pre-requisite. The exothermic nature of the chemisorption process supplies the energy required for loosening or dissociation of bonds present in the adsorbed reactant molecules. Thus, when H_2 molecules are chemisorbed on tungsten surface, the bonding between the hydrogen atoms in the H_2 molecules is disrupted, and the resulting H atoms are adsorbed. The bond between the catalytic adsorbent and the adsorbed species is usually covalent, except in case of the metal-oxygen bonds which are ionic.

It is important that the atoms should not be adsorbed on the catalytic surface so strongly that they act as a catalytic poison. The catalytic activity is usually inversely proportional to the strength of adsorption of the reactant species on the catalyst surface.

Transitional metals of Groups IV, V and VI and Fe are capable of chemisorbing gases like CO, O_2, H_2 and N_2. Cobalt and nickel also can chemisorb the above gases excepting N_2. The heat of adsorption of H_2 is relatively low for Ni, Pt, Pd, Rh, Ir and Ru, and hence they are used effectively as hydrogenation catalysts. Similarly, the heats of adsorption of N_2 on Pd, Rh, Ru, etc. are very low whereas those on metals of vanadium and chromium group are quite high, while that on Fe, it is intermediate. In case of synthesis of ammonia, the catalytic activity of Fe is found to be the best.

It is believed that for efficient catalytic activity, the crystals of transition metals should possess electronically incomplete d-bands called d-holes (or unpaired atomic d-orbitals).

According to the adsorption theory, for a **general heterogeneous** catalytic reaction such as :

$$A(g) + B(g) \xrightarrow{\text{solid catalyst}} C(g) + D(g),$$

the following steps are involved sequentially.

Step-1. Adsorption of reactant molecules on the catalytic surface by weak Vander Waals forces (Physical adsorption) or by partial chemical bonds (chemisorption).

Step-2. Formation of an unstable and intermediate activated complex, A ... B from the adsorbed reactant species adjacent to one another.

Step-3. Decomposition of the unstable activated complex to form the products.

Step-4. Desorption of the stable products formed to release the fresh catalytic surface for a fresh cycle of the above steps.

The pictorial representation of the sequential steps in a heterogeneous catalytic process as per the adsorption theory is given in Fig. 5.14.

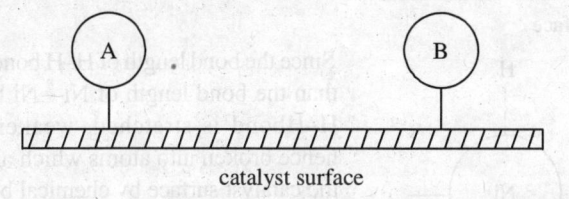

Step-1. Adsorption of reactant species on catalyst surface.

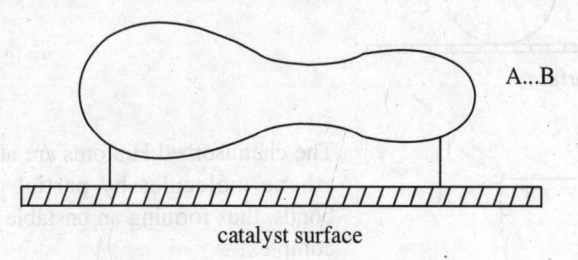

Step-2. Formation of an unstable activated complex.

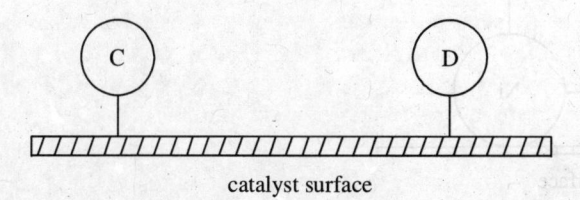

Step-3. Decomposition of the activated complex and formation of the product.

Step-4.

C   D

────────────────────
catalyst surface

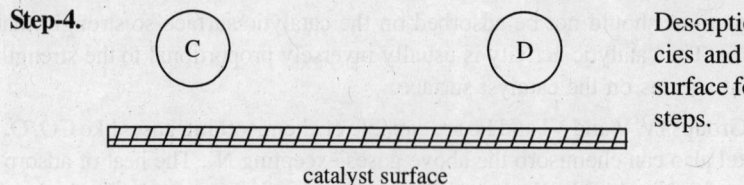

Desorption of the stable product species and release of the free catalyst surface for another cycle of the above steps.

Fig. 25.14. Pictorial representation of the sequential steps in a heterogeneous catalytic reaction.

Depending on the nature of the reactants, the mechanism may slightly alter in details. For instance, the following stepwise mechanism is contemplated for the hydrogenation of ethene to ethane in presence of Ni catalyst (Fig. 5.15)

$$\underset{\text{ethene}}{\overset{H}{\underset{H}{>}}C=C\overset{H}{\underset{H}{<}}} + H-H \xrightarrow[\text{catalyst}]{\text{Ni as}} \underset{\text{ethane}}{H-\overset{\overset{H}{|}}{\underset{\underset{H}{|}}{C}}-\overset{\overset{H}{|}}{\underset{\underset{H}{|}}{C}}-H}$$

Mechanism

Step-1.

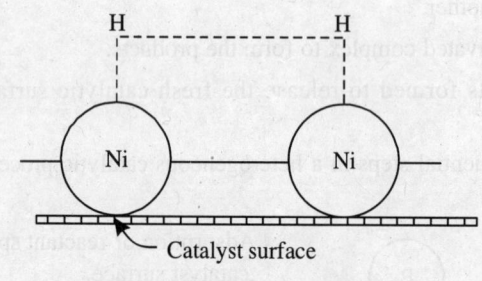

Adsorption of hydrogen molecules on the catalyst surface due to residual valence bonds of Ni atoms.

Step-2.

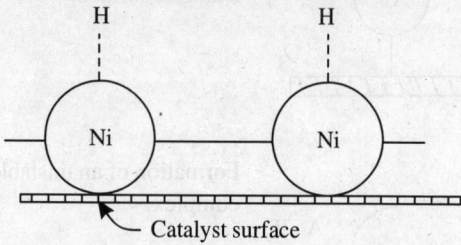

Since the bond length of H–H bond is lesser than the bond length of Ni—Ni bond, the H–H bond is stretched, weakened, and hence broken into atoms which are held to the catalyst surface by chemical bonds.

Step.3

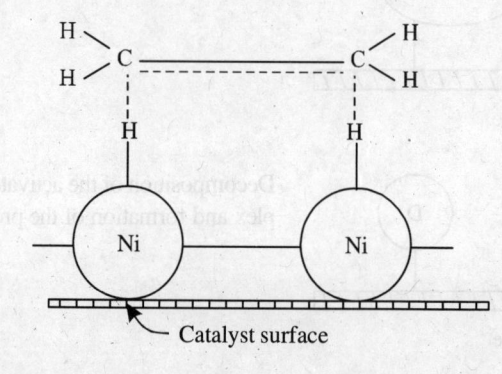

The chemisorbed H atoms are attached to ethene molecules by partial chemical bonds, thus forming an unstable activated complex.

Step-4

H₂C—ethane—CH₂ (on Ni-Ni catalyst surface)

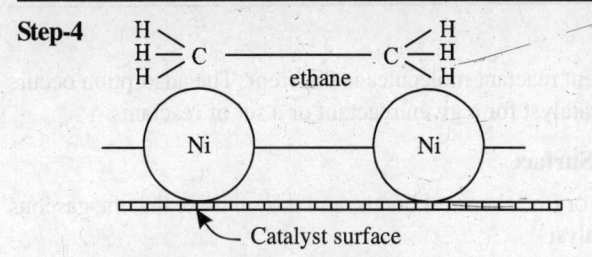

The unstable activated complex is decomposed to yield the product (ethane) and the catalyst surface is released for a fresh cycle of the above steps.

Fig. 25.15. Pictorial Representation of Mechanism of the hydrogenation of ethene to ethane in presence of Ni as catalyst.

The adsorption theory reasonably explains the various facets of catalytic activity as follows :

(a) Efficiency of Catalyst in Colloidal or Finely Divided State :

The catalyst is more efficient when present in colloidal or in finely divided state than when present in lumps.

In heterogeneous catalysis, the rate of reaction is increased because of the increase in concentration of the reactants on the surface of the solid catalyst due to adsorption at the active centres. When the catalyst is present in a finely divided state, the surface area is increased and thereby, the free valencies or active centres on the catalyst surface are increased. This accounts for the enhanced activity of the catalyst when present in finely divided state as compared to that in the lump form. The finer the catalyst is subdivided, greater will be its activity (Fig. 5.16).

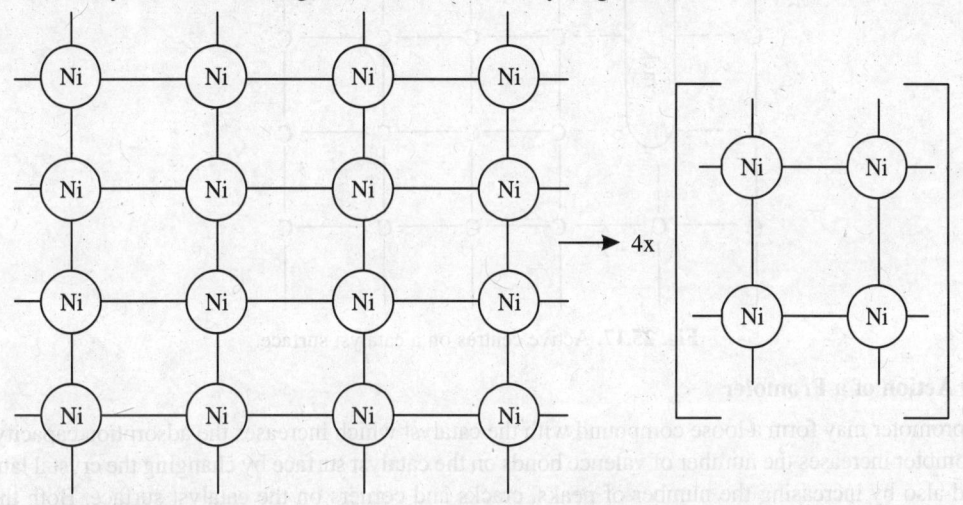

Fig. 25.16. Illustration of the increase in the free valence bonds or active centres on the surface of catalyst when it is more finely divided.

(b) Selectivity of a Catalyst

Selectivity of a catalyst usually means the efficiency of the catalyst to bring about a desired reaction whereas *Activity* of a catalyst denotes the overall conversion of the reactants into products.

In heterogeneous or contact catalysis the reaction process is initiated by the adsorption of the reactant molecules on the surface of the solid catalyst. The adsorption process results from the residual forces on the catalyst surface. These forces may lead to the formation of a chemical bond or just to polarization of the adsorbed molecule. In this process, the adsorbed molecule may undergo dissociation. The adsorbed species are in a more reactive form and hence they can undergo reactions more readily than in the absence of the catalyst. The catalytic selectivity is the consequence of the specific manner in which bond formation, polarization, etc., take place.

(c) Specificity of a Catalyst

The affinity of different catalysts for different reactant molecules is different. The adsorption occurs only when there is a strong affinity of the catalyst for a given reactant or a set of reactants.

(d) Active Centres on a Rough Catalyst Surface

A catalyst surface contains unbalanced or free chemical bonds, and it is on these that the gaseous reactant molecules are adsorbed on the catalyst.

These free valency bonds on the surface of the catalyst are not uniformly distributed. They are found to be greater in number in cracks, peaks and corners of the catalyst. The catalytic activity due to adsorption is found to be high at these spots, which are called active centres (Fig. 5.17). The active centres accelerate the rate of reaction not only by increasing the concentration of the reactants at these centres but also by activating the adsorbed reactant molecule adsorbed at two such adjacent centres by stretching.

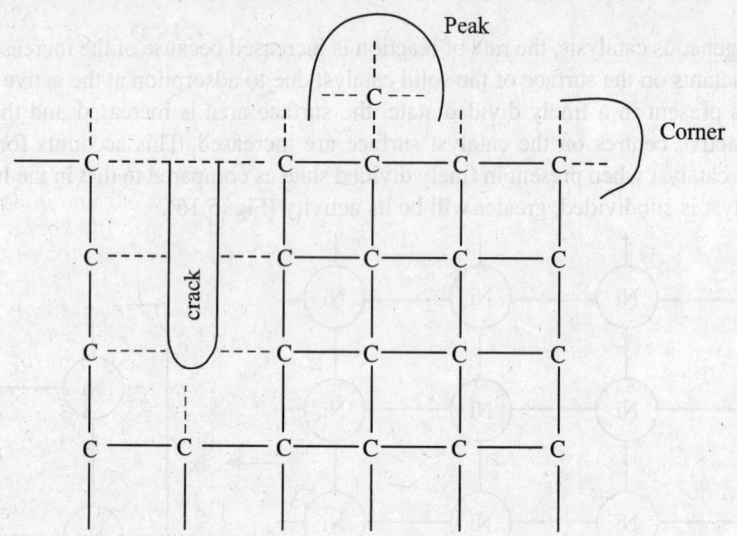

Fig. 25.17. Active centres on a catalyst surface.

(e) Action of a Promoter

A promoter may form a loose compound with the catalyst which increases the adsorption capacity. A promoter increases the number of valence bonds on the catalyst surface by changing the crystal lattice and also by increasing the number of peaks, cracks and corners on the catalyst surface. Both these aspects have been discussed earlier under "catalytic promoters" and under 'd' given above.

(f) Action of a Catalytic Poison

A catalytic poison inhibits or destroys a catalyst by blocking the free valence bonds on its surface by preferential adsorption or by chemical combination. This has been discussed in detail earlier under "catalytic poisons".

25.30. ACID-BASE CATALYSIS

Acid-base catalysis represents one of the most important category of homogeneously catalyzed reactions. A number of homogeneous catalytic reactions are known which are catalyzed by acids or bases or by both acids and bases. Such catalysts are called acid-base catalysts and the phenomenon is called acid-base catalysis.

W. Ostwald (1883-84) studied various reactions which were catalyzed by acids. Inversion of sucrose and hydrolysis of esters were some such reactions which were catalyzed by acids.

$$\underset{\text{sucrose}}{C_{12}H_{22}O_{11}} + H_2O \xrightarrow{H^+} \underset{\text{glucose}}{C_6H_{12}O_6} + \underset{\text{fructose}}{C_6H_{12}O_6}$$

$$\underset{\text{ethyl acetate}}{CH_3COOC_2H_5} + H_2O \xrightarrow{H^+} \underset{\text{acetic acid}}{CH_3COOH} + \underset{\text{ethanol}}{C_2H_5OH}$$

After the development of the electrolytic theory of dissociation of electrolytes, S. Arrhenius (1889) showed that the rates of these catalytic processes were closely proportional to the concentrations of hydrogen ions. Thus acids were found to be effective catalytic agents.

On the other hand, hydroxyl ions have been found to be particularly effective :

(a) in the mutarotation of glucose in aqueous solution,

(b) in the decomposition of nitroso triacetoneamine, and

(c) in the conversion of acetone into diacetone alcohol and vice versa.

Arrhenius (1899) further observed that the addition of neutral salts to acid solutions often had an unexpected effect in enhancing the catalytic activity greater than that based on the assumption that hydrogen ion was the only catalyst. The catalytic effect was found irrespective of whether the salt had an ion in common with the catalyzing acid or when it had not.

H.M. Dawson (1914) proposed the dual theory of catalysis in solution which postulated that both hydrogen ions and the undissociated molecules of an acid were able to act catalytically. This view has been extended by the work of J. N. Brönsted, H.M. Dawson, T.M. Lowry and their collaborators (1923) according to which it was evident that not only hydrogen and hydroxyl ions, but also undissociated molecules of acids and bases, and even cations of weak bases and anions of weak acids can have catalytic activity. According to Brönsted-Lowry definition, an acid is a substance which tends to lose a proton, whereas a base is a substance which tends to give a proton. Water is amphoteric because it behaves as an acid to substances more basic and as a base to substances more acidic than itself. Thus, if a reaction is catalysed by hydrogen ions from all acids and all Bronsted acids (*i.e.* proton donors), then this phenomenon is known as *general acid catalysis*. The general acid catalysts include H^+_{ions}, undissociated molecules of weak acids *eg.*, (CH_3COOH), cations of weak bases (*e.g.*, NH_4^+), and H_3O^+ (from water).

Similarly, if reactions catalysed by OH^- ions and all Bronsted bases (*i.e.* proton acceptors) act as base catalysts, then this phenomenon is called *general base catalysis*. The general base catalysts include OH^-, undissociated molecules of weak bases, anions of weak acids (*e.g.*, CH_3COO^-) and water.

For some processes, both acids and bases appear to be effective catalysts. For example, the decomposition of nitramide is also catalysed by both H^+ as well as acetate ions (CH_3COO^-).

$$NH_2NO_2 \xrightarrow[CH_3COO^-]{H^+ \text{ or }} N_2O + H_2O.$$

The catalysis brought about by general acids and bases is called *general acid-base catalysis*.

A reaction which is catalysed by H^+ ions (or H_3O^+ ions) but not by other proton donors (Bronsted acids) is called *specific acid catalysis*.

Example : (a) Hydrolysis of esters

$$\underset{\text{ethyl acetate}}{CH_3COOC_2H_5} + H_2O \xrightarrow{H^+} \underset{\text{acetic acid}}{CH_3COOH} + \underset{\text{ethyl alcohol}}{C_2H_5OH}$$

(b) Inversion of cane sugar

$$\underset{\text{cane sugar}}{C_{12}H_{22}O_{11}} + H_2O \xrightarrow{H^+} \underset{\text{glucose}}{C_6H_{12}O_6} + \underset{\text{fructose}}{C_6H_{12}O_6}$$

(c) Keto-enol tautomerism of acetone

$$\underset{\text{Keto form}}{CH_3-\underset{\underset{O}{\|}}{C}-CH_3} \xrightarrow{H^+} \underset{\text{Enol form}}{CH_3-\underset{\underset{OH}{|}}{C}=CH_2}$$

A reaction which is catalyzed only by OH⁻ ions and not by other proton acceptors is called *specific base catalysis*.

Example:

$$R-\underset{\underset{O}{\|}}{C}-OR' + H_2O \xrightarrow[\text{catalyst}]{OH^-} RCOOH + R'OH$$

Mechanism of Acid-Base Catalysis

(i) General Acid Catalysis

In the general acid catalysis, the H⁺ or a proton donated by Brönsted acid forms an unstable intermediate complex with reactant, which then decomposes to give the reaction product while giving back the proton.

Example : Keto-enol tautomerism of acetone

$$\underset{\text{Acetone (Keto form)}}{CH_3-\underset{\underset{O}{\|}}{C}-CH_3} \xrightarrow{H^+} \underset{\substack{\text{Intermediate}\\\text{complex}}}{CH_3-\underset{\underset{OH}{\|}}{C}-CH_3} \longrightarrow \underset{\text{Enol-form}}{CH_3-\underset{\underset{OH}{|}}{C}=CH_2} + H^+$$

(ii) General Base Catalysis

In the general base catalysis, the OH⁻ ion or any other Brönsted base accepts a proton from the reactant to give an unstable intermediate complex, which then decomposes to give the reaction product, while regenerating the OH⁻ or the Brönsted base, as the case may be.

Example : (i) (a) Decomposition of nitramide to give N₂O using OH⁻ as catalyst

$$\underset{\text{nitramide}}{NH_2\,NO_2} \xrightarrow{OH^-} \underset{\substack{\text{intermediate}\\\text{complex}}}{NH\,NO_2^-} + H_2O$$

$$NH\,NO_2^- \longrightarrow N_2O + OH^-$$

(ii) (b) Decomposition of nitramide to give NO₂ using CH₃COO⁻ as catalyst

$$NH_2\,NO_2 + CH_3\,COO^- \longrightarrow CH_3COOH + NH\,NO_2^-$$

$$NH\,NO_2^- \longrightarrow N_2O + OH^-$$

$$OH^- + CH_3\,COH \longrightarrow H_2O + CH_3COOH$$

REACTION DYNAMICS & CATALYSIS

25.31. ENZYME CATALYSIS

Enzymes are complex protein molecules which catalyze organic reactions taking place in living cells. Enzymes are capable of bringing about complex chemical and biochemical reactions which seem to be impossible under normal conditions.

Example : (i) Zymase converts glucose to ethanol.

$$C_6H_{12}O_6 \xrightarrow{zymase} 2C_2H_5OH + 2CO_2$$

(ii) Urease brings about hydrolysis of urea.

$$CO(NH_2)_2 + H_2O \xrightarrow{urease} CO_2 + 2NH_3$$

(iii) Invertase causes inversion of sucrose.

$$\underset{\text{sucrose}}{C_{12}H_{22}O_{11}} + H_2O \xrightarrow{invertase} \underset{\text{glucose}}{C_6H_{12}O_6} + \underset{\text{fructose}}{C_6H_{12}O_6}$$

(iv) Diastase catalyses starch to produce, maltose which in turn is hydrolysed by another enzyme called maltase to glucose.

$$2\underset{\text{starch}}{(C_6H_{10}O_5)_n} + nH_2O \xrightarrow{diastase} n\underset{\text{maltose}}{C_{12}H_{22}O_{11}}$$

$$\underset{\text{maltose}}{C_{12}H_{22}O_{11}} + H_2O \xrightarrow{maltase} 2\underset{\text{glucose}}{C_6H_{12}O_6}$$

The enzymes are capable of readily bringing about hydrolysis, reduction, oxidation and other complex chemical and biochemical reactions. The enzymatic action is highly specific and different enzymes can catalyze specific reactions, by lowering the activation energy for the respective reaction. The catalysis brought about by enzymes is termed as *"enzyme catalysis"*.

Enzyme catalysis takes place to bring about several reactions in life processes in human body. Enzyme catalysis is also used for the manufacture of ethyl alcohol from molasses using "mycoderma aceti".

Characteristics of Enzyme Catalysts

(1) Efficiency : Enzymes are highly efficient catalysts one molecule of an enzyme may transform one million molecules of the substrate (reactant) per minute.

Enzyme catalysts, like inorganic catalysts, function by lowering the activation energy of a reaction. For example, for the decomposition of H_2O_2, the activation energy is 18 K cal/mole. In presence of colloidal platinum catalyst, the activation energy is lowered to 6.3 K.cal/mole whereas the enzyme "catalase" lowers the activation energy to < 2 K.cal/mole.

(2) Specificity : Enzyme catalysts are highly specific. A particular enzyme catalyses only a particular reaction. For example, "urease" catalyses the hydrolysis of urea only (to form NH_3 and H_2O) and not any other amide, not even methyl urea. Similarly, *"Penicillium glaucum"* (an enzyme present in ordinary mould) when added to a mixture of d – and l – forms of tartaric acid, decomposes the d-form only.

(3) State of equilibrium : Just like inorganic catalysts, enzyme catalysts do not disturb the final state of equilibrium in a reversible reaction.

(4) Maximum efficiency at optimum temperature : The rate of enzyme catalyzed reactions is maximum at an optimum temperature. Ex :- The enzymatic reactions in human body take place with maximum efficiency at 98.6° F.

(5) Maximum efficiency at optimum pH : The rate of enzyme catalyzed reactions generally increases with pH until the optimum pH is reached and then decrease with further increase of pH. Many of the enzyme catalyzed reactions in human body are highly efficient at pH of 7.4.

(6) Increase of activity in presence of activators : The enzymatic activity is enhanced in presence of metal ions (eg., Na^+, Cu^{2+}, Co^{2+}, Mn^{2+}) that get weakly bonded to the enzyme molecules. Similarly, coenzymes eg., vitamins promote the catalytic activity of the enzymes.

(7) Inhibition of activity by poisons : The catalytic activity of enzymes is inhibited or even completely destroyed by the presence of certain inhibitors or poisons. For example, heavy metal ions eg., Hg^{2+} can react with the –SH groups of the enzymes and destroy the enzymatic activity by poisoning.

$$2\ Enz - SH + Hg^{2+} \longrightarrow 2\ Enz - S-Hg^{2+} + 2H^+$$

The physiological activity of several antibiotics is based on the inhibition of the enzymatic activity of bacteria.

Mechanism of Enzyme Catalysis

The long chains of the enzyme molecules are coiled on each other to form a rigid colloidal particle with cavities on its surface. These cavities which are of characteristic size and shape are present in plenty in active groups eg., SH, OH, NH_2, COOH etc., which are called *active centres*. Only such molecules of the substrate having complementary shape can just fit into the cavities, in the same way as a key fits into its lock. This theory is therefore known as *Lock and Key theory*. Owing to the presence of the active groups, the enzyme forms an activated complex with the substrate which at once decomposes to yield the products. Thus, the substrate molecules enters the cavities, forms the complex which decomposes quickly to yield the products and the products get out of the cavities quickly.

$$\underset{\text{enzym + substrate}}{E + S} \rightleftharpoons \underset{\substack{\text{activated}\\ \text{complex}}}{ES} \longrightarrow \underset{\text{product enzyme}}{P + E}$$

25.32. SOME IMPORTANT INDUSTRIAL PROCESSES USING CATALYSTS

S. No.	Process	Catalyst
1.	Haber's process for the manufacture of NH_3.	Fe + (Mo as promoter)
2.	Contact process for the manufacture of H_2SO_4.	Pt
3.	Ostwald's process for the manufacture of HNO_3.	Pt gauze
4.	Hydrogenation of oils for the manufacture of Vanaspati.	Ni
5.	Bosch process for the manufacture of H_2.	$Fe_2O_3 + Cr_2O_3$
6.	Manufacture of methyl alcohol from H_2 and CO.	$ZnO + Cr_2O_3$
7.	Polymerisation of alkenes.	H_3PO_4 on Kieselguhr
8.	Deacon's process for the manufacture of Cl_2.	$CuCl_2$

9.	Cracking of heavy petroleum fractions.	Silica, Alumina gel
10.	Manufacture of ethyl alcohol by fermentation of starch.	Maltose, Diastase
11.	Acetic acid from alcohol	Mycoderma aceti
12.	Ethyl alcohol manufacture by sugar fermentation.	Zymase, Invertase
13.	Acetic acid from acetaldehyde by aereal oxidation.	V_2O_5
14.	Acetaldehyde from acetylene.	$HgSO_4$

25.33. CRITERIA FOR CHOOSING A CATALYST FOR INDUSTRIAL APPLICATION

1. It should be cheap and easily avilable.
2. It should be stable.
3. It must be selective so that the desired product is the major project.
4. It should possess maximum possible activity in for the reaction under consideration in terms of the yield of the desired product.
5. It should be amenable for easy regeneration or reactivation.

QUESTIONS

1. What do you mean by activation energy ? Explain with the help of energy reaction coordinate diagram. How does the nature of the activated complex change in the presence of a catalyst.
2. Discuss the effect of a catalyst on the rate of a reaction. How can you explain the action of a catalyst ?
3. Why and how does a catalyst accelerates a reaction ?
4. Discuss the validity of the statement that "A catalyst only hastens the approach of equilibrium in a reversible reaction but it does not alter the position of the equilibrium".
5. Discuss the various theories advanced to explain the action of catalysts.
6. Discuss the important characteristics of catalysts.
7. Explain the following with examples :
   (a) positive catalysts
   (b) negative catalysts
   (c) catalytic promoters
8. Write explanatory notes on any two of the following :
   (a) Enzyme catalysts
   (b) Auto-catalysis
   (c) Induced catalysis
9. Write short notes on any three of the following :
   (a) Catalytic poisons
   (b) Negative catalysis

(c) Active centres

(d) Auto-catalysis

(e) Induced catalysis

(f) Catalytic promoters

10. Write informative notes on any two of the following :

   (a) Industrial applications of catalysts

   (b) Theories to explain the action of catalysts.

   (c) Heterogenous catalysis

   (d) Homogeneous catalysis

   (e) Enzyme catalysis

11. What are catalysts ?

    Explain homogeneous and heterogeneous catalysis with examples.

12. Explain the action of catalysts in the light of adsorption theory and intermediate compound formation theory.

13. Explain how the rate of reaction can be increased with increase in temperature or in presence of a catalyst.

14. Discuss the contribution of catalysts in industrial development.

15. Explain the role of catalysts in establishing chemical equilibrium in a reversible reaction.

16. Explain how

   (a) a promoter increases the activity of a catalyst

   (b) a catalytic poison paralyses the activity of a catalyst.

   (c) a catalyst influences a reversible reaction at equilibrium

   (d) a reactant or a product retards the rate of a surface reaction.

17. How can you say that a particular substance taking part in a reaction is not a reactant but is a catalyst. Explain the mechanism involved with examples to support your answer.

18. State whether the physical state of the catalyst influences the catalytic activity. Explain with examples.

26

Photochemistry

26.1. INTRODUCTION

Photochemistry mainly deals with chemical reactions resulting directly or indirectly from absorption of radiation. Strictly speaking, the term radiation includes the entire electromagnetic spectrum, but the radiations of photochemical importance usually lie in the visible (–400 to 800 nm) and ultraviolet (–200 to 400 nm) region. The energy range corresponding to this region is 35 Kcal to 145 Kcal/mol. From these energy values and the bond energies deduced from thermodynamic studies, it can be expected that the absorption of light in the visible and ultraviolet regions can be expected to be adequate to break a bond or at least to produce a high energy reactive (or excited) molecule.

The near infrared region, corresponding to energy range of about 3 Kcal/mol, is too small to effect any chemical changes but can bring about vibrational and rotational changes in the molecules, on the other hand, exposure of a system to radiations of very short wave lengths (*e.g.* X-rays, Y-rays, beams of electrons, α-particles etc. primarily produces ionization, although chemical reactions may take place due to secondary effects. These effects come under the regime of radiation chemistry.

Although the fact that light can bring about certain chemical changes is known for several years, systematic study of photochemistry was started only after the development of quantum chemistry and study of molecular spectra. The subject of photochemistry is important not only because of its intrinsic interest but also because of its contributions to the understanding of reaction kinetics, photosynthesis etc. The fact that solar energy can be used for diverse applications such as for heating, generation of electricity and production of hydrogen from water provided further impetus to the study of photochemistry.

26.2. PHOTOCHEMICAL REACTIONS

Diverse types of reactions such as synthesis, decomposition, hydrolysis, oxidation, reduction, polymerization and isomerization can be brought about photochemically by exposure to suitable radiations.

Absorption of electromagnetic radiation can initiate several processes. Among them, those processes that take place by capturing the radiant energy of the sun are the most important, some of these processes lead to the heating of the atmosphere during the daytime by absorption of U.V. radiation. Others include the absorption of visible radiation during photosynthesis. But for these photochemical processes, the earth would have been simply a warm, sterile rock.

Some of the common photochemical processes are as follows :

(*i*) Ionization

$$A^* \longrightarrow A^+ + e^-$$

(*ii*) Electron transfer

$$A^* + B \longrightarrow A^+ + B^- \text{ (or } A^- + B^+\text{)}$$

(*iii*) Dissociation

$$A^* \longrightarrow B + C$$
$$A^* + B-C \longrightarrow A^- + B + C$$

(*iv*) Addition

$$2A^* \longrightarrow B$$
$$A^* + B \longrightarrow AB$$

(*v*) Abstraction

$$A^* + B-C \longrightarrow A-B + C$$

(*vi*) Isosmerization or rearrangement

$$A^* \longrightarrow A'$$

(* denotes excited state)

In ordinary chemical reactions, the energy of activation is supplied by thermal energy. In photochemical reactions, the necessary activiation energy is supplied through the absorption of quanta of visible or ultraviolet radition.

Ordinary chemical reactions that take place in presence or absence of light are called thermal or "dark" reactions, on the other hand, reactions that take place only by exposure to visible or ultraviolet light are called photochemical reactions. Thermal or dark reactions are always accompanied by a decrease in free energy of the reacting system (*i.e.* negative values of ΔG). On the contrary, certain photochemical reactions are accompanied by increase in free energy of the system (*i.e.* positive value of ΔG). Examples of such reactions are the ozonisation of oxygen, photosynthesis, decomposition of ammonia, polymerization of anthracene, etc. However, such reactions, though accompanied by increase in free energy, do not take place spontaneously but occur only in presence of light. The system tends to return to its original state, when the source of light is removed.

26.3. LAWS OF PHOTOCHEMISTRY

1. Grothus-Draper Law : *According to this law, only those radiations which are absorbed by the reacting system are effective in producing a chemical change.* It means that mere exposure of a system to radiation cannot cause a chemical reactions unless and until quanta of light are absorbed by the reacting system. Further, although a photochemical reaction can result only from the absorption of light, all the light that is absorbed is not necessarily effective chemically; some of it may be converted into heat and some may be re-emitted as light of the same or different frequency. This emission of radiation is known by the general name of *"fluorescence"*.

(**Note :** The name is derived from the observation that the mineral "fluorite" emits visible light when exposed to UV radiation).

2. Stark-Einstein Law : This second law of photochemistry enunciated by Stark and Einstein is known as the law of *photochemical equivalence*. According to this law, *each molecule taking part in a chemial reaction induced by exposure of light absorbs one quantum of radiation, causing the reaction*. If the absorbing molecule decomposes or reacts immediately without further successive or side reactions, then for every quantum of light absorbed, one molecule in the reacting system should be involed in the reaction.

If v is the frequency of the absorbed radiation, then the corresponding quantum is hv, which is the amount of energy absorbed by each reacting molecule as per the law of photochemical equivalence. The energy, E, of an Avogadro number (N) of photons is referred to as one Einstein. If N is the Avogadro number, then the energy, E absorbed by 1 mole of the reacting molecules is as follows :

$$E = N \cdot hv = Nhc/\lambda$$

PHOTOCHEMISTRY

The value of E is inversely proportional to wavelength, λ. The value of E can be expressed in any energy units, viz. erg, Kcal or eV.

Substituting for,
$N = 6.02 \times 10^{23}$,
$h = 6.625 \times 10^{-27}$ erg, and
$c = 3.0 \times 10^{10}$ cm S^{-1}, (velocity of light)

$$E = N \cdot h \cdot c/\lambda \text{ erg mol}^{-1}$$

$$= \frac{6.02 \times 10^{23} \times 6.625 \times 10^{-27} \times 3.0 \times 10^{10}}{\lambda} \text{ erg mol}^{-1}$$

$$= \frac{1.196 \times 10^{8}}{\lambda} \text{ erg mol}^{-1}$$

$$= \frac{1.196 \times 10^{8}}{4.184 \times 10^{10}} \times \frac{1}{\lambda} \text{ Kcal mol}^{-1}$$

$$= \frac{2.859 \times 10^{-3}}{\lambda} \text{ Kcals mol}^{-1}$$

If λ is expressed in A°,

$$E = \frac{2.859 \times 10^{-3}}{\lambda} \text{ Kcals mol}^{-1}$$

For example, if $\lambda = 3500$ A°,
the corresponding value for 1 einstein, E

$$= \frac{1.196 \times 10^{16}}{3500} \text{ ergs.}$$

It is clear from the above equation that the shorter the wavelength (λ), the greater will be the energy absorbed per mole. In other words, the higher the wavelength, the smaller will be the energy per Einstein.

	Wave length, A°	Energy absorbed per mole, K cals
Violet light	4000	71
Red light	7500	38

obviously, radiations in the UV and violet portions of the spectrum are chemically more active than those of longer wavelength.

26.4. QUANTUM EFFICIENCY

Processes which give products directly as a result of absorption of radiation are referred to as primary processes :

$$A \xrightarrow{h\nu} A^* \longrightarrow \text{products}$$

According to the law of photochemical equivalence, one mole of an absorbing substance should decompose for every $2.859 \times 10^{5}/\lambda$ Kcal of radiation absorbed. The photochemical equivalent of 1 Kcal is thus $= \lambda/2.859 \times 10^{5}$ moles.

This relationship between the energy (number of quanta) absorbed and the number of reacting molecules enables us to test the law of photochemical equivalence experimentally. The results are expressed in terms of the *Quantum efficiency* or *the quantum yield* (ϕ), which is defined as the number of molecules decomposed (or reacted) by each quantum of radiation absorbed. In other words, quantum efficiency may be defined as the number of moles reacting per an Einstein of the light absorbed.

$$\phi = \frac{\text{Number of molecules reacting in a given time}}{\text{Number of quanta of light absorbed in the same time}}$$

$$= \frac{\text{Number of molecules reacting chemically in a given time}}{\text{Number of Einsteins of light absorbed in the same time}}$$

If Stark-Einstein's law is strictly obeyed, ϕ should be equal to unity. However, Stark and Bodenstein (1913) pointed out that the Stark-Einstein law applies only to the primary photochemical reaction, i.e. the one in which the radiation was absorbed. The resulting activated or dissociated molecule might be involved in secondary thermal reactions. Because of these secondary reactions, the quantum efficiency of the primary light-absorbing state might be obscured. If the secondary reactions are connected with the primary reaction in a simple stoichiometric manner, then the over-all quantum yield will be a simple integer, namely 1, 2 or 3.

For instance, the photochemical reaction involving decomposition of HBr proceeds as follows :

$$HBr + h\nu \longrightarrow H + Br \quad \text{(primary reaction)}$$
$$HBr + H \longrightarrow H_2 + Br \quad \text{(Secondary reactions)}$$
$$Br + Br \longrightarrow Br_2$$

Overall reaction

$$2 HBr + h\nu \longrightarrow H_2 + Br_2$$

The law of photochemical equivalence is obeyed only by the primary reaction.

Similarly, when anthracene dissolved in benzene is exposed to ultraviolet light, it is converted to dianthracene :

$$\underset{\text{Anthracene}}{2\,C_{14}H_{10}} + h\nu \longrightarrow \underset{\text{Dianthracene}}{C_{28}H_{20}}$$

The quantum yield for this reaction is obviously 2 but in practice it is actually found to be 0.5. The reason is that the reaction is accompanied by fluorescence which deactivates the excited anthracene molecules. Moreover, the above reaction is reversible too.

The following are some of the major reasons for low quantum yield :

(i) Deactivation of reacting molecules

$$R + h\nu \longrightarrow R^* \quad \text{(activation)}$$

$$R^* \longrightarrow R + h\nu' \quad \text{(fluorescence)}$$

(ii) Reversal of primary reaction and establishment of equilibrium.

$$2R \underset{\text{Thermal reaction}}{\overset{h\nu}{\rightleftharpoons}} (R)_2$$

(iii) Recombination of dissociated fragments

$$AB + h\nu \longrightarrow A + B \quad \text{(primary reaction)}$$
$$A + B \longrightarrow (AB) \quad \text{(secondary reaction)}$$

Example 1 : *Calculate the value of an Einstein of energy for a radiation of frequency 3×10^{13} per second.*

Solution.

$$E = 2.859 \times 10^5 / l$$

PHOTOCHEMISTRY

$$\lambda = \frac{c}{v} = \frac{3 \times 10^{10} \text{ cm S}^{-1}}{3 \times 10^{13} \text{ cm S}^{-1}} = 10^{-3} \text{ cm} = 10^5 \text{ A}°$$

[Here, λ = wave length in Angstrom units, c = velocity of light, v = **frequency** and E = energy absorbed (in einsteins) by one mole of the reacting substance]

$$\therefore \quad E = \frac{2.859 \times 10^5}{10^5} = 2.859 \text{ Kcal/mole}$$

$$= \frac{2.859}{23.06} \text{ eV/mole}$$

(1 eV = 23.06 Kcal/mole).

(**Note** : If we want to obtain the equation in SI units, we have to substitute $N = 6.022 \times 10^{23}$ mol^{-1}, $h = 6.626 \times 10^{-34}$ JS, $c = 3.0 \times 10^8$ m S^{-1}. Then,

$$E = \frac{0.1197}{\lambda} \text{ J mol}^{-1} = \frac{0.1197 \times 10^5}{\lambda} \text{ kJ mol}^{-1}. \text{ In this equation, l is expressed in meters)}$$

Example-2 : *Calculate the value of an Einstein of energy for radiation of wavelength 7500A°.*

Solution.
$$E = \frac{2.859 \times 10^5}{\lambda} \text{ Kcal/mole}$$

(Here λ is the wavelength expressed in Angstrom units)

$$\therefore \quad E = \frac{2.859 \times 10^5}{7500} \text{ Kcal/mole}$$

$$= 38 \text{ K cal/mole}$$

Example 3 : *Calculate the wavelength of light absorbed if the value of an Einstein is 71.5 Kcal/mole.*

Solution.
$$E = \frac{2.859 \times 10^5}{\lambda}$$

$$\therefore \quad \lambda = \frac{2.859 \times 10^5}{E} \text{ Angstroms}$$

$$= \frac{2.859 \times 10^5}{71.5} = 4000 \text{ A}°.$$

Example 4 : *(a) Calculate the energy associated with a proton of light of wavelength 8000 A°. Given that $h = 6.62 \times 10^{-27}$ erg. sec. and $c = 3 \times 10^{10}$ cm sec^{-1}.*

(b) What is the energy per einstein of the above radiation ? (Avagadro number, $N = 6.02 \times 10^{23}$)

(c) Will this be able to dissociate a bond in a diatomic molecule which absorbs this photon of light. The bond energy is 100 Kcal per molecule.

Solution.

(a) Energy associated with a photon of light

$$= \frac{hc}{\lambda} = \frac{6.62 \times 10^{-27} \times 3 \times 10^{10}}{8000 \times 10^{-8}}$$

$$= \frac{6.62 \times 3}{8} \times 10^{-12} \text{ erg}$$

$$= 2.4825 \times 10^{-12} \text{ erg}$$

$$= \frac{2.4825 \times 10^{-12}}{4.184 \times 10^7} \text{ cals.}$$

(b) Energy per Einstein of the above radiations

$$= \frac{Nhc}{\lambda} = \frac{6.02 \times 10^{23} \times 6.62 \times 10^{-27} \times 3 \times 10^{10}}{8000 \times 10^{-8}}$$

$$= \frac{6.02 \times 6.62 \times 3}{8} \times 10^{11} \text{ erg} = 1.4945 \times 10^{12} \text{ erg}$$

$$= \frac{1.4945 \times 10^{12}}{4.184 \times 10^{7}} \text{ cals.} = 35.72 \times 10^{3} \text{ cals}$$

(c) Energy required for the dissociation of one bond of the diatomic molecule

$$= \frac{100,000}{6.02 \times 10^{23}} \text{ cal} = 1.666 \times 10^{-19} \text{ cals}$$

The energy of a photon of the given radiation as calculated from (b) above = 35.72×10³ cals.

∴ Since the energy of the photon in this case is greater than the energy needed to break the bond, the bond of the said diatomic molecules will be dissociated.

Example 5 : *Calculate the quantum efficiency for the photochemical reaction*

$$X \longrightarrow Y$$

where 2×10^{-5} moles of Y are formed on absorption of 12×10^{7} ergs using light of wavelength 3600 A°. Given that

$$h = 6.0 \times 10^{-27} \text{ erg sec}$$
$$c = 3 \times 10^{10} \text{ cm/sec}$$
$$N = 6.02 \times 10^{23}$$

Solution. Wavelength of the light = 3600 A° = 3600×10^{-8} cm

Energy associated with one quantum

$$= h\nu = h\frac{c}{\lambda} = \frac{(12.0 \times 10^{-27})(3 \times 10^{10})}{3600 \times 10^{-8}} \text{ ergs}$$

$$= 10.0 \times 10^{-12} \text{ ergs}$$

∴ Number of quanta of radiation absorbed

$$= \frac{\text{Total energy absorbed}}{\text{Energy associated with each quantum}}$$

$$= \frac{12 \times 10^{7}}{10 \times 10^{-12}} = 1.2 \times 10^{19}$$

Number of moles reacted are given as

$$= 2 \times 10^{-5}$$

Avogadro number $= 6.02 \times 10^{23}$

∴ Number of molecules reacted

$$= 2 \times 10^{-5} \times 6.02 \times 10^{23} = 12.04 \times 10^{18}$$

∴ Quantum efficiency, $\phi = \frac{\text{No. of molecules reacted}}{\text{No. of quanta absorbed}}$

$$= \frac{12.04 \times 10^{18}}{1.2 \times 10^{19}} \approx 1$$

Example 6 : *In a photochemical reaction $X \to Y$, 0.004 mole of X reacted in 20 minutes. During this time 3.0×10^{6} photons of light per second were absorbed by the system. What is the quantum efficiency of the photochemical reaction ?*

Solution.

No. of molecules of X reacted

$$= 0.004 \times N \quad \text{(Where } N \text{ is Avogadro number)}$$
$$= 0.004 \times 6.02 \times 10^{23}$$

No. of photons absorbed per second

$$= 3.0 \times 10^6$$

∴ Number of photons absorbed in (20 × 60) seconds

$$= 3.0 \times 10^6 \times 1200$$

∴ Quantum efficiency $= \dfrac{\text{No. of moles reacted}}{\text{No. of Photons absorbed}}$

$$= \dfrac{0.004 \times 6.02 \times 10^{23}}{3.0 \times 10^6 \times 1200} \approx 6.7 \times 10^{11}$$

26.5. HIGH AND LOW QUANTUM YIELDS

A photochemical reaction usually takes place in two steps viz. (a) primary process and (b) secondary process.

(a) The primary process involves one of the following :

(i) activation of certain molecules by absorption of light

$$XY + h\nu \longrightarrow XY^{\bullet}$$
$$\text{activated molecule}$$

(ii) dissociation of certain molecules to produce active atoms or free radicals.

$$XY + h\nu \longrightarrow X + Y^{\bullet}$$
$$\text{active atom or}$$
$$\text{excited species}$$

(b) The secondary process involves the reaction of the activated molecules or the active products produced by the primary process with other molecules or the deactivation of the activated molecules or some other reactions.

Experimental studies have proved that the quantum yield φ is unity only in case of very few photochemical reactions. For a large number of reactions, however, φ is < 1 or very much > 1. These results show that the law of photo-chemical equivalence applies only to primary processes. A low value of f indicates that all the activated molecules do not produce photoproducts because some of them may get deactivated by other means. On the other hand, a high value of φ may be inferred due to secondary processes in which the photoproducts may be involved in thermal or "dark" reactions, thereby increasing the number of moles of the reactant consumed. For instance, a very high value of φ is inferred in case of photo-initiated chain reactions, such as formation of HCl from H_2 and Cl_2.

Further, if some of the reactions in secondary process are exothermic or endothermic, they may also affect the overall reaction resulting in high or low quantum yields.

26.6. MECHANISM OF SOME PHOTOCHEMICAL REACTIONS

The mechanism of some photochmical reactions are discussed below :

(a) **Photochemical decomposition of hydrogen iodide (HI)**

The photochemical decomposition of HI to give hydrogen and iodine occurs by the absorption of radiation in the ultraviolet region at around 2500 A°. Quantum yield measurements show that two molecules of HI are decomposed for each quantum absorbed. The following mechanism has been proposed for the photolysis of HI :

Primary process
$$HI + h\nu \longrightarrow H^\bullet + I^\bullet$$
Secondary process
$$H^\bullet + HI \longrightarrow H + I^\bullet$$
$$M + I^\bullet + I^\bullet \longrightarrow I_2 + M$$

(when M is some third body, perhaps the wall of the container, that takes up the energy liberated by bond formation). Only these two reactions are energetically reasonable. A number of other steps can be written but their contributions are unimportant.

Adding the above three reaction the overall reaction may be given as follows :
$$2HI + h\nu \longrightarrow H_2 + I_2$$

Thus, for every quantum of light absorbed, two molecules of HI are decomposed. This mechanism therefore agrees with the quantum yield value of 2.

(b) Photochemical decomposition of hydrogen bromide (HBr)

The photolytic decomposition of HBr occurs as follows :

Primary reaction
$$HBr + h\nu \longrightarrow H^\bullet + Br^\bullet$$
Secondary reaction
$$HBr + H^\bullet \longrightarrow H_2 + Br^\bullet$$
$$Br^\bullet + Br^\bullet \longrightarrow Br_2$$
Overall reaction
$$2\,HBr + h\nu \longrightarrow H_2 + Br_2$$

The quantum yield of the reaction is 2. (obviously, the law of photochmical equivalence is followed by the primary reaction only).

(c) Photochemical synthesis of HCl from hydrogen and chlorine

The photochemical reaction between H_2 and Cl_2 to form HCl is an excellent example of a photochemical chain reaction.

Primary reaction

This involves molecular dissociation resulting from the absorption of radiation by chlorine :
$$Cl_2 + h\nu \longrightarrow 2\,Cl^\bullet$$
Secondary reactions
$$Cl^\bullet + H_2 \longrightarrow HCl + H^\bullet$$
$$H_2^\bullet + Cl \longrightarrow HCl + Cl^\bullet$$

Thus, a chain reaction starts resulting in a very high quantum yield.

About a million chain steps may occur before the free radicals are destroyed by any of the following chain-termination steps occur :
$$M + 2Cl^\bullet \longrightarrow Cl_2 + M$$
or $\qquad M + 2H^\bullet \longrightarrow H_2 + M$
or $\qquad M + H^\bullet + Cl^\bullet \longrightarrow HCl + M$

where M is a third body such as the container wall. The quantum yield in such chain reactions depends on the size of the reaction vessel and this can be attributed to the chain-terminating surface reactions.

(d) Photolysis of organic compounds

Photochemical reactions are being increasingly used for the synthesis of organic compounds,

PHOTOCHEMISTRY

particularly those new compounds that cannot be easily synthesized by the conventional methods. Photolysis of some organic compounds in different types of reactions are given below :

(i) Photochemical elimination reactions

In case of ketones, these reactions usually involve the loss of carbon monoxide. The products formed during the photolysis of acetone in the vapour phase depend on temperature. At elevated temperature, the products of photolysis are ethane and carbon monoxide.

$$CH_3\overset{O}{\underset{\|}{C}}CH_3 \xrightarrow[(v.p)]{h\nu} [CH_3\overset{O}{\underset{\|}{C}}CH_3]^\bullet \longrightarrow CH_3\overset{O}{\underset{\|}{C}}{}^\bullet + CH_3{}^\bullet$$

$$CH_3\overset{O}{\underset{\|}{C}}{}^\bullet \longrightarrow CH_3{}^\bullet + CO$$

$$2CH_3{}^\bullet \longrightarrow C_2H_6$$

It may be noted that out of all the bonds present in acetone, C—C bond is the weakest.

However, at room temperature, the products of photolysis of acetone are biacetylacetaldehyde, methane and possibly hexane 2,5-dione.

$$CH_3\overset{O}{\underset{\|}{C}}CH_3 \xrightarrow{h\nu} [CH_3\overset{O}{\underset{\|}{C}}CH_3]^\bullet \longrightarrow CH_3\overset{O}{\underset{\|}{C}}{}^\bullet + CH_3{}^\bullet$$

$$CH_3\overset{O}{\underset{\|}{C}}{}^\bullet + CH_3\overset{O}{\underset{\|}{C}}CH_3 \longrightarrow CH_3\overset{O}{\underset{\|}{C}}H + {}^\bullet CH_2\overset{O}{\underset{\|}{C}}CH_3$$

$$CH_3{}^\bullet + CH_3\overset{O}{\underset{\|}{C}}CH_3 \longrightarrow CH_4 + CH_3\overset{O}{\underset{\|}{C}}CH_2{}^\bullet$$

$$2CH_3\overset{O}{\underset{\|}{C}}{}^\bullet \longrightarrow CH_3-\overset{O}{\underset{\|}{C}}-\overset{O}{\underset{\|}{C}}-CH_3$$

$$2CH_3\overset{O}{\underset{\|}{C}}CH_2{}^\bullet \longrightarrow CH_3\overset{O}{\underset{\|}{C}}CH_2CH_2\overset{O}{\underset{\|}{C}}CH_3$$

Acetone is a symmetrical ketone. In case of unsymmetrical ketones, preferential fission takes place to give the more stable alkyl radical. For example, photo-dissociation of ethyl methyl ketone gives ethyl and methyl radicals. Ethyl radical is formed predominantly.

$$CH_3\overset{O}{\underset{\|}{C}}{}^\bullet + C_2H_5{}^\bullet \xleftarrow{h\nu} CH_3\overset{O}{\underset{\|}{C}}C_2H_5 \xrightarrow{h\nu} CH_3{}^\bullet + {}^\bullet \overset{O}{\underset{\|}{C}}C_2H_5$$

Photochemical reactions of ketones (and aldehydes) which result in fission of the C—CO bond followed by elimination of carbon monoxide are classified as the "*Norrish type I process.*"

With increasing length and complexity of the alkyl groups, the products formed by photolysis become more and more complicated. For example, when a hydrogen atom in the alkyl group can enter into the formation of a six-membered cyclic transition state involving the oxygen atom as one member of the ring, abstraction of X-hydrogen and fission of the molecule take place to give an alkene and a ketone (via the enol form), e.g., hexan-2-one gives mainly propene and acetone. In this case, the elimination of carbon monoxide occurs to a lesser extent.

$$CH_3CH(CH_2)(CH_2)CCH_2(O) \xrightarrow{hv \text{ (v.p)}} \left[CH_3CH\cdots CH_2 \quad O\cdots CCH_3 \atop \cdots CH_2 \cdots H \right] \longrightarrow CH_3CH=CH_2 + \left[\underset{CH_2 \quad CH_3}{HO-C} \right] \downarrow CH_3COCH_3$$

This type of photolysis undergone by ketones (and aldehydes) is classified as the *"Norrish type II process"*.

α–β unsaturated ketones in presence of adequate light energy undergo photolysis with elimination of carbon monoxide.

Cyclic ketones eg. cyclopentanone undergo photolysis with elimination of carbon monoxide and also ring opening with the formation of an unsaturated aldehyde.

The photolysis of aldehydes is more complicated than that of ketones. The primary reaction usually results in the formation of the formyl radical :

$$RCHO \xrightarrow[\text{(vapour phase)}]{hv} R^\bullet + {}^\bullet CHO$$

This is usually followed by the secondary process :

$$R^\bullet + {}^\bullet CHO \longrightarrow RH + CO$$

All the above decarboxylations are vapour phase reactions, and the products are identical if the photochemical reactions are carried out in chemically inert solvents. However, if these reactions are carried out in protic solvents, the products are generally different.

Compounds containing nitrogen-nitrogen double bonds, such as azo-compounds in the gaseous phase undergo photolysis with elimination of nitrogen :

$$CH_3-N-CH_3 \xrightarrow[\text{(gaseous phase)}]{hv} N_2 + 2CH_3^\bullet \longrightarrow C_2H_6$$

$$\underset{\text{Diazomethane}}{CH_3N_2} \xrightarrow[\text{(gaseous phase)}]{hv} :CH_2 + N_2 \atop \text{Methylene}$$

The photolysis of nitrites that do not contain Y-hydrogen atoms generally takes place with elimination of nitric oxide and formation of hydroxy and carbonyl compounds.

$$\overset{\beta}{CH_3}\overset{\alpha}{CH_2}CH_2ONO \xrightarrow{hv} NO + CH_3CH_2CH_2O^\bullet$$

$$CH_3CH_2CH{-}O^\bullet \longrightarrow CH_3CH_2CH=O + HOCH_2CH_2CH_3$$
$$H\,{}^\bullet OCHC_2H_2CH_3$$

PHOTOCHEMISTRY

When Y-hydrogen atoms are present (for the formation of a 6-membered transition state, then an oxime or the corresponding nitroso dimer are formed. This reaction, known as Barton reaction, is synthetically very important.

(ii) Photochemical oxidations

In presence of oxygen, several photochemical reactions lead to the formation of peroxides and hydroperoxides or oxidation of functional groups. Although oxygen is a triplet (diradical) in its ground state, it doesnot add to cyclic dienes. However, in presence of a sensitizer (eg., eosin or fluorescein), the oxygen is raised to its S_1 state (all electrons paired), and in this state, the oxygen is in its reactive state. But, many linear polynuclear aromatic compounds add oxygen even in the absence of a sensitiser, probably because in these cases, the substrate molecule itself acts as sensitiser.

(iii) Photochemical reductions

In photochemical reductions, a hydrogen atom is abstracted from the solvent or from another reactant to produce two free radicals. For example, radialysis of benzophenone in an alcohol (eg. isopropanol) gives benzopinacol and oxidation products of alcohol.

Benzophenone, in presence of benzhydrol also gives benzopinacol on photochemical irradiation.

(iv) Photochemical isomerisations

Cis-trans-Isomerisation can be carried by simple photochemical irradiation, usually in solution, or in the presence of a sensitiser or a catalyst. Usually, an equilibrium mixture of the two isomers is reached, and it remains constant irrespective of the duration of irradiation. The cis-trans isomerisation of stilbene by simple irradiation or in presence of a sensitiser such as benzophenone.

Photochemical cis-trans isomerisation can also be effected in presence of bromine or iodine. Apart from photochemical cis-trans isomerisation of alkenes, isomerisation can be carried out under similar conditions with oximes, azo-compounds, and other systems too.

(v) Photochemical rearrangements

Photochemical irradiation of a-b unsaturated ketones (eg., cyclic and acyclic enones and dienones) results in various types of rearrangements. Acyclic a-b unsaturated ketones containing Y-hydrogen atom generally undergo double-bond migration provided that the light energy is not high enough to cause dissociation. The presence of Y-hydrogen is essential for the formation of a 6-membered cyclic transition state resulting in hydrogen abstraction and transfer to the oxygen atom, as in the case of 5 methylhex-3-en-2-one.

Many rearrangements that occur thermally or in the presence of catalysts, can also be effected photochemically, e.g., Orton, Fries, benzidine rearrangements.

26.7. Photosynthesis

This is an excellent example of biosynthesis. It represents the processes by which plants containing the pigment chlorophyll absorb light energy and utilize it to convert atmospheric carbohydrates, in the presence of water, to carbohydrates. The first involves a photochemical process in which light energy, absorbed by chlorophyll, is utilized to form activated compounds. These are called "*light reactions.*" The second involves the reduction of carbondioxide by the active molecules produced in the first process. The products are oxygen, carbohydrates and some other compounds. Since this process can proceed in the presence or absence of light, these reactions are called the "*dark reactions.*" Both the light and dark reactions occur only in presence of various enzymes. The overall equation for photosynthesis may be written as :

$$6 CO_2 + 6 H_2O \longrightarrow 6(CH_2O) + 6 O_2$$

The pathway of carbon dioxide in photosynthesis has been investigated using $^{14}CO_2$ as tracer.

26.8. Types of photochemical reactions

Photochemical reactions are usually divided into two main types :

(*i*) **Direct light-induced or non-sensitised reactions :** In these reactions, the molecule that actually absorbs the light undergoes various secondary processes to give the final products of the reaction.

(*ii*) **Photosensitised reactions :** In these reactions, one type of molecule, the donor (*D*) that is present absorbs the light (which is of the wavelength characteristic of the donor), becomes excited, and then transfers its excitation energy to another type of the molecule, the acceptor (*A*), that is also present in the same system. Thus, the donor molecule returns to the ground state and the acceptor molecule is raised to an excited state, which then undergoes emission or photochemical reaction. Such a transfer of energy from *D* to *A*, followed by emission of light or reaction of A is called "*sensitisation*". However, for the donor-acceptor relationship to function, the following conditions must be fulfilled.

(*a*) The donor-excited state must have a sufficiently long life-time to be able to transfer its energy to the acceptor molecule.

(*b*) The excitation energy of the donor must be higher than the excitation energy of the acceptor.

Photosensitised reactions do not normally proceed even though sufficient energy is available photochemically, but do so in presence of *sensitizers*. They absorb sufficient energy from the radiation and pass it on to the reacting molecules (acceptors). Benzophenone is a commonly used sensitiser. Photochemical reactions are being increasingly used for the synthesis of organic compounds. Some new organic compounds which cannot be obtained by the usual methods have been prepared with the help of photochemical reactions. Photosentized dissociation of hydrogen in presence of mercury is a classical example :

$$Hg + h\nu \longrightarrow Hg^*$$
$$Hg^* + H_2 \longrightarrow H + H + Hg$$

Mercury vapour also sensitizes the decomposition of water, ammonia, acetone and ethanol.

Sensitized photodecomposition of oxalic acid in presence of uranyl ion is another classical example of photosensitization in solution. When exposed to radiation of wavelength 2540 to 4350 A°, the uranyl ion absorbs the radiant energy and transfers it to oxalic acid, which then undergoes decomposition :

$$H_2C_2O_4 \xrightarrow[+ h\nu]{UO_2 \text{ ion (sensitizer)}} H_2O + CO + CO_2$$

In nature, chlorophyll acts as a sensitizer for photosynthesis in bringing about the combination of CO_2 and water to produce carbohydrates :

$$CO_2 + H_2O \longrightarrow 1/6\ C_6H_{12}O_6 + O_2 \ ; \quad \Delta H = 112\ Kcal$$

The reaction, being endothermic, requires an activation energy of atleast 112 Kcal, which will be provided by radiation of wavelength 2300 A° or less. The sunlight reaching the earth doesnot contain radiation of such short wave length. The constituents of chlorophyll absorb in the blue (4700 A°) and mainly in the red (700 A°) regions. The chlorophyll molecule then transfers the energy absorbed to CO_2 and H_2O so as to facilitate the above reaction. The energy corresponding to 7000 A° is only about 41 Kcal. Thus it indicates that nearly three photons are required to effect the direct conversion of CO_2 and H_2O into the carbohydrate. It further indicates that the reaction takes place stepwise.

There are two commonly used terms in photochemistry. They are :

(i) **Photolysis :** This term is used to describe the breaking of chemical bonds by light energy. The bonds are broken homolytically to form free radicals, and some times small molecules are eliminated while in relatively few cases, photolysis produces ions.

(ii) **Flash photolysis :** In this process, a flash of light of very high intensity, lasting for a few microseconds (10^{-6} S), is used to irradiate the reaction mixture. In this way, reactive intermediates are formed in high concentration. Their rate of decay can be measured spectroscopically and by various other methods. By using a laser pulse, th flash duration has been reduced to 15 nanoseconds (10^{-9} S) is some reactions.

A flash of light, of duration of about 10^{-5} second, can be generated with as much as 500 cal of light energy. In the visible region, this means that about 10^{-3} einstein of radiation is emitted, and therefore, if this flash is allowed to fall on a small sample of millimole size, most of the sample molecules can be brought to excited state. A typical flash photolysis assembly is shown in Fig. 26.1 to study the absorption spectrum of the species present at the time of spectroscopic flash, which can be synchronised to follow the principal flash by a fraction of a millisecond.

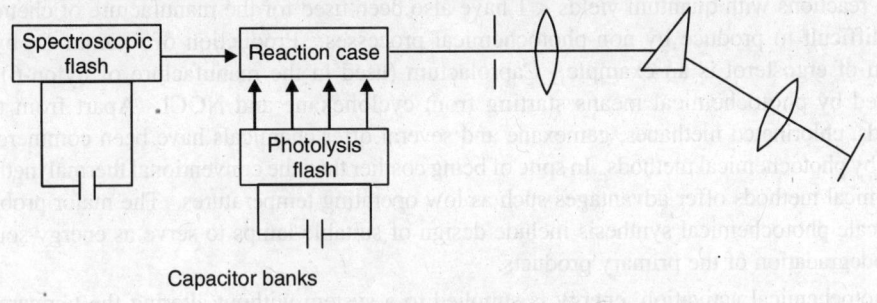

Fig. 26.1. A Flash-photolysis assembly

26.9. Apparatus for photochemical studies

Quantitative photochemistry requires the determination of the quantum yield, f, which is given as :

$$\phi = \frac{\text{number of moles reacting chemically}}{\text{number of einsteins absrobed}}$$

The energy corresponding to an Avogadro number of quanta is called one Einstein (E). The value of E is inversely dependent on wave-length, λ, one must use monochromatic radiation of suitable wavelength. The essential experimental assembly for photochemical studies is illustrated in Fig. 26.2 given below :

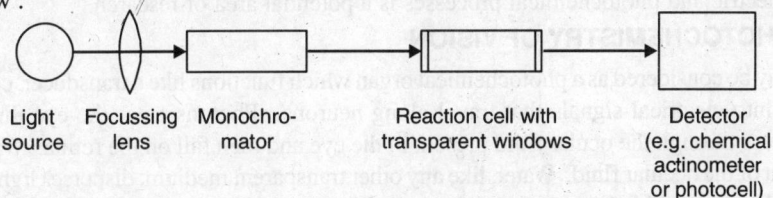

Fig. 26.2. Experimental assembly for photochemical studies.

The energy or intensity of the incident radiation is first measured without the cell, by making use of the detector. In an actinometer, the energy of the radiation is measured by titrimetrically determining the extent of chemical reaction for which the quantum yield is known. One such simple device is uranyl oxalate actinometer. It contains 0.05 M oxalic acid and 0.01 M uranyl sulphate in water. When exposed to radiation of wavelength 2540 to 4350 A°, the uranyl ion absorbs the radiant energy and transfers it to oxalic acid which subsequently decomposes as follows :

$$UO_2^{2+} + h\nu \longrightarrow (UO_2^{2+})^*$$
$$\text{uranyl ion} \qquad \text{excited uranyl ion}$$

$$(UO_2^{2+})^* + \underset{|}{\underset{\text{COOH}}{\text{COOH}}} \longrightarrow UO_2^{2+} + CO_2 + CO + H_2O^*$$
$$\text{oxalic acid}$$

The concentration of oxalic acid that remains is determined by titration with a standard $KMnO_4$ solution. The used up concentration of oxalic acid is a measure of the intensity of radiation. Thus it is possible to calculate the number of einsteins from the equation for ϕ, given because the quantum yield above, for this reaction for a number of wavelengths is known. For example, the value of ϕ for wavelength range of 2540 to 4350 A° is 0.57.

The experiment is then repeated by incorporating the reaction cell in the assembly.

26.10. APPLICATIONS OF PHOTOCHEMISTRY IN TECHNOLOGY

Photochemical reactions with high quantum yields are usually preferred for industrial applications. However, reactions with quantum yields < 1 have also been used for the manufacture of chemicals that are difficult to produce by non-photochemical processes. Production of vitamin D_2 by the irradiation of ergosterol is an example. Caprolactum (used in the manufacture of nylon-6) was synthesized by photochemical means starting from cyclohexane and NOCl. Apart from these compounds, chlorinated methanes, gamexane and several offer chemicals have been commercially produced by photochemical methods. In spite of being costlier than the conventional thermal methods, photochemical methods offer advantages such as low operating temperatures. The major problems of large-scale photochemical synthesis include design of suitable lamps to serve as energy sources and photodegradation of the primary products.

In photochemical activation, energy is supplied to a system without altering the temperature. Therefore, the equilibrium position in a photochemical reaction can be different from that which is expected in normal reactions at the reaction temperature. Further, the rate of a photochemical reaction is not appreciably affected by temperature.

The photodimerization of unsaturated organic compounds in suitable organic solvents is useful to coat materials with thin films. This finds application in the preparation of integrated circuits.

The use of photochemistry in photography is well known.

Intense monochromatic laser beams are used not only in surgery but also in cutting and welding of metals.

The use of photochemistry for the quantum conversion and chemical conversion of solar energy through photoelectric and photochemical processes is a potential area of research.

26.11. THE PHOTOCHEMISTRY OF VISION

The eye may be considered as a photochemical organ which functions like a transducer, converting radiant energy into electrical signals that travel along neurons. Photons enter the eye through the cornea. They pass through the ocular fluid present in the eye and then fall on the retina. Water is the main constituent of the occular fluid. Water, like any other transparent medium, disperses light because its refractive index varies with frequency. We know that the greater the polarizabilty, the greater is the refractive index and hence the greater will be the bending of light rays. Water is more polarizable in response to light rays of high frequency (such as blue light) than those of low frequency (such as red light) because photons of blue light possess more energy than the photons of red light. Therefore, photons of blue light can interact more readily with water molecules as compared with photons of lower frequency, which possess lower energy. Thus, different frequencies of light passing through the eye are brought to slightly different focuses. This is largely responsible for the "chromatic

PHOTOCHEMISTRY

aberration" of the eye with the consequent blurring of the image. The "chromatic aberration" is reduced to some extent by the tinted region called the "macular pigment" which covers part of the retina. The pigments in this region are xanthophylls (Fig. 26.3) which are similar to carotenes present in leaves which protect the delicate chlorophyll molecules from high-energy photons. The xanthophylls remove some of the blue light, thereby helping to sharpen the image. They also protect the photoreceptor molecules from excessive flux of potentially dangerous high energy photons. The xanthophylls have delocalized electrons spreading along the chain of conjugated double bonds, and the $\pi - \pi^*$ transitions lie in the visible region.

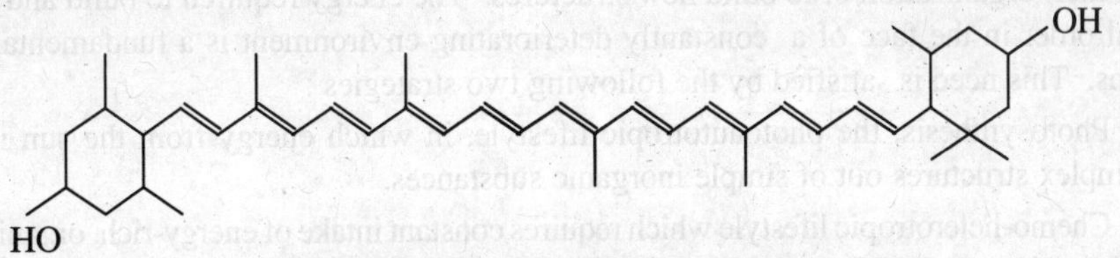

Fig. 26.3. Xanthophyll molecule

Out of the photons that enter the eye, only about 57% reach the retina, while the rest are scattered or absorbed by the ocular fluid. It is here that the primary act of vision takes place. The membrane covering the retina contains rhodopsin molecules. The chromaphore of a rhodopsin molecule absorbs a photon in a π- to -π^* transition. Rhodopsin molecule consists of an opsin protein molecule to which an 11-cis-retinal molecule (Fig. 26.4) is attached. This resembles half of a carotene molecule.

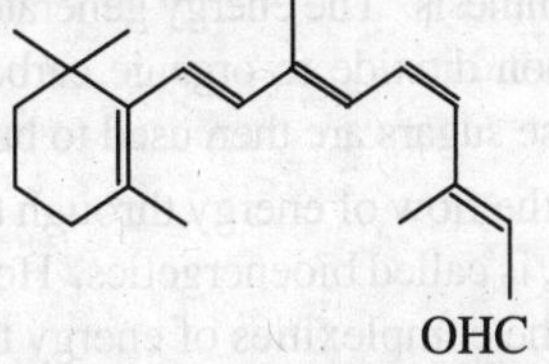

Fig. 26.4. 11-cis-retinal molecule

Although the free 11-cis-retinal molecule absorbs in the ultraviolet region, its attachment to the opsin protein molecule shifts the absorption to the visible region of the electromagnetic spectrum.

As soon as a photon is absorbed, the 11-cis-retinal molecule undergoes photoisomerization into all-trans-retinal (Fig. 26.5) in a fraction of a second.

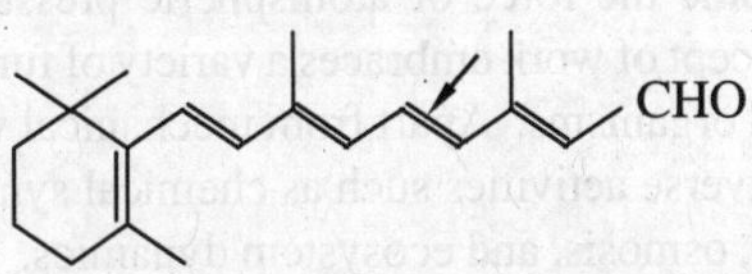

Fig. 25.5 : All-trans-retinal

The process occurs due to p- to -π^* excitation of an electron that loosens one of the p-bonds and the loss of its torsional rigidity. This ultimately leads to the activation of the rhodopsin molecule. This is followed by a sequence of biochemical events (known as the *biochemical cascade*"), which converts the energy accumulated in the activated form of rhodopsin molecule into a pulse of electric potential that travels through the optical nerve into the optical cortex. Here, it is interpreted as a signal and incorporated into the web of events, which is called *"vision"*. At the same time, a series of non-radiative chemical events powered by ATP (adenosinetriphosphate) restores the resting state of the rhodopsin molecule. The process involves the escape of all-trans-retinal as all-trans-retinol (in which –CHO is reduced to –CH$_2$OH) from the opsin molecule by a process catalysed by the enzyme rhodopsin kinase and the attachment of another protein molecule, *arrestin*. Now, the free all-trans-retinol undergoes enzyme-catalysed isomerization into 11-cis-retinol. This is then converted to 11-cis-retinal by dehydrogenation, which is then delivered back into the opsin molecule. This becomes the starting point for a fresh cycle of excitation, photoisomerization and regeneration.

26.12. PHOTOSYNTHESIS AND BIOENERGETICS

Organization is the very essence of biological world. Biological systems are complex and yet highly organized at every level viz. cell, tissue, organ or even community. However, proteins, nucleic acids and other molecules that constitute the cell are constantly under attack and subject to breakdown by hydrolysis, oxidation, etc. still, biological organisms extract materials from their environment to maintain their organization or to build new structures. The energy required to build and preserve the structural order in the face of a constantly deteriorating environment is a fundamental need of all organisms. This need is satisfied by the following two strategies :

(*i*) Photosynthesis, the photoautotropic lifestyle, in which energy from the sun is trapped to build complex structures out of simple inorganic substances.

(*ii*) Chemo-heterotropic lifestyle which requires constant intake of energy-rich, organic substances from their environment. However, even these substances originate from photosynthesis.

In the end, all life on earth derives its power from the sun through photosynthesis.

Photosynthesis occurs in green plants, the cyanobacteria and certain groups of bacteria. In higher plants, the reactions of photosynthesis occur in the chloroplast, which is simplistically akin to a thermodynamic machine. The chloroplast traps the radiant energy from the sunlight and conserves some of it in a stable chemical form. The reactions that accomplish these energy transductions are identified as the light-dependent reactions of photosynthesis. The energy generated by these reactions is subsequently used in the reduction of inorganic carbon dioxide to organic carbon in the form of sugars. Both the carbon and the energy conserved in those sugars are then used to build the order and structure.

The field of study concerned with the flow of energy through the living organisms, whether at the level of molecules, cells or ecosystems, is called bioenergetics. However, for the past several decades, bioenergetics has focussed to unravel the complexities of energy transductions in photosynthesis and respiration and also to understand how that energy is used to drive energy requiring reactions such as synthesis of ATP (adenosine triphosphate) and accumulation of ions across membranes.

Energy flow is dictated by certain fundamental rules of thermodynamics. In addition to energy transformations, thermodynamics also helps to describe the capacity of a system to do work. Work may be defined in several ways. For a physicist, work is a displacement against a force. For a chemist, work is done to overcome the force of atomspheric pressure when the volume of a gas increases. For a biologist, the concept of work embraces a variety of functions against a wide spectrum of forces encountered in cells and organisms. Apart from mechanical work such as muscular activity, the biologist is concerned with diverse activities such as chemical synthesis, the movement of solute against electrochemical gradients, osmosis, and ecosystem dynamics. All these activities and various other essential activities of living things can be described in thermodynamic terms.

Biological energy transductions are based on the first and second laws of thermodynamics. The first law, commonly known as the *law of conservation of energy,* implies that the energy of the universe is constant. That means that any apparent decrease in one form of energy will be balanced by an increase in some other form of energy. For instance, some of the energy expended in displacing an object appears as work while some appears as heated generated by friction. In a similar way, some of the chemical energy released in the combustion of glucose will also be found as bond energy in the product molecules, namely CO_2 and H_2O.

A biologist is majorly concerned with how much work can be done because all the energy is not available to do work. The brings us to the second law of thermodynamics and the concept of *entropy*. Entropy has been described variedly as a measure of randomness, chaos or disorder. However, since entropy is a thermodynamic concept,it is probably most useful to describe it in terms of thermal energy.

Temperature is defined as the mean molecular kinetic energy of matter. Thus, any molecular system above absolute zero (–273°C) contains a certain amount of thermal energy, which is the energy in the form of the vibration and rotation of its constituent molecules and their movement through space. This quantity of thermal energy is related to temperature because as the quantity of this energy increases or decreases, the temperature also varies accordingly. Since the temperature cannot be held constant when the energy is given up, it is said to be "isothermally unavailable" (In Greek, isos = equal). This isothermally unavailable energy is given by the term, TS, where T is the absolute temperature and S is entropy.

Since the isothermally unavailable energy, and consequently, entropy, are related to the energy of molecular motion, it follows that the more random or less ordered the system, the greater will be their entropy. Thus, at absolute temperature, where all molecular motion ceases, entropy is also zero. There is always a natural tendency for entropy to increase. That means, a system tends to become increasingly disordered or chaotic. This tendency is aptly summarized by R.J. Clausius in the statement: "the entropy of the universe tends towards a maximum." This, in a way, is the statement of the second law of thermodynamics. A biologists' concern with entropy is that it primarily represents that energy which is unavailable to do work. In this context, the second law of thermodynamics can be restated as follows : "The capacity of an isolated system to do work continually decreases". It implies that it is impossible to utilize all the energy of a system to do work.

Thus, it is clear that some energy will be available under isothermal conditions, and consequently, this is available to do work. This energy is called "*Gibbs free energy*"(*G*), named after J.W. Gibbs, the physical chemist who introduced this concept. If H is the total energy (also called "*enthalpy*"), and G is the free energy, and TS is the isothermally unavailable energy, then

$$H = G + TS \qquad \ldots(1)$$

This also suggests a corollary to the second law of thermodynamics. "The free energy of the universe tends towards a minimum".

It is more convenient and relevant to measure absolute energies, but changes (designated by the symbol Δ) in energy (such as heat gain or loss or work) during the course of a reaction can usually be measured with little difficulty. Thus, eqn. (1) can be restated as :

$$\Delta G = \Delta H - T \Delta S \qquad \ldots(2)$$

Changes in free energy can give very useful information about a reaction, such as the feasibility of the reaction actually taking place and the quantity of work that might be done if it takes place. The sign of ΔG indicates the feasibility of a reaction. If ΔG is negative, the reaction is considered to be spontaneous, *i.e.* it will proceed without any input of energy. Such reactions are called "exergonic" or energy yielding, because the free energy of the products is less than that of the reactants, and that is why ΔG is negative on the other hand, if ΔG is positive (that is, the free energy of the products is more than that of the reactants), such reactions are called "*endergonic*" or energy consuming. Therefore, input of energy is required for such reactions to occur. The oxidation of glucose is an example of exergonic reactions (*i.e.* reactions with a negative ΔG). Once the activation barrier is overcome, glucose will spontaneously oxidize to form CO_2 and H_2O. In spite of the fact that large quantities of CO_2 and H_2O in the atmosphere, they do not spontaneously recombine to form glucose. This is because the equilibrium constant favours the formation of CO_2 and H_2O and the ΔG for the formation of glucose is positive.

The magnitude of free energy changes depends on the particular set of conditions for that reaction. Therefore it is convenient to compare the free energy changes of reactions under standard reaction conditions. In biochemistry, the standard free energy change, $\Delta G°$, defines the free energy change of reaction that occurs at physiological pH (*i.e.* pH = 7.0) under conditions where both the reactants and products are at unit concentration (*i.e.* 1 M).

Free energy and chemical equilibria : All chemical reactions attain dynamic equilibrium under appropriate conditions. At this state, there will not be any further net change in the concentrations of reactions and products.

The relation between standard free energy change ($\Delta G°$) and equilibrium can be expressed quantitatively as follows :

$$\Delta G° = -RT \ln K = -2.3\, RT \log K \qquad ...(3)$$

where $K = \dfrac{[\text{Products}]_{eq}}{[\text{Reactants}]_{eq}}$. Here K is called the equilibrium mass action ratio.

The actual free energy change (ΔG) of a reaction not at equilibrium is given by

$$\Delta G = \Delta G° + 2.3\, RT \log \tau \qquad ...(4)$$

where τ is the observed (non-equilibrium) mass-action ratio.

Substituting eqn. (3) in eqn. (4), we get

$$\Delta G = -2.3\, RT \log K + 2.3\, RT \log \tau$$
$$\therefore \quad \Delta G = -2.3\, RT \log (K/\tau) \qquad ...(5)$$

Thus, it is clear that the value of ΔG is a function of the degree to which a reaction is displaced from equilibrium. When $\tau = K$, the reaction is at equilibrium. When $\tau < K$, the value of ΔG is negative. When $\tau > K$, the value of ΔG is positive. All these inferences have been corroborated by experimental observations too. Finally, a reaction with a negative ΔG can drive a reaction with positive ΔG, if they are biochemically coupled.

The above discussion can be summarized as below :

1. The free energy change of a reaction at equilibrium is zero.
2. The farther a reaction is poised away from equilibrium, the more free energy is available as the reaction proceeds towards equilibrium.
3. A system can do work as it moves towards equilibrium. This is because in that case, ΔG is negative and free energy is available to do work.
4. A reaction with $-\Delta G$ can drive a reaction with $+\Delta G$, if they are biochemically coupled.

Energy conservation in photosynthesis

In photosynthesis, the energy of sunlight is used to photochemically reduce CO_2, thus conserving energy in the sugars that are produced. Sugars are a form of portable energy that can be mobilized for use elsewhere in the plant.

One of the early clues regarding the redox nature of photosynthesis was provided in 1920 from the studies of C.B. Van Niel on the photosynthetic sulphur bacteria that use H_2S as reductant in place of H_2O. He found that unlike algae and higher plants, the photosynthetic sulphur bacteria do not evolve oxygen, but they deposit elemental sulphur as per the following equation :

$$CO_2 + 2H_2S \longleftrightarrow (CH_2O) + 2S + H_2O \qquad ...(6)$$

This reaction can be split into the following two partial reactions :

$$2H_2S \longleftrightarrow 4e^- + 4H^+ + 2S \qquad ...(7)$$
$$CO_2 + 4e^- + 4H^+ \longleftrightarrow (CH_2O) + H_2O \qquad ...(8)$$

These two equations describe photosynthesis in purple sulphur bacteria as straightforward oxidation-reduction reaction. Adopting a comparative biochemistry approach, Van Niel argued that the mechanism for *oxygenic* (i.e. oxygen-evolving) photosynthesis in green plants and *non-oxygenic* photosynthesis in the sulphur bacteria, both followed the general plan.

$$2H_2A + CO_2 \longleftrightarrow 2A + (CH_2O) + H_2O \qquad ...(9)$$

where A represents O or S, depending on the type of the photosynthetic organism. The O_2 released

PHOTOCHEMISTRY

in oxygenic photosynthesis would be derived from reductant water. Correct stoichiometry would therefore require the participation of four electrons and hence two molecules of water.

Photosynthesis is fundamentally an oxidation-reduction reaction, allthough it may not be obvious at the first glance. This can be seen by examining the summary equation of photosynthesis reaction:

$$6 CO_2 + 12 H_2O \longleftrightarrow C_6H_{12}O_6 + 6 O_2 + 6 H_2O \qquad ...(10)$$

In this photosynthesis is shown as a reaction between CO_2 and water to produce glucose, a six-carbon carbohydrate or hexose. Although glucose is not the first product of photosynthesis, it is a common form of accumulated carbohydrate which provides a convenient basis for discussion. Further, it can be seen from eqn. 6 that eqvi-molar quantities of CO_2 and O_2 are consumed and evolved, respectively. Hence, it is possible to measure photosynthesis in the laboratory in terms of the uptake of CO_2 or the evolution of O_2.

Equation (6) can be simplified as below:

$$CO_2 + 2 H_2O \longleftrightarrow (CH_2O) + O_2 + H_2O \qquad ...(11)$$

where (CH_2O) represents the basic building block of carbohydrate. This equation can be interpreted as a simple redox reaction involving reduction of CO_2 to carbohydrate, where H_2O is the reductant and CO_2 is the oxidant.

It can be seen that in eqn. (7), 2 molecules of water are shown in the reactants side and 1 molecule is shown in the products side. Experimental studies have corroborated that the O_2 released in oxygenic photosynthesis is derived from the reductant, H_2O. Correct stoichiometry would therefore require the participation of four electrons and hence 2 molecules of H_2O are required.

Thus, photosynthesis can be viewed as a photochemical reduction of CO_2. The energy of light is used to generate strong reducing equivalents from H_2O (strong enough to reduce CO_2 to carbohydrate. The reducing equivalents are in the form of reduced $NADP^+$ (or, $NADPH + H^+$). Additional energy for carbon reduction is required in the form of ATP, which is also generated at the expense of light. The main function of the light-dependent reactions of photosynthesis is therefore to generate the NADPH and ATP required for carbon reduction. This is achieved through a series of reactions which constitute the *"Photosynthetic electron transport chain."*

Photosynthetic electron transport chain

The photosynthetic unit of oxygenic photosynthetic organisms is organized as two separate photosystems that operate in series. This two-step series formulation is known as *"Z-Scheme"*, which caused considerable amount of excitement when proposed in the early 1960s. This idea was based on a series of experiments mostly on quantum efficiency and studies on action spectra for photosynthesis in *chlorella*.

Quantum efficiency can be expressed either as *quantum yield* or *quantum requirement*. Quantum yield (ϕ) expresses the efficiency of a process as a ratio of the yield or product to the number of photons absorbed. For example, in photosynthesis, the quantum yield is measured as the amount of CO_2 taken up or O_2 evolved. On the other hand, the quantum requirement ($1/\phi$) tells how many photons are required for every molecule of CO_2 reduced or O_2 evolved.

It has been identified from eqn. 8 that a minimum of four electrons are required for every molecule of CO_2 reduced. Also, it was established that one photon is required for each electron excited. Therefore, the minimum quantum requirement for photosynthesis is 4. However, it has been well established experimentally that the minimum quantum requirement for photosynthesis is 8 to 10 photons for every molecule of CO_2 reduced. If 8 photons are required (it is usual to absume the minimum) for 4 electrons, then each electron must have been excited twice.

Further evidence came from Emerson and Lewis (1943), during their studies to determine the action spectra for photosynthesis in *"chlorella"*, reported that the value of f was remarkably constant over most of the spectrum. This indicates that any photon absorbed by chlorophyll was more or less effective in driving photosynthesis. However, there was an unexpected drop in the quantum yield at wavelengths > 680 nm, although chlorophyll still was found to absorb in that range. This puzzling drop in quantum efficiency in the long red portion of the spectrum is called *"red drop"*.

In another experiment, Emerson and his co-workers set up two beams of light-one in the region of 650 to 680 nm and the other in the region of 700 to 720 nm. The fluence rates of these two beams were adjusted to give equal rates of photosynthesis. Surprisingly, it was found that when the two beams were applied simultaneously, the rate of photosynthesis was 2 to 3 times greater than the sum of the rates obtained with each beam separately. This phenomenon was known as the *"Emerson enhancement effect"*. This suggests that photosynthesis might involve two photochemical events or systems – one that is driven by short wave-length light (# 680 nm) and another that is driven by long wave-length light (> 680 nm). In order to achieve optimal photosynthesis, both the systems must be driven simultaneously or in quick succession.

The key to the photosynthetic electron transport chain is the presence of two large, multimolecular complexes known as photosystem I (PS I) and photosystem II (PS II). These two photosystems operate in series linked by a third multiprotein aggregate called the *cytochome complex*. A linear representation of the photosynthetic electron transport chain is given in Fig. 26.6.

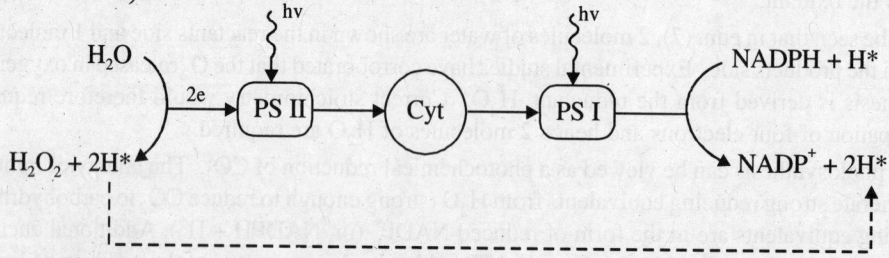

Fig. 26.6. A linear representation of photosynthetic electron transport chain.

The overall effect of the chain is to extract low-energy electrons from H_2O and using light energy trapped by chlorophyll to raise the energy level of those electrons to produce the strong reductant, NADPH.

A new model for the two-step formulation series for photosynthesis was first proposed by R. Hill and Fay Bendall in 1960 to explain the redox potential values in the studies with cytochromes. The model depicts (Fig. 26.7) two photochemical acts, both operating in series – one serving to oxidize the cytochromes and the other to reduce them.

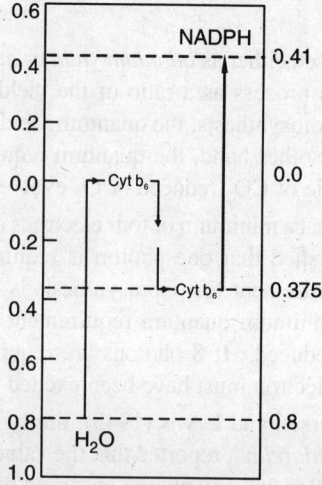

Fig. 26.7. The Z-scheme originally proposed by Hill and Bandall

This model was confirmed in the following year by L. Duysens et al. (1961) who showed that cytochromes were oxidized in the presence of long wavelength light. The effect could be reversed by short wavelength light.

Although the Hill and Bendell's Z-scheme has now been considerably modified and several details have been added, the scheme certainly helped to a large extent to evolve our present understanding of photosynthetic electron transport and oxygen evolution.

Experimental studies revealed that PS I and PS II each contain several different proteins along with a collection of chlorophyll and carotenoid molecules that absorb photons. Most of the chlorophyll in the photosystem functions as *antenna chlorophyll*. A number of different chlorophyll proteins (CP) complexes are formed by the association of chlorophyll with specific proteins. The antenna for photosystem II, for instance, consists of two chlorophyll proteins known as CP 43 and CP 47. These two CP complexes each contain 20 to 25 molecules of chlorophyll - a. Antenna pigments absorb light but do not participate directly in photochemical reactions. The antenna chlorophylls lie very close together such that excitation energy can easily pass between adjacent pigment molecules by a radiation - less transfer process. The energy of the absorbed photons thus migrates through the antenna complex, passing from one chlorophyll molecule to another until it eventually arrives at the *reaction center*.

The *reaction center chlorophyll*, which consists of 4 to 6 molecules of chlorophyll is also associated with proteins and cofactors. The reaction center is the site of the primary photochemical redox reaction and it is here that light energy is actually converted to chemical energy. The reaction centers for PS I and PS II are designated as P 700 and P 680, respectively. These designations identify the reaction center as a species of chlorophyll-a, or pigment (P) with an absorbance maximum at either 700 nm (PS I) or 680 nm (PS II). The efficiency of energy transfer through the antenna chlorophyll to the reaction is as high as about 90%.

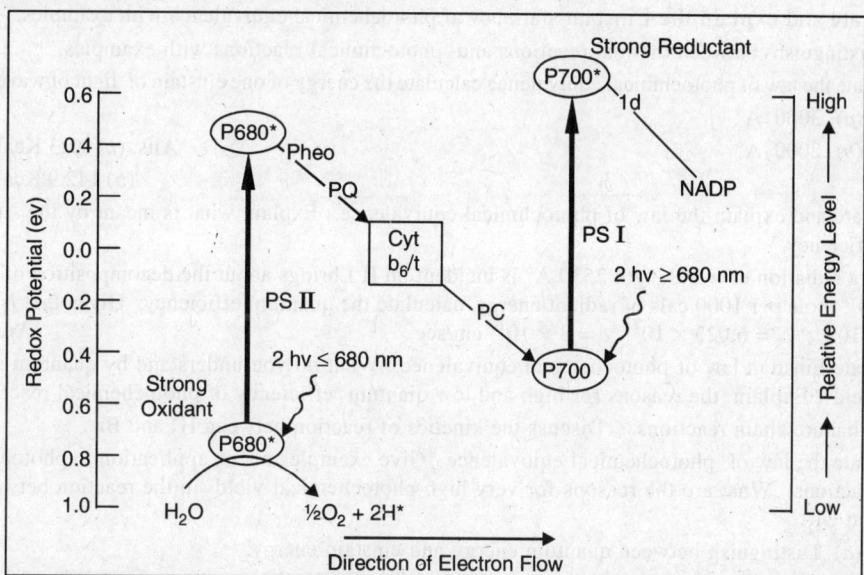

Fig. 26.8. The current version of the Z-scheme for photosynthetic electron transport.

The association of the reaction center with a large number of antenna chlorophyll molecules increases the efficiency in the collection and utilization of light energy. The advantage of the photosystem is that while the reaction center is busy in processing of one photon, the antenna chlorophyll molecules try to intercept other photons and funnel their way to the reaction center. Thus, as soon as the reaction center becomes free, more excitation energy is immediately available for a fresh cycle of processing. Plants growing under low light (or shade) augment their antenna systems with even more chlorophyll in order to ensure efficient light harvesting. A major function of this expanded antenna system is tofacilitate harvesting available photons and keep the reaction center operating at near optimal rates. The augmented antenna complexes are known as *light* harvesting complexes.

The current version of the Z-scheme for photosynthetic electron transport is illustrated in Fig. 26.8.

In this scheme, the redox components are placed at their approximate midpoint redox potential. The vertical direction indicates a change in energy level. The horizontal direction shows the direction of electron flow. The net effect of the process is to use energy of light to generate a strong reductant from low-energy electrons of water. The down-hill transfer of electrons between P 680* and P 700 represents a negative free-energy change. Some of this energy is used to establish a proton gradient which in turn drives ATP synthesis.

The ATP required for carbon reduction and other metabolic activities of the chloroplast is synthesized by a process called photophosphorylation. This is a very important process because ATP (along with NADPH) is required not only for the reduction of CO_2 but also to continually support various other metabolic activities in the chloroplast such as synthesis of protein in the stroma and the transport of proteins and metabolites across the envelope membranes.

[(For further details on bioenergetics and the light-dependent reactions of photosynthesis, readers may refer to standard treatises on the subjects such as "Introduction to plant physiology" by William G. Hopkins, John Wiley & Sons Inc., NY (1999)].

QUESTIONS

1. State and explain the Einstain-stark law af photochemical equivalence with examples.
2. Distinguish between thermal reactions and photochmical reactions with examples.
3. State the law of photochmical equivalence calculate the energy of one einstain of light of wavelength
   (a) 3000 A°
   (b) 2000 A°

   Ans. (a) 95.3 Kcal mol^{-1}
   (b) 142.9 Kcal mol^{-1}

4. State and explain the law of photochmical equivalence. Explain what is meant by the "quntum efficiency."
   If a radiation of warelength 2530 A° is incident on H I brings about the decomposition of 1.85×10^{-2} mole per 1000 cals of radiant energy, calculate the quantum efficiency. Given that $h = 6.62 \times 10^{-27}$; $N = 6.023 \times 10^{23}$; $c = 3 \times 10^{10}$ cm/sec. **Ans.** 8.74

5. State Einstian law of photochemical equivalence. What do you understand by quantum yield ? Explain the reasons for high and low qiantum efficiency of photochemical reactions.
6. What are chain reactions ? Discuss the kinetics of reaction between H_2 and Br_2.
7. State the law of photochemical eqiovalence. Give examples of its application to photochmical reactions. What are the reasons for very high photochemical yield in the reaction between H_2 and Cl_2.
   (a) Distinguish between quantum energy and einstain energy.
   (b) For a photochmical reaction X Y, 1.00×10^{-5} mole of Y is formed as a result of the absorption of 6.00×10^7 ergs at 3600 A°. Calculate the quantum yield. Given $h = 6.0 \times 10^{-27}$ erg/sec; $N = 6.02 \times 10^{22}$. **Ans.** 0.9092
8. What do you understand by the the term quantum yield of a photochmical reaction. How is it determined experimentally ?
9. Write short notes on
   (a) Primary and secondary photochemical reactions
   (b) Grothus-Draper law
   (c) Chain reactions

10. In an experiment involving the decomposition of H I into H_2 and I_2 it was found that 0.01 mole of H I were decomposed. Given that the quantum yield for the reaction is 2 and $N = 6.02 \times 10^{23}$, calculate the number of photons aborbed in the experiment.

Ans. 3.01×10^{21}

11. In a photochmical reaction system, 8.81×10^8 ergs of radiation of wave-length 2540 A°, were absorbed in a given time during which 1.12×10^{-4} moles of the irradiated substance has reacted. Calculate the quantum efficiency of the process. Given that $N = 6.023 \times 10^{23}$; $h = 6.625 \times 10^{-27}$ erg sec and $c = 3 \times 10^{10}$ cm/sec. **Ans. 0.5977**

13. Write explanatory note on any two of the following :
    (a) Photochemistry of vision
    (b) Photolysis of aldehydes and ketones
    (c) Flash photolysis
    (d) Photosensitized reactions.

14. Write explanatory note on photosynthetic electron transport with special reference to the Z-scheme.

15. Give some applications of photochemistry in technology.

QUESTION BANK

Unit 1 Chemical Bonding and Instrumentation

1. Discuss the salient features of molecular orbital theory of bonding. How does it different from valence bond theory. (10)
2. Explain velence shell electron pair repulsion theory on the basis of bond angle of methane, ammonia and water. (5)
3. Explain conductivity in conductors, semiconductors and insulators. (5)
4. What do you understand by the term "Metallic bond". Explain it on the basis of molecular orbital theory? (5)
5. What do you understand by stoichiometric and non-stoichiometric defects? (5)
6. What do you understand by Born-Haber cycle? What is its significance ? (5)
7. Predict the type of hybridisation and geometry of the following : (5)
   (i) XeO_3 (ii) NH_3 (iii) H_2O (iv) PCl_5
8. Differentiate between p-type and n-type semiconductors. (5)
9. With the help of molecular orbital diagram, calculate the bond order of : (5)
   (i) O_2^- (ii) O_2^{-2} (iii) He_2^+ (iv) NO^-
10. Derive Braggs equation for diffraction of X-rays by crystals. (5)
11. Write information notes on :
    (i) Frenkel defect
    (ii) F-centres
    (iii) Metal excess defects
    (iv) Dislocation
    (v) Schottky defects
12. What do you understand by extrinsic and intrinsic semiconductors ? (5)
13. Explain conduction in p-type and n-type semiconductors. (5)
14. Calculate the bond order in B_2, N_2, O_2, F_2 and NO. (5)
15. What do you understand by covalent bond. Give three characteristics of a covalent bond. (5)
16. What is the effect of temperature and impurities on conductivity of semiconductor? (5)
17. State the factors which favour covalency. (5)
18. What is hybridisation? Calculate % S orbital character of a hybrid orbital if bond angle between hybrid orbital is 105°. (5)
19. State differences between α (Sigma) and π (pi) bonds. (5)
20. Calculate number of α (Sigma) and π (pi) bonds in C_2H_2, C_2H_4, CO_3^{--}, NH_3, PCl_5. (10)
21. Predict shape, bond angle, dipole moment and hybridisation of the following compounds on the basis of VSEPR theory : (10)
    $Ag(CN)_2^+$, XeO_3, H_2O, PCl_5, XeF_4
22. Predict shape, bond angle dipole moment and hybridisation of the following compounds on the basis of VSEPR theory : (5)
    SO_4^{--}, NH_4^+, BF_4^-, X_3O_4, NH_3
23. Explain paramagnetic nature of O_2 and its less stability in comparison to that of N_2. (5)
24. NO_2 is coloured but N_2O_4 is colouriess. Explain (6)
25. Write order for the filling of electrons in various molecular orbitals. (5)

QUESTION BANK 949

26. What is meant by 'Bond order' ? Give a formula to calculate it. Calculate bond order for NO.
27. Explain formation of bonding and antibonding molecular orbitals on the basis of wave function.
28. State rules for LCAO method for formation of molecular orbits. (5)
29. Which of the following is paramagnetic :
$$O_2{}^-, O_2{}^-, N_2, C_6H_6, CO_3{}^-$$
30. Explain order of acidity $BF_3 < BCl_3 < BBr_3$. (5)
31. Arrange the following in the decreasing order of acidity and C—H bond length: CH_4, C_2H_6, $C_2H_4 C_2H_2$ (5)
32. Density of NaCl (at 20°C) = 2.17 gm CC^{-1}; molar mass = 58.5. Calculate the interplanar distance (d200). The first order reflection maximum of the K_α rays of Pd from the 200 planes of NaCl occurs at a glancing angle of 5.9°. Calculate the wavelength of the X-radiation. (5)
33. Distinguish between ethyl alcohol and diethyl ether on the basis of IR spectroscopy. (5)
34. How will you distinguish between the isomers CH_3COCH_3 and CH_3—CH_2—CHO on the basis of NMR spectroscopy? (5)
35. Define chemical shift. What is its significance in determination of structure of molecules? (5)
36. What do you mean by splitting of NMR signal? How many NMR signals are expected in the NMR spectrum of a acetaldehyde? Discuss their splitting. (5)
37. Why do ethene and ethyne, unlike propene and propyne, have no C to C multiple and bond stretching bands? (5)
38. How will you distinguish among aldehydes, ketones and carboxylic acids on the basis of IR spectroscopy? (5)
39. How 1, 3-pentadiene and 1, 4-pentadiene can be distinguished by UV spectroscopy? (5)
40. Discuss various types of electronic transitions. (5)
41. Name two molecules which undergo $\eta \to* \pi$ electronic transition. Discuss the effect of solvents on these transitions. (5)
42. State Beer's Law. If the path length of a beam of light through the sample is doubled and the concentration is made half, what will be the change in the value of absorbance? (5)
43. A compound contains 54.55 mass% carbon, 9.09 mass % hydrogen and 36.36 mass % oxygen. Its NMR spectrum consists of two peaks. The downfield peak is a doublet. Propose a structure for the compound. (10)
44. What are the various types of molecular vibrations? Write the stretching frequencies of the following functional groups : (10)
—OH, >C = O, —NH_2, ether (C—O—C), ≡ C —H, —C—H
45. The low resolution NMR spectrum of the compound $C_4H_8O_2$ exhibits three peaks at 8.1, 5.2 and 1.3 ppm with relative intensity ratio 1 : 1 : 6. Deduce the structure.

Ans. $\left(\begin{array}{c} CH_3 \\ CH_3 \end{array} CH-C \begin{array}{c} O \\ OH \end{array} \right)$

46. The low resolution NMR spectrum of the compound $C_5H_{10}O$ exhibits two peaks with relative intensities of 2 and 3 respectively. Deduce its structure.

(Ans. $CH_3 CH_2 . CO . CH_2 . CH_3$)

47. Assign the figures shown below to the respective isomer on the basis of the following UV spectral information:
α - Isomer shows a peak at 228 nm (ε = 1400) white β - isomer has a bant at 2.96 nm (ε = 1100)

Fig. 1

Fig. 2

Unit 2 Physical Chemistry

1. How can the rate constant k_1 and k_2 be determined for the reaction. (10)

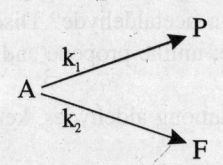

    A →k_1 P
    A →k_2 F

2. Derive the equations for determining k_1 and k_2 for the following reaction. (10)

    $$A \underset{K_2}{\overset{K_1}{\rightleftharpoons}} B$$

3. Differentiate between (any two) (10)
    (i) Order and molecularity.
    (ii) First order and pseudo first order reaction.
    (iii) Rate constant and equilibrium constant.
    (iv) Heat of reaction and activation energy.

4. Explain the reasons for the following - (10)
    (i) HCl is stronger acid than H_2S.
    (ii) NH_3 is more basic than H_2O.
    (iii) Order of acid strength in $H_2SO_4 > H_3PO_4 > H_3BO_3$.
    (iv) HI is stronger acid than HBr.
    (v) $HClO_4$ is stronger acid than HClO.

5. Write the short notes (any two). (10)
    (a) Bronsted Concept of Acid and Base.
    (b) Buffer solutions.
    (c) pH value.

6. Write a brief note on the theories of catalysis. (10)
7. Distinguish between : (10)
    (i) Homogenous and hetrogenous catalysis.
    (ii) Catalytic promotors and poisons.

QUESTION BANK

      (*iii*) Positive and negative catalyst.

8. Explain the valence bond theory of coordination complexes. (10)
9. How can the formation of coordination complexes be explained with the help of crystal field theory. (10)
10. What do you mean by energy of activation? Explain with the help of energy reaction coordinate diagram. How does the nature of activated complex change in the presence of a catalyst. (10)
11. Nitration of phenol gives the two products *o*-nitrophenol as major product and *p*-nitrophenol as side product. If *a* and *b* are initial concentration of phenol and nitric acid respectively. How can the rate constant (k_1) for the formation of *o*-nitrophenol and rate constant (k_2) for the formation of *p*-nitrophenol be obtained? (10)
12. What effect does a catalyst have on equilibrium system? Explain from the point of view of kinetics. (5)
13. Explain the action of promotors and poisons with examples. (5)
14. How does the nature of the activated complex change in the presence of a catalyst. (5)
15. How does the concept of activation energy explain the role of catalyst in a chemical reaction ? (5)
16. Discuss the modern theory of heterogenous catalysis. (5)
17. How does a catalyst act in a reversible reaction in equilibrium? Give the reasons for your answer. (5)
18. Discuss the mechanism of heterogenous catalysis. (5)
19. What are characteristics of a catalytic reaction. (5)
20. Discuss the intermediate compound formation theory of catalysis. (5)
21. Define the terms 'Catalysis', 'Inhibition' and 'Auto catalysis' with the help of examples. (5)
22. Differentiate between homogenous and heterogenous catalysis with their example. (5)
23. What do you mean by parallel and reversible reaction. Give one example of each. (5)
24. For reaction : (5)

$$A + 2B \rightarrow C + D$$

The following data are given

[A] mols l^{-1}	[B] mols l^{-1}	Rate mol $l^{-1} s^{-1}$
.01	.10	.0032
.03	0.10	.0096
.01	0.20	.0064

Find the order with respect to A, B and total order of reaction. (5)

25. For reaction

$$2NO + Cl_2 \rightarrow 2NOCl$$

It was found that on doubling the conc. of both reactants, the rate increases eight fold. But on doubling the conc. of Cl_2 alone the rate only double itself. What is the order of reaction with respect to NO? Justify your answer (5)

26. Integrate the following equation. (5)

$$-\frac{dc}{dt} = k[C]^{3/2}$$

If Co is the initial concentration find k and calculate half-life time.

27. The specific rate constant for second order neutralization of nitropropane by a base is given by (5)
$$\log k = 3163/T + 11.899$$
where concentrations expressed in moles litre^{-1} and time in min. If [Acid] = [Base] = 0.005 M, Calculate E_a and $t_{1/2}$ (half-life period).

28. Consider the acid catalysed reaction of carboxylic acid with alcohol as,
$$RCOOH + R'OH \xrightarrow{H^+} RCOOR' + H_2O$$
$$(A) \qquad (B)$$
$[A]_0 = [B]_0 = 0.2$ mol l^{-1} and pH = 3, it was also found that half life of above reaction is 50 min. Calculate K_{true} and $K_{apparent}$ for above reaction. (5)

29. What do you mean by pK value. The pKa and pKb values for formic acid and phenol are 3.75 and 9.92 respectively. Find their dissociation constant. (5)

30. Justify the statement 'water can act as an acid as well as base'.

31. Derive the Hunderson equations for buffer consisting $NH_4Cl + NH_3OH$.

32. Derive the Hunderson equation for buffer consisting. (5)
$$CH_3COOH + CH_3COONa$$

33. Calculate the pH of a solution consisting 0.05 molal lactic acid and 0.122 molal Na lactate. (5)
$$(K_a = 1.37 \times 10^{-4})$$

34. Calculate pH of solution consisting 0.1 M of NH_4OH and 0.24 M NH_4Cl. (5)
$$(K_b = 1.8 \times 10^{-5})$$

35. 10 CC of 0.1 M NaOH solution added to 50 CC of 1.1 M acetic acid solution. What will be the pH of the resulting solution. (5)
$$(K_{HAC} = 1.75 \times 10^{-5})$$

36. A solution which is 0.05 M in respect of benzoic acid and 0.1 M in sodium benzoate, has pH = 4.50. Find dissociation constant of the acid. (5)

37. In a reversible reaction
$$A \underset{K_2}{\overset{K_1}{\rightleftharpoons}} B$$
both the forward and reverse reaction are of first order with half lives of 10.0 hrs at 400 K. If one starts with 1.0M of A. How much will be left after 10 hrs at 400 K. (5)

38. Explain the formation of $Fe(CN)_6^{4-}$ complex ion. (5)
39. Explain the formation of $Fe(CN)_6^{3-}$ complex ions. (5)
40. What are the reasons for the formation of complexes by transition metalcations. (5)
41. Write the rules for naming the coordination compounds. (5)
42. pKa for acetic acid is 4.76. How will you prepare a buffer solution of pH 4.40 by mixing sodium acetate and acetic acid. (5)
43. Explain the Lewis concept of acids and bases. Identify acids and bases in the following, NH_3, H_3O^+, BF_3, $ZnCl_2$, $C_2H_5NH_2$, SO_3, H_2O, ROH (any six may be given). (5)
44. Establish the relation between pH and pOH. What is the pH of solution containing the H+ concentration of 3×10^{-7} moles per litre ? (5)
45. What are conjugate acid and base ? Give the conjugate base of the following : (5)
$$NH_4^+, HCl, H_3PO_4, H_2S, HCO_3^-$$
46. Explain Werner's coordination theory. (5)
47. What are chelating complexes. (5)
48. What factors determine the stability of the complex ? (5)

QUESTION BANK

49. Explain bonding in coordination compounds or explain crystal field theory of bonding of coordination complex. (5)
50. Write the I.U.P.A.C. names of the following (10)
    1. $K_4[Fe(CN)_6]$
    2. $K_3[Al(C_2O_4)_3]$
    3. $[Cr(en)_3]Cl_3$
    4. $[(en)_2 \, CO \underset{OH}{\overset{NH_2}{\diagdown \diagup}} CO \, (en)_2] \, (SO_4)_2$
    5. $[COCl(NH_3)_5]Cl_2$
    6. $K[PtCl_3-NH_3]$
    7. $[Au(CN)_4]^-$
    8. $K_2[NiF_6]$
    9. $(NH_4)_3[ZrF_7]$
    10. $[Pt(NH_3)_4Cl_2]^{+2}$
    11. $Na_2[CO(H_2O)_2(OH)_4]$
    12. $[Pt(NH_3)_2 \, Cl(en)NO_2]Cl_2$
    13. $[V(H_2O)_6]Cl_3$
    14. $Na_2[ZnCl_4]$
    15. $K_2[O_5Cl_5N]$
    16. $Na_3[Ag(S_2O_3)_2]$
    17. $[(N_3)_4 \, CO \underset{OH}{\overset{NH_2}{\diagdown \diagup}} CO \, (NH_3)_4]^{+4}$
    18. $Li[AlH_4]$
    19. $[Co(en) \, (NH_3)_2 \, ClBr]NO_3$
    20. $(NH_4)_3[CO(C_2O_4)_2]$
51. Explain crystal field splitting in octahedral complexes. (10)
52. What is the difference between an inner orbital complex and outer orbital complex? Discuss with one example each. (5)
53. Fill in the blanks: (5)
    (i) Crystal field theory was developed by _____.
    (ii) Unit of magnetic moment is _____.
    (iii) T_2g denotes _____.
    (iv) Δ_o denotes and Δ_t deontes _____.
54. What is primary valency, secondary valency. (5)
55. (a) Define degeneracy ? (5)
    (b) What do you understand triply degenerate and doubly degenerate orbitals? (5)
56. Give reasons for the following. (5)
    (a) $Fe(CN)_6^{+4}$ is diamagnetic and $Fe(H_2O)_6^{+3}$ is paramagnetic.
    (b) $Ni(CO)_4$ possesses titrahedral geometry, while $Pt(NH_3)_2 \, Cl_2$ square planar.
    (c) $Fe(CN)_6^{-3}$ is weakly paramagnetic while $[Fe(CN)_6]^{-4}$ is diamagnetic.
57. What do you understand by high spin complex and low spin complex. (5)
58. Explain crystal field stabilization energy. (5)

59. Write two differences between double salts and coordination compounds. (5)
60. Define the following terms: (5)
    (a) Chelates
    (b) Polynuclear complexes
    (c) Ligands (5)
61. With the help of valence bond theory, predict the geometry of the following. (5)
    (a) $[FeF_6]^{-3}$
    (b) $[Ni(CO)_4]$

Unit 3 Environmental Chemistry

1. What is corrosion of metals ? Explain the basic reason of metallic corrosion. (5)
2. Describe briefly the consequences of corrosion. (5)
3. Define metallic corrosion. Explain electrochemical theory of corrosion. (10)
4. State and explain Pilling–Bedworth rule. (5)
5. Discuss in detail about chemical corrosion. (10)
6. Differentiate chemical and electrochemical corrosion with suitable examples. (5)
7. What is stress corrosion ? Give two examples. How can it be controlled ? (5)
8. Discuss any two factors which influence the corrosion rate in detail. (5)
9. Why does corrosion occur in steel pipe connected to copper plumbing ? Explain. (5)
10. Discuss briefly (a) Soil corrosion (b) Galvanic corrosion (c) Pitting corrosion (d) Crevice corrosion (10)
11. Explain briefly the various factors influencing corrosion. (10)
12. Silver and copper metals do not undergo much corrosion like iron in moist atmosphere. Explain. (10)
13. Explain methods of prevention of corrosion. (5)
14. Discuss briefly corrosion inhibitors. (10)
15. What is corrosion ? Briefly discuss the various methods employed for Protection of metals from corrosion. (10)
16. How are metals protected against corrosion by modifying the environment. (10)
17. How will you protect an underground pipe line from corrosion by sacrificial anodic and impressed current cathodic protection methods ? (10)
18. What are cathodic and anodic protection for controlling corrosion ? Discuss their merits and demerits. (5)
19. Discuss any four methods of corrosion control. (5)
20. Distinguish between anodic and cathodic inhibitors. (5)
21. What are corrosion inhibitors ? Explain with examples how anodic and cathodic inhibitors provide protection against corrosion. (10)
22. Explain rusting of iron with the help of electrochemical theory of corrosion. (10)
23. Discuss briefly passivity and its significance. (5)
24. Discuss briefly various types of corrosion. (10)
25. Distinguish between anodic and cathodic coating. (5)
26. Explain the preliminary treatment given to a metallic surface before coating. (10)
27. Discuss the importance of design and material selection in controlling corrosion. (10)
28. Define lubricant. Discuss the classification of lubricants with suitable examples. (10)
29. What are solid lubricants ? When can they be used ? Give details about graphite. (10)
30. Discuss the important properties of lubricants and indicate the significance of these properties. (10)
31. What are the different synthetic lubricants used ? How are they superior over petroleum lubricants ? (10)

QUESTION BANK 955

32. Discuss the theory of lubrication. (10)
33. What is meant by dry corrosion ? (5)
34. Discuss briefly the nature of dry corrosion product. (5)
35. Discuss liquid metal corrosion briefly. (5)
36. How does the physical state of the metal influence corrosion rate ? (5)
37. Discuss in detail the nature of corroding environment. (10)
38. Which are the gases majorly responsible for causing green house effect and how are they released into the atmosphere ? What are the possible consequences of uncontrolled global warming? (10)
39. How is ozone formed and depleted in nature ? What are the consequences of the depletion of the O_3 layer in the atmosphere ? (10)
40. What are the pollutants responsible for causing acid rain ? What are its deleterious effects on material and terrestrial ecosystems ? (10)
41. What are the different methods adopted to limit the levels of various air pollutants in the atmosphere ? (10)
42. Name important air pollutants and mention their effects on human life. (10)
43. What is meant by the terms pollutant, contaminant, receptor, sink, reservoir. (5)
44. Discuss the sources, harmful effects and methods of control of the following Pollutants (a) CO (b) Particulate matter (c) Photochemical oxidants. (10)
45. Explain the harmful effects of air pollution with respect to
    (a) depletion of ozone layer
    (b) Photochemical smog
    (c) green house effect (10)
46. Define pollution. List various sources of air pollution and discuss their effects on the environment. (10)
47. Explain briefly (a) Acid rain (b) Photochemical smog (c) Green house effect (d) Ozone hole (e) Radiation balance of earth. (10)
48. Describe the sources and constituents of air pollution. (10)
49. What is meant by green house effect ? Describe two monitoring techniques used for green house effect. (10)
50. What is the green house effect ? How do pollutants affect this ? (5)
51. What is the effect of ozone depletion ? How does it occur ? (5)
52. How does automobile exhaust contribute to air pollution ? (5)
53. How do you account for "Ozone depletion" ? Indicate its consequences. (5)
54. Explain the sources and harmful effect of the following air pollutants.
    (a) CO (b) SO_2
    (c) Smoke (d) Particulate matter (10)
55. Give a detailed account of the various environmental segments. (10)
56. Describe characteristics of major atmosphere regions. (10)
57. Write detailed informative note on the following (a) Earth's radiation balance (b) Lapserate and temperature inversion. (10)
58. How are air pollutants classified ? What are the major effects of CO, SO_2, NO, Hydrocarbon and particulates on human health. (10)
59. What are the sources of NO and SO_2 global atmosphere ? Give an account of the global efforts to control the release of these two pollutants into the atmosphere by anthropogenic source. (10)
60. How would you broadly divide the major regions of the atmosphere ? State their respective altitudes and temperature ranges ? What are the important chemical species in each

region ? (10)
61. "Carbon dioxide, a non-pollutant, is perhaps the single most important environmental question facing us at present. Discuss in terms of the green house effect. (10)
62. "Oxygen plays a key role in the troposphere, while ozone, in the stratosphere". Elucidate. (5)
63. Discuss briefly ozone hole and its effect on human health. (5)
64. What are the primary pollutants ? Discuss their sources and relative contribution to air pollution. (10)
65. How can internal combustion engine be modified to make auto-exhausts free from pollutants? (5)
66. How do you account for Pb in particulate matter ? Can you suggest a corrective step for making air free from Pb. (5)

Unit 4 Applied Chemistry

1. Distinguish between natural and synthetic fuels. (5)
2. List the raw materials which can be utilized for biogas manufacture. How is biogas obtained from cattle dung ? (10)
3. Write the approximate composition and applications of biogas. (5)
4. Write the constituents of the amorphous conductors used in solar energy unit. Also explain the role of impurities. (5)
5. Define calorific value of a fuel and write the units. (5)
6. What are non-conventional fuels ? (5)
7. Write the mechanism of solar energy plant to convert sunlight into electricity. (5)
8. Define fuels in the light of modern concept. (5)
9. What are the merits of solid fuels ? (5)
10. What are demerits of liquid fuels ? (5)
11. What are demerits of solid fuel ? (5)
12. What are merits of liquid fuel ? (5)
13. Write the constituents responsible for temporary hardness of water. Discuss the treatment method. (5)
14. What are primary treatment methods for wastewater ? (5)
15. What are secondary fuels ? (5)
16. Write the constituents responsible for permanent hardness of water. Discuss one treatment method. (5)
17. Write the names of indicator and buffer solution used in the determination of the hardness of water by EDTA titration. (5)
18. Write the names of the ingredients which impart colour in wastewater. Explain the method for colour estimation. (5)
19. Why is it conventional to express hardness of water in terms of $CaCO_3$ at the international level? Write other units also. (5)
20. Write the structure and complete name of EDTA (di-sodium salt) and EBT. (5)
21. Write the names of the substances which impart undesirable odour in wastewater and list their removal methods. (5)
22. Write the raw materials required for biomass energy technique. Explain anerobic fermentation. (5)
23. Compare the calorific value of cattle dung cake and biogas obtained from the same amount of cattle cake. Also write limitation of biogas. (5)
24. What are the reasons for discouraging the direct use of cattle dung as a fuel ? (5)
25. List the names of secondary methods for wastewater treatment. (5)

QUESTION BANK

26. What are fuels ? Discuss their classification. (10)
27. Discuss the manufacturing of biogas. Write its merits and demerits. (10)
28. Explain the process involved in the conversion of solar energy into electrical energy and heat energy in solar plants. (10)
29. Describe the reasons for not using solar energy at domestic and industrial levels economically, inspite of the free availability of sun light. (5)
30. Explain the gross and net calorific values of fuels. (5)
31. Discuss in brief the solid fuels. (5)
32. Discuss the liquid fuels briefly. (5)
33. State the lime-soda process for the removal of hardness of water. (10)
34. What do you understand by temporary and permanent hardness of water. Which unit is used to express hardness of water (5)
35. State zeolite process for the removal of hardness of water. Discuss its merits over soda time process. (10)
36. Describe the methods to determine colour and odour in wastewaters. (10)
37. Write a short note on the wastewater treatment methods. (10)
38. Write a short note on boiler feed water. (5)
39. Describe the method to determine the net and gross calorific values of the fuels. (10)
40. Write a short note on biomass as a source of energy. How does the techniquediffer from biogas technique ? (10)
41. Describe the methods to determine temporary, permanent and total hardnesses of hard water. (10)
42. Compare the merits and demerits of solid and liquid fuels. (10)
43. What is the main purpose of secondary wastewater treatment ? Discuss the sludge treatment method. (10)
44. Discuss the characteristics imparted by impurities in wastewater. (10)
45. A sample of ground water has 150 mg/l of Ca^{2+} and 60 mg/l of Mg^{2+}. Find the total hardness expressed in milliequivalents per litre (meq/l) and mg/l as $CaCO_3$. (10)
46. Discuss the process of conversion of solar energy into electrical energy. (10)
47. Write a short note on biomass energy. (10)
48. Describe liquid and air solar energy collectors. (10)
49. Discuss p-type and n-type silicons used in solar energy plant. (5)
50. Justify that solar energy is a very good source of non-conventional energy. What do you understand by renewable and non-renewable fuels ? Categorise the following : Coal, biomass, solar energy, electrical energy. (5)
51. Discuss conventional and non-conventional sources of heat energy ? (10)
52. Give a detailed analysis of liquid fuels. (10)
53. Discuss the various methods of analysis to asses the quality of a coal sample. (10)
54. Calculate the hardness of a water sample containing per litre, 16.2 mg of $CaSO_4$; 1.4 mg $Mg(HCO_3)_2$ and 9.5 mg of $MgCl_2$. (10)
55. Differentiate between net and gross calorific values. Also discuss about correction factors. (10)
56. What are boiler troubles and what are their consequences ? (10)
57. Compare the hot lime soda process and zeolite process of water softening and discuss their merits and demerits. (10)
58. Calculate the quantities of lime and soda required for cold softening of 2,00,000 litres of water using 16.4 ppm of sodium aluminate as a coagulant. The results of the analysis of raw water and softened water are as follows. (10)

Raw water
Ca²⁺ — 160 ppm
Mg²⁺ — 72 ppm
HCO₃⁻ — 732 ppm
Dissolved CO₂ — 44 ppm

Softened water
CO₃²⁻ — 30 pp
OH⁻ — 17 ppm

59. How is the calorific value of a solid fuel determined using bomb calorimeter experiment? (10)

60. The following data is obtained in a bomb calorimeter experiment :
Weight of the crucible = 3.649 g
Weight of the crucible fuel = 4.678 g
Water equivalent of the calorimeter = 570 g
Water taken in the calorimeter = 2200 g
Observed rise in temperature = 2.3°C
Cooling correction = 0.047°C
Acids correction = 62.6 calories
Fuse wire correction = 3.8 calories
Cotton thread correction = 1.6 calories
Calculate the gross calorific value of the fuel sample. If the fuel contains 6.5% hydrogen, determine the net calorific value.

61. What are the terrestrial and aquatic plants ? (5)
62. Write the categories of biomass. (5)
63. What impurities are mixed in *n*-type and *p*-type silicon and why ? (5)
64. Write the applications of solar energy. (5)
65. Justify the statement "biomass stores solar energy". (5)
66. Write various units to express hardness of water and correlate them. (5)
67. It is conventional to express the hardness of water and in terms of $CaCO_3$. Why ? (5)
68. Write expected problems using untreated water in boiler. (5)
69. Write advantages and limitations of liquid fuels. (5)
70. Write significance of proximate analysis of a coal in determining the utility of coal for a particular purpose. (5)

Industrial Chemistry

1. (a) Identify the repeating unit in the following and determine the structure of the monomer. (5)

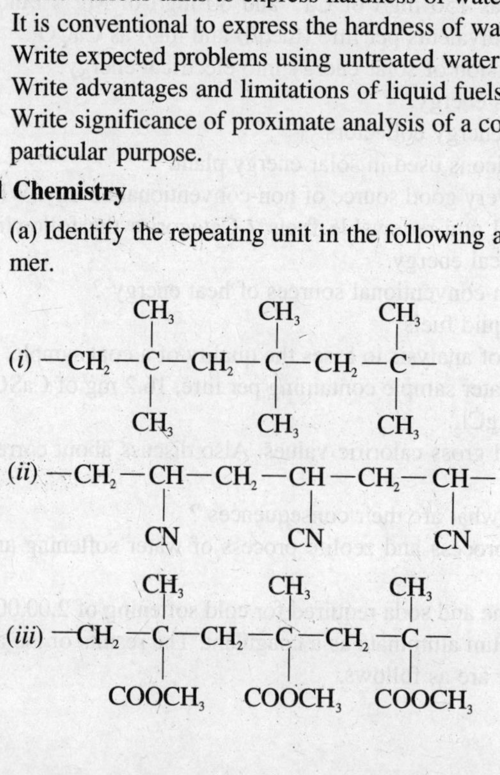

(i) $-CH_2-\underset{\underset{CH_3}{|}}{\overset{\overset{CH_3}{|}}{C}}-CH_2-\underset{\underset{CH_3}{|}}{\overset{\overset{CH_3}{|}}{C}}-CH_2-\underset{\underset{CH_3}{|}}{\overset{\overset{CH_3}{|}}{C}}-$

(ii) $-CH_2-\underset{\underset{CN}{|}}{CH}-CH_2-\underset{\underset{CN}{|}}{CH}-CH_2-\underset{\underset{CN}{|}}{CH}-$

(iii) $-CH_2-\underset{\underset{COOCH_3}{|}}{\overset{\overset{CH_3}{|}}{C}}-CH_2-\underset{\underset{COOCH_3}{|}}{\overset{\overset{CH_3}{|}}{C}}-CH_2-\underset{\underset{COOCH_3}{|}}{\overset{\overset{CH_3}{|}}{C}}-$

(iv) 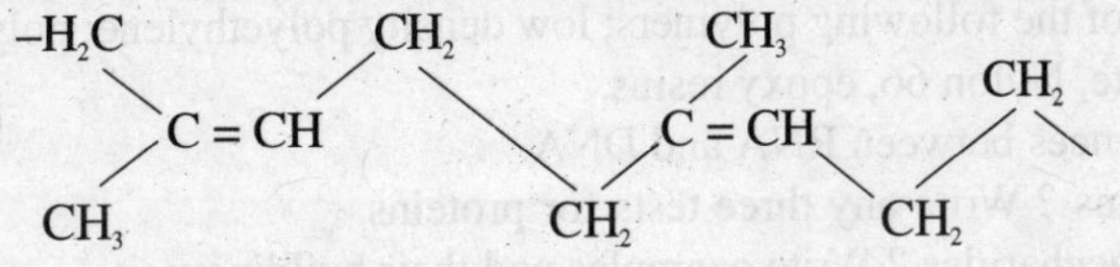
$$\begin{array}{c}-H_2C\\ \diagdown\\ C=CH\\ \diagup\\ CH_3\end{array}\begin{array}{c}CH_2\\ \diagdown\\ \\ \diagup\\ CH_2\end{array}\begin{array}{c}CH_3\\ \diagdown\\ C=CH\\ \diagup\\ CH_2\end{array}\begin{array}{c}CH_2\\ \diagdown\\ \\ \diagup\\ CH_2\end{array}$$

(v) 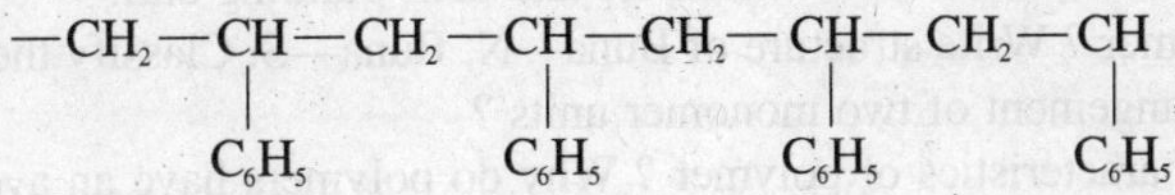 —CH₂—CH—CH₂—CH—CH₂—CH—CH₂—CH—
          | | | |
        C₆H₅   C₆H₅   C₆H₅   C₆H₅

2. If 20 gm of poly (ethylene) were completely burnt in the presence of excess of air, how many moles of CO_2 will be produced ? (5)
3. What are addition polymers ? Give five examples. (5)
4. Write four units for the polymers obtained from the following monomers (assuming head to tail structure), and name the monomeric units. (5)

(1) 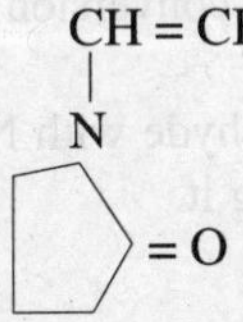 Structure: CH=CH₂ attached to N of a 5-membered ring with =O (N-vinyl pyrrolidone)

(2) CH₂ = CH — CH = CH₂

(3) CH₂ = C — CN
         |
         COOCH₃

(4) CH₂ = CF₂

(5) CH₂ = C — CH = CH₂
         |
         CH₃

5. What is the difference between a monomer and a polymer ? Classify polymers on the basis of occurance and chemical nature. (5)
6. What are graft and block copolymers ? Give their examples ? (5)
7. Give the structure of the following polymers ? (5)
   Nylon 6, 6
   Terylene
   Orlon
   Chloroprene
8. What are organic polymers ? Give the structure of five organic polymers. (5)
9. What are vinyl monomers ? Give four examples. (5)
10. Give the mechanism of free radical polymerization of vinyl monomers. (5)
11. What are initiators ? Name four free radical initiators. (5)
12. What are inhibitors ? Give examples. (5)
13. What is Zieglar Natta catalyst ? Give four examples. What is the significance of a catalyst in polymerization (5)
14. What are condensation polymers ? Give four examples of condensation polymers. (5)
15. What are bio-polymers ? Give three examples. (5)
16. Classify the polymers on the basis of stereochemistry. (5)
17. If average degree of polymerization of polymethyl methacrylate is 10^3, calculate its average molecular weight. (5)
18. State the differences between addition and condensation polymerization. (10)

19. Write two uses of the following polymers; low density polyethylene; polystyrene, polymethyl, ethacrylate, Nylon 66, epoxy resins. (10)
20. State the differences between RNA and DNA. (5)
21. What are proteins ? Write any three tests for proteins. (5)
22. What are polysaccharides ? Write examples and their building unit. (5)
23. What is copolymer ? Write structure of Buna—N, Buna—S. Classify the copolymers on the basis of arrangement of two monomer units ? (10)
24. What are the characteristics of polymer ? Why do polymers have an average molecular weight. (10)
25. State the differences between Free radical and ionic polymerization. (5)
26. What benzyne ? Draw its structure and explain the mechanism of the reaction involving it. (5)
27. Describe the mechanism of Skraup Synthesis. (5)
28. What is molecular rearrangement ? Discuss mechanism of conversion of caprolactum into nylon-6. Name the rearrangement involved. (5)
29. What cannizzaro reaction ? Discuss mechanism of formaldehyde with NaOH. (5)
30. What is nitrene ? Discuss mechanism of a reaction involving it. (5)
31. Complete followings :

(i) $\underset{H}{\overset{H}{>}}C=O$, $\underset{H}{\overset{H}{>}}C=O \xrightarrow{NaOH} ?$

(ii) $HCHO + C_6H_5CHO \xrightarrow{NaOH} ?$

(iii) benzene-1,3-dicarbaldehyde $\xrightarrow{NaOH} ?$

(iv) 2-hydroxybenzaldehyde $\xrightarrow{CHCl_3 / KOH} ? \xrightarrow{CH_3COCl: \text{ in pyridine}} / \xrightarrow{CH_3OH^⊕}$

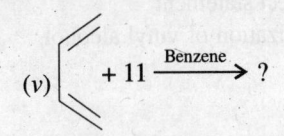

(v) [cyclopentadiene structure] + 11 $\xrightarrow{\text{Benzene}}$?

32. Write an example of :
   (i) Crossed Cannizzaro reaction
   (ii) Reimer Tiemann reaction
   (iii) Hoffmann reaction
   (iv) Diel's Alder reaction
   (v) Beckmann rearrangement. (5)
33. What is Diel's Alder reaction ? Describe the mechanism of the reaction. What is the importance of the reaction in industry ? (10)
34. Write two applications of each of the following reactions :
   Cannizzaro reaction, Hoffmann rearrangement, Reimer Tiemann reaction and Beckmann rearrangement. (5)
35. What is Beckmann rearrangement ? Discuss the mechanism and application of the reaction. (5)
36. Convert indole into indole-3-aldehyde. (5)
37. What is Hoffmann rearrangement ? Discuss the mechanism and application of the reaction. (10)
38. Write structure of Phenol formaldehyde resin, urea formaldehyde resin, polymides, polycarbonate and polyurethanes. (5)
39. (i) If two polymers of molecular weight 10,000 and 1,00,000 are mixed together in equal parts by weight, determine the number average and weight average molecular weights. (5)
   (ii) If the above polymers are mixed so that equal number of molecules are added, determine $\overline{M}_n$ and $\overline{M}_w$. (5)
40. A sample of polystyrene is composed of a series of fractions of different sized molecules.

Fraction	Weight Fraction	Molecular Weight
A	0.10	12000
B	0.19	21000
C	0.24	35000
D	0.18	49000
E	0.11	73000
F	0.08	102000
G	0.06	122000
H	0.04	146000

(5)

Calculate the numbr average and weight average molecular weights of this polymer sample.
41. Fractions of a polymer when dissolved in an organic solvent gave the following intrinsic viscosity values at 25°C. (10)
   M(g mol⁻¹) :   34000   61000   130000
   [η] :   1.02   1.60   2.73
   Determine K and α.
42. What is polydispersity index ? Give its importance write expressions for calculation of $\overline{M}_w, \overline{M}_v$ and $\overline{M}_z$ for polymers. Name the techniques to determine the above average molecular weight(s). (10)

43. State whether true or false. If false, give the correect statement
    (i) Polyvinyl alcohol can be prepared by polymerization of vinyl alcohol.
    (ii) CH_4 can be polymerised.
    (iii) C_2H_2 and aniline cannot be polymerised.
    (iv) Polymers have sharp melting point.
    (v) Ziegler-Natta catalyst is used for the preparation of syndiotactic polymer. (5)
44. What are the essential conditions for a compound to show geometrical isomerism ? (5)
45. Write three examples of geometrical isomerism. (5)
46. Explain optical isomerism of chiral organic compounds. (5)
47. Write five examples of organic compounds showing optical isomerism without chirality. (5)
48. Define the terms : Plane of symmetry, Centre of symmetry, Alternating axis of symmetry. (5)
49. State differences between enantiomers and diastereoisomers with suitable example. (5)
50. (a) An optical pure compound A, gave an $[\alpha]^{25} = +20°$. A mixture of A and its enantiomer B, gave $[\alpha]^{25} = +10°$. Calculate the ratio of A to B in the mixture. (5)
51. (a) The specific rotation of a pure enantiomer is +10°. If it is isolated from a reaction with 30% recemisation and 70% retention, what will be its observed rotation ? (5)
    (b) What do you understand by Altropisomerism ? (5)
52. (a) A optically active compound A having molecular formula C_6H_{12} gives inactive compound B of molecular formula C_6H_{14} on hydrogenation. Write the structures of A and B. (5)
    (b) The specific rotation of a solution of conine containing 0.35 g/ml measured in 10 cm tube is +16°. What will be the observed specific rotation of its enantiomer ? (5)

Appendix-1
El Nino Phenomenon and Its Effects

Oceans and seas play a vital role in global climatic variations. These variations are largely driven by fluctuations in the air-sea momentum and heat exchange. This effect was dramatically expressed in the 1982-1983 *"El Nino"* event, which sparked wide-ranging aberration in weather conditions such as flooding in south California and droughts in Australia, and is believed to be of oceanic origin.

"El Nino" event is generally characterized by warming of the coastal waters off the coasts of Peru and Ecuador. Under normal conditions, these waters are held relatively cool by the upwelling activity typical of this region of the eastern pacific. The atmospheric circulation of this area is characterized by what is commonly known as the Southern oscillation,. i.e. the seasonal shift in atmospheric surface pressure between the Australian Indian ocean region and the South-eastern pacific. This Southern Oscillation has been shown to influence surface pressure, temperature and rainfall variations over many parts of the earth. The disastrous effect of *"El Nino"* events (such as abnormal rainfall variations of a global scale) are postulated to be due to the alteration of the Southern Oscillation by the near-shore surface-warming along Peru and Ecuador.

"El Nino" effect is generally related to an anomalous weather condition which gives rise to major changes in ocean circulation and biological productivity along the coast of Peru. Under normal conditions, upwelling along the coast brings up nutrient-rich deeper water which results in high biological productivity. During *"El Nino"* event, wind patterns along the peruvia coast are changed, the upwelling is interrupted, and the coast is invaded by warm water. This is believed to be due to the reduction in trade-wind intensity following the periods of extremely strong winds which could cause warm tropical waters to pile up along the coast of South America and then further spread Southwards. Absence of upwelling results in a massive mortality of marine organisms and the collapse of the important peruvian anchoy fishery. Such a phenomenon generally occur around Christmas; hence this effect is called *"El Nino"* which is spanish, means *"the child"* (implying, the "christ's child").

"El Nino" phenomenon which is believed to occur in cycles, results in reversal of wind flows. *"El Nino"* causes drastic change from very heavy rains and floods to drought, poor crop yields and starvation. Initially, the warm Southward wind currents that appeared every December along the Coastal Ecuador and Peru, used to be described as *"El Nino"*, but now the word is used for recurrences that are exceptionally intense and persistant, causing climatic disturbances of varying severity around the globe, for more than a year.

Under normal conditions, the trade winds flow from the East to the West, pushing sun-warmed surface sea-waters west-wards and exposing cold water to the surface in the East. However, during the *"El Nino"* event, these normal trade winds collapse or even reverse. The weakening of the wind causes a modest change in Sea-surface temperature but the change in wind and pressure becomes drastic. The warm water of the pacific flows back Eastward and Sea-Surface temperature increases significantly on the western coast of South America. Due to this, the wet weather conditions normally present in the Western pacific move to the East and the arid conditions common in the East appear in the West, thereby resulting in heavy rains in South America and possible drought in South-East Asia, Indian and southern Africa. Thus, instead of bringing much needed rains to Asia, it takes away the moist air and dumps unwanted rains across the Pacific causing floods. In Asia, it leaves large tract like

Indonesia without rains and creates droughts in some South Pacific Islands or even Australia.

The year 1982-83 is supposed to be the worst year under "El *Nino*" for America causing losses to the tune of 13 billion dollars. The year 1997 turned out to be the worst under "*El Nino*" for Asia, particularly Indonesia, which had far below average rainfall. Papua New Guinea too suffered severe drought. The effect of the "*El Nino*" Phenomenon is varied, the worst damage could be the disruption to fish migration and damage to coral reefs. Those who are dependent on marine ecosystems for their livelyhood would be affected. It has also been documented that the number of births of weddell seals declined every four to six years, possibly coinciding the "*El Nino*" events.

Appendix-2
Basic Principles of Green Chemistry

Green chemistry is the use of chemistry for pollution prevention by environmentally-conscious design of chemical products and processes that reduce or eliminate the use or generation of hazardous substances.

The basic principles of green chemistry are as follows :

1. Prevention : It is better to prevent waste than to treat or clean up waste after it has been created.

2. Atom Economy : Synthetic methods should be designed to maximize the incorporation of all materials used in the process into the final product.

3. Less Hazardous Chemical syntheses : Wherever practicable, synthetic methods should be designed to use and generate substances that possess little or no toxicity to human health and the environment.

4. Designing Safer Chemicals : Chemical products should be designed to effect their desired function while minimizing their toxicity.

5. Safer Solvents and Auxiliaries : The use of auxiliary substances (e.g., solvents, separation agents, etc.) should be made unnecessary wherever possible and innocuous when used.

6. Design for Energy Efficiency : Energy requirements of chemical processes should be recognized for their environmental and economic impacts and should be minimized. If possible, synthetic methods should be conducted at ambient temperature and pressure.

7. Use of Renewable Feedstocks : A raw material or feedstock should be renewable rather than depleting whenever technically and economically practicable.

8. Reduce Derivatives : Unnecessary derivatization (use of blocking groups, protection/deprotection, temporary modification of physical/chemical processes) should be minimized or avoided if possible, because such steps require additional reagents and can generate waste.

9. Catalysis : Catalytic reagents (as selective as possible) are superior to stoichiometric reagents.

10. Design for Degradation : Chemical products should be designed so that at the end @ of their function they break down into innocuous degradation products and do not persist in the environment.

11. Real-time analysis for Pollution Prevention : Analytical methodologies need to be further developed to allow for real-time, in-process monitoring and control prior to the formation of hazardous substance.

12. Inherently Safer Chemistry for Accident Prevention : Substances and the form of a substance used in a chemical process should be chosen to minimize the potential for chemical accidents, including releases, explosions, and fires.

Bibliography

1. *Chemtech 1 to 4, First to Fourth volumes of Manual of Chemical Technology,* Chemical Engineering ; Education Development Centre, I.I.T., Madras, Editor : **D. Venkateswarlu,** S. Chand & Co; 1984.
2. *Chemistry of Engineering Materials,* **Robert B. Leigeou,** McGraw-Hill Book Company, Inc., New York.
3. *Chemistry in Engineering* by **Lloyd A. Munro,** Prentice-Hall, Inc., N.J.
4. *Chemistry for Engineers* by **Edward Cartn 1,** Butterworths, London.
5. *Water Treatment for industrial and other use* by **Eskel Nordell,** Reinhold Publishing Corporation, New York.
6. *A Text Book of Water Supply* by **A.C. Twort,** Edward Arnold Publishers Ltd., London.
7. *Applied Chemistry for Engineers* by **T.S. Gyngell.**
8. *Water Treatment* by **F.I. Belanoe**, Mir Publishers, Moscow.
9. *Petroleum processing* by **V.P. Sukhanov,** Translated from the Russian by **V. Afanasyev,** Mir Publishers, Moscow.
10. *Efficient use of fuel* by **R.C. England,** Her Majesty's Stationery Office, London, 1958.
11. *Elements of fuel technology* by **G.W. Himus,** Leonard Hill (Books) Ltd., London 1958.
12. *An introduction to Metallurgy* by **Joseph Newton,** Second Edn., John Wiley and Sons, Inc., New York.
13. *The Analysis of fuels, gas, water and lubricants* by **S.W. Parr,** McGraw Hill Book Co.Inc., New York.
14. *Engineering Metallurgy,* Part I, *Applied Physical Metallurgy* by **R.A. Higgins**. The English University Press Ltd., London.
15. *An Introduction to Physical Chemistry* by **S. Glasstone.**
16. *Fundamentals of Corrosion* by **Michael Henthorne,** Chemical Engineering, May 17,1971, p 127 to 132.
17. *Electrochemical Corrosion,* by **Michael Henthorne,** Chemical Engineering, June 14,1971. p. 102 to 106.
18. *Polarization Data Yield Corrosion Rates* by **Michael Henthorne,** Chemical Engineering, July 26,1971, p.99 to 104.
19. *Stress Corrosion* by **Michael Henthorne,** Sept. 20 1971, Chemical Engineering, p. 159 to 164.
20. *Measuring corrosion in the process plant* by **Michael Henthorne,** Chemical Engineering, Aug.23, 1971, p. 89 to 94.
21. *Control the process and control "corrosion"* by **Michael Henthorne,** Chemical Engineering Aug, Oct.18, 1971, p. 139 to 146.
22. *Good Engineering design minimize corrosion,* by **Michael Henthorne,** Chemical Engineering, Nov.15, 1971, p.163 to 166.
23. *Cathodic and anodic protection for corrosion control* by **Michael Henthorne,** Chemical Engineering, Dec. 27, 1971, p. 73-79.
24. *Solid Lubricants and Surfaces* by **E.R. Braithwaite,** Pergamon Press, New York, 1964.
25. *Industrial Lubrication Practice* by **Paul D.Hobson,** The Industrial Press, New York, 1955.
26. *Friction and Wear* by **B.Pugh,** Newnes Butterworths, London, 1973.
27. *Basic Lubrication Theory* by **A. Cameron,** Longman, London, 1971.
28. *Synthetic Lubricants* by **R.C. Gunderson** and **Andrew W. Hart,** Reinhold Publishing Corporation, New York, 1962.
29. *Motor oils, Performance and Evaluation,* **William A. Gruse,** Reinhold, New York, 1967.
30. *Principles and Applications of Tribology* by **Desmond F. Moore,** Pergamon, 1975.
31. *Practical Lubrication* by **B.Pugh,** Newnes-Butterworths, London, 1970.

BIBLIOGRAPHY

32. *Lubrication and Friction* by **Peter Freeman,** Sir Isaac Pitman and Sons Ltd., London, 1962.
33. *Introduction to Chemistry* by **Amos Turk, Herbert Meislich, Frank Brescia, and John Arents,** Academic Press, International Edition, New York and London, 1968.
34. *Physical Chemistry of Polymers,* by **A. Tager,** Translated from the Russian, Mir Publishers, Moscow, 1978.
35. *Text Book of Polymer Science* by **F.W. Billmeyer, Jr.,** Interscience Publishers, John Wiley and Sons, New York and London, 1962.
36. *Introduction to Polymer Chemistry* by **Raymond B. Seymour**, McGraw Hill, Kogakusna.
37. *Insulating Materials for Design and Engineering Practice,* by **Clark, Frank, M., John Wiley.**
38. *Encyclopedia of Chemical Technology,* Vol. 12 by **Klrk, Raymond E** and **Othmer Donald F.,** Interscience.
39. *Modern Science and Technology,* **Colborn, Robert, D. Van Nostrand.**
40. **Mc Gregor** *Silicones and their uses,* McGraw Hill.
41. *Chemistry and Technology of Silicones* by **Walter Noll,** Academic Press, New York, 1966.
42. *Inorganic Chemistry* by **F.A. Cotton and G. Wilkinson,** Wiley Eastern Ltd., 1978.
43. *Principles of Chemistry ; An introduction to theoretical concepts,* **Paul Ander and Anthony J. Sonnessa,** Collier-Macmillan ; Student Edition, 1965.
44. *Fundamental concepts of Inorganic Chemistry* by **Esmarch S. Gilreath,** International Student Edition, McGraw Hill, Kogakusha, Ltd., Tokyo 1958.
45. *Chemical Principles and Properties,* by **M.J. Sienko and R.A. Plane,** International Student Edition, McGraw Hill Kogakusha Ltd., Tokyo 1974.
46. *Fuels and combustion* by **Amir Circar,** Orient Longmans.
47. *The Chemistry of Cement and Concrete* by **F.M. Lea,** Edward Arnold (publishers) Ltd., 1970.
48. *Nature and properties of Engineering Materials* by **D. Jastrzebski,** Toppan Company Ltd., Tokyo, Japan, 1951.
49. *Fundamentals of Environmental Pollution* by **Ram Krishan Kannan,** S. Chand & Co. Ltd., New Delhi, 1991.
50. *University General Chemistry* by **C.N.R. Rao,** Mac Millian India, 1973.
51. *Thermodynamics* by **Kenneth Wark,** 3rd Edition, Mc Graw Hill Ltd., 1977.
52. *Fundamentals of Classical Thermodynamics* by **Gordon J. Van Wylen, Richard. E. Sonntyg,** John Wiley and Sons Inc. 1976.
53. *Chemical Process Industries* by **R.Norris, Shreeve,** McGraw Hill Book Co. 1967.
54. *Power Plant Technology* by **M.M. El. Wakil,** Mc-Graw Hill Book Co., New York, 1985.
55. *Nuclear Chemistry - Principles and Applications* by **G.R. Choppin and J. Rydberg,** Pergamon Press.
56. *An Introduction to Nuclear Science,* by **M.N. Sastri,** Associated East-West Press, 1984.
57. *Source Book of Atomic Energy* by **Samuel Glasstone,** D.Van Nostrand Co., Inc., 1967.
58. *Environmental Chemistry* by **A.K.De,** Wiley Eastern Ltd., New Delhi, 1989.
59. *Environmental Engineering* by **G.N. Pandey and G.C. Carney,** Tata McGraw Hill, New Delhi,1989.
60. *Corrosion Engineering* by **Mars G. Fontana and Norbert D. Greene,** International Student Edition, McGraw-Hill International Book Co., Tokyo, 1982.
61. *Materials Science & Engineering* by **V. Raghavan,** Prentice Hall of India Pvt Ltd., 1990.
62. *Polymer Science* by **V.R. Gowarikar, N.R. Viswanathan and J. Sreedhar,** Wiley Eastern Ltd. 1987.
63. *Degradable Polymers - Principles and Applications,* Gerald Scott and Dean Gilead (Editors), Chapman & Hall, London, 1995.
64. *Plastics Waste Management - Disposal, Recycling and Reuse,* Nabil Mustafa, Marcel Dekker, Inc., 1993.
65. *Materials Science and Engineering - An Introduction,* William D. Callister, Jr., John Wiley & Sons, Inc., 1994.
66. *Mechanical Properties of Polymers and Composites,* L.E. Nilsen, and R.F. Landel, Marcel Dekker, Inc., 1994.
67. *Engineering Materials and their Applications,* R.A. Flinn, Jaico Publishing House, Bombay, 1993.

68. Engineering Materials, V. John, Macmillan, London, 1990.
69. Polymers and Polymer Composites in Construction, L.C. Hollaway (Editor), Thomas Telford Ltd., London, 1990.
70. High Performance Polymers and Composites, J.I. Kroschwitz, John Wiley and Sons, New York, 1991.
71. Metal-Polymer Composites, J. Delmonte, Van Nostrand Reinhold, New York, 1990.
72. Handbook of Polymer Fibre Composites, F.R. Jones, Longman Scientific and Technical, Harlow Essex, 1994.
73. The Structure and Properties of Polymeric Materials, The Institute of Materials, London, 1993.
74. Polymer Matrix Composites, R.E. Shalin, Chapman and Hall, London, 1995.
75. Handbook of Polymer Composites for Engineers, L. Hollaway, Jaico Publishing House, Mumbai,1995.
76. Fundamental Principles of Polymeric Materials, S.L. Rosen, John Wiley and Sons Inc., 1993.
77. Electroactive Polymer Materials - State of the Art Review of Conductive Polymers, A. Wirsen, Technomic 1991.
78. Polymer Science and Technology, Joel R. Fried, Prentice Hall of Indias Private Ltd., New Delhi, 2000.
79. Wiley Encyclopedia Series in Environmental Science, Energy, Technology and Environment, Attilio Bibio and Sharon Boots, John Wirey & Sons, Inc., New York, 199.
80. Green Chemistry – Designing chemistry for the environment, Edited by Paul T. Anastas and Tracy C. Williamson, ACS Symposium Series 626, American Chemical Society, Washington, DC, 1996.
81. Introduction to Green Chemistry, Albert S. Matlock, Marcel Dekker. Inc., NY, 2001.
82. Chemistry of Waste Minimisation, Edited by J.H. Clark, Blackie, London, 1995.
83. Pollution prevention for chemical processes, David T. Allen and K.S. Rosselot, John Wiley & Sons, Inc., NY, 1996.
84. Dryden's outlines of Chemical Technology - For the 21st Century 3rd Edn., Edited by M. Gopala Rao and Marshal Sitting, Affiliated East-West Press, New Delhi, 2002.
85. Petroleum Refining Technology, Ram Prasad, Khanna Publishers, Delhi, 2002.
86. Green Chemistry – A synonym for Innovation, Dr. Rolf Froböe, Chemistry World June 2004.
87. Green Chemistry - Theory and Practice, P.T. Anastas and J.W. Warner, Oxford University Press, NY 1998.
88. Organic Chemistry, Volumes 1 & 2, I.L. Fenar, 6th Edn., Longman, ELBS, 1990.
89. Green Engineering – Environmentally Conscious design of Chemical Processes, David T. Allen and David R. Shonnart, Prentice Hall, 2003.
90. Environmental Engineering – A design approach, A. P.Sincero & G.A. Sincero, Prentice Hall, 1999.

Index

A
Acid number, 262
Acid rain, 566
Adipic acid 789
Adsorption, 235
Alkalinity, 12
Amino resins, 398
Analysis of coal, 88
Anelastic behaviour, 499
Aniline Point, 264
Anodised oxide coatings, 227
Anodic protection, 223
Apparatus for 926
Arsenic in water
 – effects, 19
Atmosphere, 551, 555
Atom economy, 797

B
Basic components, 788
Battery Technology, 699-741
Benzol, 137
Bituminous coal, 85
Bioenergetics 929
Biodegradable lubricants, 245
Blast furnace gas, 145
Blending of coal, 95
Blow-down, 63
Blue gas, 144
Boiler compounds, 62
Boiler feed water, 54
Boiler troubles, 55
Bomb Calorimeter, 76
Born-Haber Cycle, 323
Boy's Calorimeter, 81
Bragg's law, 470
Brown coal, 84
Butyl rubber, 431

C
Calgon conditioning, 62
Calorific intensity, 75
Calorific value, 73
 – gross and net, 74
Carbon residue test, 266
Carbonate Conditioning, 60
Carbonium ions, 818
Carbanions, 820
Carbonization of coal, 99

Carburetted water gas, 145
Carry over, 56
Catalytic converters, 125
Cathode reactions, 203
Cathodic protection, 220
Caustic embrittlement, 64
Cellulose acetate, 385
Cellulose acetate butyrate, 386
Cellulose nitrate, 385
Cetane number, 135
Cement, 284
 – Additives, 291
 – Chemical equations, 297
 – Constitutional compounds, 290
 – Manufacture of, 286
 – Raw materials for, 285
 – Portland, 295
 – Properties, 292
 – Testing, 294
 – Thermo-chemical changes, 295
 – Types, 299
Ceramics (511-518)
Cermets, 441
Chain reaction, 179
Characteristics of 819
Chemical bonding, 330
Chemical dip coatings, 227
Chemical equilibria, 931
Chemical modification
 – of polymers, 372
Chlorination, 29
Chlorides in water, 17
Chromatography, 767
Cloud Point, 267
Coagulation, 23
Coal gas, 140
Coal compositions, 87
Coal storage, 95
Coal technology, 94
Coke, 175
Coking coals, 97
Coke oven gas, 140
Colloidal fuels, 137
Colloidal conditioning, 62
Colour in water, 4
Colour in lubricants, 262
Combustion, 102
Complex ions, 339

Compounded oils, 257
Composite Materials, 436-448
Concentration cells, 196
Concrete, 303
 – action of chemicals on, 361
Conductive polymers, 376
Conditioning
 – of water with EDTA, 62
Conradson's test, 266
Coordination polymerisation, 370
Coolants (Reactor), 183
Cooling curves, 323
Coordinate bond, 337
Core (reactor), 183
Covalent bond, 333
Corrosion, 200, 333
 – Causes, 200
 – Classification, 201
 – Consequences, 200
 – Control, 217
 – Factors influencing, 206
 – in boilers, 62
 – of raw water, 20
 – reactions, 202
 – inhibitors, 210
 – measurement, 216
 – protection from 217
 – testing, 216
 – theories, 200
 – types, 212
Clays, 511
Cracking,
 – Catalytic, 115
 – thermal, 115
Creep, 507
 – in metals, 509
 – in amorphous solids, 509
Crevice corrosion, 213
Crystal structure, 466,
Cryogenic lubcircants, 244
Crystal systems, 468
Crystallinity,
 – of polymers, 486
Curing, 304

D
Dative bond, 337
Decomposition potential, 198

De-ionisation, 52
Delocalised orbitals, 356
Demineralisation, 52
Derivatized, 796
Disodium, 793
Dislocations, 505
Dissolved gases, 20
Dissolved solids, 20
Distillation, 53
Drag reduction, 379
Drop-point test, 253

E

Earthenwares, 516
Efficiency quantum, 916
Elastic behaviour, 491
Electroplating, 542
- applications of, 542
- baths for, 546
- equipment and operating conditions, 545
- preparation of basic materials, 544
- Wastes, characteristics of, 547
- Wastes treatment techniques, 547
- Safety precautions, 548
Elastomers, 422
Electrode potential, 194
Emulsification, 264
Electron displacement, 806
Electron pair, 810
Emulsion paints, 232
Enamels, 231
Engler's viscometer, 273
Engineering polymers, 375-379
Enthalpy, 451
Entropy, 453
Environmental aspects of Power Generation (nuclear), 188
Environmental aspects of Power Generation (thermo-nuclear), 546
Environmental segments, 549
Environmental Chemistry, 549
Epoxy resins, 399
Equivalents, 41
Erosion corrosion, 212
Ethyl cellulose, 386
Exfoliation, 213
Excess air, 156

F

Fiber - Reinforced Composites, 440-441
Filament Winding, 451
Fibre reinforced plastics, 433
Filtration, 24
Fire point, 267
Fischer-Tropsch process, 118
Flame temperature, 75
Flame Photometry, 781
Flash point, 267
Fluoride in water
 – effects, 17
Foaming, 57
Formazin turbidity units, 5
Free energy, 459
Friction, 236
Fuel oils, 137
Fuels, 73
Fuels-nuclear, 175
Functions group, 798

G

Galvanic cells, 196
Galvanic corrosion, 212
Galvanic series, 195
Gaseous fuels, 138
Gibbs-Helmholtz,
 - free energy, 457
Glazes, 517
Glass lubrication, 243
Gas lubrication, 244
Geemen synthesis, 800
Green Chemistry, 787
Greenhouse effect, 558
Greases, 251
- Cup, 251

H

Hardness of water, 6
Hazen units, 4
Hooke's law, 494
Hybridization, 352
Hybrid Composites, 443
Hydrogen bond, 339

I

Imperfections, 473
Industrial application, 801
Instrumental Techniques, 742
Inorganic non-metallic coatings, 227

Intergranular corrosion, 213
Internal energy, 451
Internal treatment, 60
Ion-exchangers, 45
Ionic coplymerisation, 370
Ionic polymers, 376
Ionic bond, 330
 (or electrovalent bond),
Intelligent polymers, 379
Iron removal
 - from water, 53

K

Kerosene, 136
Kinetics, 461
Knocking, 122
Kopper's gas, 146

L

Lacquers, 232
Lapse rate, 560
Lattice energy, 452
Laws of photochemistry, 915
Ligancy, 340
Lignite, 84
Liquid fuels, 103
Liquid crystal polymers, 377
Low quantum yield, 920
LPG, 140
Lubricants, 234
- blended, 276
- classification, 247
- emulsions, 261
- liquid, 253
- properties, 261
- selection of, 276
- semi-solid, 250
- solid, 247
- synthetic, 261
Lubrication, 238
- Boundary, 240
 (or thin film),
- extreme pressure, 242
- fluid or hydrodynamic, 239
- mechanism, 239
- methods of, 279
Lubrication
 – gas lubrication, 243
 – liquid metal, 243
 – caryogenic, 244
 – for nuclear reactor systems, 244

INDEX

- glass lubrication, 244
- environmental and health effects, 245
- for food processing, 245

Lubricants
- biodegradable, 245
- life cycle analysis, 247

Lurgi gas, 145

M

Mechanical properties, 491
Mesomeric effect, 811
Metallic bond, 352
Metal Pawders, 653, 671
Metallic coatings, 225
Metal matrix - Fiber composites, 443
Metastable equilibrium, 460
Metallic wires, 438
Methame, 829
Methods of polymerisation, 369
Microbiological corrosion, 215
Mineral oils, 256
Moderators (Reactor), 183
Molecular orbital theory, 355
Mortars, 303

N

Natural gas, 138
Natural resins, 384
Negative calalysis, 898
Neoprene, 431
Nernst theory, 193
Neutralization number, 263
Nephelometric turbidity units, 5
Nitrates in water
- effects, 17

Nitrile rubbers, 430
Nuclear energy (binding), 175
Nuclear energy (fission), 176
Nuclear energy (fusion), 189
Nuclear Power Stations in India, 169
Nuclear Reactors (Breeder), 187
Nuclear Reactors (Power), 180
Nuclear Reactors (Thermo), 544
Nucleophilic, 831
Nylons, 397

O

Octet rule, 335
Odour in water, 4
Oil gas, 141

Oil removal (from water), 58
Oiliness, 265
Onia-Geigi process, 142
Organic coatings, 228
Order of reaction, 463
Over-voltage (or over-potential), 198, 198
Oxidability, 22
Oxidability zone, 143
Ozone layer (formation, depletion), 559

P

Particle Reinforced Composites, 438-440
Passivity, 2N
Peat, 84
Petrochemicals, 147
Petroleum, 103
Phase rule, 308
- Gibs, 308
- water system, 312
- sulfur system, 315
- Ag-Pb system, 319
- Cd-Bi system, 322
- Al-Cu system, 324
- Pb-Sn system, 324
- Fe-C system, 327

Phenolic resins, 395
Plastic deformation, 502
Plastics, 380
- constituents, 381
- thermosetting, 382
- thermoplastic, 384
- fabrication of, 383

Photochemical reactions, 914, 920
Photolysis, 921
Photosynthesis, 924
Phosphate conditioning, 61
Pigment Volume Concentration (PVC), 229
Pitting corrosion, 213
Polarization, 197
Pollution control (air), (584—586)
Pollution control (water), (586—632)
Polyamides, 397, 484
Polyesters, 400, 485
Polyethylene, 260
Polymers
- Ionic polymers, 370
- Coordination polymers, 370
- Chemical modification, 372

- engineering polymers, 375 to 379
- speciality polymers, 375 to 379
- photoconductive polymers, 376
- conductive polymers, 376
- photonic polymers, 379

Polymethylmethacrylate, 389
Polystyrene, 388
Polytetrafluoroethylene (PTFE or Teflon), 387
Polyviny lacetate, 390
Polyvinyl chloride (PVC), 390
Polyurethanes, 433
Poly-saccharides, 484
Polymers 360, 482, 488, 489
- characteristics, 379
- classification, 361
- crystallinity, 486
- inorganic, 363
- organic, 362

Polymerization, 364
- Addition or chain, 364
- Condensation, 365
- Ionic, 367
- Mechanism of Chain, 365
- Methods of, 366
- Radical, 366
- Steriospecific, 369
- Step, 369
- Thermodynamics of, 371

Porcelain and Vitreous enamels, 518
Post-tensioning, 304
Pour Point, 267
Power alcohol, 137
Precipitation Hardening, 439
Precipitation particles
- (effect on dislocation motion), 495

Prepreg Production Process, 444
Pre-stressed Concrete, 303
Priming, 56
Producer gas, 142
Proteins, 484
Protecting coatings, 224
Proximate analysis (of coal), 88
Pultrusion, 443
Pulverized coal, 96

R

Ramsbottom test, 268
Rate expressions, 462
Rate laws, 462
Reaction mechanisms, 815

Redwood viscometer, 271
Reforming, 117
Refractories (519-500)
- behaviour of, 520
- classification of, 519
- industrial outlet for, 540
- manufacture of, 526
- properties of, 521
- raw materials for, 524
- requisites of a good, 519
- types of products, 530
Relative stabilities, 817
Relaxation process, 501
Resins
- ion exchange, 50
- natural, 384, 389
- Silicone, 402
Resonance, 336
Reversible cells, 197
Renewable Energy Sources, 642
Rods—control (Reactor), 184
Rubber
- crepe, 423
- foam, 426
- natural, 426
- reclaimed, 428
- silicone, 404, 432
- smoke, 423
- sponge, 426
- synthetic, 429

S

Saponification number, 263
Saybolt viscometer, 273
Scale formation, 58
- prevention, 59
- ion exchange, 50
Screw dislocations, 505
Sediments in water, 5
Sedimentation, 22

Semiconductors, 481
Significance, 785
Silica removal (from water), 58
Silica Scale, 5
Silicones, 402
Silicone rubbers, 405
Soil corrosion, 215
Soil pollution, 621
Solid fuels, 83
Solids
- types of, 477
- structure of, 479
Solid state defects, 472
Sources of Energy, 633
Standard electrode
   Potentials, 194
Stark-Einstein law, 915
Sterilization of water 27
Speciality polyolegins, 377
Spontaneous combustion, 96
Storage of coal, 95
Stress corrosion craking, 214
Structural Composites, 446
Styrene butadiene rubber
   (SBR), 431
Synthesis gas, 146
Synthetic gasoline, 118

T

Tallow and Tallow oil, 255
Tastes in water, 4
Teflon, 387
Temperature Inversion, 560
Theory collision, 872
Thermodynamics, 364
Thermoelastic effect, 500
Thiokols, 432
Turbidity (in water), 5
Twinning, 504
Types of imperfections, 473

U

Ubbelohde viscometer, 274
Ultraviolent Spectroscopy, 750
Ultimate analysis (or coal), 89
U-tube viscometer, 274
Unleaded petrol, 5

V

Valence bond theory, 344
Variable valency, 337
Varnishes, 231
Vectra, 377
Viscoelasticity, 506
Viscosity, 269
Viscosity index, 271
Vitreous enamel coatings, 227
Vulcanization (of rubber), 423

W

Water
- sources, 1
- effect on rocks and minerals, 3
- impurities, 3
- softening of, 29
- treatment of, (1—72)
- cooling of, 65
- for steam making, 54
Water gas, 126
Water-line corrosion, 214
Wear, 237
Werner's theory, 340
Whiskers, 440
White wares, 515
Winkler's gas, 146
Wood, 84
Work function, 460

Z

Zeolite process, 50

NOTES

NOTES